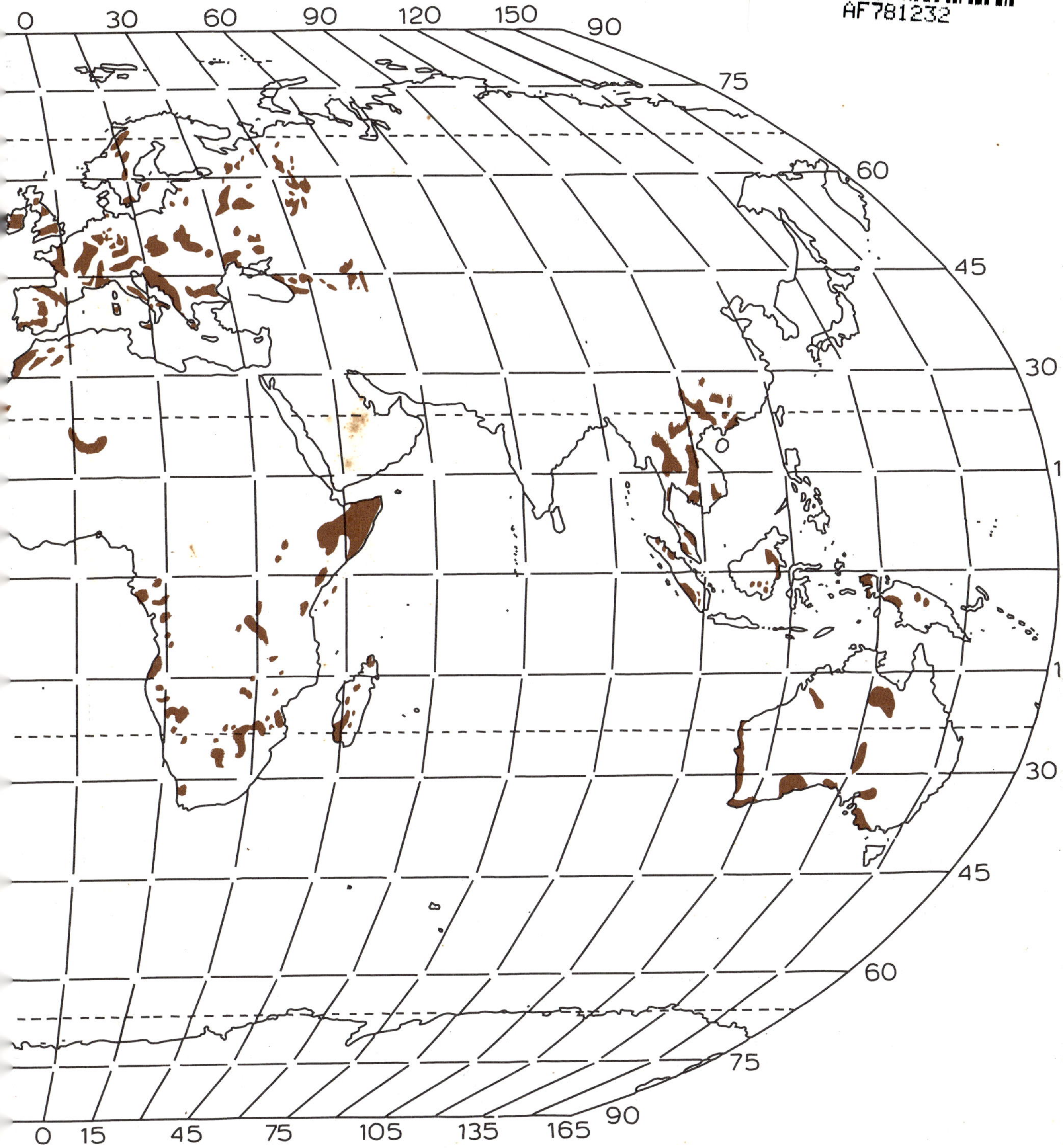

Subterranean Biota

ECOSYSTEMS OF THE WORLD 30

SUBTERRANEAN ECOSYSTEMS

ECOSYSTEMS OF THE WORLD

Editor in Chief:

David W. Goodall

Centre for Ecosystem Studies, Edith Cowan University, Joondalup, W.A. (Australia)

I. TERRESTRIAL ECOSYSTEMS

A. Natural Terrestrial Ecosystems

1. Wet Coastal Ecosystems
2. Dry Coastal Ecosystems
3. Polar and Alpine Tundra
4. Mires: Swamp, Bog, Fen and Moor
5. Temperate Deserts and Semi-Deserts
6. Coniferous Forests
7. Temperate Deciduous Forests
8. Natural Grasslands
9. Heathlands and Related Shrublands
10. Temperate Broad-Leaved Evergreen Forests
11. Mediterranean-Type Shrublands
12. Hot Deserts and Arid Shrublands
13. Tropical Savannas
14. Tropical Rain Forest Ecosystems
15. Wetland Forests
16. Ecosystems of Disturbed Ground

B. Managed Terrestrial Ecosystems

17. Managed Grasslands
18. Field Crop Ecosystems
19. Tree Crop Ecosystems
20. Greenhouse Ecosystems
21. Bioindustrial Ecosystems

II. AQUATIC ECOSYSTEMS

A. Inland Aquatic Ecosystems

22. River and Stream Ecosystems
23. Lakes and Reservoirs

B. Marine Ecosystems

24. Intertidal and Littoral Ecosystems
25. Coral Reefs
26. Estuaries and Enclosed Seas
27. Ecosystems of the Continental Shelves
28. Ecosystems of the Deep Ocean

C. Managed Aquatic Ecosystems

29. Managed Aquatic Ecosystems

III. UNDERGROUND ECOSYSTEMS

30. Subterranean Ecosystems

ECOSYSTEMS OF THE WORLD 30

SUBTERRANEAN ECOSYSTEMS

Edited by

Prof. Dr. Horst Wilkens

Zoologisches Institut und Zoologisches Museum der Universität Hamburg,
Martin-Luther-King-Platz 3,
20146 Hamburg,
Germany

Prof. David C. Culver

The American University,
4400 Massachusetts Avenue,
Washington, DC 20016-8002,
USA

Dr. William F. Humphreys

Western Australian Museum,
Francis Street, Perth,
Western Australia 6000,
Australia

ELSEVIER
Amsterdam – New York – Oxford – Shannon – Singapore – Tokyo

Elsevier Science B.V.
Sara Burgerhartstraat 25
P.O. Box 211, 1000 AE Amsterdam, The Netherlands

First edition 2000
Second impression 2001

Library of Congress Cataloging-in-Publication Data

A catalog record from the Library of Congress has been applied for.

ISBN 0 444 82299 2 (Volume)

ISBN 0 444 41702 8 (Series)

The paper used in this publication meets the requirements of ANSI/NISO Z39.48-1992 (Permanence of Paper).
Printed in the Netherlands

PREFACE

Subsurface biota extend over a very wide variety of habitats encompassing terrestrial, freshwater and marine biotopes, all of which may spatially be interconnected. From a biological point of view the subterranean ecosystem may be subdivided according to the size of the cavities colonized by animals.

Cavities large enough for people to enter are referred to as caves. Caves may have a large enough size to allow for the existence of biotopes which are also known from the lighted surface. But most are of modest proportions and are predominantly in total darkness. The subterranean fauna living here is usually characterized as troglobitic. Besides aquatic habitats like rivers, creeks, stagnant pools, and temporary puddles, various terrestrial biotopes may be found.

Such caves usually develop within soluble rock (limestone and sandstone) or are formed as lava tubes in volcanic lava. Exceptionally they are found within the gypsum cap of salt domes or in ice. Within the same substrate, the caves continue as fissures and crevices of smaller width which can only be colonized by smaller animals. This community may approach close to the surface making contact here with the so-called mesocavernous shallow stratum (MSS or superficial underground compartment), which consists of fragmented rock below the soil and which may permit the movement of subterranean fauna between adjacent limestone areas through what would otherwise be an inhospitable substrate. Submerged in seawater, the crevicular system extends along volcanic or karstic coasts, islands, and sea mounts.

The groundwater and hyporheic systems comprise additional and extensive underground habitats which are formed within unconsolidated sediments. These consist of loose gravel of varying grainsize, their open spaces filled with water ranging from freshwater to seawater. The fauna living in these interstices is aquatic and is generally called groundwater fauna or stygofauna. It has an artificial boundary line to the meiobenthic fauna, which is based on size – roughly less than 500 μm.

The principal common characteristic of true subsurface habitats is the absolute absence of light. Above all, this has two essential implications which define subsurface ecosystems. On the one hand, all structures in cave animals, the biological function of which is dependent on light as an information mediating element, get reduced. On the other, the absence of light results in a lack of photoautotrophic primary production. As a consequence, the majority of subsurface ecosystems are energetically dependent on the surface. The only exception from this second characteristic occurs where primary production is taking place in darkness, as in the case of chemoautotrophic production. Furthermore, there is a richer energy supply if the subsurface community lives in close contact to epigean biota, as for example on the roots of plants, and in this case may even directly participate in photoautotrophic primary production.

The study of subterranean ecosystems started with the scientific description of *Proteus anguinus,* the European cave salamander, by Laurent in 1768. The exploration of the cave fauna increased rapidly through the 19th century and there has been another enormous expansion of research into the subterranean realm in the second half of the twentieth century. Today we know that the hypogean habitat is the most extensive terrestrial biome on earth, being found in all regions of this planet, reaching from the polar regions to the tropics, from high altitudes to the sea and encompassing terrestrial, limnic and marine environments.

A vast amount of research has been performed on taxonomic, physiological, behavioural and ecological aspects in subsurface animals. Cave species have provided valid tools for the study of general questions of population genetics, speciation and evolution.

In this book we have brought together chapters compiled by researchers from the perspective of their own studies which exemplify different aspects of research into subsurface ecosystems. The initial chapters provide a description of the types of subsurface habitats, the characteristics of these systems (Chapters 1–3) and a description of the most characteristic taxonomic groups of troglobites (Chapters 4–6). A further set of chapters deals with the main abiotic factors being responsible for the living conditions in the dark and the resulting adaptations (Chapters 7–11). A central part of the contributions presents the different types of energetic sources hitherto discovered in subsurface habitats (Chapters 12–19). The study of evolution and speciation in subterranean forms has been the principal motivation for many studies of subterranean animals and has led to most controversy. Groundwater ecology is a rapidly growing discipline of considerable significance to the maintenance of water quality and this discipline is represented by a comprehensive characterization of the groundwater systems of the Danube and the Rhone rivers (Chapters 25–27). A series of chapters presents the contrasting characteristics of caves and cave life ranging from ice caves, through desert to the equatorial tropics (Chapters 28–32). This volume concludes with an examination of environmental issues and nature conservation in karst areas represented by consideration of Australia, Europe and North America.

Horst Wilkens
David C. Culver
William F. Humphreys
Editors

LIST OF CONTRIBUTORS

G. ALLEGRUCCI
Dipartimento di Biologia
Universitá di Roma "Tor Vergata"
Via della Ricerca Scientifica
00133 Roma
Italy

M.J. ASHMOLE
Institute of Cell, Animal and Population Biology
Edinburgh University
West Mains Road
Edinburgh EH9 3JT
Great Britain

N.P. ASHMOLE
Institute of Cell, Animal and Population Biology
Edinburgh University
West Mains Road
Edinburgh EH9 3JT
Great Britain

A. BEDOS
Laboratoire d'Ecologie Terrestre
UMR 5552 du CNRS
Université Paul Sabatier
118, route de Narbonne
31062 Toulouse Cedex 4
France

C. BOUTIN
Laboratoire de Zoologie
Université Paul Sabatier, Bat. 4R3
118 Route de Narbonne
31062 Toulouse
France

D. CESARONI
Dipartimento di Biologia
Universitá di Roma "Tor Vergata"
Via della Ricerca Scientifica
00133 Roma
Italy

L.A. CIFUENTES
Department of Oceanography
Texas A&M University
College Station, TX 77845
USA

N. COINEAU
Laboratoire Arago
Observatoire Oceanologique de Banyuls
66650 Banyuls/Mer
France

M. CREUZÉ DES CHÂTELLIERS
Université Claude Bernard, Lyon I
UMR-CNRS 5023
'Ecologie des Hydrosystèmes Fluviaux'
43 Bvd du 11 Novembre 1918
69622 Villeurbanne
France

D.C. CULVER
Department of Biology
The American University
4400 Massachusetts avenue
Washington, DC 20016-8002
USA

D.L. DANIELOPOL
Limnological Institute
Austrian Academy of Sciences
Gaisberg 116
5310 Mondsee
Austria

L. DEHARVENG
Laboratoire d'Ecologie Terrestre
UMR 5552 du CNRS
Université Paul Sabatier
118, route de Narbonne
31062 Toulouse Cedex 4
France

M.J. DOLE-OLIVIER
Université Claude Bernard, Lyon I
UMR-CNRS 5023
'Ecologie des Hydrosystèmes Fluviaux'
43 Bvd du 11 Novembre 1918
69622 Villeurbanne
France

J. DREHER
Water Resources Management
Verbundplan GmbH
Parkring 12
1011 Vienna
Austria

S. EBERHARD
Research Officer
Caveworks
Witchcliffe, Western Australia 6286
Australia

W.R. ELLIOTT
Missouri Department of Conservation,
Natural History Section
P.O. Box 180
Jefferson City, MO 65102-0180
USA

S. GALDENZI
Instituto Italiano de Speleologia
Sede Frasassi
Viale Verdi, 10
60035 Jesi, AN
Italy

M. GEIGER-KAISER
Amt der Salzburger Landesregierung
Natur- & Gewässerschutz (Referat 13/04)
Ulrich-Schreier Str. 18
5020 Salzburg
Austria

G. GENTILE
Dipartimento di Biologia
Universtà Tor Vergata
Viale della Ricerca Scientifica
00133 Roma
Italy

J. GIBERT
Université Claude Bernard, Lyon I
Laboratoire d'Ecologie des Hydrosystème Fluviaux
Equipe d'Hydrobiologie et Ecologie Souterraines
43 Bd du 11 Novembre 1918
69622 Villeurbanne Cedex
France

P. GNASPINI
Departamento de Zoologia
Universidade de São Paulo
Instituto de Biociências
Caixa Postal 11461
05422-970 São Paulo, SP
Brazil

T. GONSER
Department of Limnology
EAWAG
CH-6047 Kastanienbaum
Switzerland

A. GUNATILAKA
Water Resources Management
Berbundplan GmbH
Parkring 12
1011 Vienna
Austria

E. HAMILTON-SMITH
Chair, IUCN/WCPA Working Group on
Cave and Karst Protection
P.O. Box 36
Carlton South, Victoria 3053
Australia

M.S. HARVEY
Western Australian Museum
Francis Street
Perth, Western Australia 6000
Australia

H.H. HOBBS III
Department of Biology
Wittenberg University
P.O. Box 720
Springfield, OH 45501-0720
USA

H. HOCH
Museum für Naturkunde
Institut für Systematische Zoologie
Invalidenstr. 43
10115 Berlin
Germany

J.R. HOLSINGER
Department of Biological Sciences
Old Dominion University
Norfolk, VA 23529-0266
USA

W.F. HUMPHREYS
Museum of Natural Science,
Western Australian Museum,
Francis Street,
Perth, Western Australia 6000
Australia

K. HÜPPOP
Vogelwarte Helgoland
An der Sapskuhle
27498 Helgoland
Germany

T.C. ILIFFE
Department of Marine Biology
Texas A&M University at Galveston
Galveston, TX 77553-1675
USA

A. IPSEN
Zoologisches Institut und Zoologisches Museum der Universität Hamburg
Martin-Luther-King-Platz 3,
20146 Hamburg
Germany

E. JASINSKA
Department of Zoology
University of Western Australia
Nedlands, WA 6907
Australia

C. JUBERTHIE
Laboratoire Souterrain de CNRS
09200 Moulis
France

T.C. KANE
Department of Biological Sciences
University of Cincinnati
PO Box 210006
Cincinnati, OH 45221-0006
USA

B.K. KINKLE
Department of Biological Sciences
University of Cincinnati
PO Box 210006
Cincinnati, OH 45221-0006
USA

B. KNOTT
Department of Zoology
University of Western Australia
Nedlands, WA 6907
Australia

T.G. LANGECKER
Zoologisches Institut und Zoologisches Museum der Universität Hamburg
Martin-Luther-King-Platz 3,
20146 Hamburg
Germany

R. LAURENT
Université Claude Bernard, Lyon I
Laboratoire d'Ecologie des Hydrosystème Fluviaux
Equipe d'Hydrobiologie et Ecologie Souterraines
43 Bd du 11 Novembre 1918
69622 Villeurbanne Cedex
France

K.H. LAVOIE
Dean, Faculty of Arts and Sciences
State University of New York at Plattsburgh,
Hudson Hall 101
Plattsburgh, NY 12901-2681
USA

F. MALARD
Université Claude Bernard, Lyon I
Laboratoire d'Ecologie des Hydrosystèmes Fluviaux
Equipe d'Hydrobiologie et Ecologie Souterraines
43 Bd du 11 Novembre 1918
69622 Villeurbanne Cedex
France

P. MARMONIER
Université de Rennes I
UMR-CNRS 6553 'Ecobio'
Station Biologique de Paimpont
35380 Paimpont
France

M. MENICHETTI
Istituto di Geodinamica e Sedimentologia
Università di Urbino, Campus Universitario
61029 Urbino
Italy

F. MÖSSLACHER
EAWAG
Department of Limnology
Ueberlandstrasse 133
8600 Dübendorf
Switzerland

J. PARZEFALL
Zoologisches Institut und Zoologisches Museum der Universität Hamburg
Martin-Luther-King-Platz 3
20146 Hamburg
Germany

S. PLÉNET
Université Claude Bernard, Lyon I
UMR-CNRS 5023
'Ecologie des Hydrosystèmes Fluviaux'
43 Bvd du 11 Novembre 1918
69622 Villeurbanne
France

J.W. POHLMAN
Department of Oceanography
Texas A&M University
College Station, TX 77845
USA

P. POSPISIL
Institute of Limnology
Austrian Academy of Sciences
5310 Mondsee
Austria

T.L. POULSON
Chief Scientist of the Cave Research Foundation
The University of Illinois at Chicago
Department of Biological Sciences (M/C 066)
845 West Taylor Street
Chicago, IL 60607-7060
USA

G. RACOVITZA
Institutul de Speologie
Str. Clinicilor 5
3400 Cluj-Napoca
Romania

S.M. SARBU
Department of Biological Sciences
University of Cincinnati
821-A Rieveschl Hall
PO Box 210006
Cincinnati, OH 45221-0006
USA

V. SBORDONI
Dipartimento di Biologia
Universitá di Roma "Tor Vergata"
Via della Ricerca Scientifica
00133 Roma
Italy

W.A. SHEAR
Department of Biology
Hampden-Sydney College
Hampden-Sydney, VA 23943
USA

J.A. STANFORD
Flathead Lake Biological Station
Polson, MT 59860
USA

P. TORREITER
ACH-Siedlung 12
4950 Altheim
Austria

E. TRAJANO
Departamento de Zoologia
Universidade de São Paulo
Instituto de Biociências
Caixa Postal 11461
05422-970 São Paulo, SP
Brazil

M.J. TURQUIN
Université Claude Bernard, Lyon I
Laboratoire d'Ecologie des Hydrosystème Fluviaux
Equipe d'Hydrobiologie et Ecologie Souterraines
43 Bd du 11 Novembre 1918
69622 Villeurbanne Cedex
France

J.V. WARD
Department of Limnology
ETH/EAWAG
Überlandstrasse 133
CH-8600 Dübendorf
Switzerland

A. WEBER
Zoologisches Institut und Zoologisches Museum der Universität Hamburg
Martin-Luther-King-Platz 3,
20146 Hamburg
Germany

H. WILKENS
Zoologisches Institut und Zoologisches Museum der Universität Hamburg
Martin-Luther-King-Platz 3,
20146 Hamburg
Germany

CONTENTS

Part I

Background and Glossary

Chapter 0

BACKGROUND AND GLOSSARY

W.F. HUMPHREYS

INTRODUCTION

As with other specialized subjects, a plethora of terminology and definitions has come to surround, sometimes to obscure, the discipline of biospeleology (speleobiology, stygobiology, ...), terms sometimes so subtly different – even synonymous – as to confound even researchers within the discipline (for a fuller treatment see, for instance, Camacho, 1992).

The terminology covers six main themes:

(1) How a species came to be underground.
(2) The degree of dependence of a species on the underground realm.
(3) Whether a species shows adaptations to the underground realm.
(4) Whether a species inhabits subterranean voids filled with water or with air, and which part(s) of each it inhabits.
(5) The delimitation of the various hypogean subsystems.
(6) The amount and origin of the energy source for the ecosystem.

As a background to this volume, the main themes recurring in biospeleology are introduced below, followed by a Glossary.

The subject of caves is well introduced by Gillieson (1997), while the literature pertaining to karst hydrology, encountered especially in the context of stygofauna, is accessible in Ford and Williams (1989).

How a species came to be underground

This issue is covered from different stances in several contributions to this volume (e.g. Chapter 21 by Holsinger, Chapter 22 by Humphreys, and Chapter 31 by Deharveng and Bedos) as the topic is all-pervasive within biospeleology, and the issue will not be pursued here at length.

Troglobites were long considered to be phylogenetic or distributional relicts driven into caves by climatic change, especially in the Pleistocene refugium model. Recently, cave colonization has been viewed as an active process – the invasion of a vacant habitat. Notwithstanding, the underground realm does harbour many distributional and phyletic relicts.

Anchialine waters (inland groundwater with current marine connections) harbour a distinct (Sket, 1996) and, in some areas, a predictable fauna (Yager and Humphreys, 1996).

Stygofauna with marine ancestry are considered to colonize the shore line and to become isolated inland by marine regression following a period of marine transgression (Stock, 1977; Boutin and Coineau, 1990; Notenboom, 1991; Holsinger, 1994).

The degree of dependence of a species on the underground realm

Organisms have many pathways of interaction with subterranean spaces (Gibert et al., 1994); some are only found underground, while others move in and out of the hypogean realm. More than twenty special nomenclature schemes have been proposed to describe these relationships, based on morphological, behavioural, and ecological adaptations of animals to underground life and on their presence or absence in caves (troglofauna; see review in Camacho, 1992).

In the Schiner–Racoviţă system (Schiner, 1854; Racovitza, 1907), cave-inhabiting animals are grouped by their degree of dependence on the cave environment: *troglobites* are species which do not exist outside caves; *troglophiles* may live and reproduce underground as well as in the epigean domain; while *trogloxenes* do not normally feed underground, but may enter caves actively (regular trogloxene) or passively (accidental trogloxene). Additional criteria have subsequently entered these purely ecological definitions, so that they have become a variable mix of biological and morphological criteria (Howarth, 1973: p. 142).

The lack of detailed ecological information for most cave-inhabiting species leads to ambiguity, and this is to some extent overcome by the concept of *troglomorphy* (Christiansen, 1962), which permits classification of the cave inhabitants using a less ambiguous morphological definition (troglomorphic/non-troglomorphic species) rather than the ecological one (troglobitic/troglophilic/trogloxenic species); but this term is itself being infiltrated by behavioural and physiological attributes (Table 0.1).

Table 0.1
Characteristics of cave organisms relative to surface organisms – troglomorphies [1]

Morphological
Specialization of sensory organs (chemo-, hygro- thermo- and baro-receptors)
Elongation of appendages
Reduction of eyes, pigments and wings
Cuticle thinning (in terrestrial arthropods)
Foot modification (in Collembola and planthoppers)
Scale reduction (in fish)
Ecological and behavioural
Slowing metabolisms
Starvation resistance
Relaxation and degeneration of circadian rhythms
Lowered fecundity
Increased egg volume
Increased life span
Behavioural
Decreased aggregation (in Collembola)
Reduced reaction to alarm substances (in fish)
Increased sensitivity to vibration
Reduced intraspecific aggression

[1] From Culver et al. (1995), after Christiansen (1992).

Many species found living underground display certain characteristic traits that are thought to be adaptive to underground life. These include both the reduction or loss of characters (regressive evolution) and the enhancement of others (constructive evolution), which together produce the convergence, characteristic of cave-adapted animals, that is termed troglomorphy. These adaptations include morphological, ecological, physiological and behavioural characteristics (Table 0.1).

Whether a species shows adaptations to the underground realm

Organisms have many pathways of interaction with the subterranean realm. While some are only found underground, others move in and out of the hypogean environment as indicated in the ecological classification of species by the terms troglo(stygo)bitic, troglophilic and trogloxenic.

It is important to note that functionally these categories are more diffuse than is suggested by their ecological context. For example, while stygo(troglo)-xenes may occur underground only accidentally, they may influence happenings in the hypogean realm, for example, functioning either as predators or prey (e.g. Gibert et al. 1994: p. 11), as well as constituting a subset of the allochthonous inputs from epigean sources.

In considering troglomorphic correlates (e.g. reduced eyes, pigment, wings etc.) it is necessary to distinguish whether the traits are different from close epigean ancestors (*apomorphic*) or simply characteristics of the group or lineage (*plesiomorphic*).

Whether a species inhabits subterranean voids filled with water or with air, and which part(s) of each it inhabits

While the Schiner–Racoviţă system is often used for both terrestrial and aquatic species, Gibert et al. (1994: p. 13) have proposed stygo- (*sensu lato*) as the most ecologically descriptive prefix for groundwater fauna. Hence, *troglofauna* becomes restricted to terrestrial systems (see review in Camacho, 1992), while the aquatic fauna is referred to generally as *stygofauna* (Ginet and Decou, 1977; Botosaneanu, 1986; Gibert et al., 1994). There are complex parallel classifications in use for aquatic subterranean organisms (e.g. Botosaneanu, 1986: endpaper; Williams, 1984; Stanford and Ward, 1993).

Gibert et al. (1994) followed the main Schiner–Racoviţă system, replacing the prefix "troglo-" by "stygo-" (hence, stygoxene, stygophile, stygobite). In porous aquifers, stygophiles are subdivided into three categories: *occasional hyporheos,* which comprises essentially the larvae, especially of aquatic insects, of most of the benthos; *amphibites,* whose life cycle requires the use of both surface and groundwater systems; and *permanent hyporheos,* which comprises a diverse assemblage of organisms present in all life stages in the groundwater or in the benthic habitats – an aerial epigean stage is not required to complete the life cycle.

Stygobites are subdivided into two classes of obligatory hypogean forms: *ubiquitous stygobites* are widely present in both karstic and alluvial groundwaters and are sometimes found very close to the surface. In contrast, *phreatobite* species are stygobites that are restricted to the deep groundwater substrata of alluvial aquifers (Gibert et al., 1994).

Delimitation of the various hypogean sub-systems

The hypogean realm used to be defined in terms of caves and the degree of darkness. The realization that most caves had no openings to the surface (Curl, 1966), and that troglobitic animals may inhabit the immediate subsurface spaces even in non-karstic areas (Uéno, 1977; Juberthie et al., 1980; Juberthie and Delay, 1981), means that the limits to the hypogean realm are difficult to define. At the surface, epigean and edaphic environments may grade imperceptibly into the hypogean realm, while at the lower end of the hypogean realm the metazoans progressively fade away, leaving a microbiological world that may penetrate deep into the Earth's crust (Gold, 1992).

This complexity is apparent also at a finer scale. For example, the hyporheic zone is an ecotone between surface water and groundwater, but this ecotone is difficult to define. It has been variously interpreted in both physical and biological terms: physically by the depth to which the river water penetrates into the hyporheic zone, and biologically by the depth to which amphibites penetrate into it. In reality, it is a region of great complexity, the origin of the water being from either groundwater (upwelling zone) or surface water (downwelling zone). It is best considered as a dynamic ecotone or transition zone (see Gibert et al., 1994).

In large caves the highly troglomorphic species are generally considered to be restricted to the deep cave environments where the temperature is stable and the air is saturated with water vapour, and frequently contains elevated levels of carbon dioxide. Nonetheless, in some tropical areas (Cape Range, Australia), highly troglomorphic (and stygomorphic) species are found, under certain conditions, to be feeding in sunlight at the entrance to caves. However, even within caves, the boundaries between zones are dynamic, shifting, *inter alia,* according to the degree of biogenic carbon dioxide production (dependent on periodicity of water and energy inputs), the weather, the season and major climatic events. The extent of the zones and their vagility varies between caves according to the relative sizes of entrances and volume, and the size, shape and arrangement of the chambers (see Racovita, Chapter 28, this volume).

The amount and origin of the energy source for the ecosystem

In the absence of light subterranean ecosystems have been considered to be dependent on organic matter fixed by photosynthesis and reaching the hypogean realms by various direct and indirect pathways. Routes of entry are as diverse as logs being washed into open caves, dissolved organic matter percolating downwards through a porous medium into caves otherwise closed to the surface, and animals such as bats and crickets moving between the cave and the exterior.

Diverse subterranean ecosystems dependent upon the chemoautotrophic bacteria that rely on sulphides have recently been demonstrated (see Chapter 17 by Sarbu, Chapter 18 by Pohlman et al., and Chapter 19 by Sarbu et al.). Sulphides may be of magmatic or petroleum origin (see Chapter 12 by Poulson and Lavoie) and these caves are thus analogous to the deep-sea environments of hydrothermal vents (Childress and Fisher, 1992) and cold seeps (Arp and Fisher, 1995; Scott and Fisher, 1995), respectively.

Underground environments are typically considered to be sparsely provided with food energy – to have a low energy density – and it is hypothesized that this is a major selective pressure involved in the loss of characters, such as eyes, in troglobites. This low energy density may be evenly distributed and so difficult to harvest, or it may occur as sparsely distributed bundles of high energy and thus be difficult to find. A few systems have profuse and predictable supplies of energy usually transported into the system by animals (bats, birds and crickets) or flowing water (some chemoautotrophic systems).

TERMINOLOGY

As indicated above, biospeleology has developed a range of special terms to cover its subject matter. In addition, the terminology of speleology is frequently encountered in biospeleological literature. This vocabulary is covered well in Lowe and Waltham (1995), and the equivalent terms in English, German, French and Hungarian are listed by Attila (1996).

Specialized terms are listed and explained in the Glossary below, together with a range of terms in related disciplines which may be unfamiliar to the reader. Additional terms can be found in Lincoln et al. (1982).

GLOSSARY

Aesthetasc. Sensory or olfactory bristles on the antenna or antennules of many arthropods.

Allochthonous. Originating outside the habitat under discussion.

Allopatric. Used of species or populations occupying a geographical area different from that of another species or population.

Amensalism. An interspecific interaction in which one species population is inhibited, typically by toxin produced by the other, which is unaffected.

Amphibite. Particular type of stygophile species whose life cycle requires the use of both surface water and groundwater systems; also *amphibionts.*

Anchialine (or anchihaline). Anchialine habitats consist of bodies of haline waters, usually with a restricted exposure to open air, always with more or less extensive subterranean connections to the sea, and showing noticeable marine as well as terrestrial influences (Stock et al., 1986). They typically occur in volcanic or limestone bedrock (Iliffe, Chapter 3, this volume).

Apomorphic. Pertaining to a derived character within the taxa being considered, in contrast to the ancestral (plesiomorphic) state.

Aquiclude. Body of relatively impermeable rock or sediment acting as a boundary to an aquifer.

Autapomorphy. The possession of a unique derived character by a species or monophyletic taxon; *cf.* Synapomorphy *q.v.*

Aven. Hole in a cave roof, which may be a blind pocket or a tributary inlet shaft, and known as dome-pits in the United States; locally in France the term for pothole.

Biofilm. Accumulation of bacteria attached to solid surfaces by means of appendages and exopolymers, in which condition cross-feeding is favoured and they are protected from dislodgment, predation and adverse physico-chemical conditions (Gounot, 1994).

Blue hole. (1) Non-turbulent resurgence (Jamaica); (2) drowned sinkholes on the Bahamas Banks from which cooler water rises during high tide (Bahamas); (3) generally, shaft-like depressions deeply flooded by sea water, typically in low-lying coral islands.

Boundstone. Carbonate rock whose components were bound together during deposition and remain in position of growth, such as many reef rocks.

Calcarenite. Wackstones, *q.v.* (Gary, McAfee and Wolf, 1972).

Calcrete. Carbonate deposits that form in the soil or in the vicinity of the groundwater table as a result of the evaporation of soilwater or groundwater respectively. Groundwater calcretes form in arid climates (annual rainfall less than 200 mm) with high potential evaporation (more than 3000 mm per year: Mann and Horwitz, 1979).

Cave (*sensu lato*). Underground void, generally of a size sufficient for people to enter. However, the word means different things to different people. Cavers treat as caves natural holes large enough to enter – sometimes the term pothole is applied to vertical developments – or which extend to darkness. Archaeologists often use the term loosely to include rock-shelters. In geomorphological terms, a cave is an opening of diameter or width greater than 5–15 mm – that is, a void of sufficient dimensions to allow turbulent rather than laminar flow of water (Ford and Williams, 1989). These differences are well reconciled by Howarth (1983: p. 370) "In both soluble and volcanic rocks, therefore, a complex of interconnected voids of varying sizes anastomoses throughout the rock in a great labyrinthine system. Within this system there is a continuum of various sized voids from the microscopic to the largest caverns. The existence of these voids, their size, depth, and extent, depends on the geological history of the area. From a biological perspective, this continuum can be divided into three size classes: microcavernous (<0.1 cm), mesocavernous (0.1–20 cm) and macrocavernous (>20 cm)".

Cave lint. The accumulation on cave formations of

detritus (skin, fibres, hair, etc.) shed by human visitors to caves; mainly a problem in busy tourist caves.

Cavernicole. Pertaining to an inhabitant of caves *sensu lato,* without implying any particular dependence on, or association with the cave environment.

Cenote. Flooded collapse depressions on flat limestone plain – originally from Yucatán, Mexico.

Chaetotaxic. Pertaining to the arrangement or pattern of bristles.

Chemoautotrophs. Organisms deriving nourishment from chemical reactions of inorganic substances, as in sulphur and iron bacteria.

Chemoheterotrophs. Organisms that obtain energy from oxidation/reduction reactions and use organic electron donors; cf. chemoorganotrophs.

Chemolithoautotrophs. Organisms that obtain energy from oxidation/reduction reactions and use inorganic electron donors.

Chemotrophs. = Chemoautotrophs.

Cleitric. The fissure system in the rock matrix in contradistinction to the cave (speleic) system.

Cockpit karst. Synonym of Cone karst, referring to the depressions.

Compartiment souterrain superficiel. = Superficial underground compartment, *q.v.*

Cone karst. Karst type of tropical regions characterized by an extent of steep, conical hills and adjacent star-shaped depressions, or cockpits having steep, convex sides and slightly concave floors.

CPOM. Coarse particulate organic matter [for example, dead leaves and twigs; *cf.* FPOM, *q.v.*].

Craton. Part of the Earth's crust little deformed for a prolonged period (shield region).

Crenobionts. Inhabitants of springs and spring brooks.

Crevicular. Pertaining to the finer cracks, crevices and spaces within rock matrix.

Cryptokarst. Karst secondarily covered by sediments.

Cyclothemic. In Jurassic limestone of central Italy, pertaining to a cyclic sedimentation controlled by climatic events (Cycle of Milankovitch: Milankovitch, 1920).

Dark zone. Region of cave in perpetual total darkness.

Decoration. Speleothems, *q.v.*

Decomposer. Any organism that feeds by degrading organic matter.

Deep (zone) cave environment. Region of cave with constant temperature and near-saturated humidity; the most troglomorphic species are often restricted to this region.

Distrophic. Pertaining to caves mostly supplied with vegetal matter during floods.

Doline. Simple closed circular depression in karst area with subterranean drainage and commonly funnel-shaped, varying from a few to many metres in dimensions.

DOM. Dissolved organic matter, typically $<1\ \mathrm{mg}\,\ell^{-1}$ in pristine groundwater.

Dripstone. = Flowstone, *q.v.*

Dysaerobic. Environments in which the oxygen concentration is between 0.1 and $1.0\ \mathrm{ml}\,\ell^{-1}$.

Edaphobite. Deep soil inhabitant.

Edaphon. The biota living within the soil or substrate – *cf.* Psephon, Edaphobite, *q.v.*

Embryonic genesis (embryogenesis). Embryonic development from an ovum.

Endobenthos. Organisms living within the substratum, as, for example, burrowing species.

Endogenous zone. The zone immediately beneath the Earth's surface.

Endopsammic. Pertaining to inhabitants of water-saturated sand.

Entrance zone of cave. Area of cave where outside influences dominate both the physical and biological environment.

Epibenthic. Living on the surface of the sea bed or lake floor.

Epigean. The surface environment as opposed to the subsurface (subterranean, hypogean) environment.

Epikarst. Upper part of the percolation zone.

Epiphreatic. Inhabiting the superficial part of the Phreatic zone, *q.v*; the vadose–phreatic boundary.

Euryece. Inhabiting a wide range of habitats.

Eurytopic. Having a wide geographical distribution.

Eutrophic. Pertaining to a large amount of available organic matter, high energy density or nutrient enrichment; in caves especially that provided by bat guano or flood debris.

Exaptation. Adaptation for one function serving for another function.

Exokarst. All features that may be found on the surface of a karst landscape, ranging in size from tiny *karren* to massive *poljes.*

Exploitative competition. An interaction between

organisms in which one benefits at the expense of the other.

Flowstone. Deposit formed from trickles or films of water on walls and floors of caves, normally of calcite.

Fluviokarst. A karst landscape where the dominant landforms are valleys cut by surface rivers – the valleys become dry as karst development improves underground drainage.

Fossil cave. Cave no longer containing a stream or active speleothems.

Foul-air zone. Still-air zone of a cave, usually in a depression or pot, with high humidity and elevated carbon dioxide levels, which may result from biogenic or physical effects (see, for instance, Howarth and Stone, 1990).

FPOM. Fine particulate organic matter resulting from abrasion and mechanical disruption by biological and physical agents; *cf.* CPOM.

Gelifraction. Mechanical disintegration of rock or soil caused by the freezing of contained water (also congelifraction).

Ghyben–Herzberg principle. Expression of the relationship which occurs in near-coastal porous aquifers where fresh water overlies sea water; the depth below sea level of the fresh water/salt water interface is ~40 times the height of the water table above sea level; thus a minor drawdown of the groundwater can lead to a major intrusion of sea water into an aquifer by upward coning of the salt water in response to a thinner freshwater layer. Hence, Ghyben–Herzberg system or lens. (See Chow, 1964.).

Grainstones. Mud-free carbonate sedimentary rock.

Grieta. Crevice, fissure, crack or cleft in Spanish.

Guanobite. Species that, when in caves, inhabit guano deposits (from bats, swiftlets, etc.), and whose entire life cycle takes place in this substrate (see Gnaspini and Trajano, Chapter 13, this volume).

Guanophages. Species that feed directly on guano, or on microorganisms and fungi that grow on it (see Gnaspini and Trajano, Chapter 13, this volume).

Guanophile. Species that may inhabit and reproduce both in guano piles and in other substrates in the cave environment (see Gnaspini and Trajano, Chapter 13, this volume).

Guanoxene. A species feeding and/or reproducing in guano deposits but depending on other substrates in the cave to complete its life cycle (see Gnaspini and Trajano, Chapter 13, this volume).

Haematophagy. Feeding on blood.

Halocline. More or less abrupt change in salinity with depth in water column.

Heterochrony. Change in timing or rate of developmental events, relative to the same event in the ancestor.

Hom[o]eotic mutation. Mutation from one state to another of a structure that exist in a series of states, such as the mutation of an insect wing into a haltere.

Homoplasy. Structural resemblance due to parallelism or convergent evolution rather than to common ancestry.

Hypogean. The subsurface or subterranean environment as opposed to the surface (epigean) environment.

Hyporheos [hyporheic zone]. Interstitial spaces within the sediments of a stream bed; a transition zone between surface water and groundwater (Orghidan, 1959).

Impounded karst. Karst system in which the underground drainage is dammed, typically by another lithology, sometimes resulting in dammed springs.

Interstitial. Living in the spaces between particles, especially in alluvia.

Iridophore. Colour cells containing guanine in, for example, fishes, reptiles, spiders and butterflies; iridocytes.

Isopiestic. Regions of an aquifer where the pressure head of ground water is the same.

Karren. Synonym of Lapiaz.

Karst. Soluble-rock landscape; terrain with distinctive hydrology and landforms arising from a combination of high rock solubility and well-developed secondary porosity. The distinctive landforms above and below ground that are the hallmark of karst result from the solution of rock (mainly by carbonic acid) along pathways provided by the structure. The unusual features of the Kras (Karst in the period of the Austro–Hungarian empire) region on the Italo–Slovenian border became known as "karst phenomena". Such areas are characterized by sinking streams, caves, enclosed depressions, fluted rock outcrops and large springs (Ford and Williams, 1989).

Karstification. Production of karstic features on the landscape.

Karst window. A section of groundwater exposed by roof collapse or dissolution, especially in karstlands, as, for example, an anchialine pool.

Kipuka. Isolated protrusion of bedrock (in Hawaii) in lava flow, such as the summit of a hill; synonym: steptoe.

Lapiaz. In karst terrain, a general term for solutional furrows in a limestone surface, with depths between a few millimetres and more than a metre, usually separated by knifelike ridges; synonym of karren. (Pyrénées dialect, also *lapiés.*).

Lava tubes. Tubular caves within lava flows, subject to mechanical collapse of the roof (vulcanokarst).

Lavicole. Pertaining to an inhabitant of lava beds.

Lentic. Standing water as in swamp, lake, pond.

Limnicoid. Inhabiting lakes; also, limnicole.

Limno- (Prefix). Restricted to fresh water.

Limnostygobionts. Stygobionts restricted to fresh waters.

Lotic. Running water, as in stream or river.

Lucifuge. Intolerant of light.

Macrocavernous. Underground voids >20 cm, especially caves and underground passages in karst and volcanic substrates.

Mafic. Said of igneous rock composed of one or more ferromagnesian minerals.

Manna. Honeydew secreted by some scale insects (Coccidae).

MAST (or M.A.S.T.). The mean annual surface temperature of a region – the expected temperature of the deep cave environment in the absence of significant geothermal heat or internal climatic effects.

Meiobenthic. Small benthic (*q.v.*) organisms that pass through a 500 μm sieve but are retained by a 40 μm sieve; *cf.* meiofauna, *q.v.*

Meiofauna. Assemblage of animals that pass through a 500 μm sieve but are retained by a 40 μm sieve, often interstitial; prefix meio-, hence meiobenthic.

Mesocaverns. Underground voids in the size range 0.1–20 cm, especially in karst and volcanic substrates.

Mesopsammon. Psammon *q.v.*

Mesovoid shallow stratum. See superficial underground compartment.

Mesotrophic. Pertaining to an intermediate amount of available organic matter, medium energy density.

Messinian. Part of Miocene epoch ~6.5–5.5 million years ago. The Mediterranean area was landlocked and almost completely dried out at this time, leaving a series of hypersaline lakes which had profound implications for the aquatic fauna, hence the Messinian Salinity Crisis.

Microvoid. See mesocaverns.

Milieu souterrain superficiel. = Superficial underground compartment.

Mixohaline. Water with a salinity between 5 and 30‰ under the Venice System (equivalent to 5000 and 30 000 mg ℓ^{-1} total dissolved solids).

Mixture corrosion. The ability of two calcite-saturated water masses of different carbon dioxide content to dissolve additional calcite when mixed.

MSS. = Superficial underground compartment.

Nappes phreatiques. Pertaining to sheet flow in deep groundwater.

Nectarivory. Feeding on nectar.

Neogeoaeolian. Pertaining to an ecosystem within a newly formed landscape (lava flows, sand cays) supported by windborne allochthonous material (Howarth, 1979).

Neoteny. Paedomorphosis (*q.v.*) produced by retardation of somatic development, so that sexual maturity is attained in an organism retaining juvenile characters; *cf.* Progenesis.

Neuromast. Sensory receptor, part of the acoustico-lateralis system of aquatic chordates, to detect vibration and movement in water.

Neutral mutation. A genetic mutation that proffers no advantage or disadvantage to the organism.

Nomocoenosis. Pasture community.

Normoxic. Pertaining to habitats with the normal level of oxygen found in the atmosphere or dissolved in water in equilibrium with it.

Nycthemeral. Occurring at night.

Oligotrophic. Pertaining to low amount of available organic matter, low energy density.

Outgroup. A distinct but closely related species or lineage used as an anchor in phylogenetic and cladistic analyses.

Packstones. Sedimentary carbonate rock whose granular material is self-supporting yet contains calcareous mud.

Paedogenesis. Paedomorphosis (*q.v*) produced by precocious sexual maturity in an organism that is still at a morphologically juvenile stage; *cf*. Neoteny, Progenesis *q.v.*

Paedomorphosis. Retention of sub-adult ancestral traits in the descendant adult.

Palaeokarst. A rock or area that has been karstified and subsequently covered by sediments.

Paludicole. Pertaining to an inhabitant of marshes.

Parafluvial. Pertaining to groundwater adjacent to the river channel.

Parapatric. Pertaining to populations whose geographic ranges are contiguous but not overlapping, so that gene flow between them is possible.

Peripatric. Peripheral to the main distribution, as used in 'speciation in peripatric populations'.

Philopatric. In the context of bats, loyalty to caves (maternity or hibernal) or other roosts (Tuttle, 1976). It is applied to bats which return to the same caves on a daily and/or on a seasonal basis; thus, the bat colonies tend to be stable.

Photoautotroph. Organisms that obtain metabolic energy from light by a photochemical process; = Phototroph, *q.v.*

Photoheterotroph. Organism that utilizes radiant energy and organic electron donors (synonym: photoorganotroph).

Photo-negativity. The avoidance of light.

Phototroph. = photoautotroph, *q.v.*

Phreatic. That part of a cave system lying below the local groundwater table; below the unsaturated or vadose zone, *q.v.*

Phreatic loop. Water-filled cave tube rejoining itself.

Phreatobiological net. Net designed to sample deep groundwater, usually through bores and wells.

Phreatobite. Stygobites that are restricted to the deep groundwater substrata of alluvial aquifer (phreatic water) (Gibert et al., 1994).

Phreatomorphic. Pertaining to the convergent morphological, behavioural and physiological features of phreatobites – Stygomorphy, Troglomorphy, *q.v.*

Phreatos. Assemblage of organisms inhabiting deep groundwater (phreatic zone).

Phytophilic. Plant-eating = phytophagous, herbivorous.

Pleiotropic. Pertaining to a gene that has more than one phenotypic effect.

Pleistocene refugium model. Traditional model for isolation of species in caves including both dispersalist and vicariant elements. Ancestral populations are isolated in caves by climatic change without active invasion. The genetic bottleneck resulting from small founder population is followed by reorganization of epigenotype. Subsequent subterranean dispersal expands the geographic range as adaptation improves.

Plesiomorphic. Pertaining to an ancestral character within the taxa being considered, in contrast to the derived (apomorphic) state.

Poecilotrophic. Pertaining to variable amounts of organic matter, as in long caves with sections differing in energy density, ranging from oligotrophic to eutrophic.

Polje. Large closed depression – with a flat floor that makes a sharp break with surrounding slopes – draining underground, often containing a stream and liable to flood, becoming a lake.

Pre-adaptation. Possession by an organism of the necessary properties to permit a shift into a new niche or habitat. A structure is preadapted if it can assume a new function before it becomes modified itself.

Progenesis. Early cessation of developmental events in the descendant; produces paedomorphic traits when expressed in the adult phenotype; early sexual maturation will produce global progenesis, but early cessation in local growth fields can also produce progenesis (McNamara, 1990).

Provincialization. Progressive formation of morphologically or genetically distinct subpopulations in a region having previously a more homogeneous population.

Psammolittoral. Pertaining to sands along lake and sea shores.

Psammon. Organisms living between sand grains of freshwater or marine shores; Mesopsammon.

Psephon. Interstitial water dwellers (equivalent to Edaphon; unusual).

Pseudokarst. Karst-like landforms produced by processes other than solution or corrosion-induced subsidence and collapse; e.g. ice caves (Ford and Williams, 1989).

Pycnocline. More or less abrupt change in water density with depth, typically at the halocline or thermocline in the water column.

Regressive evolution. The loss of morphological and behavioural characters that accompanies isolation in caves.

Refuge (refugium, refugia). A region in which certain types or suites of organisms are able to persist during a period in which most of their original geographic range becomes uninhabitable because of climatic change (Morton et al., 1995).

Relics (*reliques* in French). The last survivors of an ancient radiation (Newman, 1991).

Relicts (*relictes* in French). Population of organisms separated from a parent population by some vicariant event.

Resurgence. Spring where a stream, which has a course on the surface higher up, reappears at the surface.

Riffle. Natural shallows extending across a stream bed, over which the water flows swiftly and is broken into waves by obstructions wholly or partly submerged.

Sapropel. Organisms inhabiting mud rich in decaying organic matter.

Schiner–Racoviţă system. Classification of hypogean animals on ecological grounds into troglobites, troglophiles and trogloxenes (Schiner, 1854; Racovitza, 1907).

Shaft. Passage in cave that is more or less vertical; *cf.* Vertical cave.

Sinkhole. = Doline.

Sinistral. On the left – pertaining to the direction of coiling of snail shells.

Solfatara. A type of volcanic vent (fumarole), the gases of which are characteristically sulphurous.

***Sondage* method.** Sampling by pulling a net through the water column, especially down wells, preferably using a phreatobiological net.

Spel[a]ean. Pertaining to caves.

Speleic. Pertaining to caves.

Speleifer. A cave-tolerant organism.

Speleogenesis. The origin and development of caves.

Speleothem. Secondary deposition in caves (commonly of calcite or aragonite) forming the cave decoration, including stalagmites, stalactites, flowstone, draperies, straws, helictites etc. Also known as sinter or travertine.

Stagnant-air zone. = Foul-air zone.

Stranding. The isolation inland of stygofauna owing to marine regression, especially following a period of marine transgression.

Stygal. See Stygo-.

Stygo-. Prefix referring to groundwater habitats (Gibert et al., 1994: p. 13).

Stygobiont. Animal inhabiting the various types of groundwater.

Stygobite. Species that are specialized subterranean forms, obligatorily hypogean. Some are widely distributed in all kinds of groundwater systems (both karst and alluvia), and sometimes they are found very close to the surface.

Stygobitization. = Stygogenesis.

Stygofauna. Fauna inhabiting the various types of groundwater. Numerous names are in use for subsets of this system mostly based on the habitat type (see Ginet and Decou, 1977; Botosaneanu, 1986, especially the endpaper; Camacho, 1992; Gibert et al., 1994; see also Chapter 2, this volume).

Stygogenesis. The morphological, behavioural, physiological and other changes found in species or lineages as they populate groundwater or adapt to groundwater life.

Stygophile. An organism that seeks out and exploits resources in the groundwater system (Stanford and Ward, 1993). In porous aquifers, stygophiles are subdivided into three categories: (1) the occasional hyporheos, (2) the amphibites, and (3) the permanent hyporheos. For more information see Gibert et al. (1994: p. 13).

Stygomorphic. Pertaining to morphological, behavioural and physiological characters that are convergent in subterranean aquatic populations.

Stygoxenes. Organisms that have no affinities with the groundwater systems, but occur accidentally in caves or alluvial sediments (Stanford and Ward, 1993). Nevertheless, stygoxenes can influence processes in the groundwater ecosystems – for example, by functioning either as predators or prey.

Subcutaneous zone. A zone of generally highly weathered rock that lies below the soil but above the main, relatively unweathered, rock mass of a karst aquifer.

Superficial underground compartment. The crevicular spaces deep in the soil, as found in glacially fragmented zones. By origin, the *milieu souterrain superficial* (MSS; Juberthie et al., 1980), and also the mesovoid shallow stratum, hence retaining the acronym "MSS".

Swallet. Hole into which a stream flows.

Synecomorphosis. Of the same ecological growth form.

Sympatric. Used of two or more species or populations occupying the same geographical locality.

Synapomorphy. A derived (apomorphic) state shared by two or more taxa.

Syngenetic karst development. Consolidation of calcareous dune deposits occurring concurrently with karst formation (Jennings, 1968).

Syntopically. Pertaining to populations or species that occupy the same microhabitat, are observable in close proximity, and could thus interbreed.

Tabular karst. Broad expanse of more or less horizontal karst.

Thalasso-. (Prefix) Pertaining to marine or brackish waters, hence thalassostygobionts, thalassopsammon and thalassopsephon.

Thalassoid. Having marine affinities.

Thermocline. More or less abrupt change in temperature with depth in air or water column.

Thigmotactism. The tendency of many small organisms to seek maximum surface contact, as in corners and crevices.

Tower karst. Residual karst hills with near-vertical to overhanging lower slopes, often separated by areas of alluvial deposits.

Transition zone of cave. A dynamic zone in total darkness where the microclimate is still conspicuously affected by short-term meteorological events at the surface.

Troglo-. (Prefix) For subterranean terrestrial and aquatic systems, now more commonly restricted to terrestrial systems, being replaced by "stygo-" for aquatic systems.

Troglobiont. Animal inhabiting the various types of subterranean spaces; sometimes restricted to air-filled voids; and hence terrestrial as opposed to aquatic species.

Troglobiontic stygobionts. Obligate inhabitants of underground waters found in caves.

Troglobite. Species which do not exist outside caves (Schiner, 1854); they may, however, occur in the superficial underground compartment or in the upper hypogean zone.

Troglobitization. = Troglogenesis.

Troglogenesis. The morphological, behavioural, physiological and other changes found in species/lineages as they populate/adapt to cave or hypogean life.

Troglomorphic. Pertaining to morphological, behavioural and physiological characters that are convergent in subterranean populations (Christiansen, 1962).

Troglophile. Species able to live and reproduce underground as well as in the epigean domain.

Troglos. Assemblage of organisms inhabiting underground voids, often restricted to the 'terrestrial' organisms in air-filled voids.

Trogloxene. Species that do not normally feed underground, but may enter caves actively (regular trogloxene) or passively (accidental trogloxene) (Racovitza, 1907).

Tropical winter effect. The drying of the air in tropical caves in winter when the cave temperature and the partial pressure of water vapour inside the cave exceed those outside the cave, resulting in the loss of water from the cave air (Howarth, 1980).

***Trou* technique.** Sampling method for fauna sediments involving filtering the water that has percolated into a hole dug in stream bank; synonym: Karaman–Chappuis method.

Twilight zone of cave. Region of cave extending from the limit of green vascular plants to total darkness; physical climate not stable.

Upper hypogean zone. Region immediately above the bedrock where the high porosity provides suitable hypogean habitat for troglobites (Uéno, 1987); = Superficial underground compartment, *q.v.*

Vadose zone. The zone of rock above the water table, mostly free-draining; also called the unsaturated zone, and lying above the saturated or phreatic zone, *q.v.*

Vauclusian. A spring rising from a vertical or very steep flooded bedrock cave passage directly draining the phreatic zone and emerging in daylight; named after the type example, La Fontaine de Vaucluse in southern France.

Vicariance. Speciation as a result of range disruption, typically the result of some non-biological process; also termed passive allopatric speciation.

Vicariant event. The division of the biota or taxon through the development of a natural barrier.

Wackstones. Mud-supported carbonate sedimentary rocks containing >10% grains (particles >20 μm in diameter), e.g. calcarenite.

REFERENCES

Arp, A.J. and Fisher, C.R., 1995. Introduction to the symposium: Life with sulfide. *Am. Zool.*, 35: 81–82.

Attila, K., 1996. *The Cavers' Living Dictionary.* Hungarian Speleological Society, Budapest.

Botosaneanu, L. (Editor), 1986. *Stygofauna Mundi: A Faunistic, Distributional, and Ecological Synthesis of the World Fauna inhabiting Subterranean Waters (including the marine interstitial).* E.J. Brill, Leiden, 740 pp + endpapers.

Boutin, C. and Coineau, N., 1990. "Regression Model", "Modèle Biphase" d'évolution et origine des micro-organismes stygobies

interstitiels continentaux. *Revue de Micropaléontologie,* 33: 303–322.
Camacho, A.I., 1992. A classification of the aquatic and terrestrial subterranean environment and their associated fauna. In: A.I. Camacho (Editor), *The Natural History of Biospeleology.* Monografias Museo Nacional de Ciencias Naturales, Madrid, pp. 57–103.
Childress, J.J. and Fisher, C.R., 1992. The biology of hydrothermal vent animals: physiology, biochemistry and autotrophic symbioses. *Oceanogr. Mar. Biol. Ann. Rev.,* 30: 337–442.
Chow, Van Te (Editor), 1964. *Handbook of Applied Hydrology: A Compendium of Water-resources Technology.* McGraw-Hill, New York, 1418 pp.
Christiansen, K., 1962. Proposition for the classification of cave animals. *Spelunca,* 2: 76–78.
Christiansen, K., 1992. Biological process in space and time. In: A.I. Camacho (Editor), *The Natural History of Biospeleology.* Monografias Museo Nacional de Ciencias Naturales, Madrid, pp. 453–478.
Culver, D.C., Kane, T.C. and Fong, D.W., 1995. *Adaptation and Natural Selection in Caves: the Evolution of Gammarus minus.* Harvard University Press, Cambridge, MA, 223 pp.
Curl, R.L., 1966. Caves as a measure of karst. *Journal of Geology,* 74: 798–830.
Ford, D.C. and Williams, P.W., 1989. *Karst Geomorphology and Hydrology.* Unwin Hyman, London, 601 pp.
Gary, M., McAfee, R. and Wolf, C.L. (Editors), 1972. *Glossary of Geology.* American Geological Institute, Washington, DC, 857 pp.
Gibert, J., Stanford, J.A., Dole-Olivier, M.-J. and Ward, J.V., 1994. Basic attributes of groundwater ecosystems and prospects for research. In: J. Gibert, D.L. Danielopol and J.A. Stanford (Editors), *Groundwater Ecology.* Academic Press, London, pp. 7–40.
Gillieson, D., 1997. *Caves – Processes, Development and Management.* Blackwell, Oxford, 324 pp.
Ginet, R. and Decou, V., 1977. *Initiation à la Biologie et à l'Écologie Souterraines.* J.P. Delarge, Paris, 354 pp.
Gold, T., 1992. The deep, hot biosphere. *Proc. Natl. Acad. Sci. U.S.A.,* 89(13): 6045–6049.
Gounot, A.M., 1994. Microbial ecology of groundwaters. In: J. Gibert, D.L. Danielopol and J.A. Stanford (Editors), *Groundwater Ecology.* Academic Press, London, pp. 189–215.
Holsinger, J.R., 1994. Pattern and process in the biogeography of subterranean amphipods. *Hydrobiologia,* 287: 131–145.
Howarth, F.G., 1973. The cavernicolous fauna of Hawaiian lava tubes, 1. Introduction. *Pac. Insects,* 15: 139–151.
Howarth, F.G., 1979. Neogeoaeolian habitats on new lava flows on Hawaii Island: An ecosystem supported by windborne debris. *Pac. Insects,* 20:133–144.
Howarth, F.G., 1980. The zoogeography of specialized cave animals: a bioclimatic model. *Evolution,* 34:394–406.
Howarth, F.G., 1983. Ecology of cave arthropods. *Ann. Rev. Entomol.,* 28: 365–389.
Howarth, F.G. and Stone, F.D., 1990. Elevated carbon dioxide levels in Bayliss Cave, Australia: implications for the evolution of obligate cave species. *Pac. Sci.,* 44: 207–218.
Jennings, J.N., 1968. Syngenetic karst in Australia. In: P.W. Williams and J.N. Jennings (Editors), *Contributions to the Study of Karst.* Department of Geography Publication G/5, Australian National University, Canberra, pp. 41–110.
Juberthie, C. and Delay, B., 1981. Ecological and biological implications of the existence of a Superficial Underground Compartment. *Proc. 8th International Congress of Speleology, Bowling Green, Kentucky,* 6: 203–206.
Juberthie, C., Delay, B. and Bouillon, M., 1980. Extension du milieu souterrain en zone non calcaire: description d'un nouveau milieu et de son peuplement par les Coléoptères troglobies. *Mém. Biospéol.,* 7: 19–52.
Lincoln, R.J., Boxshall, G.A. and Clark, P.F., 1982. *A Dictionary of Ecology, Evolution and Systematics.* Cambridge University Press, London, 298 pp.
Lowe, D. and Waltham, T., 1995. *A Dictionary of Karst and Caves.* Cave Studies Series No. 6, British Cave Research Association, London, 40 pp.
Mann, A.W. and Horwitz, R.C., 1979. Groundwater calcrete deposits in Australia: some observations from Western Australia. *J. Geol. Soc. Aust.,* 26: 293–303.
McNamara, K.J. (Editor), 1990. *Evolutionary Trends.* Belhaven Press, London, 368 pp.
Milankovitch, M., 1920. *Théorie Mathématique des Phénomènes Thermiques produits par la Radiation Solaire.* Gauthier-Villars, Paris, 338 pp.
Morton, S.R., Short, J. and Barker, R.D., 1995. *Refugia for biological diversity in arid and semi-arid Australia.* Biological diversity series, paper No. 4. Biological Diversity Unit, Department of the Environment, Sport and Territories, Canberra, 171 pp.
Newman, W.A., 1991. Origins of southern hemisphere endemism, especially among marine crustacea. *Mem. Queensl. Mus.,* 31: 51–76.
Notenboom, J., 1991. Marine regressions and the evolution of ground dwelling amphipods (Crustacea). *J. Biogeogr.,* 18: 437–454.
Orghidan, T., 1959. Ein neuer Lebensraum des unterirdischen Wassers, das hyporheische Biotop. *Arch. Hydrobiol.,* 55: 392–414.
Racovitza, E.G., 1907. Éssai sur les problèmes biospéologiques. *Arch. Zool. Exp. Gen.,* 6: 371–488.
Schiner, J.R., 1854. Fauna der Adelsberger, Lueger und Magdalener-grotte. In: A. Schmidl (Editor), *Die Grotten und Höhlen von Adelsberg, Lueg, Planina und Lass.* Braunmüller, Wien, pp. 231–272.
Scott, K.M. and Fisher, C.R., 1995. Physiological ecology of sulfide metabolism in hydrothermal vent and cold seep vesicomyid clams and vestimentiferan tube worms. *Am. Zool.,* 35: 102–111.
Sket, B., 1996. The ecology of anchihaline caves. *Trends Ecol. Evol.,* 11: 221–255.
Stanford, J.A. and Ward, J.V., 1993. An ecosystem perspective of alluvial rivers: Connectivity and the hyporheic corridor. *J. North Am. Benthol. Soc.,* 12: 48–60.
Stock, J.H., 1977. The taxonomy and zoogeography of the Hadziid Amphipoda, with emphasis on the West Indian taxa. *Stud. Fauna Curaçao, Amsterdam,* 55: 1–130.
Stock, J.H., Iliffe, T.M. and Williams, D., 1986. The concept 'anchialine' reconsidered. *Stygologia,* 2: 90–92.
Tuttle, M.D., 1976. Population ecology of the gray bat (*Myotis grisescens*): Philopatry, timing and patterns of movement, weight loss during migration, and seasonal adaptive strategies. *Occas. Pap. Mus. Nat. Hist. Univ. Kans.,* 54: 1–38.

Uéno, S.I., 1977. The biospeleological importance of non-calcareous caves. In: *Proc. 7th Int. Speleological Congress, Sheffield,* 407–408.

Uéno, S-I., 1987. The derivation of terrestrial cave animals. *Zool. Sci.,* 4: 593–606.

Williams, D.D., 1984. The hyporheic zone as a habitat for aquatic insects and associated arthropods. In: V.H. Resh and D.M. Rosenberg (Editors), *The Ecology of Aquatic Insects.* Praeger, New York, pp. 430–455.

Yager, J. and Humphreys, W.F., 1996. *Lasionectes exleyi*, sp. nov., the first remipede crustacean recorded from Australia and the Indian Ocean, with a key to the world species. *Invertebrate Taxon.,* 10: 171–187.

Part II

Classification, Geological Characterization, and Distribution of the Different Subsurface Biota

Chapter 1

THE DIVERSITY OF THE KARSTIC AND PSEUDOKARSTIC HYPOGEAN HABITATS IN THE WORLD[1]

Christian JUBERTHIE

HISTORY

The study of the cave fauna began in Slovenia (in what was then a part of Carinthia) with the scientific description of *Proteus anguinus* by Laurenti (1768), which previously had been reported by Valvasor (1689) to be regarded by local people as "ein junger Lindwurm [a young dragon]". The specimens were collected in a spring near Vrhnika by a postal officer. Schreibers (1801) studied the reproduction of dissected *Proteus* and made observations on living specimens in an aquarium. Cave studies developed in the region between Ljubljana and Trieste named Krst or Kras, especially following the description of the highly specialized cave beetle *Leptodirus hohenwarti* by Schmidt in 1832. After an initial faunal classification by the Danish zoologist Schiödte (1849), the Austrian naturalist Schiner (1854) established the most commonly used classification of cave fauna: troglobite (exclusively in caves), troglophile (in caves and surface habitats) and trogloxene (accidental in caves).

In an early expedition to South America, the famous European geographer Alexander von Humboldt visited the Cueva del Guacharo (Venezuela) in September 1799 and described the cave-nesting oil bird *Steatornis caripensis*.

In the middle of the 19th century, studies on cave fauna increased rapidly and extended into other countries. In Hungary, Fridvaldszky in 1841–1844 collected in the famous Baradla cave; (Fridvaldszky, 1865); Miller (1856) and Fridvaldszky and Fridvaldszky (1857) described cave beetles from the Apuseni mountains (today in Romania). In France, Delarouzée (1857) in the Pyrenees, and Lespes (1857) in the prehistoric cave of Niaux, described the first French cave beetles and a troglophile opilionid. Wankel (1856) published the first contribution to the fauna of Moravian caves in 1856. In Spain, Schaufuss (1861) described the first blind beetle from the Cantabric mountains. In Italy, the first Sicilian orthopteran *Dolichopoda palpata* was described prior to 1776, then studies were performed on *Proteus* anatomy. In Russia (Caucasus and Crimea), von Motschulsky in 1850 and 1862 collected cave fauna (von Motschulsky, 1854, 1862). In the USA, the first cave salamander and the first troglobites were described from Mammoth cave by Rafinesque (1822, 1832); De Kay (1842) published a diagnosis of the species *Amblyopsis spelaea.* Tellkampf (1844) described blind fishes and several invertebrates, Agassiz (1851) made observations on cave fishes, and the Russian coleopterist von Motschulsky (1854, 1862) collected and described blind beetles. Finally in Mexico, Bilimek (1867) discovered 11 species of arthropods in the Cuevas de Cacahuamilpa.

After the middle of the 19th century, exploration of cave fauna increased greatly. In 1907, Racoviţză and Jeannel founded in France an association named "Biospeologica" to explore caves and their fauna throughout the world with many collaborators, resulting in a massive contribution to the development of biospeology. The results appeared in *Archives de Zoologie expérimentale et générale,* and contained 77 contributions and 1825 pages.

For a long period of time, studies were limited to cave fauna to the exclusion of karstic cracks, fissures, and drains because of an anthropocentric concept of the habitat, and related to the difficulties and the lack of accurate techniques of accessing these areas.

[1] See the Glossary, p. 6, for explanations of various terms.

Since the Second World War, the results of indirect studies of populations in inaccessible parts of karstic habitat by mark–recapture techniques, the long-duration filtering of springs and cave streams, and the development of new sample techniques have enormously increased knowledge of the subject. These new sample techniques include the "phreatobiologic net" for wells (Cvetkov, 1968), the Karaman–Chappuis method (*trou* Karaman-Chappuis) for sandy-beach fauna (Chappuis, 1946), the Bou–Rouch technique for sampling groundwater fauna (*sondage* Bou–Rouch: Bou and Rouch, 1967; Bou, 1974), core-freezing sample techniques (Bretschko and Klemens, 1986) for interstitial habitats, and traps devised by Barber (1931) for terrestrial fauna in caves. In addition to these improvements in technique, the exploration of subterranean habitats in other types of rocks [lava tubes in volcanic basaltic rocks (Uéno, 1977; Howarth, 1983; Oromi et al., 1986; Peck, 1987, 1994b), caves in quartzite, diabase sill (Urbani, 1977; Martini, 1987), gneiss, granite, sandstone, and lateritic sediments in the tropical zone] has broadened knowledge of caves beyond the original concentration on karstic systems. The concept of the hypogean habitat has also been much enlarged and diversified by the exploration of anchialine and marine caves (Holthuis, 1973; Iliffe, 1981, 1991; Yager, 1981, 1994; Sket, 1986; Iliffe and Sarbu, 1990) and the inclusion of marine crevicular habitats (Iliffe, 1990), and in particular by study of interconnected microvoids and cracks near the surface [Milieu Souterrain Superficiel (MSS) (Juberthie et al., 1980; Juberthie and Delay, 1981) = Mesovoid Shallow Substratum, formerly called "Compartiment souterrain superficiel" (Juberthie et al., 1980) – the "upper hypogean zone" of Uéno (1977)] in limestone and other types of rock. In the volcanic Hawaii archipelago, it corresponds to mesovoids in lava flows (Howarth, 1981, 1983) and in the Canary Islands to the "volcanic MSS" of Oromi et al. (1986) and Medina (1991)].

In the last twenty years, in addition to studies on temperate hypogean fauna, there has been an extraordinary increase in taxonomic, ecological and biological information about hypogean fauna in the tropics, in volcanic archipelagoes, and in marine caves. Today, the hypogean environment has been more or less explored in all countries of the world.

The hypogean (subterranean) habitat represents the most extensive ecosystem on the planet. It has been recognized from the subarctic (Canada, Svalbard, etc.) to the equatorial zone, and from more than 120 m beneath sea level (limit of sea level during the Würm glaciation) to high altitudes in mountains of the tropical and equatorial zone (e.g., Mexico, Papua New Guinea), as well as in all oceanic archipelagoes, islands and atolls. Under an apparent uniformity due to the obscurity and the lack of green plants, there is a large diversity of habitats and fauna – terrestrial and aquatic (lotic, benthic, interstitial in fresh, brackish or marine water); sulphurous habitat; temperate and tropical, and with different types of food supply (organic matter, guano of bats and birds, seeds rejected by birds, feces of Mammalia, roots, chemoautotrophic production of organic mater in sulphurous caves) (Barr Jr, 1968; Camacho, 1992; Culver, 1982; Culver et al., 1995; Delamare-Deboutteville, 1960; Ford and Williams, 1989; Gèze, 1965; Ginet and Decou, 1977; Jeannel, 1926; Juberthie and Decu, 1994, 1998; Mohr and Poulson, 1966; Packard, 1888; Salomon and Maire, 1992; Racovitza, 1907; Trombe, 1952; Vandel, 1964).

Thus, a more comprehensive definition of the hypogean habitat and subterranean ecosystem is necessary, where the type of rock is only one of many factors.

The hypogean ecosystem consists of habitats located in interconnected voids of different sizes, from microvoids to large caves, isolated from the surface, with food supply coming from the surface or, exceptionally, from chemoautotrophic bacteria. All hypogean habitats present similar main characteristics: darkness and lack of photoperiod, annual temperature variations moderate without fast changes and extremes, relative humidity near saturation in the terrestrial habitat, food usually with low energy content. It is represented by natural or artificial caves in limestones or other rocks, lava tubes, interconnected cracks, fissures, drains in the rocks, microvoids in the slopes of the valleys covered by a soil or equivalent, interstitial habitats of rivers and lake beaches, phreatic, alluvial, and perched aquifers, and fresh-water Ghyben–Herzberg lenses in islands and atolls.

Features of these habitats are their long temporal scale (megascale through geologic time), and, in the case of lava tubes and the mesocavernous shallow substratum, the fast creation of new habitats available for active or passive colonization.

Biospeleology is devoted to the study of hypogean environments and all subterranean habitats. Cave ecology is only a part of this field, because caves represent only a part of the habitat of the populations. Generally, for studies of populations and communities, the scale

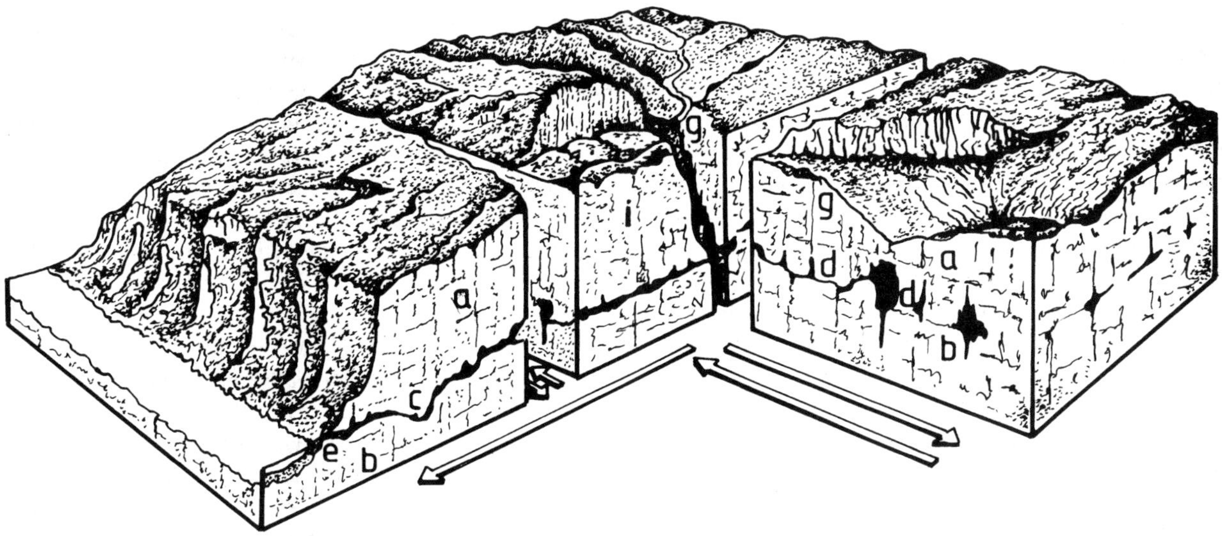

Fig. 1.1. Diagram of a karst and karstic system: (a) vadose (infiltration) zone; (b) saturated (phreatic) zone; (c) cave stream (drain); (d) annex systems; (e) main spring; (f) epikarstic aquifer; (i) sinkhole. Arrows indicate the directions of stream flow. (After Mangin, 1975.)

is the karstic system, including the mesovoid shallow substratum, or hydrogeologic system. However, caves represent the totality of the habitat for troglophilic bats, birds, and presumably several blind fishes.

KARSTIC SYSTEMS AND HABITATS

Distribution

The word "karst" has been generalized to cover all the world's limestone outcrops, plateaus and mountains, which present similar karstic morphology and processes, caves and karstic systems.

Karst covers large areas in all continents and in many countries. In Europe, karsts cover 40% of Slovenia, 25% of France and Italy, 20% of Spain, Bulgaria, Croatia, Bosnia, and Serbia, 17% of Austria, 5% of Slovakia, and small percentages of the Czech Republic, Romania and Poland. In America, it represents 15% of the USA, 12% of Canada (1 180 000 km^2), 75% of Cuba, and 15–25% of Central American countries. In Asia, it covers 20% of China (2 000 000 km^2), and is well represented in Vietnam, Thailand, Cambodia, Malaysia, Indonesia, Burma (Myanmar), India, Nepal, Iran, Turkey, and elsewhere. In Oceania, it is present around uplifted volcanic islands, and as a cap on uplifted atolls of archipelagoes (Bikini, Mururoa, Makatea, Rennell, etc.) and also in Australia, New Zealand and Papua New Guinea. In Africa it is well represented in North Africa, in South Africa, Malagasy, Somalia, Tanzania and Mozambique.

Caves are present in limestone from the oldest karsts (Cambrian, Silurian, Ordovician) to the most recent (Holocene). The number of explored caves increases every day and is related to speleological activity, the level of karstification of limestone, and the surface area of karsts. An estimate for Europe is 130 000 caves (more than 28 000 in France, around 20 000 each in Austria, Italy, Romania and the former Yugoslavia, around 10 000 each in Spain and Russia, around 3000 each in Germany, Great Britain, Greece and Hungary, more than or around 1000 each in Belgium, Norway, Poland, and Switzerland, and 700 in Finland). In America: around 10 000 each in Cuba and the USA, more than 1000 in Mexico and around 1000 in Brazil and in Jamaica. In Oceania, there are 2500 in Australia, 400 in New Zealand, 280 in the Palau Islands, and 110 in New Caledonia.

Structure of karst and karst habitats

Karst landscape (Exokarst)

The characteristic landscape of karsts (Fig. 1.1) is the result of the process of dissolution of carbonate rocks (limestone and dolomite). The process of dissolution requires carbonic acid in water, coming from the biological activity of the soil or from the atmosphere. The amount of water available, the concentration of carbon dioxide and the solubility of the rocks, primarily determine the rate of dissolution of the rock. The large regional diversity of climate, geological and structural contexts, the tectonic events, the time elapsed since the exposure to chemically aggressive water (with a high content of dissolved carbon dioxide and low pH), the presence of soil and forest, and the degree of denudation, explain the diversity of exokarst morphology. The main exokarst dissolution forms are: lapiaz, dolines, sinkhole, avens and polje.

For the ecologist, the most important factors related to the exokarst morphology are the amount of primary production, the level and the timing of precipitation, and the spatial distribution of the input of meteoric water which is the main transport agent of organic matter and vegetable debris produced on the karst surface (from the soil, or from forests, steppic and arid formations covering it).

Two other exokarst forms of great interest for biologists are: the karstic mesocavernous shallow substratum on the foot of outcrops which originates from the gelifraction of limestone and dolomite; and the river sinkholes which enter the cave and bring in vegetable detritus, animals and organic matter during flooding.

Structure of karstic habitats

The specificity of karst and karstic habitat comes from the dissolution of pre-existing interstices in the limestone or dolomite by aggressive water, a process called karstification. It provides a heterogeneous habitat of interconnected cracks, fissures and drains, filled with air or water or both; the karst process is polyphasic through geologic time, and related to the change of level of sea and land-mass; inactive (fossil) caves may be present at different elevations.

Caves are only a part of the heterogeneous, anisotropic and hierarchic system of voids inhabited by hypogean invertebrates. Thus, the term "karstic system" (Mangin, 1994) is the most accurate; it designates a functional unit involving organized and hierarchic flow pathways, storages, and voids.

For the aquatic fauna, karst covers the totality of habitats and subhabitats, and the karst system is equivalent to the karst aquifer.

For the terrestrial hypogean fauna, the subterranean habitats cannot be reduced to the hydrogeological point of view of the karstic system, and to the karstification process. Indeed, the mesovoid shallow substratum which is a part of the habitat located at the foot of karst cliffs, does not originate from karstification, but from gelifraction and accumulation of rock elements by gravity, and also develops in other types of rock, especially schists.

Thus, the karstic system includes habitats not originating from karstification, consisting of a large set of interconnected voids of different sizes.

Classically, two main zones are recognized in a karst: the vadose zone (= upper zone, infiltration zone; unsaturated zone); and the phreatic zone (= saturated zone; *zone noyée*).

Vadose zone: This zone has a vertical circulation of meteoric water, through cracks and fissures. There are two modes of infiltration (Mangin, 1994): the first is fast infiltration – runoff, through large but not numerous fissures; the second is delayed infiltration through minute and usually numerous fissures. Generally the water is mixed with air, and the flow is diphasic. Pools of water, and small cavities, can be found along these cracks and fissures. Subhabitats are aquatic, terrestrial or both, according to the morphology of the cracks, the period and amount of rainfall and the fact that some fissures are active while others are not.

Beneath the surface of the karst there is sometimes a zone several metres deep with high porosity (5–10%) and permeability, formed by the weathering of rock or a network of vertical fissures (Fig. 1.2) originating

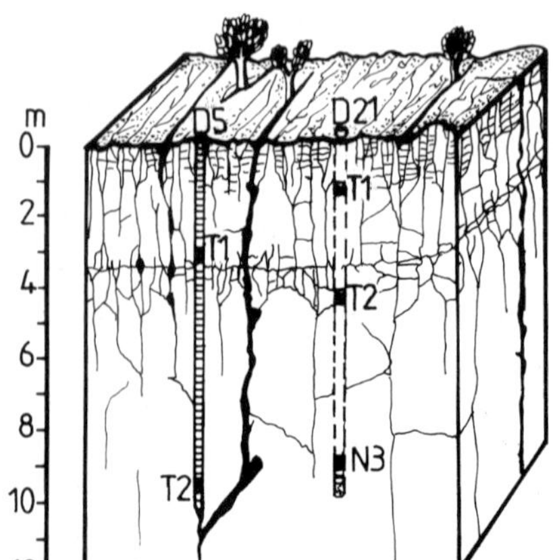

Fig. 1.2. Diagram of the epikarstic zone in the plateau of Mas de Vedel (Hérault, France), showing fissures, horizontal joints, drillings (D), traps (T) and net (N).

from surface decompression of the rock. It may contain larger or smaller permanent water reserves, because at deeper levels the permeability is lower; this is called the epikarstic zone (Mangin, 1975).

Three methods can be used for sampling terrestrial

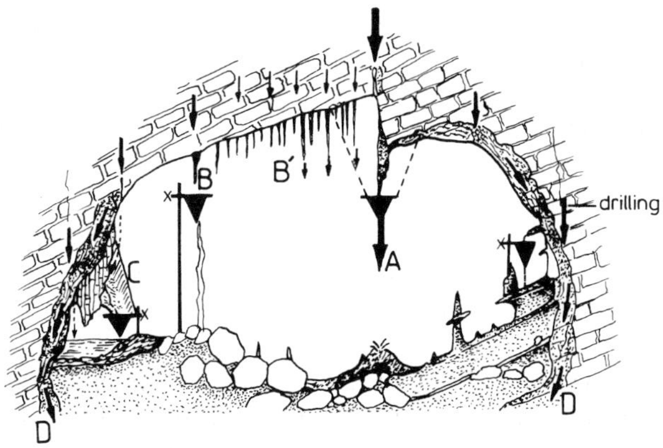

Fig. 1.3. Methods to collect aquatic fauna with nets from fissures and stalactites of the vadose zone inside a cave: (A) fast infiltration; (B, B′) delayed infiltration and dripping water from stalactites; (C) flow on calcite deposits on the wall; (D) flow under concretions covering the wall.

and aquatic fauna of the vadose zone and epikarstic aquifers.

The first to be used (Fig. 1.3) involves nets, during seasonal periods of water circulation; these filter the water dripping from stalactites and the water running on or beneath the walls covered with calcite in caves located in the vadose zone; this method was first used for stalactites by Bou in 1957, in the "Trois Cloches" cave (Bou, 1979) and by Delay (1968) in the Moulis and Liqué caves, and for a calcified wall by Rouch in the Grotte de Sainte Catherine (1968) in France.

The second uses special baited nets for aquatic environments, and baited Barber traps for terrestrial environments, in drillings from the surface, several to tens of metres deep. This method was used in a karst plateau north of Montpellier (Juberthie-Jupeau, 1966, unpublished), and on the islands of Guadeloupe and Réunion (Juberthie, 1989).

The third uses baits put in caves, combined with mark–recapture methods, to collect and estimate the size of populations; it is very accurate for cave beetles, and during the spring and the summer when beetles enter caves from fissures; it has been intensively used in the Pyrenees by Delay (1968, 1978) and many other biospeleologists in Europe (e.g., for Orthoptera by Sbordoni's team in Rome, Italy) and North America (e.g., for Ricinulei by Mitchell, 1969).

The apparent scarcity of fauna in the vadose zone is not real. The number of aquatic animals and species collected by net filtering under stalactites which drain a very small catchment basin on the karstic surface compared to the large quantity of water filtered in streams or springs is relatively high. For example, Delay (1968), by filtering 3624 litres of water running off stalactites in Liqué cave, from February to July, collected 706 individuals, 589 being hypogean crustaceans. The populations of these habitats must be higher, because only those animals carried along in the water are collected. In the case of terrestrial species, only that part of a population which is directly connected to the cave is estimated by the mark–recapture method; for example, in the French Pyrenees, the size of populations of detritus-based cave beetles (Bathysciinae = Leptodirinae), in each limited system of caves, cracks and fissures studied over a period of two to three months by Delay (1968), was around 50 000 individuals.

Saturated zone (phreatic zone, *zone noyée*): This lower part of the karstic system is a saturated zone, flooded by a considerable amount of water, generally millions of cubic metres. It is characterized by an organization of voids in two systems: large sub-horizontal drains or vauclusian drains and springs, which determine the dynamic function of the aquifer, and connected drainage systems which consist of small voids and very large cavities providing the storage function. In this zone the dissolution process is high and the porosity can reach 15%.

Two methods are used to sample the fauna of the saturated zone. The oldest is sampling in the pools inside caves with nets, baited or not; animals are collected *in situ,* but only a very small area of the habitat is sampled. Another more recent method, initiated by Chappuis (1924) then Leruth (1939) and developed by Rouch et al. (1968) in the karst system of Cent Fons (Hérault, France), and Rouch (1970) in the karst system of Baget (Ariège, France), uses nets 2 or 3 m long to filter springs, especially during flood periods. Tens of thousands of cubic metres can be filtered by this method. Rouch showed that around a million microcrustaceans a year, particularly copepods, are washed out of the karst system of Baget, which drains a catchment basin of 14 km^2. Measurement of this phenomenon (called "haemorrhage") indirectly demonstrated that the sizes of aquatic populations are high in karst systems, especially in the saturated aquifer, the volume of which was estimated to be 3 million m^3 for the Baget karst (Mangin, 1975, 1994). Studies carried out in connection with hydrogeological and water-geochemical studies of flood events offer the possibility of examining the hypogean subhabitat

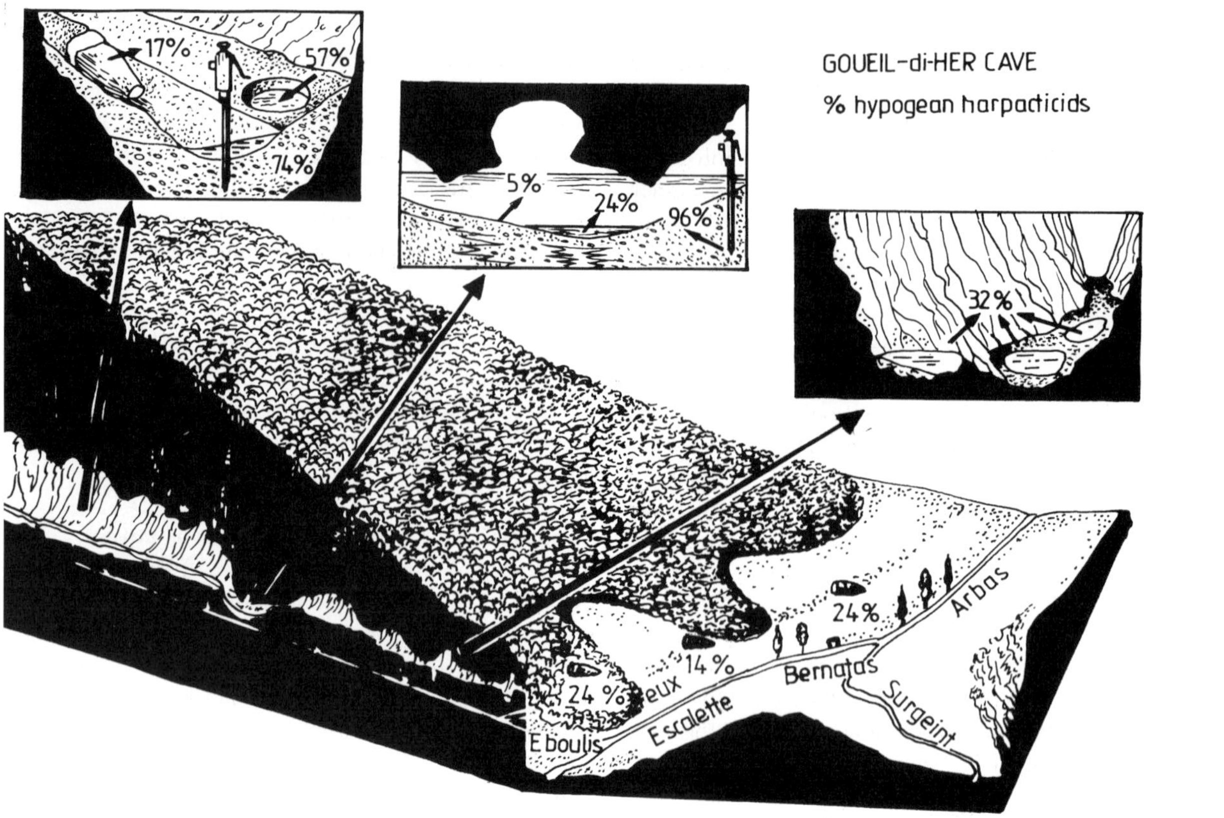

Fig. 1.4. Interstitial and stream habitats in karst; percentages of stygobitic copepods in the total copepod populations. (After Rouch, 1986.)

of harpacticids washed out of drain or annex systems. In addition, first in Cent Fons and later in Baget, it was shown that the diversity of aquatic karst communities is higher than can be estimated by *in situ* methods. Inside caves, Ginet (1960) used the filtering method for the Hautecourt cave stream, then Gibert (1986) for the Dorvan Karstic system in Jura, France. The representativity of species abundance and community structure shown by filtering methods has some limits, on account of the different capacity of aquatic invertebrates to resist an increased current as Culver (1971) demonstrated with crustaceans in a cave stream in Virginia (USA). The type of habitat of the species, whether lotic or interstitial, the location of habitat [as in the Sauve (Gard, France) karstic system: Juberthie and Juberthie-Jupeau (1975)], and the location of populations in drain or annex systems, also have an effect. Recently, high-discharge pumping in a vertical pool connected with the saturated zone of the Baget karstic system (Gouffre de la Peyrère) has shown a large population of the isopod *Stenasellus* in sediments (Rouch et al., 1993), which are not washed out, or collected by filtering the spring; a similar result was obtained with the harpacticid *Nitocrella gracilis,* which constitutes 76% of the harpacticids in this annex system, though only 7% were shown by the drain-filtering method. This is due to the fact that the level of water rises or falls slowly in a static-saturated system, and consequently the drift of animals is lower than in an active drain.

Other karstic habitats

Cave streams: An important part of the functional structure of a karstic aquifer (Fig. 1.4) is represented by the drainage axis (subterranean free rivers, cave streams), characterized by a sub-horizontal circulation and periodic flooding which can flood the galleries, totally or partially. There are two main types: endogenous cave rivers originating from the collection of water from the vadose zone, and exogenous cave rivers originating from surface rivers entering through sinkholes. For many hydrogeologists they are only the upper part of the saturated zone; for biospeleologists, they represent a special habitat for large aquatic vertebrate stygobites such as fishes and amphibia, and for crayfishes. The input of food, and the relatively high food supply necessary for their reproduction, are related to the flood factor (Hawes, 1939). In the USA, Mexico, the former Zaire, Brazil and Somalia, studies have been conducted on the ecology and biology of cave fishes and benthic forms, and in Slovenia, France and the United States on cave Amphibia (*Proteus, Typhlomolge,* etc.).

Interstitial habitat in karst: Usually the karstic habitat is considered as a heterogeneous aquifer in a series of fissures, where the speed of water circulation is high. In addition, an interstitial habitat exists in the gravel, sand and clay in the bed of some subterranean rivers; it is similar to the interstitial habitat [hyporheic habitat of Orghidan (1959)] in the bed of surface rivers. It was studied in France in the Goueil di Her cave (Haute Garonne) by Lescher-Moutoué and Gourbault (1970).

Mesovoid shallow substratum in karstic rocks: The screes at the base of limestone cliffs have a network of voids connected with cracks and fissures of the cave karstic system (Fig. 1.5). They are inhabited by troglo-

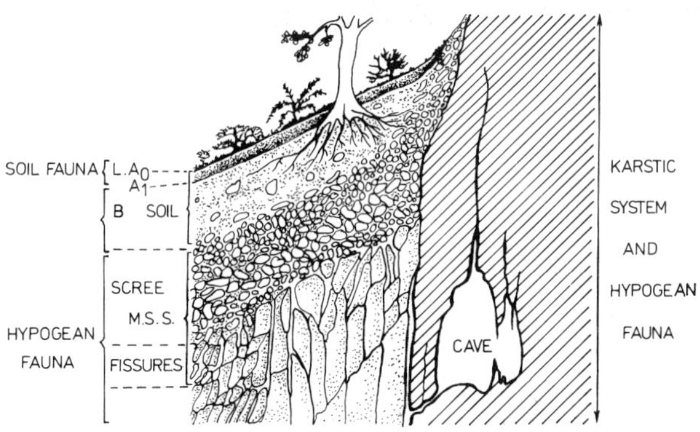

Fig. 1.5. Diagram of the mesovoid shallow substratum (MSS) in a calcareous zone in a scree at the base of a limestone cliff, connected with a cave system; soil with L, A_0, A_1, B horizons.

bitic and troglophilic beetles, Collembola, Diplura, Myriapoda and other invertebrates also recorded from caves and karstic systems. This habitat represents a type of mesovoid shallow substratum discovered in other rocks (see p. 33). Frequently, the voids of the mesovoid shallow substratum between calcareous stones are filled with clay, giving a patchy habitat.

Marine caves in carbonate rocks: anchialine (anchihaline), submarine and dissolution caves: Marine caves include any types of caves containing salt water of marine origin. Two main types are recognized: anchialine and submarine caves.

On all coasts which contain limestone, the sea water penetrates into the voids of the rock for several kilometres, and into caves.

Usually fresh water overlies the deeper sea water, and on islands the fresh water forms the Ghyben–Herzberg lens, sometimes giving brackish water at

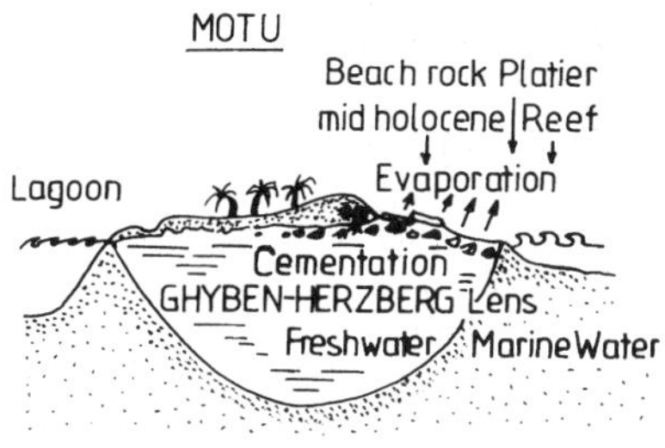

Fig. 1.6. Location of Ghyben–Herzberg freshwater lens, under an islet (motu) of an atoll.

the interface zone or halocline (Fig. 1.6). Thus, the habitat is composed of small voids and cracks, and the hypogean fauna is concentrated in the lens at the bottom of artificial wells.

Anchialine caves (see Chapter 3): The term "anchialine" was originally coined by Holthuis (1973) for coastal caves to describe "pools with no surface connection to the sea, containing salt or brackish water, which fluctuates with the tides". The definition was reconsidered by Stock and Iliffe (1986): "bodies of haline waters, usually with a restricted exposure to open air, always with more or less extensive . . . "

Several types of anchialine cave habitats are found (Iliffe, 1981, 1991, 1992; Iliffe and Sarbu, 1990; Manning et al., 1986):

The first consists of anchialine pools in tectonic faults along the coast, in limestone cracks of coastal cliffs, locally referred to as "grietas" (Cuba). Those fissures nearest to the coast may extend down below sea level and contain anchialine pools. In these pools, salinity increases with depth, from fresh water or brackish water near the surface to marine water at greater depths, often marked by a halocline. In Cuba, several anchialine grietas have been described in Seborucco Bank, 8000 years old; the first species of the new ostracod family Thaumatocyprididae were collected there (Danielopol, 1972).

A second type is found in coastal limestones of continental plateaus or archipelago carbonate platforms (Fig. 1.7) or in basaltic volcanic coasts. It occurs widely throughout the world. Access for caving may be terrestrial or marine. The Caribbean Sea is an important zone for this type of habitat, in the coastal caves of Yucatan, Florida, the Bahamas, Caicos, Turks and the Belize Cays.

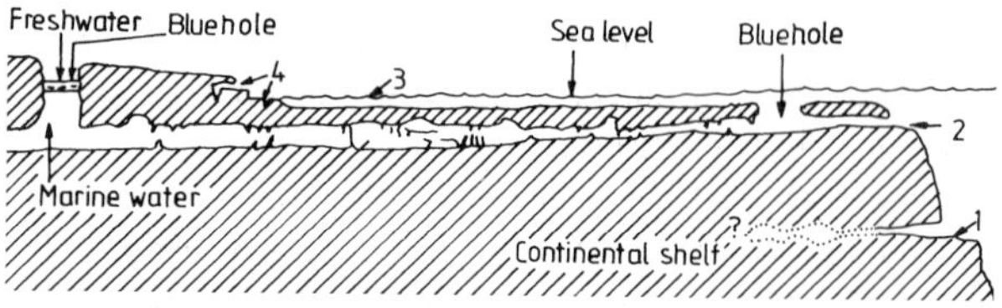

Fig. 1.7. Diagram of a long anchialine cave in limestone bedrock of the Bahamas submerged platform: (1) submerged marine bank at 120 m depth, eroded during the Würm. (2) old bank submerged since the post-Würm; (3) present sea level; (4) grietas.

Submarine caves: Submarine caves occur in coastal limestones of continental plateaus, in carbonate platforms of archipelagos, or in basaltic volcanic coasts, under the sea level (Iliffe, 1981). Access is by surface or submarine openings (inland Blue Holes or Cenotes, oceanic Blue Holes).

The most specialized fauna was found in this type of cave in the Bahamas. These caves were generated in the phreatic zone when the sea level was lower, and were secondarily submerged by marine water as a result of the post-Würmian rise of the oceanic surface level. They were presumably colonized by marine cave (Iliffe, 1990) fauna originating from crevicular habitat or submarine caves located 120 m or lower during the maximum of the Würm glaciation. Beneath the density halocline, the habitat is characterized by a low oxygen level, usually less than 1 ppm. A new group of primitive Crustacea, the Remipedia, was found in this habitat (Yager, 1981, 1994). The communities consist almost entirely of stygobite species of Crustacea: Thermosbaenacea, hadziid Amphipoda, asellot and cirolanid Isopoda, Ostracoda, Mysidacea and caridean shrimps.

Another type of submarine habitat is represented by a descending cave near sea level in the Mediterranean zone which retains cold marine water, the temperature of which is the same as that at the bottom of the Mediterranean Sea. The cave is inhabited by benthic forms from deep oceanic habitats (Boury-Esnault et al., 1993).

Dissolution caves (Fig. 1.8): Dissolution caves have the same type of habitat, but the origin of the caves differs. Marine dissolution caves are formed along the halocline, at the contact between fresh and sea water. The mixing of discharging fresh water with tide-pulsed incoming marine water under the flanks of the stable platform, as in the Bahamas, produced the necessary conditions for them to be formed (Mylroie and Carew, 1990).

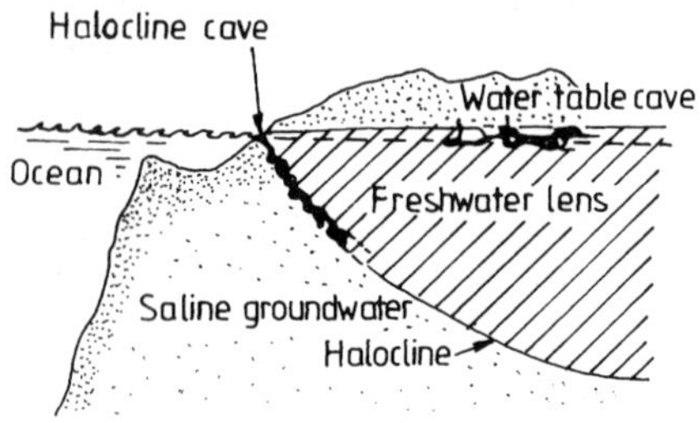

Fig. 1.8. Dissolution cave in the Bahamas: Scheme showing the two zones of dissolution: one along the halocline at the contact between the fresh and sea water, causing marine caves, another on the top of the freshwater lens. (After Mylroie and Carew, 1990.)

Artificial caves

Quarries and mines dug in limestones or other rock types have communities composed of trogloxene or troglophile species of such groups as Diptera, Lepidoptera and Orthoptera, which penetrated through the entrance.

In quarries, troglophile bats roost, hibernate or reproduce. In Europe, some of the biggest colonies are found in quarries such as the underground quarries of the Sint Pietersberg near Maastricht in the Netherlands or the artificial caves of Miedyrzecz in Poland.

Old mines of southern Europe are colonized by troglobite beetles, Collembola, Diplura, etc., which entered the deep zone of the mines through deep faults or cracks; the size of the populations is high around wet and rotten wood. For example, the Riverenert mine near St. Girons in the Pyrenees, dug in gneiss, has been colonized by the troglobite bathysciine *Speonomus hydrophilus*.

DIVERSITY OF KARST : REGIONAL KARSTS

Karst areas present a variety of geomorphological, tectonic, bioclimatic and historical conditions.

The main varieties are: karsts in sedimentary basins (the Paris and London basins, the Mississippi basin, etc.); karst in slightly raised platforms (Florida, Yucatan, Cuba, Bahama, uplifted atolls); karst in uplifted platforms (Grands Causses, Mississippi Plateau,

Guanxi in China, Darai Plateau in Papua New Guinea, Malagasy); karst in folded mountain ranges (Jura, Alps, Pyrenees, Dinaric mountains, Caucasus, the Zagros Mountains in Iran, Nepal, Yunnan, Japan, the Atlas mountains in North Africa, the Rocky Mountains, the Appalachian Mountains, the Andes); cryptokarst (Central African Republic, etc.); submerged marine karst (mainland coasts, continental plateaus, submerged part of atolls).

Types based on climatic criteria are: subarctic karst, high-altitude karst in temperate and tropical zones, temperate karst, tropical karst, sub-arid and arid karsts.

Types may also be distinguished on the origin of the food supply; in most cases it is allochthonous, but in some it is derived from chemosynthetic production *in situ* (Movile cave, and some others).

Subarctic karst

In the most northern zone of the hemisphere all the fauna was destroyed by the last glaciation (Würm in Europe = Wisconsin in America), with some exceptions for aquatic fauna. Today, a natural field experiment of recolonization of Scandinavian, Russian and Canadian caves by eurytopic terrestrial species can be observed; all these are troglophile and trogloxene species which also occur in nearby biotopes (Peck, 1994a; Sjöberg and Odell, 1994; Kejonen and Juberthie, 2000).

One of the most interesting features is the discovery of stygobitic species which survived in an ice-free refugium or under glacial ice cover in Canada during the last glaciation (Holsinger, 1980). In the Canadian Mackenzie Mountains at 61°N, the Nahami area was not ice-covered during the Wisconsin, and three types of caves have been recognized there: those that were glaciated in the Wisconsin; those that have been glaciated, but not for the past 100 000 to 400 000 years; and those that were never glaciated. Invertebrates may have colonized the caves in interglacial periods, and some aquatic species could have survived through the Wisconsinian glaciation to the present. However, most of the species were eliminated by the rigorous periglacial climate, and surveys of a cave showed that all the Coleoptera, Diptera and Lepidoptera present were post-Wisconsin troglophiles or trogloxenes (Peck, 1987).

High mountain karsts and habitats; ice caves

The high-mountain karst (alpine karst) extends into the temperate mountain zone, above the tree line. The exokarstic surface is denuded, with seasonal snow, ice and glaciers. It is characterised by numerous deep vertical shafts in the vadose zone, some caves being 1000 to 1500 m deep, and series of springs in the valleys at lower altitude.

This type of karst is well developed, and has been studied in the French, Italian, Swiss and Austrian Alps and also in Greece, Taurus (Turkey), Zagros (Iran), and the Rocky Mountains in the USA (Maire, 1990). In Dachstein, Tennengebirge and other mountains in the southern Salzburg area (Austria), in addition to vertical shafts originating from the high-altitude karstification which was accelerated during the glaciations, there are networks of horizontal conduits and sometimes large tubular galleries layered at different altitudes. The most extensive occur at altitudes between 1300 and 2000 m, and include the six largest Austrian cave systems, each with from 28 to 52 km of galleries. Their formation occurred during the end of the Miocene, under a hot and humid climate near the water table; today they are perched and inactive, following a strong uplifting of the Alps during the Pliocene (Langenscheid, 1986; Audra, 1994).

Some of these horizontal alpine caves in Austria have a large volume of ice, such as the Eisriesenwelt with ice occupying the first kilometre from the entrance (around 30 000 m^3 in volume). Endemic cave beetles (of the trechine genus *Arctaphaenops*) occur sparsely on or near the ice, but the real habitat consists of fissures in the rock, with temperatures around 5°C. The population size and the species diversity seem very low; this is related to the low level of organic matter and food supply ["oligotrophic caves" in the sense of Sbordoni et al. (1977)]. Two hypotheses can be suggested for the date of colonization: the first requires a colonization of caves under a hot and humid climate, at lower altitude, in the late Miocene or early Pliocene, and speciation by selection of psychrophile genotypes during the strong uplifting of the karst and the decrease in temperature; the second hypothesis is of a later colonization from surface trechines inhabiting the margins of the snowy zone. The same type of habitat occurs in the French Alps (Chartreuse, Vercors, etc.) with endemic species (for example the bathyscine *Isereus xambeui* in the Trou du Glas in Chartreuse), and in the subalpine zone of the Pyrenees, with endemic beetle species (of the trechine genus *Aphaenops*) or genera (the bathysciine *Josetekiella*) usually only known from one or two specimens and with small populations. The scarcity of animals, the difficulty of access, and the short period available for study, account for the fact that this habitat and its populations are not well known.

High-mountain karsts also occur in the tropical and equatorial zones [the Andes (Peru), Guatemala, Mexico, Papua New Guinea]. They occur at higher altitude, for instance around 3800–4000 m in Papua New Guinea (Maire, 1990), where a glacio-karst occurs above 3800 m, the tree line, in the Finisterre–Suarunwaged chain (Huon Peninsula); its cover is steppic with doline shafts. During the Würm glaciation (30 000–38 000 BP) the karst was covered with ice; the lower limit of snow was 3700 m and the upper limit of forest 2000 m. Around 12 000–9 000 BP, the forest went up to 3800 m again and nutrient levels increased in hypogean habitats. In Mexico and Guatemala, high-mountain karst occurs above 3000–4000 m.

An ecological feature is the discovery of troglobites similar in morphology to those found in temperate-zone caves, in tropical and equatorial high-mountain karsts and caves (3000 m in altitude or more), where temperature is around 8–10°C – for example, the beetles *Gastrogonum* and *Speogonium* (Agoninae) in Papua New Guinea. Other species, such as *Mexaphaenops* in Mexico and *Mayaphaenops* collected in a cave at 3040 m in Guatemala with temperatures of 9–10°C, were found in mountain tropical karst under forests with the same climate as in temperate karsts, and are similar to the European *Aphaenops* and American *Neaphaenops* (Vigna Taglianti, 1977).

Temperate karsts and habitats

Temperate karsts are the classical karsts and the most studied. There are two main types: plateau karsts and folded karsts in mountains. The greatest number of stygobites and troglobites are found here. This fact results directly or indirectly from the effects of glaciation, surface denudation, local or seasonal dryness in mediterranean or continental zones and the selection pressures for energy-economizing adaptations.

Tropical karsts and habitats

Many studies have been carried out on and subterranean habitats in tropical wet karsts, in Mexico (Chiapas, Sierra del Abra), Guatemala, Venezuela (Cueva del Guacharo), Brazil, Peru, the former Zaire, Guinea, Somalia, India (Siju caves), Myanmar (Burma), Malaysia (Batu Caves; Gunung Mulu National Park in Sarawak), Java (Gunung Sewu), Cambodia, Vietnam, southern China, South Korea, Taiwan, Japan, and in Oceania.

In tropical rain forests and tropical warm and humid countries, the karst landscape shows characteristic forms: cone karsts, and tower karsts.

Cone karsts are typified by that of Gunung Sewu in Java (Lehmann, 1936) and the cockpit karst in Jamaica (Lehmann, 1927). Individual cones are rarely higher than 100 m, and their upper part is rounded; valley floors between the cones are very narrow and alluvial flats are scarce. In Jamaica the cockpit karst is generally formed in areas of high rainfall, where circulation cannot be deep because the clay and insoluble particles in the soil form an impermeable cover which controls the drainage pattern. Rain water drains to the bottom of the rounded hills where it enters the horizontal galleries in the limestone. Many cockpits have vertical sinkholes in the bottom which connect with horizontal caves. For this mechanism to be continually effective, the rate of erosion has to be matched by the rate of uplift (Versey, 1972).

The tower karst also has a wide distribution in the world; classic examples are in southern China (Yangshuo area of Guanxi) and in Vietnam. Individual towers vary in height from 30 m to over 200 m, many have vertical sides, whilst others are much more irregular. Usually, towers rise from alluvial flats and may be isolated or in clusters with internal valleys and depressions. In southern China, in an area of a million square kilometres, there is a range of hill forms between the extremes of cones and towers.

Tropical cave communities

The species diversity and size of populations are higher in tropical caves with guano than in temperate caves (e.g., Chapman, 1984). A classical feature is a relative paucity of troglobites, high diversity and abundance of troglophiles, and in contrast an abundance of stygobite aquatic species (Mitchell, 1969). However, with the increase of tropical cave surveys, the number of troglobites appears higher than was claimed in the early studies. The higher energy input into tropical subterranean streams explains the diversity of stygobite fishes observed. Some of the communities consist of families which differ from those found in temperate caves; this is true, for example, for the families Loxoscelidae, Ochyroceratidae, Oonopidae,

Tetrablemmidae, Theridiidae, etc., that characterize the spider fauna of tropical caves (Brignoli, 1973, 1980).

Karsts on oceanic volcanic islands

In Oceania, archipelagoes consist of different types of islands (Fig. 1.9). In the tropical zone there are

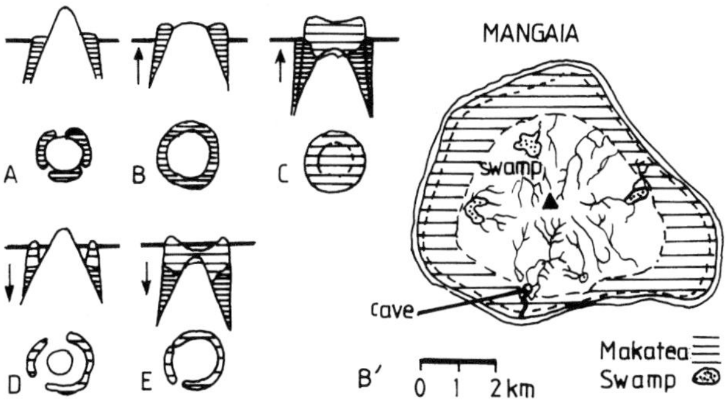

Fig. 1.9. Types of oceanic islands: (A) volcanic island with fringing reef: (B, B′) volcanic island with uplifted reef (Mangaia type); (C) uplifted atoll (Makatea type); (D) volcanic island with reef barrier (Bora-Bora type); (E) atoll on erased and submerged volcano (Bikini and Mururoa type).

islands with a volcanic core and fringing reef (the Rapa type, in the Austral Islands), islands with a core surrounded with a reef rim or a reef barrier (Bora-Bora type, Raevavae), uplifted islands with emerged reefs surrounding the core (Mangaia type in the Cook archipelago, Mariana Islands, etc.), entirely uplifted calcareous island (Makatea type in French Polynesia, Lifou, Maré, Ouvea in New Caledonia, Tongatapu in the Tonga archipelago, Nauru, Niue, Rennell Island), and atolls capping submerged volcanoes (Bikini and Mururoa are the two best known; but also the Tuamotu islands, the Marshall Islands, Kiribati).

For example, French Polynesia and the geographically related Cook Islands consist of five archipelagoes (the Austral Islands prolonged by the Cook Islands, the Society Islands, the Tuamotu Islands and the Marquesas Islands), aligned from the more ancient islands in the south-east, to the more recent islands in the northwest. Tuamotu is a massive submerged chain stretching more than 1500 km, including 80 atolls; coral limestones cap the tops of truncated volcanic cones, rising from a huge ridge at depths varying from 1500 to 3000 m; this ridge is on a wide plateau which rises 2500–3000 m above the surrounding oceanic substratum. This archipelago is much older than the others in Polynesia; in the northwest Tuamotu archipelago the beginning of the formation of the islands has been estimated to be 37.5 to 53.5 million years ago, and that of the oceanic crust to be 65 million years ago. The Society archipelago, 720 km long, which includes Tahiti and Bora-Bora, is composed of volcanic cores surrounded by fringing reef or barrier reef. The volcanic propagation from Mahetia (SE) to Bellingausen (NW), whose age is estimated to be 6.5 million years, was steady, at a rate of 12.5 cm yr^{-1}. The height of volcanoes diminishes as a function of their distance from Mahetia, which can be ascribed to decreasing hotspot activity; thus, the two youngest volcanoes are low compared to the highest altitude in Hawaii, currently an active zone. In the Austral alignment and the Cook Islands, the age of limestone cliffs on the volcanic core of Rurutu, an island between Palmerston Atoll and the submarine volcano MacDonald, dates back less than 1.85 million years, because the island has had an uplift rate of approximately +0.1 mm yr^{-1} since the early Quaternary (Chevalier, 1973; Brousse, 1985).

Two origins must be considered for islands with reef limestone: the islands built on or near the mid-Pacific ridge, and those formed on intraplate hot spots (Brousse, 1985). Tuamotu belongs to the first kind; on account of the displacement of the Pacific plate, the volcanism which formed the Tuamotu archipelago must have occurred some thousand kilometres distant from the present position, near the East Pacific ridge; this has important implications for the history of marine, and perhaps terrestrial, colonization of the islands. The other volcanic alignments in French Polynesia are situated on thermal anomalies in the intraplate mantle (second origin), as is the Hawaii archipelago.

Several types of hypogean habitats can be recognized:

(1) Sinkholes and caves through the limestone cliffs. The runoff from the volcanic core forms swamps at the foot of the interior cliff, forms subterranean galleries in the ring of limestone and gives rise to springs on the coast or beneath the sea.

(2) An anchialine habitat, below sea level, near the coast; cracks and caves in reef limestone are filled with sea water and inhabited by marine stygobites (Iliffe and Sarbu, 1990).

(3) A Ghyben–Herzberg lens of fresh water, from the coast to the interior of the reef cliff, partly below sea level, which floats on underlying denser salt water. In the islets (Motu in Polynesian) of atolls this is the only hypogean habitat present, colonized by interstitial or cave stygobites. They were colonized through the halocline by marine

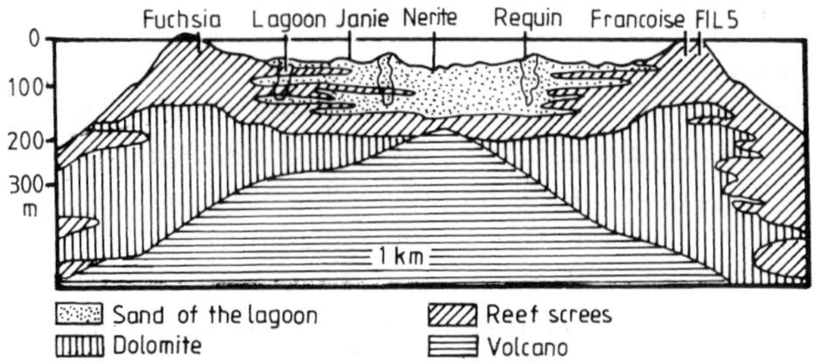

Fig. 1.10. Diagram of Mururoa atoll. (After Guilcher, 1992.)

euryhaline ancestors, secondarily adapted to fresh water.

(4) A potential crevicular marine subterranean habitat in submerged reef limestone at great depth [400 m in Mururoa (recognized through several drillings), to 1400 m in Eniwetok (one drilling) and the Maldives (one drilling)]. These rocks were karstified very early by localized sedimentation at sea level, and karstification accompanied by fluctuations of sea level along the eustatic curve (Bourrouilh-Le Jan, 1990). This type of atoll consists of a deep cap (−130 m in Mururoa, −1230 m in Eniwetok, −1400 m in Mali, an atoll of the Maldives archipelago) formed by irregular beds of coral limestone, coral sand, carbonated debris and reef conglomerates of diverse sizes (of Precambrian non-limestone rocks) with very high porosity (locally more than 50% in Mururoa between 0 and −125 m). Beneath this porous cap there is a bedrock of dolomite over the volcanic core (−125 to −435 m in Mururoa; −1230 to −1400 m in Eniwetok) with lower porosity (Fig. 1.10). However, the dolomitization does not affect the slope of the submerged high-porosity limestone in Mururoa; in Eniwetok the dolomite between −1230 and −1400 m is interbedded with coral limestone with numerous voids, which induce high porosity.

The high porosity explains the decrease of temperature to 14.5°C in drillings in Mururoa and 6.5°C in Eniwetok, similar to the ocean temperature. To explain this phenomenon and the high level of nutrients and nitrate in the Mururoa lagoon, Rougerie and Wauthy (1986) postulated upwelling of sea water and organic matter through the porous karst of the atoll (Fig. 1.11).

This potential deep crevicular habitat is unexplored, but it may be possible through drillings to take samples with baited nets.

The first colonization by marine ancestors occurred in the Miocene at the foot of the cap where there is Miocene limestone, which was near sea level as fringing reef or coral barrier.

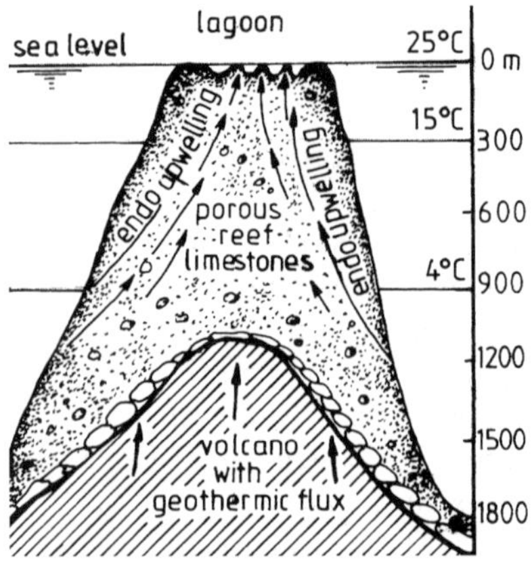

Fig. 1.11. Hypothetical upwelling through the very porous rock in an Eniwetok-type atoll. (After Rougerie and Wauthy, 1986.)

Karsts in evaporites (gypsum and salt)

Gypsum

Gypsum (hydrated calcium sulphate) is much more soluble in water than limestone, so large cave systems exist in gypsum (Nicod, 1976; Klimchouk et al., 1996). For example, in the Ukraine several labyrinthic cave systems have developed in carbonated gypsum bedrocks, and are among the longest in the world: Optimisticheskaïa cave has a development of 183 km, and Ozionaïa cave 111 km. In central Europe, Triassic (Zechstein) gypsum karsts cover large areas. In Italy, Triassic and Messinian (five million years) gypsum karsts are located in Sicily and Emilia-Romagna, where around 400 cavities are recognized. Many gypsum caves are also found in arid regions (Ford, 1976). In the subpolar zone (Svalbard, Irkutsk plateau in eastern Siberia) under permafrost or glaciers (Pulina, 1992), karsts were developed in Cambrian and Permian carbonate rocks interbedded with gypsum, with a large network of active caves and saturated aquifers. There is the possibility that psychrophile stygobites, such as copepods, may be present, but they have not been reported (see Chapter 28).

Salt

Many caves are known in salt rock (Spain, Austria, Central Asia, etc.), but usually without cave species. One of them, the Kitum Cave on Mt. Elgon in Kenya, was partly dug by elephants (*Loxodonta africana*), wanting to eat the salt (Redmond, 1986).

Sulphurous habitat in karst

A peculiar continental ecosystem based on chemoautotrophic food production by bacteria, associated with sulphurous water, has been discovered in 1986 in the southern Dobrodgea karst area of Romania; this is the Movile Cave system (Sarbu, 1990). This is the first study of a type of ecosystem possibly frequent in limestone with sulphurous springs (see Chapter 17). Several caves presumably of this type or partially of the same type have recently been recorded, such as the Cueva del Azufre in Mexico (Langecker et al., 1996).

The Movile Cave consists of a network of straight galleries partially flooded by water at 20ºC, with a concentration of dissolved hydrogen sulphide between 17 and 41 mg ℓ^{-1} (Marin and Nicolescu, 1993), and air-bells with a low oxygen concentration (10%), and a high level of carbon dioxide (2.4%) and methane (1%). A mat of bacteria develops at the air/water interface and on the foot of the walls of the air bells. The bacteria oxidize hydrogen sulphide and use the energy for carbon fixation and the synthesis of organic molecules; heterotrophic bacteria and fungi of the mat feed upon these molecules or chemoautotrophic bacteria. Microscopic terrestrial and aquatic invertebrates feed upon the mat and form a hypogean community with the predators (Sarbu and Popa, 1992).

The majority of species are troglobitic and stygobitic, endemic to the Movile system, or to subterranean water of the karstic plateau of Dobrogea which is a pontic refuge. The species have different degrees of eye regression; thus, the system has seen several colonization periods (Decu and Juberthie, 1994).

The base of the food supply of this cave system is similar to that of the deep oceanic vents, but it differs by the presence of two phases, one aquatic and one aerial, with correlated fauna.

Paleo-cryptokarsts

These consist of ancient paleokarsts secondarily covered with sediments or developed under a generally permeable cover (alluvial deposit, continental deposit, volcanic ash). One example was described by Boulvert and Salomon (1988) in the Oubangui basin in Central Africa, where old paleokarsts formed in the precambrian carbonate beds are covered by continental sediments (Fig. 1.12). On the surface, there is a pseudokarstic morphology which consists of large ponds, about 1 km in diameter, formed in sand–clay sediments, with a ring of vegetation and groundwater in the centre during the rainy season. This type of pseudokarstic morphology is reported in other tropical plains (Colombia, India, Ivory Coast, China and Malagasy).

The cryptokarst has only the saturated zone and an aquatic habitat in cracks and large voids whose fauna can be studied by baited nets in boreholes.

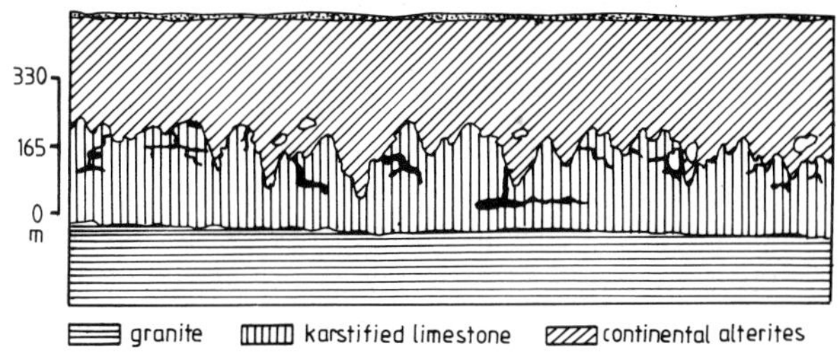

Fig. 1.12. Diagram of a paleo-cryptokarst in the Oubangui basin. (After Boulvert and Salomon, 1988.)

ABIOTIC AND BIOTIC FACTORS

In early views on subterranean environments, authors claimed that the climate of caves was constant, compared with surface environment. It is now recognized that some climatic and biotic variations occur in hypogean habitats, and that darkness is the only constant factor. In addition, the karstic system appears not to be closed, but open, receiving many inputs from the surface.

Darkness

The hypogean habitats are characterized by the total lack of light, more exactly the lack of photons and the lack of photoperiod, with consequences on eyes and rhythms of locomotory activity (see Chapter 7).

Usually three environmental zones are recognized (entrance, twilight zone, and dark zone) related to the change in environmental light conditions from the entrance to the deepest part of caves. Sometimes the dark zone is subdivided into a variable-temperature zone (transition zone) and a constant-temperature zone. In tropical karsts the dark, long horizontal passages are often interrupted by hollows with entrances and twilight zones. In the mesovoid shallow substratum, only the dark zone exists. Each zone supports a different community.

A zonation of plants has been described from the entrance to the dark zone, directly induced by the decrease of light. The leafy plants and ferns of the entrance are replaced by hepatics and mosses, and then by algae on the rocks (Maheu, 1906; Lämmermayr, 1916; Dobat, 1970).

Temperature

Generally, the dark zone in caves and karst systems has a temperature near to the mean annual surface temperature above the cave. The temperature of hypogean habitats depends on the altitude and latitude, and on the morphology of the caves. In equatorial and tropical karsts between 0 and 100 m in elevation, the caves have an average temperature near 25°C (Guinea, Somalia, Tanzania, Cuba, Venezuela, Sarawak, etc.), which decreases to 16°C at an altitude of 1600 m (East Africa), and 9–10°C at 3000 m (Guatemala). In the temperate zone, the temperature of Pyrenean caves decreases from around 15°C at sea level, to 6°C at 1300 m and around 0°C in ice caves at an altitude of 2500 m; those in the Alps decrease from 14°C to 4°C at an altitude of 1400 m, and 0°C near 3000 m. In northern Scandinavia and Canada the caves have temperatures between 0 and 5°C.

The morphology of caves can modify the temperature of a part of the habitat. This is the case in some descending caves which act as cold traps; for example, Gèze (1965) mentions the Enrico Revel Abisso in Italy where the temperature is 8°C at –25 m and 1.3°C at –304 m.

The temperature is unusually high in other caves. In some tropical caves the number of bats is so high in the rooms that the temperature increases to 30–40°C – for example, Cueva de los Majaes, Cueva del Mudo, and Cueva Bariay in Cuba; the species diversity and abundance of individuals are extraordinary in these cases. In the temperate zone, many caves or subterranean aquifers have a flow of hot thermal water; stygobitic animals are adapted to these relatively high temperatures. The most striking case is *Thermosbaena mirabilis,* which occurs in a spring in southern Tunisia at 45°C. In addition, some caves have an atmosphere with an unusually high temperature of thermal origin, for example, the Pestera lui Adam in Romania; in these caves some tropical species have survived through the glaciation in the temperate zone.

The amplitude of thermal variations is another important feature. The dark zone of static caves without air flow, and deep fissures in limestone, has an almost constant temperature (Fig. 1.13). In the variable-temperature zone, and in ventilated passages of the dark zone, the annual variation in temperature can reach five or more degrees in temperate caves. Variations may increase with latitude and altitude. In the mesovoid shallow substratum in the Pyrenees, the temperature can oscillate between 2°C and 15°C during the year. Thus, the early view that true troglobites can only inhabit zones with a constant temperature have not been confirmed. Eggs and embryonic development of troglobitic beetles need lower temperatures than imagined, and temperature becomes a limiting factor to the altitude range of these species.

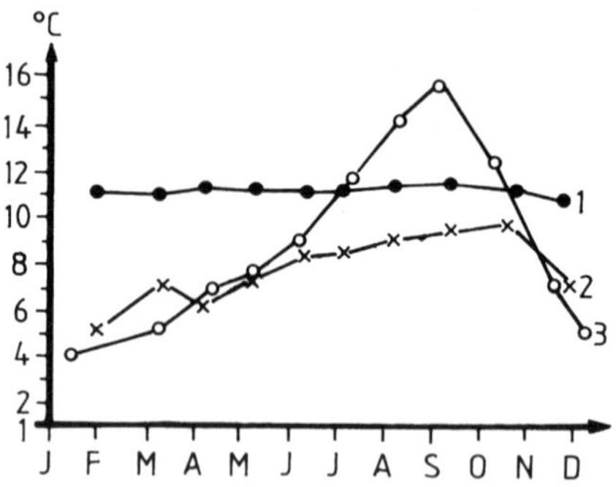

Fig. 1.13. Annual variations of temperature in (1) the static dark zone, altitude 400 m, (2) the variable dark zone, altitude 400 m, and (3) the mesovoid shallow substratum (MSS), altitude 500 m, in the northern central French Pyrenees.

Humidity

In caves, the humidity is usually very high (relative humidity >95%) and can frequently reach saturation point. This factor is important for terrestrial troglobites. Their cuticle lacks impermeable wax layers, so that the level of transpiration through the integument is high. Therefore troglobites cannot survive for long in dry air [Vannier and Juberthie (1977): cave beetles; Ahearn and Howarth (1982): cave crickets and spiders].

Microclimate

Most troglobites cannot fly. They walk on cave walls or in cracks and live in a microclimatic space characterized by thermic and hydric exchanges between the air and the wet walls (Juberthie, 1969). Thus, the climatic factors usually measured in the axis of passages are not representative of their climatic environment. In ventilated galleries, such as in the St. Catherine Cave (Ariège), the speed and the amplitude of changes in microclimatic temperature and evaporation near the walls are the main factors which explain the presence or absence of the troglobite trechine beetle *Aphaenops* in the cave, and determine their migration from the cracks

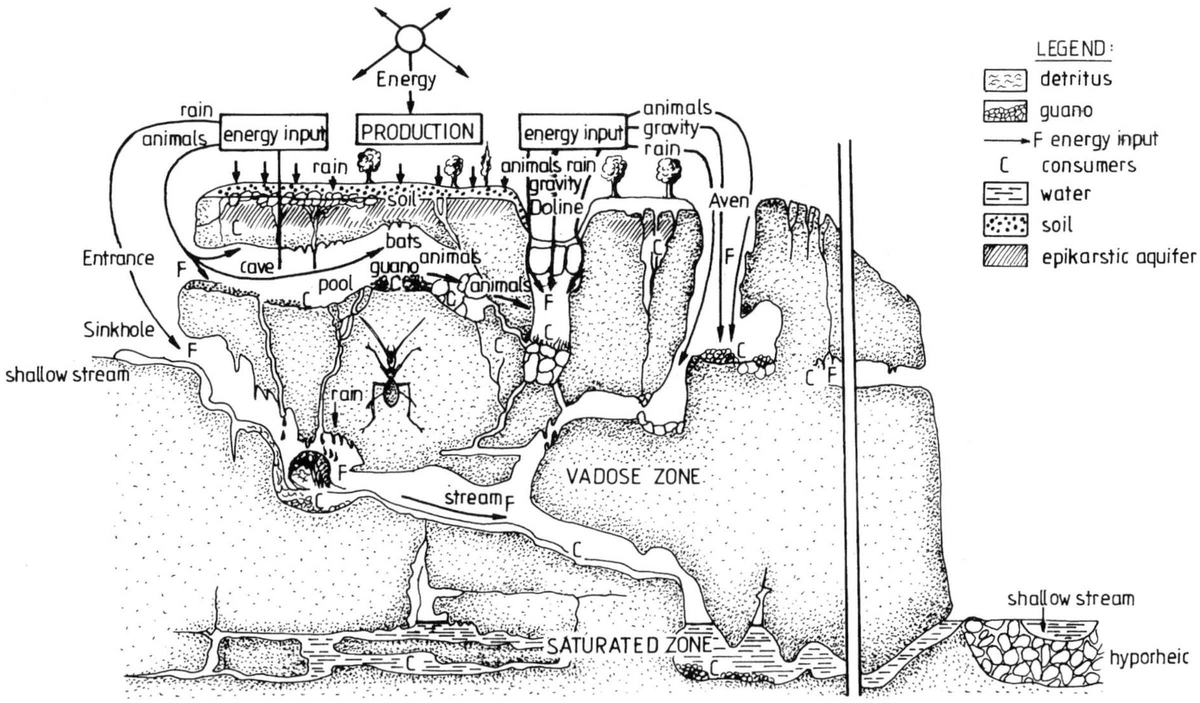

Fig. 1.14. Diagram of food input in a karstic system in the temperate zone.

to the cave in springtime and their return to the cracks in autumn.

Energy sources and trophic dynamics in karsts; flood factors

The energy sources for the karst ecosystem are allochthonous, except in the Movile cave type; they originate from the production of the surface vegetation cover, and secondarily from animals that inhabit the soil and vegetation (see Chapters 12–19).

The nutrients are carried into the underground habitat in several ways (Fig. 1.14): the groundwater percolating through the fissures and cracks of the exokarst and in the vadose zone; the streams entering the karst through sinkholes and introducing debris (vegetable, soil, and animal) into galleries, especially during floods. Hawes (1939) called this phenomenon the "flood factor". Barr Jr and Kuehne (1971) showed the importance of seasonal flooding in the Mammoth Cave (Fig. 1.15), Jegla and Poulson (1970) described the impact of the flood factor on the reproduction of crayfish, Turquin on the age pyramid of the amphipod crustacean *Niphargus*, and Rouch (1986) the impact on the washing-out of crustaceans from the karst. The flood factor acts also in the vadose zone. Nutrients are also carried into caves by air, by bats or birds (guano), and reached by deep-penetrating roots, especially in tropical caves and lava tubes (see, for instance, Jasinska and Knott in Chapter 15 of this volume).

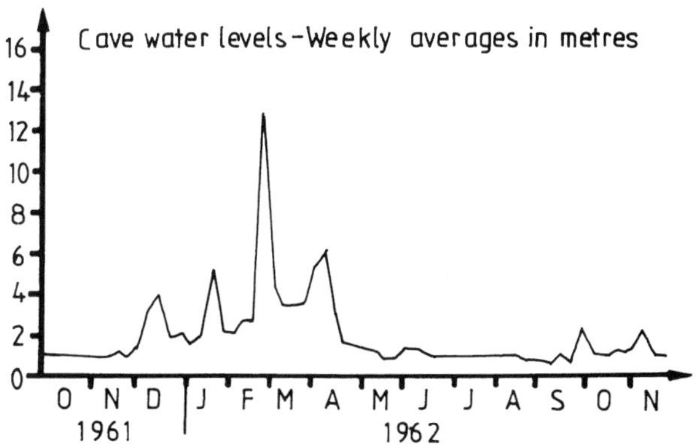

Fig. 1.15. Flood in Mammoth Cave. (After Barr Jr and Kuehne, 1971.)

Long-period changes and oscillation (Quaternary)

Recent data about oscillations in temperature and precipitation during the late Quaternary, and the best-known changes in vegetation during the Würm glaciation, suggest a scenario of long-term change in the subterranean environment, cyclic change in the size of populations, and cyclic formation of mesovoid shallow substratum with new empty habitats open to colonization, and providing the potential for cave speciation (Peck, 1980; Juberthie, 1988).

The Würm glaciation decreased cave temperatures by approximately 5–6°C in southern France. A phase of cold dryness occurs at the beginning of glacial retreat, with a change from forest to steppic cover on the karst surface, and a decrease of primary production and therefore the amount of organic matter entering the subterranean ecosystem. These oscillations in climate and food supply have induced drastic fluctuations of population size, with occasional bottlenecks and founder effects, the selection of psychrophile genotypes during a glacial period, and speciation originating from pre-existing cave species.

PSEUDOKARSTS (HYPOGEAN SYSTEM IN ROCKS OTHER THAN LIMESTONE AND DOLOMITE)

In the old Precambrian shield (Baltic, Guyana, Africa) most of the known caves are formed in primary rocks (gneiss, quartzites, diabase sills, granite, sandstones), sometimes in Quaternary boulder moraines at higher latitudes: (Sweden), and under the crust of altered primary rock (laterite = alterite) in the tropical zone.

Caves in quartzites

This type of cave (Fig. 1.16) is found in the old

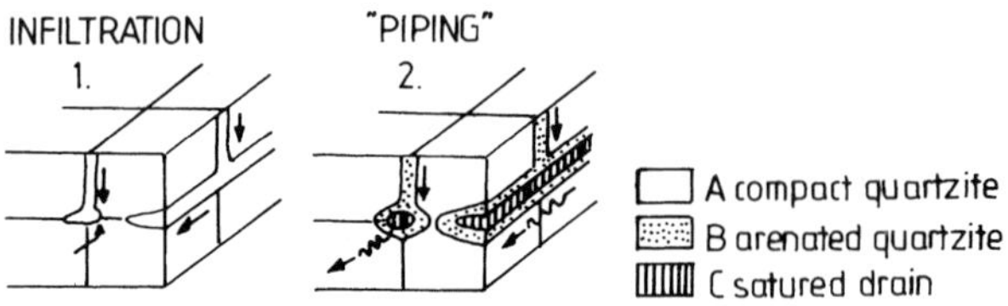

Fig. 1.16. Origin of caves in quarzites. The massive quarzite is transformed into sand (arenatization), through the dissolution of intergranular siliceous cement. The caves were formed by mechanical piping along joints and bedding planes. (After Galan, 1995 and Martini, 1982.)

Precambrian and Ordovician substratum of Africa, especially in South Africa, under a continental or temperate climate, and in the old Precambrian Guyana shield in tropical South America. More than fifty quartzite caves have been recorded in Transvaal, in a Precambrian quartzite formation immediately underlying the dolomite. About one hundred caves were surveyed in the Ordovician quartzites of Table Mountain in the Cape Province; the longest is Ronan's well with 950 m of development (Martini, 1982, 1987). Some are known in quartzite near Oubangui in the Central African Republic. Many others are known from the Precambrian quartzite of the Roraima group in Venezuela, especially in high isolated plateaus (Tepuys) which are remnants of old erosion surfaces (Urbani, 1977; Galan, 1995).

Caves in diabase sills

These caves have been developed by erosion of diabase sills within quartzite or clay resulting from weathering. The quartzite is seldom transformed into sand, and acts as a roof. Most of the passages are formed from the top of the sill, which suggests that the initial void developed here. From this initial void, in the case of capture of a surface stream, the vadose erosion excavated the cave, presumably at a fast rate. This type has been recorded from the Transvaal, where an unusually long system, Mogoto Cave, has a total of 1615 m of surveyed galleries (Martini, 1987).

Caves in laterites (alterites)

In tropical and equatorial rain forests, and the savanna zone, the primary rocks are covered with a thick layer of laterite which forms a surface crust. Along the edges of the valleys, numerous cracks disjoint the crust, and small caves are formed during the rainy season by the rainwater making pipes through the underlaying laterite (Fig. 1.17). When bats roost in these caves a

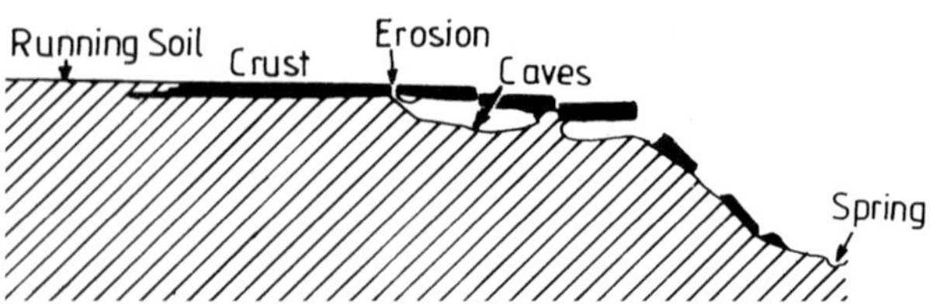

Fig. 1.17. Origin of caves in laterite (Guinea).

rich community of troglobite, and dominant troglophile and trogloxene species, is present on and around the guano. These caves are numerous in Africa (Cameroon, Guinea, Mali, Tanzania, etc.) and in South America (Kaw mountain in French Guiana, etc.).

Caves in sandstone rocks

These represent one of the most frequent cave types in occidental and oriental Africa, but most of them have not been studied biologically. They are found, for example, in the sandstone outcrop of Bandiagara, 300 km long, in Mali, where they are used by the Dogon people for burial and storage chambers; in the sandstones of the Ouadda outcrop, 1000 km long, and the Gadzi–Carnot plateau in the Central African Republic; in Cameroon around Yaoundé; and in Nigeria. They are also found in Europe, for example in the Czech Republic, where Teplice cave has more than 1 km of passages in sandstone which were eroded by a stream, in France and in Italy (Choppy, 1986).

Caves in granite and gneiss

In the Precambrian Guyana shield and in South Africa, numerous large granitic blocks scattered around the foot of inselbergs have formed small caves, where bats roost or birds nest, colonized by Orthoptera, Diptera and guanophile fauna.

Some caves are known in gneiss. For example, a cave near Znojmo (Moravia) has a system almost 2 km long and 40 m deep developed in metamorphic orthogneiss, and another cave can be seen near Yaoundé (Cameroon).

Caves in volcanic rocks

Several types of caves are recognized in volcanic rocks: lava tubes, gas bubble caves, tectonic caves, caves from shrinkage during cooling, volcanic screes caves.

Lava tubes

They are the most numerous and the most studied volcanic habitat. They are specific to basaltic volcanoes. From this type of volcano, the lava flows in two forms called a'a and pahoehoe. Lava tubes form almost exclusively in pahoehoe basalt. Pahoehoe is very fluid, very hot, and flows like a river. These flows tend to crust over as the surface cools, and the crust may thicken and become a roof. When the lave level subsides, an empty tube is formed. The pahoehoe lava can flow at great distances from the vent and form very large lava tubes. They are well known and their fauna, terrestrial or marine, is studied in oceanic archipelagoes (Hawaii, Galapagos, Canary Islands, Madera, Réunion island, Mauritius, Japan, especially the lava tubes of the Fuji Jama, Jeju islands in Korea, Fije, Tonga in Oceania) and on the mainland (U.S.A., Queensland in Australia, Kenya and Cameroun in Africa, etc.) The energy sources are organic matter deposited by percolating water and tree roots which develop on the lava tube crust. In Hawaii the main energy source is tree roots that support numerous troglobitic invertebrates (Uéno, 1971, 1977, 1995; Howarth, 1973, 1983; Peck, 1973, 1982, 1994a,b; Wilkens et al., 1986; Hoch, 1990; Oromi and Izquierdo, 1994; Borges and Oromi, 1994; Juberthie, 1995).

Mesovoid shallow substratum (in rocks other than limestone)

The mesovoid shallow substratum was discovered and described by Juberthie et al. (1980) in the central Pyrenees, in non-limestone rocks such as schists, gneiss and granite, without direct connections with caves (Fig. 1.18). Later it was found in some screes

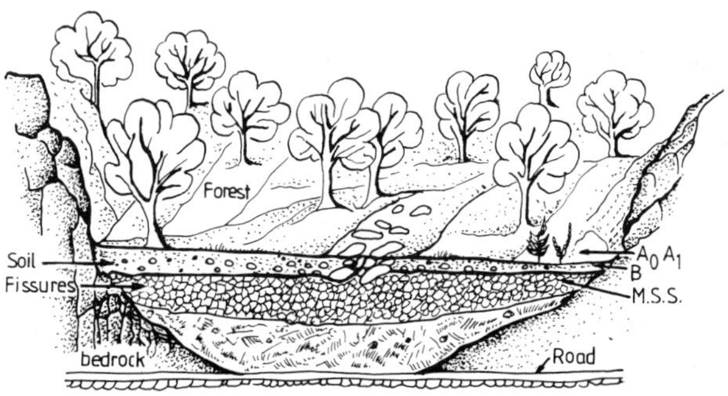

Fig. 1.18. Diagram of a mesovoid shallow substratum (MSS) ecosystem in the central Pyrenees and the Romanian Carpathians.

at the base of limestone cliffs by Juberthie et al. (1981), and in volcanic rocks in the Canary Islands by Oromi et al. (1986). Early reports on troglobites in cracks and mesovoids of volcanic rocks in Hawaii by Howarth (1981, 1983) have been interpreted as inhabitants of similar superficial underground habitats (mesovoid habitat). In Japan, Uéno (1977, 1981, 1995) found eyeless beetles from colluvial deposits under steep slopes at the sides of gullies or narrow shallow streams, in spaces between the soil layer and the bedrock; usually the thickness of the soil mantle was 50 cm or more. He called this peculiar environment, a mesovoid shallow substratum along the slopes of streams, "Upper Hypogean habitat".

At present, the mesovoid shallow substratum has been described in European mountains (Pyrenees, Alps, Carpathians, Stara Planina and Rhodope) and in Asia (Turkey, Japan, Taiwan). Presumably it is present in all mountains in the temperate zone. By contrast, it seems absent in tropical areas, where the voids are filled with laterite and clay.

The abbreviation MSS is the most useful for this habitat: it corresponds to Milieu Souterrain Superficiel in French, Mesovoid Shallow Substratum in English, Medio Subterraneo Superficial in Spanish, Medio Sotterraneo Superficiale in Italian, etc.

Structure

The mesovoid shallow substratum is composed of a network of small voids and cracks, isolated from the surface by soil or by cracks which have been filled in. Two main types of mesovoid shallow substratum are found. The first consists of screes along the slope of valleys and volcanic scoriaceous layers, covered by soils; the second consists of cracks and interconnected mesovoids in the superficial part of the rocks or in lava flows.

Origin

Mesovoid shallow substratum is formed in one of two ways:

- The first is due to frost, which fragments non-volcanic rocks into stones usually of small size. These fall down the slopes and gather in several layers of particles of various sizes on the slopes of valleys. When soil has covered the screes a new mesovoid shallow substratum is available for colonization by cave or soil invertebrates. The cracks in the superficial zone of the rocks have the same origin, and they are enlarged by mechanical expansion of the rock.
- The second is volcanic mesovoid shallow substratum formed by some types of cracked lava flow or by scoriaceous layers, sometimes intermixed, generated by volcanoes (Fig. 1.19).

Abiotic factors

The abiotic factors are similar to those of caves in the same area. The major difference consists in a greater range of seasonal variation of temperature, up to 10°C or more. In winter in the Pyrenees, at an elevation of around 1000 m, the average temperature in the mesovoid shallow substratum is 2–4°C; the troglobite bathysciine beetle *Speonomus* cannot reproduce at this temperature (Juberthie-Jupeau, 1980).

Food supply

The energy sources in the mesovoid shallow substratum are allochthonous, originating from the production of the surface vegetation cover as for karst (Fig. 1.20). However, carnivory is high compared to the karstic cave system, because many soil invertebrates enter

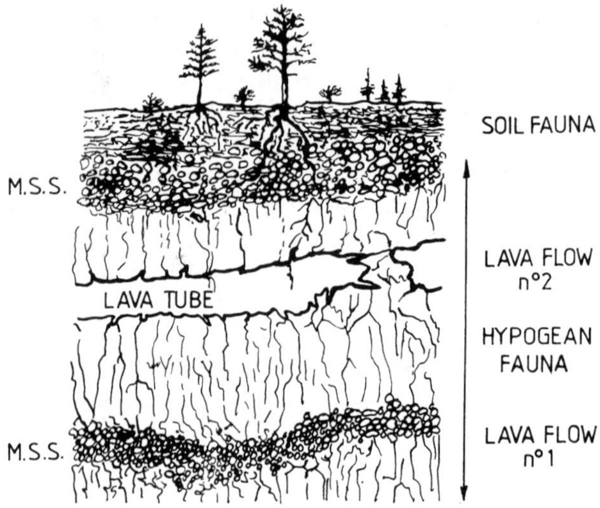

Fig. 1.19. Diagram of a volcanic mesovoid shallow substratum (MSS) in the Canary Islands. (Combined from Medina (1991) and Juberthie and Decu (1994).)

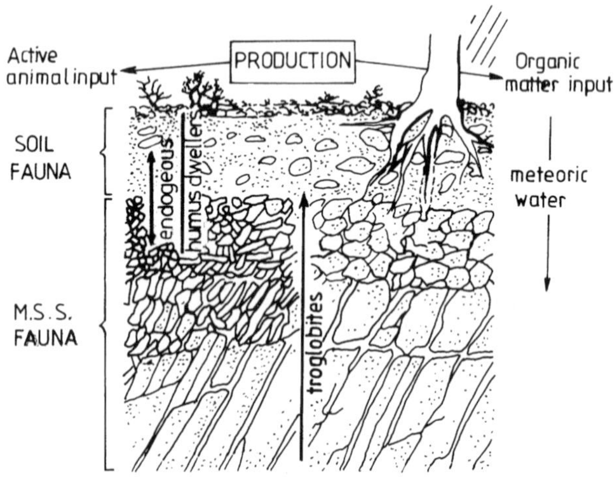

Fig. 1.20. Diagram of food chains and input in the Pyrenean mesovoid shallow substratum (MSS).

the mesovoid shallow substratum. For example, Gers (1995) showed that the diet of the trechine beetle *Aphaenops* consists in large part of larvae of the troglophilic soil insect *Tryphleba hyalinata* (Diptera: Phoridae), which also reproduces in the mesovoid shallow substratum.

Fauna

Two communities are intermixed in the mesovoid shallow substratum (Juberthie et al., 1980, 1981; Racovitza, 1984; Medina, 1991; Gers, 1992). The first community is composed of typical troglobite and troglophile species (Collembola, Diplura, Coleoptera and other insects, Chilopoda, Diplopoda, Isopoda, Pseudoscorpiones, Acari, etc.). Many species inhabit both caves and the mesovoid shallow substratum (for example the Coleoptera *Aphaenops*, *Speonomus hydrophilus,* and *S. zophosinus*), others are only found in the mesovoid shallow substratum (for example, *Speonomus colluvii*), while others again live only in caves – indeed, I have never found spiders or the coleopteran *Geotrechus* in the mesovoid shallow substratum in the Pyrenees. The most highly adapted cave beetle species are found in the mesovoid shallow substratum – for example, *Aphaenops*. The second community is composed of soil dwellers that actively enter the mesovoid shallow substratum or drift in with the flow of meteoric water. Some are prey for troglobites, for example Collembola or Diptera larvae (Gers, 1993, 1995), while others, such as Chilopoda, are predators on troglobites.

Evolution of the mesovoid shallow substratum habitat and communities

Gers (1992) showed a cyclic evolution of the mesovoid shallow substratum with three phases: the juvenile stage with colonization from soil or cave dwellers, the mature stage with an equilibrium community, and the old stage characterized by collapsed voids and disappearance of fauna. The average time of a cycle ranges from ten thousand years to several times as long. In the central Pyrenees, the most recent cycle with genesis of mesovoid shallow substratum in screes began 12 000–13 000 years ago when forests replaced the post-Würm steppic cover, following climatic changes characterized by an increase of temperatures and rainfall (Juberthie et al., 1990).

Range of troglobites and major routes of cave colonization

In the central Pyrenees, where karstic limestone areas and their cave systems are in contact with the mesovoid shallow substratum in schistose and other rocky areas, it has been demonstrated that the range of troglobitic beetles of the subfamily Bathysciinae includes both karstic and nonkarstic areas. It can be deduced that the route of cave colonization by invertebrates is not only through the karst surface and cave entrances but also through the mesovoid shallow subsystem. The soil ancestors of cave dwellers entered the hypogean habitat across the mesovoid shallow substratum as well as through the karstic surface. In volcanic areas, the colonization of new lava tubes originates from the stock of troglobites in the volcanic mesocavernous shallow subsystem and in old lava tubes.

HYPOTHETICAL CAVE ENVIRONMENTS ON OTHER PLANETS, SUCH AS MARS

Recent data suggests the hypothesis of primitive life on Mars based on possible fossil Archaebacteria.

High-resolution images and information from the US Mariner 9 and Viking missions (Carr, 1981; Greeley, 1991) suggest that the predominant volcanic rocks were extremely fluid (basaltic) and may contain lava tubes. The high-resolution images from the Viking Orbiter spacecraft may be interpreted as open channels and roofed channel segments.

The possibility of sulphurous volcanism on the Moon, the presence of small quantities of water and presumably volcanic carbon dioxide, perhaps with higher levels in the past, the discovery of Archaebacteria in sulphurous and other extreme habitats (hydrothermal vents in oceanic ridges and Movile cave in the mainland) which can develop at high temperature (120ºC), suggest the possible presence on Mars of fossil archaebacteria and bacteria related to hypogean sulphurous habitats.

The role of the roof of lava tubes in protecting against radiation, and possibly low water pressure and sulphurous vapours in these lava tubes, would be favourable for the presence of these fossil archaebacteria, but their survival is more problematic.

REFERENCES

Agassiz, L., 1851. Observation on the blind-fish of Mammoth Cave. *Am. J. Sci.,* 11: 127–128.

Ahearn, G.A. and Howarth, F.G., 1982. Physiology of Cave Arthropods in Hawaii. *J. Exp. Zool.,* 222: 227–238.

Audra, Ph., 1994. *Karst Alpins. Genèse de Grands Réseaux Souterrains. Karstologia Mém.* 5, 277 pp.

Barber, H.S., 1931. Traps for cave-inhabiting insects. *Elisha Mitchell Sci. Soc.,* 46(2): 259–266, Plate 23.

Barr Jr., T.C., 1968. Cave ecology and the evolution of troglobites. In: Th. Dobzhansky, M.K. Hecht and W.C. Steere (Editors), *Evolutionary Biology*, Vol. 2. North-Holland, Amsterdam, pp. 35–102.

Barr Jr., T.C. and Kuehne, R.A., 1971. Ecological studies in the Mammoth cave ecosystem in Kentucky. II. The ecosystem. *Ann. Spéléol.,* 26: 47–96.

Bilimek, D., 1867. Fauna der Grotte Cacahuamilpa in Mexico. *Verhandl. Zool. Bot. Ges. Wien,* 17: 901–908.

Borges, P.A.V. and Oromi, P., 1994. Azores. In: T.I.C. Juberthie and V. Decu (Editors), *Encyclopaedia Biospeologica,* Vol. I. Société de Biospéologie, Moulis–Bucarest, pp. 605–617.

Bou, C., 1974. Les méthodes de récolte dans les eaux souterraines interstitielles. *Ann. Spéléol.,* 29(4): 611–619.

Bou, C., 1979. La grotte des Trois-Cloches et son intérêt scientifique. *Bull. Féd. Tarn Spéléo-Archéol.,* 16: 3–10.

Bou, C. and Rouch, R., 1967. Un nouveau champ de recherches sur la faune aquatique souterraine. *C.R. Acad. Sci. Paris,* 265: 369–370.

Boulvert, Y. and Salomon, J.N., 1988. Sur l'existence de paléo-cryptokarst dans le bassin de l'Oubangui (République Centrafricaine). *Karstologia,* 11–12: 37–48.

Bourrouilh-Le Jan, F.G., 1990. Diagenèse des carbonate de plates-formes, récifs et mangroves en Atlantique et Pacifique. Contrôle de la diagenèse par les variations thermo-glacio-eustatique d'émersion-submersion. *Thèse Doc. Etat, Univ. P. et M. Curie, Paris VI.* 216 pp.

Boury-Esnault, N., Harmelin, J.-G. and Vacelet, J., 1993. Les abysses méditerranéennes à vingt mètres de profondeur? *Recherche,* 256(24): 849–851.

Bretschko, J. and Klemens, W.E., 1986. Quantitative methods and aspects in the study of the interstitial fauna of running waters. *Stygologia,* 2: 297–316.

Brignoli, P.M., 1973. Il popolamento di ragni nelle grotte tropicale. *Int. J. Speleol.,* 5: 325–336.

Brignoli, P.M., 1980. La valeur biogéographique des Araignées cavernicoles. *Verh. 8e Int. Arachnol. Kongr. Wien,* pp. 427–432.

Brousse, R., 1985. L'âge des îles du Pacifique : volcanisme et installation corallienne [The age of the islands in the Pacific Ocean: volcanism and coral-reef build up]. In: *Proc. 5th Int. Coral Reef Congress, Tahiti,* pp. 389–397.

Camacho, A.I. (Editor), 1992. *The Natural History of Biospeology,* Monogr. Mus. nat. Cienc. nat. Graphicas. Mar-Car, SA, Madrid, 551 pp.

Carr, M.H., 1981. The Surface of Mars. Yale University Press, New Haven, 232 pp.

Chapman, P., 1984. The invertebrate fauna of the caves of Gunung Mulu National Park. *Sarawak Mus. J.,* 30(51): 1–18.

Chappuis, P.A., 1924. Sur les Copépodes et les Syncarides des eaux souterraines de Cluj et des Monts Bihar. *Bull. Soc. Sci. Cluj,* 2(2): 157–182.

Chappuis, P.A., 1946. Un nouveau biotope de la faune souterraine. *Notes biospéol.,* 5: 1–35.

Chevalier, J.P., 1973. Geomorphology and geology of Coral Reefs in French Polynesia. In: *Biology and Geology of Coral Reefs. Tahiti.* Academic Press, N.Y., p. 113–141.

Choppy, J., 1986. Les karsts de gypse italiens. *Karstologia,* 8: 39–46.

Culver, D.C., 1971. Analysis of simple cave communities III. Control of abundance. *Am. Midl. Nat.,* 85: 173–187.

Culver, D.C., 1982. *Cave Life: Evolution and Ecology.* Harvard University Press, Cambridge, MA.

Culver, D.C., Kane, T.C. and Fong, D.W., 1995. *Adaptation and Natural Selection in Caves.* Harvard University Press, Cambridge, MA, 223 pp.

Cvetkov, L., 1968. Un filet phréatobiologique. *Bull. Inst. Zool. Mus. Acad. Bulgare Sci., Sofia,* 27: 215–215.

Danielopol, D.L., 1972. Sur la présence de *Thaumatocypris orghidani* n. sp. (Ostracoda, Myodocopida) dans les grottes de Cuba. *C.R. Acad. Sci. Paris, sér. D,* 274: 1390–1393.

De Kay, J.E., 1842. Description of *Amblyopsis spelaeus. Zool. New York,* pt 3: 187.

Decu, V. and Juberthie, C., 1994. *Un monde retrouvé: la grotte de Movilé.* Encyclopaedia Universalis, Pais, pp. 224–228.

Delamare-Deboutteville, C., 1960. *Biologie des Eaux Souterraines Littorales et Continentales.* Hermann éd., Paris, 740 pp.

Delarouzée, Ch., 1857. Trois Coléoptères nouveaux qu'il trouvé dans la caverne de Bétharram (Hautes-Pyrénées). *Ann. Soc. Entomol. Fr.,* 5: 93.

Delay, B., 1968. Données sur le peuplement de la zone de percolation temporaire. *Ann. Spéléol.,* 20(4): 705–716.

Delay, B., 1978. Milieu souterrain et ecophysiologie de la reproduction et du développement des Coléoptères Bathysciinae hypogés. *Mem. Biospel.,* 5: 349 pp.

Dobat, K., 1970. Considérations sur la végétation cryptogamique des grottes du Jura Souabe (Sud-Ouest de l'Allemagne). *Ann. Spéléol.,* 25(4): 871–907.

Ford, D.C. and Williams, P.W., 1989. *Karst Geomorphology and Hydrology.* Unwin Hyman, London, 601 pp.

Ford, T.D., 1976. The geology of caves. In: T.D. Ford and C.H. J. Cullingford (Editors), *The Science of Speleology.* Academic Press, London, pp. 11–60.

Fridvaldszky, E. and Fridvaldszky, J., 1857. Drei neue Grotten-Käfer aus Ungarn. *Verhand. Zool. Bot. Wien,* 7: 44–46.

Fridvaldszky, J., 1865. Adatok a magyarhoni barlangok faunajahoz. *Math. ès termés-zettud. Köslem.,* 3: 27–43.

Galan, C., 1995. Exploracion y estudio de cavidades en rocas siliceas precambricas del grupo Roraima, Guyana venezolana: una sintesis actual. *Karraitza,* 4: 3–12.

Gers, C., 1992. *Ecologie et Biologie des populations d'Arthropodes terrestres du milieu Souterrain Superficiel: fonctionnement et écologie évolutive, Thèse.* Université Paul Sabatier, Toulouse, 392 pp.

Gers, C., 1993. Population of *Triphleba hyalinata* (Meigen, 1830; Diptera, Phoridae) in the superficial underground compartment (mesocavernous shallow substratum or rocks fissures and voids in screes). *Eur. J. Soil Biol.,* 29(2): 35–41.

Gers, C., 1995. Stratégie alimentaire de Coléoptères troglobies du genre *Aphaenops* (Coleoptera, Trechinae). *Mém. Biospéol.,* 22: 35–45.

Gèze, B., 1965. *La Spéléologie Scientifique.* Edition du Seuil, Paris, 190 pp.

Gibert, J., 1986. Ecologie d'un système karstique jurassien; hydrogéologie, dérive animale, transits de matières, dynamique de la population de *Niphargus* (Crustacé Amphipode). *Mém. Biospéol.,* 13: 1–376.

Ginet, R., 1960. *Ecologie, Ethologie et Biologie de* "Niphargus" *(Amphipodes Gammaridés hypogés).* Maurice Declume, Lons-le-Saunier, 254 pp.

Ginet, R. and Decou, V., 1977. *Initiation à la Biologie et à l'Écologie Souterraines.* Jean-Pierre Delarge, Paris, 345 pp.

Greeley, R., 1991. Lava tubes in the Solar System. In: G. Thomas Rea (Editor), *6th International Symposium on Vulcanospeleology.* pp. 223–230.

Guilcher, A., 1992. Relations entre la circulation karstique et la dolomitisation dans la structure des atolls. In: J. Salomon and R. Maire (Editors), *Karst et Évolutions Climatiques.* Presse Universitaire Bordeaux, pp. 105–114.

Hawes, R.S., 1939. The flood factor in the ecology of caves. *J. Anim. Ecol.,* 8: 1–5.

Hoch, H., 1990. Cavernicolous Meenoplidae (Homoptera Fulgoroidae) from Australia. *Occas. Pap. Bishop Mus.,* 30: 188–203.

Holsinger, J.R., 1980. *Stygobromus canadensis,* a new subterranean amphipod crustacean (Crangonyctidae) from Canada, with remarks on Wisconsin refugia. *Can. J. Zool.,* 58: 290–297.

Holthuis, L.B., 1973. Caridean shrimp found in land locked saltwater pools at four Indo-West Pacific localities (Sinai Peninsula, Funufati Atoll, Maui and Hawaii Islands). *Zool. Verh.,* 128: 1–48.

Howarth, F.G., 1973. The cavernicolous fauna of Hawaiian lava tubes I. Introduction. *Pac. Insects,* 15: 139–151.

Howarth, F.G., 1981. Non-relictual terrestrial troglobites in the tropical Hawaiian caves. In: B.F. Beck, Americus (Editor), *Proceedings of the 8th International Congress on Speleology, Bowling Green,* Vol. 2, pp. 539–541.

Howarth, F.G., 1983. Ecology of cave Arthropods. *Ann. Rev. Entomol.,* 28: 365–389.

Iliffe, T.M., 1981. The Submarine Caves of Bermuda. In: B.F. Beck, Americus (Editor),*Proceedings of the 8th International Congress on Speleology, Bowling Green,* Vol. 1, pp. 161–163.

Iliffe, T.M., 1990. Crevicular dispersal of marine cave fauna. *Mém. Biospéol.* 17: 93–96.

Iliffe, T.M., 1991. Anchialine cave fauna of the Galapagos Islands. In: M.J. James (Editor), *Galapagos Marine Invertebrates,* Vol. XV. Plenum Press, New York, pp. 209–231.

Iliffe, T.M., 1992. Anchialine cave biology. In: A. Camacho (Editor), *The Natural History of Biospeleology.* Madrid, pp. 615–636.

Iliffe, T.M. and Sarbu, S.M., 1990. Anchialine caves and cave fauna of the South Pacific. *N.S.S. News,* 88–97.

Jeannel, R., 1926. *Faune Cavernicole de la France, avec une Étude des Conditions d'Existence dans le Domaine Souterrain.* P. Lechevalier, Paris, 334 pp.

Jegla, T.C. and Poulson, T.L., 1970. Circadian rhythms – I – Reproduction in the cave crayfish, *Orconectes pellucidus inermis. Comp. Biochem. Physiol.,* 33: 347–355.

Juberthie, C., 1969. Relations entre le climat, le microclimat et les *Aphaenops cerberus* dans la grotte de Sainte-Catherine (Ariège). *Ann. Speleol.,* 24(1): 75–104.

Juberthie, C., 1988. Paleoenvironment and speciation in the cave beetle *Speonomus delarouzeei* (Coleoptera Bathysciinae). *Int. J. Speleol.,* 17: 31–45.

Juberthie, C., 1989. Données sur la faune souterraine de la Guadeloupe et de ses dépendances. *Mém. Biospéol.,* 16: 71–85.

Juberthie, C., 1995. Les habitats souterrains, leur faune, et les modalités de colonisation de l'île de la Réunion. *Mem. Biospeol.,* 22: 73–82.

Juberthie, C. and Decu, V. (Editors), 1994. *Encyclopaedia Biospeologica,* Vols. I and II. Société de Biospéologie, Moulis (France) and Bucarest, 834 pp.

Juberthie, C. and Decu, V. (Editors), 1998. *Encyclopaedia Biospeologica,* Vol. II. Société de Biospéologie, Moulis (France)-Bucarest, 835 pp.

Juberthie, C. and Delay, B., 1981. Ecological and biological implications of the existence of a superficial underground compartment. In: B.F. Beck, Americus (Editor),*Proceedings of the 8th International Congress on Speleology, Bowling Green,* Vol. 1, pp. 203–206.

Juberthie, C. and Juberthie-Jupeau, L., 1975. La réserve biologique du Laboratoire souterrain du C.N.R.S. à Sauve (Gard). *Ann. Spéléol.,* 30(3): 539–551.

Juberthie, C., Delay, D. and Bouillon, M., 1980. Extension du milieu souterrain en zone non calcaire: description d'un nouveau milieu et de son peuplement par les Coléoptères troglobies. *Mém. Biospéol.,* 7: 19–52.

Juberthie, C., Delay, B. and Bouillon, M., 1981. Extension du milieu souterrain superficiel en zone calcaire. *Mém. Biospéol.,* 8: 77–93.

Juberthie, C., Dupré, E. and Jalut, G., 1990. *Aphaenops reberti*: espèce "endogée" du sous-gentre *Geaphaenops* présente dans le M.S.S. de la vallée glaciaire d'Ossau, daté de 12 000 B.P. *Mem. Biospeol.,* 17: 181–190.

Juberthie-Jupeau, L., 1980. Sur la présence d'un rythme de reproduction dans le milieu souterrain superficiel; étude chez les Coléoptères Bathysciinae. In: B.F. Beck, Americus (Editor), *Proceedings of the 8th International Congress on Speleology, Bowling Green,* Vol. 1, pp. 714–716.

Kejonen, A. and Juberthie, C., 2000. Finland. In: C. Juberthie and V. Decu (Editors), *Encyclopaedia Biospeologica,* Vol. III. Société de Biospéologie, Moulis (France) and Bucarest, in press.

Klimchouk, A., Lowe, D. and Sauro, U., 1996. Gypsum karst of the world. *Int. J. Speleol.,* 25(3–4): 3–307.

Lämmermayr, L., 1916. Die grüne Pflanzenwelt der Höhlen. I. Teil (Dritte Mitteilung). *Denkschr. Kaiserl. Akad. Wiss., math.–natwiss. Kl.,* 92: 107–148.

Langecker, T.G., Wilkens, H. and Parzefall, J., 1996. Studies in the trophic structure of an energy-rich Mexican cave (Cueva de las Sardinas) containing sulfurous habitat. *Mém. Biospéol.,* 23: 121–125.

Langenscheid, E., 1986. Höhlen und ihre Sedimente in den Berchtesgadener Alpen. *Forschungsbericht* 10, 95 pp.

Laurenti, J.N., 1768. *Specimen medicum, exhibens synopsin reptilium emendatam cum experimentis circa venena et antidota reptilium Austriacorum.* Viennae, in 8, 219 pp.

Lehmann, H., 1936. Morphologischen Studien auf Java. *Geogr. Abhandl.,* 9: 39–45.

Lehmann, O., 1927. Das Tote Gebirge als Hochkarst. *Mitt. Geol. Ges. Wien,* 70: 201–242.

Leruth, R., 1939. Notes d'hydrobiologie souterraine. VII. Une méthode intéressante pour l'étude de la nappe phréatique. *Bull. Soc. r. Sci. Liège,* 2: 10–15.

Lescher-Moutoué, F. and Gourbault, N., 1970. Etude écologique du peuplement des eaux souterraines de la zone de circulation permanente d'un massif karstique. *Ann. Spéléol.,* 2(4): 765–848.

Lespes, C., 1857. Notes sur quelques insectes des grottes de l'Ariège. *Ann. Sc. nat.,* 283–284, Plate XVII.

Maheu, J., 1906. Contribution à l'étude de la flore souterraine de France. *Ann. Sci. Nat., 9, ser. Bot.* 3: 1–189.

Maire, R. (Editor), 1990. La haute montagne calcaire. *Karstologia, Mém.,* Vol. 3, 731 pp.

Mangin, A., 1975. Contribution à l'étude hydrodynamique des aquifères karstiques. Troisième partie. Constitution et fonctionnement des aquifères karstiques. *Ann. Spéléol.,* 30(1): 21–124.

Mangin, A., 1994. Karst hydrogeology. In: J. Gibert, D.L. Danielopol and J.A. Stanford (Editors), *Groundwater Ecology.* Academic Press, New York, pp. 43–67.

Manning, R.B., Hart Jr., C.W. and Iliffe, T.M., 1986. Mezozoic relicts in marine caves of Bermuda. *Stygologia,* 2(1/2): 156–166.

Marin, C. and Nicolescu, ??., 1993. The geochemistry of groundwater from southeastern Dobrogea, Romania. *Trav. Inst. Speol. Emile Racovitza,* 32: 229–247.

Martini, J., 1982. Karst in Black Reef and Volkberg Group quartzite of the eastern Transvaal escarpment, South Africa. *Bol. Soc. Venez. Espeleol.,* 7: 5–12.

Martini, J., 1987. Les phénomènes karstiques des quartzites d'Afrique du Sud. *Karstologia,* 9: 45–52.

Medina, A.L., 1991. *El medio subterraneo superficial en las Islas Canarias: Caracterizacion y consideraciones sobre su fauna, Thesis.* Universita La Laguna, 205 pp.

Miller, L., 1856. Beschreibung eines neuen Grottenkäfferfauna Ungarans. *Verh. Zool. Bot. Ges. Wien,* 6: 635.

Mitchell, R.W., 1969. A comparison of temperate and tropical cave communities. *Southwestern Nat.,* 14(1): 73–88.

Mohr, C.E. and Poulson, T.L., 1966. *The Life of the Cave.* McGraw-Hill, New York, 232 pp.

Mylroie, J.E. and Carew, J.L., 1990. The flank margin model for dissolution cave development in carbonate platforms. *Earth Surface Processes and Landforms,* 15: 413–424.

Nicod, J., 1976. Le karst des gypses et évaporites associés. *Ann. Géogr.,* 471: 513–554.

Orghidan, T., 1959. Ein neuer Lebensraum des unterirdischen Wassers, das hyporheische Biotop. *Arch. Hydrobiol.,* 55: 392–414.

Oromi, P. and Izquierdo, I., 1994. Canary Islands. In: T.I.C. Juberthie and V. Decu (Editors), *Encyclopaedia Biospeologica,* Vol. I. Société de Biospéologie, Moulis (France)-Bucarest, pp. 631–639.

Oromi, P., Medina, A.L. and Tejedor, M.L., 1986. On the existence of a superficial underground compartment in the Canary Islands. *Act. IXe Congr. Int. Espeleol. Barcelona,* 2: 147–151.

Packard, A.S., 1888. The cave fauna of North America, with remarks on the anatomy and the origin of the blind species. *Mem. Nat. Acad. Sci. U.S.A.,* 4(1): 1–156.

Peck, S.B., 1973. A review of the invertebrate fauna of volcanic caves in western North America. *Bull. Natl. Soc. Speleol.,* 35(4): 99–107.

Peck, S.B., 1980. Climataaic change and the evolution of cave invertebrates in the Grand Canyon, Arizona. *Bull. Natl. Soc. Speleol.,* 42: 53–60.

Peck, S.B., 1982. Invertebrate faunas and zoogeographic significance of lava tube caves of Arizona and New Mexico. *Great Basin Nat.,* 44(3): 405–412.

Peck, S.B., 1987. A review of the cave fauna of Canada, and the composition and ecology of the invertebrate fauna of caves and mines in Ontario. *Can. J. Zool.,* 66: 1197–1213.

Peck, S.B., 1994a. Canada. In: C. Juberthie and V. Decu (Editors), *Encyclopaedia Biospeologica,* Vol. II. Société Biospéologie, Moulis (France) and Bucarest, pp. 381–388.

Peck, S.B., 1994b. Equador. In: T.I.C. Juberthie and V. Decu (Editors), *Encyclopaedia Biospelogica.* Société de Biospéologie, Moulis (France) and Bucarest, pp. 551–560.

Pulina, M., 1992. Glacio-karsts gypseux de la zone polaire et périglaciaire (Exemple du Spitzberg et de Sibérie orientale). In: J. Salomon and R. Maire (Editors), *Karst et Évolutions Climatique.* Presse Universitaire Bordeaux, pp. 267–283.

Racovitza, E.G., 1907. Essai sur les problèmes biospéologiques. Biospeologica, I. *Arch. Zool. Exp. Gén., 4e sér,* 6: 371–488.

Racovitza, G., 1984. Etude faunistique et écologique du Milieu Souterrain Superficiel de la vallée d'Ordincusa (Monts du Bihor). *Trav. Inst. Spéol. "Emile Racovitza",* 23: 27–37.

Rafinesque, C.S., 1822. Description of *Eurycea lucifuga. Kentucky Gazette, N.S.,* 1(9):3.

Rafinesque, C.S., 1832. The caves of Kentucky. *Atlantic J.,* 1(9): 22.

Redmond, I., 1986. Underground elephants. *Youth News,* 34: 3.

Rouch, R., 1968. Contribution à la connaissance des Harpacticides hypogés. *Ann. Speleol.,* 23(1): 1–167.

Rouch, R., 1970. Le système karstique du Baget. 1. Le phénomène de "hémorragie" au niveau de l'exutoire principal. *Ann. Speleol.,* 25(3): 665–709.

Rouch, R., 1986. Sur l'écologie des eaux souterraines dans le karst. *Stygologia,* 2(4): 352–398.

Rouch, R., Juberthie, L. and Juberthie, C., 1968. Peuplement de la zone noyée. Essai d'un etude du karst. *Ann. Spéléol.,* 23(4): 717–733.

Rouch, R., Pitzalis, A. and Descouens, A., 1993. Effets d'un pompage à gros débit sur le peuplement des Crustacés d'un aquifère karstique. *Ann. Limnol.,* 29(1): 15–29.

Rougerie, F. and Wauthy, B., 1986. Le concept d'endo-upwelling dans le fonctionnement des atolls-oasis. *Oceanol. Acta,* 9: 133–148.

Salomon, J. and Maire, R. (Editors), 1992. *Karst et Évolutions Climatique.* Presse Universitaire Bordeaux, 520 pp.

Sarbu, S.M., 1990. The unusual fauna of a cave with thermomineral waters containing H_2S, from southern Dobrogea, Romania. *Mém. Biospéol.,* 18: 193–197.

Sarbu, S.M. and Popa, R., 1992. A unique chemoautotrophically based cave ecosystem. In: A. Camacho (Editor), *The Natural History of Biospeleology,* Madrid, pp. 638–666.

Sbordoni, V., Argano, R., Vomero, V. and Zullini, A., 1977. Ricerche sulla fauna cavernicola del Chiapas (Messico) e delle regioni limitrophe: grotte esplorate nel 1973 e nel 1975. Criteri per una classificazione biospeologica delle grotta. In: *Subterranean fauna of Mexico, Accad. Naz. Lincei, Roma,* 3: 5–74.

Schaufuss, L., 1861. Zwei neue Silphiden-Gattungen. *Entomol. Z. Stettin,* 22: 423.

Schiner, J.R., 1854. Fauna der Adelsberger-Lueger und Magdanenen Grotte. In: A. Schmidt (Editor), *Die Grotten und Höhlen von Adelsberg, Lueg, Planina und Loos.* Braumüller, Wien, pp. 231–272.

Schiödte, J.C., 1849. *Specimen faunae-subterraneae. Bidrag til den underjordische Fauna.* Extr. de *K. Dan. Vidensk. Selsk. Skr., 5 Raekke, Nat. Math. Afd.,* 2 Bind. Copenhague, Luno.

Schmidt, F., 1832. *Leptodirus hohenwarti* n.g. n.sp. *Illyr. Bl.,* 21(3): 9.

Schreibers, C.H., 1801. A historical and anatomical description of a doubtful amphibian animal of Germany, called by Laurenti *Proteus anguinus. Philos. Trans. R. Soc. London,* 91(2): 241–264.

Sjöberg, R. and Odell, B., 1994. Sweden. In: C. Juberthie and V. Decu (Editors), *Encyclopaedia Biospeologica,* Vol. I. Société de Biospéologie, Moulis (France) and Bucarest, pp. 819–823.

Sket, B., 1986. Ecology of the mixohaline hypogean fauna along the yugoslav coasts. *Stygologia,* 2(4): 317–338.

Stock, J. and Iliffe, T.M., 1986. The concept "anchialine" reconsidered. *Stygologia,* 2: 90–92.

Tellkampf, T., 1844. Ueber die blinden fische der Mammuthöhle und in Kentucky aufgefundener Gattungen von Gliedertieren. *Müllers Arch. Anat. Physiol.,* 4: 380–394.

Trombe, F., 1952. *Traité de Spéléologie.* Payot, Paris, 376 pp.

Uéno, S.-I., 1971. The fauna of the lava caves around Mt. Fuji-san. VI. Trechinae (Coleoptera). *Bull. Nat. Sci. Mus. Tokyo,* 14(3): 337–350.

Uéno, S.-I., 1977. The biospeological importance of non-calcareous caves. In: *Proceedings of the 7th International Speleological*

Congress, Sheffield. Cave Research Assoc., Somerset, England, pp. 407–408.
Uéno, S.-I., 1981. A remarkable new trechine beetle found in a superficial subterranean habitat near Tokyo, Central Japan. *J. Speleol. Jpn.*, 6: 1–10.
Uéno, S.-I., 1995. Half a century of subterranean explorations. *Bull. Jpn. Soc. Coleopt., Spec. Bull.*, 4: 30 pp.
Urbani, P.F., 1977. Nuevos comentarios sobre estudios realizados en las formas carsicas de las cuarcitas del Grupo Roraima. *Bol. Soc. Venez. Espeleol.*, 7(14): 129–145.
Valvasor, J.W., 1689. *Die Ehre des Herzogthums Crain, IV. Buch. Von den Naturraritäten des Landes Crain.* Laybach.
Vandel, A., 1964. *Biospéologie. La Biologie des Animaux Souterrains.* Gauthier-Villars, Paris, 619 pp.
Vannier, G. and Juberthie, C., 1977. Etude dynamique de la transpiration chez *Aphaenops cerberus,* Coléoptère Trechinae troglobie. *C.R. Acad. Sci. Paris,* 285: 805–808.
Versey, H.R., 1972. Karst of Jamaica. In: M. Herak and V.T. Stringfield (Editors), *Karst.* Elsevier, Amsterdam, pp. 445–466.
Vigna Taglianti, A., 1977. Due Nuovi Trechini Troglobi del Messico meridionale e del Guatemala (Coleoptera, Carabidae). In: *Subterranean Fauna of Mexico*, Vol. 3, Accad. Naz. Lincei, Roma, pp. 326–339.
von Motschulsky, T.V., 1854. Etudes entomologiques, 3e année. Soc. Liter. Finnoise Helsingfors, 55 pp.
von Motschulsky, T.V., 1862. Etudes entomologiques, 11e année. Soc. Liter. Finnoise Helsingfors, 55 pp.
Wankel, J., 1856. Über die Fauna der mährischen Höhlen. *Schr. Zool.-botan. Verh., Abh.,* 6: 467–470.
Wilkens, H., Parzefall, J. and Iliffe, T.M., 1986. Origin and age of the marine stygofauna of Lanzarote, Canary Islands. *Mitt. Hamb. Zool. Mus. Inst.,* 83: 223–230.
Yager, J., 1981. Remipedia. A new class of Crustacea from a marine cave in the Bahamas. *J. Crustacean Biol.,* 1(3): 328–333.
Yager, J., 1994. Remipedia. In: C. Juberthie and V. Decu (Editors), *Encyclopaedia Biospeologica,* Vol. II. Société de Biospéologie, Moulis (France) and Bucarest, pp. 87–90.

Chapter 2

INTERSTITIAL AQUATIC FAUNA OF SHALLOW UNCONSOLIDATED SEDIMENTS, PARTICULARLY HYPORHEIC BIOTOPES

J.V. WARD, F. MALARD, J.A. STANFORD AND T. GONSER

THE INTERSTITIAL MILIEU

The water-filled interstitial spaces in unconsolidated sediments are inhabited by a diverse array of animals representing most metazoan phyla (Botosaneanu, 1986; Higgins and Thiel, 1988). The interstitial fauna is derived from a variety of phylogenetic lineages constituting a range of biogeographical patterns (Coineau and Boutin, 1992; Giere 1993). On ecological time scales, interstitial biotopes form interconnected dispersal pathways between different aquatic environments and between hypogean (subsurface) and epigean (surface) waters (Stanford and Ward, 1988; Gibert et al., 1990). Over evolutionary time, interstitial biotopes serve as evolutionary invasion routes between marine and fresh waters on a global scale (Ward and Palmer, 1994).

Although this chapter focuses on the unconsolidated sediments (alluvia) of running waters, the faunae of other freshwater and marine interstitial biotopes are also considered. A broad view of interstitial habitats is necessary to provide the evolutionary and biogeographic underpinnings of the ecology of interstitial fauna in alluvial sediments. The ensuing material begins with brief descriptions of environmental conditions and fauna in the various freshwater and marine interstitial biotopes. The ecology of the hyporheic biotope (*sensu lato*) constitutes the major focus of the chapter, especially faunal distribution patterns at different scales. The final section addresses the interconnections of interstitial biotopes and the evolutionary invasion of fresh waters.

The psammolittoral

The term psammal refers to the specific habitat (sand) within which the psammon, organisms inhabiting sand, reside. Psammolittoral is a general term for sand beaches, the habitat where the psammon community is best developed (Pennak, 1951). Grain size is the primary determinant of habitat conditions for interstitial animals because of its relationship with variables such as pore space, water circulation, capillary action, oxygen concentration, salinity gradients, and organic-matter retention. The interstitial volume of most beach sands typically ranges between 36 and 43% (Pennak, 1951). The best-developed psammon is associated with medium to fine sand. Below a median grain size of about 125 μm, pore spaces are too small for interstitial animals, and burrowing forms predominate.

Marine psammolittoral zone

The environmental conditions in the interstitial habitat of marine beaches have been thoroughly reviewed (e.g., Swedmark, 1964; Higgins and Thiel, 1988; Giere, 1993). The intertidal zone, where interstitial populations are best developed, is subject to wave action and to large spatio-temporal variations in temperature and salinity. However, marked changes in these factors are generally restricted to the top few centimetres of sand. In addition, surface layers of intertidal sand are subject to drying at low tide. Light is rapidly attenuated, so that active photosynthesis by microflora occurs only in the top centimetre or so of sand. Because of tidal action, water circulation and interstitial oxygen levels are higher than in freshwater beaches.

Freshwater psammolittoral zone

Most research on the interstitial habitat of freshwater beaches has been conducted on lakes, although some important work deals with river sand (Pennak, 1988, and references therein). Capillarity raises the water in sand beaches as much as 10 cm above the water table. At 8 cm below the sand surface of a typical lake beach, the extent to which the interstitial spaces are filled with water varies with distance from the water's edge as follows: 100% at 0 cm, 100% at 50 cm, 98% at 100 cm, 76% at 200 cm, and 46% at 300 cm (Pennak, 1951). In the absence of tidal action, the freshwater psammal is a more stagnant environment than on the coast, with steep oxygen gradients, although distinct oxygen gradients may not occur in high-energy beaches of very large lakes (Whitman et al., 1994). Pennak (1951) reported typical oxygen concentrations in interstitial water at 8 cm depth (when lake water contained 8.4 mg ℓ^{-1} O_2) as follows: 5.5 mg ℓ^{-1} at the water's edge, 1.0 mg ℓ^{-1} at 50 cm, and 0.4 mg ℓ^{-1} at 100 cm. The interstitial water of lake beaches, but not marine beaches, contains appreciably greater amounts of dissolved organic and inorganic matter than the adjacent open water, which Pennak (1951) regarded as the most striking chemical feature of the freshwater psammal.

Alluvial sediments

The term "alluvium" refers to sediments transported and deposited by running water. Because of unidirectional current, the potential exists for alluvial deposits to have larger mean grain sizes and interstitial spaces than are found in sediments associated with standing water bodies. Indeed, gravel deposits are a common characteristic of running waters of various sizes. Compared with sand, water-filled interstices in alluvial gravels provide more habitat space and greater circulation. The most "open" alluvial sediments, derived from crystalline rock, consist of coarse sand and gravel with silt-free interstices. Nonetheless, current and light rapidly decline with depth below the stream bed. Under the best conditions, interstitial waters remain well oxygenated to depths exceeding 60 cm beneath the bed surface (e.g., Bretschko, 1981). Temperatures within the stream bed are dampened so that interstitial animals are exposed to reduced diel and seasonal fluctuations compared to surface waters.

The term "hyporheic zone" originally described the biotope of interstitial spaces within gravel deposits beneath stream channels (Orghidan, 1959). The hyporheic biotope is an ecotone or transition zone between surface waters and true (phreatic) ground waters (Fig. 2.1). The hyporheic biotope is now regarded as a wider and more dynamic boundary zone that, under some conditions, may extend metres vertically beneath the channel and hundreds of metres or even kilometres away from the channel (Stanford and Ward, 1988; Gibert et al., 1994a). The spatial extent of the hyporheic zone may be defined by reasonably objective chemical or physical characteristics that distinguish surface water from phreatic water, or by penetration by riverine species as described later in this chapter. At a finer scale of resolution, there are local areas of upwelling and downwelling, where groundwater emerges through the gravel to mix with the surface water in the stream channel or where surface water moves into the phreatic zone (see, for instance, Shepherd et al., 1986).

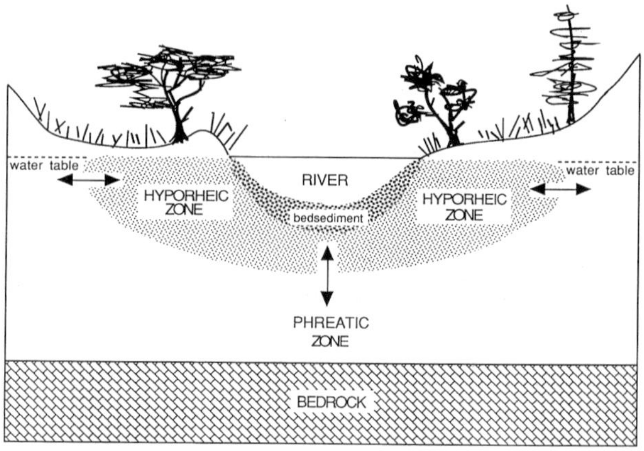

Fig. 2.1. Diagrammatic vertical structure of a river–aquifer system, showing the hyporheic and phreatic zones. Reprinted from *Biological Conservation*, Vol.83(3): J.V. Ward, Riverine landscapes: biodiversity patterns, disturbance regimes, and aquatic conservation, pp. 269–278, © 1998, with kind permission from Elsevier Science Ltd.

Riverine floodplains

The formation and structure of alluvial aquifers are inextricably linked with the dynamic processes that form riverine floodplains (e.g., Wolman and Leopold, 1957; Hey et al., 1982; Amoros and Petts, 1993; Gibert et al., 1994b; Brown, 1995). Alluvial aquifers and fringing flood plains are the subterranean and lateral extensions of river systems (Fig. 2.2). Natural river systems are interactive along longitudinal, lateral, and vertical dimensions (Ward, 1989).

Floodplains are broad valleys of alluvium formed by erosion/deposition as river channels migrate laterally. Channel migration results in a high level of habitat heterogeneity across the floodplain, reflected by a diversity of water-bodies and vegetation zones each encompassing a range of successional stages (Welcomme, 1979; Amoros and Roux, 1988; Ward and Stanford, 1995). The subterranean landscape beneath the floodplain surface, also formed by channel dynamics, is by no means homogeneous. Alluvial aquifers

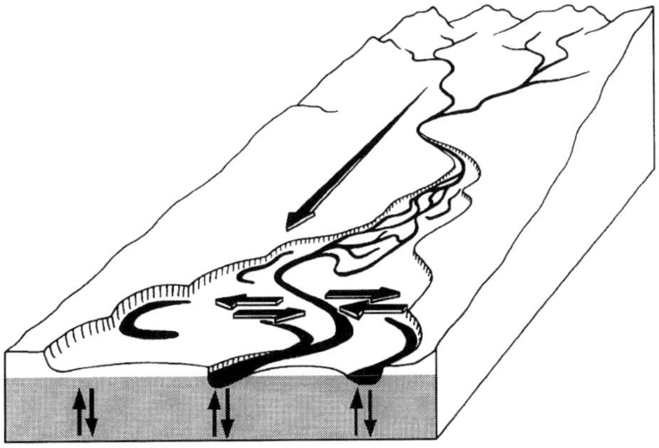

Fig. 2.2. Major interactive pathways in a riverine-floodplain/alluvial-aquifer system. After Amoros and Petts (1993).

consist of a latticework of patches differing in features such as particle size, stability, hydraulic conductivity, oxygen concentration and food resources (Castany, 1985; Stanford and Ward, 1993; Creuzé des Châtelliers et al., 1994; Gibert et al., 1994b).

THE INTERSTITIAL FAUNA

Psammon

The animals comprising the psammon (mesopsammon) community reside in a labyrinth of narrow voids where small body size is requisite. The metazoans of this biotope belong to the meiofauna, a special size class for aquatic organisms too small to be called macroinvertebrates and too large to be micro-organisms (Higgins and Thiel, 1988; Giere, 1993). Meiofaunae are represented in 22 of 33 metazoan phyla. Four of these are exclusively meiobenthic (Tardigrada, Gastrotricha, Kinorhyncha, Loricifera). Many other groups, such as Turbellaria, Oligochaeta, Copepoda, Rotifera, and Nematoda, contain many meiobenthic representatives. The young stages of some macrofaunal groups (e.g., Chironomidae) are important temporary members of the meiofauna. The following summary of adaptations to the interstitial biotope applies to the meiofauna in general, although such adaptations often attain their most characteristic form in the psammon. Given the constrained physical space of the interstitial habitat, it is not surprising that permanent members of the psammon community are either from groups in which all or most species are small, or are small species within larger-sized taxa. A good example of the latter is the hydroid *Psammohydra nana,* which is only 1 mm long (Swedmark, 1964).

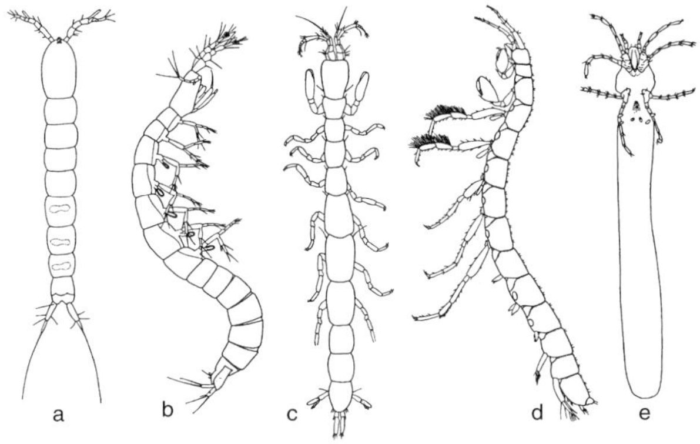

Fig. 2.3. "*Lebensformtypen*" of interstitial animals, showing the characteristic vermiform morphology. (a) *Parastenocaris* (Harpacticoida, ~0.4–0.5 mm); (b) *Leptobathynella* (Bathynellacea, ~0.6–0.8 mm); (c) *Microcerberus* (Isopoda, ~0.7–1.35 mm); (d) *Ingolfiella* (Amphipoda, ~1.5 mm); (e) *Wandesia* (Hydrachnellae, ~2.5 mm). After Husmann (1978).

A filiform body shape is a common feature of interstitial animals (Fig. 2.3). In some cases an attenuated form is characteristic of the taxon (e.g., Nematoda, Oligochaeta), whereas in others elongation is a morphological aberration found in interstitial representatives (e.g., the water mite *Wandesia*, Fig. 2.3). Other morphological adaptations to the interstitial environment (from Giere, 1993) include greater articulation of the exoskeleton (e.g., harpacticoid Copepoda), allowing greater maneuverability; dorsoventral flattening (e.g., chromadorid Nematoda), adhesion devices (e.g., the adhesive glands of many Rotifera); ciliation, not only in animals such as gastrotrichs and turbellarians, but also as locomotory devices in psammic hydroids, molluscs and others; extensible, sticky appendages used to draw the body through the interstitial spaces (e.g., *Monobryozoon*, a noncolonial mobile bryozoan); armoring with cuticular plates, rings, scales, spicules and other structures, presumably as protection from mechanical stress in shifting sand (e.g., the turbellarian *Acanthomacrostomum spiculiferum* and the minute sea cucumber *Leptosynapta*). Static organs (statocysts) are well developed among interstitial forms, which must orient in a dark, three-dimensional labyrinth, whereas eyes and pigmentation are typically reduced.

Small body size, coupled with limited water movement, are associated in interstitial animals with specialized means of reproduction and development (Giere, 1993). Many meiobenthic forms have only one ovary and produce only one or a few eggs. Sperm is often

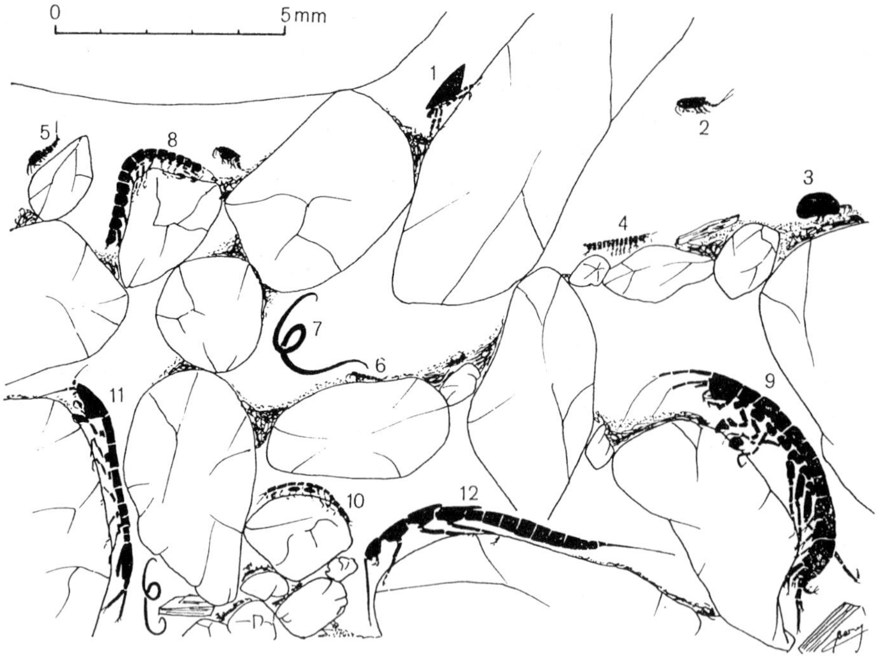

Fig. 2.4. Optical section through alluvial sediment showing typical interstitial animals moving between gravel particles, diagrammatic: (1) Hydrachnellae; (2) Cyclopoida; (3) Ostracoda; (4) *Bathynella* (Bathynellacea); (5) *Elaphoidella* (Harpacticoida); (6) *Parastenocaris* (Harpacticoida); (7) Nematoda; (8) *Balcanella* (Amphipoda); (9) *Niphargus* (Amphipoda); (10) *Microcharon* (Isopoda); (11) *Stenasellus* (Isopoda); (12) *Leuctra* (Plecoptera). After Bou (1974).

contained in spermatophores, that may be transferred directly to the partner. Copulatory organs are present in the meiobenthic representatives of groups such as polychaetes, which typically shed spermatozoa into the open water. Hermaphroditism is well developed among interstitial forms, even in groups in which this is not a general feature of the taxon. Other features that are considered adaptations to the interstitial biotope are direct development (lack of free-swimming larvae), short generation times, more or less continuous reproduction, hatching of well-developed offspring (juveniles as opposed to larvae), vivipary, parental protection of eggs and even offspring, and parthenogenesis. Special dormancy mechanisms that enhance survival during adverse conditions (e.g., drying, excessive heat or cold) are possessed by some meiobenthic forms; these are especially well developed in freshwater tardigrades and nematodes.

The selective pressures imposed by the constraints of life in the interstitial biotope appear to have resulted in convergent evolution by groups of quite different phylogenetic lineages. This convergence is perhaps best exemplified by the close relationship between morphology and biotope (*Lebensformtypen*) exhibited by the meiofauna (Remane, 1952; Fig. 2.3). According to Giere (1993), the presence of these convergent traits "indicates the ecological links connecting all (mesopsammic) meiofauna".

Hyporheos

The hyporheos is best developed in gravel-bed streams and rivers that have sediment with large interstitial spaces (Fig. 2.4). The fauna residing in the hyporheic zone is a mixed assemblage consisting of epigean forms that colonize from the surface benthos, and groundwater forms that colonize from the phreatic zone, plus species that characteristically occur in the hyporheic ecotone (hyporheos, *sensu stricto*). Because interstitial spaces tend to be larger than in the psammal, the hyporheos includes both meiofauna and macrofauna. The hyporheic zone plays important ecological roles in lotic ecosystems (Schwoerbel, 1967; Williams and Hynes, 1976). It serves as an incubation site for eggs, a nursery for juveniles, and a habitat for quiescent stages. Predation pressures are reduced within the bed sediments, and environmental conditions are less variable and more predictable. The hyporheos is buffered from adverse surface conditions such as floods, drought and extreme temperatures. The epigean component of the hyporheos forms a faunal reserve capable of rapidly recolonizing the surficial stream bed following depletion of the surface benthos by adverse environmental events.

Most studies of the hyporheos have examined faunal distribution immediately beneath the stream bed (Williams and Hynes, 1974). For example, Coleman and Hynes (1970) studied the vertical distribution of the fauna in four 7.5 cm layers within the top 30 cm of a Canadian stream bed. The proportion of total animals in each of the four layers was similar [(20% (top), 27%, 26%, 26% (bottom)] and a similar pattern was exhibited by most groups [e.g., Ephemeroptera (21%, 28%, 25%, 26%), Copepoda (23%, 37%, 19%, 22%)]. Only black flies (Simuliidae) appeared to be confined to the surface stratum.

It is clear that most species of lotic benthos are not normally confined to the bed surface. The depth to which epigean animals move downward depends on physicochemical conditions within the sediment, species-specific environmental requirements, and temporal phenomena such as season and life stage. Streams with very fine substrata or segments where bedrock is exposed lack a hyporheic biotope. At the other extreme, rivers with alluvial floodplains consisting of deep, porous gravels have expansive subterranean habitat for interstitial fauna, as discussed subsequently.

Phreatos

"Phreatos" is a collective term for the assemblage of subterranean animals that reside in alluvial sediments of the phreatic zone (see Fig. 2.1). Cave-dwelling animals ("troglos") are also subterranean, but not necessarily interstitial. Stygofauna is an all-encompassing term for animals that occur in subterranean waters. Some species have clear affinities for a specific groundwater biotope, whereas others are ubiquitous. Some epigean animals penetrate ground waters, whereas others are confined to the surface realm. Figure 2.5 (overleaf) classifies groundwater animals on the basis of habitat occurrences and epigean/hypogean affinities.

Stygoxenes are epigean animals that enter subterranean waters accidentally. Although they have no affinity for the hypogean realm, some individuals passively infiltrate alluvial sediments (or caves). Many species of stream benthos may penetrate bed sediments a short distance in this manner.

Stygophiles have clear affinities for the subterranean realm and actively enter ground waters. The **occasional hyporheos** category is for facultative species that can spend the entire aquatic phase of their life cycle in epigean waters, but that commonly reside in alluvial sediments during early life stages (eggs, early instars). Later aquatic life stages occur in epigean habitats, although they actively enter sediment interstices to avoid adverse conditions in surface waters. Most members of occasional hyporheos are insects that have a winged terrestrial stage. The **permanent hyporheos**, in contrast are fully aquatic species of microcrustaceans, mites, nematodes, oligochaetes etc. that may spend their entire lives in the subterranean realm. However, both the permanent hyporheos and the occasional hyporheos are facultative in the sense that they can also reside permanently in surface waters. They may, therefore, occur in stream reaches lacking substantial subsurface habitat. That is not the case for the third category of stygophiles, the **amphibites**, that require both surface and subsurface waters during different stages of their life cycles. Several species of stoneflies (Plecoptera) spend the majority of their nymphal existence deep within floodplain alluvia, but they must return to the river channel to emerge, mate, and oviposit (Stanford and Ward, 1988).

Stygobites are specialized for and restricted to subterranean waters. Some are ubiquitous stygobites that inhabit a variety of hypogean biotopes, including both caves and alluvial sediments. Ubiquitous forms may penetrate up to the hypogean–epigean interface (Ginet and David, 1963; Ward and Holsinger, 1981). **Phreatobites**, in contrast, are confined to deep phreatic waters. Stygobites all exhibit a common suite of adaptations that transcend taxonomic lines (Gibert et al., 1994a), including reduced pigmentation, ocular regression, well-developed chemical and mechanical receptors, reduced body size, vermiform morphology, slower metabolism and ontogenetic development, greater longevity, less frequent reproduction, and reduced reproductive output, in comparison with epigeal relatives.

THE HYPORHEIC ECOTONE

Following the preceding overview of interstitial habitats and interstitial faunas, we now focus on the hyporheic

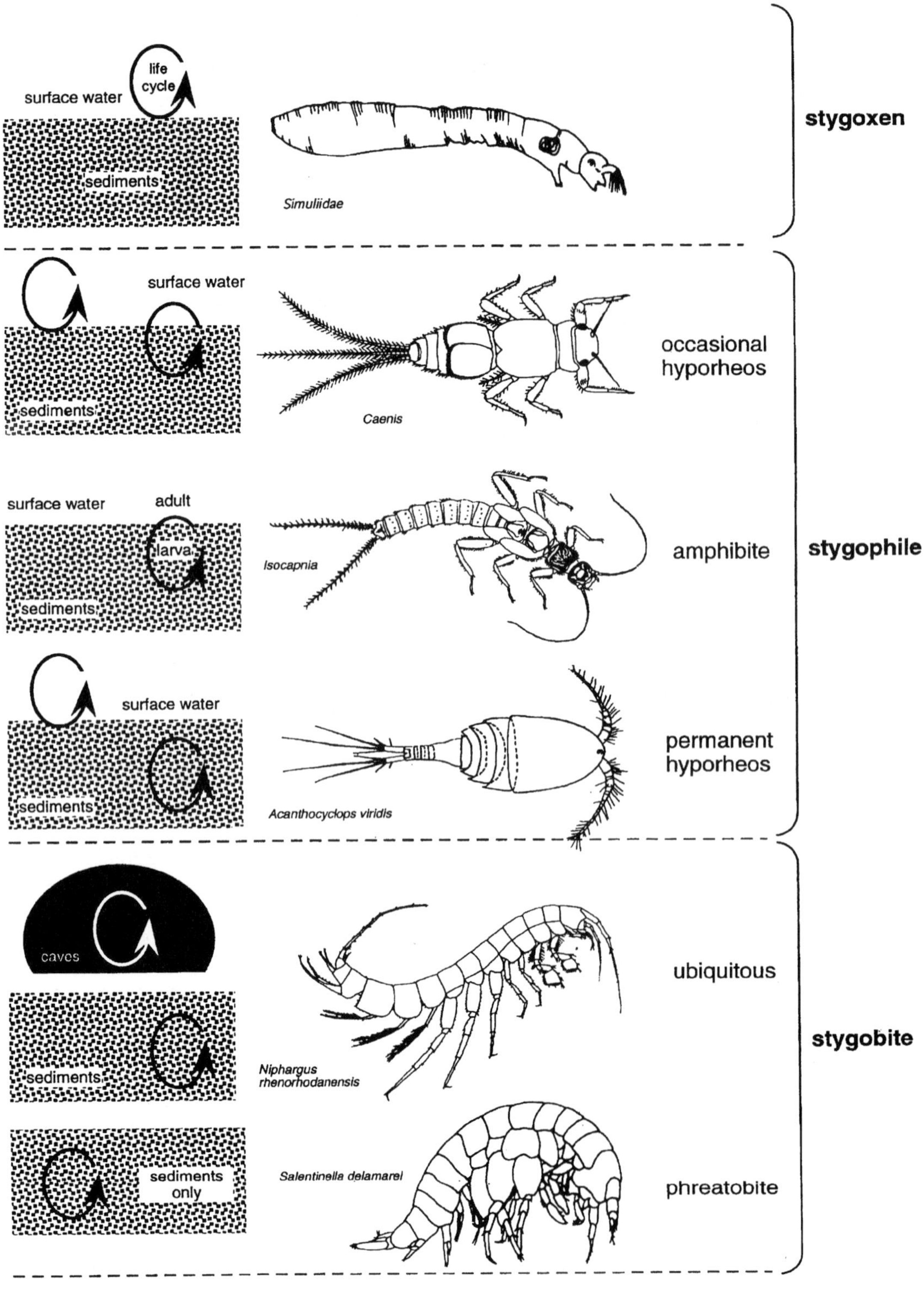

Fig. 2.5. Classification of stygofauna based on habitat occurrences and epigean/hypogean affinities. After Gibert et al. (1994a).

zone (*sensu lato*). This introduction of the ecotone properties and spatial extent of the hyporheic zone is intended to serve as a template for examining faunal distribution patterns within and across ecosystem boundaries.

Traditionally, the boundary between surface waters and ground waters was regarded as a sharp interface at the surface of the stream bed (Thienemann, 1926). However, surface waters may penetrate the bed for considerable distances. In the uppermost portion of the sediment, interstices are filled with essentially pure stream water with the same temperature regime and oxygen concentration as surface waters, although light levels and current rapidly decline. With increasing depth, interstitial water becomes a mixture of surface and phreatic water. Eventually a depth is reached where interstices are filled with true ground water, not interactive with surface water over short time scales. It is the mixing zone starting a short distance beneath the bed surface and extending downward to true ground water that defines the vertical extent of the hyporheic ecotone (Fig. 2.1). This is, however, a dynamic ecotone, the boundaries of which shift as a function of discharge and other variables.

The extent to which surface waters penetrate substratum interstices varies spatially along the longitudinal dimension, reflecting pool/riffle series and other locations of upwelling and downwelling. Therefore, the vertical extent of the hyporheic ecotone differs from one location to the next along a river's course.

The hyporheic zone is also dynamic in the lateral dimension. The distance that stream/river water penetrates beneath the banks varies over time and as a function of channel pattern, alluvial characteristics, and geomorphic features.

From this perspective the hyporheic zone is a dynamic three-dimensional ecotone of interconnected interstitial spaces. The shape and size of the hyporheic zone vary within the same stream reach over time, from one reach to the next along a river's course, and from one river system to another.

FAUNAL DISTRIBUTION PATTERNS

The distribution patterns of interstitial animals can be examined across a range of spatial scales. In a recent paper, Ward and Palmer (1994) emphasized the spatial heterogeneity of interstitial freshwater meiofauna in alluvial river–aquifer systems. They provided examples of faunal distribution patterns at the floodplain scale, reach scale and gravel-bar scale. In this section, we will describe distribution patterns and examine the prevailing factors responsible for these patterns at three other spatial scales: biogeographic scale, geomorphic (landscape) scale and microspatial scale.

The biogeographic scale

There has been a long tradition of species inventory at the geographic scale amongst groundwater ecologists. This tradition has resulted in the publication of impressive books such as Stygofauna Mundi (Botosaneanu, 1986) and more recently the first volume of the Encyclopaedia Biospeologica (Juberthie and Decu, 1994). This latter book contains no less than 87 maps showing the geographic distribution of at least 32 faunal groups. The biogeographic approach has greatly contributed to a global perspective of subterranean life. For example, Holsinger (1993), who identified regions with significant concentrations of stygobiont amphipods, showed that these animals may be found in nearly all parts of the world with suitable groundwater habitats [Fig. 2.6; but see also Holsinger (1994a,b)]. The world subterranean amphipod fauna is composed of 740 species belonging to 138 genera and 36 families. However, there are great differences in taxonomic richness between the regions, and three primary factors were put forward to explain the distributional patterns of amphipods.

As many as 573 species (i.e., 77% of the world's subterranean amphipod fauna) belonging to 17 families can be found in the central and southern Europe–Mediterranean and the eastern and southern North American–West Indian regions. The high number of families, genera and species in these two distant regions of the world were attributed to a set of similar geographical and geological factors, which include the absence of Pleistocene glaciation, exposure to marine transgressions in the Tertiary and Cretaceous, and the occurrence of a diverse array of groundwater habitats (from limnic habitats in karst terrains to marine interstitial and anchialine habitats).

At the family level, Holsinger also demonstrated that the spatial distribution pattern depended on the ecological tolerances of the amphipods. The exclusively freshwater families usually had small distribution ranges, while the euryhaline families spread over a greater number of regions. For example, the

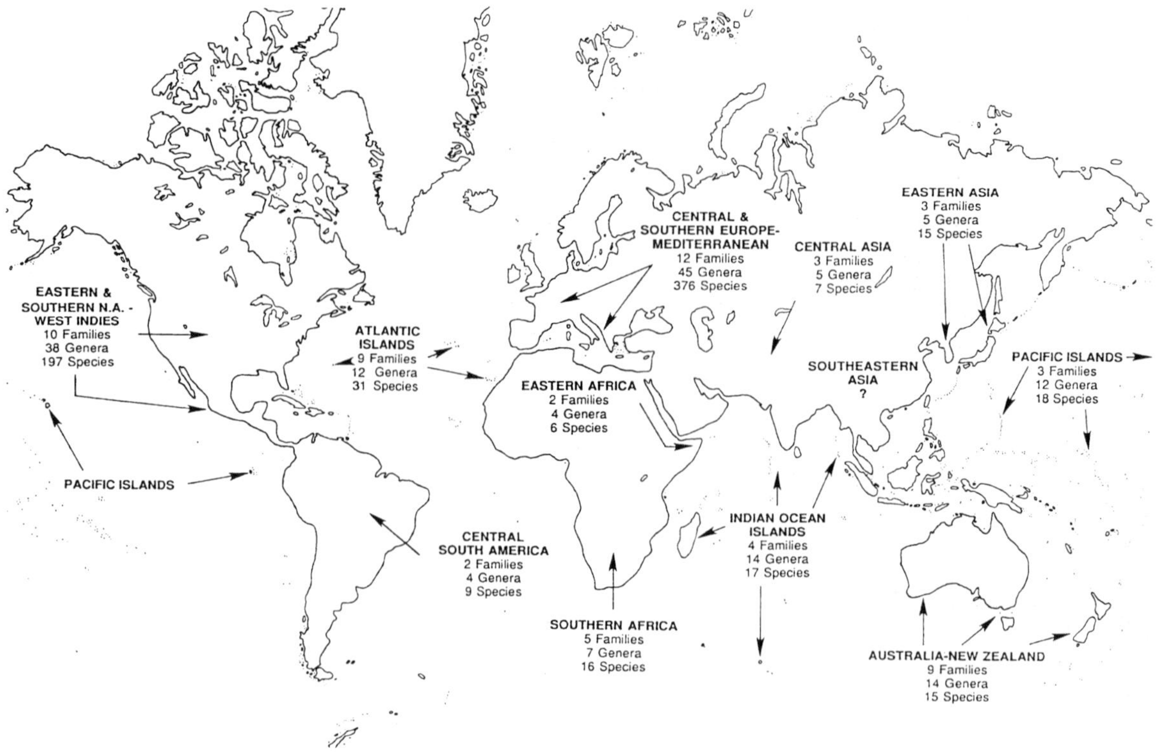

Fig. 2.6. World-wide pattern of taxonomic diversity in stygobiont amphipods. After Holsinger (1993), with kind permission from Taylor & Francis Group Ltd.

niphargids (exclusively freshwater amphipods represented by 211 species) are restricted to the central and southern Europe–Mediterranean region, while the euryhaline bogidiellids (representing by 86 species) are present in all regions except the eastern Asian region.

Finally, spatial distribution patterns of taxonomic richness were shown to originate from the combination of geographic and organismic factors. Holsinger showed that large freshwater families such as the Niphargidae or Crangonyctidae contained few genera but numerous species, while the contrary was observed for large euryhaline families. He suggested that such a trend could be due to the fact that freshwater species were distributed in continental groundwaters that show a higher degree of connectivity than island aquifers inhabited by marine/brackish-water species. Strong geographic isolation in insular species would have resulted in greater morphological changes, justifying the creation of several genera.

The geomorphic (landscape) scale

Several conceptual models have been elaborated to integrate the structure and function of the hyporheic zone on a landscape scale. Bencala (1993) stressed the fact that streams should not be considered as pipes draining subsurface water from their catchment but rather "as part of the catchment's downvalley and downgradient transport of water and solutes". In this context, the hyporheic zone is seen as the part of the landscape containing water of both subsurface and surface origin that establishes the hydrologic connection between streams and catchments. In agreement with Bencala's point of view, White (1993) predicted the physical development of the hyporheic zone along a longitudinal geomorphological–hydrologic continuum from headwaters to floodplains. Hyporheic-zone development should be weak in groundwater-dominated headwater streams where flow paths resulting from advection of channel water occur on the scale of centimetres. Mid-reach streams are characterized by pool–riffle–pool sequences that would result in infiltration of stream water and exfiltration of groundwater at the scale of metres. Finally, proportionately less surface-water–groundwater interactions are expected in large river floodplains, but the hyporheic zone increases in size

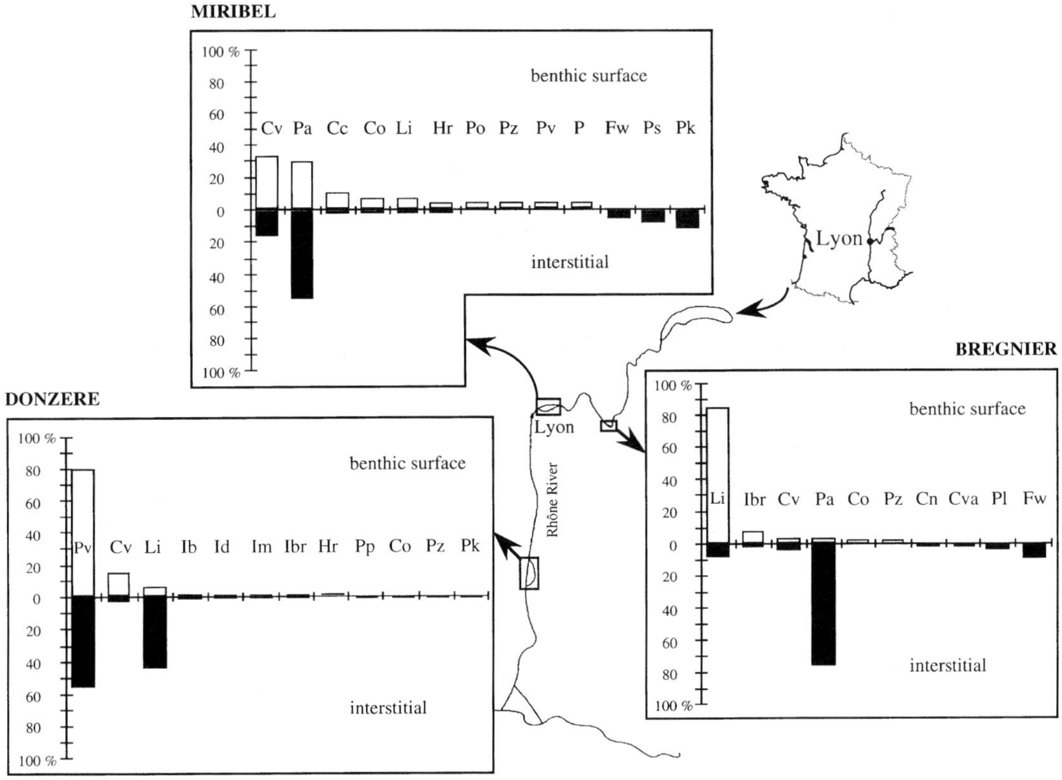

Fig. 2.7. Relative abundances of ostracod species in surface water and interstitial water in three geomorphological sectors of the Rhône river. Only the names of dominant species are indicated: Cv, *Cypridopsis vidua*; Li, *Limnocythere inopinata*; Pa, *Pseudocandona albicans*; Pv, *Potamocypris cf. variegata*. Redrawn after Creuzé des Châtelliers and Marmonier (1993, pp. 268, 272).

due to the advection of channel water on the scale of hundreds of metres.

Considering the case of large rivers, Creuzé des Châtelliers (1991a) suggested that the hyporheic layer should be viewed as a sequence of geomorphological sub-systems (e.g., constrained, braided, anastomosed and meandering sectors), thus forming a biological discontinuum. Stanford and Ward (1993) used a similar conceptualization, describing the longitudinal arrangement of the hyporheic zone as "beads on a string". These authors emphasized the role of this hyporheic corridor in the bioproduction within the channel, the transformation of materials by interstitial food-webs, and the formation of diverse riparian habitats sustaining high biodiversity within the landscape.

Although predictive models have flourished, substantial data are still very scarce. Only two studies are known to us that have examined the distribution of interstitial fauna within a landscape perspective. In the first study, Ward and Voelz (1990) investigated the spatial distribution pattern of interstitial fauna along a 475-km segment of the South Platte River in Colorado. In marked contrast with the typical longitudinal distribution of benthos, gradient analysis, classification and ordination techniques run on a "9 sites 64 taxa" data set failed to show distinct gradients in interstitial species composition. Thus, these authors suggested that the structure of interstitial communities was more likely to be determined by site-specific hydraulic or geomorphologic conditions (Ward and Voelz, 1994). The second such study was based on the extensive sampling of the benthic and interstitial communities of three geomorphological sectors (see Fig. 2.7) along the Rhône river in France (Creuzé des Châtelliers, 1991a; Creuzé des Châtelliers and Poinsart, 1991). As part of this study, Creuzé des Châtelliers and Marmonier (1993) examined more specifically the degree of similarity between benthic and interstitial ostracod assemblages (Fig. 2.7). A total of 358 samples were collected, and twenty-four species belonging to two families (Limnocytheridae and Cyprididae) were identified. No similarity between benthic and interstitial assemblages was observed in the Brégnier sector, where the dominant and subdominant species in interstitial and surface habitats differed markedly. In the other two sectors, ostracod assemblages were dominated by the same species in both surface and interstitial habitats (Fig. 2.7): *Cypridopsis vidua* and *Pseudocandona albicans* in the Miribel sector and *Potamocypris cf. variegata* and *Limnocythere inopinata* in the Donzère sector. At the latter site, the same subdominant species also occurred in interstitial and surface habitats (i.e., *Isocypris beauchampi* and three species of *Ilyocypris*). Spatial patterns of assemblage similarity were attributed to the geomorphological and hydrological attributes of the different sectors. In the Brégnier sector, vertical exchange of water was reduced due to aggradation and accumulation of fine sediments, thus preventing the colonization of interstitial layers by benthic ostracods. In the other sectors, the degradation process and the absence of fine sediments resulted in greater water fluxes between the river and the hyporheic zone, which promoted faunal exchanges between these two subsystems. Because geomorphic and hydrologic features may be the prevailing factors structuring the hyporheic communities along a watercourse, it is not then surprising that the longitudinal patterns of benthic and interstitial animals are uncoupled.

The microspatial scale

Investigations at a macro or mesoscale usually focus on the environmental factors that structure the distribution of interstitial animals. At the microscale level, the organism is the primary focus because investigators are interested in determining how the organisms actively seek their ecological preferendum and even transform their environment. In that sense, microscale study is a necessary step to understanding the chemical and biological processes in which organisms are active, and thus explaining the patch-dynamic processes (Danielopol and Pospisil, 1991). Although numerous investigations have been carried out to document the microdistribution patterns of meiofauna living in the marine psammolittoral zone (see review by Fleeger and Decho, 1987), comparable microscale studies have rarely been conducted in the hyporheic zone. In a study of the three-dimensional distribution of the hyporheos in Salem Creek (Ontario), Godbout and Hynes (1982) sampled nine quadrats (at least three interstitial samples per quadrat) regularly distributed over an area of 20 m^2. Danielopol (1983, 1989, 1991) used a regular grid of 24 fixed pipes to determine the spatial and temporal distribution of the interstitial fauna in a 30 m^2 area of the Eberschüttwasser backwater (Danube river, Vienna). His main contribution was to enhance the concept of the organism as an active subject: "meiobenthic species disperse steadily within the interstitial space, form dynamic patches and seem to perceive actively their preferred environment" (Danielopol, 1991, p. 97). More recently, Rouch and his collaborators investigated the distribution of interstitial crustaceans over a 75 m^2 area of Lachein brook in the Pyrenees (Rouch, 1988, 1991, 1992; Rouch et al., 1989; Rouch and Lescher-Moutoué, 1992). A detailed study of the physico-chemical and hydrogeological characteristics of the hyporheic zone was concomitantly conducted (i.e., porosity, sediment composition, 15 physico-chemical parameters, particulate organic matter, hydraulic conductivity, piezometric map). The study area was divided into ten quadrats and the fauna was collected (two samples of ten litres per quadrat) at a depth of 60 cm every two months during one year. A total of 63 crustacean species were identified, amongst which 31 were stygobites. Distribution patterns differed from one species to another, but with the exception of dominant epigean copepods (e.g., the cyclopoid *Eucyclops serrulatus*), all species exhibited restricted distributions (Fig. 2.8). Eurytopic species such as *Stenasellus virei boui* (Isopoda) and the amphipods *Ingolfiella thibaudi* and *Niphargus kochianus*, occurred in all habitats except the low-oxygen zone (mean oxygen saturation 16.2%). Some species (e.g., the isopod *Microcharon rouchi*) colonized exclusively the gravel bars with high hydraulic conductivity (i.e., 10^{-3} m s^{-1}) while the syncarid *Vandelibathynella vandeli* and the harpacticoid *Elaphoidella bouilloni* were restricted to the low-

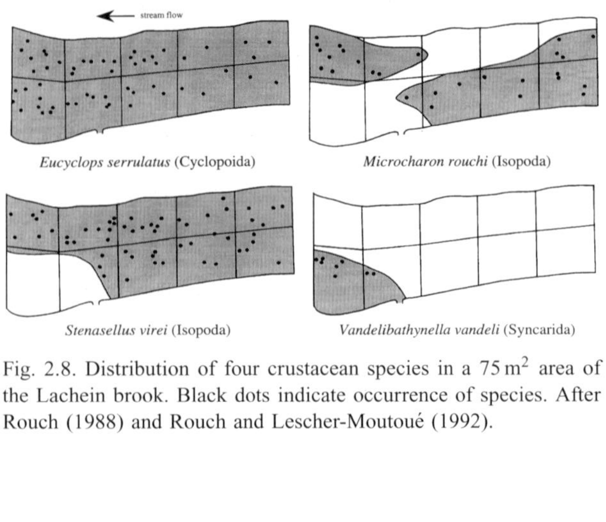

Fig. 2.8. Distribution of four crustacean species in a 75 m^2 area of the Lachein brook. Black dots indicate occurrence of species. After Rouch (1988) and Rouch and Lescher-Moutoué (1992).

permeability and poorly oxygenated zone. Although the distribution of species was primarily related to the combined effect of geomorphological and hydrogeological factors, Rouch and Lescher-Moutoué (1992) suggested that the spatial pattern of cyclopoid diversity could be due in part to competitive exclusion. Indeed, the highest abundance and species richness of cyclopoids, but the lowest diversity (as measured by the Shannon–Wiener index) was observed in an area dominated by the eurytopic species *Graeteriella* sp. Finally, the crustacean community was found to remain remarkably stable over the year, although the physico-chemical conditions and surface-water–groundwater exchanges exhibited some seasonal patterns. This result, which was in marked contrast with those of most studies, was attributed by Stanley and Boulton (1993) to the fact that the "river originated from a karstic system" (p. 80). However, while conducting another similar one-year study in a non-karstic stream, Rouch (1995) demonstrated again that floods did not modify the bed morphology, the organization of interstitial flow, or the composition of the crustacean community. He concluded that in some streams the hyporheic zone could be considered as being stable over a period of several years.

These three case studies, as well as those presented by Ward and Palmer (1994), emphasize the necessity of maintaining a spatial and temporal perspective in groundwater ecology. The structure and operation of groundwater ecosystems depend upon a number of biological and physical processes which operate and generate various patterns at different scales (Gibert et al., 1994a). For example, at the scale of the floodplain of the Rhône river, the interstitial fauna was distributed along a "physical stability" gradient linked to the hydrological regime of the river and to the circulation of flood waters (Marmonier et al., 1992). However, in the same river system, but at the scale of the gravel bar, the fauna exhibited a patchy distribution in relation to upwelling and downwelling water movement and to the geomorphological characteristics of the river bed (Dole-Olivier and Marmonier, 1992).

During the last 20 years, most studies have revealed the primacy of hydrologic and geomorphological factors, across a wide range of spatial scales, in determining the distribution of the fauna in the hyporheic zone (Godbout and Hynes, 1982; Marmonier and Dole, 1986; Creuzé des Châtelliers, 1991b; Cooling and Boulton, 1993). However, hydrology and geomorphology are synthetic factors that act upon many other covariable environmental parameters such as the porosity, solute concentrations, and particulate and dissolved organic matter. Innovative experiments and sampling designs are desirable to partition the effects of such covariable factors. For example, Rouch (1992) clearly separated the influence of the interstitial water flow and surface-water–groundwater interactions on the distribution of crustaceans in sediments of the Lachein brook in the Pyrenees by simultaneously studying the piezometric surface and its position relative to the stream level. Little is known about the effects of biotic factors such as differential migration amongst age classes, aggregation due to reproduction, resource partitioning, food selection, predation and interspecific competition. All these factors have been shown to influence the distribution of interstitial marine meiofauna (Fleeger and Decho, 1987), and there is no doubt that they will prove to have a role in structuring the hyporheos as more microscale studies are conducted.

As pointed out by Gibert et al. (1994b), it is not enough merely to recognize that ecological phenomena are scale-dependent; between-scale interactions have also to be considered. The importance of examining which large-scale patterns may originate from processes operating at a finer scale has already been expressed by some researchers (Danielopol et al., 1994; Rogulj et al., 1994). Danielopol et al. (1994) demonstrated that the world-wide distribution of six subterranean species of ostracods was not linked to any geological event but rather to "ecological processes occurring at local and/or regional scales" (e.g., ecological flexibility and tolerance). Inversely, small-scale patterns may result from a process that acts at a higher spatial scale. In the degraded part of the Miribel Canal, the interstitial communities of the upstream and downstream riffle zones differed noticeably; but such a riffle-scale heterogeneity was not observed in the aggraded part of the channel (Creuzé des Châtelliers and Reygrobellet, 1990). Long-term and multiscale ecological studies such as those conducted on the alluvial groundwater of the Rhône river (Dole-Olivier et al., 1994) and the Danube (Danielopol, 1984; Pospisil, 1994) must be encouraged, because they can provide insight into interactive processes that transfer across scales. For example, Culver and Fong (1994) examined patterns of subterranean crustacean diversity in the Virginias in the USA across five different spatial scales (rocks, riffles, streams, karst basins and river valleys). Although no true biological explanation was given for the observed fractal pattern of diversity, their results had important implications with respect to the evolutionary invasion of groundwater because they suggested that the regional cave communities were not saturated with species.

EVOLUTIONARY INVASION OF GROUNDWATER

Groundwater has long been considered as an extreme environment inhabited by only a few specialized species. This paradigm is now being revised (Marmonier et al., 1993; Stoch, 1995); these case studies clearly show that groundwater environments, which are by no means homogeneous, harbour relatively diverse communities across various spatial scales. In addition to ecological determining factors, species diversity is also related to evolutionary time. Evolutionary invasion of groundwater – that is, the colonization of groundwater by epigean species and production of stygobite lineages – has occurred for many millions of years for some taxa, only a few million years for others. For example, some ancient subterranean groups were probably established in groundwater prior to the separation of North America from Eurasia in the late Mesozoic period [*Stygobromus* (Amphipoda): Holsinger (1988)] or even prior to the breakup of Pangaea during the Jurassic [e.g., Bathynellidae (Syncarida): Schminke (1981)]. Other younger groups colonized groundwater during sea regressions in the Tertiary [Microparasellidae (Isopoda): Coineau (1992)]. Moreover, some epigean freshwater groups are known to have subterranean populations containing partly oculated and depigmented individuals, which may be considered as incipient stygobites (see Henry et al., 1986, p. 436).

Several authors have examined different aspects of the evolutionary invasion of groundwater – for instance, the colonization of groundwater environments (Rouch and Danielopol, 1987; Botosaneanu and Holsinger, 1991; Holsinger, 1994a), speciation of subterranean species (Sbordoni, 1982; Barr Jr and Holsinger, 1985; Peck and Finston, 1993), and processes involved in the acquisition of stygomorphic traits (Barr Jr, 1968; Culver, 1982; Christiansen, 1992, Wilkens, 1992). Therefore, we will not discuss each of these aspects at length in this section, rather we will address the various pathways that may account for diversity of species in groundwater. The word

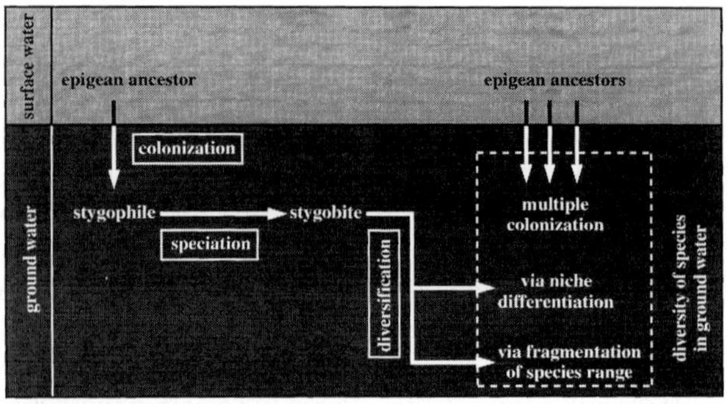

Fig. 2.9. Potential evolutionary pathways resulting in diversity of species in groundwater.

"pathway" is used here to denote the succession of processes which determine the appearance of a stygobite lineage. For example, one particular pathway may be that of a marine endobenthic species having a set of preadaptations to interstitial life. This species migrates actively in interstitial littoral waters, passively colonizes continental groundwater during a sea regression, and then speciates allopatrically and finally gives rise to further groundwater species by vicariance. The "adaptive zone model" recently developed by Stoch (1995) provides a good description of the multiplicity of these pathways (Fig. 2.9). He distinguished three different steps, but the completion of one step does not imply that another will subsequently proceed.

Colonization results from active exploration and passive migration of benthic fauna in the interstitial biotopes. It is a continuous process which does not require any environmental constraints. However, the success of an organism in entering the interstitial milieu and maintaining a viable population requires a set of preadaptations. Those particular morphological, physiological or behavioral features that may pre-exist in a number of benthic forms can include elongated body shape, euryhalinity in marine forms, and the ability to feed in darkness. Because progenetic species, characterized by morphological similarity to juvenile stages of their ancestors, are particularly suited to interstitial life, progenesis may have played an important role as a preadaptive process in the colonization of groundwater.

Speciation is probably one of the more debated questions in subterranean ecology. Geographic speciation models, either passive allopatric speciation (vicariance) or peripatric speciation (as a result of genetic revolutions in peripheral founder populations), are strongly supported by many authors (Sbordoni, 1982; Barr Jr and Holsinger, 1985; Botosaneanu and Holsinger,

1991; Notenboom, 1991). While the geographic isolation of populations unequivocally interrupts gene flow, the fact that adaptation and evolution may occur in some contiguous epigean and hypogean populations, as in *Astyanax fasciatus* (Pisces) (Wilkens and Hüppop, 1986) or troglobites of the Hawaiian archipelago (Howarth, 1987), suggests that other kinds of barriers exist. If evolution in contiguous populations leads to differences in mating behavior or competitive differences, then parapatric or sympatric speciation is also possible (Rouch and Danielopol, 1997; Stoch, 1995; Culver et al., 1995). Finally, the case when speciation does not occur (that is, not all stygophiles are stygobites in training), but the colonizer remains a stygophile, still contributes to species diversity in groundwater. Once a colonizer is established in groundwater it usually acquires specific features called stygomorphic traits [but see Danielopol and Rouch (1991) and Christiansen (1992) for cases of non-stygobitization]. These morphological, physiological and behavioural traits include both the reduction of some features (e.g., reduction or loss of eyes and pigment) and the enhancement of other features (elongation of appendages, K-strategy, resistance to starvation). If enhanced features are considered as adaptations, then causes of regressive changes remain unclear. A number of potential causes have been suggested (Barr Jr, 1968; Christiansen, 1985; Culver et al., 1995), of which selection and neutral mutation seem to be the most plausible, on the basis of studies using *Gammarus minus* (Amphipoda) (Jones and Culver, 1989; Kane and Culver, 1992; Culver et al., 1995) and *Astyanax fasciatus* (Pisces) (Wilkens, 1988, 1992).

Once the hypogean form is separated genetically from its epigean ancestor, adaptive radiation may occur following two different pathways. First, the stygobite may evolve into a wide range of related species by occupying a variety of niches. As suggested by Stoch (1995), this diversification hypothesis, which is based on non-geographic speciation processes, is open to criticism (see previous paragraph). The second hypothesis refers to the possibility of a stygobite having a wide distribution range and generating new species through vicariance (i.e., population fragmentation due to subsequent geographic barriers). An expanded species range may result from groundwater colonization on a large spatial scale or from "point" colonization followed by dispersal into newly opened groundwater areas. However, the fact that stygobites are often considered to possess low vagility and mobility severely limits the plausibility of this scenario of diversification by dispersal.

Since establishment of a stygobite in groundwater may not be followed by either diversification mechanisms, species diversity in groundwater may originate from a multiple colonization process. For example, several epigean species belonging to the same phylogenetic group, or several populations of one epigean ancestor, may give rise simultaneously or at different times to different stygobitic species.

Although some steps presented in Fig. 2.9 are still hypothetical (e.g., diversification via niche differentiation), Stoch's model clearly shows the variety of potential pathways involved in the evolutionary invasion of groundwater which may lead to the present-day diversity of groundwater taxa. The number of models developed or used by different authors to explain the occurrence and distribution of subterranean species also suggests a multiplicity of possible scenarios: the regression model (Stock, 1980), the adaptive zone model (Iliffe, 1986), the model of active colonization as an alternative to colonization under constraint (see Rouch and Danielopol, 1987; Botosaneanu and Holsinger, 1991), the dispersionist model (Danielopol et al., 1994), the "*modèle biphase d'évolution*" (Boutin and Coineau, 1990), the two-step model (Notenboom, 1991) and the three-step model (Holsinger, 1994a). The presentation of all existing models is beyond the scope of this chapter, thus, we have selected two concrete examples illustrating the evolutionary invasion of groundwater by marine forms.

The first example concerns the evolutionary biogeography of Metacrangonyctidae (Amphipoda) in Morocco (Boutin and Messouli, 1988; Messouli et al., 1991; Boutin et al., 1992). The *Metacrangonyx panousei* group includes six species which are found in groundwater of the Atlas mountains as well as the adjacent plain of the Drâa river system (Figure 2.10A). A species cladogram obtained from analysis of 22 morphological characters was congruent with a geological cladogram (Fig. 2.10B). This cladogram showed the marine origin and age of the *panousei* lineage, as well as the events that lead to species diversification. This model suggests that colonization of continental groundwater took place during the Senonian regression about 70 million years ago. Colonization was the result of a double process which included active horizontal migration of a benthic marine organism into the littoral interstitial groundwater (this interstitial organism is then called a thalassostygobiont) and passive vertical

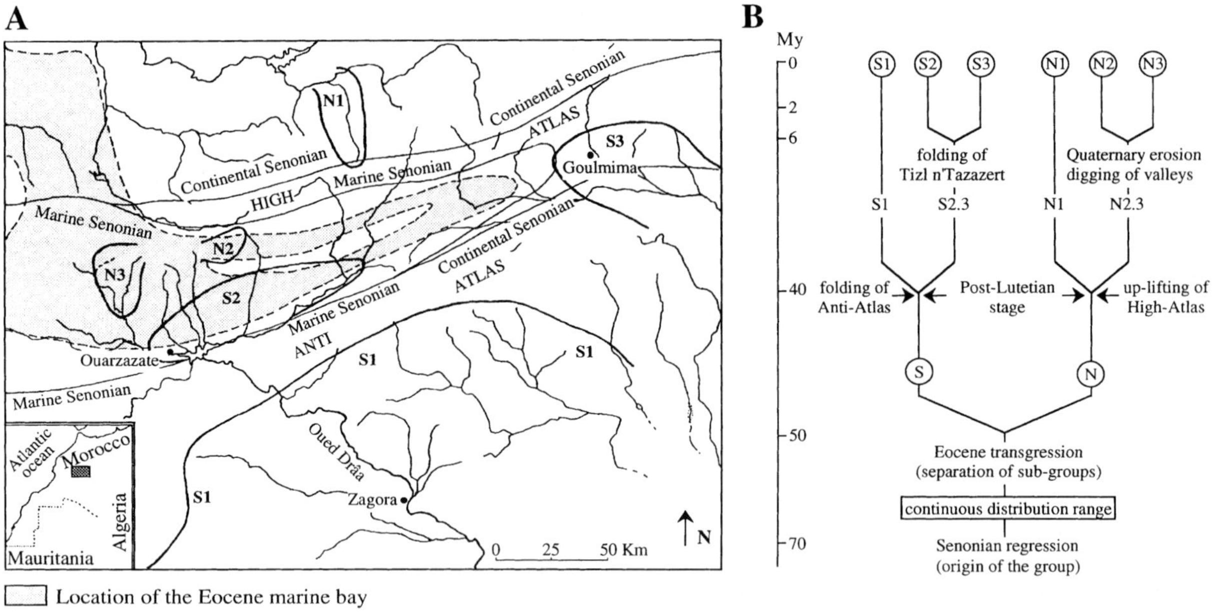

Fig. 2.10. Evolutionary biogeography of the *Metacrangonyx panousei* group (Metacrangonyctidae, Amphipoda). (A) Distribution of species in Morocco: N1, *Metacrangonyx aroudanensis*; N2, *M. ruffoi*; N3, *M. gineti*; S1, *M. panousei*; S2, *M. delamarei*; S3, *M. goulmimensis*. (B) Geographic cladogram of distribution areas of species. Redrawn after Boutin et al. (1992, pp. 172, 173).

migration of this interstitial organism into continental groundwater as the sea receded (this latter organism is then called a thalassoid stygobiont). At the end of the marine regression, probably only one species of Metacrangonyctidae inhabited groundwater. Subsequently, its distribution was fragmented by geographic barriers formed during the transgression of the Eocene Sea, the orogenic phases of the Haut Atlas and Anti-Atlas mountains, and the erosional deepening of valleys during the Quaternary. In this evolutionary scenario, which fits well with the "*modèle biphase d'évolution*" (Boutin and Coineau, 1990), the active dispersal ability of Metacrangonyctidae is considered minimal. Even the large distribution range of *Metacrangonyx panousei*, which extends into the catchment of the Drâa River (south of the Atlas mountains) is thought to result from the passive downstream drift of individuals at the surface of sediment. Finally, species diversification within the *panousei* group is exclusively dependent upon allopatric speciation events.

The historical description by Magniez (1976, 1981) of the occidental group of *Stenasellus virei* (Stenasellidae, Isopoda) is presented here as a second example because it provides a different scenario of groundwater invasion by marine forms (see also Henry et al., 1986; Coineau et al., 1994). This group contains four species which are thought to originate from a common mesogean ancestor (Fig. 2.11A). The period of invasion into groundwater by this marine form is not precisely known (Middle Cretaceous period?). This uncertainty is because colonization of inland groundwater is an active process and therefore cannot be associated with a particular transgression. As Magniez (1976) stated, no transgression in the Mesozoic and Cenozoic periods can explain the presence of stenasellids in the center of Africa (Congolian and Nigerian shields). Many species of stenasellids, which are active burrowers, can migrate and thus frequently extend their distribution range. The ancestor of the *S. virei* group probably invaded groundwater of the Tyrrhenian plate by migrating into interstitial water of the littoral and then the hyporheic zone of large rivers. A thermophilic species lived in the water-saturated superficial zone of crystalline rocks until the Middle Eocene. Diversification of the group is seen as the result of several pre-Pliocene vicariant events such as the Eocene and Oligocene transgression which isolated *Stenasellus racovitai*. The dynamism of the evolutionary scenario of the *Stenasellus* group can be further exemplified by considering the history of the polytypic species *Stenasellus virei* in the Aquitanian Basin during the Quaternary (Fig. 2.11B). The distribution range of *S. virei* was greatly reduced during glacial times (Riss, Würm) and the cold dry epoch of the Late Würm. Although Magniez (1976) suggested

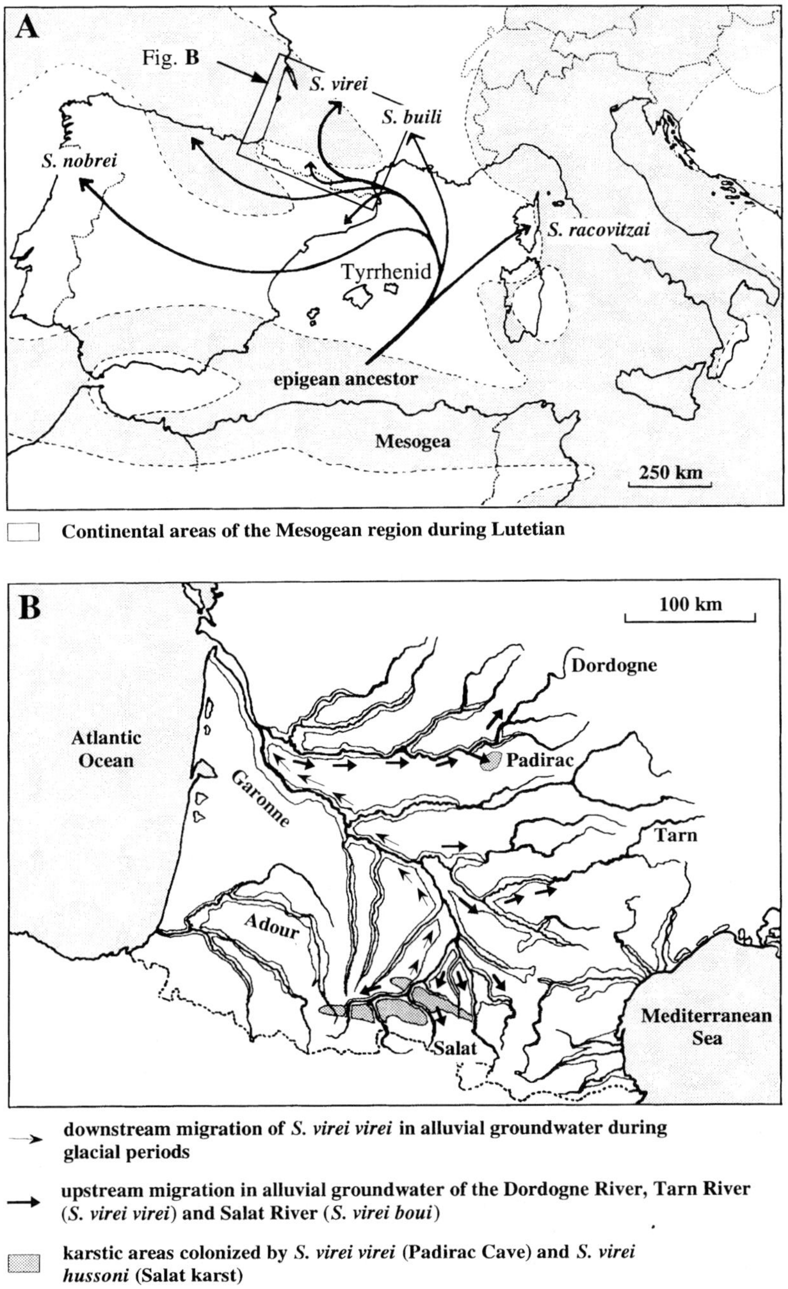

Fig. 2.11. Evolutionary biogeography of the occidental group of *Stenasellus virei* (Stenasellidae, Isopoda). (A) Active colonization of the Tyrrhenian Plate by a Mesogean ancestor. (B) Upstream and downstream migration of *S. virei* in alluvial groundwater of the Aquitanian Basin during the Quaternary. Redrawn after Magniez (1976, pp. 76, 165).

that populations might have subsisted in some places, these periods probably resulted in significant reductions in the *S. virei* stock (especially in the central Pyrenees). Then, melting of glaciers and intense aggradation during the Holocene allowed re-colonization and sub-speciation to take place in depopulated areas. Indeed, *S. virei* migrated into alluvial groundwater of the Dordogne, Tarn and Salat rivers, and also colonized some adjacent karstic basins (the Padirac cave for *S. virei virei*, the Salat Massif for *S. virei hussoni*). Ward and Palmer (1994) also emphasized the significance of alluvial aquifer systems as a long-term dispersal route for the meiobenthos between different groundwater biotopes with limited interconnection. Finally, the present geographic distribution of the polytypic species *S. virei* reflects the long history of a thermophilic marine form that invaded the groundwater of the Tyrrhenian Plate during a period of continental biostasis (Middle Cretaceous?), speciated during the Paleogene and Neogene, withstood a tremendous reduction of its distribution range in the Aquitanian Basin during the Riss and Würm, but underwent a new expansion stage and sub-speciation during the Holocene.

REFERENCES

Amoros, C. and Petts, G.E. (Editors), 1993. *Hydrosystèmes Fluviaux.* Masson, Paris, 322 pp.

Amoros, C. and Roux, A.L., 1988. Interactions between water bodies within the floodplains of large rivers: function and development of connectivity. In: K.-F. Schreiber (Editor), *Connectivity in Landscape Ecology. Münstersche Geogr. Arb.,* 29: 125–130.

Barr Jr., T.C., 1968. Cave ecology and the evolution of troglobites. *Evol. Biol.,* 2: 35–102.

Barr Jr., T.C. and Holsinger, J.R., 1985. Speciation in cave faunas. *Annu. Rev. Ecol. Syst.,* 16: 313–337.

Bencala, K.E., 1993. A perspective on stream-catchment connections. *J. N. Am. Benthol. Soc.,* 12(1): 44–47.

Botosaneanu, L. (Editor), 1986. *Stygofauna Mundi.* E.J. Brill, Leiden, 740 pp.

Botosaneanu, L. and Holsinger, J.R., 1991. Some aspects concerning colonization of the subterranean realm – especially of subterranean waters: a response to Rouch & Danielopol (1987). *Stygologia,* 6(1): 11–39.

Bou, C., 1974. Les méthodes de récolte dans les eaux souterraines interstitielles. *Ann. Spéléol.,* 29: 611–619.

Boutin, C. and Coineau, N., 1990. "Regression model", Modèle biphase d'évolution et origine des micro-organismes stybobies interstitiels continentaux. *Rev. Micropaléontol.,* 33(3/4): 303–322.

Boutin, C. and Messouli, M., 1988. *Metacrangonyx gineti* n. sp. d'une source du Haut-Atlas marocain et la famille des Metacrangonyctidae n. fam. (Crustacés Amphipodes Stygobies). *Vie Milieu,* 38(1): 67–84.

Boutin, C., Messouli, M. and Coineau, N., 1992. Phylogénie et biogéographie évolutive d'un groupe de Metacrangonyctidae, Crustacés Amphipodes stygobies du Maroc. II – Cladistique et paléobiogéographie. *Stygologia,* 7(3), 159–177.

Bretschko, G., 1981. Vertical distribution of zoobenthos in an alpine brook of the RITRODAT–LUNZ study area. *Verh. Int. Verein. Limnol.,* 21: 873–876.

Brown, A.G. (Editor), 1995. *Geomorphology and Groundwater.* Wiley, Chichester, UK.

Castany, G., 1985. Liaisons hydrauliques entre les aquifères et les cours d'eau. *Stygologia,* 1: 1–25.

Christiansen, K.A., 1985. Regressive evolution in Collembola. *N.S.S. Bull.,* 47(2): 89–100.

Christiansen, K.A., 1992. Biological processes in space and time: cave life in the light of modern evolutionary theory. In: A.I. Camacho (Editor), *The Natural History of Biospeleology.* National Museum of Natural Sciences, Madrid, pp. 453–478.

Coineau, N., 1992. Biogéographie évolutive du crustacé interstitiel *Microcharon* (Isopoda, Janiroidea) dans l'ouest du bassin Méditerranéen. *Bull. Inst. Océanogr. Monaco,* 9: 101–114.

Coineau, N. and Boutin, C., 1992. Biological processes in space and time. Colonization, evolution and speciation in interstitial stygobionts In: A.I. Camacho (Editor), *The Natural History of Biospeleology.* National Museum of Natural Sciences, Madrid, pp. 423–451.

Coineau, N., Henry, J.-P., Magniez, G. and Negoescu, I., 1994. Isopoda aquatica. In: C. Juberthie and V. Decu (Editors), *Encyclopaedia Biospeologica,* Vol. I. Société de Biospéologie, Moulis (France) and Bukarest, pp. 123–140.

Coleman, M.J. and Hynes, H.B.N., 1970. The vertical distribution of the invertebrate fauna in the bed of a stream. *Limnol. Oceanogr.,* 15: 31–40.

Cooling, M.P. and Boulton, A.J., 1993. Aspects of the hyporheic zone below the terminus of a South Australian arid-zone stream. *Aust. J. Mar. Freshwater Res.,* 44: 411–426.

Creuzé des Châtelliers, M., 1991a. *Dynamique de répartition des biocénoses interstitielles du Rhône en relation avec des caractéristiques géomorphologiques (secteurs de Brégnier-Cordon, Miribel-Jonage et Donzère-Mondragon), Ph.D. Thesis.* University Lyon 1, France, 161 pp.

Creuzé des Châtelliers, M., 1991b. Geomorphological processes and discontinuities in the macrodistribution of the interstitial fauna. A working hypothesis. *Verh. Int. Verein. Limnol.,* 24: 1609–1612.

Creuzé des Châtelliers, M. and Marmonier, P., 1993. Ecology of benthic and interstitial ostracods of the Rhône River, France. *J. Crust. Biol.,* 13(2): 268–279.

Creuzé des Châtelliers, M. and Poinsart, D., 1991. Caractéristiques des aquifères alluviaux et diversité faunistique du sous-écoulement du Rhône. *Hydrogéologie,* 3: 201–215.

Creuzé des Châtelliers, M. and Reygrobellet, J.-L., 1990. Interactions between geomorphological processes, benthic and hyporheic communities: first results on a by-passed canal of the French upper Rhône River. *Regulated Rivers: Res. & Manag.,* 5: 139–158.

Creuzé des Châtelliers, M., Poinsart, D. and Bravard, J.-P., 1994. Geomorphology of alluvial groundwater ecosystems. In: J. Gibert, D.L. Danielopol and J.A. Stanford (Editors), *Groundwater Ecology.* Academic Press, New York, pp. 157–185.

Culver, D.C., 1982. *Cave Life. Evolution and Ecology.* Harvard University Press, Cambridge, MA, 189 pp.

Culver, D.C. and Fong, D.W., 1994. Small scale and large scale biogeography of subterranean crustacean faunas of the Virginias. *Hydrobiologia,* 287: 3–9.

Culver, D.C., Kane, T.C. and Fong, D.W., 1995. *Adaptation and Natural Selection in Caves. The Evolution of Gammarus minus.* Harvard University Press, Cambridge, MA, 223 pp.

Danielopol, D.L., 1983. Der Einfluß organischer Verschmutzung auf das Grundwasser-Ökosystem der Donau im Raum Wien und Niederösterreich. *Forschungsber. Bundesminist. Gesund. Umweltschutz,* 5: 155 pp.

Danielopol, D.L., 1984. Ecological investigations on the alluvial sediments of the Danube in the Vienna area – a phreatobiological project. *Verh. Int. Verein. Limnol.,* 22: 1755–1761.

Danielopol, D.L., 1989. Groundwater fauna associated with riverine aquifers. *J. N. Am. Benthol. Soc.,* 8(1): 18–35.

Danielopol, D.L., 1991. Spatial distribution and dispersal of interstitial Crustacea in alluvial sediments of a backwater of the Danube at Vienna. *Stygologia,* 6(2): 97–110.

Danielopol, D.L. and Pospisil, P., 1991. Ecological basic research with potential application for groundwater management. *IAHS Publ.,* 202: 215–228.

Danielopol, D.L. and Rouch, R., 1991. L'adaptation des organismes au milieu aquatique souterrain. Réflexions sur l'apport des recherches écologiques récentes. *Stygologia,* 6(3): 129–142.

Danielopol, D.L., Marmonier, P., Boulton, A.J. and Bonaduce, G., 1994. World subterranean ostracod biogeography: dispersal or vicariance. *Hydrobiologia,* 287: 119–129.

Dole-Olivier, M.J. and Marmonier, P., 1992. Patch distribution of interstitial communities: prevailing factors. *Freshwater Biol.,* 27: 177–191.

Dole-Olivier, M.J., Marmonier, P., Creuzé des Châtelliers, M. and Martin, D., 1994. Interstitial fauna associated with the alluvial floodplains of the Rhône River (France). In: J. Gibert, D.L. Danielopol and J.A. Stanford (Editors), *Groundwater Ecology.* Academic Press, New York, pp. 313–346.

Fleeger, J.W. and Decho, A.W., 1987. Spatial variability of interstitial meiofauna: a review. *Stygologia,* 3(1): 35–54.

Gibert, J., Dole-Olivier, M.J., Marmonier, P. and Vervier, P., 1990. Surface water–groundwater ecotones. In: R.J. Naiman and H. Décamps (Editors), *The Ecology and Management of Aquatic-Terrestrial Ecotones.* Parthenon, Casterton Hall, England, pp. 199–225.

Gibert, J., Stanford, J.A., Dole-Olivier, M.J. and Ward, J.V., 1994a. Basic attributes of groundwater ecosystems and prospects for research. In: J. Gibert, D.L. Danielopol and J.A. Stanford (Editors), *Groundwater Ecology.* Academic Press, New York, pp. 7–40.

Gibert, J., Danielopol, D.L. and Stanford, J.A. (Editors), 1994b. *Groundwater Ecology.* Academic Press, New York, 571 pp.

Giere, O., 1993. *Meiobenthology. The Microscopic Fauna in Aquatic Sediments.* Springer, Berlin, 328 pp.

Ginet, R. and David, J., 1963. Présence de *Niphargus* (Amphipode Gammaridae) dans certaines eaux épigées des forêts de la Dombes (département de l'Ain, France). *Vie Milieu,* 14: 299–310.

Godbout, L. and Hynes, H.B.N., 1982. The three dimensional distribution of the fauna in a single riffle in a stream in Ontario. *Hydrobiologia,* 97: 87–96.

Henry, J.-P., Lewis, J.J. and Magniez, G., 1986. Isopoda: Asellota: Aselloidea, Gnathostenetroidoidea, Stenetrioidea. In: L. Botosaneanu (Editor), *Stygofauna Mundi.* E.J. Brill, Leiden, pp. 434–464.

Hey, R.D., Bathurst, J.C. and Thorne, C.R. (Editors), 1982. *Gravel-Bed Rivers.* Wiley, Chichester, UK, 875 pp.

Higgins, R.P. and Thiel, H. (Editors), 1988. *Introduction to the Study of Meiofauna.* Smithsonian Institution Press, Washington, DC, 488 pp.

Holsinger, J.R., 1988. Troglobites: the evolution of cave-dwelling organisms. *Am. Sci.,* 76: 146–153.

Holsinger, J.R., 1993. Biodiversity of subterranean amphipod crustaceans: global patterns and zoogeographic implications. *J. Nat. Hist.,* 27: 821–835.

Holsinger, J.R., 1994a. Pattern and process in the biogeography of subterranean amphipods. *Hydrobiologia,* 287: 131–145.

Holsinger, J.R., 1994b. Amphipoda. In: C. Juberthie and V. Decu (Editors), *Encyclopaedia Biospeologica,* Vol. I. Société de Biospéologie, Moulis (France) and Bucarest, pp. 147–163.

Howarth, F.G., 1987. The evolution of non-relictual tropical troglobites. *Int. J. Speleol.,* 16: 1–16.

Husmann, S., 1978. Die Bedeutung der Grundwasserfauna für biologische Reinigungsvorgänge im Interstitial von Lockergesteinen. *GWF Wasser/Abwasser,* 119: 293–302.

Iliffe, T.M., 1986. The zonation model for the evolution of aquatic faunas in anchialine caves. *Stygologia,* 2(1/2): 2–9.

Jones, R. and Culver, D.C., 1989. Evidence for selection on sensory structures in a cave population of *Gammarus minus* (Amphipoda). *Evolution,* 43(3): 688–693.

Juberthie, C. and Decu, V. (Editors), 1994. *Encyclopaedia Biospeologica,* Vol. I. Société de Biospéologie, Moulis (France) and Bucarest, 834 pp.

Kane, T.C. and Culver, D.C., 1992. Biological processes in space and time: analysis of adaptation. In: A.I. Camacho (Editor), *The National History of Biospeleology.* National Museum of Natural Sciences, Madrid, Spain, pp. 377–399.

Magniez, G., 1976. *Contribution à la connaissance de la biologie des Stenasellidae (Crustacea Isopoda Asellota des eaux souterraines), Thesis.* University of Dijon, France, 228 pp.

Magniez, G., 1981. Biogeographical and paleobiogeographical problems in Stenasellids (Crustacea Isopoda Asellota of underground waters). *Int. J. Speleol.,* 11: 71–81.

Marmonier, P. and Dole, M.J., 1986. Les amphipodes des sédiments d'un bras court-circuité du Rhône. *Sci. Eau,* 5: 461–486.

Marmonier, P., Dole-Olivier, M.J. and Creuzé des Châtelliers, M., 1992. Spatial distribution of interstitial assemblages in the floodplain of the Rhône River. *Regulated Rivers: Res. & Manag.,* 7: 75–82.

Marmonier, P., Vervier, P., Gibert, J. and Dole-Olivier, M.J., 1993. Biodiversity in ground waters. *Trends Ecol. Evol.,* 8: 392–395.

Messouli, M., Boutin, C. and Coineau, N., 1991. Phylogénie et biogéographie évolutive d'un groupe de Metacrangonyctidae, Crustacés amphipodes stygobies du Maroc. I – Les espèces du groupe *Panousei. Mém. Biospéol.,* 18: 247–262.

Notenboom, J., 1991. Marine regressions and the evolution of groundwater dwelling amphipods (Crustacea). *J. Biogeogr.,* 18: 437–454.

Orghidan, T., 1959. Ein neuer Lebensraum des unterirdischen Wassers: Der hyporheische Biotop. *Arch. Hydrobiol.,* 55: 392–414.

Peck, S.B. and Finston, T.L., 1993. Galapagos Islands troglobites: the questions of tropical troglobites, parapatric distributions with eyed-sister-species, and their origin by parapatric speciation. *Mém. Biospéol.,* 20: 19–37.

Pennak, R.W., 1951. Comparative ecology of the interstitial fauna of fresh-water and marine beaches. *Ann. Biol.,* 27: 449–480.

Pennak, R.W., 1988. Ecology of the freshwater meiofauna. In: R.P. Higgins and H. Thiel (Editors), *Introduction to the Study of Meiofauna.* Smithsonian Institution Press, Washington, DC, pp. 39–60.

Pospisil, P., 1994. The groundwater fauna of a Danube aquifer in the "Lobau" wetland in Vienna, Austria. In: J. Gibert, D.L. Danielopol and J.A. Stanford (Editors), *Groundwater Ecology.* Academic Press, New York, pp. 347–366.

Remane, A., 1952. Die Besiedlung des Sandbodens im Meere und die Bedeutung der Lebensformtypen für die Ökologie. *Verh. Dtsch. Zool. Ges. Wilhelmshaven,* 1951: 327–359.

Rogulj, B., Marmonier, P., Lattinger, R. and Danielopol, D.L., 1994. Fine-scale distribution of Ostracoda in the interstitial habitats of the Rivers Sava and Rhône. *Hydrobiologia,* 287: 19–28.

Rouch, R., 1988. Sur la répartition spatiale des Crustacés dans le sous-écoulement d'un ruisseau des Pyrénées. *Ann. Limnol.,* 24(3): 213–234.

Rouch, R., 1991. Structure du peuplement des Harpacticides dans le milieu hyporhéique d'un ruisseau des Pyrénées. *Ann. Limnol.,* 27(3): 227–241.

Rouch, R., 1992. Caractéristiques et conditions hydrodynamiques des écoulements dans les sédiments d'un ruisseau des Pyrénées. Implications écologiques. *Stygologia,* 7(1): 13–25.

Rouch, R., 1995. Peuplement des Crustacés dans la zone hyporhéique d'un ruisseau des Pyrénées. *Ann. Limnol.,* 31(1): 9–28.

Rouch, R. and Danielopol, D.L., 1987. L'origine de la faune aquatique souterraine, entre le paradigme du refuge et le modèle de la colonisation active. *Stygologia,* 3(4): 345–372.

Rouch, R. and Danielopol, D.L., 1997. Species richness of microcrustacea in subterranean freshwater habitats. Comparative analysis and approximate evaluation. *Int. Rev. Ges. Hydrobiol.,* 82(2): 121–145.

Rouch, R. and Lescher-Moutoué, F., 1992. Structure du peuplement des Cyclopoides (Crustacea: Copepoda) dans le milieu hyporhéique d'un ruisseau des Pyrénées. *Stygologia,* 7(4): 197–211.

Rouch, R., Bakalowicz, M., Mangin, A. and D'Hulst, D., 1989. Sur les caractéristiques du sous-écoulement d'un ruisseau des Pyrénées. *Ann. Limnol.,* 25(1): 3–16.

Sbordoni, V., 1982. Advances in speciation of cave animals. In: C. Barigozzi (Editor), *Mechanisms of speciation,* Alan R. Liss, New York, pp. 219–240.

Schminke, H.K., 1981. Adaptation of Bathynellacea (Crustacea, Syncarida) to life in the interstitial ("Zoea Theory"). *Int. Revue Gesamten Hydrobiol.,* 66(4): 575–637.

Schwoerbel, J., 1967. Das hyporheische Interstitial als Grenzbiotop zwischen oberirdischem und subterranem Ökosystem und seine Bedeutung für die Primär-Evolution von Kleinsthöhlenbewohnern. *Arch. Hydrobiol. Suppl.,* 33: 1–62.

Shepherd, B.G., Hartman, G.F. and Wilson, W.J., 1986. Relationships between stream and intragravel temperatures in coastal drainages, and some implications for fisheries workers. *Can. J. Fish. Aquat. Sci.,* 43: 1818–1822.

Stanford, J.A. and Ward, J.V., 1988. The hyporheic habitat of river ecosystems. *Nature,* 335: 64–66.

Stanford, J.A. and Ward, J.V., 1993. An ecosystem perspective of alluvial rivers: connectivity and the hyporheic corridor. *J. N. Am. Benthol. Soc.,* 12: 48–60.

Stanley, E.H. and Boulton, A.J., 1993. Hydrology and the distribution of hyporheos: perspectives from a mesic river and a desert stream. *J. N. Am. Benthol. Soc.,* 12(1): 79–83.

Stoch, F., 1995. The ecological and historical determinants of crustacean diversity in groundwaters, or: why are there so many species? *Mém. Biospéol.,* 22: 139–160.

Stock, J.H., 1980. Regression model evolution as exemplified by the genus *Pseudoniphargus* (Amphipoda). *Bijdr. Dierkd.,* 50(1): 105–144.

Swedmark, B., 1964. The interstitial fauna of marine sand, *Biol. Rev.,* 39: 1–42.

Thienemann, A., 1926. *Die Binnengewässer Mitteleuropas, Band I, eine limnologische Einführung.* E. Schweizerbart'sche Verlagsbuchhandlung, Stuttgart, 255 pp.

Ward, J.V., 1989. The four-dimensional nature of lotic ecosystems. *J. N. Am. Benthol. Soc.,* 8: 2–8.

Ward, J.V., 1998. Riverine landscapes: biodiversity patterns, disturbance regimes, and aquatic conservation. *Biol. Conserv.,* 83: 269–278.

Ward, J.V. and Holsinger, J.R., 1981. Distribution and habitat diversity of subterranean amphipods in the Rocky Mountains of Colorado. *Int. J. Speleol.,* 11: 63–70.

Ward, J.V. and Palmer, M.A., 1994. Distribution patterns of interstitial freshwater meiofauna over a range of spatial scales, with emphasis on alluvial river–aquifer systems. *Hydrobiologia,* 287: 147–156.

Ward, J.V. and Stanford, J.A., 1995. Ecological connectivity in alluvial river ecosystems and its disruption by flow regulation. *Regulated Rivers,* 11: 105–119.

Ward, J.V. and Voelz, N.J., 1990. Gradient analysis of interstitial meiofauna along a longitudinal stream profile. *Stygologia,* 5(2): 93–99.

Ward, J.V. and Voelz, N.J., 1994. Groundwater fauna of the South Platte River System, Colorado. In: J. Gibert, D.L. Danielopol and J.A. Stanford (Editors), *Groundwater Ecology.* Academic Press, New York, pp. 391–423.

Welcomme, R.L., 1979. *Fisheries Ecology of Floodplain Rivers,* Longman, London.

White, D.S., 1993. Perspectives on defining and delineating hyporheic zones. *J. N. Am. Benthol. Soc.,* 12(1): 61–69.

Whitman, R.L., Andrzejewski, M.C., Kennedy, K.J. and Sobat, T.A., 1994. Composition, spatial–temporal distribution and environmental factors influencing the interstitial beach meiofauna of Southern Lake Michigan. *Verh. Int. Verein. Limnol.,* 25: 1389–1397.

Wilkens, H., 1988. Evolution and genetics of epigean and cave *Astyanax fasciatus* (Characidae, Pisces) – Support for the neutral mutation theory. *Evol. Biol.,* 23: 271–367.

Wilkens, H., 1992. Neutral mutations and evolutionary progress. In: A.I. Camacho (Editor), *The Natural History of Biospeleology,* National Museum of Natural Sciences, Madrid, pp. 402–478.

Wilkens, H. and Hüppop, K., 1986. Sympatric speciation in cave fishes? Studies on a mixed population of epi- and hypogean *Astyanax* (Characidae, Pisces). *Z. Zool. Syst. Evolutionsforsch.,* 24: 223–230.

Williams, D.D. and Hynes, H.B.N., 1974. The occurrence of benthos deep in the substratum of a stream. *Freshwater Biol.,* 4: 233–256.

Williams, D.D. and Hynes, H.B.N., 1976. The recolonization mechanisms of stream benthos. *Oikos,* 27: 265–272.

Wolman, M.G. and Leopold, L.B., 1957. River flood plains: some observations on their formation. *U.S. Geol. Surv. Prof. Pap.,* 282-C: 87–109.

Chapter 3

ANCHIALINE CAVE ECOLOGY

Thomas M. Iliffe

INTRODUCTION

Reports of a variety of red-colored caridean shrimps inhabiting tropical landlocked saltwater pools prompted Holthuis (1973) to propose a new term describing this peculiar habitat. "Anchialine", derived from the Greek *anchialos,* near the sea, was defined by Holthuis (1973, p. 1) to indicate "pools with no surface connection to the sea, containing brackish water, which fluctuates with the tides". Such anchialine pools occur in both uplifted reef limestone and irregular porous lava flows. The water in these pools varies in salinity from nearly fresh to fully marine. Dampened tidal fluctuations indicate the influence of a continuous water table extending inland from the open sea. Both salinity and the degree of connection to the sea control the nature of the biota. Pools close to the sea contain typical marine species, while those farther inland have fewer but more unusual species.

Recent explorations by cave divers have resulted in the discovery of extensive, entirely submerged cave systems in which the classical anchialine pool represents only a small portion. Thus, an expanded definition of the term anchialine, more refined in the ecological sense, was proposed at the 1984 International Symposium on the Biology of Marine Caves: "Anchialine habitats consist of bodies of haline waters, usually with a restricted exposure to open air, always with more or less extensive subterranean connections to the sea, and showing noticeable marine as well as terrestrial influences" (Stock et al., 1986a, p. 91).

ANCHIALINE HABITATS

Geology and distribution

Anchialine habitats exist in a variety of forms, although they appear to be most common in tropical locales (Fig. 3.1). One type of anchialine habitat consists of water-filled depressions in the substratum. For example, anchialine pools are found in 200-year-old, highly porous lava flows on the Cape Kinau Peninsula of Maui in the Hawaiian Islands (Holthuis, 1973; Maciolek, 1986). Salinity in these pools ranges from 6 to 33.5 g ℓ^{-1} and temperature from 20 to 30.5ºC, in response to solar and tidal cycles. The biota consists of more than 140 species of algae, together with a fauna dominated by crustaceans, but also including a few species of fishes, molluscs, polychaetes, and other invertebrates (Maciolek, 1986). A pool, apparently of this type, on Funafuti Atoll (Tuvalu) is bordered by the coral limestone of the island on one side and a boulder ridge, possibly deposited by hurricanes, on the other (Holthuis, 1973).

Coastal tectonic faults that extend below sea level constitute another form of anchialine habitat. Such faults in volcanic rock in the Galapagos Islands may stretch for up to 20 km and contain pools as deep as 30 m (Iliffe, 1991). Similar faults in uplifted reef limestone on the island of Niue in the Central Pacific produce deep chasms containing anchialine pools (Iliffe and Sarbu, 1990). The Ras Muhammad's Crack in the Sinai Peninsula consists of a water-filled crack in an elevated fossil reef formed by a 1968 earthquake (Holthuis, 1973).

Many of the famed Bahamian "Blue Holes" consist of submarine faults along the platform edge that exhibit exceptionally strong, reversing tidal currents (Dill, 1977; Farr and Palmer, 1984; Smart, 1984; Palmer, 1985, 1986; Mylroie et al., 1995). Since tides on the shallow banks are out of phase by several hours or more in comparison with tides in the open sea, a tidal imbalance is created between the two water

Fig. 3.1. World map showing locations of some of the more significant limestone (circles) and volcanic (asterisks) anchialine caves.

bodies. Blue Holes serve as hydrological conduits that respond to this imbalance and, as a result, transport water between the banks and the shelf edge. During the "suck" phase of the tidal cycle when tides in the open ocean are higher than on the bank, large volumes of ocean water are drawn into these cave entrances. Conversely, on the "blow" phase when water levels on the banks are higher than the ocean, cave water is forced out.

The most extensive of the known anchialine habitats are solutionally-developed limestone caves. The largest anchialine cave explored is Systema Nohoch Nah Chich with 70.4 km of surveyed passage (Mike Madden, pers. comm.). This totally submerged cave is located on the Caribbean coast of the Yucatan Peninsula in Mexico. The Yucatan is a flat limestone plain with no surface streams or rivers. All drainage is subterranean through extensive networks of submerged cave systems. Primary orientation of these caves is perpendicular to the coast, suggesting that they serve as major drainage conduits to the sea. Other long underwater limestone caves on the Caribbean coast of the Yucatan mainland include Systema Dos Ojos (64.0 km), Systema Ox Bel Ha (45.1 km), and Systema Naranjal (18.5 km). Cueva Quebrada (7.60 km) and Areolito (6.10 km) are long anchialine caves on the island of Cozumel, separated from the Yucatan Peninsula by a deep channel.

Extensive anchialine limestone caves are also present in the Bahamas, Bermuda and other West Indian islands. Lucayan Caverns on Grand Bahama Island in the Bahamas has been explored by divers for 9.18 km (Courbon et al., 1989). Green Bay Cave in Bermuda is 2.29 km long (Iliffe, 1987). Anchialine caves in Belize, the Dominican Republic, and Bonaire also show good potential for diving exploration and biospeleological investigations.

Anchialine lava tube caves are known from the Canary Islands, Galapagos Islands, Hawaii and Western Samoa. The longest of these is the Jameos del Agua (Atlantida Tunnel) on Lanzarote in the Canary Islands, the submerged portion of which extends 1.60 km out from the coastline, reaching a depth of 50 m (Wilkens and Parzefall, 1974; Iliffe et al., 1984b; Kornicker and Iliffe, 1995).

Anchialine groundwater also exists in the form of small cracks and fissures within either volcanic or limestone bedrock. Caves and adjacent groundwater accessible through wells frequently share a very similar stygofauna (Stock et al., 1986b; Wilkens et al., 1986; Humphreys, 1993).

Exploration and investigation of anchialine caves

Specialized cave-diving techniques are required for the thorough investigation of anchialine caves, owing to their considerable length, their varied microhabitats, and the restricted occurrence of most anchialine troglobites in only the deeper, more saline waters in the far interior of the caves. Biological collections in anchialine caves typically involve use of suction bottles, diver-towed plankton nets and baited traps, as well as direct collection of specimens in individual vials. Water sampling can be done using bottles initially filled with sterilized water. This water is forced out by inverting the bottle over the exhaust tee of the diver's regulator to purge it with air, before refilling the bottle with water from the sampling site. Hypodermic syringes can also be used to collect water samples. Additionally, divers can carry *in situ* probes that are programmed to record data on depth, temperature, salinity, dissolved oxygen, pH, redox potential, turbidity and other parameters at intervals as short as one second.

Hydrology and water chemistry

Salinity

The water in anchialine pools is an extension of the surrounding groundwater. The permeability and porosity of the rock surrounding the pools determine the nature of the groundwater. In younger limestone, groundwater movement is limited to interstitial spaces within the bedrock. As slightly acid fresh groundwater moves through the rock, dissolution alters and enlarges the pore spaces. In older limestone, the pore spaces have become connected into solution channels, crevices and caverns. Since the intervening rock has become tightly cemented, water flow is largely through underground channels. Highly karstified limestone is particularly cavernous with nearly infinite permeability.

Fresh groundwater occurs as a lens-shaped body that is surrounded and underlain by saline groundwater. Much like an iceberg, the fresh groundwater floats on denser brackish to saline groundwater. The thickness of the freshwater lens increases with distance from the coast, and is a function of the porosity of the bedrock. Since the surface of the water table is elevated above sea level, overall movement of fresh groundwater is toward the sea. Inland recharge (rainfall) is thus balanced by loss of fresh groundwater at the margins of the lens.

The boundary between these fresh and marine water bodies consists of a more or less distinct transition zone or halocline produced by mixing between the two water bodies. The permeability of the rock affects the extent of mixing and thus the thickness of the halocline. At increasing distances inland from the coast, there is a dampening of tidal oscillations, thus reducing tidal mixing and thinning the halocline.

Due to the large surface area of karstic reef-derived limestones, lack of surface drainage, and a relatively high precipitation rate, the Yucatan Peninsula is underlain by an immense freshwater lens. The thickness of this freshwater lens increases with distance inland from the coast. For example, the freshwater lens at Cenote Abejas, located about 2 km inland from the coast, is 10 m thick, while 10 km inland at Cenote Vaca Ha, it is 20 m thick.

Water-filled sinkholes, called cenotes along the Caribbean coast, contain a nearly fresh (1 to $3\,g\,\ell^{-1}$) surface layer separated by a relatively sharp halocline from fully marine waters (Fig. 3.2a). Primary cave development has occurred at the halocline where mixing corrosion[1] between fresh and salt water has resulted in limestone dissolution (Back et al., 1986).

Bermuda's anchialine caves typically have a thin (0.5 to 2 m) surface layer of brackish water overlying nearly full-strength sea water (Fig. 3.3a). Surface salinity measurements in 100 cave pools ranged from 0 to $35\,g\,\ell^{-1}$ with an average of $24.5\,g\,\ell^{-1}$. In the same caves, the salinity at 1 m depth varied from 15 to $35\,g\,\ell^{-1}$ with an average of $32\,g\,\ell^{-1}$. Caves most isolated from the sea have the lowest surface salinity, on account of increased freshwater input and reduced tidal perturbation. The highly stratified water column in these cave pools results from the absence of wind- or wave-induced mixing. Seasonal changes in surface salinity may be related to lunar tidal cycles, precipitation, and water temperature variations.

[1] Mixing of two water bodies, each saturated with different concentrations of calcium carbonate, produces water that is undersaturated (Bögli, 1964). This undersaturation occurs because the relationship between CO_2 partial pressure and calcium carbonate saturation is exponential.

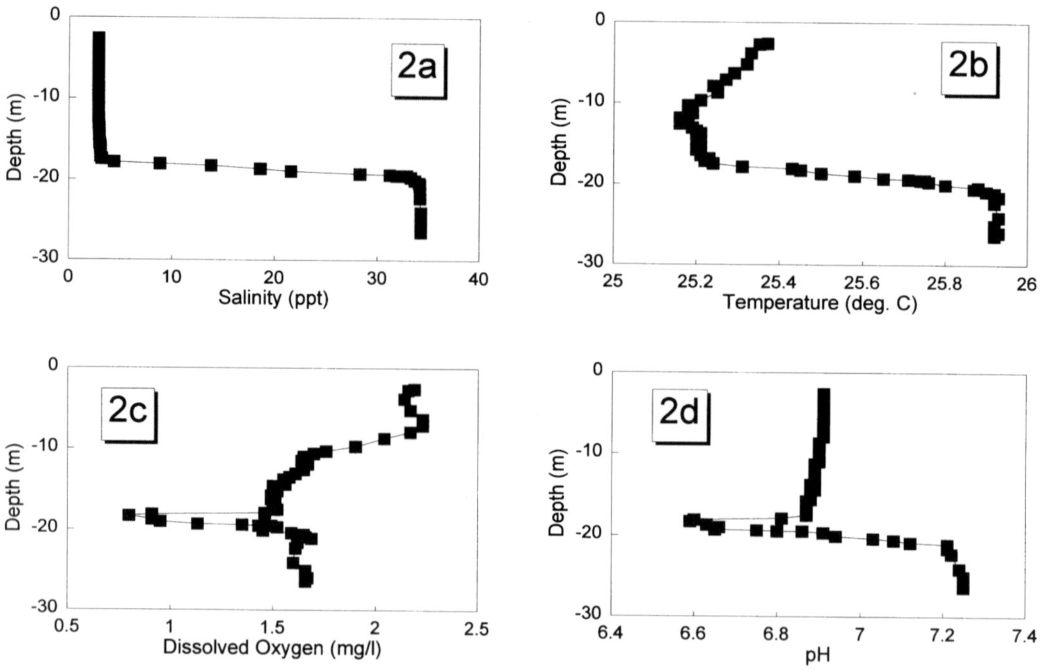

Fig. 3.2. Vertical profiles of (a) salinity, (b) temperature, (c) dissolved oxygen, and (d) pH in the B Tunnel of Maya Blue Cenote, Tulum, Quintana Roo, Mexico. Each black square in a graph represents a single datum point. A Hydrolab Datasonde III profiler was used to obtain these measurements on 18 March 1995.

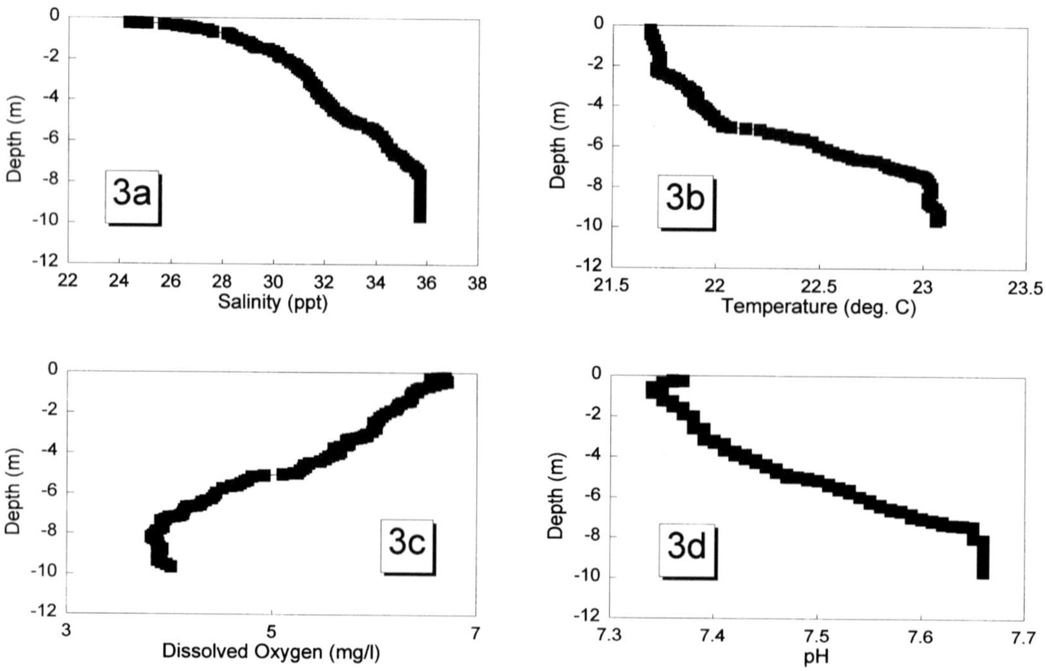

Fig. 3.3. Vertical profiles of (a) salinity, (b) temperature, (c) dissolved oxygen, and (d) pH in Wonderland Cave, Hamilton Parish, Bermuda. Each black square in a graph represents a single datum point. A Hydrolab Datasonde III profiler was used to obtain these measurements on 17 December 1995.

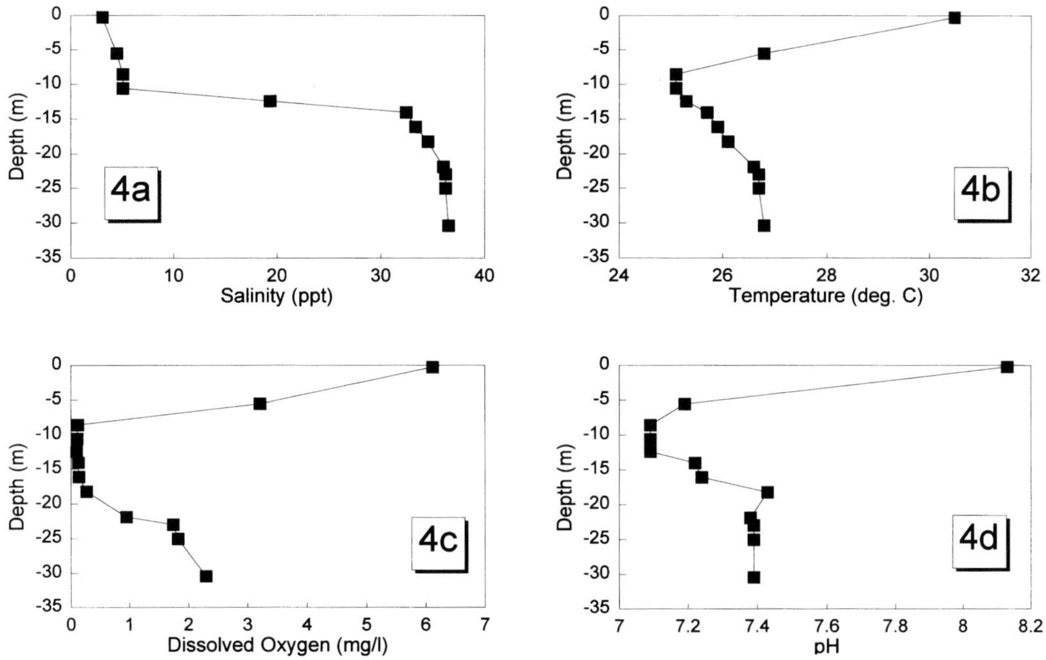

Fig. 3.4. Vertical profiles of (a) salinity, (b) temperature, (c) dissolved oxygen, and (d) pH in Cueva de los Carboneros, Playa Giron, Matanzas, Cuba. Each black square in a graph represents a single datum point. These plots are derived from data reported by Yager (1994b) using a YSI 600 Sonde.

Plotting the data supplied by Yager (1994b) for a Cuban cave shows a relatively sharp halocline similar to those observed in Yucatan caves (Fig. 3.4a).

Temperature

Water temperatures in salinity-stratified anchialine caves show similar steep gradients with depth (Figs. 3.2b and 3.3b). Temperatures are lowest at the surface, increase markedly through the halocline and then more gradually with increasing depth. The increase in temperature with depth could be due either to evaporative cooling at the surface or geothermal warming below the halocline (Iliffe et al., 1983). The Cuban cave studied by Yager (1994b) is entered through an open sinkhole pond that contained a bloom of microalgae at the time. Temperature drops sharply beneath a warm surface layer, but increases across the halocline with increasing salinity (Fig. 3.4b).

Seasonal variations in temperature and salinity were investigated in two anchialine cave pools from Bermuda. Walsingham Cave is situated about 100 m inland from the shore of Castle Harbour and is connected to it by submerged channels. The main pool in this cave is located immediately inside the entrance and is thus subject to external influences. In contrast, the pool in Crystal Cave is relatively far from the entrance and also more remote from the sea. However, the two pools are interconnected by submerged passageways. Surface water temperatures in both caves closely mirror the cave air temperatures. For most of the year, water at 1 m depth in Crystal Cave averaged one degree warmer than the surface waters (Fig. 3.5a). Only for a two-month period from July to August did surface temperatures exceed those at 1 m. Since the humidity in this cave remains constant at saturation, evaporative cooling should not be significant. Thus, an extension of the island's documented geothermal gradient into the deeper waters of the cave pools could explain this inverse temperature relationship.

The water column in Walsingham Cave was stratified by temperature and salinity for winter and early spring 1982 (Fig. 3.6). The cooler, lower-salinity surface layer abruptly disappeared on 26 May leaving a homogeneous water column that remained intact for at least the following six months. In Crystal Cave, the water column remained salinity-stratified for the entire study period, with only minor fluctuations at 1 m depth in comparison with the surface (Fig. 3.5b).

Tides

A recording tide gauge was placed in the entrance or interior pools of 27 Bermuda caves, and the

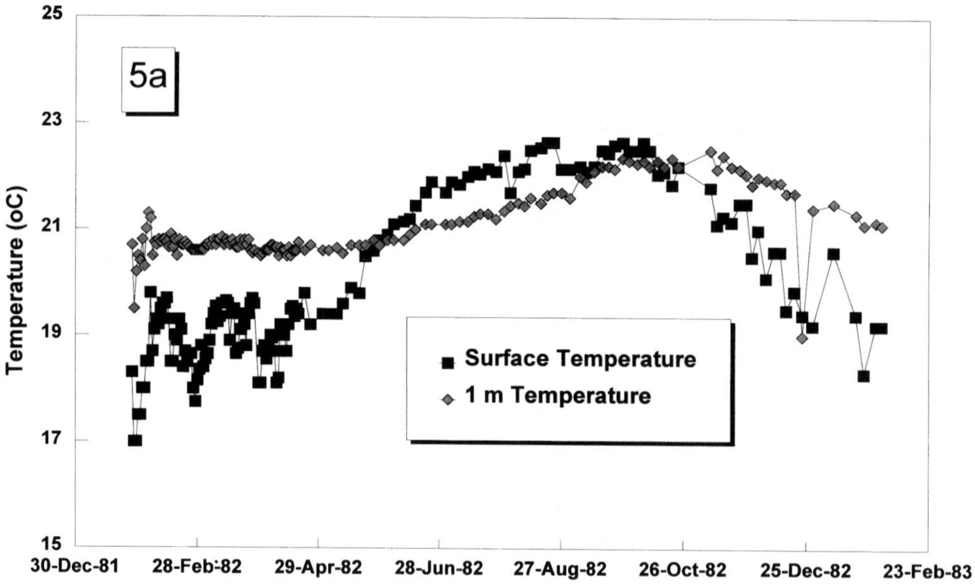

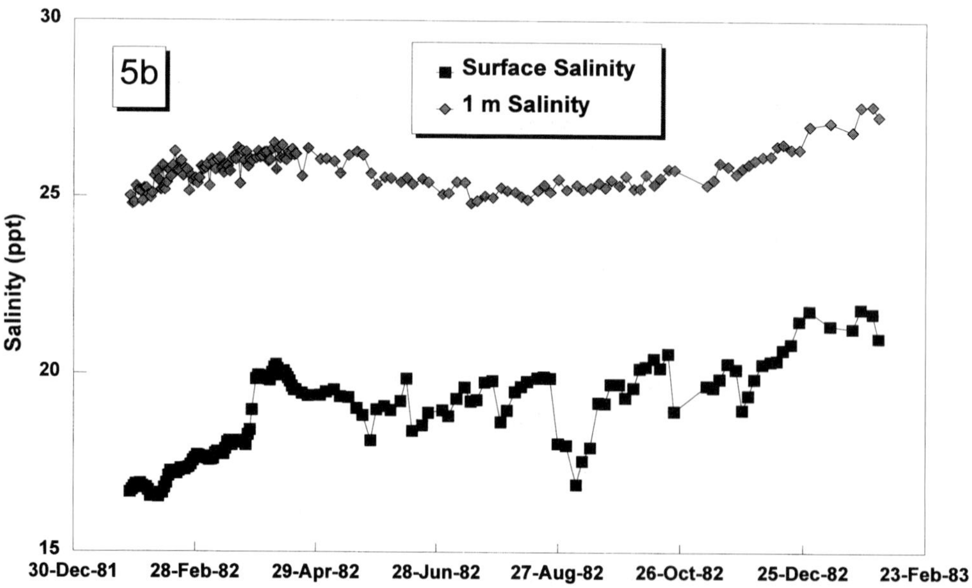

Fig. 3.5. Seasonal variations in (a) water temperature and (b) salinity, from the surface and 1 m depths of Crystal Cave, Hamilton Parish, Bermuda. Data were collected between 27 January 1982 and 31 January 1983. Temperatures of water samples collected were measured immediately with a thermometer, whereas salinities were later analyzed with a laboratory salinometer.

collected data were compared with a similar gauge simultaneously measuring tidal cycles in the open ocean. Cave tides characteristically showed a reduced amplitude and occurred up to $2\frac{1}{2}$ hours after the corresponding tide in open waters. In general, coastal caves that had a relatively direct connection to the sea had a short lag time and a tidal range nearly identical to that of the open sea. Caves located farther inland had a longer lag and smaller percent range. However, the longest lag time and smallest percent range in comparison with open ocean tides were observed in caves connecting directly to Harrington Sound. This

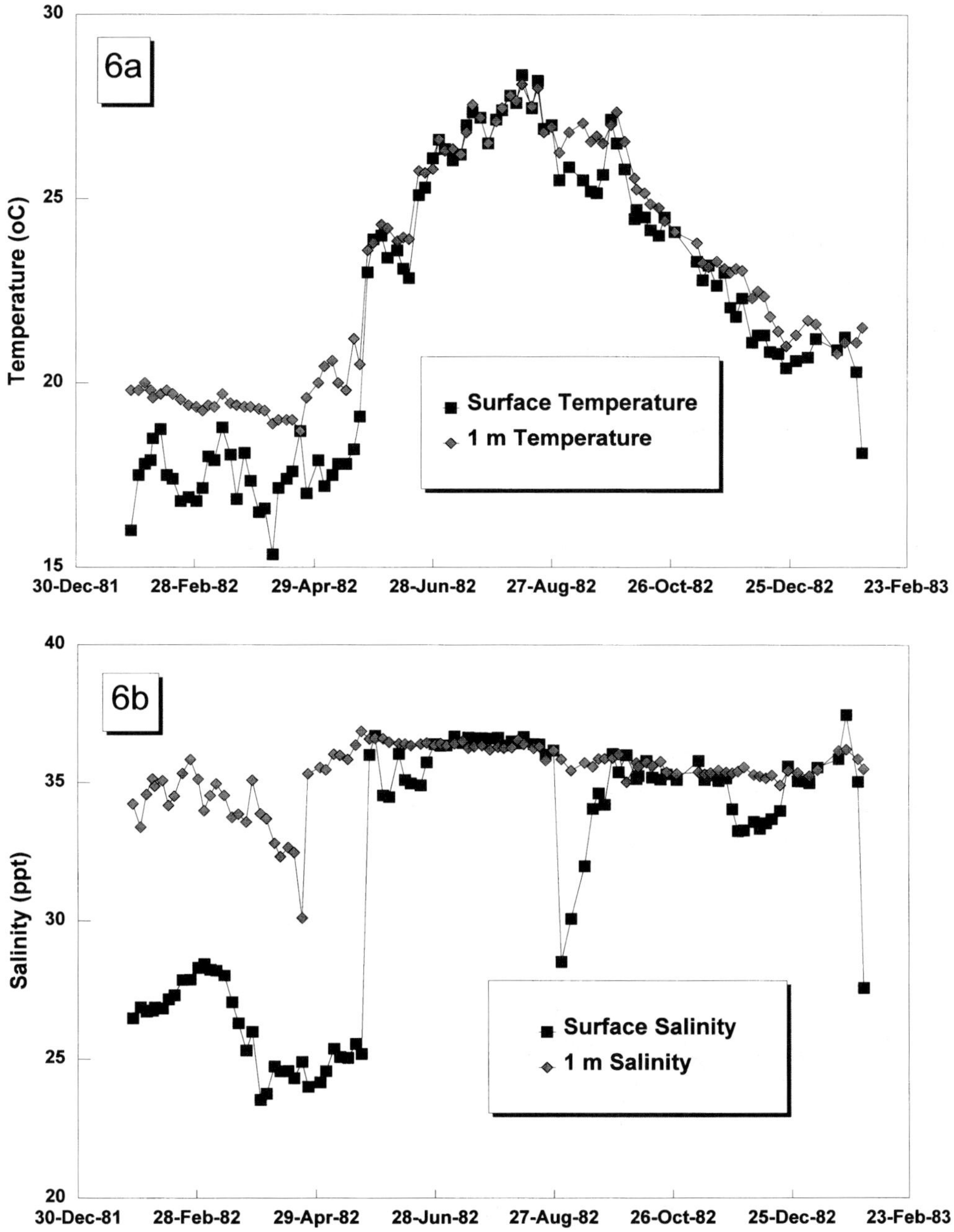

Fig. 3.6. Seasonal variations in (a) water temperature and (b) salinity from the surface and 1 m depths of Walsingham Cave, Hamilton Parish, Bermuda. Data were collected between 27 January 1982 and 31 January 1983. Temperatures of water samples collected were measured immediately with a thermometer, whereas salinities were later analyzed with a laboratory salinometer.

inshore lagoon, almost totally enclosed, had a lag time of 165 minutes and a 30% range. A linear relationship ($R^2 = 0.85$) exists between lag time and percent range for Bermuda cave pools (Fig. 3.7). Extrapolating to 0% range (i.e., no tidal oscillation) yields a lag time of 171 minutes.

Tidally generated water currents in most of Bermuda's coastal caves allow for complete exchange of water on each tidal cycle. Caves located farther inland have a reduced amplitude and relatively long delay, indicating a rather indirect connection with the sea and a correspondingly long residence time (turnover time)

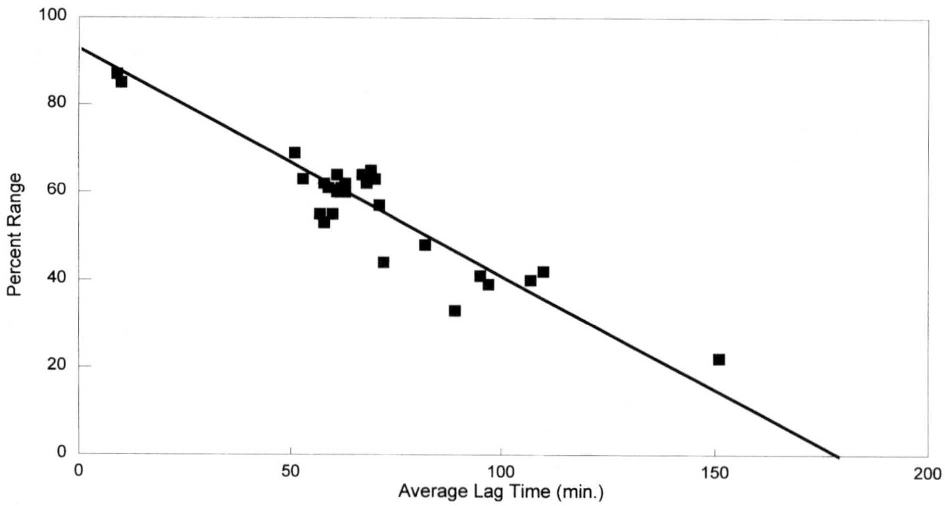

Fig. 3.7. Plot of percent range (i.e., tidal amplitude in comparison with open ocean tides) and average lag time (i.e., average delay in comparison with open-ocean tides) for 27 inland cave pools from Bermuda. A recording tide gauge was placed in each of the cave pools for 11 to 24 tidal cycles while an identical gauge simultaneously recorded data on open-ocean tides. Tide data were collected between 22 August and 30 November 1980.

for the cave waters. In such caves, undisturbed water is essentially "air-clear", with no observable particulate matter in the water column.

Water residence time was estimated for Church Cave, an inland Bermuda cave with no obvious connection to the sea (Iliffe et al., 1984a). In Church Cave, the maximum water depth was estimated as 30 m (the depth of the basalt underlying the cavernous limestone) and the tidal amplitude was determined to be 40 cm. Assuming complete mixing during each tidal cycle, the residence time would be 37.5 days [30 m/(0.4 m 2 tidal cycles per day)]. However, based on similar assumptions, residence times for Bermuda's inshore waters were underestimated by a factor of four (Barnes and Bodungen, 1978). Since wind- and wave-induced mixing is absent in caves, this underestimation may be at least twice that for the inshore waters. With these corrections, cave residence times would be about one year.

In the Bahamas, extremely strong, reversing tidal currents are observed in some oceanic blue holes, particularly those located in shallow waters along the shelf edge. Currents measured in the main cave passage of Conch Sound Blue Hole on Andros Island were as high as 0.53 m sec^{-1} (Warner and Moore, 1984). Plots of current velocity during blow and suck phases indicate that, during the two-day observation period, there appeared to be a net loss of water from the cave (i.e., the volume of water blown out was greater than the volume sucked in). Currents in these ocean blue holes are apparently due to tidal delays and phase differences between the extensive shallow water banks and the open ocean, generating hydrostatic gradients at opposite ends of the cave systems.

Dissolved oxygen and biological oxygen demand

Biological Oxygen Demand (BOD) levels in cave waters in Bermuda are very low (<0.25 mg $O_2\,\ell^{-1}$). A small maximum in BOD at the halocline is probably the result of trapping of surface-derived particulate matter at the pycnocline (Iliffe et al., 1984a). Oxygen levels are close to saturation in the surface waters, owing to exchange with the atmosphere and drip-water contributions. However, despite the low BOD, dissolved oxygen levels fall off sharply at the halocline and remain fairly constant with increasing depth (Fig. 3.3c). Although oxygen levels in deeper water vary from cave to cave, the lowest value obtained for any of seven normal Bermuda cave systems investigated by Iliffe et al. (1984a) was 55% of saturation (3.75 mg ℓ^{-1}).

Considerably lower dissolved oxygen levels have been reported from caves in Cuba, in Western Australia and along the Adriatic (Sket, 1996; Yager, 1994b; Yager and Humphreys, 1996). An oxygen minimum associated with the halocline was encountered in both Yucatan and Cuban caves (Figs. 3.2c and 3.4c). This oxygen minimum was especially sharp and well defined in the Yucatan cave. It may be generated by microbial oxidation of organic matter accumulating at the pycnocline. An anoxic layer, 7.5 m thick, in the Cuban cave could have resulted from similar microbial decomposition of organic matter generated by the phytoplankton bloom in the surface waters.

Acidity

In the Yucatan cave, a sharp and well-defined pH minimum, with values as low as 6.6, accompanied the oxygen minimum (Fig. 3.2d). A similar, but broader, pH minimum occurred in the anoxic zone of the Cuban cave (Fig. 3.4d). With the exception of a minor minimum just below the surface, pH increased along with increasing salinity in the Bermuda cave (Fig. 3.3d). In fully marine waters below the halocline, pH values were uniform at 7.25 in the Yucatan cave, 7.39 in the Cuban cave and 7.68 in the Bermuda cave.

Nutrient chemistry

The cave waters of Bermuda show nutrient characteristics similar to groundwater, with high nitrate content (up to $600\,\mathrm{mg}\,\ell^{-1}$), but low ammonia, nitrite and phosphate concentrations (Iliffe et al., 1984a). The nitrate concentrations decrease with increasing salinity in a linear manner, implying conservative mixing and a single freshwater source. Nitrate concentrations in both caves and groundwater appear to arise from cesspit seepage and fertilizer usage, since nutrient concentrations in rainwater are too low to account for the observed values. Ammonia and phosphate concentrations are very low but show maxima associated with the area of maximum BOD, indicating that these ions are released during oxidation of organic matter. Subsequent oxidation of the ammonia produces a very weak nitrite maximum slightly deeper than the ammonia maximum. Silicate concentrations in cave waters are inversely related to salinity, with soil leaching as the probable source. Nitrate and phosphate concentrations in the deeper waters of the caves are about five times higher than for inshore waters of comparable salinity, because of biological utilization in the latter.

Trace metals

In Bermuda caves, cadmium, copper, iron, manganese, nickel, and zinc showed rather similar distributions (Iliffe et al., 1984a). Elevated concentrations present in the surface waters fell sharply at the halocline, and remained fairly constant with depth at concentrations similar to Bermuda's inshore waters. Metal concentrations are probably controlled by freshwater inputs and interaction with limestone.

Water chemistry of polluted caves

In 1980, several large piles of partially burned domestic waste materials were bulldozed into the anchialine pool of a Bermuda cave (Iliffe et al., 1984a). Shortly thereafter, subsurface water samples from pools in this and several other nearby caves gave off a strong odor of hydrogen sulfide. Levels of BOD in the polluted cave waters were 10 to 60 times higher than normal, with a maximum in the region of the halocline. Depth profiles showed that dissolved oxygen levels fell rapidly to zero, and only began to recover slightly at depths below 10 m. Hydrogen sulfide was present in all samples from the anoxic zone. In oxygen-depleted subsurface waters, nitrate was removed by denitrification, while maxima of ammonia and phosphate occurred. A nitrite maximum appeared in deeper waters, possibly from partial denitrification of ammonia. Most metal concentrations were higher in the surface waters of the dumping area. Iron and manganese concentrations were 3–100 times higher than in normal cave waters, probably as a result of mobilization of their more soluble lower oxidation states from dumped material or sediments. In the anoxic zone, precipitation of insoluble cadmium, copper, nickel and zinc sulfides took place. Tidal transfer of water between caves primarily occurs at depth, so that pollutants, such as metals which are removed in the anoxic zone, are not transported. However, anoxic waters present in several nearby caves indicate that BOD loadings at least are transported. Modeling calculations for a lake in the nearby and relatively unspoiled Church Cave (35 35 30 m deep) suggest that an approximate doubling of already low natural carbon inputs ($40\,\mathrm{kg\,yr^{-1}}$) could produce anoxic conditions in the cave waters.

Anoxic conditions at or below the halocline are commonly observed in open sinkholes in which logs, leaves and other organic matter enter directly from the surface. Cenote Angelita in Yucatan is an open sink, with oxygenated fresh water overlying the halocline at 30 m depth. The halocline is marked by a 1–2 m thick, cloud-like layer of sulfide precipitates. Beneath this layer, water clarity improves, but the water is anoxic to a depth of at least 50 m. Similar anoxic conditions also exist in a number of inland blue holes[2] in the Bahamas and organically enriched caves along the Adriatic coast (Sket, 1996).

ANCHIALINE FAUNAL DIVERSITY

Virtually all species inhabiting anchialine caves are of marine origin. Since marine organisms are far more numerous and diverse than freshwater taxa, anchialine caves, in contrast to freshwater caves, should have a larger and more diverse pool of potential cave colonizers to draw from. In addition, various freshwater cavernicoles including many nematodes, triclad turbellarians, polychaetes, copepods, mysids, thermosbaenaceans, isopods, amphipods, shrimps and fishes are marine relicts (Vandel, 1965; Reddell, 1981) that may have inhabited anchialine caves before their adaptation to fresh water (Iliffe, 1986).

As recently as 30 years ago, it was thought that animals inhabiting marine caves had not undergone noticeable modification, and were therefore of little interest to the biospeleologist (Vandel, 1965; Riedl, 1966). The first faunal surveys of anchialine caves began in the early 1980s with the development of cave diving procedures that allowed biologists access to this habitat. Early surveys of anchialine caves were conducted in Bermuda (Sket and Iliffe, 1980; Iliffe et al., 1983), the Canary Islands (Iliffe et al., 1984b) and the Bahamas (Warner and Moore, 1984). Recent reviews of stygofauna, including anchialine taxa, have been prepared by Botosaneanu (1986), Iliffe (1992) and Juberthie and Decu (1994, 1996).

Bacteria: A variety of aerobic and anaerobic, heterotrophic and autotrophic bacteria are apparently present in anchialine caves. Unfortunately, few studies have attempted to identify or quantify anchialine bacteria.

Protozoa: In Bermuda, at least two species of ciliates appear to be limited to cave environments and are thus considered to be troglobites (Small et al., 1986; Hill et al., 1986).

Porifera: Sponges are common in most strongly tidal anchialine caves having direct communication with the sea. In such caves, all exposed rock surfaces are completely covered with sponges and other encrusting marine growth. Most of these sponges are probably troglophiles that can be found in suitable habitats outside caves. However, several highly unusual sponges from marine caves in the Mediterranean include one that is carnivorous (Vacelet and Boury-Esnault, 1995) and another that lacks skeletal elements (Boury-Esnault et al., 1995). Three new species of sponges apparently limited to caves were described from the anchialine pool in Lighthouse Cave, San Salvador (Van Soest and Sass, 1981).

Hydrozoa: Like sponges, troglophilic hydroids are common inhabitants of strongly tidal caves (Calder, 1988).

Oligochaeta: The tubificid oligochaete *Phallodrilus macmasterae* was described from an anchialine cave in Bermuda (Erséus, 1986). It is most closely related to a congener inhabiting waters of the continental slope off Surinam.

Polychaeta: One species of anchialine archiannelid was described from Bermuda caves (Sterrer and Iliffe, 1982), while several undescribed, but probably troglobitic, species have been found in the Bahamas and Yucatan Peninsula (Mexico). Two species of polynoid polychaetes inhabit caves in the Canary Islands (Hartmann-Schröder, 1974) and the Caicos Islands (Pettibone, 1985).

Remipedia: Six genera of remipedes are known from anchialine caves in the Bahamas, Caicos Islands, Canary Islands, Yucatan and, most recently, Cuba and Western Australia (Yager, 1994a,b; Yager and Humphreys, 1996). Remipedes are the most primitive of living crustaceans.

Ostracoda: The ostracod orders Halocyprida, Myodocopida, Platycopida and Podocopida include anchialine species. Of these, the halocyprids contain the most cave-adapted species from three primarily or exclusively anchialine genera: *Danielopolina, Deeveya* and *Spelaeoecia* (Kornicker and Iliffe, 1985, 1989a–c, 1992, 1995, 1998; Angel and Iliffe, 1987; Kornicker and Palmer, 1987; Kornicker et al., 1990). Only one species of *Danielopolina, D. carolynae,* has been found outside anchialine caves, and that was collected from a water depth of 3459 m in the South Atlantic (Kornicker and Sohn, 1976). Anchialine myodocopids, podocopids and platycopids come primarily from genera including open-water species.

Copepoda: Copepods from the orders Calanoida, Cyclopoida, Harpacticoida, Misophrioida, and Platycopioida inhabit anchialine caves. The calanoid family

[2] Inland blue holes are typically circular, water-filled shafts formed by solution or collapse. Some inland blue holes contain extensive development of solutional cave passages with abundant, subaerially-formed speleothems.

Epacticeriscidae is almost exclusively anchialine. The epacteriscid genus *Erebonectes*, inhabiting caves in Bermuda and the Caicos Islands, is one of the most primitive of known calanoids (Fosshagen and Iliffe, 1985, 1994). The cyclopoid family Speleoithonidae consists of only three species from anchialine caves in the Bahamas and Bermuda (da Rocha and Iliffe, 1991, 1993). Four genera of cyclopoids occur in anchialine habitats on western Mediterranean and eastern Atlantic islands (Jaume and Boxshall, 1996, 1997). Three harpacticoid families are limited to anchialine caves: Rotundiclipeidae with a single species from an anchialine pool in the Canary Islands (Huys, 1988a); Superornatiremidae with three genera and nine species from anchialine caves in Bermuda, the Canary Islands and the Balearic Islands (Huys, 1996; Jaume, 1997); and Novocriniidae, with one species from anchialine caves in Belize (Huys and Iliffe, 1998). Seven genera of misophrioids are found exclusively in anchialine habitats (Boxshall and Iliffe, 1986, 1987, 1990; Huys, 1988b; Jaume et al., 1998). Two anchialine genera of platycopids, *Antrisocopia* and *Nanocopia*, are known only from a single Bermuda cave (Fosshagen and Iliffe, 1985, 1988).

Mysidacea: Seventeen species of troglobitic mysids from six genera are known to inhabit anchialine caves (Bowman, 1986; Pesce et al., 1994). Most are blind and unpigmented.

Mictacea: The crustacean order Mictacea (Bowman et al., 1985b) includes three species – one from anchialine caves in Bermuda (Bowman and Iliffe, 1985) and two from deep waters of the Equatorial Atlantic (Sanders et al., 1985) and the Indo-Pacific (Just and Poore, 1988). Based on the discovery of a new anchialine species from the Bahamas, Guţu and Iliffe (1998) have proposed erecting a new order, Bochusacea, to include this anchialine species, as well as the two deep sea species previously included within the Mictacea.

Isopoda: The isopod family Cirolanidae includes ten anchialine genera (Botosaneanu et al., 1986; Botosaneanu and Iliffe, 1997; Carpenter, 1994). The anthurid genus *Curassanthura* consists of three anchialine species from Bermuda, the Canary Islands and the Netherlands Antilles (Kensley, 1981; Wägele, 1985; Wägele and Brandt, 1985). The monotypic family Atlantasellidae is known from only a single cave in Bermuda (Sket, 1979).

Amphipoda: The amphipod families Bogidiellidae, Gammaridae, Hadziidae, Hyalidae, Ingolfiellidae, Liljeborgiidae, Lysianassidae, Pardaliscidae, Phoxocephalidae and Sebidae include anchialine species (Holsinger, 1994). Of these families, the Hadziidae + Melitidae complex contains the largest number of anchialine taxa.

Phyllocarida: A blind, unpigmented leptostracan, *Speonebalia cannoni,* occurs in anchialine caves in the Caicos Islands (Bowman et al., 1985a).

Thermosbaenacea: Four genera of anchialine thermosbaenaceans include *Monodella* and *Tethysbaena* from the Mediterranean region and Caribbean, *Halosbaena* from the Canary Islands, Caribbean and Western Australia, and *Tulumella* from the Yucatan Peninsula and the Bahamas (Wagner, 1994; Boutin, 1996).

Decapoda: Anchialine shrimps include at least 18 genera from six families: Agostocarididae, Alpheidea, Atyidae, Hippolytidae, Palaemonidae, and Procarididae (Holthuis, 1986; Hobbs, 1996). The galatheid crab *Munidopsis polymorpha* inhabits an anchialine lava tube in the Canary Islands (Wilkens and Parzefall, 1974). Two species of brachyuran crabs occur in anchialine habitats in the Galapagos Islands (Guinot and Iliffe, 1991; Garth and Iliffe, 1992), while a third brachyuran inhabits anchialine limestone caves in the Solomon Islands (Ng et al., 1994).

Chaetognatha: *Paraspadella anops* is a blind, unpigmented chaetognath inhabiting caves in the Bahamas (Bowman and Bieri, 1989).

Pisces: Anchialine fish from the order Ophidiiformes, family Bythitidae, include two species of *Lucifuga* from Cuba and the Bahamas and one species of *Ogilbia* from the Galapagos Islands (Thinès and Proudlove, 1986; Weber et al., 1996). *Milyeringa veritas* (Perciformes: Eleotridae) and *Ophisternon candidum* (Synbranchiformes) are both highly troglomorphic and are found in anchialine habitats on the Cape Range Peninsula of Western Australia (Yager and Humphreys, 1996; Humphreys, pers. comm.).

CHARACTERISTICS OF ANCHIALINE TAXA

Troglomorphic adaptations

Most obligate anchialine taxa exhibit typical troglomorphic adaptations to the lightless, food-poor cave environment. These include varying degrees of eye loss and depigmentation (Fig. 3.8), increased tactile and chemical sensitivity, and larger but fewer eggs. Similar adaptations are also found in marine organisms

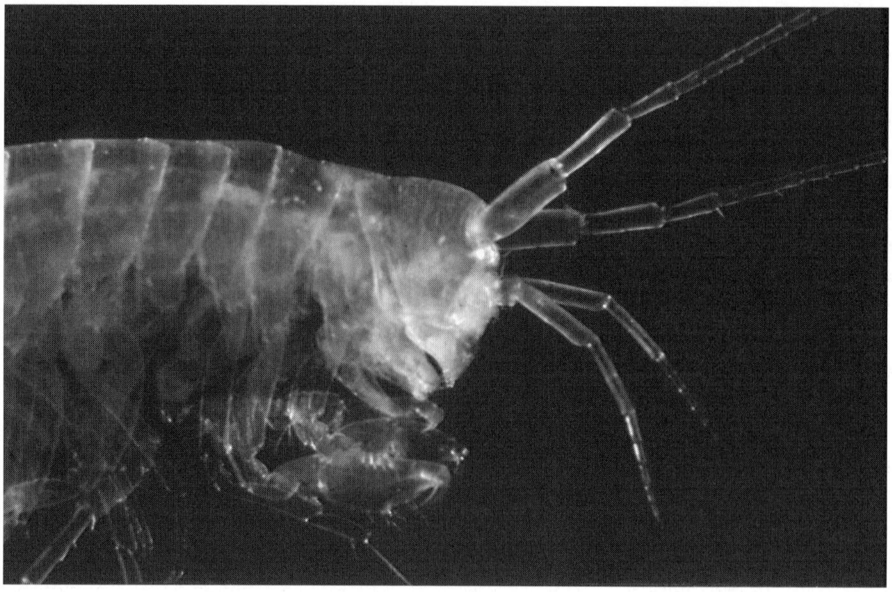

Fig. 3.8. Head region of the troglobitic amphipod *Pseudoniphargus grandimanus* inhabiting anchialine caves and groundwaters in Bermuda. Note the lack of eyes and external body pigmentation.

inhabiting other lightless environments such as the deep sea and the submarine interstitial (Danielopol et al., 1996). Total depigmentation and loss of eyes are found in anchialine polynoid polychaetes, remipedes, halocyprid ostracods, epacteriscid and misophrioid copepods, cirolanid isopods, leptostracans, thermosbaenaceans, chaeotognaths, fishes, some shrimps, and numerous amphipods. Such pronounced troglomorphic adaptations usually occur in more primitive taxa which apparently have been restricted to caves for long time periods.

As with deep-sea shrimps, many anchialine shrimps are scarlet to blood-red in color. Ten species of insular, Indo-Pacific anchialine shrimps, representing nine genera in five families, all possess red integumentary chromatophores (Maciolek, 1983). However, color display is variable between individuals within a population. In contrast, continental cave shrimps of the Indo-Pacific are colorless with the exception of *Macrobrachium cavernicola*, a red palaemonid from inland caves in India (Holthuis, 1956). The hippolytid shrimp *Barbouria cubensis* displays an intense red coloration when found in anchialine pools open to the surface, but is colorless in totally dark caves (Viña and Dávilla, 1980). If these shrimps are moved from darkness to light, a color transformation occurs within a matter of minutes.

Eyes are also present in many anchialine shrimps. Of nine species within the atyid genus *Typhlatya*, pigmented eyes are found in four (mostly brackish or marine) species, while the remaining five (mostly freshwater) species lack eye pigment. Reduced but pigmented eyes are present even in some of the most primitive anchialine shrimps such as *Procaris hawaiana* (irregular pigment spot; no distinct corneal elements), *Procaris chacei* (eye unfaceted but containing black pigment), *Agostocaris williamsi* (eyes weakly pigmented), *Yagerocaris cozumel* (small weakly pigmented area), and *Ventrocaris chaceorum* (faint scattered pigment) (Holthuis, 1973; Hart and Manning, 1986; Kensley and Williams, 1986; Kensley, 1988). The eyes of *Agostocaris bozanici* lack pigment and are not differentiated into cornea and stalk (Kensley, 1988).

Primitive characteristics

A number of anchialine crustaceans possess extremely primitive characteristics. Remipedes, along with the copepods *Antrisocopia* and *Erebonectes*, are regarded as "living fossils". Anchialine caves also contain many of the most primitive known copepods in the orders Calanoida, Misophrioida and Platycopioida (Huys and Boxshall, 1991). The platycopioid *Antrisocopia prehesilis* agrees in many ways with the theoretical ancestral copepod (Fosshagen and Iliffe, 1985). *Erebonectes nesioticus* is considered one of the most primitive calanoids because of its first antennae with 27 segments (the highest number ever recorded), essentially unmodified mouthparts, and primitive legs (Fosshagen and Iliffe, 1985). Misophrioid copepods are of considerable phylogenetic interest because they represent the first branch to diverge from the main podoplean lineage within the copepoda (Boxshall and Iliffe, 1986). Before their discovery in anchialine caves, misophrioids were considered to be an essentially deep-sea group.

Anomalous biogeographical distribution

Despite being limited to caves, many anchialine taxa have a broad and highly disjunct distribution, thus raising questions as to their origins. A variety of genera contain species inhabiting caves on opposite sides of the Atlantic Ocean. These include the remipede *Speleonectes*, the anthurid isopod *Curassanthura*, the mysid *Heteromysoides*, the ostracod *Danielopolina*, the thermosbaenacean *Halosbaena*, the polychaete *Namanereis*, the shrimp *Typhlatya*, and the amphipods *Gevegiella, Pseudoniphargus* and *Spelaeonicippe*. Such an amphiatlantic distribution suggests a Tethyan origin, with dispersal by plate-tectonic rafting (Iliffe et al., 1984b; Wilkens et al., 1986).

Other anchialine genera, previously known from caves in the Caribbean and Atlantic, are found in the Galapagos Islands, located 1000 km off the South American coast in the eastern equatorial Pacific. Included in this group are the shrimp *Typhlatya*, the ostracod *Danielopolina*, and the copepods *Enantiosis* and *Expansophria*. The Caribbean-derived anchialine fauna of the Galapagos must have entered the Pacific prior to the closure of the Panama land bridge about 3 to 5 million years ago (Iliffe, 1991).

Even more complex zoogeographic relationships exist for other anchialine taxa. The primitive shrimp *Procaris* is known only from caves and anchialine pools on Hawaii in the mid-Pacific, Ascension Island in the South Atlantic, Bermuda in the North Atlantic and Cozumel in the Caribbean (Hart and Manning, 1986). Equally perplexing is the case of the misophrioid copepod genus *Expansophria*. It is represented by cave species from Palau and the Galapagos Islands on opposite sides of the Pacific and from the Canary Islands in the Atlantic (Boxshall and Iliffe, 1987, 1990; Boxshall, 1989).

Anchialine taxa, originally thought to be limited to or known primarily from the Atlantic and Caribbean, have recently been discovered from caves in the Cape Range of Western Australia (Humphreys and Adams, 1991; Humphreys, 1993; Knott, 1994; Yager and Humphreys, 1996). These include the remipede *Lasionectes*, the isopod *Haptolana*, the thermosbaenacean *Halosbaena*, the ostracod *Danielopolina* and the amphipod *Liagoceradocus* (Humphreys, Chapter 30, this volume). The wide range and overlapping distributions of these five taxa, and many other anchialine taxa in general, suggest considerable age, coupled with similar dispersal mechanisms and simultaneous colonization of the present habitat.

Deep-sea affinities

A considerable number of anchialine species have close relatives inhabiting the deep sea. Anchialine genera with affinities to present-day bathyal taxa include the galatheid crab *Munidopsis*, the polynoid polychaetes *Gesiella* and *Pelagomacellicephala*, the mictacean *Mictocaris*, the ostracod *Danielopolina*, the amphipod *Spelaeonicippe*, and the misophrioid copepods *Boxshallia, Dimisophria, Expansophria, Misophria, Palpophria* and *Speleophria* (Hart et al., 1985; Wilkens et al., 1986). Close ties between anchialine and deep-sea taxa suggest a possible deep-sea origin for cave species.

ORIGINS OF ANCHIALINE FAUNA

Several theories have been proposed to explain the origins and anomalous distribution of anchialine taxa.

The **Vicariance Model** suggests plate tectonics as a mechanism for dispersal of interstitial and anchialine fauna (Sterrer, 1973; Rosen, 1976, 1985; Pregill, 1981; Briggs, 1987). Since most islands of truly oceanic origins have never been part of or nearer to a continental coast than they are now, a direct role of plate tectonics in the origin of such oceanic cave faunas can be ruled out. However, indirect effects (i.e., distances between these islands and land masses on other tectonic plates changing over the course of geologic history) could have played an important role. For example, the presence of Old World taxa in Bermuda could be accounted for by the island's proximity to Africa and Europe very early in its geologic history (Stock et al., 1986b; Stock, 1990). Also, the existence of an anchialine cave fauna with primarily Caribbean and West Atlantic affinities in the

Fig. 3.9. Underwater photograph in Church Cave, Bermuda, showing a diver swimming between two large columns. These columns were formed during Pleistocene periods of glaciation when sea level was 100 m or more below present levels and the cave was air-filled. Photo by Cristian Lascu.

Galapagos Islands indicates a colonization before the closure of the Panama land bridge (Iliffe, 1991).

The **Regression Model** (Stock, 1977, 1980, 1990; Boutin and Coineau, 1990; Notenboom, 1991) is based on the observation that the distribution patterns of numerous marine-derived aquatic cave organisms correlate well with the shorelines of Late Mesozoic or Tertiary seas. To explain this distribution, the Regression Model suggests that marine littoral species first colonized crevicular or interstitial habitats. During sea-level regressions, resulting from tectonic uplift or eustatic glacial lowering of sea levels, these species became "stranded" and adapted to mixohaline or limnic groundwater conditions. The Regression Model also implies very limited mobility of anchialine and interstitial species, allowing them to be stranded by the slow rate of sea-level retreat. In the case of oceanic islands such as Bermuda and the Bahamas, a rich and diverse anchialine fauna inhabits caves that were completely dry and air-filled (as evidenced by their now submarine speleothems; Fig. 3.9) less than 10 000 years ago. Since many of these are phylogenetically old species that have apparently existed in caves for long periods of time, they must have found suitable refuge in habitats beneath the low point of Pleistocene glacial sea level. As the post-Glacial sea level rose, these species followed it up to colonize the caves they presently occupy. If anchialine species can be carried up by rising sea levels, it is not clear why declining sea levels should pose a significant barrier.

A **Deep-Sea Origin** for some anchialine species has been proposed (Fuchs, 1894; Hart et al., 1985; Manning et al., 1986; Iliffe, 1990). Both caves and the deep sea are old, climatically stable and non-rigorous environments (Poulson and White, 1969; Poulson, 1971). The recent discoveries of numerous marine troglobites with close taxonomic affinities to bathyal forms (i.e., from the same genus or from a higher taxon known primarily from bathyal depths) add considerable support to this hypothesis. Along with caves, marine subterranean habitats may also consist of crevices or fissures occurring at any depth within the ocean. A continuum of such "crevicular" habitats could provide a link between shallow-water caves and the deep sea (Hart et al., 1985; Iliffe, 1990). Cladistic analysis indicates that misophrioid copepods inhabiting anchialine caves on oceanic islands in both the Atlantic and Pacific are descended from deep-sea ancestors (Boxshall, 1989). However, at least three factors have been cited by other authors as making a deep-sea origin of cave faunas unlikely (Newman, 1985; Stock, 1986; Danielopol, 1990; Danielopol et al., 1996). These include the questionable ability of deep-sea species to cross the oceanic thermocline, the relatively recent nature of deep-sea species (due to lack of oxygen in Atlantic bathyal waters during the late Oligocene) and morphological evidence.

An **Active Migration Model**, independent of climatological and geological variations, has also been suggested (Danielopol, 1980; Rouch and Danielopol, 1987). In this model, marine species of an expansionistic nature actively colonize empty niches within the range of the species. Rouch and Danielopol (1987) considered that most subterranean species are not relicts, caves do not function as biological refugia and glacial–interglacial influences are not essential to the isolation and evolution of subterranean fauna. Botosaneanu and Holsinger (1991) have countered these conclusions and cited a number of examples of passive versus active colonization.

ACKNOWLEDGEMENTS

Special appreciation is extended to Drs. M. Bacescu, T. Bowman, G. Boxshall, A. Fosshagen, J. Holsinger, L. Kornicker, R. Maddocks, J. Parzefall, C. Rocha, B. Sket, W. Sterrer, J. Stock, and H. Wilkens, among others, who have contributed to my faunal surveys of anchialine caves. Cave divers B. Dodson, S. Exley, S. Gerrard, P. Hobbs, M. Madden, J. Pohlman, R. Power and others assisted with this work. This research has been funded by grants from the National Science Foundation (870219, BSR-8700079, BSR-8417494, BSR-8215672), National Geographic Society (3412-86, 4725-92), Smithsonian Institution Caribbean Coral Reef Ecosystem Program, NOAA National Undersea Research Program, The Explorers Club and Texas A&M University at Galveston. All diving done during these investigations met cave diving standards set by the Cave Diving Section of the U.S. National Speleological Society.

REFERENCES

Angel, M.V. and Iliffe, T.M., 1987. *Spelaeocia bermudensis* new genus, new species, a halocyprid ostracod from marine caves in Bermuda. *J. Crustacean Biol.*, 7: 541–553.

Back, W., Henshaw, B.B., Herman, J.S. and Van Driel, J.N., 1986. Differential dissolution of a Pleistocene reef in the ground-water mixing zone of coastal Yucatan, Mexico. *Geology*, 14: 137–140.

Barnes, J.A. and Bodungen, B.V., 1978. *The Bermuda Marine Environment*, Vol. 2. Spec. Pub. No. 17, Bermuda Biological Station, Ferry Reach, Bermuda, 190 pp.

Bögli, A., 1964. Mischungskorrosion – ein Beitrag zur Verkarstungsproblematik. *Erdkunde*, 18: 83–92.

Botosaneanu, L., 1986. *Stygofauna Mundi. A Faunistic, Distributional, and Ecological Synthesis of the World Fauna Inhabiting Subterranean Waters (Including the Marine Interstitial)*. E.J. Brill, Leiden, 740 pp.

Botosaneanu, L. and Holsinger, J.R., 1991. Some aspects concerning colonization of the subterranean realm – especially subterranean waters: a response to Rouch & Danielopol, 1987. *Stygologia*, 6: 11–39.

Botosaneanu, L. and Iliffe, T.M., 1997. Four new stygobitic cirolanids (Crustacea: Isopoda) from the Caribbean – with remarks on intergeneric limits in some cirolanids. *Bull. Inst. R. Sci. Nat. Belg. Biol.*, 67: 77–94.

Botosaneanu, L., Bruce, N. and Notenboom, J., 1986. Isopoda: Cirolanidae. In: L. Botosaneanu (Editor), *Stygofauna Mundi*. E.J. Brill, Leiden, pp. 412–422.

Boury-Esnault, N., Muricy, G., Gallissian, M.F. and Vacelet, J., 1995. Sponges without skeleton – a new Mediterranean genus of Homoscleromorpha (Porifera, Demospongiae). *Ophelia*, 43(1): 25–43.

Boutin, C., 1996. Thermosbaenacea. In: C. Juberthie and V. Decu (Editors), *Encyclopedia Biospeologica*, Vol. II. Société de Biospéologie, Moulis (France) and Bucharest, pp. 877–888.

Boutin, C. and Coineau, N., 1990. "Regression model", "Modèle biphase" d'évolution et origine des micro-organismes stygobies interstitiels continentaux. *Rev. Micropaléontol.*, 33: 303–322.

Bowman, T.E., 1986. Mysidacea. In: L. Botosaneanu (Editor), *Stygofauna Mundi*. E.J. Brill, Leiden, pp. 405–409.

Bowman, T.E. and Bieri, R., 1989. *Paraspadella anops*, new species from Sagittarius Cave, Grand Bahama Island, the second troglobitic chaetognath. *Proc. Biol. Soc. Washington*, 102: 586–589.

Bowman, T.E. and Iliffe, T.M., 1985. *Mictocaris halope*, a new unusual peracaridan crustacean from marine caves on Bermuda. *J. Crustacean Biol.*, 5: 58–73.

Bowman, T.E., Yager, J. and Iliffe, T.M., 1985a. *Speonebalia cannoni*, n. gen., n. sp., from the Caicos Islands, the first hypogean leptostracan (Nebaliacea, Nebaliidae). *Proc. Biol. Soc. Washington*, 98: 439–446.

Bowman, T.E., Garner, S.P., Hessler, R.R., Iliffe, T.M. and Sanders, H.L., 1985b. Mictacea, a new order of Crustacea Peracarida. *J. Crustacean Biol.*, 5: 74–78.

Boxshall, G.A., 1989. Colonization of inland marine caves by misophrioid copepods. *J. Zool.*, 219: 521–526.

Boxshall, G.A. and Iliffe, T.M., 1986. New cave-dwelling misophrioids (Crustacea: Copepoda) from Bermuda. *Sarsia*, 71: 54–64.

Boxshall, G.A. and Iliffe, T.M., 1987. Three new genera of misophrioid copepods (Crustacea) from anchialine caves on Indo-West Pacific and North Atlantic Islands. *Zool. J. Linn. Soc.*, 91: 223–252.

Boxshall, G.A. and Iliffe, T.M., 1990. Three new species of misophrioid copepods from oceanic islands. *J. Nat. Hist.*, 24: 595–613.

Briggs, J.C., 1987. Biogeography and plate tectonics. *Dev. Palaeontol. Stratigr.*, 10: 1–204.

Calder, D.R., 1988. *Shallow-water hydroids of Bermuda: the Athecatae*, Life Sciences Contributions 148, Royal Ontario Museum, Toronto, 107 pp.

Carpenter, J.H., 1994. *Dodecalana yagerae,* new genus, new species, a troglobitic marine cirolanid isopod from Grand Bahama Island, Bahamas. *J. Crustacean Biol.,* 14(1): 168–176.

Courbon, P., Chabert, C., Bosted, P. and Lindsey, K., 1989. *Atlas of the Great Caves of the World.* Cave Books, St. Louis, MO, 369 p.

da Rocha, C.E.F. and Iliffe, T.M., 1991. Speleoithonidae, a new family of Copepoda Cyclopoida, from anchialine caves on the Bahama Islands. *Sarsia,* 76: 167–176.

da Rocha, C.E.F. and Iliffe, T.M., 1993. New cyclopoids (Copepoda) from anchialine caves in Bermuda. *Sarsia,* 78(1): 43–56.

Danielopol, D.L., 1980. An essay to assess the age of the freshwater interstitial ostracods of Europe. *Bijdr. Dierkd.,* 50: 243–291.

Danielopol, D.L., 1990. The origin of anchialine cave fauna – the "deep sea" versus the "shallow water" hypothesis tested against empirical evidence of the Thaumatocyrididae (Ostracoda). *Bijdr. Dierkd.,* 60: 137–143.

Danielopol, D.L., Baltanas, A. and Bonaduce, G., 1996. The darkness syndrome in subsurface-shallow and deep-sea dwelling Ostracoda (Crustacea). In: F. Uiblein, J. Ott and M. Stachowitsch (Editors), *Biosystematics and Ecology Series,* Vol. 11. Austrian Academy of Sciences, Vienna, pp. 78–99.

Dill, R.F., 1977. The Blue Holes – geologically significant submerged sink holes and caves off British Honduras and Andros, Bahama Islands. In: D.L. Taylor (Editor), *Proceedings of the Third International Coral Reef Symposium.* University of Miami, FL, pp. 237–242.

Erséus, C., 1986. A new species of *Phallodrilus* (Oligochaeta, Tubificidae) from a limestone cave on Bermuda. *Sarsia,* 71: 7–9.

Farr, M. and Palmer, R., 1984. The Blue Holes: description and structure. *Cave Sci.,* 11: 9–22.

Fosshagen, A. and Iliffe, T.M., 1985. Two new genera of Calanoida and a new order of Copepoda, Platycopioida, from marine caves on Bermuda. *Sarsia,* 70: 345–348.

Fosshagen, A. and Iliffe, T.M., 1988. A new genus of Platycopioida (Copepoda) from a marine cave on Bermuda. *Hydrobiologia,* 167/168: 357–361.

Fosshagen, A. and Iliffe, T.M., 1994. A new species of *Erebonectes* (Copepoda, Calanoida) from marine caves on Caicos Islands, West Indies. *Hydrobiologia,* 292/293: 17–22.

Fuchs, T., 1894. Über Tiefseethiere in Höhlen. *Ann. K.K. naturhist. Hofmus.,* 9: 54–55.

Garth, J.S. and Iliffe, T.M., 1992. *Guitonia troglophila,* a new genus and species of anchialine crab from the Galápagos, Isla Santa Cruz, Grieta de Caleta la Torta (Crustacea, Decapoda, Brachyura). *Proc. Biol. Soc. Washington,* 105: 310–316.

Guinot, D. and Iliffe, T.M., 1991. *Garthiope anchialina* sp. nov., espèce anchialine des Galapagos, île Isabela, Cueva de la Cadena, avec des remarques sur la faune carcinologique anchialine (Crustacea Decapoda Brachyura). *Bull. Mus. Natl. Hist. Nat., Paris,* 4e sér., 12, sect. A, (3–4), 1990(1991): 607–621.

Guţu, M. and Iliffe, T.M., 1998. Description of a new hirsutiid (n.g., n.sp.) and reassignment of this family from Order Mictacea to the new Order, Bochusacea (Crustacea, Peracarida). *Trav. Mus. Hist. Nat. Grigore Antipa,* 40: 93–120.

Hart Jr., C.W. and Manning, R.B., 1986. Two new shrimps (Procarididae and Agostocarididae, new family) from marine caves of the Western North Atlantic. *J. Crustacean Biol.,* 6: 408–416.

Hart Jr., C.W., Manning, R.B. and Iliffe, T.M., 1985. The fauna of Atlantic marine caves: Evidence of dispersal by sea floor spreading while maintaining ties to deep waters. *Proc. Biol. Soc. Washington,* 98: 288–292.

Hartmann-Schröder, G., 1974. Die Unterfamilie Macellicephalinae Hartmann-Schröder, 1971 (Polynoidae, Polychaeta). Mit Beschreibung einer neuer Art, *Macellicephala jameensis,* n. sp., aus einen Höhlengewässer von Lanzarote (Kanarische Inseln). *Mitt. Hamb. Zool. Mus. Inst.,* 71: 75–85.

Hill, B.F., Small, E.B. and Iliffe, T.M., 1986. *Euplotes iliffei* n. sp.: A new species of *Euplotes* (Ciliophora, Hypotrichida) from the marine caves of Bermuda. *J. Washington Acad. Sci.,* 76: 244–249.

Hobbs III, H.H., 1996. Decapoda. In: C. Juberthie and V. Decu (Editors), *Encyclopedia Biospeologica,* Vol. II. Société de Biospéologie, Moulis (France) and Bucharest, pp. 891–911.

Holsinger, J.R., 1994. Amphipoda. In: C. Juberthie and V. Decu (Editors), *Encyclopaedia Biospeologica,* Vol. I. Société de Biospéologie, Moulis (France) and Bucarest, pp. 147–163.

Holthuis, L.B., 1956. An enumeration of the Crustacea Decapoda Natantia inhabiting subterranean waters, *Vie Milieu,* 7: 43–76.

Holthuis, L.B., 1973. Caridean shrimp found in land locked saltwater pools at four Indo-West Pacific localities (Sinai Peninsula, Funafuti Atoll, Maui and Hawaii Islands), with the description of one new genus and four new species. *Zool. Verh.,* 128: –48.

Holthuis, L.B., 1986. Decapoda. In: L. Botosaneanu (Editor), *Stygofauna Mundi.* E.J. Brill, Leiden, pp. 589–615.

Humphreys, W.F., 1993. Stygofauna in semi-arid tropical Western Australia: A Tethyan connection?, *Mém. Biospéol.,* 20: 111–116.

Humphreys, W.F. and Adams, M., 1991. The subterranean aquatic fauna of the North West Cape peninsula, Western Australia. *Rec. West. Aust. Mus.,* 15: 383–411.

Huys, R., 1988a. Rotundiclipeidae, a new family (Copepoda, Harpacticoida) from an anchialine cave on Tenerife, Canary Islands. *Stygologia,* 4: 42–63.

Huys, R., 1988b. Stygofauna of the Canary Islands: 11. *Boxshallia bulbantennulata,* new genus new species (Copepoda: Misophrioida) from an anchialine lava pool on Lanzarote, Canary Islands. *Stygologia,* 4: 138–154.

Huys, R., 1996. Superornatiremidae fam. nov. (Copepoda: Harpacticoida): an enigmatic family from North Atlantic anchialine caves. *Sci. Mar.,* 60(4): 497–542.

Huys, R. and Boxshall, G.A., 1991. *Copepod Evolution.* Ray Society, London, 468 pp.

Huys, R. and Iliffe, T.M., 1998. Novocriniidae, a new familiy of harpacticoid copepods from anchialine caves in Belize. *Zool. Scr.,* 27(1): 1–15.

Iliffe, T.M., 1986. The zonation model for the evolution of aquatic faunas in anchialine waters. *Stygologia,* 2: 2–9.

Iliffe, T.M., 1987. Observations on the biology and geology of anchialine caves. In: H.A. Curran (Editor), *Proceedings of the Third Symposium on the Geology of the Bahamas.* CCFL Bahamian Field Station, San Salvador, Bahamas, pp. 73–80.

Iliffe, T.M., 1990. Crevicular dispersal of marine cave faunas. *Mém. Biospéol.,* 17: 93–96.

Iliffe, T.M., 1991. Anchialine cave fauna of the Galápagos Islands. In: M.J. James (Editor), *Galápagos Marine Invertebrates.* Plenum Press, New York, pp. 209–231.

Iliffe, T.M., 1992. Anchialine Cave Biology. In: A.I. Camacho (Editor), *The Natural History of Biospeleology.* National Museum of Natural Sciences, Madrid, pp. 613–636.

Iliffe, T.M. and Sarbu, S.M., 1990. Anchialine caves and cave fauna of the South Pacific. *Natl. Speleol. Soc. News,* 48: 88–96.

Iliffe, T.M., Hart Jr., C.W. and Manning, R.B., 1983. Biogeography and the caves of Bermuda. *Nature,* 302: 141–142.

Iliffe, T.M., Jickells, T.D. and Brewer, M.S., 1984a. Organic pollution of an inland marine cave from Bermuda. *Mar. Environ. Res.,* 12: 173–189.

Iliffe, T.M., Wilkens, H., Parzefall, J. and Williams, D., 1984b. Marine lava cave fauna: composition, biogeography, and origins. *Science,* 225: 309–311.

Jaume, D., 1997. First record of Superornatiremidae (Copepoda: Harpacticoida) from Mediterranean waters, with description of three new species from Balearic anchialine caves. *Sci. Mar.,* 61(2): 131–152.

Jaume, D. and Boxshall, G.A., 1996. Two new genera of cyclopinid copepods (Crustacea) from anchialine caves on western Mediterranean and eastern Atlantic islands. *Zool. J. Linn. Soc.,* 117(3): 283–304.

Jaume, D. and Boxshall, G.A., 1997. Two new genera of cyclopinid copepods (Cyclopoida: Cyclopinidae) from anchialine caves of the Canary and Balearic Islands, with a key to genera of the family. *Zool. J. Linn. Soc.,* 120(1): 79–101.

Jaume, D., Boxshall, G.A. and Iliffe, T.M., 1998. Two new genera of misophrioid copepods (Crustacea) from an anchialine cave in the Bahamas. *J. Nat. Hist.,* 32(5): 661–681.

Juberthie, C. and Decu, V., 1994. *Encyclopaedia Biospeologica,* Vol. I. Société de Biospéologie, Moulis (France) and Bucarest, 834 pp.

Juberthie, C. and Decu, V., 1996. *Encyclopaedia Biospeologica,* Vol. II. Société de Biospéologie, Moulis (France) and Bucharest, pp. 835–1373.

Just, J. and Poore, G.C.B., 1988. Second record of Hirsutiidae (Peracarida: Mictacea): *Hirsutia sandersetalia,* new species, from southeastern Australia. *J. Crustacean Biol.,* 8: 483–488.

Kensley, B., 1981. Amsterdam expeditions to the West Indian Islands, Report 10. *Curassanthura halma,* a new genus and species of interstitial isopod from Curaçao, West Indies (Crustacea: Isopoda: Paranthuridae). *Bijdr. Dierkd.,* 51: 131–134.

Kensley, B., 1988. New species and records of cave shrimps from the Yucatan Peninsula (Decapoda: Agostocarididae and Hippolytidae). *J. Crustacean Biol.,* 8: 688–699.

Kensley, B. and Williams, D., 1986. New shrimps (families Procarididae and Atyidae) from a submerged lava tube on Hawaii. *J. Crustacean Biol.,* 6: 417–437.

Knott, B., 1994. Stygofauna from Cape Range peninsula, Western Australia: tethyan relicts. In: W.F. Humphreys (Editor), *Biogeography of Cape Range, Western Australia.* Western Australia Museum, Perth, pp. 109–128.

Kornicker, L.S. and Iliffe, T.M., 1985. Deeveyinae, a new subfamily of Ostracoda (Halocyprididae) from a marine cave on the Turks and Caicos Islands. *Proc. Biol. Soc. Washington,* 98: 476–493.

Kornicker, L.S. and Iliffe, T.M., 1989a. New Ostracoda (Halocyprida: Thaumatocyprididae and Halocyprididae) from anchialine caves in the Bahamas, Palau and Mexico. *Smithson. Contrib. Zool.,* 470: 1–47.

Kornicker, L.S. and Iliffe, T.M., 1989b. Troglobitic Ostracoda (Myodocopa: Thaumatocyprididae) from anchialine pools on Santa Cruz Island, Galapagos Islands. *Smithson. Contrib. Zool.,* 483: 1–38.

Kornicker, L.S. and Iliffe, T.M., 1989c. Ostracoda (Myodocopina, Cladocopina, Halocypridina) from anchialine caves in Bermuda. *Smithson. Contrib. Zool.,* 475: 1–88.

Kornicker, L.S. and Iliffe, T.M., 1992. Ostracoda (Halocypridina, Cladocopina) from anchialine caves in Jamaica, West Indies. *Smithson. Contrib. Zool.,* 530: 1–22.

Kornicker, L.S. and Iliffe, T.M., 1995. Ostracoda (Halocypridina, Cladocopina) from anchialine lava tube in Lanzarote, Canary Islands. *Smithson. Contrib. Zool.,* 568: 1–32.

Kornicker, L.S. and Iliffe, T.M., 1998. Myodocopid Ostracoda (Halocypridina, Cladocopina) from anchialine caves in the Bahamas, Canary Islands, and Mexico. *Smithson. Contrib. Zool.,* 599: 1–93.

Kornicker, L.S. and Palmer, R.J., 1987. *Deeveya bransoni,* a new species of troglobitic halocyprid ostracode from anchialine caves on South Andros Island, Bahamas (Crustacea, Ostracoda). *Proc. Biol. Soc. Washington,* 100: 610–623.

Kornicker, L.S. and Sohn, I.G., 1976. Phylogeny, ontogeny and morphology of living and fossil Thaumatocypridacea (Myodocopa: Ostracoda). *Smithson. Contrib. Zool.,* 219: 1–124.

Kornicker, L.S., Yager, J. and Williams, D., 1990. Ostracoda (Halocyprididae) from anchialine caves in the Bahamas. *Smithson. Contrib. Zool.,* 495: 1–51.

Maciolek, J.A., 1983. Distribution and biology of Indo-Pacific insular hypogeal shrimps. *Bull. Mar. Sci.,* 33(3): 606–618.

Maciolek, J.A., 1986. Environmental features and biota of anchialine pools on Cape Kinau, Maui, Hawaii. *Stygologia,* 2: 119–129.

Manning, R.B., Hart Jr., C.W. and Iliffe, T.M., 1986. Mesozoic relics in marine caves of Bermuda. *Stygologia,* 2: 156–166.

Mylroie, J.E., Carew, J.L. and Moore, A.I., 1995. Blue Holes – definition and genesis. *Carbonates & Evaporites,* 10(2): 225–233.

Newman, W.A., 1985. The abyssal hydrothermal vent invertebrate fauna: A glimpse of antiquity? *Biol. Soc. Washington Bull.,* 6: 231–242.

Ng, P.K.L., Guinot, D. and Iliffe, T.M., 1994. *Sesarmoides ultrapes* new species, a remarkable sesarmine crab from caves in the Solomon Islands (Decapoda: Brachyura: Grapsidae). *Crustacean Res.,* 23: 12–22.

Notenboom, J., 1991. Marine regressions and the evolution of groundwater dwelling amphipods (Crustacea). *J. Biogeogr.,* 18: 437–454.

Palmer, R.J., 1985. The Blue Holes of Eastern Grand Bahama. *Cave Sci.,* 12: 85–92.

Palmer, R.J., 1986. The Blue Holes of South Andros. *Cave Sci.,* 13: 3–6.

Pesce, G.L., Juberthie-Jupeau, L. and Passelaigue, F., 1994. Mysidacea. In: C. Juberthie and V. Decu (Editors), *Encyclopaedia Biospeologica,* Vol. I. Société de Biospéologie, Moulis (France) and Bucarest, pp. 113–119.

Pettibone, M.H., 1985. Polychaete worms from a cave in the Bahamas and from experimental wood panels in deep water of the North Atlantic (Polynoidae: Macellicephalinae, Harmothoinae). *Proc. Biol. Soc. Washington,* 98: 127–149.

Poulson, T.L., 1971. Biology of cave and deep sea organisms: A comparison. *Bull. Natl. Speleol. Soc.,* 33: 51–61.

Poulson, T.L. and White, W.B., 1969. The cave environment. *Science,* 165: 971–981.

Pregill, G.K., 1981. An appraisal of the vicariance hypothesis of Caribbean biogeography. *Syst. Zool.,* 30(2): 147–155.

Reddell, J.R., 1981. *A review of the Cavernicole Fauna of Mexico, Guatemala, and Belize,* Bulletin 27. Texas Memorial Museum, Austin, TX, 327 pp.

Riedl, R., 1966. *Biologie der Meereshöhlen.* Verlag Paul Parey, Hamburg, 636 pp.

Rosen, D.E., 1976. A vicariance model of Caribbean biogeography. *Syst. Zool.,* 24: 431–464.

Rosen, D.E., 1985. Geological hierarchies and biogeographic congruence in the Caribbean. *Ann. Mo. Bot. Gard.,* 72: 636–659.

Rouch, R. and Danielopol, D.L., 1987. L'origine de la faune aquatique souterraine, entre le paradigme du refuge et le modéle de la colonisation active. *Stygologia,* 3: 345–372.

Sanders, H.L., Hessler, R.R. and Garner, S.P., 1985. *Hirsutia bathyalis,* a new unusual deep-sea benthic peracaridan crustacean from the tropical Atlantic. *J. Crustacean Biol.,* 5: 30–57.

Sket, B., 1979. *Atlantasellus cavernicolus* n. gen., n. sp. (Isopoda Asellota, Atlantasellidae n. fam.) from Bermuda. *Biol. Vestn., Ljubljana,* 27: 175–183.

Sket, B., 1996. The ecology of anchihaline caves. *Trends Ecol. Evol.,* 11(5): 221–225.

Sket, B. and Iliffe, T.M., 1980. Cave fauna of Bermuda. *Int. Rev. Gesamten Hydrobiol.,* 65: 871–882.

Small, E.B., Heisler, J., Sniezak, J. and Iliffe, T.M., 1986. *Glauconema bermudensis,* a troglobitic ciliophoran from Bermudian marine caves. *Stygologia,* 2: 167–179.

Smart, C.C., 1984. The hydrology of the inland blue holes, Andros Island. *Cave Sci.,* 11: 23–29.

Sterrer, W., 1973. Plate tectonics as a mechanism for the dispersal and speciation in interstitial sand fauna. *Neth. J. Sea Res.,* 7: 200–222.

Sterrer, W. and Iliffe, T.M., 1982. *Mesonerilla prospera,* a new archiannelid from marine caves in Bermuda, *Proc. Biol. Soc. Washington,* 95: 509–514.

Stock, J.H., 1977. The taxonomy and zoogeography of the hadziid Amphipoda, with emphasis on the West Indian taxa. *Stud. Fauna Curacao,* 177: 1–130.

Stock, J.H., 1980. Regression model evolution as exemplified by the genus *Pseudoniphargus* (Amphipoda). *Bijdr. Dierkd.,* 50: 105–144.

Stock, J.H., 1986. Deep sea origin of cave faunas, an unlikely supposition. *Stygologia,* 2: 105–111.

Stock, J.H., 1990. Insular groundwater biotas in the (sub)tropical Atlantic: A biogeographic synthesis. *Atti Conv. Lincei (Accad. Naz. Lincei Roma),* 85: 695–713.

Stock, J.H., Iliffe, T.M. and Williams, D., 1986a. The concept "anchialine" reconsidered. *Stygologia,* 2: 90–92.

Stock, J.H., Holsinger, J.R., Sket, B. and Iliffe, T.M., 1986b. Two new species of *Pseudoniphargus* (Amphipoda) in Bermudian groundwaters. *Zool. Scripta,* 15: 237–249.

Thinès, G.L. and Proudlove, G., 1986. Pisces. In: L. Botosaneanu, (Editor), *Stygofauna Mundi.* E.J. Brill, Leiden, pp. 709–733.

Vacelet, J. and Boury-Esnault, N., 1995. Carnivorous sponges. *Nature,* 373(6512): 333–335.

Van Soest, R.W.M. and Sass, D.B., 1981. Marine sponges from an island cave on San Salvador Island, Bahamas. *Bijdr. Dierkd.,* 51: 332–344.

Vandel, A., 1965. *Biospeleology: The Biology of Cavernicolous Animals.* Pergamon Press, Oxford, 524 pp.

Viña, N. and Dávilla, Z., 1980. Coloración de *Barbouria cubensis* (Martens) (Crustacea Decapoda Natantia). In: N. Viña (Editor), *Cuevas de Cuba.* Editorial Oriente, Santiago de Cuba, pp. 45–53.

Wägele, J.W., 1985. On the Tethyan origin of the stygobiont Anthuridea *Curassanthura* and *Cyathura* (*Stygocyathura*), with a description of *Curassanthura canariensis* n. sp. from Lanzarote (Crustacea, Isopoda). *Stygologia,* 1: 258–269.

Wägele, J.W. and Brandt, A., 1985. New West Atlantic localities for the stygobiont paranthurid *Curassanthura* (Crustacea, Isopoda, Anthuridea) with descriptions of *C. bermudensis* n. sp. *Bijdr. Dierkd.,* 55: 324–330.

Wagner, H.P., 1994. A monographic review of the Thermosbaenacea (Crustacea: Peracarida). *Zool. Verh.,* 291: 1–338.

Warner, G.F. and Moore, C.A.M., 1984. Ecological studies in marine Blue Holes of Andros Island, Bahamas. *Cave Sci.,* 11: 30–44.

Weber, A., Proudlove, G.S. and Nalbant, T.T., 1996. Morphology, systematic diversity, distribution, and ecology of stygobitic fishes. In: C. Juberthie and V. Decu (Editors), *Encyclopaedia Biospeologica,* Vol. II. Société de Biospéologie, Moulis (France) and Bucharest, pp. 1179–1190.

Wilkens, H. and Parzefall, J., 1974. Die Ökologie der Jameos del Agua (Lanzarote). Zur Entwicklung limnischer Höhlentiere aus marinen Vorfahren. *Ann. Spéléol.,* 29: 419–434.

Wilkens, H., Parzefall, J. and Iliffe, T.M., 1986. Origin and age of the marine stygofauna of Lanzarote, Canary Islands. *Mitt. Hamb. Zool. Mus. Inst.,* 83: 223–230.

Yager, J., 1994a. Remipedia. In: C. Juberthie and V. Decu (Editors), *Encyclopaedia Biospeologica,* Vol. I. Société de Biospéologie, Moulis (France) and Bucarest, pp. 87–90.

Yager, J., 1994b. *Speleonectes gironensis,* new species (Remipedia: Spelonectidae), from anchialine caves in Cuba, with remarks on biogeography and ecology. *J. Crustacean Biol.,* 14(4): 752–762.

Yager, J. and Humphreys, W.F., 1996. *Lasionectes exleyi,* sp. nov., the first remipede crustacean recorded from Australia and the Indian Ocean, with a key to the world species. *Invertebr. Taxon.,* 10: 171–187.

Part III

Subterranean Organisms

Chapter 4

ONYCHOPHORA, ARACHNIDA, MYRIAPODS AND INSECTA

Mark S. HARVEY, William A. SHEAR and Hannelore HOCH

INTRODUCTION

Subterranean ecosystems provide suitable habitats for a number of terrestrial and aquatic arthropod groups, some of which dominate the faunal assemblages in such habitats. Relatively constant environmental parameters in the subterranean environment, including temperature and humidity, provide a somewhat uniform microclimate that has been exploited over and over again, for a number of reasons, by such groups.

In this chapter, we treat four major invertebrate groups: Onychophora, Arachnida, the 'myriapods' and Hexapoda. Although the latter three groups are usually regarded as members of the Phylum Arthropoda (along with Crustacea and some other groups), alternative classifications have been proposed in which the arachnids are placed along with the marine Xiphosura and Pycnogonida in the Phylum Chelicerata, and the myriapods and Insecta are placed in the Phylum Uniramia. Debates over the status of problems such as whether the Arthropoda are monophyletic, the composition and status of the Uniramia, and the position of the Onychophora, abound in the literature, and we choose to leave these major groups unranked at present.

ONYCHOPHORA

All Recent onychophorans, or velvet worms, are included in a single order [and often phylum, e.g., Harvey and Yen (1989); Nielsen et al. (1996)], and are divided into two families, Peripatidae and Peripatopsidae. They are soft-bodied and somewhat worm-shaped, with pseudo-segmented legs and razor-like mouthparts that are used to dismember prey, which is captured with a sticky exudate sprayed from the mouthparts. Onychophorans are found primarily in the southern continents, but some species occur in Asia and central America. Several species have been reported from caves in South Africa, Jamaica and Australia; these species are often extremely pale (Fig. 4.1).

Fig. 4.1. Illustration of a troglobitic onychophoran, *Speleoperipatus spelaeus* (Peripatidae) (redrawn after Peck, 1975).

ARACHNIDA

The class Arachnida is conventionally divided into 11 recent orders, although some authorities recognize differing numbers of mite orders (see p. 84), thus increasing the total. Arachnids comprise one of the dominant groups of living organisms, with some 100 000 species currently recognized and many times that number yet to be described in the scientific literature. They have a fossil record dating back to the Silurian, when the first scorpions appeared. These early forms were probably aquatic, the first terrestrial species emerging during the Devonian when spiders, pseudoscorpions and mites first appeared. Most other extant orders appeared in the fossil record during the Carboniferous, with three extinct orders known from the Silurian to the Permian (Selden, 1993; Dunlop, 1996).

Most arachnids are primarily predators; they utilize a variety of techniques to secure prey, ranging from direct capture with anteriorly situated mouthparts, pedipalps and legs, or with the use of trapping devices such as silk snares. Some mite groups are secondarily detritivorous, parasitic or herbivirous. Arachnids have successfully colonized most terrestrial ecological zones, including aquatic systems where numerous mite groups have invaded freshwater and marine environments, even occurring in benthic sediments in the deepest oceans.

Therefore, it is not surprising that arachnids have been recorded from many subterranean environments, including a wide range of terrestrial cavernous spaces, and from aquatic systems. The following review is designed to provide a brief overview of the major groups of arachnids known to occur in subterranean environments, with an introduction to the primary literature.

Subclass Cteidophora

This subclass comprises the Scorpiones and the Eurypterida. The latter became extinct during the late Permian.

Recent scorpions (order Scorpiones) are currently arranged into nine families (Sissom, 1990), and can be easily distinguished from all other arachnids by their elongate metasoma (tail) which bears a terminal sting, and by the presence of pectines on the ventral surface of the abdomen. Over 1100 species are currently recognized, but troglobitic species have only been reported from Mexico, Ecuador and Sarawak (Lourenço, 1994); thirteen troglobitic species in four families are known. These include a number of chactid genera known primarily from cave ecosystems in Mexico (*Alacran*, *Sotanochactas*, *Troglocormus* and *Typhlochactas*) and Ecuador (*Troglotayosicus*) in which the eyes are completely lacking or where there is only a faint indication of lateral eyes. Most also possess extremely attenuated appendages and pale coloration. *Alacran tartarus* (Fig. 4.2a) from caves in Oaxaca, Mexico, occurs at depths of 720–812 m below ground level (Francke, 1982). From the fact that a blind species of *Typhlochactas* had been found living in litter in Mexico, Mitchell and Peck (1978) speculated that this might have been the original habitat for members of the genus, later species of which have adapted to cave systems.

Subclass Lipoctena

This subclass contains the remainder of the arachnids, including several orders thought to have become extinct by the end of the Carboniferous (Selden, 1993).

Superorder Haplocnemata

Pseudoscorpions (order Pseudoscorpiones) are small arachnids (less than 1 cm) with long, chelate pedipalps. They superficially resemble scorpions but lack the long metasoma (tail) and associated sting found in the latter group. Over 3000 species are currently recognized in 23 families (Harvey, 1992): most occur in epigean habitats such as under tree bark, under stones, in litter and in soil. Numerous representatives occur in caves, and many have significant troglobitic adaptations. Most families possess at least some representatives with troglobitic facies, although notable exceptions include the Tridenchthoniidae, Feaellidae, Olpioidea (Olpiidae and Menthidae), Garypidae and Geogarypidae. Very few genera are fully restricted to cave environments, but these include *Troglochthonius* (Chthoniidae) from the Mediterranean region, most genera of the Central American family Bochicidae, *Hyella* (Hyidae) from Western Australia (Fig. 4.2b), several genera of Neobisiidae from North America and the Mediterranean region (e.g., *Alabamocreagris, Balkanoroncus* and *Insulocreagris*), the southern European genera *Hadoblothrus, Pseudoblothrus* and *Troglobisium* (Syarinidae), and *Troglochernes* (Chernetidae) from southern Australia.

Solifugids or sun-spiders (order Solifugae) are active predators with greatly enlarged chelicerae, which mostly occur in Africa, Asia, southern Europe and the Americas. Over 1000 species have been described in 12 Recent families, but none have been reported from subterranean ecosystems.

Superorder Megoperculata

Spiders (order Araneae) are one of the most successful terrestrial animal groups, and have colonized virtually every available habitat. They differ from all other arachnids by the presence of abdominal spinnerets producing a proteinaceous secretion which hardens after being extruded, and by the presence of venom glands discharging through the fangs.

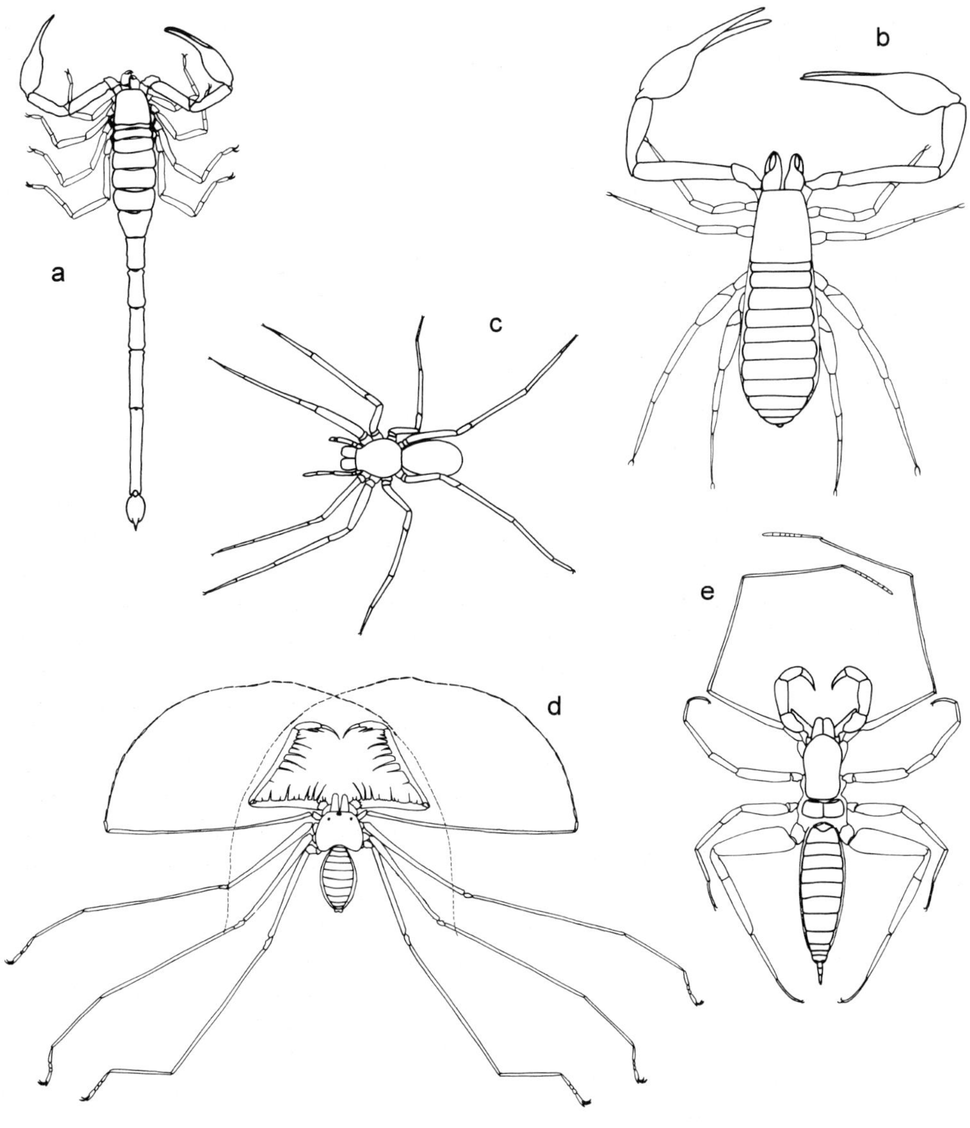

Fig. 4.2. Illustrations of troglobitic arachnids: (a) Scorpiones, *Alacran tartarus* (Chactidae) (redrawn after Francke, 1982); (b) Pseudoscorpiones, *Hyella humphreysi*; (c) Araneae, Ctenidae, new species from Western Australia; (d) Amblypygi, *Paraphrynus emaciatus* (Phrynidae) (redrawn after Mullinex, 1975); (e) Schizomida, *Draculoides vinei* (Hubbardiidae) (redrawn after Harvey, 1988).

Two suborders are recognized. The first, the Mesothelae, are relatively stocky spiders which retain dorsal abdominal tergites, have seven or eight spinnerets, and possess paraxial fangs. The group comprises a single Recent family Liphistiidae with two genera restricted to Southeast Asia. All spend the bulk of their lives in deep, silk-lined burrows, and only a few species have been reported from caves. These include *Liphistius batuensis* from the Batu Caves, Malaysia.

The second suborder, the Opisthothelae, is divided into two infraorders, the Mygalomorphae and the Araneomorphae. The former comprises some 15 families (Raven, 1985), which differ from the araneomorphs by the presence of paraxial fangs and the retention of two pairs of book lungs. Most representatives spend very little of their lives outside of their retreats, which are usually deep burrows in the soil. Very few troglobitic representatives of the Mygalomorphae have been reported, but these include the genera *Troglothele* (Barychelidae) from Cuba, *Troglodiplura* (Dipluridae) from southern Australia, *Speloctenіza* (Nemesiidae) from Ecuador and *Spelopelma* (Theraphosidae) from

Mexico. In representatives of each of these genera, the eye-group is either lost or substantially reduced, but there are very few other significant troglobitic adaptations.

The infraorder Araneomorphae comprises the remainder of the spiders, which are currently divided into some 90 families with over 34 000 described species. They are easily distinguished from the preceding two groups by their diaxial fangs (which pinch together). Those araneomorph spiders with paraxial fangs possess only one pair of book-lungs (as opposed to two pairs in mesotheles and mygalomorphs). Although most species are primarily epigean, large numbers of troglobitic species have been described. Many completely lack eyes, possess elongate appendages and are extremely pale in coloration (Fig. 4.2c). Numerous other species are restricted to cave entrances and similar habitats and, as expected, possess no obvious adaptations to the cave environment or display only partial adaptation. Gertsch (1992) has shown that the North American troglobitic spider fauna is diverse, with a large radiation of species of *Cicurina* in Texas and neighboring regions, with eyes ranging from fully formed to completely absent.

The following three orders are primarily tropical and constitute a natural, monophyletic group based primarily upon their elongate, subsegmented first legs which function and appear like antennae.

Amblypygids or whip spiders (order Amblypygi) are fast-moving arachnids with a flattened appearance and large, raptorial pedipalps. Most species are found under or amongst rocks where they utilize their elongate, antenniform first legs to detect prey. The order has been divided into two suborders, Pulvillata and Apulvillata, based upon the presence or absence, respectively, of tarsal pulvillae (Quintero, 1986), but Weygoldt (1994) preferred not to recognize these categories. The number of families recognized also varies, from three (Weygoldt, 1994) to five (Quintero, 1986). Several species have been reported from caves in Africa, Asia, Australia and the Americas, although most are known from shallow cave systems only and lack modifications associated with cavernicolous life. The most cave-adapted species occur in Central America, where many have reduced eyes and elongated appendages and are pale in colour. Of these species, most belong to the genera *Phrynus* and *Paraphrynus* (Phrynidae) (Fig. 4.2d), while some belong to *Charinus* (Charinidae). Several species of *Phrynichus* (Phrynichidae) have been reported from caves in Southeast Asia.

Schizomids (order Schizomida) are a uniform group of arachnids divided into two Recent families, Protoschizomidae and Hubbardiidae (Reddell and Cokendolpher, 1995). The Protoschizomidae currently includes only two genera, *Agastoschizomus* and *Protoschizomus*, all described species being from Mexico [a juvenile protoschizomid has been taken from a cave in Texas (U.S.A.), but the species is unnamed]. *Agastoschizomus* comprises four named troglobitic species, while *Protoschizomus* is currently known from four troglobitic and three epigean species. The Hubbardiidae (formerly Schizomidae) are much more widespread, and epigean species have been reported from most tropical regions with sufficient habitat, mostly rainforest. Cavernicolous hubbardiids have been reported from various parts of the world, and at present over 40 species have been described. The vast majority have been reported from the intensively explored Central American region, and Mexico and Cuba are rich in troglobitic or troglophilic hubbardiids. Others have been reported from Australia, Burma, China, Congo, India, Japan and Tanzania. Several genera are currently thought to be restricted to cave environments, including *Draculoides* from Western Australia (Fig. 4.2e), *Sotanostenochrus* from Mexico, *Stewartpeckius* from Jamaica and *Tayos* from Ecuador.

Whip scorpions (order Uropygi) are robust, heavily sclerotized arachnids with an elongate tail (flagellum) the base of which is equipped with a gland that delivers an acidic spray, providing an effective defensive mechanism. Only two Recent families are recognized, Thelyphonidae and Hypoctonidae, and approximately 100 species have been described, mostly from tropical and subtropical zones. Very few species have been reported from caves and no species is known to possess any troglomorphic facies. The epigean *Mastigoproctus giganteus* has been recorded from cave entrances in central America (e.g., Reddell and Mitchell, 1971).

Superorder Apulmonata

Palpigrades (order Palpigradi) constitute one of the most obscure arachnid groups, with considerable divergence of opinion as to their relationships. Some authors place them amongst the preceding three orders, while others consider them to be related to the Acari. A conservative approach is adopted here in which they are placed in their own superorder. Two Recent families, Eukoeneniidae and Prokoeneniidae, are recognized (Condé, 1996), and of the 80 or so

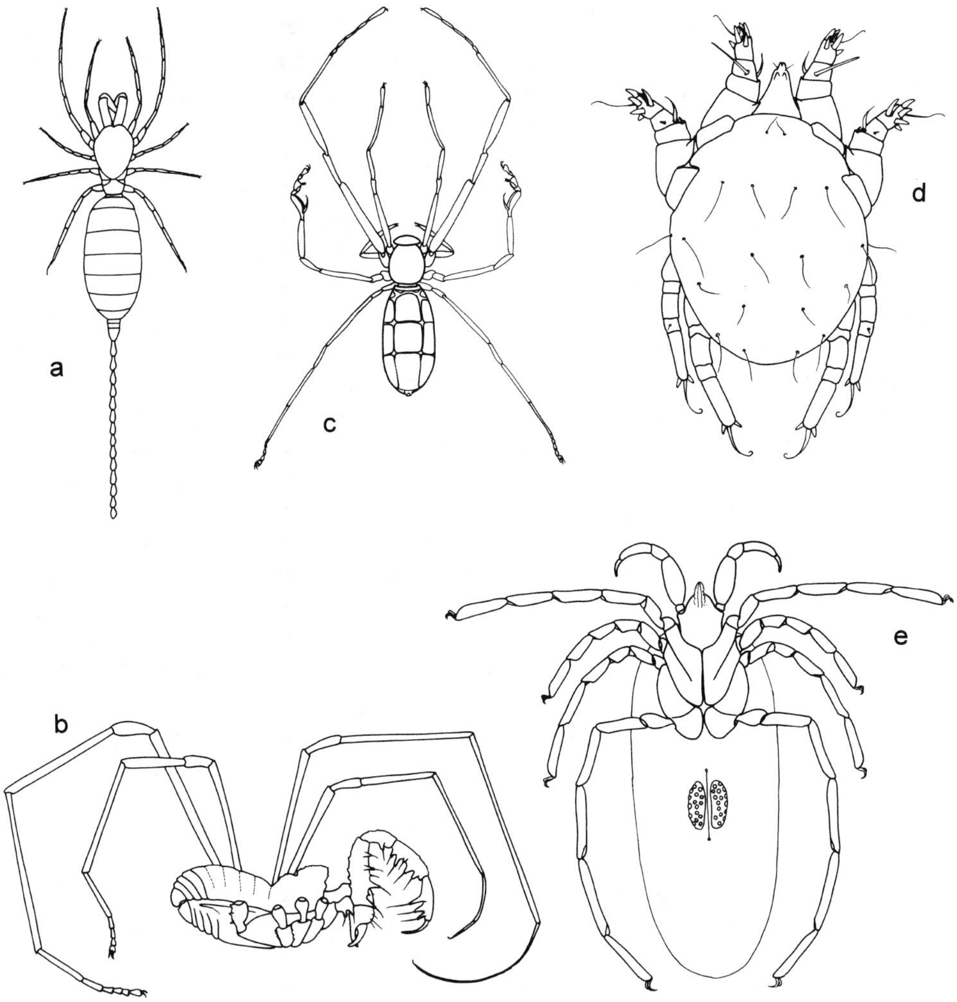

Fig. 4.3. Illustrations of troglobitic or hyporheic arachnids: (a) Palpigradi, *Eukoenenia* sp. (Eukoeneniidae) (redrawn after Harvey and Yen, 1989); (b) Opiliones, *Maiorerus randoi* (Phalangodidae), lateral (redrawn after Rambla, 1993); (c) Ricinulei, *Pseudocellus reddelli* (Ricinoididae) (redrawn after Gertsch, 1971); (d) Acari, *Neoguanolichus panamensis* (Guanolichidae) (redrawn after Fashing, 1979); (e) Acari, *Omartacarus elongatus* (Omartacaridae), ventral (redrawn after Cook, 1963).

described species, 60 are accommodated within the genus *Eukoenenia.* Most species occur in leaf litter and soil, and one species has been collected from interstitial beach deposits. Cavernicolous representatives have been reported from subterranean systems in many parts of the world, mostly from Europe, but none possess any remarkable troglomorphies (Fig. 4.3a).

Superorder Cryptoperculata

Harvestmen (order Opiliones) are distinctive arachnids with variable body shape. Some are short-legged, but others possess immensely elongated legs. They are currently classified into two or three suborders and 26 families (Martens, 1986) and over 5000 species are named. Troglobitic representatives occur in most families (Rambla and Juberthie, 1994), often showing blindness, elongate appendages and pale coloration (Fig. 4.3b).

Ricinuleids (order Ricinulei) are heavily sclerotized, thick-set arachnids; combined with their slow, deliberate gait, their morphology gives them an extremely primitive appearance. They first appear in the fossil record during the Carboniferous, but are at present only found in the tropical rainforests of western Africa (*Ricinoides*) and in rainforests and caves of Central America and northern South America (*Cryptocellus* and *Pseudocellus*). All extant species are placed in a single family, Ricinoididae, and adult males possess a unique structure on leg III which is used in sperm transfer to the female (Fig. 4.3c). Eleven cavernicolous species have been described, mostly in the genus *Pseudocellus,* from Cuba, Guatemala, Mexico and Venezuela. Like most of their epigean counterparts, these species have low vagility and are usually restricted to very small areas, such as single cave systems. None possess truly troglobitic facies such as complete blindness or extremely pale coloration, but most have elongate appendages (e.g., Gertsch, 1971; Fig. 4.3c). Most Mexican species prey on other invertebrates, usually those associated with guano deposits.

Traditionally, mites are placed in a single order (Acari), but most recent authors divide them into at least three orders, Opilioacarida, Parasitiformes and Acariformes. Mites are an extremely abundant group, occurring in virtually every terrestrial and aquatic habitat. Many species occur in soil and leaf litter. Only a few cave-dwelling mites are known to possess traditional troglobitic facies such as elongate appendages, blindness or pale coloration.

The Opilioacarida are considered to be the most primitive of the living mites and retain many plesiomorphic features such as abdominal segmentation. Although found in many disparate regions around the world, only 19 species have been described in a single family, Opilioacaridae. The only known troglophilic species occur in Cuba and Thailand.

Amongst the remaining two orders, Parasitiformes and Acariformes, many species are known to inhabit bat guano (Fig. 4.3d) where they either feed on particulate matter or are predators of other cohabiting animals.

Two major groups of acariform mites, the Halacaridae and the Hydracarina, are represented in aquatic subterranean systems. The Halacaridae are primarily a marine group which have made many independent incursions into freshwater environments (Bartsch, 1996). Conversely, the Hydracarina are mostly restricted to freshwater environments, although one family is secondarily marine. Many limneic water mites possess swimming setae on one or more legs, which are invariably lost in hyporheic and hypogeic species (Fig. 4.3e).

Adaptations of cave arachnids

Subterranean arachnids often display marked morphological features which distinguish them from their epigean relatives. These include:

(1) Increased size. Cavernicoles are often larger in size than epigean species.
(2) Depigmented cuticle. Paleness is a feature of many troglobitic and hyporheic arachnids.
(3) Loss or reduction of eyes. The standard number of eyes for each arachnid group is relatively fixed, and epigean arachnids rarely show reductions from that number. Examples include spiders and whip spiders with eight eyes, pseudoscorpions with four eyes and schizomids with two eyes or eye-spots. Endogean arachnids such as palpigrades lack eyes. Many subterranean arachnids show reduction in eye size, and many lack eyes altogether (Figs. 4.2, 4.3).
(4) Attenuate appendages. The chelicerae, pedipalps and legs of many subterranean arachnids are greatly elongated compared with their epigean relatives (Fig. 4.3c).
(5) Elongate body. Many hyporheic and hypogeic Hydracarina possess elongated bodies, presumably facilitating movement through the interstitial medium.

"MYRIAPODS"

The term "myriapod", while not formally designating a taxon, covers four classes of atelocerate arthropods related to insects. The two largest classes, Chilopoda (centipedes) and Diplopoda (millipedes) are familiar to most zoologists and to the general public, but the two smaller classes, Pauropoda and Symphyla, have mostly escaped notice, except for the occasional damage done to crops by a few species of Symphyla. Similarly, only the large classes are of any significance in subterranean ecosystems (symphylans have been reported from caves, but no troglobitic species have been described). Therefore this brief review will focus on chilopods and diplopods.

Centipedes (class Chilopoda) can be recognized by their soft bodies and prominent grappling poison claws just behind the head, and by having only one pair of legs on each segment (however, one dorsal plate may cover more than one segment). Very few cave-adapted chilopods have been reported, a somewhat surprising result given that their usual habitat – soil and litter in moist environments – would preadapt them for life in caves. Millipedes (class Diplopoda) usually have a hard, crusty cuticle and all lack poison claws (Hopkin and Read, 1992). Their segments are obviously fused in pairs so that each has two pairs of legs; the segments of the trunk are often ringlike. Most species are able to produce evil-smelling or irritating chemicals from pores on the segments. Scores of species of highly adapted troglobites are known amongst the diplopods, and in some cave ecosystems they are the dominant organisms. This may be attributed to the fact that millipedes, as feeders on organic detritus and debris, find more sustenance in most caves than would the predatory centipedes.

Class Chilopoda

The class Chilopoda (centipedes) contains four orders, including 18–20 families; centipede systematics are in a chaotic state at present and there are few active specialists (Lewis, 1981). The total number of centipede species, including undescribed ones, has been estimated at about 10 000.

Centipedes of the order **Scutigerimorpha** (with a single family: Scutigeridae) are easily recognized as having 15 pairs of extremely elongate legs, long, filamentous antennae, and large multifaceted eyes. On close examination, it can be seen that a large breathing pore, or spiracle, occurs at the rear of each dorsal plate (tergite). The most familiar species is *Scutigera coleoptrata,* a Mediterranean native that has become established as an inhabitant of houses in many parts of the world. Scutigerimorphs are found in caves in eastern Asia, Africa, Australia and South America, the distribution of the order being primarily tropical. While some of the Asian species are extraordinarily large (with leg spans to six inches), none are known to show troglobitic adaptations.

The order **Lithobiomorpha** also includes centipedes with 15 pairs of legs, but the legs are stout and the tarsi are not multi-articulated. The antennae are usually short, and at most only a few ocelli are present. The spiracles are located on the sides of the segments and generally not very conspicuous. Lithobiomorphs are most abundant and diverse in the temperate northern hemisphere (Lithobiidae) and in Australia (Henicopidae). A few troglobitic species have been recorded from caves in Europe and the United States. These species are typically pale, eyeless, and lack pigment; their legs are proportionally longer than those of their surface-dwelling relatives (Fig. 4.4a). Four families, Anopsobiidae, Cermatobiidae, Henicopidae and Lithobiidae are usually recognized.

Members of the order **Scolopendromorpha** have either 21 or 23 pairs of stout legs and relatively short antennae, and are blind or have only a few ocelli. Spiracles are on the sides of the segments, but vary in distribution. They are most abundant and diverse in the tropics, especially in dry climates, and include the largest known centipedes. Species of the genus *Scolopendra* can reach 30–35 cm in length, and are much feared as venomous. While often found in tropical caves, few troglobitic species have been reported. There are two families, Scolopendridae and Cryptopidae.

The order **Geophilomorpha** is the largest centipede order in number of species, and members of the order are common on all continents (except Antarctica) and in nearly all climates. They are long, thin centipedes with more than 35 pairs of legs – some species may have more than 100 pairs, making them true "hundred-leggers". The antennae are short, consisting of 18 segments in all species, and all known forms are eyeless. Lateral spiracles are found on each segment. These centipedes inhabit the soil, from the superficial leaf litter to considerable depth, and are often found in caves as "accidentals". No troglobitic species have been recorded. Thirteen families were recognized in a recent treatment (Hoffman, 1982).

The order **Craterostigmatomorpha** (with one family, Craterostigmatidae) is known from two soil-dwelling species found in Tasmania and New Zealand. They have 15 pairs of legs and a peculiar posterior end like a clam shell; neither species has been found in caves.

Class Diplopoda

A recent classification (Hoffman, 1979) recognized 15 orders of millipedes (class Diplopoda), containing 115 families and 1733 genera; a good guess at the total number of species (described and undescribed) is about 80 000. Only a sampling of orders can be discussed here.

The order **Polyxenida** consists of small millipedes, all less than 5 mm long, with soft bodies densely beset with clusters of peculiar hairs. There are from 13 to 17 pairs of legs. While sometimes found in caves, no definitively troglobitic species have been named. Their usual habitat is under stones and bark.

Pill millipedes, which can roll up into an almost

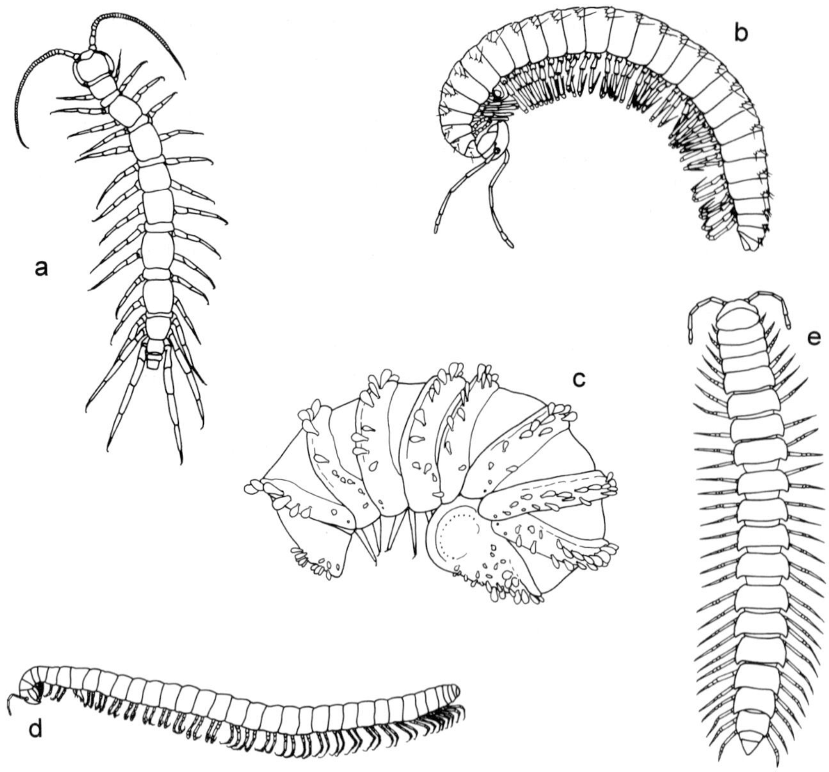

Fig. 4.4. Illustrations of troglobitic myriapods: (a) Lithobiomorpha, *Lithobius anophthalmus* (redrawn after Serra, 1980); (b) Chordeumatida, new species from China (redrawn after M. Bertoncini); (c) Glomerida, *Trachysphaera orientalis* (redrawn after Golovatch, 1990); (d) Julida, *Antrokoreana arcuata* (redrawn after Murakami, 1969); (e) Polydesmida, *Epanerchodus fontium* (redrawn after Murakami, 1969).

perfect sphere, are members of the order **Glomerida**. The second segment behind the head and the terminal segment are enlarged to complete the rolling-up process (Fig. 4.4c). In males, the last pair of legs is many times larger than the other pairs and is used to grasp the female during mating. Species are abundant in Europe but rare in North America; a distinctive group of families is found in southeast Asia, Africa and Australia. Cave-adapted species are known from Europe, and Mexico and Central America.

The following orders are often referred to as helminthomorphs, or worm-like millipedes. In all, the males have strongly modified legs on the seventh and eighth body segments, used to transfer sperm to the females. The form of these gonopods, which can be extraordinarily baroque, is the main criterion for species recognition. Even the distinctions between genera and families are sometimes made on the basis of these appendages. Hence it is important to collect mature male individuals for identification.

The **Polydesmida** is the largest millipede order and may contain as many as 30 families. They are easily recognized because in most species the segments are fused into complete rings, and bear more or less prominent wing-like extensions called paranota (Fig. 4.4e). All species are blind. Typically there are 20 segments behind the head, but a few species with 18, 19 or 21 occur. The usual colour is black or brown; but this is often enlivened with bright spots and bands of yellow, orange, or red; one family common in Mexico (Rhachodesmidae) includes blue and green species. Many troglobitic species have been described, especially from Europe, Japan, and Mexico. Polydesmids appear to dominate the cave habitat in tropical regions.

The order **Spirobolida**, comprising 9 or 10 families, contains the largest species of millipedes (up to 12 inches long). The millipedes usually pictured in general textbooks, or studied in introductory zoology laboratories, belong here. The body is elongated and cylindrical, usually brown or black (though some highly colourful tropical species are known). They can be unpleasant to handle since their defensive secretions can cause damage to human skin. There are a few small, troglobitic species known from Mexico, all in the family Typhlobolellidae, and a single troglobitic species is found in Western Australia.

The order **Spirostreptida** is diverse. While some species closely resemble spirobolids, others are quite small, with long, thin bodies and often the trunk segments are ornamented with various kinds of crests. Members of the genus *Cambala* are commonly found in caves in the southern United States and Mexico, and the genus *Nannolene* has diversified to an extraordinary degree on Hawaii, where some species are troglobites in lava tubes.

Members of the order **Julida** are restricted to the northern hemisphere, except where they have been transported by humans to other regions. The body is cylindrical, and as in the Polydesmida, individual segments are fused to form complete rings (Fig. 4.4d). The family Julidae is the dominant millipede family in Europe, and numerous troglobitic species are known. Likewise, the family Blaniulidae includes troglobites from North America and Europe.

Callipodid millipedes (order **Callipodida**) are elongate animals with crested segments; the sterna (to which the legs are attached) are not fused to the other plates of the segments, and there is a median dorsal suture running down the body. These millipeds are highly active, can withstand dry conditions, and include the only carnivores known from the entire class. Troglobites and troglophiles are common in Europe and the Near East, and the genus *Tetracion* consists entirely of troglobites from the southern United States.

Undoubtedly the most important millipede order in temperate-zone subterranean habitats is the **Chordeumatida**. Like callipodids, the sterna are not fused with the rest of the trunk segments, and there is a dorsal median suture. However, chordeumatids rarely have crested segments and entirely lack chemical defences. Each trunk segment bears six prominent macrosetae on the dorsal surface (Fig. 4.4b). Most chordeumatids are adapted to high humidity and cool temperatures, and are found in the soil; this preadapts them to invade cave habitats. Scores of troglobitic species are known, most prominently from North America, Europe and Japan. In the United States, species of *Pseudotermia* (Cleidogonidae) and *Trichopetalum* (Trichopetalidae) are common and often very abundant in caves (Causey, 1964), as are species of *Cleidogona* (Cleidogonidae) in Mexico (Shear, 1972). Many of the European troglobites are important to studies of biogeography and evolution because of their relict nature. In Japan, *Diplomaragna* (Diplomaragnidae) has many troglobitic species. This order is generally rare in tropical regions and in the southern hemisphere.

A group of unusual and evidently rather primitive millipede orders are combined in the superorder **Colobognatha**, though some authorities argue that they are not in fact closely related. These millipedes are predominantly tropical, though some species are common in the forested regions of the temperate zone. They are distinguished by having reduced mouth parts, and in at least some families (Siphonophoridae, Siphonorhinidae) the head projects forward in a distinct beak. The body is usually flattened, and may be relatively short or enormously elongated – the millipede with the largest number of segments is a member of this superorder. While often collected in tropical caves, no species of colobognaths are much modified for troglobitic life.

Geographic distribution of cave myriapods

As with most other troglobitic organisms, the best-studied karst regions have yielded the most completely known faunas. Among these areas are, of course, the United States, Europe and Japan. Somewhat unexpectedly, Mexico must also be included in this list and may eventually prove to harbour as many troglobitic millipede species as the other regions combined. Other tropical and subtropical countries with karst regions remain poorly explored for cave millipedes. Preliminary data suggest that the karst areas of China are probably rich in undescribed genera and species of cave millipedes.

Troglobitic centipedes have only been reported from Europe, United States (Shear, 1969), and possibly Cuba.

Abundant and diverse faunas of cave millipedes are found in the temperate regions of North America and Europe. Subtropical and tropical karsts, such as those in Mexico, are in general lower in diversity at high taxonomic levels but may harbour many congeneric species.

Adaptations of cave myriapods

Morphological adaptations of cave myriapods are those typical of troglobitic arthropods of other groups. The legs and antennae tend to become elongate, and the multisegmented body may also be attenuated when compared with surface-dwellers. The cuticle of millipedes is strengthened by calcium carbonate, like that of crustaceans. In highly adapted troglobitic species, the organic fraction of the cuticle may be greatly reduced, producing a very brittle cuticle. Pigmentation is reduced or entirely lost, as are eyes, if they are usually present; but in one of the largest millipede orders, Polydesmida, all known species, troglobitic or otherwise, are eyeless. Sensory setae may be more numerous and longer in troglobitic species. Many troglobitic chordeumatids have the characteristic six tergal setae much elongated, and some appear to be glandular.

If the number of segments in the family to which a troglobitic species belongs is not genetically fixed, troglobitic millipedes tend to have fewer segments, shorter bodies, and probably abbreviated life cycles.

Helminthomorph millipede males transfer sperm to females using modified legs located on the seventh and/or eighth body segment. These gonopods are often extremely complicated, but in troglobitic species they may be of reduced complexity when compared to surface-dwelling relatives.

Among the unique adaptations of a few species are the modified mouthparts of the Italian troglobites *Trogloiulus binii* and *T. vailatii* (Enghoff, 1985). Unsuited for chewing, the mandibles of these millipedes are used to comb bacteria from surfaces or to filter food particles from water; they are often collected submerged in cave pools or streams.

HEXAPODA

The superclass Hexapoda comprises all six-legged arthropods [Protura + Collembola + Diplura + "true" insects (Archaeognatha + Thysanura s.str. + Pterygota)]. The relationships between these taxa are still not completely resolved. The classification presented here reflects the view of Kristensen (1991).

The oldest known hexapods are Collembola from Lower Devonian strata of Scotland (Tillyard, 1928). The monophyly of the Hexapoda is inferred primarily from the unique type of tagmosis shared by its members: successive body segments unite morphologically and functionally into three distinct tagmata: head, thorax and abdomen. The head consists of 6 segments and the acron, the thorax of 3 segments, and the abdomen of 11 segments and the telson. Within the hexapods, this ground pattern remains largely conserved, apart from reductions of abdominal segment numbers.

Nevertheless, hexapods are by far the most diverse and dominant group of organisms, including about 750 000 described species, but with estimates of the number of existing species ranging from 3–4 million to tens of millions. Thirty orders of Hexapoda are recognized today, the majority of which are regarded as monophyletic – that is, descending from a single ancestral species.

Following the evolutionary history of the hexapods, it becomes evident that comparatively few major "evolutionary innovations" account for the biggest success stories – that is, in the evolution of remarkably species-rich taxonomic groups: evolution of flight (the Pterygota); the development of a basal wing articulation that allows folding the wings against the abdomen in a resting position (the Neoptera); a holometabolous life cycle – the introduction of a pupal stage into postembryonic development (the Holometabola or Endopterygota); and the (independent) development of true social behaviour (eusociality) in Isoptera (termites) and within the Hymenoptera (bees, wasps and ants).

Additionally, and superimposed on the above trends, modifications leading to specializations of virtually all aspects of hexapod morphology (e.g., mouthparts, respiratory organs, locomotory organs, to mention only a few), development, physiology and behaviour have allowed colonization of nearly all conceivable habitats and exploration of virtually all kinds of organic matter as an energy resource. Hexapods even occur in seemingly extremely hostile habitats: in hot springs, with water temperature close to boiling, on the snow surface on high mountain-tops, in extremely arid regions such as deserts, in highly acid environments like vertebrate intestines, in pools of petroleum in oil fields, on the open ocean, skating the surface or hidden in the nostrils of sea lions (Berenbaum, 1995), and beneath the earth's surface, in the permanent darkness of limestone caves and lava tubes. An engraving of a cave cricket on a fragment of bison's bone in a cave in the French Pyrenees is the most ancient evidence we have of real cave fauna (Richards, 1961).

Many hexapod orders contain species that are cavernicolous in a broad sense – occurring in caves but also in other environments. These "troglophiles" do not show distinct morphological specializations for cave life. As there is often uncertainty or even controversy in respect to the biology of these species, they are not considered here. In contrast to troglophilic taxa, troglobites (organisms *restricted* to cave life) are usually highly specialized and – compared to epigean representatives of the same group – often display profound morphological alterations (troglomorphies), such as the reduction and loss of eyes, wings, and body pigmentation. Troglobitic hexapods occur in limestone caves and lava tubes. The cave faunas of temperate regions, especially of Europe [see chapters in the section "Europe" in: Juberthie and Decu (1994)] and North America (Barr Jr et al., 1994) including Mexico (Palacios-Vargas, 1994) have long been relatively well known. In the past 25 years, however, highly specialized cave faunas have been discovered in tropical regions and on oceanic islands – in Australia (Howarth, 1988), Central and South America [see various chapters in the sections "Amerique Centrale, Antilles" and "Amerique du Sud" in: Juberthie and Decu (1994)], Hawaii (Howarth, 1991), Galapagos (Peck, 1994) and the Canary Islands (Oromi and Izquierdo, 1994).

The following account aims to introduce hexapod orders which are commonly represented in caves. In each case, one or a few prominent troglobitic species have been selected to illustrate morphological and ecological diversity in underground habitats. More detailed information of the cavernicolous species of many of the groups treated here has recently been compiled by Juberthie and Decu (1994).

Class Ellipura

The two orders included in this class, **Protura** and **Collembola**, like other primitive hexapods, are wingless and soft-bodied. Springtails (order Collembola) are small hexapods with a forked jumping organ derived from modified abdominal segments which enable many species to jump considerable distances when disturbed (Fig. 4.5a). They feed on micro-organisms associated with the rhizosphere and decomposing organic matter (Greenslade, 1994).

Class Diplura

Diplurans (class and order **Diplura**) are small (less than 5 mm) to comparatively large (~50 mm) wingless hexapods with narrow, usually elongate and unpigmented bodies (Fig. 4.5b). The cerci are either filiform or forceps-like, and diplurans are herbivorous or carnivorous (including cannibalism), the prey being caught by the maxillae or cerci (Condé and Pagés, 1994). They are usually found in damp soil under logs and stones, and some troglobitic species have been named.

Class Insecta

This class contains the bulk of the Recent hexapods and includes two subclasses, Archaeognatha, with only a single order of the same name, and Dicondylia for the remaining orders. The latter contains two infraclasses, Thysanura with a single order, and Pterygota, whose primary feature is the presence of wings on the meso- and metathorax in the adults.

Infraclass Thysanura

The infraclass Thysanura contains only a single order of the same name (order Thysanura). Silverfish are wingless insects with a dorsoventrally flattened body, and with filiform cerci and terminal filament (Fig. 4.5c). Many species are covered by scales, and the body may be pigmented or unpigmented. Silverfish are free-living and agile, but invariably inhabit cryptozoic habitats (e.g., under bark, in leaf litter, inquilines in ant nests). Although they are usually omnivorous, all species in the subfamily Nicoletiinae, which is principally subterranean and cavernicolous, are vegetarian (Smith and Watson, 1994).

Infraclass Pterygota

This taxon comprises all primarily winged and secondarily apterous hexapods. Pterygotes are characterized by a variably united meso- and metathorax (pterothorax), well-developed pleura, 8–11 abdominal segments, a well-developed tracheal system with spiracles primarily with muscular closing apparatus, mating by copulation, and no adult ecdysis (except for the subimago in Ephemeroptera).

The following taxa (orders) are included: Ephemeroptera (mayflies), Odonata (dragonflies and damselflies), Plecoptera (stoneflies), **Blattodea** (cockroaches), Isoptera (termites), Mantodea (praying mantids), Grylloblattodea, **Dermaptera** (earwigs), **Orthoptera** (grasshoppers, crickets, katydids), Phasmatodea (stick-insects), Embioptera (foot-spinners), Zoraptera, Psocoptera (psocids, booklice), Phthiraptera (lice), **Hemiptera** (Heteroptera: bugs; and Homoptera: leafhoppers, treehoppers, froghoppers, planthoppers, cicadas, aphids, scale insects), Thysanoptera (thrips), Strepsiptera, **Coleoptera** (beetles), Megaloptera (alder-flies, dobsonflies), Raphidioptera (snake-flies, camel-neck-flies), Neuroptera (lacewings), Mecoptera (scorpion-flies, hanging-flies), Siphonaptera (fleas), **Diptera**

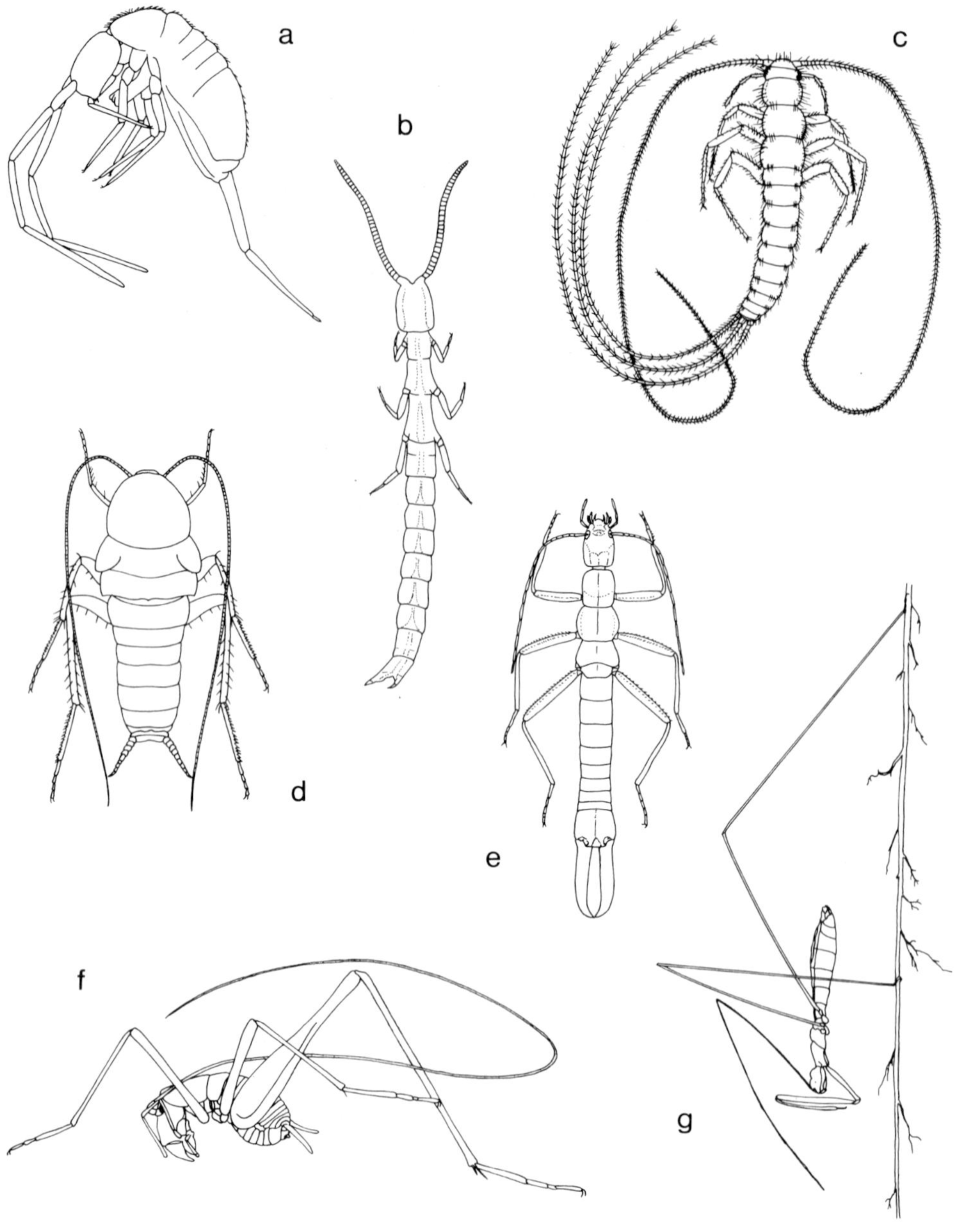

Fig. 4.5. Illustrations of troglobitic hexapods: (a) Collembola, *Troglopedetes delamarei* (Paronellidae) (redrawn after Thibaud and Deharveng, 1994); (b) Diplura, *Troglojapyx hauseri* (Japygidae) (redrawn after Pagés and Bareth, 1994); (c) Thysanura, *Cubacubana negreai* (Nicoletiidae) (redrawn after Mendes, 1994); (d) Blattodea, *Loboptera troglobia* (Blattellidae) (redrawn after Izquierdo and Oromi, 1994); (e) Dermaptera, *Anataelia troglobia* (Pygidicranidae) (redrawn after Brindle and Oromi, 1994); (f) Orthoptera, *Spelaiacris tabulae* (Rhaphidophoridae) (redrawn after Samways, 1994); (g) Hemiptera, Heteroptera, *Nesidiolestes ana* (Reduviidae: Emesinae) (redrawn after Gagné and Howarth, 1975).

(flies), Trichoptera (caddisflies), Lepidoptera (butterflies and moths), Hymenoptera (wasps, bees, ants, sawflies). We treat in more detail only those orders (in bold type) which contain obvious troglobitic species.

Cockroaches (order **Blattodea**) are primitive relatives of the Orthoptera with a dorsoventrally flattened body, and the head usually reflexed under the pronotum (Fig. 4.5d). Most species are nocturnal, and usually found near the ground, hiding under logs, bark and stones during daylight (Roth, 1994). They feed on decaying vegetation. Blindness and depigmented cuticle are common features in troglobitic cockroaches, which occur in many cave systems.

Earwigs (order **Dermaptera**) possess elongate bodies, may be winged or apterous, and possess cerci which are modified into terminal forceps (Fig. 4.5e). They primarily occur in tropical and warm-temperate zones, and are usually nocturnal. Earwigs occur in cryptozoic habitats, such as in crevices, under bark of dead trees, in rotting logs. Most species are presumably omnivorous, but some are herbivorous, saprophagous or carnivorous, with the forceps used to catch prey (Rentz and Kevan, 1994).

Grasshoppers and their relatives (order **Orthoptera**) are medium to large-sized insects (up to 11.5 cm in body length), which usually possess saltatorial hind legs (Fig. 4.5f). They are found worldwide, especially in warmer climates and in the tropics, and are carnivorous or phytophagous (most species); some species are myrmecophilous, burrowing, or cavernicolous (Rentz, 1994).

With more than 70 000 species, the **Hemiptera** (Heteroptera and Homoptera) are the dominant group among the insects with gradual development (hemimetabolous). They range in body size from less than 1 mm to 11 cm. The mouthparts are modified into a rostrum, with stylet-like mandibles and maxillae which are enclosed in an anteriorly grooved rostrate labium; the suction canal and salivary canal lie between the maxillary stylets. These unusual mouthparts enable hemipterans to feed by suction; they extract liquid contents from plants or from animal prey. They are extraordinarily diverse in morphology and ecology, but most species are terrestrial and phytophagous (all Homoptera). However, the Heteroptera contain many carnivorous and aquatic taxa, including some marine and even fully oceanic species (Carver et al., 1994). Troglobitic species are found in numerous regions of the world, and include long-legged predatory heteropterans (Fig. 4.5g), and micropterous or apterous homopterans (Figs. 4.6a–c). All cavernicolous Homoptera feed by sucking sap from roots of surface vegetation which reach down into the caves (Hoch, 1994).

With about 300 000 described species, the beetles (order Coleoptera) clearly are the dominant group not only among hexapods (ca. 40%) but also among all animals (30%). Their most prominent morphological feature are the forewings which are modified into more or less hardened, rigid elytra, in repose meeting each other medially in a straight line, thus protecting the wings and abdomen (Figs. 4.6d,e). They are cosmopolitan, in virtually every available terrestrial and freshwater habitat, including soil, humus, leaf litter, flood debris and decomposing organic matter, and with feeding habits ranging from phytophagous and saprophagous to carnivorous and even parasitic (Lawrence and Britton, 1994).

With at least 150 000 species, the flies (order **Diptera**) are one of the major insect orders, and are characterized by a pair of membranous wings on the mesothorax only, and the metathoracic pair being modified into club-like haiteres (Fig. 4.6f). The mouthparts of adults are suctorial and often adapted for piercing. Adult Diptera are free-living and ubiquitous; they occur in practically all terrestrial habitats and climates, but the majority of species are associated with forest or water. Adults of most species are polyphagous, concentrating on liquid food (e.g., animal and plant secretions, decomposing organic matter, including honeydew, and the body fluids of prey and hosts), but some species also eat pollen. Larvae usually inhabit moist situations (e.g., mud, soil, decomposing organic matter, plant or animal tissues, free water). Many species are of economic importance, either as pests and pathogen-vectors (e.g., of malaria) or – beneficially – as pollinators or in biological control (Colless and McAlpine, 1994). Few troglobitic species have been recorded; examples include *Mormotomyia hirsuta* (Mormotomyiidae) (Fig. 4.6f), which is known only from a vertical pit-like cave in Kenya. The larvae feed on bat guano, and adults may either be guanophiles or feed on skin secretions of the bats (Matile, 1994).

Adaptations of cave hexapods

Looking at the ecology of epigean relatives of troglobitic hexapods, it becomes evident that most of them, at least during one developmental stage, have developed strategies to cope with environmental factors that are also common within caves – high humidity, permanent darkness, food scarcity, etc. Species at such a level of preadaptation are likely to adopt a permanently subterranean way of life. This is excellently demonstrated by the Homopteran family of planthoppers Cixiidae. The nymphal instars of cixiids generally live within the soil, where they feed on roots. Only the adult emerges to the surface, where it feeds on the epigean parts of vegetation and where courtship and mating take place.

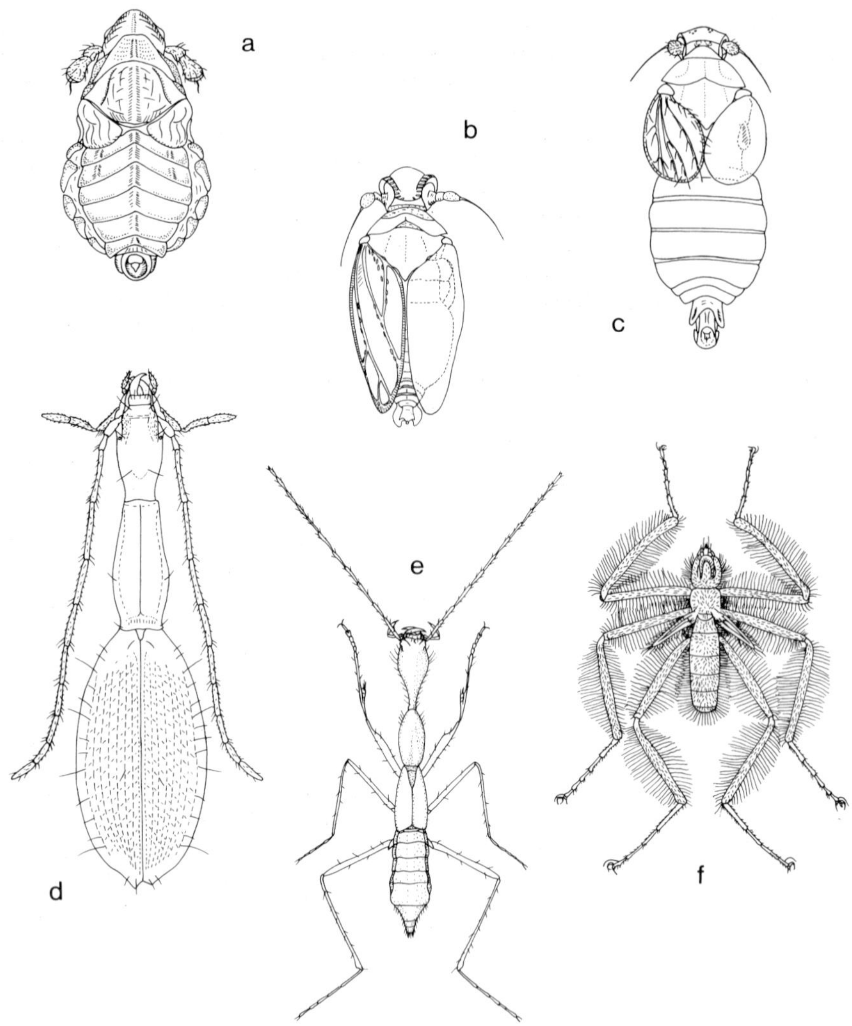

Fig. 4.6. Illustrations of troglobitic hexapods: (a) Hemiptera, Homoptera, *Notuchus larvalis* (Delphacidae) (redrawn after Hoch, 1994); (b) Hemiptera, Homoptera, *Meenoplus cancavus* (Meenoplidae) (redrawn after Remane and Hoch, 1988); (c) Hemiptera, Homoptera, *Cixius ariadne* (Cixiidae) (redrawn after Hoch and Asche, 1993); (d) Coleoptera, *Rhadine austinica* (Carabidae) (redrawn after Barr Jr, 1974); (e) Coleoptera, *Domene vulcanica* (Staphylinidae) (redrawn after Oromi and Hernandez, 1986); (f) Diptera, *Mormotomyia hirsuta* (Mormotomyiidae) (redrawn after Matile, 1994).

Eggs are deposited on the soil surface (Müller, 1942). In troglobitic planthoppers, the nymphal habitat and mode of nutrition are retained also in the adult stage, and the whole life cycle is completed underground (Remane and Hoch, 1988).

Thus, reductive evolutionary trends are not only observed in cavernicolous species, but also in inhabitants of leaf litter and soil, nocturnal or otherwise cryptic species. The most striking alterations concern the loss of eyes and pigmentation, as well as the reduction of tegmina and wings. These reductive trends are usually most strongly pronounced in troglobitic species. Besides these obvious troglomorphies, other adaptations in cave hexapods have been recognized: prolonged antennae to enhance sensory function, specialization of body form (either compressed or depressed) to aid in moving along small voids and cracks, specialized praetarsi for walking on rock or wet surfaces (Howarth, 1983), mechanisms to cope with abnormally high concentrations of carbon dioxide (Howarth and Stone, 1990), modifications of water balance and metabolic rates in response to water availability and energy utilization (Ahearn and Howarth, 1982), as well as alterations of life-history strategies (e.g., a prolonged life cycle, fewer eggs) (Howarth, 1983) and reproductive behaviour in response to the modified conditions of the cave habitat (see Chapter 10, this volume).

REFERENCES

Ahearn, G.A. and Howarth, F.G., 1982. Physiology of cave arthropods in Hawaii. *J. Exp. Zool.*, 222: 227–238.

Barr Jr., T.C., 1974. Revision of *Rhadine* LeConte (Coleoptera, Carabidae). I. The *subterranea* group. *Am. Mus. Novit.*, 2539:1–30.

Barr Jr., T.C., Culver, D.C. and Kane, T.C., 1994. United States. In: C. Juberthie and V. Decu (Editors), *Encyclopaedia Biospeologica*, Vol. I. Société de Biospéologie, Moulis (France) and Bucarest, pp. 403–416.

Bartsch, I., 1996. Halacarids (Halacaroidea, Acari) in freshwater. Multiple invasions from the Paleozoic onwards? *J. Nat. Hist.*, 30: 67–99.

Berenbaum, M.R., 1995. *Bugs in the System. Insects and their Impact on Human Affairs.* Addison-Wesley, Reading, MA, 377 pp.

Brindle, A. and Oromi, P., 1994. Dermaptera. In: C. Juberthie and V. Decu (Editors), *Encyclopaedia Biospeologica*, Vol. I. Société de Biospéologie, Moulis (France) and Bucarest, pp. 289–294.

Carver, M., Gross, G.F. and Woodward, T.E., 1994. Hemiptera. In: I.D. Naumann (Editor), *Systematic and Applied Entomology. An Introduction.* Melbourne University Press, Melbourne, pp. 316–329.

Causey, N.B., 1964. New trichopetalid millipeds from caves in North and Central America. *Proc. La. Acad. Sci.*, 27: 63–66.

Colless, D.H. and McAlpine, D.K., 1994. Diptera. In: I.D. Naumann (Editor), *Systematic and Applied Entomology. An Introduction.* Melbourne University Press, Melbourne, pp. 374–387.

Condé, B., 1996. Les Palpigrades, 1885–1995: acquisitions et lacunes. *Rev. Suisse Zool., volume hors série,* pp. 87–106.

Condé, B. and Pagés, J., 1994. Diplura. In: I.D. Naumann (Editor), *Systematic and Applied Entomology. An Introduction.* Melbourne University Press, Melbourne, pp. 240–241.

Cook, D.R., 1963. Omartacaridae, a new family of water mites from the ground waters of North America. *Entomol. News,* 74: 37–43.

Dunlop, J., 1996. Systematics of the fossil arachnids. *Rev. Suisse Zool., volume hors série,* pp. 173–184.

Enghoff, H., 1985. A new species of *Trogloiulus* with modified mouthparts, with a revised key to the species and new records of the genus (Diplopoda, Julida, Julidae). *Lav. Soc. Veneziana Sci. Nat.*, 10: 19–77.

Fashing, N.J., 1979. *Neoguanoichus panamensis,* a new species of Guanolichidae from bat caves in Panama (Acarina: Sarcoptiformes). *Acarologia,* 21:108–116.

Francke, O.F., 1982. Studies of the scorpion subfamilies Superstitioninae and Typhlochactinae, with description of a new genus (Scorpiones, Chactoidea). *Bull. Assoc. Mex. Cave Stud.*, 8: 51–61.

Gagné, W.C. and Howarth, F.G., 1975. The cavernicolous fauna of Hawaiian lava tubes, 7. Emesinae or thread-legged bugs (Heteroptera: Reduviidae). *Pac. Insects,* 16: 399–413.

Gertsch, W.J., 1971. Three new ricinuleids from Mexican caves (Arachnida, Ricinulei). *Bull. Assoc. Mex. Cave Stud.*, 4:127–135.

Gertsch, W.J., 1992. Distribution patterns and speciation in North American cave spiders with a list of the troglobites and revision of the cicurinas of the subgenus *Cicurella. Tex. Mem. Mus. Speleol. Monogr.*, 3: 75–122.

Golovatch, S., 1990. Diplopoda of the Caucasus, 3. Trachysphaeridae, with contributions to the fauna of Turkey. *Senckenbergiana Biol.*, 70: 331–358.

Greenslade, P.J.M., 1994. Collembola. In: I.D. Naumann (Editor), *Systematic and Applied Entomology. An Introduction.* Melbourne University Press, Melbourne, pp. 233–237.

Harvey, M.S., 1988. A new troglobitic schizomid from Cape Range, Western Australia (Chelicerata: Schizomida). *Rec. West. Aust. Mus.*, 14: 15–20.

Harvey, M.S., 1992. The phylogeny and classification of the Pseudoscorpionida (Chelicerata: Arachnida). *Invertebr. Taxon.*, 6: 1373–1435.

Harvey, M.S. and Yen, A.L., 1989. *Worms to Wasps: an illustrated Guide to Australia's Terrestrial Invertebrates.* Oxford University Press, Melbourne, 201 pp.

Hoch, H., 1994. Homoptera (Auchenorrhyncha Fulgoroidea). In: C. Juberthie and V. Decu (Editors), *Encyclopaedia Biospeologica,* Vol. I. Société de Biospéologie, Moulis (France) and Bucarest, pp. 313–325.

Hoch, H. and Asche, M., 1993. Evolution and speciation of cave-dwelling Fulgoroidea in the Canary Islands (Homoptera: Cixiidae and Meenoplidae). *Zool. J. Linn. Soc.*, 109: 53–101.

Hoffman, R.L., 1979. *Classification of the Diplopoda.* Museum d'Histoire Naturelle, Genève, 237 pp.

Hoffman, R.L., 1982. Chilopoda. In: S.P. Parker (Editor), *Synopsis and Classification of Living Organisms,* Vol. 2. McGraw Hill, New York, pp. 681–688.

Hopkin, S.P. and Read, H.J., 1992. *The Biology of Millipedes.* Oxford University Press, New York, 233 pp.

Howarth, F.G., 1983. Ecology of cave arthropods. *Annu. Rev. Entomol.*, 28: 365–389.

Howarth, F.G., 1988. Environmental ecology of North Queensland caves: or why there are so many troglobites in Australia. In: L. Pearson (Editor), *Preprints of the 17th Biennial Conference of the Australian Speleological Federation, Tropicon Conference, Tinaroo, Queensland (December 1988).* Australian Speleological Federations, Cairns, pp. 76–84.

Howarth, F.G., 1991. Hawaiian cave faunas: macroevolution on young islands. In: E.C. Dudley (Editor), *The Unity of Evolutionary Biology.* Dioscorides Press, Portland, OR, pp. 285–295.

Howarth, F.G. and Stone, F.D., 1990. Elevated carbon dioxide levels in Bayliss Cave, Australia: implications for the evolution of obligate cave species. *Pac. Sci.*, 44:207–218.

Izquierdo, I. and Oromi, P., 1994. Dictyoptera – Blattaria. In: C. Juberthie and V. Decu (Editors), *Encyclopaedia Biospeologica,* Vol. I. Société de Biospéologie, Moulis (France) and Bucarest, pp. 295–300.

Juberthie, C. and Decu, V. (Editors), 1994. *Encyclopaedia Biospeologica,* Vol. I. Société de Biospéologie, Moulis (France) and Bucarest, 834 pp.

Kristensen, N.P., 1991. Phylogeny of extant hexapods. In: *The Insects of Australia,* 2nd edn. CSIRO/Melbourne University Press, Melbourne, pp. 141–179.

Lawrence, J.F. and Britton, E.B., 1994. Coleoptera. In: I.D. Naumann

(Editor), *Systematic and Applied Entomology. An Introduction.* Melbourne University Press, Melbourne, pp. 345–359.

Lewis, J.G.E., 1981. *The Biology of Centipedes.* Cambridge University Press, London, 476 pp.

Lourenço, W., 1994. Scorpions. In: C. Juberthie and V. Decu (Editors), *Encyclopaedia Biospeologica,* Vol. I. Société de Biospéologie, Moulis (France) and Bucarest, pp. 181–184.

Martens, J., 1986. Die Grossgliederung der Opiliones und die evolution der Ordnung (Arachnida). In: J.A. Barrientos (Editor), *Actas X Congreso Internacional de Aracnologia, Jaca.* Imprenta Juvenil, Barcelona, pp. 289–310.

Matile, L., 1994. Diptera. In: C. Juberthie and V. Decu (Editors), *Encyclopaedia Biospeologica,* Vol. I. Société de Biospéologie, Moulis (France) and Bucarest, pp. 341–357.

Mendes, L.F., 1994. Thysanura. In: C. Juberthie and V. Decu (Editors), *Encyclopaedia Biospeologica,* Vol. I. Société de Biospéologie, Moulis (France) and Bucarest, pp. 285–287.

Mitchell, R.W. and Peck, S.B., 1978. *Typhlochactas sylvestris,* a new eyeless scorpion from montane forest litter in Mexico (Scorpionida, Chactidae, Typhlochactinae). *J. Arachnol.,* 5: 159–168.

Müller, H.J., 1942. Über Bau und Funktion des Legeapparates der Zikaden (Homoptera Cicadina). *Z. Morphol. Ökol. Tiere,* 38: 534–628.

Mullinex, C.L., 1975. Revision of *Paraphrynus* Moreno (Amblypygida: Phrynidae) for North America and the Antilles. *Occas. Pap. Calif. Acad. Sci.,* 116: 1–80.

Murakami, Y., 1969. Myriapods found in limestone caves of northern Honshu, Japan. *Bull. Nat. Sci. Mus. Tokyo,* 12: 557–582.

Nielsen, C., Scharft, N. and Eibye-Jacobsen, D., 1996. Cladistic analyses of the animal kingdom. *Biol. J. Anim. Kingdom,* 57: 385–410.

Oromi, P. and Hernandez, J.J., 1986. Dos nuevas especies cavernicolas de *Domene* de Tenerife (Islas Canarias) (Coleoptera, Staphylinidae). *Fragm. Entomol.,* 19: 129–144.

Oromi, P. and Izquierdo, I., 1994. Canary Islands. In: C. Juberthie and V. Decu (Editors), *Encyclopaedia Biospeologica,* Vol. I. Société de Biospéologie, Moulis (France) and Bucarest, 1: 631–639.

Pagés, J. and Bareth, C., 1994. Les Diploures japygides. In: C. Juberthie and V. Decu (Editors), *Encyclopaedia Biospeologica,* Vol. I. Société de Biospéologie, Moulis (France) and Bucarest, 1: 280–283.

Palacios-Vargas, J.G., 1994. Mexique. In: C. Juberthie and V. Decu (Editors), *Encyclopaedia Biospeologica,* Vol. I. Société de Biospéologie, Moulis (France) and Bucarest, pp. 391–401.

Peck, S.B., 1975. A review of the New World Onychophora with the description of a new cavernicolous genus and species from Jamaica. *Psyche,* 82: 341–358.

Peck, S.B., 1994. Ecuador. In: C. Juberthie and V. Decu (Editors), *Encyclopaedia Biospeologica,* Société de Biospéologie, Moulis (France) and Bucarest, pp. 551–560.

Quintero Arias, D., 1986. Revision de la clasification de Amblypygidos pulvinados: creacion de subordenes, una nueva familla y un nuevo genero con tres nuevas especies (Arachnida: Amblypygi). In: W.G. Eberhard, Y.D. Lubin and B.C. Robinson (Editors), *Proceedings of the Ninth International Congress of Arachnology, Panama, 1983.* Smithsonian Institution Press, Washington, DC, pp. 203–212.

Rambla, M., 1993. *Maiorerus raucindoi* n. gen., n. sp., the first laniatorid from a Canary Island cave (Opiliones, Phalangodidae). *Mem. Biospéol.,* 20: 177–182.

Rambla, M. and Juberthie, C., 1994. Opiliones. In: C. Juberthie and V. Decu (Editors), *Encyclopaedia Biospeologica,* Vol. I. Société de Biospéologie, Moulis (France) and Bucarest, pp. 215–230.

Raven, R.J., 1985. The spider infraorder Mygalomorphae (Araneae): cladistics and systematics. *Bull. Am. Mus. Nat. Hist.,* 182: 1–180.

Reddell, J.R. and Cokendolpher, J.C., 1995. Catalogue, bibliography, and generic revision of the order Schizomida (Arachnida). *Tex. Mem. Mus. Speleol. Monogr.,* 4: 1–170.

Reddell, J.R. and Mitchell, R.W., 1971. A checklist of the cave fauna of Mexico. I. Sierra de El Abra, Tamaulipas and San Luis Potosi. *Bull. Assoc. Mex. Cave Stud.,* 4: 137–173.

Remane, R. and Hoch, H., 1988. Cave-dwelling Fulgoroidea (Homoptera: Auchenorrhyncha) from the Canary Islands. *J. Nat. Hist.,* 22: 403–412.

Rentz, D.C.F., 1994. Orthoptera. In: I.D. Naumann (Editor), *Systematic and Applied Entomology. An Introduction.* Melbourne University Press, Melbourne.

Rentz, D.C.F. and Kevan, D.K.McE., 1994. Dermaptera. In: I.D. Naumann (Editor), *Systematic and Applied Entomology. An Introduction.* Melbourne University Press, Melbourne, pp. 286–289.

Richards, A.M., 1961. Some observations on New Zealand cave-wetas. *Tuatara,* IX(2): 80–83.

Roth, L.M., 1994. Blattodea. In: I.D. Naumann (Editor), *Systematic and Applied Entomology. An Introduction.* Melbourne University Press, Melbourne, pp. 266–270.

Samways, M.J., 1994. *Insect Conservation Biology.* Chapman and Hall, London, 358 pp.

Selden, P.A., 1993. Fossil arachnids – recent advances and future prospects. *Mem. Queensl. Mus.,* 33: 389–400.

Serra, A., 1980. Chilopoda, Lithobiomorpha. In: Contribucion al conocimiento de la fauna cavernicola del Pais Vasco. *Kobie Bilbao* 10: 545–551.

Shear, W.A., 1969. A synopsis of the cave millipeds of the United States, with an illustrated key to genera. *Psyche,* 76: 126–143.

Shear, W.A., 1972. Studies in the milliped order Chordeumida (Diplopoda): a revision of the family Cleidogonidae and a reclassification of the order Chordeumida in the New World. *Bull. Mus. Comp. Zool.,* 141:151–352.

Sissom, W.D., 1990. Systematics, biogeography, and paleontology. In: G.A. Polis (Editor), *The Biology of Scorpions.* Stanford University Press, Stanford, CA, pp. 64–160.

Smith, G.B. and Watson, J.A.L., 1994. Thysanura. In: I.D. Naumann (Editor), *Systematic and Applied Entomology. An Introduction.* Melbourne University Press, Melbourne, pp. 245–247.

Thibaud, J.M. and Deharveng, L., 1994. Collembola. In: C. Juberthie and V. Decu (Editors), *Encyclopaedia Biospeologica,* Vol. I. Société de Biospéologie, Moulis (France) and Bucarest, pp. 267–276.

Tillyard, R.J., 1928. Some remarks on the Devonian fossil insects from the Rhynie chert beds, Old Red Sandstone. *Trans. Entomol. Soc. London,* 76: 65–71.

Weygoldt, P., 1994. Amblypygi. In: C. Juberthie and V. Decu (Editors), *Encyclopaedia Biospeologica,* Société de Biospéologie, Moulis (France) and Bucarest, pp. 241–247.

Chapter 5

CRUSTACEA

H.H. HOBBS III

INTRODUCTION

The subterranean environment historically has been considered simplistic, stable, and predictable. Stereotypical descriptions of the communities associated with this environment are generally made in terms of low diversity, low competition, relict species, high specialization, uniformity, and a high degree of convergence in numerous regressive characters of stygobionts and troglobites. In fact, the subterranean environment (not merely "caves") is indeed complex and is made up of a mosaic of microhabitats divisible into microcavernous (<1 mm diameter), mesocavernous (1–200 mm diameter), and macrocavernous (>200 mm diameter) categories of hole size (Howarth, 1983). Obviously, the two smaller divisions are not readily accessible by humans, yet they are commonly inhabited by numerous small biota. Many crustaceans are permanent residents of microcaverns (in part, "the interstitial fauna"); ostracods, calanoid and harpacticoid copepods, and syncarids are representative inhabitants of this immensely important, extensive, yet small habitat. This, and other parts of the subterranean complex are not isolated, but are connected to surface environments, and constitute a highly integrated, dynamic karst drainage system (see White et al., 1995) which may be strictly freshwater or, near or along coastal areas, brackish or saltwater. Near-shore subterranean biota may be found in a distinct habitat, called **anchialine** by Holthuis (1973), and defined to be those pools, occurring typically in volcanic and limestone bedrocks, that have no surface connection to the sea, yet contain brackish water and are under tidal influence. Pools are the obvious surface indicator of these habitats, which can contain a significant subterranean component and often have a lens of fresh or brackish water overlying salt water, and are inhabited, for example, by troglomorphic crustaceans such as remipedes, some copepods and mysids, as well as numerous caridean decapods.

Intensive study for more than a century, and particularly during the last two decades, indicates clearly that the global diversity of subterranean fauna, and specifically crustaceans, is very high. In spite of the characteristic limiting conditions common to most hypogean environments (see Holsinger, 1988; Hobbs, 1992), which include absence of light energy, low amplitude of variation for abiotic factors, and impoverished trophic resources (see Chapter 1 of this volume, and elsewhere), the subterranean biota is rich, crustaceans being particularly successful, with approximately 2700 stygobiont species occupying caves and other subterranean habitats.

Crustaceans are a diverse group of arthropods (approximately 40 000 recognized species) that are geographically abundant in epigean habitats but also have invaded the hypogean environment with great success (see Table 5.1). Five of the ten classes have representatives dwelling in caves as well as other hypogean habitats (Fig. 5.1) and, like numerous other cavernicoles, many species are stygobionts and demonstrate adaptations for existing in the nutrient-poor spelean environment (e.g., loss or reduction to rudiments of eyes and pigments, attenuated appendages, provision of sensory setae, lowered metabolic rate, K-reproductive strategy). The classes Branchiopoda, Remipedia, Copepoda, Ostracoda, and Malacostraca and their representative subterranean inhabitants are treated herein. The primitive Class Mystacocarida, considered to be morphologically conservative stygobionts (in the sense of Botosaneanu, 1986), is not included in this discussion since it does not inhabit "caves". Instead, it is a meiobenthic group, restricted to interstitial waters of intertidal sand beaches and subtidal sand substrates, which shares closest affinities

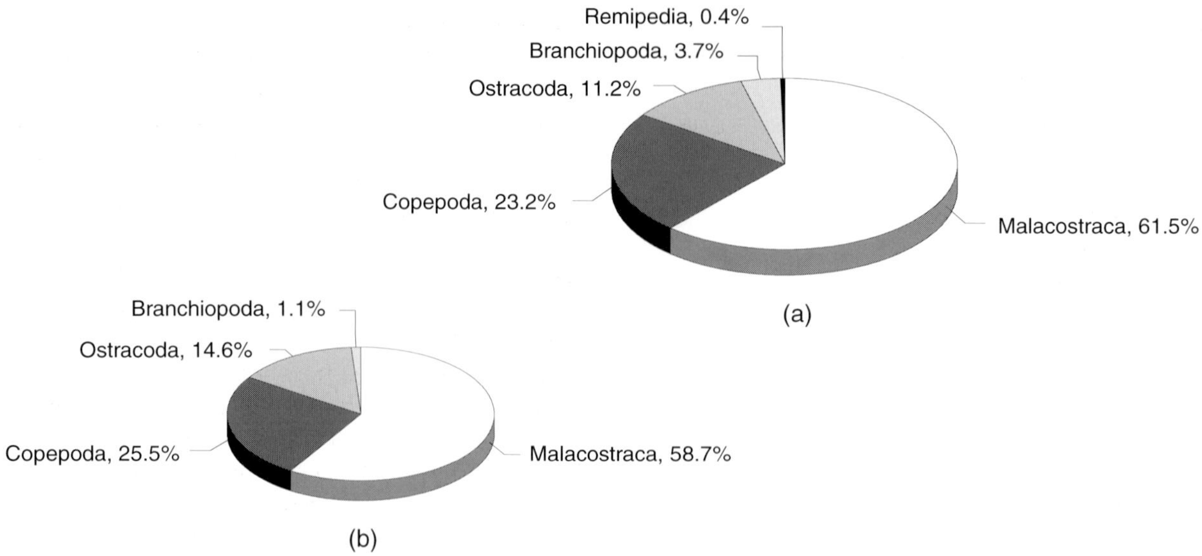

Fig. 5.1. Frequency distribution of species of the five classes of crustaceans having subterranean representatives. (a) Subterranean species; (b) all species. (Remipedia omitted from (b), since they represent only 0.028% of the total.)

to copepods, and includes 13 species and subspecies in two genera found in Africa, Europe, North and South America (Zinn, 1986), and Australia (Bartsch, 1993).

Although the subterranean crustacean fauna of the world has been studied for well over a century, many gaps exist in our understanding of their distribution, ecology, and evolution. It is generally accepted that most stygobionts evolved from epigean ancestors that had some degree of pre-adaptation for existence in subterranean environments. Although many stygobiont crustaceans are relicts, numerous obligates are recent invaders of actively developing karst. For the most part, speciation occurs as the result of selection pressure acting on the phenotype of the genotype of an isolated population, and this process is particularly evident on oceanic islands. However, it may be equally manifest, or nearly so, on "ecological islands" within a continental landmass, where caves in many ways resemble islands or archipelagos. The regressed troglomorphic state of obligate cavernicoles has and continues to be debated in terms of either natural selection (adaptation) or neutral theory (non-selection – genetic drift) (see Culver, 1982; Wilkens, 1992) and is a feature of crustaceans worldwide.

Most cavernicolous crustaceans are aquatic, but there are certainly terrestrial representatives (e.g., some isopods). Invasion of groundwaters by ancestral stocks came from shallow and deep marine and freshwater environments; the similarities of known species of various crustacean groups (e.g., *Typhlatya*) suggest a common origin, but their relationships are often obscured by convergence. Both active and passive dispersal of ancestral faunas, coupled with geologic structure and climatic conditions, played varying roles in the current distribution of subterranean crustaceans. Although many parts of the world are inhabited by these organisms, the central and southern European and Mediterranean regions as well as the southern North American and the West Indian regions support the greatest diversity. This is attributed in part to the southern portions of both regions being probable remnants of the Tethys Sea, much of the world's unglaciated, humid temperate and subtropical karst being located here, and it is likely that subterranean environments of these regions have long been available for founding populations of marine and freshwater ancestors. In part, because there are fewer and much less extensive karst areas, the taxonomic diversity of crustaceans is significantly lower in the Southern Hemisphere; but the stygofauna of northwestern Australia has a close affinity with those found on either side of the North Atlantic (Humphreys, Chapter 22, this volume). For more in-depth coverage of all crustaceans the reader is referred to the informative volumes edited by Botosaneanu (1986), and Juberthie and Decu (1994), and, for further discussion of decapods, to Hobbs et al. (1977) and Hobbs (1994).

Table 5.1
Classification of the Phylum Crustacea listing only those groups dwelling in cave or other subterranean habitats

CLASS Branchiopoda (850)[1] – mostly fresh water – benthos, plankton
    ORDER Cladocera (100 hypogean)
CLASS Remipedia (11) – anchialine caves
    ORDER Nectiopoda
CLASS Copepoda (10 000) – marine and fresh water – plankton, commensal parasitic, interstitial
    ORDER Platycopioida (2)
    ORDER Misophrioida (2)
    ORDER Calanoida (17)
    ORDER Cyclopoida (153)
    ORDER Harpacticoida (450+)
CLASS Ostracoda (5700) – mostly marine, some fresh water, few terrestrial; stygobionts (300+) mostly in the order Podocopida
CLASS Malacostraca (23 000) – marine, lotic and lentic fresh water, terrestrial
  SUPERORDER Syncarida
    ORDER Anaspidacea (3)
    ORDER Stygocaridacea (7)
    ORDER Bathynellacea (145)
  SUPERORDER Pancarida
    ORDER Thermosbaenacea (34)
  SUPERORDER Peracarida
    ORDER Mysidacea (17)
    ORDER Spelaeogriphacea (2)
    ORDER Mictacea (3)
    ORDER Isopoda (575+)
    ORDER Amphipoda (700+)
  SUPERORDER Eucarida
    ORDER Decapoda
      INFRAORDER Caridea (87)
      INFRAORDER Astacidea (40)
      INFRAORDER Anomura (2)
      INFRAORDER Brachyura (38)

[1] The numbers in parentheses are the numbers of species.

CLASS BRANCHIOPODA

The Class Branchiopoda is comprised of three subclasses, the Calmanostraca, Sarsostraca, and Diplostraca. The Calmanostraca and Sarsostraca are both known from the fossil record; extant species of Calmanostraca are placed in the order Notostraca and those of Sarsostraca in the Anostraca. The Recent subclass Diplostraca includes the Conchostraca and **Cladocera** and it is to this latter group that species occupying subterranean habitats (the hyporheos, the phreatic zone, and caves) are assigned. Approximately 11 families, 70 genera, and 450 species are within the order, yet fewer than 100 species are known to have invaded subterranean waters. Those that have are laterally compressed and quite small (0.2–2 mm), and are included in seven families, 24 species being considered stygobionts. Of these, the chydorids have the greatest number of stygobiont species, nl., four (Negrea, 1994).

Cladocera are known from subterranean waters (particularly the hyporheos) on all the continents; currently, their distribution is best known in Belgium, France, Hercegovina, Romania and Spain. These interstitial and cave-dwelling organisms have been studied only partially in Africa, the Americas, Asia, and Australia. Morphological adaptations to a subterranean existence are very minor except in only a few species, and most are considered to be of recent origin. The cavernicolous species *Alona hercegovinae* (Fig. 5.2a) and *A. sketi* lack eyes, and the carapace is translucent and sparsely pigmented. One may note that the epigean ancestor of these two species is limited for the most part to tropical and sub-tropical habitats; thus, these subterranean species are much older than other subterranean Cladocera, and probably entered the cave environment during Quaternary glaciation (Negrea, 1994).

CLASS REMIPEDIA

The remarkable Class Remipedia (Fig. 5.2b) includes 11 species, restricted to anchialine caves in the Bahamas, Cuba, the Yucatán Peninsula, the Canary Islands, and Western Australia (see Yager and Humphreys, 1996). The density interface between oligohaline waters and the underlying salt water appears to act as a barrier to these unique crustaceans, since *Speleonectes benjamini* has been observed in Sagittarius Cave on Grand Bahama Island only below the interface in completely dark, poorly oxygenated waters (as low as $0.08\,\mathrm{mg}\,\ell^{-1}$ at ~18 m depth) (Yager, 1994). The animals were seen to swim on their "backs" as a result of metachronal beating of the trunk appendages. These blind, elongate, centipede-like crustaceans, up to 45 mm in total length, with translucent bodies, are predators and are found associated typically with other stygobiont crustaceans (ostracods, mysids, thermosbaenaceans, cirolanid isopods, haziid

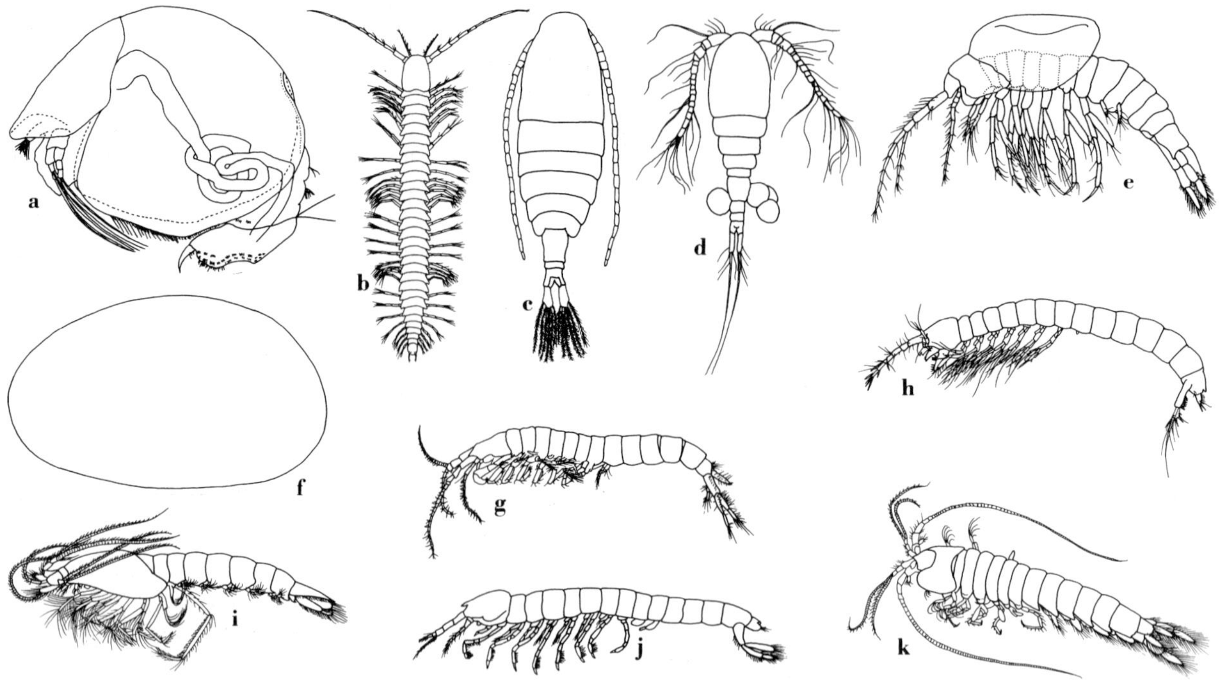

Fig. 5.2. Line drawings of lateral [l] or dorsal [d] views of representative unpigmented, blind, subterranean crustaceans: (a) cladoceran branchiopod [l] *Alona hercegovinae* (modified from Negrea, 1994; length 540 μm); (b) Remipedia [d] *Speleonectes lucayensis* (after Yager, 1986; length 19.8 mm); (c) calanoid copepod [d] *Stygodiaptomus kieferi* (after Rouch, 1994; length 3 mm); (d) cyclopoid copepod [d] *Kieferiella delamarei* (after Rouch, 1994; length 740 μm); (e) Thermosbaenacea [l] *Monodella sanctaecrucis,* female with empty brood pouch (modified from Stock, 1976; length 2 mm); (f) podocopid ostracod [l] *Sagittocythere barri* (after Hart and Hobbs, 1961; shell length 470 μm); (g) anaspidacean syncarid [l] *Psammaspides williamsi* (modified from Schminke, 1986; length 6.7 mm); (h) bathynellacean syncarid [l] *Notobathynella williamsi* (modified from Schminke, 1986; length 1.46 mm); (i) Mysidacea [l] *Spelaeomysis longipes* (modified from Bowman, 1986; length 5.8 mm); (j) stygocaridacean syncarid [l] *Stygocaris* sp. (after McLaughlin, 1980; length 500 μm); (k) spelaeogriphacean peracarid [l] *Spelaeogriphus lepidops* (modified from Botosaneanu, 1986; length 6.0 mm).

amphipods, and caridean shrimps). This little-known group was first discovered in 1979 in Lucayan Cavern in the Bahamas (Yager, 1981) and is among the most primitive of all extant crustaceans. Their closest known relative is an extinct Carboniferous species, *Tesnusocaris goldichi* (Schram, 1983, 1986).

CLASS COPEPODA

Most authors recognize some 10 000 copepod species assigned to ten or twelve orders, four of which contain species that are mostly parasitic, at least during some stage in their life history. The three remaining orders are generally free-living (**Calanoida**, **Cyclopoida**, and **Harpacticoida**), are distributed globally, and are the focus of this section, although some representative cyclopoids and harpacticoids also are parasitic (Williamson, 1991). The orders Platycopioida and Misophrioida have anchialine representatives. An additional order, **Gelyelloida**, was proposed by Huys (1988) to receive two species, *Gelyella droguei* and *G. monardi,* restricted to caves in France and Switzerland, respectively.

Order Calanoida

Approximately 2300 species are assigned to about 250 genera, yet only some 17 species in six families are known from subterranean waters (eight freshwater, nine brackish-anchialine-marine) in Madagascar, southeastern China, the Krym (Crimean Peninsula) in the Black Sea, Hercegovina, the western part of the Istra Peninsula on the Adriatic Sea, southern France, Philippines, central Mexico, Belize, Cuba, Barbuda, and Bermuda (Fosshagen and Iliffe, 1985, 1988; Bowman, 1986). Some are troglomorphic (mostly diaptomids), demonstrating morphological adaptations such as loss of pigmentation and eye color (Fig. 5.2c);

most are planktonic filter-feeders, yet some spend time in or on the benthos.

Order Cyclopoida

There are approximately 670 species and subspecies of freshwater cyclopoid copepods and, although they have not been well studied, Lescher-Moutoué (1986) reported 153 species from subterranean interstices, caves, and wells in Africa, Madagascar, Asia, Europe, and North and South America. Most of the species are comparatively recent invaders of the subterranean environment, yet some are descendants of tropical epigean ancestors dispersed during the Quaternary glaciation. Numerous cyclopoids are found in marine and brackish interstitial habitats, yet only a few of these species have been observed in caves (Herbst, 1986). Iliffe (1992) indicated that two speleoithonids are known from caves in the Bahamas, and that *Neocyclops stocki* has been observed in anchialine groundwater and caves in Bonaire, Columbia, and Bermuda. Absence of pigments and eyes are common traits of subterranean cyclopoids (Fig. 5.2d).

Order Harpacticoida

Some 375 genera containing nearly 3000 species in 40 families of harpacticoids are recognized, and at least 10 families have stygobiont species. Most freshwater stygobionts are assigned to the families Ameiridae, Canthocamptidae, and Parastenocarididae, and inhabit the interstitial habitats of caves as well as anchialine environments [Rouch (1994) reported 441 freshwater subterranean species]. They are vermiform in body shape, and primarily scrape food off the surface of a diversity of substrata. The distribution of this rich order and its successful invasion of subterranean environments are summarized well by Rouch (1994).

CLASS OSTRACODA

The Class Ostracoda is a very successful group of crustaceans made up of about 5700 planktonic and benthic, laterally compressed bivalve species; they are mostly marine, but are known from virtually all aquatic habitats. Numerous inland species are confined to the interstices of lentic and lotic substrates. At least 300 species of ostracods are known to be restricted to subterranean waters, most belonging to the Order Podocopida. Stygobiont ostracods demonstrate various morphological specializations, including reduction in eye structure, pigments, and the number of setae, and some have a very elongate or trapezoidal-shaped carapace. Most stygobiont species live in interstitial habitats where populations can become quite dense. However, within the podocopid family Entocytheridae, one subfamily (Sphaeromicolinae) lives exclusively in freshwater cave habitats (Hart and Hart, 1974) where the species are commensal with isopods (sphaeromatids in France, Italy, and Croatia and cirolanids in Mexico); another subfamily (Entocytherinae) has representatives that are ectocommensal on cambarid decapods (Figs. 5.2f, 5.3) inhabiting caves in North America

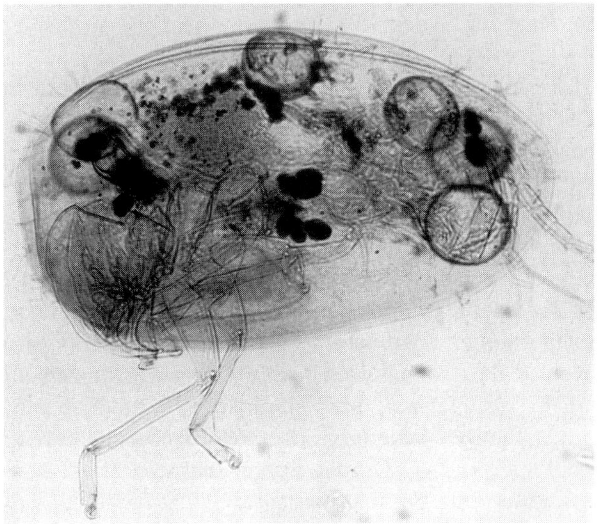

Fig. 5.3. Photograph of lateral view of the eyeless, unpigmented entocytherid ostracod, *Sagittocythere barri* infested with protozoans (male; anterior to right; shell length 450 μm).

(Hart and Hobbs, 1961). Stygobiont ostracods are known from Africa (Somalia, Zaire), North America (Cuba, Haiti, Mexico, United States), South America (Venezuela), Japan, South Korea, Afghanistan, India, Java and Western Malaysia, and from central and southern Europe. The origin of most inland hypogean ostracods (>120 species) can be traced to surface limnetic groups; only a small number have marine ancestors that adapted to subterranean freshwater environments (Danielopol and Hartman, 1986). The phylogeny of marine interstitial ostracods (about 200 species) is not as well understood. Iliffe (1992) summarized the biogeography of the anchialine species of ostracods (see also Baltanas and Danielopol, 1995), of which 17 are halocyprids, eight are myodocopids, 56 are podocopids, and one is a platycopid; some are clearly of deep-sea origin.

CLASS MALACOSTRACA

This is by far the largest class of crustaceans, with some 23 000 described species, the majority being assigned to the peracarids and eucarids (see below). Six major groups of orders (superorders – see Table 5.1) can be differentiated: Phyllocarida, Hoplocarida, Syncarida, Pancarida, Peracarida, and Eucarida. Of these, only the first two are not known to be represented by hypogean species.

Superorder Syncarida

The Syncarida is a freshwater group that evolved from marine stock. It is a relatively small taxon (probably because the group has received so little study) showing the total absence of any vestige of a carapace or cephalic shield. The superorder is divided into three orders: Anaspidacea, Stygocaridacea, and Bathynellacea. The **anaspidaceans** (Fig. 5.2g) are restricted to Tasmania and Australia, where they are found in surface streams and lakes as well as springs, caves, and even in the groundwater in the burrows of parastacid crayfish (pholeteros) (see Schminke, 1986). All species of the family Psammaspididae are stygobionts and are troglomorphic. The order **Stygocaridacea** (Fig. 5.2j) is known from South America and New Zealand, and has only four genera and seven species. They lack eyes and are inhabitants of interstitial fresh waters in both epigean and subterranean environments. The order **Bathynellacea** (Fig. 5.2h) was the first to move into the groundwater and originated by neoteny from surface ancestors. It is by far the richest of the three orders, with some 145 species known from the freshwater interstitial medium in epigean, well, and cave hypogean habitats in Europe, Africa, Madagascar, Asia, Japan, Australia, New Zealand, and North and South America. Of interest, *Bathynella* demonstrates the extreme in K-reproductive strategy by producing a single, large yolky egg.

Superorder Pancarida

This group (order **Thermosbaenacea**) has strong affinities with the Peracarida (see below), and is characterized by a short carapace (fused with the first thoracic somite) which, in females, provides a dorsal brood pouch (Fig. 5.2e). These small, blind crustaceans ($\leqslant$4 mm) are K-strategists and entirely stygobionts, inhabiting freshwater, oligohaline, to hypersaline waters in thermal springs and cisterns, wells, cenotes, caves, springs, interstices in coral and sand, and anchialine habitats. These sites are typically near coastal areas (but not, however, limited to them) and have been recorded from wells, caves, and anchialine habitats in Cambodia, Croatia, France, Spain, Italy, Greece, Balearic Islands (Mallorca), Sicily, the Canary Islands (Lanzarote), the British and U.S. Virgin Islands, Culebra (Costa Rica), Puerto Rico, Haiti, Cuba, Mexico (Yucatán Peninsula), Bahamas, the United States (Texas - Edwards Aquifer) and northwestern Australia. Four families, seven genera, and some 34 species are recognized in this group (Wagner, 1994), and one genus (*Halosbaena*) has a wide distribution (Poore and Humphreys, 1992) corresponding to the Miocene extent of the Tethyan basin (Stock, 1986; Yager, 1987; Bowman and Iliffe, 1988).

Superorder Peracarida

This successful group of crustaceans has a carapace that is never fused to all the thoracomeres (the carapace in isopods and amphipods has been lost secondarily), forms brood chambers from outgrowths of the pereiopods, and includes seven orders: Mysidacea, Cumacea, Spelaeogriphacea, Tanaidacea, Mictacea, Isopoda, and Amphipoda; the cumaceans and tanaidaceans have no known subterranean species.

Order Mysidacea

This rich order (about 800 species) is widely distributed in marine environments (open and coastal), brackish, freshwater, and various groundwater habitats, including land-crab burrows. Because the known distribution of most groundwater and surface freshwater species (except for the widely distributed relict *Mysis relicta*) is in close proximity to various shorelines, this suggests that the freshwater/anchialine mysids, as a whole, evolved recently from marine ancestors. Of interest, most carcinologists agree that the mysids and amphipods (see below) are more closely related to each other than to the isopods (Schram, 1986). Two suborders are recognized: the Lophogastrida are strictly pelagic and have no stygobiont representatives, and the Mysida are grouped into four families, three of which have stygobiont species (Lepidomysidae, Stygiomysidae, and Mysidae) which are blind and unpigmented.

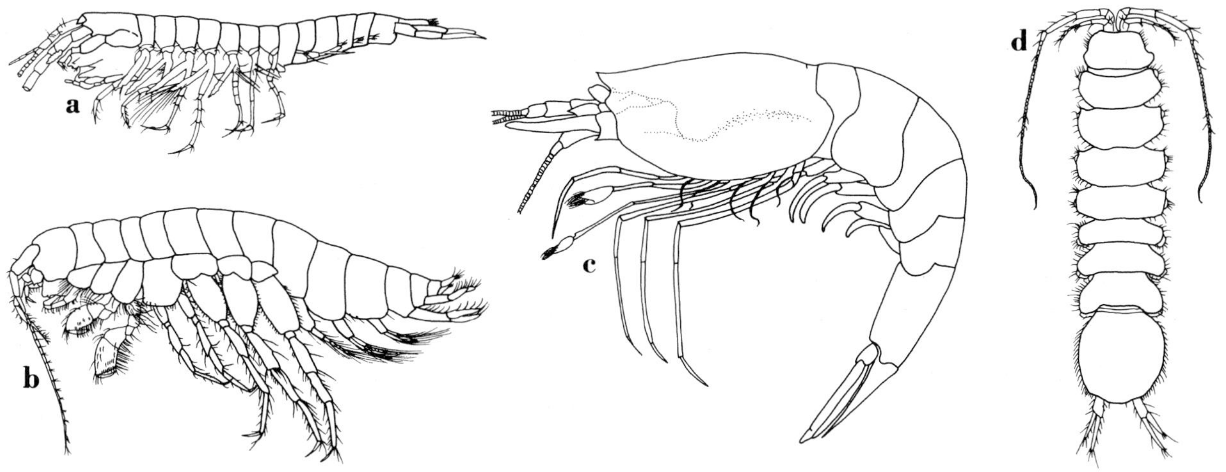

Fig. 5.4. Line drawings of lateral [l] or dorsal [d] views of representative unpigmented, blind, subterranean crustaceans: (a) mictacean peracarid [l] *Mictocaris halope* (after Bowman and Iliffe, 1985; length 3.0 mm); (b) amphipod peracarid [l] *Stygobromus exilis* (modified from Holsinger, 1994; length 6.0 mm); (c) caridean decapod [l] *Typhlatya mitchelli* (after Hobbs et al., 1977; postorbital carapace length 4.8 mm); (d) isopod peracarid [d] *Proasellus valdensis* (modified from Coineau et al., 1994; length 8.0 mm).

The primitive genus *Spelaeomysis* (Fig. 5.2i), is widely distributed in fresh coastal cave and anchialine waters, and shows varying degrees of eye regression and depigmentation. Stygophilic representatives also are known within the Lepidomysidae as well as from the Mysidae (see Pesce et al., 1994).

At least 17 stygobiont species are known from coastal caves, wells (phreatic water), blue holes, and anchialine caves in southern Italy, the Balearic Islands, Bosnia/Hercegovina, India, the Canary Islands, the Lesser Antilles, Cuba, Jamaica, Puerto Rico, Bahamas, Turks and Caicos Islands, and Mexico (including the Yucatán Peninsula). Pesce et al. (1994) have summarized what is known of the ecology and behavior of the cave mysids.

Order Spelaeogriphacea

Until recently only a single species of this order was known, *Spelaeogriphus lepidops* (Fig. 5.2k), known only from the type locality (freshwater stream in Bats Cave, Table Mountain, South Africa); a second genus (*Potiicoara*) has now been found in Brazil (Pires, 1987). These blind, unpigmented stygobionts have a small saddle-like carapace; this, coupled with the reduced nature of the thoracopods and pleopods, suggests paedomorphosis as a factor in spelaeogriphacean evolution. In the type locality *S. lepidops* was observed swimming swiftly using rapid undulations of the body, and it may be a detritivore (Gordon, 1957). Although its evolutionary development is unclear, this crustacean shows affinities with Tanaidacea, Syncarida, and also with Thermosbaenacea.

Order Mictacea

This is the most recently described order of peracarids (Bowman et al., 1985), and contains three known species in two families (all lacking gills). Two deep-sea benthic species are placed in the Hirsutiidae and a single blind, unpigmented stygobiont, *Mictocaris halope* (Fig. 5.4a), was described by Bowman and Iliffe (1985) from marine caves on Bermuda and assigned to the family Mictocarididae. It is usually observed swimming in the water column, rarely resting or walking on the substrata of anchialine caves. Although the food of *M. halope* is unknown, the mouth-parts suggest that it probably is not a predator, but may be a scraper.

Order Isopoda

These dorsoventrally compressed crustaceans lack a carapace, have pleopods adapted for respiration, and are extremely successful, with some 11 000 known species. They are grouped into 11 suborders, eight of which have subterranean representatives (Anthuridea, Microcerberidea, Cymothoidea, Sphaeromatidea, Asellota, Calabozoidea, Phreatoicidea, and Oniscidea), the oniscids being terrestrial (in part) and the others predominantly aquatic (see Juberthie and Decu, 1994). The four families of **anthurideans** are primarily marine; about 20 blind species in two families (Anthuridae and Paranthuridae) have been reported (especially from the interstitial media of sediments) from rivers, bays, beaches, wells, anchialine habitats, and caves in Indonesia, the Indian Ocean islands, the Canary Islands, the Caribbean, Mexico, and South America. The slender, blind, translucent **Microcerberidea** are stygobionts inhabiting interstices of marine beaches as well as caves and wells. About 60 species within six genera are placed in a single family, Microcerberidae, found mostly near coastal areas in the Mediterranean, southern Africa, islands in the Indian Ocean, southeast Asia, Japan, and the Caribbean. The continental species are the most primitive, suggesting that during the lower Cretaceous certain groups regressed to the sea. Among **cymothoideans**, the predominantly marine Cirolanidae are the only one of ten families with subterranean representatives. There are about 300 species recognized in the Cirolanidae, yet only 50 of these are subterranean, clearly derived from marine forms. They are blind, without pigments, and are somewhat convex dorsally (some stygobionts are capable of rolling into a ball when disturbed), and most are benthic, opportunistic scavengers. The subterranean species (some occupying anchialine habitats), have a peri-Caribbean, a peri-Ponto-Mediterranean, and a small East African distribution. In addition, disjunct species in northwestern Australia (see Bruce and Humphreys, 1993), Bermuda, and the Appalachians in North America are recognized. The mostly marine **sphaeromatideans** have some 36 stygobiont representatives in the family Sphaeromatidae that are restricted to karst waters in southern Europe. They are detritus feeders and/or scrapers and are considered to be marine relicts in cave waters.

The **Asellota** are divided into five superfamilies: Aselloidea, Stenetrioidea, Gnathostenetroidoidea, Protojanoidea, and Janiroidea. The family **Asellidae**, one of three families of the Aselloidea, is a large benthic group (about 260 species in 15 genera) that is found primarily in caves, wells, and springs (Fig. 5.4d) in Japan, Europe, northwestern Africa, and North and Central America. The origin of most species is from various epibenthic marine forms, some of which moved into epigean, and later, as surface and subterranean hydrologic systems developed, into hypogean freshwater environments. The **Atlantasellidae** consists of a single species, *Atlantasellus cavernicolus*, which inhabits anchialine caves in Bermuda; its marine origin is evident. The **Stenasellidae** represents a very old group (with plesiomorphic characters) that are blind and often are orange or rose in color. Species of the Stenasellidae inhabit thermal springs, wells, hyporheos, springs, and caves in western, southern, and eastern Europe, much of Africa, central and southeastern Asia, Malaysia, Indonesia, and Mexico and the Edwards Plateau in North America. It is likely that the ancestors of the West African and American forms invaded continental waters prior to or at the time of the breakup of the two continents. Various preadapted stocks moved through the channels and alluvia of streams, invading hypogean waters of interstices, crevices, and ultimately caves. In some areas, recolonization of subterranean habitats becoming available after glaciation occurred during the past 10 000–15 000 years. The superfamilies **Gnathostenonetroidoidea** and **Stenetrioidea** are small isopods of marine origin, the latter from mixohaline waters in a cave on Curaçao and the former showing a disjunct distribution in the Bahamas (anchialine cave), Japan (intertidal zone), and a small coastal island west of Naples, Italy. The **protojaniroideans** are composed of three genera and six species that are quite separate from other members of the Asellota. From surface fresh water they invaded subterranean waters in southern Africa and southern India. The **janiroideans** are mostly marine, but two of the 20 families include marine-derived stygobionts: **Janiridae** are known from spring, well, and cave waters in eastern Asia, Japan, Italy, western Europe, and California (U.S.A.). The **Microparasellidae** are widely known from interstitial waters in Polynesia, Indonesia, Asia, the Middle East, western, central, and eastern Europe and some of the Mediterranean islands, northern and western Africa, Madagascar, southern India, islands of the Indian Ocean, Galapagos Islands, Bonaire, Cuba, and Belize.

The monotypic suborder **Calabozoidea** is known only from a phreatic system in northern Venezuela. This blind isopod shows a mixture of plesiomorphic and apomorphic characters, and is considered to be most closely related to the suborder Asellota.

The suborder **Phreatoicidea** are blind, albinistic crustaceans known from wells, caves, and the pholeteros as well as epigean environments in New Zealand, Australia, India, and southern Africa. This is the most ancient group of isopods, dating back to the Carboniferous. They invaded the surface fresh waters before the breakup of Gondwana and, in fact, most species are epigean.

The suborder **Oniscidea** is essentially a terrestrial group, although some stygobiont species of the family Trichoniscidae inhabit cave waters. They have their closest affinities with the Calabozoidea and 34 families and approximately 3500 species are assigned to this group of isopods; however, only 14 families have subterranean representatives. These are considered to be relicts of an ancient fauna once widely distributed. As glaciers advanced during the Pleistocene, holarctic species invaded the subterranean environment, and remain as relict stygobionts. They are global in distribution but occur particularly in North America, the Greater Antilles, all of southern Europe, the Middle East, and northwestern Africa.

Order Amphipoda

These laterally compressed crustaceans, like the isopods, lack a carapace and, although a few (e.g., Talitridae) dwell in terrestrial habitats, they are particularly abundant in many marine, brackish, and freshwater ecosystems. The order is composed of some 6700 species placed in three suborders: Gammaridea, Caprellidea, and Hyperiidea.

The **Gammaridea**, with 95 families, over 1000 genera, and approximately 5700 species is clearly the largest amphipod suborder, and is the only one including subterranean species. Most of those inhabiting groundwater ecosystems are obligate cavernicoles, showing troglomorphic adaptations that evolved from epigean ancestors preadapted to life in subterranean waters. The ancestral invasion of groundwater undoubtedly involved active and passive dispersal from shallow marine and various freshwater environments. Habitats of extant stygobiont species include the sediment interstices, open waters in meso- and micro-caverns (including lotic and lentic environments), as well as marine and brackish waters of anchialine caves. Like so many cave-dwelling crustaceans, the primary nutrients of cavernicolous amphipods are derived from allochthonous materials and associated microbial communities; no parasitic amphipods are known from hypogean waters.

Stygobiont amphipods (>700 species) are assigned to some 140 genera in 36 families (or family groups). However, 12 families contain more than 90% of these obligate subterranean crustaceans, and the families Bogidiellidae, Crangonyctidae (Fig. 5.4b), the Hadziidae complex and Niphargidae show the greatest species richness (Holsinger, 1993, 1994). The two largest genera, *Niphargus* (mostly European) and *Stygobromus* (mostly North American) are exclusively stygobionts, and each has more than 100 recognized species. One may note that the holarctic stygobionts in the Gammaridae are probably derived from freshwater stocks, which in karst regions migrated upstream into caves from surface springs during Pleistocene and Recent times (Holsinger, 1986). Culver et al. (1995) examined the role of evolutionary history and natural selection in molding the morphology of the amphipod *Gammarus minus,* found in springs and caves. They suggested that genetic isolation in Appalachian cave populations is quite pronounced, and that invasions of caves in distinct hydrologic systems were independent of each other.

The **Niphargidae** includes eight genera, almost all of which are freshwater stygobionts; their ancestors probably originated in shallow marine basins of the Paratethys during the Tertiary, and spread north from southeastern Europe through hypogean waters of central and western Europe. *Niphargus* has the widest geographic distribution and the greatest number of species in the family, and is found in much of central and southern Europe. Central Europe also is inhabited by bogidiellids and crangonyctids; yet in southern Europe the niphargids are sympatric with bogidiellids, gammarids, hadziids, ingolfiellids, metacrangonyctids, pseudoniphargids, and salentinellids. In northern Africa the niphargids are absent and the metacrangonyctids and pseudoniphargids are the major families represented.

The majority of the species of **Crangonyctidae** (derived from epigean freshwater ancestors) occur in North America (mostly U.S.A., but at least three species from Canada); but the family is also represented in many parts of the Holarctic region. Six genera are recognized; only *Bactrurus, Stygobromus,* and *Stygonyx* are strictly stygobionts, whereas epigean and hypogean species are assigned to the genera *Crangonyx* and *Synurella.* In Mexico the hypogean amphipods are made up of bogidiellids and **hadziids**, but on the Yucatán Peninsula only hadziids occur. In the Caribbean, and particularly throughout the Greater Antilles, the hadziids are most diverse and abundant. The hadziids are mostly marine, and have invaded interstitial anchialine habitats, and freshwater inland karst areas formerly exposed to shallow marine embayments. Currently, 42 genera and 105 species are recognized. The family **Bogidiellidae** is probably an ancient group that is widely distributed globally in continental fresh waters, embayment areas, islands and coastal zones, and even in the interstices of sublittoral sands. It is made up solely of stygobiont species occurring in the interstitial of many groundwater habitats.

Hypogean amphipods are known from at least 31 additional families or family groups, and are found in virtually all parts of the world with suitable groundwater habitats: South America (e.g., bogidielloids), islands in the Atlantic (e.g., bogidiellids, pseudoniphargids) and in the Indo-Pacific (e.g., bogidiellids, hadziids, melitids), eastern Asia (e.g., pseudocrangonyctids), eastern Africa (e.g., bogidiellids, hadziids), southern Africa (e.g., ingolfiellids, sternophysingids), southern Australia (e.g., paramelitids), northwestern Australia (bogidiellids, haziids, melitids), and New Zealand (e.g., paracrangonyctids, phreatogammarids). Yet two regions stand out as supporting the greatest taxonomic diversity and richness of stygobiont amphipods: central and southern Europe and the Mediterranean region; and eastern and southern North America, as well as the Caribbean region. Geographically and ecologically both areas have been available optimally for colonization by populations from marine and freshwater environments. It is of interest that a recent collection of 11 individuals on Barrow Island, Western Australia, included nine species and three families – 1% of the world's known stygobiont amphipods! More have been found since (see Humphreys, Chapter 30, this volume; Bradbury and Williams, 1996a, b).

Superorder Eucarida

The superorder Eucarida is the most highly evolved group of malacostracans, and is characterized by a carapace that is fused to all the thoracomeres to form a cephalothorax. Two of the three orders, **Euphausiacea** (lacking maxillipeds) and **Amphionidacea**, are worldwide open-water marine crustaceans, the former represented by about 85 known epipelagic planktonic species and the latter by a single circumglobal species.

The third order, **Decapoda**, constitutes the largest and most diverse order, with nearly 10 000 species recognized. Generally, members of this eucarid order possess three pairs of maxillipeds, which leaves five pairs of functional pereiopods (hence the name Decapoda), and the group is split into two suborders: **Dendrobranchiata** (mostly penaeid and sergestid shrimps) with dendrobranchiate gills (synapomorphy), and **Pleocyemata** (all remaining decapods) never possessing dendrobranchiate gills. This latter suborder is broken generally into seven infraorders, but only the carideans, astacideans, anomurans, and brachyurans are found in subterranean environments. The infraorder Palinura is represented by a single possibly stygoxenic palinurid, *Jasus edwardsii*. The stygoxenic *Penaeus indicus*, a penaeid shrimp inhabiting an East African cave, is the only known cave-dwelling dendrobranchiate decapod (Holthuis, 1956).

Nearly 2000 species of natant, pleocyemate decapods are assigned to the infraorder **Caridea** (shrimps; Fig. 5.4c). Of these, some 87 obligate, troglomorphic species in 35 genera and six families inhabit subterranean waters (springs, wells, caves, blue holes, cenotes, and anchialine habitats) in North America (southern U.S.A. and Mexico); Caribbean and western Atlantic islands (Bermuda, Bahamas, Caicos, Greater and Lesser Antilles); Ascension Island; north, east, and central Africa; Madagascar; Middle East; southern Europe; Georgia; India; Indian Ocean islands; Japan; Philippine Islands; Indonesia; Western Australia; Micronesia; and Melanesia. Because of the vast size of the Pacific and its many islands, the caridean fauna (and the crustacean fauna generally, for that matter) is still incompletely known. For years disjunct distributions of many closely related species of caridians have posed questions to crustacean biogeographers, but proposed explanations have not been tested rigorously (see Hobbs, 1994). Many of these hypogean species are completely blind and lack any pigmentation (e.g., *Typhlatya mitchelli*; Fig. 5.4c), whereas some subterranean shrimps have functional eyes and are orange to red in body coloration (e.g., *Barbouria cubensis*). An additional 38 species inhabit global subterranean waters but are not obligate cave species with troglomorphic adaptations. The caridians are a polyphyletic group, being distributed globally with many obligate cavernicolous species derived from marine ancestors, that are concentrated in tropical latitudes (see above).

The infraorder **Astacidea** consists of crayfishes and clawed lobsters, some 900 species having trichobranchiate gills and chelate pereiopods (first three pairs). Only the astacidean crayfishes occupy subterranean waters, and their lineage can be traced back to the Late Permian. Widespread radiation during the Mesozoic and current distribution patterns suggest dispersal associated with continental movements. For instance, the superfamily Astacoidea is primarily Laurentian and the Parastacoidea primarily Gondwanan in distribution. Of these, only the astacoideans have invaded the subterranean environment, and include 40 obligate, troglomorphic species (Fig. 5.6) assigned to four genera (*Cambarus, Orconectes, Procambarus, Troglocambarus*) in one family, the **Cambaridae**

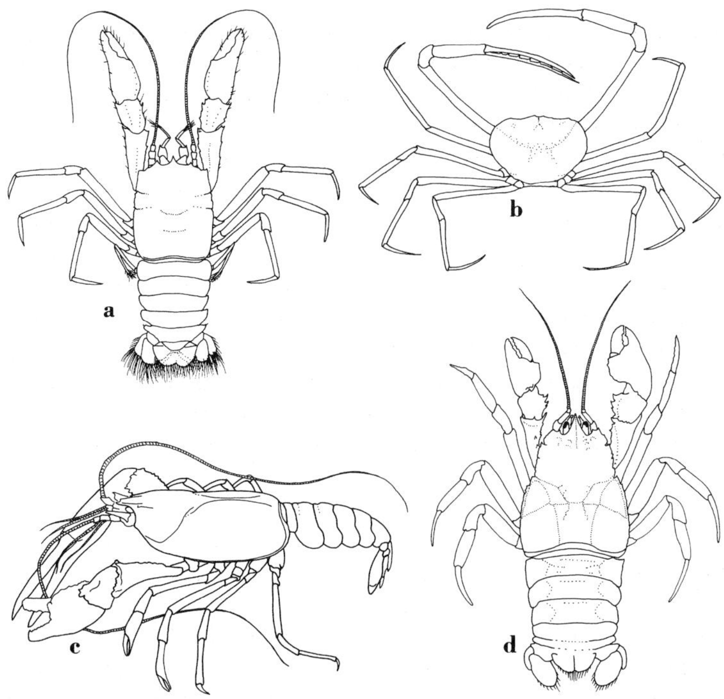

Fig. 5.5. Line drawings of dorsal [d] or dorsolateral [dl] views of representative unpigmented, subterranean decapod crustaceans: (a) galatheid anomuran [d] *Munidopsis polymorpha* (modified from Holthuis, 1986); (b) brachyuran [d] *Typhlopseudothelphusa acanthochela* (modified from Hobbs, 1986; carapace width 37.5 mm); (c) astacidean [dl] *Cambarus (Jugicambarus) setosus* (carapace length 40.5 mm); (d) aeglid anomuran [d] *Aegla cavernicola* (modified from Hobbs et al., 1977; carapace length 15 mm).

Fig. 5.6. Photograph of the stygobiont crayfish *Orconectes (Orconectes) inermis inermis*. (carapace length 31.2 mm).

(Fig. 5.5c). Some 37 additional crayfishes are inhabitants of caves, yet show no troglomorphic adaptations. Sixteen stygobionts belonging to the genera *Cambarus, Procambarus,* and *Troglocambarus* are known from the state of Florida (U.S.A.). This represents the most diverse stygobiont crayfish fauna in the world, and results from subterranean invasion by at least six different epigean species initiated as far back as the late Miocene or Pliocene (see Franz and Lee, 1982).

The infraorder **Anomura** includes hermit crabs, porcelain crabs, mole and sand crabs, as well as a few additional crab-like decapods, most of which are marine and have a reduced fifth pereiopod. One species has been observed in caves in Saipan, and another on Aruba, Bonaire, and Curaçao; neither shows any degree of troglomorphy. The superfamily Galatheoidea includes two families having subterranean species: Aeglidae and Galatheidae. Very little is understood about the aeglid *Aegla cavernicola* (Fig. 5.5d), known only from Grutas das Areias, São Paulo, Brazil where it was collected 300 m from the entrance. There are about 30 species in the family, all restricted to South America; but only *A. cavernicola* is stygobiont. *Munidopsis polymorpha* (Fig. 5.5a) is the only galatheid to occupy the subterranean environment. It is the most abundant organism in a 5 km long anchialine lava tube (Jameos del Aqua) on Lanzarote in the Canary Islands, and is clearly of marine stock, probably descendant of a widely distributed deep-water ancestor. Feeding on diatoms occurs primarily during the night, indicating a light-driven circadian rhythm, and males actively seek females using rhythmic movements of their chelipeds as "communication". It is an aggressive species and, while resting or feeding, individuals maintain a minimal distance (depending on the length of the second antenna) from one another; their agonistic behavior is demonstrated by the extended cheliped attack (Parzefall and Wilkens, 1975).

The infraorder **Brachyura**, the so-called true crabs, are distinctly flattened dorsoventrally and have phyllobranchiate gills. At least 4500 extant species are recognized, of which nine families, 23 genera, and 38 species are associated with subterranean waters (Fig. 5.5b) in the Philippines, Thailand, Indonesia, Melanesia, Madagascar, Jamaica, Guatemala, southern Mexico, Belize, northwestern South America, and the Galapagos Islands. These stygobionts have multiple origins; some are traced to marine ancestors that invaded fresh waters, others moved into anchialine habitats from the marine environment, and many moved into hypogean freshwaters from inland epigean sources. They show variations in degree of troglomorphy, ranging from well-developed eyes and orange carapace coloration, indicating probable Holocene invasion of subterranean waters (e.g., *Phaibulamon stilipes* in Thailand), to complete lack of pigments, blindness and gross reduction of eyes, and elongated pereiopods – regressive traits implying long-term hypogean existence (e.g., *Typhlopseudothelphusa* in Guatemala and *Cancrocaeca xenomorpha* in Sulawesi, Indonesia). Guinot (1994) presented graphic illustrations showing the regressive character of the development of eyes in representative stygobiont crabs. In addition, at least nine crabs occupy anchialine habitats; 57 species are inhabitants of hypogean waters as well as subterranean terrestrial environments, but are not obligate cavernicoles.

REFERENCES

Baltanas, A. and Danielopol, D.L., 1995. Cladistic analysis of *Danielopolina* species (Ostracoda: Thaumatocyprididae) and the origin of anchialine fauna. *Mitt. Hamb. Zool. Mus. Inst.*, 92: 315–324.

Bartsch, I., 1993. *Ctecheilocaris echra*, a new mystacocarid (Crustacea) from Western Australia. In: F.E. Wells, D.I. Walker, H. Kirkman and R. Lethbridge (Editors), *The Marine Flora and Fauna of Rottnest Island, Western Australia*, Vol. 1. Western Australia Museum, Perth, pp. 105–111.

Botosaneanu, L. (Editor), 1986. *Stygofauna Mundi. A Faunistic, Distributional, and Ecological Synthesis of the World Fauna Inhabiting Subterranean Waters (Including the Marine Interstitial)*. E.J. Brill, Leiden, vi + 740 pp.

Bowman, T.E., 1986. Copepoda: Calanoida. In: L. Botosaneanu (Editor), *Stygofauna Mundi*. E.J. Brill, Leiden, pp. 295–298.

Bowman, T.E. and Iliffe, T.M., 1985. *Mictocaris halope*, a new unusual peracaridan crustacean from marine caves on Bermuda. *J. Crustacean Biol.*, 5: 58–73.

Bowman, T.E. and Iliffe, T.M., 1988. *Tulumella unidens*, a new genus and species of thermosbaenacean crustacean from the Yucatan Peninsula, Mexico. *Proc. Biol. Soc. Washington*, 101: 221–226.

Bowman, T.E., Garner, S.P., Hessler, R.H., Iliffe, T.M. and Sanders, H.L., 1985. Mictacea, a new order of Crustacea Peracarida. *J. Crustacean Biol.*, 5: 74–78.

Bradbury, J.H. and Williams, W.D., 1996a. Freshwater amphipods from Barrow Island, Western Australia. *Rec. Aust. Mus.*, 48(1): 33–74.

Bradbury, J.H. and Williams, W.D., 1996b. Two new species of anchialine amphipod (Crustacea: Hadziidae: *Liagoceradocus*) from Western Australia. *Rec. West. Aust. Mus.*, 17: 395–409.

Bruce, N.L. and Humphreys, W.F., 1993. *Haptolana pholeta* sp. nov., the first subterranean flabelliferan isopod crustacean (Cirolanidae) from Australia. *Invertebr. Taxon.*, 7: 875–884.

Coineau, N., Henry, J.-P., Magniez, G. and Negoescu, I., 1994. Isopoda Aquatica. In: C. Juberthie, and V. Decu (Editors), *Encyclopaedia Biospeologica*, Vol. I. Société de Biospéologie, Moulis (France) and Bucarest, pp. 123–140.

Culver, D.C., 1982. *Cave Life: Evolution and Ecology*. Harvard University Press, Cambridge, MA, 189 pp.

Culver, D.C., Kane, T.C. and Fong, D.W., 1995. *Adaptation and Natural Selection in Caves: The Evolution of Gammarus minus*. Harvard University Press, Cambridge, MA, 223 pp.

Danielopol, D.L. and Hartman, G., 1986. Ostracoda. In: L. Botosaneanu (Editor), *Stygofauna Mundi*. E.J. Brill, Leiden, pp. 265–295.

Fosshagen, A. and Iliffe, T.M., 1985. Two new genera of Calanoida and a new order of Copepoda, Platycopioida, from marine caves on Bermuda. *Sarsia*, 70: 345–348.

Fosshagen, A. and Iliffe, T.M., 1988. A new genus of Platycopioida (Copepoda) from a marine cave on Bermuda. *Hydrobiologia*, 167/168: 357–361.

Franz, R. and Lee, D.S., 1982. Distribution and evolution of Florida's stygobiont crayfishes. *Bull. Fla. State Mus., Biol. Sci.*, 28: 53–78.

Gordon, I., 1957. On *Spelaeogriphus*, a new cavernicolous crustacean from South Africa. *Bull. Brit. Mus. Nat. Hist. Zool.*, 5: 29–47.

Guinot, D., 1994. Decapoda Brachyura. In: C. Juberthie and V. Decu (Editors), *Encyclopaedia Biospeologica*, Vol. I. Société de Biospéologie, Moulis (France) and Bucarest, pp. 165–179.

Hart Jr., C.W. and Hobbs Jr., H.H., 1961. Eight new troglobitic ostracods of the genus *Entocythere* (Crustacea, Ostracoda) from the eastern United States. *Proc. Acad. Nat. Sci. Philadelphia*, 113(8): 173–185.

Hart, D.G. and Hart Jr., C.W., 1974. The ostracod family Entocytheridae. *Monogr. Acad. Nat. Sci. Philadelphia*, 18: 1–239.

Herbst, H.V., 1986. Copepoda: Cyclopoida aus dem Meeres- und Brackwasser-Interstitial. In: L. Botosaneanu (Editor), *Stygofauna Mundi*. E.J. Brill, Leiden, pp. 313–320.

Hobbs III, H.H., 1992. Caves and Springs. In: C.T. Hackney, S.M. Adams and W.H. Martin (Editors), *Biodiversity of the*

Southeastern United States: Aquatic Ecosystems. Wiley, New York, pp. 59–131.
Hobbs III, H.H., 1994. Biogeography of subterranean decapods in North and Central America and the Caribbean region (Caridea, Astacidea, Brachyura). *Hydrobiologia,* 287: 95–104.
Hobbs Jr., H.H., 1986. A new troglobitic crab (Crustacea: Decapoda: Pseudothelphusidae) from Belize. *Tex. Mem. Mus. Speleol. Monogr.,* 1: 1–4.
Hobbs Jr., H.H., Hobbs III, H.H. and Daniel, M.A., 1977. A review of the troglobitic decapod crustaceans of the Americas. *Smithson. Contrib. Zool.,* 244: 1–183.
Holsinger, J.R., 1986. Zoogeographic patterns of North American subterranean amphipod crustaceans. In: R.H. Gore and K.L. Heck (Editors), *Crustacean Biography (Crustaceans Issues 4).* Balkema, Rotterdam, pp. 85–106.
Holsinger, J.R., 1988. Stygobionts: the evolution of cave-dwelling organisms. *Am. Sci.,* 76: 146–153.
Holsinger, J.R., 1993. Biodiversity of subterranean amphipod crustaceans: global patterns and zoogeographic implications. *J. Nat. Hist.,* 27: 821–835.
Holsinger, J.R., 1994. Amphipoda. In: C. Juberthie and V. Decu (Editors), *Encyclopaedia Biospeologica,* Vol. I. Société de Biospéologie, Moulis (France) and Bucarest, pp. 147–163.
Holthuis, L.B., 1956. An enumeration of the Crustacea Decapoda Natantia inhabiting subterranean waters. *Vie Milieu,* 7: 43–76.
Holthuis, L.B., 1973. Caridean shrimp found in land locked saltwater pools at four Indo-West Pacific localities (Sinai Peninsula, Funafuti Atoll, Maui, and Hawaii Islands), with the description of one new genus and four new species. *Zool. Verh.,* 128: 1–48.
Holthuis, L.B., 1986. Decapoda. In: L. Botosaneanu (Editor), *Stygofauna Mundi.* E.J. Brill, Leiden, pp. 589–615.
Howarth, F.G., 1983. Ecology of cave arthropods. *Annu. Rev. Entomol.,* 28: 365–389.
Huys, R., 1988. Gelyelloida, a new order of stygobiont copepods from European karstic systems. *Hydrobiologia,* 167/168: 485–495.
Iliffe, T.M., 1992. Anchialine cave biology. In: A.I. Camacho (Editor), *The Natural History of Biospeleology.* National Museum of Natural Sciences, Madrid, pp. 613–636.
Juberthie, C. and Decu, V. (Editors), 1994. *Encyclopaedia Biospeologica,* Vol. I. Société de Biospéologie, Moulis (France) and Bucarest, 834 pp.
Lescher-Moutoué, F., 1986. Copepoda Cyclopoida Cyclopidae des Eaux Douces Souterraines Continentales. In: L. Botosaneanu (Editor), *Stygofauna Mundi.* E.J. Brill, Leiden, pp. 299–312.
McLaughlin, P.A., 1980. *Comparative Morphology of Recent Crustacea.* W.H. Freeman, San Francisco, CA, 177 pp.
Negrea, S., 1994. Cladocera, In: C. Juberthie and V. Decu (Editors), *Encyclopaedia Biospeologica,* Vol. I. Société de Biospéologie, Moulis (France) and Bucarest, pp. 99–104.
Parzefall, J. and Wilkens, H., 1975. Zur Ethologie augenreduzierter Tiere. Untersuchungen an *Mundiopsis polymorpha* Koelbel (Anomura, Galatheidae). *Ann. Spéleol.,* 30: 325–335.
Pesce, G.L., Juberthie-Jupeau, L. and Passelaigue, F., 1994. Mysidacea, In: C. Juberthie and V. Decu (Editors), *Encyclopaedia Biospeologica,* Vol. I. Société de Biospéologie, Moulis (France) and Bucarest, pp. 113–119.
Pires, A.M.S., 1987. *Potiicoara brasiliensis*: a new genus and species of Spelaeogriphacea (Crustacea: Peracarida) from Brazil, with a phylogenetic analysis of the Peracarida. *J. Nat. Hist.,* 21: 225–238.
Poore, G.D.B. and Humphreys, W.F., 1992. First record of Thermosbaenacea (Crustacea) from the Southern Hemisphere: a new species from a cave in tropical Western Australia. *Invertebr. Tax.,* 6: 719–725.
Rouch, R., 1994. Copepoda, In: C. Juberthie and V. Decu (Editors), *Encyclopaedia Biospeologica,* Vol. I. Société de Biospéologie, Moulis (France) and Bucarest, pp. 105–111.
Schminke, H.K., 1986. Syncarida. In: L. Botosaneanu (Editor), *Stygofauna Mundi.* E.J. Brill, Leiden, pp. 389–404.
Schram, F.R., 1983. Remipedia and crustacean phylogeny, In: F.R. Schram (Editor), *Crustacean Phylogeny.* A.A. Balkema, Rotterdam, pp. 23–28.
Schram, F.R., 1986. *Crustacea.* Oxford University Press, Oxford, xii + 606 pp.
Stock, J.H., 1976. A new genus and two new species of the Crustacean order Thermosbaenacea from the West Indies. *Bijdr. Dierkd.,* 46: 47–70.
Stock, J.H., 1986. Thermosbaenacea. In: L. Botosaneanu (Editor), *Stygofauna Mundi.* E.J. Brill, Leiden, pp. 585–588.
Wagner, H.P., 1994. A monographic review of the Thermosbaenacea (Crustacea: Peracarida). *Zool. Verhand.,* 291: 1–338.
White, W.B., Culver, D.C., Herman, J.S., Kane, T.C. and Mylroie, J.E., 1995. Karst lands. *Am. Sci.,* 83: 450–459.
Wilkens, H., 1992. Neutral mutations and evolutionary progress. In: A.I. Camacho (Editor), *The Natural History of Biospeleology.* National Museum of Natural Sciences, Madrid, pp. 401–422.
Williamson, C.E., 1991. Copepoda. In: J.H. Thorp and A.P. Covich (Editors), *Ecology and Classification of North American Freshwater Invertebrates.* Academic Press, San Diego, CA, pp. 787–822.
Yager, J., 1981. Remipedia, a new class of Crustacea from a marine cave in the Bahamas. *J. Crustacean Biol.,* 1: 328–333.
Yager, J., 1986. Remipedia. In: L. Botosaneanu (Editor), *Stygofauna Mundi.* E.J. Brill, Leiden, pp. 382–384.
Yager, J., 1987. *Tulumella grandis*, new species and *Tulumella bahamensis*, new species of thermosbaenacean crustacean (Monodellidae) from anchialine caves in the Bahamas. *Stygologia,* 3: 373–382.
Yager, J., 1994. *Speleonectes gironensis*, new species (Remipedia: Speleonectidae) from anchialine caves in Cuba, with remarks on biogeography and ecology. *J. Crustacean Biol.,* 14: 752–762.
Yager, J. and Humphreys, W.F., 1996. *Lasionectes exleyi,* sp. nov., the first remipede crustacean recorded from Australia and the Indian Ocean, with a key to the world species. *Invertebr. Taxon.,* 10: 171–187.
Zinn, D.J., 1986. Mystacocarida, In: L. Botosaneanu (Editor), *Stygofauna Mundi.* E.J. Brill, Leiden, pp. 385–388.

Chapter 6

FISH AND AMPHIBIA

Axel WEBER

INTRODUCTION

Despite the fact that vertebrates represent only a small fraction of the presently known troglobitic fauna, they were the first obligate cave-dwellers to excite the curiosity of man, certainly because of their comparatively large bodies (Vandel, 1964) and their bizarre appearance. A transparent fish (apparently *Sinocyclocheilus hyalinus*) inhabiting the Alu limestone caves in Yunnan, China, was reported in ancient books published in 1541 (Chen et al., 1994). A cave-dwelling salamander is mentioned by Baron Johann Weichard Valvasor (1689, p. 597) in his book dealing with the natural rarities of the duchy of Carniole (nowadays a part of Slovenia) (Durand and Vandel, 1968). This salamander was formally described by J.N. Laurenti in 1768 as *Proteus anguinus,* and merits the designation as the first scientifically recognized troglobite. The first troglobitic invertebrate was not described until 1832 (Vandel, 1964). The earliest recognized cave fish is *Amblyopsis spelaea* from the Mammoth Cave in Kentucky (U.S.A.) which was presented to the Academy of Natural Sciences of Philadelphia in 1842 by W.T. Craige and described in the same year by J.E. DeKay (Putnam, 1872).

A number of vertebrates spend their life, or at least some part of it, in the subterranean environment; but not all are considered as true cavernicoles. Many gymnophiones (= Apoda) (e.g., Caeciliidae, Ichthyophiidae, Scolecomorphidae), lizards (e.g., some in each of the families Amphisbaenidae, Anniellidae, Dibamidae and Scincidae), snakes (e.g., Anomalepidae, Atractaspinae, Leptotyphlopidae, Typhlopidae and Uropeltidae) and mammals (e.g., Bathyergidae, Chrysochloridae, Ctenomyidae, Geomyidae, Myospalacidae, Notoryctidae, Rhyzomyidae, Spalacidae, and Talpidae) are adapted to a burrowing way of life in the soil. Some of them even show a remarkable reduction of their eyes, normally a characteristic feature of troglobites. But these burrowing vertebrates are restricted to the food-rich upper layer of the pedosphere. Most of them leave the subterranean habitat from time to time, as indicated by the body pigmentation, which is usually retained. Some fishes are known for their cryptic way of life in rock crevices or coarse gravel of river rapids (Roberts and Stewart, 1976), in deep river channels with turbid water (Lundberg and Rapp Py-Daniel, 1994), or in the marine littoral (Thinès, 1969). Some of them not only have reduced eyes but also have little or no body pigmentation. However, they are usually not considered to be true troglobites (but compare Thinès and Proudlove, 1986), because their habitats differ from the subterranean environment.

Caves are frequently visited by vertebrates for shelter in times of unfavourable environmental conditions at the surface. Well-known examples are the retreat of amphibians into caves to escape desiccation during dry periods, and the hibernation of bats and bears in temperate latitudes. The Chinese cyprinid fish *Varicorhinus (Scaphesthes) macrolepis* hibernates for about six months in the comparatively warm and stable cave waters within its harsh mountainous distribution range (Zhang, 1986). Avoidance of predation is another reason for cave frequentation. The North American pack rats of the genus *Neotoma* (Cricetidae), for example, like to spend the days in the safety of a cave wherever possible, and leave at night to search for food (Mohr and Poulson, 1966). This holds true for oilbirds (Steatornithidae) and cave bats, too. Caves are also used as breeding sites by some vertebrates – for example some salamanders, some frogs (Diesel et al., 1995), oilbirds, some cave swiftlets (Apodidae) and some bats and other mammals (Mohr and Poulson, 1966).

According to the ecological classification of cavernicoles, as proposed by Barr Jr (1968), most of the vertebrates mentioned as encountered in caves have to be judged either as accidentals or as trogloxenes. The sporadic accidentals are usually restricted to the entrance zone, or they are trapped in sinkholes. Normally, they are not able to survive in the subterranean environment for long, and their influence on the cave ecosystem is comparatively low. The life cycles of trogloxenes include regular occupation of the cave environment, but these animals have to return periodically to the surface for food. Trogloxenic vertebrates are able to inhabit even remote parts of the cave because they can navigate in complete darkness. Bats (Chiroptera) are famous for their highly developed echolocation system which is used for the detection of obstacles and prey. Some birds are able to make use of echolocation, too. The oilbird *Steatornis caripensis* (Steatornithidae) from northern South America, and some species of cave swiftlets of the genus *Collocalia* (Apodidae) from South-east Asia, nest in caves. But unlike the ultra-sound signals emitted by bats, the sound signals of these birds are within the range of human ears (Griffin and Luthers, 1970; Konishi and Knudsen, 1979). Hence, the efficiency of the birds' echolocation is restricted. It is only used within the cave. Outside, the oilbird relies on its keen eyesight and perhaps olfaction during its nocturnal search for fruits and seeds. The cave swiftlets prey on flying insects in bright daylight only. Trogloxene vertebrates can play an important role in cave ecology because their faeces and dead bodies can be a main source of food for cavernicoles (Barr Jr, 1968; Poulson, 1972). Especially, perennial bat colonies are known to produce immense deposits of guano on the cave floor.

There are only two vertebrate groups in which troglophilic and troglobitic species have evolved: fishes (Table 6.1; pp. 111–115) and salamanders (Table 6.3; p. 122). Hence they are the focus of this chapter. Figures 6.2A–M show the range of body form shown by fishes in caves (see pp. 116–121), and Figs. 6.3A–C the range for amphibians (see pp. 121–123).

CAVE COLONIZATION BY VERTEBRATES

The transition from life above ground to life in subsurface habitats confronts the colonizing vertebrates with a peculiar ecological regimen. The access is limited mainly by three factors:

(1) darkness,
(2) scarcity of food (in most cases),
(3) dimensions of cavities.

Successful colonists have appropriate morphological, physiological and ethological adaptations that enable them to survive and reproduce in the cave habitat.

The requirements for a continual cave existence evidently explain the restriction of troglophilic and troglobitic vertebrates to fishes and amphibians. Birds and mammals, as homoiothermic animals, must eat more or less continuously to maintain their high activity and body temperature. In the usually food-poor cave environment this requirement cannot be covered (Mohr and Poulson, 1966). The poikilothermic reptiles should therefore be preadapted as cave colonists because they need less food than homoiotherms. But reptiles are mainly terrestrial, and food scarcity is stronger in the terrestrial than in the aquatic cave habitat (Culver, 1982; Hüppop, 1985; see also Chapter 8, this volume). So it seems likely that, even for reptiles, the food supply in caves is too low (Mohr and Poulson, 1966). The two remaining poikilothermic vertebrate groups, fishes and amphibians, exhibit a broad norm of reaction to environmental conditions in many developmental processes (Barr Jr, 1968). They are able to cope with food scarcity. One adaptation to the low food supply is the body size of troglobitic fishes and salamanders, which are usually small in comparison to epigean relatives (Bertin, 1958; Greenwood, 1967). The total length rarely exceeds 15 cm. Nevertheless, troglobitic vertebrates are the largest animals of the cave communities. Therefore, they usually represent the highest trophic level and are the top predators in the cave food pyramid (Culver, 1982, 1985) – or at least scavengers (Culver, 1994). Hence, to a degree that is extreme even for troglobites, their food is usually diffuse and scarce. Only very food-rich caves, for example those with bat colonies roosting over the water, can maintain large populations of cave-dwelling vertebrates.

Troglobitic vertebrates are potentially at risk owing to restricted distribution, usually small population sizes, low reproductive potential, and top position in the food pyramid (Sweet, 1986). Only a few species are legally protected. Miller (1977) listed several species as vulnerable or even as endangered. Nevertheless, the degradation of their environment due to pollution and pumping still constitutes a serious threat.

Table 6.1
List of troglobitic fish species (**Teleostei**), their synonyms and distribution[1]

	Species	Distribution
	Ostariophysi	
	Characiformes	
	Characidae	
1	*Astyanax fasciatus* (Cuvier, 1819) cave populations *Astyanax mexicanus* Fillippi, 1853 *Anoptichthys jordani* Hubbs & Innes, 1936 (Chica Cave) *Anoptichthys antrobius* Alvarez, 1946 (Pachon Cave) *Anoptichthys hubbsi* Alvarez, 1947 (Sabinos Cave)	Sierra de El Abra and adjacent regions (Huastecan Province) in Tamaulipas and San Luis Potosi, Mexico
2	*Stygichthys typhlops* Brittan & Böhlke, 1965	subterranean waters near Jaiba, Minas Gerais, Brazil
	Cypriniformes	
	Cyprinidae	
3	*Barbopsis devecchii* Di Caporiacco, 1926 *Eilichthys microphthalmus* Pellegrin, 1929 *Barbopsis stefaninii* Gianferrari, 1930 *Zaccarinia stefaninii* Gianferrari, 1934	valley of the Nugaal River, Eyl (Eil), Kalis (Callis), and Taleex (Taleh), Nugaal County, Somalia
4	*Barbus speleops* Roberts, 1991	Tham Phu Khieo and Tham Pathewada, Phu Khieo, Thailand (tham = cave)
5	*Caecobarbus geertsi* Boulenger, 1921	limestone caves in the Kwilu Basin south of Mbanza-Ngungu (Thysville), Bas-Congo, Congo
6	*Caecocypris basimi*[2] Banister & Bunni, 1980	sinkhole at the Sheik Hadid Shrine near Al Hadithah, Al Anbar Province, Iraq
7	*Garra barreimiae* Fowler & Steinitz, 1956 cave population	cave in the south flank of the Jabal Akhdar, 9 km east–south-east of Al-Hamra in the north of Oman
8	*Garra dunsirei*[2] Banister, 1987	sink hole at Tawi Atair in the Jabal Qara or Jabal Samhan mountains, Dhofar, Oman
9	*Gibbibarbus cyphotergous* Dai, 1988	Daxiao Dong Cave, Dajing Village, Luodian County, Guizhou, China
10	*Iranocypris typhlops* Bruun & Kaiser, 1943	Kajee-Ru Source at the Baq-e-Loveh Oasis near Tang-e-Haft in the Zagros Mountains, Khorramabad Province, Iran
11	*Phreatichthys andruzzii* Vinciguerra, 1924	subterranean waters in the region around Bud Bud and Dirri, Galguduud County, Somalia
12	*Sinocyclocheilus anatirostris* Lin & Luo, 1986	subterranean waters in Yunnan, China
13	*Sinocyclocheilus anophthalmus* Chen & Chu, 1988	subterranean river, Yiliang County, Yunnan, China
14	*Sinocyclocheilus hyalinus* Chen & Yang, 1994	Alu Limestone caves, Luxi County, Yunnan, China
15	*Sinocyclocheilus microphthalmus* Li, 1989	karst cave at Luo-luo Village, Lingyun County, Guangxi, China
16	*Typhlobarbus nudiventris* Chu & Chen, 1982	near Yangjieba, Jianshui County, Yunnan, China
17	*Typhlogarra widdowsoni* Trewavas, 1955	sinkholes near Al Hadithah, Al Anbar Province, Iraq
	Cobitidae	
18	*Protocobitis typhlops* Yang, Chen & Lan, 1994	limestone cave near Xiaao, Duan County, Guangxi, China
	Balitoridae	
19	*Homaloptera thamicola* Kottelat, 1988	Tham Susa, Mae Hong Son Province, Thailand
20	*Indoreonectes evezardi*[2] (Day 1878) Kottelat, 1990 *Noemacheilus evezardi* Day 1878 *Oreonectes evezardi* (Day 1878)	Kotumsar Cave, Bastar District, Madhya Pradesh State, India

continued on next page

Table 6.1, *continued*

	Species	Distribution
21	*Nemacheilus troglocataractus* Kottelat & Géry, 1989	Tham Sai Yok Noi, Nam Tok, and Tham Wang Ba Dan, Ban Sai Yok, Kanchanaburi Province, Thailand
22	*Oreonectes anophthalmus* Zheng, 1981 *Nemacheilus anophthalmus* in Kottelat, 1988	karst cave in the Quifeng Hill, Wuming County, Guangxi, China
23	*Oreonectes* sp.[2] in Yang, Chen & Lan, 1994	limestone cave near Xiaao, Duan County, Guangxi, China
24	*Paracobitis longibarbatus* Chen, Yang, Sket & Aljançiç, 1998	subterranean waters, Guizhou, China
25	*Paracobitis smithi* (Greenwood 1976) *Noemacheilus smithi* Greenwood 1976	Kajee-Ru Source at the Baq-e-Loveh Oasis near Tang-e-Haft in the Zagros Mountains, Khorramabad Province, Iran
26	*Paracobitis starostini* (Parin 1983) *Noemacheilus* (*Troglocobitis*) *starostini* Parin 1983	subterranean water flow, Khrebet Kugitantau (Mountains), Chardzhou Province, Turkmenistan
27	*Schistura jarutanini*[2] Kottelat 1990	Tham Ba Dan, Mae Nam Khwae Nai basin, Amphoe Sri Sawat, Kanchanaburi Province, Thailand
28	*Schistura oedipus* (Kottelat, 1988) Kottelat & Géry 1989 *Nemacheilus oedipus* Kottelat, 1988	Tham Nam Lang, Mae Hong Son Province, Thailand
29	*Sundoreonectes tiomanensis*[2] Kottelat 1990	cave in the Mount Gunung Kajang, Tioman Island, Malaysia
30	*Triplophysa gejiuensis* (Chu & Chen 1979) Chen & Yang 1992 *Noemacheilus gejiuensis* Chu & Chen 1979 *Schistura gejiuensis* in Zhu, 1989	Bajiaoquing near Quiafang, Gejiu County, Yunnan, China
31	*Triplophysa shilinensis* Chen & Yang 1992	cave at Weiboyi Village, Lunan County, Yunnan, China
32	*Triplophysa xiangxiensis* (Yang, Yuan & Liao, 1986) Chen & Yang, 1993 *Noemacheilus xiangxiensis* Yang, Yuan & Liao, 1986 *Schistura xiangxiensis* in Zhu, 1989	subterranean waters of Huoyanxiang, Lonshan County, Hunan, China
Siluriformes		
	Siluridae	
33	*Pterocryptis* or *Silurus* sp.[2] in Kottelat & Géry, 1989	Tham Sai Yok Noi, Nam Tok, Kanchanaburi Province, Thailand
34	*Silurus cucphuongensis*[2] Mai 1978[2]	subterranean waters in the Cuc Phuong National Park, northern Vietnam
	Clariidae	
35	*Clarias cavernicola* Trewavas, 1936	Aigamas Cave north of Otavi and west of Grootfontein, Namibia
36	*Horaglanis krishnai* Menon, 1950	wells at Kottayam, Kerala State, India
37	*Uegitglanis zammaranoi* Gianferrari, 1923	Waajid (Uegit), Bakool (locus typicus) and Baydhabo (Baidoa), subterranean waters of the plateau between Jubba and Webi Shabeelle, Somalia
	Ictaluridae	
38	*Prietella lundbergi* Walsh & Gilbert, 1995	Nacimiento de San Rafael de Los Castros, Ciudad Mante, Tamaulipas, Mexico
39	*Prietella phreatophila* Caranza, 1954	El Potrero de Doña Mariana, Municipio de Muzquiz, Coahuila, Mexico; several other locations in northern Mexico (Hendrickson, pers. comm.)
40	*Satan eurystomus* Hubbs & Bailey, 1947	phreatic waters near San Antonio, Bexar County, Texas, USA
41	*Trogloglanis pattersoni* Eigenmann, 1919	phreatic waters near San Antonio, Bexar County, Texas, USA

continued on next page

Table 6.1, *continued*

	Species	Distribution
	Pimelodidae	
42	*Imparfinis* sp. nov. or gen. nov.[2] in Trajano, 1997	caves in the Chapada Diamantina, Itaete County, Bahia State, Brazil
43	*Phreatobius cisternarum*[2] Goeldi, 1904	phreatic waters near the lower Amazonas River, Amazonia State, Brazil
44	*Pimelodella kronei* (Ribeiro 1907) Pavan, 1946 *Typhlobagrus kronei* Ribeiro, 1907 *Pimelodella lateristriga* var. *kronei* Haseman, 1911 *Caecorhamdella brasiliensis* Borodin, 1927	karstic caves of the upper Ribeira Valley, Iporanga County, São Paulo State, Brazil
45	*Rhamdia laticauda typhla* Greenfield, Greenfield & Woods, 1982	subterranean waters of Las Cuevas, Millionario, Mountain Pine Ridge, Belize
46	*Rhamdia quelen urichi* (Norman 1926) Mees, 1974 *Caecorhamdia urichi* Norman, 1926 *Caecorhamdella urichi* in Hubbs, 1938 *Rhamdia urichi* in Miller, 1984	Guacharo or Oropuche Cave, St. Andrew, Northern Range, Trinidad
47	*Rhamdia reddelli* Miller, 1984	Cueva del Nacimiento del Rio San Antonio, Acatlan, Oaxaca, Mexico
48	*Rhamdia macuspanensis* Weber & Wilkens, 1998	Grutas de Agua Blanca, near Macuspana, Tabasco, Mexico
49	*Rhamdia zongolicensis* Wilkens, 1993	Cueva del Ostoc, Sierra de Zongolica, Veracruz, Mexico
50	*Taunayia* sp.[2] in Trajano, 1997	Toca do Goncalo Cave, Campo Formoso County, Bahia State, Brazil
	Trichomycteridae (= Pygiidae)	
51	*Trichomycterus chaberti* Durand, 1968	Grutas de Umayalanta, Charcas, Torotoro Massiv, south of Cochabamba, Bolivia
52	*Trichomycterus guianense*[2] (Eigenmann 1909) cave population	Guacharo Cave, Rio San Juan drainage, Caripe, Monagas State, Venezuela
53	*Trichomycterus itacarambiensis* Trajano & Pinna, 1996	Gruta Olhos d'Agua, Itacarambi County, Minas Gerais State, Brazil
54	*Trichomycterus* sp.[2] in Trajano & Souza, 1994	São Mateus and São Vicente cave system, Sierra de São Domingos, Goias State, Brazil
55	*Trichomycterus* sp.[2] in Trajano, 1997	Buraco das Abelhas Cave, Bonito County, Mato Grosso do Sul State, Brazil
	Astroblepidae	
56	*Astroblepus pholeter* Collette, 1962	Cueva de Jumandi, 4 km north of Archidona, Napo Province, Ecuador
57	*Astroblepus riberae* Cardona & Guerao, 1994	Grutas de Ninabamba, Cajamarca region, Peru
	Loricariidae	
58	*Ancistrus cryptophthalmus* Reis, 1987	São Vicente cave system and Angelica-Bezzera cave system, São Domingos County, Goias State, Brazil
59	*Ancistrus formoso* Sabino & Trajano, 1997	cave waters of the Bodoquena karstic region, Bonito County, Mato Grosso do Sul State, Brazil
60	*Ancistrus galani* Perez & Viloria, 1994	El Saman cave system, Sierra de Perija, Zulia State, Venezuela
Gymnotiformes		
	Sternopygidae	
61	*Eigenmannia vicentespelaea*[2] Triques, 1996	São Vicente cave system, São Domingos County, Goias State, Brazil

continued on next page

Table 6.1, *continued*

	Species	Distribution
Paracanthopterygii		
Amblyopsiformes (= Percopsiformes)		
	Amblyopsidae	
62	*Amblyopsis rosae* (Eigenmann 1898) *Typhlichthys rosae* Eigenmann 1898 *Troglichthys rosae* Eigenmann 1899	subterranean waters of the Springfield Plain (Ozark Plateau), southwest Missouri, northwest Arkansas, and northeast Oklahoma, USA
63	*Amblyopsis spelaea* De Kay, 1842 after Woods & Inger, 1957 *Amblyopsis spelaeus* DeKay 1842	from East Fork of White River in southern Indiana south to the Mammoth Cave system in west central Kentucky (Western Mississippian Plateaus), USA
64	*Speoplatyrhinus poulsoni* Cooper & Kuehne, 1974	Key Cave, Lauderdale County, northwest Alabama, USA
65	*Typhlichthys subterraneus* Girard, 1859 *Typhlichthys osborni* Eigenmann, 1905 *Typhlichthys wyandotte* Eigenmann 1905 *Typhlichthys eigenmanni* Hubbs 1938	from southern Kentucky (Mississippian Plateaus), central Tennessee and north Alabama (Interior Low Plateaus) to north Arkansas, northeast Oklahoma, and southern central Missouri (Ozark Plateau), USA
Ophidiiformes		
	Bythitidae	
66	*Lucifuga dentatus* Poey 1856 *Stygicola dentatus* (Poey 1856) Gill 1863	anchialine caves in western Cuba (Pinar del Rio, Matanzas, and Habana Province) and in eastern Cuba (Holguin Province)
67	*Lucifuga simile* Nalbant, 1981	anchialine caves at the northern coast of Cuba (Matanzas Province)
68	*Lucifuga spelaeotes* Cohen & Robins, 1970	limestone sinkholes on New Providence Island, Grand Bahama, and Andros, Bahamas
69	*Lucifuga subterraneus* Poey, 1856	anchialine caves at the northwestern and southwestern coast of Cuba
70	*Lucifuga teresinarum* Diaz Perez, 1988	anchialine caves in Cuba
71	*Ogilbia pearsei* (Hubbs 1938) Cohen & Nielsen, 1978 *Typhlias pearsei* Hubbs 1938 *Typhliasina pearsei* (Hubbs 1938) Whitley, 1951	caves and sinkholes south of Merida, Yucatan Peninsula, Yucatan State, Mexico
72	*Ogilbia galapagosensis* (Poll & Leleup 1965) Cohen & Nielsen, 1978 *Caecogilbia galapagosensis* Poll & Leleup 1965	anchialine cave waters on Santa Cruz Island, Galapagos Archipelago, Ecuador
Acanthopterygii		
Cyprinodontiformes		
	Poeciliidae	
73	*Poecilia mexicana* Steindachner, 1863 cave population *Poecilia sphenops* Valenciennes, 1846 *Mollienesia sphenops* Regan, 1913 *Mollienisia mexicana* Hubbs, 1961	Cueva de las Sardinas, near Tapijulapa Village, about 26 km south of Teapa, Tabasco, Mexico
Synbranchiformes		
	Synbranchidae	
74	*Monopterus eapeni* Talwar, 1991 *Monopterus indicus* Eapen, 1963	wells at Kottayam, Kerala State, India
75	*Monopterus roseni* Bailey & Gans, 1998	well at Periyam, Kerala State, India

continued on next page

Table 6.1, *continued*

	Species	Distribution
76	*Ophisternon candidum* (Mees 1962) Rosen & Greenwood, 1976 *Anommatophasma candidum* Mees 1962	subterranean waters at the coast of the North-West Cape Range National Park, Western Australia
77	*Ophisternon infernale* (Hubbs 1938) Rosen & Greenwood, 1976 *Pluto infernalis* Hubbs 1938 *Furmastix infernalis* (Hubbs 1938) Whitley, 1951 *Synbranchus infernalis* (Hubbs 1938) Mees, 1962	subterranean waters on the Yucatan Peninsula, Yucatan State, Mexico
Perciformes		
	Gobiidae	
78	*Caecogobius cryptophthalmus* Berti & Ercolini, 1991	Calbiga cave system, Samar Island, Philippines
79	*Glossogobius ankaranensis* Banister, 1994	subterranean waters near Ankarana, Madagascar
80	*Luciogobius pallidus* Regan, 1940 *Luciogobius guttatus guttatus* Gill, 1859 after Tomiyama, 1936	underground waters near the sea in southern Honshu and in Chugoku, Japan
81	*Luciogobius albus* Regan, 1940 *Luciogobius guttatus guttatus* Gill, 1859 after Tomiyama, 1936	underground waters near the sea in southern Honshu, Chugoku, Shikoku, Kyushu, and the Goto Islands, Japan
	Eleotridae	
82	*Milyeringa veritas* Whitley, 1945	subterranean waters at the coast of the North-West Cape Range National Park, Western Australia
83	*Oxyeleotris caeca* Allen, 1996	caves at the upper Kikori River, Papua New Guinea
84	*Typhleotris madagascariensis* Petit, 1933	phreatic and cave waters of the Mahalafy Limestone Plateau near the Tsimanampetsotsa Lake, south-west Madagascar
85	*Typhleotris pauliani* Arnoult, 1959	Andranomaly Cave in the Andolambezo region near Morombe, south-west Madagascar

[1] Discarded synonyms are listed after the numbered accepted species.
[2] Species names thus marked designate forms of uncertain taxonomic or ecological recognition.

REDUCTION OF EYES AND BODY PIGMENTATION IN CAVE VERTEBRATES

It is a well-known fact that features which have lost their biological significance are subject to regression. In the cave environment with its constant darkness this holds true especially for all characteristics influenced by light stimuli under epigean conditions (see Chapter 7, this volume). At a first glance, the most impressive differences of subterranean vertebrates from epigean ones are the reduced eyes and body pigments. Although this regression is the result of convergent evolution and consequently each species shows a distinctive developmental pattern, cave vertebrates reveal some common fundamental characteristics of this process. The degree of structural degeneration depends on the ontogenetic age of the individual (Langecker et al., 1993, 1995) and the phylogenetic age of the troglobitic species (Wilkens, 1982; Wilkens et al., 1989).

Eye regression includes reduced eye growth and impaired histogenesis. In phylogenetically old cave vertebrates the normal histological organization of the eye has almost completely broken down. In some cave fish species the optic nerve may lose its contact with the brain, clearly revealing that no light-induced stimuli can be transmitted from the optic cyst in this case. The eye reduction is accompanied by a reduction of the optic centres of the brain (Fig. 6.1). The optic cysts may remain near the neurocranium during body growth, which leads to the impression that the eye remnants are sunken into the head. In some cases, a cord of dermal cells or a pit connecting the optic cyst and the body surface indicates the former position of the eye. However, in some phylogenetically old cave vertebrates the skin in the region over the eye exhibits the same

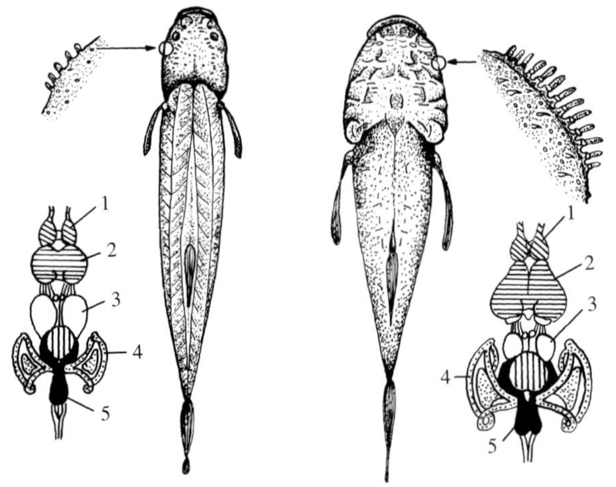

Fig. 6.1. Schematic comparison of the troglobitic *Amblyopsis spelaea* (right) with its troglophilic relative *Chologaster agassizi* (left). 1, Olfactory; 2, higher order brain integration; 3, visual; 4, equilibrium; 5, lateral line. *A. spelaea* has a larger head and stouter body. Its pectoral fins are elongated. Eyes are lacking. The free neuromasts are more numerous and project further out into the water. The chemoreceptive and mechanoreceptive centres of the brain are enlarged, whereas the optic centres are reduced. The enlarged brain parts for higher-order integration indicate an increasing processing of sensory inputs (redrawn from Poulson and White, 1969).

structural morphology as the surrounding skin, and the eye remnants can no longer be detected externally.

The Ozark cave salamander *Typhlotriton spelaeus* (Fig. 6.3C) shows a special case of eye reduction that is unique among cave vertebrates. The larvae are pigmented and have functional eyes. But with the onset of metamorphosis the body pigmentation disappears and the eyes begin to degenerate. Enhanced growth of the eyelids is induced, and frequently the eye remnants are completely covered (Besharse and Brandon, 1974).

The colour patterns of epigean fishes are caused by pigment cells scattered in the skin and structural colours generated by reflection, refraction, and interference. Beside the dominating black-brown melanophores, there are yellowish xanthophores, red erythrophores, and reflecting guanophores. The chromatophores are subject to physiological and morphological colour change. The physiological colour change is a rather fast process involving aggregation or dispersal of pigment particles within each chromatophore. During the morphological colour change the number of chromatophores varies. This process requires some time. According to these different modes of colour variation, the reduction of the body pigmentation in cave fishes is achieved by a decrease in the pigment content of the chromatophores as well as by a decrease in chromatophore number itself (Wilkens, 1988). Phylogenetically old cave populations of *Astyanax* are still able to effect physiological colour change, although it is hardly recognizable because of the reduced pigment content of the few chromatophores. On the other hand, morphological colour change is no longer possible. Thus, when maintained in daylight, phylogenetically old cave fishes do not become darker. Usually, even in the skin of phylogenetically old cave-fish species a few scattered melanophores can still be found. However, albinism has also been observed in some cave fish populations, for instance in *Triplophysa xiangxiensis* (Durand et al., 1995), in the *Astyanax* forms from the Pachon and Yerbaniz caves in the Mexican Sierra de El Abra (Wilkens, 1988) and in *Trichomycterus itacarambiensis* (Trajano and de Pinna, 1996). In the latter species about one-third of the population are albinotic. Albinism is indicated by the complete lack of pigmentation even in the eye remnants. Since albinism is not negatively selected in the dark cave environment, it might be fixed in the population by genetic drift (Peters, 1992; Weber, 1996).

Like cave fishes, troglobitic salamanders show a more or less reduced body pigmentation, apparently depending on the degree of cave adaptation. In life, pigmentless cave vertebrates usually show a pale pink or reddish coloration because of the blood in the capillaries shining through the translucent skin. Although the regression of the body pigmentation and of the eyes are simultaneous events in the course of cave adaptation, they are not genetically linked. The two characteristics show independent inheritance (Wilkens, 1988).

SYSTEMATIC DIVERSITY, EXTERNAL MORPHOLOGY, AND DISTRIBUTION OF CAVE VERTEBRATES

Cave fishes

All known troglophilic and troglobitic fishes belong to the Teleostei (i.e., higher bony fishes), the most species-rich and diversified vertebrate group comprising approximately 23 600 species (Nelson, 1994). The exact number of troglophilic fishes is unknown, but it is justifiable to assume that they are quite numerous. Barbs (Cyprinidae), loaches (Cobitidae, Balitoridae),

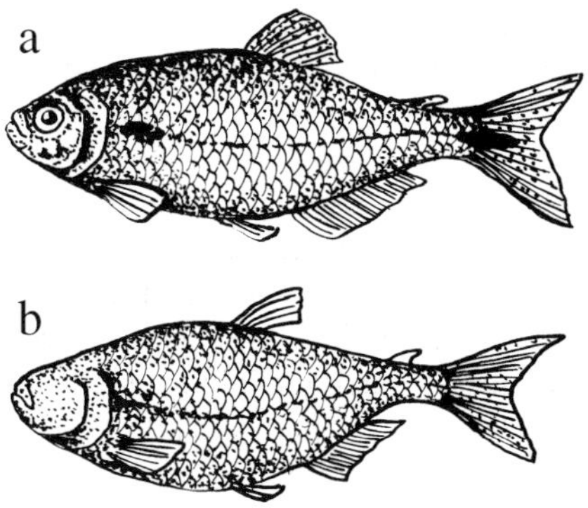

Fig. 6.2A. *Astyanax fasciatus* (Characidae); (a) epigean form; (b) troglobitic form (redrawn from Wilkens, 1988).

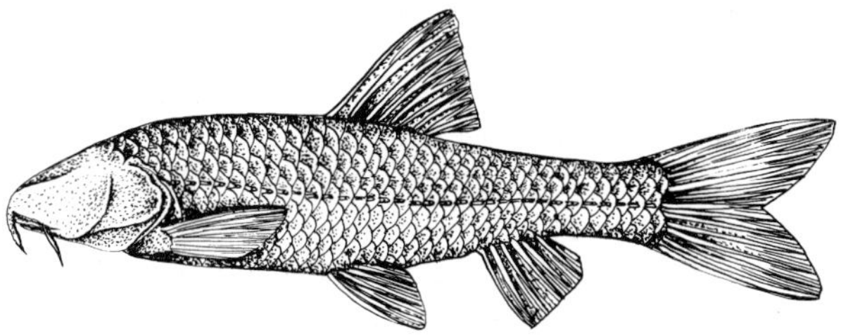

Fig. 6.2B. *Caecobarbus geertsi* (Cyprinidae) (redrawn from T. Nalbant).

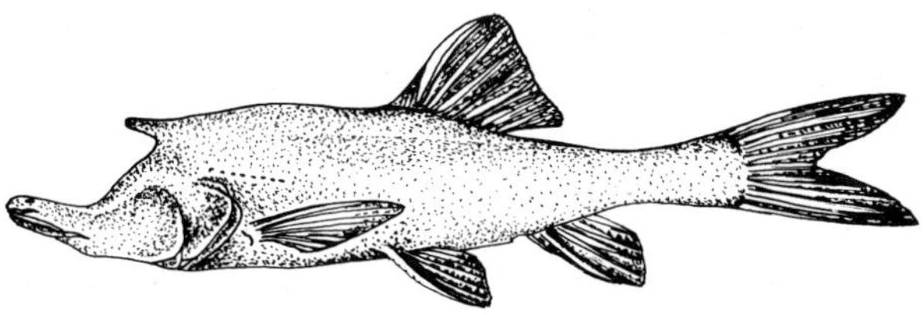

Fig. 6.2C. *Sinocyclocheilus hyalinus* (Cyprinidae) (redrawn from Chen, Yang and Zhu, 1994).

catfishes (Siluriformes), and sculpins (Cottidae) of epigean habitus are commonly encountered in caves. At present (1997), 70 teleost species, subspecies or populations restricted to caves are known; the status of 14 more is not resolved through lack of information about their ecology and taxonomic status (Table 6.1). The troglobitic taxa (comprising only about 0.27% of all teleost species) are distributed among 8 orders (21%), 16 families (4%), and 45 genera (1%). The comparatively numerous orders involved indicate the remarkably extensive systematic diversity of cave-adapted fishes. The high number of genera in comparison to the number of species is explained by the large proportion of monospecific genera (seventeen). This could be attributed in part to the practice of former authors in creating a new genus simply

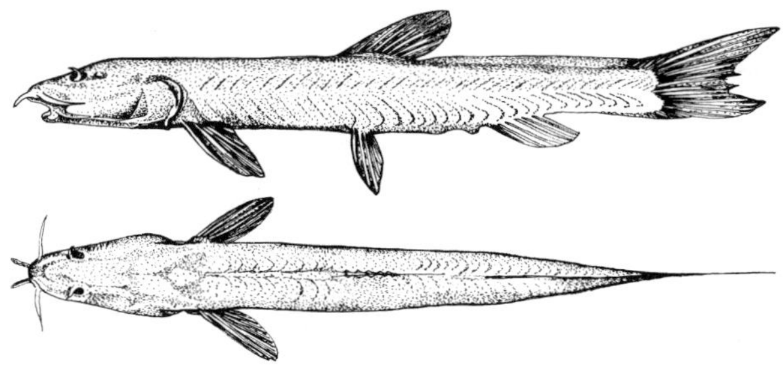

Fig. 6.2D. *Paracobitis smithi* (Balitoridae), lateral and dorsal view (redrawn from Greenwood, 1976).

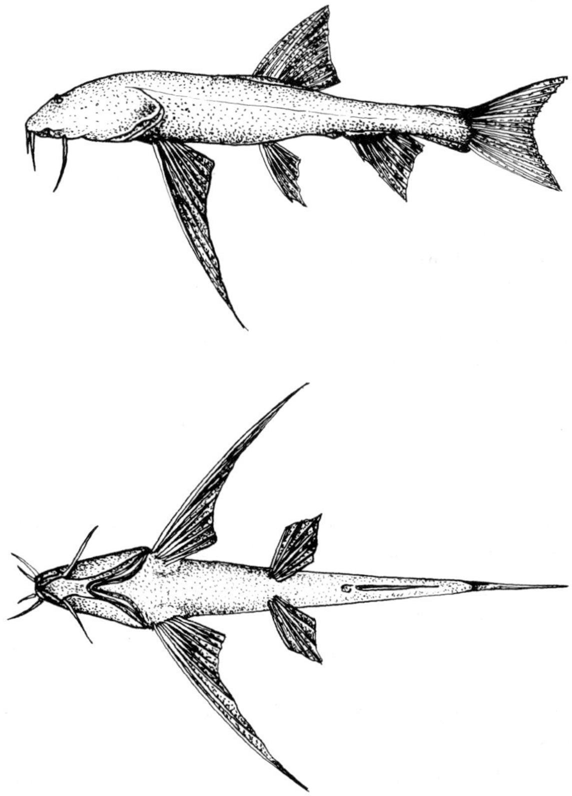

Fig. 6.2E. *Triplophysa xiangxiensis* (Balitoridae), lateral and ventral view (redrawn from Yang, Yuan and Liao, 1986).

because the newly-discovered cave-fish population lack functional eyes and pigmentation. A typical example is Norman's (1926) description of *Caecorhamdia urichi* (Pimelodidae), which is nowadays included in the genus *Rhamdia*. In recent years most of these cases have been corrected. A well-known example is the Mexican cave characids. Three populations from different caves have been described as distinct species within the genus *Anoptichthys* (Hubbs and Innes, 1936; Alvarez, 1946, 1947). However, laboratory experiments

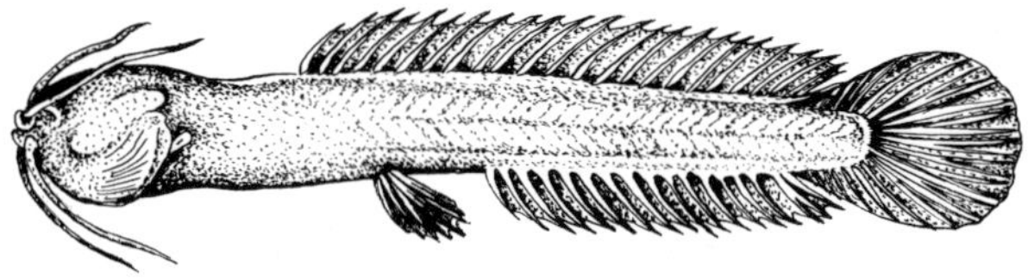

Fig. 6.2F. *Horaglanis krishnai* (Clariidae) (redrawn from Menon, 1950).

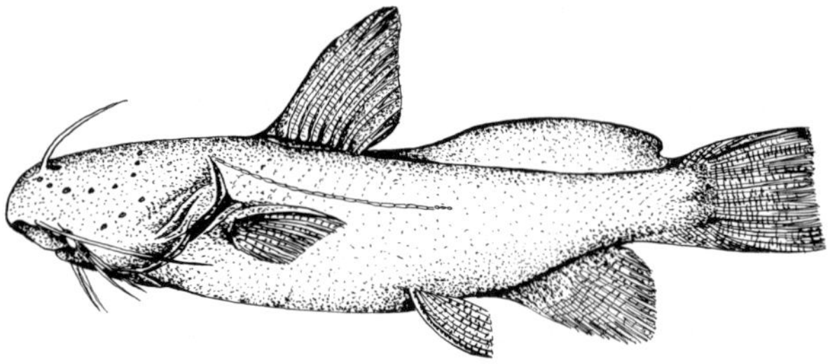

Fig. 6.2G. *Trogloglanis pattersoni* (Ictaluridae) (redrawn from Lundberg, 1982).

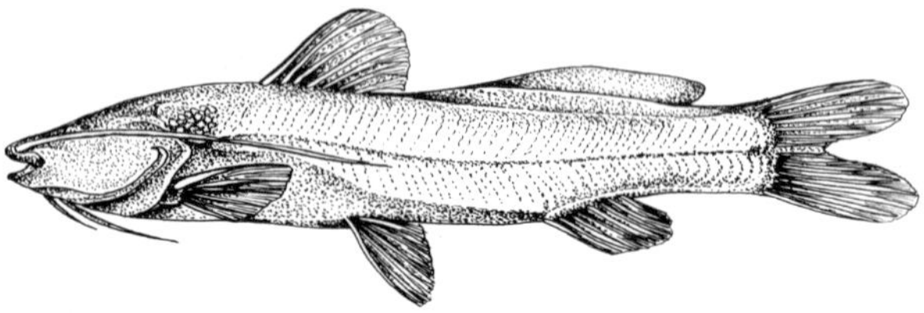

Fig. 6.2H. *Rhamdia zongolicensis* (Pimelodidae) (taken from Wilkens, 1993).

revealed a complete interfertility among the different troglobitic "species" and even with the closest epigean relative *Astyanax fasciatus.* The population of the Chica Cave, formerly described as a separate species, *Anoptichthys jordani,* even turned out to be a natural hybrid population between the epigean species and a troglobitic form (Wilkens, 1988). Nowadays, all cave populations of the Mexican cave characids are considered by most authors to be conspecific with epigean *Astyanax fasciatus* (Fig. 6.2A) [but compare Trajano (1993) for a contrary view]. Another reason for creating a new genus for a cave fish is that the generic allocation could not yet be established. Indeed, phylogenetically old cave fishes may have evolved sufficient derived characteristics to merit a generic recognition of their own. Another possibility may be the extinction of the closest epigean relatives. In this case, the cave-fish species represents a phylogenetically and geographically isolated relict.

Despite their systematic diversity, as many as 46 of the cave-fish species (66%) belong to the Ostariophysi. This seems not surprising, since this order comprises

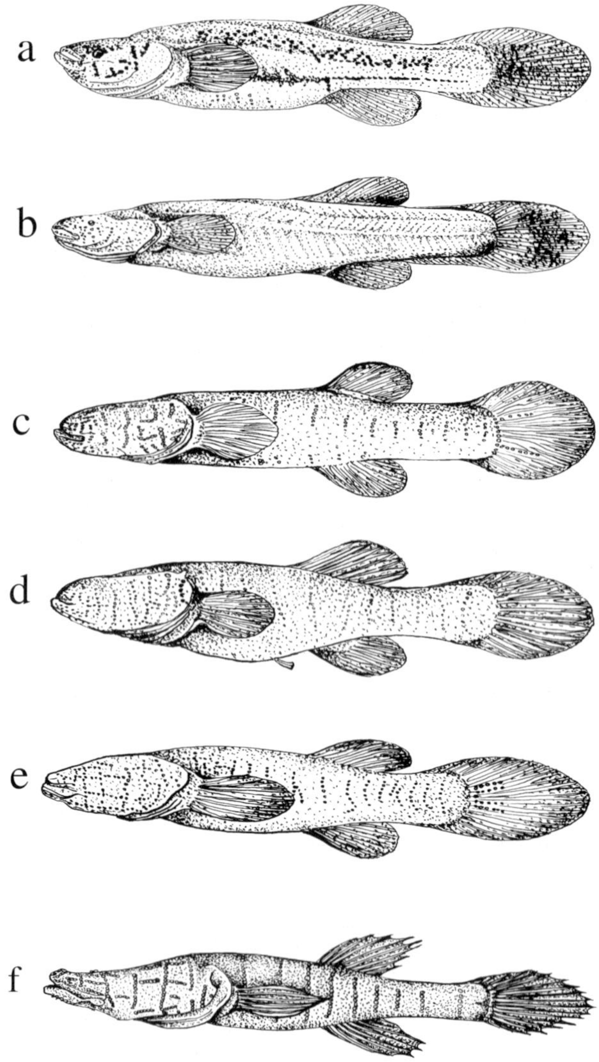

Fig. 6.2I. Amblyopsidae: (a) *Chologaster cornutus* (epigean); (b) *Chologaster agassizi* (troglophilic); (c) *Typhlichthys subterraneus* (troglobitic); (d) *Amblyopsis spelaea* (troglobitic); (e) *Amblyopsis rosae* (troglobitic); (f) *Speoplatyrhinus poulsoni* (troglobitic) (a–e redrawn from Woods and Inger, 1957; f redrawn from Cooper and Kuehne, 1974).

the majority of the primary freshwater fishes. Higher teleost taxa, especially the Perciformes, are poorly represented among the cave fishes. Only 20 cave-fish species (29%) are either secondary freshwater fishes (Poeciliidae, Synbranchidae) or of marine origin (Bythitidae, Gobiidae, Eleotridae). Secondary freshwater fishes have some tolerance for normal seawater. Hence, they may be able to use the sea for dispersal, either across open sea barriers or along coasts. Primary freshwater fishes are exclusively restricted to fresh water; they therefore rely on occasional connections

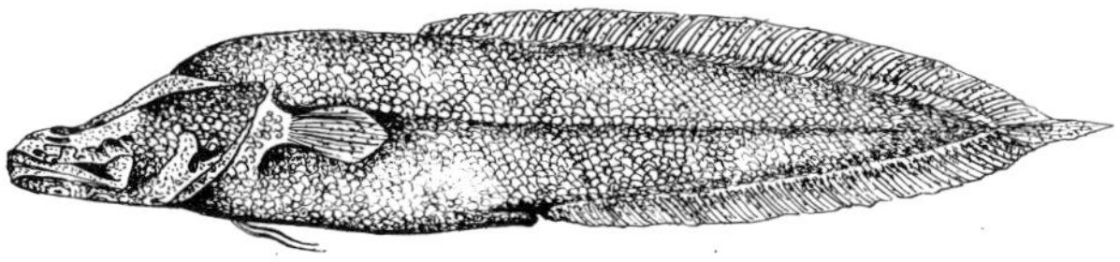

Fig. 6.2J. *Lucifuga subterraneus* (Bythitidae) (redrawn from Eigenmann, 1902).

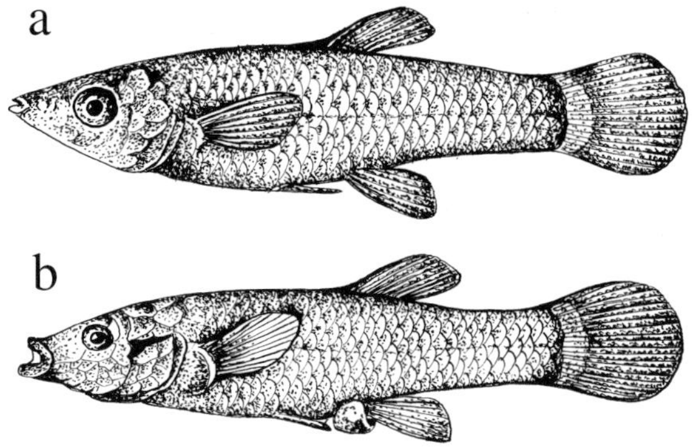

Fig. 6.2K. *Poecilia mexicana* (Poeciliidae); (a) epigean female; (b) troglobitic female (redrawn from Peters et al., 1973).

Fig. 6.2L. *Ophisternon infernale* (Synbranchidae) with detail of the head (redrawn from Hubbs, 1938).

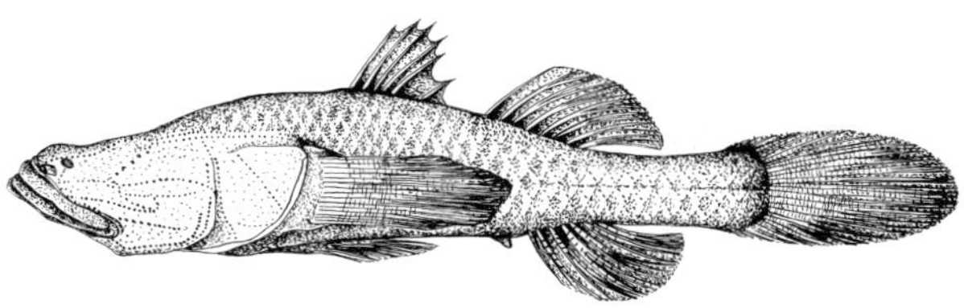

Fig. 6.2M. *Caecogobius cryptophthalmus* (Gobiidae) (redrawn from Berti and Ercolini, 1991).

between drainage systems and need land bridges for dispersal (Miller, 1966; Myers, 1966).

Certainly, many more cave fishes will be discovered in the future, especially in remote and poorly investigated karst areas in South America, Africa, and East Asia. Since 1980, 33 new species of troglobitic fishes have been described, nearly half of the total number of species recognized at present. Eighteen were from Asia, and eleven from South and Middle America.

The most conspicuous difference of subterranean fishes from epigean ones are the convergently reduced eyes and body pigments. However, troglobitic vertebrates have no unique traits which are solely restricted to them and cannot be found in colonizers of other habitats (Vandel, 1964). They are merely characterized by a set of regressive and constructive traits. However, these traits show a high level of convergent evolution, leading to a typical "troglomorphic" appearance. Figure 6.2 shows troglobitic representatives of different fish orders. However, according to developmental constraints reflected by the characteristic morphological organization of every group involved, not every illustrated example exhibits the complete set of typical troglomorphic traits. The degree of adaptation to the cave environment is also important for the superficial perceptibility of these characteristics. But thorough investigations usually reveal common troglomorphic trends. Most cave fishes have comparatively slender, elongated bodies, like *Paracobitis smithi* (Fig. 6.2D), *Triplophysa xiangxiensis* (Fig. 6.2E), *Horaglanis krishnai* (Fig. 6.2F), and *Lucifuga subterraneus* (Fig. 6.2J). The synbranchid eel *Ophisternon infernale* (Fig. 6.2L) even has a whip-shaped body. In fact, the Australian synbranchid *Ophisternon candidum* seems to be the longest cave-fish species, with a total length of about 370 mm (Mees, 1962) (but with a diameter less than 10 mm). The less slender *Amblyopsis spelaea* (Fig. 6.2Id.) reaches a length of 200 mm (Greenwood, 1967). The heads of troglobitic fishes are comparatively large and often distinguished by a depressed, shovel-like snout, as in *Sinocyclocheilus hyalinus* (Fig. 6.2C), *Speoplatyrhinus poulsoni* (Fig. 6.2If), *Lucifuga subterraneus* (Fig. 6.2J) and *Ophisternon infernale* (Fig. 6.2L). Some cave fish forms show a remarkable elongation of body appendages. For instance, *Rhamdia zongolicensis* (Fig. 6.2H) has very long maxillary barbels and *Triplophysa xiangxiensis* (Fig. 6.2E) possesses extremely elongated pectoral fins. In many species, a marked development of the mechanosensory lateral line system can be observed. The bythitid brotulas, like *Lucifuga subterraneus*, have a very wide head-canal system. Additionally, protruding free neuromasts are scattered over the body surface. Numerous elongated free neuromasts, sometimes elevated on dermal ridges, can also be seen on the skin of the amblyopsid and gobiid cave fishes (Figs. 6.2I, 6.2M.), especially on their heads. A common troglomorphic trend is the reduction of scaling, culminating in its complete absence in cave fishes like *Sinocyclocheilus hyalinus* (Fig. 6.2C). Most cave fishes belong to families which are adapted to a bottom-dwelling way of life. Exceptions are the troglobitic forms of the Characidae and Poeciliidae. The epigean representatives of these families normally are active swimmers in the whole water column, and often show schooling behaviour (Parzefall, 1993).

The majority of the troglobitic fishes so far known live in subterranean waters of the tropics and subtropics. Nevertheless, some species occur in the transition zone to the temperate latitudes, for example the amblyopsids in the United States. However, troglobitic vertebrates are absent from areas that have been subject to glaciation during the Pleistocene. The biogeographic distribution of cave fishes can be seen in Table 6.2.

Table 6.2
Biogeographic distribution of troglobitic fishes

Biogeographic region	Number of species	Families represented
Nearctic	9	Characidae, Ictaluridae, Amblyopsidae
Neotropical	24	Characidae, Pimelodidae, Trichomycteridae, Astroblepidae, Loricariidae, Sternopygiidae, Bythitidae, Poeciliidae, Synbranchidae
Palearctic	7	Cyprinidae, Balitoridae, Gobiidae
Oriental	18	Cyprinidae, Cobitidae, Balitoridae, Clariidae, Synbranchidae, Gobiidae, Eleotridae
Ethiopian	6	Cyprinidae, Clariidae
Madagascan	3	Eleotridae
Australian	3	Synbranchidae, Eleotridae

The Neotropical and Oriental regions possess the greatest diversities of cave fishes. In the Neotropical region most of the families involved are obviously primary freshwater fishes. The majority belong to the siluriforms. The pantropical, secondary freshwater synbranchid eels have troglobitic representatives not only in the Neotropical region but in the Oriental and Australian regions, too. It seems quite astonishing that troglobitic Bythitidae inhabiting anchialine or freshwater caves are at present known only from the Neotropical region. Doubtlessly they are derived from marine ancestors, and the family is widely distributed in the Atlantic, Pacific and Indian Oceans. In fact, a troglobitic representative of the genus *Ogilbia*, *O. pearsei*, lives in the caves of the Yucatan Peninsula on the Atlantic side of the Neotropical region, whereas the congeneric *O. galapagosensis* is restricted to the Galapagos Islands in the Pacific (Cohen and Nielsen, 1978). The troglobitic genus *Lucifuga* comprises four described species living in anchialine caves of Cuba, and another species in the "blue holes" of the Bahamas (Nalbant, 1981). Rumours of other blind, cave-dwelling Ophidiiformes in Jamaica were mentioned by Eigenmann (1909) and again by Cohen and Robins (1970). It seems reasonable to expect the discovery of additional troglobitic Ophidiiformes in other parts of the world.

The cave-fish assemblage of the Old World is dominated by cyprinids of barbid habitus and balitorid loaches. The number of troglobitic siluriforms is surprisingly low in comparison to the Neotropical region. The cave-fish fauna of the Ethiopian main region seems relatively poor. This impression is enhanced by the fact that half of the species are restricted to Somalia, and a troglobitic population of *Garra barreimiae* is only known from Oman on the Arabian peninsula near the transition to the Oriental region. From the major part of the African continent only two cave fishes have been reported. Three troglobitic sleepers (Eleotridae) can be found in the Madagascan subregion. Eleotrid and gobiid cave species are restricted to Australia and to islands like Madagascar, Samar (Philippines), Japan and New Guinea. Australia is known for the low diversity of its freshwater fish fauna, so it is not surprising that one of the two known troglobitic species is a secondary freshwater fish, the synbranchid eel *Ophisternon candidum*; the other is an eleotrid of marine origin, *Milyeringa veritas*. In accordance with these origins, they can only be found in subterranean waters near the coast in their distribution area in the Northwest Cape Range (Mees, 1962; Humphreys and Adams, 1991).

The cave-fish fauna of the Nearctic region is dominated by two families which can only be found there: the Ictaluridae and the Amblyopsidae. This restriction is remarkable since the Cyprinidae, which constitute a major part of the cave-fish assemblage of the Old World, occur in North America too; but obviously they did not evolve troglobitic representatives there. The Amblyopsidae represent a unique teleost family in that the majority of its species are troglobitic. The swampfish *Chologaster cornutus* is the only epigean species of the family. The springfish *Chologaster agassizi* is troglophilic. It inhabits springs and caves and emerges at night to exploit epigean food resources. The remaining four species of the Amblyopsidae are truly troglobitic (Fig. 6.2I). In all other teleost families with troglobitic representatives, the few cave-dwellers constitute merely a marginal group (Greenwood, 1967). Another part of the Nearctic cave-fish fauna are the troglobitic populations of *Astyanax fasciatus* (Characidae) in northern Mexico (Mitchell et al., 1977). Nevertheless, the Nearctic representatives of the characids are relatively recent invaders of Neotropical origin.

Holarctic cave fishes are restricted to southern regions which were not covered by glaciers during the Pleistocene. In the Palearctic region, most cave-fish populations live near the transition to the Oriental region. In Europe there are no troglobitic fishes at all (Greenwood, 1967), possibly because most of the primary freshwater fishes became extinct during the Pleistocene since the retreat to southerly refugia was severely restricted by the geological barriers of the Alps and the Pyrénées. The major part of the present European fish fauna consists of relatively recent invaders of southeastern origin (Karaman, 1923). Possibly, there has not been enough time for the evolution of troglobitic species.

Cave salamanders

Some groups of amphibians are more likely to colonize caves than others. No troglophilic or troglobitic frogs (Anura) have been found, probably because they cannot complete their life cycle in the cave. As an explanation, it has been proposed that the tadpoles of many species rely on green algae for food, and green plants cannot grow in the darkness of the cave (Mohr and Poulson, 1966). Additionally, the mode of locomotion of adult frogs might be disadvantageous in the dark cave environment (Eigenmann, 1909). Nevertheless, not all anuran species have exclusively herbivorous tadpoles, and the adults of many species are nocturnal. Hence, the lack of troglobitic anurans is still unexplained. The same holds true for the Gymnophiona, since some of them seem to be quite well preadapted for a troglobitic existence. In contrast, a number of troglophilic and troglobitic salamanders (Urodela) are known (Cooper, 1961; Clergue-Gazeau, 1974; Sweet, 1986).

Of the 4522 amphibian species of the world, only 392 (11.5%) are Urodela (= Caudata) (Duellman, 1993). Among the salamanders which undergo complete metamorphosis, a few members of the families Ambystomatidae, Dicamptodontidae, Hynobiidae and Salamandridae are reported as troglophiles (Sweet, 1986). The salamandrid *Euproctus asper* frequently can be found in caves of the Pyrénées. But truly

Fig. 6.3A. *Proteus anguinus* (Proteidae) (redrawn from Sweet, 1986).

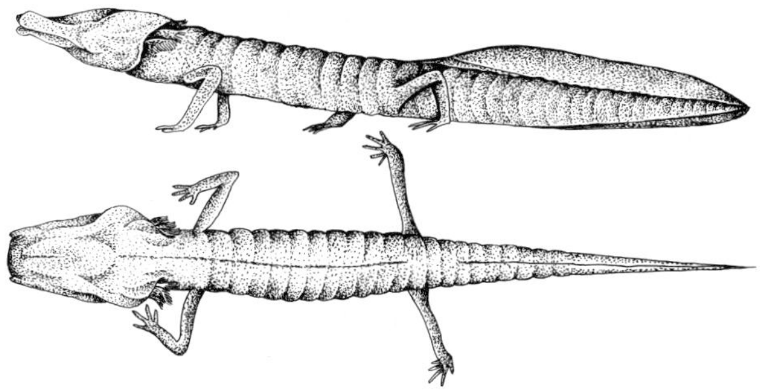

Fig. 6.3B. *Typhlomolge rathbuni* (Plethodontidae); lateral and dorsal view (redrawn from Eigenmann, 1909).

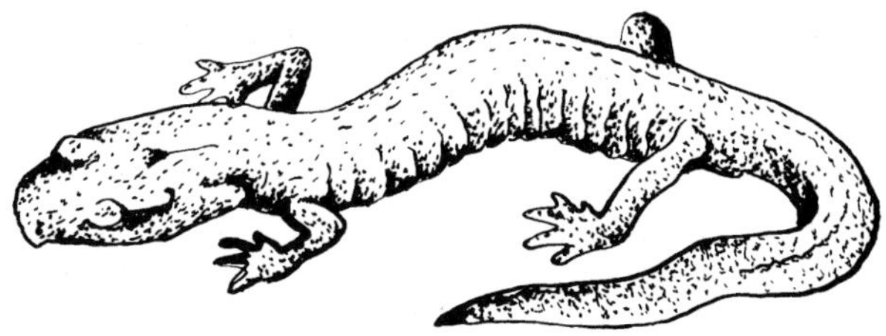

Fig. 6.3C. *Typhlotriton spelaeus* (Plethodontidae); metamorphosed adult (redrawn from Sweet, 1986).

troglobitic urodelan species are restricted to two families: the Plethodontidae and the Proteidae. Both families show a disjunct distribution pattern. They have some representatives in the Palearctic region, but their main distributions are in the New World. The Proteidae comprise six species, but only one is troglobitic. It is the famous grotto olm *Proteus anguinus,* which might

Table 6.3
List of troglobitic salamanders (**Amphibia**), their synonyms and distribution[1]

	Species	Distribution
Urodela		
	Proteidae	
1	*Proteus anguinus* Laurenti, 1768 *Hypochthon* spp. Fitzinger, 1850 (7 species) *Proteus anguinus anguinus*[2] in Sket & Arntzen, 1994	dinaric karst from Trieste, Italy in the northwest over Slovenia and Croatia to the Trebisnica River in Bosnia-Herzegovina in the southeast
2	*Proteus anguinus parkelj* Sket & Arntzen, 1994	springs near Crnomelj in the Bela Krajina, southern Slovenia
	Plethodontidae	
3	*Eurycea tridentifera* Mitchell & Reddell, 1965 *Eurycea latitans* Smith & Potter, 1946 (in part) *Eurycea troglodytes* Baker, 1957 (in part) hybrid forms: *E. tridentifera* *E. neotenes*	caves in the Cibolo sinkhole plain and adjacent regions, Comal and Bexar Counties, Edwards Plateau, Texas, USA
4	*Gyrinophilus gulolineatus* (Brandon 1965) in Brandon et al., 1986 *Gyrinophilus palleucus gulolineatus* Brandon 1965	caves at the east bank of the Tennessee River Valley (Appalachian Valley), Tennessee, USA
5	*Gyrinophilus palleucus necturoides* Lazell & Brandon, 1962	Big Mouth Cave, Elk River drainage, Grundy County, southern Cumberland Plateau, southern Tennessee, USA
6	*Gyrinophilus palleucus palleucus* McCrady, 1954	caves of the southern Cumberland Plateau, southern Tennessee and northern Alabama, USA
7	*Gyrinophilus subterraneus* Besharse & Holsinger, 1977	General Davis Cave near Alderson, Greenbrier County, (Greenbrier Valley), West Virginia, USA
8	*Haideotriton wallacei* Carr, 1939	caves of the Dougherty Plain of southwestern Georgia and adjacent northwestern Florida, USA
9	*Typhlomolge rathbuni* Stejneger, 1896	San Marcos pool of the Balcones Aquifer, Hays County, Texas, USA
10	*Typhlomolge robusta* Longley, 1978	Balcones Aquifer to the north and east of the Blanco River, Hays County, Texas, USA
11	*Typhlotriton spelaeus* Stejneger, 1892	southwestern Ozark Plateaus (Springfield and Salem Plateaus), of southwest Missouri, southeast Kansas, northeast Oklahoma, and northwest Arkansas, USA

[1] Discarded synonyms are listed after the numbered, accepted species.
[2] Species names thus marked designate forms of uncertain taxonomic or ecological recognition.

be subdivided into two subspecies (Sket and Arntzen, 1994; but compare Grillitsch and Tiedemann, 1994). *Proteus* (Fig. 6.3A) is the only Palaearctic representative of the family, and the only cave salamander of the Old World. Additionally, it is the only troglobitic vertebrate of any group in Europe (Vandel and Bouillon, 1959). *Proteus* is distributed in the Dinaric karst, from Trieste in Italy in the northwest across Slovenia and Croatia to Bosnia-Hercegovina in the southeast. The family Plethodontidae comprises 246 species and is distributed mainly in the Nearctic region and in the adjacent Neotropical Middle America and northern South America. But the genus *Hydromantes* has members in southwestern North America (California) as well as in southeastern France, in northern and central Italy, and on Sardinia (Wake et al., 1978). Many members of the family Plethodontidae are troglophilic (e.g., species of *Aneides, Bolitoglossa, Chiropterotriton, Desmognathus, Plethodon, Pseudoeurycea* and *Thorius*), and some are rarely found far from caves (e.g., *Eurycea longicauda, E. lucifuga, E. multiplicata, E. neotenes, E. tynerensis, Gyrinophilus porphyriticus* and *Hydromantes* spp.) but only nine species or subspecies are troglobitic (Table 6.3) and they are restricted to North America (Sweet, 1986). As a whole, only about 0.24% of all amphibians and about 2.8% of the Urodela are troglobites.

Cave salamanders share some troglomorphic trends

with cave fishes. Advanced cave-adapted forms, like *Typhlomolge rathbuni* (Fig. 6.3B), also have reduced eyes and body pigments, and exhibit a slender body with a large, anteriorly depressed, shovel-like head. *Proteus anguinus* seems to be the longest troglobitic salamander. It may reach more than 40 cm in total length (Briegleb, 1962). The longest Nearctic cave salamander is probably *Gyrinophilus palleucus necturoides,* with a maximal total length of 18.4 cm (Lazell and Brandon, 1962). The smallest one may be *Haideotriton wallacei,* reaching a total length of only 7.55 cm (Pylka and Warren, 1958).

With the exception of *Typhlotriton spelaeus* (Fig. 6.3C) and *Gyrinophilus subterraneus,* all other cave salamanders are usually paedomorphic – that is, they attain sexual maturity without metamorphosis. Adult paedomorphic salamanders retain some larval characteristics, like the feathery external gills, a fin-shaped tail and a weak ossification of the skeleton. Consequently, these species are restricted to the aquatic cave environment. Because of the blood in the well-developed capillary system, the external gills are brightly red and in sharp contrast to the whitish body of the troglobitic species. Beside the external gills, *Proteus anguinus* possesses lungs. Although Vandel and Bouillon (1959) stated that they could not observe air-breathing behaviour in adult *Proteus,* Briegleb (1962) reported that adult individuals periodically swim to the water surface to gasp for breath. Possibly, the lungs and an air-filled buccal cavity may function additionally as transducers of sound waves in underwater conditions (Bulog, 1989). In contrast, all Plethodontidae are lungless. Completely metamorphosed terrestrial specimens use their skin and probably the epithelium of the buccal cavity and pharynx for respiration. Plethodontid cave salamanders have a reduced number of costal grooves and trunk vertebrae in comparison to the closest epigean relatives. Additionally, the troglobitic plethodontids show a trend to a weakly developed muscular system and a markedly elongation of the thin legs, giving them a fragile appearance (Fig. 6.3B). Interestingly, this is not true of the European troglobitic proteid *Proteus anguinus* (Fig. 6.3A). Its legs are reduced to short, thin appendages with a reduced number of toes.

NEOTENY IN CAVE VERTEBRATES

Neotenic trends in cave fishes are less obvious than in cave salamanders (see below). The comparatively larger head, larger fins, and a longer caudal peduncle, which are frequently observed in cave fishes, may be correlated with the trend to a reduced body size (Poulson, 1964). Thus, larval characteristics retain a certain influence on the body shape. The reduction in scaliness of many cave fishes possibly also represents a neotenic trait (Banister, 1984). A deep adipose ridge between the dorsal and the caudal fin of *Paracobitis smithi* has been considered to represent a rudiment of the larval median fin fold (Greenwood, 1976; Banister, 1984). Strong evidence for a neotenic trait has been found in the amblyopsid *Speoplatyrhinus poulsoni.* This highly cave-adapted species lacks bifurcate fin rays even in the adult stage (Cooper and Kuehne, 1974). Elsewhere among the Amblyopsidae, this feature has been observed in very young *Typhlichthys subterraneus.* Additionally, *S. poulsoni* resembles juvenile *T. subterraneus* in head shape and body proportions (Culver, 1982). The ictalurid catfishes of the artesian waters in Texas, *Satan eurystomus* and *Trogloglanis pattersoni,* show a reduced ossification of the skeleton, a number of paedomorphic skeletal traits, and a retained connection of head and trunk kidney, normally only seen in juvenile ictalurids (Langecker and Longley, 1993b).

Since the metamorphosis in amphibians is usually connected with a shift from the aquatic to the terrestrial habitat and with distinct morphological and behavioural changes, the absence of this process and hence the retention of the larval characteristics in the mature animal (paedomorphosis) is a very conspicuous feature. It is an interesting fact, that most troglobitic salamanders are paedomorphic. Approximately half of all known paedomorphic salamander species are troglobites (Sweet, 1986). It seems likely that the food scarcity of the terrestrial cave habitat favours the retention of larval characteristics in the adult, since the aquatic cave habitat is more reliable (Bruce, 1979; Culver, 1982). Consequently, nearly all troglobitic salamanders are strictly aquatic, and even the adults of the few species that undergo metamorposis are never found far from water. Another explanation for paedomorphosis in cave salamanders might be the generally low food supply in the subterranean environment (Brandon, 1971b). The energetic costs of metamorphosis might outweigh any benefits, and selection would act towards prolongation of the larval phase. A good example is *Gyrinophilus subterraneus,* a species approaching paedomorphosis as witnessed by its very large larval size relative to adult size. This indicates that the approach to paedomorphosis in cave salamanders has been through retardation of somatic development – that is, neoteny (Bruce, 1979). There is no evidence for the alternative approach through an acceleration of sexual development – that is, paedogenesis (Culver, 1982). However, the reasons for paedomorphosis in distinct phyletic lineages of cave salamanders may be different. Metamorphosis of the epigean species *Gyrinophilus porphyriticus* in the eastern United States involves a dietary shift from aquatic invertebrates to other salamander species like *Eurycea lucifuga* or *Plethodon glutinosus* (Bruce, 1979). In the cave environment, troglophilic salamanders may be too rare as prey to yield a reliable and sufficient food supply. Hence, selection may favour the retention of the larval stage as in *Gyrinophilus palleucus,* which allows a continued exploitation of aquatic invertebrates as the more reliable food source. However, some exclusively cavernicolous populations of *Gyrinophilus porphyriticus,* which still show complete metamorphosis, are known. Transformed individuals of these populations are apparently able to rely mainly on terrestrial invertebrate food (Culver, 1982). However, the caves they inhabit seem to be quite food-rich. On the other hand, transformed individuals of the epigean forms of the genus *Eurycea* in Texas normally feed on riparian invertebrates, but have to cope with very dry terrestial conditions and small habitat ranges. This may favour the retention of the larval stage to avoid the rigour of the terrestial habitat. The case of the Texas species of *Eurycea* is discussed in more detail below. Only four cave salamander species are known to undergo metamorphosis in their natural habitat, and only in *Gyrinophilus subterraneus* and *Typhlotriton spelaeus* is it a regular process. The populations of *G. gulolineatus* and *G. palleucus palleucus* are normally neotenic, but metamorphosed individuals can occur rarely (Brandon et al., 1986). Generally, these specimens are very emaciated, which underlines the disadvantageous effect of metamorphosis in food-poor cave environments. In fact, *Gyrinophilus subterraneus* and *Typhlotriton spelaeus* can thrive only in caves with a rich food supply, where they can afford metamorphosis (Mohr and Poulson, 1966; Besharse and Holsinger, 1977).

In some neotenic salamanders metamorphosis can be induced artificially by application of the hormone thyroxine. But in neotenic cave salamanders which are probably phylogenetically old, like *Eurycea tridentifera*, *Haideotriton wallacei*, *Proteus anguinus* and *Typhlomolge rathbuni,* complete metamorphosis can no longer be released by such a treatment (Durand and Vandel, 1968; Sweet, 1977). Their tissue seems to have lost its sensitivity for thyroxine. Nevertheless, these cave salamanders still have a thyroid gland system (Gorbman, 1957; Vandel and Bouillon, 1959), and *Proteus anguinus* is known to produce thyroxine. This indicates that there are other important functions of this hormone which are not related to metamorphosis (for a further discussion see Chapter 7, this volume).

It has been stressed that neoteny may be an important preadaptation for cave colonization by salamanders (Mitchell and Smith, 1971). This opinion is confirmed by the existence of *Proteus anguinus* which belongs to the "perennibranchiates", a group that never exhibits metamorphosis. Nevertheless, there is no conclusive evidence that the ancestor of any troglobitic species of the Plethodontidae was necessarily already neotenic. The examples of *Gyrinophilus subterraneus* which approaches neoteny, and the likewise cave-adapted congeners *G. gulolineatus* and *G. palleucus* which have already abandoned metamorphosis, seem to indicate that neoteny is an adaptation which evolved within the cave environment. This might explain why there are so many troglobitic species among the neotenic salamanders. However, the relation of paedomorphosis and cave adaptation might be more complicated, because some salamander species can show a remarkable variability in their life cycle depending on the prevailing environmental conditions. An example is *Eurycea neotenes,* a mainly epigean salamander with a disjunct distribution in the arid Balconian Biotic Province of the Edwards Plateau in central Texas (Raun, 1971). Since salamanders need a mesic environment, *E. neotenes* can only be found as a relict in damp headwater canyons, springs and caves within its distribution range. The large number of allopatric but phenotypically similar populations living under comparable environmental conditions complicate a systematic assessment. Although there is a compact morphocline (Hendricks, 1973), *E. neotenes* may rather represent a complex of sibling species subject to parallel evolution (Sweet, 1984). Most populations are paedomorphic, but a few have been found which contain a proportion of metamorphosed individuals (Sweet, 1977). Some exclusively cavernicolous populations are known, and even in two of these metamorphosis has been observed. The occurrence of transformed individuals in some populations seems to depend on the reliability of the spring discharge (Sweet, 1977).

Apparently, only temporally stable mesic conditions allow the process of metamorphosis to develop as a successful strategy within the life cycle. Habitats with episodic fluctuations in spring discharge are inhabited solely by paedomorphic populations. During a dry period of weeks or months the resident salamander population is obliged to retreat underground through the spring openings. As already mentioned, the subterranean environment is disadvantageous for transformed individuals. Hence, paedomorphosis in *E. neotenes* seems to be a response to selection for the ability to pass dry periods in subterranean aquatic refugia (Sweet, 1977). If the epigean habitat becomes permanently uninhabitable, the local salamander population is subject either to extinction or to further adaptation to the subterranean environment. Obviously, the process of adaptation to the cave environment within the genus *Eurycea* is still currently in progress (Baker, 1957; Sweet, 1986). *Eurycea neotenes* is known to hybridize occasionally with the troglobitic and paedomorphic *E. tridentifera*. Two hybrid populations have even been described as distinct species: *E. latitans* and *E. troglodytes*. These taxa are now considered as invalid (see Sweet, 1984). However, the status of the parental forms remains unresolved because the hybrid specimens seem to be fertile as indicated by the normally developed gonads. Hence, the parental forms apparently do not show complete reproductive isolation as would be required for their continued recognition as distinct species. However, Sweet (1984) preferred to retain the original species names because *E. tridentifera* continues to be morphologically quite distinct from even the most highly cave-adapted specimens of *E. neotenes*, although the distribution range of *E. tridentifera* is completely contained within the range of *E. neotenes*. Where they live in sympatry, they normally exhibit assortative mating. The few habitats in which hybridization occurs seem to be atypical in several respects. However, the case of these two salamander forms resembles that of the epigean and troglobitic forms of the characid fish *Astyanax fasciatus* in many ways. Accordingly, the continued recognition of *E. neotenes* and *E. tridentifera* as two distinct species seems questionable despite their morphological distinctness.

ECOLOGICAL NOTES

The usually low food supply and the top position in the food pyramid has important implications for the population ecology of cave vertebrates. However, only a few studies have been concerned with this topic, probably on account of the complications and difficulties of the cave habitat. The ecology of cave salamanders, especially, is largely unknown. Most papers are concerned with taxonomic and systematic problems in the first place, and may contain only sporadic notes about ecological and behavioural observations.

One of the most influential investigations concerned with the ecology of a troglobitic vertebrate is T.L. Poulson's study of a population of the cave fish *Amblyopsis spelaea* in the Upper Twin Cave, Indiana. The fishes inhabit a small watercourse with consolidated mud-rock substrates in shoals and silt-sand substrates in pools (Poulson, 1963). Larger fishes are found in deeper water, especially along underwater ledges. Each individual seems to occupy a limited area. It moves upstream along a ledge for about 20 m, searching for prey, predominantly aquatic isopods, under rocks and in crevices. The time a fish spends under a rock is directly related to the number of prey organisms to be found there. Finally, the fish swims to the surface, where the current is swifter, and floats down to the starting point where it begins to move upstream again (Mohr and Poulson, 1966). This behaviour indicates a well-developed topographic memory. The population shows density-dependent size regulation related to the food supply. Although there are some fluctuations in age distribution and population size, the total biomass remains relatively stable (Poulson, 1969). As the population reaches its size peak, there is such competition for food that the adult fishes get barely enough to stay alive. For the production of eggs, a female would need about three times as much food as usual. Hence, reproduction is suppressed. On the other hand, when the number of adult individuals is low, there may be enough food available for a few females to produce eggs. The other less fortunate females do not breed. In fact, despite increased longevity only a very small proportion of females reproduce at all in their lifetime. When a low population of adults coincides with a very favourable food supply, more females than normal are able to reproduce. Subsequently, when the food supply decreases the numerous juveniles compete with the adults for food. However, *A. spelaea* exhibits a density-dependent cannibalism. Cannibalism as a means of regulation may be adaptive in the simple food web of the cave, since it adds another link which may help to stabilize the community (Poulson, 1969).

The cave salamander *Gyrinophilus palleucus necturoides* also shows a pronounced cannibalistic behaviour (Lazell and Brandon, 1962). However, this behaviour may be not only related to the regulation of population density and size in the cave environment. *Gyrinophilus palleucus* is probably derived from a *G. porphyriticus*-like ancestor. Since metamorphosed individuals and large larvae of *G. porphyriticus* feed mainly on other salamanders, the cannibalism exhibited by *G. palleucus* may be a reminiscence of the former dietary specialization. Nevertheless, the troglophilic salamanders *Eurycea lucifuga* and *Plethodon glutinosus* have also been observed in the caves inhabited by *G. palleucus,* and the occasionally occurring larvae of the former species may be part of the diet of *G. palleucus* (Cooper and Cooper, 1968).

The reduced growth rate of cave vertebrates implies a reduced energy demand per unit time. In consequence, more individuals can survive on a defined amount of food. Hence, through a lower metabolic rate, the reduced energy demand per unit time allows an increase of population density with increased cave adaptation. For example, cavernicolous populations of the troglophilic amblyopsid fish *Chologaster agassizi* have only about 50 individuals per hectare, whereas populations of *Amblyopsis spelaea* have 520 animals per hectare (Poulson, 1969). This allows an increase in the number of breeding females. In consequence, the genetic variability within the population may be enhanced. Some features of the population ecology of troglobitic amblyopsids have also been observed in *Pimelodella kronei,* a troglobitic catfish from Brazil (Trajano, 1991). Its growth rate is reduced and the lifetime increased. Only a small proportion of the animals investigated had mature gonads, indicating infrequent reproduction. The population density is comparable to that of *A. spelaea.* The individuals seem to occupy a territory, and defending of food sources against conspecifics has been observed. In summary, *P. kronei* has a K-selected life style like the troglobitic amblyopsids.

Among cave-dwelling salamanders the life strategy of *Typhlotriton spelaeus,* is unique and its assessment as a true troglobite or merely a troglophile is controversial. The larvae have well-developed, functional eyes and a dark body pigmentation. They live mainly in the twilight of the cave entrance where food is relatively abundant. There they prey on insect larvae, crustaceans, and snails. But when food is scarce, they may even be found outside the cave. However, larvae may occasionally be found in completely dark parts of the cave, too (Brandon, 1971a). With the onset of metamorphosis after two to three years of larval development, the animals retreat into the deeper parts of the cave. The individuals become progressively paler and the eyes degenerate. Even adult individuals stay near the water. Often they can be observed on moist vertical walls which rise out of the water. When disturbed, they readily drop into the water and swim away. The food consists mainly of riparian invertebrates and guanobites. In some caves, heliomyzid flies constitute the major part of their diet. When food within the cave becomes inadequate, however, the adult individuals will travel to the entrance zone or even out of the cave, where they stay hidden in crevices near the entrance during the daylight hours and emerge at night to search for food (Nicholas, 1971). Such behaviour has been reported for the neotenic *Gyrinophilus gulolineatus,* too (Cooper and Cooper, 1968). The reproductive activities of *Typhlotriton* seem to be correlated with the wettest part of the year in spring and summer, when food is most abundant. The occurrence of adult females which had already bred at least once in their life, as indicated by anatomical hints, with only tiny, unyolked ova in their ovaries indicate that these individuals are in a nonbreeding year. Apparently, the accumulation of yolk requires more than one year and may depend on the amount of availiable food (Brandon, 1971a).

Although *Proteus anguinus* is neotenic and possesses only weak limbs, it can occasionally be found out of the water on flat and moist banks, where it has a snake-like mode of locomotion. In this way, *Proteus* is able to prey on invertebrates in small isolated pools (Briegleb, 1962). Probably, this behaviour is also important for dispersal. One of the most interesting recent discoveries with regard to cave salamanders was the existence of another form of *Proteus* in Slovenia, formally described as *Proteus anguinus parkelj*. It is quite different from the troglobitic *Proteus anguinus anguinus* in having a nearly black dorsal pigmentation, externally visible eyes with a conjunctiva, a stouter head, a smaller number of teeth, more trunk vertebrae, a shorter tail, and shorter legs (Sket and Arntzen, 1994). With the exception of the shorter legs, all other characteristics indicate a markedly lower degree of cave adaptation in *P. a. parkelj*. All specimens collected have been caught in springs. Stomach contents consisted predominantly

Table 6.4
Sympatrically occurring troglobitic fish species

Species	Distribution
Amblyopsis spelaea (Amblyopsidae) *Typhlichthys subterraneus* (Amblyopsidae)	Mammoth Cave, Kentucky, USA
Satan eurystomus (Ictaluridae) *Trogloglanis pattersoni* (Ictaluridae)	Edwards Aquifer, Texas, USA
Lucifuga dentatus (Bythitidae) *Lucifuga simile* (Bythitidae) *Lucifuga subterraneus* (Bythitidae)	anchialine waters of western Cuba
Ophisternon infernale (Synbranchidae) *Ogilbia pearsei* (Bythitidae)	subterranean waters of Yucatan, Mexico
Trichomycterus sp. (Trichomycteridae) *Ancistrus cryptophthalmus* (Loricariidae) *Eigenmannia vicentespelaea* (Sternopygidae)	São Vicente cave system, Goias State, Brazil
Caecocypris basimi (Cyprinidae) *Typhlogarra widdowsoni* (Cyprinidae)	sinkholes near Al Haditah, Al Anbar Province, Iraq
Iranocypris typhlops (Cyprinidae) *Paracobitis smithi* (Balitoridae)	Kajee-Ru Source, Zagros Mountains, Iran
Nemacheilus troglocataractus (Balitoridae) *Pterocryptis* sp. (Siluridae)	Tham Sai Yok Noi, Kanchanaburi Province, Thailand
Protocobitis typhlops (Cobitidae) *Oreonectes* sp. (Balitoridae)	limestone cave near Xiaao, Guangxi, China
Ophisternon candidum (Synbranchidae) *Milyeringa veritas* (Eleotridae)	subterranean waters of the North-West Cape Range, Australia

of epigean invertebrates. These observations suggest that *P. a. parkelj* is merely a troglophile which takes shelter in the cave during the day and emerges at night to search for food in the epigean environment. The existence of a troglophilic form of *Proteus* sheds a new light on the phylogeny of the genus. Perhaps the troglobitic form does not represent an old relict, as which it has commonly been regarded. This is corroborated by morphological hints like the only moderately reduced eyes and the retained ability to reconstitute a dark body pigmentation in light.

The status of vertebrates as food generalists and top predators of the food pyramid in the usually oligotrophic subterranean environment readily explains why there is normally only a single species of cave vertebrate within a distinct cave system. Obviously, troglobitic populations already established would readily exclude later colonization by less adapted forms. Nevertheless, a number of exceptions are known (Table 6.4). However, on account of a lack of ecological information, most of these examples of syntopic occurrence remain unexplained. Possibly, some sites may represent rare places where the distribution ranges of the normally allopatric species overlap. This is probably the case in the amblyopsid species of the Mammoth Cave (Woods and Inger, 1957). However, an example of niche segregation within the same subterranean habitat is known: the ictalurid catfishes *Satan eurystomus* and *Trogloglanis pattersoni* of the artesian waters of the Edwards Aquifer in southwestern Texas. Adult specimens of *Trogloglanis* (Fig. 6.2G) are characterized by a unique, ventrally placed and toothless sucker mouth, whereas *Satan* has a wide, terminal mouth with teeth in villiform bands like most ictalurids (Langecker and Longley, 1993b). The unusual mouth structure of *Trogloglanis* is related to the mode of food ingestion. Probably, *Trogloglanis* is a grazer and takes up detritus. Its stomach contents consist only of mud-like substances. In contrast, the stomach of *Satan* frequently contains parts of crustacean exoskeletons. These differences in diet are also reflected by the different development of the sensory structures. In *Trogloglanis*, the head canals of the lateral line system are frequently interrupted, narrow, and provided with small pores. But the fleshy lips are densely covered with numerous taste buds and the nares are large. These observations indicate a mainly chemosensory-guided feeding behaviour in *Trogloglanis*. In contrast, *Satan* has a well-developed lateral line canal system with enlarged pores. This mechanosensory system seems well suited for the detection of moving prey. The sympatric situation of the Edwards Aquifer is possibly paralleled by that of the São Vicente cave system in Brazil. The loricariid catfish *Ancistrus cryptophthalmus* is equiped with a sucker mouth and feeds mainly on detritus, whereas an undescribed species of *Trichomycterus* with a mouth more terminally situated feeds preferentially on invertebrates (Trajano and de Souza, 1994). A comparable difference in mouth structure can also be found in sympatric cave fishes from Iraq. *Typhlogarra widdowsoni* has a small, ventrally directed mouth opening, whereas *Caecocypris basimi* has a wide, terminal mouth (Banister and Bunni, 1980). Unfortunately, their feeding behaviour is largely unknown. The sympatric troglobitic fishes in Australia, *Milyeringa veritas* and *Ophisternon candidum*, both feed opportunistically on every invertebrate available. However, stomach content analyses revealed that the synbranchid *O. candidum* prefers to pick up small stygofauna from the bottom, whereas the eleotrid *M. veritas* feeds preferentially on larger accidentals, like insect larvae or terrestrial isopods in the free water or at the surface (Humphreys and Feinberg, 1995). *Gyrinophilus palleucus palleucus* seems to be the only troglobitic salamander that occasionally lives sympatrically with a cave fish. In Shelta Cave, Alabama, and in McKinney Pit it occurs together with *Typhlichthys subterraneus* (Cooper and Cooper, 1968). Sympatry of two troglobitic salamander species has not yet been observed.

In general, caves can be judged as food-poor environments, and the implications of this for colonizing vertebrates have already been discussed. However, some food-rich caves are known too, and some of them are inhabited by vertebrates. For example, the Cueva del Azufre (Cueva de las Sardinas) in Tabasco, Mexico, is the habitat of a cavernicolous form of *Poecilia mexicana* (Fig. 6.2K) (Gordon and Rosen, 1962). The water has a high concentration of hydrogen sulphide, but *P. mexicana* is apparently able to tolerate these normally unfavourable environmental conditions (Peters et al., 1973). However, they are ideal for some bacteria, which results in an abundant production of organic material. This seems to be the basis of the food web in the cave. The Poeciliidae appear not to be preadapted for a cave existence, because their feeding and courtship behaviour usually depend on visual signals, and accordingly the eyes are well developed. However, *P. mexicana* is able to reproduce in complete darkness, an unusual feature among the Poeciliidae. Apparently, this behavioural preadaptation together with the exceptional food abundance enabled the colonization of this particular cave. Troglomorphic traits are faintly developed in the cave specimens. The size of the eyes is only slightly reduced, and they remain functional. Specimens from the deeper parts of the cave are almost colourless. Interestingly, a continuous morphological gradient from the epigean habitus at the cave entrance to the most troglomorphic habitus in the deepest part of the cave can be observed. This is exceptional among cavernicolous fishes with a high variability in troglomorphic features, since normally every morphological variant can be found throughout the range of cave habitats (Weber, 1996).

It has been claimed that troglobitic vertebrates have largely lost their ablity for regulatory adaptation in the constancy of the cave environment (Heuts, 1951). However, numerous observations prove that troglobites can tolerate even wider fluctuations than occur in their natural environment (Poulson, 1964). The cave salamanders *Proteus anguinus* (Briegleb, 1962) and *Typhlomolge rathbuni* (Uhlenhuth, 1921) proved to be able to tolerate extreme temperature variations, at least for a short period of time.

Normally, there are no predators of troglobitic vertebrates in the cave, but occasional accidentals may constitute a threat. However, endoparasitic Plathelminthes have sometimes been found in cave fishes, for example in *Lucifuga* (Eigenmann, 1910) and *Trogloglanis pattersoni* (Langecker and Longley, 1993a).

Cave waters share some enviromental conditions with the deep sea – for instance, the permanent darkness, relatively constant temperature, and food scarcity in the free water (or bathypelagial) envionment (Mohr and Poulson, 1966). However, the fishes of these two biotas show some differences. Bathypelagic fishes are normally neither blind nor pigmentless. They show weak ossification and loosely packed muscles. The swim bladder is reduced or filled with lipids. In many deep-sea fishes functional eyes are retained because of the phenomenon of bioluminescence – the biogenic production of light. A gas-filled swim bladder is ineffective and metabolically expensive under conditions of high hydrostatic pressure. Neutral buoyancy is achieved by the storage of fat, and reduced ossification of bones and scales. The accumulation of fat represents also an adaptation to the food-poor environment. The same holds true for the weakly developed muscular system, related to energy economy. Beside the parallel need of energy economy, the other constraints of the deep-sea environment normally are not apparent in the cave. Usually, cave fishes are not exposed to high hydrostatic pressure. Bioluminescence has never been observed in cave vertebrates (Vandel, 1964). Normally, the skeleton is well ossified. However, a trend to a diminution in size of the swim bladder has been confirmed for troglobitic *Astyanax*. Probably, this reduction is correlated with the enhanced storage of fat in the tissue. Remarkable exceptions are troglobitic fishes living in artesian waters, like the ictalurid catfishes of the Edwards Aquifer. *Satan eurystomus* and *Trogloglanis pattersoni* are exposed to a high hydrostatic pressure in their natural environment. In consequence, they exhibit some morphological parallels to deep-sea fishes. A swim bladder is lacking, the bones are poorly ossified, and the muscular system is weakly developed (Langecker and Longley, 1993a). However, these trends are not as advanced as in deep-sea fishes. This reflects the different ages of the two biotas. The deep sea is a very old environment and many of its inhabitants have been subject to a long period of adaptive evolution. In comparison, the cave environment is relatively young. Probably, *Satan* and *Trogloglanis* may represent examples of the oldest living troglobitic fishes, with a supposed Miocene origin (Lundberg, 1982).

ACKNOWLEDGEMENTS

I am indebted to E. Trajano and M. Kottelat for valuable informations about South American and Asiatic cave fishes. I want to thank H. Wilkens and T.G. Langecker and J. Hallermann for their help in the completion of the manuscript. The figures have been made by M. Hänel. The Konrad-Adenauer-Stiftung was so kind as to support me with a grant.

REFERENCES

Alvarez, J., 1946. Revision del genero *Anoptichthys* con descripcion de una especie nueva (Pisc., Characidae). *An. Esc. Nac. Cienc. Biol. Mexico City,* 4(2-3): 263–282.

Alvarez, J., 1947. Descripcion de *Anoptichthys hubbsi* caracinido ciego de la Cueva de Los Sabinos, S.L.P. *Rev. Soc. Mex. Hist. Nat.,* 8(1-4): 215–219.

Baker, J.K., 1957. *Eurycea troglodytes*: a new blind cave salamander from Texas. *Tex. J. Sci.,* 9: 328–336.

Banister, K.E., 1984. A subterranean population of *Garra barreimiae* (Teleostei: Cyprinidae) from Oman, with comments on the concept of regressive evolution. *J. Nat. Hist.,* 18: 927–938.

Banister, K.E. and Bunni, M.K., 1980. A new blind cyprinid fish from Iraq. *Bull. Brit. Mus. Nat. Hist. Zool.,* 38(3): 151–158.

Barr Jr., T.C., 1968. Cave ecology and the evolution of troglobites. In: T. Dobzhansky, M.K. Hecht and W.C. Steere (Editors), *Evolutionary Biology,* Vol. 2. North-Holland, Amsterdam, pp. 35–102.

Berti, R. and Ercolini, A., 1991. *Caecogobius cryptophthalmus* n. gen. n. sp. (Gobiidae Gobiinae), the first stygobic fish from Philippines. *Trop. Zool.,* 4: 129–138.

Bertin, L., 1958. Écologie. Poissons Cavernicoles. In: P.-P. Grassé (Editor), *Traité de Zoologie, 13, (3). Agnathes et Poissons. Anatomie, Éthologie, Systématique.* Masson, Paris, pp. 1918–1923; pp. 2660–2662.

Besharse, J.C. and Brandon, R.A., 1974. Postembryonic eye degeneration in the troglobitic salamander *Typhlotriton spelaeus*. *J. Morphol.,* 144(4): 381–406.

Besharse, J.C. and Holsinger, J.R., 1977. *Gyrinophilus subterraneus,* a new troglobitic salamander from southern West Virginia. *Copeia,* (4): 624–634.

Brandon, R.A., 1971a. Correlation of seasonal abundance with feeding and reproductive activity in the grotto salamander (*Typhlotriton spelaeus*). *Am. Midl. Nat.,* 86(1): 93–100.

Brandon, R.A., 1971b. North American troglobitic salamanders: some aspects of modification in cave habitats, with special reference to *Gyrinophilus palleucus. Nat. Speleol. Soc. Bull.,* 33: 1–22.

Brandon, R.A., Jacobs, J., Wynn, A. and Sever, D.M., 1986. A naturally metamorphosed Tennessee cave salamander (*Gyrinophilus palleucus*). *J. Tenn. Acad. Sci.,* 61: 1–2.

Briegleb, W., 1962. Zur Biologie und Ökologie des Grottenolms (*Proteus anguinus* Laur. 1768). *Z. Morphol. Ökol. Tiere,* 51: 271–334.

Bruce, R.C., 1979. Evolution of paedomorphosis in salamanders of the genus *Gyrinophilus. Evolution,* 33(3): 998–1000.

Bulog, B., 1989. Differentiation of the inner ear sensory epithelia of *Proteus anguinus* (Urodela, Amphibia). *J. Morphol.,* 202: 325–338.

Chen, Y.-R., Yang, J.-X. and Zhu, Z.-G., 1994. A new fish of the genus *Sinocyclocheilus* from Yunnan with comments on its characteristic adaptation (Cypriniformes: Cyprinidae). *Acta Zootaxon. Sin.,* 19(2): 246–253. In Chinese, with English abstract.

Clergue-Gazeau, M., 1974. Urodèles cavernicoles d'Amérique du Nord. Analyse critique des travaux effectués sur les trogloxènes. *Ann. Biospéléol.,* 29(3): 435–457.

Cohen, D.M. and Nielsen, J.G., 1978. *Guide to the identification of genera of the fish order Ophidiiformes with a tentative classification of the order.* National Oceanic and Atmospheric Administration (NOAA) Technical Report, National Marine Fisheries Service (NMFS) Circular 417, 72 pp.

Cohen, D.M. and Robins, C.R., 1970. A new ophidioid fish (genus *Lucifuga*) from a limestone sink, New Providence Island, Bahamas. *Proc. Biol. Soc. Washington,* 83(11): 133–144.

Cooper, J.E., 1961. Cave records for the salamander *Plethodon r. richmondi* Pope, with notes on additional cave-associated species. *Herpetologica,* 17(4): 250–255.

Cooper, J.E. and Cooper, M.R., 1968. Cave-associated Herpetozoa II:

salamanders of the genus *Gyrinophilus* in Alabama caves. *N.S.S. Bull.*, 30: 19–24.

Cooper, J.E. and Kuehne, R.A., 1974. *Speoplatyrhinus poulsoni*, a new genus and species of subterranean fish from Alabama. *Copeia*, (2): 486–493.

Culver, D.C., 1982. *Cave Life. Evolution and Ecology.* Harvard University Press, Cambridge, MA, 189 pp.

Culver, D.C., 1985. Trophic relationships in aquatic cave environments. *Stygologia*, 1: 43–53.

Culver, D.C., 1994. Species interactions. In: J. Gibert, D.L. Danielopol and J.A. Stanford (Editors), *Groundwater Ecology.* Academic Press, San Diego, CA, pp. 271–285.

Diesel, R., Bäuerle, G. and Vogel, P., 1995. Cave breeding and froglet transport: a novel pattern of anuran brood care in the Jamaican frog, *Eleutherodactylus cundalli*. *Copeia*, (2): 354–360.

Duellman, W.E., 1993. Amphibian species of the world: additions and corrections. *Spec. Pub. Univ. Kansas Mus. Nat. Hist.*, 21: 1–372.

Durand, J.P. and Vandel, A., 1968. *Proteus*: an evolutionary relic. *Sci. J. London*, 4: 44–49.

Durand, J.P., Keller, N., Laumont, C., Delprat, B. and Pouliquen, Y., 1995. Depigmented and degenerated eye of the Chinese blind dwelling: *Triplophysa xiangxiensis* Y. (abstract). *Résumés du XII. Colloque International de Biospéologie, Moulis, 28. Août–2. Septembre 1995.* Société de Biospéologie, Moulis.

Eigenmann, C.H., 1902. The fresh-water fishes of western Cuba. *U.S. Fish Commission Bulletin*, pp. 211–236, plates 19–21.

Eigenmann, C.H., 1909. Cave vertebrates of North America. A study in degenerative evolution. *Carnegie Inst. Washington Publ.*, 104: 1–241.

Eigenmann, C.H., 1910. The Cuban blind fishes. In: *Proceedings of the 7th International Zoological Congress.* Boston Meeting, August 19–24, 1907. Cambridge, Mass.

Gorbman, A., 1957. The thyroid gland of *Typhlomolge rathbuni*. *Copeia*, (1): 41–43.

Gordon, M.S. and Rosen, D.E., 1962. A cavernicolous form of the poeciliid fish *Poecilia sphenops* from Tabasco, Mexico. *Copeia*, (2): 360–368.

Greenwood, P.H., 1967. Blind cave fishes. *Stud. Speleol.*, 1(5): 262–274, plates XL A-B.

Greenwood, P.H., 1976. A new and eyeless cobitid fish (Pisces, Cypriniformes) from the Zagros Mountains, *Iran. J. Zool. London*, 180: 129–137.

Griffin, D.R. and Luthers, R.A., 1970. Sensitivity of echolocation in cave swiftlets. *Biol. Bull.*, 139: 495–501.

Grillitsch, H. and Tiedemann, F., 1994. Die Grottenolm-Typen Leopold Fitzingers (Caudata: Proteidae: *Proteus*). *Herpetozoa*, 7(3-4): 139–148, plates 1–2.

Hendricks, F.S., 1973. Systematic reevaluation of the central Texas paedogenetic plethodontid salamanders by utilizing multivariate analyses. Abstracts of the Summer Meeting of the Herpetologists' League and the Society for the Study of Amphibians and Reptiles, contributed papers. *H.I.S.S. News-J.*, 1(4): 111.

Heuts, M.J., 1951. Ecology, variation and adaptation of the blind African cave fish *Caecobarbus geertsii* Blgr. *Ann. Soc. R. Zool. Belg.*, 82(2): 155–230.

Hubbs, C.L., 1938. Fishes from the caves of Yucatan. *Carnegie Inst. Washington Publ.*, 491: 261–295, plates 1–4.

Hubbs, C.L. and Innes, W.T., 1936. The first known blind fish of the family Characidae: a new genus from Mexico. *Occas. Pap. Mus. Zool. Univ. Mich.*, 342: 1–7, plate I.

Humphreys, W.F. and Adams, M., 1991. The subterranean aquatic fauna of the North West Cape peninsula, *Western Australia. Rec. West. Aust. Mus.*, 15(2): 383–411.

Humphreys, W.F. and Feinberg, M.N., 1995. Food of the blind cave fishes of northwestern Australia. *Rec. West. Aust. Mus.*, 17: 29–33.

Hüppop, K., 1985. The role of metabolism in the evolution of cave animals. *Bull. Natl. Speleol. Soc.*, 47: 78–85.

Karaman, S., 1923. Über die Herkunft der Süßwasserfische unseres Karstes. *Glas. Zem. Mus. Bosne Hercegovine, Sarajevo*, 35: 163–176.

Konishi, M. and Knudsen, E.I., 1979. The oilbird: hearing and echolocation. *Science*, 204: 425–427.

Kottelat, M., 1988. Two new cave fishes from northern Thailand in the genera *Nemacheilus* and *Homaloptera* (Osteichthyes: Homalopteridae). *Rec. West. Aust. Mus.*, 40: 225–231.

Kottelat, M. and Géry, J., 1989. *Nemacheilus troglocataractus*, a new blind cavefish from Thailand (Osteichthyes, Balitoridae). *Spixiana*, 11(3): 273–277.

Langecker, T.G. and Longley, G., 1993a. Blind catfish (*Trogloglanis pattersoni* and *Satan eurystomus*) from deep artesian waters: a study on convergent adaptations to cave and deep sea biota. In: J.H. Schröder, J. Bauer and M. Schartl (Editors), *Trends in Ichthyology. An International Perspective*, GSF-Bericht 7/92. Blackwell Scientific Publications, Oxford, pp. 239–255.

Langecker, T.G. and Longley, G., 1993b. Morphological adaptations of the Texas blind catfishes *Trogloglanis pattersoni* and *Satan eurystomus* (Siluriformes: Ictaluridae) to their underground environment. *Copeia*, (4): 976–986.

Langecker, T.G., Schmale, H. and Wilkens, H., 1993. Strukturelle und molekulargenetische Untersuchungen zur Augenreduktion beim Höhlenfisch *Astyanax fasciatus* (Teleostei, Characidae). *Mitt. Hamb. Zool. Mus. Inst.*, 90: 363–374.

Langecker, T.G., Wilkens, H. and Schmale, H., 1995. Developmental constraints in regressive evolution: studies of the expression of the β_s-crystallin gene in the developing lens of cave-dwelling *Astyanax fasciatus* (Cuvier, 1819) (Teleostei, Characidae) by in situ hybridization. *J. Zool. Syst. Evol. Res.*, 33: 123–128.

Laurenti, J.N., 1768. Specimen medicum, exhibens synopsin reptilium emendatam cum experimentis circa venena et antidota reptilium Austriacorum. Vienna, 214 pp.

Lazell, J.D. and Brandon, R.A., 1962. A new stygian salamander from the Southern Cumberland Plateau. *Copeia*, (2): 300–306.

Lundberg, J.G., 1982. The comparative anatomy of the toothless blindcat, *Trogloglanis pattersoni* Eigenmann, with a phylogenetic analysis of the ictalurid catfishes. *Misc. Publ. Mus. Zool. Univ. Mich.*, 163, 85 pp.

Lundberg, J.G. and Rapp Py-Daniel, L., 1994. *Bathycetopsis oliveirai*, gen. et sp. nov., a blind and depigmented catfish (Siluriformes: Cetopsidae) from the Brazilian Amazon. *Copeia*, (2): 381–390.

Mees, G.F., 1962. The subterranean freshwater fauna of Yardie Creek Station, North West Cape, Western Australia. *J.R. Soc. West. Aust.*, 45: 24–32.

Menon, A.G.K., 1950. On a remarkable blind siluroid fish of the family Clariidae from Kerala (India). *Rec. Indian Mus.*, 48: 59–66, plate I.

Miller, R.R., 1966. Geographical distribution of Central American freshwater fishes. *Copeia,* (4): 773–802.

Miller, R.R., 1977. *Red Data Book No. 4: Fishes.* International Union for the Conservation of Nature and Natural Resources, Gland, Switzerland.

Miller, R.R., 1984. *Rhamdia reddelli,* new species, the first blind pimelodid catfish from Middle America, with a key to the Mexican species. *Trans. San Diego Soc. Nat. Hist.,* 20(8): 135–144.

Mitchell, R.W. and Smith, R.E., 1971. Some aspects of the osteology and evolution of the neotenic spring and cave salamanders (Eurycea, Plethodontidae) of central Texas. *Tex. J. Sci.,* 23: 343–362.

Mitchell, R.W., Russell, W.H. and Elliott, W.R., 1977. Mexican eyeless characin fishes, genus *Astyanax*: Environment, distribution, and evolution. *Spec. Pub. Mus. Tex. Tech. Univ.,* 12: 1–89.

Mohr, C.E. and Poulson, T.L., 1966. *The Life of the Cave.* McGraw-Hill, New York, 232 pp.

Myers, G.S., 1966. Derivation of the freshwater fish fauna of Central America. *Copeia,* (4): 766–773.

Nalbant, T.T., 1981. A study of the subterranean species of *Lucifuga* from Cuba, with the description of *Lucifuga simile* sp. n. (Pisces, Ophidiiformes, Bythitidae). In: T. Orghidian, A. Nuñez Jimenez, V. Decou, S. Negrea, and N. Viña Bayes (Editors), *Résultats des Expéditions Biospéologiques Cubano-Roumaines à Cuba,* Vol. 3. Editura Academiei Republicii Socialiste România, Bucuresti, pp. 185–190.

Nelson, J.S., 1994. *Fishes of the World.* 3rd edn. Wiley, New York, 600 pp.

Nicholas, G., 1971. Entrance, twilight, and dark. *Nat. Hist.,* 80: 30–35.

Norman, J.R., 1926. A new blind catfish from Trinidad, with a list of the blind cave-fishes. *Ann. Mag. Nat. Hist., Ser. 9,* 18: 324–331.

Parzefall, J., 1993. Schooling behaviour in population-hybrids of *Astyanax fasciatus* and *Poecilia mexicana* (Pisces, Characidae and Poeciliidae). In: J.H. Schröder, J. Bauer and M. Schartl (Editors), *Trends in Ichthyology. An International Perspective,* GSF-Bericht 7/92. Blackwell Scientific Publications, Oxford, pp. 297–302.

Peters, N., 1992. Mehr Zwang als Zufall: Über die zeitliche Parallelität regressiver und rekonstruktiver Evolutionsprozesse bei Höhlenfischen. *Mitt. Hamb. Zool. Mus. Inst.,* 89, Ergänzungsband 1: 97–113.

Peters, N., Peters, G., Parzefall, J. and Wilkens, H., 1973. Über degenerative und konstruktive Merkmale bei einer phylogenetisch jungen Höhlenform von *Poecilia sphenops* (Pisces, Poeciliidae). *Int. Rev. Gesamten Hydrobiol.,* 58(3): 417–436.

Poulson, T.L., 1963. Cave adaptation in amblyopsid fishes. *Am. Midl. Nat.,* 70(2): 257–290.

Poulson, T.L., 1964. Animals in aquatic environments: animals in caves. In: D.B. Dill (Editor), *Adaptation to the Environment,* Vol. 47, Section 4 of *Handbook of Physiology.* American Physiological Society, Washington, DC, pp. 749–771.

Poulson, T.L., 1969. Population size, density, and regulation in cave fishes. *Proc. 4th Int. Congr. Speleol.,* (4-5): 189–192.

Poulson, T.L., 1972. Bat guano ecosystems. *Bull. Natl. Speleol. Soc.,* 34(2): 55–59.

Poulson, T.L. and White, W.B., 1969. The cave environment. *Science,* 165(3897): 971–981.

Putnam, F.W., 1872. The blind fishes of the Mammoth Cave and their allies. *Am. Nat.,* 6: 6–30, plates 1–2.

Pylka, J.M. and Warren, R.D., 1958. A population of *Haideotriton* in Florida. *Copeia,* (4): 334–336.

Raun, G.G., 1971. Herpetology of Texas caves. In: E.L. Lundelius and B.H. Slaughter (Editors), *Natural History of Texas Caves.* Gulf Natural History, Dallas, pp. 100–107.

Roberts, T.R. and Stewart, D.J., 1976. An ecological and systematic survey of fishes in the rapids of the lower Zaïre or Congo River. *Bull. Mus. Comp. Zool.,* 147(6): 239–317.

Sabino, J. and Trajano, E., 1997. A new species of blind armoured catfish, genus *Ancistrus,* from caves of Bodoquena Region, Mato Grosso do Sul, southeastern Brazil. in: *Proceedings of the 1997 Meeting of the Neotropical Ichthyological Association, July 20–26, 1997, Porto Alegre, Brazil.* Abstract.

Sket, B. and Arntzen, J.W., 1994. A black, non-troglomorphic amphibian from the karst of Slovenia: *Proteus anguinus parkelj* n. ssp. (Urodela: Proteidae). *Bijdr. Dierkd.,* 64(1): 33–53.

Sweet, S.S., 1977. Natural metamorphosis in *Eurycea neotenes,* and the generic allocation of the Texas *Eurycea* (Amphibia: Plethodontidae). *Herpetologica,* 33(3): 364–375.

Sweet, S.S., 1984. Secondary contact and hybridization in the Texas cave salamanders *Eurycea neotenes* and *E. tridentifera. Copeia,* (2): 428–441.

Sweet, S.S., 1986. Caudata. In: L. Botosaneanu (Editor), *Stygofauna Mundi. A Faunistic, Distributional, and Ecological Synthesis of the World Fauna inhabiting Subterranean Waters (including the Marine Interstitial).* E.J. Brill, Leiden, pp. 734–740.

Thinès, G.L., 1969. *L'Évolution Régressive des Poissons Cavernicoles et Abyssaux.* Masson, Paris, 394 pp.

Thinès, G.L. and Proudlove, G., 1986. Pisces. In: L. Botosaneanu (Editor), *Stygofauna Mundi. A Faunistic, Distributional, and Ecological Synthesis of the World Fauna inhabiting Subterranean Waters (including the Marine Interstitial).* E.J. Brill, Leiden, pp. 709–733.

Tomiyama, I., 1936. Gobiidae of Japan. *Jpn. J. Zool.,* 7: 37–112.

Trajano, E., 1991. Population ecology of *Pimelodella kronei,* troglobitic catfish from southeastern Brazil (Siluriformes, Pimelodidae). *Environ. Biol. Fishes,* 30: 407–421.

Trajano, E., 1993. Considerations about systematics and the evolution of behaviour in siluriform cave fishes. *Mém. Biospéol.,* 20: 257–260.

Trajano, E., 1997. Synopsis of Brazilian troglomorphic fishes. *Mém. Biospéol.,* 24: 119–126.

Trajano, E. and de Pinna, M.C.C., 1996. A new cave species of *Trichomycterus* from eastern Brazil (Siluriformes, Trichomycteridae). *Rev. Fr. Aquariol.,* 23(3-4): 85–90.

Trajano, E. and de Souza, A.M., 1994. Behaviour of *Ancistrus cryptophthalmus,* an armoured blind catfish from caves of central Brazil, with notes on syntopic *Trichomycterus sp.* (Siluriformes, Loricariidae, Trichomycteride). *Mém. Biospéol.,* 21: 151–159.

Uhlenhuth, E., 1921. Observations on the distribution and habits of the blind Texan cave salamander, *Typhlomolge rathbuni. Biol. Bull. Woods Hole, Mass.,* 40(2): 73–104.

Valvasor, J.W., 1689. *Die Ehre des Herzogthums Crain, IV. Buch. Von den Naturraritäten des Landes Crain.* Laybach.

Vandel, A., 1964. *Biospéologie. La Biologie des Animaux Cavernicoles.* Gauthier-Villars, Paris, 619 pp.

Vandel, A. and Bouillon, M., 1959. Le Protée et son intérêt biologique. *Ann. Spéléol.,* 14: 111–127.

Wake, D.B., Maxson, L.R. and Wurst, G.Z., 1978. Genetic

differentiation, albumin evolution, and their biogeographic implications in plethodontid salamanders of California and southern Europe. *Evolution,* 32(3): 529–539.

Weber, A., 1996. Cave dwelling catfish populations of the genus *Rhamdia* (Pimelodidae, Siluroidei, Teleostei) in Mexico. *Mém. Biospéleol.,* 23: 73–85.

Wilkens, H., 1982. Regressive evolution and phylogenetic age: The history of colonization of freshwaters of Yucatan by fish and crustacea. *Bull. Assoc. Mex. Cave Stud.,* 8: 237–243; *Bull. Tex. Mem. Mus.,* 28: 237–243.

Wilkens, H., 1988. Evolution and genetics of epigean and cave *Astyanax fasciatus* (Characidae, Pisces). Support for the neutral mutation theory. In: M.K. Hecht and B. Wallace (Editors), *Evolutionary Biology,* Vol. 23, Plenum, New York, pp. 271–367.

Wilkens, H., 1993. A new species of *Rhamdia* (Pisces: Pimelodidae) from a cave in the Sierra de Zongolica (Veracruz, Mexico). *Mitt. Hamb. Mus. Inst.,* 90: 375–378.

Wilkens, H., Strecker, U. and Yager, J., 1989. Eye reduction and phylogenetic age in ophidiiform cave fish. *Z. Zool. Syst. Evolutionsforsch.,* 27: 126–134.

Woods, L.P. and Inger, R.F., 1957. The cave, spring, and swamp fishes of the family Amblyopsidae of central and eastern United States. *Am. Midl. Nat.,* 58: 232–256.

Yang, G.-J., Yuan, F. and Liao, Y., 1986. A new blind Cobitidae fish from the subterranean water in Xiangxi, China. *J. Huazhong Agric. Univ.,* 5(3): 219–223. In Chinese, with English abstract.

Yang, J.-X., Chen, Y.-R. and Lan, J.-H., 1994. *Protocobitis typhlops,* a new genus and species of cave loach from China (Cypriniformes: Cobitidae). *Ichthyol. Explor. Freshwaters,* 5(1): 91–96.

Zhang, C., 1986. On the ecological adaptation and geographical distribution of the Barbinae fish *Varicorhinus* (*Scaphesthes*) *macrolepis* (Bleeker). *Acta Zool. Sin.,* 32(3): 266–272. In Chinese, with English abstract.

Zhu, S.-Q., 1989. [The loaches of the subfamily Nemacheilinae in China (Cypriniformes: Cobitidae).] Jiangsu Science and Technology Publishing House, Nanjing, China, 150 pp. In Chinese.

Part IV

Adaptation to the Main Abiotic Factors

Chapter 7

THE EFFECTS OF CONTINUOUS DARKNESS ON CAVE ECOLOGY AND CAVERNICOLOUS EVOLUTION

Thomas G. LANGECKER

Introduction

No other natural environment demonstrates the tremendous modeling influence of light on the ecology and evolution of organisms more spectacularly than subterranean biotopes. It is the lack of light that basically characterizes subterranean habitats, and most particularities of their ecology are directly or indirectly related to this feature. In contrast to the deep sea, which parallels the conditions of permanent darkness, caves provide structurally and geographically repeating habitats that are isolated one from the other. They occur in a series of different geographic areas, and they are inhabited by terrestrial and aquatic animals of various systematic groups. Therefore, caves provide many independent evolutionary scenarios, all linked by one factor: the lack of light.

This chapter will summarize the large variety of ecological and evolutionary consequences of permanent darkness for cave animals. Under particular consideration are the direct consequences of permanent darkness, whereas indirect effects, such as food scarcity, will only briefly be noticed; they are described in more detail in separate chapters of this volume (see chapter 8, this volume).

LIGHT AS A TRANSPORT OF ENERGY AND INFORMATION

Light is electromagnetic radiation in the wavelength range that can be detected by the eye (from about 400 nm to 750 nm). In the following, this term will also be applied to adjacent wavelength ranges, the ultraviolet light (below 400 nm), and infrared radiation (above 750 nm). In biological systems, light has not only a fundamental function as a transport of energy, but also in the transport of information (Fig. 7.1). Both aspects shape the basic ecological characteristics of the cave.

Like all electromagnetic radiation, light is the transmission of energy in the form of waves. Although the earth itself supplies some energy, most organisms directly or indirectly depend on solar energy. The radiation of the sun provides the heat that is responsible for the spectrum of temperatures appropriate for organic life on earth. It is also the source for the process of photosynthesis, which fixes the light energy in organic compounds and thus converts light energy into food. Lack of light means lack of photoautotrophic primary production. Consequently, many subterranean biotopes are characterized by severe food scarcity; they depend entirely on the limited food input from surrounding epigean habitats.

An important aspect in the ecology of epigean organisms is the periodical fluctuation of solar radiation. Cave biotopes are not directly exposed neither to diurnal nor to seasonal alterations of abiotic factors. Visible light cannot penetrate the surface; the rock and land masses that encompass the cave absorb heat during warm intervals and slowly release it at colder periods. The result is an environment that is relatively stable in its climatic characteristics.

Most obvious to anyone who ever entered a natural cave, the lack of light results in the absence of any visual information. This not only concerns the loss of visual orientation, but also all kinds of visual communication. Finally, the primarily light-entrained circadian and seasonal time-keeping mechanisms, that coordinate physiological and behavioral activities with seasonal and daily rhythms, are out of control in deep subterranean habitats.

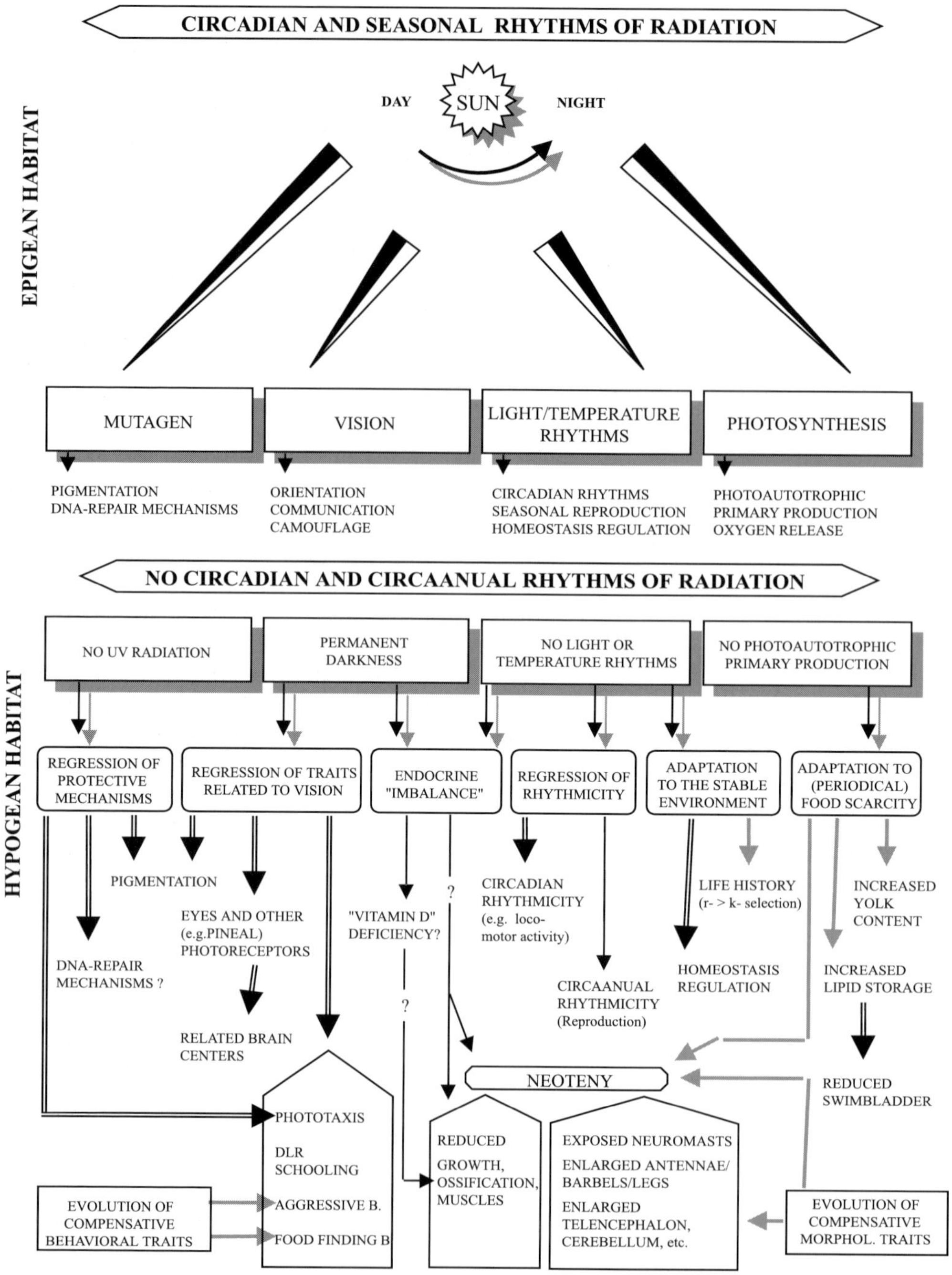

Fig. 7.1. Suggested scheme of the role of continuous darkness for the evolution of troglomorphic traits in cave animals (gray arrows, constructive traits; black arrows with double line, regressive traits; black arrows with single lines, speculative relationships. B, behavior; DLR, dorsal light reaction).

Preadaptations for cave life

The subterranean fauna originated from epigean animals. Some species actively colonize aphotic environments driven by competition or behavioral preadaptations, or in search for food and/or refuge; others are accidentally captured in caves and develop to troglobites without an intermediate, troglophilic stage (see Rouch and Danielopol, 1987; Botosaneanu and Holsinger, 1991). In any case, the successful foundation of a new cave population requires preadaptations that allow orientation, reproduction and the search for food in complete darkness.

Hence, it is not surprising that many cave invaders belong to scotophilic or nocturnal groups. Among vertebrates, only fishes and salamanders developed troglobitic populations. These taxa appear to be especially well preadapted because they possess an effective non-visual distant sense organ, the lateral-line system, and some species can even rely on electroperception for orientation. Weber (1995) suggested that the exceptional success of catfish in the colonization of caves may partially be caused by their electric sense, which unfortunately attracted little attention in cave biology until today. Terrestrial animals like crustaceans and insects depend entirely on tactile or chemical sensory organs for orientation and locating food. Besides the ability of non-visual orientation and foraging, facultative cave dwellers need to be independent of visual signals for communication. Most important, mating behavior must be controlled solely by chemical or mechanical stimuli.

It is a well-known phenomenon that several species with the capacity to orientate, to forage and to reproduce in the darkness of the cave nevertheless fail to develop a true troglobitic population (see Christiansen, 1992). It must thus be assumed that the evolution of a troglobitic lineage requires further preadaptations that have not yet been analyzed. These may be related to physiological and endocrinological adaptations to food scarcity and continuous darkness.

ORGANISMIC EFFECTS OF CONTINUOUS DARKNESS

Morphology

Typical cave animals are pale and blind. These genetically fixed characters convergently evolved in all true troglobites of many different systematic groups which have been exposed to absolute and continuous darkness for a sufficiently long period. The apparent loss of biological function results in a surprisingly fast regression of structures that directly depend on light. Even cave animals considered as phylogenetically young show a conspicuous regression of the eyes (Anderßon, 1992). The catalogue of so-called troglomorphic traits, however, includes many more entries (Table 7.1).

Pigmentation

It is commonly believed that melanin pigmentation evolved in epigean animals as a protective mechanism against solar ultraviolet radiation and as an element of visually guided behavior. Among the spectrum of UV-radiation, from 200 nm to 400 nm, only those wavelengths that match the absorption spectrum of biomolecules result in potentially harmful biological effects. Proteins and especially nucleic acids, which show absorption maxima around 280 nm and 260 nm, respectively, are of importance. The absorption of UV radiation in nucleic acids can result in gene inactivation, most significantly by the induction of pyrimidine dimers (Diffey, 1991). Melanin pigmentation in the skin provides some protection against damaging sunlight, but it is apparently not very effective. Morison (1985) suggested that, at least in humans, melanin did not primarily evolve as a protective mechanism against deleterious UV radiation but for camouflage and to keep the body warm in a forest environment. Important tools for the protection of nucleic acids from UV-induced mutations are several molecular mechanisms, among them photoreactivation, excision repair, postreplication repair and SOS repair[1], which evolved to repair damaged molecules (Diffey, 1991). Unless these mechanisms are not pleiotropically linked with other essential functions, it can be predicted that they will degenerate in cave animals.

Most likely, the regression of melanin and other body pigments in cave animals primarily derives from the uselessness of all visually guided behavior. Minimizing visual detection (camouflage) and mimicry, and maximizing visual detection and recognition (signaling), are no longer successful strategies in an aphotic environment. Deprived of its biological function, pigmentation becomes subject to regressive evolution in cave animals.

The most common regression of the pigment system concerns melanin pigmentation but other pigments, such as guanin, can be affected as well. Especially well studied is the regression of pigmentation in *Astyanax*.

[1] "SOS-repair" is a well-established term in molecular biology that describes a particular mechanism which is induced by cell-damage, primarily ultraviolet radiation. The SOS response results in an increased activation of several genes that are involved in DNA repair, among them *rec*A and *lex*A.

Table 7.1
Characteristic genetically based morphological (troglomorphic) traits of cave animals

Traits directly related to the absence of light

1 *Pigmentation*

- Reduced melanin content in the melanophores of skin, peritoneum and meninges
- Reduced number of melanophores
- Albinism
- Reduced guanine pigmentation in the skin

2 *Photoreceptors*

- Regression of the eyes, ocelli
- Degeneration of pineal photoreceptors

3 *Non-visual sensory organs (sensory compensation)*

- increased number of tactile (neuromasts, sensorial hairs, etc.) and chemical (taste buds, olfactory organs, etc.) sensory structures
- elongated sensory hairs, neuromast cupulae
- elongated antennae, barbels, appendages
- changed pattern of distribution of sensory organs

4 *Brain*

- Regression of brain centers related to vision (in fish: regression of the external layers of the optic tectum)
- Hypertrophy of brain centers related to non-visual sensory organs, memory, learning and spatial orientation (in fish: enlarged telencephalon, cerebellum, etc.)

Traits indirectly related to the absence of light (food scarcity, stable environment)

1 *Tendency to paedomorphosis and neoteny*

- in fish and amphibians: small body size, enlarged head, weak muscles, reduced ossification, reduced and partially exposed lateral line system
terrestrial: small body size, changes is allometric development, elongated body

2 *Lipid storage*

- in fish: increased lipid content in the subcutis, muscles, viscera and cranium; compensative reduction of the swimbladder to maintain bouancy

3 *Skin*

- in fish: reduced scales
- terrestrial: reduced cuticula

The epigean freshwater fish *Astyanax fasciatus* is a common and widespread species in Middle America. The fish has large eyes and a body coloration that is typical for fish of free water: a dark melanin pigmentation on the dorsal surface and a bright, silvery guanin pigmentation on the ventral surface (counter-shading), which makes the fish difficult to recognize for potential prey and predators. Regression of both pigment systems can be observed in the different cavernicolous populations of this species that evolved independently in a limited area of Central Mexico (Mitchell et al., 1977). The guanin pigmentation is still present in most populations but partially reduced in some cases. The melanin pigmentation is largely reduced in most blind populations. Only the phylogenetically young population from Micos cave, which is characterized by a high degree of variability in eye regression (from small and largely degenerate to almost normal eyes) still has a dark melanin pigmentation phenotypically similar to that of the epigean specimens. A genetic analysis, however, revealed an incipient regression of this system (Wilkens, 1993a). All other populations show the typical pale cave phenotype, which results from a reduced melanin content in the melanophores and additionally from a reduced number of melanophores (Wilkens, 1993a). In at least two populations (in Pachon cave and Yerbaniz cave, but see Langecker et al., 1991) the melanin depigmentation is absolute because of the evolution of albinism. This mutation occurred additionally to those that resulted in a reduced melanin content and a reduced number of the melanophores.

The examples show that the loss of biological function results in potential mutations on several gene loci that are directly related to the pigment system. Because mutations occur randomly, the resulting cave phenotype is especially variable in those cave populations that recently colonized caves. Phylogenetically old cave forms, however, are rather uniform in their pale appearance. Some confusion arises from the fact that several blind and pale cave species begin to develop a darkened skin when they are kept under light conditions in the laboratory. This effect is due to the morphological and/or physiological mechanisms for color change that may not always degenerate in cave animals. In the cave salamander *Proteus anguinus* the morphological color change still functions, and results in a darker appearance of cave specimens kept under light conditions. The grade of this effect depends on the degree of regression of the melanin content and of the number of melanophores, it does not indicate that the reduced pigmentation might only be environmental.

Eyes

Most animal phyla convergently developed light-sensitive organs; but they differ greatly in morphology, physiology and development (Salvini-Plawen and Mayr, 1977; Land and Fernald, 1992; Nilsson, 1996). Light sensitivity might originally have evolved for the control of circadian rhythms that are virtually ubiquitous in eukaryotes and even exist in some prokaryotes (Rosbash, 1995). Eyes for spatial vision originated in the Cambrian explosion, with the evolution of fast-moving bilateral animals requiring an effective distant sense for orientation (Nilsson, 1996). The invention of vision was then accompanied by the evolution of visually guided behavior and morphological traits like pigmentation and body shape that take part in camouflage and signaling behavior. Photoreception is thus related to a large complex of traits which became functionless in the permanent darkness of the cave.

The morphology and positioning of photoreceptors is highly variable in animals of different systematic groups. In invertebrates, eyes and photoreceptors can be found at the most unusual locations of the body (Nilsson, 1996). Generally, photoreceptors are either associated with nervous or ectodermal tissues. Aside from the primary photoresponse of melanophores, photoreception in most vertebrates is limited to the diencephalon (see Foster et al., 1993), the pineal complex and the lateral eyes. It has been suggested (Wolken and Mogus, 1979) that a "dermal light sense", which is common in invertebrates, is present in some amphibians. In fish and higher vertebrates, dermal light sensitivity is obviously restricted to embryonic or immature tissues (Wolken and Mogus, 1979). The fate of photoreceptors during cavernicolous evolution reflects their origin and function: eyes – that is, specialized organs for spatial vision – rapidly degenerate in cave-adapted species, pineal photoreceptors show a remarkably retarded regression, and diencephalic or dermal light sensitivity may even persist in some phylogenetically old troglobites (Langecker, 1992a).

The regression of eyes for spatial vision is the most conspicuous troglomorphic trait. It has occurred convergently in all taxonomic groups of cave animals, but has been most thoroughly studied in vertebrates. The degree of regression varies from species to species, but can be traced back to a common pattern:

(1) Slightly microphthalmic species with a low variability of eye morphology. Among this group are many troglophiles that do not obligatorily live in caves. These species have a nocturnal and/or scotophilic life style and do not primarily depend on visual orientation. Troglophiles with small eyes have often been referred to as "troglobites in training" because it was believed that they represent intermediate stages from an epigean to a hypogean life style. This may not necessarily be the case (see Culver, 1982). The Mexican catfish *Rhamdia guatemalensis* frequently invades caves in southern Mexico but never developed a troglobitic form, whereas the closely related *Rhamdia laticauda* developed troglobitic populations in Mexico several times. The same applies to well-adapted troglophilic cave crickets (Gryllidae) in Hawaii, which do not tend to develop obligatory cave forms (Howarth, 1981). The eyes of troglophiles may thus be small and of subordinate importance in the sensory equipment, but they are still under control of stabilizing selection. They are well differentiated and do not show high variability.

(2) Slightly or moderately microphthalmic species with a high variability of eye morphology. This group represents species that are totally or largely exposed to continuous darkness and that are in the initial stages of cavernicolous evolution. Contrary to the troglophiles, they are true troglobites *in statu nascendi*. The *Astyanax* population from the Micos cave and the pimelodid catfish *Rhamdia reddelli* from southern Mexico belong to this category. Typically,

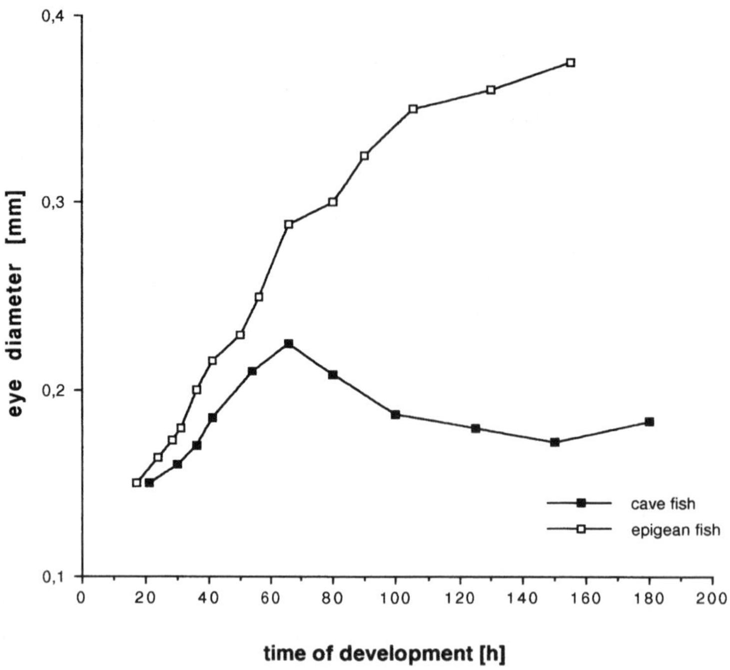

Fig. 7.2. Eye growth during early development of epigean and cave-dwelling *Astyanax fasciatus*.

these cave populations are very heterogeneous. The eye may still be functioning in some individuals, but is more or less reduced in size and degree of differentiation in others. Because the eye has definitely lost its biological function in continuous darkness, it is no longer controlled by stabilizing selection and, consequently, begins to degenerate.

(3) Extreme microphthalmic or anophthalmic species. The majority of true troglobites living in continuous darkness belongs to this group, because eye regression does not proceed at a constant rate but is especially fast in initial stages (Peters, 1990). Externally, they have a uniform eyeless appearance, but the eye rudiment sunken beneath the skin can still show a high morphological variability. Only the final stages of regression are characterized by a low variability of the remaining small rudiments. The unusual case of complete anophthalmia was reported for the Somalian troglobitic cyprinid *Phreatichthys andruzzii* (Ercolini and Berti, 1975).

The classification presented above concerns the phylogenetic aspects of eye regression in cave animals. Generally, the degree of regression increases and the degree of variability decreases with phylogenetic age. Studies on several troglobitic vertebrates also revealed that the regression of the eye follows distinct rules during ontogeny (Langecker et al., unpublished). Regardless of the phylogenetic age, the eye appears to develop normally during very early stages of ontogeny because the eye anlage is required as an inductor for craniofacial development (Durand, 1971; Wilkens, 1988; Langecker et al., 1993). Even the extremely cave-adapted *Phreatichthys andruzzii,* which is reported to be completely anophthalmic in adult stages, develops a presumably normal eye anlage during early development (Berti, pers. comm.). The decline of eye development occurs early in phylogenetically old species (*Amblyopsis spelaea*; Eigenmann, 1909) and late in troglobites of early origin (*Rhamdia reddelli*; Anderßon, 1992). The initial stages of developmental regression are characterized by a slightly reduced rate of growth and diminished mitotic activity, which is associated with the first appearance of sporadic cell death. In the following, eye growth ceases for a limited period, or the eye size may even decrease, while cell death is observed in all retinal layers. Thereafter, a continuous ontogenetic regression takes place – that is, the growth of the eye is greatly retarded and already differentiated structures continue to degenerate (Fig. 7.2). The regression of the previously differentiated structures generally begins with the lens tissue and those retinal layers that differentiated last during development; the final degree of degeneration depends on the ontogenetic age of the cave species.

The genetics of eye regression have been most thoroughly studied in *Astyanax*. It was found that the regression is caused by the inactivation of several genes that take part in developmental control, whereas structural eye genes appear to be well conserved in cave fishes (Wilkens, 1988; Langecker et al., 1993, 1995b; Yokoyama et al., 1995). The nature of the inactivated "eye genes" of cave animals is still a subject of research. Remarkably, the developmental control genes *pax2* and *pax6*, which are known to be essential for the development of the eye in vertebrates, are unchanged in cave fish (Behrens et al., 1996), whereas factors at a lower level of developmental control (growth factors) appear to be involved in the degeneration of the eye (Langecker et al., unpublished).

Present knowledge on the mechanisms of eye regression in cave fish as outlined above, demonstrates that the observed heterogeneity of regressive traits does not necessarily indicate divergent environmental conditions reflecting different selective pressures. Indeed much heterogeneity of regressive traits is primarily caused by different phylogenetic and ontogenetic age of the species studied.

Pineal and extraocular, extrapineal photoreception

Another prominent site of photoreception in fish and amphibians is the pineal organ. It develops as a dorsal evagination of the diencephalon and contains photosensitive cells that are morphologically similar to the photoreceptors of the eye (Dodt, 1973). However, the pineal photoreceptors do not only perceive light; they are also neuroendocrine cells. They synthesize and release the hormone melatonin under the control of light; melatonin secretion is high at night and low during the daytime. Although the biological function of rhythmic melatonin secretion is still not fully understood, it is likely that it plays a key role in the synchronization of internal rhythms through external *Zeitgeber* (time-keepers), especially light and temperature (Falcon et al., 1994).

The evolution of the pineal complex in cave animals is enigmatic. Even cave species with completely reduced visual photoreceptors can have well differentiated, light-sensitive and actively secretory pineal cells, which was demonstrated by behavioral, ultrastructural and endocrinological studies on *Astyanax fasciatus* (Langecker, 1992a,b; Wilkens et al., 1993; Missal et al., 1994). Nevertheless, there is a clear trend to degeneration of the outer segments of pineal photoreceptors from troglobites of early (Langecker and Wilkens, 1992) to moderate (Langecker, 1992a) and late (McNulty, 1978) origin, which indicates that the regression of the light-sensitive structures is greatly retarded compared to the eye. This phenomenon was explained by developmental constraints (Langecker, 1992b), in that the pineal photoreceptors do not become totally functionless like those of the eyes, but retain an obviously essential secretory function even in phylogenetically old troglobites (McNulty, 1978). Hence, they cannot freely degenerate, but must be converted from photoneuroendocrine to exclusively endocrine cells. The problem, however, remains unsolved; it is not clear why cave animals that lose rhythmic behavior benefit from a continuous, rhythmic secretion of melatonin at relatively high levels (Wilkens et al., 1993; Missal et al., 1994). The adaptive value of the persisting melatonin secretion and its behavioral and ecological implications (see below) are still a subject of research.

Several cave animals have been shown to exhibit extraocular, extrapineal photoreception. This phenomenon, however, appears to be of minor significance. A rather diffuse dermal light sensitivity of invertebrates, as well as diencephalic light sensitivity of vertebrates, is likely to be non-selected and to be caused by photoreactive proteins that primarily serve non-sensory functions (see Langecker, 1992a). Foster et al. (1993), who studied deep diencephalic photoreceptors in the lizard *Anolis carolinensis,* summarized the few studies demonstrating that gonadal development of lower vertebrates may be entirely mediated by deep diencephalic photoreceptors. Extraocular, extrapineal photoreception may thus be selected in some vertebrate species, but certainly not in cave animals, which show a functioning gonadal induction under aphotic conditions.

Non-visual sensory organs

It is an obvious supposition that the loss of vision should be compensated by the constructive evolution of non-visual sensory organs in cave animals. Although sensory compensation takes place in troglobites, it is usually less uniform and less accentuated than the regression of eyes and pigmentation. The causes for this phenomenon will be considered below.

It has already been stated that particular preadaptations ensuring survival and reproduction under continuous darkness are obligate prerequisites for epigean specimens that colonize caves. Whether caves are actively colonized or accidentally, with or without adaptive, transitional stages, in any case the colonizers must be able to reproduce and to orientate with the help of their existing non-visual sensory equipment. Consequently, there is little need for a radical improvement of non-visual sensory organs. Nevertheless, competition at the beginning within the initially small founder population certainly exerts a selection pressure for several improvements that essentially increase fitness in the subterranean environment. However, among them are obviously adaptations regarding life history, metabolic rate, development and resistance to starvation (see Chapter 8, this volume), whereas a refined non-visual orientation appears to take a subordinate part. In the Mexican cave fish *Astyanax fasciatus,* which accidentally colonized caves (Mitchell et al., 1977; Langecker, 1992a), there is hardly any improvement in the morphology of the lateral-line system, taste buds and olfactory organ, but lipid storage, starvation resistance, egg size and larval development show a remarkable adaptive development in different cave populations (Hüppop, 1988). The slightly refined non-visual spatial orientation of cave fish (Hahn, 1960) is probably caused by an increased number of free neuromasts (Schemmel, 1967) and the elongation of neuromast cupulae (Teyke, 1990). The improved food-finding ability of cave specimens, on the other hand, is probably not caused by a lowering of tasting thresholds (Hüppop, 1987) but by behavioral adaptations in foraging strategy and an associated change in the distribution of taste buds (Schemmel, 1967, 1980). The best documented case of significant sensory compensation in troglobitic fishes is in a group of amblyopsids (Poulson, 1963, 1985) that inhabit the Mammoth Cave in North America. Compared to their trogloxene relative *Chologaster cornutus,* the troglobitic species, *Amblyopsis rosae, A. spelaea, Speoplatyrhinus poulsoni* and *Typhlichthys subterraneus* have hypertrophied olfactory receptors, neuromast receptors and tactile receptors. A thorough analysis of degenerative (eye regression, pigmentation) and compensative traits in the phylogenetic lineage of the amblyopsids revealed the interesting result that the degree of regression increases continuously from the trogloxene to the most troglobitic species, whereas constructive traits (sensory compensation, metabolism, etc.) may increase, stop, or even reverse within this lineage (Poulson, 1985). The observed heterogeneity of compensative traits in troglobites is largely caused by the factors considered below.

Phylogenetic age: The evolution of new or improved traits requires sufficient time regardless of all other underlying mechanisms that shape or limit the outcome. Elaborated troglomorphic features can thus only be expected in troglobites of at least moderate phylogenetic age; the overall degree of cave adaptation should increase with the time of cavernicolous evolution.

Preadaptations: The selection pressure for improved non-visual sensory organs is certainly less intense on cave invaders that are well preadapted to non-visual orientation, foraging and mating. Especially, nocturnal and scotophilic colonizers can be expected to require little sensory improvement.

Life style and ecological niche: Troglobites inhabiting different niches are under the regime of different selection pressures regarding their sensory organs. In fact, the sensory endowment of a given species largely reflects its life style and ecological niche, especially its foraging mode. Predators require sensory capabilities different from those needed by detritus-consuming bottom-feeders, which might depend on distance sense organs only for mating. The two ictalurid blindcats *Satan eurystomus* and *Trogloglanis pattersoni* which coexist in the San Antonio region of the Edwards Aquifer, Texas, are both extreme troglomorphic species, but they differ considerably in the development of their lateral-line system. The top carnivore *S. eurystomus* has a hypertrophied canal system with large pores, which may serve as a very efficient distance sense organ to locate crustaceans and other potential prey. The detritus-feeding *T. pattersoni,* on the other hand, has a poorly developed lateral-line system. Instead, the latter has developed a very specialized sucker mouth densely equipped with taste buds in order to search for organic compounds on the bottom (Langecker and Longley, 1993).

Developmental constraints: Despite the large variety of form and function in living nature it is evident that evolution is quite conservative and is largely confined by developmental constraints of fundamental body plans (*Baupläne*). Developmental constraints do not only explain some peculiarities in the regression of eyes and pineal photoreceptors (Langecker, 1992b; Langecker and Wilkens, 1992, Langecker et al.,

1995b), they may also be responsible for the almost monotonous constructional evolution of troglobites. Aside from the regressive traits (eyes, pigmentation, cuticule, scales, etc.), the repertoire of cavernicolous evolution of morphological traits is limited to a few principles, particularly the increase in number and the spatial distribution of sensory receptors, elongated appendages, and the trend to paedomorphosis and neoteny (heterochrony). The complex cascades of epigenetic interactions during development allow only slight changes in those features of adaptive form and function that differentiate last during formation of the body (Hall, 1992). Cave-dwelling arthropods typically have elongated appendages, allowing the extension of sensory organs because of an increased surface. For the same reason some troglobitic catfish developed elongated barbels (Wilkens, 1993b); the trend to paedomorphosis in vertebrates, typically bringing about an enlarged head, leads to the same result – an expanded sensory area. Enlargement of features equipped with sensory receptors is a typical troglomorphic trait because it is not in conflict with the formation of the basic body plan; it is developmentally barely constrained and highly adaptive. Culver (1982) discussed the adaptive value of increased sensory organs, and gave several examples where there was no selection for appendage elongation in cave-dwelling species. Taking into account the different ecological niches troglobites may occupy and the divergent environmental conditions (especially the availability and density of food), it is obvious that selection pressure cannot be uniform. The Texas blindcat *Trogloglanis pattersoni* has almost minute barbels, which is unusual for ictalurids, apparently because its foraging strategy does not require elongated sensory structures (Langecker and Longley, 1993). Typical troglomorphic traits are thus the realization of slight developmentally constrained changes in features of adaptive form and function, which evolve in response to distinctive life styles and selection pressures.

Brain

The brain of lower vertebrates shows marked structural variations that can be correlated with the ecology of species (Bath, 1962; Rao, 1967). Because several clearly discernible brain areas are known to process primarily certain sensory modalities, brain morphology can also be used to estimate the degree of sensory regression and sensory compensation in cave vertebrates.

There is a clear trend to convergent evolution of particular brain centers in the cave species studied up to today (see Table 7.2). The fact that not all cerebral structures show a parallel development corresponds with the findings on sensory compensation discussed above. Because the development of the brain is not only determined by genetic factors, which may be intrinsic (neural genes) or extrinsic (growth factors), but is also largely modified by sensory input (Calof, 1995; Cramer and Sur, 1995), it is not clear whether the changes in brain morphology are just a consequence of the evolution of sensory organs or whether genetically fixed changes occurred in the brain itself. Especially well studied is the reduction of the optic tectum, which evolved convergently in all microphthalmic cave fishes. In the case of the Mexican troglobite *Astyanax fasciatus*, it has been suggested that the regression of the tectum is primarily an epigenetic consequence of eye regression (Pfeiffer, 1967) but to some degree caused by changes in neural genes (Schmatolla, 1972; Voneida and Fish, 1984). Regardless of this question, it is well established that all cerebral structures of troglobites that are directly related to visual input show a defective development (hypoplasia). In vertebrates these are the outer layers of the optic tectum and the pineal organ (see above). The innermost layers of the tectum, which also process input from non-visual sensory organs (Guthrie, 1990; Voneida and Fish, 1984) show little or no regression in cave fish (Schmatolla, 1972; Langecker and Longley, 1993); in blind *Astyanax fasciatus* the thickness of the inner nuclear layer and the density of neurons even increase compared to the epigean fish (Langecker et al., unpublished). These findings are in accordance with studies of their electrophysiology (Voneida and Fish, 1984) and behavior (Langecker et al., 1995a). Deep tectal neurons that receive signals from both visual and somatosensory organs in epigean specimens may retain their function in spatial orientation of cave fish.

Hypertrophy of the telencephalon is the most homogeneous trend in the cerebral development of cave fishes besides the regression of the optic lobes. Even cave fishes of moderate phylogenetic age (*Astyanax fasciatus*) or species with especially well preadapted ancestors (*Pimelodella kronei*) show a significant enlargement of the telencephalon compared to their epigean relatives (Peters et al., 1993; Trajano, 1994). Although the telencephalon is closely related to olfaction and taste (Nieuwenhuys, 1982; Rao, 1967) and is often referred to as "olfactory lobes", its

Table 7.2
Regressive and constructive changes of the brain in cave fishes

Brain structures	Related to	Development in cave species		
		Increase/ Decrease	Species	Literature
Tectum opticum (optic lobes)				
Overall size	vision and mechanical senses	––	*Astyanax fasciatus*	Peters et al. (1993)
		––	*Typhlichthys subterraneus*	Poulson (1985)
		––	*Amblyopsis spelaea*	Poulson (1985)
		––	*Amblyopsis rosae*	Poulson (1985)
		––	*Speoplatyrhinus poulsoni*	Poulson (1985)
		–	*Pimelodella kronei*	Trajano (1994)
		––	*Phreatichthys andruzzii*	Ercolini and Berti (1975)
		––	*Uegitglanis zammaranoi*	Ercolini and Berti (1977)
Outer layers	vision	–	*Astyanax fasciatus*	Schmatolla (1972)
		––	*Trogloglanis pattersoni*	Langecker and Longley (1993)
		––	*Satan eurystomus*	Langecker and Longley (1993)
		––	*Amblyopsis rosae*	Eigenmann (1909)
		––	*Typhlogarra widdowsoni*	Marshall and Thinès (1958)
Inner layers	vision and mechanical senses	+/–	*Astyanax fasciatus*	Schmatolla (1972)
		+/–	*Trogloglanis pattersoni*	Langecker and Longley (1993)
		+/–	*Satan eurystomus*	Langecker and Longley (1993)
		–	*Amblyopsis rosae*	Eigenmann (1909)
		–	*Typhlogarra widdowsoni*	Marshall and Thinès (1958)
Telencephalon	Olfaction, gustation, spatial memory, integration of complex behavior, etc.	+	*Astyanax fasciatus*	Peters et al. (1993)
		++	*Trogloglanis pattersoni*	Langecker and Longley (1993)
		++	*Satan eurystomus*	Langecker and Longley (1993)
		+	*Typhlichthys subterraneus*	Poulson (1985)
		++	*Amblyopsis spelaea*	Poulson (1985)
		++	*Amblyopsis rosae*	Poulson (1985)
		+	*Speoplatyrhinus poulsoni*	Poulson (1985)
		(+)	*Pimelodella kronei*	Trajano (1994)
Cerebellum	mechanical senses (lateral line system, free neuromasts, semicircular canals)	+/–	*Astyanax fasciatus*	Peters et al. (1993)
		+	*Trogloglanis pattersoni*	Langecker and Longley (1993)
		+	*Satan eurystomus*	Langecker and Longley (1993)
		+	*Typhlichthys subterraneus*	Poulson (1963)
		+	*Amblyopsis spelaea*	Poulson (1963)
		+	*Amblyopsis rosae*	Poulson (1963)
		+/–	*Speoplatyrhinus poulsoni*	Cooper and Kuehne (1974)
		–	*Pimelodella kronei*	Trajano (1994)
Vagal lobe	taste buds of the mouth	–	*Trogloglanis pattersoni*	Langecker and Longley (1993)
		––	*Satan eurystomus*	Langecker and Longley (1993)
Fascial lobe	taste buds of the barbels and the skin	+/–	*Trogloglanis pattersoni*	Langecker and Longley (1993)
		––	*Satan eurystomus*	Langecker and Longley (1993)
Acusticolateralis lobe	auditory and equilibratory functions	+/–	*Trogloglanis pattersoni*	Langecker and Longley (1993)
		+	*Satan eurystomus*	Langecker and Longley (1993)
Eminentia granularis, cristae cerebellum	input paths for mechanical senses	+	*Typhlichthys subterraneus*	Poulson (1963)
		+	*Amblyopsis spelaea*	Poulson (1963)
		+	*Amblyopsis rosae*	Poulson (1963)

convergent enlargement in cave fishes is probably not primarily caused by improved olfactory receptors. This is especially evident from cave species that have a reduced number of olfactory lamellae (*Astyanax fasciatus*, Schemmel, 1967; *Pimelodella kronei*, Trajano, 1994) or reduced olfactory bulbs (*Satan eurystomus*, Langecker and Longley, 1993), but still exhibit a conspicuously enlarged telencephalon. The enlarged forebrain of the amblyopsids, however, is accompanied with a better developed olfactory rosette. Riedel (1993) found no histological differences between the enlarged telencephalon of a cave-dwelling *Astyanax* of undefined origin and its epigean relative. His behavioral studies provided evidence that the telencephalon of the cave fish is involved in all kinds of learning because it takes part in short-term memory, general arousal, selective inhibition of behavioral responses, and the induction of motivation. Poulson (1963) assumed the adaptive value of a hypertrophied telencephalon in sensory compensation, better spatial memory and a better integration of sensory information, Peters et al. (1993) suggested a close relationship to an improved sense of taste.

The cerebellum is enlarged in several cave fish studied, but not in all. The degree of hypertrophy cannot be correlated with phylogenetic age. The cerebellum of the extreme troglomorphic *Speoplatyrhinus poulsoni* is not significantly enlarged (Cooper and Kuehne, 1974), and in the phylogenetically young *Pimelodella kronei* its size is even reduced compared to its epigean relative (Trajano, 1994). The cerebellum receives input from the lateral line system, free neuromasts, semicircular canals and also from several neural pathways originating in the optical tectum (Nieuwenhuys, 1982). It is therefore suggested that it is a coordinating center of distance sense organs, and its development should reflect the locomotor activity of cave species (Rao, 1967). Predators that require well-coordinated locomotor powers especially may be expected to develop a hypertrophied cerebellum in compensation for the loss of vision. However, this view does not fit all cases, as is especially evident in *Trogloglanis pattersoni.* This blind catfish shows no evidence of compensatory development of distance sense organs (Langecker and Longley, 1993), but displays a significantly enlarged cerebellum. The development of other cerebral structures like vagal and fascial lobes, eminentia granularis, etc., can well be related to the habits and life styles of the species concerned.

ENDOCRINOLOGY

About forty years ago, Rasquin and Rosenbloom (1954) concluded from their work on endocrinological aspects of cave colonization that the "central problem in the evolution of blind cave fish is one of endocrinology". In an extensive study, based on a total of 270 epigean *Astyanax fasciatus,* they demonstrated radical consequences of permanent darkness on the morphology and physiology of fishes. However, their results could not be verified in later studies. Franck (1964) and Wilkens (unpublished) found no profound effects of continuous darkness on epigean *Astyanax* that could clearly be related to endocrinological imbalance. Consequently, this subject attracted little attention in later research. Apart from studies on the role of melatonin synthesis for circadian rhythms and thyroxine for metamorphosis (see below), few efforts were made to investigate endocrinological adjustments of cave animals until recently.

Vitamin D_3 and calcium metabolism

The control of calcium metabolism by vitamin D might be a limiting factor for several potential cave colonizers. In vertebrates, calcium metabolism is regulated at the blood-plasma level by the action of the active form of vitamin D_3 (calcitriol) and the hormones calcitonin and parathyrin. Physiologically speaking, vitamin D is not a vitamin but rather a hormone, because it is not a dietary constituent required for normal body functions (Hurst, 1982). It can be synthesized in the skin through the action of sunlight (Fig. 7.3), which is the only known beneficial effect of ultraviolet radiation on the skin of higher animals (Diffey, 1991).

Calcitriol plays a key role in the control of calcium metabolism, because it stimulates the absorption of calcium in the intestines. Depending on the calcium influx mediated by calcitriol, the antagonists calcitonin and parathyrin are then able to regulate the calcium level of the blood (Fig. 7.3). Low levels result in the mobilization of calcium from the bones, which are the predominant site of calcium storage. On the other hand, high levels will cause the mineralization of bones and/or an increased calcium excretion by the kidneys (Hurst, 1982). In view of the role of light in calcitriol synthesis, it is to be expected that continuous life in aphotic environments would result in vitamin D deficiency in most vertebrates and thus in the demineralization of the bones, a defect known as

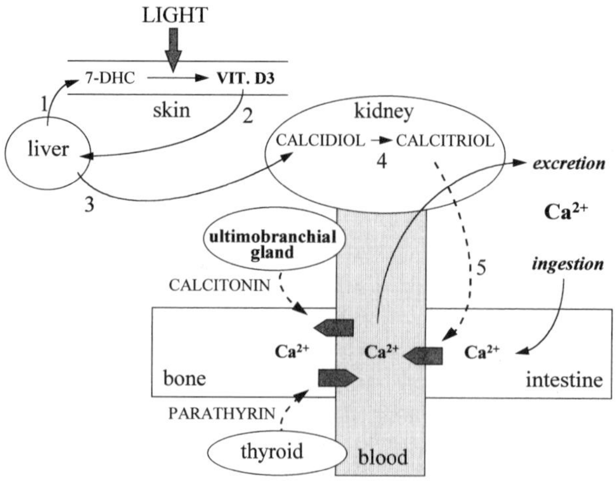

Fig. 7.3. General scheme of calcium metabolism in vertebrates. Numerals indicate successive reactions of the metabolic pathway, the arrows indicate the transport of metabolites. Large amounts of 7-dehydrocholesterol (7-DHC) are deposited in the skin and converted to vitamin D3 by ultraviolet radiation. This is metabolized in the liver to calcidiol, transported to the kidney and transformed to calcitriol, the most active form of vitamin D, which mediates the absorption of calcium in the intestines.

rickets (skeletal malformation) in vertebrate pathology. It is not likely that cave vertebrates can compensate for this deficiency by taking up sufficient quantities of vitamin D with their diet, because it is rare in nature. It is virtually absent in invertebrates, which do not require vitamin D for their calcium metabolism, and is only sparsely found in algae (Amlacher, 1986).

Obviously, vertebrate troglobites must posses adaptive mechanisms that allow a sufficient absorption of calcium without the stimulating action of calcitriol, which can only be synthesized through the action of ultraviolet radiation. While the regulatory mechanisms outlined above are well established for tetrapods, the role of vitamin D for fishes is still a controversial subject (Reinboth, 1980). Schäperclaus (1979) assumed that vitamin D might not be essential for fishes, because they are able to take up large amounts of calcium through the gill membranes, and thus do not depend on the major function of calcitriol. If the absorption of calcium through the gills indeed is the mechanism that makes fishes independent from vitamin D, it might be one additional explanation for the relatively large diversity of troglobitic fishes compared to other vertebrates. Interestingly, among the many vertebrate groups only fishes and salamanders have been able to colonize aphotic environments, and most troglobitic salamanders are neotenic (see Chapter 6, this volume). Because these forms attain sexual maturity without metamorphosis, they retain several larval traits – among them external gills. Although the implications of calcium metabolism for cave colonization are at present purely speculative, it is worthwhile to consider this aspect in future studies.

Endocrinological imbalance

Hormones are chemical signals that regulate many developmental and physiological functions of the body, such as metabolism, growth, metamorphosis and sexual maturation (Reinboth, 1980). The secretion of hormones that stimulate or inhibit particular body functions is controlled by the hypothalamus, which itself is influenced by several neural factors (Fig. 7.4). Most interesting for cave ecology and evolution is the fact that the levels of several hormones have been shown to change in direct response to light and temperature (Matty, 1985). More generally, in most vertebrates studied the release of hormones from the pituitary (e.g., prolactin, gonadotropin, corticotropin) shows circadian rhythms that are primarily entrained by light and temperature. It can be concluded that the colonization of continuously dark and arrhythmic environments should cause some endocrinological imbalance in species that are not only behaviorally but also physiologically adapted to the circadian rhythmicity of their former epigean habitat.

Rasquin and Rosenbloom (1954) described several, ultimately lethal, effects of continuous darkness on the epigean ancestor of the cave fish *Astyanax fasciatus*. Besides the reduced ossification discussed above they found a decreased growth rate, degeneration of the kidney, disturbed gametogenesis, accumulation of adipose tissue and reduced musculature, which was associated with the atrophy of the thyroid, adrenal glands and gonads. Although the authors cited a series of prior studies that revealed similar effects in other animals subjected to continuous darkness, none of them showed such destructive results. It is thus likely that Rasquin and Rosenbloom's observations were not basically related to the effects of darkness but might primarily have been caused by other experimental conditions, such as diet and water quality. Because endocrine regulatory mechanisms are not only influenced by many endogenous and exogenous factors (light and temperature, diet, levels of feeding or starvation, salinity changes in the environment of aquatic species, age, maturation, stress, etc.) but also vary to some degree between different species (Matty, 1985), it is

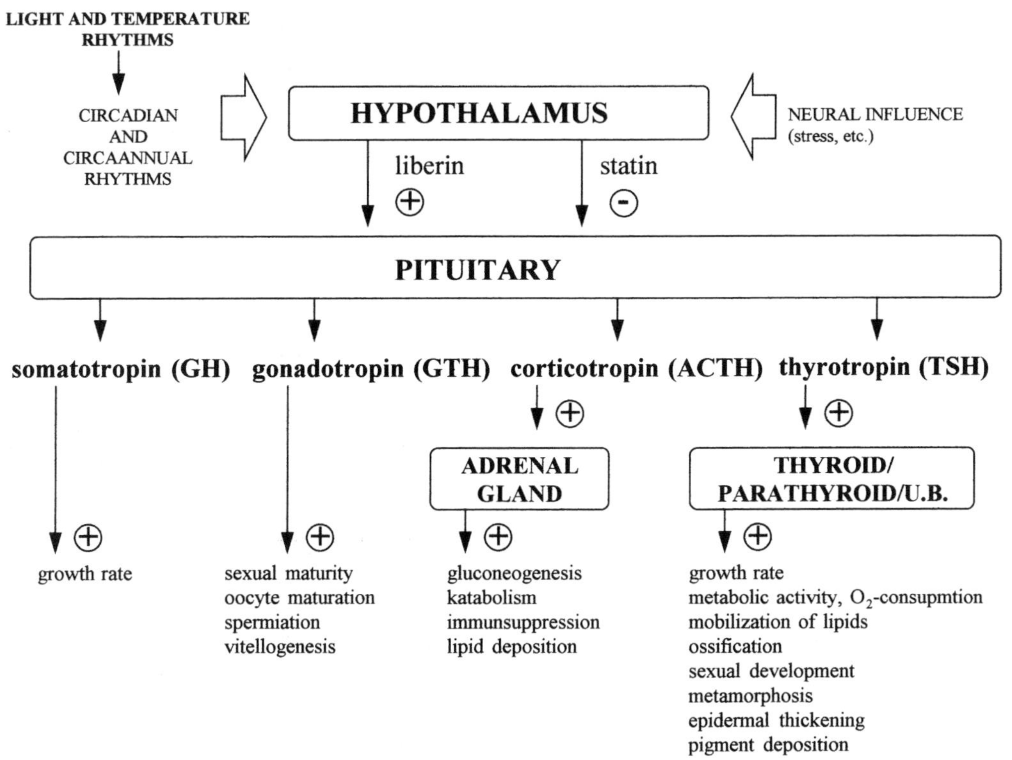

Fig. 7.4. Simplified scheme of hormonal control in vertebrates. Several troglomorphic traits appear to be related to thyroid activity.

not justifiable to draw far-reaching conclusions from scanty experimental data. Nevertheless, the information available indicates that continuous darkness can result in significant disturbances of endocrine regulatory mechanisms.

Most promising for an analysis of endocrine adjustments in cave vertebrates is a comparative analysis of thyroid function in troglobites and their epigean relatives. The thyroid is controlled by the pituitary hormone thyrotropin and promotes a series of essential metabolic and developmental functions (Fig. 7.4). Hypoplasia of the thyroid was not only found in epigean *Astyanax fasciatus* exposed to continuous darkness (Rasquin and Rosenbloom, 1954) but also in the phylogenetically-old cave fish *Caecobarbus geertsi* (Olivereau and Francotte-Henry, 1954) and *Uegitglanis zammaranoi* (Olivereau, 1960) and the cave salamander *Typhlomolge rathbuni* (Gorbman, 1957). Hypothyroidism is characterized by a reduced growth rate, reduced metabolic rate and oxygen consumption, accumulation of lipids, reduced ossification, retarded sexual development and hindered metamorphosis in amphibians (Reinboth, 1980; Kaplan and Gass, 1982; Matty, 1985). These physiological effects closely match some of the most prominent changes of troglobites that are believed to have evolved convergently in response to the conditions of food scarcity and stable environment – that is, the trend to paedomorphosis combined with a reduced metabolic rate, reduced growth rate and increased lipid storage (see Chapter 8, this volume).

The only studies available on the role of the thyroxine hormone for the evolution of the cave phenotype concerned the neotenic development of cave salamanders. Because it is known that metamorphosis can be induced in the neotenic axolotl (*Ambystoma mexicanum*) by the injection of thyroxine, several efforts were made to induce metamorphosis in the neotenic cave salamander *Proteus anguinus* by immersion of larvae, juveniles and adults into solutions of thyroxine or by injection of the hormone into the body (Vandel and Bouillon, 1959; Durand and Vandel, 1968), but none of the experiments gave positive results. Recently, an ultrastructural study (Magnan, 1995) showed that the thyroid of *Proteus* appears to be well-developed and endocrinologically active. It can be concluded that the neotenic state of the cave salamander is not caused by hypotrophy of the thyroid, but that the tissues became insensitive to thyroxine during cavernicolous evolution. Durand and Vandel (1968), however, argued that *Proteus* belongs to an ancient group of primitive salamanders that have never undergone metamorphosis during development, but lived a permanently aquatic life. The lack of metamorphosis could thus not be related to cave life in this species. Interestingly, Gorbman (1957) reported that the thyroid tissue of the Texas blind salamander *Typhlomolge rathbuni* appears to be inactive, and Kezer (1952) found that closely related neotenic salamanders (*Eurycea neotenes*, *E. tynerensis*) are sensitive to thyroxine. In this case, the absence of metamorphosis could thus result from a failure of thyroxine secretion by the thyroid.

Future studies will have to show whether the characteristics described above are indeed related to a hypoplastic development of the thyroid in different cave species, and whether continuous darkness can result in thyroid atrophy. In any case it is obvious that cave colonization requires preadaptations of the endocrine system, and that possible adaptive adjustments of the endocrine system can result in complex changes of physiology and development. Some presumably adaptive characters of troglobites may thus not themselves be selected but in fact represent side effects of endocrine adjustments.

Rhythmicity

Circadian rhythms generated by an endogenous clock are apparently an ubiquitous feature of eukaryotes (Bünning, 1973; Rosbash, 1995). The inner clock can be located within a single cell (Welsh et al., 1995); higher animals can have multiple clocks which all show rhythmic oscillations with a periodicity near 24 hours under constant environmental conditions (Rosbash, 1995). Melatonin has been shown to play an important role in the synchronization of these oscillators with environmental rhythms, primarily the change of day and night (Underwood, 1989). Light and temperature are the major zeitgeber that entrain the inner clock with environmental conditions.

Recent studies have provided much insight into the molecular control of circadian rhythms [see Rosbash (1995) for review]. Mechanisms of clock-controlled gene expression have been described, as well as several rhythm mutants that allowed different clock genes to be characterized. Nevertheless, the precise biological function of the inner clock is still not fully understood. It is evident that circadian rhythms provide the advantage of adapting behavior to external conditions; most animals forage primarily during either day, night or twilight periods (Helfman, 1993). The timed activity does not only serve to avoid predators (cover-seeking, daily migrations) but also regulates locomotor activity, interactions between individuals, or the distribution pattern of complex communities, for example. By measurement of the length of the day, internal rhythms can also adjust to seasonal cycles. These can influence the physiological and metabolic status of individuals, for instance gonadal development, and their behavior (reproductive behavior, seasonal migration, etc.). Circadian and seasonal rhythms are thus certainly selected for behavioral and physiological adaptations to temporally altering environments. However, it is not clear whether circadian rhythms are also essential for the control and synchronization of internal metabolic reactions that need to be regulated regardless to external, environmental conditions (Lamprecht and Weber, 1992). If this is the case, cave animals living in a relatively stable environment without light and dark cycles should keep free-running circadian rhythms; otherwise they would lose their rhythmicity, just like other behavioral traits that have lost their biological function.

The studies available on the rhythmicity of cave animals [see Lamprecht and Weber (1985, 1992) for review] provide much evidence for the ecological significance of behavioral rhythms. Rhythms in locomotor activity are still selected in troglophilic animals or in cave species that inhabit the twilight zone, but they degenerate in phylogenetically old troglobites inhabiting the deep cave zone. Lamprecht and Weber (1992) listed several vertebrate and invertebrate troglobites that were definitely shown to have lost circadian elements in their behavioral activity. The Mexican cave fish *Astyanax fasciatus* has been especially well studied. Although this cave form is only of moderate phylogenetic age, Thinès and Weyers (1978) could not find a free-running periodicity of locomotor activity under constant conditions (permanent darkness). A more thorough study by Erckens and Martin (1982a,b), on the other hand, revealed only a partial regression of circadian-activity rhythms in cave-dwelling *Astyanax*. Activity rhythms could still be induced by light and dark changes of any period length, but immediately disappeared in permanent darkness. However, if the induced rhythm had a period length of 24 h, a free-running circadian rhythm could even be demonstrated.

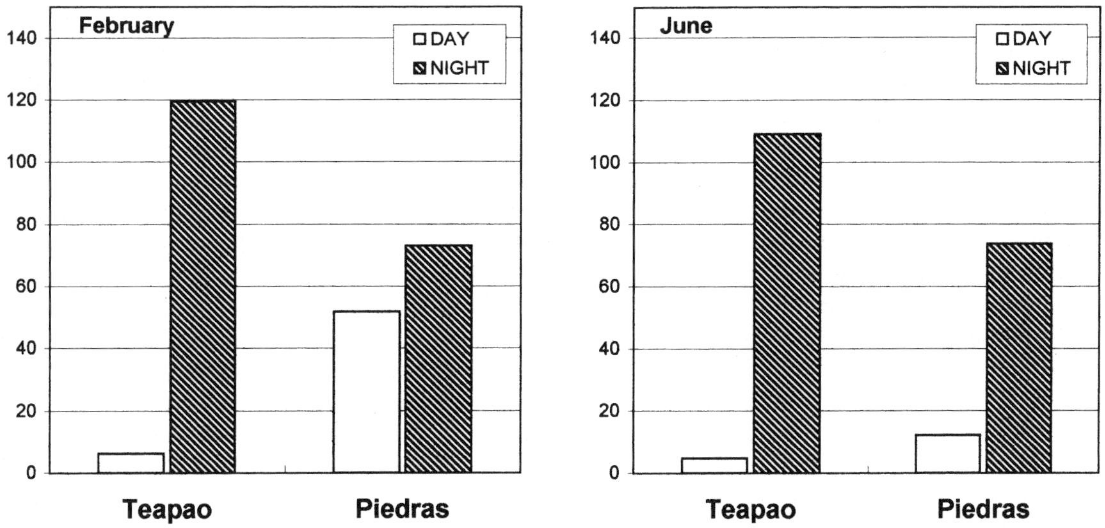

Fig. 7.5. Circadian changes of mean melatonin content (in pg) in the pineal organ of epigean (Teapao) and cave-dwelling (Piedras) *Astyanax fasciatus* in February and June.

Erckens and Martin concluded that only a period of 24 h is able to stabilize the degenerate circadian system, possibly by the synchronization of multiple circadian clocks.

The examples show that circadian locomotor activity is not under control of stabilizing selection in the aphotic and relatively stable cave environment. Yet, one does not know whether the inner clock of cave animals itself is subject to regressive evolution or whether some clock-controlled functions, such as locomotor activity patterns, are just uncoupled from the clock.

More basic functions of the circadian system of cave animals have only recently been studied in *Astyanax fasciatus* (Wilkens et al., 1993; Missal et al., 1994). The melatonin content of the pineal organ of epigean and cave specimens was measured during daytime and night in two seasons of the year by radio immuno-assay (RIA) (Wilkens et al., 1993). The epigean fish showed the expected pattern: the melatonin content was very low during the day and high at night (Fig. 7.5); there were no significant differences in the values in February and June. The cave fish still showed a rhythmic pattern of melatonin content, but the amplitude of low and high levels was significantly diminished. Light did not suppress the melatonin production as dramaticaily as in the epigean form, and the night value was notably reduced (Fig. 7.5). Contrary to the epigean specimens, the cave fish showed a marked difference between the two seasonal samples. In June, if a light/dark cycle of 18/6 h was imposed, the day value was much lower than when a cycle of 12/12 h was imposed in February. Following this, a long-term superfusion system was developed to study melatonin release profiles of epigean and cave fishes *in vitro*. Pineal organs were cultured individually in a light/dark cycle of 12:12 h, and under permanent darkness for up to 5 days, and the melatonin synthesized was automatically sampled every hour (Missal et al., 1994). Again, both populations showed a statistically significant diurnal rhythm of melatonin synthesis when exposed to a light/dark cycle of 12:12 h. Most interestingly, recent experiments (Langecker et al., unpublished) showed that a free-running rhythm of melatonin secretion is present in epigean and cave specimens of the phylogenetically young Micos population, but not in the phylogenetically old form in Pachon cave. The Pachon fishes show a steady, high rate of melatonin secretion under conditions of permanent darkness. These results are in accordance with the ultrastructural studies on the pineal organ of cave-dwelling *Astyanax* (Langecker, 1992b) that demonstrated a persisting secretory activity in the pineal organ of the Pachon fish. They furthermore provide evidence that the initial regression of locomotor activity rhythms in cave fish (Erckens and Martin, 1982a,b) might not just be a consequence of the uncoupling of behavioral rhythms from the clock, but could involve the circadian system itself.

These data are in accordance with the assumption of Lamprecht and Weber (1985, 1992) that the circadian system of phylogenetically old troglobites degenerates without any residue, and may be replaced by a stochastic control of activity and rest; but more information about the nature of the inner clock in cave animals is needed to verify their hypothesis. It is probably not useful to perform additional studies on the locomotor activity of further cave species; it is well established that there is a clear trend to the regression of circadian rhythms of behavior. In spite of this, it is necessary to prove whether the regression of free-running rhythms of melatonin synthesis can be confirmed for other cave species; if so, it would be exciting to see whether cultures of individual cells from the suprachiasmatic nucleus, which is a prominent site of clock cells, still show circadian oscillations. Those experiments could definitely answer the question whether the regression of the circadian system of cave animals is confined to some clock-dependent functions, or whether it involves the control mechanisms of the clock itself.

Besides controlling and separating temporally incompatible metabolic reactions, circadian rhythmicity could also exist in cave animals to establish seasonal rhythms that are based on circadian clocks. Slight rhythmic alterations in temperature, bat flight or other external zeitgeber could entrain the inner clock of cave species with day and night rhythms and thus provide a measure for seasonal changes. However, there is little information available about seasonal rhythmicity of cave animals. The amblyopsid cave fishes *Amblyopsis rosae* and *A. spelaea* show a well-defined yearly reproductive cycle (Poulson, 1963). Likewise, spawning of the cave fishes *Astyanax* is believed to occur twice in the year during the rainy season (Parzefall, 1983); but other cave animals, for example the shrimp *Palaemonetes antrorum* from Texas, do not show any seasonal pattern but reproduce continually throughout the year (Strenth and Longley, 1990). Circadian and seasonal rhythms of cave animals thus provide a wide-open field for future research.

Behavior

The behavioral ecology of cave fishes and general behavioral aspects of animals living in caves were most recently reviewed by Parzefall (1992, 1993); a thorough analysis of the role of agonistic behavior for cave communities is also available (see Chapter 11, this volume).

Behavioral changes in cave animals are primarily a story of regressive evolution. Deprived of their obligate visual stimuli, many behavior patterns are not manifested in cave animals. Consequently, they are no longer controlled by stabilizing selection and become a subject to regressive evolution. In some cases it was possible to prove that behavioral regression in cave animals is genetically based (Fricke, 1988; Langecker, 1993). Behavior that is not solely triggered by visual stimuli but also depends on mechanosensory input can partially degenerate, but in parallel develop constructively with new, non-visual triggers. Hausberg (1995) found a regressive evolution of the visual elements and a constructive evolution of the non-visual elements in the aggressive behavior of the cave fish *Astyanax fasciatus.* A purely constructive evolution of new behavior patterns in cave animals is rarely reported. The best- known case is a changed feeding behavior that evolved convergently in different cave fish.

Choice–preference behavior to light

The orientation of animals to light is commonly referred to as phototactic behavior. The classical taxis concept of Kuehn (1919) only applies to a positive or negative orientation to abiotic factors, such as light. However, the natural behavior of animals is better approached by the concept of optimal conditions that assigns a distinctive range of preferred and tolerated abiotic conditions to each species (Schwertfeger, 1963; Schöne, 1980). Some misleading assumptions on the role of phototactic behavior in cave animals and their epigean relatives arose from the erroneous usage of the simple taxis concept.

Photonegative behavior of epigean *Astyanax fasciatus* was discussed to support the concepts of active cave colonization (Eigenmann, 1909; Breder Jr and Rasquin, 1947) and of the acceleration of adaptation to cave life by a genetic coupling of increased scotophilia with the loss of eyes and pigmentation (Ludwig, 1942). Recent studies could not confirm these assumptions. A thorough case study on epigean and cave-dwelling *Astyanax fasciatus* analysed the choice–preference behavior to light within its ecological context (Langecker, 1989, 1992a). It was found that the photonegative behavior of the ancestral epigean fishes is by no means a behavioral preadaptation to cave entry. On the contrary, epigean fishes avoid complete darkness just as much as bright light, and show a preference for low-light conditions (Fig. 7.6). This does not indicate that visually guided behavior had minor significance but instead that it evolved from visual demands. Hovering in dark and shaded places, the

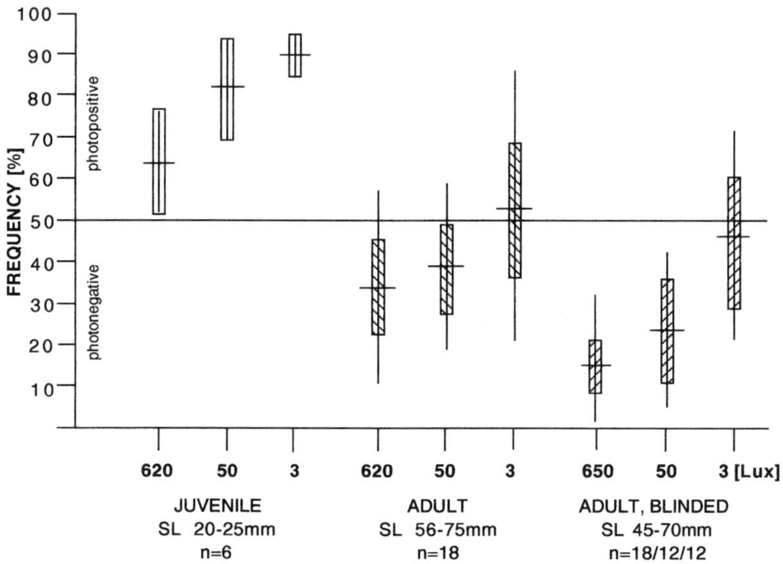

Fig. 7.6. Phototactic behavior in epigean *Astyanax fasciatus* depends on the light intensity and the phylogenetic age. The increased photonegative behavior in blinded specimens demonstrates the influence of visual demands on this behavior. SL, Standard Length – a biometric measure of the size of the organism.

fish is not only inconspicuous to potential prey but is also better able to see approaching objects (Helfman, 1981) – that is, the fish gains a visual advantage. So-called scotophilia is thus not a sufficient indicator for behavioral preadaptation of epigean individuals but can simply refer to the avoidance of bright light. It was further shown that phototactic behavior is largely influenced by changes in motivation and ontogenetic age. The photopositive behavior of young specimens can be correlated with a different vertical distribution of juveniles and adults; the latter prefer deeper and darker zones of the water whereas the juveniles are frequently found close to the surface. Although *Astyanax* is not representative of the majority of cave-dwelling species, which are mostly related to nocturnal or photophobic species, it is questionable whether the reaction to light plays a significant part in cave colonization at all. The Mexican troglophilic catfish *Rhamdia guatemalensis*, for example, frequently invades the aphotic subterranean streams of the Yucatan peninsula (Wilkens, 1979) but never developed a troglobitic population. In any case the term "scotophilia" does not sufficiently characterize the orientation to light, and has to be analysed in its behavioral and ecological context if its role for cavernicolous evolution is to be evaluated.

The discovery that cave-adapted troglobites living in continuous darkness, are still sensitive to light leads to the doubtful belief that it might have a persisting function in most cave forms. Breder Jr and Rasquin (1947) suggested that photonegative behavior should be selected in cave-adapted troglobites to prevent them from leaving their environment because they could no longer compete in the surrounding epigean habitats. On the contrary, all comparative studies on fishes performed until today have shown that cave-adapted individuals have a significantly reduced photonegative reaction to light compared with their closest relatives [see Langecker (1992a) for review] and cave-dwelling *Poecilia mexicana* even have photopositive behavior (Krauss, 1992). These findings indicate that there is generally not much need for a protective phototactic mechanism to keep aquatic troglobites in the environment to which they are adapted. The light reaction of troglobitic fish is rudimentary; the light-dependent behavior degenerates in the course of cavernicolous evolution just as the photoreceptors do (see above). However, terrestrial cave communities living close to open connections with epigean habitats may be selected for photophobic behavior, whereas species inhabiting the deep-cave zone are unlikely to show any light-dependent behavior.

Camouflage and dorsal light reaction

The convergent regression of the pigment system in troglobites evolved primarily because pigmentation had lost its biological function for camouflage and visual signaling in the aphotic environment (see above). In parallel, it can be predicted that all related behavior patterns will degenerate in cave species. One example of such a behavior was genetically analysed in epigean and cave-dwelling *Astyanax fasciatus*. The dorsal light reaction (DLR) is a mechanism in fish that helps to maintain an appropriate orientation of the body axis in free water (von Holst, 1935). Both gravity and the angle of incident light determine the position of the body axis with respect to the vertical. When the angle of the incoming light deviates from the vertical, the fish tilt their body axis to make a species-specific angle with the light. The advantage of this postural mechanism influenced by two different forces is not clear, but one can suggest that it takes part in camouflage, because the DLR may increase the inconspicuousness of free-swimming fish. Munz and McFarland (1977) provided evidence that intense momentary flashes of light reflected from the silvery ventral and lateral sides of countershaded fish might be an important factor in locating pelagic prey or predators. Although this effect cannot totally be avoided in turbid water with ripples and waves that produce rapid fluctuations in light intensities, it can be reduced when the fish orientate their dark dorsal surface towards the light. Regardless of the significance of this behavior, it is strongly selected for in epigean populations of *Astyanax fasciatus*. When illuminated from the side (angle 90°, 500 Lux), the fish tilt their body axis with an angle of approximately 40° towards the light (Langecker, 1993). Because this genetically-fixed behavior is mediated from the eyes it cannot be tested in cave specimens. However, a genetic analysis can be performed with phylogenetically young cave specimens (Micos fish) that were artificially selected for large and functioning eyes, and with the F1 and F2 hybrids between epigean and phylogenetically old cave populations that also were selected for large eyes. Such experiments showed that the DLR, which is not manifested in the aphotic cave, is significantly reduced in cave-dwelling fish (Fig. 7.7). It was calculated that the behavioral regression was caused by the inactivation of at least three genetic factors (Langecker, 1993).

Signaling and alarm reaction

Members of the Ostariophysi, which include the large group of cave-dwelling cyprinid, characid and siluriform fishes, contain the so-called alarm-substance in their skin, which is released into the water when the skin is damaged (Pfeiffer, 1977). It is suggested that this substance induces an alarm reaction in conspecifics, which can result in cover-seeking, closer crowding of conspecifics, increased locomotor activity, 'zigzag'-swimming and immobility. A regression of this behavior was found in cave-dwelling *Astyanax fasciatus* (Pfeiffer, 1963) and *Caecobarbus geertsi* (Thinès and Legrain, 1973), and was explained by the lack of predators in the cave. A detailed study by Fricke (1988), however, revealed a persisting but modified reaction of cave-dwelling *Astyanax fasciatus* to alarm-substance. Cave specimens still showed an increased locomotor activity after the release of alarm-substance and significantly avoided the site of release. Compared to their epigean relatives, however, they have lost visually guided elements of the alarm reaction – optical signaling of the detected threat to conspecifics by a characteristic mode of swimming, zigzag movements, and cover-seeking at the ground. It is thus likely that the alarm reaction still has a selective advantage in the continuous darkness of the cave because it results in the avoidance of hazardous areas, but it has lost those elements that are related to vision.

Territoriality and social groups

Epigean species are often territorial, or associated in social groups. In both cases, visually-guided behavior generally plays an important role for orientation and communication. It is thus not surprising that most troglobites, which usually occur widely scattered and in low population densities, are solitary animals with a wide home range, which only meet for mating or in competition for food (terrestrial troglobites especially are often clustered around sparse food sources).

Groups of fish that remain together for social reasons were termed shoals by Pitcher (1983). Pitcher and Parrish (1993) emphasized that schooling, the polarized and synchronized swimming of groups, is only one category of shoals which primarily evolved as an antipredator tactic. In the absence of predators, shoals are not homogeneous and individuals may leave, join or stay with the group. The behavior of epigean *Astyanax fasciatus* fits this concept very well (my own field observations). In their natural habitat the fishes are commonly associated in shoals that do not display a homogeneous or synchronized behavior. As soon as a predator (or the observer) is detected, they form a strictly polarized school that again returns to the shoaling pattern if the expected attack fails to come. Under exceptional conditions, for instance in small and secluded pools, few specimens become territorial. Cave fishes can be expected to

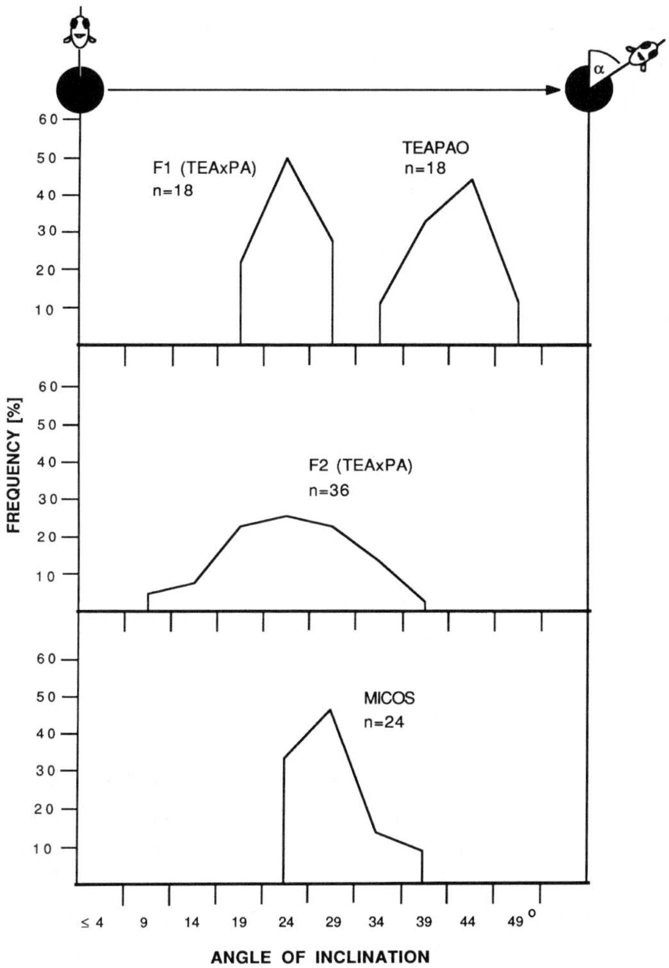

Fig. 7.7. Frequency distribution of the mean angle of tilt of epigean specimens (Teapao), F1 and F2 hybrids between epigean and phylogenetically-old cave specimens (Teapao Pachon) and of phylogenetically young cave specimens (Micos) of *Astyanax fasciatus*.

lose shoaling behavior because it is primarily visually guided, and because predators are absent in most caves. Field studies confirmed this assumption (Parzefall, 1983; and my own observations). Cave fish display no group structure, but randomly swim through the water searching for food on the bottom. Even sudden mechanical stimuli (rapid water movement) do not trigger any protective behavior. Instead, the fishes often approach the location of the stimulus, probably because they expect a potential food source (e.g., bat guano). The ability to school, which requires the exact, visual coordination of movement, also degenerated in cave fish (Parzefall, 1993). Furthermore, the alternative behavioral strategy to shoaling – that is, territoriality – was never observed in the cave (but see Hausberg, 1995). Quinn (1980) found that juvenile cave fishes are attracted by the odors of conspecifics; this might have evolved in epigean populations as a secondary mechanism for maintaining group cohesion. The reaction in cave specimens, however, is so weak that he suggests this behavior to be a behavioral vestige of epigean existence, because long-term behavioral studies revealed no apparent social interactions in cave-dwelling *Astyanax* (Quinn, 1980).

Similar behavior was observed in the carnivorous amblyopsid cave fishes *Amblyopsis spelaea* and *Typhlichthys subterraneus* which are irregularly distributed over the cave area. They move over long distances in search for food, but never defend territories and generally exhibit reduced aggressive behavior (Poulson, 1963; Mohr and Poulson, 1966; Bechler, 1983). Other cave fish are known to exhibit a complex intraspecific aggressive behavior – for instance, *Pimelodella kronei* (Trajano, 1991) or *Uegitglanis zammaranoi* (Berti et al., 1982) – which is mediated from mechanical senses. Only the blind cyprinid *Caecobarbus geertsi* and the cave-dwelling synbranchid eel *Ophisternon infernale* have been reported to defend territories (Parzefall, 1992). However, these studies were performed in the laboratory, which rarely reflects the conditions in the natural habitat, especially regarding agonistic behavior. There is a great demand for field studies about the natural behavior of cave animals; the information available is insufficient.

Mating

The reproductive behavior of cave species is necessarily non-visually guided but controlled by chemical and mechanical stimuli (see Parzefall, 1992). Because cave colonizers must possess those obligate preadaptations for successful reproduction, mating behavior shows little adaptive change in troglobites. Parzefall (1992) described the case of the cave fish *Poecilia mexicana,* in which aggressive elements of the reproductive behavior were reduced under selection pressure. Epigean males frequently fight for attractive females, but cave specimens prefer a quick copulation when they make contact with attractive females. The advantage of the latter strategy in the cave is obvious, because fighting males will readily lose contact with the female without visual orientation.

Feeding behavior

Cave-adapted animals are generally more effective in locating food in darkness than their epigean relatives. In amblyopsid cave fishes this is achieved by somatosensory adaptations, and also by more effective swimming behavior (Poulson, 1963; Poulson and White, 1969). The improved food-finding ability of the cave fish *Astyanax* (Hüppop, 1987), however, is primarily due to behavioral changes. Epigean fishes hover motionless in the shade and observe the open water for approaching prey, or they search on the bottom with a steep angle of their body axis to the horizontal. This posture optimizes the visual detection of food particles, but it has the disadvantage that only a small area of the snout equipped with chemoreceptors can participate in the location of food (Fig. 7.8). Cave fishes do not search for food in the open water, but continuously move close to the ground with a significantly reduced angle of the body axis to the horizontal. As a consequence of this changed posture, a larger area of the lower jaw, which is equipped with an increased number of taste buds (Fig. 7.8), is in contact to the ground (Schemmel, 1980). Cave fishes thus have a tremendous advantage in locating food in the darkness compared to their eyed relatives (Hüppop, 1987). A similar feeding mode is realized in the Texas blindcat *Trogloglanis pattersoni*, which has a very specialized mouth for chemical location of food on the bottom (Langecker and Longley, 1993). The cave cyprinid *Phreatichthys andruzzii,* finally, exhibits the same trend. Compared to its epigean relative *Barbus filamentosus,* the cave fish has a greater affinity to the bottom, a smaller angle of the body axis to the horizontal, and, as a result, a significantly increased efficiency in locating food (Berti and Masciarelli, 1993).

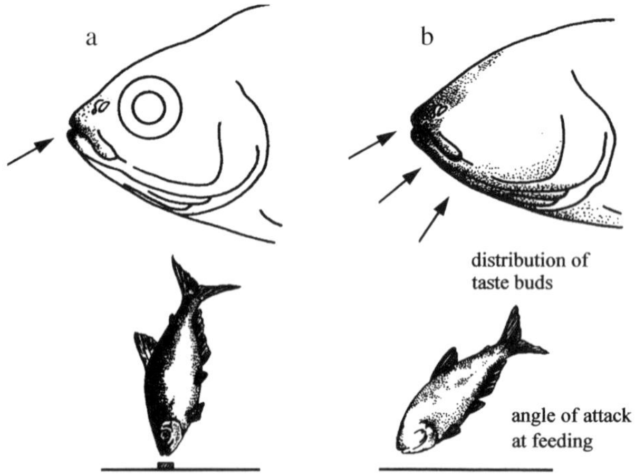

Fig. 7.8. Cave specimens of *Astyanax fasciatus* show a decreased angle of the body axis to the horizontal compared with epigean specimens, and an increased area and number of chemoreceptors (arrows) especially on the lower jaw. Both features improve the chemical detection of food in the darkness. (a) Epigean fish; (b) cave fish.

ACKNOWLEDGMENTS

I am very grateful to N. Peters and H. Wilkens for many valuable suggestions and critical reading of the manuscript.

REFERENCES

Amlacher, E., 1986. *Taschenbuch der Fischkrankheiten.* VEB Gustav Fischer Verlag, Jena.

Anderßon, O., 1992. *Vergleichende Untersuchungen der Augenentwicklung bei oberirdischen und Höhlenformen der Gattung Rhamdia (Pimelodidae, Teleostei). Diploma Thesis.* Fachbereich Biologie, Universität Hamburg, pp. 60.

Bath, H., 1962. Vergleichende biologisch-anatomische Untersuchungen über die Leistungsfähigkeit der Sinnesorgane für den Nahrungserwerb, ihre gegenseitige Abhängigkeit und ihre Beziehungen zum Bau des Gehirns bei verschiedenen Knochenfischarten. *Z. Wiss. Zool.*, 166/167: 238–290.

Bechler, D.L., 1983. The evolution of agonistic behaviour in amblyopsid fishes. *Behav. Ecol. Sociobiol.*, 12: 35–42.

Behrens, M., Langecker, T.G., Wilkens, H. and Schmale, H., 1996. Comparative analysis of *Pax-6* sequence and expression in the eye development of the blind cave fish *Astyanax fasciatus* and its epigean conspecific. *Mol. Biol. Evol.*, 14(3): 299–308.

Berti, R. and Masciarelli, L., 1993. Comparative performances of non-visual food search in the hypogean cyprinid *Phreatichthys andruzzii* and in the epigean relative *Barbus filamentosus*. *Int. J. Speleol.*, 22: 121–130.

Berti, R., Ercolini, A. and Cianfanelli, A., 1982. The aggressive behavior repertoire of an anophthalmic phreatic fish from Somalia. *Experientia*, 39: 217–219.

Botosaneanu, L. and Holsinger, J.R., 1991. Some aspects concerning colonization of the subterranean realm – especially of subterranean waters: a response to Rouch & Danielopol, 1987. *Stygologia*, 6(1): 11–39.

Breder Jr., C.M. and Rasquin, P., 1947. Comparative studies in the light sensitivity of blind characins from a series of Mexican caves. *Zoologica N.Y.*, 26(16): 123–131.

Bünning, E., 1973. *The Physiological Clock.* Springer, Berlin, 258 pp.

Calof, A.L., 1995. Intrinsic and extrinsic factors regulating vertebrate neurogenesis. *Curr. Op. Neurobiol.*, 5: 19–27.

Christiansen, K.A., 1992. Biological processes in space and time. Cave life in the light of modern evolutionary theory. In: A.I. Camacho (Editor), *The Natural History of Biospeleology.* National Museum of Natural Sciences, Madrid, pp. 453–478.

Cooper, J.E. and Kuehne, R.A., 1974. *Speoplatyrhinus poulsoni*, a new genus and species of subterranean fish from Alabama. *Copeia*, 486–493.

Cramer, K.S. and Sur, M., 1995. Activity-dependent remodeling of connections in the mammalian visual system. *Curr. Op. Neurobiol.*, 5: 106–111.

Culver, D.C., 1982. *Cave life. Evolution and Ecology.* Harvard University Press, Cambridge, MA, pp. 189.

Diffey, B.L., 1991. Solar ultraviolet radiation effects on biological systems. *Rev. Phys. Med. Biol.*, 36(3): 299–328.

Dodt, E., 1973. The parietal (pineal and parietal organs) eye of lower vertebrates. *Handbook of Sensory Physiology*, VII/3B. pp. 113–137.

Durand, J.P., 1971. Recherches sur l'appareil visuel du Protee *Proteus anguinus*. Laurenti, urodele hypoge. *Ann. Speleol.*, 26/3: 497–824.

Durand, J.P. and Vandel, A., 1968. Proteus: an evolutionary relic. *Sci. J. R. Coll. Sci.*, 44–49.

Eigenmann, C.H., 1909. Cave vertebrates of America. A study in degenerative evolution. *Carnegie Inst. Washington Publ.*, 104: 1–241.

Erckens, W. and Martin, W., 1982a. Exogenous and endogenous control of swimming activity in *Astyanax mexicanus* (Characidae, Pisces) by direct light response and by a circadian oscillator. I. Analysis of the time-control system of an epigean river population. *Z. Naturforsch.*, 37: 1253–1265.

Erckens, W. and Martin, W., 1982b. Exogenous and endogenous control of swimming activity in *Astyanax mexicanus* (Characidae, Pisces) by direct light response and by a circadian oscillator. II. Features of time-controlled behavior of a cave population and their comparison to an epigean ancestral form. *Z. Naturforsch.*, 37: 1266–1273.

Ercolini, A. and Berti, R., 1975. Light sensitivity experiments and morphology studies of the blind phreatic fish *Phreatichthys andruzzii* Vinciguerra from Somalia. *Monit. Zool. Ital. Suppl.* VI(2): 29–43.

Ercolini, A. and Berti, R., 1977. Morphology and response to light of *Uegitglanis zammaranoi* Gianferrari, anophthalmic phreatic fish from Somalia. *Monit. Zool. Ital. Suppl.* IX(8): 183–199.

Falcon, J., Bolliet, V., Ravault, J.P., Chesneau, D., Ali, M.A. and Collin, J.P., 1994. Rhythmic secretion of melatonin by the superfused pike pineal organ: thermo- and photoperiod interaction. *Neuroendocrinology*, 60: 535–543.

Foster, R.G., Garcia-Fernandez, J.M., Provencio, I. and DeGrip, W.J., 1993. Opsin localization and chromophore retinoids identified within the basal brain of the lizard *Anolis carolinensis*. *J. Comp. Phys. A*, 172: 33–45.

Franck, A., 1964. Vergleichende Untersuchungen am Höhlenfisch *Anoptichthys antrobius* und seinem oberirdischen Vorfahren *Astyanax mexicanus*. *Zool. Anz.*, 172: 95–107.

Fricke, D., 1988. Reaction to alarm substance in cave populations of *Astyanax mexicanus* (Characidae, Pisces). *Ethology*, 76: 305–308.

Gorbman, A., 1957. The thyroid gland of *Typhlomolge rathbuni*. *Copeia*, 1957: 41–43.

Guthrie, S.D.M., 1990. The physiology of the teleostean optic tectum. In: R.H. Douglas and M.B.A. Djamgoz (Editors), *The visual system of fish.* Chapman and Hall, London, New York, pp. 279–343.

Hahn, G., 1960. Ferntastsinn und Strömungssinn beim augenlosen Höhlenfisch *Anoptichthys jordani* im Vergleich zu anderen Teleosteern. *Naturwissenschaften*, 47: 611.

Hall, B.K., 1992. *Evolutionary Developmental Biology.* Chapman & Hall, London, N.Y., 491 pp.

Hausberg, C., 1995. *Das Aggressionsverhalten von Astyanax fasciatus (Cuvier 1819; Characidae, Teleostei): Ontogenie, Genetik und Evolution bei der epigäischen und hypogäischen Form.* Ph.D. thesis, Fachbereich Biologie, Universität Hamburg, 139 pp.

Helfman, G.S., 1981. The advantage to the fishes of hovering in shade. *Copeia*, 392–400.

Helfman, G.S., 1993. Fish behavior by day, night and twilight. In: T.J. Pitcher (Editor), *Behavior of teleost fishes.* Chapman & Hall, London, N.Y., pp. 479–512.

Howarth, F.G., 1981. Non-relictual terrestrial troglobites in the tropical Hawaiian caves. *Proc. 8th Intern. Cong. Speleol.*, Nat. Speleol. Soc., Huntsville, Al., pp. 539–541.

Hüppop, K., 1987. Food-finding ability in cave fish (*Astyanax fasciatus*). *Int. J. Speleol.*, 16 (1–2): 59–66.

Hüppop, K., 1988. *Phänomene und Bedeutung der Energieersparnis beim Höhlensalmer Astyanax fasciatus. Ph.D. thesis.* Fachbereich Biologie, Universität Hamburg.

Hurst, J.G., 1982. Vitamin D. In: G.H. Gass and H.M. Kaplan (Editors), *Handbook of Endocrinology.* CRC Press, Inc., Boca Raton, Florida, pp. 297–316.

Kaplan, H.M. and Gass, G.H., 1982. The thyroid gland. In: G.H. Gass and H.M. Kaplan (Editors), *Handbook of Endocrinology.* CRC Press, Inc., Boca Raton, Florida, pp. 241–296.

Kezer, J., 1952. Thyroxine-induced metamorphosis of the neotenic salamanders *Eurycea tynerensis* and *Eurycea neotenes*. *Copeia*, 1952: 234–237.

Krauss, Ch., 1992. *Zur Biologie sowie zum Verhalten der phylogenetisch jungen Höhlenpopulation von Poecilia mexicana (Pisces, Poecilidae) unter besonderer Berücksichtigung phototaktischen Verhaltens*. Thesis, Fachbereich Biologie, Universität Hamburg, 61 pp.

Kuehn, A., 1919. *Die Orientierung der Tiere im Raum*. Gustav Fischer Verlag, Jena.

Lamprecht, G. and Weber, F., 1985. Time-keeping mechanisms and their ecological significance in cavernicolous animals. *Bull. Natl. Speleol. Soc.*, 47(2): 147–162.

Lamprecht, G. and Weber, F., 1992. Spontaneous locomotion behavior in cavernicolous animals – the regression of the endogenous circadian system. In: A.I. Camacho (Editor), *The Natural History of Biospeleology*. Monografias del Museo Nacional de Ciencias Naturales, Madrid, pp. 225–262.

Land, M.F. and Fernald, R.D., 1992. The evolution of eyes. *Annu. Rev. Neurosci.*, 15: 1–29.

Langecker, T.G., 1989. Studies on the light reaction of epigean and cave populations of *Astyanax fasciatus* (Characidae, Pisces). *Mem. Biospeol.*, 16: 169–176.

Langecker, T.G., 1992a. Light sensitivity of cave vertebrates. Behavioral and morphological aspects. In: A.I. Camacho (Editor), *The Natural History of Biospeleology*. Monografias del Museo Nacional de Ciencias Naturales, Madrid, pp. 295–326.

Langecker, T.G., 1992b. Persistence of ultrastructurally well-developed photoreceptor cells in the pineal organ of a phylogenetically old cave-dwelling population of *Astyanax fasciatus* Cuvier, 1819 (Teleostei, Characidae). *Z. Zool. Syst. Evolutionsforsch.*, 30(4): 287–296.

Langecker, T.G., 1993. Genetic analysis of the dorsal light reaction in epigean and cave-dwelling *Astyanax fasciatus* (Teleostei, Characidae). *Ethol. Ecol. Evol.*, 3(5): 357–364.

Langecker, T.G. and Longley, G., 1993. Morphological adaptations of the Texas blind catfishes *Trogloglanis pattersoni* and *Satan eurystomus* (Siluriformes: Ictaluridae) to their underground environment. *Copeia*, 976–986.

Langecker, T.G. and Wilkens, H., 1992. Comparative ultrastructural studies on the pineal organ of the Mexican catfish *Rhamdia laticauda* Heckel, 1858 and one of its cave-dwelling derivates (Pimelodidae, Teleostei). *Acta Zool.*, 73(4): 247–253.

Langecker, T.G., Wilkens, H. and Junge, P., 1991. Introgressive hybridization in the Pachon cave population of *Astyanax fasciatus* (Characidae, Pisces). *Ichthyol. Explor. Freshwaters*, 2(3): 209–212.

Langecker, T.G., Schmale, H. and Wilkens, H., 1993. Transcription of the opsin gene in degenerate eyes of cave-dwelling *Astyanax fasciatus* (Teleostei, Characidae) and of its conspecific epigean ancestor during early ontogeny. *Cell Tissue Res.*, 273: 183–192.

Langecker, T.G., Neumann, B., Hausberg, C. and Parzefall, J., 1995a. Evolution of the optical releasers for aggressive behavior in cave-dwelling *Astyanax fasciatus* (Teleostei, Characidae). *Behav. Process.*, 34: 161–168.

Langecker, T.G., Wilkens, H. and Schmale, H., 1995b. Developmental constraints in regressive evolution: studies on the expression of the γ_s-crystallin gene in the developing lens of cave-dwelling *Astyanax fasciatus* (Cuvier, 1819) (Teleostei, Characidae) by *in situ* hybridization. *J. Zool. Syst. Evol. Res.*, 33: 123–128.

Ludwig, W., 1942. Zur evolutorischen Erklärung der Höhlentiermerkmale durch Allelelimination. *Biol. Zentralbl.*, 62: 447–482.

Magnan, G., 1995. *Le Système thyroïdien d'un vertébré cavernicole (Proteus anguinus Laur.)*. Thesis Univ. Claude Bernard, Lyon; Univ. Aix-Marseille II; Univ. Jean Monet, Saint Etienne, 25 pp.

Marshall, N.B. and Thinès, G.L., 1958. Studies of the brain, sense organs and light sensitivity of a blind cave fish (*Typhlogarra widdowsoni*) from Iraq. *Proc. Zool. Soc. London*, 131: 441–456.

Matty, A.J., 1985. *Fish endocrinology*. Croom Helm, London, Sydney, 267 pp.

McNulty, J.A., 1978. Fine structure of the pineal organ in the troglobitic fish, *Typhlichthys subterraneus* (Pisces: Amblyopsidae). *Cell Tissue Res.*, 195, 535–245.

Missal, J., Wilkens, H., Langecker, T.G. and Olcese, J., 1994. Studies on melatonin rhythmicity in a species of cave fish and its surface ancestor *Astyanax fasciatus* (Cuvier, 1819) (Characidae). In: N. Elsner and H. Breer (Editors), *Göttingen Neurobiology Report 1994*. Georg Thieme Verlag, Stuttgart, N.Y., p. 317.

Mitchell, R.W., Russell, W.H. and Elliott, W.R., 1977. Mexican eyeless characin fishes, genus *Astyanax*. Environment, distribution, and evolution. *Spec. Pub. Mus. Texas Tech. Univ.*, 12: 1–89.

Mohr, C.M. and Poulson, T.L., 1966. *The Life of the Cave*. McGraw-Hill, N.Y., 232 pp.

Morison, W.L., 1985. What is the function of melanin? *Arch. Dermatol.*, 121: 1160–1163.

Munz, F.W. and McFarland, W.N., 1977. Evolutionary adaptations of fishes to the photic environment. In: F. Crescitelli (Editor), *Handbook of Sensory Physiololgy VII/5*. Springer Verlag Berlin, Heidelberg, N.Y., pp. 193–274.

Nieuwenhuys, R., 1982. An overview of the organization of the brain of actinopterygian fishes. *Am. Zool.*, 22: 287–310.

Nilsson, D.-E., 1996. Old genes for new eyes. *Curr. Biol.*, 6: 39–42.

Olivereau, M., 1960. Etude anatomique et histologique de la glande thyroïde d'*Uegitglanis zammaranoi* Gianferrari, poisson aveugle et cavernicole, comparaison avec un clariidae voisin, *Claris buthupogon* A. Dum. *Ann. Soc. R. Zool. Belg.*, 90: 99–116.

Olivereau, M. and Francotte-Henry, M., 1954. Etude histologique et biometrique de la glande thyroïde de *Caecobarbus geertsi* Blgr. *Ann. Soc. R. Zool. Belg.*, 86: 129–150.

Parzefall, J., 1983. Field observation in epigean and cave population of Mexican characid *Astyanax fasciatus* (Pisces, Characidae). *Mem. Biospeol.*, 10: 171–176.

Parzefall, J., 1992. Behavioral aspects in animals living in caves. In: A.I. Camacho (Editor), *The Natural History of Biospeleology*. Monografias del Museo Nacional de Ciencias Naturales, Madrid, pp. 327–376.

Parzefall, J., 1993. Behavioural ecology of cave-dwelling fishes. In: T.J. Pitcher (Editor), *Behaviour of Teleost Fishes*. Chapman and Hall, London, N.Y., pp. 573–606.

Peters, N., 1990. Evolution without selection: quantitative aspects of the eye rudimentation in cave fishes. *Mem. Biospeol.*, 17: 43–48.

Peters, N., Schacht, V., Schmidt, W. and Wilkens, H., 1993. Gehirnproportionen und Ausprägungsgrad der Sinnesorgane von *Astyanax mexicanus* (Pisces, Characinidae). *Z. Zool. Syst. Evolutionsforsch.*, 31: 144–159.

Pfeiffer, W., 1963. Vergleichende Untersuchungen über die Schreckreaktion und den Schreckstoff der Ostariophysen. *Z. Vgl. Physiol.*, 47: 111–147.

Pfeiffer, W., 1967. Die Korrelation von Augengröße und

Mittelhirngröße bei Hybriden aus *Astyanax Anoptichthys* (Characidae, Pisces). *Roux Arch. Entwicklungsmech.,* 159: 365–378.

Pfeiffer, W., 1977. The distribution of fright reaction and alarm substance cells in fishes. *Copeia,* 653–665.

Pitcher, T.J., 1983. Heuristic definitions of shoaling behaviour. *Anim. Behav.,* 31: 611–613.

Pitcher, T.J. and Parrish, K., 1993. Functions of shoaling behaviour in teleosts. In: T.J. Pitcher (Editor), *Behaviour of Teleost Fishes.* Chapman and Hall, London, N.Y., pp. 363–439.

Poulson, T.L., 1963. Cave adaptation in amblyopsid fishes. *Am. Midl. Nat.,* 70: 257–290.

Poulson, T.L., 1985. Evolutionary reduction by neutral mutations: plausibility arguments and data from amblyopsid fishes and linyphiid spiders. *Bull. Natl. Speleol. Soc.,* 47: 109–117.

Poulson, T.L. and White, W.B., 1969. The cave environment. *Science,* 165: 971–981.

Quinn, T.P., 1980. Locomotor responses of juvenile blind cave fish, *Astyanax jordani,* to the odors of conspecifics. *Behav. Neural Biol.,* 29: 123–127.

Rao, P.D.P., 1967. Studies on the structural variations in the brain of teleosts and their significance. *Acta Anat.,* 68: 379–399.

Rasquin, P. and Rosenbloom, L., 1954. Endocrine imbalance and tissue hyperplasia in teleosts maintained in darkness. *Bull. Am. Mus. Nat. Hist.,* 104: 359–426.

Reinboth, R., 1980. *Vergleichende Endokrinologie.* Georg Thieme Verlag, Stuttgart, New York, 437 pp.

Riedel, G., 1993. *Die Rolle des Telencephalon beim Orientierungsverhalten. Untersuchungen am blinden Höhlenfisch (Astyanax hubbsi).* Haag und Herchen, Frankfurt am Main, 200 pp.

Rosbash, M., 1995. Molecular control of circadian rhythms. *Curr. Op. Gen. Dev.,* 5: 662–668.

Rouch, R. and Danielopol, D.L., 1987. L'origine de la faune aquatique souterraine, entre le paradigme du refuge et le modèle de la colonisation active. *Stygologia,* 3(4): 345–372.

Salvini-Plawen, L. and Mayr, E., 1977. On the evolution of photoreceptors and eyes. *Evol. Biol.,* 10: 207–263.

Schäperclaus, W., 1979. *Fischkrankheiten. Teil 2.* Akademie - Verlag, Berlin, pp. 512–1087.

Schemmel, C., 1967. Vergleichende Untersuchungen an den Hautsinnesorganen ober- und unterirdisch lebender *Astyanax* - Formen. *Z. Morphol. Tiere,* 61: 255–316.

Schemmel, C., 1980. Studies on the genetics of feeding behaviour in the cave fish *Astyanax mexicanus* f. *Anoptichthys.* An example of apparent monofactorial inheritance by polygenes. *Z. Tierpsychol.,* 53: 9–22.

Schmatolla, E., 1972. Dependence of tectal neuron differentiation on optic innervation in teleost fish. *J. Embryol. Exp. Morphol.,* 27: 555–576.

Schöne, H., 1980. *Orientierung im Raum: Formen und Mechanismen der Lenkung des Verhaltens im Raum bei Tier und Mensch.* Wissenschaftliche Verlagsgesellschaft, Stuttgart.

Schwertfeger, F., 1963. *Autökologie, die Beziehung zwischen Tier und Umwelt.* Paul Parey Verlag Hamburg, Berlin, 461 pp.

Strenth, N.E. and Longley, G., 1990. Reproductive patterns of the subterranean shrimp *Palaemonetes antrorum* Benedict (Decapoda, Palaemonidae) from Central Texas. *Stygologia,* 5(4): 221–224.

Teyke, T., 1990. Morphological differences in neuromasts of the blind cave fish *Astyanax hubbsi* and the sighted river fish *Astyanax mexicanus. Brain Behav. Evol.,* 35: 23–30.

Thinès, G.L. and Legrain, J.M., 1973. Effects of alarm substance on the behavior of the cave fish *Anoptichthys* and *Caecobarbus geertsi. Ann. Speleol.,* 28: 291–297.

Thinès, G.L. and Weyers, M., 1978. Réponses locomotrices du poisson cavernicole *Astyanax jordani* (Pisces, Characidae) a des signaux périodiques et apériodiques de lumière et de température. *Int. J. Speleol.,* 10: 35–55.

Trajano, E., 1991. Agonistic behaviour of *Pimelodella kronei,* a troglobitic catfish from Southeastern Brazil (Siluriformes, Pimelodidae). *Behav. Process.,* 23: 113–124.

Trajano, E., 1994. Comparative study of the brain and olfactory organ of the troglobitic catfish, *Pimelodella kronei* (Ribeiro 1907), and its putative ancestor, *P. transitoria* (Ribeiro 1912) (Siluriformes, Pimelodidae). *Trop. Zool.,* 7: 145–160.

Underwood, H., 1989. The pineal and melatonin: regulators of circadian function in lower vertebrates. *Experientia,* 45: 914–922.

Vandel, A. and Bouillon, M., 1959. Le Protee et son intérêt biologique. *Ann. Speleol.,* 14: 111–127.

von Holst, E., 1935. Über den Lichtrückenreflex bei Fischen. *Pubbl. Stn. Zool. Napoli,* 15: 143–158.

Voneida, T.J. and Fish, S.E., 1984. Central nervous system changes related to the reduction of visual input in a naturally blind fish (*Astyanax hubbsi*). *Am. Zool.,* 24: 775–782.

Weber, A., 1995. The lateral line system of epigean and cave-dwelling catfishes of the genus *Rhamdia* (Pimelodidae, Teleostei) in Mexico. *Mem. Biospeol.,* 12: 215–225.

Welsh, D.K., Logothetis, D.E., Meister, M. and Reppert, S.M., 1995. Individual neurons dissociated from rat suprachiasmatic nucleus express independently phased circadian firing rhythms. *Neuron,* 14: 697–706.

Wilkens, H., 1979. Reduktionsgrad und phylogenetisches Alter: Ein Beitrag zur Besiedlungsgeschichte der Limnofauna Yukatans. *Z. Zool. Syst. Evolutionsforsch.,* 17: 262–272.

Wilkens, H., 1988. Evolution and genetics of epigean and cave *Astyanax fasciatus* (Characidae, Pisces). *Evol. Biol.,* 23: 271–367.

Wilkens, H., 1993a. Neutrale Mutationen und evolutionäre Fortentwicklung. *Z. Zool. Syst. Evolutionsforsch.,* 31: 98–109.

Wilkens, H., 1993b. A new species of *Rhamdia* (Pisces: Pimelodidae) from a cave in the Sierra de Zongolica (Veracruz, Mexico). *Mitt. Hamb. Zool. Mus. Inst.,* 90: 375–378.

Wilkens, H., Langecker, T.G. and Olcese, J., 1993. Circadian rhythms of melatonin synthesis in the pineal organ of cave-dwelling *Astyanax fasciatus* (Teleostei: Characidae). *Mem. Biospeol.,* 20: 279–282.

Wolken, J.J. and Mogus, M.A., 1979. Extra-ocular photosensitivity. *Photochem. Photobiol.,* 29: 189–196.

Yokoyama, S., Meany, A., Wilkens, H. and Yokoyama, R., 1995. Initial mutational steps toward loss of opsin gene function in cavefish. *Mol. Biol. Evol.,* 12(4): 527–532.

Chapter 8

HOW DO CAVE ANIMALS COPE WITH THE FOOD SCARCITY IN CAVES?

Kathrin HÜPPOP

INTRODUCTION

Most obligate cave species share similar morphological, physiological and behavioral features, indicating the presence of similar strong selection forces in their subterranean environments. The comparatively low food supply is one of the main limiting factors in caves (Eigenmann, 1909; Heuts, 1953; Barr, 1968; Poulson, 1963; Poulson and White, 1969; Culver, 1982; Sbordoni, 1982; Howarth, 1983, 1993). Potential cave-dwellers, many of them nocturnal, are said to be preadapted to cave life (Jeannel, 1943; Heuts, 1951; Poulson, 1963; Vandel, 1964; Barr, 1968; Culver, 1982; Sbordoni, 1982; Barr and Holsinger, 1985; Danielopol and Rouch, 1991; Parzefall, 1992), if they have the sensory equipment and the foraging behavior to find food in total darkness, and they are able to reproduce in caves. However, such preadaptations are not sufficient in food-scarce caves.

Four main characters can evolve under the aspect of limited food supply and will be dealt with in following sections:

(1) an increased food-finding ability based on increased length of appendages and an increased number of sensory organs;
(2) an increased starvation resistance;
(3) a reduced energy demand as a result of reduced metabolic rates; and
(4) life-history changes towards more K-selected features (including lower growth rate, increased longevity, delayed reproduction, fewer and bigger eggs).

Further adaptations can be observed: feeding generalism, dietary shift and increased feeding efficiency. Many of these characteristics have evolved together, depending on the kind of food scarcity. As a consequence, most cave animals show no or only minor signs of malnutrition despite the food scarcity in the cave environment (e.g., Heuts, 1951, 1953; Poulson, 1963; Wilkens and Hüppop, 1986; Brancelj, 1992).

The degree of adaptation is dependent on the intensity and quality of food scarcity and on the duration of this selection pressure. The difficulty in quantifying the selection pressure of food scarcity on the development of troglobitic features aggravates the interpretation of these features in general, and the comparison of different cave species with each other and with (often lacking) surface relatives. For understanding the adaptations of cave animals one needs a definition of the different types of food input and food scarcity in caves.

THE NATURE OF FOOD SUPPLY IN CAVES

The basic food resource in most caves is organic matter from external sources: wind, percolating rainwater, floodings and streams provide input of dissolved organic matter, micro-organisms, detritus, feces and accidental or dead animals. Some caves are visited by epigean animals which use the cave for shelter or reproduction, but normally feed outside. Such caves are much richer in food than are isolated ones, because the visitors provide an additional food input in the form of their feces. Bat guano can present an immense source of food for guanobionts (e.g., Breder Jr, 1942). Bacteria and above all microfungi decompose detritus. Numerous ciliates, aquatic isopods and snails, and terrestrial animals seem to feed directly on these fungi and bacteria, as well as on chemoautotrophic forms (Sarbu and Popa, 1992). An example of a food web in a generalized limestone cave in southeastern Brazil has been described in detail by Trajano and Gnaspini-Netto (1991).

An example of food input in caves is the cave cricket *Hadenoecus subterraneus* from the Central Kentucky karst, which leaves the cave only at night to feed. Cricket guano deposited inside the cave is an important food source for many species. Additionally, the female crickets oviposit inside the cave in sandy habitats. Their eggs are the major diet for some cave beetles (Kane et al., 1975; Norton et al., 1975; Kane and Poulson, 1976; Poulson and Kane, 1981). Another source of food input can be observed in certain lava tubes. They can be rich in food as a result of exudates from roots growing through the ceilings into the caves (Howarth, 1981a, 1983).

Caves are furthermore characterized by an almost total absence of primary producers, except for a few species of chemoautotrophic bacteria (Caumartin, 1963) which are quantitatively important in a few exceptional caves (Sarbu and Popa, 1992). However, the long survival of several forms of Collembola on cave clay with no external source of organic matter suggests a major role of chemosynthetic bacteria living on the clay surface as food for Collembola in caves in general (Christiansen, 1970).

The degree of food supply in caves is not only dependent on the nature of the caves but also on their connections to the surface and on their geographic location. Mitchell (1969) specified that food supply in tropical and subtropical caves is higher than in temperate ones because the biomass in the tropical epigeum is greater and its production is uninterrupted. As a consequence, selection pressure can be expected to be weaker, the evolutionary rate slower and the appearance of troglobites not as fast in such caves as in caves with low energy input such as those in the temperate zone (Mitchell, 1969). In fact, troglobites are far more abundant in temperate zones than in the tropics (Barr, 1968), and species diversity in caves is often correlated with the amount of available energy (Howarth, 1993). More information on the nature of energy supply in caves has been provided by Vandel (1964), Barr (1968), Ginet and Decou (1977), Culver (1982, 1985) and Howarth (1983) and in Chapter 12 of this volume.

Generally, compared to epigean environments, less food is available in caves. The food scarcity can be considered under three aspects:

(1) **General food scarcity** occurs in caves with no or low but continuous food input.

(2) **Periodic food supply** characterizes caves which are flooded periodically, normally several times during the rainy season, or caves where food input by animals visiting the caves exhibits periodicity. In fact, caves generally are less variable than surface environments with respect to abiotic factors such as darkness, high moisture, temperature and water chemistry, as well as with regard to biotic factors such as lack of predators (Heuts, 1951; Poulson, 1964; Barr, 1968; Culver and Poulson, 1971; Culver, 1982; Howarth, 1983). As a consequence, many cave animals have lost the ability to compensate for temperature changes: they are stenothermic (e.g., Dérouet, 1953b; Poulson, 1964, 1985; Ginet and Mathieu, 1968). However, in many cases the cave environment is clearly not stable. Seasonally flooded caves are subjected to severe changes regarding food input, water quality, oxygen content, temperature and competitors or predators (Hawes, 1939; Breder Jr, 1942; Heuts, 1951; Poulson, 1964, 1972; Howarth, 1987). During the wet season the food supply can be very high and even abundant for some weeks or months. After exhaustion of the food reserves the animals in these caves suffer food scarcity, like animals in generally food-poor caves. Periodicity of food input can also be called **patchiness of food in time**, or **temporal heterogeneity**. A good example of this is the obligate cave-dwelling carabid beetle *Neaphaenops tellkampfi* from Kentucky. Its life history pattern represents an adaptation to seasonal food supply (Norton et al., 1975). *Neaphaenops* prefers the eggs and nymphs of the common cave cricket (*Hadenoecus subterraneus*), the availability of which is patchy in time. Cave crickets lay their eggs in sandy cave habitats which support few alternate prey. As a result *Neaphaenops* is forced to switch diet, habitat and foraging mode (Kane and Poulson, 1976).

(3) Some cave animals have to cope with **patchiness of food in space** or **spatial heterogeneity**. For cave animals living under such conditions food may not be limited, but difficult to find and to exploit (Howarth, 1993). For this situation the North American carabid beetle *Pseudanophthalmus menetriesii,* inhabiting many caves of the Central Kentucky karst, can be taken as an example. It preys on Collembola which are confined to habitats in which organic matter is patchily distributed but continuously available. By moving between patches, *P. menetriesii* can utilize this even temporal distribution of prey items (Kane and

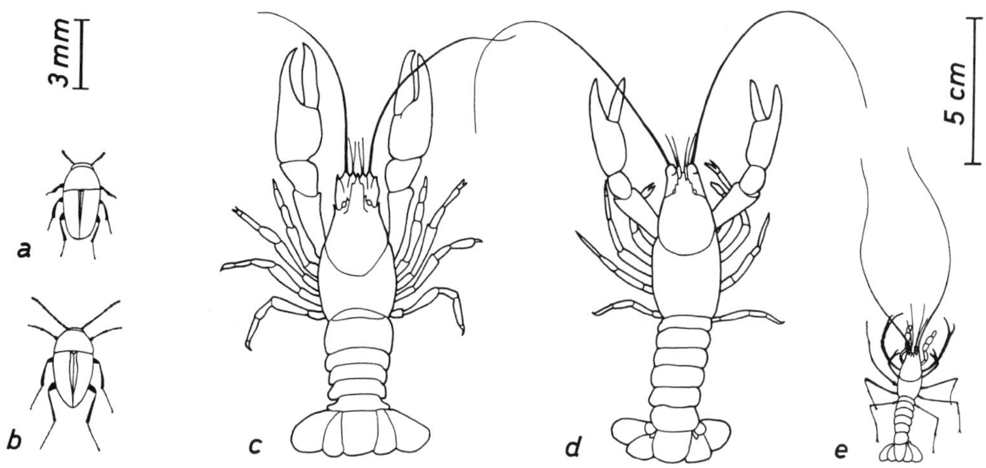

Fig. 8.1. Prolonged antennae and legs in cave beetles and cave crayfish compared to related epigean species: (a) *Ptomaphagus brevior,* an epigean forest species; (b) *Ptomaphagus longicornis,* a highly evolved troglobite; (c) *Cambarus bartoni,* an epigean crayfish; (d) *Orconectes pellucidus,* a hypogean species; and (e) *Troglocambarus maclanei,* a highly cave-adapted crayfish. (a, b) after Peck (1986), (c–e) after Mohr and Poulson (1966).

Poulson, 1976). Not surprisingly, other cave organisms have also been observed to aggregate at patchy food resources (Barr, 1968; Peck, 1976; Howarth, 1983).

FOOD FINDING ABILITY

It is generally held that selection favors increased feeding efficiency. That means increased food-finding ability; this leads to decreased energy requirements for foraging in cave animals living in low continuous or patchy food-scarce cave environments (Poulson, 1963, 1964; Mohr and Poulson, 1966; Poulson and White, 1969; Holsinger and Culver, 1970; Brandon, 1971; Peck, 1973b, 1986; Durand et al., 1981, 1982; Culver, 1982, 1985; Jones and Culver, 1989; Culver et al., 1990, 1995; Parzefall, 1992). For cavernicoles with epigean ancestors that foraged nocturnally, food scarcity rather than darkness is the main selection factor for the development of a superior ability to locate food (Cooper, 1969). For cavernicoles with diurnal ancestors, both factors act together. As a result some morphological, ethological and physiological differences in cavernicoles, compared to surface relatives, can be observed. They may include longer appendages such as antennae, legs or fins (Fig. 8.1), changes in head form, an increased number of sensory buds or organs, enlarged associated brain parts, increased sensitivity to chemical and mechanical stimulants, and changes in food preference and foraging behavior.

A higher efficiency of food utilization in cave animals as adaptation to the food scarcity, as suggested Poulson and White (1969), is very likely but has never been investigated in cave animals up to now. In the trogloxene cave cricket *Hadenoecus subterraneus* Studier et al. (1986) found a calorific assimilation efficiency of 80.4 to 82.5%, obviously higher than that of the unspecified "house cricket" they studied (71 to 73%) or other insects, and they observed exceptionally low calorific contents in mixed waste of the cave cricket, although it forages outside caves. Similarly low calorific contents were also found by Poulson and White in the mixed waste of the cave cricket.

In addition, most cave animals take a wide range of food (Barr, 1968; Peck and Richardson, 1976; Culver, 1985). In the food-poor environment of caves those animals that have a broad diet, or are able to shift their diet if one food source becomes scarce, will be favored by selection (Banta, 1907; Culver, 1973, 1985). For example, Hobbs (1973) observed a dietary shift to detritus by cave crayfish.

Appendages and sensory equipment

Compared to epigean relatives, longer legs, antennae, fins and barbels, or enlarged or flattened heads, have been observed in cave animals. Their advantages for more efficient and energy-saving foraging in the darkness and food scarcity of caves have been discussed in respect of amphipods (Holsinger and Culver, 1970), crayfish (Cooper, 1969), isopods (Banta,

1910), spiders (Poulson, 1981), beetles (Peck, 1973a, 1986), adult and larvae of fish (Miller, 1984; Langecker and Longley, 1993; Wilkens, 1993; Junge, 1993), and salamanders (Brandon, 1971). Often, these body parts bear sensory organs, and their enlarged surface enables the development of an increased number of chemosensitive or mechanosensitive organs. Cave fish have been intensively studied in this respect. In *Caecobarbus geertsi,* a cave barbel from Africa, numerous tactile or taste buds are present on the head, while they are absent in the epigean relative *Barbus holotaenia* (Heuts, 1953). Cave forms of *Astyanax fasciatus,* a Mexican characid fish, have an increased number of taste buds, and an extension of them at the front of the head compared to the surface form, but no difference was found in larvae (Breder Jr and Rasquin, 1943; Schemmel, 1967). In contrast, in the larvae of *Rhamdia guatemalensis,* a Mexican cave catfish, taste buds which cover the front of the mouth and the barbels, are spread over a wider area than in the epigean form. The increase of the average barbel length and the distinctly wider mouth can be interpreted as a trend to enlarge the range covered by taste buds (Junge, 1993).

The North American cave-fish family, the Amblyopsidae, show positive trends in several troglomorphic features. In comparison with nocturnal surface species of *Chologaster,* the least adapted *Typhlichthys subterraneus* shows an enlargement of its head, pectoral and caudal fins, cerebellum and telencephalon. It also has a more developed olfactory rosette and greater body mass per unit length (reflecting larger head and thicker, deeper and shorter body). These features are expressed even more in the highly cave adapted *Amblyopsis rosae, A. spelaea* and *Speoplatyrhinus poulsoni* (Eigenmann, 1909; Poulson, 1963; Cooper and Kuehne, 1974). The increase in relative head size with increased cave adaptation is associated with increased importance of neuromasts and tactile receptors (Poulson, 1963). Broadening and flattening of the head in troglobitic salamanders, which also occurs in epigean bottom-dwelling species, may result in more efficient foraging movements and probably also in a more efficient functioning of the neuromasts. The increased number of premaxillary and prevomerine teeth in the more specialized troglobites is probably correlated with increased efficiency in catching and holding prey (Brandon, 1971).

For cave salamanders the most likely function of the elongated limbs is to raise the body and particularly the head above the substrate (Poulson, 1964; Wake, 1966; Brandon, 1971). *Gyrinophilus porphyriticus,* which otherwise shows little sign of adaptation to cave life, rises up on its legs during prey detection for more efficient operation of the lateral-line system, and thus has a high feeding efficiency (Culver, 1973). Larger limbs also permit the salamander to search a larger area per unit of energy expended (Poulson, 1964). *Typhlomolge rathbuni,* a cave salamander from Texas, not only has a well-developed lateral line but also long legs, which enable it to search wide areas with little effort and help to keep its head high so that it can sense disturbance in the water over a wider area. Its slow stealthy stalking motions ensure better hunting, and disturbance of the water is so slight that there is little interference with the function of its lateral line (Poulson, 1963; Mohr and Poulson, 1966; Brandon, 1971). This advantage of elongated appendages in cave animals probably also holds for spiders and insects, and obviously does for cave fish. For example, the longer fins of the more cave-adapted amblyopsid fish species allow them to move a greater percentage of their body length per pectoral-caudal fin stroke in quiet water: *Amblyopsis rosae* needs 1.8 strokes to move a body length compared to 4.6 strokes in *Chologaster agassizi.* By this a higher swimming efficiency and velocity are achieved, which enable the cave fish to cover more territory in searching for food (Poulson, 1963). *Amblyopsis* covers ten times as much territory and searches 30 times as much water as its troglophilic relative *C. agassizi* (Poulson and White, 1969). The amblyopsid cave fish adopted a different swimming behavior, called glide-and-rest swimming. This behavior, facilitated by the larger fins, not only conserves energy but also results in the reduction of interference noise for neuromast receptors, as mentioned above for salamanders, thus improving prey detection (Poulson, 1963).

Sometimes, improvement of one feature may have more than one positive effect on the cave animals. For example, the elaboration of the antennae of *Gammarus minus,* a highly variable amphipod found in springs and caves in the eastern United States, not only enhances food-finding ability, and thus survivorship, but also improves mate-finding ability, since populations may often have low densities (Culver et al., 1990).

Food-finding success

Many morphological changes of cave animals are the

basis for more efficient foraging and increased food-finding success in the darkness and food scarcity of caves. They enable cave animals to detect the availability of food faster and at a greater distance from their bodies. Because of this they need less energy for food-finding. For example, *Procambarus* cave crayfish from Florida appear to have evolved a more efficient foraging habit than a related surface species. If both are in darkness, the cave crayfish senses its prey faster and locates it more precisely, but it takes much longer to capture it (Mohr and Poulson, 1966). *Eurycea lucifuga* can successfully exclude other salamander species from food-poor terrestrial habitats when competing in caves. These other species have been found to be comparatively undernourished when coexisting with *E. lucifuga* (Peck and Richardson, 1976). The crayfish *Orconectes inermis,* which is restricted to caves, locates injured worms twice as fast as blinded *Orconectes limosus,* a closely related epigean species (Cooper, 1969). The performance of the cave salamander *Proteus anguinus* in detecting prey by substances transported by water – that is, by chemoreception – is superior to that of *Necturus maculosus,* an epigean member of the same family (Proteidae). *Proteus anguinus* probably is better adapted to find rare prey in its habitat than *N. maculosus* (Durand et al., 1981). Other feeding experiments showed that the ability to discriminate different densities of prey items is higher in *P. anguinus* than in the facultative cave-dweller *Euproctus asper* (Engelke, 1994). When compared in darkness, *P. anguinus* approaches living chironomids significantly faster than *E. asper,* and clearly prefers living to dead prey in choice experiments, whereas *E. asper* does not discriminate significantly between living and dead prey (Durand et al., 1982; Uiblein et al., 1992; Uiblein, 1992; Uiblein and Parzefall, 1993). A similar choice behavior has been found in a cave-dwelling form of the Mexican characid fish *A. fasciatus* (Klimpel and Parzefall, 1990).

Prey density can have an important influence on food-finding abilities (Poulson and White, 1969). With high prey density the troglophilic amblyopsid fish *Chologaster agassizi* finds prey sooner than a troglobiotic *Amblyopsis* species, whereas with only one water flea in a 100-liter aquarium *Amblyopsis* finds the prey hours before *C. agassizi.* The more developed lateral-line system of *Amblyopsis* allows it to locate prey farther from its body and to achieve greater success in capturing the prey after locating it (Poulson, 1963). Even the cave form (the "Pachon-fish") of the Mexican surface fish *A. fasciatus,* can find much more food when competing with surface conspecific fish under conditions similar to those in caves. The cave fish not only reacts much faster to food, but also its food-finding ability is four times higher (Hüppop, 1987). The faster reaction to a chemical or a mechanical stimulus does not necessarily imply a lowered threshold for these functions. It may be that the interpretation of non-optic sensory information in the cave fish was changed in comparison with the epigean ones (Hüppop, 1987), which find their food predominantly by optic orientation (Parzefall, 1983). However, in food-scarce caves, where the cave fish may occur together with epigean fish that have been washed in, as in Micos cave (Mitchell et al., 1977), the cave fish have an advantage (Wilkens and Hüppop, 1986). The inferiority of the epigean fish under conditions resembling those of caves explains their undernourished appearance in caves.

One has to differentiate carefully between adaptations of the food-finding ability to food scarcity or to darkness. Gordon and Rosen (1962) observed the cave form of *Poecilia sphenops* from Mexico, living in a cave where food does not seem to be limiting. They mentioned differences in the cave form, compared to the surface form, with respect to mouth position, lip thickness, angle of gape, belly profile and caudal peduncle attitude. Walters and Walters (1965) added, as differences in the cave form, a broader head with a blunter snout, increased numbers of cutaneous sensory structures, enlarged sensory pores, and cephalic lateralis canals less completely roofed over, which should enable the cave form to feed more efficiently on the bottom. These alterations in the cave form of *P. sphenops* can be interpreted to be adaptive for a life in darkness. In caves, a food-searching behavior which is concentrated on the two-dimensional bottom is much more effective in time and cost than food-searching in a three-dimensional space, as the surface animals do in the light, and also try to do in darkness.

Foraging behavior

Behavioral changes in foraging, more or less based on morphological alterations, also can increase food-finding ability. The spidery, small cave crayfish *Troglocambarus maclanei* (Fig. 8.1e) adopted a special feeding behavior. Most of the time these crayfish hang nearly motionless beneath a floating stick or the underside of a rock ledge. Their legs and antennae are covered with tiny hairs which continually accumulate debris carrying numerous bacteria and protozoans, probably the main food objects of *T. maclanei* (Mohr and Poulson, 1966). This feeding habit, in addition to its preferred place of residence and its small and spidery body shape, supports the hypothesis that *T. maclanei* evolved as a result of competition with *Procambarus* species which are much larger and feed on the bottom of the caves (Mohr and Poulson, 1966).

The ventral side of the head of the cave fish *Astyanax fasciatus* is covered with many additional taste buds (Schemmel, 1967) which are used very efficiently, as a result of a smaller angle between body and bottom when searching for food (Schemmel, 1980). As a consequence of darkness in the cave biotope, cave fish gave up the schooling behavior of the surface form in the epigean habitat and adopted a continuous swimming mode. They compensate for loss of the optic-orientated and spatially limited food-finding mode of the epigean fish by covering a greater area using chemosensors and mechanosensors. The same has been shown for hypogean isopods (Danielopol et al., 1994), for the cave salamander *Proteus anguinus,* and even for the facultative cave-dwelling salamander *Euproctus asper* (Uiblein et al., 1992). Hypogean isopods give up grouping and a sit-and-wait behavior, in favor of solitary high exploration activity in habitats patchy in food (Danielopol et al., 1994). At least for adult cave animals for which food supply is restricted, this behavior seems to be advantageous in general.

STARVATION RESISTANCE

Besides a general food scarcity, many cave animals are faced with temporal periodicity of food. Hence, they need an improved ability to survive long periods of starvation (Vandel, 1964; Poulson and White, 1969; Christiansen, 1970; Gagné and Howarth, 1975a,b; Peck, 1975; Poulson, 1981; Culver, 1982; Howarth, 1983). Kammerer (1912) and Gadeau de Kerville (1926) reported that the cave urodele *Proteus anguinus* could survive three to eight years in captivity without food. This is extraordinary even if one assumes some feeding on mud in the aquarium water.

Seasonality in caves, as already mentioned, is based on periodic floodings or on animals visiting the cave periodically. Normally, the results are annual cycles. Many bat caves, especially in temperate zones, show a seasonal cycle and are characterized by a large and fast increase of food input starting again each year when bats return to a maternity cave (Poulson, 1972). Poulson also presented the hypothesis, still to be tested, that the rate of change in food supply is as important as the amount of food, or even more so, in influencing a species' biology through natural selection. Sometimes the life history of a species itself is the reason for a periodicity in feeding, although the cave in which the species lives may be characterized by a generally poor food supply. One famous example is that of the French cave beetle *Speonomus longicornis* (Deleurance-Glaçon, 1963). It lays only one large egg with a huge and concentrated food-rich yolk. The larvae of related surface cave beetles, hatching from small eggs, cannot survive in food scarce caves because they would have to travel long distances searching for the small widely dispersed particles of food. Beetle larvae do not have very good legs for crawling and they would have exhausted themselves and starved before they could locate and eat enough food. The larvae of the cave beetle *S. longicornis* do not feed. As soon as they hatch from the eggs, they encase themselves in clay and soon molt to the pupa stage. Only adults are required to forage in the cave (Deleurance-Glaçon, 1963). The phenomenon that cave species resist starvation periods for their young by producing bigger eggs will be described in more detail later.

In general, three ways to achieve an increased starvation resistance, acting alone or together, can be differentiated:

(1) Many cave animals are able to engorge large amounts of food at a single meal (Trimmel, 1968; Mitchell, 1971; Gagné and Howarth, 1975a; Howarth, 1981b, 1983).
(2) A lower energy consumption can be achieved through a reduction of activity and metabolic rate.
(3) Many cave animals store large amounts of fat.

Fat content and starvation survival

The main way to increase survival capacity in cave environments with intermittent energy input may be an improved ability to accumulate large amounts of adipose tissue. This may be achieved through excessive feeding, increased feeding efficiency, and improved metabolic pathways favoring lipid deposition during the food-rich season. The energy content per gram of lipids is roughly twice that of proteins or carbohydrates, therefore fat accumulation is the best way to store energy. High lipid contents have been observed not only in cave animals but also in many surface species subjected to seasonal changes in food supply. In addition to some endotherms such as hibernating mammals or some birds, e.g. penguins and tubenoses (Procellariiformes), many ectotherms are known to tolerate periodic starvation. For example Junk (1985) showed that various fish from the Amazon river, which are subjected to a strong seasonality of food caused by annual water-level fluctuations, are able to store large amounts of fat. After a starvation period of several months this fat is replenished within a few weeks when food resources are abundant again.

Gibert and Mathieu (1980) and Mathieu and Gibert (1980) compared the lipid content and starvation ability of two hypogean amphipods, the interstitial *Niphargus rhenorhodanensis* and the cavernicolous *Niphargus virei*. When fed *ad libitum,* the cave amphipod *N. virei* is able to accumulate fat up to 6.3% of wet body mass. The interstitial amphipod *N. rhenorhodanensis* builds up only 3% fat in wet body mass, which is comparable to epigean crustaceans. However the decrease in fat, protein and carbohydrate contents is much slower during starvation in *N. rhenorhodanensis* than in other crustaceans, indicating that even interstitial forms are relatively better adapted to food scarcity than epigean forms (Mathieu and Gibert, 1980; Mathieu and Taveau, 1984). Although *N. virei* seems to be able to build up more fat reserves than *N. rhenorhodanensis,* it depletes its lipid reserves during starvation faster than *N. rhenorhodanensis.* The reduction in lipids was 63% in *N. virei* during six months of starvation compared to 42% in *N. rhenorhodanensis.* However, protein and carbohydrates were reduced to a greater extent in interstitial *N. rhenorhodanensis* (53% and 94%, respectively) compared to *N. virei* (35% and 51%, respectively). The decrease of protein and carbohydrates is proportionally much more important in *N. rhenorhodanensis* than in the karstic *N. virei.* The two *Niphargus* species therefore present two different metabolic types, perhaps in adaptation to their environment. However, all comparisons of these species should be taken with caution because of the big difference in mass (they differ by a factor of over ten – *N. virei* about 130 mg, *N. rhenorhodanensis* about 8 mg).

The hepatopancreas of decapod crustaceans is a major storage organ and in all decapod species investigated thus far, reserve lipids are stored intracellularly as oil globules. In the cave-dwelling shrimp *Troglocaris schmidtii* from Slovenia, the intracellular lipid accumulation observed in the hepatopancreas cells already exceeds that of other decapods. Additional lipid reserves are deposited extracellularly in specialized compartments of the hepatopancreas, called oleospheres. The total amount of pure lipid accumulated in the oleospheres in the individual investigated (about 10 mm^3) is enormous for an animal with a maximal length of only 3 cm (Vogt and Strus, 1992). Yager (1991) observed huge lipid-like droplets in the remipede *Godzilliognomus frondosus* living in Bahamian anchialine caves.

The cave-dwelling collembolan *Tomocerus problematicus* stores an exceptional bulk of lipid compared to its surface-living but troglophilic relative *Tomocerus minor,* and the water content shows the inverse relationship (Thibaud, 1977). In the troglobitic catopid beetles of the subfamily Bathysciinae the adipose tissue is well developed and the animals are rich in lipid and protein reserves, but not so with troglophilic catopid beetles of the subfamily Catopinae (Strambi, 1970).

Fat content and/or starvation resistance have been studied in many cave fish. In the African cave barbel *Caecobarbus geertsi* subdermal fat layers and large fat deposits have been observed. The fish continued to live and reproduce without showing any signs of starvation (Heuts, 1951, 1953). The cave fish *Trogloglanis pattersoni* and *Satan eurystomus* (Ictaluridae, Siluriformes) are found only in the Edwards Aquifer, a subterranean water system in southeast Texas. Sensory compensation appears to be a minor subject of the cavernicolous evolution in both these blind catfish (Langecker and Longley, 1993). The fat deposits of some individuals were considerably reduced and the guts were empty. They were apparently starving. Most individuals of both species, however, appeared to be well fed. They had conspicuously large deposits of adipose tissue in the hypodermis and in association with the mesenteries. The swim bladder was totally reduced in all individuals examined, and its place was filled with prominent fat deposits (Langecker and Longley, 1993). This effect may be due to large differences in hydrostatic pressure in the aquifer system. Another explanation may be given by the body size: for animals living in the crevices of an aquifer it may be advantageous not to get thick when accumulating fat reserves. The substitution of the swim bladder by the same volume of fat allows the advantageous small body size to be preserved. Additionally, a high lipid content results in a decrease of specific gravity. Therefore, the findings of Iles and

Wood (1965) and Bone and Marshall (1985), that there is a clear correlation between the lipid content and the volume of the swimbladder in fish, may be taken into account as well.

The catfish genus *Rhamdia* is represented by several species in Mexico; some of them are troglophilic or troglobitic. Wilkens (1993) described a new species, *Rhamdia zongolicensis,* from Cueva del Ostoc. It is characterized by a well-developed adipose fin, located on a ridge consisting of fat tissue. In the holotype and the paratypes of *R. zongolicensis* this fat-containing ridge was almost completely reduced. The condition factor (CF),

$$CF = 100wl^{-3},$$

where w is the wet body mass in grams, and l the body length in centimeters, is a good measure of the nutritional state of animals within one species or within closely related species. This value was very low in the holotype (0.87), the paratypes (0.92, $n=3$) and 13 other specimens (0.93) caught in the cave in 1988 (Wilkens, 1993), when compared to the mean CF of laboratory-fed individuals (1.74, $n=19$) (Wilkens, pers. comm.). Recent starvation experiments showed that the specimens of *R. zongolicensis* caught in the cave must have been starving or living in permanently severe food scarcity for at least one year. After an artificial starvation period in the laboratory of 320 days *R. zongolicensis* had a mean CF of 1.23, corresponding to a reduction by about 30%, whereas epigean *Rhamdia laticauda,* closely related, had a mean CF of 0.99, corresponding to a reduction by about 47% (Wilkens, pers. comm.). These data not only demonstrate the reduced fat depletion in hypogean *R. zongolicensis* during starvation compared to epigean *R. laticauda,* but also indicate their capability to survive periods in apparently bad body condition.

Perhaps the genus *Rhamdia* is generally preadapted to resist long starvation periods which result in very bad body condition. A few epigean *R. laticauda* when re-fed after the starvation experiment survived CFs of 0.8 to 0.9, corresponding to a reduction of 55 to 48% in body weight (Wilkens, pers. comm.).

Poulson's (1961, 1963) comparison of the starvation resistance of the North American amblyopsid fish revealed great differences in loss of body mass between the nocturnal but epigean swamp fish *Chologaster cornutus,* the troglophilic spring cave fish *Chologaster agassizi* and the troglobitic fishes *Amblyopsis rosae, A. spelaea* and *Typhlichthys subterraneus,* (Table 8.1).

Table 8.1
Effects of starvation on amblyopsid fish acclimated in the laboratory to 13 3.5°C, under aphotic conditions [1]

Species	Body mass [2]	Activity [3]	Fat [4]
Chologaster cornutus	34	46	1
Chologaster agassizi	24	116	1–2
Typhlichthys subterraneus	15	230	2–4
Amblyopsis spelaea	13	280	3
Amblyopsis rosae	9	400	1–4

[1] From Poulson (1961).
[2] Loss of body mass (%) after 20 days.
[3] End of activity after the number of days specified in the column.
[4] Amount of fat (subjective degrees).

Although the amount of fat was not measured, it was obvious that the troglobitic species were fatter than the *Chologaster* species and that they were able to survive starvation much better, as is shown in the time before spontaneous activity ceased in an open-ended starvation experiment (Table 8.1). Poulson (1963, 1964) ranked the five amblyopsid species according to their characters increasingly adapted to cave environments, and correlated this order with the time the species had been isolated in caves. The increasing ability to cope with food scarcity in the amblyopsids follows the same order as the length of isolation (Table 8.1) and by this supports the high adaptive value of starvation resistance in cave animals.

A case study: Mexican cave fish

Another cave fish intensively studied for decades is the Mexican characid fish *Astyanax fasciatus.* One cave form, the hybrid Chica-fish from La Cueva Chica, had early been observed to have large fat reserves and high resistance to starvation (Breder Jr, 1944; Rasquin, 1947, 1949). More recent investigations were made by Rose and Mitchell (1982) on a phylogenetically old cave form of *A. fasciatus,* the Yerbaniz-fish from El Sotano de Yerbaniz. They found a lipid content of 55.4% fat in dry body mass in the Yerbaniz-fish compared to 22.1% fat in dry body mass in the epigean conspecific form of *A. fasciatus.* The fat stores were visible in the abdominal cavity, between the integument and the muscle mass, intramuscularly and within the orbital sockets of the reduced eye. Little fat was observable in the epigean fish as discrete deposits. Lipid mass linearly increased with lean dry body mass in both cave and surface forms. Additionally, the

Table 8.2
Condition of epigean and hypogean fish of *Astyanax fasciatus* before and after starvation [1]

	Days of starvation				
	0		109		174
	surface	cave	surface	cave	cave
n	54	56	4	7	4
body mass (%)	100	100	72.2	84.6	68.4
body mass (g)	3.5–17.5	3.6–9.5			
standard length (%)	100	100	99.1	99.5	96.3
standard length (cm)	5.5–9.1	5.0–6.6			
CF ($100\,g\,\ell^{-3}$)	2.03 (1.68–2.54)	2.95 (2.32–4.04)	1.55	2.45	1.89
n	7	7	4	7	4
length of intestine (% standard length)	74.4	72.9	50.5	69.2	70.9
dry body mass (% wet body mass)	30.0	51.3	22.7	44.8	44.0
fat content (% wet body mass)	8.6	36.6	2.2	28.5	27.1
fat content (% dry body mass)	27.1	71.0	8.1	63.3	61.8

[1] From Hüppop (1988).

slope of the relationship in the cave fish was steeper, indicating that with increasing body mass the difference in fat content between cave and surface fish is even greater.

My own investigations regarding fat content and starvation resistance in the cave fish *A. fasciatus* were restricted to the phylogenetically oldest known cave form, the Pachon-fish from La Cueva de El Pachon (Hüppop, 1988, 1989). Surface fish from the Rio Teapao (Tabasco, southern Mexico) and Pachon-fish of the same age, both forms reared in the laboratory and fed *ad libitum,* were compared regarding body mass, body length, CF (condition factor), proximate body composition, length of intestine, muscle fiber histology, swim bladder and general activity before and during starvation. The animals were held in groups of seven in 25°C water in a dark and silent room, thus simulating cave conditions. Observations were made using an infrared night-sight apparatus. Measurements of body mass and body length and samplings for analyses during a starvation period of 109 days were done in white light at intervals increasing from three to about 13 days.

At the egg stage, there was no difference in proximate composition between the cave and surface populations, but the cave fish began to accumulate visible lipid reserves at an age of about 19 days; none was observable in the surface fish. At the same age of one year, cavernicolous Pachon-fish were significantly smaller than the epigean Teapao-fish, but were heavier and had a higher condition factor at comparable sizes (Table 8.2). The high condition factor in the cave fish reflects the high fat content accumulated under the *ad libitum* feeding conditions. Also the slope of the relation of fat content to wet body mass was significantly steeper in the cave fish, indicating that if the cave fish had grown as big as surface fish the difference in fat content and condition factor would be even higher. Most of the lipid reserves were accumulated in the abdominal cavity. In the cave fish three-quarters of the cavity was filled with fat. Subcutaneous and inter- and intramuscular fat reserves (Hüppop, 1991), and accumulation in the orbit of the reduced eyeball could be observed. In support of observations by Hubbs and Bailey (1947) and Langecker and Longley (1993) on the Texas cave catfish, the Pachon Cave fish *A. fasciatus* had obviously smaller swim bladders than the surface fish. A correlation with the high fat content seems plausible for this species as well.

During starvation the decrease of body mass, the standard length (body length from the tip of the mouth to the base of the tail fin), the condition factor and the fat content were significantly lower in the cave fish than in the epigean ones. The water content was lower in unstarved cave fish, and increased in both forms during starvation – even more so in the surface fish (Table 8.2). This corresponds with earlier observations by Gibert and Mathieu (1980) on cave amphipods, and by Thibaud (1977) on cave Collembola. Due to the increase in water content the real loss of fat reserves is higher than the data show and even higher in the surface fish than in the cave fish. Even after 174 days of starvation the condition factor of cave fish was significantly higher than that of surface fish starved for 109 days, and only a little lower than that of unstarved surface fish (Fig. 8.2). The length of intestine was not

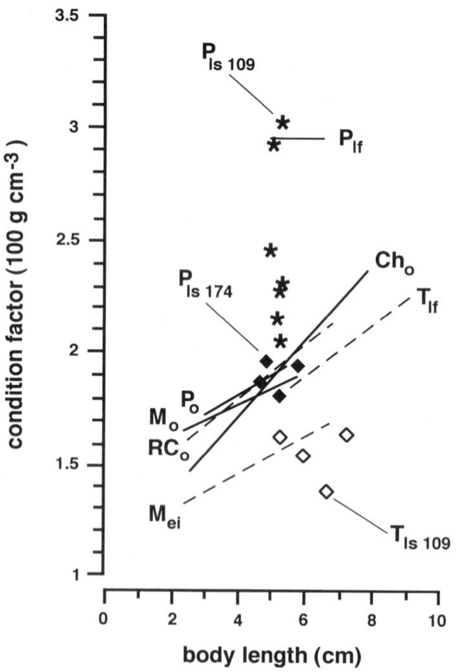

Fig. 8.2. Relation between condition factor and body length in different cave and surface forms of the characid fish *Astyanax fasciatus*: RC_o, epigean Rio Coy fish, original ($n = 46$, $y = 1.31 + 0.12x$, $r = 0.69$, $p < 0.01$). T_{lf}, epigean Teapao fish, laboratory, fed ($n = 54$, $y = 1.22 + 0.114x$, $r = 0.501$, $p < 0.001$). $T_{ls\,109}$, epigean Teapao fish, laboratory, starved for 109 days ($n = 4$). P_o, hypogean Pachon fish, original ($n = 10$, $y = 1.44 + 0.09x$, $r = 0.30$, $p > 0.1$). P_{lf}, hypogean Pachon fish, laboratory, fed ($n = 56$, $y = 2.95 + 0.0001x$, $r = 0.0001$). $P_{ls\,109}$, hypogean Pachon fish, laboratory, starved for 109 days ($n = 7$). $P_{ls\,174}$ = hypogean Pachon fish, laboratory, starved for 174 days ($n = 4$). Ch_o, hypogean Chica fish, original ($n = 24$, $y = 1.02 + 0.17x$, $r = 0.86$, $p < 0.01$). M_o, hypogean Micos fish, original ($n = 46$, $y = 1.48 + 0.07x$, $r = 0.27$, $p < 0.1$). M_{ei}, epigean inwashed Micos fish, original ($n = 72$, $y = 1.15 + 0.08x$, $r = 0.35$, $p < 0.01$).

different in unstarved fish; after starvation, it showed a significant atrophy in the surface fish, but not in the cave fish even after 174 days. Therefore the cave fish can be assumed to be able to feed again immediately floodings bring new food into the cave at the beginning of the rainy season. Regarding protein content no differences between the two forms of *A. fasciatus* could be observed, and no changes during starvation, when the influence of the fat and the water content on body mass were considered. During starvation some surface fish died. They were very lean with no visible fat reserves, with degenerated and autolytic entrails and with very large swollen gall bladders sometimes coloring the liver green.

Thus, the epigean form of *A. fasciatus* seems to be only slightly adapted to resist long starvation periods. The freeliving surface fish from a river near the cave as investigated by Rose and Mitchell (1982) had 22% fat in dry body mass, which was even less fat than the laboratory-bred fish from the Rio Teapao (27%). However, compared to other similar-sized freshwater fish the loss of body mass during starvation, even at the high water temperatures of 25°C in the tropics, are relatively low in epigean *A. fasciatus* (see Love, 1970; Pandey and Singh, 1984). Even the availability of surface food in the tropics, in the area of Mexico this investigation relies on, shows periodicity (Mitchell et al., 1977). Hence, the epigean *A. fasciatus* can be characterized as preadapted for the cave environment, with regard to starvation resistance, in addition to other preadaptations like the lateral line system, auditory and olfactory capacity, and sexual behavior (Wilkens, 1988).

The Pachon Cave form of *A. fasciatus* is obviously adapted to resist long starvation periods. It is able to accumulate large fat reserves during food-rich times and to survive absolute starvation over at least six months without any effects on the body. Pachon Cave fish caught in the cave at the end of the dry season – that is, at the end of a starvation period of several months – had a condition factor of about 2 (Wilkens and Hüppop, 1986), like the surface fish in an unstarved condition (Table 8.2, Fig. 8.2). The slow decrease of the fat content, in the laboratory as well as in the field, suggests an even higher resistance to starvation in this cave fish, perhaps for one year or more. Assuming that animals die when the total loss of body mass approaches about half the initial body mass, survival times can be calculated from metabolic rate (MR) and body size, or from extrapolating the body-mass curve from starvation experiments (Peters, 1983). From the latter the ability of Pachon Cave fish to survive starvation can be calculated to be about one year. The fact that Pachon Cave fish were even fatter than those from Yerbaniz Cave (see Rose and Mitchell, 1982) may be the result of laboratory *ad libitum* feeding conditions for the former. Or it may indicate that the Pachon Cave fish, which are relatively old phylogenetically, have a higher degree of adaptation to periodical food scarcity.

Micos Cave, located in the same area as Pachon Cave in northern Mexico, is populated by a phylogenetically young form of *A. fasciatus* (Wilkens and Burns, 1972; Peters et al., 1975; Wilkens, 1976), and is regularly invaded by epigean conspecifics during the rainy season (Wilkens and Burns, 1972; Mitchell et al., 1977). Nevertheless, hybridization does not occur very often. It happens to any extent only in Chica Cave, which is characterized by food abundance from a bat roost (Mitchell et al., 1977). The comparison of the condition-factor values shows that the epigean fish caught in Micos Cave at the end of the dry season had a lower body mass in relation to standard length than fish from Micos Cave, from other phylogenetically older cave populations, hybrid Chica Cave fish, or surface fish caught in the epigean environment (Fig. 8.2; see also Wilkens and Hüppop, 1986). The surface fish obviously had been starving for a long time, most likely during the whole dry season. Only a few small epigean individuals had been caught in Micos Cave, indicating that most of the smaller fish had already perished by starvation or cannibalism. The occasional event of cannibalism in the cave may be the reason why fish survive in the cave with a very low condition factor compared to the laboratory, where no fish with a condition factor less than about 1.4 survived. The relationship between preadaptations, and adaptation and competition in conditions of food scarcity, are discussed in more detail in the context of sympatric speciation by Wilkens and Hüppop (1986).

A comparison of the starvation resistance and the point of no return (days from hatching, after which 50% or more larvae die even when re-fed; Blaxter and Hempel, 1963) for larvae of cave and surface populations of *A. fasciatus* shows no superiority for Pachon Cave fish larvae. They seem not to be able to survive (short) starvation periods. This can be clearly attributed to conditions in the cave. The seasonal input of food, and changes of abiotic factors often initiate the reproduction of many cave animals (Hawes, 1939; Poulson, 1963, 1964, 1971, 1975; Jegla and Poulson, 1970; Poulson and Smith, 1969; Poulson and White, 1969; Kane et al., 1975; Kane, 1981; Poulson and Kane, 1981; Howarth, 1983). This holds for *A. fasciatus* as well (Breder Jr, 1942; Mitchell et al., 1977). The larvae start feeding only a few days after spawning. At this time, it can be assumed that there is still enough food for them, so that they are not forced to be adapted to food scarcity. The adaptation of the larvae is on another level, regarding food particle size. This will be discussed below.

ENERGY DEMAND

The term "energy economy" was coined by Poulson (1963), and describes the reduction in energy consumption observed in many cave animals (summaries in Poulson, 1964; Vandel, 1964; Barr, 1968; Culver, 1982; Hüppop, 1985). A generally low food supply is often one main limiting factor in caves. Where this is so, reduced metabolic rates in cave animals have a high selective value. These reduced metabolic rates may reflect resistance not only to starvation during periodic food limitation or to general food scarcity, but also to food patchiness, darkness, low oxygen supply or other abiotic factors in the cave environment (Heuts, 1951, 1953; Caine, 1978; Schmidt-Nielsen, 1984; Withers, 1992; Howarth, 1993).

Parallels between the deep sea and the cave environment with respect to darkness, environmental stability and food scarcity, have been the object of several studies (e.g., Poulson, 1971; Thinès, 1979; Culver, 1982). Generally, it was accepted that the reduced metabolic rates observed in some deep-sea animals are an adaptation to food scarcity, as in many cave animals. In a recent synopsis Childress (1995), who earlier agreed, gave convincing arguments **against** this hypothesis and argued instead that low metabolic rates in some deep sea animals were correlated with darkness.

The energy demand of troglobites was first studied by Gal (1903) and Jeannel (1929) in amphipods and by Eigenmann (1909) in amblyopsid fish. Their suggestion that cave animals have a low metabolic rate in comparison to their epigean relatives was taken up by many biologists working with a variety of phyla (Fage, 1931; Jeannel, 1943; Heuts, 1951, 1953; Vandel, 1950, 1964; Poulson, 1964; Barr, 1968; Ginet and Decou, 1977; Culver, 1982). Experimental papers on metabolic rate are numerous. Those up to 1985 have been summarized by Hüppop (1985) and will be reconsidered in this chapter, in addition to more recent investigations. The most meaningful information on metabolic rates is given by the measurement of oxygen consumption of the entire organism or parts of it. In addition, parameters such as respiratory frequency, resistance to anoxia, activity, ability to survive starvation periods, body composition, growth rate, gill area, or rate of turnover of adenosine triphosphate (ATP) give indirect information on the metabolic rate. In most comparisons, the metabolic rate of the hypogean species was found to be lower than that of their epigean relatives. Only in a few cases were similar or even higher metabolic rates found in the cave animals, sometimes interpreted as related to food abundance in the specific caves (Schlagel and Breder Jr, 1947; Korzhuev, 1950; Vandel and Bouillon, 1959; Dresco-Dérouet, 1967; Culver and Poulson, 1971; Mathieu and Gibert, 1980; Ahearn and Howarth, 1982; Mathieu and Taveau, 1984; Hüppop, 1986, 1989; Humphreys and Collis, 1990). That activity may have an important influence on energy demand is demonstrated by hypogean *Asellus aquaticus,* which have a higher routine metabolic rate (which includes spontaneous activity) than surface *A. aquaticus,* as a result of high movement activity in searching for food (Danielopol et al., 1994; Mösslacher and Creuzé des Châtelliers, 1996). Sometimes the very different body sizes of the animals compared were not taken into account. This can be important, since maintenance energy expenditure per unit mass decreases with increasing body size (Hemmingsen, 1950; Zeuthen, 1953; Phillipson, 1981). Finally, some comparisons of metabolic rates were based on species of different genera or even families.

A brief overview: aquatic cave animals

Not only caves but also interstitial habitats play an important role for subterranean aquatic animals. Gal (1903) was the first to measure oxygen consumption of a cave animal, an amphipod (*Niphargus* sp.). Jeannel (1929) compared the survival times of the hypogean *Niphargus longicaudatus* and an epigean species of *Gammarus* in a closed bottle. *Niphargus longicaudatus* survived 48 hours whereas *Gammarus* died after a few hours. Dérouet (1949, 1952, 1953a,b), Dresco-Dérouet (1959), Troiani (1954) and Wautier and Troiani (1960) found a much lower respiration rate in hypogean *Niphargus orcinus* and *Niphargus virei* than in epigean *Gammarus pulex.* Their findings of slightly lower metabolic rate in *Niphargus rhenorhodanensis* compared to *Gammarus pulex* or other epigean crustaceans was confirmed by Mathieu and Gibert (1980) and Mathieu and Taveau (1984) on the basis of body composition and starvation experiments. However, this method revealed no differences in energetics between a cavernicolous form of *N. virei* and an interstitial form of *N. rhenorhodanensis* in contrast to the studies mentioned above, which is probably due to differences in measuring techniques; only Mathieu and his colleagues measured activity as well. The comprehensive work of Mathieu and his team [summarized by Mathieu and Turquin (1992)] on *N. rhenorhodanensis* unfortunately does not permit comparisons between its metabolic rate and those of surface animals.

Camiciottoli and Messana (1994) observed very low mean hourly rates of oxygen consumption in the cave isopod *Stenasellus racovitzai*, compared to data from the literature for hypogean and epigean isopods. Oxygen consumption rates of *S. racovitzai* were constant through time, and were not influenced by level of starvation.

Several decapods have been shown to have a lower metabolic rate than surface forms (Jeannel, 1929; Burbanck et al., 1948; Eberly, 1960; Jegla, 1964; Weingartner, 1977; Caine, 1978; Franz, 1978). An important investigation on metabolic rates of cave animals was that of Dickson and Franz (1980) on *Procambarus* species. Using excised gills to exclude the influence of animal movement [like Biswas (1991), see below] they determined a lower rate of oxygen consumption by gill tissue in the cave than in the surface species. The difference between the two cave species was remarkable: the rate of oxygen consumption in *Procambarus pallidus* from a low-energy cave was only one-sixth of that observed in the epigean *Procambarus clarkii,* and one-seventh to one-eighth of the value for the epigean *Procambarus pictus,* whereas the hypogean *Procambarus franzi* from a comparatively high-energy cave had a metabolic rate one-half as great as the two epigean species. Turnover times for ATP in the gill tissue showed the opposite relationship, thus indicating a decreasing rate of energy turnover with increasing food scarcity. Comparison of these *Procambarus* species shows very clearly the effect of the intensity of selection pressure on the degree of metabolic adaptation (Fig. 8.3).

Other indirect methods to determine metabolic rates are starvation experiments and growth-rate determinations. Biswas (1991), in the Indian cave fish *Nemacheilus evezardi,* observed a reduced loss of body mass under conditions of starvation or restricted

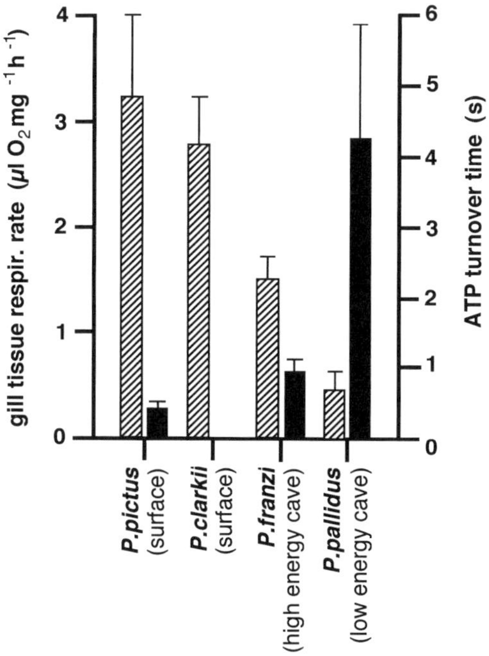

Fig. 8.3. Mean gill oxygen consumption rate (striped columns) and ATP turnover times (black columns) (+1 s.d.) of four *Procambarus* species from different habitats (after Dickson and Franz, 1980).

feeding, in comparison with the epigean form. Heuts (1951, 1953) provided evidence that the metabolic rate of the African cave fish *Caecobarbus geertsi* was only one-third that of its epigean relative *Barbus conchonius,* as was also found by Koch (unpublished observations, cited in Heuts, 1951, 1953 and in Olivereau and Francotte-Henry, 1956), on the basis of observations of resistance to starvation.

The most detailed analysis of cave adaptation in fish was by Poulson (1963, 1964). Following observations of Eigenmann (1909), who reported a respiration frequency of only 19 min^{-1} in *Amblyopsis spelaea* compared to 80 min^{-1} in the epigean *Chologaster papilliferus,* Poulson (1963) demonstrated a decreasing trend in the metabolic rate from the epigean *Chologaster cornutus* (Cc) through the troglophilic *Chologaster agassizi* (Ca), and the troglobitic *Typhlichthys subterraneus* (Ts) to the most cave-adapted *Amblyopsis spelaea* (As) and *Amblyopsis rosae* (Ar) (Table 8.3). The amblyopsids seem to be extraordinarily preadapted to cave life, since even the epigean *Chologaster cornutus* has a standard metabolic rate only 0.9 times that of the goldfish *Carassius auratus*, which has the lowest known metabolic rate of non-amblyopsid epigean teleosts (Poulson, 1963). The trend in the metabolic rate from Cc to Ar is negatively correlated with the ability

Table 8.3
Metabolic rates (MR) and foraging activity in five fish species[1,2]

	Cc	Ca	Ts	As	Ar
Standard MR ($ml\ O_2\ g^{-1}\ h^{-1}$)	0.060	0.028	0.016	0.018	0.011
Routine MR ($ml\ O_2\ g^{-1}\ h^{-1}$)	0.116	0.087	0.061	0.022	0.023
Routine MR per fish ($ml\ O_2\ h^{-1}$)	0.140	0.121	0.092	0.036	0.058
Foraging activity (body length d^{-1})	1000	1854	5454	10980	8082

[1] The epigean *Chologaster cornutus* (Cc), the troglophile *Chologaster agassizi* (Ca) and the troglobites *Typhlichthys subterraneus* (Ts), *Amblyopsis spelaea* (As) and *Amblyopsis rosae* (Ar).
[2] From Poulson (1985).

to survive starvation (see above). High fat reserves, together with low standard and routine metabolic rates explain the long survival time of the most troglobitic amblyopsid species under starvation.

Terrestrial cave animals

The metabolic rates of terrestrial cave animals have been little studied. Although food scarcity generally is even greater in the terrestrial than in the aquatic cave environment, only a few studies of differences between cave and surface forms have been reported. The North American cave cricket *Haedenoecus subterraneus*, also feeding outside the cave, was estimated to have a metabolic rate (4.52 $J\,h^{-1}$, Studier et al., 1986) only half of that expected for insects of similar mass (8.08 $J\,h^{-1}$, Kayser and Heusner, 1964). Metabolic rate was calculated from loss of body mass during "starvation", in this study meaning the mass loss in the crickets while the crop was being emptied. However, the research of Ahearn and Howarth (1982) on cave crickets from Hawaii, and that of Humphreys and Collis (1990) on some cave arthropods from Western Australia, revealed no reduction in respiratory rates compared to surface species. The oxygen consumption rates of the cave species are at the lower end of values from the literature for the respective taxa, but are not exceptionally low. From differences in loss of body mass during food scarcity, Poulson (1981) found that the metabolic rate in the troglobitic spider *Anthrobia monmouthia* was lower than in the troglophilic *Phanetta subterranea*, and attributed this to isolation time in the cave, life-history traits and their different habitats. The metabolic rate of the cave spider

Lycosa howarthi is about 70% of that of spiders of the same genus in lava flows, but both values fall within the range of metabolic rates reported for arachnids (Hadley et al., 1981). Since these are typically lower than those of other ectotherms of comparable size, arachnids seem to be well preadapted for energy economy, developed as a result of the potential problem faced even in the epigean environment of an inconstant food supply as a consequence of an usually predatory life (Hadley et al., 1981). Dérouet (1953a) and Dresco-Dérouet (1960, 1967) found that rates of oxygen consumption in European cave spiders were only one-fifth to one-seventh of those in surface species, but in opilionids no difference in metabolic rate was observed between cave and surface species.

Hypoxic conditions

Many cave or interstitial animals have to cope with temporary or permanent conditions of hypoxia in their environments. A prominent example is provided by the animals living in the recently discovered Movile Cave in Romania. Sarbu and Popa (1992) reported a high number of species living in severely hypoxic gas chambers. Those cave animals are able to survive in low partial pressures of oxygen (P_{O_2}), as a result of generally or temporarily reduced metabolic rates, by the limitation of energy expenditures linked to locomotion and ventilation, and/or through anaerobic metabolism (Hochachka and Somero, 1980; Hervant et al., 1995).

In the hypogean Indian fish *Nemacheilus evezardi* the survival rate in an oxygen-limited environment was three times as long as that of its epigean conspecific counterpart, and the rate of oxygen consumption of isolated muscle tissue was three-quarters of the corresponding figure (Biswas, 1991).

Ercolini et al. (1987) reported that the rate of oxygen consumption of hypogean cyprinids from Somalia, *Barbopsis devecchii* and *Phreatichthys andruzzii,* was only about half that of the related epigean species *Puntius lateristriga,* the lowest rate being shown by the species more adapted to the cave environment, *P. andruzzii.*

Caine (1978) not only observed an oxygen consumption rate in several hypogean species of *Procambarus* only one-half to one-third of that in surface species, but also interpreted the low metabolic rate of *Procambarus horsti* as an adaptation not only to low food availablity but also to the low oxygen tension sometimes found in caves. Caine's attempt to determine the critical oxygen level for the seven crayfish species compared was only successful in the epigean forms, which had a discontinuity in the curve relating oxygen uptake to partial oxygen pressure. This implied that the surface crayfish could withstand depletion of oxygen tension only down to a specific (critical) oxygen concentration in the water, after which they showed asphyxia, whereas hypogean species had no sharp break in the oxygen-uptake curve.

This phenomenon of the absence of a discontinuity in the oxygen-uptake curve, as in the cave animals mentioned above, is called oxyregulation and is considered to be adaptive in environments characterized by variable oxygen conditions. Also some European groundwater (interstitial) isopods are defined as oxyregulators. The hypogean species *Proasellus slavus* and *Proasellus strouhali* and a hypogean form of *Asellus aquaticus* are able to cope with decreasing oxygen concentrations. The critical P_{O_2} increases in these three isopod species according to the sequence in the previous sentence. In contrast, the epigean *A. aquaticus* and another epigean isopod species *Proasellus meridianus,* which have a critical P_{O_2}, are classified as oxyconformer (Danielopol et al., 1994). *Proasellus slavus* even adopted a special dorsally arched position of the body and an increased operculum opening under low oxygen conditions. Energy may also be saved in all hypogean isopods by a decrease of mobility during low oxygen conditions (Danielopol et al., 1994).

Two hypogean species of *Niphargus,* the interstitial *N. rhenorhodanensis* and the karstic *N. virei,* had survival times during anoxia longer than those of epigean *Asellus aquaticus, Gammarus fossarum,* and most other epigean crustaceans (Hervant and Mathieu, 1995). *Niphargus virei* had the greatest resistance to anoxia, surviving 53 hours compared to about 6 hours in *G. fossarum.* Hervant et al. (1995) compared the hypogean *N. rhenorhodanensis,* which has to cope with severe hypoxia conditions during about 6 months a year, and the epigean *G. fossarum,* which lives continuously in well oxygenated water. The lethal times for 50% of the animals were 46.7 and 6.3 hours, respectively, in very severe hypoxia. The main reasons for their different resistance are probably increased stores of phosphagen and glycogen as anaerobic fuels, and reduced behavioral and ventilatory activities in the hypogean species. During recovery, *G. fossarum* excreted a great part of the lactate accumulated during severe hypoxia, whereas *N. rhenorhodanensis* remetabolized glycogen from lactate, as a result of a higher capacity for gluconeogenesis (Hervant et al., 1995). In the sense of energy economy again this can also be interpreted as an adaptation to food scarcity in subterranean environments.

Influence of activity

Every activity increases an animal's energy consumption. Unfortunately, in some studies on cave animals the influence of the animal's activity on the metabolic rates during the experiments was not considered. The standard metabolic rate (that is, the lowest oxygen consumption rate recorded during a test) excludes motion activity, and is one measure for the physiological adaptation of cave animals to food scarcity. Although the often very different motion patterns of cave and surface animals (Schlagel and Breder Jr, 1947; Poulson, 1963) make comparisons difficult, the routine metabolic rate (that is, the mean metabolic rate over 24 h, including spontaneous activity) is a more appropriate index of the actual energy expenditure in nature. It is sometimes reduced in cave animals, allowing the animals to save energy without a reduction in the standard metabolic rate. In *Anthrobia monmouthia,* a troglobitic spider, the energy economy from a routine metabolic rate has the highest rank, among other adaptations, reducing energy demand (Poulson, 1985); in most other cave animals, too, acitivity is reduced (Verrier, 1929; Burbanck et al., 1948; Dresco-Dérouet, 1959; Eberly, 1960; Ginet, 1960; Thinès, 1969; Thinès and Piquemal, 1978; Erckens and Martin, 1982; Mathieu, 1983).

An increase in food-finding ability in cave animals often seems to go along with an increase in food-finding activity. Poulson (1963) stated that the most highly specialized amblyopsid troglobites were most active. Nevertheless they had the lowest routine metabolic rate, quite apart from very low standard metabolic rates (Table 8.3). Fewer strokes to move a body length are required in the most adapted amblyopsid species, as a result of morphological adaptations. Although their activity is higher, the savings from reduction of routine metabolic rate make up over 90% of the total energy savings by adaptations in the most cave-adapted amblyopsids (Poulson, 1985).

However, in some subterranean animals increased motion activity for food searching clearly increases the routine metabolic rate when compared to surface forms, as has been observed in the hypogean isopod *Proasellus slavus* (Danielopol et al., 1994; Mösslacher and Creuzé des Châtelliers, 1996). In this case, the interstitial *P. slavus* and an interstitial *Asellus aquaticus* form are both much smaller than the surface *A. aquaticus.* In consequence, their higher mass-specific routine metabolic rate corresponds to a smaller routine metabolic rate per individual. In addition to their adaptive oxyregulatory character, the interstitial forms are adapted to their habitat in two ways: firstly, they fit better into small crevices; and secondly, from the energetic point of view they are able to cope better with food scarcity (Danielopol et al., 1994; Mösslacher and Creuzé des Châtelliers, 1996).

Energy economy is sometimes realized by reduction in other activities. Energy may be saved in the darkness of caves where aggressive and territorial behavior are no longer displayed (e.g., Poulson, 1963, 1964; Parzefall, 1982, 1983, 1992; Burchards et al., 1985) or are restricted (Hausberg, 1995), and by changed efficiency of motion, as has been mentioned.

Excitement definitely elevates the metabolic rate. Interior activity, meaning excitement without expression in motion, elevates the standard metabolic rate, whereas external restlessness with high activity metabolic rate increases the routine metabolic rate. This underlines the importance of the great conservation of energy which results from the increase in resistance to disturbance with increase in adaptation to the cave environment, as in the amblyopsids (Poulson, 1963). The generally low standard and routine metabolic rates of the cave amblyopsids and their resistance to disturbance are interpreted not only as adaptations to the poor food supply, which is reduced by a factor of about 100 compared to the surface, but also as a result of relatively stable cave conditions and the general lack of predators in the amblyopsid cave environment.

Especially in the interstitial-like environment a reduction in activity in cave animals, compared to animals in more open water, may sometimes not have been for energy economy but rather was a consequence of their cramped habitat (Culver and Poulson, 1971; Mathieu, 1983).

Influence of fat content

The Mexican cave fish *Astyanax fasciatus* is a very interesting object of research because the existing surface ancestor lives side by side, and sometimes even syntopically (Wilkens and Hüppop, 1986), with several distinguishable cave populations. The interfertility of the animals of all localities not only allows one to study many regressive and constructive features in cave populations of different phylogenetical ages in comparison to the epigean ancestor, but also their genetics (summary in Wilkens, 1988). Only a few years after the first description of a cave form, formerly named *Anoptichthys jordani* (Hubbs and Innes, 1936) in Chica Cave in Mexico, Schlagel and Breder Jr (1947) investigated oxygen consumption and found even higher rates in Chica Cave fish than in the epigean ones. However, their closed-bottle experiments gave doubtful results and, furthermore, Chica Cave is food-rich, suggesting no selection pressure for energy economy on the fish there.

A repetition of the experiments with modern techniques (flow-through respirometer) and with animals from three different caves showed a reduction of the metabolic rate in the cave animals. But only in the phylogenetically oldest form of *A. fasciatus* from Pachon Cave, where food restrictions were most severe, was the reduction significant – to about three-quarters that of the surface form (standard metabolic rate = $0.23\,mg\,O_2\,g^{-1}\,h^{-1}$ compared to $0.30\,mg\,O_2\,g^{-1}\,h^{-1}$: Hüppop, 1986). Studies of *A. fasciatus* in respect of body composition and starvation resistance revealed very high fat content in the fish from Pachon Cave (37% fat in wet body mass compared to 9% fat in wet body mass in the epigean fish: Hüppop, 1988). Fat tissue is known to have a relatively low maintenance metabolism compared to other tissues or organs and it is preferable to use lean body mass as a metabolic reference since high fat contents of animals lead to misinterpretations of the metabolic rate (Ball and Jungas, 1965; Hayward, 1965; Schmidt-Nielsen, 1984; Kersten and Piersma, 1987). A recalculation of the metabolic rate of hypogean *A. fasciatus* based on lean body mass no longer showed a significant difference in the oxygen consumption rates between cave and surface fish (standard metabolic rate = $0.37\,mg\,O_2\,g^{-1}\,h^{-1}$ and $0.33\,mg\,O_2\,g^{-1}\,h^{-1}$, respectively: Hüppop, 1988, 1989). However, total exclusion of the fat content for the calculation of the metabolic rate is also not correct, since even fat tissue requires some energy under basal conditions (Hayward, 1965), and in consequence the fat-free metabolic rate of the cave fish appears to be even higher than that of surface ones. The metabolic rate of Pachon Cave fish, calculated again on a basis of 9% fat as found in the surface fish (equivalent to body mass minus 28% as a calculation basis for the cave fish) resulted in nearly identical metabolic rates for the cave and the surface fish (standard metabolic rate = $0.32\,mg\,O_2\,g^{-1}\,h^{-1}$ and $0.30\,mg\,O_2\,g^{-1}\,h^{-1}$, respectively). Although obviously adapted to a periodically low-energy environment, as can be seen from the high fat content, the hypogean *A. fasciatus* shows no reduction in metabolic rate. It seems to not have the biological ability to reduce the metabolic rate – which, besides genetic variability, is one fundamental prerequisite for adaptation and optimization according to Culver (1982).

Character reduction

Many features become reduced during the evolution of cave animals. This "regressive evolution", is generally described as the reduction of "functionless" characters in cave animals and concerns not only structural but also behavioral and physiological characters. For example, the reduction of the metabolic rate due to lowered levels of interior and motion activity and perhaps other mechanisms, observed in many cave animals, could be interpreted as a phenomenon of regressive evolution. Several theories for the explanation of regressive evolution have been proposed (Eigenmann, 1909; Barr, 1968; Culver, 1982; Christiansen, 1985; Kane and Richardson, 1985). One of the most important is the **energy economy theory** which favors, as a reason for their reduction, the saving of material by not building and/or not maintaining characters which are useless in the darkness of the caves (e.g., Park, 1951; Poulson, 1964; Mitchell, 1969; Poulson and White, 1969; Brandon, 1971; Poulson, 1981; Sket, 1985; Howarth, 1993). Beside the **pleiotropy theory** (Wright, 1964; Barr, 1968), meaning that regressive evolution happens as a side effect of adaptive evolution as a result of interrelated (pleiotropic) genes, the **neutral mutation theory**, that is, the reduction of characters by accumulated neutral mutations, is favored by the opponents of the energy economy theory (e.g., Wilkens, 1971, 1987, 1992; Wilkens et al., 1979; Culver, 1982; Poulson, 1985; Peck, 1986).

It obviously would be an advantage for cave animals to use the energy saved by not building and/or not maintaining useless characters, when living in strongly food-limited cave environments, by transferring it to the development or support of other characters or to growth, reproduction or survival during starvation periods.

However, there exist only few experimental data supporting the theory of energy economy by character reduction. Park (1951) observed that female *Batrisodes* beetles have smaller eyes than males, which can be explained by the fact that energy demands on females are greater than on males (Culver, 1982). Mitchell (1969) found rhadinid beetles in caves with extensive guano usually had eyes, but those in food-limited non-guano caves usually had none. Poulson (1981), in the cave spider *Anthrobia monmouthia,* attributed web-spacing with less silk, a thinner exoskeleton, a thinner egg case, a reduced body size and a reduced body mass per unit length to selection for energy economy, whereas eyes, pigment or epicuticular wax, which require little energy for development or maintenance, may not be affected by selection for energy economy. Poulson (1981) therefore suggested that both selection for energy economy and neutral mutations are involved in regressive evolution in general, and that their relative importance depends on the structure involved, selection being more important for structures involving great amounts of energy. Culver (1982) and Christiansen (1985) also defended the coexistence of two, three or even more theories of regressive evolution, depending on the character, its genetic basis and the selection pressure on it.

Sometimes the reduction of characters in cave animals is interpreted as due to selection pressure other than energy economy (e.g., Parzefall, 1985; Jones et al., 1992; Culver et al., 1995). Parzefall explained the reduction of aggressive behavior in the cyprinodont fish *Poecilia mexicana* as a consequence of the risk of losing contact with the female in darkness during an aggressive encounter against an opponent. In the North American amphipod *Gammarus minus,* the selective advantage of smaller eyes in cave populations without food restriction (Culver and Poulson, 1971) is likely to be at the neuronal level, where optic connections to the brain may be co-opted or used also by extra-optic sensory systems such as antennae (Jones and Culver, 1989; Jones et al., 1992; Culver et al., 1995).

Poulson (1985) suggested that any energetic basis for selection against useless characters is during or just after development, when savings are comparatively greatest. Also Sket (1985) expected a higher quantity of energy and material to be needed for development and growth than for maintenance. In this sense my own investigations on the Mexican cave fish, *Astyanax fasciatus*, do not support the energy economy theory, at least regarding eye and pigment reduction in the hypogean Pachon form. Epigean and hypogean larvae of *A. fasciatus* show no difference in growth rate and in yolk-consumption rate, but a reduced growth rate of the eye in the cave fish. Thus, the "underdevelopment" of the eye does not result in a measurably slower rate of yolk consumption, and thus appears to be unimportant for energy economy (Hüppop, 1988; see also Wilkens, 1988, fig. 49). To test the energy economy theory, Christiansen (1985) suggested that one should measure metabolic efficiency in a variety of species showing different levels of eye reduction, and see whether increased efficiencies are associated with reduced eyes. Because of the interfertility of cave and epigean *A. fasciatus* I had the chance to make this experiment with F_2-hybrids between surface and Pachon Cave fish. These animals show a high variability in eye size and pigmentation, ranging from an eye of normal shape to nearly total extinction, and both features are inherited independently. The lack of correlation of the characters "eye size" and the "degree of pigmentation" with an energy-relevant character ("improved ability to store fat") in F_2-hybrids of cave and surface fish of *A. fasciatus* evidently supports the neutral mutation theory. Eye and pigmentation seem not to be reduced for energy economy, since their absence does not result in a lower energy demand in F_2-hybrids (Hüppop, 1989, fig. 2).

In summary, it is clear that each character that is reduced in cave animals has to be investigated separately. It is indisputable that energy economy results from not building or not maintaining certain functionless characters in some cave species. This has to be accepted as one "strategy" of cave animals to adapt to a food-restricted cave environment. It is also just as obvious that certain characters in the same or in other cave species are reduced by mutation pressure, or possibly even by pleiotropy or by any other of the mechanisms described in theories for understanding regressive evolution in cave animals. Each theory may have its equivalents in nature, depending on the character and its energetic significance, on the species and its genetic background, and on the specific cave environment and its degrees of selection pressures. Any discussion about which theory provides the main or universal explanation for regressive evolution is as useless as are the reduced characters in cave animals.

Ectothermy

Another question of interest in the context of energy economy is why troglobites are exclusively ectotherms. Three characteristics of ectotherms are most important in the colonization of the cave environment:

(1) Ectotherms generally have very low metabolic rate (only 10–20% of that of birds or mammals of similar size, or even less: Pough, 1980). Low energy demand is the basis for the success of ectotherms in zones characterized by a limited resource supply like shortages in food, oxygen or water. Ectotherms resist starvation very well because the energy in their tissues is mobilized very slowly (Pough, 1980; Peters, 1983). Energy metabolism at comparable temperatures is several times higher in endotherms than in ectotherms (Withers, 1992).

(2) Ectotherms can utilize energy for reproduction, which endotherms use for thermoregulation (Pough, 1980; Peters, 1983).

(3) Ectotherms can also exploit a world of small body sizes that is unavailable to endotherms. Body sizes less than 2 grams are not feasible for endotherms, because the curve relating metabolism to body mass becomes asymptotic to the metabolism axis at body masses lower than 2 grams (Pough, 1980; Withers, 1992). However, within each metabolic group the energy reserves of larger animals last longer and are more resistant to food shortage than in small animals because the metabolic rate is related to mass by the power equation:

$$\mathrm{MR} = aM^b,$$

where a is the intercept, M is the body mass, and b is the mass exponent or slope, smaller than one (about 0.7 to 0.8) (Phillipson, 1981; Peters, 1983). Consequently, cave animals have to resolve the conflict between two advantages: to be preferably small to need less energy per animal, and also to be able to live in crevices, and to be larger to need less energy per unit mass, but more per individual. This question leads into another "strategy" of adaptation of cave animals in respect of life history.

LIFE HISTORY

There is evidence of an upper limit of the cumulative energy turnover per unit mass of tissue, the "metabolic span". That means that the total metabolic turnover in a lifetime not only of endotherms but also in ectotherms is the product of the energy turnover rate and the duration of life (Schmidt-Nielsen, 1984; Calder, 1985). Therefore, metabolic reduction should be inversely proportional to life-span extension, and their product should be approximately constant. The total reproductive fitness of an animal is the product of rate and time, not only the average natality per breeding season, but its lifetime reproduction. Reproductive success might then be defined by the ability to live long enough to survive the gap between good years (Calder, 1985). From this it is easy to understand that a reduction of the metabolic rate increases life-time and that this is advantageous for cave animals living in a generally food-scarce environment or in caves where floods occur irregularly (see also Culver and Poulson, 1971; Poulson, 1971).

The extremes of the spectrum of life-cycle adaptations are characterized as r- and K-selection (MacArthur and Wilson, 1967). Whereas r-selection means a trend towards high population growth rate in relatively unpredictable and changing habitats (r is the slope of the population growth curve), K-selection can be realized only in more predictable and stable habitats, and the appropriate fitness measure is the maximum lifetime fecundity. K-selected species are characterized by low or no population growth; they have reached a high K (= carrying capacity of the habitat). This is connected with fewer but larger and more nutrient-rich eggs, increased time required for hatching, prolonged larval stage, generally decreased growth rate, delayed and perhaps infrequent reproduction, increased longevity and parental care.

Many cave animals show a couple of these characters, demonstrating the trend towards more K-selection in cave species. The life history of cave animals has been reviewed by Poulson (1964), Vandel (1964), Mohr and Poulson (1966), Barr (1968), Poulson and White (1969), Dickson and Holsinger (1981), Culver (1982), Howarth (1983) and Peck (1986). The main investigations were done by Ginet (1960) on amphipods, Rouch (1968) on copepods, Cooper (1975) and Cooper and Cooper (1978a,b) on decapods, Husson (1959) on isopods, Deleurance (1955) and Poulson (1981) on spiders, Izquierdo et al. (1990) on cockroaches, Deleurance (1959), Deleurance-Glaçon (1963), Peck (1973a, 1986) and Rusdea (1994) on beetles, Eigenmann (1909), Heuts (1951, 1953), Poulson (1961, 1963, 1964), Hüppop (1988), Hüppop and Wilkens (1991) and Junge (1993) on fish and by Poulson (unpubl. obs. cited in Poulson, 1964) and Brandon (1971) on salamanders. Some of the results are described below to demonstrate the dominance of K-selection in cave animals.

Egg size

In advance, it is important to know that conclusions from egg comparisons should be based rather on yolk diameter, or even better yolk volume, than on egg diameter only. Firstly, the yolk size is more appropriate since the perivitelline space can be very variable within and among species. Secondly, the yolk volume is more meaningful because a slight increase in yolk diameter, though appearing unimportant, can cause a significant increase in yolk volume. The adaptive value of fewer but bigger eggs, not only for cave fish but for all animals living in food-poor habitats, is obvious. Bigger eggs, with more and/or energy-richer yolk, release bigger larvae. These larvae have a bigger head with a larger mouth and by this can start external feeding on a larger spectrum of food particles and have a better chance to survive. Furthermore, bigger larvae may have a higher resistance to starvation and a higher mobility for food-searching and for effective escape reactions (Poulson, 1963, 1964, 1969, 1981, 1985; Poulson and White, 1969; Culver, 1982; Hüppop, 1988; Hüppop and Wilkens, 1991; Junge, 1993).

A comparison of some European ostracods revealed an increasing tendency for egg size, in proportion to carapace length, to increase, along with a decreasing number of eggs, in the sequence from epigean through troglophile to troglobite species (Danielopol, 1980).

Recently, Izquierdo et al. (1990) investigated cockroaches on the Canary Islands. The exclusive epigean cockroach *Loboptera canariensis* had 16 to 18 ovarioles, three hypogean *Loboptera* species had about 12 ovarioles, whereas a troglobitic *Loboptera* species from Tenerife had only 6 ovarioles. This last species can be considered as the most troglobitic cockroach in the Canaries. In addition to a high degree of troglomorphism, the very low reproductive potential is comparable only to that of *Alluaudellina cavernicola,* whose ootheca contained no more than 4 to 5 embryos (Ginet and Decou, 1977; Izquierdo et al., 1990).

As already mentioned, some French cave beetles produce very few big eggs (*Isereus colasi* and *I. serrulazi*: Leiodidae) or even only one very big and yolk-rich egg (*Speonomus longicornis*: Catopidae), with the larvae developing very slowly and never feeding (Deleurance, 1959; Deleurance-Glaçon, 1963). Culver (1982) interpreted the suppression of larval feeding in cave beetles as an adaptation not only to food scarcity but also for predator escape. The advantage from being protected in the pupal cell during the whole larval development should exceed the loss from the lack of food.

In amblyopsid fish a large difference in egg size was found between the surface species and the different cave species (Poulson, 1963). The surface and the troglophilic *Chologaster* species produce ova of 0.9 to 1.2 mm diameter. The troglobitic *Typhlichthys subterraneus* and the *Amblyopsis* species have ova sizes of 1.8 to 2.3 mm diameter. The increasing egg size with increasing phylogenetical age or time in the cave is related to an increasing developmental rate.

Hüppop and Wilkens (1991) compared the surface form of *Astyanax fasciatus* with five subterranean forms. The fish of all cave forms had bigger eggs with more yolk, embryos were larger at hatching, and larvae were longer at the time of first external feeding. The difference was greatest between the phylogenetically oldest cave form from Pachon Cave and the epigean form from the Rio Teapao (Fig. 8.4). The epigean fish had a mean egg diameter of 1.03 mm and a mean yolk volume of 0.13 mm^3 12 hours after fertilization. The Pachon eggs had a mean diameter of 1.15 mm and a yolk volume of about 0.20 mm^3. The average length of the hatched embryos, 24 hours after fertilization, was 2.75 mm and 3.00 mm, respectively. First external feeding started at 3 days in the surface fish and at 4 days in the cave fish, corresponding to a body length of 4.0 mm and 4.7 mm, respectively (Fig. 8.5). The delayed start of external feeding in the cave-fish larvae results in a larger head at this time (Fig. 8.5). The mouth opening, enlarged by about 64% in the cave larvae (in area, calculated from head size) enables them to feed not only on small food particles but on bigger ones as well. Feeding on a larger food spectrum may increase their survival in the cave. An increasing trend in yolk volume, embryo length and corresponding advantages for the cave fish could be observed from the hybrid Chica Cave fish through the phylogenetically young Micos Cave fish to the phylogenetically oldest Pachon Cave fish. The enlargement of the yolk volume in Pachon Cave fish is not very prominent, compared to Poulson's data on amblyopsid fish. However, because cave and surface forms of *Astyanax fasciatus* are still conspecific, the result is especially convincing.

The eggs of two hypogean catfish species in Mexico, *Rhamdia zongolicensis* (Wilkens, 1993) from the Sierra de Zongolica and *R. reddelli* from Oaxaca, have the same size as their epigean ancestor *Rhamdia laticauda* (Wilkens, pers. comm.). The former investigation of Junge (1993), who compared *R. zongolicensis* with the

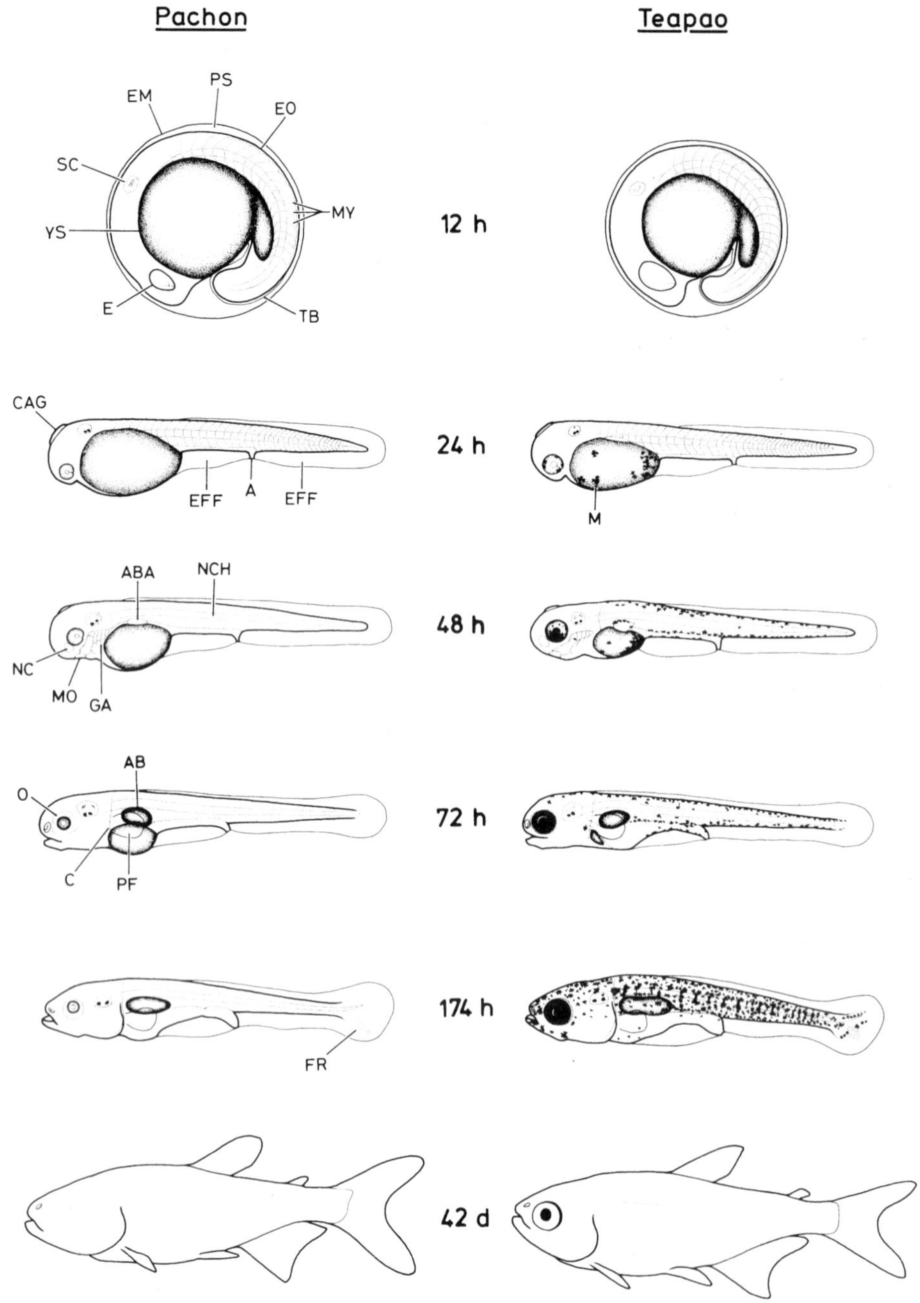

Fig. 8.4. Developmental stages of hypogean Pachon fish and epigean Teapao fish of *Astyanax fasciatus*. The sizes of the embryos can be derived from Fig. 8.5. A, anus; AB, air bladder; ABA, air bladder anlage; C, cleithrum; CAG, cephalic adhesive gland; E, eye; EFF, embryonic fin fold; EM, egg membrane; EO, embryo; FR, fin rays; GA, gill arches; M, melanophore; MO, mouth opening; MY, myomeres; NC, nasal cavity; NCH, notochord; O, orbit; PF, pectoral fin; PS, perivitelline space; SC, statocyst; TB, tail bud; YS, yolk sac.

epigean *R. guatemalensis*, has to be taken with caution since *R. guatemalensis* is not the ancestor as formerly assumed. However, in the hypogean *R. zongolicensis* and in the epigean *R. laticauda* the egg yolk volume is three times larger than in the epigean *R. guatemalensis*. Whereas *A. fasciatus* forms have the same yolk consumption rates, and in consequence start external feeding at different sizes and different ages (Fig. 8.5),

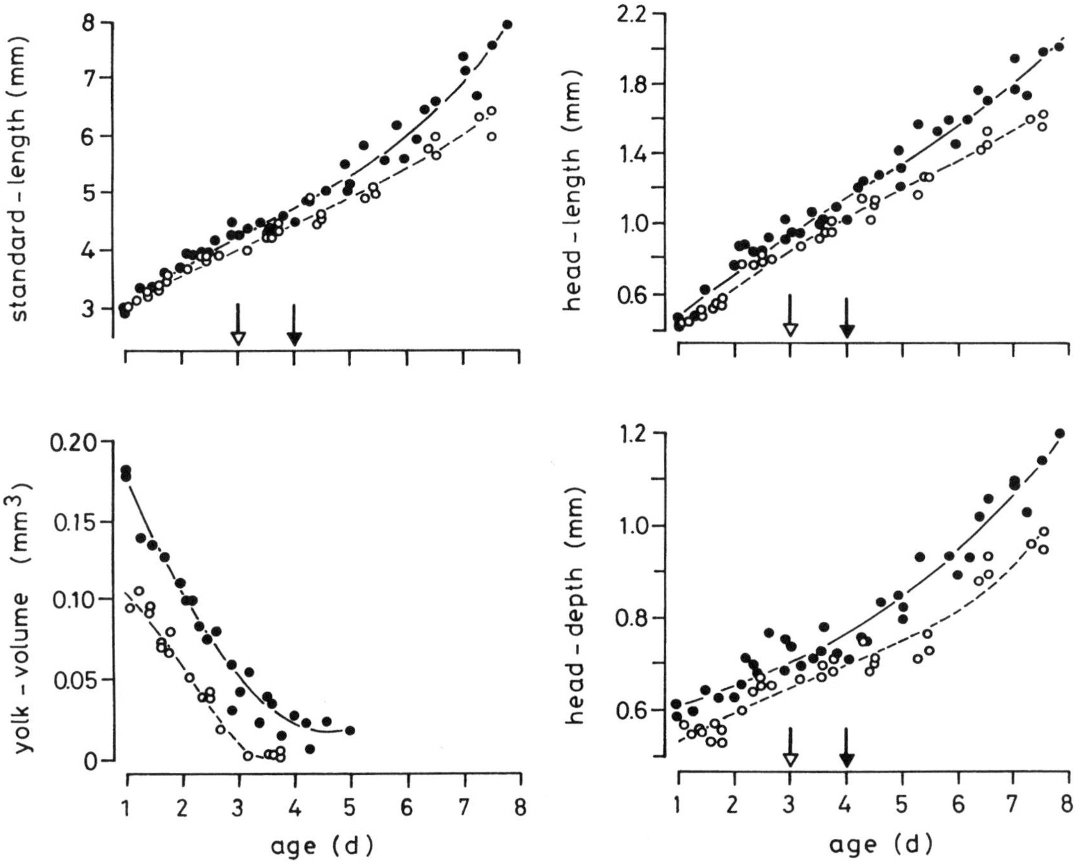

Fig. 8.5. Growth of standard length, head length and head depth and depletion of yolk volume in hypogean Pachon fish (solid symbols) and epigean Teapao fish (open symbols) of *Astyanax fasciatus* in the first eight days of life. Each point is the mean of mostly ten individuals. Lines developed from polynomial regressions (see Hüppop, 1988). Arrows indicate the beginning of external feeding.

the *R. zongolicensis* cave larvae consume their big yolk in about the same time as those of the surface *R. guatemalensis,* or only slightly longer, and by this time have grown larger. At the end of the yolk stage the area of the mouth opening of the cave larvae of *R. zongolicensis* is wider by about 80% compared to the surface species (Junge, 1993).

Growth rates

Compared to epigean and troglophile European ostracod species with short developmental times and high ovogenetic activity, troglobite species have longer developmental spans and lower ovogenetic activity, a species of *Mixtacandona* being the most extreme (Danielopol, 1980). However, in *Ptomaphagus* cave beetles Peck (1986) could demonstrate that low embryonic developmental rates do not necessarily lead to low postembryonic growth rates. Larval adaptations are not strongly coevolved with adult cave-adapted characters (Rouch, 1968; Peck, 1986).

But generally, cave animals are characterized by slower growth rates compared to surface relatives, this being connected with a delay in maturity and increased longevity. A reduced growth rate in cave animals is adaptive to food scarcity, because it means a reduced energy need per unit time. In consequence, more animals can live on a defined amount of food, or a defined group of animals can survive longer on it. A reduction in energy demand per unit time through a lower metabolic rate, together with a reduction of absolute and relative costs of reproduction, made possible an increase of population density in amblyopsid cave fish, and hence an increase in the number of females actually breeding in a given year (Poulson, 1969).

The increased longevity of cave animals, together with a trend from semelparity to iteroparity, means that the population is less likely to disappear in years when food supply is too low to allow females to produce offspring. That food supply can be a problem in this sense of reproduction is shown by the relatively high fluctuation of population size in amblyopsid cave fish (Poulson, 1969). The number of reproductions per lifetime for the troglobitic *Amblyopsis rosae* is only 0.6. This indicates that selection is potentially strong, and that the environment is so food-poor that many individuals never reproduce (Poulson, 1963). The many investigations of Poulson (1961, 1963, 1964, 1969, 1985) are very prominent examples of how cave animals adapt their life history to food scarcity. It is obvious how cave animals with increasingly slow energy turnover rates have increasingly prolonged life cycles, connected with many increasingly K-selected features, such as bigger and fewer eggs with prolonged developmental time, prolonged branchial incubation time (=parental care), bigger larvae at first external feeding, reduced growth rate, delayed maturity and multiplied chance to reproduce with increasing cave adaptation (Table 8.4). Population growth rate and population density decrease with phylogenetical age of the cave species, and the population structure shifts towards predominance of adults (Poulson, 1963).

Table 8.4
Some life-history characters of the epigean amblyopsid fish *Chologaster cornutus* compared to the troglobitic *Amblyopsis rosae* [1]

	C. cornutus	*A. rosae*
egg size (mm)	0.9 to 1.2	1.9 to 2.2
developmental time	few weeks	5/6 months
larval size when hatching (mm)	3	5
larval size when leaving gill cavity (mm)	8	12
adult maturity (yr)	1	3
adult size (mm)	23 to 55	36 to 48
female body mass (g)	0.93	1.25
maximal life span (yr)	1	5 to 6
proportion (%) of females breeding per year	100	20
number of mature ova per female	98	23
average number of reproductions per lifetime	1	0.6
maximum number of reproductions per lifetime	1	3

[1] From Poulson (1961, 1963, 1964, 1985), Poulson and White (1969) and Culver (1982).

In some other cave-fish species too, a reduced growth rate connected with an increased longevity, and delayed and infrequent reproduction, have been described – for instance, in the viviparous ophidiid cave fish *Lucifuga* and *Stygicola* (Eigenmann, 1909), in the African cave barbel *Caecobarbus geertsi* (Heuts, 1951, 1953), in a troglobitic catfish from southeastern Brazil, *Pimelodella kronei* (Trajano, 1991), and in the Mexican characid cave fish *Astyanax fasciatus* (Hüppop, 1988).

In most cases growth-rate observations have been made in the natural habitat. However, some characters which appear to be K-selected, like growth rates or birth rates, can occur as a result of ecological compensation. Hence, in habitat comparisons observed differences might be due to this rather than to selection. Hence, the separation of genetic from purely ecological effects through breeding studies under controlled conditions is required (Sibly and Calow, 1985). Epigean *A. fasciatus* and hypogean conspecifics from Pachon Cave have been reared under comparable laboratory conditions with *ad libitum* feeding (Hüppop, 1988). The results are contradictory. The growth rate of the cave-fish larvae was higher in the first days of life compared to the surface fish (Fig. 8.5). This supports the adaptiveness of having more yolk, thus growing larger until external feeding starts, and being able to feed on a larger food spectrum. However, at a later stage the cave fish had a slightly lower growth rate than the epigean ones and did not grow bigger than 5 to 6 cm and 7 to 8 g, whereas the surface fish reached 8 to 9 cm and 10 to 15 g [see also Breder Jr and Rasquin (1947) for Chica Cave fish; Nomura (1975) for epigean *Astyanax*; Hüppop and Wilkens (1991)].

Reduced adult size and reduced growth rate are both adaptive in the sense of energy economy in the cave environment. Limitation of food must be considered in relation to body size. Generally, reduced levels of resource supply may lead to communities with smaller individual size (Thiel, 1975). Although small individuals need more food per unit of body mass, one individual needs less energy per unit time and per total life (Sibly and Calow, 1985). For example, the interstitial forms of isopods are adapted to their habitat by size reduction with two effects. They need less energy per individual and they fit better into small crevices (Danielopol et al., 1994). However, with respect to egg size the smaller size of cave species is disadvantageous, since usually smaller animals have smaller eggs. If cave species were as large as their surface relatives, the enlargement in egg size would be even more prominent (Hüppop, 1988; Junge, 1993).

Data on extremely prolonged lifetimes were presented by Cooper (1975) and Cooper and Cooper (1978a,b) for some cave crayfish species from Alabama. Published information on growth of epigean crayfish shows rates ranging from 2.8 to 5.2 mm carapace length per month. In contrast, the maximum growth recorded for any cave crayfish in these studies represented a rate of 0.38 mm increase in carapace length per month. The age of the largest individuals of *Orconectes australis australis* was estimated at over 37 yr, assuming growth rates observed at various sizes, and at 175 yr, assuming average growth rates (Cooper and Cooper, 1978b). In addition these authors observed that few reproductive cycles were completed, resorption of oocytes being a common phenomenon in *Cambarus jonesi.* Another *Cambarus* species had the largest and fewest oocytes (8 to 12) and larger young at recruitment. Sexual maturity was delayed beyond the minimum adult size to 35 yr under average conditions and to 16 yr under optimal conditions; the mean age at reproduction was 105 and 29 yr, respectively. The authors interpreted these characters not only as an adaptation to an energy-limited system, but also as related to differential predation by the amblyopsid cave fish *Typhlichthys subterraneus.*

Sometimes, the reduced absolute growth rates in cave animals and their higher relative growth rates have been held responsible for increased appendage length and its advantages through allometric growth effects (Heuts, 1953; Poulson, 1963, 1964; Barr, 1968; Cooper, 1969). It is likely that allometric effects of selection for sensory compensation and lower developmental rates and growth rates, in response to food scarcity, are responsible for some apparent neotenic features (Poulson,1964). Actually, neoteny has been observed in cave animals, and attributed to the food scarcity (Poulson, 1963, 1964; Brandon, 1971; Cooper and Kuehne, 1974; Langecker and Longley, 1993); it has been proposed as a mechanism in the evolution of cave animals, notably salamanders (Culver, 1982), or regarded as a consequence of "high stress" in caves (Howarth, 1993).

SYNTHESIS

Adaptations of cave animals to the food scarcity of their environments are numerous. They build a causally interrelated network (Fig. 8.6). White arrows in the figure **allow** changes of characteristics, black arrows **require** evolution of particular characteristics in the cave habitats in question. Given preadaptations, genetic variation and the biological possibility for adaptation, cave animals can evolve. The high environmental stability, including darkness and sometimes predator scarcity, on the one hand allows the evolution of characters such as increased fright resistance (insensitivity to disturbance), life-history changes towards more K-selection and character reduction. On the other hand it requires character changes such as improved sensory orientation. In fact, most characteristics of adaptation to food scarcity can only be realized in stabilized, and above all, predator-poor caves. However, the stabilized cave environment which allows the evolution of cave animals, often becomes their prison (Heuts, 1951).

As a selective force in caves, food scarcity is divisible into general, periodic and patchy scarcity. The most important characters which can be observed in food-limited cave animals are reduced energy demand, improved food-finding ability, increased ability to survive starvation, and changes in life history with a trend towards more K-selected characters. Additionally, dietary shift, feeding generalism and higher efficiency of food utilization have evolved in some cave animals.

Reduced energy demand can be achieved under two aspects. The reduction of metabolic rates, through increased fright resistance and improved food-finding ability, and character reduction may reduce the energy consumption **per unit time**. A reduced standard metabolic rate might be leading back not only to a lowered interior activity, based on increased fright resistance and lowered aggressive behavior, but also to changes in biochemical mechanisms, such as rates of ATP turnover. A reduced routine metabolic rate, based additionally on the low level of standard metabolic rate, is the consequence of fewer escape reactions away from predators and fewer movements for food-finding. In the end, reduced metabolic rates result in a higher availability of energy for growth, and/or a greater resistance to starvation. A reduced growth rate can especially effect a reduction in energy demand per unit time as well, whereas character reduction and a reduced adult size may also lower the total energy demand **per individual life**, even if connected with increased energy demand per unit body mass.

With respect to the metabolic span, a reduced

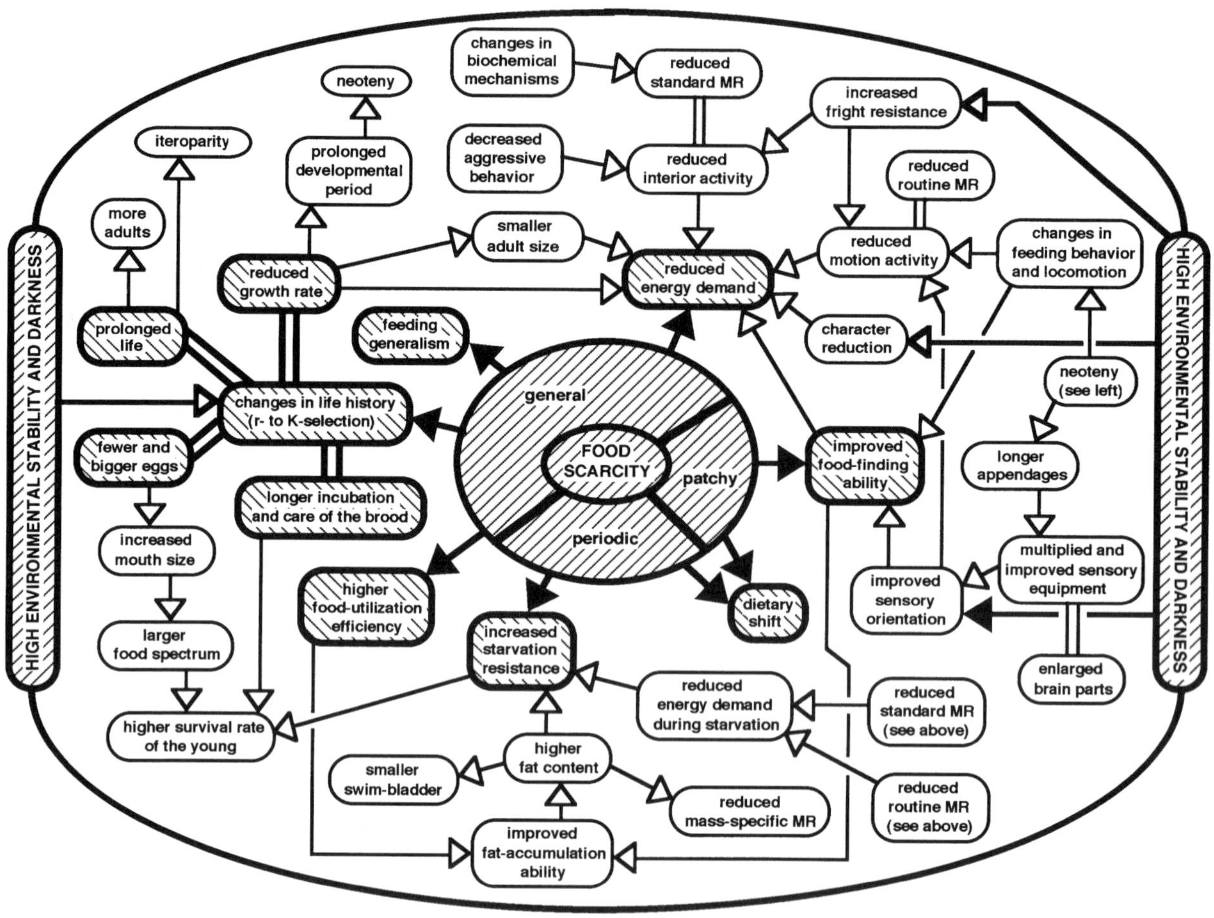

Fig. 8.6. Network of characters observed in many cave animals, their causes, consequences and interrelations (solid arrowheads require something, open arrowheads allow something).

energy demand per unit time is correlated with a prolonged lifetime. This may also be an important effect of life-history changes towards more K-selected characters, and results in the chance of iteroparity. Additionally, more K-type selection may lead to bigger and fewer eggs and more incubation and brood care, giving the offspring a higher chance of survival. A reduced growth rate and a smaller adult size may save energy. A prolonged developmental period may result in neoteny, which in turn can have an influence on food-finding ability through appendage length and changes in foraging behavior.

An improved food-finding ability is adaptive predominantly in patchy food-limited cave habitats, and may be realized through changes in foraging behavior and improved ability for sensory orientation. The latter includes improvements not only in the senses of taste and smell but also in spatial orientation, as required in the darkness of caves. Improved food-finding ability in turn may have an influence on the fat content of cave animals. A higher fat content is highly desirable in cave animals confronted with periodic food scarcity, increasing their resistance to starvation. It also can be increased by an improved ability to accumulate fat, due to a higher efficiency of food utilization. Increased starvation resistance possibly has an influence on the survival rate of the young.

ACKNOWLEDGEMENTS

I am very indebted to D.C. Culver, D.W. Goodall, H. Wilkens and J. Yager for critically reading the manuscript. They improved this chapter by many important suggestions. J. Yager also was a great help in linguistically correcting most parts of the text.

REFERENCES

Ahearn, G.A. and Howarth, F.G., 1982. Physiology of cave arthropods in Hawaii. *J. Exp. Zool.*, 222: 227–238.

Ball, E.G. and Jungas, R.L., 1965. Net gas exchange and oxygen consumption. In: A.E. Renold and G.F. Cahill Jr (Editors), *Adipose Tissue. Handbook of Physiology, Section 5.* American Physiological Society, Washington D.C., pp. 355–361.

Banta, A.M., 1907. The fauna of Mayfield's cave. *Carnegie Inst. Washington Publ.*, 67: 1–114.

Banta, A.M., 1910. A comparison of the reactions of a species of surface isopod with those of a subterranean species. *J. Exp. Zool.*, 8: 243–310; 439–488.

Barr Jr., T.C., 1968. Cave ecology and the evolution of troglobites. In: T. Dobzhansky, M.K. Hecht and W.G. Steere (Editors), *Evolutionary Biology, Vol. 2.* North-Holland Publ. Company, pp. 35–102.

Barr Jr., T.C. and Holsinger, J.R., 1985. Speciation in cave faunas. *Annu. Rev. Ecol. Syst.*, 16: 313–337.

Biswas, J., 1991. Metabolic efficiency and regulation of body weight: a comparison between life in hypogean and epigean ecosystems. *Int. J. Speleol.*, 20: 15–22.

Blaxter, J.H.S. and Hempel, G., 1963. The influence of egg size on herring larvae (*Clupea harengus* L.). *J. Cons. Perm. Int. Explor. Mer*, 28: 211–240.

Bone, Q. and Marshall, N.B., 1985. *Biologie der Fische.* Gustav Fischer Verlag, Stuttgart, New York, 236 pp.

Brancelj, A., 1992. *Alona sketi* sp.n. (Cladocera: Chydoridae), the second cave-inhabiting cladoceran from former Yugoslavia. *Hydrobiologia,* 248: 105–114.

Brandon, R.A., 1971. North American troglobitic salamanders: some aspects of modification in cave habitats, with special reference of *Gyrinophilus palleucus. Natl. Speleol. Soc. Bull.*, 33: 1–22.

Breder Jr., C.M., 1942. Descriptive ecology of La Cueva Chica, with especial reference to the blind fish *Anoptichthys. Zoologica N. Y.*, 27: 7–16.

Breder Jr., C.M., 1944. Ocular anatomy and light sensitivity studies on blind fish from Cueva de Los Sabinos, Mexico. *Zoologica N. Y.*, 29: 131–144.

Breder Jr., C.M. and Rasquin, P., 1943. Chemical sensory reactions in the Mexican blind characins. *Zoologica N. Y.*, 28: 169–199.

Breder Jr., C.M. and Rasquin, P., 1947. Evidence for the lack of a growth principle in the optic cyst of Mexican cave fish. *Zoologica N. Y.*, 32: 29–33.

Burbanck, W.D., Edwards, J.P. and Burbanck, M.P., 1948. Toleration of lowered oxygen tension by cave and stream crayfish. *Ecology,* 29: 360–367.

Burchards, H., Dölle, A. and Parzefall, J., 1985. Aggressive behaviour of an epigean population of *Astyanax mexicanus* (Characidae, Pisces) and some observations on three subterranean populations. *Behav. Process.*, 11: 225–235.

Caine, E.A., 1978. A comparative ecology of epigean and hypogean crayfish (Crustacea, Cambaridae) from Northwestern Florida. *Am. Midl. Nat.*, 99: 315–329.

Calder III, W.A., 1985. The comparative biology of longevity and lifetime energetics. *Exp. Gerontol.*, 20: 161–170.

Camiciottoli, L. and Messana, G., 1994. Some aspects of respiratory metabolism in *Stenasellus racovitzai* Razzanti, 1925 (Isopoda, Asellota, Stenasellidae). In: C. Juberthie (Editor), *XI Int. Symp. Biospeleol. Firenze,* Abstract 44. Société de Biospéology.

Caumartin, V., 1963. Review of the microbiology of the underground environments. *Nat. Speleol. Soc. Bull.*, 25: 1–14.

Childress, J.J., 1995. Are there physiological and biochemical adaptations of metabolism in deep-sea animals? *Trends Ecol. Evol.*, 10: 30–36.

Christiansen, K.A., 1970. Survival of collembola on clay substrates with and without food added. *Ann. Speleol.*, 25: 849–852.

Christiansen, K.A., 1985. Regressive evolution in collembola. *Natl. Speleol. Soc. Bull.*, 47: 89–100.

Cooper, J.E., 1975. *Ecological and behavioral studies in Shelta Cave, Alabama, with emphasis on decapod crustaceans.* Ph.D. Thesis, Univ. of Kentucky, 422 pp.

Cooper, J.E. and Cooper, M.R., 1978a. Comparative reproductive strategies of troglobitic crayfishes in Shelta Cave, Alabama. *ABS Bull.*, 25: 44.

Cooper, J.E. and Cooper, M.R., 1978b. Growth, longevity, and reproductive strategies in Shelta Cave crayfishes. *Natl. Speleol. Soc. Bull.*, 40: 97.

Cooper, J.E. and Kuehne, R.A., 1974. *Speoplatyrhinus poulsoni,* a new genus and species of subterranean fish from Alabama. *Copeia,* 486–493.

Cooper, M.R., 1969. Sensory specialization and allometric growth in cavernicolous crayfishes. *Proc. 4th Int. Congr. Speleol., Ljubljana.* 1965: 203–208.

Culver, D.C., 1973. Feeding behavior of the salamander *Gyrinophilus porphyriticus* in caves. *Int. J. Speleol.*, 5: 369–377.

Culver, D.C., 1982. *Cave Life. Evolution and Ecology.* Harvard Univ. Press, Cambridge, MA. and London, England, 189 pp.

Culver, D.C., 1985. Trophic relationships in aquatic cave environments. *Stygologia,* 1: 43–53.

Culver, D.C. and Poulson, T.L., 1971. Oxygen consumption and activity in closely related amphipod populations from cave and surface habitats. *Am. Midl. Nat.*, 85: 74–84.

Culver, D.C., Kane, T.C., Fong, D.W., Jones, R., Taylor, M.A. and Sauereisen, S.C., 1990. Morphology of cave organisms – is it adaptive? *Mem. Biospeleol.*, 17: 13–26.

Culver, D.C., Kane, T.C. and Fong, D.W., 1995. *Adaptation and Natural Selection in Caves. The Evolution of Gammarus Minus.* Harvard Univ. Press, Cambridge, MA. and London, England, 223 pp.

Danielopol, D.L., 1980. Sur la biologie de quelques ostracodes Candoninae épigés et hypogés d'Europe. *Bull. Mus. Natl. Hist. Nat. A,* 4: 471–506.

Danielopol, D.L. and Rouch, R., 1991. L'adaptation des organismes au milieu souterrain. Réflexions sur l'apport des recherches écologiques récentes. *Stygologia,* 6: 129–142.

Danielopol, D.L., Creuzé des Châtelliers, M., Mösslacher, F., Pospisil, P. and Popa, R., 1994. Adaptation of crustacea to interstitial habitats: a practical agenda for ecological studies. In: J. Gibert, D.L. Danielopol and J.A. Stanford (Editors), *Groundwater Ecology.* Academic Press, New York, pp. 217–243.

Deleurance, S.G., 1955. Sur la biologie de *Leptoneta microphthalma* (araignées cavernicoles). *C. R. Acad. Sci.*, 241: 1865–1867.

Deleurance, S.G., 1959. Contribution à l'etude des coléoptères troglobiés: sur la biologie des Bathysciinae. *Ann. Speleol.*, 14: 101–110.

Deleurance-Glaçon, S., 1963. Recherches sur les coléoptères

troglobiés de la sous-famille des Bathysciinae. *Ann. Sci. Nat. Zool. Paris,* 12e série 5: 1–172.

Dérouet, L., 1949. Comparaison des échanges respiratoires chez *Gammarus pulex* L. et *Niphargus virei* Chevreux. *C. R. Acad. Sci.,* 228: 1054–1055.

Dérouet, L., 1952. Influence des variations de salinité du milieu extérieur sur des crustacés cavernicoles et épigés. I: Etude de l'intensité des échanges respiratoires. *C. R. Acad. Sci.,* 234: 473–475.

Dérouet, L., 1953a. Métabolisme comparé de deux araignées, l'une troglophile, l'autre epigée obscuricole. Influence des variations brusques de temperature et d'humidité. *Publ. Prem. Congr. Int. Speleol.,* 3: 237–240.

Dérouet, L., 1953b. Etude comparée du métabolisme respiratoire chez certaines éspèces de crustacés cavernicoles et épigés. *Notes Biospeleol.,* 8: 103–109.

Dickson, G.W. and Franz, R., 1980. Respiration rates, ATP turnover and adenylate energy charge in excised gills of surface and cave crayfish. *Comp. Biochem. Physiol.,* 65A: 375–379.

Dickson, G.W. and Holsinger, J.R., 1981. Variation among populations of the troglobitic amphipod crustacean *Crangonyx antennatus* Packard (Crangonyctidae) living in different habitats. III. Population dynamics and stability. *Int. J. Speleol.,* 11: 33–48.

Dresco-Dérouet, L., 1959. Contribution à l'étude de la biologie de deux crustacés aquatiques cavernicoles; *Caecosphaeroma burgundum* D. et *Niphargus orcinus virei* ch. *Vie Milieu,* 10: 321–346.

Dresco-Dérouet, L., 1960. Etude biologique comparé de quelques éspèces d'araignés lucicoles et troglophiles. *Arch. Zool. Exp. Gen.,* 98: 271–354.

Dresco-Dérouet, L., 1967. Biologie et métabolisme respiratoire d'*Ischyropsalis luteipes* Simon (Opiliones) adulte, au laboratoire. *Ann. Speleol.,* 22: 537–541.

Durand, J.P., Parzefall, J. and Richard, B., 1981. Proteidae prey detection and the sensory compensation problem. In: B.F. Beck (Editor), *Proc. 8th Int. Congr. Speleol., Bowling Green,* 31–34.

Durand, J.P., Parzefall, J. and Richard, B., 1982. Etude comparé de la détection chimique des proies par *Proteus anguinus,* cavernicole, et son parent de surface *Necturus maculosus* (Proteidae, Urodela). *Behav. Process.,* 7: 123–134.

Eberly, W., 1960. Competition and evolution in cave crayfishes of Southern Indiana. *Syst. Zool.,* 9: 29–32.

Eigenmann, C.H., 1909. *Cave Vertebates of America. A Study in Degenerative Evolution.* Carnegie Institute, Washington, 104: 241 pp.

Engelke, S., 1994. Prey choice behaviour in the blind *Proteus anguinus* (Proteidae) in comparison to the facultative cave-dweller *Euproctus asper* (Salamandridae). In: C. Juberthie (Editor), *XI Int. Symp. Biospeleol., Firenze.* Société de Biospéléologie, p. 46.

Erckens, W. and Martin, W., 1982. Exogenous and endogenous control of swimming activity in *Astyanax mexicanus* (Characidae, Pisces) by direct light response and by a circadian oscillator. *Z. Naturforsch.,* 37C(I): 1253–1265; 37C(II): 1266–1273.

Ercolini, A., Berti, R., Chelazzi, L. and Messana, G., 1987. Oxygen consumption in hypogean and epigean cyprinids (Pisces). *Ital. J. Zool.,* 4: 23–30.

Fage, L., 1931. Biospéléologica LV. Araneae, précédée d'un essai sur l'évolution souterraine et son déterminisme. *Arch. Zool. Exp.,* 70: 99–291.

Franz, R., 1978. Ecological strategies of closely-related surface and troglobitic Florida crayfishes. *Bull. Ecol. Soc. Am.,* 59: 70.

Gadeau de Kerville, H., 1926. Note sur un protée anguillard (*Proteus anguinus* Laur.) ayant vécu sans aucune nourriture. *Bull. Soc. Zool. France,* 51: 81–84 [cited by Vandel and Bouillon (1959)].

Gagné, W.C. and Howarth, F.G., 1975a. The cavernicolous fauna of Hawaiian lava tubes, 6. Mesoveliidae or water treaders (Heteroptera). *Pac. Insects,* 16: 399–413.

Gagné, W.C. and Howarth, F.G., 1975b. The cavernicolous fauna of Hawaiian lava tubes, 7. Emesinae or thread-legged bugs (Heteroptera: Reduviidae). *Pac. Insects,* 16: 415–426.

Gal, J., 1903. *Niphargus* et *Caecosphaeroma.* Observations physiologiques. *Bull. Soc. Sci. Nat. Nîmes,* 31: 48–51.

Gibert, J. and Mathieu, J., 1980. Relations entre les teneurs en protéines, glucides et lipides au cours du jeûne expérimental chez deux éspèces de *Niphargus* peuplant des biotopes différents. *Crustaceana,* Suppl. 6: 137–147.

Ginet, R., 1960. Ecologie, éthologie et biologie de *Niphargus. Ann. Speleol.,* 15: 127–376.

Ginet, R. and Decou, V., 1977. *Initiation à la Biologie et à l'Ecologie Souterraines.* J.P. Delarge, Paris, 345 pp.

Ginet, R. and Mathieu, J., 1968. Comparaison des températures létales supérieures de *Niphargus longicaudatus* (Crustacé, Amphipode) hypogés et épigés. *Ann. Speleol.,* 23: 426–440.

Gordon, M.S. and Rosen, D.E., 1962. A cavernicolous form of the poeciliid fish *Poecilia sphenops* from Tabasco, Mexico. *Copeia,* 360–368.

Hadley, N.F., Ahearn, G.A. and Howarth, F.G., 1981. Water and metabolic relations of cave-adapted and epigean lycosid spiders in Hawaii. *J. Arachnol.,* 9: 215–222.

Hausberg, C., 1995. *Das Aggressionsverhalten von Astyanax fasciatus (Cuvier 1819; Characidae, Teleostei): Ontogenie, Genetik und Evolution bei der epigäischen und hypogäischen Form.* Ph. D. Thesis, Univ. Hamburg, 139 pp.

Hawes, R.S., 1939. The flood factor in the ecology of caves. *J. Anim. Ecol.,* 8: 1–5.

Hayward, J.S., 1965. Metabolic rate and its temperature-adaptive significance in six geographic races of *Peromyscus. Can. J. Zool.,* 43: 309–323.

Hemmingsen, A.M., 1950. The relating of standard (basal) energy metabolism to total fresh weight of living organisms. *Rep. Steno Mem. Hosp. Nordisk Insulinlab.,* 4: 1–58.

Hervant, F. and Mathieu, J., 1995. Ventilatory and locomotory activities in anoxia and subsequent recovery of epigean and hypogean crustaceans. *C. R. Acad. Sci.,* 318: 585–592.

Hervant, F., Mathieu, J., Garin, D. and Fréminet, A., 1995. Behavioral, ventilatory, and metabolic responses to severe hypoxia and subsequent recovery of the hypogean *Niphargus rhenorhodanensis* and the epigean *Gammarus fossarum* (Crustacea: Amphipoda). *Physiol. Zool.,* 68: 223–244.

Heuts, M.J., 1951. Ecology, variation and adaptation of the blind fish *Caecobarbus geertsi* Blgr. *Ann. Soc. R. Zool. Belg.,* 82: 155–230.

Heuts, M.J., 1953. Regressive evolution in cave animals. *Symp. Soc. Exp. Biol.,* 7: 290–309.

Hobbs III, H.H., 1973. *The population dynamics of cave crayfishes and their commensal ostracods from Southern Indiana.* Ph.D. Thesis, Indiana Univ., 247 pp.

Hochachka, P.W. and Somero, G.N., 1980. *Strategien biochemischer Anpassung.* Thieme Verlag, 403 pp.

Holsinger, J.R. and Culver, D.C., 1970. Morphological variation in *Gammarus minus* Say (Amphipoda: Gammaridae), with emphasis on subterranean forms. *Postilla,* 146: 1–24.

Howarth, F.G., 1981a. Community structure and niche differentiation in Hawaiian lava tubes. In: D. Mueller-Dombois, K.W. Bridges and H.L. Carson (Editors), *Island Ecosystems: Biological Organization in Selected Hawaiian Communities.* Hutchinson Ross. Publ. Comp., Stroudsburg, PA; US/IBP Synthesis, 15: 318–336.

Howarth, F.G., 1981b. Non-relictual terrestrial troglobites in the tropical Hawaiian caves. In: B.F. Beck (Editor), *Proc. 8th Int. Congr. Speleol., Bowling Green,* 539–541.

Howarth, F.G., 1983. Ecology of cave arthropods. *Annu. Rev. Entomol.,* 28: 365–389.

Howarth, F.G., 1987. The evolution of non-relictual tropical troglobites. *Int. J. Speleol.,* 16: 1–16.

Howarth, F.G., 1993. High-stress subterranean habitats and evolutionary change in cave-inhabiting arthropods. *Am. Nat.,* Suppl. 142: S65–S77.

Hubbs, C.L. and Bailey, R.M., 1947. Blind catfishes from artesian waters of Texas. *Occas. Pap. Mus. Zool. Univ. Mich.,* 499: 1–15.

Hubbs, C.L. and Innes, W.T., 1936. The first known blind fish of the family Characidae: a new genus from Mexico. *Occas. Pap. Mus. Zool. Univ. Mich.,* 342: 1–7.

Humphreys, W.F. and Collis, G., 1990. Water loss and respiration of cave arthropods from Cape Range, Western Australia. *Comp. Biochem. Physiol.,* 95A: 101–107.

Hüppop, K., 1985. The role of metabolism in the evolution of cave animals. *Nat. Speleol. Soc. Bull.,* 47: 78–85.

Hüppop, K., 1986. Oxygen consumption of *Astyanax fasciatus* (Characidae, Pisces): a comparison of epigean and hypogean populations. *Environ. Biol. Fish.,* 17: 299–308.

Hüppop, K., 1987. Food-finding ability in cave fish (*Astyanax fasciatus*). *Int. J. Speleol.,* 16: 59–66.

Hüppop, K., 1988. *Phänomene und Bedeutung der Energieersparnis bei dem Höhlenfisch Astyanax fasciatus (Characidae).* Ph. D. Thesis, Univ. Hamburg, 142 pp.

Hüppop, K., 1989. Genetic analysis of oxygen consumption rate in cave and surface fish of *Astyanax fasciatus* (Characidae, Pisces) – Further support for the neutral mutation theory. *Mem. Biospeleol.,* 16: 163–168.

Hüppop, K., 1991. Muskulatur als Fettspeicher beim mexikanischen Höhlensalmler *Astyanax fasciatus. Verh. Dtsch. Zool. Ges.,* 84: 379–380.

Hüppop, K. and Wilkens, H., 1991. Bigger eggs in subterranean *Astyanax fasciatus* (Characidae, Pisces). *Z. Zool. Syst. Evolutionsforsch.,* 29: 280–288.

Husson, R., 1959. Les crustacés péricarides des eaux souterraines: considérations sur la biologie de ces cavernicoles. *Bull. Soc. Zool. France,* 84: 219–231.

Iles, T.D. and Wood, R.J., 1965. The fat/water relationship in North Sea herring (*Clupea harengus*), and its possible significance. *J. Mar. Biol. Ass. U.K.,* 45: 353–366.

Izquierdo, I., Oromi, P. and Bellés, X., 1990. Number of ovarioles and degree of dependence with respect to the underground environment in the Canarian species of the genus *Loboptera* Brunner (Blattaria, Blattellidae). *Mem. Biospeleol.,* 17: 107–111.

Jeannel, R., 1929. Le vivarium du jardin des plantes pendant l'année 1928. *Rev. Hist. Nat.,* 10: 73–92.

Jeannel, R., 1943. *Les Fossiles des Vivants des Cavernes.* Avenir de la Science, N.S. I. Gallimard, Paris, 321 pp.

Jegla, T.C., 1964. *Studies of the Eyestalk, Metabolism, and Molting and Reproductive Cycles in a Cave Crayfish.* Ph. D. Thesis, Univ. Illinois, 137 pp.

Jegla, T.C. and Poulson, T.L., 1970. Circannian rythms. I. Reproduction in the cave crayfish, *Orconectes pellucidus inermis. Comp. Biochem. Physiol.,* 33: 347–355.

Jones, R. and Culver, D.C., 1989. Evidence for selection on sensory structures in a cave population of *Gammarus minus* Say (Amphipoda). *Evolution,* 43: 688–693.

Jones, R., Culver, D.C. and Kane, T.C., 1992. Are parallel morphologies of cave organisms the result of similar selection pressures? *Evolution,* 46: 353–365.

Junge, P., 1993. Egg size and larval development of epigean and cave forms of *Rhamdia guatemalensis* (Pimelodidae, Pisces). In: J.H. Schröder, J. Bauer and M. Schartl (Editors), *Trends in Ichthyology. An International Perspective.* Blackwell Sci. Publ., Oxford, pp. 165–173.

Junk, W.J., 1985. Temporary fat storage, an adaptation of some fish species related to the waterlevel fluctuations and related environmental changes in the Amazon river. *Amazoniana,* 9: 315–351.

Kammerer, P., 1912. Experimente über Fortpflanzung, Farbe, Augen und Körperreduktion bei *Proteus anguinus* Laur. *Arch. Entwicklungsmech.,* 33: 349–361 [cited by Vandel and Bouillon (1959)].

Kane, T.C., 1981. Demographic characteristics of carabid cave beetles. In: B.F. Beck (Editor), *Proc. 8th Int. Congr. Speleol., Bowling Green,* 451–452.

Kane, T.C. and Poulson, T.L., 1976. Foraging by cave beetles: spatial and temporal heterogeneity of prey. *Ecology,* 57: 793–800.

Kane, T.C. and Richardson, R.C., 1985. Regressive evolution: An historical perspective. *Nat. Speleol. Soc. Bull.,* 47: 71–77.

Kane, T.C., Norton, R.M. and Poulson, T.L., 1975. The ecology of a predacous troglobitic beetle, *Neaphaenops tellkampfii* (Coleoptera: Carabidae, Trechinae). I. Seasonality of food input and early life history stages. *Int. J. Speleol.,* 7: 45–54.

Kayser, C. and Heusner, A., 1964. Etude comparative du métabolisme énergétique dans la série animale. *J. Physiol. Paris,* 56: 489–524.

Kersten, M. and Piersma, T., 1987. High levels of energy expenditure in shorebirds; metabolic adaptations to an energetically expensive way of life. *Ardea,* 75: 175–187.

Klimpel, B. and Parzefall, J., 1990. Comparative study of predatory behaviour in cave and river populations of *Astyanax fasciatus* (Characidae, Pisces). *Mem. Biospeleol.,* 17: 27–30.

Korzhuev, P.A., 1950. Consumption of oxygen by erythrocyts of some amphibia (*Molge cristata* and *Proteus anguinus*). *C. R. Acad. Dokl. Sci. URSS.,* 72: 979–980 (in Russian).

Langecker, T.G. and Longley, G., 1993. Morphological adaptations of the Texas blind catfishes *Trogloglanis pattersoni* and *Satan eurystomus* (Siluriformes: Ictaluridae) to their underground environment. *Copeia,* 976–986.

Love, R.M., 1970. *The Chemical Biology of Fishes.* Academic Press, New York, 547 pp.

MacArthur, R.H. and Wilson, E.O., 1967. *The Theory of Island Biogeography.* Princeton Univ. Press, Princeton, New Jersey, 203 pp.

Mathieu, J., 1983. Activité locomotrice de *Niphargus rhenorhoda-*

nensis en fonction de différentes condition expérimentales. *Mem. Biospeleol.,* 10: 401–405.

Mathieu, J. and Gibert, J., 1980. Evolution des teneurs en protéines, glucides et lipides de *Niphargus rhenorhodanensis* Schellenberg comparé entre l'évelage en milieu naturel reconstitué et le jeûne expérimental. *Crustaceana,* Suppl. 6: 126–136.

Mathieu, J. and Taveau, M., 1984. Variation de la composition biochimique de *Niphargus rhenorhodanensis* (amphipode hypogé) provenant du domaine naturel ou maintenus en élevage. *Bull. Soc. Zool. France,* 109: 211–222.

Mathieu, J. and Turquin, M.J., 1992. Biological processes at the population level. II. Aquatic populations. *Niphargus* (stygobiont amphipod) case. In: A.I. Camacho (Editor), *The Natural History of Biospeleology.* Monografias. Mus. Nat. Cien. Nat., CSIC Madrid, pp. 263–293.

Miller, R.R., 1984. *Rhamdia reddelli*, new species, the first blind pimelodid catfish from Middle America, with a key to the Mexican species. *Trans. San Diego Soc. Nat. Hist.,* 20: 135–144.

Mitchell, R.W., 1969. A comparison of temperate and tropical cave communities. *Southwest. Nat.,* 14: 73–88.

Mitchell, R.W., 1971. Food and feeding habits of the troglobitic carabid beetle *Rhadine subterranea. Int. J. Speleol.,* 3: 349–370.

Mitchell, R.W., Russell, W.H. and Elliott, W.R., 1977. Mexican eyeless characin fishes, genus *Astyanax*: environment, distribution and evolution. *Spec. Publ. Mus. Texas Tech. Univ.,* 12: 1–89.

Mohr, C.E. and Poulson, T.L., 1966. *The Life of the Cave.* McCrow-Hill Book Comp. and The World Book Encyclopedia, New York, Toronto, London, 232 pp.

Mösslacher, F. and Creuzé des Châtelliers, M., 1996. Physiological and behavioural adaptations of an epigean and a hypogean dwelling population of *Asellus aquaticus* (L.) (Crustacea, Isopoda). *Arch. Hydrobiol.,* 138: 187–198.

Nomura, H., 1975. Alimentacao de tres especies de peixes do genero *Astyanax* (Osteichthyes, Characidae) do Rio Mogi Guacu, S.P. *Rev. Bras. Biol.,* 35: 595–614.

Norton, R.M., Kane, T.C. and Poulson, T.L., 1975. The ecology of a predaceous troglobitic beetle, *Neaphaenops tellkampfii* (Coleoptera: Carabidae, Trechinae). II. Adult seasonality, feeding and recruitment. *Int. J. Speleol.,* 7: 55–64.

Olivereau, M. and Francotte-Henry, M., 1956. Etude histologique et biométrique de la glande thyroide de *Caecobarbus geertsi. Ann. Soc. R. Zool. Belg.,* 86: 129–150.

Pandey, H.S. and Singh, R.P., 1984. Lipid metabolism under the influence of starvation in khostifish, *Colisa fasciatus* (Bl. and Schn.). *Zool. Jahrb.,* 88: 49–58.

Park, O., 1951. Cavernicolous pselaphid beetles of Alabama and Tennessee, with observations on the taxonomy of the family. *Geol. Surv. Alabama Mus. Pap.,* 31.

Parzefall, J., 1982. Sexual and aggressive behaviour in cave animals. In: D. Mossakowski and G. Roth (Editors), *Environmental Adaptation and Evolution.* Gustav Fischer Verlag, Stuttgart, New York, pp. 179–195.

Parzefall, J., 1983. Field observation in epigean and cave populations of the Mexican characid *Astyanax mexicanus* (Pisces, Characidae). *Mem. Biospeleol.,* 10: 171–176.

Parzefall, J., 1985. On the heredity of behavior patterns in cave animals and their epigean relatives. *Natl. Speleol. Soc. Bull.,* 47: 128–135.

Parzefall, J., 1992. Behavioural aspects in animals living in caves. In: A.I. Camacho (Editor), *The Natural History of Biospeleology.* Monografias. Mus. Nac. Cien. Nat., CSIC Madrid, pp. 325–376.

Peck, S.B., 1973a. A systematic revision and the evolutionary biology of the *Ptomaphagus* (*Adelops*) beetles of North America (Coleoptera; Leiodidae; Catopinae) with emphasis on cave-inhabiting species. *Bull. Mus. Comp. Zool. Harv. Univ.,* 145: 29–162.

Peck, S.B., 1973b. Feeding efficiency in the cave salamander *Haideotriton wallacei. Int. J. Speleol.,* 5: 15–19.

Peck, S.B., 1975. The life cycle of a Kentucky cave beetle, *Ptomaphagus hirtus* (Coleoptera; Leiodidae; Catopinae). *Int. J. Speleol.,* 7: 7–17.

Peck, S.B., 1976. The effect of cave entrances on the distribution of cave-inhabiting terrestrial arthropods. *Int. J. Speleol.,* 8: 309–321.

Peck, S.B., 1986. Evolution of adult morphology and life-history characters in cavernicolous *Ptomaphagus* beetles. *Evolution,* 40: 1021–1030.

Peck, S.B. and Richardson, B.L., 1976. Feeding ecology of the salamander *Eurycea lucifuga* in the entrance, twilight, and dark zones of caves. *Ann. Speleol.,* 31: 175–182.

Peters, N., Scholl, A. and Wilkens, H., 1975. Der Micos-Fisch, Höhlenfisch in statu nascendi oder Bastard? Ein Beitrag zur Evolution der Höhlentiere. *Z. Zool. Syst. Evolutionsforsch.,* 13: 110–124.

Peters, R.H., 1983. *The Ecological Implications of Body Size.* Cambridge Stud. Ecol., Cambridge Univ. Press, U.S.A., 329 pp.

Phillipson, J., 1981. Bioenergetic options and phylogeny. In: C.R. Townsend and P. Calow (Editors), *Physiological Ecology: an Evolutionary Approach to Resource Use.* Blackwell Sci., Oxford, pp. 20–45.

Pough, F.H., 1980. The advantages of ectothermy for tetrapods. *Am. Nat.,* 115: 92–112.

Poulson, T.L., 1961. *Cave adaptation in amblyopsid fishes.* Ph.D. Thesis, Univ. Michigan, Microf. 61–2787.

Poulson, T.L., 1963. Cave adaptation in amblyopsid fishes. *Am. Midl. Nat.,* 70: 257–290.

Poulson, T.L., 1964. Animals in aquatic environments: animals in caves. *Am. Physiol. Soc.,* 47: 749–771.

Poulson, T.L., 1969. Population size, density and regulation in cave fishes. *Proc. 4th Int. Congr. Speleol., Ljubljana,* 1965: 189–192.

Poulson, T.L., 1971. Biology of cave and deep sea organisms: a comparison. *Natl. Speleol. Soc. Bull.,* 33: 51–61.

Poulson, T.L., 1972. Bat guano ecosystems. *Bull. Natl. Speleol. Soc.,* 34: 55–59.

Poulson, T.L., 1975. Symposium on life histories of cave beetles: an introduction. *Int. J. Speleol.,* 7: 1–5.

Poulson, T.L., 1981. Variations in life history of linyphiid cave spiders. In: B.F. Beck (Editor), *Proc. 8th Int. Congr. Speleol., Bowling Green,* 60–62.

Poulson, T.L., 1985. Evolutionary reduction by neutral mutations: plausibility arguments and data from amblyopsid fishes and linyphiid spiders. *Natl. Speleol. Soc. Bull.,* 47: 109–117.

Poulson, T.L. and Kane, T.C., 1981. How food type determines community organization in caves. In: B.F. Beck (Editor), *Proc. 8th Int. Congr. Speleol., Bowling Green,* 56–59.

Poulson, T.L. and Smith, M., 1969. The basis for seasonal growth and reproduction in aquatic cave organisms. *Proc. 4th Int. Congr. Speleol., Ljubljana,* 197–201.

Poulson, T.L. and White, W.B., 1969. The cave environment. *Science,* 165: 971–981.

Rasquin, P., 1947. Progressive pigmentary regression in fishes associated with cave environments. *Zoologica N. Y.,* 32: 35–42.

Rasquin, P., 1949. The influence of light and darkness on thyreoid and pituitary activity of the characin *Astyanax mexicanus* and its cave derivates. *Bull. Am. Mus. Nat. Hist.,* 93: 522–531.

Rose, F.L. and Mitchell, R.W., 1982. Comparative lipid values of epigean and cave-adapted *Astyanax. Southwest. Nat.,* 27: 357–358.

Rouch, R., 1968. Contribution à la connaissance des harpacticides hypogés (crustacés - copepodes). *Ann. Speleol.,* 23: 5–167.

Rusdea, E., 1994. Stabilization of a *Laemostenus schreibersi* population (Carabidae) as an effect of adult longevity. *XI Int. Symp. Biospeleol., Firenze,* Abstract 8–9. Société de Biospéologie.

Sarbu, S.M. and Popa, R., 1992. A unique chemoautotrophically based cave ecosystem. In: A.I. Camacho (Editor), *The Natural History of Biospeleology.* Monografias Mus. Nac. Cien. Nat., CSIC Madrid, pp. 637–666.

Sbordoni, V., 1982. Advances in speciation of cave animals. *Progr. Clin. Biol. Res.,* 96: 219–240.

Schemmel, C., 1967. Vergleichende Untersuchungen an den Hautsinnesorganen ober- und unterirdisch lebender *Astyanax*-Formen. *Z. Morphol. Tiere,* 61: 255–316.

Schemmel, C., 1980. Studies on the genetics of feeding behaviour in the cave fish *Astyanax mexicanus* f. *anoptichthys*. An example of apparent monofactorial inheritance by polygenes. *Z. Tierpsychol.,* 53: 9–22.

Schlagel, S.R. and Breder Jr., C.M., 1947. A study of the oxygen consumption of blind and eyed cave characins in light and darkness. *Zoologica N. Y.,* 32: 17–27.

Schmidt-Nielsen, K., 1984. *Scaling: Why is Animal Size so Important?* Cambridge Univ. Press, Cambridge, 241 pp.

Sibly, R. and Calow, P., 1985. Classification of habitats by selection pressures; a synthesis of life-cycle and r/K theory. In: R.M. Sibly and R.H. Smith (Editors), *Behavioral Ecology.* Blackwell Sci. Publ., Oxford, pp. 75–90.

Sket, B., 1985. Why all cave animals do not look alike – a discussion on adaptive value of reduction processes. *Natl. Speleol. Soc. Bull.,* 47: 78–85.

Strambi, C., 1970. Etude histologique et histochimique du tissu adipeux imaginal chez quelques coléoptères troglophiles et troglobiés de la famille des Catopidae. *Arch. Biol.,* 8: 309–360.

Studier, E.H., Lavoie, K.H., Wares II, W.D. and Linn, J.A.-M., 1986. Bioenergetics of the cave cricket, *Hadenoecus subterraneus. Comp. Biochem. Physiol.,* 84A: 431–436.

Thibaud, J.M., 1977. Relationship between size, biomass, water and fat content in two species of collembola according to their distribution in the cave of Sainte Cathérine, district of Ariège, South of France. *Proc. 7th Int. Congr. Speleol., Sheffield,* 1977: 393–394.

Thiel, H., 1975. The size structure of the deep-sea benthos. *Int. Rev. Ges. Hydrobiol.,* 60: 575–606.

Thinès, G.L., 1969. *L'Evolution Régressive des Poissons Cavernicoles et Abyssaux.* Masson et Cie, Paris, pp. 394.

Thinès, G.L., 1979. Vergleich der Regressionsphänomene bei Höhlenfishen und bei Tiefseefishen. *Mitt. Hamb. Zool. Inst. Mus.,* 76: 493–496.

Thinès, G.L. and Piquemal, M., 1978. Observations sur le comportement de *Lucifuga subterranea* Poey (Pisces, Ophidiidae), poisson cavernicole de Cuba. *Int. J. Speleol.,* 10: 195–203.

Trajano, E., 1991. Population ecology of *Pimelodella kronei,* troglobitic catfish from Southeastern Brazil (Siluriformes, Pimelodidae). *Environ. Biol. Fish.,* 30: 407–421.

Trajano, E. and Gnaspini-Netto, P., 1991. Notes on the food webs in caves of southeastern Brazil. *Mem. Biospeol.,* 18: 75–79.

Trimmel, H., 1968. *Höhlenkunde.* Friedrich Vieweg Verlag, Braunschweig, 300 pp.

Troiani, D., 1954. La consommation d'oxygène de quelques gammaridae. *C. R. Acad. Sci.,* 239: 1540–1542.

Uiblein, F., 1992. Prey choice behaviour in light and darkness in a facultative cave dweller, the Pyrenean salamander *Euproctus asper. Alytes,* 10: 131–136.

Uiblein, F. and Parzefall, J., 1993. Does the cave salamander *Proteus anguinus* detect mobile prey by mechanical cues? *Mem. Biospeleol.,* 20: 261–264.

Uiblein, F., Durand, J.P., Juberthie, C. and Parzefall, J., 1992. Predation in caves: the effects of prey immobility and darkness on the foraging behaviour of two salamanders, *Euproctus asper* and *Proteus anguinus. Behav. Process.,* 28: 33–40.

Vandel, A., 1950. Le laboratoire souterrain de C.N.R.S. *Bull. Soc. Zool. France,* 75: 151–158.

Vandel, A., 1964. *Biospéologie: La Biologie des Animaux Cavernicoles.* Gauthiers-Villars Editeurs, Paris, 619 pp.

Vandel, A. and Bouillon, M., 1959. Le protée et son interêt biologique. *Ann. Speleol.,* 14: 111–127.

Verrier, M.L., 1929. Observations sur le comportement d'un poisson cavernicole: *Typhlichthys osbornii* Eigenmann. *Bull. Mus. Hist. Nat. Ser. 2,* 1: 82–84.

Vogt, G. and Strus, J., 1992. Oleospheres of the cave-dwelling shrimp *Troglocaris schmidtii*: a unique mode of extracellular lipid storage. *J. Morphol.,* 211: 31–39.

Wake, D.B., 1966. Comparative osteology and evolution of lungless salamanders, family Plethodontidae. *Mem. South. Calif. Acad. Sci.,* 4: 1–111.

Walters, L.H. and Walters, V., 1965. Laboratory observations on a cavernicolous poeciliid from Tabasco, Mexico. *Copeia,* 214–223.

Wautier, J. and Troiani, D., 1960. Contribution à l'étude du metabolisme de quelques gammaridae. *Ann. Stn. Cent. Hydrobiol. Appl.,* 8: 7–50.

Weingartner, D.L., 1977. *Production and Trophic Ecology of Two Crayfish Species Cohabiting an Indiana Cave.* Ph.D. Thesis, Michigan State Univ., 348 pp.

Wilkens, H., 1971. Genetic interpretation of regressive ecolutionary processes: studies on hybrid eyes of two *Astyanax* cave populations (Characidae, Pisces). *Evolution,* 25: 530–544.

Wilkens, H., 1976. Genotypic and phenotypic variability in cave animals. Studies on a phylogenetically young population of *Astyanax mexicanus* (Filippi) (Characidae, Pisces). *Ann. Speleol.,* 31: 137–148.

Wilkens, H., 1987. Genetic analysis of evolutionary processes. *Int. J. Speleol.,* 16: 33–58.

Wilkens, H., 1988. Evolution and genetics of epigean and cave *Astyanax fasciatus* (Characidae, Pisces) – support for the neutral mutation theory. In: M.K. Hecht and B. Wallace (Editors), *Evolutionary Biology, Vol. 23.* Plenum Publ. Corp., New York, pp. 271–367.

Wilkens, H., 1992. Neutral mutations and evolutionary progress.

In: A.I. Camacho (Editor), *The Natural History of Biospeleology.* Monografias Mus. Nac. Cien. Nat., CSIC Madrid, pp. 401–421.

Wilkens, H., 1993. A new species of *Rhamdia* (Pisces: Pimelodidae) from a cave in the Sierra de Zongolica (Veracruz, Mexico). *Mitt. Hamb. Zool. Mus. Inst.,* 90: 375–378.

Wilkens, H. and Burns, R.J., 1972. A new *Anoptichthys* cave population (Characidae, Pisces). *Ann. Speleol.,* 27: 263–270.

Wilkens, H. and Hüppop, K., 1986. Sympatric speciation in cave fishes? Studies on a mixed population of epi- and hypogean *Astyanax* (Characidae, Pisces). *Z. Zool. Syst. Evolutionsforsch.,* 24: 223–230.

Wilkens, H., Peters, N. and Schemmel, C., 1979. Gesetzmäßigkeiten der regressiven Evolution. *Verh. Dtsch. Zool. Ges.,* 72: 123–140.

Withers, P.C., 1992. *Comparative Animal Physiology.* Saunders Coll. Publ., Fort Worth, 949 pp.

Wright, S., 1964. Pleiotropy in the evolution of structural reduction and of dominance. *Am. Nat.,* 98: 65–70.

Yager, J., 1991. The Remipedia (Crustacea): Recent investigations of their biology and phylogeny. *Verh. Dtsch. Zool. Ges.,* 84: 261–269.

Zeuthen, E., 1953. Oxygen uptake as related to body size in organisms. *Q. Rev. Biol.,* 28: 1–12.

Chapter 9

ADAPTATIONS TO INTERSTITIAL GROUNDWATER LIFE

Nicole COINEAU

INTRODUCTION

Inland subterranean interstitial organisms live in the groundwater filling the interstices between the grains of the aquifer sediments and alluvia which lie along and below rivers and streams in their floodplains, in lakes, and in granular cave sediments. Pebbles, gravel, sand and clays of these deposits constitute an heterogenous granular, unconsolidated and porous medium with interconnected interstices (Zilliox, 1994). Based on recent ecological research, the interstitial milieu, which was previously considered as homogeneous, has to be viewed as a mosaic of microenvironments (Danielopol and Rouch, 1991) inhabited by different groups of species with spatial gradients (Dole, 1983; Rouch, 1988; Marmonier and Dole, 1986; Danielopol, 1991; Camacho et al., 1992; Dole-Olivier et al., 1993; Marmonier et al., 1993; Danielopol et al., 1996), and vertical, horizontal and temporal heterogeneity (Gibert et al., 1994a,b; Creuzé des Châtelliers et al., 1994; Dole-Olivier et al., 1994).

The main abiotic factors characterizing the porous aquifers are:

- the absence of light;
- limited energetic sources (oxygen and trophic supply), as in caves;
- a relative stability increasing with depth (temperature, water flow, physical and chemical variables), as compared to surface water with seasonal cycles and daily fluctuations;
- and the grain size of the substrate.

Mainly on account of the last parameter, the interstitial milieu is more constrained than cave and karstic environments, because grain size affects the permeability and porosity, the distribution of other variables (dissolved oxygen, organic matter, etc.) and the habitat space available for stygobites. Furthermore, the effect of depth has also to be considered. Nevertheless, a large number of the major invertebrate groups inhabit the porous aquifers (Delamare-Deboutteville, 1960; Vandel, 1964; Ginet and Decou, 1977; Botosaneanu, 1986; Camacho et al., 1992; Juberthie and Decou, 1994; Gibert et al., 1994a,b). Thus, one of the most important topics in interstitial biology, ecology and evolution is to understand why and how so many organisms have adapted to such a severe environment.

As troglobionts, the interstitial organisms display a set of morphological, physiological and ethological characteristics, some of which are strictly related to the porous-habitat life-style. For over 100 years, adaptation of organisms to cave life has received much attention (Packard, 1888; Racovitza, 1907; Eigenmann, 1909; Jeannel, 1923, 1926, 1943; Fage, 1931; Daum, 1954; Poulson, 1963; Vandel, 1964; Barr Jr, 1968; Culver, 1982, 1985; Sket, 1985; Barr Jr and Holsinger, 1985; Culver and Fong, 1986; Holsinger, 1988; Jones and Culver, 1989; Culver et al., 1990, 1995; Jones et al., 1992; Camacho, 1992; Christiansen, 1992; Wilkens, 1992; Kane and Culver, 1992; Howarth, 1993). In contrast, adaptation to interstitial life is sparsely documented, probably because of the small size of the fauna (Remane, 1956 for marine interstitial; Delamare-Deboutteville, 1960; Rouch, 1968; Magniez, 1976, 1985; Henry, 1976; Schminke, 1981; Coineau, 1971, 1984; Marmonier and Danielopol, 1988; Danielopol and Rouch, 1991; Camacho et al., 1992; Coineau and Boutin, 1992; Rogulj et al., 1993; Danielopol et al., 1994, 1996).

As underlined by Danielopol and Rouch (1991), most of these works refer mainly to descriptive traits of interstitial organisms which reflect their adaptation to the environment. Evolutionary processes and mechanisms are rarely considered.

"A une même époque, sous un même mot, il y a des

concepts si différents! Ce qui nous trompe, c'est que le même mot à la fois désigne et explique [There are such different concepts implied by the same word, in the same period! What deceives us is that the same word is used simultaneously both to designate and to explain]" (Bachelard, 1938). Thus, the word "adaptation" has at least two facets: it covers the adapted state of a trait, or the process leading to this state and to the fitness of the organism having the trait (Génermont and Lamotte, 1986; Brandon, 1990; Lamotte, 1994). For Stearns (1992), adaptation is a modification of the phenotype in response to an environment signal. In discussing adaptation, I shall here deal with the "product of the process of evolution by natural selection", as Brandon (1990) puts it; both descriptive features (form, behavior and function) and mechanisms and causal history will be addressed as far as possible. However, it is important to remember that "how organisms function, behave or evolve new structures is dependent on morphology" (Nitecki, 1990).

Many inland interstitial organisms are derived from surface marine ancestors. They invaded the subterranean sediments according to the "two step model of colonization and evolution" (Coineau and Boutin (1992), first proposed by Boutin and Coineau (1990) [see also Coineau (1985) and Notenboom (1991). Stock (1980) had earlier outlined the second step of the model (Botosaneanu and Holsinger, 1991). Holsinger (1994a,b) proposed a similar step-wise model, and indeed distinguished three steps]. The surface epibenthic or meroplanktonic ancestors of the thalassoid forms actively colonized the marine littoral sands during the first stage (i.e., vertical transition) and settled passively thereafter into interstitial freshwater during marine regressions (i.e., second stage); the ancestors of the limnicoid forms dispersed first in surface fresh waters, and then actively colonized the interstitial waters (i.e., vertical transition). During these stages, the evolutionary change of the organisms is marked by different steps from the surface to the interstitial environment, and involve different processes. Only generalist and unspecialized euryece forms (with a wide range of ecological distribution) showing morphological, physiological and behavioral preadaptations (Cuénot, 1909; Barr Jr, 1968), exaptations or nonaptations (Gould and Vrba, 1982) have been successful in the ecological change of environment involved in vertical colonization [Step B in Holsinger's (1994) model; (Rouch and Danielopol, 1987; Danielopol and Rouch, 1991; Coineau and Boutin, 1992)]. Some preadaptations such as small size, benthic reproduction without pelagic larvae, or unspecialized feeding may sometimes be an advantage in integrating into the new granular biotope.

Settlement in the interstitial system is correlated with processes of adaptation termed "stygobitization" (*cf*. troglobitization in caves) involving both regressive evolution and adaptive new features. Regressive evolution leads mainly to loss of eyes and pigment in response to darkness. Small size, elongated shape, elongated appendages and sensorial structures, specialized structures, slow rates of reproduction and metabolism are evolutionary novelties (Coineau, 1985; Notenboom, 1991; Coineau and Boutin, 1992). Many hypotheses have been proposed to explain the evolutionary mechanisms leading to regressive and to elaborated characters of cave organisms (reviews in Culver, 1982; Kane and Richardson, 1985; Culver et al., 1995). Poulson (1985) and Wilkens (1988, 1992) have explained regressive features by accumulation of neutral mutations, whereas elaborated structures result from selection. Others have demonstrated, in the amphipod *Gammarus minus* model, that both reduced eye and pigment and elaborated structures (extraoptic sensory structures such as large antennae) and large body size result from directional selection and neurological trade-off (Jernigan et al., 1994; Culver et al., 1994, 1995; Fong et al., 1995).

The aim of this chapter is broadly to address patterns of organism adaptation to the constraints of interstitial aquatic life, as well as the evolutionary mechanisms involved. The latter are discussed and compared, as far as possible, to evolution of cave organisms. The chapter focuses attention on crustaceans which are diversified within interstitial waters, referring mainly to thalassoid syncarids, isopods and amphipods. Discussion is limited to strict interstitial phreatobites, excluding burrowers. Both traits and whole organisms are considered, as recommended by Gould and Lewontin (1982).

PHREATOMORPHIC TRAITS AND EVOLUTIONARY PROCESSES

After Racovitza (Racovitza, 1907), who listed a set of characteristic traits of cave organisms, Christiansen (1962, 1992) coined the term "troglomorphic" to cover the convergent morphological, behavioral and physiological features of troglobites. The equivalent term "phreatomorphic" should be applied to phreatobites. Since deep-sea organisms also share these convergent pecularities, Danielopol et al. (1996) have proposed a more general concept, the darkness syndrome.

With regard to the topic of adaptation to abiotic characteristics of the aquatic subterranean milieu, Danielopol and Rouch (1991) and Coineau and Boutin (1992) stressed the supposedly adaptive features which appeared before settlement into the stygohabitat, that is, exaptations. Preadaptations and adaptive novelties are distinguished below. Exaptations may be advantageous and favour colonization of the sediment.

It has sometimes been reported that species belonging to the same genus, or to closely related genera, exhibit different features in caves and in the interstitial milieu. On the other hand, adaptive similarities have also been observed. Close attention is paid to these particular points.

RESPONSES TO DARKNESS

Lack of light is common to caves and interstitial environment. Like permanent karstic organisms, interstitial crustaceans are also blind and devoid of pigment. Their integument is thin and transparent. Most of these phreatomorphic traits appeared after or during the settlement into the stygohabitats (mainly reduction or loss of eyes and pigment, and increase of other compensatory receptor structures, generally slender appendages), and they are shared by both subterranean and deep-sea animals of many phylogenetic lines.

The monophyletic group of anaspidacean syncarids shows a remarkable evolutionary pattern from surface to interstitial forms: the large surface Anaspididae have compound stalked eyes and are pigmented; the burrowing Koonungidae display reduced sessile eyes and are depigmented, while the interstitial Psammaspididae and Stygocarididae are eyeless and without body pigment (Fig. 9.1). In respect of phylogenetic relationships, the group exhibiting the most plesiomorphic characters is the surface-dwelling oculated Anaspididae, whereas the eyeless interstitial Psammaspididae and Stygocarididae have numerous derived traits (Schminke, 1975; Schram, 1984; Coineau, 1996). Thus, there is a good correlation between the stages of eye reduction and the steps of transition from the less advanced surface groups to the most derived interstitial forms.

Regressive evolution and morphology of cave animals have been the subjects of a long debate concerning the alternative role of selection, energy economy, pleiotropy, neutral mutations and genetic drift (Barr Jr, 1968; Culver, 1982; Romero, 1985; Kane and Richardson, 1985; Sket, 1985; Christiansen, 1992). After the obsolete organicist theory (Vandel, 1964), two main schools emerged which still persist. In the neutralist school, based on the study of the fish families Amblyopsidae and Characidae, Wilkens (1988, 1992, 1993), Poulson (1985) and Parzefall (1992) have invoked accumulation of neutral mutations resulting in regressive evolution. For Wilkens, selection of constructive features is not correlated with eye loss, and stabilizing selection is reduced in the early stages of cave colonization. In Poulson and White's (1969) view, in contrast, elaboration of extraoptic sensory structures are considered to be the result of indirect selection under pressure of energy economy. On the other hand, explanations of reduction and loss of eyes and pigment, as well as increased body size and larger antennae, in the cave amphipod *Gammarus minus* involve directional selection (Culver et al., 1995). Regression of optic structures is coupled with improvement of constructive extraoptic equipment through neurological trade-offs within the central nervous system; the decrease in eye function is correlated with increased functions of other sensory organs, through compensatory innervations formerly used for optic purposes. Finally, Fong et al. (1995) concluded that "extra optic sensory structures increase in size as a result of directional selection" while "optic structures decrease in size as a result of directional selection and neutral mutation".

Multivariate visualization of natural selection, though powerful, provides only a description of the relation between phenotype and fitness (Brodie et al., 1995). Selection is statistically probable, however, it does not imply cause and effect.

In interstitial crustaceans, the situation is more complex due to the selective pressure of the porous medium, which results in both small size and elongation, especially for antennae (see discussion, pp. 198–200).

The hypothesis of the effects of accumulation of neutral mutations may provide a plausible scenario in regressive evolution of the eye and pigment in interstitial organisms, as in those of caves. Among the numerous mutations occurring in subterranean populations, only some of them result in reduction or loss of eyes; these are not deleted. They will be inherited, since there is no positive selection for

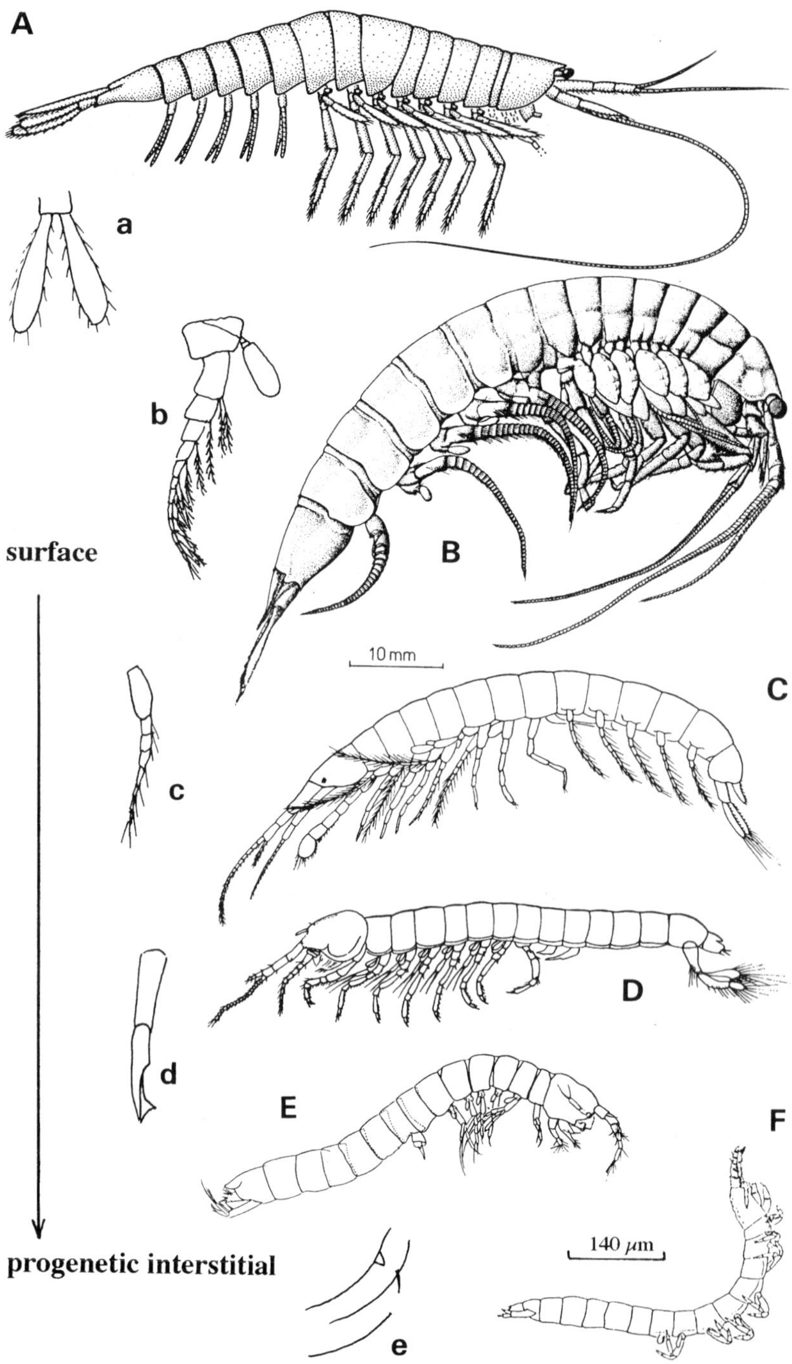

Fig. 9.1. Fossil and Present evolutionary series from surface to interstitial life in syncarids. (A) Fossil surface *Palaeocaris typus* (Carboniferous Palaeocaridacea) (total length: 1.8 cm). (B) Surface swimming *Anaspides tasmaniae* (Present Anaspidacea) (3–6 cm) with stalked eyes, pigmentation, full appendage set and a tailfan. (C) *Koonunga* (Anaspidacea) (7–9 mm); burrower, reduced sessile eyes, no pigmentation. (D) *Parastygocaris* (small interstitial anaspidacean); eyeless and depigmented. (E, F) Dwarfish interstitial progenetic bathynellaceans, *Hexabathynella knoepffleri* (1 mm) and *Acanthobathynella*; eyeless, small antennae, reduced appendage number. (a–e) Increasing reduction of the pleopod structure from surface syncarid to minute progenetic interstitial representatives. After Schram (1984) (A), Schminke (1978) (B), Noodt (1964) (C, D), Coineau (1965, 1967) (E, F).

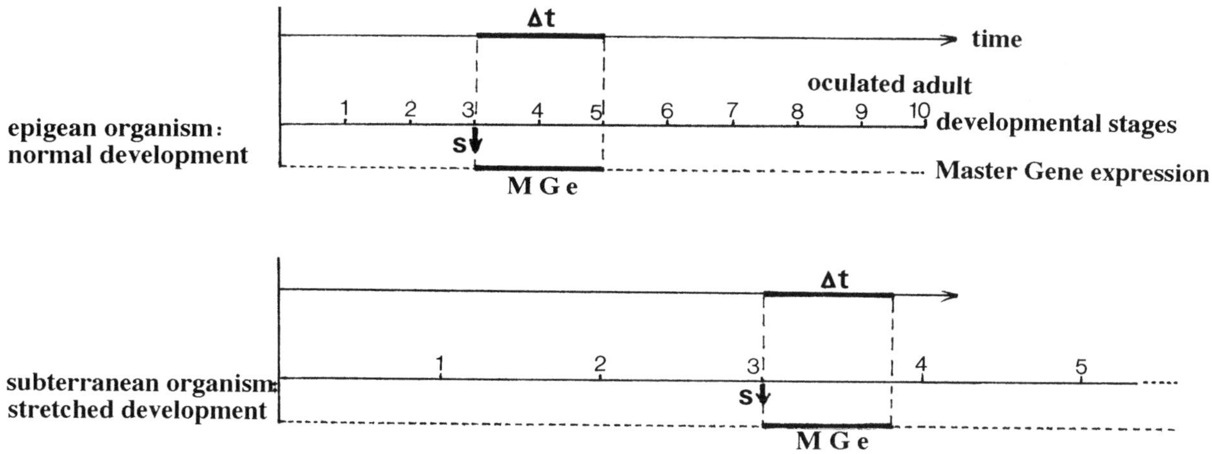

Fig. 9.2. Shift in time of the expression of the eyeless master gene expression (MG e) relative to the progress of the slower ontogenetic development in a subterranean organism, as compared with that of an epigean organism.

the maintenance of functional ocular structures in darkness. Ocular structures become gradually smaller and less numerous. Some authors (e.g., Barr Jr, 1968) have argued that the fixing of mutations may need more time than has been available for troglobitization of organisms which recently colonized lava tubes and oceanic islands. However, many interstitial organisms are very old. Their ancestors colonized the granular system at different geologic epochs from 4 up to 80 million years ago; this transition into sediments was probably a slow phenomenon, allowing enough time for fixation of the character of eye loss.

Following the studies of the *Pax-6* gene, another hypothesis may be considered. It is commonly assumed that the early developmental stages are governed by genes, especially homeotic genes. Mutational events occurring on these developmental genes result in different heterochronies (Gould, 1977). An eyeless master gene, occurring from arthropods to vertebrates and controlling the elaboration of the eye complex, has recently been highlighted (Quiring et al., 1994). Even if the functioning of this eyeless master control gene is not modified by subterranean evolution, the slower ontogenetic development of subterranean organisms suggests a hypothesis for reduction and loss of eyes. The expression program of this master gene might be triggered by an ontogenetic signal (e.g., Stage 3, Fig. 9.2) and work only during a set period (Δt) of the developmental sequence. Within an epigean organism, this period Δt covers perhaps Stages 3 to 5 – the complete development of the eye, which appears after the Stage 5. For a subterranean organism, the signal of the master gene expression acts later on, because Stages 1, 2, 3 are prolonged. At the end of Δt (=end of the expression of the master gene), the organism has hardly reached Stage 4; the eye is more or less reduced, not even visible from outside, and its morphogenesis stops. The adult is microphthalmic or eyeless (Fig. 9.2). Finally, for a subterranean organism with slow development, the shift in time of the expression of the master gene relative to the slower developmental stages might result in eyelessness as compared to a close surface relative with normal development.

For marine and freshwater interstitial ostracods, reduced ocular structures or eye loss are viewed, according to the group considered, either as adaptations (species group *arcturi* of the genus *Xestoleberis*), or as exaptations (e.g., Candonidae) (Danielopol et al., 1996).

RESPONSES TO THE PORE SYSTEM

Small size and miniaturization

The generally small dimensions of the intra-granular spaces in the sediments is a major constraint on morphology. It determines the size class of the interstitial fauna. Strictly interstitial organisms cannot be wider than the interstice diameter, and small body size is a characteristic trait of interstitial stygobites. Furthermore, there is often a strong correlation between the particle size and the animal size. Thus, many interstitial crustaceans (isopods, amphipods, anaspidacean

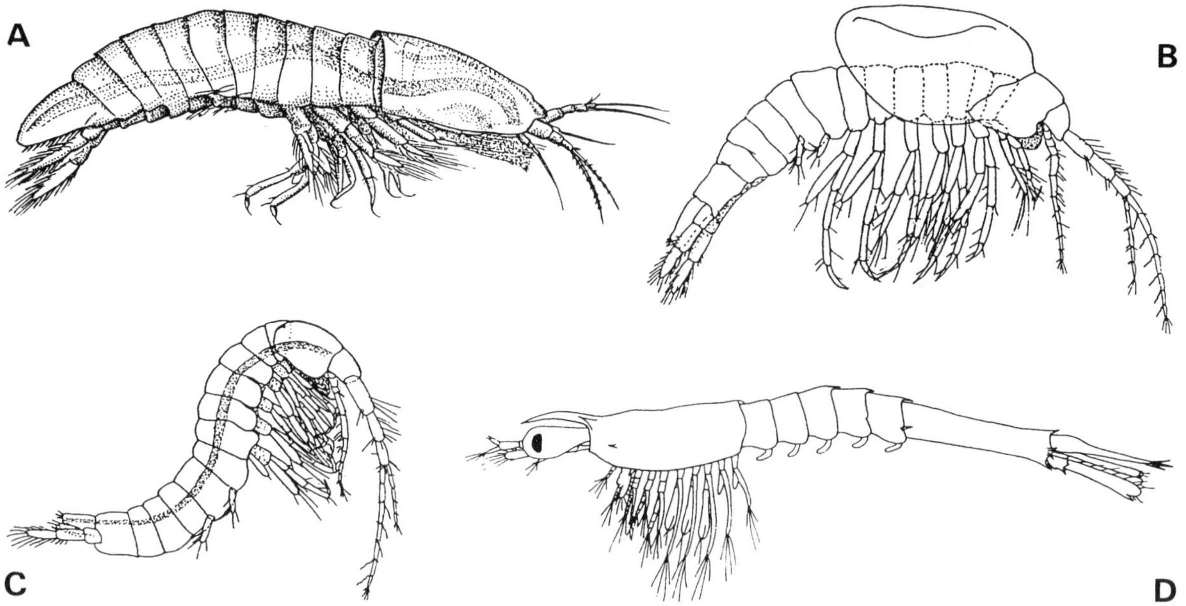

Fig. 9.3. Habitus of two genera of Thermosbaenacea, illustrating their zoea-like morphology: lateral views. (A) Adult *Thermosbaena mirabilis.* (B) Ovigerous female of *Tethysbaena* showing the dorsal brood-pouch. (C) *Tethysbaena.* (D) Larval stage of a decapod penaeid, *Penaeus duorarum,* zoea 2. After Monod in Delamare-Deboutteville (1960) (A), Stock (1986) (B, C), Dobkin (1961) (D).

syncarids, thermosbaenaceans) in coarse well-sorted sandy gravels are 1–3 mm long, whereas those occurring in fine sands (harpacticoid copepods, bathynellacean syncarids, ostracods) are less than 1 mm long. Others, such as the microcerberids, *Bogidiella* and *Typhlocirolana,* generally require mixed very coarse and fine elements; the tiny amphipod *Metacrangonyx knidirii,* which shows the most derived traits, has a narrower range of granulometry than *M. spinicaudatus* (Coineau, 1971; de Bovée et al., 1995).

Within monophyletic groups, such as syncarids, the smallest specialized species have fewer extremities than more primitive species; they lack segments or rami of normally biramous appendages and the telson. Spines and bristles are less numerous; many structures are simpler, and become simpler from surface-dwelling species to the subterranean representatives, which are the most derived (Schminke, 1981; Coineau, 1996) (Fig. 9.1). In several species of Parabathynellidae, the 7th pereiopod is absent, whereas the 8th pereiopod is present. The thermosbaenacean *Thermosbaena mirabilis* possesses only five pereiopods (instead of seven) and five free pleomeres (instead of six). The number of pleopods, oostegites, and articles in the antennae is complete neither in the Microparasellidae (Isopoda) nor in the amphipod families Bogidiellidae, Ingolfiellidae and Metacrangonyctidae. The Bogidiellidae and Ingolfiellidae also have reduced coxal plates, and a reduced number of spines and setation of appendages compared to surface relatives.

It is known that progenesis is one of the evolutionary processes responsible for miniaturization. According to Gould (1977), this type of heterochrony is defined as precocious sexual maturation of an organism resulting in an adult descendant exhibiting the larval or juvenile morphology of its ancestor. The repeated abbreviation of developmental sequences results in new small paedomorphic body forms and types (Raff et al., 1990). The progenetic nature of some interstitial groups has been demonstrated (marine psammodrilid and dorvilleid polychaetes, Lobatocerebridae, ascidians, freshwater syncarid and thermosbaenid crustaceans), and it has been considered as possible for others (copepods, opisthobranch molluscs, kinorhynchs) (Swedmark, 1964; Monniot, 1971; Rieger, 1980; Westheide, 1987; Schminke, 1981). Adult Syncarida (Schminke, 1981) and Thermosbaenacea (Coineau and Boutin, 1995) display structural as well as functional similarities shared with free-swimming larvae of primitive decapods, especially the protozoea and zoea (Fig. 9.3). The lack of cephalization of the first thoracic somite of bathynellaceans is also a retained larval condition. In these groups, development is interrupted before metamorphosis as compared with the primitive decapod sequences. Therefore, such small crustaceans, the development of which is today holobenthic, may have evolved from larger free-living epibenthic or merobenthic marine ancestors with swimming larvae. It can be assumed that, in order to get protection against predators or currents, the adult laid its eggs beneath the gravel or sand, just under the substrate surface, as *Anaspides* does today. The larvae of these ancestors actively colonized the interstitial of littoral sediments. Progenesis seems to be correlated with this vertical transition in marine sand – that is, during the first phase of the two-step model for colonization of inland interstitial by thalassoid stygobionts mentioned above (Boutin and Coineau, 1990; Notenboom, 1991; Coineau and Boutin, 1992). The resulting smaller progenetic forms reproduce without leaving the sediment, and with a shortened development may have a selective advantage within the sediment, producing a new, strictly interstitial, species. In this case, progenesis is an adaptive novelty selected by the interstitial milieu. It would be one of the processes leading from a large surface ancestor to a minute permanent interstitial stygobite (Fig. 9.4).

This is corroborated by the existence of extant syncarid Anaspidacea and Bathynellacea which exhibit a series of these adaptive stages (see Fig. 9.1): the Anaspididae swim and walk in open surface water of ponds, lakes and rivers in mountains, are 3–5 cm long, and have eyes, pigmentation and a full set of appendages. The Koonungidae burrow in muddy bottoms, are 7–9 mm long, lack pigment, and have small sessile eyes. The blind Psammaspididae walk within coarse sediments. Finally, the small Stygocarididae (1.4–4.0 mm) are truly blind and transparent members of the interstitial fauna. Interstitial Anaspidacea live only in coarse sediment. The minute Bathynellacea, the most fully adapted to interstitial life, occur only in fine substrates. At the same time, the body becomes smaller from Anaspididae to Parabathynellidae, and structures become increasingly simpler. The adult looks like the juveniles of the closest more primitive sister group (Schminke, 1975; Coineau, 1996). A similar but less complete and progressive transition can be observed in isopods, from the surface Janiridae with eyes, pigmentation and wide body segments, to the tiny eyeless Microparasellidae living in coarse and fine sediments (Fig. 9.5).

It has been suggested by Wägele (1983) that the derived isopod family Microcerberidae, which exhibits the juvenile structures of the Stenasellidae (primitive

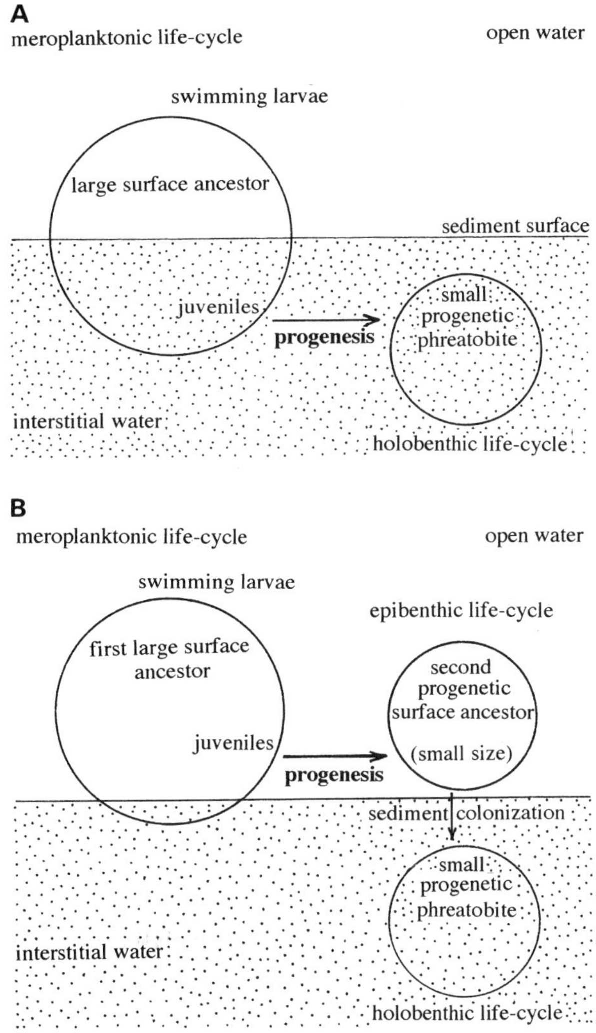

Fig. 9.4. (A) Progenesis as an adaptation resulting from selection by the interstitial milieu. The development of the surface ancestor included swimming larvae and interstitial juveniles. The development of the progenetic interstitial species is abbreviated and holobenthic, without swimming larvae. (B) Progenesis in surface species as an exaptation favouring the colonization of the interstitial habitat.

sister group), is progenetic. Henry (1976) reported also that some eyeless species of *Proasellus* retain a juvenile morphology in the adult. In the same way, the tiny amphipod family Ingolfiellidae, and some Bogidiellidae such as *Pseudingolfiella,* have paedomorphic pleopods and telson, and also have a progenetic origin (Fig. 9.6).

Paedomorphic limnicoid as well as thalassoid phreatobites may result from progenesis acting during colonization of the sediments and selected by this milieu (Fig. 9.4).

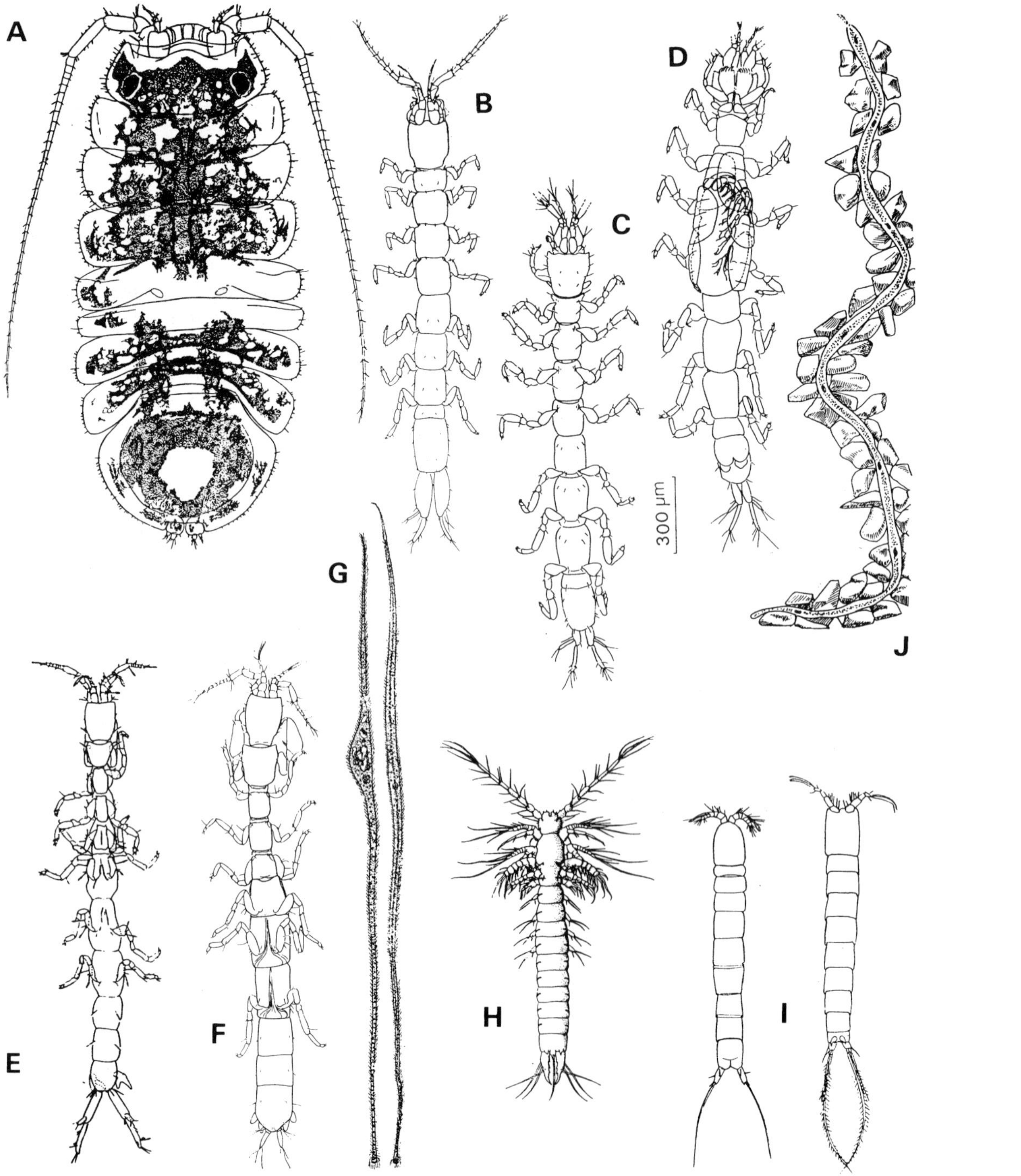

Fig. 9.5. Elongation of the body and small size in different interstitial groups. (A) *Jaera marina albifrons,* 5–8 mm pigmented surface isopod (Janiridae) with eyes, long antennae and short uropods. (B–J) Eyeless, slender and elongated typical interstitial representatives: (B) *Microcharon* sp., 1.2 mm, isopod; (C) *Angeliera phreaticola* (1.3 mm), isopod; (D) Female *A. phreaticola* (isopod) with one embryo; (E) *Protocerberus schminkei kruegeri* (isopod), 1.2 mm.; (F) *Yvesia striata* (isopod); (G) *Trachelocerca* sp., ciliate. (H) *Derocheilocaris remanei* (Mystacocarida) <0.7 mm. (I) Harpacticoid copepods. (J) *Marenda nematoides* (Foraminifera). After Bocquet (1953) (A), Coineau (1971) (B–D), Wägele (1983) (E), Coineau and Botosaneanu (1973) (F), Dragesco (1954) (G), Delamare-Deboutteville (1960) (H–J).

In other groups, the acquisition of a small size through progenesis is viewed as an exaptation (or a nonaptation: Gould and Vrba, 1982), having occurred in epibenthic or meroplanktonic surface populations (Fig. 9.4). The resulting small new species is preadapted to life in the substrates, and can actively invade the sediment. Small body size is an exaptation favouring vertical movement into the sediment. This case is documented in some ostracod crustaceans (Danielopol and Bonaduce, 1990; Marmonier and Danielopol, 1988; Rouch and Danielopol, 1987; Danielopol et al., 1994; Danielopol and Marmonier, 1994). The paedomorphic features in *Nannocandona faba* (Candoninae), such as small size, the rectangular valve, and the reduced number of antennula segments and setae, are in fact exaptations, as these characters may also be observed in epigean populations. Similar arguments have been proposed for the origin of some small epibenthic copepods (Newman, 1983; Boxshall, 1983; Rouch, 1994).

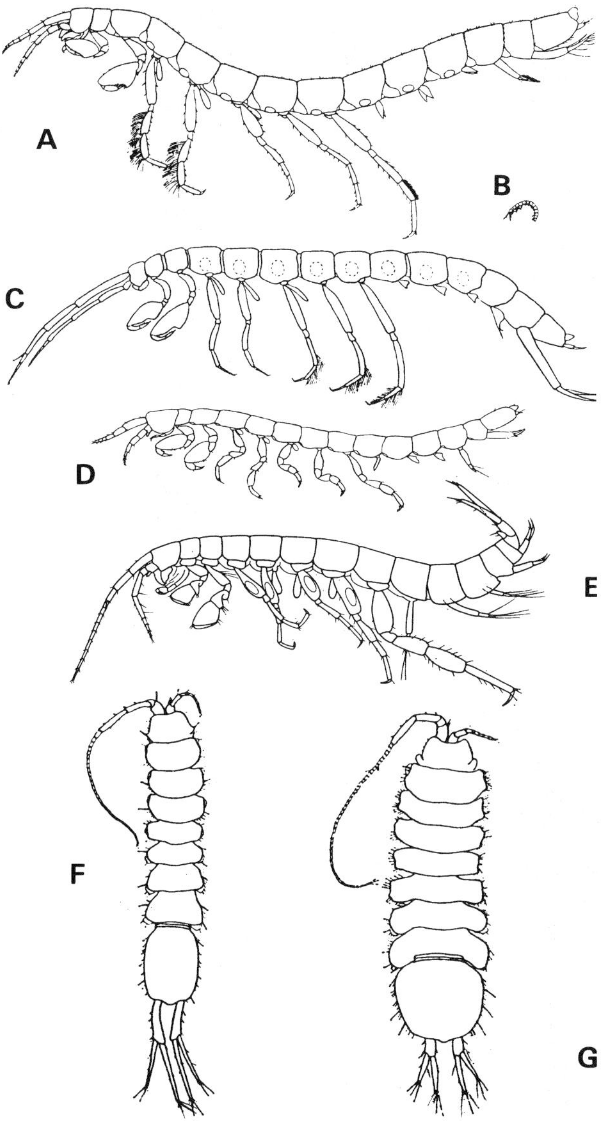

Fig. 9.6. Troglobitic and interstitial amphipods and isopods. (A) *Trogloleleupia leleupi* (cave ingolfiellid), 12–20 mm. (B) *Ingolfiella (Gevgeliella)* sp. (interstitial ingolfiellid) at the same scale. (C) *Trogloleleupia opisthodorus* (cave ingolfiellid), 24–28 mm, with long antennae and appendages. (D) *Ingolfiella (Gevgeliella) petkovskii,* progenetic interstitial Ingolfiellidae, 1 mm. (E) *Bogidiella (Orchestigidiella) orchestipes* (interstitial bogidiellid). (F) *Proasellus albigensis* (interstitial isopod: Asellidae). (G) *Proasellus vandeli* (cave species). After Ruffo (1951) (A), Barnard (1966) (B, C), Karaman (1959) (D), Ruffo and Vigna Taglianti (1977) (E), Henry and Magniez (1983) (F, G).

However, small size is not always the result of progenesis (Coineau and Boutin, 1992). In numerous interstitial groups, progenetic evolution either allowed colonization of the new milieu (exaptation) or more often acted within the interstitial milieu (selection) (Fig. 9.4). But exaptation may also be selected, and thus become an adaptation (Brandon, 1990).

In karst, where dimensional constraints do not exist in open water, close relatives of minute interstitial forms may exhibit a considerably larger range of body size – for instance, species of the amphipod *Trogloleleupia,* some bogidiellids, and isopods (*Proasellus*). Species of the genus *Trogloleleupia* in caves are 14–23 mm long, while the forms living in microporous substrates are small (1–3 mm) (Fig. 9.6). The proportions as well as the external organization of the body are identical; only details of the morphology appear at the species and generic level, and numerous setae for swimming can be seen on cave specimens. Elongation of the antennae can be observed in several cave species, whereas all interstitial ingolfiellids have very short antennae, pereiopods and uropods. Based on the clear progenetic nature of this worm-like amphipod, the primary adaptation was probably to interstitial life, and secondarily to karstic habitats (increased body size). Larger cave species are derived. Strictly interstitial forms such as bathynellids, *Microcharon* and bogidiellids often drift passively on subterranean currents from the interstitial milieu to karsts, and may live within the bottom of patchy lakes or along cave fissures. In the latter case, it is a temporary habitat, and subsequent floods will again drift these interstitial members towards springs. Selective forces are quite opposite in cave and interstitial habitats: larger and increased size is selected in caves whereas small size is selected in the interstitial milieu.

Elongation

Other common adaptive traits involved in phreatomorphology and related to the porous habitat are larger body length, and lengthening of appendages and sense organs, when compared to open-water relatives. Thread-like species are reported in different taxonomic groups (Delamare-Deboutteville, 1960; Coineau, 1984; Notenboom, 1991; Coineau and Boutin, 1992). In the isopod *Microcharon,* the elongation of the body, despite the reduced number of abdominal segments, is enhanced by the parallel sides and the very long uropods. In the freshwater species the pereiopods are very slender with long claws (Figs. 9.5, 9.7); the general aspect of the stygobiont is quite different from the surface relatives – for instance in *Iais* and *Jaera,* with a short and wide body (pelstatism) and short uropods. Like *Microcharon,* microcerberids have a long and slender body, and inland species also have slender pereiopods and an elongation of the distal claws. It is thought that these appendages are more elongated in stable environments. The stenasellids, syncarid Bathynellacea and Stygocarididae are also long and narrow. In interstitial amphipods, the coxal and epimeral plates as well as pleopods are often reduced, so that the body is less high and more vermiform (Bogidiellidae, Ingolfiellidae); another phreatomorphic feature is the lengthening of the 6th and 7th pereiopods and of the third uropods in several genera: *Medigidiella, Niphargus, Pseudoniphargus.* Harpacticoid copepods also have a long body shape, with elongation of the furca (Fig. 9.5). Such elongation, often significantly more developed in interstitial crustaceans than in their epigean relatives (see Fig. 9.1: epigean *Anaspides* versus interstitial species; or Fig. 9.5: *Jaera* versus *Microcharon*) gives the animal a better ability to thread its way and move through the labyrinth of the narrow channels within granular sediment. Long posterior appendages and claws may be an advantage for pulling, pushing and walking among grains. However, the antennae of interstitial crustaceans are not often very long; they are sometimes short, on account of the low number of segments (sometimes only 2 or 3 articles, see Figures) as a result of progenesis. Progenesis acts towards small size, whereas there is an adaptive elongation of the remaining segments, especially in the basal part of the flagellum and on the aesthetascs (Fig. 9.7). When walking in the sediment, interstitial stygobionts direct their antennae forward, and the appendages move and constantly examine the channels between the grains; longer sensorial receptors such as aesthetascs provide a better detection of prey, food and congeners, and recognition of the size of interstices.

Obviously advantageous features such as elongation of the body, appendages and sensory structures, involved in phreatomorphology might be regarded as a "product of constructive evolutionary processes" as suggested by Notenboom (1991). Furthermore, a chronology of the evolutionary changes can be observed in some cases: historical biogeographic studies of some monophyletic groups of isopods and amphipods show that more primitive species which colonized subterranean freshwater in the Turonian or Eocene display more elongate and slender body and appendages than those species colonizing interstitial freshwater later (Tortonian: Pliocene), or extant derived species still living in more instable littoral and marine interstitial biotopes.

Similar elaborations are reported in caves. Thus, the troglobitic *Asellus aquaticus cavernicolus* has long appendages, in contrast to its epigean relatives (Sket, 1985). Longer antennae can sometimes already be seen in epigean relatives (Barr Jr, 1968; Camacho et al., 1992). The cave amphipod *Gammarus minus* has longer antennae, seventh pereiopods and uropods (Holsinger and Culver, 1970; Culver et al., 1990). After tentative explanations involving allometry (Barr Jr, 1968; Camacho, 1992), more general elongation (Holsinger and Culver, 1970), tendencies of the structural group, and exaptations (Marmonier and Danielopol, 1988; Danielopol and Marmonier, 1994), direct selection was proved to have effected the elongation of the body and the antennae in the cave *G. minus.* Furthermore, a similar pattern of selection responsible for the parallel evolution of cave morphologies has been tested on a series of isolated and independent cave populations (Jones and Culver, 1989; Culver et al., 1990, 1994, 1995; Jones et al., 1992). Jones et al. (1992) concluded that "parallel patterns of selection in caves are responsible for larger body and the elaboration of antennal size". Antennae are regarded as extraoptic sensory structures, with an evolutionary trade-off in neurological connections of the brain (Fong et al., 1995).

Are "parallel patterns of selection" also responsible

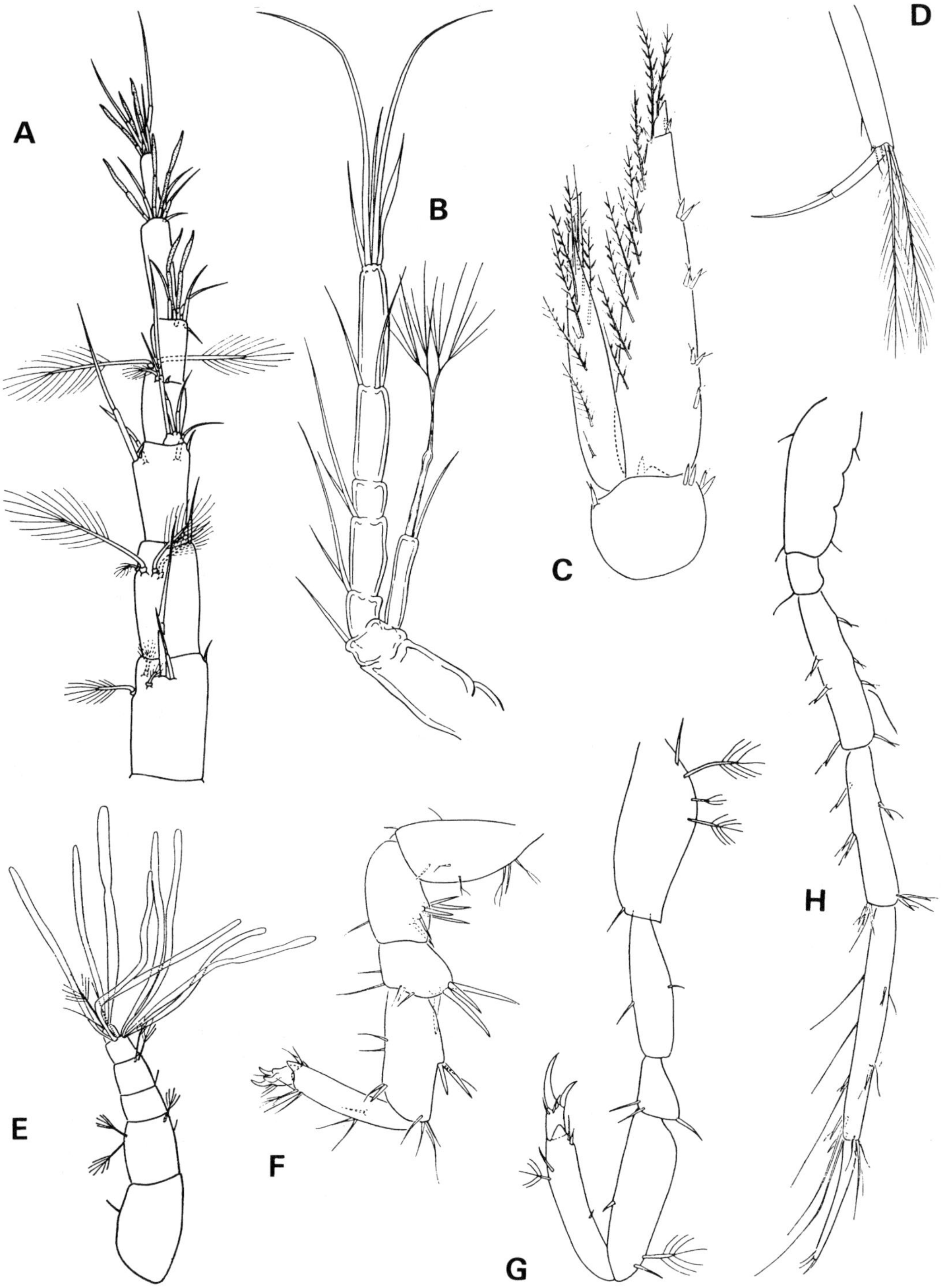

Fig. 9.7. (A) Antenna 1 of the syncarid *Iberobathynella lusitanica* showing numerous sensorial setae. (B) The sensitive setae on antenna 2 of *Transvaalthynella coineaui.* (C) Sensorial setae on uropod 3 of the amphipod *Salentinella lescherae.* (D) Elongated dactylus, claw and plumose setae on pereiopod 7 of *Salentinella petiti.* (E) Aesthetasc bunch on the male antenna 1 of the isopod *Microcharon boui.* (F, G) Short pereiopod 6 of the surface isopod *Iais aquilei* as compared to the elongated one in the isopod *Microcharon reginae*. (H) Elongated pereiopod 7 of the amphipod *Bogidiella (Bog.) skopljensis.* After Schminke (1973) (A), Serban and Coineau (1975) (B), Coineau (1968) (C, E, H), Coineau (1963) (D), Coineau (1977) (F), Dole-Olivier and Coineau (1987) (G).

for the elongation of the body, sensory structures, appendages, and claws in the interstitial milieu, as in caves? Is it possible to extrapolate? Experiments need to be done. Some constructive adaptive features are not sensory organs, and should not be considered to be the result of trade-offs against eye loss (see next section).

Antennae are elongated in caves, but are rather short in interstitial habitats; while elongation acts on all appendages in caves, there is evidence for a differential elongation in interstitial habitats, only the posterior appendages being elongated. Selection seems to work in opposite ways in caves and in the interstitial habitat: increased size in caves (cf. giant ingolfiellids), small size allied to elongation of body and posterior appendages, resulting in slenderness, in porous biotopes.

In a few other cases (isopods, amphipods such as *Pseudoniphargus*), uropods display hypertrophy, resulting in the elongation of the peduncle or the endopodite in comparisons between closely related species in the same geographic area. This type of phenotypic change generally appeared gradually and seems to have no advantage or disadvantage, and thus no indication for positive or negative selection. Such different specific characters in isolated populations might result from neutral mutations with phenotypic effects (Lynch, 1990) and be consistent with Kimura's (1990) neutral theory of evolution.

Profound modifications of the internal organization in interstitial crustaceans also result in simpler structures – even the lack of organs – as compared to surface relatives. In the isopod *Angeliera phreaticola* the gut is linear without stomodeal sinuosity, and some caeca are lacking; the number of testicular utricula is reduced, only one ovary persists; the cerebron shows only two small lateral expansions instead of the optic lobes, and elongation of all units. Similar but less marked modifications exist in *Microcharon* (Coineau and Renaud-Mornant, 1977; Renaud-Mornant and Coineau, 1978). In the syncarid *Bathynella,* the digestive caeca and the right ovary are lacking, and the germinative cells are reduced to some cells of the left female gonad; *B. chappuisi* lays only one egg from the left ovary. In the ostracod *Mixtacandona,* only three testicular lobes can be seen (four in epigean species: Danielopol, 1980). Such anatomical reductions are probably ontogenetic modifications due to developmental gene modifications, mainly related to the elongation and the slender shape of the body inter-alia.

Evolutionary innovations

Evolutionary innovations may include processes as well as the resulting structures, behavior and physiology. Functions and processes are important, but only morphological changes allow one to infer the evolutionary story (Nitecki, 1990). Moreover, Nitecki asks if "new" is "more than reshuffling and rearranging". Mayr (1963) defined evolutionary novelty as "any newly acquired structure or property that permits the performance of a new function, which in turn, will open a new adaptive zone". More recently, Müller (1990) has indicated that "new functional properties have been suggested as indicators of novelty" and "that the change of function precedes the alteration of the structure", in agreement with Futuyma (1986). Interstitial organisms display a set of evolutionary innovations related to the porous habitat, compared to surface relatives:

(1) In addition to the elongation of sensory structures, interstitial crustaceans often display numerous aesthetascs on antennae, especially in the male, or plumose setae on pereiopods and uropods (Fig. 9.7). These sensory receptors may be compensatory elaborations used for detection of the female, chemical gradients, prey, or predators within the intragranular water. In psammaspidid syncarids (Schminke, 1974), the number of aesthetascs is higher than in the epigean anaspidids (personal observations of the latter, compared with Schminke's drawings). In *Bathynella,* the antennula exhibits three categories of sensory organs: long aesthetascs, long plumose setae and "poils en lancette" (Serban, 1972). There are also "sensitive" setae on antenna 2. In the isopod *Microcharon,* the male antennae have 1 to 6 long aesthetascs on the same article (Fig. 9.7).

(2) In syncarids, the end of the body in interstitial walking groups also shows evolutionary elaborations as adaptations to the granular medium. The furca is a novelty in bathynellids which have no free telson. In contrast, surface swimming relatives have a wide and flat entire telson associated with two flat uropods; the three pieces form the "tail fan" used for swimming. Conversely, uropods of the phreatobites are styliform, and more elongated; moreover, strong curved spines arm the uropod margin and tips, allowing the crustacean to push and pull on the substrate grains (Fig. 9.8). These appendages also provide an anchorage, allowing the pleon to bend beneath the thorax and propel the animal forward or to turn it within the sediment channel. Schminke (1976, 1981)

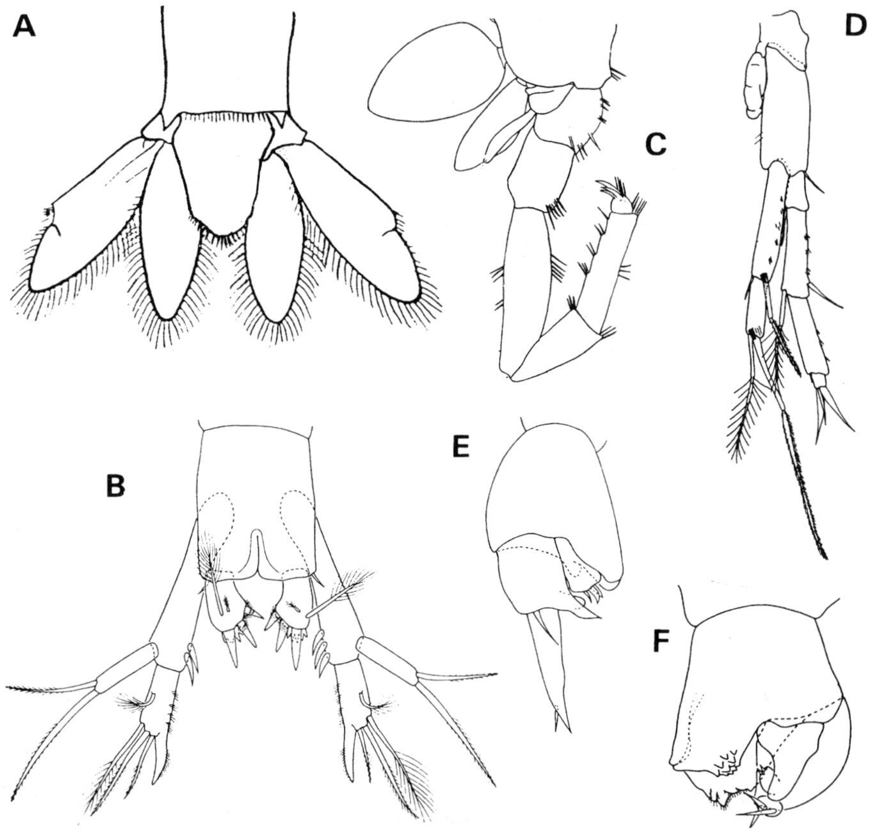

Fig. 9.8. (A) Flat swimming tail fan of the surface syncarid *Anaspides tasmaniae.* (B) Pleotelson, styliform uropods and furca of the interstitial *Cteniobathynella noodti.* (C) Thoracopod 8 of *Anaspides tasmaniae.* (D) Pereiopod 7 of *Iberobathynella lusitanica.* (E) Pereiopod 8 (male) of *Hexabathynella knoepffleri.* (F) Pereiopod 8 (male) of *Iberobathynella lusitanica.* After Noodt (1964) (A) modified, Schminke (1973) (B, F), Smith (1908) (C), Camacho (1987) (D), Coineau (1965) (E).

indicated the new sensory function of the furca, which is also used to clean the mouthparts. It can be added that the furca is involved in the new mode of locomotion. As regards its origin, the furca may be a modification of the forked larval telson retained from the ancestral zoea: in bathynellaceans, the two parts of the fork were separated and segmented, and became moveable. Thus, the swimming tail fan of the surface fossils and extant relatives is replaced by the furca and styliform uropods in interstitial forms, used mainly for moving and cleaning: this it is both a restructuring and a change of function. Furthermore, phylogenetic relationships highlight the fact that the flat swimming telson and uropods is the plesiomorphic condition shown by primitive groups.

(3) Another novelty in phreatobites is a strong thigmotactism which keeps the organism closely applied to elements of the substrate. Deprived of this contact with the grains, the animal loses coordination; it cannot swim in open water, and walks awkwardly on the bottom. Elongation of the body, posterior appendages and sense organs undoubtedly increases the contact with the channel walls. Thigmotactism may be viewed as a compensatory behavior, linked with the short antennae in the interstitial milieu (versus long antennae in caves).

(4) Interstitial bathynellaceans, isopods and amphipods have a strong muscular system allowing specific movements, body flexibility and better contacts within the micropores. Such elaborations may also be regarded as selected.

(5) Another constructive evolutionary process is shown by the copulatory organ of the bathynellaceans. In the sister group the anaspidaceans, which have surface relatives, the sperm contained in spermatophores is transmitted through the petasma resulting from pleopod modifications. In the zoea-like bathynellaceans, pleopods are not present because of a shortened

development. Since the ancestral genital male openings were in the thoracal segment 8, bathynellaceans elaborated a new small copulatory appendage arising from thoracopod 8, so that the primary locomotion function was replaced by the sexual one (Fig. 9.8). And it was not a gradual process, as emphasized by Serban (1972) and Schminke (1981); in a first step, endopodites, the ancestral function of which was mainly locomotion, remained reduced and incomplete, due to progenesis; in a second step, they were specialized into a copulatory appendage, with the *vas deferens* opening in a penial complex, and a modified articulation of the paedomorphic thoracopod 8.

(6) Ovigerous female thermosbaenaceans have a dorsal brood-pouch (Fig. 9.3) which corresponds to an enlargement of the carapace. Such a dorsal brood chamber, very rare within crustaceans, may be a conservative trait resulting from a progenetic evolution of the peracaridan ancestor.

(7) Other adaptations are peculiar evolutionary elaborations, such as the dorsal insertion of the pereiopods in isopod angelierids and microcerberids. This derived feature is correlated with body elongation. Moreover, the coxae have spines and a triangular shape in coastal microcerberids and may perhaps serve to increase contact with the grains in unstable sandy shores; in a relatively more stable environment, freshwater species have simple straight coxae.

The interstitial environment has thus promoted evolutionary innovations as: 1, modifications of pre-existing structures for a change of function in the new milieu (male pereiopod 8 in syncarids); 2, constructive features (furca) for a new function within the interstitial habitat; 3, new behavior (thigmotactism). Progenesis is partly involved in some products (copulatory organ). Most novelties probably require modifications of ontogeny and are realized at later stages of development. The integration of new structures into the organisms is probably facilitated by a relative developmental plasticity. "Development and its mechanisms are central to the problem of novelty" (Müller, 1990).

RESPONSE TO STABILITY AND LOW ENERGY SOURCES

Other evolutionary elaborations have arisen in response to low energy sources and the relative stability in the interstitial environment – characteristics elaborated relative to surface species.

Life-history adaptations

Most interstitial crustaceans display a K-reproductive strategy compared to closely related surface representatives which are r-strategists. They have a slower or lower reproductive potentiality. Precopulation is lacking in asellid, stenasellid isopods and amphipods. The sex ratio favours females, whereas demographic structures are "skewed" toward large size classes with a low number of reproducing females (Henry, 1976; Magniez, 1976; Coineau et al., 1994; Holsinger, 1994a,b). The number of eggs is reduced: only one in *Angeliera* and bathynellids, two in *Microcharon,* three in *Medigidiella,* 6–10 in *Proasellus walteri*; 4 ovocytes in the ostracod *Mixtacandona,* and less than ten per ovigerous sack in harpacticoid copepods (Rouch, 1968; Coineau, 1971, 1984; Rogulj et al., 1993). Correlatively, egg volume is increased with more yolk. Egg-laying rhythm is very slow; from once or twice a year in microparasellids to biennial in stenasellids.

In contrast to the short epigean life spans, in interstitial species both embryonic and post-embryonic development are much longer, though the number of stages is reduced. Growth is also lower, and intermoult periods are longer. There is no season for reproduction; interstitial organisms exhibit only an attenuated seasonal maximum, often related to high water periods (= more food supply) (Turquin, 1975; Juberthie, 1975; Mathieu and Turquin, 1992). The increased longevity may range from 7–10 months (asellids) to 2 years (microparasellids, bogidiellids) and up to 12, 15 or 20 years in *Niphargus* and large stenasellids (Henry, 1976; Magniez, 1976; Coineau, 1971, 1984; Halley, 1986; Holsinger and Culver, 1988; Holsinger, 1994a,b; Coineau et al., 1994).

At hatching, the young already has a morphology very close to that of the adult. Reproductive effort in the interstitial habitat is relatively low and energy loss is limited: increased parental care, with the embryo spending a long period in the egg or protected within the marsupium (brood pouch), low juvenile mortality, protected gametes, a seminal receptacle in the female, precocious fertilization of the post-embryonal female Stage 2 in *Angeliera* (Coineau, 1971), and viviparity in *Typhlocirolana* (Boutin and Boulanouar, pers. comm.).

This convergent evolution towards energy economy is reported in most other groups, such as the polychaetes (Westheide, 1984), and is common to most troglobites (Vandel, 1964; Turquin, 1975; Culver, 1982; Camacho et al., 1992). The responses may vary according to the habitat (Wilkens, 1992). K-reproductive strategies are new constructive features, and may have been selected by the relatively stable interstitial milieu and the underground oligotrophic environment. These specializations are also considered by Culver (1982) to be adaptations to the characteristics of the subterranean environment. However, Camacho and Valdecasas (1992) attempted to separate convergent evolution (adaptation) and inheritance from ancestors in harpacticoid copepods. From a preliminary phylogenetic analysis and ecological data, as well as from consideration of the reproduction effort indicated by the number of eggs, they demonstrated that the causal force is not "exclusively due to food". Wilkens (1992) considered the enhanced yolk content in the cave fish *Astyanax* as probably a "constructive adaptation to the limited food supply" with a genetic basis. This increase in yolk content evolves rapidly, and the energy economy resulting from eye loss is probably too low to be involved in this reproductive cost (Hüppop and Wilkens, 1991).

Ecological strategies and other physiological adaptations

As highlighted during the last decade, the interstitial environment appears to be more heterogeneous, complex and variable than had generally been considered earlier (Danielopol and Rouch, 1991; Dole-Olivier et al., 1993; see also review in Gibert et al., 1994a,b). The diversity of the "mosaic" of habitats may result in numerous and underestimated adaptive responses of the phreatobites (Camacho et al., 1992; Mathieu and Turquin, 1992; Danielopol et al., 1994). Nevertheless, from a general point of view, the deepest interstitial water is characterized by more stable conditions (physical and chemical factors) than nearer the surface, where it is subject to surface hydrodynamism and to strong variations of the fluxes and the physical and chemical quality of the groundwater.

Dole-Olivier et al. (1993) showed that "a differential sensitivity of stygobite species to environmental fluctuations" is expressed by repeated gradients in distribution along the three spatial dimensions of the flood plain of the Rhône River – transversally, vertically and longitudinally. This distribution reflects a gradient of habitat stability, and strict phreatobites inhabit the most stable area in the three dimensions (*Bathynella* sp., *Microcharon reginae, Salentinella* sp.). This distribution may be regarded as a more or less marked adaptation to a stable milieu, reflecting the life-history requirements of the species. Thus, despite their seemingly convergent morphology, phreatobite species exhibit different physiological responses integrated in this gradient.

The oxygen content of interstitial water is generally low; it decreases from the surface to greater depths where hypoxic conditions prevail; fluctuations are higher near the top of the sediment layers, and from site to site over small or large scales (Rouch, 1991; Pospisil, 1994; Strayer, 1994).

Danielopol et al. (1994) have attempted to achieve an ecological research program dealing with adaptation. Following their "practical agenda", they demonstrated that the interstitial isopod *Proasellus slavus vindobonensis* is able to live in well-oxygenated deposits ($>1\,mg\,\ell^{-1}\,O_2$) as well as in poorly oxygenated and hypoxic sediments ($<0.5\,mg\,\ell^{-1}\,O_2$) of the superficial alluvia in the Danube wetland. Oxygen consumption is low (1000 to $3500\,\mu\ell\,O_2\,g^{-1}\,d.w.h^{-1}$) and independent of the oxygen concentration. Experiments conducted in varying oxygen concentration (from 9.1 to $0.1\,mg\,\ell^{-1}\,O_2$) show that *P. slavus* has a high ventilating activity and rapid respiratory pleopod beats irrespective of oxygen values. In hypoxic conditions exploratory movement decreases, whereas a special arched position of the posterior part of the body, with the operculum widely opened, allows a wider pleopod pulsation; oxygen consumption is thus more effective. In the same experimental conditions, the epigean *Asellus aquaticus* shows the opposite behavior. Among three evolutionary scenarios (adjusters, regulators and extraregulators: Danielopol et al., 1994), *P. slavus* appears to be a regulator: the organism is able to regulate respiration rate, varying the ventilatory activity of its pleopods and its motility according to oxygen availability – it is able to "compensate" in energy allocation. Such regulation exists in several isopods investigated, and Danielopol et al. (1994) considered this pattern as an adaptation of subterranean crustaceans living under low and varying oxygen conditions. These adaptive physiological mechanisms and ecological strategies have been selected by the heterogeneous interstitial environment with limited and fluctuating oxygen content. They might also only represent ecological accommodation which disappears as soon as the new situation disappears, and thus consequently contributes to the niche range of the species.

Similarly, *Niphargus rhenorhodanensis* in interstitial waters has a greater capacity for physiological responses (oxygen consumption, locomotory activity) to environmental fluctuations than do karstic populations which live in more stable conditions. Moreover, in drastic low water and drained conditions, the amphipod migrates into deeper sediments and survives on its own carbohydrate and lipid reserves. As soon as the piezometric level increases, the organisms again accumulate reserves for reproduction (Mathieu, 1983; Mathieu and Turquin, 1992). Thus, the phreatobiontic populations display different strategies related to environmental fluctuations, and are also able to compensate for energy allocation.

In contrast, microparasellids and bogidiellids, as well as mystacocarids of unstable marine sandy shores, are located in highly oxygenated areas and consume relatively high quantities of oxygen [e.g., *Angeliera* (10–48 $\mu\ell\ O_2\ g^{-1}$ d.w.h^{-1}), *Microcharon marinus* (8–52 $\mu\ell\ O_2\ g^{-1}$ d.w.h^{-1}) and *Medigidiella chappuisi* (5–52 $\mu\ell\ O_2\ g^{-1}$ d.w.h^{-1}) (unpublished results), compared with *Niphargus rhenorhodanensis* (0.7–3.4 $\mu\ell\ O_2\ g^{-1}$ d.w.h^{-1}) and *Proasellus slavus* (0.7–3.0 $\mu\ell\ O_2\ g^{-1}$ d.w.h^{-1}) (Danielopol et al., 1994; Mathieu, 1983)]. These relatively high values are related to their small size, and to varying and higher temperature and unstable littoral characteristics.

For some of the crustaceans, the caloric values (*Angeliera*: 4.80 to 7.02 cal.mg^{-1} ash-free d.w.; *Medigidiella*: 3.52 to 6.92 cal.mg^{-1} ash-free d.w.) are high compared with surface representatives; the fluctuations are correlated with that of the dry weight, and are related to the reproductive season (Coineau, 1971, 1983). The highest value is found in June–July when ovigerous females are most abundant; furthermore, *Angeliera* lays only one very large egg. This high energy content is due to storage of energy sources in the form of lipids. The part played by the males in the caloric increase is noticeable because of their sexual maturation during the same period; after copulation with the juvenile females, a decrease takes place by energy loss through gametes and respiratory cost. The high caloric values reflect good trophic conditions as well as instability due to wave action and floods in spring. The ash content of interstitial crustaceans is relatively low as compared to other isopods and amphipods, with a low proportion of calcium carbonate, since the integument is very thin in phreatobites.

Continental interstitial waters are very poor in food, and the distribution of resources is patchy. There is no photosynthesis, and chemoautotrophy is very low (Ginet and Decou, 1977; Camacho et al., 1992; Culver et al., 1995; Strayer, 1994; Notenboom et al., 1994). Most of the energy input is allochtonous, and it is mainly composed of detritus providing small-size particulate organic matter. Responses to the limited and homogeneous resources are the low densities of the phreatobites, and simple food webs with few trophic levels (Notenboom et al., 1994; Culver et al., 1995). Interstitial stygobites represent the higher-level consumers in the microporous milieu. The trophic behavior of phreatobites is poorly documented. Some experiments show that microparasellids and bogidiellids are non-selective deposit feeders, grazing, sucking and rasping the epistratum that cover the grains, and using the bristles, spines and teeth covering the buccal appendages. They ingest fungi, particulate organic deposits, biofilms, animal and plant microdetritus, bacteria covering detritus; bogidiellids and bathynellids also ingest small prey such as worms or copepods, as well as faecal pellets (N. Coineau, unpublished). *Proasellus slavus* ingests fine sediment, organic detritus and bacteria (Danielopol et al., 1994). Typhlocirolanids are detrivorous and carnivorous. Phreatobites of different groups often exhibit similar feeding strategies involving similar selection forces. According to Culver et al. (1994), two opposite evolutionary pressures and adaptive responses to the above-mentioned characteristics of the interstitial milieu may coexist: limited resources result in a "strong selection for omnivory and broadening of diet", and there is also a "strong selection for increased efficiency at food finding, including predation". Energy economy represents an effective selective pressure.

CONCLUSION

Adaptations of crustaceans to life in the interstitial groundwater, compared with close relatives on the surface and in caves, have been discussed. The main abiotic factors to be faced are darkness, grain size, limited energetic sources and, despite the biotope heterogeneity, a relative stability which increases with depth and distance from the river channel. The most significant variable is obviously grain size, which results in a more constraining milieu than the karstic environment.

Small size is a characteristic adaptation to aquatic interstitial life. It results generally from progenesis. The progenetic origin of several taxa is strongly supported by larval or juvenile morphology, the early abbreviated development and extant evolutionary series. Modifications of developmental timing represent a major way of morphological evolution associated with size modifications (Raff et al., 1990). Progenesis is correlated with colonization of the interstitial habitat by surface ancestors. In syncarid and peracarid crustaceans, this type of heterochrony is stimulated and positively selected by the porous ecosystem, and thus is an adaptation. In maxillopod crustaceans (ostracods, copepods), progenesis may be viewed as an exaptation already acting at the surface. As there is selection for small size in the interstitial milieu, this exaptation is thereafter selected.

Both elongation and slenderness, as responses to the porous system, are selected novelties. Elongation mainly acts on the body size, and a differential increase in size mainly affects the posterior appendages (pereiopods 6 and 7, uropods). Progenesis results in decreased size of body and appendages; the combined effect is to produce a slender, elongated, small body. In caves, selection acts in the opposite direction, since there are less dimensional constraints – larger body size and general increased appendage size are favoured. According to Wilkens, increased size is not coupled with regressive evolution and is a selected constructive trait. In the view of Culver et al. (1995), increased size of elaborated extraoptic sensory structures (larger antennae and body in *Gammarus minus*) result from directional selection and neurological trade-off. However, it is possible that such a trade-off is not general, and that, in the porous milieu, elongation may be an independent and selected evolutionary innovation.

Phreatomorphology involves regressive evolution as a response to darkness. As in caves, eye loss may be explained by different hypotheses: 1, accumulation of neutral mutations; 2, both mutation and directional selection correlated with a trade-off of neuronal connections, resulting in increased size of extraoptic sensory elaborations also directionally selected; 3, poor synchronization of the slower ontogenetic development in subterranean organisms with the stage of development for expression of the eyeless master gene, as compared with normal development in epigean animals. Eyelessness may sometimes be an exaptation which may also be selected in darkness (some ostracod lineages).

The increased number of sensory setae, the furca, the styliform uropod shape, the copulatory organ of the syncarid Bathynellacea as well as some dimorphic characters in males, thigmotactism and strong musculature are evolutionary novelties resulting from selection. Some of these adaptive innovations involve restructuring and transformation within a given structure, others are new constructive traits for a new function realized during morphogenesis (late development).

Responses to low energy allocation and stability are more or less similar to adaptation to cave life: K-strategies, capacity to regulate the metabolism rate in the isopod *Proasellus slavus*, efficiency at food finding, unspecialized diet, and spatial gradients in distribution are constructive features resulting from selection. Though energy economy is a driving selection force in these life histories and in ecophysiological adaptive traits (i.e., symecomorphosis: Liem, 1990) in the interstitial milieu, it does not seem to be involved in regressive evolution.

Evolutionary convergence in the interstitial habitat, as in caves, might be due to the limited phenotype possibilities provided by genotype expression, associated with developmental constraints and the epigenetic effects of abiotic factors. However a relative developmental plasticity allows the integration of novelties into organisms.

Phylogenies show that the chronology of evolutionary change in some cases (syncarids, amphipods) is correlated with colonization of the interstitial habitat by surface ancestors, and phreatomorphology appears to be derived.

Finally, the most common process by which adaptation to aquatic interstitial environment occurs is natural selection (Brandon, 1990).

Phreatobites may provide good tools and experimental systems in studies of environmental change and evolution. Only matching of evolutionary genetical, developmental, morphological, biological, ecological, ecophysiological and phylogenetical results will allow the resolution of adaptive patterns and mechanisms, both in subterranean and surface organisms – since, according to Nitecki (1990), structure and function must be well integrated with ecological factors for innovations to initiate evolutionary diversification.

ACKNOWLEDGEMENTS

The author is indebted to C. Boutin for kindly reading the manuscript and valuable suggestions. Thanks are due to S. Boletzky for helpful discussions and to D.C. Culver, J. Gibert and D. Danielopol who supplied me with their recent publications. I am grateful to Drs Wilkens, Culver and Humphreys for suggesting that I should contribute a chapter to this volume of "Ecosystems of the World". Research was partly supported by G.D.R. 88 "Evolution des organismes microscopiques et environnement" and ACC-SV n°7: Systématique et Biodiversité: "Diversité Fonctionnelle et Evolution en Milieu Souterrain".

REFERENCES

Bachelard, G., 1938. *La Formation de l'Esprit Scientifique.* In: J. Vrin (Editor), coll. Librairie scientifique, 15e éd., Bibliothèque des textes philosophiques (1993), 256 pp (repr.)

Barnard, K.H., 1966. The occurrence of the genus *Ingolfiella* (Crustacea, Amphipoda) in South Africa, with description of a new species. *Ann. Mag. Nat. Hist. London,* 9(100–102): 189–197.

Barr Jr., T.C., 1968. Cave ecology and the evolution of troglobites. *Evol. Biol.,* 2: 35–102.

Barr Jr., T.C. and Holsinger, J.R., 1985. Speciation in cave faunas. *Ann. Rev. Ecol. Syst.,* 16: 313–337.

Bocquet, C., 1953. Recherches sur le polymorphisme naturel des *Jaera marina* (Fabr.) (Isopodes Asellotes). *Arch. Zool. Exp. Gen.,* 90(4): 187–450.

Botosaneanu, L., 1986. *Stygofauna Mundi.* E.J. Brill/Dr W. Backhuys, Leiden, 740 pp.

Botosaneanu, L. and Holsinger, J.R., 1991. Some aspects concerning colonization of the subterranean realm-especially of subterranean waters: a response to Rouch and Danielopol, 1987. *Stygologia,* 6: 11–39.

Boutin, C. and Coineau, N., 1990. "Regression Model", Modèle biphase d'évolution et origine des micro-organismes stygobies interstitiels continentaux. *Rev. Micropal.,* 33(3/4): 303–322.

Boxshall, G.A., 1983. A comparative functional analysis of the major maxillopodan groups. In: F.R. Schram (Editor), *Crustacean Issues, 1. Crustacean Phylogeny.* Balkema, Rotterdam, pp. 121–143.

Brandon, R.N., 1990. *Adaptation and Environment.* Princeton University Press, 214 pp.

Brodie III, E.D., Moore, A.J. and Jansen, F.J., 1995. Visualizing and quantifying natural selection. *Trends Ecol. Evol.,* 10(8): 313–318.

Camacho, A.I., 1987. *La familia Parabathynellidae (Crustacea, Syncarida Bathynellacea) en la Peninsula Iberica. Taxonomia, Filogenia y Biogeografia.* Thesis University Madrid, 889 pp.

Camacho, A.I., 1992. A classification of the aquatic and terrestrial environment and their associated fauna. In: A.I. Camacho (Editor), *The Natural History of Biospeleology.* Monogr. 7, Mus. Nac. Cien. Nat., C.S.I.C., Madrid, pp. 57–103.

Camacho, A.I. and Valdecasas, A.G., 1992. The subterranean environment. *Limnetica,* 8: 167–174.

Camacho, A.I., Bello, E., Becerra, J.M. and Vaticon, I., 1992. A natural history of the subterranean environment and its associated fauna. In: A.I. Camacho (Editor), *The Natural History of Biospeleology.* Monogr. 7, Mus. Nac. Cienc. Nat., C.S.I.C., Madrid, pp. 171–197.

Christiansen, K.A., 1962. Proposition pour la classification des animaux cavernicoles. *Spelunca Mem.,* 2: 76–78.

Christiansen, K.A., 1992. Biological processes in space and time. Cave life in the light of modern evolutionary theory. In: A.I. Camacho (Editor), *The Natural History of Biospeleology.* Monogr. 7, Mus. Nac. Cien. Nat., C.S.I.C., Madrid, pp. 453–478.

Coineau, N., 1963. Etude sur les Amphipodes. II. *Salentinella petiti* n. sp. *Vie Milieu,* 14(1): 107–122.

Coineau, N., 1965. Recherches sur la faune des îles méditerranéennes. I. le genre *Parabathynella* (Crust. Syncar.) en Corse. *Vie Milieu,* 15(4): 993–1005.

Coineau, N., 1967. *Acanthobathynella,* nouveau genre de Syncaride africain (Côte d'Ivoire). *C. R. Acad. Sci. Ser. D,* 265: 1988–1990.

Coineau, N., 1968. Contribution à l'étude de la faune interstitielle. Isopodes et Amphipodes. *Mem. Mus. Natl. Hist. Nat. Ser. A Zool.,* 55(3): 145–216.

Coineau, N., 1971. Les isopodes interstitiels. Documents sur leur écologie et leur biologie. *Mem. Mus. Natl. Hist. Nat. Ser. A Zool.,* 64: 1–170.

Coineau, N., 1977. Isopodes aquatiques. La Faune Terrestre de l'île de Sainte Hélène. 4. 2. *Ann. Mus. R. Afrique Centrale, Tervuren, Sc. Zool.,* 220: 427–447.

Coineau, N., 1983. Energetic values in interstitial isopods and amphipods from sandy beaches as a function of body size and season (Western Mediterranean). In: A. McLachlan and T. Erasmus (Editors), *Proceedings of the International Symposium on "Sandy Beaches as Ecosystems", Port Elizabeth.* Junk, The Hague, pp. 687–692.

Coineau, N., 1984. Quelques aspects de l'adaptation des Crustacés Isopodes, Amphipodes et Syncarides à la vie dans le milieu interstitiel. *Mem. Biospeol.,* 11: 7–15.

Coineau, N., 1985. Colonisation du milieu souterrain continental par des organismes d'origine marine. Moyens d'accès par le milieu interstitiel. *Bull. Soc. Biospeol.,* 8: 26–30.

Coineau, N., 1996. Sous Classe des Eumalacostracés (Eumalacostraca Grobben, 1892), Super Ordre des Syncarides (Syncarida Packard, 1885). In: J. Forest (Editor), *Traité de Zoologie, 7(2) Crustacés.* Masson, Paris, pp. 897–954.

Coineau, N. and Botosaneanu, L., 1973. Isopodes interstitiels de Cuba. In: T. Orghidan, A. Nunez Jimenez, L. Botosaneanu, V. Decou, S. Negrea and V. Viña Bayès (Editors), *Résultats des expéditions biospéologiques cubano-roumaines à Cuba: I.* Academei Republicii Socialiste România Editura, Bucarest, pp. 191–220.

Coineau, N. and Boutin, C., 1992. Biological processes in space and time: Colonization, evolution and speciation in interstitial stygobionts. In: A.I. Camacho (Editor), *The Natural History of Biospeleology.* Monogr. 7, Mus. Nac. Cien. Nat., C.S.I.C., Madrid, pp. 423–451.

Coineau, N. and Boutin, C., 1995. Progenesis as an evolutionary process in some interstitial crustaceans. In: A. McLachlan and T. Erasmus (Editors), *Proc. Ninth Int. Meiofauna Conf., Perpignan.* Junk, The Hague, p. 32.

Coineau, N. and Renaud-Mornant, J., 1977. Etude anatomique des Isopodes interstitiels. II. Morphologie fonctionnelle et évolution saisonnière de l'appareil génital femelle d'*Angeliera phreaticola* Chappuis et Delamare Deboutteville, 1952. *Arch. Zool. Exp. Gen.,* 118(3): 349–366.

Coineau, N., Henry, J.P., Magniez, G. and Neogescu, I., 1994. Isopoda Aquatica. In: C. Juberthie and V. Decou (Editors), *Encyclopaedia*

Biospeologica, Vol. I. Société de Biospéléologie, Moulis, Bucarest, pp. 123–140.

Creuzé des Châtelliers, M., Poinsard, M. and Bravard, J.P., 1994. Geomorphology of alluvial groundwater ecosystems. In: J. Gibert, D.L. Danielopol and J.A. Stanford (Editors), *Groundwater Ecology.* Academic Press, New York, pp. 157–185.

Cuénot, L., 1909. Le peuplement des places vides dans la nature et l'origine des adaptations. *Rev. Gen. Sci. Paris,* 20: 8–14.

Culver, D.C., 1982. *Cave Life, Evolution and Ecology.* Harvard Univ. Press, 189 pp.

Culver, D.C., 1985. Regressive evolution. Caves and karst studies. *Natl. Soc. Speleol. Bull.,* 47(2): 1–162.

Culver, D.C. and Fong, D.W., 1986. Why all cave animals look alike. *Stygologia,* 2(3): 208–216.

Culver, D.C., Kane, T.C., Fong, D.W., Jones, R., Taylor, M.A. and Sauereisen, S.C., 1990. Morphology of cave organisms – is it adaptive? *Mem. Biospeol.,* 17: 13–26.

Culver, D.C., Jernigan, R.W., O'Connell, J. and Kane, T.C., 1994. The geometry of natural selection in cave and spring populations of the amphipod *Gammarus minus* Say (Crustacea Amphipoda). *Biol. J. Linn. Soc.,* 52: 49–67.

Culver, D.C., Kane, T.C. and Fong, D.W., 1995. *Adaptation and Neutral Selection in Caves. The Evolution of Gammarus minus.* Harvard Univ. Press, 223 pp.

Danielopol, D.L., 1980. Sur la biologie de quelques Ostracodes Candoninae épigés et hypogés d'Europe. *Bull. Mus. Natl. Hist. Nat. Ser. D,* 2: 471–506.

Danielopol, D.L., 1991. Spatial distribution and dispersal of interstitial Crustacea in alluvial sediments of a backwater of the Danube at Vienna. *Stygologia,* 6: 97–110.

Danielopol, D.L. and Bonaduce, G., 1990. The colonization of subsurface habitats by the Loxoconchidae Sars and the Psammocytheridae Klie. In: R. Watley and C. Maybury (Editors), *Ostracoda and Global Events.* Chapman and Hall, London, pp. 437–458.

Danielopol, D.L. and Marmonier, P., 1994. Ostracoda. In: C. Juberthie and V. Decou (Editors), *Encyclopaedia Biospeologica, Vol. I.* Société de Biospéléologie, Moulis, Bucarest, pp. 91–97.

Danielopol, D.L. and Rouch, R., 1991. L'adaptation des organismes au milieu aquatique souterrain. Réflexions sur l'apport des recherches écologiques récentes. *Stygologia,* 6(3): 129–142.

Danielopol, D.L., Creuzé des Châtelliers, M., Mösslacher, F., Pospisil, P. and Popa, R., 1994. Adaptation of Crustacea to interstitial habitats: a practical agenda for ecological studies. In: J. Gibert, D.L. Danielopol and J.A. Stanford (Editors), *Groundwater Ecology.* Academic Press, New York, pp. 218–243.

Danielopol, D.L., Baltanas, A. and Bonaduce, G., 1996. The darkness syndrome of subsurface-shallow and deep-sea dwelling Ostracoda. In: F. Uiblein, J. Ott and M. Stachchowitsch (Editors), *Deep-Sea and Extreme Shallow-Water Habitats: Affinities and Adaptation.* Biosystematics and Ecology Series, Austrian Academy of Sciences, Vienna, 11: 123–143.

Daum, J., 1954. Zur Biologie einer Isopodenart unterirdischen Gewässer: *Caecosphaeroma (Vireia) burgundum* Dollfuss. *Ann. Univ. Sarav. Sci.,* 3(1/2): 104–160.

de Bovée, F., Yacoubi-Khebiza, M., Coineau, N. and Boutin, C., 1995. Influence du substrat sur la répartition des Crustacés stygobies interstitiels du Haut-Atlas occidental. *Int. Rev. Ges. Hydrobiol.,* 80(3): 453–468.

Delamare-Deboutteville, C., 1960. Biologie des eaux souterraines littorales et continentales. *Vie Milieu,* Suppl. 9: 1–740.

Dobkin, S., 1961. Early developmental stages of pink shrimp, *Penaeus duorarum* from Florida waters. *Fish. Bull.,* 61(190): 321–349.

Dole, M.J., 1983. *Le domaine aquatique souterrain de la plaine alluviale du Rhône à l'est de Lyon: écologie des niveaux supérieurs de la nappe.* Thèse Doct., 3e cycle, Univ. Lyon I, 168 pp.

Dole-Olivier, M.J. and Coineau, N., 1987. L'Isopode *Microcharon* (Crustacé) abondant dans les eaux interstitielles de l'Est lyonnais. *M. reginae* n. sp. Ecologie et biogéographie. *Stygologia,* 3(3): 200–216.

Dole-Olivier, M.J., Creuzé des Châtelliers, M. and Marmonier, P., 1993. Repeated gradients in subterranean landscape – Example of the stygofauna in the alluvial floodplain of the Rhône River (France). *Arch. Hydrobiol.,* 127(4): 451–471.

Dole-Olivier, M.J., Marmonier, P., Creuzé des Châtelliers, M. and Martin, M., 1994. Interstitial fauna associated with the alluvial floodplains of the Rhône River (France). In: J. Gibert, D.L. Danielopol and J.A. Stanford (Editors), *Groundwater Ecology.* Academic Press, New York, pp. 313–346.

Dragesco, J., 1954. Diagnoses préliminaires de quelques ciliés nouveaux des sables de Banyuls-sur-Mer. *Vie Milieu,* 4(4): 627–632.

Eigenmann, C.H., 1909. Cave vertebrates of America; a study in degenerative evolution. *Carnegie Inst. Washington Publ.,* 104: 1–241.

Fage, L., 1931. Essai sur l'évolution souterraine et son déterminisme. Aranee. *Arch. Zool. Exp. Gen. Biospeol.,* 71: 99–291.

Fong, D.W., Kane, T.C. and Culver, D.C., 1995. Vestigialization and loss of nonfunctional characters. *Annu. Rev. Ecol. Syst.,* 26: 249–268.

Futuyma, D.J., 1986. *Evolutionary Biology. Second Edition.* Sinauer Associates, Inc., Publishers, Sunderland, Massachussetts, 599 pp.

Génermont, J. and Lamotte, M., 1986. Place et rôle de l'adaptation dans l'évolution des organismes. *Boll. Zool.,* 53: 215–237.

Gibert, J., Danielopol, D.L. and Stanford, J.A. (Editors), 1994a. *Groundwater Ecology.* Academic Press, New York, 571 pp.

Gibert, J., Stanford, J.A., Dole-Olivier, M.J. and Ward, J.V., 1994b. Basis attributes of groundwater ecosystems and prospects for research. In: J. Gibert, D.L. Danielopol and J.A. Stanford (Editors), *Groundwater Ecology.* Academic Press, San Diego, pp. 7–40.

Ginet, R. and Decou, V., 1977. *Initiation à la Biologie et à l'Ecologie Souterraines.* Delarge, Paris, 345 pp.

Gould, S.J., 1977. *Ontogeny and Phylogeny.* Belkmap Press, Harvard Univers. Press, Cambridge, MA, 501 pp.

Gould, S.J. and Lewontin, R.C., 1982. L'adaptation biologique. *Recherche,* 139: 1494–1502.

Gould, S.J. and Vrba, E.S., 1982. Exaptation – a missing term in the science of form. *Paleobiology,* 8: 4–15.

Halley, E., 1986. *Etude de la Dynamique d'une Population Psammique de Proasellus walteri (Crustacé Isopode souterrain).* D.E.A. Univers. Dijon, 47 pp.

Henry, J.-P., 1976. *Recherches sur les Asellides hypogés de la lignée cavaticus (Crustacea, Isopoda, Asellota).* Thèse, Univers. Dijon, 270 pp.

Henry, J.-P. and Magniez, G., 1983. Crustacés Isopodes (principalement Asellotes). In: Introduction pratique à la systématique des organismes des eaux continentales françaises. 4. *Bull. Soc. Linn. Lyon,* 52(10): 319–357.

Holsinger, J.R., 1988. Troglobites: the evolution of cave dwelling organisms. *Am. Sci.,* 76: 146–153.

Holsinger, J.R., 1994a. Pattern and process in the biogeography of subterranean amphipods. In: D.C. Culver and J.R. Holsinger (Editors), *Biogeography of Subterranean Crustaceans: the Effects of Different Scales. Hydrobiologia,* 287(1): 131–145.

Holsinger, J.R., 1994b. Amphipoda. In: C. Juberthie and V. Decou (Editors), *Encyclopaedia Biospeologica, Vol. I.* Société de Biospéologie, Moulis, Bucarest, pp. 145–163.

Holsinger, J.R. and Culver, D.C., 1970. Morphological variation in *Gammarus minus* Say (Amphipod, Gammaridae) with emphasis on subterranean forms. *Postilla,* 146: 1–24.

Holsinger, J.R. and Culver, D.C., 1988. The invertebrate cave fauna of Virginia and a part of eastern Tennessee: zoogeography and ecology. *Brimleyana,* 14: 1–162.

Howarth, F.G., 1993. High-stress subterranean habitats and evolutionary change in cave-inhabiting arthropods. *Am. Nat.,* 142: 565–577 (supplement).

Hüppop, K. and Wilkens, H., 1991. Bigger eggs in subterranean *Astyanax fasciatus* (Characidae, Pisces). Their significance and genetics. *Z. Zool. Syst. Evolutionforsch.,* 29: 280–288.

Jeannel, R., 1923. Sur l'évolution des Coléoptères aveugles et le peuplement des grottes dans les monts du Bihar, Transylvanie. *C. R. Acad. Sci. Paris,* 176: 1070–1673.

Jeannel, R., 1926. *Faune Cavernicole de France, avec une Etude des Conditions d'Existence dans le Milieu Souterrain.* P. Lechevallier Ed, Paris, 344 pp.

Jeannel, R., 1943. *Les Fossiles Vivants des Cavernes.* Gallimard, Paris, 321 pp.

Jernigan, R.W., Culver, D.C. and Fong, D.W., 1994. The dual role of selection and evolutionary history as reflected in genetic correlations. *Evolution,* 48(3): 587–596.

Jones, R. and Culver, D.C., 1989. Evidence for selection on sensory structures in a cave population of *Gammarus minus* Say (Amphipoda). *Evolution,* 43: 688–693.

Jones, R., Culver, D.C. and Kane, T.C., 1992. Are parallel morphologies of cave organisms the result of similar selection pressures? *Evolution,* 46: 353–365.

Juberthie, C., 1975. Vie souterraine et reproduction. *Bull. Soc. Zool. France,* 100(2): 177–201.

Juberthie, C. and Decou, V. (Editors), 1994. *Encyclopaedia Biospeologica, Vol. I.* Société de Biospéologie, Moulis, Bucarest, 834 pp.

Kane, T.C. and Culver, D.C., 1992. Biological processes in space and time: Analysis of adaptation. In: A.I. Camacho (Editor), *The Natural History of Biospeleology.* Monogr. 7, Mus. Nac. Cienc. Nat., CSIC, Madrid, pp. 377–399.

Kane, T.C. and Richardson, R.C., 1985. Regressive evolution: an historical perspective. *Natl. Soc. Speleol. Bull.,* 47(2): 71–77 (special issue).

Karaman, S.L., 1959. Über die Ingolfielliden Jugoslawiens. *Biol. Glas.,* 12: 63–79.

Kimura, M., 1990. *Théorie Neutraliste de l'Evolution.* Flammarion, Paris, 472 pp.

Lamotte, M., 1994. *Théorie Actuelle de l'Evolution.* Coll. Histoire et Phylosophie des Sciences, Hachette, Paris, 444 pp.

Liem, K.F., 1990. Key evolutionary innovations, differential diversity, and symecomorphosis. In: M.H. Nitecki (Editor), *Evolutionary Innovations.* University of Chicago Press, pp. 148–170.

Lynch, M., 1990. The rate of morphological evolution in mammals from the standpoint of the neutral expectation. *Am. Nat.,* 136(6): 727–741.

Magniez, G., 1976. *Contribution à la Connaissance de la Biologie des Stenasellidae (Crustacea Isopoda Asellota des Eaux Souterraines).* Thèse, Univers. Dijon, 228 pp.

Magniez, G., 1985. Regressive evolution in stenasellids (Crustacea Isopoda Asellota of underground waters). *Natl. Soc. Speleol. Bull.,* 47(2): 118–122.

Marmonier, P. and Danielopol, D.L., 1988. Découverte de *Nannocandona faba* Eckman (Ostracoda, Candoninae) en Basse Autriche. Son origine et son adaptation au milieu interstitiel. *Vie Milieu,* 38(1): 35–48.

Marmonier, P. and Dole, M.J., 1986. Les Amphipodes des sédiments d'un bras court-circuité du Rhône: logique de répartition et réaction aux crues. *Sci. Eau,* 5: 461–486.

Marmonier, P., Vervier, P., Gibert, J. and Dole-Olivier, M.J., 1993. Biodiversity in ground waters. *Trends Ecol. Evol.,* 8(11): 392–394.

Mathieu, J., 1983. *Le Métabolisme Respiratoire de Niphargus (Amphipode hypogé). Déterminisme de sa Variabilité par la Comparaison de deux Populations de N. rhenorhodanensis.* Thèse Doc. Etat, Lyon I, 55 pp.

Mathieu, J. and Turquin, M.J., 1992. Biological processes at the population level. II. Aquatic populations *Niphargus* (stygobiont amphipod) case. In: A.I. Camacho (Editor), *The Natural History of Biospeleology.* Monogr. 7, Mus. Nat. Cie. Nat., CSIC, Madrid, pp. 263–293.

Mayr, E., 1963. *Animal Species and Evolution.* Belknap Press, Havard Univ. Press, Cambridge, MA.

Monniot, F., 1971. Les Ascidies littorales et profondes des sédiments meubles. *Smithson. Contrib. Zool.,* 76: 119–126.

Müller, G.B., 1990. Developmental mechanisms at the origin of morphological novelties: a side-effects hypothesis. In: M.H. Nitecki (Editor), *Evolutionary Innovations.* University of Chicago Press, pp. 100–130.

Newman, W.A., 1983. Origin of the Maxillopoda; Urmalacostracan ontogeny and progenesis. In: F.R. Schram (Editor), *Crustacean Issues 1, Crustacean Phylogeny.* A.A. Balkema, Rotterdam, pp. 105–119.

Nitecki, M.H., 1990. The plurality of evolutionary innovations. In: M.H. Nitecki (Editor), *Evolutionary Innovations.* University of Chicago Press, pp. 3–18.

Noodt, W., 1964. Natürliches System und Biogeographie der Syncarida (Crustacea: Malacostraca). *Gewässer Abwässer,* 37–38: 77–186.

Notenboom, J., 1991. Marine regressions and the evolution of groundwater dwelling amphipods (Crustacea). *J. Biogeogr.,* 18: 437–454.

Notenboom, J., Plénet, S. and Turquin, M.J., 1994. Groundwater contamination and its impact on groundwater animals and ecosystems. In: J. Gibert, D.L. Danielopol and J.A. Stanford (Editors), *Groundwater Ecology.* Academic Press, New York, pp. 477–504.

Packard, A.S., 1888. The cave fauna of North America, with remarks on the anatomy of the brain and the origin of blind species. *Mem. Natl. Acad. Sci. USA,* 4: 1–156.

Parzefall, J., 1992. Biological processes. Behavioural aspects in animals living in caves. In: A.I. Camacho (Editor), *The Natural*

History of Biospeleology. Monogr. 7, Mus. Nac. Cie. Nat., CSIC, Madrid, pp. 327–376.

Pospisil, P., 1994. The groundwater fauna of a Danube aquifer in the "Loban" Wetland in Vienna, Austria. In: J. Gibert, D.L. Danielopol and J.A. Stanford (Editors), *Groundwater Ecology.* Academic Press, New York, pp. 347–366.

Poulson, T.L., 1963. Cave adaptation in amblyopsid fishes. *Am. Midl. Nat.,* 70: 257–290.

Poulson, T.L., 1985. Evolutionary reduction by neutral mutations: plausibility arguments and data from amblyopsid fishes and limyphiid spiders. *Natl. Soc. Speleol. Bull.,* 47 (2): 109–117.

Poulson, T.L. and White, W.B., 1969. The cave environment. *Science,* 165: 971–981.

Quiring, R., Walldorf, U., Kloter, K. and Gehring, W., 1994. Homology of eyeless gene of *Drosophila* to the small eye gene in mice and Aniridia in humans. *Science,* 265: 785–789.

Racovitza, E.G., 1907. Essai sur les problèmes biospéologiques. Biospeologica I. *Arch. Zool. Exp. Gen.,* 4: 371–488.

Raff, A.R., Parr, B.A., Parks, A.L. and Wray, G.A., 1990. Heterochrony and other mechanisms of radical evolutionary change in early development. In: M.H. Nitecki (Editor), *Evolutionary Innovations.* University of Chicago Press, pp. 71–98.

Remane, A., 1956. *Die Grundlagen des natürlichen Systems der Vergleichen den Anatomie und der Phylogenetik.* Akad. Verlag Geest. Portig, Leipzig, 400 pp.

Renaud-Mornant, J. and Coineau, N., 1978. Etude anatomique des isopodes interstitiels. I. *Angeliera phreaticola* Chap. et Del., 1952. Données gérérales. *C. R. Acad. Sci. D,* 286(171): 1249–1252.

Rieger, R.M., 1980. A new group of interstitial worms, Lobatocerebridae nov. fam. (Annelida) and its significance for metazoan phylogeny. *Zoomorphologie,* 95: 41–84.

Rogulj, B., Danielopol, D.L., Marmonier, P. and Pospisil, P., 1993. Adaptive morphology, biogeographical distribution and ecology of the species group *Mixtacandona bravensis* (Ostracoda Candoninae). *Mem. Biospeol.,* 20: 195–207.

Romero, A., 1985. Can evolution regress? *Natl. Soc. Speleol. Bull.,* 47(2): 86–88.

Rouch, R., 1968. Contribution à la connaissance des Harpacticides hypogés (Crustacés-Copépodes). *Ann. Limnol.,* 23(1): 1–167.

Rouch, R., 1988. Sur la répartition spatiale des Crustacés dans le sous-écoulement d'un ruisseau des Pyrénées. *Ann. Limnol.,* 24: 213–234.

Rouch, R., 1991. Structure et peuplement des Harpacticides dans le milieu hyporhéïque d'un ruisseau des Pyrénées. *Ann. Limnol.,* 27: 227–234.

Rouch, R., 1994. Copepoda. In: C. Juberthie and V. Decou (Editors), *Encyclopaedia Biospelogica, Vol. I.* Société de Biospéologie, Moulis, Bucarest, pp. 107–111.

Rouch, R. and Danielopol, D.L., 1987. L'origine de la faune aquatique souterraine, entre le paradigme du refuge et le modèle de colonisation active. *Stygologia,* 4: 345–372.

Ruffo, S., 1951. *Ingolfiella Leleupi* n.sp. nuovo troglobio del Congo Belgo. *Rev. Zool. Bot. Afr.,* 44: 189–209.

Ruffo, S. and Vigna Taglianti, A., 1977. Secondo contributo alla conoscenza del genere *Bogidiella* in Messico e Guatemala (Crustacea, Amphipoda, Gammaridae). *Accad. Nazion. Lincei,* 374(171): 125–172.

Schminke, H.K., 1973. Evolution, System und Verbreitungsgeschichte der Familie Parabathynellidae (Bathynellacea: Malacostraca). *Akad. Wiss. Lit. Mainz Math. Kl. Mikrofauna Meeresbodens,* 24: 1–192.

Schminke, H.K., 1974. *Psammaspides williamsi* gen. n., sp. n., ein Vertreter einer neuen Familie mesopsammaler Anaspidacea (Crustacea, Syncarida). *Zool. Scr.,* 3(4): 177–183.

Schminke, H.K., 1975. Phylogenie und Verbreitungsgeschichte der Syncarida (Crustacea, Malacostraca). *Verh. Dtsch. Zool. Ges.,* 1974: 384–388.

Schminke, H.K., 1976. The ubiquitous telson and deceptive furca. *Crustaceana,* 30(3): 292–300.

Schminke, H.K., 1978. Die Phyllogenetische Stellung der Stygocarididae (Crustacea: Syncarida). Unter besonderer Berücksichtigung morphologischer Ähnlichkeiten mit Larven-formen der Eucarida. *Z. Zool. Syst. Evolutionforsch.,* 16: 225–239.

Schminke, H.K., 1981. Adaptation of Bathynellacea (Crustacea Syncarida) to the life in the interstitial ("Zoea Therory"). *Int. Rev. Ges. Hydrobiol.,* 66(4): 575–637.

Schram, F.R., 1984. Relationships within Eumalacostracan Crustacea. *Trans. San Diego Soc. Nat. Hist.,* 20(16): 301–312.

Serban, E., 1972. Bathynella. *Trav. Inst. Speologie Emile Racovitza,* 11: 10–225.

Serban, E. and Coineau, N., 1975. Sur les Bathynellidae (Podophallocarida: Bathynellacea) d'Afrique du Sud. Les genres *Transvaalthynella* nov. et *Transkeithynella* nov. *Ann. Speleol.,* 30(1): 137–165.

Sket, B., 1985. Why all cave animals do not look alike – a discussion on adaptative value of reduction process. *Natl. Soc. Speleol. Bull.,* 47(2): 78–85.

Smith, G., 1908. On the Anaspidacea, living and fossil. *Q. J. Microsc. Sci.,* 53: 489–578.

Stearns, S.C., 1992. *The Evolution of Life Histories.* Oxford Univ. Press, 249 pp.

Stock, J.H., 1980. Regression Model Evolution as examplified by the genus *Pseudoniphargus* (Amphipoda). *Bijdr. Dierkd.,* 50: 105–144.

Stock, J.H., 1986. Thermosbanacea. In: L. Botosaneanu (Editor), *Stygofauna Mundi.* E.J. Brill/Dr W. Backhuys, Leiden, pp. 585–588.

Strayer, D.L., 1994. Limits to biological distributions in groundwater. In: J. Gibert, D.L. Danielopol and J.A. Stanford (Editors), *Groundwater Ecology.* Academic Press, San Diego, pp. 287–310.

Swedmark, B., 1964. The interstitial fauna of marine sand. *Biol. Rev. Cambridge Philos. Soc.,* 39(1): 1–42.

Turquin, M.J., 1975. Cycle biologique et rythme de ponte chez les Crustacés aquatiques troglobies. *Bull. Soc. Zool. Fr.,* 100(2): 169–176.

Vandel, A., 1964. *Biospéologie, la Biologie des Animaux Cavernicoles.* Gauthier-Villars, Paris, 619 pp.

Wägele, J.W., 1983. *Protocerberus* gen. n. und *Afrocerberus* gen. n., neue limnische Microcerberidea aus Africa (Crustacea: Isopoda). *Bull. Zool. Mus. Amsterdam,* 9(8): 65–74.

Westheide, W., 1984. The concept of reproduction in polychaetes with small body size: adaptations in interstitial species. In: *Fortschritte der Zoologie, Bd 29: Polychaete Reproduction.* Gustav Fischer Verlag, Stuttgart, pp. 265–287.

Westheide, W., 1987. Progenesis as a principle in meiofauna evolution. *J. Nat. Hist.,* 21: 843–854.

Wilkens, H., 1988. Evolution and genetics of epigean and cave *Astyanax fasciatus* (Characidae, Pisces). Support for the neutral theory. In: M.K. Hecht and B. Wallace (Editors), *Evol. Biol.*, 23: 271–367.

Wilkens, H., 1992. Neutral mutations and evolutionary progress. In: A.I. Camacho (Editor), *The Natural History of Biospeleology*. Mongr. 7, Mus. Nat. Cie. Nat., CSIC, Madrid, pp. 401–422.

Wilkens, H., 1993. Neutral Mutationen und evolutionäre Fortentwicklung. *Z. Zool. Syst. Evolutionsforsch.*, 31: 98–109.

Zilliox, L., 1994. Porous media and aquifer systems. In: J. Gibert, D.L. Danielopol and J.A. Stanford (Editors), *Groundwater Ecology*. Academic Press, San Diego, pp. 69–96.

Chapter 10

ACOUSTIC COMMUNICATION IN DARKNESS

Hannelore HOCH

INTRODUCTION

In arthropods, several studies have been devoted to morphological, physiological, and developmental adaptations to the cave environment (Culver, 1982). Comparatively little, however, is known about the behavioral alterations which accompany a drastic ecological shift such as that from a surface-dwelling to a cavernicolous way of life. One of the main questions concerning the behavioral traits of cavernicolous arthropods is: how do obligate cavernicoles communicate? How do they cope with permanent darkness, not only in regard to their own orientation, but especially in regard to recognition and location of conspecific partners?

To ensure survival and reproductive success of a given species, it is essential for conspecific organisms to communicate – to interact, so as to pass on information to potential partners. Generally, mobile, sexually reproducing organisms have developed specific behavioral patterns which serve to bring together conspecific males and females for mating. These **specific mate-recognition systems** (SMRS: Paterson, 1985) may consist of optical, chemical, tactile or acoustic signals, or a combination thereof. The comparative analysis of specific mate-recognition systems in epigean and cavernicolous taxa will essentially contribute to understanding of how communication among individuals of a given species works in a particular environment. It may also provide insight into the evolution of communication systems and their role in cave adaptation and speciation. It is evident that specific mate-recognition systems of obligate cavernicolous species are dependent on selective forces different from those of species in epigean habitats. My own research on small insects of the order Homoptera (Auchenorrhyncha: Fulgoroidea) or **planthoppers** have revealed a specific mate-recognition system which is nearly ideal for studying its function as well as its evolution. In this chapter, I shall focus on the specific mate-recognition system of planthoppers and its alteration during cave adaptation, and outline some evolutionary inferences that arose from my studies.

Epigean planthoppers (Fulgoroidea) comprise about 15 000–20 000 species worldwide. Since the first record of a cavernicolous planthopper species from Madagascar (Synave, 1953), more than 60 cavernicolous species (from four out of 18 families: Cixiidae, Delphacidae, Kinnaridae and Meenoplidae) have been described from many parts of the world. They are known from Galápagos, Hawaii, Western Samoa, New Caledonia, New Guinea, New Zealand, Australia, southern France, the Balearic Islands, the Canary Islands, the Azores, Jamaica and Mexico (Hoch, 1994) and Réunion (unpublished data).

Cave-dwelling planthoppers differ strongly from their epigean relatives by displaying troglomorphies which are usually associated with cave-adaptation in arthropods, such as the reduction of compound eyes, ocelli, tegmina, wings and bodily pigmentation (Figs. 10.2–10.4), the most highly modified taxa being completely eyeless, flightless and unpigmented. Besides these reductive characters, other morphological specializations in response to their subterranean habitat have been recognized, such as optimization of body shape to facilitate movement through small voids and cracks (Hoch and Howarth, 1989b), waxy exudations along tegminal margins, perhaps to repel potential predators (Fig. 10.4), and special configuration of leg spinulation and pretarsal claws which might facilitate walking on wet, rocky surfaces (Howarth, 1981; Hoch and Howarth, 1989b). Most troglobitic species seem to be restricted to the deep cave zone, where high humidity and constant temperature prevail (Howarth, 1983); some even appear to be especially adapted to

Fig. 10.1. Root curtains of *Metrosideros polymorpha* in Lanikai Cave, Hawaii Island.

environments with abnormally high concentrations of carbon dioxide in the air (Howarth and Stone, 1990). Like their epigean relatives, cavernicolous planthoppers are phytophagous, but have switched their sap-sucking mode of feeding from the green parts of living plants to roots which penetrate the cave ceiling (Howarth, 1973, 1983), and can be abundant in caves, especially in lava tubes where passages often run parallel to the surface (Fig. 10.1). It has been hypothesized that exploitation of this previously unutilized food resource might have been the driving force for the evolution of cave planthoppers (adaptive shift hypothesis: Howarth, 1986). All cavernicolous species of Fulgoroidea belong to groups where the immature stages live close to the soil (e.g., under the dead bark of rotting logs, in leaf litter or moss) or even within the soil, feeding on roots or perhaps on fungi (Remane and Hoch, 1988). From this plateau of ecological preadaptation, it seems to be a small evolutionary step to switch to a permanent life underground (Hoch, 1994). Within the cave ecosystem, rhizophagous Fulgoroidea are primary consumers (Howarth, 1973, 1981).

THE SPECIFIC MATE RECOGNITION SYSTEM OF SURFACE-DWELLING PLANTHOPPERS

Well-known relatives of the planthoppers are the cicadas (Cicadidae), which are not only much bigger in body size, but are also famous for the loud, air-borne signals emitted by male individuals only. In contrast, the majority of planthoppers are less than a few millimetres long. It has been shown that, because of physical limitations of their size, small insects would need to use very high frequency (or ultrasonic) sounds (Mebes, 1978; Michelsen, 1983) to communicate over longer distances. These, however, are not suitable for communication in structurally complex habitats characterized by dense vegetation (Michelsen, 1983) such as grasslands. Planthoppers – along with other small insects – have thus developed an alternative way to pass on information, namely by the use of low-frequency vibrations (100–2000 Hz, with maximum energy at 100–500 Hz: de Vrijer, 1984). These vibrations are transmitted through a dense substrate which usually is the host plant (Ichikawa, 1976; Bell, 1980; Michelsen et al., 1982). Vibrational "songs" apparently have an air-borne component which can be detected by suitable microphones, for instance, a particle-displacement microphone (Bennet-Clark, 1984), with an intensity of 25 to 35 dB at a distance of only 1 cm in air from the emitter (Claridge, 1985). Such low-intensity signals are estimated not to stimulate even sensitive receptor organs over distances greater than a few centimetres, whereas the vibrational components in the plant substrate may be detected over distances of at least 1–2 m (Michelsen et al., 1982). Although only comparatively few species have been studied in detail, it is clear that the vibrational signals play an important role in recognition and location of sexual partners. The sexual behaviour patterns of planthoppers can be rather complicated (Booij, 1982), involving a repertoire of call structures according to the particular situation, for instance, distant communication between potential mates by spontaneous calling (initiated by either male or female), searching, and courtship in the strict sense. The acoustic signals of males are usually more differentiated and characteristic than in those of females (Booij, 1982). The acoustic signalling is often accompanied by wing-fluttering of the male (Strübing,

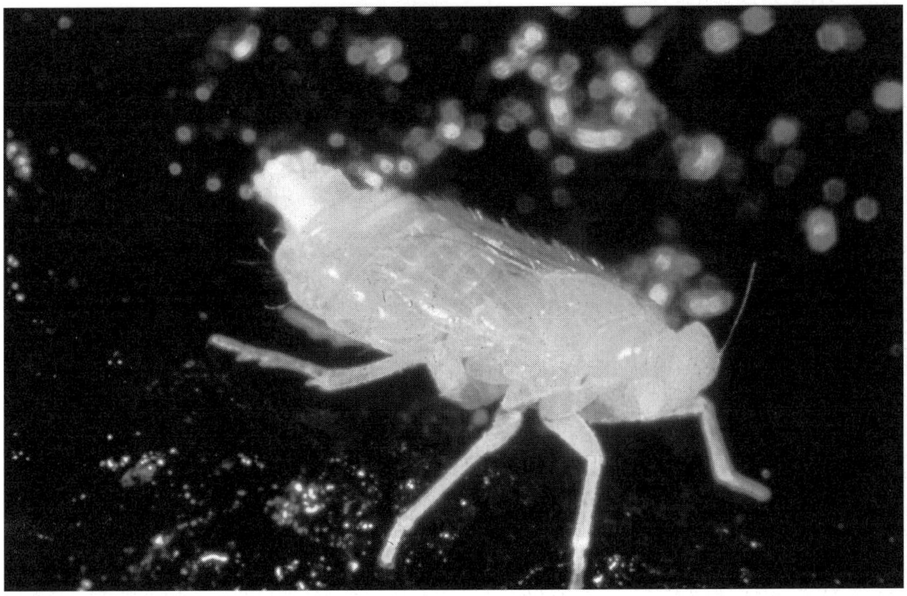

Fig. 10.2. *Oliarus polyphemus,* adult female (body length ~4 mm).

1960; Ichikawa, 1976; Booij, 1982), which perhaps provides an additional optical stimulus for the female.

THE SPECIFIC MATE RECOGNITION SYSTEM OF CAVERNICOLOUS PLANTHOPPERS

Lessons from Hawaiian and Australian cave planthoppers

Recordings of substrate-borne vibratory signals of cave-dwelling planthoppers were obtained from various populations of *Oliarus polyphemus* (Cixiidae) in lava tubes on the Island of Hawaii (Hoch and Howarth, 1993), and from two *Solonaima* species (Cixiidae) from limestone and lava caves in northeastern Australia. *Oliarus polyphemus* individuals display characteristic troglomorphies of obligate cavernicolous insects: the compound eyes are entirely missing, and tegmina and wings as well as bodily pigmentation are strongly reduced (Fig. 10.2). The two *Solonaima* species studied differ considerably in their degree of troglomorphy. *Solonaima pholetor* is considered facultatively cavernicolous; it shows only a slight reduction of the compound eyes (Fig. 10.3). Tegmina, wings and bodily pigment are well developed. In contrast, *Solonaima baylissa* is a highly modified, troglobitic species which is completely eyeless, flightless and unpigmented (Fig. 10.4). *Solonaima pholetor* is found in several isolated karst towers around Chillagoe, 130 km west of Cairns, while *S. baylissa* is restricted to the Undara lava tube system, ~170 km southwest of Cairns (Hoch and Howarth, 1989b).

General courtship pattern

Here, I refer mainly to observations on Hawaiian species of *Oliarus,* since the information on reproductive behaviour of the Australian cave planthoppers is still too fragmentary to draw general conclusions.

All surface-dwelling *Oliarus* species possess well-developed compound eyes, therefore it seems likely that visual clues are a component of courtship behaviour. In contrast, *O. polyphemus* is completely eyeless, thus excluding any visual information from courtship. Also, there is no evidence of chemical communication (e.g., by pheromones) in planthoppers. Thus, one can assume that, in troglobitic cixiids, substrate vibrations are the most important (if not the only) means by which potential mates can recognize each other (Hoch and Howarth, 1993).

Recording the vibratory communication signals of cave planthoppers

Recordings were made in the field (Hawaii, Australia) and in the laboratory (Hawaii). To record the vibrational signals, a magneto-dynamic system (Strübing

Fig. 10.3. *Solonaima pholetor*, adult female (body length ~8 mm).

Fig. 10.4. *Solonaima baylissa*, adult female (body length ~4.5 mm).

and Rollenhagen, 1988) was used, which operates on the principle of electromagnetic induction; with the help of a small magnet, mechanical oscillations (caused by the vibrating insect and transferred to the substrate via its legs) are transformed by an induction coil into electrical oscillations, which are then recorded on tape. In the process, these low-intensity signals are amplified by a factor of ~1000, and thus become audible to the human ear. For testing the localization and efficiency of vibrational communication in the natural habitat, the recording system was set up in a prehistoric lava tube of Kilauea volcano on Hawaii Island (Pahoa Cave) (Fig. 10.5) which is known to house a large and stable population

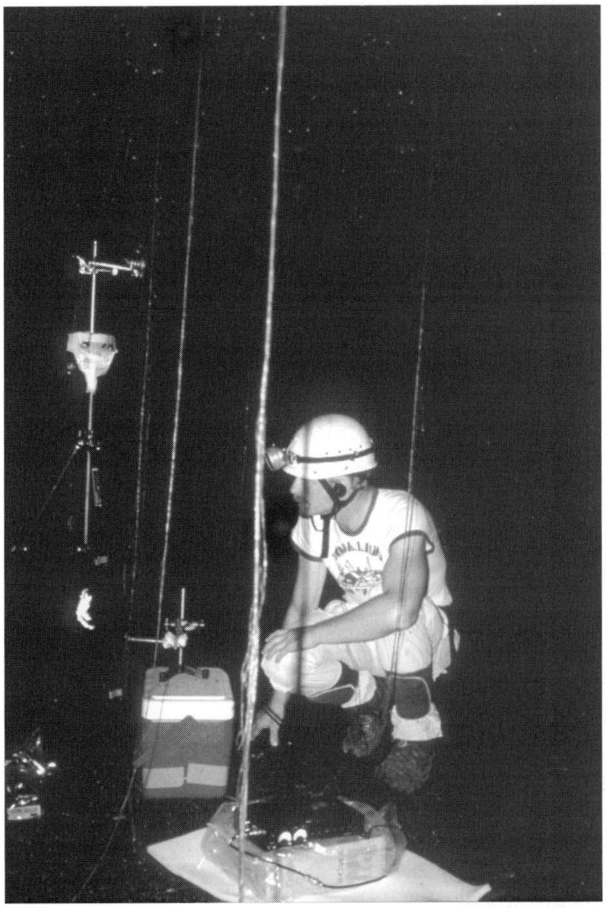

Fig. 10.5. Testing the magneto-dynamic recording system in Pahoa Cave, Hawaii Island.

of *Oliarus polyphemus*. Roots of the endemic tree *Metrosideros polymorpha* (Myrtaceae), the host plant of *O. polyphemus,* penetrate the cave ceiling and dangle into the cave, forming "root curtains". Young-instar nymphs (I–IV) of *O. polyphemus* are mostly found feeding on these roots, whereas V. instar nymphs (ready to moult) and adults (as the dispersal stage) are often also seen wandering on the rock substrate of ceiling, walls and floor (Hoch and Howarth, 1993). Thus, both roots and rock surface were conceivable transmission substrates.

In the experiment, both a living *Metrosideros* root of ~2.50 m length (average diameter ~3–4 mm) and the rock of the cave wall served as test substrates. Signals of an *Oliarus polyphemus* individual previously recorded in the laboratory were played back as vibrations to the two substrates with the help of a modified loudspeaker; a pin attached to the membrane ensured transmission of purely vibrational signals. To determine maximal transmission distance, recordings were made at different distances between loudspeaker (source of played-back signal) and magnet/coil (2 cm–2.50 m).

Our studies revealed that cave planthoppers maintain the communication system of their epigean relatives during cave adaptation, communicating by substrate-borne vibrations. The field experiments showed that communication by vibratory signals is extraordinarily efficient within the cave environment; living root material is an especially well-suited substrate for transmitting low-frequency vibrations (Hoch and Howarth, 1993). In our experiment in Pahoa Cave, signals could be detected at distances up to 2.50 m from the vibration source, and may be transmitted along root curtains through interconnecting rootlets branching off the main stems. In contrast, rocky substrates proved to have extremely poor transmitting characteristics – even intense banging with a forceps was not detectable a few centimetres away from the source.

ALTERATION OF REPRODUCTIVE BEHAVIOUR PATTERN IN CAVERNICOLOUS PLANTHOPPERS

Hawaii

Structure of calls

Compared to courtship signals of epigean planthoppers (e.g., Claridge, 1985; de Vrijer, 1986), the calls of troglobitic *Oliarus* individuals appeared drastically reduced in complexity, their time–amplitude pattern consisting of more or less homogeneous pulse trains (Fig. 10.6). Moreover, in contrast to the few epigean species studied, both sexes displayed similar call structures.

Within each population inter-individual variation of the vibrational signals was found to be larger than intra-individual variation, although overall variation was comparatively small. Amazingly, however, comparison of various populations of *Oliarus polyphemus* from different lava tubes showed that variation in song pattern between these populations was extraordinarily high. Consequently, Hoch and Howarth (1993) assumed that *O. polyphemus,* previously considered a widely distributed species, has to be regarded as a complex of reproductively isolated populations – that is, separate species – and developed a hypothesis to explain evolutionary divergence in dynamic habitats (Hoch and Howarth, 1993; Hoch, 2000).

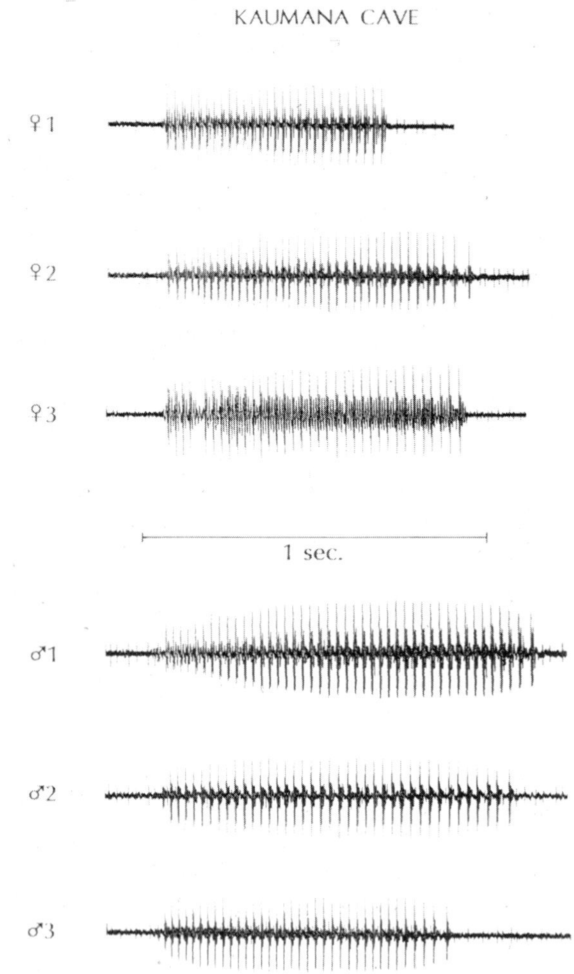

Fig. 10.6. Oscillograms showing variation of male and female calls within the *Oliarus polyphemus* population from Kaumana Cave, Hawaii Island.

General pattern of reproductive behaviour

In all populations of *Oliarus polyphemus* studied, the following general pattern of sexual behaviour was observed: in the majority of male–female interactions, the female emitted spontaneous calls – that is, the female performed the initial step of courtship behaviour. In cases where these initial calls remained unanswered, the female eventually stopped calling. In cases where a nearby male answered these calls, the female would remain stationary, emitting calls at regular intervals. The male would then orientate itself towards the calling female, but calling less often than the female and at irregular intervals. In all observed cases courtship persisted for about an hour before copulation commenced. Copulations lasted from 36 to 57 minutes. While the animals were in copula, no calls were emitted. In none of the observed male–female interactions displaying the courtship pattern described, was a song-active female found to reject a responding male, or to show escape reaction during male mating attempts (Hoch and Howarth, 1993).

Australia

Structure of calls

In the facultative cavernicolous *Solonaima* species, *S. pholetor* from Carpentaria Cave, the structure of the vibrational signal was observed to be extremely rudimentary: calls of males and females consisted of simple "clicks" (Fig. 10.7, lower). Amazingly, these signals contained sufficient information to lead eventually to successful courtship and copulation. In contrast, males of the highly troglomorphic, obligate cavernicolous *S. baylissa* have much more differentiated calling signals, consisting of three distinguishable elements: two initial chirps, a longer trill, and a final pulse (Fig. 10.7, upper).

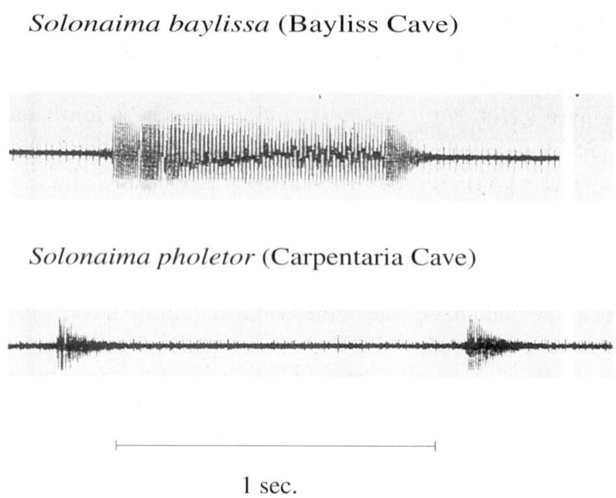

Fig. 10.7. (Upper) male call structure of *Solonaima baylissa*, Undara, Queensland; (lower) male call structure of *Solonaima pholetor*, Chillagoe, Queensland.

General pattern of sexual behaviour

From both cavernicolous *Solonaima* species, observations and recordings were obtained under field conditions. For each species, only few male–female interactions were observed. Thus, information is too sporadic to recognize a specific behaviour pattern.

POSSIBLE ADAPTIVE VALUE OF OBSERVED BEHAVIORAL ALTERATIONS DURING CAVE ADAPTATION AND EVOLUTIONARY INFERENCE

Reduction of signal complexity

It seems remarkable that, although substrate-borne vibratory signals are apparently the most important component of the specific mate-recognition system of the cavernicolous planthoppers, the signal complexity is greatly reduced in comparison with epigean taxa. Two alternative hypotheses to explain this phenomenon are conceivable:

1. Energy-economy hypothesis. Energy economy is most commonly invoked as a selective agent involved in reductive evolution of cave animals. Many cave organisms are strongly food-limited, so that a mutation causing the reduction of structures which are apparently "useless" in the cave environment (such as eyes, wings and pigment), or of energy-consuming behavioral traits, will have a selective advantage (Culver, 1982).

2. Ethological-release hypothesis. When organisms are released from the competitive pressure of related species, selection on highly specific sounds may be less intense (Booij, 1982). "When we have a great variety of allied sympatric species there must be an emphasis on very accurate mating barriers which is unnecessary where virtually no sympatric allies occur" (Hutchinson, 1958).

The cavernicolous planthoppers from Australia and Hawaii provide an excellent model to test these hypotheses by displaying different degrees of troglomorphy and by different numbers of competing species (i.e., species using the same communication system) in the same habitat.

Both in Australia and in Hawaii, the assumption of limited food resources does not hold true for the troglobitic planthoppers; roots of surface-living plants are abundant all the year round, and constitute an ample energy source which is utilized by rhizophagous troglobites. These are the primary consumers of the food web within the corresponding cave ecosystem. Thus, I conclude that the special quality of the habitat in regard to energy content is not likely to be a major selective agent involved in reduction of signal complexity. Instead, the scenario described by Hutchinson (1958) seems to provide a more plausible way of explaining the observed reduction of signal complexity. In Hawaii, cavernicolous *Oliarus* species occur largely allopatrically, and in any given cave the species present is the only cavernicolous planthopper (with one exception – see Hoch and Howarth, 1993). Consequently, competitive pressure by sympatric allies to maintain high signal complexity in order to bring together the sexes of the same species does not exist. Thus, maintaining a signal with minimal complexity may have been sufficient to serve the purpose of mate recognition and location.

The situation is different in the *Solonaima* species from Australian lava tubes and limestone caves. *Solonaima pholetor,* the less troglomorphic species, displays a much higher degree of signal reduction than the strongly troglomorphic, obligate cavernicolous *S. baylissa.* In the cave where *S. pholetor* is found, it is the only cave-dwelling cixiid species, whereas the *S. baylissa* population studied coexists with another cavernicolous cixiid species of a different taxon (*Undarana*: Hoch and Howarth, 1989a). Individuals of the two species have been found on the same root. Compared to signals of the other cave planthoppers studied, *S. baylissa* maintains a comparatively complex song structure with little variation. Although *Undarana* is not a close relative of *S. baylissa,* maintaining this complexity might be necessary to ensure specific mate recognition in an environment where mate recognition is apparently totally dependent on acoustic clues.

While these results provide evidence that the reduction of a complex call structure may be attributed to ethological release, the factors responsible for maintaining the small degree of call variability (e.g., Fig. 10.6) observed in all populations of cavernicolous planthoppers studied so far, remain to be studied. Clues to our understanding of the evolutionary history of the specific mate-recognition system of cavernicolous planthoppers may lie in the populations' past (e.g., founder effects, genetic drift, bottleneck events) and/or in specific factors relating to reproductive behaviour (e.g., sexual selection).

Change of courtship pattern

An important question in regard to the evolutionary history of cavernicolous taxa is whether and how patterns of mating behaviour change during adaptation to a permanent life in subterranean habitats. As mentioned previously, visual and chemical clues on potential partners are not available to troglobitic cixiids; thus all critical information must be contained in the vibratory communication signals. It is remarkable that, in nearly all cases observed, *Oliarus polyphemus* females initiated calling, remained the more song-active partner throughout the courtship, and were never observed to reject the responding male that had been successful in locating her (Hoch and Howarth, 1993). This behaviour differs markedly from observations made in epigean planthoppers (e.g., in the family Delphacidae). There the males usually initiate calling, and in the genus *Javesella* were observed to emit a few successive calls; then they stopped calling for a variable amount of time, walked off the grass stem, and produced a new series of calling signals when they were placed back on the grass stem. In the field, this strategy is probably used by males to explore acoustically the usually dense vegetation (e.g., grass tussocks, etc.) of their natural habitat, to locate receptive females (de Vrijer, 1986).

To confirm whether the switch to female calling occurred before or during cave-adaptation in Hawaiian *Oliarus* one needs to record the courtship and communication behaviour of epigean relatives. In any case, the observed behaviour of cave-dwelling *O. polyphemus* could be a strategy to economize in mate-location. The fact that song-active females readily accepted responding males for copulation indicates that song activity in females is strongly correlated with receptiveness. The female is actively advertising a mating opportunity to searching males. Consequently, males are aware of and respond to females only when the latter are ready to mate, and in this way invest less energy in exploring the environment acoustically in search for a receptive female, but may instead be able to cover a wider area. In the natural habitat, the female's signals are perceptible for males within an area with a radius of at least 2.50 m. The first male to reach the female will be the most likely to perform a successful mating. The apparent low motility of the female during courtship may facilitate the male's attempts to locate her. Both low song-activity in the male and low motility of the female may also minimize the risk of predation (Hoch and Howarth, 1993) by other troglobitic arthropods, such as the spiders and crickets which also colonize the lava tubes (Howarth, 1981).

ACKNOWLEDGMENTS

I would like to express my sincere thanks to all colleagues and friends who have supported my work on cavernicolous planthoppers over the years for their continuing enthusiasm, their patient help in the field under sometimes rather uncomfortable conditions and their stimulating discussions. A particularly warm *mahalo* goes to Dr. F.G. Howarth, Bishop Museum, Honolulu, for introducing me to the underground realm of Hawaiian and Australian caves. I am especially indebted to Dr. Manfred Asche, my long-time life and field companion, for his everlasting logistic and emotional support. Without him, my work on cave planthoppers would not have been possible. Finally, I also thank the editors for inviting me to contribute to this volume. Financial support was provided by the Deutsche Forschungsgemeinschaft and the Hawaii Conservation Biology Initiative of The Nature Conservancy of Hawaii.

REFERENCES

Bell, P., 1980. Transmission of vibrations along plant stems: implications for insect communication. *J. N. Y. Entomol. Soc.*, 88: 210–216.

Bennet-Clark, H.C., 1984. A particle velocity microphone for the songs of small insects and other acoustic measurements. *J. Exp. Biol.*, 108: 459–463.

Booij, C.J.H., 1982. Biosystematics of the *Muellerianella* complex (Homoptera, Delphacidae), interspecific and geographic variation in acoustic behaviour. *Z. Tierpsychol.*, 58: 31–52.

Claridge, M.F., 1985. Acoustic signals in the Homoptera: behavior, taxonomy, and evolution. *Annu. Rev. Entomol.*, 30: 297–317.

Culver, D.C., 1982. *Cave life. Evolution and Ecology*. Harvard University Press, Cambridge, MA, 189 pp.

de Vrijer, P.W.F., 1984. Variability in calling signals of the planthopper *Javesella pellucida* (F.) (Homoptera: Delphacidae) in relation to temperature, and consequences for species recognition during distant communication. *Neth. J. Zool.*, 34(3): 388–406.

de Vrijer, P.W.F., 1986. Species distinctiveness and variability of acoustic calling signals in the planthopper genus *Javesella* (Homoptera: Delphacidae). *Neth. J. Zool.*, 36(1): 162–175.

Hoch, H., 1994. Homoptera (Auchenorrhyncha: Fulgoroidea). In: C. Juberthie and V. Decu (Editors), *Encyclopedia Biospeologica*. Société de Biospéologie, Moulis-Bucarest, pp. 313–325.

Hoch, H., 2000. The Hawaiian cave planthoppers (Homoptera: Cixiidae) – a model for rapid subterranean speciation? *Int. J. Speleol.*, in press.

Hoch, H. and Howarth, F.G., 1989a. Reductive evolutionary trends in two new cavernicolous species of a new Australian cixiid genus (Homoptera: Fulgoroidea). *Syst. Entomol.*, 14: 179–196.

Hoch, H. and Howarth, F.G., 1989b. Six new cavernicolous cixiid planthoppers in the genus *Solonaima* from Australia (Homoptera: Fulgoroidea). *Syst. Entomol.*, 14: 377–402.

Hoch, H. and Howarth, F.G., 1993. Evolutionary dynamics of behavioral divergence among populations of the Hawaiian cave-dwelling planthopper *Oliarus polyphemus* (Homoptera: Fulgoroidea: Cixiidae). *Pac. Sci.*, 47(4): 303–318.

Howarth, F.G., 1973. The cavernicolous fauna of Hawaiian lava tubes. 1. Introduction. *Pac. Insects,* 15: 139–151.

Howarth, F.G., 1981. Community structure and niche differentiation in Hawaiian lava tubes, In: D. Mueller-Dombois, K.W. Bridges and H.L. Carson (Editors), *Island Ecosystems.* US/IBP Synthesis Series 15, pp. 318–336.

Howarth, F.G., 1983. Ecology of cave arthropods. *Annu. Rev. Entomol.,* 28: 365–389.

Howarth, F.G., 1986. The tropical cave environment and the evolution of troglobites. In: *Proceedings of the 9th International Congress of Speleology, Vol. II, Barcelona, Spain.* pp. 153–155.

Howarth, F.G. and Stone, F.D., 1990. Elevated carbon dioxide levels in Bayliss Cave, Australia: implications for the evolution of obligate cave species. *Pac. Sci.,* 44(3): 207–218.

Hutchinson, G.E., 1958. Concluding remarks. *Cold Spring Harbor Symp. Quant. Biol.,* 22: 415–427.

Ichikawa, T., 1976. Mutual communication by substrate vibration in the mating behavior of planthoppers (Homoptera: Delphacidae). *Appl. Entomol. Zool.,* 11: 8–23.

Mebes, H.D., 1978. Zur Biophysik der Schallerzeugung bei Kleinzikaden. *Forma Functio,* 7: 95–118.

Michelsen, A., 1983. Biophysical basis of sound communication. In: B. Lewis (Editor), *Bioacoustics: a Comparative Approach.* Academic Press, London, pp. 3–38.

Michelsen, A., Fink, F., Gogala, M. and Traue, D., 1982. Plants as transmission channels for insect vibrational songs. *Behav. Ecol. Sociobiol.,* 11: 269–281.

Paterson, H.E.H., 1985. The recognition concept of species. In: E.S. Vryba (Editor), Species and Speciation. *Transvaal Mus. Monogr.,* 4: 21–30.

Remane, R. and Hoch, H., 1988. Cave-dwelling Fulgoroidea (Homoptera: Auchenorrhyncha) from the Canary Islands. *J. Nat. Hist.,* 22: 403–412.

Strübing, H., 1960. Paarungsverhalten und Lautäußerung von Kleinzikaden, demonstriert an Beispielen aus der Familie der Delphacidae (Homoptera: Auchenorrhyncha). *Verh. XI Int. Kongr. Entomol. Wien,* 3: 12–14.

Strübing, H. and Rollenhagen, T., 1988. Ein neues Aufnehmersystem für Vibrationssignale und seine Anwendung auf Beispiele aus der Familie Delphacidae (Homoptera: Cicadina). *Zool. Jahrb. Abt. Allg. Zool. Physiol. Tiere,* 92: 245–268.

Synave, H., 1953. Un cixiide troglobie découvert dans les galeries souterraines de Namoroka (Hemiptera: Homoptera). *Nat. Malgache,* 5(2): 175–179.

Chapter 11

ECOLOGICAL ROLE OF AGGRESSIVENESS IN THE DARK

Jakob PARZEFALL

INTRODUCTION

Animals living in darkness have to compete for sources like food, mates and space for an undisturbed reproduction, just as their epigean counterparts do in epigean habitats. But there is one striking difference: in light animals can use visual signals. Thus, one important part of aggressiveness, visual signals of threatening behaviour, effective at greater distance to prevent escalating fights, cannot be used in darkness. Therefore the question arises how cave dwellers interact in complete darkness. I shall compare the situation in cave-dwelling populations with that in closely related epigean populations.

AGGRESSIVE BEHAVIOUR

Aggressive behaviour consists of different patterns: threatening postures and attacks followed by fights. The threatening animal in general enlarges his body shape, shows striking colours or sends impressive calls. During fights the normal way is to bite, strike or beat the opponent. To prevent an injury, fights can be ritualized to techniques which allow each to measure how strong the opponent is. The weaker participant then has the chance to switch to submissive behaviour without being injured. Both patterns will be combined under the term agonistic behaviour. I examine the different functional aspects of agonistic behaviour and their changes under cave conditions. Table 11.1 gives a list of cave dwellers in which aggression has been studied.

Competition for food

Food competition may result in territories for groups, pairs or solitary animals. Within a group food limitation can also lead to aggressive encounters. In general, defending of food resources is only convenient when the costs are not higher then the incoming energy obtained from the food.

Most data are available for aggression in cave and epigean populations of the characid fish *Astyanax fasciatus*. The epigean form is widely distributed in Mexican waters, living in rivers, ponds and even in cenotes on Yucatan. When undisturbed the epigean fish defend small territories, ten to twenty centimetres in diameter depending on body size, by fin-spreading, snake-swimming and ramming. It is very difficult to observe this territorial behaviour in the field because the animals switch immediately to shoaling or schooling (see Langecker, Chapter 7, this volume, for more detail) when disturbed by an observer. I have observed territoriality in the field only in the Rio Naranjos population (Parzefall, 1983).

In the laboratory, epigean fish show the complete aggressive pattern: fin-spreading, snake-swimming and ramming attempts are followed by ramming, biting and circling during longer, escalated fights. The subdominant fish demonstrates submission by a head-up-position and tries to hide or to escape. In smaller aquaria without any possibility of hiding, escalated fights may lead to the death of the subdominant fish. There are no differences between males and females in respect of agonistic behaviour.

Aggressive behaviour depends on optical releasers. Langecker et al. (1995), using dummies of different types, demonstrated that "natural shape" and "locomotion" are visual signals. Tests using infra-red video have shown that epigean fish are not able to perform aggressive patterns in complete darkness (Burchards et al., 1985; Hausberg, 1995). They do not establish territories at all. From these data it is concluded that

Table 11.1
Aggression in cave dwellers

Species	Terr.[1]	Rank order	Remarks	References
AMPHIBIA				
Euproctus asper	+	?	males territorial	Schulz (1991)
Proteus anguinus	+		males territorial; females egg-guarding	Parzefall (1976)
Typhlomolge rathbuni	+	?	both sexes territorial	Bechler (1986)
TELEOSTEI				
AMBLYOPSIDAE				
Amblyopsis rosae		+	tail-beat only	Bechler (1983)
Amblyopsis spelaea		+	bite, tail-beat, chase, jaw-lock	Bechler (1983)
Chologaster agassizi		+	bite, tail-beat, chase, jaw-lock	Bechler (1983)
Typhlichthys subterraneus		+	jaw-lock reduced	Bechler (1983)
CHARACIDAE				
Astyanax fasciatus				
Epigean fish	+		feeding territories in both sexes	Burchards et al. (1985)
Micos cave fish	?		quantitative reduction	Hofmann (1990); Hoffmann and Hausberg (1993)
Pachon cave fish	?		quantitive reduction, new type of aggression	Hausberg (1995)
Piedras cave fish	?		quantitative reduction	Burchards et al. (1985)
CLARIDAE				
Uegitglanis zammaranoi	–	+	biting	Ercolini et al. (1981)
CYPRINIDAE				
Caecobarbus geertsi	?	?	attacks in both sexes	Thinès (1969)
Garra barreimiae	+		attacks in both sexes	Fausel (1990)
PIMELODIDAE				
Pimelodella kronei		+	chasing, biting	Trajano (1991)
Rhamdia guatemalensis	–	+	biting, tail-beating	Bormann (1989)
POECILIIDAE				
Poecilia mexicana		+	quantitative reduction	Parzefall (1979)
OPHIIDIDAE				
Ogilbia pearsei		+	ramming	Liebherz, unpublished
SYNBRANCHIDAE				
Ophisternon infernale	?		biting	Parzefall, unpublished
CRUSTACEA: ANOMURA				
GALATHEIDAE				
Munidopsis polymorpha	+		feeding territories, both sexes	Parzefall and Wilkens (1975)

[1] Territoriality.

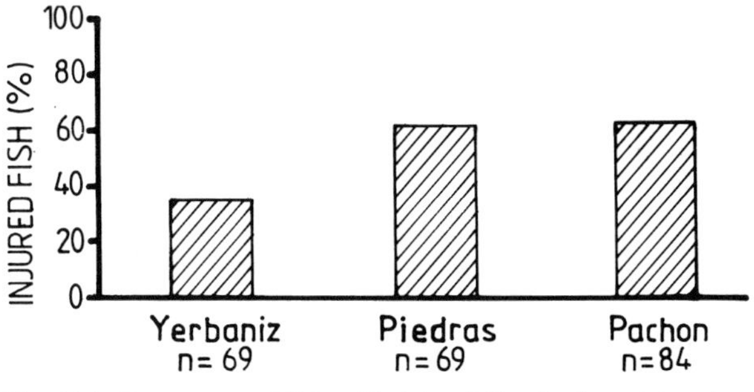

Fig. 11.1. Percentage of fish with injured fins in three laboratory stocks of *Astyanax fasciatus* from different caves (after Hausberg, 1995).

epigean fish in the cave environment are no longer able to perform aggressive encounters. However, Hausberg (1995) noticed in the eyeless populations of the Pachon, Piedras and Yerbaniz Caves a high percentage of fish with injuries on fins and scales (Fig. 11.1). In an experiment with the Pachon population the absence of food increased the proportion of injured fish (Fig. 11.2).

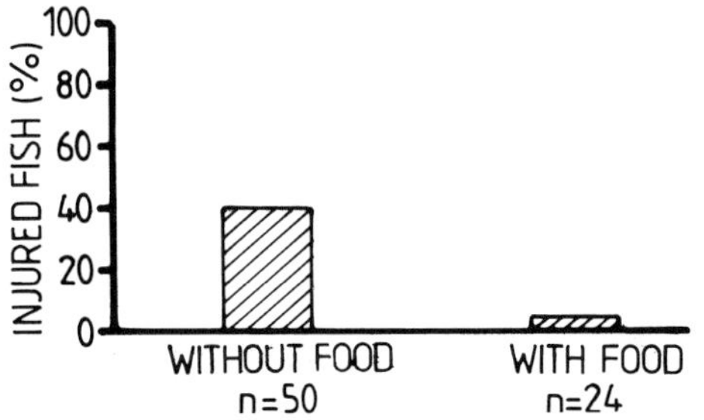

Fig. 11.2. Comparison of the degree of injuries in *Astyanax fasciatus* of the Pachon cave kept in groups of two fish for one week with or without food (after Hausberg, 1995).

In this situation, aggressive behaviour was observed: the fish defended small territories a few centimetres across by biting, circling and tail-beating. In these experiments a striking difference in the swimming behaviour was also observed. Fishes which were regularly fed showed the typical slow gliding through the water of the whole aquarium without aggressive encounters against conspecifics. After food was absent for some days the locomotory activity decreased. The fish mostly remained stationary at the bottom with a rythmical flicking of the fins, and fish entering the small area of a few centimetres were attacked. The territory size was positively correlated with the aggressiveness of the fish, and the aggressive pattern differs from the pattern shown in the epigean fish. Only signals effective in close body contact exist. Biting, circling and tail-beating have been observed, but not fin-spreading, snake-swimming and ramming. Hausberg (1995) argued that the aggressive behaviour of the cave fish is a new type of aggressive pattern which has evolved in adaptation to lightless habitat after the reduction of the visually guided aggression in the epigean ancestor.

Trajano (1991) observed comparable agonistic interactions in both sexes of the blind pimelodid catfish *Pimelodella kronei* from caves in southeastern Brazil. At food baits, smaller catfish were chased away by the larger ones. Behaviours used during the aggressive encounters included oscillations, shoves and bites (Fig. 11.3). The epigean ancestor *Pimelodella*

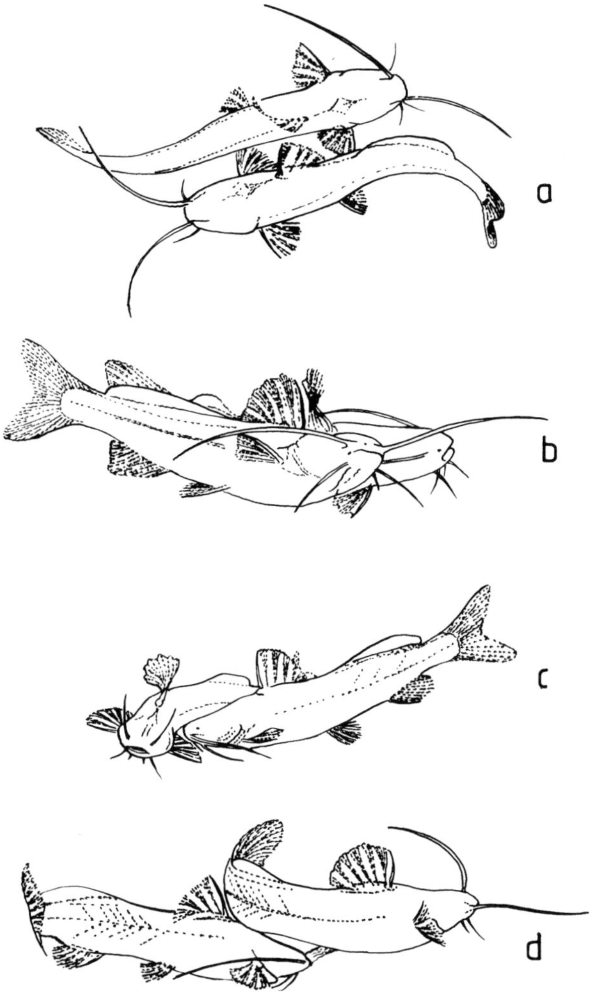

Fig. 11.3. Aggressive behaviour in *Pimelodella kronei*: (a) zig-zagging in head-to-tail position; (b) zig-zagging in head-to-head position; (c) ventral shove; (d) fin biting (after Trajano, 1991. Reprinted from Behavioural Processes, Vol. 23, 1991, pp. 113–124: "Agonistic behaviour of *Pimelodella kronei,* a troglobitic catfish from southeastern Brazil", with permission from Elsevier Science).

transitoria, which inhabits epigean and subterranean streams, shows more or less the same type of aggressive behaviour. The only difference from the cave fish is the use of dens for hiding and their defence. In contrast to *Astyanax fasciatus,* the epigean nocturnal catfish seems preadaptated for life in darkness, and does not need to change aggressive behaviour in the dark habitat.

In amblyopsid cave fish, males and females of the four species that behaved agonistically did not appear different (Bechler, 1983). The complete aggressive repertoire consists of six aggressive and two submissive acts (Table 11.2). In some of the species studied only an incomplete repertoire has been observed. The two submissive acts "freeze" and "escape" have been observed in all the species studied. All these species seem to have feeding territories. The *Amblyopsis* species are a mobile constantly swimming species; the other species (Table 11.2) are more sedentary, resting for long periods of time under debris and rocks (Bechler, 1983; Poulson, 1963).

Table 11.2
Aggressive repertoire for each amblyopsid species [1]

Acts	*Chologaster*		*Typhlichthys subterraneus*	*Amblyopsis*	
	cornutus	*agassizi*		*spelaea*	*rosae*
Aggressive acts					
Tail-beat	–	+	+	+	+
Head-butt	–	+	+	+	–
Attack	–	+	+	+	–
Bite	–	+	+	+	–
Chase	–	+	+	+	–
Jaw-lock	–	+	–	+	–
Submissive acts					
Freeze	–	+	+	+	+
Escape	–	+	+	+	+

[1] +, act observed in a species; –, not observed. From Bechler (1983).

In *Typhlomolge rathbuni,* the Texas blind cave salamander, tail-beating and biting have been reported for both sexes by Bechler (1986), without giving any functional interpretation. The fact that both sexes react in the same way can most likely be interpreted as competition for space and food.

Among cave-living invertebrates, the galatheid crab *Munidopsis polymorpha* of the Jameos del Aqua on Lanzarote (Canary Islands) feeds mainly on diatoms on lava rocks. The animals keep at least a minimal distance from one another, which depends on the length of their second antenna. Within this distance *Munidopsis* attacks with extended chelipeds, and by snapping its pincers (Fig. 11.4). This behaviour does not depend on optical releasers but on water movements (Parzefall and Wilkens, 1975). The aggressive pattern of *Munidopsis* is very similar to that described by Berrill (1970) for the deep-water bottom-living epigean galatheid *Munida sarsi.* Berrill assumed that, despite their naturally dim environment, vision was the primary sense involved in the aggressive behaviour of this species. So it can be concluded that aggressive behaviour in the galatheids seems to be effective both in light and in darkness.

Fig. 11.4. Aggressive interaction in *Munidopsis polymorpha* (after Parzefall and Wilkens, 1975).

Competition for mates

To compete for females, males can defend territories which only females are allowed to enter. In such territories normally monogamous pairs reproduce. The other possibility is that the males compete within a social group of conspecifics by establishing a rank order dependent on body length. In such cases, only the high-ranking male has access to females. The subdominant males counteract by sneaking behaviour (Taborsky, 1994). In all these cases vision is an important sensory system that is used by males to compete for females. None of the visual signals can be used in darkness. So male cave-dwellers must use other signals when competing for females.

Territoriality

For the blind cave salamander *Proteus anguinus* from the Adriatic karst in Slovenia, Croatia and some other parts of the former Yugoslavia it is known from observations in the Laboratoire Souterrain in Moulis (France) that the males only show aggressive behaviour and territoriality for a very short reproductive period. Normally, the animals rest under stones in groups of both sexes without any aggressive reaction. When a male becomes sexually active it starts to control his conspecifics by contacts with the snout and allows only females in the reproductive state to remain in the hiding place. All these females had well-developed eggs visible in the oviducts. Intruders are attacked by tail-beating, ramming and biting (Fig. 11.5)

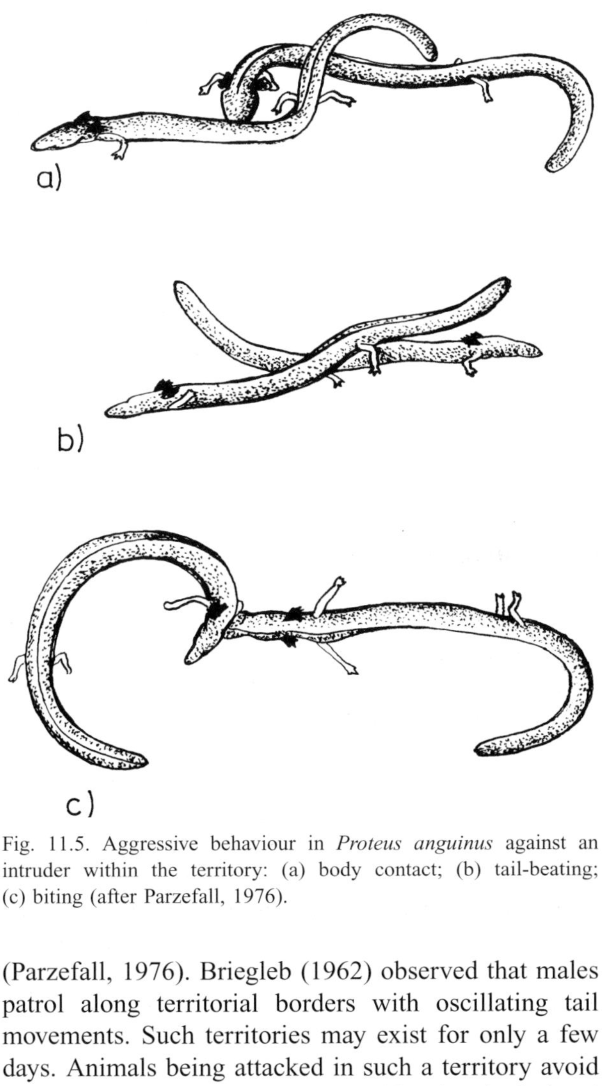

Fig. 11.5. Aggressive behaviour in *Proteus anguinus* against an intruder within the territory: (a) body contact; (b) tail-beating; (c) biting (after Parzefall, 1976).

(Parzefall, 1976). Briegleb (1962) observed that males patrol along territorial borders with oscillating tail movements. Such territories may exist for only a few days. Animals being attacked in such a territory avoid it on the basis of a substrate-specific chemical signal (Parzefall, 1976). This avoidance reaction also persists for only a few days (Parzefall, 1976).

For an epigean member in the family Proteidae, the genus *Necturus* from North America, there are only a few comparable data. In aquaria, *Necturus maculosus* defended food items by tail-beating and biting against conspecifics (Parzefall, unpubl.). However, *Proteus* has never been seen fighting when food was offered (Parzefall, 1976; Parzefall et al., 1980). Despite functional differences, these aggressive patterns seem similar in both species. But data on reproduction and territoriality in *Necturus* are not available.

Only in *Proteus* has egg-guarding and defending by females been observed (Briegleb, 1962; Parzefall, unpubl.). The females show the same aggressive pattern as the males during territory defence. The water coming from such egg-clutches has been avoided by conspecifics in choice experiments (Durand et al., 1983).

During spawning behaviour in *Astyanax fasciatus,* which does not show differences in cave and epigean populations, no aggressive encounters have been observed (Wilkens, 1972).

Rank order

In the poeciliid fish *Poecilia mexicana,* the males establish a size-dependent rank order within a mixed school by aggressive pattern. The females have a reproductive cycle of about 28 days. They are attractive to males within the first three days of the cycle. The dominant male controls the females by nipping, and responds to an attractive female by following behaviour and dark body coloration. In the field, the pair separate from the school and become more or less sedentary. The male nips and tries to copulate. Other males are chased away from the female. During aggressive encounters with more or less equal-sized males, small males coloured like females sneak copulations (Parzefall, 1973, 1979). The population of *P. mexicana* which colonized a limestone cave in Tabasco (Mexico) does not school and the males do not fight. In laboratory studies with epigean fish and cave fish having functional eyes a quantitative genetically-based reduction of the aggressive pattern (Fig. 11.6) and schooling has been demonstrated (Parzefall, 1974, 1979, 1989, 1993). The reduction of the aggressive pattern is highly variable within the population. Some of the cave fishes tested in light do not show an aggressive pattern, and it seems that they do not "understand" attacks by other males. Instead, they answer aggression by nipping and copulation attempts against the aggressor. Attractive females can be seen followed by several males without any aggression (Parzefall, unpubl.). There is no other example in cave dwellers with a quantitative reduction

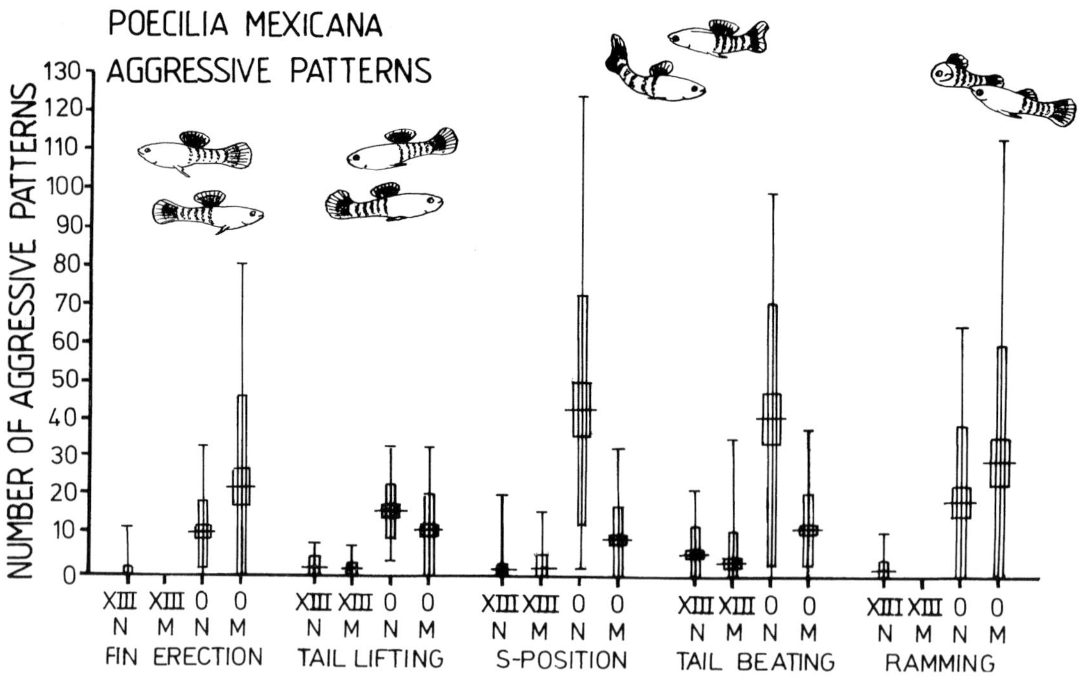

Fig. 11.6. Aggressive patterns in males of *Poecilia mexicana*: XIII, cave population; O, epigean population; N, standard test of a male against an opponent of the same size and the same population; M, male teted against an opponent of the other population. The total variability, mean, standard error and standard deviation are shown (after Parzefall, 1976).

of the whole aggressive system and a reduction of the rank order.

DISCUSSION

The data on aggression in cave dwellers demonstrate that, during cave colonization in some species like *Poecilia mexicana,* aggressive behaviour undergoes a reduction process. It is possible that this species does not need aggressive behaviour in the cave habitat. But in comparison to the data presented on *Astyanax fasciatus* one can also argue that the phylogenetically young cave fish *P. mexicana* has not reached the final stage like *A. fasciatus,* which has reduced the aggressive pattern of the epigean ancestor and replaced it by a new type of aggressive behaviour. The reduction of the aggressive pattern and the rank order in the cave population of *Poecilia mexicana* could also be caused directly by the absence of optical releasers for some aggressive pattern, which is connected with dark body coloration. This means that these elements are not effective in darkness, and therefore no longer controlled by stabilizing selection. But it is also possible to argue that fights in darkness are too expensive and disadvantageous. Therefore males would do better to retain contact with a female they meet in darkness than to fight with an opponent. Following this argument, the final stage should be a complete reduction of aggressive behaviour in *P. mexicana.*

On the basis of the other examples presented one can state that aggressive behaviour exists also in lightless habitats. But there are some differences from the aggressive behaviour in epigean habitats. Territories for food and reproduction seem to be smaller. This is probably caused by the fact that nonvisual signals like water waves are effective only at shorter distances. Acoustic signals have been described in many fishes (Hawkins, 1993) during agonistic behaviour [in the Centrarchidae (Ballantyne and Colgan, 1978)] or during reproduction [in the Pomacentridae (Myrberg et al., 1986)]. This type of signal, which can be transmitted at greater distances, has been found in lightless deep-sea habitats (Parzefall, 1996), but it has not been detected in cave-dwellers.

During fights in darkness the animals act in distances of only a few centimetres. This allows the perception of water waves caused by tail-strokes, or direct contact by biting, head-butting or tail-beating. Submissive positions seem to be rare in the dark habitat, which facilitates flight and hiding. Only in the cave amblyopsids has freezing been described as submissive behaviour.

Social groups probably exist in the dark only in slow-moving species like salamanders but not in permanently swimming fishes. One cannot argue that the chemical signals which the salamanders use are lacking in fishes. However, Quinn (1980) and De-Fraipont (1992) have also found that *Astyanax fasciatus* releases chemical signals in the water which attract conspecifics. But in comparison to the epigean fish the reaction in the cave fish was weak. It is hypothesized that the aggregation tendency is maintained as a vestige in the cave population (Quinn, 1980).

ACKNOWLEDGEMENTS

I should like to thank the Deutsche Forschungsgemeinschaft (Pa 148) for financial support and the Mexican Government for the permits (242.-2190 276/36, 210696-213-03) which allowed me to study the Mexican cave fish in the laboratory and the field.

REFERENCES

Ballantyne, P.K. and Colgan, P.W., 1978. Sound production during agonistic and reproductive behaviour in the pumpkinseed (*Lepomis gibbossus*), the bluegill (*L. macrochirus*), and their hybrid sunfish. 2. Recipients. *Biol. Behav.*, 3: 207–220.

Bechler, D., 1983. The evolution of agonistic behaviour in Amblyopsid fishes. *Behav. Ecol. Sociobiol.*, 12: 35–42.

Bechler, D., 1986. Pheromonal and tactile communication in the subterranean salamander *Typhlomolge rathbuni*. *Proc. Ninth Int. Cong. Speleol., Barcelona, Spain,* 2: 120–122.

Berrill, M., 1970. The aggressive behaviour of *Munida sarsi* (Crustacea: Galatheidae). *Sarsia,* 43: 1–11.

Bormann, J., 1989. *Das Sozialverhalten von Rhamdia guatemalensis (Pimelodidae, Pisces). Vergleich der Höhlenpopulation mit der oberirdisch lebenden Population.* Unpubl. Diploma thesis, Universität Hamburg, 126 pp.

Briegleb, W., 1962. Zur Biologie und Ökologie des Grottenolms (*Proteus anguinus* LAUR., 1768). *Z. Morph. Ökol. Tiere,* 51: 271–334.

Burchards, H., Dölle, A. and Parzefall, J., 1985. Aggressive behaviour of an epigean population of *Astyanax mexicanus* (Characidae, Pisces) and some observations of three subterranean populations. *Behav. Process.,* 11: 225–235.

DeFraipont, M., 1992. Reponse d'*Astyanax mexicanus* aux stimulation chimiques provenant de groupes de congeneres a different stades du developement. *Mem. Biospeol.,* 19: 209–213.

Durand, J.P., Bouillon, M. and Parzefall, J., 1983. Reponse de *Proteus anguinus* L., amphibien cavernicole, à des stimuli chimiques provenant de la ponte. *Mem. Biospeol.,* 10: 395–399.

Ercolini, A., Berti, R. and Cianfanelli, A., 1981. Aggressive behaviour in *Uegitglanis zammaronoi* Gianferrari (Clariidae, Siluriformes) an anophthalmic phreatic fish from Somalia. *Monit. Zool. Ital.,* 5: 39–56.

Fausel, P., 1990. *Das Sozialverhalten der Höhlenbarbe Garra barreimiae (Cyprinidae, Pisces).* Unpubl. Diploma thesis, Univ. Hamburg, 92 pp.

Hausberg, C., 1995. *Das Aggressionsverhalten von Astyanax fasciatus (Characidae, Teleostei): Zur Ontogenie, Genetik und Evolution der epigäischen und hypogäischen Form.* Unpubl. Diss., Univ. Hamburg, 139 pp.

Hawkins, A.D., 1993. Underwater sound and fish behaviour. In: T.J. Pitcher (Editor), *Behaviour of Teleost Fishes.* Chapman & Hall, London, pp. 129–171.

Hoffmann, S. and Hausberg, C., 1993. The aggressive behaviour of the Micos cave population (*Astyanax fasciatus,* Characidae, Teleostei) after selection for functional eyes in comparison to an epigean one. *Mem. Biospeol.,* 20: 101–103.

Hofmann, S., 1990. *Vergleichende Untersuchung zum Aggressionsverhalten einer oberirdischen und einer unterirdischen Population von Astyanax fasciatus (Characidae, Pisces).* Unpubl. Diploma thesis, Univ. Hamburg.

Langecker, T.G., Neumann, B., Hausberg, C. and Parzefall, J., 1995. Evolution of optical releasers for aggressive behaviour in cave dwelling *Astyanax fasciatus* (Teleostei, Characidae). *Behav. Process.,* 34: 161–168.

Myrberg, A., Mohler, M. and Catala, J.D., 1986. Sound production by males of a coral reef fish (*Pomacentrus partitus*): its significance to females. *Anim. Behav.,* 34: 913–923.

Parzefall, J., 1973. Attraction and sexual cycle of Poeciliidae. In: J.H. Schröder (Editor), *Genetics and Mutagenesis of Fish.* Springer, Berlin, pp. 177–183.

Parzefall, J., 1974. Rückbildung aggressiver Verhaltensweisen bei einer Höhlenform von *Poecilia sphenops* (Pisces, Poeciliidae). *Z. Tierpsychol.,* 35: 66–84.

Parzefall, J., 1976. Die Rolle der chemischen Information im Verhalten des Grottenolms *Proteus anguinus* LAUR. (Proteidae, Urodela). *Z. Tierpsychol.,* 42: 29–49.

Parzefall, J., 1979. Zur Genetik und biologischen Bedeutung des Aggressionsverhaltens von *Poecilia sphenops* (Pisces, Poeciliidae). Untersuchungen an Bastarden ober- und unterirdisch lebender Populationen. *Z. Tierpsychol.,* 50: 399–422.

Parzefall, J., 1983. Field Observation in epigean and cave population of the mexican characid *Astyanax mexicanus* (Pisces, Characidae). *Mem. Biospeol.,* 10: 171–176.

Parzefall, J., 1989. Sexual and aggressive behaviour in species hybrids of *Poecilia mexicana* and *Poecilia velifera* (Pisces, Poeciliidae). *Ethology,* 82: 101–115.

Parzefall, J., 1993. Schooling behaviour in population-hybrids of *Astyanax fasciatus* and *Poecilia mexicana* (Pisces, Characidae and Poeciliidae). In: H. Schröder, J. Bauer and M. Schartl (Editors), *Trends in Ichthyology.* GSF-Bericht 7/92, GSF-Forschungszentrum, Neuherberg, pp. 297–303.

Parzefall, J., 1996. Behavioural and morphological changes caused by light conditions in deep sea and shallow water habitats. In: F. Uiblein, and M. Stachowitsch (Editors), *Deep Sea and Shallow Water Habitats: Affinities and Adaptations.* Biosystematics and Ecology Series 11, The Austrian Academy of Sciences, Vienna, pp. 91–122.

Parzefall, J. and Wilkens, H., 1975. Zur Ethologie augenreduzierter

Tiere: Untersuchungen an *Munidopsis polymorpha* Koelbel (Anomura, Galatheidae). *Ann. Speleol.*, 30: 325–335.

Parzefall, J., Durand, J.P. and Richard, B., 1980. Chemical communication in *Necturus maculosus* and his cave-living relative *Proteus anguinus* (Proteidae, Urodela). *Z. Tierpsychol.*, 53: 133–138.

Poulson, T.L., 1963. Cave adaptation in amblyopsid fishes. *Am. Midl. Nat.*, 70: 257–290.

Quinn, P.T., 1980. Locomotor response of juvenile blind cave fish, *Astyanax jordani,* to the odors of conspecifics. *Behav. Neural Biol.*, 29: 123–127.

Schulz, W., 1991. *Visuelle Reaktion von Euproctus asper Duges im Hinblick auf eine Augenverpflanzung bei Proteus anguinus Laurenti (Proteidae).* Unpubl. Diploma thesis, Univ. Hamburg, 80 pp.

Taborsky, M., 1994. Sneakers, satellites, and helpers: parasitic and cooperative behaviour in fish reproduction. *Adv. Stud. Behav.*, 23: 1–100.

Thinès, G.L., 1969. *L'Evolution Régressive des Poissons Cavernicoles et Abyssaux.* Masson et Cie, Paris, 394 pp.

Trajano, E., 1991. The agonisitc behaviour of *Pimelodella kronei,* a troglobitic catfish from Southeastern Brazil (Siluriformes, Pimelodidae). *Behav. Process.*, 23: 113–124.

Wilkens, H., 1972. Über Präadaptation für das Höhlenleben, untersucht am Laichverhalten ober- und unterirdischer Populationen des *Astyanax mexicanus. Zool. Anz.*, 188: 1–11.

Part V

The Trophic Basis of Subsurface Ecosystems

Chapter 12

THE TROPHIC BASIS OF SUBSURFACE ECOSYSTEMS

Thomas L. POULSON and Kathleen H. LAVOIE

INTRODUCTION

In this chapter we try to give a conceptual framework for the case-study chapters that follow in this section. We first show how to determine whether a system is food-limited. In this context we then describe the differences in spatial distribution of food expected in aquatic and terrestrial systems. We finally compare food input by abiotic and biotic agents into aquatic and terrestrial systems with particular emphasis on feces of trogloxenes.

Terminology

Because usage differs we wish to be clear about our terminology. Organisms can be grouped on the basis of their source of energy as **chemotrophs**, using chemical bond energy, or **phototrophs**, using radiant energy. Adding the prefixes **hetero-** or **auto-** describes the source of carbon as, respectively, "other feeders" that use organic carbon, or "self feeders" that use fixation of carbon dioxide. Thus, all **decomposers**, like fungi and most bacteria, which break down organic matter, are **chemoheterotrophs**, as are all animals and protozoans. Because of lack of sunlight in subsurface ecosystems, **photoautotrophic** green plants do not grow. However, many **chemoautotrophs** depend on **photoautotrophs** indirectly, for example by using oxygen as a terminal electron receptor, or directly, for example by using organic matter produced in food chains that start with photosynthesis.

Chemoautotrophic bacteria are important in a few below-ground (**hypogean**) ecosystems; but in most hypogean ecosystems that we will discuss, lack of primary productivity by autotrophs requires that organic nutrients be introduced from above-ground (**epigean**) sources. By definition, hypogean systems do not include lightless environments of the deep sea, large lakes, and large rivers. We consider four hypogean habitats. Lava tubes and limestone caves are large enough to accommodate humans, whereas mesocaverns or mesovoids (~10–100 cm maximum dimensions) and interstitial spaces or microvoids (~1–10 mm) are too small.

Caveats

Except in passing, we do not consider interstitial spaces or mesocaverns because we know of no data on trophic input or community organization. However, both are potentially important as intermediate habitats during the evolutionary origin of troglobites. The fauna of epigean interstitial spaces among gravels and sands along and under rivers and lakes somewhat overlaps the fauna in limestone caves (Rouch and Danielopol, 1987). And the fauna of mesocaverns, also known as the *compartiment surficiel,* overlaps even more with that in caves (Juberthie and Delay, 1980) and lava tubes (Howarth, 1987).

Because of the confounding effects of differences in evolutionary time we will be cautious in using differences among regions in diversity of species or specialization to the cave environment in order to infer difference in degree of energy limitation. However, differences in diversity and specialization within regions or within cave systems give strong indirect evidence for the relative importance of energy limitation. We briefly consider two examples to illustrate potential difficulties of relating energy available to diversity and specialization.

Considering two systems based on the use of sulfides by chemoautotrophic bacteria for energy, deep-sea vents have been available for colonization for thousands of millions of years (Jannasch, 1984), whereas the

Movile Cave system in Romania has been available only for several million years (Sarbu and Popa, 1992). The Movile Cave system has fewer species and they are less specialized than those at deep-sea vents. Many macroscopic consumers at deep-sea vents depend on an intimate mutualism with sulfide-using bacteria which provide carbon and energy in the protected closed intestines of their hosts. By contrast, macroscopic consumers in Movile Cave have typical open intestines with no mutualistic symbionts.

Between temperate and tropical latitudes it was once erroneously believed that there was a large difference in diversity of troglobites, and that this was due to differences in opportunities for isolation of troglophiles which then evolve into troglobites (Poulson, 1964; Barr, 1968; Poulson and White, 1969; Howarth, 1987; Rouch and Danielopol, 1987). Opportunities for evolution of troglobites were certainly great in temperate periglacial regions with multiple advances and retreats of continental glaciers [see Barr (1967, 1968) for carabid beetles and Peck (1981) for leiodid beetles]. In the lowland, wet tropics there was no continental glaciation and so no rigors of cold, wet glacial or warm, dry interglacials that would have made epigean survival problematic for the troglophilic ancestors of troglobites. In addition, most tropical caves have many entrances and abundant abiotic input of detritus, as well as biotic input of feces by many bats. With so much food one might think that most species in caves would be troglophiles. While there is truth to this argument, it now appears that some tropical caves and many tropical lava tubes have many species of troglobites. One reason is that climatic changes in montane and arid parts of the tropics were greater than was once thought (e.g., Reddell, 1981). In lava tubes, and perhaps tropical limestone caves as well, another hypothesis is that there was active invasion, allowed in part by high food input and intermediate mesocavern habitats. Under this scenario, speciation occurs by an "adaptive shift" associated with "crash and flush" fluctuations in population size (Howarth, 1987).

Another example of confounding regional differences is at the continental scale for intermediate epigean–hypogean habitats of mesocaverns and interstitial spaces. For reasons which are not clear, Europe has more of these intermediate habitats than North America, a higher diversity of troglobites, and greater morphological (i.e, troglomorphic) and physiological specialization to the cave environment. Interstitial habitats are widespread in gravels of rivers with headwaters in the Alps and Pyrénées, and there is broad overlap of a diverse interstitial fauna and cave fauna (Rouch and Danielopol, 1987); but this habitat has few or no specialized interstitial species or troglobitic relatives in North America. Mesocaverns are common in Europe, especially where karst is interrupted and/or overlain by montane scree slopes of poorly sorted rocks, and many species in non-limestone mesocaverns are also found in limestone caves (Juberthie and Delay, 1980). However, using the same collecting technique as in Europe, Juberthie found no mesocavern fauna in North America. Part of the reason is that there appears to be more soil over the karst in North America, and this fills the mesocavern-sized voids at the epikarst border between the soil and bedrock.

In caves, workers have documented species with greater specialization to the cave environment in Europe than in North America. For example, compared to North American species, European bathysciine beetles include species with much greater evolutionary reduction of larval stages, and entomobryine springtails have more troglomorphic traits [reviewed by Poulson and White (1969)]. We suggest that these contrasts are due not to differences in selection intensity based on food limitation but rather to differences in geological setting and chances for initial isolation in hypogean habitats. In Europe, karst and non-karst montane terrains are closely interspersed, and the east–west orientation of mountains to the south served as barriers to migration during severe glacial or interglacial climates. Consequently, opportunities for evolutionary isolation may have occurred earlier and/or more often, and led to more speciation, in Europe than in North America. However it is not clear how the greater prevalence of interstitial spaces and mesocaverns in Europe might help to explain the observations. On the one hand, both could have been transition habitats which decreased the rate of extinction of troglophiles, but on the other hand the higher food supplies in both should not select for extreme specializations to the cave environment.

Our final caveat is that our experience is mainly with limestone caves in temperate North America. Thus, our general perspectives should be viewed as working hypotheses to be tested in other parts of the world and in different kinds of hypogean ecosystems. We believe, nonetheless, that the patterns we suggest are general and should be useful as a conceptual framework for all hypogean ecosystems.

A hypothetical and conceptual framework

Rigor, variability, and predictability of nutrient input

One important concept is the scale in time and space at which nutrients are introduced into or produced in hypogean ecosystems. Nutrients vary in their caloric density (rigor), and their renewal is not always predictable on temporal scales ranging from days, to seasons, and to decades. Nutrient input also varies greatly at spatial scales ranging from local to regional to latitudinal to continental. On all scales variability of nutrient input is affected by differences in rock type and vertical extent of each rock type, number and sizes of openings to the hypogean environment, age of the epigean landscape and its connectivity to the hypogean environment, average and peak rainfall events, climate, and epigean vegetation.

Tradeoffs

A second important concept is ecological tradeoffs – that is, advantages (+) and disadvantages (–). For example, caves with large sinking streams may have huge concentrations of large woody detritus (+), but the water velocity and scouring associated with large streams may be problematic (–) for survival of many organisms. And, even if such abiotic rigor is not a problem there may be biotic rigor. For example, leached woody debris may be high in quantity (+) but low in quality (–), as measured by caloric density and by very low rates of degradation by decomposers.

Within these frameworks we consider the general questions of whether all hypogean systems are energy-limited, what the differences are between aquatic and terrestrial systems, and what the differences are between abiotic and biotic input.

HYPOGEAN SYSTEMS ARE TYPICALLY ENERGY LIMITED

Both direct and indirect evidence is consistent with the conclusion that all hypogean ecosystems are food-limited compared to photosynthetically-based epigean systems. Measurements of primary chemosynthetic productivity or secondary productivity needed to directly assess energy limitation have rarely been made in hypogean systems, since the practical and technical difficulties are greatly compounded. However, there are substantial data from hypogean ecosystems on readily observable correlates of different levels of productivity in epigean ecosystems. These data, on correlations of food chain or web and organism biology with high versus low productivity in epigean ecosystems, lead us to hypothesize that hypogean systems, whether driven autotrophically or heterotrophically, are relatively energy-limited.

Epigean models of high vs. low productivity

In epigean aquatic ecosystems, highly pulsed productivity is associated with high dominance of a few species in each trophic level, short food chains, large size of organisms at the lowest trophic level, and small size and short lifespan of the dominant species toward the end of food chains. In contrast, continuously low productivity is associated with a number of species in each trophic level and no dominance, small size of organisms at the lowest trophic level, and large size and long lifespans of those at the highest trophic level. In principle these contrasts apply to terrestrial systems as well, but our examples here are from aquatic systems.

A clear example is provided by the contrast between aquatic food chains for coastal upwelling and the open ocean (e.g., Russell-Hunter, 1970; Paulik, 1971). Along the coast of Peru, upwelling results in fast renewal and high concentrations of mineral nutrients needed by phytoplankton. Often there are only two aquatic trophic levels. The few species of phytoplankton are large colonial species (10^{-1} to 10^{0} cm) which can be eaten directly by a small clupeid fish (10^{1} cm), the anchovy, which occurs in immense schools and lives about two years. Average net primary productivity is $\geqslant 800\,g\,C\,m^{-2}\,yr^{-1}$ with little spatial variation but considerable temporal variation associated with El Niño events. In contrast, in the open ocean of the tropical Pacific, maximum net primary productivity averages $<50\,g\,C\,m^{-2}\,yr^{-1}$ with extreme spatial variation but little overall temporal variation. Six trophic levels are present and a "microbial loop" conserves limiting nutrients. The many species of non-colonial phytoplankton are too minute (10^{-6} cm) to be eaten by the huge bluefin tuna (*Thunnus thunnus*) ($>10^{3}$ cm) at the top trophic level, which live for decades and eat smaller fish and squid. Areas with adequate prey are so unpredictable in space and time that the schools of tuna must search immense areas of the ocean. Consequently the average density of tuna in the ocean is extremely low.

"Freshwater" examples of simplified high-productivity systems are thermal-vent communities and hypersaline lakes. In both cases the rigors of high temperature or high salinity limit the number of species (–) but the very high concentrations of mineral nutrients allow a few autotrophic species to do extremely well (+). At Yellowstone National Park in Wyoming, there is a strong zonation with distance from thermal vents, starting with extremely simplified communities with only autotrophs and progressing to more complex communities with autotrophs and heterotrophs. Starting at 100°C there are only bacteria, then Cyanobacteria are added as the flowing waters start to cool, and then eukaryotic algae and their consumers are found (Wiegert and Fraleigh, 1972). In hypersaline lakes, as in East Africa and the Bahamas, temperatures are moderate and zonation based on salinity and nutrient concentrations is not so striking but the communities are greatly simplified. Only filter-feeding flamingos (Pheonicopteridae) and a few species of fish can eat the filamentous Cyanobacteria that are virtually the only autotrophs and contain a variety of toxins.

Compared to the above examples, there are virtually no other epigean systems and no hypogean systems where such complete data on different aspects of community and food chain or food web relationships exist, but data on diversity and dominance are readily available for many systems, so we shall use these to infer the levels of productivity and energy limitation. As a complement to these community-level data we shall consider direct measurements and inferred rates of input of energy into different systems.

Hypogean systems directly based on photosynthesis

Lava tubes become more energy-limited with time

Howarth (1987) implicitly showed that the amount and spatial predictability of production and density of organisms decrease with age of lava tubes. On recent lava flows, tree roots and aeolian particulate organic matter penetrate directly into young lava because there is little soil to fill the innumerable mesovoids. Tree roots grow deeply, including into lava tubes, in "search" of water and nutrients, and aeolian detritus works its way quickly into the interconnected voids in the lava. This relative openness of young lava flows and lava tubes also allows troglophilic species to invade actively from the inhospitable surface of the lava flow and troglobitic species to move in from older lava tubes. As a result there are relatively high frequencies and densities of both troglophilic and troglobitic species. As lava flows age, over short periods of tens to thousands of years, soil fills the mesovoids around the lava tubes and retains water, and so new tree roots need not grow extensively into the tubes. Accumulating soil also restricts input of particulate organic matter and movement of animals. The frequency and density of animals declines, and the restriction of organic input puts strong selection on troglophiles to evolve the efficiency of finding and utilizing food that are characteristic of troglobites in food-poor environments.

An evolutionary consequence of this scenario, about which Howarth (1987) was very clear, is that the model of troglobite evolution based on isolation of hypogean populations of troglophiles through extinction of their epigean populations (e.g., Barr, 1968) does not apply to lava tubes in volcanically active areas. Rather, Howarth espoused a model of speciation by "adaptive shift", facilitated by a crash of populations as lava tubes age, and a flush of populations as new lava flows occur over old lava flows, and allow renewed input of allochthonous organic matter into the lava tubes. This is an excellent model for lava tubes, but we are not convinced that Howarth's extension of the model to tropical limestone caves is reasonable, because we expect the opposite trends of energy input with age in limestone caves in general.

Limestone caves become less energy-limited with time

Caves in limestone karst terrains may become less energy-limited because they become more connected to the surface as they develop over thousands to millions of years. With time, slow diffuse input of water and nutrients continues and fast concentrated input increases. Under well-developed soils at the epikarst surface a rich decomposer community increases carbon dioxide output, and this results in faster dissolution of limestone. Tiny new openings to the incipient caves below are formed as fast as older openings become filled with soil. The largest openings in the limestone carry most of the water, and so are preferentially increased in size. Thus a cave is "born". As time goes on, flowing epigean waters carry organic matter into the largest caves without passing through the soil and epikarst. Thus, we hypothesize, in contrast to lava tubes, that the connectivity of caves to the surface and openness of caves to input of organic matter and movement of animals should increase over geological time. A corollary is that this process proceeds faster in the tropics. In addition, caves in the tropics are often older than in temperate zones because many caves in temperate zones were destroyed by past glaciation. And tropical caves have many more entrances than temperate caves. All of this suggests that tropical caves should be less energy-limited than temperate caves.

Mesocaverns and interstitial spaces may be intermediate in energy limitation

We suggest that interstitial spaces and mesocaverns lie between the extremes of young lava tube and old limestone caves in age, connectivity, input of particulate organic matter, and movement of organisms. Study of food-chain correlates of productivity should elucidate the relative degree of food limitation in these systems, but we are not aware of any relevant data.

Chemoautotrophically-based ecosystems

The deep-sea vents that depend on methane and inorganic sulfides have extremely high primary productivity, high biomass per unit area, simple food chains, and striking zonation of species (Jannasch, 1984). The zonation of species is along sharp local gradients both of energy availability and of abiotic rigor for temperature, oxygen, and sulfide concentration. The zonation is most manifest in the 100% occupancy of space by the largest primary consumers, especially clams and plumed tube worms (*Riftia*), which are tolerant of high sulfide concentrations because they have mutualistic sulfide-utilizing bacteria that live in their closed intestines, and so can use the energy source that others species cannot use, or cannot use efficiently.

In comparison to deep-sea vents, a cave ecosystem based on chemoautotrophic use of sulfides has lower primary productivity, lower biomass per unit area, more complex food chains, and less striking zonation of species (Lisowski et al., 1985; Sarbu and Popa, 1992; see also Chapter 17, this volume). We suggest that these differences are due to contrasts in the abiotic rigor and rate of input of sulfide to the two systems. Progressing away from vents energy supply and abiotic rigor decrease rapidly due to decreasing sulfide concentrations, decreasing temperature, and increasing oxygen concentration. In contrast, in Movile Cave in Romania there is a lower energy supply and no abiotic rigor of high temperature or low oxygen, because the magmatic sources of sulfide are many kilometers away. Unlike the deep-sea vents, species diversity is high on a scale of meters, and local zonation, based on competitive dominance and/or physiological tolerance, is absent. Overall there are 28 terrestrial and 18 aquatic endemic species of troglobites. These depend on mats of mainly colonial bacteria which cover the water surface and adjacent walls. The aquatic food chains in the cave appear to be complex, with a high diversity of protozoa, nematodes, microcrustaceans, and macrocrustaceans between the bacteria and the top predators, which include a leech, a water scorpion, and a diving beetle. Although a few species are locally quite dense, none of the largest aquatic or terrestrial consumers show the striking zonation seen around deep-sea vents. This zonation is based on differential tolerance of abiotic rigor and ability to use high concentrations of sulfide (Sarbu and Popa, 1992; Sarbu, personal communication).

We suggest three interrelated alternative hypotheses for the lack of local dominance in Movile Cave, all related to the long distance from the study site to multiple presumed magmatic sources of sulfide-rich waters. First, a low gradient in concentration of inorganic sulfide in the cave means that monopolization is unlikely. Second, the multiple sources converging from many kilometers upstream in the aquifer makes the energy supply downstream predictable in time and space, even if any one upstream source is not. And, third, there are no sharp chemical or temperature gradients associated with gradients of the energy source, and so there is no zonation based on differential tolerance of abiotic rigor. At a much smaller scale, the microzonation of a few centimeters, through and just below bacterially dominated mats on the water surface, is based on oxygen concentration and not on sudden pulses of input or sharp local gradients of sulfides.

Chemoheterotrophically-based ecosystems

Compared to photoautotrophically-based epigean ecosystems, chemoheterotrophically-based aquatic systems are more energy-limited. One line of evidence is from metabolic rates, growth rates, sense-organ elaboration, and food-finding efficiency. When appropriate controls for evolutionary relatedness are made (Culver and Poulson, 1972), hypogean species have lower metabolic rates than epigean species, lower growth rates, greater sense-organ elaboration, and greater food-finding efficiency (Poulson 1963, 1985; Poulson and White, 1969; Riser-Cooper, 1969; Culver and Poulson, 1972; Cooper, 1975; Culver, 1982; Culver et al.,

1995). A second line of evidence is from the relation of productivity to community organization and food chain/food web structure. This line of evidence has not been considered heretofore so we will concentrate on it in some detail.

Organic enrichment in epigean and hypogean systems

With extreme organic enrichment of easily biodegradable organic matter, the biological oxygen demand of aerobic decomposers is great and so oxygen concentrations are much reduced. Even wood can cause sudden organic enrichment and depletion of dissolved oxygen to near-lethal levels if it is not well leached and has a high ratio of surface area to volume. One documented example is related to sawdust from a lumber mill (Culver et al., 1992). We have observed a loss of troglobites with appearance of fungal-bacterial bloom when the leached black layer on the surface of old construction wood was suddenly broken up during removal to expose the relatively unleached brown interior. In all massive organic enrichment the result, at best, is that a few troglophilic species dominate and there is a much reduced species diversity. At worst, oxygen is completely depleted and only anaerobic decomposers occur – as is always the case with easily decomposable organic matter like urea fertilizer and human or animal wastes.

If the introduction of easily decomposed organic matter is not too fast, then secondary productivity increases dramatically and the indigenous fauna is replaced by a much simplified community of species that can quickly utilize the high densities of bacteria and tolerate low oxygen concentrations. The typical community is the same as in organically enriched epigean systems with dense mats of colonial "sewage" bacteria (e.g., *Sphaerotilus*) as the first trophic level. If sufficient oxygen remains then these bacteria are consumed by small, short-lived tubificid worms. These worms exist in extremely high densities and may cover 100% of available space in shallow water by the shore, where diffusion can replace oxygen used up by the bacteria as they decompose the organic matter. And, because concentrations of oxygen near to zero have a high abiotic rigor, there is no third trophic level. If the input of organic matter is stopped, or if it is slow, then a community of intermediate simplicity is favored. Troglobitic species are replaced by or dominated by troglophilic species. It is not clear why this occurs since the troglobites, with a low metabolic rate, have a refuge from competition with the troglophiles, which have a higher metabolic rate. Hypothetically, the troglophilic species gain a numerical advantage when food supply is high, because they have higher growth rate and reproductive rate.

Moderately energy-limited epigean streams provide a model against which to compare hypogean streams

The extreme of low energy input in flowing epigean waters occurs in shaded streams with slow degradation of dead leaves and twig litter (coarse particulate organic matter: CPOM). Despite strong seasonality of input in temperate zones, the food quality of CPOM is so low that, as in the open-ocean example discussed earlier, there is little species dominance, long food chains, and large, long-lived top consumers.

The sequential degradation of leaves in streams was first established by Kaushik and Hynes (1971). As leaves fall in autumn they are quickly wetted and leached. In the first 24 hours up to 25% of their initial mass is converted to dissolved organic matter (DOM). This readily useable DOM is sequestered by decomposers which become a biofilm on surfaces. The remaining leached CPOM is broken down into fine particulate organic matter (FPOM) by mechanical disruption, and is colonized by decomposers. The decomposers "condition" the CPOM and FPOM, often by increasing its nitrogen concentration relative to its carbon concentration. This makes CPOM more attractive to the shredder guild of invertebrates, such as isopods and amphipods, whose feces and shredding are another source of FPOM. In addition, FPOM comes from runoff into the catchment, and from aggregation of DOM by decomposers. Whatever the source, FPOM is in turn used by the strainer guild, such as larvae of caddis flies (e.g., Hydropsychidae) and blackflies (Simuliidae), and by the collector–scraper guild, such as mayfly larvae (Ephemeroptera) and snails. All of these organisms are in turn eaten by a variety of invertebrate carnivores and fish. Going downstream, the collective effect of continued recycling of FPOM is an efficient use of all organic matter.

A major reason why energy is limited in headwater epigean streams is that much of the nutrients are scarcely useable. In fact the bulk of organic carbon in running waters is refractory POM, such as lignins and hemicelluloses, with such high carbon/nitrogen ratios that they are not quickly useable by decomposers.

Severe energy limitation in most hypogean streams

Animals in most cave streams experience tradeoffs. If organic input is of high quantity and quality (+) then there is high risk of injury due to high water velocity (–). But if high water velocity is not a problem (+) then the quantity and quality of organic input is low and relatively unpredictable (–). Refractory DOM is the only abiotic input that occurs every year. Large quantities of CPOM and FPOM are washed in only by unusually great spring floods that occur unpredictably on time scales of decades to centuries. Nonetheless, aquatic troglobites reproduce seasonally since subtle increases in DOM input occur during spring rains when the ground is saturated (Poulson and Smith, 1969). The ubiquity of this reproductive response to subtle increases in energy availability is indirect evidence for extreme energy limitation in aquatic cave ecosystems.

The exact timing of seasonal food renewal varies by many months in different years (Poulson and Smith, 1969) and so troglobites should be prepared. Jegla and Poulson (1970) have argued that circannual rhythms allow troglobitic crayfish to be reproductively ready to lay eggs at any time during the winter and spring when a subtle pulse of organic renewal occurs. Poulson and Smith (1969) have pointed out that failure to recognize seasonal reproduction in aquatic troglobites by other investigators is due to their having lumped data among years and developmental stages. Among years and locales, timing of organic-matter renewal may vary by 3–6 months. And the duration of development from eggs to free-living immatures ranges from 2 to 6 months in different troglobites. Therefore arguments about seasonality must be based on within-year studies of the timing of the occurrence of the earliest developmental stages in each species.

On those rare occasions when CPOM or FPOM is washed into caves, the material is usually highly leached and can be only very slowly decomposed. This means that POM cannot be monopolized, and explains why neither high densities nor local dominance of aquatic troglobites are seen in unpolluted systems. Typically, any FPOM in cave streams is black and, even when it accumulates as centimeter-thick layers over areas of 1–2 m^2 in quiet pools and backwaters, it supports only very low densities of troglobites. Rather, biofilms of decomposers on rocks are a much more important food for organisms such as isopods and amphipods. Biofilms that are well enough developed to be slippery to the touch are encountered only with organic enrichment from sources such as septic tanks and the spreading of manure on farm fields.

Least energy limitation in caves fed by sinking streams

By far and away the best-studied and least energy-limited hypogean cave systems are fed by sinking streams. In these streams either food input and use, or biotic interactions, have been quantified, but not both in one system.

A small stream in Shiloh Cave in southern Indiana is the only system where a year-long assessment of organic input and secondary production has been made on epigean and hypogean portions of the same stream (Weingartner, 1975). Weingartner's data show that the hypogean system is quite energy-limited compared to the epigean stream that flows out of the cave. The caloric density of mud, and the caloric density and mass flux of CPOM and FPOM, differed by a factor of from four to seven between the epigean and the hypogean stream, whereas the biomass per unit area of all invertebrates was from 10 to 20 times greater in the epigean stream. The biomass of benthic organisms in the cave stream (annual mean: 77 mg dry m. m^{-2}), was about two orders of magnitude lower than in the stream flowing outside of the cave (4884 mg dry m. m^{-2}). Put another way, the biomass per unit area of the most abundant species was about 300 times that of the least abundant species in the epigean stream, whereas the corresponding ratio for the hypogean stream was only about 45. This reflects less dominance by the most successful species in the cave. In comparisons among dominants a similar contrast is seen. In the cave, the most dominant group of metazoan primary consumers (various oligochaetes, 45 m^{-2}) was 2.7 times as numerous as the second group (the isopod *Caecidotea stygia*), of which there were 17 m^{-2}; but there was much more dominance in the stream leaving the cave, with a ratio of 24.1 (4427 amphipods to 184 isopods m^{-2}).

Why were troglophilic amphipods and crayfish so highly dominant in the epigean portion of the stream? It appears to us that the troglophilic amphipod (*Gammarus* sp.) and the crayfish (*Cambarus laevis*), both shredder omnivores, were able to utilize the high amounts of CPOM, and so dominate the epigean stream. Inside the cave CPOM was absent, and only troglobitic amphipods, both grazer–collectors, were present. A troglobitic species of crayfish (*Orconectes inermis*) could coexist in the cave with the troglophilic species because neither could monopolize the food of FPOM, oligochaetes, and isopods on which both groups depend. The troglobite was probably more efficient at finding and utilizing the food, but the troglophile was probably better at eating large quantities in a given time.

To us, Weingartner's data show that Shiloh Cave is a sink habitat for the troglophilic crayfish, with survival dependent on immigration from outside. The density of the troglophilic crayfish in the cave was only one-hundredth of that outside, mainly because of poorer reproductive success and poorer survival of young. The troglophile was five times less numerous and, unlike the troglobite, had many fewer small than large individuals. This indicates that the troglophile had much poorer yearly success in reproduction and survival of young in the cave than the troglobite. Despite these disadvantages and a three-fold higher metabolic rate than the troglobite, the few troglophilic crayfish inside the cave had twice the growth rate of the troglobites, and the same standing crop. Compared to sinking streams in thick, flat limestones of the Interior Lowland Plateau geological province, such as that studied by Weingartner, the sinking streams in thin, steeply tilted limestones of the Appalachian Valley and Ridge province are potentially even less energy-limited. Even though no quantitative studies of organic-matter inputs have been made, we infer high energy input based on the fact that streams regularly sink where insoluble rocks of the ridges meet soluble limestone on the valley slopes. Some enter openings large enough to carry abundant FPOM and CPOM in the form of leaf and twig litter into caves. However, there is a tradeoff between this high food supply (+) and risk of injury and wash-out (–). Even in streams with no in-wash of CPOM, the tradeoff is that FPOM comes to amphipods (*Gammarus minus* and *Stygonectes* sp.) and isopods (*Caecidotea* spp.) in the rocky riffles (+), but the swift currents in springtime can result in wash-out and injury (–) to them.

The biotic interactions associated with this tradeoff have been exhaustively studied (Culver, 1982; Culver et al., 1991). Using an artificial stream, Culver showed that, despite infrequent encounters associated with low densities, the ability to avoid being displaced from a rock refuge in a riffle by direct competitive interactions explained much of the pattern of distribution and abundance of several species of isopods and amphipods. In such situations an evolutionarily young troglobitic amphipod, *Gammarus minus*, dominates the system because it is large and least likely to be displaced from the safety of rocks during encounters with more specialized troglobites of smaller size, lower metabolic rate, and longer lifespan.

In slower-moving streams in the same region, with no input of CPOM and less input of FPOM, the biotic interactions are much more complicated and there is no local dominance. These streams have more silt–sand pools with predators, mainly salamanders and crayfish, but fewer rocky riffles as refuges from predation. Culver et al. (1991) used a series of artificial stream and field transplant experiments with different combinations of species to show both expected direct effects and unexpected indirect effects. Direct effects included predation and interference competition, while indirect effects included exploitation competition, apparent competition, amensalism, and mutualism. We suggest that this complexity helps to explain the high species diversity and lack of dominance. In addition it suggests that energy input is either limited and/or highly predictable in space and time.

Contrasts of very high and very low energy availability

We contrast overall density and diversity of species in two groundwater systems. Both have slowly moving water and high diversity of species, and both have been adversely affected by or are threatened by anthropogenic impacts. In the case of Shelta Cave, Alabama, energy input was historically marginal, and anthropogenic disturbance of bat guano input is the most likely reason for nearly complete loss of the community. In the case of the artesian Edwards Aquifer, Texas, energy availability is still high but pollution from upstream, and upstream movement of petroleum in the "badwater" zone due to excess withdrawal of water, both threaten the community. Comparison of these two systems is complicated by inability to examine directly the densities and relative abundances of the species in the artesian system of the Edwards Aquifer. Nets placed on the outflow of deep wells provide the only sampling of the fauna below.

In the huge Edwards Aquifer in Texas the extraordinary diversity of troglobitic species (Longley, 1981) is due both to a long and complicated evolutionary history and to a reliable food supply based on petroleum. We suggest that this artesian system presents tradeoffs for the troglobites. At the upstream end of the 'badwater' zone fungal decomposition of the fossil petroleum results in high biological oxygen demand which precludes high densities of aerobic consumers at higher trophic levels. Upstream from the reservoir of petroleum, oxygen concentrations are high enough to support troglobites with a low metabolic rate. Here the energy availability is reduced to the point that logically no one species of consumers should dominate. Nobody has seen the communities *in situ,* so we suggest that the inferred high densities of fungal decomposers and of species that consume the decomposers could be due to accumulation of organisms filtered by nets from a high volume of artesian waters drawn to the sampling well from a great area at high rates of flow. Our hypothesis could be partially tested by making calculations, based on volume flow rates and presumed volume of the voids in the limestone, as to the area and volume sampled by these nets. Whatever the case, if human withdrawal of groundwater continues to accelerate, the "badwater zone" will move upgradient and the "goodwater zone" will decrease in area. If this happens, the diverse troglobitic communities will first be simplified as the fastest growing species become dominant, and finally all aerobic organisms will be lost.

We suggest that the "goodwater zone" in the relatively unimpacted parts of the Edwards Aquifer has a low dominance of macroscopic organisms because either they are energy-limited, or resources cannot be monopolized, or both. In the Edwards Aquifer the petroleum reservoir is extensive, and the system is so huge that sharp gradients of energy supply or oxygen concentrations are absent. Thus, the bacterial and fungal decomposers are evenly dispersed, and no consumer species can monopolize this food base. Our hypothesis is consistent with the high species diversity and lack of dominance by one or two of the many species in samples from artesian wells.

In contrast to the Edwards Aquifer, historically the Shelta Cave "window" to the top of the Madison Aquifer in Alabama was extremely energy-limited. Dissolved organic matter of low energy availability was the only abiotic input of energy, and it was sparse. Little or no dripping water and no streams enter the cave, whose groundwater level fluctuates by 2–10 m seasonally. In the 1960s and 70s, deductions from gradients of organism densities indicate that the important energy input was from a small colony of grey bats (*Myotis grisescens*) (Cooper, 1975; and personal observations). From extensive data on size distributions, growth and reproduction, especially of marked individuals of a troglobitic crayfish *Orconectes australis* (Cooper, 1975), we infer that food was extremely limited even when grey bats were present. Cooper was able to deduce that *O. australis* had a lifespan of a hundred years, with an age at maturity of about 45 years. In five years he saw three adult females carrying young, of which only the largest female successfully "fledged" her young.

This apparently extreme energy limitation in Shelta Cave is emphasized by comparison to Shiloh Cave, Indiana, which has a relatively high energy input of POM from a sinking stream. In Shiloh a species of crayfish (*Orconectes inermis*), closely related to the one in Shelta Cave, had about a lifespan of about eight years, with a high proportion of adult females reproducing annually (Weingartner, 1975). Consistent with its much lower food supply, the aquatic fauna of the Shelta Cave was much more diverse than for Shiloh and included no troglophiles. Just considering top troglobitic consumers, Shelta had three species of crayfish, an endemic shrimp, and a fish in addition to consumers at lower trophic levels such as amphipods, copepods and isopods (Cooper, 1975). On the other hand, though the food supply was low it could be partially monopolized, because most occurred under one or two small bat roosts. In these areas the largest species of crayfish dominated in numbers and, especially, in total biomass. Most individuals of other species were relegated to areas with scattered feces deposited by flying bats entering or leaving the cave.

Toxins and energy resources

Toxins can either be the potential energy resources themselves, or be produced by decomposers of the resources. In either case they would make the system more energy-limited than if there were no toxic effects.

Bacterial and fungal decomposers are usually beneficial, but can sometimes produce toxins. They convert recalcitrant compounds, like lignin and chitin, to bacterial and fungal biomass which can be consumed by larger consumers. However, they often also produce compounds that make a rotting organic resource unpalatable or poisonous to larger consumers as a strategy for monopolizing resources for themselves (Janzen, 1977).

We hypothesize that chemoheterotrophic Actinomycetes have toxins that preclude their use as food by most organisms. Our rationale is that some Actinomycetes are sources of antibiotics, such as streptomycin, and we seldom observe animals on them. Actinomycetes are common to abundant in damp caves and are apparently responsible for the characteristic cave mud odor that all cavers know but cannot describe. In many Hawaiian lava tubes Actinomycetes are particularly abundant, and colonies cover many square meters of smooth rock walls (personal observations). They certainly show no obvious evidence of being grazed, and few organisms are seen on them. Growth rates of Actinomycetes seem to be so slow that they could not withstand significant grazing. In White Cave in Mammoth Cave National Park (central Kentucky, U.S.A.), Lavoie (unpublished data) has been unable to detect any growth of marked colonies of about 0.5 cm diameter over a period of 12 years!

Many organisms of the deep-sea vents have specialized symbionts that metabolize sulfides; but sulfides are toxic to most organisms that have not been associated with sulfides over evolutionary time. Certainly toxic reactions occur when aquatic species are suddenly exposed to high concentrations of sulfides associated with brines from oil exploration. It is not known whether the organisms in Movile Cave in Romania have evolved sulfide resistance, but the great distance from the magmatic sources to the cave suggests that sulfide concentrations may be low enough to be non-toxic.

It is known that fossil hydrocarbons, such as gas and oil, can be energy sources for hypogean communities; but highly refined hydrocarbons like gasoline appear to be toxic. Apparently generalized decomposers cannot metabolize gasoline that is introduced by accidental spills into caves, and so the hypogean fauna is invariably killed.

AQUATIC/TERRESTRIAL CONTRASTS

Epigean and hypogean ecosystems

Whether epigean or hypogean, aquatic habitats change more slowly over longer distances than terrestrial habitats. The fast mixing and dilution and rapid leaching in water result in only subtle gradients of nutrients, with distance from discrete points of input. Readily available nutrients are quickly lost by leaching, especially by flowing water, and so successional change in species is minimal. Because water has great thermal stability, the range of temperature and its rate of change are slow. Thus, aquatic organisms have little opportunity to choose different thermal microhabitats but can physiologically acclimate to slow temperature change. Thus, in water abiotic and biotic gradients of energy availability are usually not sharp and one does not see local zonation of species and niche differentiation.

A major consequence of reduced local species diversity in aquatic systems is that often food chains have relatively few species. As a result there is often clear "top-down" control by predators and "bottom-up" control through energy input. The large size and high feeding impact of top and intermediate adult predators like fish and crayfish apparently overrides the fact that their young feed only at lower trophic levels, whereas adults feed at several trophic levels.

With the exception of resources that can be monopolized by large consumers in the absence of predation, terrestrial food webs tend to be complicated with many species at each trophic level, and trophic cascades from bottom-up or top-down are hard to demonstrate. There are great spatial changes in abiotic conditions over short distances, and great changes over short periods of successional time. Further blurring of food chains occurs because species often have distinct feeding niches at different ontogenetic stages. This includes species that grow over a large size range from young to adult, like cave crickets, and species with distinct larval and adult stages, like cave beetles. With distinct feeding niches there can be high local species diversity and complicated food webs. There are great spatial changes in abiotic and biotic conditions over short distances and great temporal changes over short periods of successional time.

Hypogean ecosystems

In the hypogean environment, some of the epigean differences between aquatic and terrestrial systems discussed above are accentuated and others are reduced. The aquatic–terrestrial contrasts seen in epigean systems are still seen in hypogean systems; but greater energy limitation results in trophic dynamics that are simpler with fewer species. In addition, hypogean streams often have few tributaries; organic input occurs only at the headwaters, and decomposable organic matter and associated secondary production decrease as one goes downstream. In contrast, epigean streams have the same or increasing production as one goes downstream. As production based on decomposition of leaf and twig litter decreases with widening of the stream there is more light and photoautotrophic production increases.

In terrestrial hypogean systems, slight differences in abiotic factors affect distribution and abundance of species as much as great differences in epigean systems. We suggest that hypogean species are evolutionarily adjusted to relatively little abiotic rigor or variability, and so are quite sensitive to slight differences. The analogue in epigean systems is between temperate-zone and tropical species. In tropical epigean and temperate hypogean environments, the variability between years in population sizes is fully as great as in temperate epigean environments [e.g., Leigh et al. (1982) for the tropics and Poulson et al. (1995) for caves]. Thus, the old paradigms of constant tropical and constant cave environments (Poulson, 1964) are faulted. It is also now known that rare events of unusual abiotic conditions, such as El Niño events in the tropics and especially wet and cool years for caves, can be very important in both epigean and hypogean environments.

INPUT BY ABIOTIC vs. BIOTIC AGENTS TO HYPOGEAN ENVIRONMENTS

Background

Generalizations

In addition to the general energy limitation of hypogean systems and the contrasts between terrestrial and aquatic systems with respect to spatial and temporal variation just discussed, we list three generalizations about inputs by abiotic and biotic agents to hypogean ecosystems:

(1) Great variation often occurs between years in abiotic input by wind and water. Owing to vagaries in weather, abiotic renewal of organic matter is much less predictable than biotic input in the form of feces of trogloxenic and troglophilic animals.

(2) Abiotic input is usually of dead plant parts which have uniformly low caloric availability whereas biotic input is usually of feces and carcasses which have highly variable caloric availability. As a consequence, terrestrial successional decomposition can be striking for such inputs as bat feces and carcasses, and is not detectable for such inputs as leached wood washed in by floods.

(3) Within a year abiotic input is seasonal but localized and is diluted with distance from input points. In contrast, biotic input can be high even at some distance from "entrances" because of the mobility of trogloxenes and troglophiles.

Number and size of openings

The number and size of openings strongly influence all epigean inputs, whether abiotic or biotic, to the hypogean environment. Curl (1966) has estimated that as few as five percent of limestone caves in temperate zones have "entrances" large enough to admit humans and flying bats. Many more caves have openings large enough for small trogloxenes such as pack rats, salamanders, and crickets. But most have openings only large enough for bacteria and recalcitrant dissolved organic matter carried by flowing and seeping water.

Terrestrial communities based on different kinds of input

Poulson and Kane (1981) and Poulson (1992) have presented a framework for considering the kinds of terrestrial community that can be supported by various abiotic and biotic inputs. The most important variable is the range in caloric density per unit area per unit time for both the initial input and subsequent stages of successional decomposition. They also considered how abiotic rigor of too much or too little moisture and low digestibility could decrease the useability of different kinds of organic matter. The general conclusion is that inputs of low useable energy per unit area per unit time and without successional heterogeneity, like leached leaf fragments and beetle feces, support communities of long-lived troglobites with low species diversity, including a few generalists, with none dominant. In contrast, inputs with high useable energy per unit area per unit time and great successional heterogeneity, like carcasses, large feces of raccoons (*Procyon lotor*) with high energy per fecal unit, and bat guano in large volumes, support communities with more species including a few short-lived opportunistic troglophiles or troglobites that are highly dominant at any given time or place.

Input by abiotic agents

Nitrogen limitation

Most abiotic input is of low quality because high carbon to nitrogen ratios greatly slow the rates of utilization by bacterial and fungal decomposers. For example, wood and leaf litter have carbon to nitrogen ratios about 100-fold greater than live animals or carcasses and 10- to 30-fold greater than various kinds of feces. If the material is leached by water, then nitrogen limitation of decomposition is even greater. If ratios of carbon to nitrogen are not too high, then leached particulate organic matter may be colonized and thus "conditioned" by decomposers. In essence the ratio of carbon to nitrogen is decreased by addition of the decomposer biomass. Thus, consumers can gain some nutrition by eating the organic matter along with the decomposer biofilm on the organic matter. The same principle is used by many consumers that re-ingest their own low-energy feces after they are colonized by bacteria and fungi.

Dissolved and particulate organic matter

Here we consider how organic matter gets into the hypogean environment, why it usually supports communities with low species diversity, and why it supports terrestrial communities of higher diversity than aquatic communities.

In contrast to aeolian input to young lava tubes, and seeping, dripping, and flowing water input to limestone caves, terrestrial abiotic input by gravity at some vertically oriented entrances to caves is minimal in frequency, amount, and distance carried into the cave. One reason is that vertically oriented entrances are rare relative to other orientations of entrances. The only positive aspect is that terrestrial input into vertical entrances may not be thoroughly leached. Larger animals rarely fall in and die, but their carcasses are a bonanza for opportunistic decomposers and scavengers.

Usually, any water-borne input of leaf and twig litter is so thoroughly leached that the only compounds that remain are recalcitrant to decomposition, particularly blackened humic acids, lignins, and hemicelluloses. Based on known times of construction of wooden stairs for tourism in wet caves, the rate of decomposition of initially non-leached wood, whose surface becomes black with time, is extremely slow. It is as little as a cubic centimeter per 25–50 years, whether in aquatic or terrestrial environments. It is certainly even slower for wood that is leached before being washed into the cave.

The food webs that depend on diffuse and continual input of very small amounts of low-quality dissolved organic matter in flat-lying limestone terrains have few species and they tend to be long-lived troglobites. Input of dissolved organic matter is diffuse and more or less continuous, through many tiny voids in the epikarst. However, input of significant amounts of particulate organic matter is quite localized in space and rare in time – on the scale of decades or centuries during 10–100 year floods that flush in accumulations of debris at normally dry input locations.

On the other hand, cave ecosystems that depend on concentrated and episodic input of high amounts of relatively unleached particulate organic matter have a moderate number of troglobitic species. Some species can attain high densities in rocky streams with near-yearly renewal of input by sinking streams. Examples are the *Gammarus minus* and *Styonectes* amphipods and the *Caecidotea* isopods in caves of the Valley–Ridge Province in the Appalachians. However, species diversity is low because the material is too leached to have successional decomposition, with different consumer species at different stages of succession.

Compared to aquatic habitats, terrestrial habitats have less leached particulate organic matter and higher species diversity. Partly leached leaf and twig litter deposited by floods at intervals of 1 to 10 years in terrestrial habitats has some successional heterogeneity of species. In several karst regions of the United States such flood-deposited leaf and twig debris supports oligochaetes and millipedes as the consumers with highest frequency, density and body size. When several species of millipedes (Diplopoda) are present, then the most troglomorphic species are associated with the successionally oldest organic matter – that is, at the highest elevation above normal stream level in cracks in the mud-silt stream banks (Culver, 1982: p. 133).

The best-studied mud stream bank community is in a small cave in Indiana where there is a gradient of decreasing flood rigor and decreasing food supply from streamside to tops of mud banks (Keith, 1975). Keith studied these mud banks for two years, using a combination of visual census and both baited and unbaited pitfall trapping for all species, and mark-recapture studies for the troglobitic carabid beetle (*Pseudophthalmus tenuis*) that was the top predator. A sminthurid collembolan (*Arrhopalites pygmaeus*) comprised 85% of all individuals, with an overall density of $1.5\,m^{-2}$ and a biomass of $0.23\,mg\,m^{-2}$. The carabid beetle predator had the highest average biomass of $0.80\,mg\,m^{-2}$ but a density of only $0.20\,m^{-2}$. Oligochaete worms in the flood zone were also important, but could not be censused accurately. All other species had numbers and biomass lower by 1 to 2 orders of magnitude. The main consumers included a myriapod (*Pseudotermia indianae*), a dipluran (*Litocampa* sp.), two oligochaete worms

(Enchytraeidae and Lumbricidae), and two Collembola (*Arrhopalites pygmaeus* and *Sinella cavernarum*). Predators included the carabid beetle mentioned, a linyphiid spider (*Phanetta subterranea*), a rhagidid mite (*Rhagidia cavicola* Banks), and a pseudoscorpion (*Kleptochthonius packardi*). Except for the worms these are all troglobites.

Keith used multiple regression and principal components analysis to assay the effects of abiotic factors and fine particulate organic matter on species diversity and organism densities. He concluded that abiotic rigor of flooding was the primary controlling factor whereas food supply was secondary, and that there was a tradeoff between these two factors. At the stream side, fine particulate organic matter was renewed regularly by floods (+) but flooding rigor was high (–). Overall species diversity was high, but overall density of organisms was very low. In contrast, on the tops of the mud banks flooding rigor was almost nonexistent (+), but food supply was low and rarely renewed (–). Overall species diversity was low, but density of those species present was high. At the tops of the mud banks all predators fed on the abundant sminthurid springtails, whereas at the stream side the only common predator was the carabid beetle, and it fed both on oligochaete worms, in the short time that they are active as flood waters recede, and on a patchily distributed and seasonally abundant entomobryine springtail.

Input by biotic agents

In temperate zones, trogloxenic bats and rhaphidophorine "crickets" provide the most important biotic input of energy to hypogean environments, because they have high frequencies and high densities, and high impacts per individual. Other groups are too infrequent or too sparse, or of too low impact per individual, for their carcasses or feces to support significant communities. For example, in the Mammoth Cave region, cave salamanders (*Eurycea lucifuga*) are in a few entrances only, and even in entrances where they attain moderate densities they collectively produce few feces despite their large size because their metabolic rate is low. At the other extreme a pack rat (*Neotoma magister*) is large and has a high metabolic rate, so a single individual deposits abundant feces in its spatially predictable latrine. When rat latrines are in damp areas they support a diverse community with a few dominant consumers and predators (Richards, 1992). These latrine communities are in jeopardy because pack rats are becoming restricted to places where they cannot be infected by a raccoon nematode of undetermined species which is lethal to the rats. But even where pack rats are still doing well in the mesic eastern United States the latrine community is not common, because most pack rats select areas for their latrines that are too dry for decomposers to use their feces. In the dry western United States, perfectly preserved plant fragments and pollen in pack-rat latrines in caves date from previous interglacial epochs.

Carcasses

Except under maternity roosts of bats, input of dead animals is rare and unpredictable in time. Animals sometimes fall into the openings of vertical shafts and die, but the microclimate at the bottom of open shafts in temperate regions is seasonally too cold or too wet for most troglophilic or troglobitic consumers, and open vertical shafts are rare overall. In parts of the tropics open vertical shafts can be common, and microclimate at the bottom is moderate, but we know of no studies on rates of input of carcasses of animals which fall into shafts in the tropics and die.

Where moisture conditions are suitable for troglophiles and troglobites, the fate of dead animals depends on their size and predictability of input. Small invertebrates are quickly scavenged, especially by large orthopterans such as cockroaches and cave crickets. If these large orthopterans are injured or killed then they are quickly scavenged by individuals of their own species.

Baby bats are the largest predictably occurring type of carcass. Maternity roosts of free-tailed bats (*Tadarida* spp.) include millions of individuals, and baby bats regularly fall to the floor. They are often reduced to skeletons by hordes of dermestid beetles in less than an hour, so that successional decomposition does not exist (personal observations). Martin (1980) has shown that successional decomposition does occur over several months where maternity colonies of grey bats (*Myotis grisescens*) number only in the thousands. Because fallen babies under these maternity roosts are rare, scavengers like dermestid beetles are rare or absent. In fact, the species responsible for the successional decomposition are similar to those that decompose the large piles of feces onto which baby bats fall (see below).

Nutrient composition of dead animals varies, but this variation is not the basis for different hypogean communities. Input of carcasses is rare and unpredictable, so that scavengers are all opportunistic food generalists. Thus, it does not matter whether a dead animal has a high or low caloric density through differences in fat content. [Caloric density of animals is a direct function of fat content, since its energy density of about 9.4 Kcal g^{-1} dry mass is markedly higher than that of protein at 4.3 or carbohydrate at 4.0. Kcal g^{-1} dry mass]. Regardless of protein and carbohydrate concentrations, an animal that dies while in reproductive condition or after storing fat for winter hibernation will have a high caloric density, whereas an animal that dies of starvation will have a low caloric density. In either extreme, the water content is similar at 60–70%, and the protein and carbohydrate concentrations are similar.

Feces

Unlike those dependent on carcasses, communities supported by feces of trogloxenes differ greatly, depending on the size of fecal units or aggregate amount, predictability of renewal in time and space, heterogeneity of material in the feces, the substrate on which the feces are deposited, and the chances of leaching by water. We will briefly consider each of these in general, and then illustrate the generalities by considering bats of various feeding habits and the detailed studies of the use by decomposers (Fletcher, 1982) of feces of insectivorous grey bats and arthropods that eat the decomposers (Martin, 1980). Examples of other kinds of fecal communities in the Mammoth Cave region have been recently reviewed by Poulson (1992).

Feces deposited in water are generally leached and dispersed before they can be eaten by aquatic organisms. Thus, the feces are similar to leached plant organic material in having low energy availability, and they cannot be monopolized. Nonetheless, there are significant aquatic communities under large maternity bat roosts in the temperate zone, and in many tropical caves where many bats roost year-round. For example (personal observations), under a grey bat maternity roost the community consists, in order of decreasing abundance, of troglophilic copepods (Cyclopoida) > troglophilic snails (*Physa* sp.) > troglophilic crayfish (*Orconectes* sp.) > larvae of a troglobitic salamander (*Typhlotriton spelaeus*). This community extends downstream many hundreds of meters, with slowly decreasing abundance of all species. Upstream this largely troglophilic community is replaced within 10 meters by a largely troglobitic community of amphipods (*Stygobromus ozarkensis*) and a tiny snail (*Antrobia culveri*) which have densities lower by orders of magnitude than the troglophiles under the bat roost. And in La Cueva Chica in Mexico the relative abundance of different morphs of a cave fish (*Astyanax mexicanus*) changes between locations under a bat colony and those upstream from it (Mitchell et al., 1977). There are abundant surface, cave, and hybrid forms in the cave, and the ratio of cave to surface morphs increases as one moves upstream from the bat colony. Presumably, the high and constant input of energy below the bat colony decreases the advantage of cave adaptation.

The size of fecal units and associated differences in ratio of surface area to volume are important determinants of the kinds of decomposers (Lavoie, 1982). In terrestrial areas with humidity below saturation, decomposers of small feces risk drying out. Hypothetically, phycomycete fungi can dominate small feces, because their hyphae need only penetrate several millimeters to obtain enough energy to produce their tiny fruiting bodies during their short life cycle. At the other extreme, basidiomycete fungi cannot use small feces unless they are in piles. Their hyphae must penetrate many centimeters in order to obtain enough energy to produce their large fruiting bodies ("mushrooms") during their longer life cycle. Ascomycetes and Fungi Imperfecti are intermediate in their requirements. It follows from these facts that large fecal units or piles of small fecal units could have a typical fungal succession from phycomycetes through ascomycetes and Fungi Imperfecti to basidiomycetes. Of course, metazoan consumers, such as larvae of fungus gnats (Mycetophilidae) and scavenger beetles, could prevent succession if they used the fecal resource before basidiomycete fungi could garner enough energy from their extensive hyphal networks to produce their large fruiting bodies. These hypotheses have been tested and supported by experiments (Lavoie, 1982; Poulson, 1992) which manipulated the size of the fecal unit while keeping the nutritional composition constant, and using fine-mesh netting to exclude metazoan consumers.

In terrestrial habitats, feces that are more heterogeneous in organic matter will have more successional heterogeneity. Feces of carnivores and omnivores often have much hair during the molting season, and this is not easily decomposed. The combination of labile and recalcitrant organic matter results in successional heterogeneity, as we have observed for raccoon feces (Poulson, 1992).

The terrestrial substrate on which feces are deposited has a great effect on the moisture in the feces and, thus, on the community of decomposers, especially for pack-rat feces (Poulson, 1992), but also for cave-cricket feces (unpublished). If pack-rat feces are deposited on rock, the feces–rock interface will be wet and often low in oxygen. Leached nutrients will be concentrated at the interface, and favor bacteria, nematodes, and saprovore mites. At the other extreme, on sand, feces are much drier, and leached nutrients will be lost into the sand. Under these circumstances fungi will be more common. Typical fungivores expected are Collembola, and sciarid or mycetophilid fly larvae. In terrestrial passages with actively developing dripstone formations, dripping water can disperse fresh cave-cricket feces and leach useable nutrients so quickly that the feces are useable only by a greatly restricted number of species. In terrestrial areas with humidity well below saturation there is no leaching, but these small feces dry out quickly and decomposers are at risk from desiccation.

If feces are frequently renewed, then they should be able to support specialized rather than just opportunistic decomposers. Frequent and/or predictable renewal of feces is most likely with species that regularly use the cave like obligate trogloxenes. These are mainly bats and either cockroaches in the tropics or rhaphidophorine orthopterans in the temperate zone. If these organisms use the same roosting areas near cave entrances year-round, then guano renewal is predictable in space as well as in time.

Bat feces

Bat feces illustrate general principles about fecal use by species in hypogean communities, since bats show great differences in predictable use of caves from the temperate zone to the tropics, and great differences in feeding habits in the tropics. Often, in temperate-zone caves, only one species of bat is present, so that the same kind of feces occurs over a wide range of densities and renewal rates and on a wide variety of substrates. This allows one clearly to test the hypothesis that low and steady rates of organic input are not associated with dominance by a few species, but that pulsed input leads to dominance.

Essentially, all bats in temperate zones are insectivorous, so that the nutrient composition of guano varies only in subtle ways; but there can be great differences in the amount deposited per unit time and per unit area. The kinds of insects available, and whether they have high caloric concentrations, vary with seasonality of reproduction (Studier et al., 1994), so that this should affect fecal caloric density. In addition, the rate of input of fresh bat feces has great consequences for the guano community, since the rate varies tremendously with cave type and season. Rate of fecal deposition is highest and most predictable in caves with warm traps which are occupied during the late spring and early summer by maternity colonies, and are abandoned during the fall when bats migrate. Rate of deposition is lowest and least predictable in caves that are cold traps, occupied only during the winter by hibernating colonies, and abandoned in spring as epigean insects become available again. On account of the very low metabolic rate of hibernating bats, there is no fecal deposition in winter, and only a little occurs when bats enter the cave in fall and evacuate their guts in preparation for hibernation. When bats periodically arouse from hibernation in spring, they shift their roosting areas and guano is deposited slowly and irregularly due to periodic torpor associated with cold weather and erratic availability of epigean insects. Thus, guano in hibernacula is scattered and in low amounts. In contrast, in caves with bat maternity roosts guano accumulations range from scattered with moderate rates of deposition to concentrated with high rates of deposition. In late spring and summer under maternity roosts, guano deposition increases with food consumption by females during pregnancy and lactation. Hyperphagy associated with fat deposition in fall may also result in increased rates of guano deposition.

The predictability of bat guano availability is much less in temperate-zone caves than in the tropics. In the temperate zone most caves do not have bats, and where they do occur bats use different caves for prehibernation swarming, for maternity, and for hibernation. In the tropics the amount of guano deposition is great and continuous, because food for bats is available all year. Almost all tropical caves have bats year-round and the several species have quite different diets.

As a working hypothesis we suggest that dietary differences between species in the tropics affect the nutritional composition of bat feces much more than differences between seasons, for a given species, in the temperate zone. We hypothesize that tradeoffs for consumers are associated with their use of feces of blood-eating (sanguivorous), fruit-eating (frugivorous), and insect-eating (insectivorous) bats. The tradeoffs are based on digestibility of the bats' food and heterogeneity of the feces both in nutrients and moisture content.

Vampires (*Desmodus*) eat blood which is highly nutritious, high in protein, and highly digestible. However, vampires must urinate copiously to get rid of the great amount of water in blood (McFarland and Wimsatt, 1969). For potential decomposers of vampire feces, the result is a high urea content (+) but little fat or carbohydrate (–), a viscous consistency, and high iron content (–). We predict that it can be used by only a few bacteria that are highly dominant, and that they may generate a high biological oxygen demand, lowering the oxygen tension and so making the feces unusable for all macroscopic consumers.

The feces of large fruit bats, and also oil-birds (*Steatornis caripensis*) that roost in caves, have a bimodal distribution of nutrients because the seeds in the feces are largely indigestible. The pulp of fruit is highly digestible, so that little of its carbohydrate and fat is left in the feces (–), but the pulp is also high in water content, so that dilute bat urine with little urea is a major component of the guano (+/–). Feces of frugivorous birds and mammals are not attractive to dung beetles because they cannot use the seeds (Poulson, unpublished data). Thus, we deduce that the seeds are not a likely source of food for hypogean consumers. The seeds are adapted to be resistant to digestion and to pass quickly through the gut so that the bats act as a dispersal agent for the plant (Howe and Westley, 1988), but the seeds may also have toxic compounds as a defense against seed predators and this might be expected to reduce or preclude the useability of the feces by decomposers (Janzen, 1977). Consistent with this scenario are "forests" of etiolated seedlings under roosts of oil-birds in caves. Thus, the seedlings may be the most important source of food for communities of consumers associated with feces of frugivorous birds and bats in tropical caves.

The feces and urine of insectivorous bats are likely to support the highest diversity of decomposers because the components of the bat guano are quite heterogeneous. Thus, overall species diversity is increased with successional decomposition. Under roosts, decomposers very quickly metabolize large amounts of urea from urine (+), slowly metabolize small amounts of fat and protein (+/–), and very slowly metabolize large amounts of chitin (–). Combining these different rates of nutrient use with Poulson's (1972) early hypotheses about the effects of the rate of input of bat feces on community structure, we expect that early successional stages will be dominated by a few fast-growing bacterial species that monopolize urea, and that a low diversity of consumers will be present due to potentially low oxygen concentrations (–) and ability of a few fast-growing consumers to monopolize the resources. Late successional stages will have a variety of slow-growing fungal species that can metabolize chitin, and the consumers will have high diversity.

These expectations have been tested by Martin (1980) and Fletcher (1982) in Tumbling Creek Cave (Missouri, USA), where dispersion and rates of renewal of feces of insectivorous bats (*Myotis grisescens*) vary greatly, and where the decomposer communities are correspondingly diverse. Overall, the consumers include 13 carnivorous mites, 7 fungivorous mites, 5 saprovorous mites, 7 springtails, 5 flies, 3 beetles, 2 millipedes, 2 spiders and one species each of bristletail, pseudoscorpion, and harvestman. The data are from a combination of observational counts, pitfall traps, and guano cores (Martin, 1980). The greatest community contrasts are under grey bat maternity roosts, where fecal input is seasonally pulsed. On the guano pile communities change dramatically over a few meters from the top of the pile, where the rate of fecal and urine deposition is high, to the periphery where the rate of deposition is low and feces are scattered. When mothers are lactating and baby bats are growing most rapidly, the top of the pile has an abundant fecal input, and so much urine that the feces are very wet and near neutral in pH. At a microscopic scale enteric bacteria are abundant, at about 10^8 g^{-1}, and bacteriophagous and predatory nematodes are also abundant. And at a macroscopic scale saprovorous and predatory mites make up a seething mass.

To a depth of 5–10 centimeters at the center of the pile, and 2–3 meters laterally from the area of highest input of feces and urine, succession is evident. Moisture levels are lower and pH has dropped to 3–4. Here, fungi have become the dominant primary decomposers. The number of mite species and their abundance have declined dramatically and springtails and fly larvae support a high frequency of a predatory pseudoscorpion (*Hesperochernes occidentalis*). Succession is not so evident at the edge of the pile, where fungivorous fly larvae (*Bradysia* sp.) are more common and only one mite species and two species of springtails (*Folsomia candida* and *Pseudosinella argentea*) are present. The larger predatory species are coequal in abundance and include a small carabid beetle (*Bembidion* sp.), a large pseudoscorpion (*Hesperochernes occidentalis*), a tiny staphylinid beetle (*Atheta* sp.), and a large fungus gnat larva (*Macrocera nobilis*).

Just away from the pile edge, feces cover is no longer continuous and no succession is evident. Overall species abundance and diversity is still lower, on account certainly of the low energy availability and possibly the great variety of predators. In addition to the same predators as at the edge of the pile there are two species of spiders (*Cicurina cavealis* and *Islandiana* sp.). Two species of flies (*Leptocera tenebrarum* and *Psychoda* sp.) and a millipede (*Antriadesmus* sp.) are added as mid-level consumers. The community is similar farther away from active pile edges and in areas of scattered feces far from guano piles. Overall dominance is low and diversity is low. Flies, springtails, and a millipede are the most common large consumers.

We close by considering how these new data of Martin (1980) and Fletcher (1982), and our overall conclusions about nutrients in this review, relate to the hypotheses induced by Poulson in his 1972 review of temperate bat-guano ecosystems. His general framework (Poulson and Culver, 1969) was that rigor, variability, and predictability of abiotic factors and nutrient input determine species biology, which in turn affects community organization, such as dominance–diversity patterns. The data still seem to support these predictions but some of his hypotheses about bat guano need to be modified.

(1) Compared to other temperate-zone hypogean communities, the input of energy into bat-guano communities is more variable and less predictable in time and space (Poulson, 1992). One now knows that the issue of scale is quite important. Pulses of abiotic input of organic matter into caves can be quite variable and unpredictable because of rare floods at the scale of decades. It can also be understood that shallow gradients in abiotic variables and in food concentration, in stable microclimates and with great energy limitation, can be as important to the troglobitic species as steep gradients on piles of bat guano are to troglophilic species.

(2) One species "strategy" would be a specialist guanobite; such a species would be small with a narrow feeding niche (Poulson, 1972). It would survive the seasonal absence of bats by having dormant stages, and show explosive reproduction when guano input resumes. "Zones" with different pH and urea concentration near the top of a guano pile would each be dominated numerically by one or two species adapted to the local abiotic conditions. Martin's data largely support all of these hypotheses, except for the question of dormancy which she did not study. What was not expected, with the high dominance of one or two species, was the high number of relatively rare species in each pitfall trap or guano-core sample.

(3) The contrasting species "strategy" would be a generalist troglophile. These would be large and mobile species with a broad feeding and habitat niche (Poulson, 1972). Such species would survive the seasonal absence of bats by subsisting on lower-energy foods, and would opportunistically switch back to fresh bat feces and show reproductive pulses when the bats return. A few of these species would comprise most of the biomass in the guano. We now know that this scenario fits for some species of camel cricket (*Ceuthophilus*) but not the other species of large and mobile troglophiles. These other troglophiles live and reproduce continuously on the scattered feces at the edge of the guano pile where low amounts of low-quality food are always present, like hair for a tineid moth larva, or dry powdery guano for a tenebrionid beetle larva. It is also now known that troglobites occur on scattered feces away from the edges of guano piles. We believe that they avoid the guano piles because they cannot sustain the high predation rates that would be incurred from the high densities of generalist predators on the piles. The hypothesis is that troglobites have low reproductive rates, but they must have repeated reproduction so that each female will have at least one female offspring to survive and replace her mother when she dies (Poulson, 1981).

REFERENCES

Barr Jr., T.C., 1967. Ecological studies in the Mammoth Cave System of Kentucky. I. The biota. *Int. J. Speleol.*, 3: 147–203.

Barr Jr., T.C., 1968. Cave ecology and the evolution of troglobites. In: T. Dobzhansky, M.C. Hecht, and W.C. Steere (Editors), *Evolutionary Biology*, Vol. 2. Plenum Press, NY, pp. 35–202.

Cooper, J.E., 1975. *Ecological and Behavioral Studies in Shelta Cave, Alabama with Emphasis on Decapod Crustaceans. Ph.D. Dissertation in Zoology.* University of Kentucky, Lexington, Kentucky, 364 pp.

Culver, D.C., 1982. *Cave Life: Evolution and Ecology.* Harvard University Press, Cambridge, Massachusetts, 189 pp.

Culver, D.C. and Poulson, T.L., 1972. Oxygen consumption and activity in closely related amphipod populations from cave and surface habitats. *Am. Midl. Nat.*, 85: 74–84.

Culver, D.C., Fong, D.W. and Jernigan, R.W., 1991. Species interactions in cave stream communities: Experimental results and microdistribution effects. *Am. Midl. Nat.,* 126: 364–379.

Culver, D.C., Jones, W.K. and Holsinger, J.R., 1992. Biological and hydrological investigations of the Cedars, Lee County, Virginia: An ecologically significant and threatened karst area. In: J.A. Stanford and J.J. Simons (Editors), *Proceedings of the First International Conference on Groundwater Ecology.* American Water Resources Association, Bethesda, Maryland, pp. 281–290.

Culver, D.C., Kane, T.C. and Fong, D.W., 1995. *Adaptation and Natural Selection in Caves: The Evolution of Gammarus minus.* Harvard University Press, Cambridge, Massachusetts, 223 pp.

Curl, R., 1966. Caves as a measure of karst. *J. Geol.,* 74: 798–830.

Fletcher, M.W., 1982. *Microbial Ecology of a Bat Guano Community. M.S. Thesis in Life Sciences.* Southwest Missouri State University, Springfield, Missouri, 425 pp.

Howarth, F.G., 1987. The evolution of non-relictual tropical troglobites. *Int. J. Speleol.,* 16: 1–16.

Howe, H.H. and Westley, L.C., 1988. *Ecological Relationships of Plants and Animals.* Oxford University Press, New York, 273 pp.

Jannasch, H.H., 1984. Chemosynthesis: The nutritional basis for life at deep-sea vents. *Oceanus,* 27: 73–78.

Janzen, D.H., 1977. Why fruits rot, seeds mold, and meat spoils. *Am. Nat.,* 111: 691–713.

Jegla, T.C. and Poulson, T.L., 1970. Circannian rhythms I. Reproduction in the cave crayfish *Orconectes pellucidus inermis. Comp. Biochem. Physiol.,* 33: 347–356.

Juberthie, C. and Delay, B., 1980. Extension du milieu souterrain en zone non-calcaire: Description d'un nouveau milieu et de son peuplement par les Coleopteres troglobies. *Mem. Biospeol.,* 7: 19–52.

Kaushik, N.K. and Hynes, H.B.N., 1971. The fate of dead leaves that fall into streams. *Arch. Hydrobiol.,* 68: 465–515.

Keith, J.H., 1975. *Biotic and abiotic studies of a terrestrial cave ecosystem in Indiana. Ph.D. Dissertation in Zoology.* University of Indiana, Bloomington, Indiana, 197 pp.

Lavoie, K.H., 1982. *The Influence of Biotic and Abiotic Factors on the Successional Decomposition of Cave Rat Dung. Ph.D. Disseration in Biological Sciences.* University of Illinois at Chicago, Chicago, Illinois, 224 pp.

Leigh, E.G., Rand, A.S. and Windsor, D.M. (Editors), 1982. *The Ecology of a Tropical Forest: Seasonal Rhythms and Long-term Changes.* Smithsonian Institution Press, Washington, D.C., 468 pp.

Lisowski, E.A., Olson, R.A., Roy, W.R. and Thompson, W.J., 1985. Geochemistry and biology of Sulphur River, Parker's Cave, Kentucky. In: *Cave Research Foundation Annual Report for 1985, Cave Research Foundation.* Cave Books, Saint Louis, Missouri, pp. 26–27.

Longley, G., 1981. The Edwards Aquifer: Earth's most diverse groundwater ecosystem? *Int. J. Speleol.,* 11: 123–128.

Martin, B.J., 1980. *The Community Structure of Arthropods on Bat Guano and Bat Carcasses in Tumbling Creek Cave, M.S. Thesis in Biological Sciences.* University of Illinois at Chicago, Chicago, Illinois, 178 pp.

McFarland, W. and Wimsatt, W., 1969. Renal function and its relation to the ecology of the vampire bat. *Comp. Biochem. Physiol.,* 28: 985–1006.

Mitchell, R.W., Russell, W.H. and Elliott, W.R., 1977. Mexican eyeless Characin fishes, genus *Astyanax*: Environment, distribution, and evolution. *Spec. Publ. Mus. Texas Tech Univ.,* 12: 1–89.

Paulik, G.J., 1971. Anchovies, birds, and fishermen in the Peruvian Current. In: W.W. Murdoch (Editor), *Environment.* Sinauer, Stamford, Connecticut, pp. 156–185.

Peck, S.B., 1981. Evolution of cave Cholevinae in North America (Coleoptera: Leiodidae). In: B.F. Beck (Editor), *Proceedings of the Eighth International Congress of Speleology, Bowling Green, Kentucky,* 2: 503–507.

Poulson, T.L., 1963. Cave adaptation in amblyopsid fishes. *Am. Midl. Nat.,* 70: 257–290.

Poulson, T.L., 1964. Animals in aquatic environments: Animals in caves. In: D.B. Dill (Editor), *Handbook of Physiology 4: Adaptation to the Environment.* American Physiological Society, Washington, D.C., pp. 749–771.

Poulson, T.L., 1972. Bat guano ecosystems. *Natl. Speleol. Soc. Bull.,* 34: 56–60.

Poulson, T.L., 1981. Variations in life history of Linyphiid cave spiders. In: B.F. Beck (Editor), *Proceedings of the Eighth International Congress of Speleology, Bowling Green, Kentucky,* 1: 60–62.

Poulson, T.L., 1985. Evolutionary reduction by neutral mutations: Plausibility arguments and data from amblyopsid fishes and linyphiid spiders. *Natl. Speleol. Soc. Bull.,* 47: 109–117.

Poulson, T.L., 1992. The Mammoth Cave Ecosystem. In: A.I. Camacho (Editor), *The Natural History of Biospeleology.* Monograph of the National Museum of Natural Sciences, Madrid, Spain, pp. 569–611.

Poulson, T.L. and Culver, D.C., 1969. Diversity in terrestrial cave communities. *Ecology,* 50: 153–158.

Poulson, T.L. and Kane, T.C., 1981. How food type determines community organization in caves. In: B.F. Beck (Editor), *Proceedings of the Eighth International Congress of Speleology, Bowling Green, Kentucky,* 1: 56–59.

Poulson, T.L. and Smith, P.M., 1969. The basis for seasonal growth and reproduction in aquatic cave organisms. *Proc. Fourth Int. Congr. Speleol. Ljubjana, Yugoslavia (1965),* 4–5: 197–201.

Poulson, T.L. and White, W.B., 1969. The cave environment. *Science,* 165: 971–981.

Poulson, T.L., Lavoie, K.H. and Helf, K., 1995. Long-term effects of weather on the cave cricket (*Hadenoecus subterraneus,* Orthoptera, Rhaphidophoridae) guano community in Mammoth Cave National Park. *Am. Midl. Nat.,* 134: 226–236.

Reddell, J.R., 1981. A review of the cavernicole fauna of Mexico, Guatemala, and Belize. *Bulletin 27 of the Texas Memorial Museum.* Texas Memorial Museum, Austin, Texas, USA, 327 pp.

Richards, P.M., 1992. *Predatory Effects, Life Cycle, and Husbandry of the Staphylinid Beetle Quedius erythrogaster in the Cave Rat Latrine. M.S. Thesis in Biological Sciences.* University of Illinois at Chicago, Chicago, Illinois, 131 pp.

Riser-Cooper, M., 1969. Sensory specialization and allometric growth in cavernicolous crayfishes. *Proc. Fourth Int. Congr. Speleol. Llubjana, Yugoslavia (1965),* 4–5: 203–208.

Rouch, R. and Danielopol, D.L., 1987. The origin of subterranean aquatic fauna: Between the refuge paradigm and the active colonization model. *Stygologia,* 3: 345–372.

Russell-Hunter, W.D., 1970. *Aquatic Productivity.* MacMillan, New York, 306 pp.

Sarbu, S.M. and Popa, R., 1992. A unique chemoautotrophically based cave ecosystem. In: A.I. Camacho (Editor), *The Natural History of Biospeleology.* Monograph of the National Museum of Natural Sciences, Madrid, Spain, pp. 637–666.

Studier, E.H., Sevick, S.H., Ridley, D. and Wilson, D.E., 1994. Mineral and nitrogen concentrations in guano of some neotropical bats. *J. Mammal.,* 75: 674–680.

Weingartner, D.L., 1975. *Studies of Productivity in a Cave Stream in Indiana. Ph.D. Dissertation in Zoology.* Michigan State University, Lansing, Michigan, 261 pp.

Wiegert, R.G. and Fraleigh, P.C., 1972. Ecology of Yellowstone thermal effluent systems: Net primary production and species diversity of a successional blue-green algal mat. *Limnol. Oceanogr.,* 17: 215–228.

Chapter 13

GUANO COMMUNITIES IN TROPICAL CAVES

Pedro GNASPINI and Eleonora TRAJANO

INTRODUCTION

Animal matter such as droppings and carcasses (from both cavernicolous and accidental animals) have been widely considered to be an important energy input in the subterranean biotope. The most important source of animal droppings in caves throughout the world is bat guano (Fig. 13.1). In tropical caves, bird droppings such as those from guácharos (*Steatornis caripensis*) (northern South America and Caribbean) and swiftlets (*Aerodramus* sp.) (tropical Oriental caves) may be an important food source as well. On the other hand, in temperate caves, guano of rhaphidophorid crickets represents a large amount of food input for cavernicoles.

Cavernicolous guano communities are an important field for ecological studies, because they are somewhat isolated and, in general, present a simple structure. Moreover, the physical and nutritional properties of the guano vary according to the kind of animals producing the piles – namely, bats, birds or insects. The diversity of feeding habits observed among bats provides an additional field for comparison of community structures. This is true especially for Neotropical bats, which show the highest diversity of feeding habits among all bats in the world. Because of this, Neotropical caves constitute important laboratories for describing, understanding and comparing guano communities.

Thus, several faunistic surveys have focused on bat-guano communities and the relationships among guano inhabitants. The chemical analysis of bat-guano deposits was extensively reviewed by Hutchinson (1950), who included data from all over the world. Here we present some considerations about guano communities, especially those from tropical caves.

DEFINITION OF TERMS: ECOLOGICAL–EVOLUTIONARY CLASSIFICATION OF GUANO CAVERNICOLES

Communities associated with guano have specific attributes that require further classification in addition to the standard trogloxene, troglophile and troglobite categories. Theoretical definitions about guano inhabitants are based mainly on papers by Decu and colleagues (focusing on eastern European caves), and Poulson and colleagues (relative to North American caves). Both groups of researchers started their field and theoretical work on guano fauna in the 1960's, and their ideas are summarized in papers by Poulson (1972) and Decu (1986). This work was discussed by Gnaspini (1992), who reviewed the theoretical definitions and added data from Brazil and other Neotropical countries.

Decu, Poulson and their colleagues did not consider bat-guano inhabitants to be real cavernicoles, because they were supposed to depend on caves only from a biological (search for food) and not an ecological point of view (e.g., Decu, 1986). This would mean that (1) guano inhabitants colonize caves only because of guano, and would not be there if that food supply was missing; and (2) they do not depend on the cave ecosystem nor are subjected to cave environmental constraints. However, in order to reach and colonize the guano piles, frequently deposited far from known cave entrances, organisms should be able to orient themselves and also survive inside the cave. Therefore, guano piles should simply be treated as a substrate, and, as with any substrate inside a cave, it would be colonized by cavernicoles. Hence, guano inhabitants may be classified according to their ecological and evolutionary relationships both as regards the cave environment and the guano deposits (Gnaspini, 1992).

Decu (1986) classified guano inhabitants according

Fig. 13.1. General view of a deposit of guano from haematophagous bats in a cave in São Paulo State. Photo: J. Allieri.

to their ecological relationship to the substrate as guanobites, guanophiles and guanoxenes, following the same rationale of "affinity to the environment" used in the classification of cavernicoles. Actually, he called guanobites "troglobites-guanobites" considering that "the guanobite notion must be considered to be equivalent to that of troglobite". However, these cave and guano relationships are not necessarily interdependent. The degree of fidelity to, or dependence on, the cave environment is neither influenced by nor influences the relationships of the cavernicoles to the guano (i.e., their degree of fidelity to the guano environment). For example, an animal restricted, in caves, to guano piles (thus, a guanobite) may occur in the epigean environment as well, in this case not being a troglobite, as expected if the two classifications were correlated.

Therefore, the guano inhabitants should be classified independently according to their cave and guano relationships. When analyzing the cave relationships of an organism, one must take into account its possible occurrence in the epigean environment, as well as the possible modifications acquired during colonization and isolation in the hypogean environment, in order to classify it as a trogloxene, a troglophile or a troglobite. Independently, one may also consider whether or not an organism inhabits substrates other than guano inside the cave.

The following classification, based on the guano relationships (Gnaspini, 1992), agrees with that proposed by Decu (1986), but with a new sense for guanobites:

Guanobites are organisms that, when in caves, inhabit exclusively guano deposits, and whose entire biological cycle takes place in this substrate. One should bear in mind that guano deposits are discontinuous in caves, and hence guanobites may be found on any cave substrate as they move through the cave environment to colonize guano piles. Guanobites, however, do not reproduce or feed in these substrates. **Guanophiles** may inhabit and reproduce both in guano piles and in other substrates of the cave environment. **Guanoxenes** may be found feeding and/or reproducing on guano deposits but depend on other substrate(s) in the caves to complete their biological cycle.

The entire biological cycle of a guanobite takes place in the guano piles inside caves, and, therefore, entirely in the cave environment. Thus, a guanobite may be either a troglobite or a troglophile, but never a trogloxene. This applies also to guanophiles, because at least the populations observed in the guano piles complete their life cycles inside the cave. Finally, guanoxenes can be trogloxenes, troglophiles, or troglobites. It is noteworthy that a classification combining troglo- and guano-related classifications is very useful because it gives a precise notion of the ecological–evolutionary relationships of the organisms studied. Moreover, bat-guano fauna consists of guanophages and their predators. **Guanophages** include animals that feed directly on the guano (i.e., the faecal matter), and animals feeding on microorganisms and fungi that grow on it. Decu (1986) stated that guanophages may be fitted into any of the categories whereas all their predators are guanophiles, but this does not seem to be true for Brazilian animals. For instance, at least one species of chilopod (in the order Lithobiomorpha) found exclusively in frugivorous bat guano – where it probably preys on Heteroptera (in the families Cydnidae and Lygaeidae, also guanobites) – shows apparent troglomorphisms (Gnaspini-Netto, 1989a).

DIVERSITY OF FEEDING HABITS IN NEOTROPICAL BATS

Bats are the most conspicuous and common trogloxenes throughout the world, and their guano is one of the most important food sources for other cavernicoles – in some cases, the most important. The occurrence, distribution and amount of guano is a consequence of bat biology, and they determine the relevant features of cave ecology.

Among mammalian orders, Chiroptera is second in diversity to Rodentia, reaching its highest diversity in the Neotropical and Southern Oriental regions. The Neotropical region is outstanding not only for the great abundance of bats, which belong to nine families and about 250 species (Koopman, 1993), but also for the highest diversity of food habits in the group: aerial and arboreal insectivory, carnivory, frugivory, nectarivory, omnivory, and haematophagy, are found among the endemic phyllostomids, piscivory is typical of noctilionids, whereas strict insectivory, as in temperate bats, is observed in other, usually less common, families.

For mammalogists, Neotropical bats represent an excellent opportunity to study biological processes such as niche-partitioning, plant–animal interactions, extreme feeding specializations (as observed in haematophages), and social behaviour. For speleobiologists, Neotropical bat faunas provide an unique opportunity of comparing cave communities associated with strikingly different kinds of guano.

Here, we use a speleobiological approach. Nevertheless, to understand the role of bat guano in Brazilian cave ecosystems and the functioning of cave communities associated with such substrates, it is necessary to analyse the ecology of cave bats.

BRAZILIAN TROGLOXENES: CAVE-BAT ECOLOGY

Except for guácharos on the border between Brazil and Venezuela and Guyana (C. Galán, pers. commun., 1995), guano deposition in Brazilian caves is practically restricted to bats. Other trogloxenes such as otters, rodents (rats, agoutis), and didelphids, also leave their faeces in caves, but as scattered occurrences representing subsidiary food sources for cavernicoles. The only Brazilian invertebrate unequivocally shown to be a trogloxene is the harvestman *Goniosoma spelaeum,* from southeastern Brazil (Gnaspini, 1996). It forms large populations near cave entrances, but guano deposits were not identified.

Up to now, about 35 bat species, representing 25% of those recorded for the country, have been found in Brazilian caves. Since many karst areas have not been surveyed for bats, the proportion of bat species using caves is probably higher. The presence of caves favours those bats with preference for rocky shelters, mainly the insectivorous taxa, such as the phyllostomines (including large carnivores like *Chrotopterus auritus, Phylloderma stenops,* and *Trachops cirrhosus*), natalids (*Natalus stramineus*), furipterids (*Furipterus horrens*), and some emballonurids (*Peropteryx* spp.), mormoopids (*Pteronotus parnellii*), and vespertilionids (*Myotis nigricans*), the diversity of which increases in karstic regions. However, these species are generally present in small populations, and their importance as guano producers is, with some exceptions [*P. parnellii* in Amazonia (Trajano and Moreira, 1991); *F. horrens* in the Ibiapaba region, Ceará State (Uieda et al., 1980)], relatively limited.

The most common bat in inhabited areas is, by far, the common vampire bat, *Desmodus rotundus* (Phyllostomidae). Owing to the high availability of prey in the form of domestic animals, and roosts not being a limiting factor in cave areas, the populations of these haematophages are high, sometimes outnumbering all other bats together. Thus, in most of the Brazilian karstic regions studied which are situated near livestock areas, haematophagous bat guano is the most abundant type. This is the case for limestone caves in the Ribeira Valley, São Paulo and Paraná States (Fig. 13.1), and in the neighbourhood of Brasília, Distrito Federal, and of sandstone caves in the Serra Geral, São Paulo State (Trajano, 1985; Bredt et al., 1999).

Second to *D. rotundus, Carollia perspicillata* (Phyllostomidae) is frequently found in Brazilian caves. This ecologically flexible bat, a frugivore that opportunistically feeds on other items such as insects, is common throughout South America, outnumbering *D. rotundus* in uninhabited areas such as the Altamira–Itaituba region, Pará State (Trajano and Moreira, 1991). Frugivorous stenodermines (Phyllostomidae) may also be frequent in caves. This is the case for *Artibeus fimbriatus,* the fourth commonest cave bat in the Ribeira valley, and *Platyrrhinus lineatus,* with small cave populations found in several Brazilian karstic areas (Trajano, 1985; Campanhã and Fowler, 1993; Bredt et al., 1999).

Glossophagine bats (Phyllostomidae), specialized in nectarivory, are relatively frequent in Brazilian caves – *Anoura caudifer* is the third commonest species in the Ribeira valley, and several other species have been recorded throughout the country. However, we were unable to identify their guano piles, probably because the colonies are in general small and, on account of its liquid nature, the guano would mix with the substrate. The piscivorous noctilionids were not found in the caves studied, hence we have no information on communities associated with this kind of guano.

Thus, the available data refer mainly to invertebrate communities associated with haematophagous, carnivorous, insectivorous, and frugivorous bat guano in Brazilian caves.

In most of the Neotropical region, there is no selective pressure for thermoregulatory bat aggregations in roosts, and most species seem not to be highly gregarious. Therefore, when a great number of suitable shelters are available, as in many karstic areas, the local populations tend to spread throughout them, forming small to medium-sized colonies. Moreover, the degree of fidelity to particular roosts is variable. In the Ribeira Valley, phyllostomines like *Chrotopterus auritus* and *Trachops cirrhosus* show particularly high fidelity to their roosts (phylopatry), whereas glossophagines such as *Anoura caudifer* tend to be more nomadic, frequently moving between caves – hence the lack of recognizable piles of nectarivorous guano. *Artibeus fimbriatus* and *Carollia perspicillata* are intermediate in this aspect (Trajano, 1996). In addition to moving between caves, these bats frequently move inside the caves, forming itinerant colonies that change their location in an irregular, non-seasonal fashion. This behaviour was well documented for *Desmodus rotundus* in several areas such as the Ribeira Valley (São Paulo State), Bodoquena region (Mato Grosso do Sul State), and northern Minas Gerais State (Trajano, 1996; Trajano and Gimenez, 1998).

On the other hand, in areas with a low cave density, bats with a preference for rocky shelters would have no choice but to concentrate on the few available roosts, forming larger colonies and showing a higher degree of roost fidelity. Such concentrations were observed for *Desmodus rotundus* using sandstone caves (usually smaller and less abundant than limestone caves) in Serra Geral, São Paulo State (Campanhã and Fowler, 1993), and small, rather isolated limestone caves such as those located close to the limestone border in the Ribeira Valley (Trajano, 1996) and in the Brasília region, Goiás State. However, these populations usually do not exceed hundreds of individuals.

The only reported case of typical "bat caves" in Brazil are the sandstone caves of the Altamira–Itaituba region, Pará State, which harbour huge colonies of bats, specially the insectivorous *Pteronotus parnellii* (the most abundant), *Myotis nigricans, Natalus stramineus,* and *Peropteryx* sp., besides the omnivorous *Carollia perspicillata,* and, to a lesser extent, the haematophagous *Desmodus rotundus* (Trajano and Moreira, 1991). Two non-exclusive hypotheses may explain such concentrations of bats. First, the low density of caves in sandstone rocks would compel the bats to share the same shelters. Second, bat concentration may be due to a high degree of sociability in *P. parnellii,* which would, in this aspect, be rather an exception among Brazilian bats.

BAT GUANO DEPOSITION AND EUTROPHY IN TROPICAL CAVES

Five types of caves can be distinguished according to the kind and amount of trophic resources (e.g., Forestiero and Sbordoni, 1978). **Oligotrophic** caves are characterized by low amounts of available organic matter derived from animals and/or plants. **Eutrophic** caves have high amounts of available organic matter, mainly from animals – especially bat guano. **Dis-trophic** caves are those mostly supplied with plant organic material during floods. **Mesotrophic** caves are intermediate among the three types above, and are characterized by a moderate amount of animal and/or plant material. Finally, **poecilotrophic** caves are those of extended length with sections of differing energy supplies – ranging from oligotrophic to eutrophic sections.

It is widely held in the literature that tropical caves are mostly eutrophic, whereas temperate caves are oligo- or mesotrophic (e.g., Decu, 1986). This view is based on the great amount of food, represented by large guano deposits, found in some caves due to the continuous activity of tropical bats. According to Decu (1981), all caves in Cuba are eutrophic, and the trophic chain is totally dependent on bat guano. On the other hand, Sbordoni et al. (1980) observed that some caves in Mexico were not eutrophic, and concluded that tropical caves may be eutrophic or poecilotrophic (Forestiero and Sbordoni, 1978).

In the Brazilian sandstone caves of Pará State, the input of bat guano is indeed high due to the presence of particularly numerous bat colonies, mainly *Pteronotus parnellii,* and these caves are actually eutrophic (Trajano and Moreira, 1991). However, as already mentioned, the situation is quite different in most Brazilian limestone caves. Usually, the small to medium deposits of bat guano are scattered, and do not constitute the main or only base of the trophic chain. Furthermore, there are generally other important food sources, such as plant debris and sediment banks. Thus, these caves are mesotrophic, oligotrophic, or, in some cases, poecilotrophic. We estimate that no more than 5% of the Brazilian caves can be considered to be eutrophic. This is supported by analysis of the composition of Brazilian cave communities, which shows that a substantial part of the fauna is not closely associated with guano.

SUCCESSION IN BAT-GUANO PILES IN TROPICAL CAVES

In temperate regions, the seasonal activity of cave bats leads to seasonal changes in the structure of guano piles and, consequently, in the associated fauna (Poulson, 1972; Decu, 1986). In general, these bats are active during the summer and hibernate (in the same or in other caves) in winter. When bats are present and active, fresh guano accumulates continuously. As bats reduce their activity or leave the sites, there is a progressive desiccation of the deposits, promoting chemical and microclimatic changes. The microclimatic properties of the guano depend on the depth and the distance from the centre of the deposit. Because several taxa show microclimatic preferences, a faunistic succession is observed in the guano deposits. Guano piles are initially alkaline owing to the concentration of ammonia in the faeces. They subsequently become acid as a result of the decomposition of alkaline compounds and fermentation of the piles. Although this process has seldom been studied (e.g., Hill, 1969; Negrea and Negrea, 1971; Martin, 1980), it seems to be widespread in guano communities worldwide. On the other hand, Decu (1986), based on the literature and his own observations in Cuba, stated that bats are permanently active in tropical caves. As the guano input is continuous, there would be no seasonal differences in the structure and composition of the guano deposits and the associated fauna. In both cases guano deposition is predictable; but, whereas in temperate bat caves deposition is seasonal, in tropical bat caves such deposition is continuous.

In the Brazilian karstic areas so far studied, such as the Ribeira valley, continuous guano deposition usually does not occur because roost availability is high and the bats form itinerant colonies, frequently moving between and inside caves. Therefore, guano piles show temporal changes in these caves, passing through the same phases of deposition, and subsequent desiccation after the bat colony leaves the sites. In contrast to temperate caves, succession in Brazilian caves is not necessarily (if at all) seasonal. In fact, the temporal and spatial deposition of guano may be highly unpredictable. Neotropical bats show regular activity throughout the year, but can leave a specific roost at any time, causing the interruption of guano deposition and consequent succession. From whichever cave region, piles of bat guano are "qualitatively and microclimatically non-uniform ... [and] ... look like a mosaic of microhabitats that shelter a mosaic of zoological communities in various evolutional stages" (Decu, 1986).

BAT-GUANO COMMUNITIES: CASE STUDY IN BRAZIL

Here we present two examples of food webs based

Fig. 13.2. Brazilian speleological provinces. Limestone provinces: 1, Ribeira valley; 2, Bambuí; 3, Serra da Ibiapaba; 4, Serra da Bodoquena; 5, Alto Paraguay; 6, Rio Pardo. Sandstone provinces: 7, Furnas; 8, Itararé; 9, Serra Geral; 10, Altamira–Itaituba; 11, Alto Urubu; 12, Espinhaço. From Trajano and Sánchez (1994).

on guano in Brazilian caves and discuss the main taxonomic groups which constitute these communities. In Fig. 13.2, we present a map showing the main Brazilian speleological provinces (including sandstone and limestone provinces). Figures 13.3 and 13.4 show, respectively, the food webs from bat caves of the sandstone speleological province Altamira–Itaituba, Pará State, and from poecilotrophic limestone caves of the Ribeira valley, São Paulo State. In the following we discuss the main taxa and their relationships with the different kinds of bat guano based on both unpublished and published data (e.g., Gnaspini-Netto, 1989a,b; Trajano and Gnaspini-Netto, 1991b; Trajano and Moreira, 1991; Pinto-da-Rocha, 1993; Gnaspini and Trajano, 1994). The taxonomic data are summarized in Table 13.1, which provides a comparative view of the main taxa found in different tropical continents and in different kinds of guano.

As expected, mites (Acari) are by far the most diversified group of guano inhabitants. This is well illustrated by a pile of frugivorous guano in São Paulo State, where eleven species (ten families) were collected in the same vial. Guano-inhabiting mites include guanophages and predators. Owing to the lack of taxonomists covering several Neotropical groups, taxonomic identification remains mostly at the family level, and consequently it is almost impossible to determine the cave relationships of many species. Mites colonize all kinds of guano and belong to several groups such as Acarida, Actinedida, Oribatida and Gamasida.

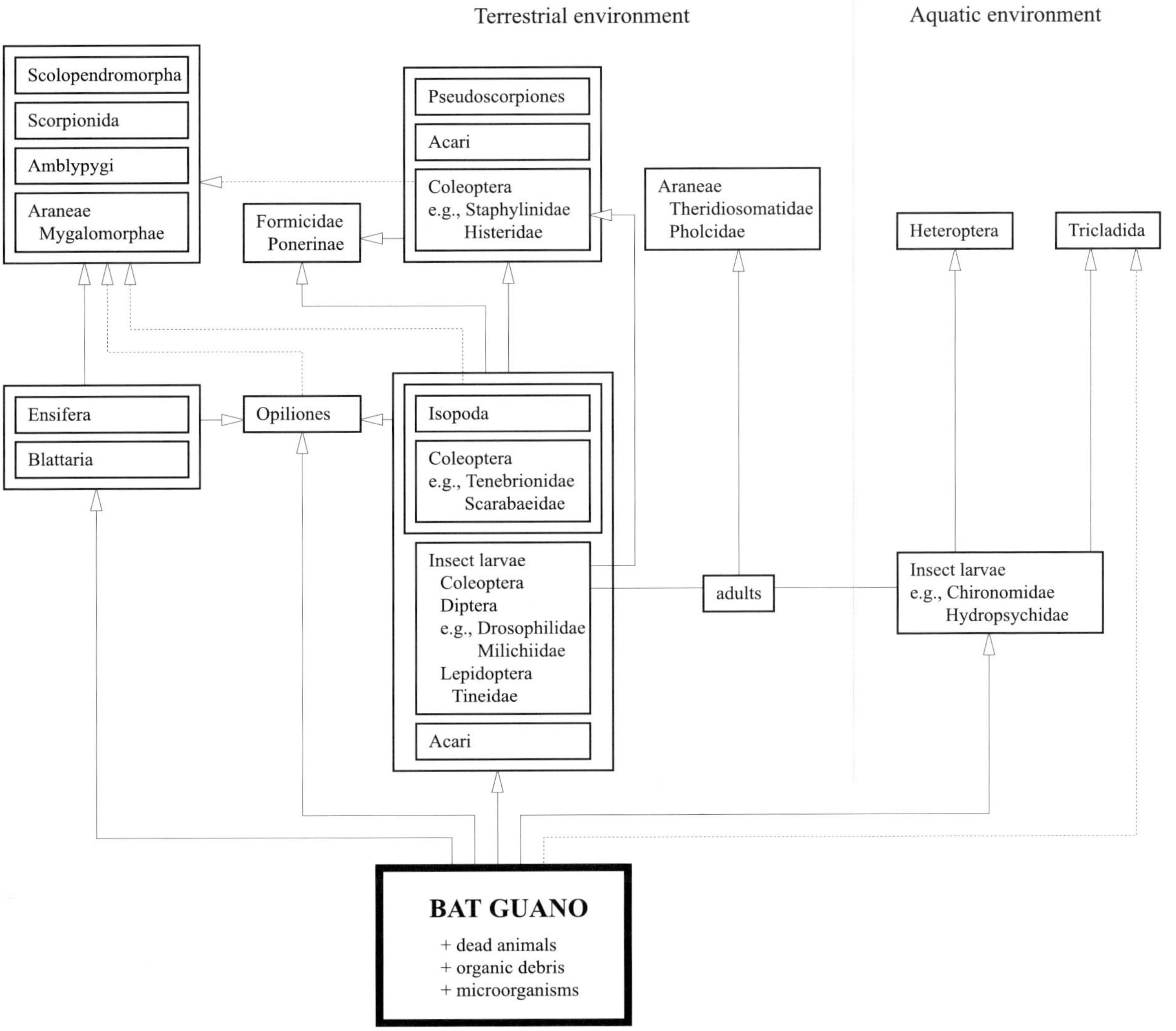

Fig. 13.3. Hypothetical food web for bat-guano communities from Pará bat caves, northern Brazil (modified from Trajano and Moreira, 1991). Arrows indicate direction of energy flow.

The last is the most diversified taxon in guano piles and includes many families, the commonest being the Macrochelidae, Phytoseiidae, Rhodacaridae, and Uropodidae. The Macrochelidae (mainly *Macrocheles* and *Macrolaspis* spp.) are predatory mites that frequent both guano piles and other kinds of organic debris in caves, preying on several mesoinvertebrates. Ticks (mainly *Ornithodoros* spp.) may occasionally be found on guano, but they should not be considered to be true guano-inhabitants because they drop from their hosts (bats) when well fed, and stay there for a while before starting to look for new hosts.

Springtails (Collembola) are, together with mites, the most speciose invertebrates in guano piles, and may form extremely large populations. Among the most common taxa, Cyphoderidae, Entomobryidae and Isotomidae occur in low populations in almost all kinds of guano. Paronellids are common in many types of substrates inside caves, and, when associated with guano piles, they seem to have preference for those of haematophagous species. Hypogastrurids may form large populations associated with bat guano in several countries. The Brazilian species *Acherontides eleonorae* is one of the best examples of Brazilian guanobites and is noteworthy in view of the extremely large populations (hundreds of individuals per cm^2)

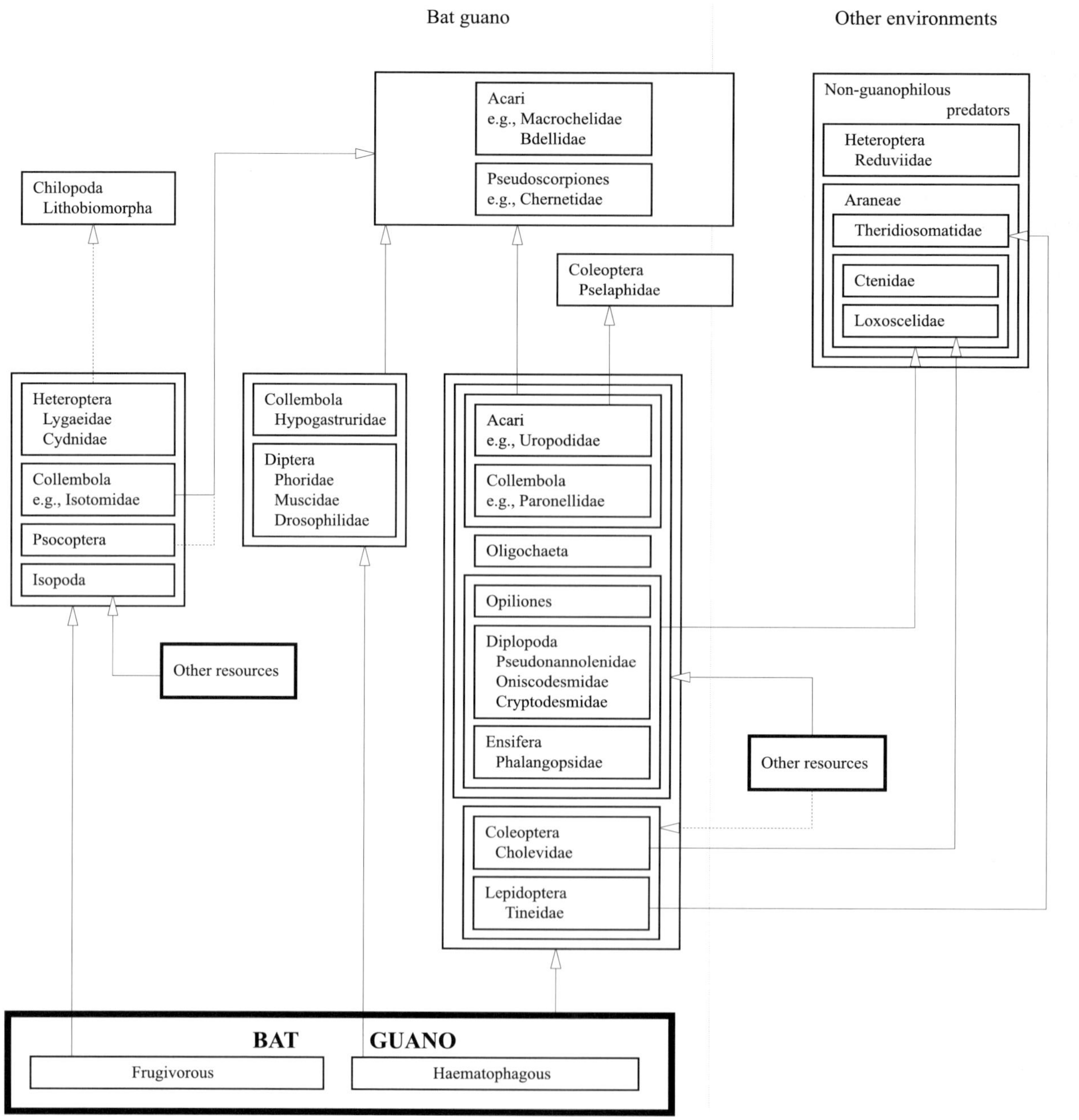

Fig. 13.4. Hypothetical food web for bat-guano communities from São Paulo caves, southeastern Brazil (modified from Trajano and Gnaspini-Netto, 1991a). Arrows indicate direction of energy flow.

found on old, dry piles of haematophagous bat guano from São Paulo caves (Fig. 13.5). Species of this genus occur in guano in other countries, but never in association with guano from haematophagous species, even when it is available (as in Mexican caves). Moreover, they survive and breed when kept in closed vials with this substrate, in contrast to other taxa also found in this type of guano which cannot stand the highly ammonia-rich atmosphere inside closed containers. Arrhopalitids may also establish very large populations in guano of frugivorous species in the São Paulo caves.

Pseudoscorpions may be found in any kind of guano, preying on the abundant springtails and mites. In some

Table 13.1
Distribution of the main taxa reported from tropical bat-guano piles in relation to the main continents and guano types (see text for references)

Taxon	Continent				Guano type[1]				
	New World	Africa	Australia	Asia	carn.	haemat.	insect.	frugi.	bird
PLATHELMINTHES (Dugesiidae)	+								
NEMATODA	+					+	+		
MOLLUSCA Gastropoda (Subulinidae and Physidae)	+								
ANNELIDA Oligochaeta	+					+		+	
ARTHROPODA									
ARACHNIDA									
Scorpiones	+								
Schizomida				+					
Amblypygi	+			+					+
Araneae	+	+		+		+	+	+	+
Ricinulei	+								
Pseudoscorpiones (mainly Chernetidae)	+		+	+		+	+	+	+
Opiliones	+	+				+	+	+	
Acari	+	+	+	+	+	+	+	+	+
–Gamasida (mainly Macrochelidae, Uropodidae, Phytoseiidae, Rhodacaridae)	+		+			+	+	+	
–Actinedida	+		+			+	+	+	
–Oribatida	+	+				+	+	+	
–Acaridida	+		+			+	+	+	
CRUSTACEA **Isopoda**	+	+	+	+	+	+	+	+	+
–Phylosciidae	+							+	
–Scleropactidae	+						+		
–Armadillidae	+		+	+	+		+		
MYRIAPODA									
Chilopoda (Lithobiomorpha and Geophilomorpha)	+			+	+			+	+
Diplopoda	+	+		+		+	+	+	+
Symphyla	+								
INSECTA									
Collembola	+	+	+	+	+	+	+	+	+
–Hypogastruridae	+			+	+	+		+	
–Onychiuridae	+							+	
–Isotomidae	+							+	
–Entomobryidae	+			+		+	+	+	
–Paronellidae	+			+	+	+	+	+	
–Cyphoderidae	+								

continued on next page

Table 13.1, *continued*

Taxon	Continent				Guano type[1]				
	New World	Africa	Australia	Asia	carn.	haemat.	insect.	frugi.	bird
–Oncopoduridae	+								
–Sminthuridae	+								
–Arrhopalitidae	+					+		+	
–Dicyrtomidae	+								
Thysanura	+	+						+	
Ensifera	+	+	+	+		+	+	+	+
Blattaria	+	+	+	+		+	+	+	+
Dermaptera	+			+					+
Psocoptera	+		+	+		+	+	+	
Thysanoptera	+						+		
Heteroptera	+	+		+			+	+	+
–Veliidae	+	+					+	+	
–Reduviidae	+	+		+			+	+	
–Cydnidae and Lygaeidae	+							+	+
Coleoptera	+	+	+	+	+	+	+	+	+
–Carabidae	+		+				+		+
–Hydrophilidae	+	+		+		+	+		
–Histeridae	+		+			+	+		+
–Ptiliidae	+								
–Cholevidae	+		+		+	+	+	+	+
–Scydmaenidae	+								
–Staphylinidae	+			+	+	+	+	+	
–Pselaphidae	+					+		+	+
–Trogidae			+	+			+		+
–Scarabaeidae	+					+	+	+	
–Elateridae	+								+
–Lampyridae	+					+			
–Dermestidae	+			+			+		
–Ptinidae		+	+				+	+	
–Nitidulidae	+								
–Cerylonidae	+								
–Tenebrionidae	+	+	+				+	+	+
–Aderidae				+					+
–Curculionidae	+								+
Lepidoptera	+	+	+	+	+	+	+	+	+
–Tineidae	+		+	+		+	+	+	+
–Lyonetiidae		+					+		
–Pyralidae	+								+

continued on next page

Table 13.1, *continued*

Taxon	Continent				Guano type[1]				
	New World	Africa	Australia	Asia	carn.	haemat.	insect.	frugi.	bird
Diptera	+	+	+	+	+	+	+	+	+
–Ceratopogonidae	+								
–Psychodidae	+	+					+		
–Scatopsidae	+								
–Sciaridae	+					+			
–Stratiomyidae	+			+			+	+	
–Scenopinidae	+						+		
–Empididae	+								
–Dolichopodidae	+								
–Phoridae	+		+	+	+	+	+		
–Sphaeroceridae	+	+		+		+	+		+
–Drosophilidae	+					+	+		
–Milichiidae	+			+		+	+	+	
–Fanniidae	+		+		+	+			
–Muscidae	+				+	+			
Hymenoptera (Formicidae)	+	+			+	+		+	+

[1] Guano types: carn., carnivorous bat guano; haemat., haematophagous bat guano; insect., insectivorous bat guano; frugi., frugivorous bat guano; bird, bird guano.

cases, they may form large populations of some tens of individuals per m^2 (Fig. 13.6). Chernetids (mainly species of *Chelanops* and of an undescribed genus) are the most common pseudoscorpions in Brazilian caves and are found in different kinds of guano piles; they are probably mostly guanophiles.

After Collembola and Acari, larvae of tineid Lepidoptera are among the most ubiquitous guano-inhabitants in Brazil, occurring in any kind of bat guano, whereas the adults have been collected in other substrates or during flight. The larvae make their cases of silk and organic matter from guano and are likely to be guanophages.

Beetles (Coleoptera) of the families Dermestidae, Histeridae, Hydrophilidae, Pselaphidae, Scarabaeidae, Staphylinidae, and Tenebrionidae are occasionally found associated with guano piles, sometimes in large populations, as is the case for histerids in guano of insectivorous species in Pará State and hydrophilids in guano of haematophagous bats in São Paulo State. However, as observed in other countries, cholevids are the commonest guano-inhabiting beetles in Brazil. Representatives of the genera *Adelopsis* and *Dissochaetus* were found in all kinds of guano and other decaying animal matter in caves, although some specific preferences were observed – for instance, *A. leo* was always found associated with guano of haematophagous species, whereas *D. vanini* and *D. villosus* form very large populations, including adults and larvae, in guano of carnivorous bats (although they occur in other kinds of guano) (Fig. 13.7). The case of lampyrid beetles of the genus *Photuris* is also noteworthy because the larvae occasionally form large populations in guano of haematophagous bats in São Paulo caves; although lampyrid larvae are known as predatory, those from guano piles probably ingest this substrate.

The most important Brazilian guano flies (Diptera) belong to the families Drosophilidae, Fanniidae, Milichiidae, Muscidae and Phoridae, many of which are probably guanobites. Phorids (mainly species of *Conicera, Dohrniphora* and *Megaselia*) may inhabit any kind of guano, but show a preference for piles of haematophagous guano. Although some other species may be occasionally found in caves (and in guano), *Drosophila eleonorae* is by far the most frequent

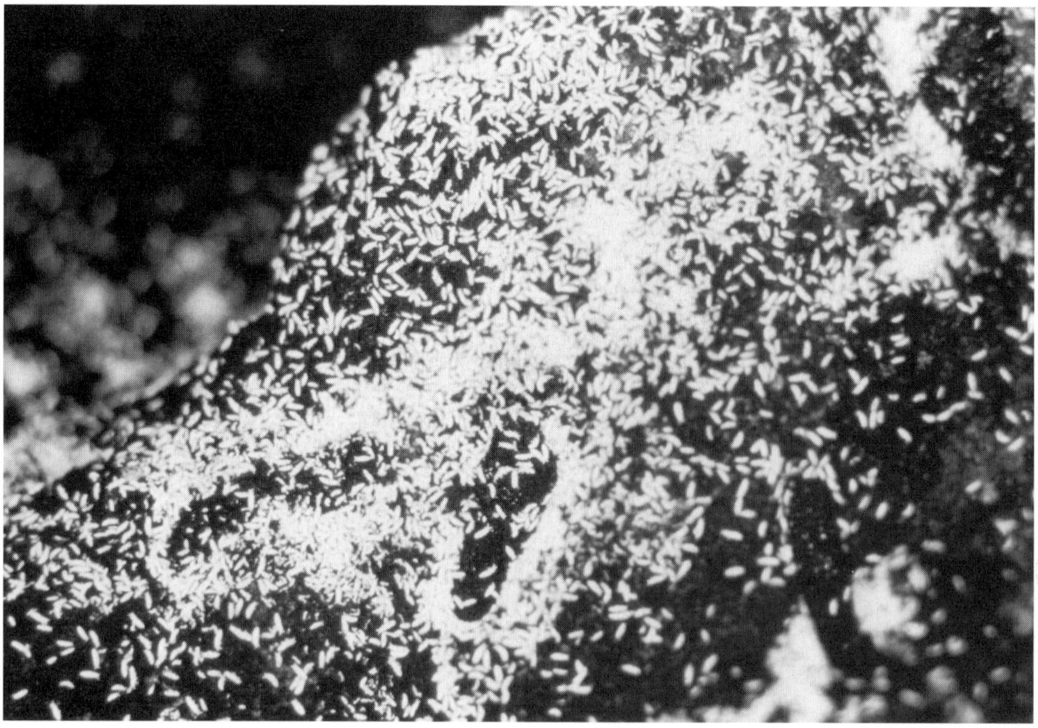

Fig. 13.5. Large population of *Acherontides eleonorae* springtails dwelling in a pile of guano from haematophagous bats. Length of each individual is ~1.5 mm. Photo: J. Allieri.

Fig. 13.6. The pseudoscorpion *Maxchernes iporangae*, which forms large populations of tens of individuals in piles of guano from frugivorous bats. Body length is ~2 mm. Photo: P. Gnaspini.

Fig. 13.7. Larvae of *Dissochaetus* beetles (Cholevidae) dwelling in a pile of guano from carnivorous bats. Length of each individual is ~4 mm. Photo: P. Gnaspini.

guano-inhabiting drosophilid. These flies have been observed in caves of several limestone and sandstone speleological provinces (from Paraná to Pará State) breeding in guano of haematophagous bats, and constitute a typical case of troglophile-guanobites. Milichiids (*Pholeomyia* spp.), faniids (*Fannia* spp.) and muscids (*Psilochaeta* spp.) occur almost exclusively in guano from haematophagous bats in several cave areas (Fig. 13.8). The latter two are occasionally found associated with guano from carnivorous bats in São Paulo and Paraná caves. Psychodids, scenopinids, sciarids, sphaerocerids and stratiomyids are also occasionally observed in guano.

Bugs (Heteroptera) of the families Cydnidae and Lygaeidae are associated exclusively with guano from frugivorous bats, where they may establish very large breeding populations, such as those observed in São Paulo. In caves these insects are restricted to guano, and thus should be considered to be guanobites. Representatives of these families are in general plant-eaters (and some are detritivores), which may explain why the cave species occur only in guano from frugivorous bats.

Pigmented lithobiomorphan chilopods were found occasionally in Brazilian caves. However, in some São Paulo caves with large piles of guano from frugivorous bats, large breeding populations of unpigmented lithobiomorphans are observed. Since they have not been found in other substrates, and in view of their apparent troglomorphisms, these chilopods might be classified as troglobite-guanobites. These animals probably feed on the common bugs of the families Cydnidae and Lygaeidae referred to above.

Cockroaches (Blattaria) are typical tropical cave-dwellers, which are rare in caves to the south of São Paulo State, probably owing to temperature constraints. These highly opportunistic insects are, as expected, found in several substrates, including different types of guano. In Pará bat caves, they form huge populations which, in certain rooms, cover the floor and walls. Crickets (Ensifera) are omnivorous cavernicoles observed throughout the country, and concentrate near and on guano piles. In Pará caves, they may also attain exceptionally high population densities.

Although isopods are not widespread in Brazilian guano piles, they sometimes form large populations in this substrate – for instance, philosciids in guano of frugivorous bats from São Paulo caves, scleropactids in guano of insectivorous bats from one Pará cave, and armadillids in guano of carnivorous bats from Mato

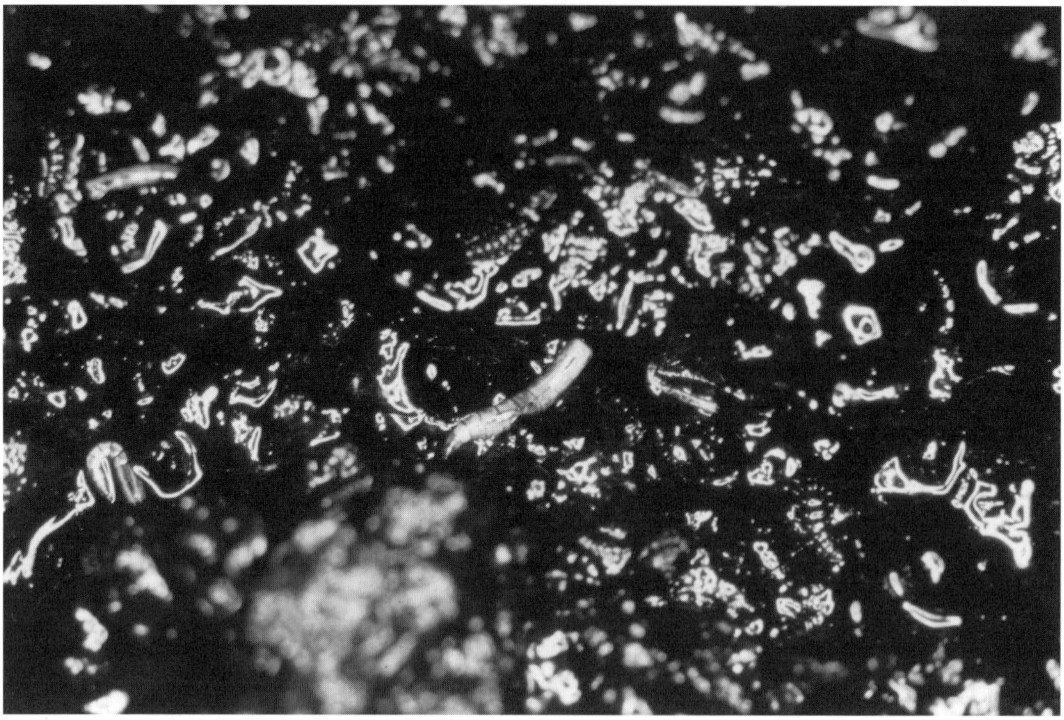

Fig. 13.8. Fanniid and muscid larvae (Diptera) dwelling in a pile of guano from haematophagous bats. Length of muscid larvae (at centre) is ~9 mm. Photo: J. Allieri.

Grosso do Sul. It is not known if the kind of guano with which the isopods are mostly associated is related to the biology of these taxa and/or to biogeographic factors.

Diplopods are found mainly associated with guano of haematophagous bats, the Cryptodesmidae (e.g., *Peridontodesmella alba,*) and Oniscodesmidae (*Crypturodesmus* spp.) being common in São Paulo, Paraná, and Mato Grosso do Sul caves, and Pseudonannolenidae (*Pseudonannolene* spp.) widespread in Brazilian caves as a whole. It is noteworthy that the latter two families include some troglomorphic taxa, which are possibly troglobite-guanobites.

In addition, other higher taxa were occasionally collected in or near guano piles. Earthworms (Oligochaeta) may use guano as an extension of their endogean habitat; they need a moist and nutrient-rich substrate, such as that found in guano from haematophagous and frugivorous species (where large populations of enchytraeid worms were found). Harvestmen (Opiliones) may also be found near the piles and are observed occasionally feeding on guano. Spiders (Araneae) such as *Blechroscelis* spp., *Nesticoides* sp. and *Loxosceles* spp. sometimes make their webs over or near guano piles, opportunistically preying on guanophages (Ferreira and Martins, 1998). Psocopterans (e.g., Psyllipsocidae) feed on fungi and may be found in guano.

Summarizing, some taxa may be found in any kind of guano (e.g., tineid larvae), whereas some are restricted to one kind (e.g., flies in guano from haematophagous bats, and heteropterans and lithobiomorphans in that from frugivorous species).

BIRD-DROPPING COMMUNITIES

Several birds in Brazil, such as doves, owls, parakeets, swiftlets, and vultures, nest near cave entrances, and their droppings may be used as a food supply in the twilight zone. On the other hand, guácharos (oilbirds, *Steatornis caripensis*) in northern South America and swiftlets (*Aerodramus* sp.) in Southeast Asia colonize the dark zones, and their droppings form piles which are an important food supply for cavernicoles, including troglobites.

Guácharos are common in northern South American caves, occurring in northern Bolivia, Colombia, Guyana, Peru, and Venezuela, besides the island of Trinidad. They feed mainly on palm fruits, which they carry inside the caves. Consequently, their guano comprises both rejected seeds (with some pulp adhering) and droppings, and the guano fauna includes animals feeding on the seeds and on the organic matter, and their predators. The taxa recorded in guácharo guano communities were: isopods, pseudoscorpions, spiders, mites, diplopods, springtails, a large number of cockroaches, crickets, cydnid and lygaeid bugs, carabid, curculionid, elaterid, histerid, and pselaphid beetles, a large number of tenebrionid beetles, several fly larvae, and pyralid and tineid moth larvae (Dourojeanni and Tovar, 1974; Decu et al., 1987). Guácharo guano is similar to that of frugivorous bats as regards both physical and nutritional characteristics (both are granulated and consist of seeds and decaying plant matter), and the associated communities (e.g., the presence of cydnid and lygaeid bugs in both kinds of guano and their absence in other kinds of animal droppings).

Swiftlets feed on insects and are common in Southeast Asia, where their nests are exploited for a traditional dish. Their droppings are available for a guano community based mainly on Amblypygi, rhaphidophorid crickets, aderid beetles, several fly larvae and tineid moths (Deharveng and Leclerc, 1989), cockroaches, dermapterans, isopods, geophilomorphans, and scutigerans (E. Trajano, pers. obs.). Brazilian swiftlets were reported to fly in the dark zone of a few caves in Minas Gerais State and are supposed to use echolocation, but the biology of these birds and their guano has not been studied.

Other birds nest and roost at cave entrances, and their droppings are available for the twilight community. Richards (1971) recorded cockroaches, carabid, tenebrionid and trogid beetles, and tineid moths in droppings from falcons, owlet-nightjars (*Aegotheles cristata*: Podargidae), swallows, and *Tyto* owls from southern Australian caves. *Tyto alba* is one of the birds which enter deeper into Brazilian caves, being frequently observed in dark zones, but always where the light from entrances is visible. The pellets regurgitated by these owls are used by some animals such as mites, cholevid beetles (e.g., *Dissochaetus*) and ants (e.g., *Solenopsis*). The frequent and ancient use of Brazilian caves by *T. alba* is attested by the accumulations of bones of their prey (mainly small mammals), thus representing important paleontological sites (Lund, 1950).

It can be seen that bird-guano communities do not differ much from those of bat guano (see also Table 13.1). Whether the observed differences result from biogeographical or taxonomic differences will be discussed below.

GUANO COMMUNITIES IN THE TROPICS

Guano communities have been the subject of ecological studies in several caves around the world, but more commonly in tropical than in temperate caves, contrary to what would be expected from the higher development of temperate speleobiology. Unfortunately, many faunal lists do not discriminate between the inhabitants of guano and those of other substrates. Some, however, do include food-web sketches from which to draw comparative data. In addition, tropical bat-guano communities were used as a base for mathematical simulation by Mazanov and Harris (1971), who studied a population of mites in Australia. We present here a synopsis of tropical guano communities based mainly on faunal lists, which is summarized in Table 13.1.

Guano communities of tropical American caves, from Barbados, Cuba, Jamaica, Panama, and Puerto Rico (Peck, 1971, 1975, 1981a,b; Decu [Decou], 1983), include subulinid gastropods, chernetid pseudoscorpions, several mites, isopods (the platyarthrid *Trichorhina* sp. being the main guanophage), cyphoderid, entomobryid, isotomid, hypogastrurid, onychiurid, paronellid, and sminthurid springtails, earwigs, cockroaches, phalangopsid crickets, cydnid bugs, carabid, cholevid, histerid, hydrophilid, ptiliid, and tenebrionid beetles, drosophilid, milichiid, muscid, phorid, and sphaerocerid flies, pyralid and tineid moths, and ants. The main regional differences are the occurrence of scorpions, thysanurans, and a large number of arachnids of the order Ricinulei in Cuba; henicopid chilopods in Puerto Rico; and flatworms, earthworms and symphylans in Barbados. Guano communities from Argentina (Trajano, 1991) included earthworms, arrhopalitid, entomobryid, and isotomid springtails, cholevid beetles and tineid moths. Those from Ecuador (Peck, 1985) included diplurid spiders, diplopods, oncopodurid and paronellid springtails, cydnid bugs, and carabid and cholevid beetles. Finally, Hill (1969, 1981) compared the fauna of guano from insectivorous and frugivorous bats in Trinidad, recording nematodes, guanophage and predatory mites, cockroaches, ptiliid beetles and tineid moths in guano piles from insectivorous species, and nematodes, guanophage and predatory mites, a chthoniid pseudoscorpion and an entomobryid springtail in those from frugivorous species.

In African caves, communities in guano from insectivorous and frugivorous bats were also surveyed. Braack (1989) recorded thysanurans, a large number of cockroaches, ptinid and tenebrionid beetles, ants, and reduviid bugs in guano from frugivorous bats in South Africa. In guano from insectivorous species in Nigeria, Edington (1984) found springtails, crickets and spiders, but stated that guano was not the main food source for the cave community studied; Hiernaux and Villiers (1955) recorded ctenizid spiders, cockroaches, phalangopsid crickets, hydrophilid beetles, psychodid and sphaerocerid flies, and lyonetiid moths from Guinea; and Irish (1991) found diplopods, cockroaches, crickets and reduviid bugs in the piles, and harvestmen and veliid bugs on floating guano particles from Namibia.

Southern Australian communities in guano from insectivorous bats included armadillid isopods, cheiridiid and chernetid pseudoscorpions, several mites, springtails, psyllipsocid psocopterans, rhaphidophorid crickets, carabid, cholevid, histerid, ptinid, and tenebrionid beetles, phorid flies, and tineid moths (Richards, 1971).

Based on a review of Southeast Asian cave fauna (Deharveng and Leclerc, 1989, and references therein), the taxa associated with guano include mites, spiders, atemnid, cheiridiid, cheliferid, and chernetid pseudoscorpions, schizomids, large populations of armadillid isopods and of cambalopid diplopods, geophilomorphan chilopods, entomobryid, hypogastrurid, and paronellid springtails, psocopterans, aderid, dermestid, hydrophilid, staphylinid, and trogid beetles (among others), and tineid moths, besides reduviid bugs near the piles. Although they mentioned the existence of faunistic differences, those authors did not identify the kind of guano (swiftlet, insectivorous bat or frugivorous bat) with which each taxon was associated.

GENERAL DISCUSSION

Comparison between different kinds of guano communities showed some differences in their composition and structure for caves in Brazil (as shown above), Southeast Asia (Deharveng and Leclerc, 1989), and southern Australia (Richards, 1971) (Table 13.1). Richards compared the fauna associated with bat guano, other mammalian faeces and bird droppings, and observed a much more diversified community in bat guano than in the other two substrates.

It is notable that typical guano-inhabitants may also be found in faeces of other mammals, when available. This is the case of trombidiid mites, and ptinid and tenebrionid beetles [in faeces of cats (*Felis catus*), dingos (*Canis familiaris dingo*), foxes (*Vulpes vulpes*), and rats (*Rattus* sp.)] from southern Australia (Richards, 1971); and Brazilian cavernicoles such as nematodes [in faeces of puma (*Felis concolor*)], glycyphagid, macrochelid, and sejid mites [agouti (*Agouti paca*) and puma], staphylinid [otter (*Lutra longicaudis*)], cholevid and tenebrionid beetles (puma and rat, respectively), and pseudoscorpions (puma).

On the other hand, apart from some differences based on the kind of guano and/or faunal regionalism (e.g., rhaphidophorid crickets in Australian caves, phalangopsid crickets in African and South American caves), the cavernicolous guano communities in the tropics are quite similar, with most guanophages belonging to the same families. The main differences occur among the predators associated with such communities. There are several similarities between tropical and temperate regions, and again the differences are mainly related to biogeographical factors. For instance, guano communities from Romanian caves (where specific studies were carried out – e.g., Negrea and Negrea, 1971; Decu and Tufescu, 1976) included earthworms, macrochelid, oribatid, trombiculid, and uropodid mites (among others), spiders, pseudoscorpions, diplopods, chilopods, isopods, entomobriid, hypogastrurid, isotomid, and onychiurid springtails, psocopterans, pyralid and tineid moths, carabid, histerid and staphylinid beetles, and fanniid, helomyzid, phorid, and sciarid flies – taxa also found in tropical guano piles.

At the family level, some taxa are widespread in piles all over the world, and constitute the basis of most guano communities. These "basic" taxa are macrochelid, oribatid, and uropodid mites, chernetid pseudoscorpions, entomobriid, hypogastrurid, and isotomid springtails, tineid moths, carabid, cholevid, histerid, and staphylinid beetles, and drosophilid, fanniid, milichiid, and phorid flies. Besides them, there are some less common but still frequent groups, which could be called "accessory" taxa (*sensu,* for instance, Negrea and Negrea, 1971), such as earthworms, harvestmen, spiders, isopods, and psocopterans.

It is noteworthy that the same structure and basic taxa occur in epigean guano communities (e.g., Bernath and Kunz, 1981). Because subterranean communities are in general simpler than those above ground, one could expect simpler hypogean guano communities resulting from a kind of "subterranean-related filter effect". Such simplicity observed both in epigean and cave guano communities reveals a dependence on the same structural, microclimatical and nutritional constraints.

In summary, the composition of tropical guano communities show some differences related to the kind of guano (e.g., bird vs. bat guano, guano of frugivorous vs. insectivorous vs. haematophagous bats, etc.). Even some of the "basic" taxa may be missing from some types of guano, whereas most of these "basic" taxa are present in guano communities of northern temperate caves, where only insectivorous bat guano is available. In contrast, in tropical caves, where this ecological constraint is relaxed on account of the widespread availability of other kinds of guano, the taxa have the opportunity to select a particular kind of guano.

REFERENCES

Bernath, R.F. and Kunz, T.H., 1981. Structure and dynamics of arthropod communities in bat guano deposits in buildings. *Can. J. Zool.,* 59: 260–270.

Braack, L.E.O., 1989. Arthropod inhabitants of a tropical cave "island" environment provisioned by bats. *Biol. Conserv.,* 48(2): 77–84.

Bredt, A., Uieda, W. and Magalhães, E.D., 1999. Morcegos cavernícolas da refião do Distrito Federal, centro-oeste do Brasil (Mammalia, Chiroptera). *Rev. Bras. Zool.,* 16(3): 731–770.

Campanhã, R.A.C. and Fowler, H.G., 1993. Roosting assemblages of bats in arenitic caves in remnant fragments of Atlantic forest in Southeastern Brazil. *Biotropica,* 25(3): 362–365.

Decu, V., 1981. Quelques aspects de la biospéologie tropicale résultant des expéditions biospéologiques cubano-roumaines à Cuba. *Résultats des expéditions biospéologiques cubano-roumaines à Cuba,* 3: 9–15.

Decu, V., 1983. Sur la bionomie de certaines espèces d'animaux terrestres qui peuplent les grottes de Cuba. *Résultats des expéditions biospéologiques cubano-roumaines à Cuba,* 4: 9–17.

Decu, V., 1986. Some considerations on the bat guano synusia. *Trav. Inst. Spéologie Emile Racovitza,* 25: 41–51.

Decu, V. and Tufescu, M.V., 1976. Sur l'organisation d'une biocénose extrême: la biocénose du guano de la grotte "Pestera lui Adam" de Baile Herculane (Carpates Méridionales, Roumanie). *Trav. Inst. Speologie Emile Racovitza,* 15: 113–132.

Decu, V., Bordon, C. and Linares, O., 1987. Sinopsis de los invertebrados citados de las cuevas de Venezuela. In: V. Decu, T. Orghidan, D. Dancau, C. Bordon, O. Linares, F. Urbani, J. Tronchoni and C. Bosque (Editors), *Fauna hipogea y hemiedáfica de Venezuela y de otros paises de América del Sur. v. 1.* Editura Academiei Republicii Socialiste România, Bucaresti, pp. 47–60.

Deharveng, L. and Leclerc, P., 1989. Recherches sur les faunes cavernicoles d'Asie du Sud-Est. *Mem. Biospeol.,* 16: 91–110.

Dourojeanni, M.J. and Tovar, A., 1974. Notas sobre el ecosistema y la conservación de la Cueva de las Lechuzas (Parque Nacional de Tingo Maria, Perú). *Rev. For. Perú,* 5(1–2): 28–45.

Edington, M.A., 1984. Biological observations on the Ogbunike cave system, Anambra State, Nigeria. *Stud. Speleol.,* 5: 31–38.

Ferreira, R.L. and Martins, R.P., 1998. Diversity and distribution of spiders associated with bat guano piles in Morrinho cave (Bahia State, Brazil). *Diversity and Distributions,* 4: 234–241.

Forestiero, S. and Sbordoni, V., 1978. Biospeleologia. In: Società Speleologica Italiana (Editor), *Manuale di Speleologia.* Longanesi & C., Milano, pp. 219–273.

Gnaspini, P., 1992. Bat guano ecosystems. A new classification and some considerations, with special references to Neotropical data. *Mem. Biospeol.,* 19: 135–138.

Gnaspini, P., 1996. Population ecology of *Goniosoma spelaeum,* a cavernicolous harvestman from south-eastern Brazil (Arachnida: Opiliones: Gonyleptidae). *J. Zool.,* 239(3): 417–435.

Gnaspini, P. and Trajano, E., 1994. Brazilian cave invertebrates, with a checklist of troglomorphic taxa. *Rev. Bras. Entomol.,* 38(4): 549–584.

Gnaspini-Netto, P., 1989a. Análise comparativa da fauna associada a depósitos de guano de morcegos cavernícolas no Brasil. Primeira aproximação. *Rev. Bras. Entomol.,* 33(2): 183–192.

Gnaspini-Netto, P., 1989b. Fauna associated with bat guano deposits from Brazilian caves (a comparison). In: *Proceedings 10th International Congress of Speleology, Budapest,* pp. 52–54.

Hiernaux, C.R. and Villiers, A., 1955. Speologica africana. Étude préliminaire de six cavernes de Guinée. *Bull. Inst. Fr. Afrique Noire,* 17(3): 926–946.

Hill, S.B., 1969. *The Ecology of Bat Guano in Tamana Cave, Trinidad, W.I., Ph.D. Thesis.* University of West Indies, Trinidad, 311 p.

Hill, S.B., 1981. Ecology of bat guano in Tamana Cave, Trinidad, W.I. In: *Proceedings 8th International Speleological Congress, Bowling Green,* pp. 243–246.

Hutchinson, G.E., 1950. Survey of contemporary knowledge of biogeochemistry. 3. The biogeochemistry of vertebrate excretion. *Bull. Am. Mus. Nat. Hist.,* 96: 1–554.

Irish, J., 1991. Cave investigations in Namibia. I. Biospeleology, ecology and conservation of Dragon's Breath Cave. *Cimbebasia,* 13: 59–67.

Koopman, K.F., 1993. Order Chiroptera. In: D.E. Wilson and D.M. Reeder (Editors), *Mammal Species of the World.* Smithsonian Institution Press, Washington, pp. 137–241.

Lund, P.W., 1950. *Memórias Sobre a Paleontologia Brasileira.* Instituto Nacional do Livro, Rio de Janeiro, 589 pp.

Martin, B.J., 1980. *The Community Structure of Arthropods on Bat Guano and Bat Carcasses in Tumbling Creek Cave. M.Sc. Thesis.* University of Illinois, Chicago.

Mazanov, A. and Harris, J.A., 1971. Simulation of a cave microcosm: A trip in eco-mathematical reality. In: H.A. Nix (Editor), *Proc. Ecol. Soc. Aust.,* 6: 116–134.

Negrea, A. and Negrea, S., 1971. Sur la synusie du guano des grottes du Banat (Roumanie). *Trav. Inst. Speologie Emile Racovitza,* 10: 81–122.

Peck, S.B., 1971. The invertebrate fauna of tropical American caves, part I: Chilibrillo Cave, Panama. *Ann. Speleol.,* 26(2): 423–437.

Peck, S.B., 1975. The invertebrate fauna of tropical American caves, part III: Jamaica, an introduction. *Int. J. Speleol.,* 7(4): 303–326.

Peck, S.B., 1981a. Zoogeography of invertebrate cave faunas in southwestern Puerto Rico. *Bull. Natl. Speleol. Soc.*, 43: 70–79.

Peck, S.B., 1981b. Community composition and zoogeography of the invertebrate cave fauna of Barbados. *Fla. Entomol.*, 64(4): 519–527.

Peck, S.B., 1985. The invertebrate faunas of tropical American caves, part 6: Jumandi Cave, Ecuador. *Int. J. Speleol.*, 14(1–4): 1–8.

Pinto-da-Rocha, R., 1993. Invertebrados cavernícolas da porção meridional da Província Espeleológica do Vale do Ribeira, Sul do Brasil. *Rev. Bras. Zool.*, 10(2): 229–255.

Poulson, T.L., 1972. Bat guano ecosystems. *Bull. Natl. Speleol. Soc.*, 34(2): 55–59.

Richards, A., 1971. An ecological study of the cavernicolous fauna of the Nullarbor Plain, southern Australia. *J. Zool.*, 164: 1–60.

Sbordoni, V., Vomero, V., Argano, R. and Zullini, A., 1980. The cave fauna of Chiapas (Mexico): Tropical and temperate cave communities and their evolution. In: J.I. Furtado (Editor), *Proceedings 5th International Symposium of Tropical Ecology, Malaysia.* The International Society of Tropical Ecology, pp. 687–692.

Trajano, E., 1985. Ecologia de populações de morcegos cavernícolas em uma região cárstica do sudeste do Brasil. *Rev. Bras. Zool.*, 2(5): 255–320.

Trajano, E., 1991. Notas biológicas sobre cavernas argentinas (Resultados de la Primera Expedición Espeleológica Argentino–Brasileña, Neuquén - Mendoza). *Spelaion,* 2(2): 3–8.

Trajano, E., 1996. Movements of cave bats in southeastern Brazil, with emphasis on the population ecology of the common vampire bat, *Desmodus rotundus* (Chiroptera). *Biotropica,* 28(1): 121–129.

Trajano, E. and Gimenez, E.A., 1998. Bat community in a cave from eastern Brazil, including a new record of *Lionycteris* (Phyllostomdae, Glossophaginae). *Stud. Neotrop. Fauna Environ.*, 33(2–3): 69–75.

Trajano, E. and Gnaspini-Netto, P., 1991a. Notes on the food webs in caves from southeastern Brazil. *Mem. Biospeol.,* 18: 75–79.

Trajano, E. and Gnaspini-Netto, P., 1991b. Fauna cavernícola brasileira, com uma análise preliminar da distribuição dos táxons. *Rev. Bras. Zool.*, 7(3): 383–407.

Trajano, E. and Moreira, J.R.A., 1991. Estudo da fauna de cavernas da Província Espeleológica Arenítica Altamira-Itaituba, Pará. *Rev. Bras. Biol.*, 51(1): 13–29.

Trajano, E. and Sánchez, L.E., 1994. Brésil. In: C. Juberthie and V. Decu (Editors), *Encyclopaedia Biospeologica. Tome I.* Société de Biospéologie, Moulis, pp. 527–540.

Uieda, W., Sazima, I. and Storti-Filho, A., 1980. Aspectos da biologia do morcego *Furipterus horrens* (Mammalia, Chiroptera, Furipteridae). *Rev. Bras. Biol.*, 40(1): 59–66.

Chapter 14

FALLOUT OF DISPERSING ARTHROPODS SUPPORTING INVERTEBRATE COMMUNITIES IN BARREN VOLCANIC HABITATS

N. Philip ASHMOLE and Myrtle J. ASHMOLE

INTRODUCTION

A few years before we went to work on the Canary Islands in 1984, Howarth (1979) had described from Hawaii a population of endemic crickets (*Caconemobius fori*) supported by windblown debris on recently formed and totally barren lava. On Tenerife in the Canary Islands the extensive areas of lava and tephra[1] resulting from eruptions within historical times appeared equally lacking in primary production. Previous experience on Scottish mountain snowfields (Ashmole et al., 1983; Ashmole, 1983) had impressed us with the importance of fallout of dispersing arthropods as a resource for scavenger communities in barren places. Finding that arthropod fallout also occurred high up on Tenerife, and, remembering that nature abhors an ecological vacuum, we predicted the presence of an arthropod community living in the barren volcanic terrain and utilizing the fallout.

We therefore developed a research program to investigate the arthropod fallout on mountain snowfields and barren lava flows on Tenerife (Ashmole and Ashmole, 1988), and also to study populations of invertebrates resident on some of the most recent lava flows and the cinder cones (Ashmole and Ashmole, 1987). Further samples were obtained from recent lava flows on Lanzarote (Ashmole and Ashmole, 1987; Ashmole et al., 1990), and a parallel study was carried out on El Hierro by Martín et al. (1987). Our later work in the Canaries was done cooperatively with Dr Pedro Oromi and his group of cave biologists at the Universidad de La Laguna (Ashmole et al., 1992), and two joint expeditions were mounted to investigate comparable habitats in the Azores (Menier and Constantin, 1989; Oromi et al., 1990; Eason and Ashmole, 1992; Ashmole et al., 1996). More recent work of ours on Ascension Island (Ashmole and Ashmole, 1997) has shown that even this young and extremely isolated oceanic island has endemic arthropods adapted to life in barren volcanic terrain, including some troglomorphic species living underground in tephra as well as in caves.

NEOGEOAEOLIAN HABITATS

Animals in subterranean habitats have two distinct potential sources of energy: either they depend on autotrophic microorganisms that use chemical energy in the process of carbon fixation (Sarbu et al., 1996); or they depend on the input of plant and animal material from the surface, and thus ultimately on solar energy. In most terrestrial volcanic areas the latter source seems to be of prime importance, although recent data from the crater of Vulcano in the Isole Lipari off Sicily suggest that a sulphur-based trophic system may operate there (Ashmole and Ashmole, unpublished observations).

In vegetated volcanic terrain, resources for subterranean animal communities are derived mainly from local primary production. In barren areas, with no vascular plants and negligible primary production, the energetic and nutrient base for the resident heterotrophs – including those living underground – is provided by organic material arriving from elsewhere. Terrestrial ecosystems in which this incoming material is transported by the wind were termed "aeolian" by Swan (1963, 1968), and have been discussed for instance by Ashmole et al. (1983), Howarth (1987a),

[1] The term tephra refers to all particles or fragments ejected into the air from a volcano, irrespective of size, shape, or composition.

Edwards (1987), Thornton et al. (1988) and New and Thornton (1988).

The present discussion relates mainly to the windborne material, termed "manna" or "biological fallout" by Ashmole and Ashmole (1987), which derives mainly from terrestrial primary production elsewhere. It should be noted, however, that resident arthropods in barren coastal volcanic habitats may receive more resources from the sea than from the air. As indicated below (p. 277), coastal lava on Lanzarote has higher invertebrate densities than areas further inland which have more lichens and, thus, presumably higher primary production. The area has prevalent onshore winds, so that input of arthropod fallout is likely to be small. Resources from the marine environment, however, may reach terrestrial communities by several routes, as discussed by Ashmole and Ashmole (1987), Ashmole et al. (1990) and Martín (1991), and in different contexts by Heatwole (1971), Lindroth et al. (1973) and Heatwole et al. (1981). Possibilities include input of marine bacteria and other organic material from the sea-salt aerosol (Wilson, 1959; Swan, 1963; Blanchard, 1983); input of marine biomass by nesting or roosting seabirds; input from carrion, drifted marine algae and flotsam; and transfer into terrestrial habitats of organic material derived from primary production in the intertidal zone. We have little idea of the relative importance of these sources, but their combined ecological effect is clearly substantial. Polis and Hurd (1995) have elegantly shown that, on small arid islands in the Gulf of California, transfer of marine resources to terrestrial food webs results in exceptionally high densities of predators of arthropods (spiders, scorpions and lizards).

Windborne biological fallout in barren volcanic areas has diverse origins. It may include pollen, spores, seeds, plant fragments, nutrient particles derived from the sea or more productive land, and a wide variety of aerially dispersing arthropods. The proportions of the different components doubtless vary enormously, both in time and space. Plant fragments are likely to be dominant in aeolian habitats close to vegetated areas, and marine input in areas near the coast, while flying or ballooning arthropods are relatively important in more remote places, and especially on islands (Farrow, 1984; Thornton et al., 1988; Edwards and Sugg, 1993). Edwards (1986) and Edwards et al. (1986) presented data indicating that, at their study site on Mount St. Helens (Washington State) in 1984, arthropods comprised almost exactly 50% and organic material of plant origin the other 50% of total dry weight biomass in the fallout. Spores and pollen were presumably not collected by their traps.

Although windborne fallout occurs on all parts of the earth's surface, its ecological significance is greatest in areas where it forms a high proportion of the total biomass available to consumers. Aeolian ecosystems occur mainly at very high latitudes or altitudes, and in extreme deserts (Swan, 1968). Since none of these are normally rich in subterranean cavities they are not considered further here. Young volcanic substrates of lava or tephra, however, also form aeolian ecosystems, for which Howarth (1979) coined the adjective "neogeoaeolian". In these, arthropod fallout not only provides resources for consumers, but also constitutes a significant source of nutrients for pioneer microorganisms and plants (Edwards and Sugg, 1993).

Raw volcanic substrates contain a labyrinth of interconnected cracks and spaces of varied sizes. The "microcavernous" habitats, with spaces less than 1 mm in width, are utilized by soil organisms. Cavernicoles occur mainly in the "mesocavernous" habitats, with spaces ranging from 1 mm to 200 mm in width, although larger "macrocavernous" passages (caves) are also used (Howarth, 1983). All over the world, these labyrinths prove to contain distinctive faunas of arthropod scavengers and predators (Howarth, 1979; Martín et al., 1987; Ashmole and Ashmole, 1987, 1997; New and Thornton, 1988; Thornton et al., 1988; Peck, 1990; Ashmole et al., 1990, 1992). The data of Thornton et al. (1988), Edwards and Sugg (1993) and our own observations suggest that aerially dispersing arthropods provide a crucial resource for these resident animals.

Volcanic aeolian ecosystems tend to be of short duration unless they are in polar, high-altitude or desert regions. Elsewhere, weathering and soil formation on the surface is followed by vegetational succession and increase in primary production (Whittaker et al., 1989; del Moral and Wood, 1993), with accompanying faunal succession (Ashmole et al., 1992). At the same time the subterranean cavities typically either collapse or become blocked by a variety of geological and biological processes (Howarth, 1980; Ashmole et al., 1992).

While they are barren and full of cavities, however, volcanic substrates tend to blur the distinction between epigean (surface) and hypogean (subterranean) habitats. Some of the resident animals live entirely on the surface, though often seeking shade or hiding under superficial rocks. Others spend the day in cool and relatively moist subterranean spaces and emerge to forage on the surface at night. Active diurnal predators such as lycosid spiders catch many vagrant arthropods while they are still alive, whereas nocturnally foraging lavicoles are likely to be feeding mainly on moribund or dead animals. Together, these scavengers doubtless intercept a high proportion of the resources derived from fallout. They, thus, ensure that any troglobites living permanently at lower levels have few resources and only sparse populations.

Nocturnally foraging scavengers of volcanic terrain are analogues of the well-known "habitual trogloxenes" (Chapman, 1993) associated with limestone cave systems. Trogloxenes often live in substantial caves and emerge in large numbers at night to forage in the area around the caves; they are thus conspicuous and well-documented components of the subterranean fauna. The scavengers on recent lava and tephra, however, spend the day dispersed in cracks and crevices and probably move relatively short distances during their nightly search for fallout deposited on the surface. Since the substrate is typically rough and pitted, most of the vagrant insects landing on it quickly become lodged in crevices, where they are found and scavenged by the resident fauna. On such terrain – often jagged and dangerous to walk on at night – the scavenging process is much less obvious to human observers than it is on summer snowfields. The scavengers are thus rarely encountered, and since they also lack obvious specializations for subterranean life they have – until recently – attracted little attention from either entomologists or cave biologists. The discovery of populations of mildly troglomorphic scavengers and predators living just below the surface in neogeoaeolian tephra habitats on Ascension is discussed later in this chapter.

AERIAL DISPERSAL AND FALLOUT BY ARTHROPODS

In recent decades much attention has been paid to aerial dispersal (or "migration") of arthropods. Many studies have focused on economically significant migrant pests, but others have analysed the adaptive nature of dispersal of arthropods from their natal habitat patch (Drake and Gatehouse, 1995), or considered the zoogeographic implications of dispersal (Peck, 1994a; Ashmole and Ashmole, 1997). The review by Southwood (1962) led to general agreement that long-distance one-way migration is primarily an adaptation to environmental instability, favoured by selection when there is a high risk that the natal habitat will become unsuitable before the individual is ready to reproduce. Habitats are often ephemeral because of seasonal or successional changes in the environment, but migration may also be stimulated by population changes leading to local crowding and intensified intraspecific competition. Although aerial dispersal is presumably adaptive in those species that show it, the associated risks for individuals must always be high. One hazard is that of predation by aerial-feeding birds and bats. Another risk is the element of indeterminacy in trajectory (Edwards, 1986) and consequent likelihood of landing in a hostile environment. The scale of this problem varies greatly. For instance, species living on islands or in patchy habitats run a risk of landing in the surrounding sea or in an inappropriate habitat, and host-specific herbivores may land in a suitable area but on the wrong kind of plant. In extreme cases, the former may lead to selection against aerial dispersal (Gressitt, 1964; Ashmole and Ashmole, 1988), while the latter may lead to evolution of appropriate trial-and-error behaviour (Kennedy, 1975).

Most dispersing insects generally drift downwind, since they are outside the "boundary layer" of relatively still air close to the ground where they would retain control of their flight direction. Even in relatively strong-flying insects such as locusts it has been shown that, although the individual insects in part of a swarm may be uniformly orientated at an angle to the wind, the overall displacement of the swarm is directly downwind (Rainey, 1978). Because migration is generally downwind, the abundance of arthropod fallout on a given barren area depends on the distance from a vegetated source area in the upwind direction.

Fallout also varies with the number of individuals in the source population that are at a migratory stage in the life cycle (Johnson, 1969). On a day-to-day basis, weather factors have major effects on the density of arthropods in the air (review by Pedgley, 1982). For insects, temperature has a threshold above which flight is not inhibited (Taylor, 1963). At high and moderate latitudes the threshold for many insects is often reached only in summer, so that fallout is highly seasonal in its occurrence. There is also a strong diurnal pattern, with the numbers of insects in the air generally being low at night (Johnson, 1969).

Even small insects will normally remain aloft only while they continue flying. For instance, the bean aphid *Aphis fabae* sinks at a rate of about 80 cm sec^{-1} if it stops beating its wings (Kennedy, 1975); further data are provided by Pedgley (1982). Strong convection, however, doubtless sometimes keeps passive insects in the air, and many spiders routinely "balloon" suspended from a thread of silk.

Arthropod fallout occurs when aerially dispersing animals make a landing. They may choose a landing place, as in the case of dragonflies or butterflies, or they may simply stop flying so that they fall to the ground. The journey may also end by accident, as when insects are caught in down draughts or in eddies to leeward of a ridge (Lewis and Dibley, 1970; Pedgley, 1982; Antor, 1994), or when a weather change leads to a loss of lift, for instance for ballooning spiders. Often, landing is followed quickly by another flight. Some substrates, however, act as traps, water and snow being the most obvious examples. Landing on water often leads to death, though some insects can take off again, and others can survive for some time on the surface (Peck, 1994b).

Landing on snow risks cooling of the body below the threshold temperature for flight, and there will also usually be a lack of convective lift immediately above a snow surface. Although some insects manage to take off again from snow (Antor, 1994), accumulations of dead and moribund arthropods are a common sight on summer snowfields on mountains (Antor, 1994, 1995) and in the arctic (Edwards, 1972; Edwards and Banko, 1976). In and around such snowfields, fallout provides the energetic and nutrient base for communities of resident scavengers and predators (Ashmole, 1983; Edwards, 1987; Antor, 1995).

Measurement of the amount of arthropod fallout is always difficult, but some information about the input of wind-borne arthropods has been provided by studies on summer snowfields and mountain tops (Ashmole et al., 1983; Edwards, 1986; Antor, 1994). A temptation to assume that arthropod fallout derived from distant sources is negligible in quantity, should be countered by analyses such as those by Johnson (1969) and Pedgley (1982) on long-distance migration of pest insects under particular weather systems. That even remote islands may receive significant quantities of fallout is shown by the data of Farrow (1984). He analysed insect catches made with a light trap on 351 successive nights on Willis Island in the Coral Sea, which is a minimum of 450 km from continental source areas and measures only about 400 m 150 m. Farrow showed that a minimum of several hundred migrant insects belonging to nearly one hundred species reached the island during the year. Similarly, in one of the earliest and most striking demonstrations of fallout following wind drift, Elton (1925) found large numbers of a forest aphid *Dilachnus piceae* and smaller numbers of a syrphid *Syrphus ribesii* high on the ice cap in Nordaustlandet in the Svalbard group; these animals were deduced to have come from spruce forests in the Kol'sky Poluostrov (Kola Peninsula), about 1300 km away across the Barents Sea.

More relevant to subterranean communities in volcanic terrain are the data collected in the blast zone of Mount St. Helens in the years after the 1980 eruption, and on Anak Krakatau. The research on Mount St. Helens involved use of a special fallout trap mimicking the desert pavement in which it was sunk, which would in principle allow arriving arthropods to leave again from the surface (Edwards, 1986; Edwards et al., 1986; Edwards and Sugg, 1993). The trap consisted of an array of close-packed golf balls in a square frame backed by nylon mesh. During the summers of 1983 and 1984 the investigators found a net dry weight input of arthropods amounting to ~6–10 mg m^{-2} day^{-1} on the surface at a representative site in the blast zone of the volcano. This input is of the order of 1 kg ha^{-1} during the season, providing a major resource for resident scavengers and predators (Edwards, 1987; Antor, 1995).

In the work on Anak Krakatau, Thornton et al. (1988) used water traps on bare ash-covered lava near the windward coast, in an attempt to monitor fallout of animals on to the island. Excluding probable members of the ground fauna, they caught dispersing arthropods at a rate of about 20 individuals m^{-2} day^{-1}. It was not known to what extent the traps were attractive rather than passive receivers of fallout, but the results clearly indicated a massive input of biomass to the resident scavengers, including the lavicolous cricket *Speonemobius* sp. (New and Thornton, 1988).

Composition of arthropod fallout naturally varies according to the distance and character of the source area. In the blast zone of Mount St. Helens the fallout collected by Edwards et al. (1986) included at least 76 families of insects from 13 orders, and 9 families of spiders. The authors commented that their catches were dominated by nematoceran Diptera originating from fresh water, and the presence of water bodies nearby doubtless had a major effect on the composition of the fallout. The arthropods caught in traps on Anak Krakatau by Thornton et al. (1988) were also taxonomically varied. The most numerous groups, after exclusion of probable members of the resident community, were Diptera (especially Chloropidae), Coleoptera (especially Curculionidae), Araneae and Aphididae. In a parallel study, New and Thornton (1988) examined the gut contents of 80 lavicolous crickets, *Speonemobius* sp. Only four contained plant material, while 62 contained arthropod material. Much of the latter was unrecognizable, but the most abundant identifiable arthropods in the diet were Coleoptera.

ARTHROPOD DISPERSAL ON TENERIFE

Introduction

Our original plan of research on Tenerife was based on an evolutionary hypothesis and concerned the incidence of aerial dispersal in endemic and other resident arthropods presumably adapted to island life. We have presented the data elsewhere (Ashmole and Ashmole, 1988), and a summary is provided here only of data relevant to the role of arthropod fallout in supporting resident lava communities.

Tenerife, the largest (>2000 km^2) and highest of the Canary Islands, lies in the subtropical Atlantic Ocean at 28°N and 16°W; it is nearly 300 km from the coast of Africa. Like the other western Canary Islands, it is oceanic in origin and composed entirely of volcanic rocks. The island is dominated by the central stratovolcano of Pico de Teide (3718 m), which rises from the caldera of Las Cañadas at ~2000 m. The volcano is still active, and there have been a series of eruptions within historic times, the last in 1909. On the island of La Palma, where work was also carried out, eruptions occurred as recently as 1949 and 1971.

On Tenerife the northeast trade winds predominate for much of the year, and the high relief of the island often results in build-up of cloud on the northern slopes below a temperature inversion at less than 1800 m. Under these conditions few dispersing insects are likely to reach Las Cañadas or El Teide from the north. During relatively calm conditions, however, there is less cloud and there are opportunities for transport of insects from the lowlands by diurnal upslope (anabatic) winds.

Fallout on snowfields on El Teide

The mountain El Teide has a substantial aeolian zone where vegetation is almost entirely lacking. Snow lies on the peak (Pico de Teide) for a few weeks in most winters. During these periods arthropods are conspicuous on the snow surface, but seem to arrive mainly on days when air temperatures rise well above freezing (Ashmole and Ashmole, 1988).

In 1984, substantial snowfalls during January provided an opportunity to make several collections of fallout between mid-February and early March; late snowfalls on 10–11 and especially 19–20 March allowed sampling during the latter part of that month, and snow patches remained in a few places until the beginning of May. Collections were made at an altitude of about 3550 m near the peak of Teide, and at 2500 m on Montaña Blanca, an area on the eastern slopes of El Teide where snow tends to persist. Though the latter site was not strictly aeolian, vegetation was very sparse and largely snow-covered, so that input of local insects engaged in trivial flights was undoubtedly low.

A general feature of arthropod fallout on snow is that it is spatially (as well as temporally) extremely variable: groups of particular kinds of animals tend to be concentrated in certain parts of the snowfields. The best way of assessing the general composition of the fallout – and thus of the dispersing arthropod fauna – seems to be to obtain the largest samples possible and to collect over a considerable area. Accordingly, the main collecting method used was a 30-minute "walking transect" in which the observer walked slowly across the snow in a series of approximately straight lines, turning whenever the edge of a patch of snow was approached, and collecting (with an aspirator) all live arthropods that could be reached without taking more than one pace to the side. This technique ensures that some specimens are obtained even when fallout is at very low density; it is less satisfactory when fallout is abundant, since the number of animals collected becomes asymptotic with increasing density as a result of the limitation imposed by "handling time".

Estimates of the absolute density of fallout were obtained by means of series of 20 randomly placed quadrat samples of area 0.5 m^2. Details of the methods are given in Ashmole and Ashmole (1988). Quadrat sampling was used on three days near Pico del Teide between 3550 and 3600 m, and on five occasions on Montaña Blanca at 2500 m. The amount of fallout was extremely variable. On 13 March none was seen at either level, and on 22 March only scattered insects were seen (none in the quadrats). In contrast, fallout at Montaña Blanca on 24 February amounted to almost 18 insects m^{-2}, which is among the highest values recorded anywhere (Ashmole et al., 1983). Although a few of the insects may have survived from previous days, the great majority evidently represented that day's fallout. On the one day (30 March) when quadrat samples were taken at the same time near Pico del Teide and on Montaña Blanca, the density was much greater at the lower site.

Because quadrats of 0.5 m^2 give many zero values unless fallout is very heavy, transect data are more useful for assessing the role of the weather. The sparse data (in Ashmole and Ashmole, 1988) suggested the existence of a threshold air temperature somewhere between −1°C and 6°C, above which substantial fallout occurs. Fallout is evidently also influenced by wind direction and by incident radiation, which when strong may enable some insects alighting on the snow to take off again rather than being trapped at the snow surface.

Fallout on snow provides one of the least biased indications of the kinds of arthropods likely to become available to scavengers in barren lava habitats. Ashmole and Ashmole (1988) provided lumped data from walking transects and quadrat samples from the aeolian zone of Pico del Teide on five dates between mid-February and early May 1984. The catch of 536 arthropods was dominated at the ordinal level by Homoptera, Aphidoidea constituting 39% and Psylloidea 14% of the catch. The abundance of homopterans suggests that many of the dispersing insects came from croplands at much lower levels on the island. Diptera were next in abundance, with Cyclorrhapha 12% and Nematocera 11%, and were followed by Hymenoptera 10% and Thysanoptera 5%.

High variability of composition is characteristic of catches of dispersing arthropods. On the only occasion when it was feasible to sample simultaneously high on Pico de Teide and ~1000 m lower on Montaña Blanca there was a striking contrast in the catches, with Psocoptera forming 12% of the catch and Heteroptera 5% of the catch at the lower level, but much lower proportions higher up. Furthermore, large cyclorrhaphan flies were found on the snow near the peak, whereas at the lower level the only ones seen were making controlled flights and did not become part of the fallout.

The snowfield samples demonstrated the great taxonomic diversity of dispersing insects on Tenerife, even in winter, nine orders and nearly 40 families being represented. Among the major orders, the Diptera included ten families of Cyclorrhapha and six of Nematocera (Sciaridae being the commonest). The Hymenoptera included eight families, with Braconidae dominant.

Fallout on lava flows on Tenerife

In seasons other than winter and at levels where snow does not lie, other methods of sampling arthropod fallout must be used. Many kinds of traps are liable to attract windborne animals, rather than catching a sample landing at random. Even if the traps are not attractive there is the problem that animals arriving in the absence of traps may often take off again and continue their journey, while trapping denies them this option and may thus give misleading quantitative results. Edwards (1986) has offered an ingenious solution to these problems, as already described (p. 272).

On Tenerife simple water traps were placed in areas of barren lava (Fig. 14.1). Traps consisted of circular brown plastic trays of area 0.1 m^2 containing water together with a little formalin and detergent. These were used over long periods to determine the taxa involved in aerial dispersal, but on one occasion a specific attempt was made to quantify fallout. Two of the traps were left for two days (22–24 July, 1984) on a barren and jagged lava flow at one of the study sites (Narices del Teide, altitude ~2075 m). In analysing the catch, Diptera and other strong-flying insects which could reasonably be assumed to be capable of taking off from the lava after they had landed were excluded from consideration. During the two-day period, 26 other insects were caught. These were air-dried and weighed. The weights obtained indicated fallout at the rate of 40 mg m^{-2} day^{-1}, with an average weight of insect of 0.58 mg. Although this is a very limited study, it indicates that a significant quantity of allochthonous arthropod fallout is available to support the scavenger community on the recent lava. The rate of input is comparable to that observed in a volcanic desert on Mount St. Helens (see above, p. 272).

Fig. 14.1. Water trap used to sample biological fallout on the Narices del Teide lava flow of 1798 on Tenerife, Canary Islands.

RESIDENT INVERTEBRATES OF LAVA IN THE CANARY ISLANDS

Background

Lava flows derived from eruptions within the historical period in the Canary Islands give the appearance of being some of the most inhospitable environments on earth. The jagged rocks, lack of soil or vegetation, intense heat by day and – at high altitude – severe cold at night combine to convey an impression of sterility. Search of the lava surface for animals by day is likely to produce only an occasional salticid spider and a few flies. In this context, the recent demonstration – by trapping – of the presence of diverse arthropod communities in such habitats is as remarkable as the discovery long ago of analogous communities in desert regions.

No investigations seem yet to have been carried out in barren volcanic terrain of the behavioural and physiological adaptations permitting the various species that spend time at the surface to withstand the rigours of their environment. It is clear, however, that diurnal species (with the possible exception of flies) depend on the presence of shade. Salticid spiders make brief forays into sunlit areas before retreating into the shade of rocks, while pholcids remain permanently in shaded cavities just below the surface; other species spend the day in deeper crevices, but forage on the surface at night.

Our initial investigations of these communities were carried out on Tenerife in 1984, where some of the most barren areas of recent lava and cinders on the island were selected for study. A visit to Lanzarote in the same year permitted comparable work on several sites within the enormous "mar de lava" produced by the eruptions in the 18th and 19th centuries. Further investigations on Lanzarote were carried out with Dr Pedro Oromi in spring 1985, and in later years joint expeditions were made to a number of other islands, as explained above (p. 269).

Methods

Throughout the research our primary objective has always been to detect the widest possible variety of invertebrates that might be present in the lava habitats, taking account of the likelihood that the populations are patchy and sparse. To this end, a combination of timed hand collecting (by day) and trapping has been used. By standardizing techniques within projects it was also possible to obtain some indication of relative abundance of different taxa at particular sites, and to compare abundances in different places. Because much of the work had to be carried out in remote places and on difficult terrain, it was essential to miniaturize equipment and simplify field procedures; various improvements were introduced during the successive phases of the work. The methods described here are those currently employed; those used during the earlier projects were similar and have been explained in the relevant papers.

At each site we set traps and whenever possible also carried out three person-hours (sometimes less) of visual searching and collecting in the vicinity of the traps; ideally this is done by two or more people to minimize the effect of individual differences in collecting technique. We have preferred to use eight pairs of traps at each site, and to set them in a variety of situations within the site. The members of each pair are set 1–5 m apart and normally at least 10 m from the next pair. One trap in each pair is a pitfall trap consisting of a straight-sided white polypropylene jar of 110 ml capacity, 82 mm high, internal diameter 42 mm. If possible the pitfall is dug in flush with the substrate, but otherwise small rock fragments are used to build up an artificial surface around it. Rocks are used to shade the trap from direct sunlight. Each pitfall contains ~50 ml of modified Turquin's liquid (Turquin, 1973). The form that we use consists of 10 g chloral hydrate, 5 ml formalin, 5 ml glacial acetic acid, 1 ml liquid detergent, plus beer (preferably including one-third stout) to make 1 litre. The second trap in each pair is a 250 ml clear polyvinyl chloride bottle with internal neck diameter 21.5 mm; it is set at an angle of 45–60 degrees and small rocks are placed around it to facilitate the entry of animals. Each bottle contains 25 ml of Turquin's liquid (as above) but also ~.2 g of "Danish Blue" cheese, contained in a small plastic cup (the inverted screw-top of a specimen tube) impaled on a galvanized nail that supports it above the level of the liquid. Traps are preferably left for four days and nights, and are then closed with screw caps and taken back to base for sorting and preservation of specimens. In some projects records are kept of the contents of individual traps, but specimens of particular taxa are then lumped. A special type of trap used recently in sampling tephra habitats is described in a later section.

Results

In the first phase of the investigation on Tenerife, trapping was carried out on a transect of four stations crossing the boundary between recent unvegetated lava and older rocks at a later stage of succession. The resulting data provided the crucial demonstration that several invertebrate taxa were lava specialists, having significant populations at stations on the barren lava but being absent or rare at the stations on older rocks only a few hundred metres away (Ashmole and Ashmole, 1987). The most important lava specialists at high altitudes on Tenerife were lepismatid Thysanura, with much smaller numbers of pygidicranid Dermaptera, *Anataelia canariensis*; there was a hint that the phalangiid *Bunochelis spinifera* (Opiliones) also favoured the barren lava.

The presence of barren-lava specialists conformed with the earlier discovery by Howarth (1979) of a cricket *Caconemobius fori* on Hawaii that occurred only on recent barren lava. A similar situation involving another cricket was described by Thornton et al. (1988) on Krakatau. Lava specialists of this type were termed "lavicoles" by Martín et al. (1987), who studied the aeolian ecosystem of a recent lava flow on El Hierro in the Canary Islands, finding that the fauna there was dominated by a species of pygidicranid dermapteran, *Anataelia lavicola.*

During the 1984 and 1985 studies on Lanzarote (Ashmole and Ashmole, 1987; Ashmole et al., 1990), which were followed up by more extensive work by

Fig. 14.2. The 'islote' or 'kipuka' of Halcones, on Lanzarote, Canary Islands. This site provided an opportunity to compare the invertebrate community on partly vegetated volcanic rocks with that on the surrounding barren 'mar de lava' only two hundred and fifty years old.

Martín (1991); see also Martín and Oromi (1990), sampling was carried out at many barren lava sites; controls were available at a few sites with some vegetation, such as the Islote Halcones (Fig. 14.2). Clearly lavicolous distribution was shown by a thysanuran, *Ctenolepisma longicaudata,* and a new species of cricket *Hymenoptila lanzarotensis* (Kevan and Hsiung, 1992). Lavicolous but also halophilic distribution was shown by two melyrid beetles, the remarkable *Gietella fortunata* (Constantin and Menier, 1987) and *Ifnidius petricola,* and also by a new species of entomobryid collembolan, *Seira dinizi* (da Gama, 1988).

The work on Lanzarote, as well as confirming the presence of a complex community of lavicoles in appropriate habitats, also showed an intriguing tendency for animal abundance on recent lava to be higher near the coast (where the only vegetation consisted of scattered crustose lichens) than further inland (where the fruticose *Stereocaulon vesuvianum* was sparsely distributed). The effect was first noted in 1984 (Ashmole and Ashmole, 1987), and was confirmed during another visit to the island in 1985 by use of a set of four collecting stations forming a transect from the coast to 4 km inland. The catches (especially of Collembola) were much higher on the beach and 200 m inland than 500 m and 4 km inland (Ashmole et al., 1990). Similar results were obtained in the same area by Martín (1991). Some possible reasons for this effect were mentioned above (p. 270).

Comparison of the fauna of lava and adjacent caves

As the various studies on the Canary Islands progressed, it became apparent that the populations of resident scavengers sampled at the surface of the lava and cinders were closely linked to those inhabiting the lava tubes in the same areas. The fauna of caves in the Canary Islands was already under study by cave biologists and entomologists at the Universidad de La Laguna (e.g., Martín and Oromi, 1986; Oromi et al., 1991). To explore the relationship between the lava and cave populations, a joint study was undertaken of adjacent surface and subterranean habitats at six sites on the islands of La Palma and El Hierro. This study used standardized trapping and searching techniques to characterize lava and cave communities, together with various analytical techniques to compare them (Ashmole et al., 1992). As on Tenerife and Lanzarote, the most recent volcanic substrates on La Palma and

Fig. 14.3. Setting traps outside the entrance to Cueva Todoque in the pahoehoe lava flow of 1949 at San Juan, La Palma, Canary Islands. The dermapteran family Pygidicranidae was represented in the catch by sister species of the genus *Anataelia*: *A. lavicola* was trapped both on the surface and in the shallow cave, while a single individual of *A. troglobia* was obtained in the cave.

El Hierro had little primary production and – with the exception of a shore site on La Palma – the only available resource appeared to be biological fallout, within which dispersing arthropods doubtless formed the dominant component.

In spite of the limited energy base, trapping on the barren lava sites showed fairly abundant arthropod populations, though low community diversity. Samples from caves had much smaller numbers of individuals than adjacent lava, but rather similar taxonomic composition. Significantly, this similarity was strongest at the San Juan lava flow on La Palma, erupted in 1949 and containing Cueva Todoque, the youngest cave studied (Fig. 14.3). The collections from the surface of the lava and the cave within it were both dominated by a pygidicranid dermapteran and entomobryid collembolans. The dermapteran *Anataelia lavicola* is clearly a pioneer on new lava, occurring both on the surface and in several young caves; in the cave at San Juan one individual of the sister species *Anataelia troglobia* (Martín and Oromi, 1988) was also found. Of the Collembola, one abundant species was confined to the surface but another, *Pseudosinella canariensis*, was found both on the surface and in the cave.

These results led to a key conclusion of the study – that the animals colonizing a lava flow in the period shortly after an eruption form a single community, dominated by a few species of generalized predators and scavengers and occupying both the surface of the flow and cavities within it. These animals probably harvest the great bulk of the arthropod fallout and other incoming organic material on the surface at night, although some fallout may drift directly into caves and mesocavernous spaces and be harvested there.

Some of the lavicolous consumers remain on the surface throughout the diurnal cycle (salticid spiders and perhaps some lepismatid Thysanura), but others (mainly entomobryid Collembola, pygidicranid Dermaptera and gryllid Orthoptera) exploit fallout on the lava surface by night, while taking refuge underground during the day. In doing so, they make use of the fact that new volcanic substrates are penetrated by a range of cracks and cavities that permit easy transit from the surface to deeper levels. These animals are thus in a sense "habitual trogloxenes", seeking shelter in underground cavities but leaving them regularly to feed on the surface (Chapman, 1993).

Successional changes

As in the original work on Tenerife, the fauna of the barren lava sites on La Palma and El Hierro was compared with that in geologically older terrain on the same islands. The results showed that vegetational succession on the surface was associated with disappearance of the pioneer lavicolous species and colonization by a variety of herbivores and predators. The reasons for the changes are only partly understood. While the arrival of new trophic groups is clearly linked to the establishment of plants, the loss of the lavicoles may be caused partly by the arrival of new predators such as ants. Also relevant, however, are physical changes in the substrates.

As lava flows age, weathering and the input of windblown dust leads to filling up of cracks, while later eruptions nearby may deposit tephra on the surface. As F.G. Howarth first pointed out to us, the vertical movements of trogloxenic lavicoles must be progressively hampered by these processes. Ultimately, they are prevented completely by the filling up of larger cavities, the collapse of lava tubes and the development – in all but the driest climates – of a soil layer with its own characteristic endogean fauna, which blocks the route to any remaining mesocavernous subterranean spaces. In parallel with these physical changes the development of plant communities on the surface gradually reduces the relative importance of biological fallout in the energy base of subterranean communities, which become increasingly dominated by specialized troglobitic species. Many of the latter are physiologically unsuited for excursions to the surface, even at night, and their scarce resources are overwhelmingly derived from the local plant community on the surface.

These studies on the Canary Islands, therefore, provide a view of a relatively brief stage in the development of the lava fauna, before the establishment of clear distinctions between epigean, endogean and hypogean communities. During this stage, arthropod fallout plays a key role in providing resources for the lava community, within which many species utilize both surface and subterranean environments. Later, fallout is doubtless fully exploited by epigean predators and scavengers, which also harvest prey and detritus derived from local plant growth.

SUCCESSION AS A CAUSE OF SPECIATION IN VOLCANIC TERRAIN

The fact that lavicoles occupy both surface and subterranean habitats in new volcanic substrates, but disappear from the surface during the course of succession, seems to provide a special opportunity for isolation and speciation. Howarth (1987b), Oromi et al. (1991), Peck and Finston (1993), Ashmole (1993), and others, have emphasized that the occurrence of a pair of species – respectively epigean and hypogean – on the same island at low latitude presents a challenge to conventional views of allopatric speciation. Martín et al. (1989) suggested that alternating wet and dry periods might have led to speciation on the islands. In the following simple model we suggest that successional changes alone could have had a similar effect.

The process starts with the colonization of a new volcanic habitat by a relatively widespread lavicolous species. In due course, as indicated above, it will be excluded from the surface of this habitat by successional changes, while surviving elsewhere by colonizing newer substrates or persisting in unvegetated terrain near the coast (cf. Martín and Oromi, 1990; Martín, 1991). Meanwhile, a part of the original colonizing population might survive underground, occupying caves or the smaller spaces of the Mesocavernous Shallow Substratum (MSS)[2]. Provided that this population persisted for long enough in isolation, it would become specialized for subterranean life and would be likely to develop reproductive isolation from the parental form, thus achieving separate species status.

This scenario seems almost certainly to have been followed on the island of La Palma, where the lavicolous pygidicranid dermapteran *Anataelia lavicola* (which also occurs on El Hierro) has apparently given rise to *A. troglobia* (Martín and Oromi, 1988; Oromi et al., 1991). The two species now sometimes occur sympatrically, and indeed were found together during sampling in the cave in the lava flow of San Juan on La Palma, as mentioned above; only *A. lavicola*, however, was present on the surface (Ashmole et al., 1992).

Subsequent to the studies of new lava habitats on the Canary Islands, comparable investigations were made during joint Edinburgh/La Laguna expeditions to the Azores archipelago in 1987 and 1989 (Oromi et al.,

[2] The acronym MSS originally stood for Milieu Souterrain Superficiel (Juberthie et al., 1980); in English this is sometimes referred to as the Superficial Underground Compartment, but we use the free translation suggested by Ashmole et al. (1990). The designation of this habitat is further discussed by Juberthie in Chapter 1 of this volume.

Fig. 14.4. Pedro Oromi and Myrtle Ashmole setting traps for arthropods on the Mistério Negro on Terceira, Azores, a barren volcanic dome formed in an eruption in 1761. The mildly troglomorphic endemic subspecies of centipede, *Lithobius obscurus mediocris,* was the dominant predator here but was replaced by *Lithobius pilicornis* on adjacent vegetated areas and by the more highly troglomorphic *L. obscurus azoreae* in caves a few kilometres away.

1990; Borges and Oromi, 1991; Eason and Ashmole, 1992). The much higher rainfall in that archipelago leads to higher rates of successional change, and it was not easy to find barren volcanic habitats for study. However, the lithobiomorph centipedes on the Azores provided evidence supporting the possibility of speciation by the mechanism just discussed (Eason and Ashmole, 1992), although it is doubtful whether reproductive isolation has been achieved in this case.

A new troglobitic centipede, *Lithobius obscurus azoreae,* which is substantially modified for cavernicolous life, was found in totally dark lava tubes and in nearly dark volcanic pits on Pico, Faial and Terceira; later work by P.A.V. Borges showed that it also occurs in the Mesocavernous Shallow Substratum. A second form, *L. obscurus mediocris,* was found on Terceira at the Mistério Negro, a barren and deeply dissected lava dome dating from 1761 (Fig. 14.4). This form is apparently a lavicolous trogloxene and is morphologically intermediate between *L. o. obscurus* from continental areas and the troglobitic *L. o. azoreae*. For instance, *L. o. obscurus* has 24–28 antennal articles, *L. o. mediocris* 30–35 and *L. o. azoreae* 39–44, indicating progressive adaptation to cavernicolous life.

Intriguingly, traps on vegetated ground only 200 m

Fig. 14.5. Five pipe traps and associated equipment assembled for field work on St. Helena and Ascension. Myrtle Ashmole is inserting three vials mounted on a 'ladder' into a pipe. The vertical trap is arranged to demonstrate extraction of the pipe using a scaffolding joint, car jack and foot plate. The traps are readily transported in boxes on the car roof-rack.

away from the barren site dominated by *L. obscurus mediocris* caught none of that species but abundant *Lithobius pilicornis,* and a few individuals of the latter species were also found on the barren lava. It is not clear whether *L. pilicornis* reached the Azores naturally during the Pleistocene, or later with the help of man. It seems certain that, as weathering, soil formation and vegetational succession proceeds on the barren lava dome, *L. pilicornis* will become dominant at the surface and the population of *L. obscurus mediocris* will become restricted to subterranean habitats; it will then presumably experience directional selection for more extreme troglomorphic characteristics and will converge with *L. o. azoreae*, which is already present in a cave under agricultural land a few kilometres away. If the two forms then meet, they will either merge or behave as distinct species. In the absence of *L. pilicornis*, it seems likely that speciation of the two forms of *obscurus* would have given rise to one lavicolous and one troglobitic species.

FAUNA OF TEPHRA ON ASCENSION

Howarth (1987a) pointed out that, in the aeolian ecosystem at high altitudes on Hawaii, the resident arthropods "are segregated by substrate type into three communities: on cinders, ash and lava surface", but we have not seen an account of these communities. Recent work on Ascension (Ashmole and Ashmole, 1997) has included a preliminary investigation of the subterranean fauna of barren tephra habitats including drifts of cindery pyroclasts and piles of rubbly scoria, and also one that appeared to be a'a lava with some scoria.

Sampling of underground habitats was by means of pipe traps derived from a kind used by Juberthie (pers. comm.) while investigating the mesocavernous superficial substratum in talus slopes. In our version (Fig. 14.5) the trap consists of a medium-wall steel pipe ~2 m long and 48 mm external diameter, 42 mm internal. The lower end is closed by a solid cylindrical steel plug which also extends beyond the pipe by 30 mm at a diameter of 51 mm (to reduce friction on the pipe itself) and which terminates in a cone with apical angle 50°. Four equally spaced lines of 9 mm holes (50 mm apart) start 200 mm above the tip and finish half way up the pipe. The top of the pipe is reinforced by an external steel collar, and just below this is a lug to which a steel cap (shaped like an inverted mug) can be padlocked. The pipe is driven more or less vertically into the ground using a sledge hammer. If an excessive amount of powder falls in during insertion it can be removed with a narrow metal tube, after moistening if necessary to improve adhesion. Three plastic vials are then lowered into the lower half of the pipe, spaced on a metal ladder-like strip of a type used with tabs to support shelves. The vials contain modified Turquin's liquid (see above, p. 276) together with a solid bait which is held above the level of the liquid and which can be varied to taste; we generally use Danish Blue cheese. Animals entering the pipe through the holes in the wall either fall into the vials or are attracted to them.

The traps are left baited for a period varying from a few days to several weeks, after which the vials can be emptied and – if desired – replaced without disturbing the pipe. Eventual removal of the pipe is by means of a car jack engaging with a scaffolding clamp which is attached to the pipe and moved down as necessary during extraction. The base of the jack fits loosely into a socket on a separate steel foot-plate that prevents the jack from digging into the substrate; a hole at one side of the plate enables it to be threaded over the top of the pipe, thus keeping the jack aligned with the pipe. The pipe traps can be used in cinders, scoria, and loose and fragmented lava, though not in massive flows.

Although the pipe traps were used on Ascension only briefly, and in the intervals of a project relating to seabirds and alien mammal predators, they show that barren tephra habitats have fauna including at least Araneae, Pseudoscorpiones, Collembola, Psocoptera, Coleoptera and Phoridae (Diptera). Some of the animals captured (e.g., the presumably introduced nitidulid beetle *Urophorus humeralis*) may be epigean species, which detected the underground baits by smell and travelled down through the spaces between the cinders to reach them. Others (e.g., the probably indigenous scytodid spider *Scytodes* sp.) may spend time both underground and on the surface. Three of the species captured, however, are moderately troglomorphic and seem to be genuinely cavernicolous animals. The endemic pseudoscorpion *Apocheiridium cavicola* is a taxonomically isolated member of its genus considered to be adapted to subterranean life (Mahnert, 1993); it was discovered by us in 1990 in a semi-dark lava tube, but in 1995 an individual was captured in a pipe trap fully inserted in barren scoria on the other side of Ascension, implying wide subterranean distribution. The endemic entomobryid collembolan *Pseudosinella lava* is the dominant collembolan in samples from caves but also from pipe traps; it has unusual troglomorphic features (Christiansen, 1998) and is related to another endemic species apparently confined to the surface of lava flows, and to a third found only close to the shore. Finally, the liposcelidine psocopteran *Troglotroctes ashmoleorum,* which was discovered in 1990, represents a new genus and is apparently also adapted to subterranean life (Lienhard, 1996); it was originally found in caves, but in 1995 many individuals were caught in a pipe trap in barren volcanic rubble (Fig. 14.6), suggesting that it is widespread underground. However, a single female was also trapped in a crevice on barren lava in 1990, so that individuals must sometimes come to the surface, probably at night.

The finding of these troglomorphic species in widely separated caves and pipe traps implies that the barren volcanic habitats in the arid western part of Ascension are populated not only by epigean lavicoles, but also by hypogean species that live both in caves and in the mesocavernous spaces between fragments of tephra, or in cracks in lava, a metre or more below the surface.

Although alien plants are now spreading in the arid parts of Ascension, there is still little primary production in the areas sampled. However, scattered plants were probably present even in the distant past. The animal communities in these habitats, dominated by scavengers and predators, seem analogous to faunas of extreme deserts, in that many species are presumably dependent on plant and animal remains concentrated by interactions of wind and topography. Some of the remains are eaten on the surface, but there appears to be a sufficient downward energy flow to support populations of primarily subterranean animals.

Communities of this type seem not to have been much studied elsewhere, perhaps because appropriate sampling equipment has not been available. Studies on the Canary Islands have shown that a form of mesocavernous shallow substratum occurs in some areas where there is soil at the surface and there are mesocaverns between the soil layer and the bedrock (Medina and Oromi, 1990; Oromi et al., 1991). This habitat is occupied by a number of highly troglomorphic species, but they are isolated from the surface by the soil layer. On the other hand, trapping on the surface of barren lava and tephra in the Canaries has detected the presence of significant communities of scavengers and predators, as described above, but in general has not led to the finding

Fig. 14.6. Myrtle Ashmole beside a pipe trap 2 m long fully inserted into lava rubble on the slopes of South Gannet Hill, Ascension. This trap caught individuals of a troglomorphic entomobryid collembolan, *Pseudosinella lava,* and of a troglomorphic liposcelidine psocopteran, *Troglotroctes ashmoleorum*, both of which were also abundant in caves.

of troglomorphic species. One exception, however, is an entomobryid collembolan, *Pseudosinella trioculata* (da Gama, 1988), which seems to be somewhat troglomorphic; it occurs regularly in cave habitats on Lanzarote (Martín and Oromi, 1990), and has also been trapped at the surface of barren lava habitats (Ashmole et al., 1990). An analogous case from the Azores is provided by the mildly troglomorphic centipede *Lithobius obscurus mediocris* discussed in the previous section, though here the caves were inhabited by a distinct, more highly troglomorphic form.

Fuller understanding of the fauna of barren tephra habitats will require sampling – at the same sites and by day and by night – of animals living at the surface and in the subterranean spaces at various depths below the surface. We suspect that deep banks of barren scoria, or drifts of finer cinders lacking a surface layer of soil, constitute a special category of mesocavernous shallow substratum. At a depth of a metre or so in such substrates, light is lacking and there is presumably a relatively equable temperature regime. Furthermore, there is a tendency for condensation to occur at night in surface layers of cinders or pumice, when hot days are succeeded by cold, clear nights. This effect is exploited for agriculture in the Canaries (Anonymous, 1984; Ashmole et al., 1992) and probably maintains higher humidity in tephra mesovoids than is typical in cracks in more massive volcanic rocks at comparable depths. We suggest, therefore, that in barren tephra habitats selection for mild troglomorphy may occur even in populations where individuals have access to the surface and may sometimes forage there at night. It is likely, however, that the kinds of animals living in different tephra habitats depend critically on the typical size of the tephra fragments, since this will determine the dimensions of the mesovoids.

CONCLUSIONS

Studies in the Canary Islands and elsewhere indicate that communities of scavengers and predators in barren volcanic terrain are supported to a substantial extent by fallout of arthropods – the "derelicts of dispersal" of Edwards (1986). Originating in more productive places, these insects and arachnids are drifted with the wind to inappropriate barren localities, where they succumb to desiccation, starvation or predatory residents.

Many new volcanic substrates, formed of lava or tephra, contain extensive systems of cracks and cavities that permit free movement of scavengers from the surface into the subterranean environment. In such situations the distinction between surface and subterranean communities is blurred, with some animals – the lavicolous trogloxenes – foraging for fallout on the surface (mainly at night) but retreating underground when conditions outside become unfavourable. Recent work in these habitats on Ascension, however, shows that some species with troglomorphic features live in a zone just below the surface, which thus forms a kind of mesocavernous shallow substratum even though there is no soil at the surface.

As volcanic substrates age, processes of weathering, soil formation and vegetational succession gradually eliminate the superficial cavities and obstruct the passage of organic debris downwards, and of scavengers from the surface to refuges below. Trophically more specialized taxa invade the surface habitat as primary production increases, the soil layer is colonized by specialized soil organisms, and the cavities remaining at greater depths are occupied mainly by troglobites that have no access to arthropod fallout. It is suggested that this successional process sometimes leads to speciation by splitting an originally generalized lavicolous species into separate forms confined respectively to surface and subterranean habitats.

ACKNOWLEDGEMENTS

Studies on the Canaries were supported by the University of Edinburgh, the Nuffield Foundation, the British Council and the "Acciones Integradas" program of the Spanish and British Governments. Expeditions to the Azores were financed by the National Geographic Society (USA). Recent work on Ascension was supported by the Carnegie Trust for the Universities of Scotland, the British Ecological Society and the Percy Sladen Memorial Fund. Much of the work has been carried out in cooperation with colleagues at the Universidad de La Laguna, Tenerife, whose friendship, ideas and practical help are gratefully acknowledged. We also wish to thank all the taxonomists who have worked on our material and whose expertise is so vital in ecological, evolutionary and biogeographic studies.

REFERENCES

Anonymous, 1984. *Congreso Internacional de Suelos Volcánicos. Guia de Campo.* Departamento de Edafología, Universidad de La Laguna, Tenerife, pp. 250–256.

Antor, R.J., 1994. Arthropod fallout on high alpine snow patches of the Central Pyrenees, Northeastern Spain. *Arct. Alp. Res.,* 26: 72–76.

Antor, R.J., 1995. The importance of arthropod fallout on snow patches for the foraging of high-alpine birds. *J. Avian Biol.,* 26: 81–85.

Ashmole, M.J. and Ashmole, N.P., 1987. Arthropod communities supported by biological fallout on recent lava flows in the Canary Islands. *Entomol. Scand.* 32(suppl): 67–88.

Ashmole, N.P., 1983. *Spiders and Beetles (and other Insects) at High Levels on the Cairngorms.* Unpublished report to the Nature Conservancy Council.

Ashmole, N.P., 1993. Colonization of the underground environment in volcanic islands. *Mem. Biospeol.,* 20: 1–11.

Ashmole, N.P. and Ashmole, M.J., 1988. Insect dispersal on Tenerife, Canary Islands: high altitude fallout and seaward drift. *Arct. Alp. Res.,* 20: 1–12.

Ashmole, N.P. and Ashmole, M.J., 1997. The land fauna of Ascension Island: new data from caves and lava flows, and a reconstruction of the prehistoric ecosystem. *J. Biogeogr.,* 24: 549–589.

Ashmole, N.P., Nelson, J.M., Shaw, M.R. and Garside, A., 1983. Insects and spiders on snowfields in the Cairngorms, Scotland. *J. Nat. Hist.,* 17: 599–613.

Ashmole, N.P., Ashmole, M.J. and Oromi, P., 1990. Arthropods of recent lava flows on Lanzarote. *Vieraea,* 18: 171–187.

Ashmole, N.P., Oromi, P., Ashmole, M.J. and Martín, J.L., 1992. Primary faunal succession in volcanic terrain: lava and cave studies on the Canary Islands. *Biol. J. Linn. Soc.,* 46: 207–234.

Ashmole, N.P., Oromi, P., Ashmole, M.J. and Martín, J.L., 1996. The invertebrate fauna of early successional volcanic habitats in the Azores. *Bol. Mus. Munic. Funchal,* 48(264): 5–39.

Blanchard, D.C., 1983. The production, distribution and bacterial enrichment of the sea-salt aerosol. In: W.G.N. Slinn and P.S. Liss

(Editors), *Air–Sea Exchange of Gases and Particles*. D. Reidel Publishing Company, Dordrecht, pp. 407–454.

Borges, P. and Oromi, P., 1991. Cave-dwelling ground beetles of the Azores (Col., Carabidae). *Mem. Biospeol.*, 18: 185–191.

Chapman, P., 1993. *Caves and Cave Life*. HarperCollins, London, 224 pp.

Christiansen, K.A., 1998. New species of *Pseudosinella* (Collembola) from Ascension Island. *J. Nat. Hist.*, 32: 149–156.

Constantin, R. and Menier, J.J., 1987. Etude d'un remarquable *Melyridae* aptère des Iles Canaries: *Gietella fortunata*, n. gen., n. sp., type d'une sous-famille nouvelle *Gietellinae (Coléoptera, Cleroidea)*. *Rev. Fr. Entomol.*, 9(2): 53–63.

da Gama, M.M., 1988. Colembolos das Canarias. *Acta III Congréso Ibérico de Entomología, Granada*: 73–90.

del Moral, R. and Wood, D.M., 1993. Early primary succession on a barren volcanic plain at Mount St. Helens, Washington. *Am. J. Bot.*, 80: 981–991.

Drake, V.A. and Gatehouse, A.G., 1995. *Insect Migration: Tracking Resources Through Space and Time*. Cambridge University Press, Cambridge, 478 pp.

Eason, E.H. and Ashmole, N.P., 1992. Indigenous centipedes (Chilopoda: Lithobiomorpha) from Azorean caves and lava flows. *Zool. J. Linn. Soc.*, 105: 407–429.

Edwards, J.S., 1972. Arthropod fallout on Alaskan snow. *Arct. Alp. Res.*, 4: 167–176.

Edwards, J.S., 1986. Derelicts of dispersal: arthropod fallout on Pacific Northwest volcanoes. In: W. Danthanarayana (Editor), *Insect Flight: Dispersal and Migration*. Springer-Verlag, Berlin, pp. 196–203.

Edwards, J.S., 1987. Arthropods of alpine aeolian ecosystems. *Annu. Rev. Entomol.*, 32: 163–179.

Edwards, J.S. and Banko, P.C., 1976. Arthropod fallout and nutrient transport: a quantitative study of Alaskan snowpatches. *Arct. Alp. Res.*, 8: 237–245.

Edwards, J.S. and Sugg, P., 1993. Arthropod fallout as a resource in the recolonization of Mount St. Helens. *Ecology*, 74: 954–958.

Edwards, J.S., Crawford, R.L., Sugg, P.M. and Peterson, M.A., 1986. Arthropod recolonization in the blast zone of Mount St. Helens. In: S.A.C. Keller (Editor), *Mount St. Helens: Five Years Later*. Eastern Washington University Press, Cheney, pp. 329–333.

Elton, C., 1925. The dispersal of insects to Spitsbergen. *Trans. Entomol. Soc. London*, 1925: 289–299.

Farrow, R.A., 1984. Detection of transoceanic migration of insects to a remote island in the Coral Sea, Willis Island. *Aust. J. Ecol.*, 9: 253–272.

Gressitt, J.L., 1964. Insects of Campbell Island. Summary. *Pac. Insects Monogr.*, 7: 531–600.

Heatwole, H., 1971. Marine-dependent terrestrial biotic communities on some cays in the Coral Sea. *Ecology*, 52: 363–366.

Heatwole, H., Done, T. and Cameron, E., 1981. *Community Ecology of a Coral Cay. A Study of One-Tree Island, Great Barrier Reef, Australia*. Monographiae Biologicae Volume 43, W. Junk, The Hague, 379 pp.

Howarth, F.G., 1979. Neogeoaeolian habitats on new lava flows on Hawaii Island: an ecosystem supported by windborne debris. *Pac. Insects*, 20: 133–144.

Howarth, F.G., 1980. The zoogeography of specialized cave animals: a bioclimatic model. *Evolution*, 34: 394–406.

Howarth, F.G., 1983. Ecology of cave arthropods. *Annu. Rev. Entomol.*, 28: 365–389.

Howarth, F.G., 1987a. Evolutionary ecology of aeolian and subterranean habitats in Hawaii. *Trends Ecol. Evol.*, 2: 220–223.

Howarth, F.G., 1987b. The evolution of non-relictual tropical troglobites. *Int. J. Speleol.*, 16: 1–16.

Johnson, C.G., 1969. *Migration and Dispersal of Insects by Flight*. Methuen, London, 763 pp.

Juberthie, C., Delay, B. and Bouillon, M., 1980. Sur l'existence d'un milieu souterrain superficiel en zone non calcaire. *C. R. Acad. Sci. Ser. D*, 290: 49–52.

Kennedy, J.S., 1975. Insect dispersal. In: D. Pimentel (Editor), *Insects, Science and Society*. Academic Press, New York, pp. 103–119.

Kevan, D.K.McE. and Hsiung, C.-C., 1992. A summary of orthopteroid insects known from Lanzarote, Canary Islands, with description of a new cricket species of the genus *Hymenoptila* Chopard, 1943 (Grylloptera; Grylloidea; Gryllidae). *Vieraea*, 21: 77–100.

Lewis, T. and Dibley, G.C., 1970. Air movement near windbreaks and a hypothesis of the mechanism of accumulation of airborne insects. *Ann. Appl. Biol.*, 66: 477–484.

Lienhard, C., 1996. Psocoptères nouveaux ou peu connus de quelques îles atlantiques (Canaries, Madère, Açores, Ascension) et de l'Afrique du Nord (Insecta: Psocoptera). *Bol. Mus. Munic. Funchal*, 48(267): 87–151.

Lindroth, C.H., Andersson, H., Bodvarsson, H. and Richter, S.H., 1973. Surtsey, Iceland. The development of a new fauna, 1963–1970. Terrestrial invertebrates. *Entomol. Scand.*, 5(Suppl.), 280 pp.

Mahnert, V., 1993. Pseudoskorpione (Arachnida: Pseudoscorpiones) von Inseln des Mittelmeers und des Atlantiks (Balearen, Kanarische Inseln, Madeira, Ascension), mit vorwiegend subterraner Lebensweise. *Rev. Suisse Zool.*, 100: 971–992.

Martín, J.L., 1991. *Fauna Invertebrada del Parque Nacional de Timanfaya (Lanzarote - I. Canarias)*. Servicio de Publicaciones de la Caja General de Ahorros de Canarias (Número 145), 166 pp.

Martín, J.L. and Oromi, P., 1986. An ecological study of the Cueva de los Roques lava tube (Tenerife, Canary Islands). *J. Nat. Hist.*, 20: 375–388.

Martín, J.L. and Oromi, P., 1988. Dos nuevas especies de *Anataelia* Bol. (Dermaptera, Pygidicranidae) de cuevas y lavas recientes del Hierro y La Palma (Islas Canarias). *Mem. Biospeol.*, 15: 49–59.

Martín, J.L. and Oromi, P., 1990. Fauna invertebrada de las lavas del Parque Nacional de Timanfaya (Lanzarote, Islas Canarias). *Ecol. (ICONA, Madrid)*, 4: 297–312.

Martín, J.L., Oromi, P. and Izquierdo, I., 1987. El ecosistema eólico de la colada volcánica de Lomo Negro en la isla de El Hierro (Islas Canarias). *Vieraea*, 17: 261–270.

Martín, J.L., Izquierdo, I. and Oromi, P., 1989. Sur les relations entre les troglobies et les espèces epigées des isles Canaries. *Mem. Biospeol.*, 16: 25–34.

Medina, A.L. and Oromi, P., 1990. First data on the superficial underground compartment in La Gomera (Canary Islands). *Mem. Biospeol.*, 17: 87–91.

Menier, J.J. and Constantin, R., 1989. Description complémentaire et remarques sur la morphologie de *Gietella faialensis* Menier & Constantin, 1988. Notes sur la phylogénie, la biologie at la distribution du genre (Coleoptera, Melyridae, Gietellinae). *Rev. Fr. Entomol.*, 11(2): 79–84.

New, T.R. and Thornton, I.W.B., 1988. A pre-vegetation population of crickets subsisting on allochthonous aeolian debris on Anak Krakatoa. *Philos. Trans. R. Soc. London Ser. B,* 322: 481–485.

Oromi, P., Martín, J.L., Ashmole, N.P. and Ashmole, M.J., 1990. A preliminary report on the cavernicolous fauna of the Azores. *Mem. Biospeol.,* 17: 97–105.

Oromi, P., Martín, J.L., Medina, A.L. and Izquierdo, I., 1991. The evolution of the hypogean fauna in the Canary Islands. In: E.C. Dudley (Editor), *The Unity of Evolutionary Biology.* Dioscorides Press, Portland, Oregon, pp. 380–395.

Peck, S.B., 1990. Eyeless arthropods of the Galapagos Islands, Ecuador: composition and origin of the cryptozoic fauna of a young, tropical, oceanic archipelago. *Biotropica,* 22: 366–381.

Peck, S.B., 1994a. Aerial dispersal of insects between and to islands in the Galápagos Archipelago, Ecuador. *Ann. Entomol. Soc. Am.,* 87: 218–224.

Peck, S.B., 1994b. Sea-surface (pleuston) transport of insects between islands in the Galápagos Archipelago, Ecuador. *Ann. Entomol. Soc. Am.,* 87: 576–582.

Peck, S.B. and Finston, T.L., 1993. Galapagos Islands troglobites: the questions of tropical troglobites, parapatric distributions with eyed-sister-species, and their origin by parapatric speciation. *Mem. Biospeol.,* 20: 19–37.

Pedgley, D.E., 1982. *Windborne Pests and Diseases. Meteorology of Airborne Organisms.* John Wiley, New York, 250 pp.

Polis, G.A. and Hurd, S.D., 1995. Extraordinarily high spider densities on islands: Flow of energy from the marine to terrestrial food webs and the absence of predation. *Proc. Natl. Acad. Sci. USA,* 92: 4382–4386.

Rainey, R.C., 1978. The evolution and ecology of flight: the "oceanographic" approach. In: H. Dingle (Editor), *Evolution of Insect Migration and Diapause.* Springer-Verlag, New York, pp. 33–48.

Sarbu, S.M., Kane, T.C. and Kinkle, B.K., 1996. A chemoautotrophically based cave ecosystem. *Science,* 272: 1953–1955.

Southwood, T.R.E., 1962. Migration of terrestrial arthropods in relation to habitat. *Biol. Rev.,* 37: 171–214.

Swan, L.W., 1963. Aeolian zone. *Science,* 140: 77–78.

Swan, L.W., 1968. Alpine and aeolian regions of the world. In: H.E. Wright Jr and W.H. Osburn (Editors), *Arctic and Alpine Environments.* Indiana University Press, Bloomington, pp. 29–54.

Taylor, L.R., 1963. Analysis of the effect of temperature on insects in flight. *J. Anim. Ecol.,* 32: 99–117.

Thornton, I.W.B., New, T.R., McLaren, D.A., Sudarman, H.K. and Vaughan, P.J., 1988. Air-borne arthropod fall-out on Anak Krakatau and a possible pre-vegetation pioneer community. *Philos. Trans. R. Soc. London B,* 322: 471–479.

Turquin, M.J., 1973. Une biocenose cavernicole original pour le Bugey; le Puits de Rappe. *C. R. 96e Congr. Natl. Soc. Savantes, Toulouse 1971, Sciences,* Tome 3: 235–256.

Whittaker, R.J., Bush, M.B. and Richards, K., 1989. Plant recolonization and vegetation succession on the Krakatau Islands, Indonesia. *Ecol. Monogr.,* 59: 59–123.

Wilson, A.T., 1959. Surface of the ocean as a source of air-borne nitrogenous material and other plant nutrients. *Nature,* 184: 99–101.

Chapter 15

ROOT-DRIVEN FAUNAS IN CAVE WATERS

Edyta J. JASINSKA and Brenton KNOTT

INTRODUCTION

In the dark zones of most caves, food is a limiting resource for the establishment and development of cave faunas. Due to the absence of light and consequently the lack of photosynthesis, food supply in caves tends to be low or sporadic in quantity and poor or variable in quality (Culver, 1982). Common examples of energy sources in caves include cadavers, bat and cricket guano, and debris washed in from the surface, which can comprise organic particulates and litter, dissolved organic matter, bacteria, protozoans and other organisms. The allochthonous material is decomposed and transformed by microflora, which themselves become a source of energy when detritus is consumed [Hobbs (1992) and references therein]. Primary productivity in caves, achieved through bacterial chemosynthesis, is very low compared to photosynthetic productivity (Culver, 1982), and to date there is only one location known where a highly diverse fauna is sustained mainly by energy derived from chemosynthetic bacteria in non-thermal cave environments (Popa and Sarbu, 1991; Sarbu, 1990; Sarbu et al., 1996; Chapter 17, this volume). The lack of a nutritious, substantial and reliable food source is one of the principal reasons why cave faunas are commonly depauperate in both biomass and diversity when compared to the faunas of epigean ecosystems from the same area. However, in shallow sections of caves to a maximum depth of approximately 30 m, tree roots can provide a food source which is substantial and stable in supply.

Tree roots are not uncommon in shallow limestone caves and lava tubes; in tropical caves especially, aerial roots descending from cave ceilings have been reported to sustain populations of invertebrates, but of limited species diversity (Howarth, 1983). In temperate limestone caves in southwestern Australia, only two species of oniscoid isopod and a species of moth (with caterpillar stages which both feed and pupate on fine aerial tree roots) similarly inhabit aerial roots (E.J. Jasinska, unpublished). In some aquatic cave environments, submerged tree roots support a much greater number of species than observed for aerial root mats, often combined with high animal abundance (Jasinska et al., 1996). In these habitats the food supply clearly is no longer a major limiting factor for the establishment of cave faunas.

There are remarkably few observations in the literature of aquatic root habitats in caves, and those comments which have been made are generally included as habitat notes accompanying species descriptions of cave animals, although Notenboom (1981) did describe the distribution of isopods and amphipods within *Ficus* root mats from caves in the Caribbean. The presence of aquatic root mats has often been overlooked in the past by speleologists; but in recent years, with the growing awareness of their significance, root mats are noted to be reasonably common in subterranean waters of shallow caves. Cavernicoles have now been collected from aquatic root mats in caves in three Australian States: New South Wales, Tasmania, and in both the southwest and northwest of Western Australia (Fig. 15.1) (S.M. Eberhard, pers. commun., 1995; B. Knott, unpublished).

Occurrence and distribution of cave waters containing root mat habitats

Not all caves which are penetrated by tree roots and which contain streams or lakes also support the growth of aquatic root mats. The roots may produce aerial mats which remain suspended above the water surface and, although some rootlets may descend to within millimetres of the water surface, none actually grow into it.

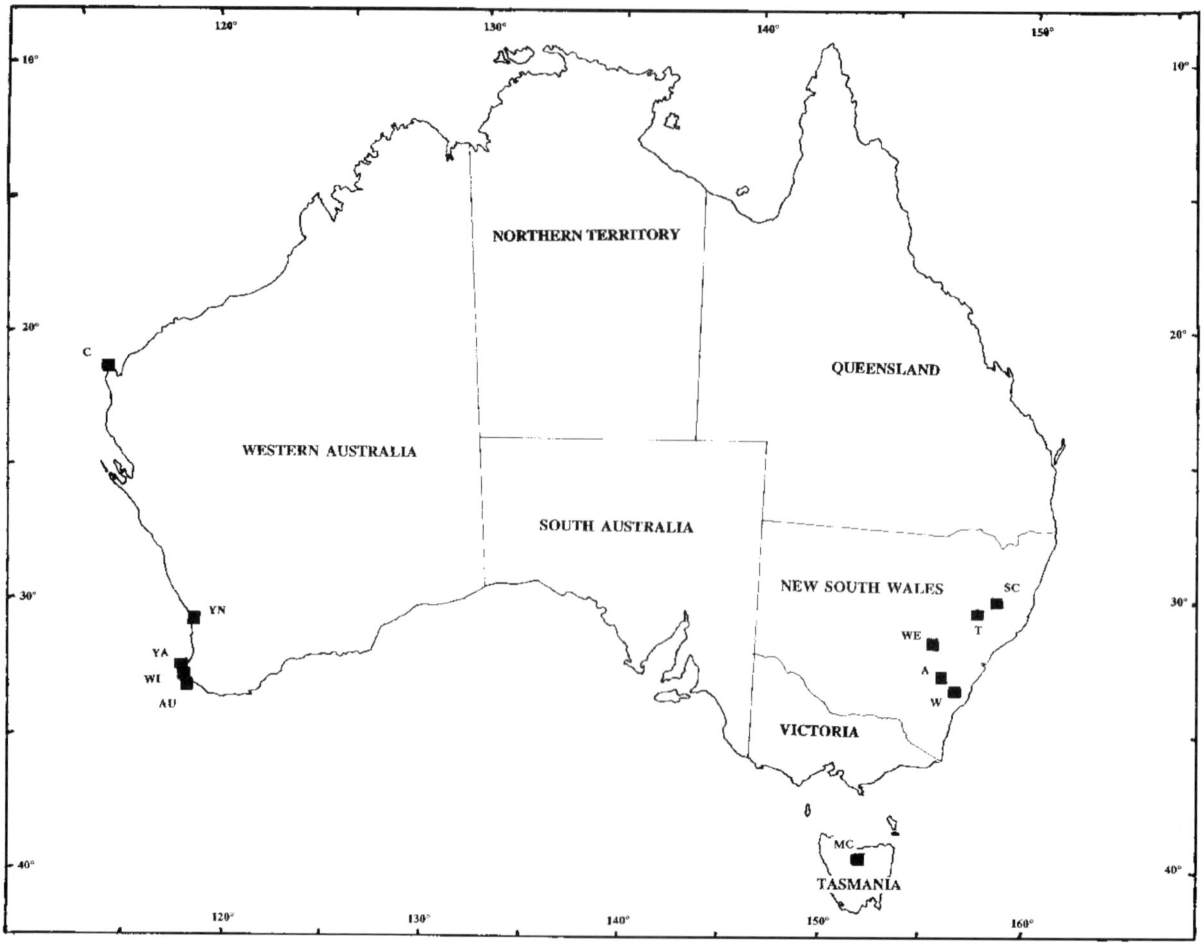

Fig. 15.1. Cave areas in Australia where cavernicoles have been collected from aquatic root mats in caves. A, Abecrombie; AU, Augusta; C, Cape Range; MC, Mole Creek; SC, Stockyard Creek; T, Timor; W, Wombeyan; WE, Wellington; WI, Witchcliffe; YA, Yallingup; YN, Yanchep.

Based on the Australian examples of aquatic root mats in caves, the features in common [E.J. Jasinska (unpublished) and S.M. Eberhard (pers. commun., 1995) *re* eastern Australian caves; W.F. Humphreys (pers. commun., 1997) *re* northwestern Australian caves] which appear to be necessary for the growth of aquatic root mats in caves are listed below, together with our interpretations of their functional significance.

(1) Presence of trees above the caves. Trees from a number of unrelated species produce aquatic root mats in caves, including several species of eucalypt (*Eucalyptus calophylla, E. diversicolor*, *E. gomphocephala* in Western Australia, and a number of species in southeastern Australia), eastern Australian she-oaks (*Casuarina* spp.), peppermint (*Agonis flexuosa*), Australian figs (*Ficus* spp.), exotic trees such as the peppercorn tree (*Schinus* sp.), and unidentified species from a dry subtropical forest in southeastern Australia. Plants with smaller growth habits, such as shrubs, usually have shallower root systems which, even in arid areas, tend not to reach deeper than 3.5 m below ground (Dodd et al., 1984) and thus are unlikely to reach to the depth of the majority of cave waters.

(2) Cavernous rock with fissures or solution channels. The rock needs to be penetrable by roots either along solution pipes, fissures or collapses, and/or by being prone to dissolution by a mixture of root exudates and water.

(3) Depth to cave waters approximately 35 m or less below the ground surface. This is likely to be a

reflection of the maximum depth to which tree roots in cavernous areas can grow.

(4) Arid conditions in the cave atmosphere and in soil above the cave water levels for extended periods of time during the year. Trees with roots growing in water tend to suffer from waterlogging problems, stemming mainly from a decreased availability of dissolved oxygen (Davison and Tay, 1985; Drew, 1992); in ground waters especially, the ratio of carbon dioxide to oxygen is much higher than in air because carbon dioxide is more soluble in water than oxygen (Allan, 1994). Therefore, trees would be expected to grow their roots, preferentially, in moist but not water-saturated conditions. Aquatic root mats presumably are produced where there is insufficient moisture in the soil and/or cave air to meet the tree's water requirements.

(5) Permanent streams or lakes in caves. In southwestern Australia, we have observed the disappearance of root mats from once-permanent cave waters which in recent years have become subject to drying-out in summer due to progressive lowering in the regional water table.

Thus, the salient requisite factors for the formation of aquatic root mats in caves are: (i) permanent water bodies at (ii) shallow depth, in (iii) fissured cavernous rock which supports (iv) the growth of trees and (v) where, at least for part of the year, the local climate and soil structure create arid conditions, forcing the trees to grow roots in cave waters in order to meet their water requirements.

It would be valuable to have detailed documentation of the occurrence of root mats in cave waters worldwide in order to describe in greater detail their patterns of distribution and to elucidate further the factors contributing to their development. Here we focus on describing root mat habitats and associated fauna from cave waters in two areas of karst along the southwestern coast of Western Australia: Yanchep National Park and the Leeuwin–Naturaliste Ridge (Fig. 15.2A).

Both Yanchep National Park and the Leeuwin–Naturaliste Ridge are located within the southern half of the Perth Basin (Fig. 15.2A), separated from each other by nearly 300 km. The Perth Basin is a major north–south trending sedimentary structure extending from just east of Shark Bay to the southwestern corner of the continent. The Basin is demarcated from the Archaean rocks of the Yilgarn Craton by the Darling Fault which in the vicinity of Perth forms the Darling Scarp. The oldest known sediments in the Basin are

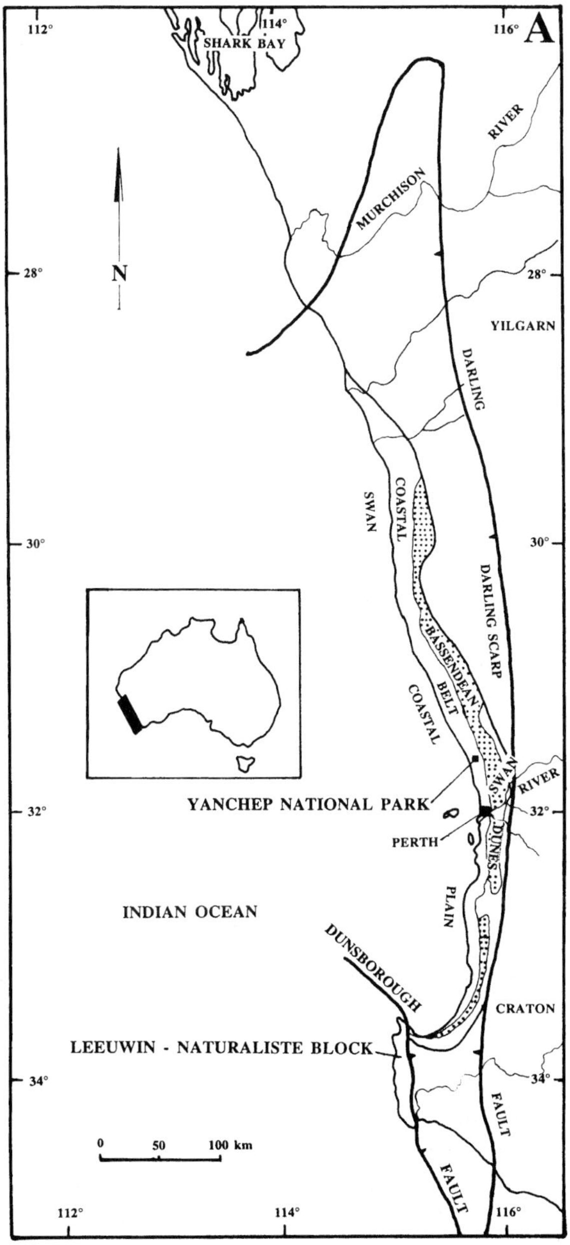

Fig. 15.2A. Locality maps of cave areas in southwestern Australia. Major features of the Perth Basin showing the location of the Yanchep National Park on the Swan Coastal Plain and of the Leeuwin–Naturaliste Ridge situated on the Leeuwin Block. Insert: Australia with the Perth Basin indicated.

of Silurian age (Cockbain, 1990), but the caves of both areas occur in strata of Quaternary age. The Quaternary superficial strata comprise an extensive system of dunes formed approximately parallel to the coast and increasing in age (Holocene to early

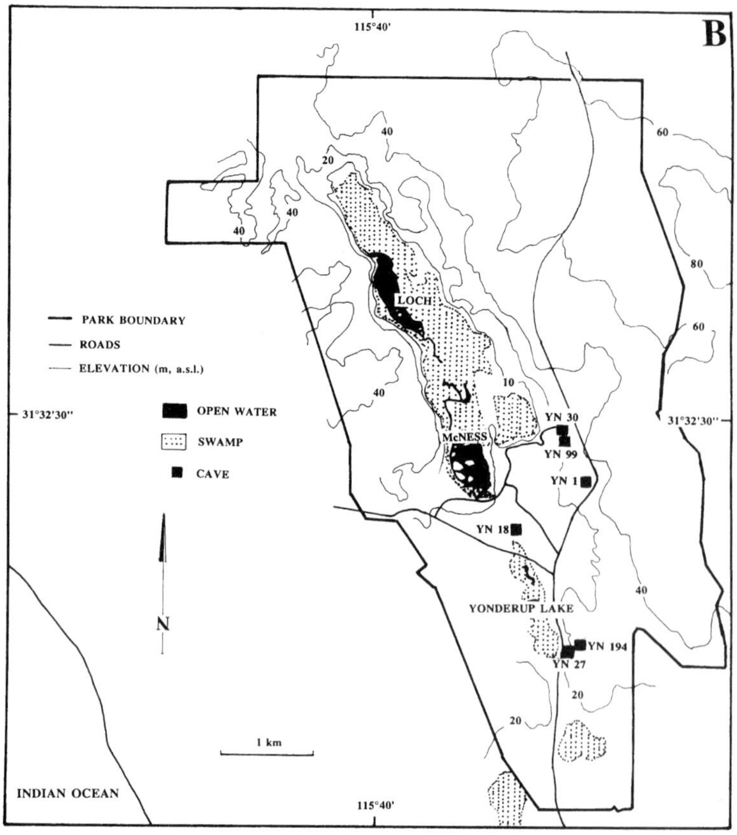

Fig. 15.2B. Locality maps of cave areas in southwestern Australia. Topography of Yanchep National Park. YN 1, Crystal Cave; YN 18, Carpark Cave; YN 27, Gilgie Cave; YN 30, Cabaret Cave; YN 99, unnamed cave; YN 194, Twilight Cave.

Pleistocene) eastwards from the Indian Ocean, and rising to 110 m above sea level. The oldest sequence abuts the Darling Scarp, which itself reaches to about 350 m above sea level. The gently undulating dune sequences, together with belts of alluvium in front of the Darling Scarp and along the river valleys comprise the Swan Coastal Plain (Fig. 15.2A) (Playford et al., 1976).

The caves which are the subject of this chapter occur within 20 km of the coastline in consolidated calcareous dune deposits termed the Tamala limestone [or "Coastal Limestone" of earlier usage (Playford et al., 1976)], in places reaching a thickness of more than 150 m. Within Yanchep National Park and the Leeuwin–Naturaliste Ridge the Tamala limestone is of mid-Pleistocene age (Kendrick et al., 1991). Jennings (1968) deduced that the processes of consolidation of the dunal sediments and of karst formation in the Tamala limestone were concurrent and consequently termed this type of karst "syngenetic". In the region of the Yanchep National Park the Tamala limestone overlies early to mid-Pleistocene noncalcareous dune deposits, the Bassendean sands (Kendrick et al., 1991; Playford et al., 1976).

The Leeuwin–Naturaliste Ridge, rising to 220 m above sea level, is a prominent surface feature of the Leeuwin Block. The Leeuwin Block is an upthrust inlier of the mid-Proterozoic crystalline rocks which form the basement of the Perth Basin (Palfreyman, 1984; Cockbain, 1990). The crystalline rocks, metamorphosed into gneisses and granulites, and mostly capped by laterite, are mantled by the Tamala Limestone and superficial sands (Playford et al., 1976; Jongsma and Petkovic, 1977; Cockbain, 1990).

The caves from both areas have provided important palaeontological finds (Archer et al., 1984), remains of Late Pleistocene extinct marsupials dated at >37 000 yrs having been recovered from Mammoth Cave (WI 38) of the Leeuwin–Naturaliste Ridge (Merrilees, 1968).

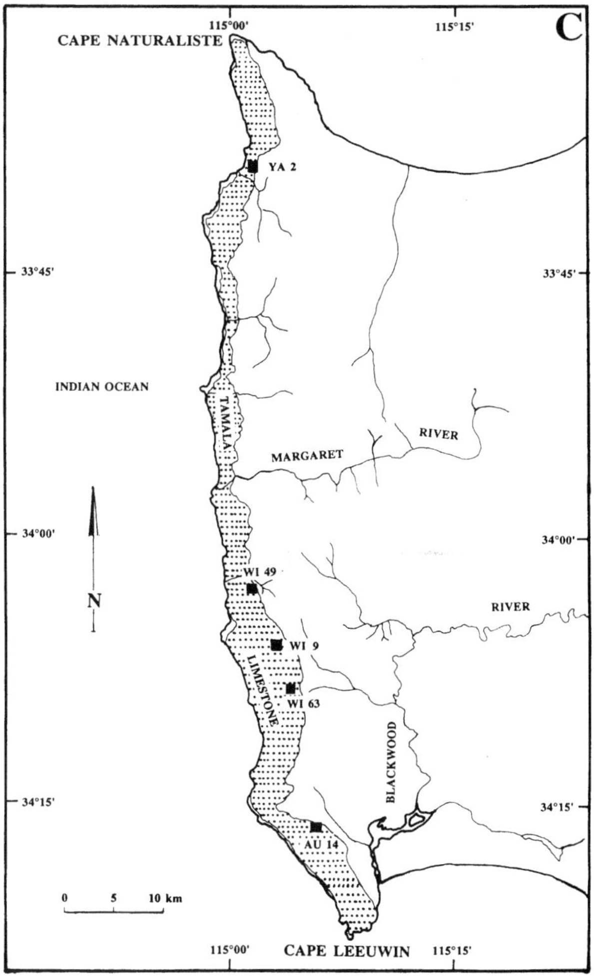

Fig. 15.2C. Locality maps of cave areas in southwestern Australia. Leeuwin–Naturaliste Ridge showing the extent of Tamala limestone and major drainage systems. AU 14, Easter Cave; WI 9, Kudjal Yolgah; WI 49, Calgardup Cave; WI 63, Strongs Cave; and YA 2, Northcote Grotto.

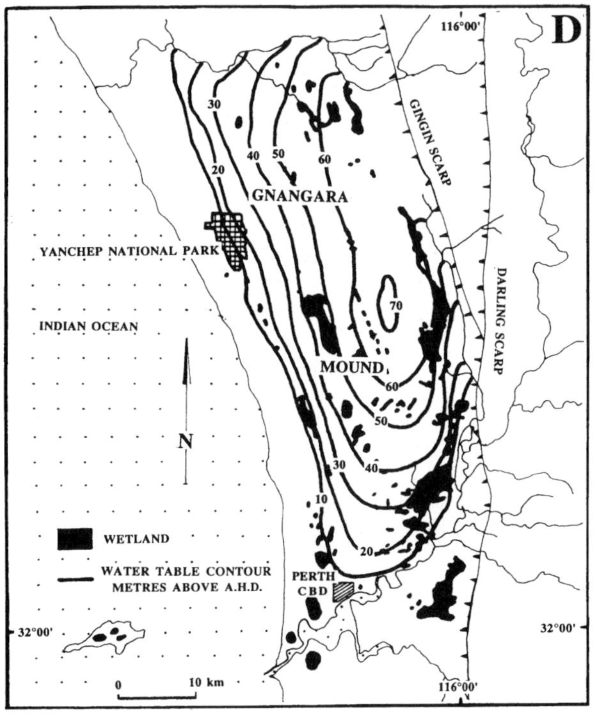

Fig. 15.2D. Locality maps of cave areas in southwestern Australia. Location of Yanchep National Park in relation to the Gnangara Mound aquifer.

CASE STUDY: ROOT MATS IN CAVE WATERS OF SOUTHWESTERN AUSTRALIA

Study area

In Western Australia root mat habitats in cave waters have been studied intensively since 1990 in two areas of karst situated on the southwestern coast of the continent. The karst areas comprise the Yanchep National Park (31°30′S to 31°35′S and 115°39′E to 115°43′E) (Fig. 15.2B) and the Leeuwin–Naturaliste Ridge (33°31′S to 34°23′S and 114°59′E to about 115°15′E) (Fig. 15.2C). Caves of the Yanchep area are given the prefix "YN" in the Australian Karst Index (Matthews, 1985), while those on the Leeuwin–Naturaliste Ridge occur in three areas with the prefix "AU", "WI" and "YA" denoting Augusta, Witchcliffe and Yallingup, respectively.

The climate is mediterranean, characterized by hot summer droughts (December to February) and cool winters with successions of abundant rains (June to August) (Gentilli, 1972). Yanchep National Park has similar weather patterns to Perth, which lies 50 km south of the Park. The average annual rainfall for Perth is 866 mm with an average of 120 raindays per year. The Leeuwin–Naturaliste Ridge experiences more rain-days per year, the annual average ranging from 154 days at Cape Naturaliste to 183 days at Cape Leeuwin. The average annual rainfall for the Leeuwin–Naturaliste Ridge varies from 823 mm at Cape Naturaliste to 1001 mm at Cape Leeuwin (data from the Bureau of Meteorology, 1995).

The topography within the Yanchep National Park ranges from 10 to 80 m above sea level, the lowest elevations occurring about Loch McNess (Fig. 15.2B), a lake fed by groundwater. In the Yanchep National Park, the uppermost Tamala limestone, 10 to 20 m

Fig. 15.3. Tuart (*Eucalyptus gomphocephala*) forest above Carpark Cave (YN 18), shortly after a fire in Yanchep National Park, showing partially exposed Tamala limestone development over the cave (by Edyta Jasinska).

Fig. 15.4A. Aquatic tuart root mats. Fringing the epiphreatic stream in Cabaret Cave (YN 30) (by Tadeusz Jasinski).

thick, rests on the siliceous Bassendean sands (Playford et al., 1976; Kendrick et al., 1991). The occurrence of calcified tree roots suggests that forests covered the calcareous dunes during the formation of the limestone, and Jennings (1968) argued that they had an important role in the process of syngenetic karst formation – as points of active dissolution and reprecipitation of carbonates. The limestone is not highly concreted, and cave collapses are a common occurrence. The underlying sands contain a major groundwater reservoir, the Gnangara Mound (Fig. 15.2D).

The Gnangara Mound is an extensive unconfined aquifer, its highest point reaching 70 m above sea level east of Yanchep National Park (Allen, 1981). Its water-table levels depend predominantly on a balance between recharge through winter rainfall, groundwater abstraction for human uses and evapotranspiration, as well as discharge from springs, especially on the eastern flanks of the aquifer (Davidson, 1995). In the area of the Yanchep National Park, the water table of the Gnangara Mound coincides with the boundary between the calcarenite and the underlying sands, where it forms a system of shallow, epiphreatic streams. A number of these streams [e.g., those in caves YN 18, YN 27, YN 30, YN 99 and YN 194 (Fig. 15.2B)] are permanent and, for the most part, only 2 cm deep but with deeper channels, up to 20 cm, along the banks and in the narrowest stream sections. The overall direction of drainage in this part of the Gnangara Mound is westerly. Along these groundwater streams there is an extensive development of caves. Nevertheless, the caves in the Yanchep National Park are not large, having a vertical range of less than 20 m, and the most extensive, Crystal Cave (YN 1), comprises only 340 m of passages (Matthews, 1985). The caves with aquatic root mats are particularly small, the length of accessible stream chambers ranging from 3 to 25 m.

The karst terrain within the Yanchep National Park is

Fig. 15.4B. Aquatic tuart root mats. Close-up of the root mats (by Stan Hopwood).

largely covered by a dry, sclerophyllous forest. Of particular relevance to the cave ecosystems is the growth of tuart trees (*Eucalyptus gomphocephala*). Where, within the Park, there is little soil development over the limestone, tuart trees (Fig. 15.3) stand about 20 m tall, dominant or co-dominant with the smaller tree candle banksia (*Banksia attenuata*). In five shallow caves [Carpark Cave (YN 18), Gilgie Cave (YN 27), Cabaret Cave (YN 30), an unnamed cave (YN 99), and Twilight Cave (YN 194)], where the epiphreatic waters flow only 10 to 15 m below the ground surface, dense aquatic root mats fringe the stream edges (Figs. 15.4A,B), typically extending up to 15 cm into the stream. Apart from the roots of tuart trees no other photosynthetic plants, or parts thereof, grow in the cave waters, even though small sections of some of the streams are exposed to low levels of light filtering through collapsed sections of cave roofs (E.J. Jasinska, unpublished). This is probably because such cave streams are fast-flowing and there are few stable surfaces suitable for periphyton establishment. Furthermore, these cave streams are produced by ground water issuing up from the stream bed all along the water course (Jasinska and Knott, 1991) and consequently are unlikely to carry spores of photosynthetic plants such as ferns, mosses, liverworts or algae which might be able to grow under twilight conditions.

The caves of the Leeuwin–Naturaliste Ridge are developed in the superficial, mid-Pleistocene Tamala limestone which, in this region, achieves its greatest elevations of over 220 m above sea level and, in places, a thickness of more than 150 m. The cave streams flow at the base of the limestone sometimes directly over the Precambrian rocks but, more often, over a sandy or lateritic layer. The cave streams form part of a westward-flowing drainage system and are either of groundwater origin or, especially in the case of temporary streams, a continuation of surface creeks which flow into the karst. Cave streams which originate directly from surface creeks are typically tea-coloured due to high tannin content (e.g., WI 49, WI 51), while groundwater cave streams are invariably colourless (e.g., WI 9, WI 63). Many caves have developed along the subterranean water courses, and the caves overall are much larger [e.g., Easter Cave (AU 14) has over 7 km of passages surveyed and a vertical range of 35 m], more stable and better decorated than those at Yanchep, probably as a result of the greater thickness of the limestone layer and wetter climate which results in greater calcification of the aeolian calcarenite. Again, the occurrence of calcified tree roots indicates that forests covered the calcareous dunes during the formation of the limestone.

The cave streams of the Leeuwin–Naturaliste Ridge

Table 15.1
Physicochemical comparison of cave waters from Yanchep National Park (sampled in August 1994) and the Leeuwin–Naturaliste Ridge (sampled in August 1995)

Caves	Relative cations[1] (%)	Total cations[2] (mg ℓ^{-1})	Water temperature (°C)	pH
YANCHEP NATIONAL PARK				
Cabaret Cave (YN 30)	Ca (49) > Na (41) > Mg (8) > K (2)	109.3	19.0	6.7
Carpark (YN 18)	Ca (57) > Na (35) > Mg (7) > K (1)	163.6	18.7	7.1
Gilgie (YN 27)	Ca (57) > Na (35) > Mg (7) > K (1)	186.0	18.1	7.1
Twilight (YN 194)	Ca (58) > Na (34) > Mg (7) > K (1)	193.7	18.0	7.3
Cave YN 99	Ca (48) > Na (42) > Mg (8) > K (2)	122.7	17.3	6.8
LEEUWIN–NATURALISTE RIDGE				
Calgardup Cave (WI 49)	Na (74) > Ca (16) > Mg (9) > K (1)	153.9	14.0	5.3
Easter Cave (AU 14)	Na (73) > Ca (19) > Mg (6) > K (2)	632.2	16.5	7.2
Kudjal Yolgah (WI 9)	Ca (50) > Na (42) > Mg (7) > K (1)	141.4	15.6	6.9
Northcote Grotto (YA 2)	Na (84) > Mg (9) > Ca (5) > K (2)	357.2	15.6	6.0
Strongs Cave (WI 63)	Na (50) > Ca (44) > Mg (5) > K (1)	158.9	15.5	6.8

[1] Relative concentrations of major cations (%).
[2] Total concentration of the major cations (mg ℓ^{-1}).

are up to 1 m deep. However, the depth of cave streams with root mats in the Leeuwin–Naturaliste Ridge varies by 10 cm or more annually in three out of the five caves sampled. These pronounced water-level fluctuations contrast with the situation in the shallow cave streams (with root mats) of the Yanchep National Park where the water levels usually vary by less than 2 cm during the year (Jasinska and Knott, 1991; unpublished data from the Western Australian Water and Rivers Commission). Furthermore, in the southernmost cave (Easter Cave) especially, there has been a progressive drop in water levels since 1965, by 1.5 m during the 1970s (measurements recorded by the Western Australian Speleological Group); at the last visit in 1995, it was at an all-time low. Unlike the Yanchep cave streams which are all derived from one source (the Gnangara Mound), the five groundwater streams [Easter Cave (AU 14), Strongs Cave (WI 63), Kudjal Yolgah (WI 9), Calgardup Cave (WI 49) and Northcote Grotto (YA 2)] sampled on the Leeuwin–Naturaliste Ridge appear to originate from different sources (only WI 63 and WI 9 possibly sharing the same source), based on water-quality information (Table 15.1).

Presently, where land has not been developed or cleared for agriculture on the Leeuwin–Naturaliste Ridge, dry sclerophyllous forests still cover much of the karst. In the southern cave areas of the Ridge, karri

Fig. 15.5. Karri (*Eucalyptus diversicolor*) forest above Strongs Cave (WI 63), Leeuwin–Naturaliste Ridge (by Edyta Jasinska).

Fig. 15.6. Detail of aquatic karri root mat from a pool 50 cm deep in Easter Cave (AU 14) (by Edyta Jasinska).

(*Eucalyptus diversicolor*) forms tall forests with trees reaching heights of 70 to 90 m (Fig. 15.5). The shorter trees of marri (*Corymbia calophylla*) and peppermint (*Agonis flexuosa*) also occur in karri forests. In the central and northern cave areas of the Ridge, mixed forests of marri, jarrah (*Eucalyptus marginata*), peppermint, bull banksia (*Banksia grandis*), candle banksia (*B. attenuata*), and she-oak (*Casuarina fraseriana*) stand 10 to 30 m tall. The deepest caves with aquatic root mats where the streams flow at depths to 35 m underground (Matthews, 1985; E.J. Jasinska, unpublished), are found beneath the karri forests. In the shorter mixed forests, aquatic root mats have been observed at depths of 10 to 18 m underground (Matthews, 1985; E.J. Jasinska, unpublished). Only a few caves on the Ridge contain submerged roots; these are produced by karri, marri or peppermint. Karri (Fig. 15.6) and peppermint produce root mats which are looser in structure than the mats in the Yanchep National Park, but which typically fill sections of the stream channel. The marri root mats are distinctive in resembling long tassels hanging into the water in discrete loose bunches, which retain practically no root debris because of the very limited formation of short lateral rootlets. All of the aquatic root mats sampled on the Leeuwin–Naturaliste Ridge occur in the dark zones of caves.

Root mat structure

Healthy root mats in caves at Yanchep National Park are compact structures composed of finely branching rootlets typically less than 2 mm in diameter. The rootlets all grow in mycorrhizal association with fungi (Jasinska, 1990). The outer surface of the rootlets appears smooth, there are neither root hairs nor obvious mycelial strands or tufts (which would be expected for mycorrhizal roots) (Fig. 15.7). The spaces between the densely interwoven rootlets are packed with fine detritus originating from faeces of root-grazers, sloughed-off outer layers of rootlets, and other mycorrhizal fragments.

Histology of the tuart rootlets (Jasinska, 1990) revealed that they were ecto-endomycorrhizal with a number of fungi – Zygomycetes, Ascomycetes and Basidomycetes – the hyphae of which are frequently

Fig. 15.7. Microscopic structure of root mats from the stream in Cabaret Cave showing smooth mycorrhizal rootlets and fine root debris within the root mat (by Edyta Jasinska).

interwoven, intercellularly, within the same area of a rootlet. Some short lateral rootlets were enclosed by fungal mantles (Fig. 15.8A), while others supported mainly inter- and intra-cellular fungal growths (Figs. 15.8B–D). Free mycelial growth extending out from the roots was uncommon, but occasionally did occur especially within dying root mats.

All mycorrhizal rootlets (only some new shoots and thick mother roots were non-mycorrhizal) contained heavy deposits of tannins, both in the hyphae and in the outer cortical and epidermal root cells. It is likely that fungi contribute nutritional value to the tree rootlets, but the presence of fungi also coincides with heavy tannin deposition. Tannins tend to impede digestion in animals in at least three ways: by blocking the action of digestive enzymes; by binding to proteins being digested; and/or by interfering with protein activity in the gut wall (Howe and Westley, 1988). Nevertheless, the root-grazing cavernicoles are clearly able to cope with a high tannin diet, although the mechanisms by which they achieve this remain to be elucidated.

The morphology of roots in cave waters on the Leeuwin–Naturaliste Ridge varies both with the tree species involved and with the depth and/or flow rate of the stream (Figs. 15.9A–D). In the deeper (0.5 m or more in depth) slow-flowing or still cave waters, the root mats tend to have a very open structure which traps little organic debris (Fig. 15.6); also, the individual rootlets are longer and more sparsely branched than in root mats of the same species which grow in shallower and/or faster-flowing waters. In sections of Strongs Cave extensive fungal mycelia of various textures and colours co-occur with karri root mats (Figs. 15.10A,B). The mycelia may be used for food by some of the fauna, but we have not yet observed any animal associations specifically with the fungi.

Fauna

The most diverse and abundant aquatic fauna in any of the caves studied to date in southwestern Australia (Jasinska and Knott, 1991; Burt, 1982; Jasinska, unpublished data) occurs in the aquatic root mats of the Yanchep National Park caves (Fig. 15.11). The aquatic fauna of Cabaret Cave (Table 15.2), being the best known, is the focus of the present account. The groundwater fauna of Australia has been poorly studied to date, so that there is a dearth of taxonomic knowledge of these animals; consequently the identification to species and descriptions of many taxa relevant to this study have yet to be completed.

The largest aquatic cavernicoles in Cabaret Cave attain body lengths of just over 10 cm, and include the carnivorous night fish (*Bostockia porosa*: Percichthyidae) and *Cherax quinquecarinatus* (Decapoda: Parastacidae), an omnivorous crayfish known locally as the gilgie. Predatory leeches (Hirudinea: Erpobdellidae), about 5 cm long when extended, are often

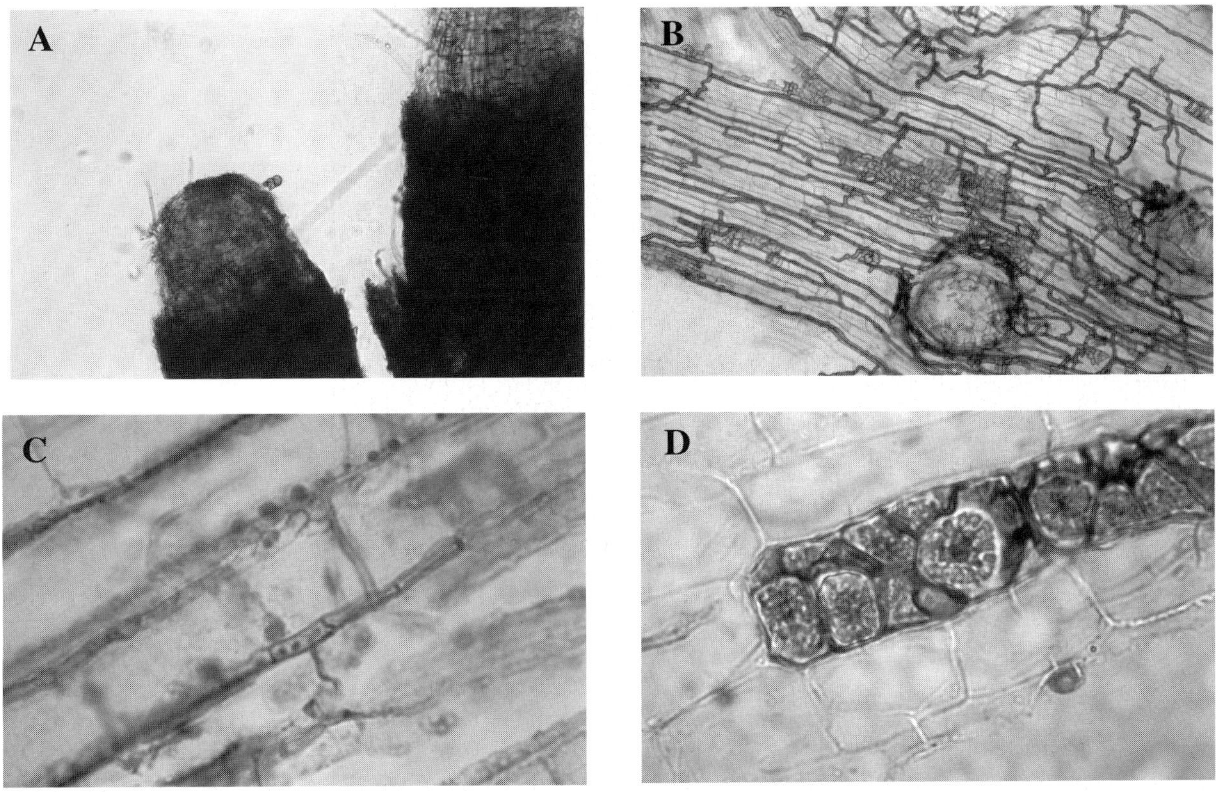

Fig. 15.8. Tuart mycorrhizae from Cabaret Cave stream cleared in potassium hydroxide and stained with Pianese IIIB (stains tannin-free fungi pink); in the preparation, fungal hyphae appear brown due to accumulation of tannins. (A) Rootlets both enclosed by fungal mantles and penetrated by hyphae; (B) whole mount of a rootlet showing fungal-hyphae forming pseudoparenchyma within the cortex; (C) intercellular cortical hyphae with intracellular projections (?haustoria); (D) intracellular (root cortex) fungal structures (?spores) (by Edyta Jasinska).

seen swimming in the open water, often along the root mats. Occasionally, predatory amphipods *Perthia* sp. (Perthidae), reaching a body length of nearly 1 cm, are collected from the root mats; they are nonpigmented and with rudimentary eyes. The remaining fauna comprises animals less than 1 cm in length. Twenty two species inhabit the root mats, with total animal densities usually exceeding 200 individuals per gram dry weight of root mat. The janirid isopods, four species of acarines, gomphodellid ostracods, naidid oligochaetes, the occasional trichopteran larva (in a case constructed from rootlets) and minute, Collembola-like hexapods (rare), are essentially non-swimming species which crawl on the submerged rootlets. The amphipod *Austrochiltonia subtenuis* (Ceinidae), cyclopoid copepods (*Paracyclops* sp. nov.), harpacticoid copepods, dytiscid beetle larvae (*Sternopriscus* sp.), each of the five species of microturbellarians [*sensu* Schwank (1986)], aeolosomatid oligochaetes and nematodes will swim short distances from one rootlet to another within the root mat, but more often they are observed to move over the rootlets. Chironomid larvae [*Polypedilum* (*Pentapedilum*) sp.] use the root detritus to build their cases against the rootlets.

The remaining aquatic cavernicoles include crangonyctoid amphipods (*Hurleya* sp.: Paramelitidae), which, together with nematodes, cyclopoid copepods, at least two species of rotifers and some protozoans, reside in shallow miniature puddles present in some places on the shore side of the root mat. The puddles have a bottom of loose, silty root detritus, and a few, finely-branched living rootlets all covered by less than a centimetre of water; when disturbed, the amphipods "dive" into the loose silty sediments. Similar deposits in deeper, quiet sections of the stream contain nematodes and cyclopoid copepods.

In the open water there are only small numbers of animals: sweeps of the length of the stream with a 90 μm-mesh net typically capture only 2 to 5 cyclopoid copepods. The interstices of the sandy

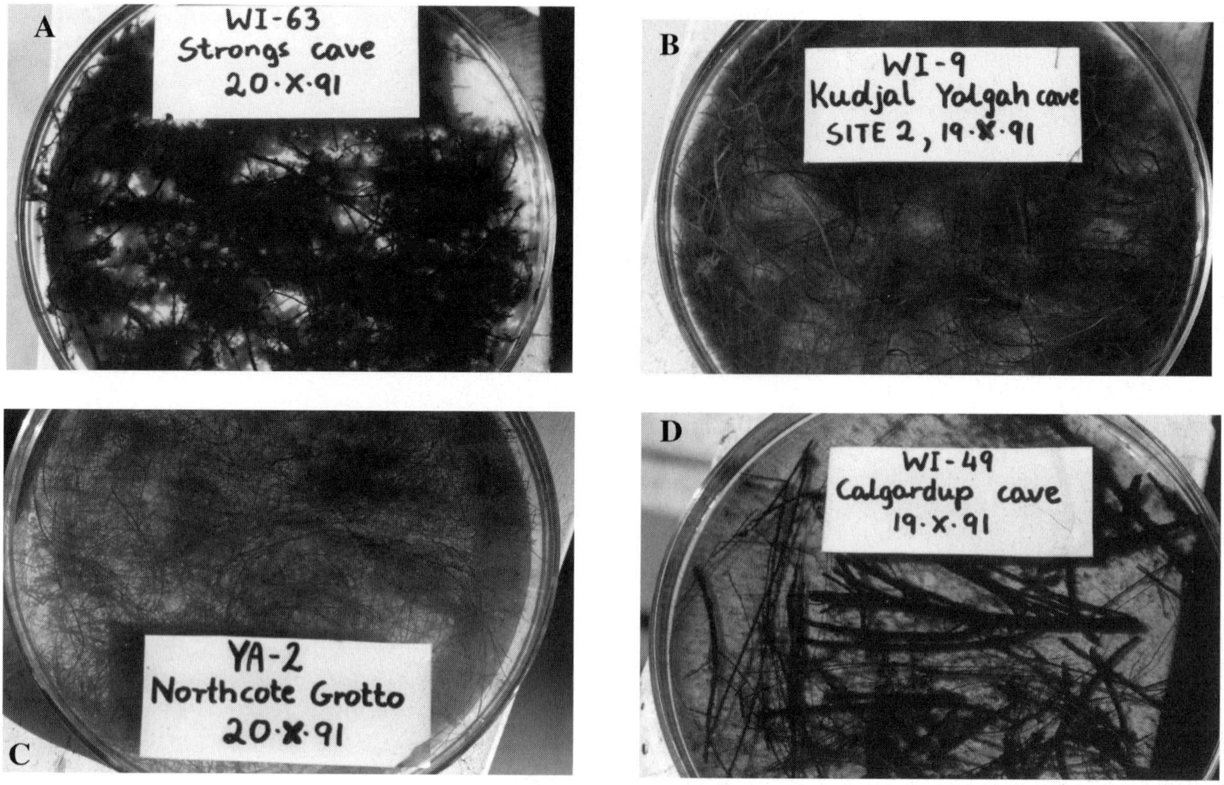

Fig. 15.9. Gross morphology of root mats in cave waters on the Leeuwin–Naturaliste Ridge. (A) Karri mycorrhizae with dark mycelial tufts from nearly still, deep (~40 cm) section of Strongs Cave stream; (B) smooth karri mycorrhizae from a shallow (~10 cm) fast-flowing stream in Kudjal Yolgah cave; (C) smooth peppermint tree mycorrhizae from a slow-flowing stream in Northcote Grotto; (D) Marri mycorrhizae showing very little fine branching, from deep (~60 cm) nearly still waters in Calgardup Cave (by Edyta Jasinska).

stream bottom contain acarines (*Lobohalacarus weberi*), nematodes, turbellarians (*Macrostomum* sp. 1), oligochaetes, harpacticoids, bathynellid syncarids and Collembola-like hexapods. Unless the sand contains a large proportion of root detritus, animals are sparsely distributed in the stream bed. Of six core samples of 50 cm^3 taken from the top 4 cm of the stream bed, five contained fewer than 10 animals; but one core, which included unusually large quantities of root detritus, contained 195 animals (193 being nematodes). Only the bathynellid syncarid, although rare in the sediment cores, appears to be restricted to the interstitial habitat of the sandy stream bottom.

Cabaret Cave has a well-developed trophic system which includes primary consumers, predators and decomposers (Jasinska et al., 1996). In studies during the past five years, we have established that: the amphipod *A. subtenuis* feeds primarily by grazing on living rootlets, though not on the new pale or white growth; the gilgie, *Cherax quinquecarinatus,* also feeds on the roots but inadvertently collects the microinvertebrates together with the root mat; the janirid isopods, the amphipod *Hurleya* sp., gomphodellid ostracods, chironomid larvae and possibly the trichopteran larvae (Williams, 1980) feed either on the outermost layers of the root epidermis (which is otherwise sloughed off) or on fine organic detritus. All species of mites found in root mats appear to feed either on fine organic detritus or on the root sap, and nematodes of at least one type are often found with their bodies rigid and the stylet deeply embedded in rootlets. The five species of microturbellarians are predatory, and have been observed to ingest juvenile ostracods, chironomid larvae, rotifers and, in Carpark Cave (YN 18), tardigrades. The dytiscid larvae (adults are very rarely seen) may be expected to be predators on microinvertebrates in general (Williams, 1980). The night fish and the leech feed on small invertebrates. Rotifers, cyclopoid copepods and the nematode species which lack a stylet, and are commonly found in association with

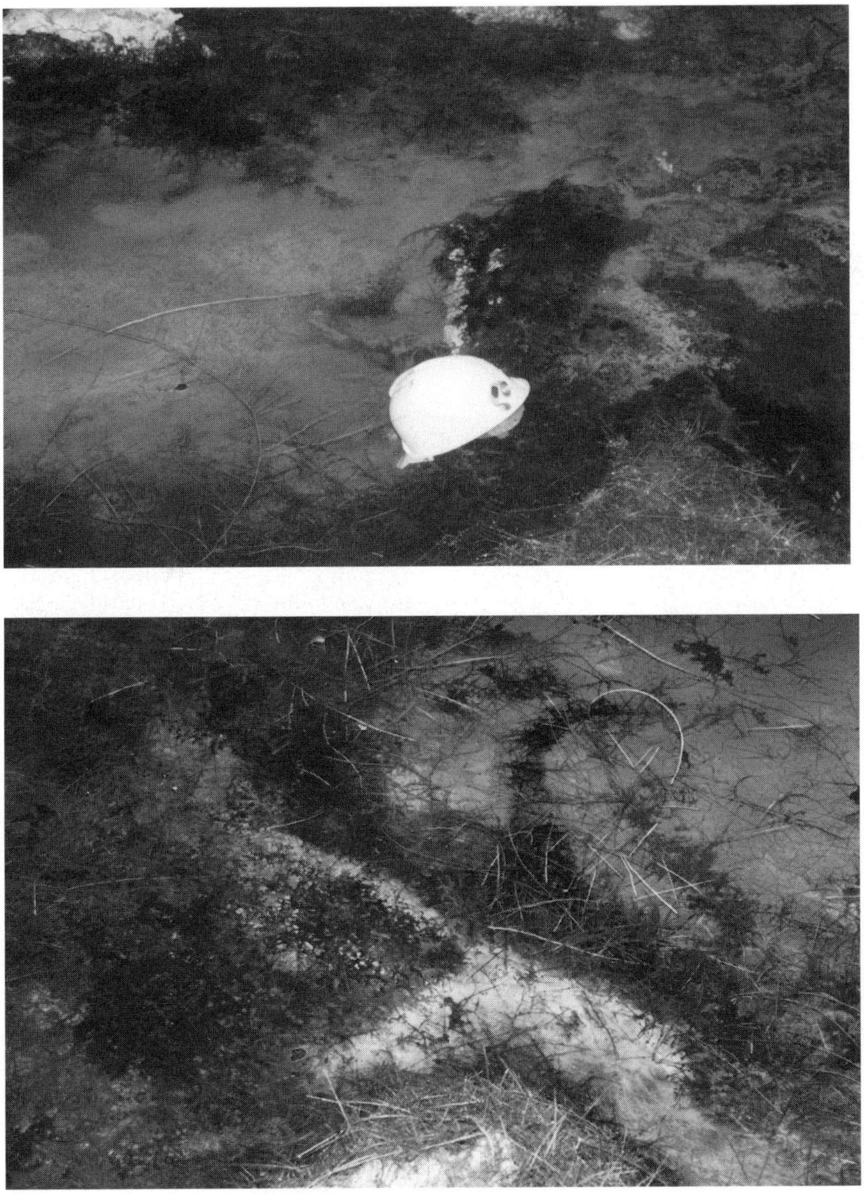

Fig. 15.10. Mycelia amongst karri root mats in slow-flowing sections of Strongs Cave. (top) View across the Strongs Cave stream with a half-submerged helmet shown for scale; (bottom) close-up of the root mat and mycelia (by Edyta Jasinska).

fine organic detritus, most likely feed on bacteria and protozoans which, in turn, probably break down some components of the organic detritus. Both the aeolosomatid and naidid oligochaetes, together with protozoans, are typically found "cleaning out" the inside of dead isopods, amphipods or insect larvae. The syncarids and interstitial harpacticoids most likely all feed on bacteria adhering to sand grains, since they are more common in sands with minimal amounts of organic debris. Such bacterial growths are likely to occur only in small amounts, since analyses by High-Pressure Liquid Chromatography (HPLC) of Cabaret water, with a limit of detectability, 1 nmol ℓ^{-1}, failed to detect any dissolved amino acids, sugars or organic acids (Jasinska, 1990; Jasinska et al., 1996). Tannins are present in detectable quantities levels in the Cabaret stream, but such compounds are utilized extremely slowly as food by bacteria (Münster and Chróst, 1990).

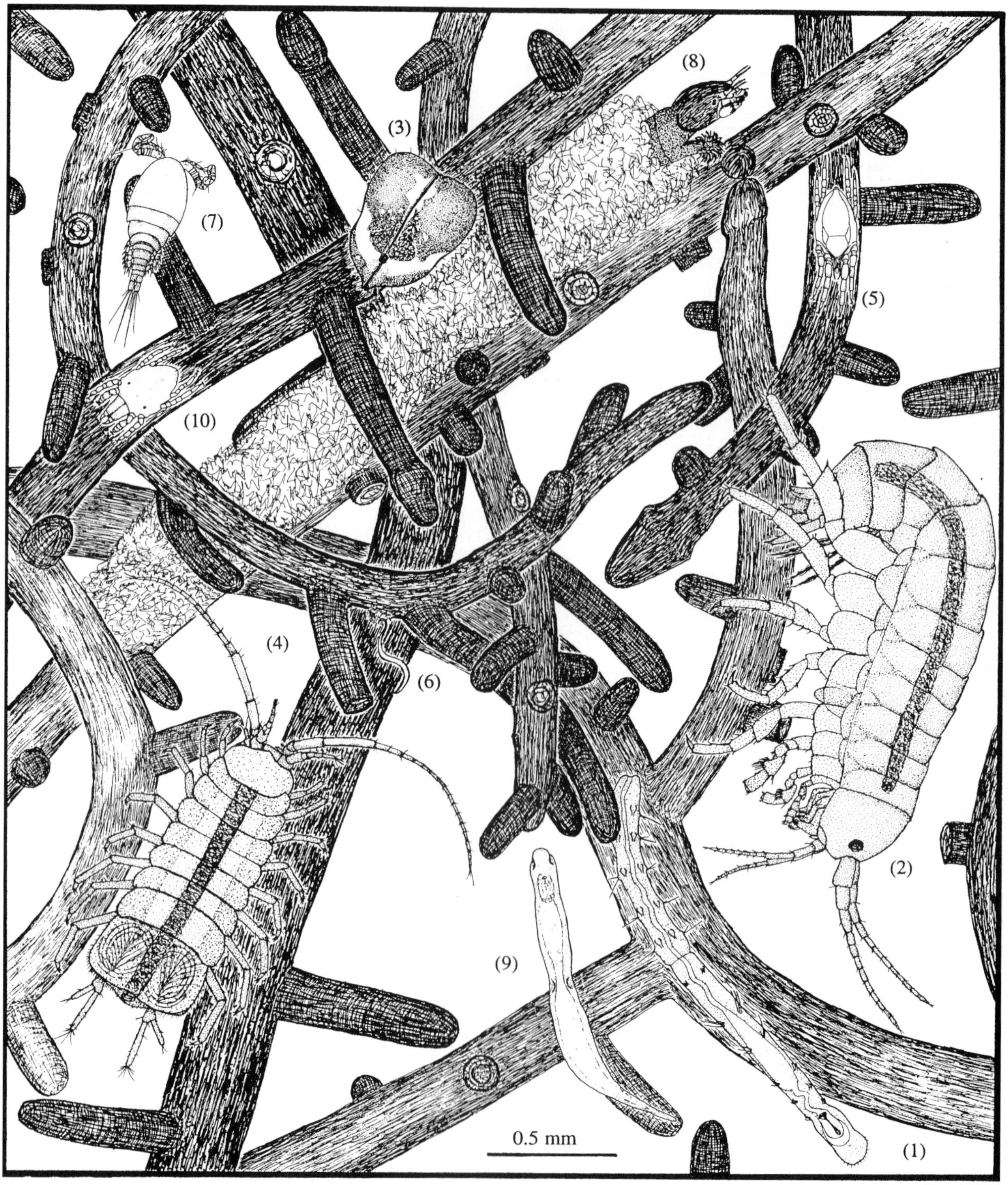

Fig. 15.11. Aquatic cavernicoles within root mats in Cabaret Cave stream. Fine root detritus which normally fills the smaller spaces in tuart root mats has been omitted for better clarity. (1) Aeolosomatid oligochaete; (2) *Austrochiltonia subtenuis* (female); (3) *Gomphodella* aff. *maia* (female); (4) janirid isopod (male); (5) *Lobohalacarus weberi*; (6) nematode grasping fungal hyphae with its tail; (7) *Paracyclops* sp. (male); (8) *Polypedilum* (*Pentapedilum*) sp.; (9) *Stenostomum* sp.; (10) *Soldanellonyx monardi* (by Edyta Jasinska).

Table 15.2
Aquatic fauna of the Cabaret Cave stream, Yanchep National Park

Animal taxa	Morphology
VERTEBRATES	
OSTEICHTHYES: **Percichthyidae**	
Bostockia porosa	pigmented, eyes present
INVERTEBRATES	
ANNELIDA	
HIRUDINEA: **Erpobdellidae**	
sp. 1	pigmented, eyes present
OLIGOCHAETA: **Aeolosomatidae**	
sp. 1	non-pigmented, eyeless
Naididae	
Pristina sp.	non-pigmented, eyeless
Pristina longiseta	non-pigmented, eyeless
Phreodrilidae	
sp. 1	non-pigmented, eyeless
Tubificidae	
?*Antipodrilus* sp.[1]	non-pigmented, eyeless
ARTHROPODA	
CHELICERATA: ACARINA	
Halacarida	
Lobohalacarus weberi	non-pigmented, eyeless
Soldanellonyx monardi	non-pigmented, eye spots present
Oribatida	
sp. 1	body and appendages tinged faint brown, eye spots present
sp. 2	body and appendages dark brown, eye spots present
sp. 3	body and appendages dark brown, postero-dorsal body margin silvery, eyes present
CRUSTACEA: AMPHIPODA	
Ceinidae	
Austrochiltonia subtenuis	reduced pigmentation: faint red pigment dorsally, rest non-pigmented, eyes present
Paramelitidae	
Hurleya sp.	non-pigmented, eyeless
Perthidae	
Perthia sp.[1]	non-pigmented, eyeless
COPEPODA: **Cyclopoida**	
Paracyclops sp. nov.	non-pigmented, eye distinguishable but non-pigmented
Eucyclops linderi	non-pigmented, eye distinguishable but non-pigmented
Harpacticoida	
sp. 1	non-pigmented, eye if present then non-pigmented
DECAPODA: **Parastacidae**	
Cherax quinquecarinatus	lacking patterns present in epigean specimens from nearby populations, individuals range from pigmented to non-pigmented, eyes present

continued on next page

Table 15.2, *continued*

Animal taxa	Morphology
ISOPODA: **Janiridae**	
gen. and sp. nov.	non-pigmented, eyeless
OSTRACODA: **Darwinulidae**	
Darwinula sp.	non-pigmented, eye present
Limnocytheridae	
Gomphodella aff. *maia*	reduced violet/brown pigmentation, asymmetrical patterns, eye present
SYNCARIDA: **Bathynellidae**	
sp. 1	non-pigmented, eyeless
INSECTA	
entognathous hexapod sp.	non-pigmented, eyeless
DIPTERA: **Chironomidae**	
Polypedilum sp. (larvae)	head capsule brown, body red from haemoglobin, eyes present
Tipulidae	
sp. 1 (larvae)	colourless or slightly tinged yellow, eyeless
COLEOPTERA: **Dytiscidae**	
Sternopriscus sp.	pigmented, eyes present
TRICHOPTERA: **Leptoceridae**	
sp. 1 (larvae) [1]	faintly pigmented, eyes present
NEMATODA	
six distinct species	non-pigmented
PLATHELMINTHES	
TURBELLARIA: **Catenulida**	
Stenostomum sp.	non-pigmented, eyeless
sp. 1	non-pigmented, eyeless
MACROSTOMIDA	
Macrostomum sp. 1	non-pigmented, eye spots present
Macrostomum sp. 2	non-pigmented, eye spots present
TYPHLOPLANOIDA	
sp. 1	non-pigmented, eyeless
ROTATORIA	
BDELLOIDEA	
sp. 1	non-pigmented, eyeless
MONOGONONTA: PLOIMA	
sp. 1	non-pigmented, eyeless

[1] Only rarely encountered.

A possible food web relating the aquatic fauna of Cabaret Cave is presented in Fig. 15.12.

Although some inhabitants of the aquatic root mats in Cabaret Cave feed directly upon the rootlets, others do not; these may use the root mat mainly as a solid substrate within which they can forage and shelter. This possibility was investigated by observing the degree of colonization of artificial root mats (made from tenikalon, a split fibre of polyethylene tape) compared to living root mats of similar dimension

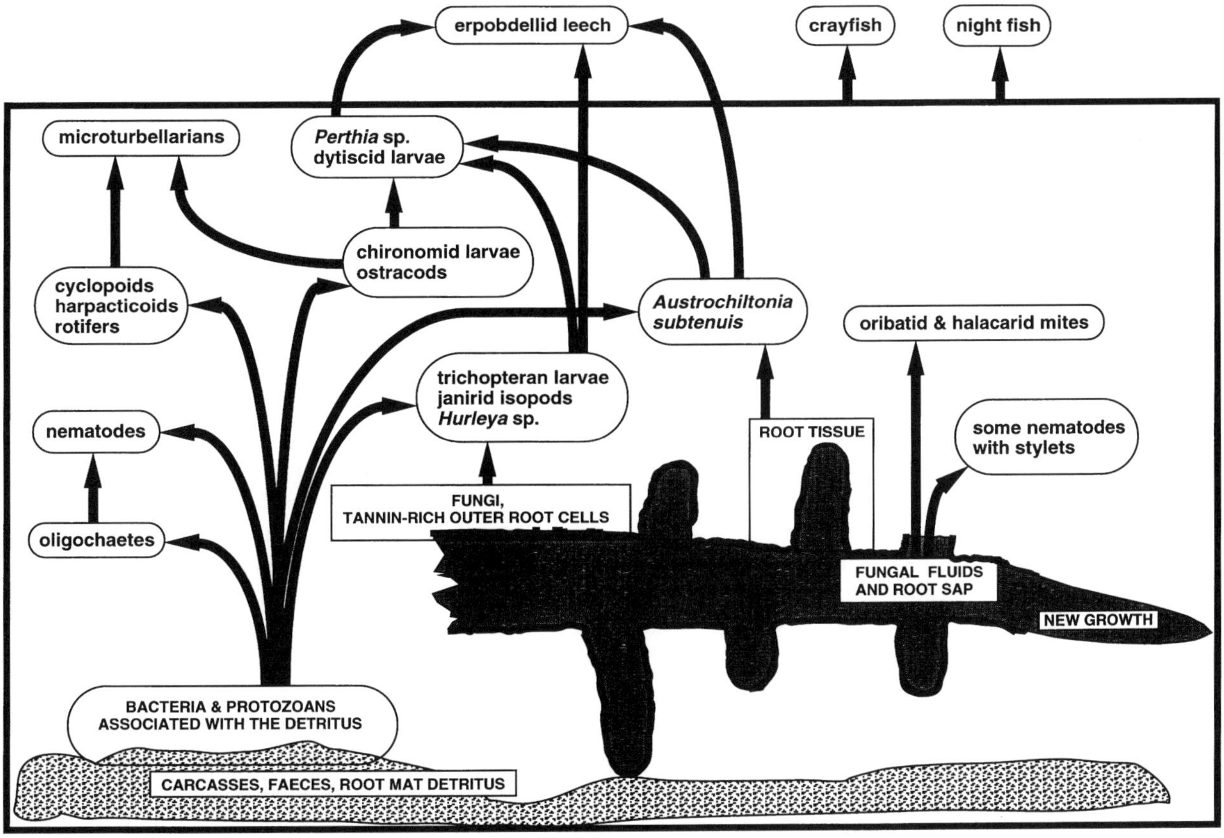

Fig. 15.12. Food web of the fauna from the stream in Cabaret Cave, Yanchep National Park; the interstitial fauna is not included. All root-mat animals together with debris from the mycorrhizal rootlets contribute to the root-mat detritus. The pattern of tannin deposition within the rootlets is highly variable but most deposition occurs in the outer epidermal layers; minimal mycorrhizal development occurs at the growing tip of rootlets (by Edyta Jasinska).

cleared of animals and fine root debris. The natural and artificial root mats were of similar dimensions, but the artificial root mats had much greater dry weight so the abundances of animals were not standardized per unit dry weight of material. After three months, a total of 14 species of animals had recolonized the living rootlets and, of these 14 species, 8 colonized the artificial root mats (Table 15.3). Both the natural root mats and the artificial root mats were colonized by large numbers of nematodes, which are usually fairly uncommon in intact root mats. Because of their great numbers, nematodes are excluded from the following abundance figures. In the living root mats, the numbers of animals ranged from 10 in the smallest root mat (0.135 g dry weight) to 396 in the largest root mat (1.9 g dry weight). In the artificial root mats the numbers of animals ranged from zero for the smallest substrate to 34 for the largest substrate. The artificial root mats trapped considerable amounts of root-mat pieces and fine-root debris. The crayfish *Cherax quinquecarinatus*, when feeding in Yanchep caves, breaks off many pieces of root mat and these, together with their invertebrate fauna, are carried by the stream current; this debris would have constituted a major source of the animals and root mat material that settled on the artificial substrates. The results support the idea that animals colonize tree roots mainly to obtain food rather than for shelter or to maintain their position in the stream. The fauna which remained on the artificial substrate may have utilized the entrapped pieces of root mat and/or the fine root debris as their main sources of nutrients.

The other four caves at Yanchep which contain aquatic root mats (Gilgie, Carpark, Twilight and YN 99) are also inhabited by similar species assemblages, the animals always occurring in great abundance within the root mats – typically about 350 animals per gram dry weight of root material, which is equivalent to a fistful of fresh root mat. Curiously, no two caves have the same total assemblage of species; only a limited number of species are found in all the caves: the janirid isopods, halacarid acarines, the ostracod *Gomphodella* aff. *maia,* the turbellarian *Stenostomum* sp., one species of aeolosomate oligochaetes, some of the nematode species, and the crayfish *C. quinquecarinatus.* The aquatic cavernicoles include pigmented and eyed species which have surface conspecifics (Table 15.2) such as *G.* aff. *maia, C. quinquecarinatus,* and *Austrochiltonia subtenuis*; non-pigmented species which also occur in springs on the eastern side of the Gnangara Mound, such as *Paracyclops* spp. and halacarid acarines; and also species restricted to the epiphreatic waters at Yanchep which exhibit classical troglobitic features (elongate antennae, lack of eyes and pigmentation), such as the janirid isopods and *Hurleya* sp., together with two other undescribed troglobitic amphipods.

Table 15.3
Fauna settling on natural root mats (previously cleaned of animals) and artificial root mats in the Cabaret Cave stream over a period of three months

Tenikalon "roots"	Natural roots
Austrochiltonia subtenuis	**acarines** [1]
harpacticoids	*Austrochiltonia subtenuis*
janirid isopods	**entognathous hexapod**
microturbellarians	***Gomphodella* aff. *maia***
naidid oligochaetes	harpacticoids
nematodes	janirid isopods
Paracyclops sp.	**leptocerid trichopteran larva**
Polypedilum sp.	microturbellarians
	naidid oligochaetes
	nematodes
	Paracyclops sp.
	Polypedilum sp.
	***Sternopriscus* sp. larvae**
	tipulid larva

[1] Taxa recovered only from one type of substrate are in bold type.

At present we see no obvious pattern to the distribution of animals within Yanchep cave waters. This lack of obvious pattern is perhaps surprising since all the caves receive water from the same source, the Gnangara Mound; all are in close proximity to one another; all lie within the same geomorphic unit; and all contain tuart root mats. Part of the reason behind the lack of obvious distribution patterns may be multiple invasions of the cave streams starting in the mid- to late-Pleistocene during the karst syngenesis, especially during the wetter interglacials in southwestern Australia (Hopper, 1979). Aquatic animals then may have been able to move from the Darling Scarp on to the Swan Coastal Plain. In fact, the genus *Hurleya* was described from a specimen of *H. kalamundae* collected from groundwaters on the Darling Scarp (Straskraba, 1966). Some inhabitants of the cave root mats, such as the halacarid acarines, appear to have originated from the Gnangara Mound interstices, since we have collected the same species from groundwater bores further east on the aquifer. The pigmented aquatic cavernicoles which have surface conspecifics in nearby Lake Yonderup and Loch McNess must have invaded the caves relatively recently; allozyme analysis of the genetic structure of and surface populations of *Austrochiltonia subtenuis* revealed that the cave population has high genetic diversity, comparable with that of the epigean lake populations, and is not genetically distinct from the local surface conspecifics (Stuckey, 1991). Nevertheless, the animals show morphological and biological divergence; the cave animals have a smaller eye area, grow to smaller body size, have very little pigmentation, and produce fewer but larger eggs than their surface conspecifics (Burt, 1982; Stuckey, 1991). Similar observations were made on *Gammarus minus* in North American caves by Fong (1989) and Jones et al. (1992), for example.

Presently, there is no direct surface drainage into any of the Yanchep caves considered here, and the only way aquatic fauna could move from epigean lakes into the caves would be to travel upstream, either interstitially or through rock piles, to reach the particular caves. Therefore, of the aquatic Yanchep cavernicoles, only the flying insects might not complete their life cycle within the cave environment. Ceratopogonid larvae, for example, appear in the root mats of three of the Yanchep caves for short periods of the year only, although the caves with chironomid larvae also seem to support sizeable populations of adult chironomids, and the larvae are always present in the root mat samples, regardless of the season.

On the Leeuwin–Naturaliste Ridge, the cave waters with root mats [Easter Cave (AU 14), Strongs Cave (WI 63), Kudjal Yolgah (WI 9), Calgardup Cave (WI 49)] also support rich faunas. The single exception is Northcote Grotto (YA 2), where root mat samples yielded no animals in 1992 and 1993, and in 1994 and 1995 only tubificid oligochaetes (one species) were obtained from them. In Northcote Grotto, the root mats are formed by the peppermint *Agonis flexuosa,* and we have tested whether the root grazers from Yanchep would also eat these roots. Over a period of three weeks the amphipod *A. subtenuis* and the janirid isopod consumed both tuart and peppermint roots at the same rate. The animals fed peppermint tree roots behaved no differently from those fed tuart roots, and some individuals from each species produced offspring during the three weeks of feeding trials. Northcote Grotto had been a tourist cave, and some of the electrical wiring (for lights) has been ripped out and thrown into the water; judging from the containers found in the cave, it has been subjected to acid, oil and alcohol spills. It may be that the chemical pollution of the cave waters in Northcote Grotto was at one stage great enough to destroy the resident aquatic cavernicoles, or that as a result of some as yet unidentified factors, the environment of Northcote Grotto until recently was unsuitable for the establishment of aquatic faunas.

On the Leeuwin–Naturaliste Ridge the most species-rich cave is Kudjal Yolgah (WI 9) with about fifteen species of invertebrates, including a bathynellid syncarid different from that in Cabaret Cave, and animal abundances in root mats similar to those in Yanchep National Park caves. The tree root mats in Kudjal Yolgah include both karri and peppermint roots. Of all caves on the Leeuwin–Naturaliste Ridge, the system most similar to the Yanchep caves with root mats occurs in Kudjal Yolgah. There the root mats are dense, the stream is only 10–20 cm deep, fast-flowing and less than 1 m across, and the water levels are very stable.

The caves with aquatic root mats on the Leeuwin–Naturaliste Ridge each have an endemic fauna, the crangonyctoid amphipod *Perthia* sp. (a different species from those found in Yanchep caves) being the only species identified as being in common between three of the five caves (AU 14, WI 63 and WI 49) studied on the Leeuwin–Naturaliste Ridge. The animal taxa represented in groundwater streams with root mats in the Leeuwin–Naturaliste caves are: Coleoptera larvae (Helodidae), Amphipoda (Perthidae and Paramelitidae), parastacid Decapoda (*Cherax crassimanus*), Ostracoda, Syncarida (Bathynellacea), Copepoda (Cyclopoida and Harpacticoida), Acarina (Halacarida and Oribatida), Tardigrada, Oligochaeta, Nematoda, Rotifera and Turbellaria. The densities of animals in the karri root mats of the southernmost cave, Easter Cave (AU 14), are low compared to Kudjal Yolgah and the Yanchep caves, and only seven species are present. It seems, therefore, that the presence of aquatic root mats by itself does not necessarily result in the development of highly diverse and abundant faunas.

The groundwater caves of the Leeuwin–Naturaliste Ridge appear to have only one recent colonizer (i.e., animals which are eyed and pigmented), the helodid beetle larvae in Kudjal Yolgah which have small, barely visible eyes and some pigmentation on their head and legs. The same beetle larvae also appear to be the only direct root-grazers in the cave waters of Leeuwin–Naturaliste; the remaining cavernicoles include fungivores, detritivores, scavengers, sap-feeders and predators all of which are rare or absent in cave-stream habitats other than root mats or associated root debris.

It is apposite to know whether one of the factors controlling the distribution of aquatic faunas in the syngenetic caves of southwestern Australia is the differentiated ability to utilize root matter derived from different tree species. To test this possibility, the amphipod *Austrochiltonia subtenuis* and the janirid isopod from Cabaret Cave were placed, for a period of three weeks in either karri or peppermint roots, with water from four of the Leeuwin–Naturaliste caves. Both the amphipods and the isopods survived on the roots and water from the Leeuwin–Naturaliste caves for the duration of the experiment, and some animals of each species produced young. This further supports the observation stated in the Introduction that aquatic cavernicoles evidently can utilize root mats from different tree species.

CONSERVATION

The two cave areas, Yanchep National Park and the Leeuwin–Naturaliste Ridge, are subject to significant threats to their long-term maintenance through their proximity to human activity. The major threat to the persistence of the aquatic cave ecosystems is the progressive lowering of the water table, mainly due to groundwater abstraction for urban, rural and industry needs. The human populations and consequently the demand for water are growing rapidly, both in the Leeuwin–Naturaliste and Perth regions. The Gnangara Mound, for example, currently meets approximately 40% of the freshwater requirements for the rapidly expanding suburbs of Perth; but, as the surface catchments on the Darling Scarp become fully utilized, the groundwater abstraction will increase even more rapidly (Davidson, 1995). Management of the groundwater resource relies heavily on computer modelling (Davidson, 1995); but, so far, the models do not function at a finer resolution than 0.5 m, which clearly is inadequate for the maintenance of epiphreatic streams 2 cm deep. During the last decade especially, many aquatic habitats fed by the Gnangara Mound groundwater either have been altered from permanently wet to seasonally dry or have dried up completely, the water table remaining well below the bottom of some past wetlands even towards the end of winter. These adversely affected aquatic habitats include some wetlands lying over the high points of the Gnangara Mound (e.g., Lake Pinjar); a series of, once permanent, springs on the eastern flanks of the aquifer (Jasinska and Knott, 1994); and a few stream caves at Yanchep National Park, notably Crystal Cave (YN 1), which contains an endemic, undescribed species of troglobitic amphipod (Paramelitidae). Should the "greenhouse" predictions of a continuing decrease in rainfall in the southwest of Australia (Pittock, 1988) be realized, the preservation of the Yanchep National Park and Leeuwin–Naturaliste cave-stream faunas will present a great challenge indeed.

CONCLUSION

From the foregoing it is apparent that tree roots which grow into cave waters constitute an important source of nutrition and shelter for aquatic invertebrates. This substantial and reliable food source has enabled the development of diverse and abundant faunas in the groundwater streams of a number of caves in mid-Pleistocene syngenetic karst of southwestern Australia.

ACKNOWLEDGMENTS

We wish to acknowledge the kind assistance of Prof. A.R. Main who critically reviewed an early draft of this manuscript and provided constructive comments and Dr. W.F. Humphreys and Mr. Stefan Eberhard who kindly provided information. Taxonomic advice from a number of specialists is fully acknowledged in E.J. Jasinska's Ph.D. Thesis. Field studies were facilitated by rangers at Yanchep National Park and a number of cavers, especially Lex Bastian, Norm Poulter, Anne and Peter Wood, Barry Loveday and Rob Foulds. We also acknowledge funding from the Australian National Parks and Wildlife Service through the Endangered Species Programme, Water Authority of Western Australia (organised by Mr. Jeff Kite), Murdoch University, The University of Western Australia and from the South West Development Authority of Western Australia.

REFERENCES

Allan, J.D., 1994. *Stream Ecology, Structure and Function of Running Waters.* Chapman and Hall, London, 400 pp.

Allen, A.D., 1981. Groundwater resources of the Swan Coastal Plain, near Perth, Western Australia. In: R. Whelan (Editor), *Groundwater Resources of the Swan Coastal Plain.* CSIRO, Perth, pp. 29–74.

Archer, M., Clayton, G. and Hand, S., 1984. A checklist of Australasian fossil mammals. In: M. Archer and G. Clayton (Editors), *Vertebrate Zoogeography and Evolution in Australasia.* Hesperian Press, Perth, pp. 1027–1087.

Burt, J.S., 1982. *Aquatic Cavernicoles of the Yanchep Caves.* Unpublished Honours thesis, Department of Zoology, The University of Western Australia, Perth, 64 pp.

Cockbain, A.E., 1990. Perth Basin. Geology and Mineral Resources of Western Australia. *Geol. Surv. West. Aust. Mem.*, 3: 495–524.

Culver, D.C., 1982. *Cave Life: Evolution and Ecology.* Harvard University Press, Cambridge, 189 pp.

Davidson, W.A., 1995. Hydrogeology and groundwater resources of the Perth Region, Western Australia. *Geol. Surv. West. Aust. Bull.*, 142: 1–257.

Davison, E.M. and Tay, F.C.S., 1985. The effects of waterlogging on seedlings of *Eucalyptus marginata. New Phytol.*, 101: 743–753.

Dodd, J., Heddle, E.M., Pate, J.S. and Dixon, K.W., 1984. Rooting patterns of sandplain plants and their functional significance. In: J.S. Pate and J.S. Beard (Editors), *Kwongan, Plant Life of the Sandplain.* The University of Western Australia Press, Perth, pp. 146–147.

Drew, M.C., 1992. Soil aeration and plant root metabolism. *Soil Science,* 154: 259–268.

Fong, D.W., 1989. Morphological evolution of the amphipod *Gammarus minus* in caves: quantitative genetic analysis. *Am. Midl. Nat.*, 121: 361–378.

Gentilli, J., 1972. *Australian Climate Patterns.* Thomas Nelson (Australia) Ltd., Adelaide, 285 pp.

Hobbs III, H.H., 1992. Caves and springs. In: C.T. Hackney, S.M. Adams and W.H. Martin (Editors), *Biodiversity of the Southern United States: Aquatic Communities.* John Wiley and Sons, Inc., New York, pp. 59–131.

Hopper, S.D., 1979. Biogeographical aspects of speciation in the southwest Australian flora. *Annu. Rev. Ecol. Syst.*, 10: 399–422.

Howarth, F.G., 1983. Ecology of cave arthropods. *Annu. Rev. Entomol.*, 28: 365–389.

Howe, H.F. and Westley, L.C., 1988. *Ecological Relationships of Plants and Animals.* Oxford University Press, New York, 273 pp.

Jasinska, E.J., 1990. *Root Mat Ecosystem in an Epiphreatic Stream: Cabaret Cave, Yanchep National Park.* Unpublished

Honours thesis, Environmental Science, Murdoch University, Perth, 170 pp.

Jasinska, E.J. and Knott, B., 1991. *Stability of Root Mat Ecosystems in a Groundwater Stream: Cabaret Cave, Yanchep National Park, Western Australia.* Report to Australian National Parks and Wildlife Service, Endangered Species Project No. 46, 154 pp.

Jasinska, E.J. and Knott, B., 1994. *Aquatic Fauna in Gnangara Mound Discharge Areas of the Ellen Brook Catchment, Western Australia.* Report to the Water Authority of Western Australia, 24 pp.; Appendices 1/2, 22 pp.

Jasinska, E.J., Knott, B. and McComb, A.J., 1996. Root mats in groundwater: a fauna-rich cave habitat. *J. North Am. Benthol. Soc.*, 15: 508–519.

Jennings, J.N., 1968. Syngenetic karst in Australia. In: P.W. Williams and J.N. Jennings (Editors), *Contributions to the Study of Karst.* Department of Geography Publication G/5, The Australian National University, Canberra, pp. 41–110.

Jones, R., Culver, D.C. and Kane, T.C., 1992. Are parallel morphologies of cave organisms the result of similar selection pressures? *Evolution,* 46: 353–365.

Jongsma, D. and Petkovic, P., 1977. The structure of the Naturaliste Plateau and Trough. *APEA J.,* 17: 3–12.

Kendrick, G.W., Wyrwoll, K.H. and Szabo, B.J., 1991. Pliocene–Pleistocene coastal events and history along the western margin of Australia. *Quat. Sci. Rev.,* 10: 419–439.

Matthews, P.G., 1985. *Australian Karst Index.* Australian Speleological Federation Inc., Melbourne, Australia, pp. 6-0–6-29.

Merrilees, D., 1968. Man the destroyer: Late Quaternary changes in the Australian marsupial fauna. *J. R. Soc. West. Aust.,* 51: 1–24.

Münster, U. and Chróst, R.J., 1990. Origin, composition and microbial utilization of dissolved organic matter. In: J. Overbeck and R.J. Chróst (Editors), *Aquatic Microbial Biology: Biochemical and Molecular Approaches.* Springer-Verlag, New York, pp. 8–46.

Notenboom, J., 1981. Amsterdam expedition to the West Indian Islands, Report 12. The new hypogean cirolanid isopod crustaceans from Haiti and Mayaguana Bahamas. *Bijdr. Dierkd.,* 51: 313–331.

Palfreyman, W.D., 1984. Guide to the geology of Australia. *Bur. Miner. Resour. Geol. Geophys. Bull.,* 181: 1–111.

Pittock, A.B., 1988. Actual and anticipated changes in Australia's climate. In: G.I. Pearman (Editor), *Greenhouse, Planning for Climate Change.* CSIRO Division of Atmospheric Research, E.J. Brill, Leiden, pp. 35–51.

Playford, P.E., Cockbain, A.E. and Low, G.H., 1976. Geology of the Perth Basin Western Australia. *Geol. Surv. West. Aust. Bull.,* 124: 1–311.

Popa, R. and Sarbu, S.M., 1991. Trophic structure in an aquatic troglobitic community based upon chemoautotrophic carbon fixation. *Trav. Inst. Speologie "Emile Racovitza",* 30: 51–58.

Sarbu, S.M., 1990. The unusual fauna of a cave with thermomineral waters containing hydrogen sulfide from southern Dobrogea, Romania. *Mem. Biospeol.,* 17: 191–196.

Sarbu, S.M., Kane, T.C. and Kinkle, B.K., 1996. A chemoautotrophically based cave ecosystem. *Science,* 272: 1953–1955.

Schwank, P., 1986. Microturbellaria from subterranean freshwater habitats. In: L. Botosaneanu (Editor), *Stygofauna Mundi a Faunistic, Distributional and Ecological Synthesis of the World Fauna Inhabiting the Subterranean Waters (Including the Marine Interstitial).* E.J. Brill/Dr. W. Backhuys, Leiden, pp. 47–56.

Straskraba, M., 1966. *Hurleya kalamundae* n.g.n.sp. (Amphipoda: Gammaridae) from subterranean waters of Western Australia. *Int. J. Speleol.,* 2: 291–295.

Stuckey, M.L., 1991. *Differences between Cave and Surface Populations of Austrochiltonia subtenuis (Sayce), 1902 (Amphipoda: Ceinidae) – Selection or Developmental Plasticity?* Unpublished Honours thesis, Department of Zoology, The University of Western Australia, Perth, 71 pp.

Williams, W.D., 1980. *Australian Freshwater Life: The Invertebrates of Australian Inland Waters.* MacMillan Australia Pty Ltd., South Melbourne, 321 pp.

Chapter 16

CHEMOLITHOAUTOTROPHIC MICRO-ORGANISMS AND THEIR POTENTIAL ROLE IN SUBSURFACE ENVIRONMENTS

Brian K. KINKLE and Thomas C. KANE

INTRODUCTION

All ecosystems are ultimately based on inputs of carbon and energy provided by autotrophic organisms. It is generally accepted that the food base of most ecosystems, and certainly the best-studied ones, consists of autotrophic algae or vascular plants harvesting solar radiation – hence the description of these primary producers as photoautotrophs. This photosynthetic food base then supports heterotrophic organisms throughout the food web. With few exceptions, non-photosynthetic micro-organisms in most ecosystems are relegated to the role of decomposers of dead plant or animal material, thus mineralizing and recycling important plant nutrients, including inorganic forms of carbon, nitrogen, sulfur, and phosphorus. Obvious exceptions to this detrital role are disease-causing micro-organisms which parasitize plants and animals, as well as beneficial micro-organisms such as the bacteria which reduce atmospheric nitrogen to forms useable by plants. Recent evolutionary and ecological studies, however, are describing a more varied and complete picture of microbial diversity and activities, including those micro-organisms found in subsurface environments.

SUBSURFACE MICRO-ORGANISMS

Until recently, the numbers and activities of micro-organisms in subsurface terrestrial environments were thought to be limited by the low amounts of organic carbon available for heterotrophic activity in pristine aquifers (Winogradsky, 1890; Chapelle, 1993; Madsen and Ghiorse, 1993). In addition, many of the micro-organisms recovered were believed to be contaminants from the drilling process itself, rather than indigenous inhabitants of groundwater (Ghiorse and Wilson, 1988). Improved methodology used during deep-drilling studies, however, has unequivocally demonstrated the presence and activity of many different types of bacteria, including chemolithoautotrophs, in subsurface environments (Fredrickson et al., 1989; Pedersen and Ekendahl, 1992; Balkwill and Boone, 1997). The predominance of heterotrophic Bacteria and Protozoa, especially flagellate and amoeboid forms, has been indicated in these studies. Much of the emphasis on groundwater microbiology has been on heterotrophs, especially in the context of the use of biological means to restore contaminated environments (Hazen, 1997).

Much less is known about the microbiology of caves, a distinct type of subsurface environment. There has been some descriptive work identifying and enumerating the indigenous micro-organisms of cave systems (Coumartin, 1963; Thompson and Olson, 1988). The presence and activity of autotrophs in cave systems has been investigated, especially regarding the potential role of nitrifiers in the formation of saltpeter (KNO_3) (Fliermans and Schmidt, 1977; Hill, 1981). Even with these efforts, the microbiology of cave and groundwater environments is still poorly understood (Gounot, 1994).

CHEMOLITHOAUTOTROPHIC MICRO-ORGANISMS

Despite the anthropocentric focus of most microbiology research, concentrated on heterotrophic organisms involved in disease, nutrient-cycling, or fermentation, chemolithoautotrophic bacteria were first described by Winogradsky over 100 years ago (Winogradsky, 1890;

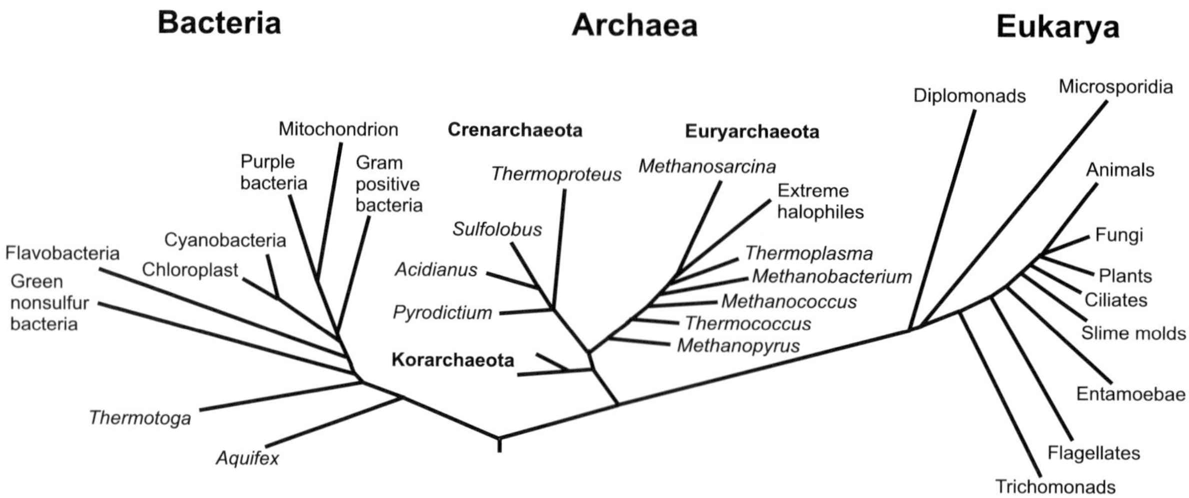

Fig. 16.1. Phylogeny of living organisms based on small subunit ribosomal RNA.

Schlegel and Bowien, 1989). These "chemosynthetic" micro-organisms obtain the majority of their cellular carbon from carbon dioxide, and their energy from the oxidation of inorganic substances (Shively and Barton, 1991). Since their original description, however, chemolithoautotrophic micro-organisms have been largely treated as marginal. The best-studied groups are those that directly affect human interests, including the sulfur-oxidizing *Thiobacillus* spp. with their positive role in ore leaching and negative environmental impacts during mining operations, as well as nitrifiers with their critical role in nitrogen cycling, especially in agricultural and forestry settings.

Since Winogradsky, a handful of microbiologists have carried out basic research on some of the more obscure chemolithoautotrophs. This approach is exemplified by the work by Wolfe's group on the biochemistry of methanogens, a group of micro-organisms with an often-central role in the anaerobic degradation of organic matter (Wolfe, 1993). This basic research on methanogens played a pivotal role in the early molecular evolution studies of Woese (1987), which revolutionized biology by discarding the paraphyletic division of all organisms into the prokaryote and eukaryote groups. There is now compelling evidence that there are three domains of life: Bacteria, Eukarya, and Archaea (Fig. 16.1), with the prokaryotic domain Archaea more closely related to the Eukarya than to the other prokaryotic domain Bacteria. The Archaea include the methanogens, as well as several other groups of micro-organisms, many of which are adapted to thermal, acidic, hypersaline, or anaerobic environments.

All known chemolithoautotrophs are micro-organisms from the bacterial and archaeal domains. In addition, with the evidence from molecular and cytological studies that Cyanobacteria (formerly known as blue-green algae) are in fact Bacteria, and that eukaryotic mitochondria and chloroplasts originated as endosymbiotic Bacteria, it is clear that all autotrophic processes, both chemo- and phototrophic, ultimately rely on the carbon-fixing capabilities of Bacteria and Archaea (Margulis, 1981). Interestingly, the obligate, symbiotic associations of several animals with chemolithoautotrophic bacteria (described below) in marine environments has resulted in the evolution of chemolithoautotrophic animals in parallel to photoautotrophic plants. The evolutionary success of eukaryotes appears to be partially due to their capabilities to harness directly the metabolic capabilities of Bacteria through obligate symbiotic associations (Maynard-Smith and Szathmary, 1995).

Despite the known diversity of morphology and life history, eukaryotic photoautotrophs are relatively similar in their photosynthetic machinery: they are all aerobic, use the Calvin cycle for carbon dioxide fixation, and split water for use as an electron donor (hence the term oxygenic photosynthesis). This is in contrast with the phylogenetic, metabolic, and ecological diversity of photo- and chemolithoautotrophs which are distributed across the archaeal and bacterial domains. Their metabolic diversity is such that there is a need to label these autotrophic organisms as to

Table 16.1
Examples of chemolithoautotrophs

Genera	Phylogenetic domain	Electron donor	Electron acceptor	Carbon fixation pathway [1]
AEROBES				
Thiobacillus, Beggiatoa, Thiomicrospira	Bacteria	HS^-, S^0, $S_2O_3^{2-}$	O_2	Calvin
Nitrosomonas, Nitrosolobus, Nitrosococcus	Bacteria	NH_4^+	O_2	Calvin
Nitrobacter, Nitrospira, Nitrococcus	Bacteria	NO_2^-	O_2	Calvin
Gallionella, Leptospirillum	Bacteria	Fe^{2+}	O_2	Calvin
Alcaligenes, Pseudomonas, Bacillus	Bacteria	H_2	O_2	Calvin
Hydrogenobacter, Aquifex	Bacteria	H_2	O_2	RCA
Sulfolobus, Acidianus	Archaea	HS^-, S^0	O_2	ACA or RCA
ANAEROBES				
Thiobacillus	Bacteria	HS^-	NO_3^-	Calvin
Acetobacterium, Clostridium	Bacteria	H_2	CO_2	ACA
Desulfococcus, Desulfovibrio	Bacteria	H_2	SO_4^{2-}	ACA or RCA
Methanococcus, Methanobacterium	Archaea	H_2	CO_2	ACA
Acidianus, Thermoproteus, Pyrodictium	Archaea	H_2	S^0, $S_2O_3^{2-}$	ACA or RCA
Pyrobaculum, Stygiolobus, Pyrolobus	Archaea	H_2	NO_3^-	ACA or RCA
Archaeoglobus	Archaea	H_2	SO_4^{2-}, $S_2O_3^{2-}$	ACA

[1] Calvin, Calvin or reductive pentose phosphate cycle; ACA, acetyl–CoA pathway; RCA, reductive citric acid cycle.

energy source (photo vs. chemo), electron donor (litho vs. organo), and oxygen requirements (aerobic vs. anaerobic). Moreover, there are at least three distinct pathways for carbondioxide assimilation in prokaryotic autotrophs, as described below.

In addition to the ultimate reliance of all ecosystems on carbon fixation by prokaryotic autotrophs, either free-living organisms or endosymbiotic chloroplasts, all aerobic organisms are dependent on oxygen produced by oxygenic Cyanobacteria, again either free-living or as endosymbiotic chloroplasts. It has been suggested that oxygen formation by Cyanobacteria may have allowed the development and subsequent radiation of plants and Metazoa (Madigan et al., 2000).

CHEMOLITHOAUTOTROPHIC DIVERSITY

Many, but not all, micro-organisms that use light or inorganic chemicals as an energy source are autotrophs. As stated previously, all known chemolithoautotrophs are either Bacteria or Archaea. These organisms fix carbon through the Calvin, acetyl–Coenzyme A, or reverse citric acid cycles. The Calvin, or reductive pentose phosphate cycle, is found in all autotrophic eukaryotes and many autotrophic Bacteria, but has not, to date, been found in any autotrophic Archaea. This cycle is nearly ubiquitous in aerobic chemolithoautotrophic Bacteria (see Table 16.1), one notable exception being the thermophilic hydrogen-oxidizing aerobe *Hydrogenobacter thermophilus* (Shiba et al., 1985). In the Calvin cycle the dual-function enzyme ribulose bisphosphate carboxylase/oxygenase (RuBisCO) incorporates a molecule of carbon dioxide into ribulose 1,5-bisphosphate, leading eventually to fructose-6-phosphate which can be assimilated into biomass.

An alternative pathway for autotrophic carbon fixation is the acetyl–Coenzyme A pathway. This pathway involves the joining and reduction of two molecules of carbon dioxide, forming a molecule of acetate, which is then incorporated into biosynthetic pathways. This pathway is present in most acetogenic and sulfate-reducing bacteria, autotrophic micro-organisms that produce acetate and hydrogen sulfide respectively. In addition, the acetyl–CoA pathway is employed by several archaeal autotrophs, including the methanogens and the sulfate-reducing *Archaeoglobus*. Most of the archaeal and bacterial autotrophs that use the acetyl–CoA pathway are obligate anaerobes.

A third pathway for carbon fixation in autotrophic

micro-organisms is the reductive citric acid cycle. This pathway reverses the citric acid cycle to form acetate, which is then incorporated into biosynthetic pathways, similar to the acetyl–CoA pathway. The reductive citric acid cycle is found in the previously mentioned *Hydrogenobacter, Aquifex,* some sulfate-reducers (e.g., *Desulfobacter*), and several thermophilic archaeal strains, both aerobes and anaerobes (see Table 16.1).

There has been a tendency to distinguish sharply between autotrophic and heterotrophic micro-organisms, and perhaps to focus research efforts on the extremes of each category, the obligate heterotrophic or obligate autotrophic strains. Many autotrophic strains, in fact, are facultative and thus able to switch to heterotrophic pathways and incorporate significant amounts of organic carbon, if it is present (Whittenbury and Kelly, 1977). In addition, some strains are mixotrophic – capable of simultaneously using both autotrophic and heterotrophic pathways (Karl, 1995). Moreover, most autotrophs are studied only in pure cultures, thus restricting the isolation of strains which grow under natural conditions in multi-species consortia.

MAJOR CHEMOLITHOAUTOTROPHIC GROUPS

Many of the best-known bacterial chemolithoautotrophs are in the phylum Proteobacteria. This phylum contains a wide variety of bacteria, ranging from aerobic to anaerobic, autotrophic to heterotrophic, and photoautotrophic to chemoheterotrophic. The ancestral strain of this group is believed to have been photosynthetic, using the purplish pigment bacteriochlorophyll, from which the common name of "purple" bacteria (Fig. 16.1) is derived (Woese, 1987). Within the Proteobacteria, the autotrophic bacteria (as with many other bacteria) are often grouped physiologically, rather than phylogenetically – for example, the hydrogen-oxidizing "*Knallgas*" bacteria, and the sulfur-oxidizing thiobacilli. Unfortunately, this has resulted in several genera, such as *Thiobacillus,* which are not monophyletic.

Compared to the Bacteria, much less is known about the biology of many of the archaeal chemolithoautotrophs, and the overall diversity of the domain Archaea is probably far greater than is currently recognized. Historically, most of the Archaea have been categorized as "extremophiles" (e.g., acidophiles, thermophiles, halophiles, or anaerobic). In fact, many Archaea fit more than one of these categories, such as the acidophilic thermophilic *Sulfolobus* spp. Yet, despite this reputation as "extremophiles", increasing numbers of Archaea are being detected in non-extreme locations such as soil and marine environments by ribosomal ribonucleicacid (rRNA) gene sequencing (DeLong, 1992; Bintrim et al., 1997). Three main groups of Archaea have been well characterized: anaerobic methanogens, halophiles, and hyperthermophiles. The kingdom Euryarchaeota (Fig. 16.1) includes the methanogens and halophiles, while the kingdom Crenarchaeota includes the hyperthermophiles, most of which are obligate anaerobes involved in sulfur transformations. The potential new kingdom Korarchaeota, a third group of Archaea only recently reported and as yet uncultured, has been identified directly from natural environments through determination of rRNA sequences amplified by the polymerase chain reaction (Barns et al., 1996). There are many known chemolithoautotrophic strains in both the Euryarchaeota and Crenarchaeota.

One common end-product formed during the anaerobic degradation of organic materials is hydrogen. There are several different groups of anaerobic, hydrogen-oxidizing autotrophs in the archaeal and bacterial domains that have evolved the capability to use hydrogen as an energy source, including anaerobes that use electron acceptors other than oxygen and produce reduced end products such as hydrogen sulfide, methane, or acetate. These organisms are grouped according to the electron acceptors that they use, and the subsequent waste-products formed. The methanogens, for example, are a phylogenetically related group of Archaea that include autotrophic strains capable of fixing carbon dioxide using the acetyl–CoA pathway, fueled by the oxidation of hydrogen, which is coupled to the reduction of carbon dioxide to methane (Boone et al., 1993). Methanogens are commonly found in habitats where electron acceptors such as oxygen, nitrate, or sulfate are limiting, including anaerobic sediments, flooded soils, and gastrointestinal tracts. The presence of electron acceptors other than carbon dioxide inhibits methanogenesis by allowing other anaerobic organisms to outcompete methanogens for reduced substrates (Lovley et al., 1982).

The acetogens, in contrast, are a phylogenetically diverse group of facultative autotrophic Bacteria, which also use the acetyl–CoA pathway for carbon dioxide fixation, but form acetate during the oxidation of hydrogen. The anaerobic acetogens are found in many of the same environments as methanogens, but are much more versatile in the range of electron donors used than most other groups of strict anaerobes. Despite this metabolic versatility, relatively little is known about their ecology or their contribution to anaerobic food webs (Diekert, 1992).

The dissimilatory sulfate-reducers consist primarily of facultative autotrophic Bacteria capable of using hydrogen, as well as small-molecule organic acids, coupled with the reduction of sulfate to hydrogen sulfide. Most of the strains in this physiological group consist of a phylogenetically coherent clade of Gram-negative bacteria in the phylum Proteobacteria. Two major exceptions are the Gram-positive bacterium *Desulfotomaculum* and the hyperthermophilic archaean *Archaeoglobus* (Stetter, 1992). All known sulfate-reducing organisms are anaerobes, using either the acetyl–CoA pathway or the reductive citric acid cycle for carbon dioxide fixation. The sulfate-reducers are commonly found in anaerobic habitats, including marine and freshwater sediments, flooded soils, and gastrointestinal tracts (Widdel, 1988).

Among the mostly anaerobic, hyperthermophilic Crenarchaeota there are several Archaea that prefer to use elemental sulfur or thiosulfate as electron acceptors during autotrophic growth on hydrogen. This group includes strains of *Acidianus, Pyrodictium* and *Thermoproteus*. There is also a group of anaerobic, hyperthermophilic Crenarchaeota that oxidize hydrogen using nitrate as a terminal electron acceptor, including strains of *Pyrobaculum, Pyrolobus* and *Stygiolobus* (Madigan et al., 2000).

If either hydrogen or any of the other reduced end-products of anaerobic bacterial metabolism, such as methane, hydrogen sulfide, or acetate, are transported to an aerobic environment by diffusion, mass flow, or degassing from water they can then support the growth of a wide range of aerobic or facultative anaerobic micro-organisms that use oxygen or nitrate as a terminal electron acceptor, including many autotrophic strains.

Most aerobic hydrogen-oxidizing bacteria are facultative autotrophs and use the Calvin cycle to assimilate carbon dioxide. The primary exceptions to this use of the Calvin cycle are the phylogenetically deep-rooted bacterial genera *Aquifex* and *Hydrogenobacter*, which use the reductive citric acid cycle. The aerobic hydrogen-oxidizers are a phylogenetically diverse group including over 20 different genera of bacteria, including both Gram-positive and Gram-negative strains, which are widespread in terrestrial, freshwater, and marine environments (Aragno and Schlegel, 1992). Since most of these strains are facultative autotrophs, they are also found in environments which lack sufficient hydrogen to support autotrophic growth.

The sulfur-oxidizing autotrophs, which oxidize reduced forms of sulfur such as HS^-, S^0, and $S_2O_3^{2-}$ using either O_2 or NO_3^- as a terminal electron acceptor, include both bacterial and archaeal species (Kelly, 1992). All of the known bacterial strains assimilate carbon dioxide through the Calvin cycle. The most-investigated group of sulfur-oxidizing bacteria, the species of *Thiobacillus*, are clearly not a monophyletic group. This genus has been used as a convenient category for any Gram-negative rod-shaped bacterium capable of deriving energy from the oxidation of reduced sulfur compounds. Since this sulfur oxidation often results in the production of sulfuric acid, it is not surprising that many strains of *Thiobacillus* spp. are acidophilic, yet some acid-producing strains are neutrophilic. *Thiobacillus* species range from obligate autotrophs – to mixotrophs – to strains which are primarily heterotrophic; they are ubiquitous in soil, freshwater, and marine environments (Kelly, 1992).

Morphologically distinct, sulfur-oxidizing bacteria include the spiral-shaped *Thiomicrospira* which are prevalent at the deep-sea hydrothermal vents, both in free-living forms and as unculturable endosymbionts associated with tube worms (*Riftia pachyptila*) and giant white clams (*Calyptogena magnifica*) (Robertson and Kuenen, 1992). Large filamentous sulfur-oxidizers include *Beggiatoa, Thioploca,* and *Thiothrix*. These aquatic micro-organisms are all capable of gliding motility, and many of them are micro-aerophilic – that is, aerobes that prefer reduced levels of oxygen. Most of these filamentous organisms are difficult to cultivate, and many have not been isolated in pure culture or been well characterized. Interestingly, they each appear to inhabit different ecological niches and are rarely found together. *Beggiatoa* species are found at the interface between aerobic and anaerobic aquatic environments, including marine and freshwater sediments, and near the deep-sea hydrothermal vents. *Thiothrix* species, in contrast, occur as attached filaments in flowing waters which contain hydrogen sulfide and oxygen. Only marine strains of *Beggiatoa* have been demonstrated to grow autotrophically (Nelson and Jannasch, 1983).

Most of the archaeal sulfur-oxidizers are acidophilic hyperthermophiles capable of growth as facultative autotrophs. The mechanism of carbon dioxide fixation is unclear in several of these hyperthermophilic Archaea, but is thought to be related to the reductive citric acid cycle. In a sense they function as an ecological replacement for the more mesophilic *Thiobacillus* spp. in thermal environments such as solfatara fields. *Sulfolobus* species are restricted to aerobic freshwater environments, whereas *Acidianus* species are also found in marine and anaerobic environments (Segerer and Stetter, 1992).

Several strains of aerobic Archaea and Bacteria are also capable of autotrophic growth during the oxidation of Fe^{2+}, including the bacteria *Gallionella, Leptospirillum,* and *Thiobacillus ferrooxidans,* as well as the archaean *Sulfolobus.* There are many other strains of bacteria that precipitate iron extracellularly, either enzymatically or non-enzymatically, but it has been difficult to determine, for many of these strains, if energy is gained, or if growth is autotrophic (Ehrlich, 1996). The existence of autotrophic manganese-oxidizers has also been suspected, but never conclusively demonstrated.

Ammonium is another compound often released during the degradation of organic materials. The nitrifiers, a group of aerobic bacteria, are capable of autotrophic growth by coupling the oxidation of reduced nitrogen compounds to carbon dioxide fixation through the Calvin cycle. The nitrifiers are divided into two distinct groups: the ammonium-oxidizers which produce nitrite, and the nitrite-oxidizers which then convert this nitrite to nitrate. These two processes, and the associated nitrifier populations, are usually tightly coupled, resulting in the overall production of nitrate from ammonium (Bock et al., 1992). The nitrifiers are very slow-growing strains, since the energy yields from these reactions are quite small, but they are widely distributed and active in soil, marine, and freshwater environments.

Methane is commonly generated from the activities of methanogenic Archaea in anaerobic environments, and transported to aerobic habitats where it supports the growth of methane-oxidizing methanotrophs. Strictly speaking, however, these Bacteria are not autotrophs since they are using an organic molecule, methane, as a carbon source instead of carbon dioxide. The methanotrophs are phylogenetically a relatively coherent group of aerobic Proteobacteria, similar to the nitrifiers. In fact, methane monooxygenase, a key enzyme in the oxidation of methane, appears to be evolutionarily related to ammonia monooxygenase, a key enzyme used by the ammonium-oxidizing nitrifiers (Hanson and Hanson, 1996). Methane oxidizers are widely distributed in aerobic soil and aquatic environments, often above anaerobic regions where methane is produced.

SUBSURFACE ECOSYSTEMS

Lotic (i.e., flowing-water) ecosystems have historically been viewed as systems that derive the majority of their energy base from dead organic material produced outside the stream boundaries, rather than from *in situ* primary (i.e., photosynthetic) production (Allan, 1995). In this regard, one type of subterranean habitat, cave streams, represent an extreme, since the absence of light in these systems precludes even the possibility of *in situ* photosynthetic production (Culver, 1982). Thus, subterranean systems in general, and cave ecosystems in particular, are viewed as open systems that rely on transport of organic matter from the surface as an energy base for the system (Poulson and White, 1969; White et al., 1995). This view emphasizes the intimate association between surface and subsurface environments, especially in karst terrains. The dependence of cave ecosystems on allochthonous inputs of organic matter has led ecologists to argue that these environments are food-limited, and that food limitation has been an important selection pressure in the evolution of cave-limited species (Culver, 1982).

Despite the prevalence of allochthonously based cave communities, there is growing evidence that autotrophic micro-organisms may contribute more to the food base of cave ecosystems than has been previously thought (White et al., 1995). *In situ* primary production is possible in caves that harbor chemolithoautotrophic bacteria capable of using reduced inorganic compounds (e.g., H_2S, NH_4^+) as energy sources. For caves, the extreme example of chemolithoautotrophy is the Movile Cave ecosystem in Romania which is based completely on chemolithoautotrophic production (Sarbu and Kane, 1995; Sarbu et al., 1996; Sarbu, Chapter 17, this volume) and appears, at present, to receive no organic inputs from the surface. However, significant autotrophic production has been reported for other systems such as the Parker Cave system in Kentucky (Thompson and Olson, 1988), which is predominantly based on allochthonous inputs. Two recent studies on marine cave systems in the Mediterranean (Southward et al., 1996) and the Caribbean (Pohlman et al., this volume) have also demonstrated the presence of significant primary production by sulfide-oxidizing bacteria in the former case and by nitrifying bacteria in the latter. Using an analysis based on the ratio of stable isotopes, these investigators have been able to determine that this bacterial production is passed through the food webs of the biological communities in these cave systems.

FOOD QUALITY

The presence of chemolithoautotrophy in cave systems potentially increases both food quantity and food quality relative to systems based completely on allochthonous production. Although detailed quantitative estimates of chemolithoautotrophic productivity in these systems are still needed, the observations of extensive microbial mats in Movile Cave (Sarbu et al., 1996) and in submarine caves of Italy (Southward et al., 1996) support the concept of an increased food base in these systems. Food quality is a measure of the nutritional value of the food source. The nutritional value of a food source can be determined in large degree on the basis of its total nitrogen content and its carbon:nitrogen (C/N) ratio (Panadian and Marian, 1986; McMahon, 1975). For microbial mats in Movile Cave, C/N = 5.7 (Sarbu and Kane, unpublished), which is within the range of C/N = 4–8 reported for periphyton in surface streams (Gregory, 1983), and the range expected for high-quality food sources (McMahon, 1975). Unlike periphyton, however, the Movile Cave microbial mats have higher nitrogen content, 7.3% (Sarbu and Kane, unpublished), than algae (4–5%: Trama, 1957; Stockner, 1971), diatoms (4%: McCullough et al., 1979) or leaves (0.4–3.1%: Aber and Melillo, 1980), the typical food sources for surface and subsurface streams.

SUBSURFACE FAUNA

The presence of chemolithoautotrophic production in caves is usually associated with an increased diversity of the cave fauna (Sarbu and Kane, 1995; Southward et al., 1996). Movile Cave, for example, contains 48 species of cave-limited invertebrates, 33 of which are endemic to the system (Sarbu and Kane, 1995). The increased biodiversity, and the increased abundance of species, in caves with chemolithoautotrophic production may be a straightforward ecosystem consequence of increased food quantity and food quality in these systems. The number of trophic levels in an ecosystem, and to some degree the number of species supported, is a function of the absolute size of the energy base and the efficiency of trophic transfer. One important trophic efficiency measure is assimilation efficiency, the rate at which food ingested by consumers is converted into consumer biomass (Calow and Fletcher, 1972). Generally, assimilation efficiencies in streams are higher for predators and lower for herbivores and detritivores (Allan, 1995). The assimilation efficiencies of primary consumers, therefore, may be critical in determining the overall diversity of these systems. Assimilation efficiency is positively correlated with food quality (Panadian and Marian, 1986). The high nitrogen content and low C/N ratio of the Movile Cave mat suggests that grazers (primary consumers) in this system may have assimilation efficiencies in the range of 70%, which is in the upper range of values for primary consumers and detritivores (Panadian and Marian, 1986). Thus, high biodiversity in such systems may be a consequence of both an increased food base and a high trophic transfer efficiency on the part of primary consumers.

OTHER POTENTIAL CHEMOLITHOAUTOTROPHICALLY-BASED ECOSYSTEMS

Descriptions of the communities of deep-sea hydrothermal vents were the first reports of ecosystems based on chemolithoautotrophic production (Jannasch and Wirsen, 1979), although there is still some debate regarding the significance of allochthonous inputs of particulate matter to these systems (Karl, 1995). Much of the primary production supporting these marine communities appears to come from the activities of autotrophic sulfur-oxidizers, both free-living forms and, perhaps most remarkably, as obligate endosymbionts associated with a range of marine invertebrates, including tube worms (*Riftia pachyptila*), mussels (*Bathymodiolus thermophilus*), and giant clams (*Calyptogena magnifica*). The significant geothermal inputs of reduced metals to the communities of the deep-sea vents also suggests the potential for autotrophic oxidizers of iron and manganese, although this has not been conclusively demonstrated.

There are an increasing number of reports describing chemolithoautotrophic symbiotic associations in non-hydrothermal marine environments, including hydrocarbon seeps in the Gulf of Mexico (Kennicutt et al., 1985; Paull et al., 1984) and in shallow-water sediments (Conway et al., 1994). In addition to sulfur-oxidizing bacterial autotrophs, symbiotic associations between marine invertebrates and methanotrophs have been reported in these environments (Cavanaugh et al., 1987). All of the ecosystems described above based on chemolithoautotrophic production are aerobic and therefore rely on photosynthetically-produced oxygen, which restricts these systems to cave or marine environments where gas exchange or water flow provides sufficient oxygen (Stevens, 1997). A very different system is the anaerobic groundwater microbial community, which appears to be based on hydrogen produced chemically from mafic basalt (Stevens and McKinley, 1995). It was suggested that these microbial communities, which are completely independent of photosynthetic carbon fixation or oxygen production, are based on chemolithoautotrophic production by methanogenic Archaea.

Recently, microbial communities were reported in deep marine sediments, where high microbial numbers (greater than $10^6\,ml^{-1}$) were recovered in cores at depths greater than 500 m below the sea floor, with no indication that the relatively small rate of decrease in bacterial populations with depth would change below 500 m (Parkes et al., 1994). This report supports the previous hypothesis that subterranean microbial biomass, with perhaps much of it based on chemolithoautotrophic production, may actually exceed total surface biomass (Gold, 1992; Gould, 1996).

Much more speculative is the possibility of life on planetary bodies other than Earth, such as Mars and Europa, based on chemolithoautotrophic production. There is increasing evidence that the first forms of life on Earth may have been chemolithoautotrophs, not heterotrophs developing in a warm pond as first suggested by Charles Darwin (Darwin, 1887, Vol. 3, p. 18; see also Waldrop, 1990). Phylogenetic and geological data support the hypothesis that the first organisms on Earth were anaerobic thermophiles (Pace, 1991). There is also speculation that subsurface environments were at least an early refuge for early forms of life on Earth, if not the site of their actual origin, because of the presence of catalytic sites on mineral surfaces, protection from intense solar radiation before the development of the protective ozone layer, and perhaps even protection from frequent meteor showers (Stevens, 1997). Significant amounts of liquid water on Mars are probably currently restricted to the subsurface, as groundwater resulting from melting of the polar caps (Boston et al., 1992). Previous attempts to look for life on Mars only examined the surface of planet. If life is restricted to subsurface environments on Mars, on account of the availability of liquid water, it would necessarily be based on microbial chemolithoautotrophic production.

CONCLUSIONS

The increasing exploration of both marine and terrestrial subsurface environments has forced a reassessment of earlier views that subterranean environments were low in microbial numbers and activity. Unless substantial amounts of allochthonous organic matter are being transported into these subsurface environments, which studies with stable isotopes have often shown not to be the case, many of the micro-organisms present must be relying on chemolithoautotrophic production. In subsurface systems based partially or completely on chemolithoautotrophic production, the microbial ecology is the critical feature, determining the trophic structure and controlling the biodiversity of the system. It seems clear then that investigations of subterranean systems at the ecosystem level must pay significant attention to the microbiology of the systems. Certainly there may be more subsurface ecosystems based on chemolithoautotrophic production yet to be discovered. However, it is also undoubtedly the case that further studies in known subterranean ecosystems will reveal the presence of important and previously unrecognized microbial activities.

REFERENCES

Aber, J.D. and Melillo, J.M., 1980. Litter decomposition: measuring relative contributions of organic matter and nitrogen to forest soils. *Can. J. Bot.*, 58: 416–421.

Allan, J.D., 1995. *Stream Ecology.* Chapman and Hall, New York, 388 pp.

Aragno, M. and Schlegel, H.G., 1992. The mesophilic hydrogen-oxidizing (knallgas) bacteria. In: A. Balows, H.G. Truper, M. Dworkin, W. Harder and K.-H. Schleifer (Editors), *The Prokaryotes.* Springer Verlag, New York, pp. 344–384.

Balkwill, D.L. and Boone, D.R., 1997. Identity and diversity of micro-organisms cultured from subsurface environments. In: P.S. Amy and D.L. Haldeman (Editors), *The Microbiology of the Terrestrial Deep Subsurface.* CRC Press, Boca Raton, Florida, pp. 105–117.

Barns, S.M., Delwiche, C.F., Palmer, J.D. and Pace, N.R., 1996. Perspectives on archaeal diversity, thermophily and monophyly from environmental rRNA sequences. *Proc. Natl. Acad. Sci. USA,* 93: 9188–9193.

Bintrim, S.B., Donohue, T.J., Handelsman, J., Roberts, G.P. and Goodman, R.M., 1997. Molecular phylogeny of Archaea from soil. *Proc. Natl. Acad. Sci. USA,* 94: 277–282.

Bock, E., Koops, H.-P., Ahlers, B. and Harms, H., 1992. Oxidation of inorganic nitrogen compounds as energy source. In: A. Balows, H.G. Truper, M. Dworkin, W. Harder and K.-H. Schleifer (Editors), *The Prokaryotes.* Springer Verlag, New York, pp. 414–430.

Boone, D.R., Whitman, W.B. and Rouviere, P., 1993. Diversity and taxonomy of methanogens. In: J.G. Ferry (Editor), *Methanogenesis: Ecology, Physiology, Biochemistry, and Genetics.* Chapman and Hall, New York, pp. 35–80.

Boston, P.J., Ivanov, M.V. and McKay, C.P., 1992. On the possibility of chemosynthetic ecosystems in subsurface habitats on Mars. *Icarus,* 95: 300–308.

Calow, P. and Fletcher, C.R., 1972. A new radiotracer technique involving ^{14}C and ^{51}Cr for estimating the assimilation efficiencies of aquatic, primary consumers. *Oecologia,* 9: 155–170.

Cavanaugh, C.M., Levering, P.R., Maki, J.S., Mitchell, R. and Lidstrom, M.E., 1987. Symbiosis of methylotrophic bacteria and deep-sea mussels. *Nature,* 325: 346–348.

Chapelle, F.H., 1993. *Groundwater Microbiology and Geochemistry.* Wiley, New York, 424 pp.

Conway, N.M., Kennicutt, M.C. and Van Dover, C.L., 1994. Stable isotopes in the study of marine chemosynthetic-based ecosystems. In: K. Lajtha and R.H. Michener (Editors), *Stable Isotopes in Ecology and Environmental Sciences.* Blackwell Scientific, Oxford, U.K., pp. 158–186.

Coumartin, V., 1963. Review of the microbiology of underground environments. *Nat. Spel. Soc. Bull.,* 25: 1–14.

Culver, D.C., 1982. *Cave Life.* Harvard University Press, Cambridge, MA, 185 pp.

Darwin, F., 1887. *The Life and Letters of Charles Darwin.* John Murray, London, 348 pp.

DeLong, E.F., 1992. Archaea in coastal marine environments. *Proc. Natl. Acad. Sci. USA,* 89: 5685–5689.

Diekert, G., 1992. The acetogenic bacteria. In: A. Balows, H.G. Truper, M. Dworkin, W. Harder and K.-H. Schleifer (Editors), *The Prokaryotes.* Springer Verlag, New York, NY.

Ehrlich, H.L., 1996. *Geomicrobiology.* Marcel Dekker, New York.

Fliermans, C.B. and Schmidt, E.L., 1977. *Nitrobacter* in Mammoth Cave. *Int. J. Speleol.,* 9: 1–19.

Fredrickson, J.K., Garland, T.R., Hicks, R.J., Thomas, J.M., Li, S.W. and McFadden, K.M., 1989. Lithotrophic and heterotrophic bacteria in deep subsurface sediments and the relationship to sediment properties. *Geomicrobiol. J.,* 7: 53–66.

Ghiorse, W.C. and Wilson, J.T., 1988. Microbial ecology of the terrestrial subsurface. *Adv. Appl. Microbiol.,* 33: 107–172.

Gold, T., 1992. The deep, hot biosphere. *Proc. Natl. Acad. Sci. USA,* 89: 6045–6049.

Gould, S.J., 1996. *Full House: The Spread of Excellence from Plato to Darwin.* Random House, New York, NY, 244 pp.

Gounot, A.M., 1994. Microbial ecology of groundwaters. In: J. Gibert, D. Danielopol and J.A. Stanford (Editors), *Groundwater Ecology.* Academic Press, San Diego, CA, pp. 189–215.

Gregory, S.V., 1983. Plant–herbivore interactions in stream ecosystems. In: J.R. Barnes and G.W. Minshall (Editors), *Stream Ecology.* Plenum Press, New York, pp. 157–190.

Hanson, R.S. and Hanson, T.E., 1996. Methanotrophic bacteria. *Microbiol. Rev.,* 60: 439–471.

Hazen, T.C., 1997. Bioremediation. In: P.S. Amy and D.L. Haldeman (Editors), *The Microbiology of the Terrestrial Deep Subsurface.* CRC Press, Boca Raton, Florida, pp. 247–266.

Hill, C.A., 1981. Origin of cave saltpeter. *Natl. Soc. Speleol. Bull.,* 43: 110–126.

Jannasch, H.W. and Wirsen, C.O., 1979. Chemosynthetic primary production at East Pacific sea floor spreading centers. *BioScience,* 29: 592–598.

Karl, D.M., 1995. *The Microbiology of Deep-sea Hydrothermal Vents.* CRC Press, Boca Raton, Florida, 299 pp.

Kelly, D.P., 1992. The chemolithotrophic prokaryotes. In: A. Balows, H.G. Truper, M. Dworkin, W. Harder and K.-H. Schleifer (Editors), *The Prokaryotes.* Springer Verlag, New York, pp. 331–343.

Kennicutt, M.C., Brooks, J.M., Bidigare, R., Fay, R.R., Wade, T.L. and McDonald, T.J., 1985. Vent-type taxa in a hydrocarbon seep region on the Louisiana slope. *Nature,* 317: 351–353.

Lovley, D.R., Dwyer, D.F. and Klug, M.J., 1982. Kinetic analysis of competition between sulfate reducers and methanogens for hydrogen in sediments. *Appl. Environ. Microbiol.,* 43: 1373–1379.

Madigan, M.T., Martinko, J.M. and Parker, J., 2000. *Brock Biology of Micro-organisms.* Prentice-Hall, Upper Saddle River, NJ, 991 pp.

Madsen, E.L. and Ghiorse, W.C., 1993. Groundwater microbiology: subsurface ecosystem processes. In: T.E. Ford (Editor), *Aquatic Microbiology.* Blackwell Scientific Publishers, Cambridge, MA, pp. 167–213.

Margulis, L., 1981. *Symbiosis in Cell Evolution.* Freeman, San Francisco, 419 pp.

Maynard-Smith, J. and Szathmary, E., 1995. *The Major Transitions in Evolution.* W.H. Freeman, New York, 346 pp.

McCullough, D.A., Minshall, G.W. and Cushing, C.F., 1979. Bioenergetics of lotic filter-feeding insects *Simulium* spp. (Diptera) and *Hydropsyche occidentalis* (Trichoptera) and their function in controlling organic transport in streams. *Ecology,* 60: 585–596.

McMahon, R.F., 1975. Growth, reproduction and bioenergetic variation in three natural populations of a freshwater limpet *Laevapex fuscus* (C.B. Adams). *Proc. Malacol. Soc. London,* 41: 331–342.

Nelson, D.C. and Jannasch, H.W., 1983. Chemoautotrophic growth of a marine *Beggiatoa* in sulfide-gradient cultures. *Arch. Microbiol.,* 136: 262–269.

Pace, N.R., 1991. Origin of life – facing up to the physical setting. *Cell,* 65: 531–533.

Panadian, T.J. and Marian, M.P., 1986. An indirect procedure for the estimation of assimilation efficiency of aquatic insects. *Freshwater Biol.,* 16: 93–98.

Parkes, R.J., Cragg, B.A., Bale, S.J., Getliff, J.M., Goodman, K., Rochelle, P.A., Fry, J.C., Weightman, A.J. and Harvey, S.M., 1994. Deep bacterial biosphere in Pacific Ocean sediments. *Nature,* 371: 410–413.

Paull, C.K., Hecker, B., Commeau, R., Freeman-Lynde, R.P., Neumann, C., Corso, W.P., Golubic, S., Hook, J.E., Sikes, E. and Curray, J., 1984. Biological communities at the Florida escarpment resemble hydrothermal vent taxa. *Science,* 226: 965–967.

Pedersen, K. and Ekendahl, S., 1992. Assimilation of CO_2 and introduced organic compounds by bacterial communities in groundwater from southeastern Sweden deep crystalline bedrock. *Microb. Ecol.,* 23: 1–14.

Poulson, T.L. and White, W.B., 1969. The cave environment. *Science,* 165: 971–981.

Robertson, L.A. and Kuenen, J.G., 1992. The colorless sulfur bacteria. In: A. Balows, H.G. Truper, M. Dworkin, W. Harder and K.-H. Schleifer (Editors), *The Prokaryotes.* Springer Verlag, New York, pp. 385–413.

Sarbu, S.M. and Kane, T.C., 1995. A subterranean chemoautotrophically based ecosystem. *Natl. Soc. Speleol. Bull.,* 57: 91–98.

Sarbu, S.M., Kane, T.C. and Kinkle, B.K., 1996. A chemoautotrophically based cave ecosystem. *Science,* 272: 1953–1955.

Schlegel, H.G. and Bowien, B., 1989. *Autotrophic Bacteria.* Science Tech Publishers, Madison, WI, 528 pp.

Segerer, A.H. and Stetter, K.O., 1992. The order Sulfolobales. In: A. Balows, H.G. Truper, M. Dworkin, W. Harder and K.-H. Schleifer (Editors), *The Prokaryotes.* Springer Verlag, New York, pp. 684–701.

Shiba, H., Kawasumi, T., Igarashi, Y., Kodama, T. and Minoda, Y., 1985. The CO_2 assimilation via the reductive tricarboxylic acid cycle in an obligately autotrophic aerobic hydrogen-oxidizing bacterium, *Hydrogenobacter thermophilus. Arch. Microbiol.,* 141: 198–203.

Shively, J.M. and Barton, L.L., 1991. *Variations in Autotrophic Life.* Academic Press, San Diego, CA, 346 pp.

Southward, A.J., Kennicutt, M.C., Alcala-Herrera, J., Abbiati, M., Airoldi, L., Cinelli, F., Blanchi, C.N., Morri, C. and Southward, E.C., 1996. On the biology of submarine caves with sulphur springs: appraisal of $^{13}C/^{12}C$ ratios as a guide to trophic relations. *J. Mar. Biol. Ass. U.K.,* 76: 265–285.

Stetter, K.O., 1992. The genus *Archaeoglobus*. In: A. Balows, H.G. Truper, M. Dworkin, W. Harder and K.-H. Schleifer (Editors), *The Prokaryotes.* Springer Verlag, New York, pp. 707–711.

Stevens, T.O., 1997. Subsurface microbiology and the evolution of the biosphere. In: P.S. Amy and D.L. Haldeman (Editors), *The Microbiology of the Terrestrial Deep Subsurface.* CRC Press, Boca Raton, Florida.

Stevens, T.O. and McKinley, J.P., 1995. Lithotrophic microbial ecosystems in deep basalt aquifers. *Science,* 270: 450–454.

Stockner, J.H., 1971. Ecological energetics and natural history of *Hedriodiscus truqui* (Bellard) (Diptera) in two thermal spring communities. *J. Fish. Res. Bd. Can.,* 28: 73–94.

Thompson, D.B. and Olson, R., 1988. A preliminary survey of the protozoa and bacteria from Sulphur River, in Parker Cave, Kentucky. *Natl. Soc. Speleol. Bull.,* 50: 42–46.

Trama, F.B., 1957. *The Transformation of Energy by an Aquatic Herbivore Stenonema pulchellum (Ephemeroptera).* Ph.D. Thesis, Univ. Michigan, Ann Arbor.

Waldrop, M.M., 1990. Goodbye to the warm little pond? *Science,* 250: 1078–1080.

White, W.B., Culver, D.C., Herman, J.S., Kane, T.C. and Mylroie, J.E., 1995. Karst lands. *Am. Sci.,* 83: 450–459.

Whittenbury, R. and Kelly, D.P., 1977. Autotrophy: a conceptual phoenix. In: B.A. Haddock and W.A. Hamilton (Editors), *Microbial Energetics.* Cambridge University Press, Cambridge, U.K., pp. 121–149.

Widdel, F., 1988. Microbiology and ecology of sulfate- and sulfur-reducing bacteria. In: A.J.B. Zehnder (Editor), *Biology of Anaerobic Micro-organisms.* Wiley, New York, pp. 469–585.

Winogradsky, S., 1890. Recherches sur les organismes de la nitrification. *Ann. Inst. Pasteur,* 4: 213–231.

Woese, C.R., 1987. Bacterial evolution. *Microbiol. Rev.,* 51: 221–271.

Wolfe, R.S., 1993. An historical overview of methanogenesis. In: J.G. Ferry (Editor), *Methanogenesis: Ecology, Physiology, Biochemistry, and Genetics.* Chapman and Hall, New York, pp. 1–32.

Chapter 17

MOVILE CAVE: A CHEMOAUTOTROPHICALLY BASED GROUNDWATER ECOSYSTEM

Serban M. SARBU

INTRODUCTION

Chemoautotrophic systems

Nearly all biological communities derive their energy and organic carbon from photosynthesis by green plants. Before 1976, no one had formulated an alternative pathway of primary energy flow in the biosphere (Karl, 1995). In the late 1970s, however, different communities were discovered near deep-sea hydrothermal vents (Ballard, 1977; Longsdale, 1977; Tunnicliffe, 1992). Similar communities were later discovered around submarine cold seeps (Childress et al., 1986; Cavanaugh et al., 1987). The food base for organisms in these aphotic habitats is supplied by chemoautotrophic bacteria deriving their energy from reduced inorganic compounds (Corliss et al., 1979; Jannasch, 1985; Fisher et al., 1994; Karl, 1995). Organic matter is produced *in situ* by sulfide-oxidizing and methane-oxidizing bacteria. Chemoautotrophic microorganisms occur in the water as bacterioplankton, and also form microbial mats that cover the substratum around the vents. Endosymbiotic chemoautotrophic bacteria are also very common in several species of bivalves, nematodes, and vestimentiferan tube-worms (e.g., *Riftia pachyptila*) living in these environments (Cavanaugh, 1983, 1993; Fisher et al., 1994; Nelson and Fisher, 1995). More recently, hydrothermal vents have been discovered in the deep waters of Lake Baikal in Siberia, one of the largest freshwater lakes of the world (Crane et al., 1991; Gracev et al., 1995). Adjacent to the vents, thick continuous microbial mats cover the sediments and are associated with a rich fauna consisting of sponges, flatworms and amphipods. Little is known to date regarding the source of energy for these hydrothermal-vent ecosystems.

Deep subterranean research identified an unusual lithoautotrophic microbial ecosystem in a deep basalt aquifer in the Columbia River basin, Washington State. Autotrophic microorganisms outnumber heterotrophs, and measurements of stable carbon isotopes indicate that lithoautotrophic methanogenesis dominates this ecosystem. Hydrogen, a potential energy source for bacteria, appears to be produced by reactions between basalt and anaerobic water (Stevens and McKinley, 1995).

Subterranean redox interfaces

The deep aphotic recesses of caves are similar to deep-sea habitats, in that photosynthetic primary producers are excluded by the absence of light. Communities of obligate cave-dwelling organisms rely entirely on organic material transported from the surface as a food base, and are thus ultimately dependent on photosynthetic production by green plants (Culver, 1982). The scarcity of food, one of the most important characteristics of cave habitats, is implicated in the low species diversity and low population densities observed in cave communities (Poulson, 1971, 1972; Culver, 1982).

However, during the last two decades several caves have been discovered that contain redox interfaces. Oxygen is usually the oxidized compound while the reduced compound is most often sulfide, but may also be methane, ammonium, iron, and manganese. Chemoautotrophic and methanotrophic microorganisms can take advantage of redox interfaces, and use the electrons resulting from the oxidation of the reduced compounds to reduce inorganic carbon and to form organic molecules (Fig. 17.1).

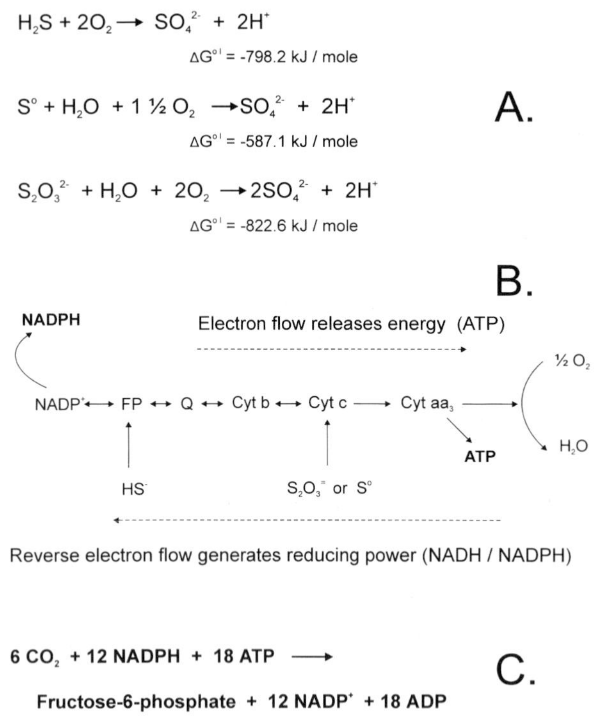

Fig. 17.1. Energetics and carbon metabolism in sulfur chemolithotrophs (after Brock et al., 1994). The most common sulfur compounds used as energy sources by sulfur bacteria are hydrogen sulfide (H_2S), elemental sulfur (S^0), and thiosulfate ($S_2O_3^{2-}$) (A) Electron flow in *Thiobacillus* grown on sulfur, thiosulfate, and sulfide releases energy (ATP). (B) Reverse electron flow generates reducing power (NADH/NADPH). (C) Overall stoichiometry of the Calvin cycle: for each six molecules of CO_2 incorporated, one molecule of fructose 6-phosphate is produced.

Lower Kane Cave, Wyoming, a cave with a small stream of sulfidic water (Egemeier, 1981), contains a rich sulfide-oxidizing microbiota consisting of filamentous bacteria (M. Buck, pers. commun., 1995). Filamentous mats also cover the walls of Grotta Azzura and Grotta Sulfurea at Capo Palinuro, Italy – two submarine caves at the bottom of which are thermal sulfidic springs (Southward et al., 1996). Extensive microbial mats consisting primarily of sulfide-oxidizing bacteria (e.g., *Beggiatoa*, *Thiothrix*) are also present in other caves with sulfide-rich waters: Grotte di Frasassi and Aquasanta Terme, near Ancona, Italy (S. Galdenzi, pers. commun., 1996; see also Chapter 19, this volume), Cesspool Cave (Virginia, USA) (Culver and R. Popa, pers. commun., 1996), Sulfur River in Parker Cave, Kentucky (N. Pace, pers. commun., 1994), and Cueva del Azuffre, Tabasco, Mexico (Pisarowicz, 1994; J. Parzefall, pers. commun., 1995; Taylor, 1999). Unfortunately, most of these caves are not large

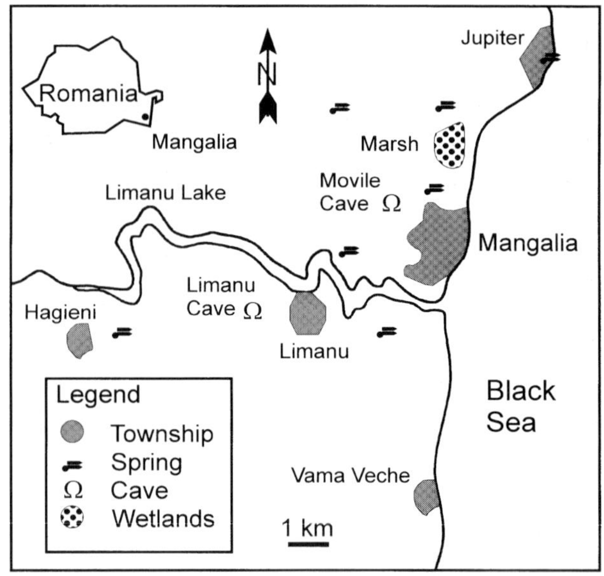

Fig. 17.2. The location of Movile Cave.

enough and not sufficiently isolated from the surface, or are affected too often by flooding, to permit the development of communities of obligate cave-dwelling organisms relying solely on the food produced by the chemoautotrophic microbiota.

Movile Cave

In June 1986 a natural cave passage was discovered at the bottom of an artificial shaft dug for geological investigations (Lascu, 1989) in the proximity of Mangalia, southeastern Dobrogea, Romania (Fig. 17.2). The narrow cave passage provided access to an unusual groundwater ecosystem located at a depth of approximately 20 m below the surface.

The artificial shaft was open until August 1990, when two gates were installed to protect the unusual ecosystem. A strong gate was installed at the surface to stop unwanted visitors from entering the cave. In addition, an air-tight gate was mounted at the bottom of the artificial shaft to prevent the exchange of air between cave and surface, as well as to stop the surface and soil fauna from invading the groundwater ecosystem.

This underground system, known as Pestera Movile (Movile Cave) (Fig. 17.3), is developed in oolitic and fossil-rich limestone of Sarmatian age (i.e., late Miocene, about 12.5 million years ago) containing numerous mollusc fossils (Lascu, 1989). It consists of a maze of upper dry cave passages 200 m long, and a lower cave level 40 m long partially flooded by

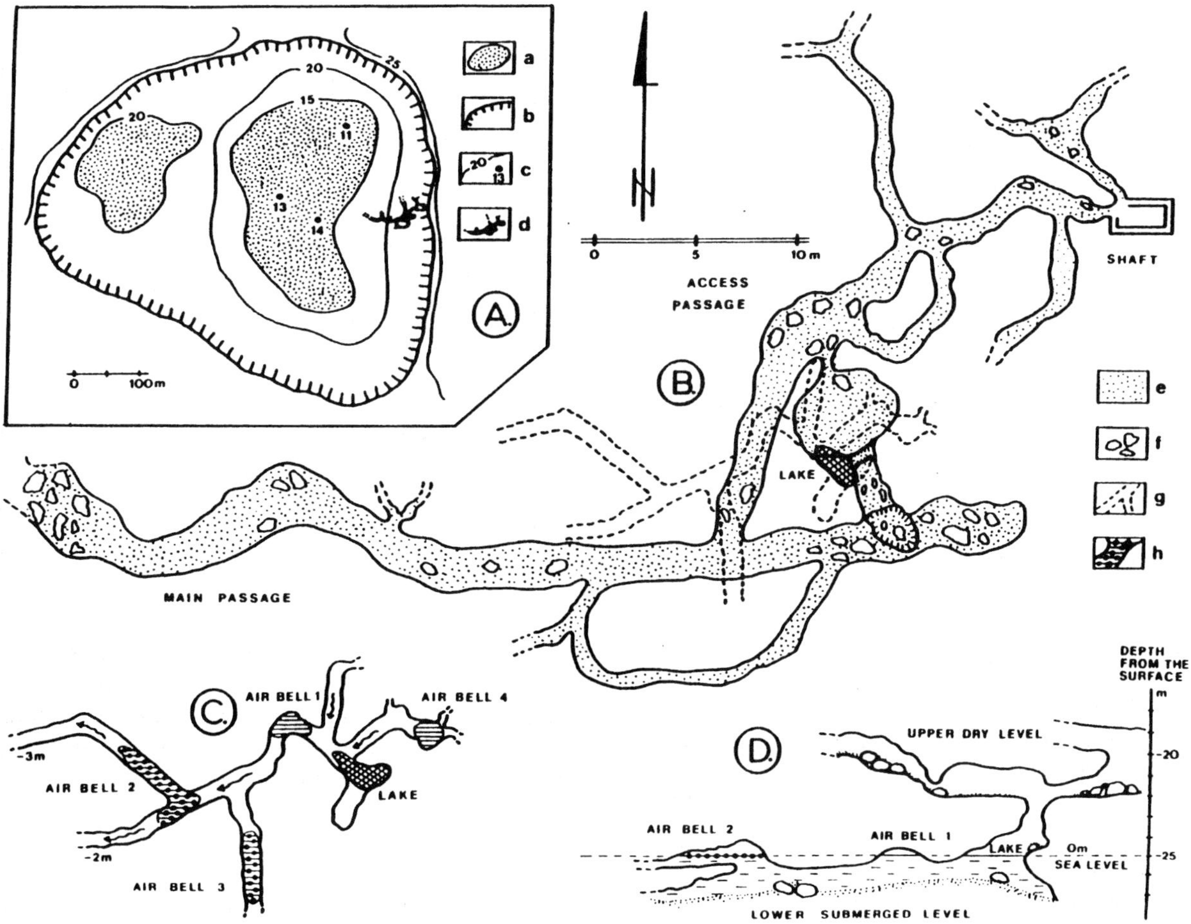

Fig. 17.3. Movile Cave. (A) Location of the cave in the Movile sinkhole (a) bottom of the sinkhole; (b) edge of the sinkhole; (c) altitude in meters; (d) cave; (B) Plan of the upper level of the cave: (e) clay; (f) breakdown; (g) lower level; (h) air bells containing floating mats. (C) Plan of the submerged level of the cave. (D) Profile of the Lake Room area.

thermal sulfidic waters (Sarbu and Popa, 1992). Water chemistry and stable isotope data indicate that the geographic extent of the sulfidic groundwater aquifer is approximately 50 to 70 square kilometers.

After examining the morphology of the cave, Constantinescu (1989) concluded that it had a phreatic origin. He also pointed out that the upper level of the cave was extremely dry as a consequence of the lack of water infiltration from the surface, and suggested that the lack of speleothems was an effect of the dry climate in the region. The peculiar characteristics of the cave atmosphere and of the water flooding the lower cave level cause accelerated rates of limestone dissolution. Sulfuric acid corrosion, active in the lower, partially submerged, cave passages, and condensation corrosion, a process affecting the cave walls in the upper dry cave passages, are the two limestone dissolution processes currently shaping the morphology of Movile Cave.

The unusual physicochemical characteristics of the water and of the atmosphere in Movile Cave have produced of a peculiar groundwater ecosystem (Sarbu and Kane, 1995). Air pockets in the submerged portion of the cave contain microbial mats composed of fungi and bacteria which float on the water surface and grow on the limestone walls (Sarbu et al., 1994a). Incubation of mat samples with radiolabeled [^{14}C]bicarbonate resulted in incorporation of radioactive carbon into microbial lipids, indicating chemoautotrophic carbon fixation (Sarbu et al., 1991). Further, ribulose 1,5-bisphosphate carboxylase/oxygenase (RuBisCO), an enzyme unique to certain types of autotrophic organisms, is present and active in homogenates made from the mat and in crude cell lysates of bacteria isolated from cave water and cultured in the laboratory (Sarbu et al., 1994b). Data for stable isotopes of carbon and nitrogen provide conclusive evidence that the groundwater ecosystem associated with the thermal sulfidic groundwaters at Mangalia is isolated from the surface and is based exclusively on autochthonous carbon fixed chemoautotrophically.

The lower level of the cave contains an abundant and diverse fauna, with a terrestrial community composed of 30 species of obligate cave-dwelling invertebrates, including 23 endemic species, and an aquatic community of 18 species, with 11 endemics (Sarbu, 1990, 1991; Sarbu and Popa, 1992; Sarbu and Kane, 1995). Many of the aquatic and terrestrial species inhabiting Movile Cave have also been found in old hand-excavated wells in villages nearby and in sulfidic springs in the Mangalia region (Sarbu and Kane, 1995). No other cave ecosystem in the world contains as many endemic species as Movile Cave. Some of the invertebrate species inhabiting this chemoautotrophically-based ecosystem presumably became isolated from the surface as early as the Upper Miocene (5.2 million years ago), whereas many others appear to have invaded the system much more recently.

GEOGRAPHIC EXTENT OF THE SULFIDIC GROUNDWATER AQUIFER

Introduction

Thermal sulfidic springs have been known in the Mangalia region of Romania for over 2000 years, at which time the Greek colony of Callatis was established (Sarbu and Popa, 1992). Because all springs and hand-excavated wells in this area contain sulfidic water, the inhabitants of the old Greek colony constructed an aqueduct which brought fresh water from a large spring located eight kilometers west of Callatis (present-day Mangalia). This is the oldest evidence of an extensive sulfidic groundwater aquifer in the Mangalia region. Some springs are a few kilometers inland, but most are along the shore of the Black Sea. There are also numerous submarine springs in shallow waters on the continental platform of the Black Sea at Mangalia (Lascu et al., 1995). Data for stable isotopes of sulfur have shown that the sulfide in the groundwaters at Mangalia has an origin different from that of the sulfide from the deep anoxic waters of the Black Sea (Sarbu and Kane, 1995). The discovery of a positive geomagnetic anomaly in southeastern Dobrogea (Airinei, 1980) supports the hypothesis that the reduced sulfur may be of magmatic origin (Sarbu and Kane, 1995). Data collected from the numerous deep boreholes dug in southern Dobrogea have shown that the thermal sulfidic groundwaters at Mangalia ascend along geological faults from a confined aquifer 200 to 400 m deep, and flood the superficial carbonate bedrock (Feru and Capota, 1991). Isotopic data showed that the sulfidic waters at Mangalia have residence times up to 25 000 years underground (Ţenu et al., 1987).

The study of the physicochemical characteristics and stable-isotope signatures of the groundwaters in southeastern Dobrogea was initiated to determine the geographic extent of the thermal sulfidic aquifer at Mangalia. Stable-isotope ratios of hydrogen and oxygen are often used in hydrogeological research as they are fairly conservative, and on a short geological time scale they do not tend to equilibrate with the rocks in which the aquifer is lodged (Faure, 1986). Differences in the hydrogen and oxygen isotope composition of the groundwaters are likely to reflect atmospheric conditions at the time of precipitation and can be used to determine the origin of the groundwater (Faure, 1986). In addition, the chemical composition of the atmosphere was determined in the system of deep limestone fractures and in the small cave passages associated with the groundwaters. Sampling and analytical procedures used are described in detail in the Appendix (pp. 339–340).

Conditions in the cave

The water that floods the lower level of the cave is thermal ($T = 21$°C), 8°C above the mean annual temperature of this region. Other sulfidic springs around Mangalia are also thermal, with temperatures ranging between 15 and 25°C. The content of total dissolved solids in the water of the various sources in the Mangalia region ranges between 1.1 and 1.5 g ℓ^{-1} – essentially fresh water (Table 17.1). The redox potential is generally low, with average values of –341 mV ($pE = -5.76$) in Movile Cave. Given the pH of 7.29, H_2S and HS^- will be the dominant sulfur species and will occur in equal concentrations in the Movile Cave water (W. White, pers. commun., 1996). Oxidized chemical species are nearly absent (Table 17.1), and the water contains large amounts of reduced compounds

Table 17.1
Physicochemical characteristics of the water in Movile Cave

Physicochemical parameter	Measured value
Ca^{2+}	1.19 mol ℓ^{-3}
Mg^{2+}	1.14 mol ℓ^{-3}
Na^+	8.78 mol ℓ^{-3}
K^+	0.22 mol ℓ^{-3}
Cl^-	6.48 mol ℓ^{-3}
HCO_3^-	3.8 mol ℓ^{-3}
NH_4^+	0.28 mol ℓ^{-3}
H_2S	0.25 mol ℓ^{-3}
CH_4	0.02 mol ℓ^{-3}
SO_4^{2-}	0.01 mol ℓ^{-3}
NO_3^-	0 mol ℓ^{-3}
Temperature	20.9°C
pH	7.29
Redox potential	~341 mV

such as sulfide (0.2–0.3 mmol ℓ^{-1}), ammonium (0.2–0.3 mmol ℓ^{-1}) and methane (0.02 mmol ℓ^{-1}). The pH of the water is neutral to slightly basic because of the buffering capacity of the carbonate bedrock. Dissolved oxygen ranges between 9 and 16 μmol ℓ^{-1} at the surface of the water and decreases to less than 1 μmol ℓ^{-1} at a depth of 5 cm, making the deep water in Movile Cave essentially anoxic.

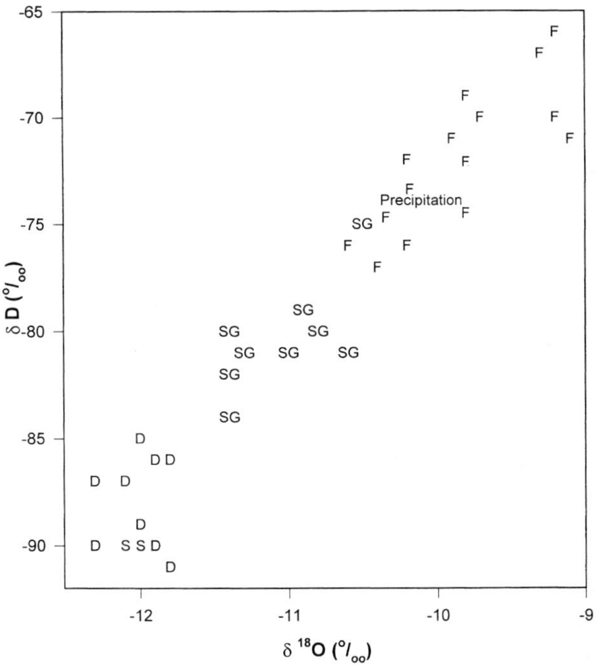

Fig. 17.4. Various categories of water sources sampled for stable isotope analysis (δD and $\delta^{18}O$): D, deep boreholes; S, warm sulfidic springs located close to the geological faults; F, freshwater (non-sulfidic) wells in villages around Mangalia; SG, sulfidic groundwater samples from Movile Cave and from wells and springs in the Mangalia region.

Stable isotope signatures of the groundwater

Data[1] for stable isotopes of oxygen and hydrogen (Fig. 17.4) show that the thermal sulfidic waters in the deep confined aquifer (D in Fig. 17.4) are isotopically much lighter for both oxygen and hydrogen than the present precipitation input, thus supporting the hypothesis that these waters are of considerable subterranean age. They also show that some of the warmest sulfidic springs (S in Fig. 17.4) also produce the isotopically lightest waters indicating that they are located in close proximity to the geological fault that brings sulfidic waters to the surface. The non-sulfidic cold wells from villages located at distances of approximately 10 km to the north, west and south of Mangalia (F in Fig. 17.4) are isotopically heavier than the present precipitation input, indicating that their origin is in the recent precipitation. The wells in the old town of Mangalia and in the village of Arsa located eight kilometers west of Mangalia (SG in Fig. 17.4), are intermediate between the isotopically light sulfidic waters and the present precipitation input, indicating that a certain degree of mixing occurs between the isotopically light sulfidic waters ascending from the deep confined aquifer and the more recent and isotopically heavier precipitation input.

Characteristics of the cave atmosphere

The air temperature in Movile Cave ranges between 19°

[1] The proportions of stable isotopes are expressed by delta (δ) values, where delta is the proportional difference of the ratio in the sample in question from an arbitrary standard, multiplied by one thousand – that is,

$$X = \frac{R_{\text{sample}} - R_{\text{standard}}}{R_{\text{standard}}} \quad 10^3, \quad \text{where } X \text{ is } \delta^{13}C, \delta D, \delta^{18}O, \text{ or } \delta^{35}S \text{ and } R \text{ is } {}^{13}C/{}^{12}C, D/H, {}^{18}O/{}^{16}O, \text{ or } {}^{35}S/{}^{32}S.$$

Standards are Pee Dee belemnite (PDB) for carbon, Standard Mean Ocean Water (SMOW) for hydrogen and oxygen, and Canon Diablo troilite (CDT) for sulfur. Precision of measurements was 0.2‰ or better.

and 21°C, with the lowest temperatures in the upper dry level of the cave and the highest in the proximity of the thermal water. Temperatures of 22.6 to 22.8°C were measured in the atmosphere of air bells #2 and #3, the highest recorded in Movile Cave. The relative humidity ranges between 98 and 100%.

Small amounts of hydrogen sulfide are present in the air in the lower level of the cave, near the water (Table 17.2). Compared to the surface, the atmosphere of Movile Cave is poor in oxygen and very rich in carbon dioxide. The upper dry level of the cave contains 20 to 21% oxygen and 1 to 1.2%, carbon dioxide, while the air bells in the lower level of the cave contain 7 to 10% oxygen and up to 2.5% carbon dioxide, as well as small amounts of methane (1–2%) (Table 17.2). The depletion in oxygen in the cave's air-bells may be the effect of biological respiration by the cave biota. The enrichment in carbon dioxide of the atmosphere of Movile Cave could be the effect of aerobic oxidation of methane in the thermal sulfidic waters, of aerobic respiration by the cave biota, and of sulfuric acid dissolution of carbonate.

Table 17.2
Composition of the Movile Cave atmosphere

Locations	O_2 (%)	CO_2 (%)	CH_4 (%)	H_2S (ppm)
Upper dry level	20.1–20.6	1.2–1.3	0	0
Lake Room	19.5–20.1	1.3–1.5	0	1
Airbell #1	16.6–18.9	1.3–1.6	0	3
Airbell #2	7.2–10.7	2.4–3.5	0–1.4	5

Discussion

Data for stable isotopes of oxygen and hydrogen analysed herein, in conjunction with data regarding the physicochemical characteristics of the groundwaters, indicate that the thermal sulfidic aquifer extends over approximately 50 to 70 square kilometers in the Mangalia region. Its northern limit is difficult to identify as there are no wells between Mangalia and the villages of 23 August, Dulcesti and Pecineaga, which are all located beyond the sulfidic aquifer. To the west, the sulfidic aquifer appears to extend as far as the village of Arsa (10 km west of Mangalia). The southern limit appears to be the Limanul Mangalia lake. All of the wells located south of this lake contain cold non-sulfidic water.

After ascending along geological faults from the deep confined aquifer, the thermal sulfidic waters reach the groundwater level and disperse into the network of fractures and small cave passages. The water temperature decreases, and the hydrogen sulfide and methane are oxidized at the air–water interface. Oxidation of the reduced compounds also takes place at the edge of the sulfidic aquifer through mixing with superficial oxygenated waters.

Although overall the environmental characteristics of Movile Cave resemble those of most other caves (absence of light, constancy of temperature and relative humidity), there are certain peculiarities that are uncommon in caves in general. Impermeable layers of clays within the limestone layers above the cave prevent surface waters from infiltrating into the cave, and reduce air exchange between the cave and the surface. The mean temperature of the cave, which is very constant throughout the year (20.9°C 0.3), is significantly higher than the mean annual temperature of caves located in temperate climates (approximately 4° to 14°C) which makes the temperature regime of Movile Cave resemble those of tropical cave environments. The characteristics of the water and of the air in the cave show no daily or seasonal fluctuations, suggesting that the aquatic and terrestrial habitats associated with the thermal sulfidic groundwaters are fairly well isolated from the surface.

FORMATION OF MOVILE CAVE

Hydrogeological setting

Southern Dobrogea, Romania, is part of the Moesian platform, the northernmost extension of the Pre-Balkan plateau (Vucev, 1974). It is located west of the Black Sea, and it is bordered by the Danube to the west and the Casimcea Valley to the north. The carbonate deposits reach a thickness of up to 800 meters in the Dobrogean plateau, and consist of marine deposits of ages ranging from Jurassic (205 to 135 10^6 yr) through Cretaceous (135 to 65 10^6 yr) and Eocene (53 to 34 10^6 yr) to Sarmatian (12.5 10^6 yr), overlying a thick layer of Paleozoic sediments (aged over 2500 10^6 yr) (Sarbu and Popa, 1992). Although some thin and discontinuous layers of impermeable rocks exist between the limestone strata, the carbonate rocks can be considered to form a unified karstifiable complex (Lascu, 1989). Paleogeographical and geological evidence (Hsü, 1978; Lascu, 1989; Lascu et al., 1994, 1995) suggest that an extensive network of subterranean conduits was first formed in southern Dobrogea as early as the Upper Miocene, approximately 5.5 million years ago.

The Black Sea, as well as the Mediterranean Sea, are remnants of the ancient Tethys Sea which separated Africa from Europe and connected the Atlantic to the Indian Ocean. About 20 million years ago, the movement of the African plate to the north divided Tethys into a southern part occupying today's Mediterranean basin and a northern part known as the Paratethys. The latter extended from the Hungarian basin in the west to the Aral Sea in the east, and included southeastern Europe, the Black Sea and the Caspian Sea. About 5.5 to 5.2 10^6 years ago, the Mediterranean basin underwent one of the most spectacular changes in its history: the "Messinian Crisis". The northward movement of Africa caused the uplifting of the Gibraltar area, closing the communication of the Mediterranean with the Atlantic Ocean. During the following millennia, the large amounts of water lost by evaporation led to a severe drop in the level of the Mediterranean Sea, which was reduced to several small lakes containing hypersaline waters. Deep drillings performed by the Glomar Challenger expedition in the Black Sea in 1970 uncovered deep sedimentary deposits, dated at 5.2 million years, typical of shallow waters. Based on these findings, Hsü (1978) concluded that the water level in the Black Sea had dropped by several hundred meters compared with today's level. At this stage, the waters of the Dacian Lake, another remnant of the Paratethys located in the west of southern Dobrogea, were drained toward the empty Black Sea basin through karst conduits developed in the deep Dobrogean carbonate deposits. Consequently, numerous large cave passages were formed at depths of about 200 m below the surface. Later oscillations in the level of the Black Sea led to several other episodes of karstification. Many such large water-filled voids have been intercepted by deep boreholes dug in the Mangalia region for geological and hydrogeological investigations (Lascu et al., 1994, 1995). Thus, what is known today as Movile Cave is a small portion of the uppermost karstification level in southern Dobrogea, the most recently formed part of an extensive three-dimensional network of subterranean conduits transporting water from west to east, from the present day River Danube into the Black Sea (Lascu et al., 1995).

The lack of significant inputs of water from the surface and the presence of highly corrosive sulfidic groundwaters in the Mangalia region (Sarbu and Kane, 1995) support the hypothesis that the water in Movile Cave has a hypogenic origin (Palmer, 1991). Recent examination of the cave sediments in the upper dry cave passage showed that the cave was open to the surface and communicated with the Movile sinkhole for a relatively short period of time during the Upper Pleistocene (Ştiucă and Ilinca, 1995). The evidence supports the hypothesis that the morphology of Movile Cave is primarily shaped by two relatively unusual limestone dissolution processes: dissolution of limestone by sulfuric acid and condensation corrosion (Sarbu and Lascu, 1997).

Sulfuric acid corrosion

Since[2] its description by Egemeier in 1973, corrosion of limestone by sulfuric acid has been reported from several caves containing sulfidic waters, including Lower Kane Cave (Wyoming, USA) (Egemeier, 1973, 1981), Parker Cave (Kentucky, USA) (N. Pace, pers. commun., 1994), Grotta di Frasassi, Italy (Galdenzi, 1990; Cucchi and Forti, 1990; see also Chapter 19, this volume); and Cueva del Azufre, Tabasco, Mexico (Pisarowicz, 1994; M. Buck, pers. commun., 1995). It is also believed that this mechanism was responsible for the formation of large caves such as Carlsbad Caverns and Lechuguilla Cave, New Mexico, where the emission of hydrogen sulfide has ceased (Hill, 1987, 1990; Buck et al., 1994). The existence of an oxygen–sulfide interface in a limestone cave establishes conditions favoring accelerated dissolution of the bedrock (Egemeier, 1981).

Hydrogen sulfide evaporates at the air–water interface and redissolves in the water film covering the limestone cave walls as well as in the small water droplets in the moist cave atmosphere. In the aqueous phase, it dissociates into various ionic species depending on the pH of the solution. At pH values close to 7, as on the surface of limestone walls,

[2] Sulfuric acid speleogenesis, also known as sulfuric acid dissolution of limestone, represents the process of carbonate dissolution under the influence of protons resulting from the oxidation of hydrogen sulfide at redox interfaces in carbonate environments such as caves. Sulfur in hydrogen sulfide can be oxidized to various degrees, resulting in $S_2O_3^{2-}$, S^0, SO_3^{2-}, SO_4^{2-}, etc.

H_2S and HS^- occur in approximately equal amounts. Oxygen in the cave atmosphere also dissolves in the water and reacts with HS^- to form elemental sulfur (Eq. 17.1). Sulfide-oxidizing microorganisms (e.g., *Beggiatoa, Thiobacillus*, *Thiothrix*) can oxidize HS^- through several steps to sulfate (Eq. 17.2).

$$2HS^- + \tfrac{1}{2}O_2 \rightarrow 2S^0 + H_2O \qquad 17.1$$

$$HS^- \xrightarrow{\text{oxidation}} S^0 \xrightarrow{\text{oxidation}} SO_3^- \xrightarrow{\text{oxidation}} SO_4^{2-} \qquad 17.2$$

The protons resulting from the dissociation and oxidation of hydrogen sulfide in water are responsible for the low pH values measured on the surface of microbial mats covering cave walls in the vicinity of sulfidic waters (Sarbu and Popa, 1992; Pisarowicz, 1994; N. Pace, pers. commun., 1994; see also Chapter 19, this volume).

The sulfate ions and protons react with the carbonate bedrock to form calcium sulfate and carbonic acid (Eq. 17.3).

$$SO_4^{-2} + 2H^+ + CaCO_3 \rightarrow Ca^{2+} + SO_4^{2-} + 2H^+ + CO_3^{2-} \qquad 17.3$$

When the water film reaches saturation with respect to calcium sulfate, gypsum precipitates ($CaSO_4$ $2H_2O$) forming crystals and crusts that cover the cave walls.

Condensation corrosion

Condensation corrosion is the dissolution of carbonate by acidic vapors condensing above the water table (Ford and Williams, 1989). This carbonate dissolution process is rarely noted and receives little attention in the mainstream cave literature (Jameson, 1989). The effects of condensation corrosion were reported from several non-thermal caves (Jameson, 1989, 1995), as well as from many hydrothermal caves which exhibit specific dissolution patterns (Collingnon, 1983; Bakalovicz et al., 1987; Cigna and Forti, 1986; Palmer and Palmer, 1989). It appears that condensation corrosion is a quantitatively important force of speleogenesis (i.e., cave formation) in semi-desert karst areas of Central Asia according to A. Maltsev (unpublished) as well as in the Mediterranean region (Pasquini, 1973).

The mechanism proposed for condensation corrosion is simple: water vapors condense when the temperature decreases to the dew point; drops of water or thin water films result when condensation occurs on walls; water drips from walls and ceiling if condensation rates are sufficiently large (Jameson, 1995); carbon dioxide from the cave atmosphere dissolves in the condensate to form carbonic acid (Eq. 17.4); this reacts with the carbonate bedrock forming bicarbonate (Eq. 17.5) which is transported by the condensate as it drains down the cave walls.

$$CO_2 + H_2O \rightarrow 2H^+ + CO_3^{2-} \qquad 17.4$$

$$2H^+ + CO_3^{2-} + CaCO_3 \rightarrow 2HCO_3^- + Ca^{2+} \qquad 17.5$$

High concentrations of carbon dioxide in the cave atmosphere favor the production of large quantities of carbonic acid thus accelerating the process of limestone dissolution (James, 1994).

Corrosion in Movile Cave

Sulfuric acid corrosion

The mechanism of sulfuric acid corrosion which shapes the morphology of the lower level of Movile Cave is shown in Fig. 17.5.

Hydrogen sulfide was detected in the cave atmosphere only in proximity to the air–water interface in concentrations of 1–5 p.p.m. No hydrogen sulfide was detected in the upper level of Movile Cave.

The water entering Movile Cave is much lower in sulfate concentration (maximum 0.01 mmol ℓ^{-1}) than the surface of the lake in Movile Cave (maximum 0.1 mmol ℓ^{-1}). The presence of sulfate only at the surface of the water indicates that hydrogen sulfide is oxidized only at the water–air interface.

The surface of the limestone walls in the Lake Room and in Air Bell #1 (see Fig. 17.3) is covered by gypsum deposits up to 6 cm thick. Diaconu and Morar (1993) showed that the saccharoidal crusts covering the walls of the Lake Room are composed solely of gypsum. The small concentrations of sulfate in the cave lake cannot account for the presence of the thick gypsum deposits on the cave walls. The amount of gypsum deposited on the experimental limestone slices placed 5 cm above the surface of the water was on average 2.67 g m^{-2} yr^{-1}, which is the equivalent of 3.1 g m^{-2} yr^{-1} of limestone dissolved as a result of sulfuric acid corrosion.

The difference in the stable isotope ratios of sulfur between hydrogen sulfide in the water ($\delta^{35}S_{CDT} = +4‰$) and gypsum on the walls ($\delta^{35}S_{CDT} = +1‰$) may be the effect of a biological fractionation by sulfide-oxidizing microorganisms living on the surface of the cave walls.

The non-carbonate residue, mostly clay, contained in the limestone is not affected by sulfuric acid corrosion. Thus, in the lower level of the cave, thick layers of clay

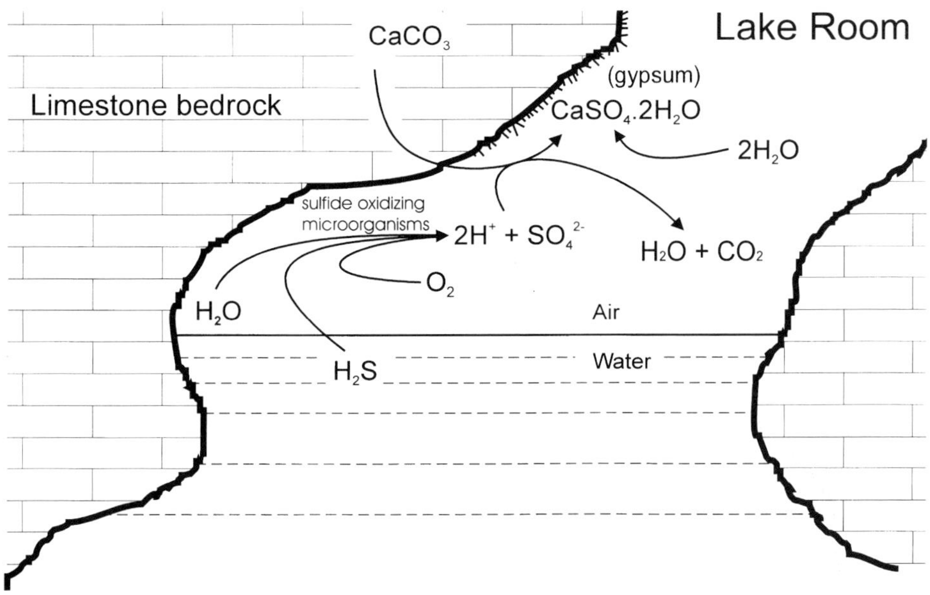

Fig. 17.5. Sulfuric acid corrosion in Movile Cave, a limestone dissolution mechanism currently shaping the morphology of the lower partially flooded level of Movile Cave.

are present on the cave walls, sometimes exposed, but more often covered by gypsum crusts.

In the remote air bells (#2 and #3), pH values of 3.8 to 4.5 were measured on the surface of the microbial mats covering the limestone walls. Protons and sulfate ions formed at the surface of the mat diffuse through the mat and react with the carbonate wall to form calcium sulfate which is deposited as gypsum crystals under the microbial mats.

When the water level rises in the groundwater aquifer, the gypsum deposits dissolve, exposing the bedrock to new episodes of sulfuric acid corrosion. The thick gypsum deposits on the cave walls in the Lake Room and in Air Bell #1 indicate that there has been no major increase in the groundwater level in the Movile region within the last several thousand years.

The effects of sulfuric acid dissolution of limestone can be seen especially in the Lake Room where the bedrock is severely corroded near the water level (Fig. 17.6). These corrosion patterns are similar to those found in other sulfide-rich caves such as Grotta di Frasassi (Galdenzi, 1990; see also Chapter 19, this volume). No effects of sulfuric acid speleogenesis are encountered in the upper dry level of Movile Cave due to the absence of hydrogen sulfide in the atmosphere.

Condensation corrosion

The temperature of the cave walls ranges from 20.7 to 20.9°C in the lower level of the cave and from 19.0 to 20.1°C in the upper level. Additionally, the floor of the dry cave passages is 0.2 to 0.3°C warmer than the ceiling, because of the thermal flux (Fig. 17.7) generated by the warm sulfidic reservoir. The warm vapor produced at the surface of the sulfidic lake (Fig. 17.7) ascends along the cave passages into the upper cave level where it condenses on the cooler cave walls and especially on the ceiling. Amounts of condensate of 0.1–0.15 ml cm^{-2} $month^{-1}$ were collected in several locations in the upper cave passage. Due to the decreased wall temperature, condensation is greater in the upper cave level and is lacking almost completely in the proximity of the lake where the cave walls are

Fig. 17.6. Limestone bedrock severely corroded near the water level, under the influence of sulfuric acid.

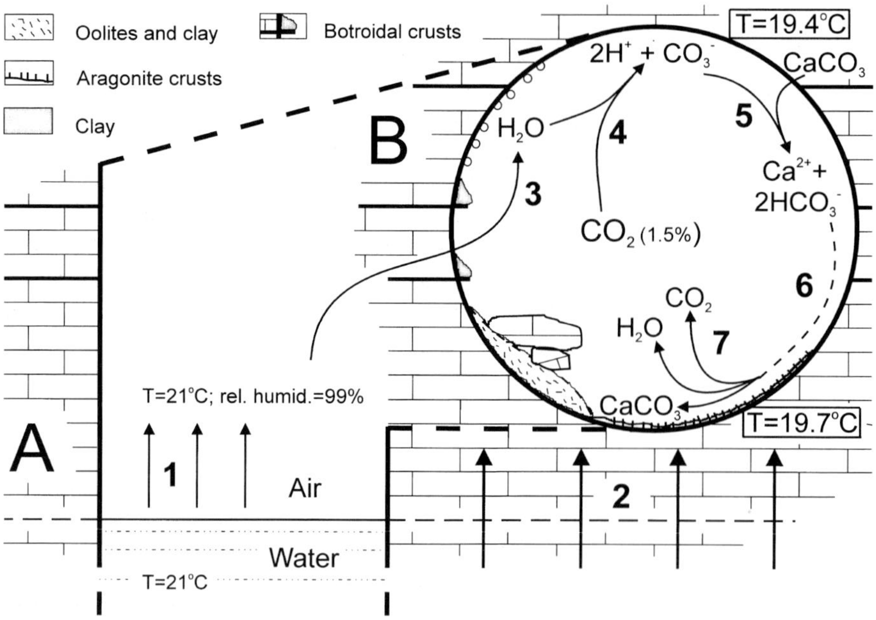

Fig. 17.7. Condensation corrosion in Movile Cave, a limestone dissolution mechanism shaping the morphology of the upper dry level of the cave. Schematic profile view of the two cave passages: (A) Lower passage; (B) Upper dry passage. (1) Evaporation of water at the surface of the thermal sulfidic lake. (2) Thermal flux through the bedrock. (3) Condensation of water vapors onto the colder cave walls and ceiling in the upper cave passage. (4) Dissolution of carbon dioxide into the condensate and formation of carbonic acid. (5) Reaction between carbonic acid and the carbonate bedrock with formation of bicarbonate. (6) Transport of bicarbonate along the walls. (7) Precipitation of aragonite at the base of the cave walls and on the floor of the cave passage.

warmer. There is no evidence of allochthonous surface water infiltrating the cave (Sarbu and Kane, 1995). However, condensation of water vapor on the cave walls appears to provide sufficient moisture for the process of condensation corrosion to take place. Through capillary movement the condensate drains down along the walls at rates of up to 10 cm $month^{-1}$.

Compared to the surface atmosphere, the atmosphere in the upper dry level of Movile Cave is depleted in oxygen and enriched in carbon dioxide. The high levels of carbon dioxide in the atmosphere of Movile Cave accelerate dissolution by the condensed water on the surface of the limestone walls. Carbon dioxide in the cave atmosphere is isotopically light ($\delta^{13}C = -22$ to -24‰; $\delta^{18}O = +4.9$ to 5.7‰) compared to surface carbon dioxide ($\delta^{13}C = -7.8$‰), indicating that methane oxidation (L. Vlasceanu, unpublished) and metabolic activity of the cave biota (Sarbu et al., 1996) are the main sources of carbon dioxide in Movile Cave. Additional carbon dioxide is produced from the sulfuric acid dissolution of the carbonate bedrock (Diaconu and Morar, 1993; Sarbu and Kane, 1995).

The superficial corroded layer of bedrock reaches a thickness of up to 8 cm and consists of uncemented oolites. Their spherical shape offers a higher resistance to corrosion compared with the matrix which is readily removed by corrosion. As a result of this preferential dissolution of the carbonate matrix (Horoi, 1994), oolites become detached from the wall and accumulate on the floor of the cave passage, primarily at the base of the walls. Not being affected by condensation corrosion, the non-carbonate component of the bedrock accumulates on the surface of the corroded walls or may become detached and accumulate at the base of the walls along with oolites (Fig. 17.7).

The numerous botryoidal calcite crusts (Diaconu and Morar, 1993) that fill the fractures within the bedrock appear to be more resistant to corrosion than the bedrock. They often protrude from the cave walls forming small ledges on which the uncorroded residue of the limestone (i.e., clays) accumulates (Fig. 17.7). Examination of the edges of these crusts shows that they are eventually affected by corrosion which transforms them into a fine white powder often found on the passage floor.

The deep limestone bedrock is isotopically heavy both for carbon and oxygen ($\delta^{13}C_{PDB} = -2.6$ to -0.8‰;

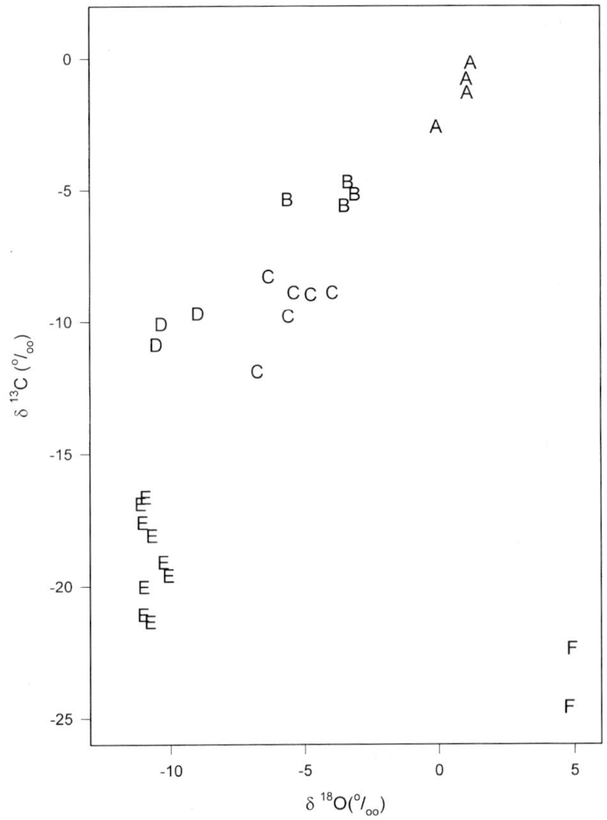

Fig. 17.8. Carbon and oxygen stable isotope ratios in samples of carbonate and carbon dioxide from Movile Cave. A, limestone bedrock not affected by corrosion. B, corroded surface of passage walls and detached oolites. C, boulders of limestone bedrock laying on the passage floor. D, botroidal calcite crusts filling the fractures within the bedrock. E, aragonite crusts covering the passage floor and the base of the walls. F, carbon dioxide from Movile Cave.

$\delta^{18}O_{PDB} = 0$ to 1.2‰)[3] (A's in Fig. 17.8). The surface of the cave walls affected by corrosion and the detached oolites, as well as the boulders of bedrock that lie on the passage floor, are isotopically lighter than the limestone bedrock (B's and C's in Fig. 17.8). The botryoidal calcite crusts protruding from the walls are also isotopically lighter than the bedrock (D's in Fig. 17.8).

The floor of the cave passage as well as the base of the walls in the upper dry level of Movile Cave are covered by carbonate crusts (Sarbu and Kane, 1995). These deposits consist of aragonite as the main mineral (Diaconu and Morar, 1993; Engel and Lascu, 1996). The water film covering the cave wall is enriched in bicarbonate when it reaches the floor of the cave passage. Additionally, due to upward capillary movement, bicarbonate may also ascend through the limestone bedrock from the lower level of the cave (Diaconu and Morar, 1993). Upon reaching warmer temperatures and evaporative conditions at the base of the walls and on the floor of the cave passage, the capillary water approaches saturation with respect to calcite. Evaporation can actually drive the solution to supersaturation with respect to the more soluble aragonite. Magnesium ions present in the water inhibit precipitation of calcite, causing aragonite to be the main precipitate, even though aragonite is more soluble. Bicarbonate dissociates as a consequence of calcium carbonate precipitation (Eq. 17.6).

$$2HCO_3^- + Ca^{2+} \rightarrow H_2O + CO_2 + CaCO_3 \quad 17.6$$

The aragonite crusts are isotopically light with respect to both carbon ($\delta^{13}C_{PDB} = -16.6$ to -21.4‰) and oxygen ($\delta^{18}O_{PDB} = -10.9$ to -11.1‰) (E's in Fig. 17.8). This suggests that the bicarbonate generating the crusts results from the isotopically light carbon dioxide in the cave atmosphere (F's in Fig. 17.8), the water vapor condensing on the cave walls ($\delta^{18}O_{PDB} = -41$‰), and the isotopically light limestone bedrock. Additionally, the isotopically very light carbon in the crusts suggest a possible epigenetic interaction between the aragonite crusts and the light carbon dioxide in the cave atmosphere.

Conclusions

The presence of gypsum deposits on the limestone walls in the vicinity of the sulfidic water (Sarbu, 1990; Sarbu and Popa, 1992; Lascu et al., 1993; Diaconu and Morar, 1993; Sarbu and Kane, 1995), the low pH values measured on the surface of the microbial mats covering the walls in the air bells, the difference between the $\delta^{35}S$ values in sulfide and sulfate, and the intensely corroded carbonate bedrock in the lower level of the cave suggest that sulfuric acid corrosion of limestone plays an important role in shaping the morphology of the lower, partially submerged, level of Movile Cave.

The presence of corroded walls in the upper cave passages despite the absence of water infiltration from the surface, the difference in temperature between the water in the lower cave level and the walls in the upper level, the high carbon dioxide concentration throughout the cave, and the isotopically light carbonate crusts

[3] The suffix "PDB" indicates that Pee Dee Belemnite was used as the standard.

covering the floor of the cave passage suggest that condensation corrosion is the major limestone dissolution mechanism currently affecting the morphology of the upper dry level of Movile Cave.

CHEMOAUTOTROPHIC FOOD WEB IN MOVILE CAVE

Introduction

Cave communities generally are dependent on input of allochthonous organic material of photosynthetic origin from the surface (Culver, 1982). As a result of this limited food base, caves usually support communities of few species and low biomass (Poulson, 1971). Unlike communities of non-thermo mineral caves, the aquatic and terrestrial communities of Movile Cave support high densities of species and a large biomass (Sarbu, 1991; Sarbu and Kane, 1995). The lack of a natural entrance and the absence of water infiltration from the surface exclude the input of photosynthetic food into this ecosystem. The unusual abundance of organisms in this hypogean environment and the apparent absence of food inputs from the surface suggest that the aquatic and terrestrial communities depend on an autochthonous non-photosynthetic food source. A layer of elemental sulfur, 10–25 μm thick, was identified by electron dispersion analysis within the floating mats collected from the remote air bells of Movile Cave (Sarbu and Popa, 1992; Sarbu et al., 1994a). Sulfur is a byproduct of sulfide oxidation and accumulates in the mat at the sulfide–oxygen interface. This discovery first suggested that chemoautotrophic sulfide-oxidizing microorganisms are present and active within the cave microbiota.

Assays performed for Ribulose-1, 5-bisphosphate-Carboxylase/Oxygenase (RuBisCO) (Sarbu et al., 1994b) showed an increase in acid-stable reaction products compared with the controls. This indicates that RuBisCO is present and active in the cave microbiota. RuBisCO is an enzyme involved in carbon dioxide fixation and its presence is an indicator of autotrophy. Recent microbiological investigations have shown that *Thiobacillus thioparus,* a chemoautotrophic sulfur-oxidizing microorganism, is present in the floating microbial mats in Movile Cave (Vlasceanu et al., 1997).

The absence of artificial gamma-emitting radionuclides (^{137}Cs and ^{134}Cs) in the water and in the sediment from the submerged level of the Movile Cave (F. Baciu, pers. commun., 1993) were the first data suggesting that no significant amounts of water and sediment are transported into the cave from the surface. As a result of the Chernobyl accident (April, 1986), surface waters and sediments in Dobrogea contain high concentrations of these nuclides. The absence of diatoms and fecal coliforms in the sulfidic groundwaters at Mangalia (Sarbu et al., 1994b) also indicate the isolation of the groundwater ecosystem from the surface. Additional evidence supporting the isolation of the Movile Cave ecosystem is the absence of dehydrogenase activity in samples of clay collected from the limestone strata covering Movile Cave, in contrast with the high activity levels at surface locations and near the cave lake (Sarbu et al., 1994b). Dehydrogenase activity is a general indicator of microbial activity, and the very low (undetectable) levels indicate the scarcity of metabolically active microorganisms in the subsoil above the cave and in most of the upper cave passages, and suggest that little if any organic material seeps in from the surface.

The objective of this study was to demonstrate chemoautotrophic production in the microbial mats observed in the Movile Cave system, and to determine whether it is also the food base for the invertebrate species inhabiting this ecosystem (Sarbu et al., 1995, 1996). Incubations with [^{14}C]bicarbonate were used to demonstrate fixation of carbon by the Movile Cave microbiota. Stable isotope ratio analysis (SIRA) of carbon and nitrogen was done to determine the relationship of selected species to the chemoautotrophic production in the cave. Stable-isotope ratio analysis is a valuable method for studying food webs because organisms fractionate isotopes of carbon (^{13}C/^{12}C) and nitrogen (^{15}N/^{14}N) in predictable ways. Consumers are typically enriched by approximately 1–2‰ in ^{13}C and by 3–5‰ in ^{15}N, relative to their food source (Paull et al., 1985; Ehleringer et al., 1986; Peterson and Fry, 1987). Techniques used in the Movile Cave studies are described in detail in the Appendix, pp. 339–340.

The food web – results and discussion

Increase in the radioactivity of the microbial lipids during a period of incubation indicated uptake of radiolabeled bicarbonate, suggesting chemoautotrophic carbon fixation. The increase in labeled carbon dioxide fixed in lipids is correlated with the length of the incubation. The radioactivity of lipids increases from

0.216 to 4.264 10^{-9} DPM nmol^{-1} of lipid phosphate for an incubation time of four hours (Sarbu et al., 1991).

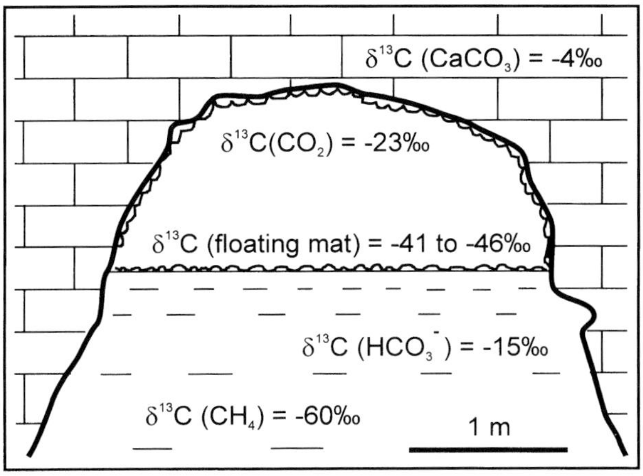

Fig. 17.9. Cross-sectional view of an air bell in Movile Cave. The atmosphere in the air bells is depleted in oxygen (7–10%), enriched in carbon dioxide (2.5–3.5%) and contains 0.5–1.0% methane.

All forms of inorganic carbon in the cave are isotopically light (Fig. 17.9). Carbon dioxide in the cave atmosphere ($\delta^{13}C = -22$ to -24‰) is a mixture of heavier carbon dioxide from the sulfuric acid dissolution of limestone ($CaCO_3$) ($\delta^{13}C = -4$‰) (Diaconu and Morar, 1993) and lighter carbon dioxide presumably resulting from oxidation of isotopically light methane ($\delta^{13}C = -60$‰; Fig. 17.9) ascending from the deep aquifer. Respiration of the cave biota is also a source of isotopically light carbon dioxide. Bicarbonate in the water ($\delta^{13}C = -15$‰; Fig. 17.9) is at isotopic equilibrium with carbon dioxide in the cave atmosphere (Goericke et al., 1994). Assuming a fractionation of approximately -25‰ between inorganic carbon and the biomass of chemoautotrophs (Degens, 1969; Ruby et al., 1987; Vlasceanu et al., 1997), the carbon isotope values for the microbial mat ($\delta^{13}C = -41$ to -46‰; Fig. 17.9) are as expected. Atmospheric carbon dioxide at the surface is isotopically heavier ($\delta^{13}C = -7.6$‰) and the $\delta^{13}C = -28$‰ observed in aquatic macrophytes from the surface aquatic habitat (point A_4, Fig. 17.10) is consistent with an expected isotopic fractionation of carbon of -20‰ by C_3 plants (O'Leary, 1981, 1988).

Ammonium in the cave water is isotopically heavy ($\delta^{15}N = 19.9$‰). However, nitrogen in the microbial mat is isotopically lighter (mean $\delta^{15}N = -9.1$‰). A fractionation of stable isotopes of nitrogen of -0 to -20‰ has been reported for ammonium uptake by microbes when ammonium is not limited (Macko and Ostrum, 1994). Further, nitrification (i.e., $NH_4^+ \rightarrow NO_3^-$) can result in a depletion in ^{15}N on the order of -10 to -30‰ (Yoshinari and Koike, 1994). Whether either of these factors is the cause of the observed differences in nitrogen isotopes between ammonium in the water and nitrogen in the microbial mats requires a better understanding of the nitrogen cycle in Movile Cave.

Organisms sampled from Movile Cave are isotopically lighter in both carbon and nitrogen than organisms sampled from the three surface habitats (Fig. 17.10). Isotopic ratios observed in surface-dwelling organisms (Fig. 17.10) reflect dependence on photosynthetic production, whereas values for organisms from Movile Cave suggest dependence on the chemoautotrophic production occurring in the cave (Fig. 17.10). This includes three pairs of cave/surface congeners (*Cryptops* T_4 and C_5; *Haemopis* A_3 and C_2; and *Nepa* A_2 and C_1; Fig. 17.10) for which isotopic values reflect the habitats in which the species occur.

If the Movile Cave ecosystem were dependent on allochthonous input of organic material from the surface, organisms in the cave should be isotopically heavier for both carbon and nitrogen than comparable organisms from the surface. This is the case for the two organisms sampled in Limanu Cave (Fig. 17.10, L_1 and L_2) which are isotopically heavier than all (nitrogen) or nearly all (carbon) of the species sampled from the surface habitats. Limanu Cave does not communicate with the sulfidic aquifer and is non-thermal (temperature approximately 14°C). None of the endemic species of Movile Cave has ever been observed or collected in Limanu Cave. There is organic matter of surface origin in Limanu Cave including wood, plant detritus, and vertebrate feces and the isotope data suggest that these provide the food base for organisms in this cave.

Two species of amphipods, *Niphargus stygius* and *Pontoniphargus racovitzai,* and one species of aquatic isopod *Asellus aquaticus infernus,* collected from the three hand-excavated sulfidic wells are also known from Movile Cave. All three samples are enriched in the heavy isotopes of both carbon and nitrogen (Fig. 17.10) relative to samples from Movile Cave, including *N. stygius* (Fig. 17.10, G_3 and W_1), which was collected from both locations. Carbon and nitrogen isotope values from the wells are intermediate between the values observed in Movile Cave and values for samples from surface habitats. Surface invertebrates were often noted on the well walls and debris, such

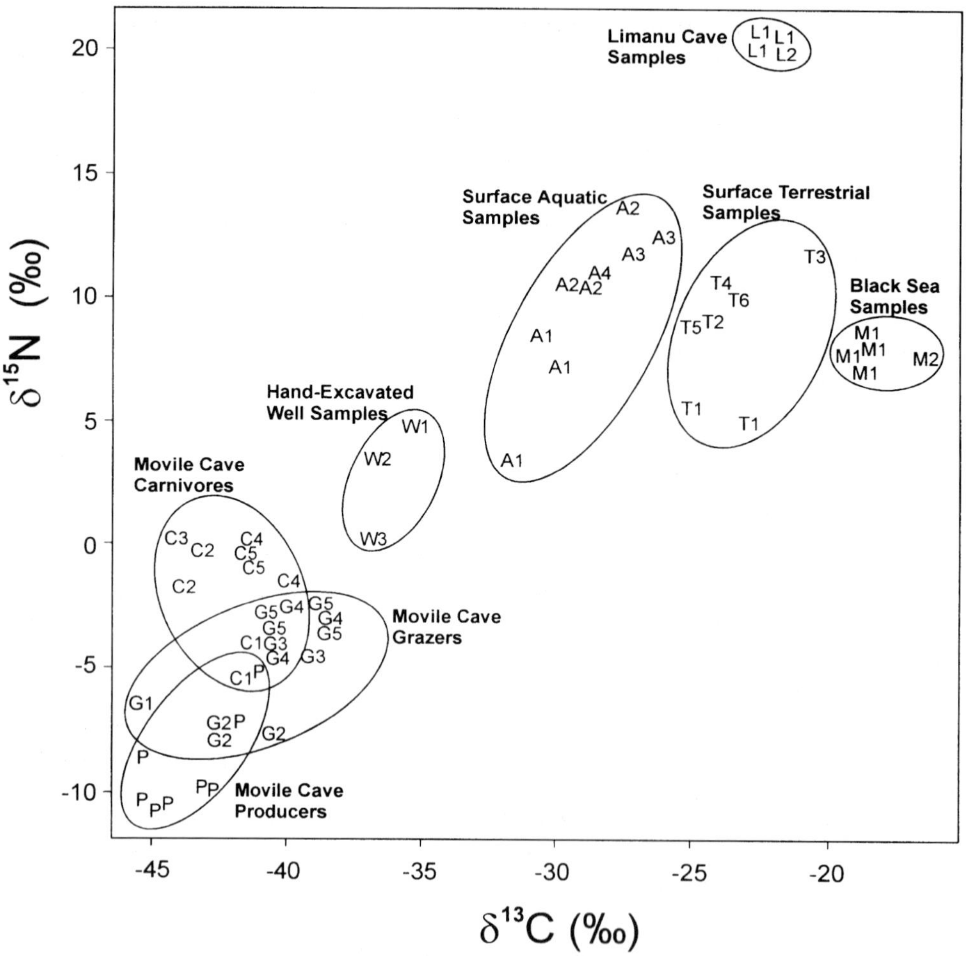

Fig. 17.10. Stable isotope ratios of carbon and nitrogen for organic samples. Movile Cave samples: Producers: P, floating microbial mats. Grazers: G_1, *Allolobophora* sp. (Oligochaeta, Lumbricidae); G_2, *Archiboreoiulus* sp. (Diplopoda, Julidae); G_3, *Niphargus stygius* (Amphipoda, Gammaridae); G_4, *Trachelipus troglobius* (Isopoda, Trachelipidae); G_5, *Armadillidium tabacarui* (Isopoda, Armadilliidae). Carnivores: C_1, *Nepa anophthalma* (Hemiptera, Nepidae); C_2, *Haemopis caeca* (Hirudinea, Haemopidae); C_3, *Medon dobrogicus* (Coleoptera, Staphylinidae); C_4, *Agraecina cristiani* (Arachnida, Liocranidae); C_5, *Cryptops anomalans* (Chilopoda, Cryptopidae). Limanu Cave samples: L_1, staphylinid beetle (Coleoptera, Staphilinidae); L_2, clubionid spider (Arachnida, Clubionidae). Surface aquatic samples: A_1, *Asellus aquaticus aquaticus* (Isopoda, Asellidae); A_2, *Nepa cinerea* (Hemiptera, Nepidae); A_3, *Haemopis sanguisuga* (Hirudinea, Haemopidae); A_4, aquatic macrophytes. Surface terrestrial samples: T_1, *Armadillidium* sp. (Isopoda, Armadillidiidae); T_2, *Trachelipus* sp. (Isopoda, Trachelipidae); T_3, *Ronchus* sp. (Pseudoscorpiones, Neobisiidae); T_4, *Cryptops anomalans* (Chilopoda, Cryptopidae); T_5, *Scolopendra* sp. (Chilopoda, Scolopendridae); T_6, clubionid spider (Arachnida, Clubionidae). Black Sea marine samples: M_1, *Mytilus* sp. (Mollusca, Bivalvia, Mytilidae); M_2, marine amphipods (Crustacea, Amphipoda, Gammaridae). Well samples: W_1, *Niphargus stygius* (Amphipoda, Gammaridae); W_2, *Pontoniphargus racovitzai* (Amphipoda, Gammaridae); W_3, *Asellus aquaticus infernus* (Isopoda, Asellidae).

as wood, was found in the water. The isotopic values for the wells suggest, therefore, that the organisms sampled use a mixed diet of chemoautotrophically produced organic material and organic material of surface origin.

Isotopic differences among Movile Cave samples correspond with the presumptive trophic status of the organisms sampled. Mean $\delta^{15}N$ values for the mat, which includes the primary producers (P's of Fig. 17.10), is –9.11‰ ($n=8$), for grazers, presumably feeding on the mat (G's of Fig. 17.10), the value is –4.6‰ ($n=13$) and for presumed predators (C's of Fig. 17.10) $\delta^{15}N=-1.58$‰ ($n=9$). These differences in $\delta^{15}N$ between successive trophic levels of 4.51‰ (G's vs. P's) and 3.02‰ (C's vs. G's) are statistically significant [Tukey's "honestly significant difference" (hsd) test – see, for instance, Steele and Torrie (1960), $p < 0.01$] and the differences in $\delta^{15}N$ were within the range of 3 to 5‰ expected. Results for the carbon isotopes are less clear. Grazers (mean $\delta^{13}C = -40.66$‰, $n = 13$) are significantly enriched in ^{13}C by 3.08‰ relative to the mat (mean $\delta^{13}C = -43.74$‰, $n = 8$), which is greater than expected (Tukey's hsd test, $p < 0.01$). However, the mean for predators [$\delta^{13}C = -42.16$‰ ($n = 9$)] is intermediate between, and not significantly different from, the mat and grazer values, which is not as expected. Many cave-dwelling organisms are dietary generalists (Culver, 1982) and thus it is possible that these presumptive "predators" are in fact omnivores, perhaps including mat material in their diets. Were this the case, one would also expect it to be reflected in the isotope data for nitrogen, which it is not. Cave-dwelling organisms also tend to have attenuated appendages and a general increase in surface-to-volume ratio (Culver, 1982). For invertebrates this can increase the proportion of exoskeleton and chitin in the individual. Chitin may be depleted in ^{13}C by approximately 2‰ relative to the rest of the organism (Macko et al., 1990; Rau et al., 1990). Thus, the carbon values for predators may reflect this 'chitin effect'.

The results of the incubations with radiolabeled bicarbonate and the stable isotope data show that the Movile Cave ecosystem derives its organic carbon from *in situ* chemoautotrophic production (Sarbu and Popa, 1992; Sarbu et al., 1996). All other limestone cave ecosystems studied require allochthonous inputs of organic material of photosynthetic origin from the surface. In addition, the rich terrestrial community in Movile Cave appears to be the first such community with a chemoautotrophic energy base. All other chemoautotrophically-based communities have been described from marine habitats (Conway et al., 1994) or fresh water (Stevens and McKinley, 1995).

THE FAUNA

Introduction

In the ten years since the discovery of Movile Cave, 33 new taxa have been identified, most of which are endemic to the system of limestone fractures and small cave passages associated with the thermomineral sulfidic waters at Mangalia. The terrestrial community consists of 30 obligate cave-dwelling invertebrate species, including 24 endemic species, and the aquatic community consists of 18 species, with nine endemics (Table 17.3). Many of the aquatic and terrestrial species have also been found in hand-excavated wells in villages nearby and in sulfidic springs located within the geographic extent of the sulfidic groundwater aquifer at Mangalia (Lascu et al., 1993; Sarbu and Kane, 1995; Sarbu et al., 1996; Turk et al., 1996).

The terrestrial fauna

Several of the terrestrial invertebrates inhabiting Movile Cave are related to species that today inhabit areas with warmer climates (i.e., the Mediterranean region, northern Africa). With few exceptions, most of the species exhibit a high degree of troglomorphy (i.e., reduced pigmentation and eyes, and elongated appendages) suggesting that these forms have been isolated underground for a considerable period (Sarbu and Kane, 1995).

Three species of pseudoscorpions have been found in Movile Cave. *Chthonius monicae* is present in the remote air bells as well as in the Lake Room at densities of 3–5 individuals m^{-2} (Table 17.4). It is depigmented and is the only blind species of this genus. *Roncus ciobanmos* and *R. dragobete* are two new species described initially from Movile Cave. They were collected in the upper dry cave passage between 1990 and 1992, before and shortly after the bottom of the artificial shaft was hermetically sealed, thus stopping the invasion of the surface and soil fauna into Movile Cave. After 1992 no additional specimens of the two species were found in Movile Cave, but they were instead collected from soil samples (Georgescu and Căpuşe, 1994). This suggests that small populations of these two species entered Movile Cave through the artificial shaft and invaded the network of cave passages associated with the sulfidic groundwater between 1986 and 1992, prior to the installation of the gates.

Five new species of spiders were described from Movile Cave. They are all endemic to this chemoautotrophically based groundwater ecosystem. *Mariana mihaili* is a new genus of Theridiidae and the first troglobitic species of this family discovered in Romania. *Lepthyphantes constantinescui* is the first blind species of the genus. The absence of eyes and the elongation of the appendages suggest that this species has a long evolutionary history in the

Table 17.3
Terrestrial cave-limited species in Movile Cave[1]

ANNELIDA

OLIGOCHAETA

Lumbricidae

Allolobophora sp.[2]

CRUSTACEA

ISOPODA

Trichoniscidae

Caucasonethes n. sp.[2]

Haplophtalmus n. sp.[2]

Trachelipodidae

Trachelipus troglobius[2]

Armadillidiidae

Armadillidium tabacarui[2]

ARACHNIDA

PSEUDOSCORPIONES

Chthoniidae

Chthonius monicae[2]

Neobisiidae

Roncus dragobete[2]

Roncus ciobanmos[2]

ARANEAE

Theridiidae

Marianana mihaili[2]

Linyphiidae

Lepthyphantes constantinescui[2]

Clubionidae

Agraecina cristiani[2]

Nesticidae

Nesticus n. sp.[2]

Hahniidae

Hahnia caeca[2]

ACARINA

Nicoletiellidae

Labidostoma motasi[2]

CHILOPODA

GEOPHILOMORPHA

Geophilidae

Geophilus sp. cfr. *insculptus*

Clinopodes trebevicensis

SCOLOPENDROMORPHA

Cryptopidae

Cryptops anomalans

SYMPHYLA

Scolopendrellidae

Symphynella sp.

DIPLOPODA

JULIDA

Julidae

Archiboreoiulus n. sp.[2]

INSECTA

COLLEMBOLA

Onychiuridae

Deuteraphorura movilae[2]

Entomobryidae

Heteromurus cfr. *nitidus*

Cyphoderidae

Onchopodura vioreli[2]

DIPLURA

Campodeidae

Plusiocampa isterina[2]

Plusiocampa euxina[2]

COLEOPTERA

Staphylinidae

Medon dobrogicus[2]

Pselaphidae

Tychobythinus n. sp.[2]

Decumarellus sarbui[2]

Clivinidae

Clivina subterranea[2]

[1] Based on the publications of Bernasconi (1991, 1994), Boghean (1989), Condé (1993, 1996), Ćurčić et al. (1993), Decu et al. (1994a,b), Decu and Georgescu (1994), Georgescu (1989, 1994), Georgescu and Căpuşe (1994), Georgescu and Decu (1994), Georgescu and Sarbu (1992), Grossu and Negrea (1989), Gruia (1989), Gruia et al. (1994), Iavorschi (1992), Karaman (1990), Karaman and Sarbu (1995, 1996), Manoleli et al. (1998), Negoescu (1989), Negrea (1993), Pleşa (1989), Poggi (1994), Poinar and Sarbu (1994), Sarbu (1990, 1991), Sarbu and Kane (1995), Sarbu and Popa (1992), Tabacaru and Boghean (1989), Turk et al. (1996), Turk-Prevorčnik and Blejec (1998) and Weiss and Sarbu (1994, 1996).

[2] Species which are endemic for the chemoautotrophically based groundwater ecosystem at Mangalia.

Table 17.4
Abundance of selected fauna in Air Bells 1, 2 and 3

Taxon	Air Bell 1	Air Bells 2, 3
GASTROPODA		
Heleobia dobrogica	very abundant	very rare
HIRUDINEA		
Haemopis caeca	present	present
PSEUDOSCORPIONES		
Chthonius monicae	3–5 m^{-2}	3–5 m^{-2}
ARANEAE		
Agraecina cristiani	3–6 m^{-2}	0.2 m^{-2}
Hahnia caeca	present	absent
Nesticus n. sp.	present	absent
Lepthyphantes constantinescui	present	absent
ISOPODA		
Asellus aquaticus infernus	present	absent
Armadillidium tabacarui	3–5 m^{-2}	100–200 m^{-2}
Trachelipus troglobius	15 m^{-2}	absent
Haplophthalmus sp.	100–200 m^{-2}	absent
Caucasonethes n. sp.	absent	present
DIPLOPODA		
Archiboreoiulus sp.	present	absent
CHILOPODA		
Cryptops anomalans	1–2 m^{-2}	0.2 m^{-2}
COLLEMBOLA		
Heteromurus nitidus	rare	>100 m^{-2}
Onchopodura vioreli	very abundant	rare
HETEROPTERA		
Nepa anophthalma	present	absent
COLEOPTERA		
Medon drobogicus	1 m^{-2}	5–10 m^{-2}
Clivina subterranea	1–2 m^{-2}	0.2 m^{-2}
Decumarellus sarbui	present	absent
Trychobytinus n. sp.	present	absent

subterranean environment (Georgescu, 1989). It was found only in the Lake Room area and in Air Bell #1 (Table 17.4). *Agraecina cristiani* (Georgescu, 1989) is the only known blind species of clubionid spider. It is endemic to the sulfidic groundwater ecosystem at Mangalia (Weiss and Sarbu, 1996), and was recently found in several wells in the town of Mangalia, at distances of over two kilometers from Movile Cave (Sarbu and Kane, 1995). Another species of this genus, *A. canariensis,* inhabits lava tubes in the Canary Islands (Wunderlich, 1991). *Hahnia caeca* is the only blind species described from the family Hahniidae. The closest relatives of both *A. cristiani* and *H. caeca* are distributed in the Mediterranean region. *Nesticus* n. sp. is the only troglobitic species of the genus found in Dobrogea (Georgescu, 1994). All the spider species are found exclusively in the Lake Room and in Air Bell #1 (Table 17.4) where the concentrations of carbon dioxide are lower and the concentrations of oxygen are higher than in the upper cave passage.

Four new species of terrestrial isopods are endemic to Movile Cave. *Armadillidium tabacarui* is present in large numbers in the remote air bells rich in carbon dioxide and poor in oxygen (Table 17.4). *Trachelipus troglobius* is the largest terrestrial isopod in Movile Cave and is only present in the Lake Room and in Air Bell #1 (Table 17.4). *Haplophthalmus* n. sp. occurs in large numbers only in Air Bell #1 living on the surface of the cave walls (Table 17.4). *Caucasonethes* n. sp. is the smallest terrestrial isopod inhabiting Movile Cave. It lives on the surface of the microbial mats covering the walls in the remote air bells (Table 17.4).

Archiboreoiulus sp., the only diplopod inhabiting Movile Cave, is closely related to a species living in France (I. Tabacaru, pers. commun., 1993). It is present only in the Lake Room area and in Air Bell #1 (Table 17.4). The species has also been found in several of the wells in the town of Mangalia.

The centipede *Cryptops anomalans* is the largest terrestrial species inhabiting Movile Cave. It occurs only in the Lake Room area and in Air Bell #1. Surface studies have identified this species under stones in the Movile sinkhole and throughout the Mangalia region (Negrea, 1993). No morphological differences exist between the surface and the Movile Cave populations. It is not known whether the Movile Cave population is isolated genetically from the surface populations.

Three species of collembolans have been found in Movile Cave. *Onchopodura vioreli,* a new species described from Movile Cave, is present mainly in the Lake Room and in Air Bell #1 (Table 17.4). Its closest relatives appear to live in Asia and in the Mediterranean region (Gruia, 1989). *Onychiurus movilae* is also a Movile Cave endemic and inhabits the remote air bells of the cave. *Heteromurus nitidus* occurs primarily in the remote air bells (Table 17.4). Individuals of *H. nitidus* living in Movile Cave are only about half the size of individuals from populations living at the surface (M. Gruia, pers. commun., 1995).

Decumarellus sarbui, one of two new species of pselaphid beetles from Movile Cave, belongs to a new

Table 17.5
Aquatic cave-limited species in Movile Cave

PLATHELMINTHES

Turbellaria

Dendrocoelidae

Dendrocoelum sp.

ASCHELMINTHES

Nematoda

Rhabditidae

Protorhabditis n. sp.[1]

Panagrolaimidae

Panagrolaimus n. sp.[1]

Leptolaimidae

Chronogaster troglodytes[1]

Rotatoria

Habrotrochidae

Habrotrocha rosa

Habrotrocha bidens

ANNELIDA

Hyrudinea

Haemopidae

Haemopis caeca[1]

Aphanoneura

Aeolosomatidae

Aeolosoma hyalinum

Aeolosoma litorale

MOLLUSCA

Gastropoda

Moitessieriidae

Heleobia dobrogica[1]

CRUSTACEA

Ostracoda

Cyprididae

Pseudocandona sp. cfr. *eremita*[1]

Copepoda

Cyclopidae

Eucyclops subterraneus scythicus[1]

Tropocyclops prasinus

Harpacticidae

Parapseudoleptomesochra italica

Amphipoda

Gammaridae

Niphargus stygius[1]

Pontoniphargus racovitzai[1]

Isopoda

Asellidae

Asellus aquaticus infernus

INSECTA

Heteroptera

Nepidae

Nepa anophthalma[1]

[1] Species that are endemic for the chemoautotrophically based groundwater ecosystem at Mangalia.

genus closely related to the genus *Marellus* known from northern and eastern Africa (Poggi, 1994).

The aquatic fauna

The aquatic fauna inhabiting the sulfidic groundwater ecosystem at Mangalia consists of 18 species (Table 17.5). Oxygen from the cave atmosphere diffuses only in a very thin superficial layer of water less than a millimeter deep (Riess et al., 1999); in consequence, the fauna is present only in the superficial section of the water column. Some of the species inhabiting this aquifer display behavioral and morphological adaptations for this environment.

Haemopis caeca is the only species of troglobitic leech known. It is blind and lacks epidermal chromatophores. It occurs throughout the cave (Table 17.4) and feeds on oligochaetes (*Allolobophora* sp.) found on the banks of the cave lake. Its closest relative is *H. sanguisuga*, the only other palearctic species of the genus.

Heleobia dobrogica is the only snail inhabiting Movile Cave. It crawls on the sediments in shallow water or it "crawls" on the surface of the water.

It is only present in the oxygen-rich section of the cave (Table 17.5). The species was recently collected from several wells in the town of Mangalia containing sulfidic water.

A blind and unpigmented population of *Asellus aquaticus infernus* was discovered in Movile Cave (Negoescu, 1989; Turk et al., 1996). Additional blind and unpigmented populations were discovered later in several sulfidic springs and sulfidic hand-excavated wells in the Mangalia region (Turk et al., 1996). Eyed and pigmented populations of *A. aquaticus aquaticus* inhabit the surface waters in the Mangalia region. *Asellus aquaticus* is apparently one of the species more recently isolated in the Dobrogean groundwater ecosystem. It originated from southeast Asia, and its area of distribution was extended during the Pleistocene when extensive stagnant-water habitats were formed as glaciers retreated. Populations of *A. aquaticus* exhibiting various degrees of depigmentation and eye reduction have also been found in groundwaters in Slovenia and France (Henry and Magniez, 1983).

Niphargus stygius and *Pontoniphargus racovitzai* are the two species of amphipods inhabiting the sulfidic groundwaters in the Mangalia region. *Niphargus stygius* is present in large numbers in all sections of the cave, in wells, and in sulfidic springs located along the shore of the Black Sea at Mangalia. Investigations of wells in the village of Vama Veche, ten kilometers south of Mangalia, led to the discovery of a new species, *Niphargus decui,* which supports the hypothesis that the sulfidic aquifer does not extend much further south than Mangalia. *Pontoniphargus racovitzai* is a new genus of gammarid first described from wells in Mangalia by Dancau (1970). Unlike most gammarids, *Pontoniphargus* sp. do not swim on their side but rather "walk" on the under surface of the water. *P. racovitzai* is present in Movile Cave and in many of the sulfidic wells in the town of Mangalia. A second species of the genus, *P. ruffoi* was discovered in a small sulfidic spring near the village of Hagieni located approximately ten kilometers west of Mangalia (Karaman and Sarbu, 1996). Water chemistry and the stable-isotope data indicate that the sulfidic spring at Hagieni is presently not connected to the sulfidic groundwater aquifer at Mangalia. This could explain the presence of two different species of *Pontoniphargus* in the two aquifers.

The blind water scorpion *Nepa anophthalma* is perhaps one of the most interesting of the new species discovered in Movile Cave. It is the first known troglobitic aquatic heteropteran and the first species of cave-adapted heteropteran discovered in temperate caves. Completely blind and lacking pigmentation, it lives along the banks of the sulfidic lake in the cave and feeds on amphipods and isopods. *Nepa anophthalma* and *Haemopis caeca* are the top predators in the aquatic community of the sulfidic groundwaters at Mangalia.

DISCUSSION

The analysis of the food-web structure in the groundwater ecosystem at Mangalia shows that, in both aquatic and terrestrial communities, the number of predator species exceeds the number of grazers. The grazers however are present in very high population densities of up to several hundred individuals per square meter. Unlike most subterranean ecosystems which are characterized by scarcity of food, this groundwater ecosystem contains extensive microbial mats, a rich food supply for the large populations of aquatic and terrestrial grazers. Preliminary carbon/nitrogen ratio analysis of samples of microbial mats collected in Movile Cave show that the nutritional quality of the food is relatively high ($C/N = 5.7$). In addition, the mat is also high in nitrogen content ($N = 7.3\%$). The abundance and the high quality of the chemoautotrophic food supply could explain the abundance of the invertebrate communities inhabiting this groundwater ecosystem.

Although the geographic extent of the sulfidic aquifer at Mangalia does not exceed 70 square kilometers, the ecosystem associated with the thermal sulfidic groundwater is uniquely rich and diverse. No other cave ecosystem in the world contains so many endemic species. The mechanisms responsible for maintaining such a high biodiversity in a relatively small-size ecosystem are yet to be understood. Paleogeographical studies have shown that some of the fauna inhabiting the cave passages associated with the sulfidic groundwaters at Mangalia became isolated as early as 5.2 million years ago. Other species, however, invaded the groundwater ecosystem during the late Pleistocene, probably about 10 000 to 20 000 years ago. Additionally, the vast network of cave passages partially flooded by sulfidic groundwater offers a large variety of microhabitats which can be occupied by different species of invertebrates. The differences in abundance and species composition between the invertebrate communities of the Lake Room and the remote air bells has been attributed to differences in the chemical composition of the atmosphere. Differences in community structure were also observed in the Lake Room and in Air Bell #1, where some species occur only very close to the water surface while others are found only on the gypsum crusts covering the ceiling. Episodes of colonization over time, the potential for niche separation, and the availability of a rich and high-quality food source produced *in situ* are possible explanations for the high biodiversity found in this ecosystem.

GENERAL CONCLUSIONS

A peculiar ecosystem occurs in association with sulfidic groundwaters at Mangalia, southern Dobrogea, Romania. Thermal waters containing sulfide, ammonium and methane, and originating in deep captive aquifers, ascend toward the surface through deep faults and invade the superficial Sarmatian limestones. This groundwater aquifer extends over a 50 to 70 square kilometer area in the Mangalia region. Movile Cave, a 240 m long network of small cave passages located 20 m beneath the surface, is a window of access to this groundwater ecosystem.

The physicochemical characteristics of the groundwater accelerate the limestone dissolution at the air–water interface. The protons and sulfate ions produced by sulfide-oxidizing microorganisms dissolve the carbonate bedrock enlarging small fractures within the limestone. This leads to the development of a vast network of small cave passages at groundwater level. Additionally, the carbonate bedrock is further dissolved by acidic vapors condensing above the water table. Condensation corrosion is made possible by the temperature gradient between the groundwaters and the surface, and by the high concentration of carbon dioxide in the cave passages. These two mechanisms cause increased rates of limestone dissolution relative to other cave systems.

The redox interface occurring in the groundwater permits the existence of abundant microbial communities consisting of chemoautotrophic and methanotrophic microorganisms. The incorporation of radiolabeled carbon dioxide by the cave microbiota, the presence of RuBisCO activity in the microbial mats, and data for stable isotopes of carbon and nitrogen show that the groundwater ecosystem derives its organic carbon from *in situ* chemoautotrophic production. The rich terrestrial community in this groundwater ecosystem appears to be the first cave community with a chemoautotrophic energy base.

Thirty four of the 48 invertebrate species collected from the Movile Cave ecosystem are endemic, and densities of some of the populations reach several hundred individuals per square meter. In addition to being energetically unusual, the Movile Cave ecosystem also has an extraordinarily high diversity for a cave ecosystem. Although the causes for the high diversity and abundance of these organisms are not yet well understood, three characteristics of this ecosystem have the potential to contribute to high biodiversity:

(1) Colonization of this ecosystem occurred over extensive periods of time in several distinct episodes, permitting accumulation of species in evolutionary time;

(2) The system has considerable habitat heterogeneity, due largely to the physicochemical characteristics of the environment;

(3) The energy base of this ecosystem, produced *in situ* by chemoautotrophic microorganisms, is abundant and of high nutritional quality, and thus has the potential to support a large biomass.

ACKNOWLEDGMENTS

This work was funded in part by the Cave Research Foundation, the Department of Biological Sciences of the University of Cincinnati, the Fulbright Program, the National Geographic Society (Grants 4639-91 and 5460-95), the National Science Foundation (DEB-9420033), the National Speleological Society, the Romanian Academy of Sciences (Grant 3919-95), the Volkswagen Foundation, and the NSF-NATO Postdoctoral Program (DGE-9633928).

During the course of this study I received valuable help from the members of my Ph.D. research advisory committee, from the researchers of the Institutul de Speleologie "Emil Racovitza" in Bucharest, from the members of the GESS team, and from many other individuals. Valuable support for the water chemistry work was received from the "Dr. Radu Haller" Institute in Stuttgart, Germany. Special thanks to all who provided help during the field and laboratory work, and who did not spare any efforts for the protection and preservation of Movile Cave. This study would not have been possible without all this generous help.

Appendix A. METHODS

Field measurements and sampling

Samples of water and air were collected in Movile Cave at three-month intervals between 1993 and 1996 for chemical analysis. Temperature, pH, and redox potential were measured *in situ* using a portable EXTECH meter with a measurement precision of 0.1%. Temperature measurements of air and cave walls were also performed with this instrument. Measurements of the concentration of oxygen, carbon dioxide, hydrogen sulfide and water vapor in the atmosphere in various sectors of Movile Cave were also performed *in situ* using Draeger tubes and a Draeger hand pump. The limit of detection for hydrogen sulfide concentration was 1 ppm. Gas samples were also collected in 50 ml glass vials sealed with Teflon-lined rubber septa; they were later analysed by gas chromatography. Using a flat-surface electrode, pH values were measured on the surface of the microbial mats covering the limestone cave walls in air bells #2 and #3 (see Fig. 17.3).

Rates of sulfuric acid corrosion of the cave walls were determined by measuring the amount of gypsum deposited during a period of 12 months on limestone slices with a surface area of 25 cm^2, and then calculating the amount of calcium carbonate necessary for the production of the gypsum. Glass plates with an area of 200 cm^2 were suspended 1 m above the passage floor in various locations in the cave to determine the amount of liquid condensing on the cave walls. Plastic funnels placed under the plates collected the condensed water in containers for subsequent measurement. Fluorescein grains were applied to the cave walls to make visible the movement of the condensate along the cave walls. Observations were performed at 30-day intervals. The depth of penetration by a stainless steel wire 3 mm in diameter under fixed pressure was used to measure the thickness of the soft, corroded layer of bedrock throughout the entire cave.

Water samples for determinations of stable isotopes were collected in 20 ml glass vials sealed with Teflon-lined rubber septa. No head space was left in the vials to avoid changes in the isotopic signatures due to selective evaporation. Gas samples for stable-isotope determinations were collected in 20 ml glass vials or in 2 l Tedlar bags.

Chemical analyses

Samples of water were sent to the Research-Extension Analytical Laboratory, Ohio State University, for chemical analysis of dissolved constituents. Several water-chemistry parameters were measured *in situ* or at the nearby field research station in Mangalia: chloride (Cl^-) was measured by titration with silver nitrate; bicarbonate (HCO_3^-) was determined by titration with hydrochloric acid to pH 3.7 (Wetzel and Likenns, 1990); ammonium (NH_4^+), sulfate (SO_4^{2-}), nitrate (NO_3^-), and hydrogen sulfide (H_2S) were determined using a HACH DR2000 portable spectrophotometer. Methane concentrations in water were determined at the Department of Biology, Pennsylvania State University, using a specially adapted gas chromatograph.

Concentrations of methane in the atmosphere were determined by gas chromatography in the Department of Civil Engineering, University of Cincinnati.

Isotopic analysis

Water samples were sent to the Department of Geological Sciences, University of Michigan, for analysis of stable isotopes of oxygen, hydrogen, and carbon (in bicarbonate). Some samples were analysed at the Institute for Isotopic Technology, Cluj, Romania, at the Department of Geology, McMaster University, Hamilton, Canada, and at the Geochron Laboratories, Krueger Inc., Cambridge, Massachusetts, U.S.A. The only form of biologically available nitrogen detected in the waters of Movile Cave was ammonium (NH_4^+) (Sarbu and Popa, 1992). The ratios of stable isotopes of nitrogen in the ammonium of cave water were analysed at the Department of Environmental Sciences, University of Virginia. All the results are expressed as delta (δ) values (see footnote, p. 323).

Samples of atmospheric carbon dioxide were analysed at the Department of Geological Sciences, University of Illinois. Samples of bedrock limestone and aragonite crusts collected in Movile Cave were sent to the Department of Geological Science, University of Michigan, for analysis of stable isotopes of carbon and oxygen in carbonate. Samples of gypsum and of sulfide fixed with cadmium acetate were collected in the lower level of the cave and sent to the Biogeochemical Laboratories, Indiana University, for analysis of stable isotopes of sulfur.

Samples of microbial mat and of 10 of the 48 species of invertebrates taken from Movile Cave were sent to the Water Research Center, University of Alaska, Fairbanks, for isotopic analysis. Some of these cave-limited species were also collected from three hand-excavated wells which intercept the sulfidic aquifer in the area. For purposes of comparison, samples of plant material and invertebrates, ecologically similar to those sampled from the cave system, were taken in the vicinity from a surface terrestrial community, from a non-sulfidic surface aquatic community, and from the Black Sea. Further, samples of invertebrates were taken from Limanu Cave, a nearby cave system (Fig. 17.2) which receives inputs of organic matter from the surface and does not connect with the sulfidic aquifer. Stable-isotope ratios were also analysed in samples of carbon dioxide from the surface and in samples of CO_2, HCO_3^-, CH_4 and $CaCO_3$ (limestone) from the cave. Inorganic carbon samples were analysed at Geochron Laboratories, Krueger Enterprises, Cambridge, Massachusetts. Results are expressed as delta (δ) values.

Incubations with [^{14}C]bicarbonate

Samples of floating mat from Air Bell #2 in Movile Cave were collected in sterile plastic bags, homogenized and mixed with sulfidic water from the cave lake. Samples of 1.2 ml of the resulting homogenate were taken up in 5 ml plastic syringes, each sample containing approximately 70 mg of biomass. Eighty microliters of [^{14}C]bicarbonate solution were added, representing 0.7207 μCi ml^{-1} of sample. The incubation was done in the cave at 20°C in total darkness for 0, 0.5, 2, 4, and 12 hours. After the incubation, total lipids were extracted with chloroform–methanol (Bligh and Dyer, 1959). The microbial biomass in the extract of the surface mat was measured by analysis of the phospholipid component of the extracted lipids (Findlay et al., 1989). The amount of fixed carbon was estimated by measuring the radioactivity of the microbial lipids extracted at the end of the incubation time (McKinley et al., 1982). The ratio between radioactive bicarbonate incorporated in the microbial lipids and biomass was used to characterize the chemoautotrophic microbial activity.

REFERENCES

Airinei, S., 1980. *Radiografia Geofizică a Subsolului României.* Editura Ştiinţifică şi enciclopedică, Bucharest, 228 pp.

Bakalovicz, W.J., Ford, D.C., Miller, T.E., Palmer, A.N. and Palmer, M.V., 1987. Thermal genesis of dissolution caves in the Black Hills, South Dakota. *Geol. Soc. Am. Bull.,* 99: 729–738.

Ballard, R.D., 1977. Notes on a major oceanographic find. *Oceanus,* 20: 34–44.

Bernasconi, R., 1991. Sur un gasteropode prosobranche de la grotte Peştera Movile, Mangalia Roumanie: *Semisalsa dobrogica* (Grossu, Negrea, 1989) comb. nov. (Hydrobiidae, Littoridininae, Heleobiini). *Mem. Biospeol.,* 18: 237–243.

Bernasconi, R., 1994. Mollusca. In: C. Juberthie and V. Decu (Editors), *Encyclopaedia Biospeologica.* CNRS, Moulis, pp. 53–58.

Bligh, E.G. and Dyer, W.J., 1959. A rapid method of total lipid extraction and purification. *Can. J. Biochem. Physiol.,* 37: 911–917.

Boghean, V., 1989. Sur un pseudoscorpion cavernicole nouveau, *Chthonius (C) monicae* n.sp. (Arachnida, Pseudoscorpionida, Chthoniidae). *Misc. Speol. Romanica,* 1: 77–83.

Brock, T.D., Madigan, M.T., Martinko, J.M. and Parker, J., 1994. *Biology of Microorganisms,* 7th edition. Prentice Hall, Englewood Cliffs, 909 pp.

Buck, M.J., Ford, D.C. and Schwarcz, H.P., 1994. Classification of cave gypsum deposits derived from oxidation of H_2S. In: I.D. Sasowsky and M.V. Palmer (Editors), *Breakthroughs in Karst Geomicrobiology and Redox Geochemistry.* Karst Water Institute, Charles Town, WV, pp. 5–9 (special publication).

Cavanaugh, C.M., 1983. Symbiotic chemoautotrophic bacteria in marine invertebrates from sulfide rich habitats. *Nature,* 302: 58.

Cavanaugh, C.M., 1993. Methanotrophic–invertebrate associations: symbioses from deep-sea cold seeps and hydrothermal vents. In: R. Guerrero and C. Pedros-Alio (Editors), *Trends in Microbial Ecology.* Spanish Society for Microbiology, Barcelona, pp. 227–230.

Cavanaugh, C.M., Levering, P.R., Maki, J.S., Mitchell, R. and Lidstorm, M.E., 1987. Symbiosis of methylotrophic bacteria and deep-sea mussels. *Nature,* 325: 346.

Childress, J.J., Fisher, C.R., Brooks, J.M., Kennicutt II, M.C., Bidigare, R. and Anderson, A.E., 1986. A methanotrophic marine mulluscan (Bivalvia, Mytillidae) symbiosis: mussels fueled by gas. *Science,* 233: 1306–1308.

Cigna, A.A. and Forti, P., 1986. The speleogenetic role of airflow caused by convection. *Int. J. Speleol.,* 15: 41–52.

Collingnon, B., 1983. Spéléogenèse hydrothermale dans les Bibans (Atlas Tellien-nord de l'Algerie). *Karstologia,* 2: 45–54.

Condé, B., 1993. Une ligne danubiene de genre *Plusiocampa* (Diploures, Campodeides). *Rev. Suisse Zool.,* 100: 735–745.

Condé, B., 1996. Diploures Campodeides de la Peştera de la Movile (Movile Cave), Dobroudja meridionale (Roumanie). *Rev. Suisse Zool.,* 103: 101–114.

Constantinescu, T., 1989. Considerations sur la zone karstique de "La Movile" (Mangalia, Dobrogea de Sud, Roumanie). *Misc. Speol. Romanica,* 1: 7–12.

Conway, N.M., Kennicutt II, M.C. and Van Dover, C.L., 1994. Stable isotopes in the study of marine chemosynthetic-based ecosystems. In: K. Lajtha and R.H. Michener (Editors), *Stable Isotopes in Ecology and Environmental Science.* Blackwell, Oxford, pp. 158–186.

Corliss, J.B., Dymond, J., Gordon, L.I., Edmond, J.M., Von Herzen, R.P., Ballard, R.D., Green, K., Williams, D., Bainbridge, A., Crane, K. and Van Andel, Tj.H., 1979. Submarine thermal springs on the Galapagos rift. *Science,* 203: 1073–1983.

Crane, K., Hecker, B. and Golubev, V., 1991. Hydrothermal vents in Lake Baikal. *Nature,* 350: 281.
Cucchi, F. and Forti, P., 1990. Evolutione speleogenetica del complesso carsico Grotta del Fiume – Grotta del Vento (S. Vittore Gegna – Marche). In: S. Galdenzi and M. Menichetti (Editors), *Il Carsismo della Gola di Frasassi.* Memorie Instituto Italiano di Speologia, Costacciaro-Perugi Costacciaro-Perugia, pp. 143–153.
Culver, D.C., 1982. *Cave Life.* Harvard University Press, Cambridge, 189 pp.
Ćurčić, B.P.M., Poinar Jr., G.O. and Sarbu, S.M., 1993. New and little known species of Chthoniidae and Neobisiidae (Pseudoscorpiones, Arachnida) from Movile Cave in southern Dobrogea, Romania. *Bijdr. Dierkd,* 63: 221–241.
Dancau, D., 1970. Sur un nouvel amphipode souterrain de Roumanie, *Pontoniphargus racovitzai,* n. g., n. sp. In: *Livre de Centenaire Emile Racovitza 1868–1968.* Académie de la République Socialiste de Roumanie, pp. 275–285.
Decu, V. and Georgescu, M., 1994. Deux espèces nouvelles de *Medon* (*M. dobrogicus* et *M. paradobrogicus*) (Coleoptera, Staphylinidae) de la grotte "Peştera Movile", Dobrogea Meridionale, Roumanie. *Mem. Biospeol.,* 21: 47–51.
Decu, V., Gruia, M., Keffer, S.L. and Sarbu, S.M., 1994a. A stygobiotic waterscorpion, *Nepa anophthalma* n. sp. (Hemiptera, Nepidae), from Movile Cave, Romania. *Ann. Entomol. Soc. Am.,* 87: 755–761.
Decu, V., Nitzu, E. and Juberthie, C., 1994b. *Clivina subterranea* (Caraboidea, Scaritidae) nouvelle espèce de la grotte "Peştera de la Movile", Dobrogea Meridionale, Roumanie. *Mem. Biospeol.,* 21: 41–45.
Degens, E.T., 1969. Biochemistry of stable carbon isotopes. In: G. Elington and M.T.J. Murphy (Editors), *Organic Geochemistry, Methods and Results.* Springer, New York, pp. 304–329.
Diaconu, G. and Morar, M., 1993. Analyses mineralogiques dans la grotte "Peştera de la Movile" (Mangalia-Dobrogea de Sud). *Karstologia,* 22: 15–20.
Egemeier, S.J., 1973. *Cavern Development by Thermal Waters with a Possible Bearing on Ore Deposition.* Ph.D. Thesis, Stanford Univ., California, 88 pp.
Egemeier, S.J., 1981. Cavern development by thermal waters. *Natl. Speleol. Soc. Bull.,* 43: 31–51.
Ehleringer, J.R., Rundel, P.W. and Nagy, K.A., 1986. Stable isotopes in physiological ecology and food web research. *Trends Ecol. Evol.,* 1: 42–45.
Engel, A.S. and Lascu, C., 1996. Clays of Movile Cave, Southern Dobrogea, Romania: speleogenetic indicators of cavern formation. *Program and Abstracts of the 33rd Annual Meeting of the Clay Minerals Society, Gatlinburg, Tennessee,* pp. 53.
Faure, G., 1986. *Principles of Isotope Geology.* Wiley, New York, 589 pp.
Feru, M.U. and Capota, A., 1991. Les eaux thermominérales karstiques de la zone de Mangalia (Roumanie). *Theor. Appl. Karstol.,* 4: 143–157.
Findlay, R.H., King, G.M. and Watling, L., 1989. Efficacy of phospholipid analysis in determining microbial biomass in sediments. *Appl. Environ. Microbiol.,* 55: 2888–2893.
Fisher, R.C., Childress, J.J., Macko, S.A. and Brooks, J.M., 1994. Nutritional interactions in Galapagos rift hydrothermal vent communities: inferences from stable carbon and nitrogen isotope analyses. *Mar. Ecol. Progr. Ser.,* 103: 45–55.
Ford, D. and Williams, P., 1989. *Karst Geomorphology and Hydrology.* Unwin and Hyman, London, 601 pp.
Galdenzi, S., 1990. Un modello genetico per la grotta del vento. In: S. Galdenzi and M. Menichetti (Editors), *Il Carsismo della Gola di Frasassi.* Memorie Instituto Italiano di Speologia, Costacciaro-Perugia, pp. 123–141.
Georgescu, M., 1989. Sur trois taxa nouveaux d'Araneides troglobies de Dobrogea (Roumanie). *Misc. Speol. Romanica,* 1: 85–102.
Georgescu, M., 1994. Sur la présence d'une espèce troglobionte de *Nesticus* (Araneae, Nesticidae) dans la grotte "Pestera de la Movile" (Dobrogea, Roumanie). *Trav. Inst. Speol. "Emile Racovitza",* 33: 85–87.
Georgescu, M. and Căpuşe, I., 1994. Sur les pseudoscorpions de la région de la grotte Movile (Mangalia, Dobrogea du Sud, Roumanie). *Trav. Inst. Speol. "Emile Racovitza",* 33: 63–78.
Georgescu, M. and Decu, V., 1994. Sur la présence de deux espèces d'*Eukoenenia* dans le souterrain et l'euédaphique du midi de la Dobrogea (Roumanie). *Trav. Inst. Speol. "Emile Racovitza",* 33: 79–84.
Georgescu, M. and Sarbu, S.M., 1992. Description d'un nouveau taxon *Iberina caeca* de la grotte "Peştera de la Movile" (Araneae: Hahniidae). *Mem. Biospeol.,* 19: 319–141.
Goericke, R., Montoya, J.P. and Fry, B., 1994. Physiology of isotopic fractionation in algae and cyanobacteria. In: K. Lajtha and R.H. Michener (Editors), *Stable Isotopes in Ecology and Environmental Science.* Blackwell, Oxford, pp. 187–221.
Gracev, M., Flalkov, L., Nakamura, T., Ohta, T. and Kawal, T., 1995. Extant fauna of ancient carbon. *Nature,* 374: 123–124.
Grossu, A.V. and Negrea, A., 1989. *Paladilhia (Paladilhiopsis) dobrogica* – une nouvelle espèce de la famille Moitessieriidae (Gastropoda, Prosobranchia). *Misc. Speol. Romanica,* 1: 33–37.
Gruia, M., 1989. Nouvelles espèces troglobiontes des Collemboles de Roumanie. *Misc. Speol. Romanica,* 1: 103–111.
Gruia, M., Iavorschi, V. and Sarbu, S.M., 1994. *Armadillidium tabacarui* (Isopoda: Oniscidea: Armadillidiidae), a new troglobitic species from a sulfurous cave in Romania. *Proc. Biol. Soc. Washington,* 107: 699–706.
Henry, J.P. and Magniez, G., 1983. Introduction practique à la systématique des organismes des eaux continentales françaises. 4. Crustacés Isopodes (Principalement Asellotes). *Bull. Soc. Linn.,* 52: 312–357.
Hill, C.A., 1987. Geology of Carlsbad Cavern and other caves in the Guadalupe Mountains, New Mexico and Texas. *N. M. Bur. Mines Miner. Resour. Bull.,* 117: 1–150.
Hill, C.A., 1990. Sulfuric acid speleogenesis of Carlsbad Cavern and its relationship to hydrocarbons, Delaware Basin, New Mexico and Texas. *Am. Assoc. Pet. Geol. Bull.,* 74: 1685–1694.
Horoi, V., 1994. The corrosion process in "Pestera de la Movile" Cave (Southern Dobrugja - Romania). *Theor. Appl. Karstol.,* 7: 187–191.
Hsü, J.K., 1978. When the Black Sea was drained. *Sci. Am.,* 238: 12–14.
Iavorschi, V., 1992. *Labidostoma motasi* n. sp. (Nicoletiellidae) a new species of the mite of Romania. *Trav. Inst. Speol. "Emile Racovitza",* 31: 47–51.
James, J.M., 1994. Microbially produced carbon dioxide and studies on its effects on speleogenesis. In: I.D. Sasowsky and M.V. Palmer (Editors), *Breakthroughs in Karst Geomicrobiology*

and Redox Geochemistry. Karst Waters Institute, Charles Town, WV, pp. 28–30 (special publication).

Jameson, R.A., 1989. Features of condensation corrosion in caves of the Greenbrier karst, West Virginia. *Program 1989 NSS Convention, Sewanee, Tennesee.* National Society of Speleology, p. 56.

Jameson, R.A., 1995. Condensation, condensation corrosion, and associated features in Snedegars and Greenville Saltpeter Caves. *Program 1995 NSS Convention, Blacksburg, Virginia.* National Society of Speleology, pp. 122–125.

Jannasch, H.W., 1985. The chemosynthetic support of life and the microbial diversity at deep-sea hydrothermal vents. *Proc. R. Soc. London,* B225: 227–297.

Karaman, G., 1990. New data of Genus *Pontoniphargus* Dancau, 1970 (Fam. Niphargidae) from Romania (Contribution to the Knowledge of the Amphipoda 1990). *Glas. Repub. Zavoda Zastitu Prir. Prir. Muz. Titogradu,* 22: 79–94.

Karaman, G.S. and Sarbu, S.M., 1995. *Niphargus decui* (Amphipoda, Gammaridea, Niphargidae) a new species from Romania. *Trav. Inst. Speol. "Emile Racovitza",* 34: 77–87.

Karaman, G.S. and Sarbu, S.M., 1996. A new species of the genus *Pontoniphargus* Dancau, 1970 (Amphipoda, Gammaridea, Family Niphargidae) from Romania, *P. ruffoi,* n. sp. *Boll. Mus. Civ. Stor. Nat. Verona,* 20: 569–582.

Karl, D.M., 1995. *The Microbiology of Deep-sea Hydrothermal Vents.* CRC Press, Boca Raton, 299 pp.

Lascu, C., 1989. Paleogeographical and hydrogeological hypothesis regarding the origin of a peculiar cave fauna. *Misc. Speol. Romanica,* 1: 13–18.

Lascu, C., Popa, R., Sarbu, S.M., Vlasceanu, L. and Prodan, S., 1993. La grotte de Movilé: une faune hors du temps. *Recherche,* 24: 1092–1098.

Lascu, C., Popa, R. and Sarbu, S.M., 1994. Le karst de Movile (Dobrogea de Sud) (I). *Rev. Roum. Geogr.,* 38: 85–94.

Lascu, C., Popa, R. and Sarbu, S.M., 1995. Le karst de Movile (Dobrogea de Sud) (II). *Rev. Roum. Geogr.,* 39: 31–40.

Longsdale, P., 1977. Clustering of suspension-feeding macrobenthos near abyssal hydrothermal vents at oceanic spreading centers. *Deep-Sea Res.,* 24: 857–863.

Macko, S.A. and Ostrum, N.E., 1994. Pollution studies using stable isotopes. In: K. Lajtha and R.H. Michener (Editors), *Stable Isotopes in Ecology and Environmental Science.* Blackwell, Oxford, pp. 45–62.

Macko, S.A., Helleur, R., Hartley, G. and Jackman, P., 1990. Diagenesis of organic matter – a study using stable isotopes on individual carbohydrates. *Org. Geochem.,* 16: 1129–1137.

Manoleli, D., Klemm, D.J. and Sarbu, S.M., 1998. *Haemopis caeca* sp. nov. (Annelida, Hirudine, Haemopidae) a troglobitic leech, from a chemoautotrophically based groundwater ecosystem in Romania. *Proc. Biol. Soc. Washington,* 111: 222–229.

McKinley, V.L., Federle, T.W. and Vestal, J.R., 1982. Effects of hydrocarbons on plant litter microbiota of an Arctic lake. *Appl. Environ. Microbiol.,* 43: 129–135.

Negoescu, I., 1989. Sur une population troglobie d'*Asellus aquaticus* (L.) (Isopoda, Asellota, Asellidae). *Misc. Speol. Romanica,* 1: 13–18.

Negrea, S., 1993. Sur une population troglobionte de *Cryptops anomalans* Newport, 1844 (Chilopoda, Scolopendromorpha) trouvée dans la grotte "Pestera de la Movile" (Dobrogea, Roumanie). *Trav. Inst. Speol. "Emile Racovitza",* 32: 87–94.

Nelson, D.C. and Fisher, C.R., 1995. Chemoautotrophic and methanotrophic endosymbiotic bacteria at deep-sea vents and seeps. In: D.M. Karl (Editor), *The Microbiology of Deep-sea Hydrothermal Vents.* CRC Press, Boca Raton, pp. 126–167.

O'Leary, M.H., 1981. Carbon isotope fractionation in plants. *Phytochemistry,* 20: 553–567.

O'Leary, M.H., 1988. Carbon isotopes in photosynthesis. *BioScience,* 38: 328–336.

Palmer, A.N., 1991. Origin and morphology of limestone caves. *Geol. Soc. Am. Bull.,* 103: 1–21.

Palmer, A.N. and Palmer, M.V., 1989. Geologic history of the Black Hills caves, South Dakota. *Natl. Speleol. Soc. Bull.,* 51: 72–99.

Pasquini, G., 1973. Aggressive condensation. *Int. J. Speleol.,* 3: 315–318.

Paull, C.K., Jull, A.J.T., Toolin, L.J. and Linick, T., 1985. Stable isotope evidence for chemoautotrophy in an abyssal seep community. *Nature,* 317: 709–711.

Peterson, B.J. and Fry, B., 1987. Stable isotopes in ecosystem studies. *Annu. Rev. Ecol. Syst.,* 18: 293–320.

Pisarowicz, J.A., 1994. Cueva de Villa Luz – an active case of H_2S speleogenesis. In: I.D. Sasowsky and M.V. Palmer (Editors), *Breakthroughs in Karst Geomicrobiology and Redox Geochemistry.* Karst Waters Institute, Charles Town, WV, pp. 60–62 (special publication).

Pleşa, C., 1989. Etude preliminaire des cyclopides (Crustacea, Copepoda) de la grotte "Peştera de la Movile", Mangalia (Roumanie). *Misc. Speol. Romanica,* 1: 13–18.

Poggi, R., 1994. Descrizione di un nuovo pselafide rumeno, primo rappresentante cavernicolo della tribù Tyrini. *Boll. Soc. Entomol. Ital.,* 125: 221–228.

Poinar Jr., G.O. and Sarbu, S.M., 1994. *Chronogaster troglodytes* sp.n. (Nematoda, Chronogasteridae) from Movile Cave, with a review of cavernicolous nematodes. *Fundam. Appl. Nematol.,* 17: 231–237.

Poulson, T.L., 1971. Comparison of cave and deep sea organisms. *Natl. Speleol. Soc. Bull.,* 33: 51–61.

Poulson, T.L., 1972. Bat guano ecosystems. *Natl. Speleol. Soc. Bull.,* 34: 55–59.

Rau, G.H., McHugh, C.M., Harrold, C., Baxter, C., Hecker, B. and Embley, R.W., 1990. $\delta^{13}C$, $\delta^{15}N$ and $\delta^{18}O$ of *Calyptogena phaseoliformis* (Bivalve Mollusc) from the Ascension Fan-Valley near Monterey, California. *Deep-Sea Res.,* 37: 1669–1676.

Riess, W., Giere, O., Kohls, O. and Sarbu, S.M., 1999. Anoxic thermomineral cave waters and bacterial mats as habitat for freshwater nematodes. *Aquat. Microb. Ecol.,* 18: 157–164.

Ruby, E.G., Jannasch, H.W. and Deuser, W.G., 1987. Fractionation of stable carbon isotopes during chemoautotrophic growth of sulfur-oxidizing bacteria. *Appl. Environ. Microbiol.,* 53: 1940–1943.

Sarbu, S.M., 1990. The unusual fauna of a cave with thermomineral waters containing H_2S, from southern Dobrogea, Romania. *Mem. Biospeol.,* 17: 191–195.

Sarbu, S.M., 1991. Contributions to the biological investigation of the "Movile Cave": the species composition and the trophic structure of the cave community and the origin of the fauna. *Mem. Biospeol.,* 18: 193–197.

Sarbu, S.M. and Kane, T.C., 1995. A subterranean chemoautotrophically based ecosystem. *Natl. Speleol. Soc. Bull.,* 57: 91–98.

Sarbu, S.M. and Lascu, C., 1997. Condensation corrosion in Movile Cave. *J. Karst Cave Stud.*, 59: 99–102.

Sarbu, S.M. and Popa, R., 1992. A unique chemoautotrophically based cave ecosystem. In: A.I. Camacho (Editor), *The Natural History of Biospeleology.* Monografias Museo Nacional de Ciencias Naturales, Madrid, pp. 637–666.

Sarbu, S.M., Popa, R. and Goliat, I., 1991. Chemoautotrophic production in a thermomineral sulfurous cave. *Trav. Inst. Speol. "Emile Racovitza"*, 30: 59–61.

Sarbu, S.M., Vlasceanu, L., Popa, R., Sheridan, P., Kinkle, B.K. and Kane, T.C., 1994a. Microbial mats in a thermomineral sulfurous cave. In: L.J. Stal and P. Chaumette (Editors), *Microbial Mats: Structure, Development and Environmental Significance.* Springer-Verlag, Berlin, pp. 45–50.

Sarbu, S.M., Kinkle, B.K., Vlasceanu, L., Kane, T.C. and Popa, R., 1994b. Microbiological characterization of a sulfide-rich groundwater ecosystem. *Geomicrobiol. J.*, 12: 175–182.

Sarbu, S.M., Gheorghe, C., Popescu-Jarnea, V., Vlasceanu, L., Popa, R. and Lascu, C., 1995. Stable isotope studies in Movile Cave. *Trav. Inst. Speol. "Emile Racovitza"*, 34: 99–102.

Sarbu, S.M., Kane, T.C. and Kinkle, B.K., 1996. A chemoautotrophically based groundwater ecosystem. *Science*, 272: 1953–1955.

Southward, A.J., Kennicutt II, M.C., Alcala-Herrera, J., Abbiati, M., Airoldi, L., Cinelli, F., Bianchi, C.N., Morri, C. and Southward, E.C., 1996. On the biology of submarine caves with sulfur springs: Appraisal of 13C/12C ratios as a guide to trophic relations. *J. Mar. Biol. Assoc. U.K.*, 76: 265–285.

Steele, R.G.D. and Torrie, J.H., 1960. *Principles and Procedures of Statistics.* McGraw-Hill, New York, 481 pp.

Stevens, T.O. and McKinley, J.P., 1995. Lithoautotrophic microbial ecosystems in deep basalt aquifers. *Science*, 270: 450–454.

Ştiucă, E. and Ilinca, H., 1995. Pleistocene mammals (Rodentia) from Movile Cave (Romania). *Theor. Appl. Karstol.*, 8: 157–161.

Tabacaru, I. and Boghean, V., 1989. Découverte, en Dobrogea (Roumanie), d'une espèce troglobie du genre *Trachelipus* (Isopoda, Oniscoidea, Trachelipidae). *Misc. Speol. Romanica*, 1: 53–75.

Taylor, M.R., 1999. A trip to the Lighted House. *Natl. Speleol. Soc. N.*, 57: 36–48.

Ţenu, A., Davidescu, F.D. and Slăvescu, A., 1987. Recherches isotopiques sur les eaux souterraines des formations calcaires dans la Dobroudja méridionale (Roumanie). In: *Isotope Techniques in Water Resources Development.* International Atomic Energy Agency, Vienna, pp. 439–453.

Tunnicliffe, V., 1992. Hydrothermal-vent communities of the deep sea. *Am. Sci.*, 80: 336–349.

Turk, S., Sket, B. and Sarbu, S.M., 1996. Comparison between some epigean and hypogean populations of *Asellus aquaticus* (Crustacea: Isopoda: Asellidae). *Hydrobiologia*, 337: 161–170.

Turk-Prevorčnik, S. and Blejec, A., 1998. *Asellus aquaticus infernus*, new subspecies (Isopoda: Asellota: Asellidae), from Romanian hypogean waters. *J. Crust. Biol.*, 18: 763–773.

Vlasceanu, L., Popa, R. and Kinkle, B.K., 1997. Characterization of *Thiobacillus thioprus* LV43 and its distribution in a chemoautotrophicaly based ground ecosystem. *Appl. Environ. Microbiol.*, 63: 3123–3127.

Vucev, V.T., 1974. Black Sea studies in Bulgaria – a brief survey. In: E.T. Degens and D.A. Ross (Editors), *The Black Sea. Geology, Chemistry and Biology.* American Association of Petroleum Geologists, Tulsa, Oklahoma, pp. 90–96.

Weiss, I. and Sarbu, S.M., 1994. Die Höhlenspinne *Agraecina cristiani* (Georgescu, 1989) n. comb. (Arachnida, Araneae, Liocranidae). *Verh. Naturwiss. Ver. Hamburg*, 36: 163–171.

Weiss, I. and Sarbu, S.M., 1996. Das unbekannte Männchen der Höhlenspinne *Hahnia caeca* n. comb., mit Hinweisen zur Funktion der Tibial-Apophyse des männlichen Pedipalpus (Arachnida, Araneae, Hahniidae). *Verh. Naturwiss. Ver. Hamburg*, 36: 163–171.

Wetzel, R.G. and Likenns, G.E., 1990. *Limnological analyses*, 2nd edition. Springer, New York, 391 pp.

Wunderlich, J., 1991. Die Spinner-Fauna der Makaronesischen Inseln. Taxonomie, Ökologie, Biogeographie und Evolution. *Beitr. Araneol.*, 1: 619.

Yoshinari, T. and Koike, I., 1994. The use of stable isotopes for the study of gaseous nitrogen species in marine environments. In: K. Lajtha and R.H. Michener (Editors), *Stable Isotopes in Ecology and Environmental Science.* Blackwell, Oxford, pp. 114–137.

Chapter 18

FOOD WEB DYNAMICS AND BIOGEOCHEMISTRY OF ANCHIALINE CAVES: A STABLE ISOTOPE APPROACH

John W. POHLMAN, Luis A. CIFUENTES and Thomas M. ILIFFE

INTRODUCTION

Partially or totally submerged cave systems occur along many tropical karstic and volcanic coastlines (Iliffe, 1992). Such caves, technically known as anchialine, are characterized by highly stratified fresh and marine waters that fluctuate with the tides (Fig. 18.1). The depth of the halocline is a function of the amount of rainfall an area receives and the distance of the cave from the coastline. In areas like the Yucatan Peninsula (Mexico), where surface rivers and streams are absent and rainfall averages around 1000–1500 mm yr^{-1}, the freshwater lens is extensive (Back et al., 1986). For example, at Cenote Abejas (2 km inland) the freshwater lens is 10 m thick, while at Cenote Vaca Ha (10 km inland) it is 20 m thick (see Chapter 3, this volume). The halocline, which is underlain by seawater which penetrates through the cracks, fissures and passages that permeate the limestone bedrock, becomes progressively deeper with increasing distance from the coast. The cave systems are accessible to investigation through open pools known as "cenotes", which are surface lesions formed by the collapse of the cave ceiling during the geologic past.

A phylogenetically and biogeographically unique assemblage of troglobitic organisms inhabit these caves. Research addressing the taxonomy and biogeography of the anchialine troglobites is extensive (e.g., Yager, 1981; Iliffe et al., 1984; Bowman et al., 1985; Hart et al., 1985; Manning et al., 1986), yet fundamental questions regarding the trophic ecology and nutrient biogeochemistry of these systems have been investigated only recently. Most information about the food web of anchialine caves comes from incidental behavioral observations by divers, or from anatomical characteristics of the animals. We here report on the use of stable isotopes to trace biotic interactions and fundamental biogeochemical processes that uniquely define the anchialine habitat. Our studies, conducted in the Yucatan Peninsula (Mexico), provide insights into events that may be typical anchialine phenomena. New perspectives on the biogeochemical dynamics of this ecosystem and the trophic interactions of anchialine fauna should provide a basis for understanding how better to protect these valuable and sensitive coastal groundwater habitats.

PRIOR OBSERVATIONS ON TROPHIC INTERACTIONS AMONG CAVE FAUNA

Observations of gut contents have provided some information on the diet of anchialine fauna. For instance, the halocyprid ostracod *Danielopolina wilkensi* from the Atlantida lava tube in the Canary Islands apparently feeds on detritus (Kornicker and Iliffe, 1995). Gut-content analysis revealed minute amber or brown colored organic particles, none of which were recognizable as animal fragments. Likewise, unrecognizable amber particles were present in the gut of *Danielopolina styx* and the myodopid ostracod *Skogsbergia galapagensis* from anchialine fissures in the Galapagos Islands (Kornicker and Iliffe, 1989a). A large ovoid fecal pellet from the halocyprid *Spelaeoecia bermudensis* from limestone caves in Bermuda contained abundant broken, hollow, transparent and colorless tubes of unknown origin (Kornicker and Iliffe, 1989b). Finally, setae and fragments of apparent crustacean origin were found in the gut of the halocyprid *Spelaeoecia cubensis* (Kornicker and Yager, 1996).

Behavioral observations of living specimens have

Fig. 18.1. Diagrammatic representation of a typical anchialine cave system along the coast of the northeastern Yucatan Peninsula. These coastal systems have limited access to open air and subterranean connections to the sea, and exhibit both terrestrial and marine influences (Stock et al., 1986).

also provided information on feeding methods. The anchialine calanoid copepod *Erebonectes nesioticus* from Bermuda holds its second maxillae and maxillipeds in a downward position, forming a basket-like structure probably used for catching prey (Fosshagen and Iliffe, 1985).

The galatheid crab *Munidopsis polymorpha* inhabits a partially illuminated tidal lake near the entrance of a coastal lava tube in the Canary Islands (Parzefall and Wilkens, 1975). They rest in crevices in the lava rocks during the day, and at night they come out to feed on diatoms that grow on the rocks.

Speculation on the food-web structure of Sagittarius Cave, an anchialine limestone cave in the Bahamas, has been proposed by Cunliffe (1985). The cave fish *Lucifuga speleotes* and the remipede *Speleonectes lucayensis* are considered to be the top-level carnivores in this cave. They probably feed on smaller ostracods, thermosbaenaceans and copepods as well as larger cirolanid isopods and amphipods. The mysids were considered to be omnivores, while the cirolanid isopods were viewed as benthic omnivores primarily feeding on detritus. Ostracods and thermosbaenaceans were also probably detritivores.

Finally, morphological features can provide details on feeding strategies. Remipedes have a terminal fang at the distal margin of the first maxilla, which probably functions in prey capture (Yager, 1987). A pore, typically visible near the tip of the fang, implies that captured prey may be envenomated. Yager (1987) concluded that the morphological variations in the raptoral feeding appendages of several remipede species from one cave allow each species to select a specific prey and avoid competition.

APPLICATIONS OF STABLE ISOTOPES TO CAVE ECOLOGY

A major impediment to conducting ecological research in anchialine caves stems from time and depth limitations inherent to cave diving. These limitations obviously preclude long-duration surveys. With respect to food-web studies, visual observations of feeding strategies only identify what the organism is ingesting, which is not necessarily what the organism is digesting. Techniques like gut-content analysis are of limited value because of the small size of the fauna and the associated material found in the digestive tract. Furthermore, the degraded status of the gut content often makes identification of the original source impossible. Introduced radioactive tracers have been effective in other systems (e.g., White and Roman, 1991), but potentially threaten the sensitive ecological balance of cave systems. In contrast, the proportions of naturally occurring stable isotopes can constitute a non-intrusive, natural tracer, which has commonly been used by terrestrial and aquatic ecologists to overcome the inadequacies of traditional approaches (Peterson and Fry, 1987). This provides information on the origins of organic food sources and on how they are transferred through an ecosystem (Macko and Ostram, 1994). Stable isotopes have been successfully used to this end in estuarine, salt-marsh, deep-sea, freshwater and terrestrial ecosystems (e.g., Fry and Sherr, 1984;

Peterson et al., 1985, 1986; Boutton, 1991; Fry, 1991; Macko, 1993), but our application in anchialine cave ecology is novel.

Background on stable isotopes

Understanding the utility of stable isotopes as an ecological tool begins with the awareness that the major atomic components of organic matter (C, N, O, S, H) exist in several isotopic forms. In this study, we will focus on carbon and nitrogen. There are two stable isotopes of carbon: ^{12}C and ^{13}C. Carbon-12 is more abundant than carbon-13, which varies between 1.00 and 1.12% in natural systems. Similarly, there are two stable isotopes of nitrogen, the more abundant ^{14}N and less abundant ^{15}N (0.36–0.40%). By looking at the ratio of the heavier to the lighter isotope in organic matter, one can gain information about the nature of its source and the transformations that have occurred since it was formed (Fogel and Cifuentes, 1993). These ratios are expressed as δ (delta) values and are calculated by the following equation:

$$\delta X = \quad R_{\text{sample}}/R_{\text{standard}} \quad - 1 \qquad 10^3 \tag{18.1}$$

δX is expressed in parts per thousand (‰), and is either $\delta^{13}C$ or $\delta^{15}N$; R is either $^{13}C/^{12}C$ or $^{15}N/^{14}N$. The standards for carbon and nitrogen are PeeDee Belemnite limestone fossil, and atmospheric dinitrogen gas, respectively. A stable-isotope mass spectrometer is used to measure the isotopic ratios of a sample (Peterson and Fry, 1987). Below, we will compare values from different samples and will refer to those with larger δ-values as "heavier" and those with smaller δ-values as "lighter".

Stable carbon isotope applications

Stable carbon isotope determinations are particularly useful in identifying sources of organic matter. Organic material is characterized by specific $\delta^{13}C$ values depending upon the autotrophic pathway responsible for fixing the carbon and the isotopic signature of the carbon source. For example, $\delta^{13}C$ values for upland C_3 plants, phytoplankton, and the C_4 plant *Spartina alterniflora,* in a New England salt marsh are all distinctively different (Peterson et al., 1985). Upland plants utilizing the C_3 photosynthetic pathway had average $\delta^{13}C$ values of –28.6‰, while *S. alterniflora,* a C_4 plant, averaged –13.1‰, and plankton averaged –21.3‰. These distinctive isotopic signatures arise from differential discrimination against ^{13}C during carbon fixation in the three photosynthetic pathways (O'Leary, 1981). In addition to distinguishing photosynthetic sources, chemoautotrophic sources are also easily discerned. Organic matter of chemoautotrophic origin has "lighter" values for stable carbon isotopes than photoautotrophic organic material. Kennicutt et al. (1992) have reported such values from mussels at deep-sea hydrothermal vents harboring endosymbiotic chemoautotrophic bacteria. The tissue of those containing thiotrophic bacteria had a $\delta^{13}C$ value of –34.7 2.1‰, while tissue with methanotrophic bacteria had a $\delta^{13}C$ value of –49.5 6.3‰.

Degradation of organic material has only a minimal effect on its stable carbon isotope ratio (Fry and Sherr, 1984); thus, the source of organic matter (e.g., vascular plants or algae) can be identified even when transformed to soil, particulate or sedimentary organic matter. This fact is particularly valuable for studying aquatic cave ecosystems, which have been traditionally regarded as detrital-based (Mohr and Poulson, 1966).

Values for stable carbon isotopes can also be used to trace the flow of organic carbon through the food web. The adage "you are what you eat" is particularly suitable, because the isotopic value of consumers resembles that of their food source. For example, $\delta^{13}C$ values of the polychaete *Capitella capitata* cultured in an isotopically homogeneous material displayed $\delta^{13}C$ values enriched by about 1‰ relative to the food source (Haines and Montague, 1979). This characteristic 1‰ enrichment (also referred to as fractionation) occurs with each trophic level advance, thus making it possible to link consumers to their food sources.

Stable nitrogen isotope applications

Stable nitrogen isotopes are even more useful for defining the hierarchy of the food web. With each trophic level transfer, the ratio in the consumer becomes 2–3‰ "heavier" than in its food source (DeNiro and Epstein, 1981). The ^{14}N in the form of ammonia or urea is preferentially excreted, thus resulting in ^{15}N enrichment in the consumer. This relationship has been confirmed by numerous studies from a variety of ecosystems (Rau, 1981; Macko et al., 1982; Checkly and Entzeroth, 1985; Peterson and Fry, 1987).

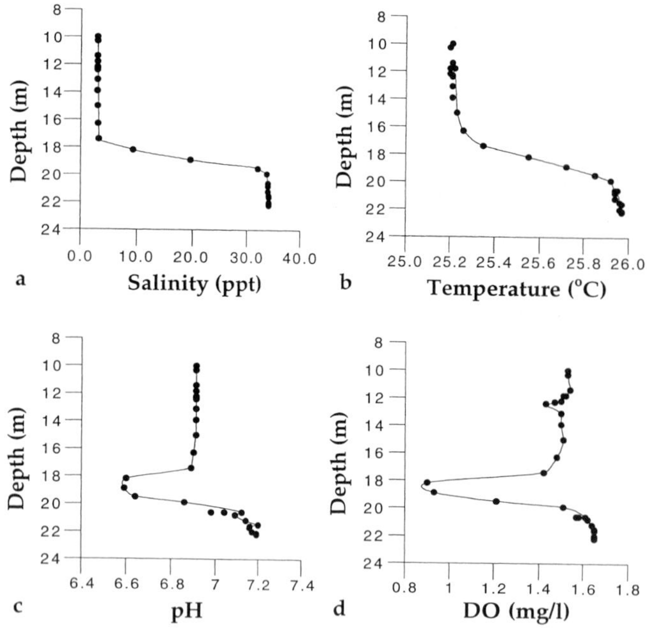

Fig. 18.2. Vertical profiles of (a) salinity, (b) temperature, (c) pH and (d) dissolved oxygen from the Mayan Blue cave system.

STUDY SITE

We have applied this stable-isotope approach and other biogeochemical techniques to investigate anchialine cave systems of the northeastern Yucatan Peninsula, Mexico. Along the Caribbean coast of the Yucatan, the world's most extensive anchialine cave systems meander through a limestone plain toward the sea. Our primary study site was Mayan Blue Cenote near the ancient Mayan city of Tulum. Supporting materials were collected from the nearby Naharone, Carwash, Temple of Doom and Angelita Cenotes. Cenotes are open-water pools that formed when the ceiling of an underlying cave passage collapsed. Temple of Doom Cenote is a jug-shaped cenote while the other four are open-pool cenotes. Cenote Angelita, although classified as an open-pool cenote, differs in that it extends to depths greater than 50 m and lacks cave passages leading off from the shaft. All other cenotes are connected to cave networks by passages large enough for divers to explore.

Water column characteristics

Anchialine caves are distinguished from their freshwater counterparts by the presence of a well-defined, distinct halocline separating freshwater above from fully marine waters below (Fig. 18.2). Associated with this interface is a slight increase in temperature, lower pH, and decrease in dissolved oxygen (DO). The salinity and temperature changes are physical phenomena resulting from the mixing of two water types, while variations in the pH and dissolved oxygen probably result from microbial activity. With the exception of Cenote Angelita, other anchialine caves in Yucatan have basically similar vertical profiles, suggesting that similar physical, chemical and biological processes are operating in these cave systems. Due to restricted water circulation and the presence of large amounts of organic debris, salt water below the halocline in Cenote Angelita is completely anoxic.

Sources of organic matter

A conceptual model (Fig. 18.3) was created from the Mayan Blue stable-isotope data to describe the dynamics of the carbon cycling into and within that system. The potential sources of organic matter available to the cave fauna included forest soil above the cave, freshwater algae in the cenote entrance pool and chemoautotrophic bacteria in the interior of the cave. Samples from the canopy and understory of the tropical forest had average $\delta^{13}C$ values of −28.7 0.5 ($n=4$). This value is typical for C_3 vegetation (Fry and

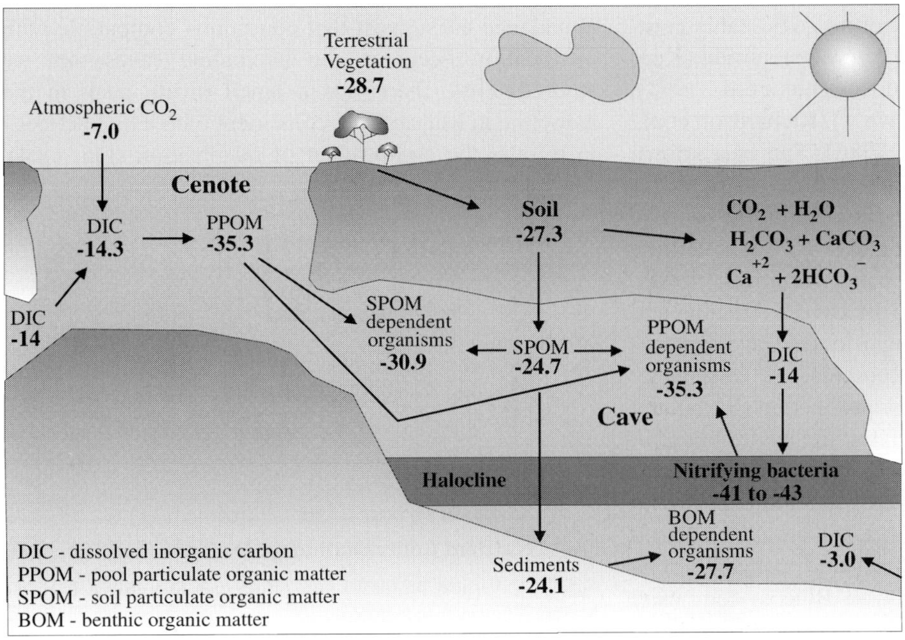

Fig. 18.3. A model for carbon cycling in an anchialine cave system.

Sherr, 1984), and resembles that of the soil organic matter generated from vegetative detritus (–27.3 0.4; $n=8$). Because of the high precipitation in this region of the Yucatan (1500 mm y^{-1}) (Back and Hanshaw, 1974) and the porosity of the limestone, the soil organic matter percolates through the porous bedrock into the cave. To reflect its likely source, particulate organic matter in the cave water column is henceforth referred to as soil particulate organic matter (SPOM). The mean $\delta^{13}C$ value for the SPOM was –24.7 2.0‰ ($n=7$). Continued fractionation of carbon isotopes in the soil detritus may be responsible for the SPOM being isotopically "heavier" than the soil material. Organic matter becomes isotopically "heavier" with increasing depth in the soil as vegetative detritus decomposes (Stout et al., 1981; Boutton, 1991). Finally, benthic organic matter (BOM) in the cave sediments had a mean $\delta^{13}C$ value of –24.1 2.8‰ ($n=10$). This value was indistinguishable from that of the SPOM, but this is to be expected considering that soil particulates in the water column settle into the sediments.

The second source of organic carbon entering the cave ecosystem was algal and vegetative detritus from the cenote. This material is hereafter referred to as pool particulate organic matter (PPOM), and had a mean $\delta^{13}C$ value of –35.3 2.6‰ ($n=4$). The PPOM probably included suspended benthic algae (–36.8 4.5‰; $n=7$) in addition to phytoplankton. Values similar to these have been found in the algae of freshwater lakes with high respiration in the water column and soil (Rau, 1978; Quay et al., 1986). Biogenic carbon dioxide produced during respiration is isotopically "lighter" than atmospheric carbon dioxide; thus, respiration in a confined lake or groundwater system produces isotopically "light" dissolved inorganic carbon (DIC). The mean $\delta^{13}C$ value of the dissolved inorganic carbon in the pool was –14.4‰ ($n=2$), whereas water saturated with atmospheric carbon dioxide has a $\delta^{13}C$ value of –7.9‰ (Boutton, 1991). Fractionation of an additional –20‰ during photosynthesis (O'Leary, 1981) explains the unusually light values measured in the pool algae.

The isotopic signature of pool organic matter was significantly different from that found in the caves below. This is surprising considering that the pool water apparently enters the caves. Either the amount of soil particulate organic matter (SPOM) overwhelmed the contribution of the PPOM, or the PPOM was selectively removed from the water by suspension feeders in the cave, or both processes contributed to the result.

The third possible source of organic matter is *in situ* chemoautotrophic production. Chemoautotrophy is a bacterially-mediated process whereby reduced compounds provide the energy needed to support bacterial growth and metabolism. A variety of oxidizable compounds can be used, including hydrogen sulfide (Kennicutt et al., 1992), methane (de Angelis et al., 1991), reduced metals (e.g., Fe^{+2} and Mn^{+2}) (Richardson et al., 1988) and ammonium (Ward, 1986). The best known chemoautotrophic communities are those of the deep-sea hydrothermal vents where endosymbiotic bacteria harbored within tube worms oxidize hydrogen sulfide to support themselves and their host (Felbeck, 1981). Chemoautotrophy also occurs in caves. Aquatic and terrestrial organisms inhabiting Movile Cave, an isolated thermomineral cave in Romania, are completely dependent on chemoautotrophic production (Sarbu and Popa, 1992; Sarbu et al., 1996), but the environment in that cave is radically different from anchialine caves. Concentrations of dissolved hydrogen sulfide in Movile Cave range between 500 and 1300 mmol ℓ^{-1} (Sarbu and Popa, 1992), whereas measured concentrations of hydrogen sulfide in the Mayan Blue system were below detection limits (2 mmol ℓ^{-1}). These differences, however, do not preclude the possibility that chemoautotrophs also utilize hydrogen sulfide in anchialine caves. In Cenote Angelita, an anchialine sinkhole lacking significant water circulation, degradation of organic debris fallen from the surrounding tropical forest results in the production of hydrogen sulfide which is utilized by thick mats of bacteria on the wall and clouds of bacteria floating above a layer of 250 mmol ℓ^{-1} hydrogen sulfide at the halocline. In any case, chemoautotrophic bacteria can be active at concentrations below 1 mmol ℓ^{-1} hydrogen sulfide (Dando et al., 1985).

Anchialine caves appear to be particularly suited for nitrification, a chemoautotrophic process whereby ammonium is sequentially oxidized to nitrite and then to nitrate (Ward, 1986):

$$NH_4^+ \rightarrow NO_2^- \rightarrow NO_3^-. \quad (18.2)$$

This process has been well documented in oceanic environments where both light and dissolved oxygen (DO) levels are low (Gundersen, 1966; Carlucci and McNally, 1969). The role of nitrification in these environments is important for the cycling of nitrogen, but considered negligible in terms of carbon production. Anchialine cave habitats share the environmental characteristics conducive to nitrification. Furthermore, because of the minute amounts of organic carbon naturally present in caves, nitrification may prove to be an important source of carbon in the cave food web.

Data for nutrients, dissolved oxygen and bacterial abundance all suggest that conditions compatible with nitrification occur in the anchialine caves that we studied. First, there was a small nitrite peak in the halocline in Mayan Blue coincident with a distinct peak in nitrate, the end-product of nitrification (Fig. 18.4).

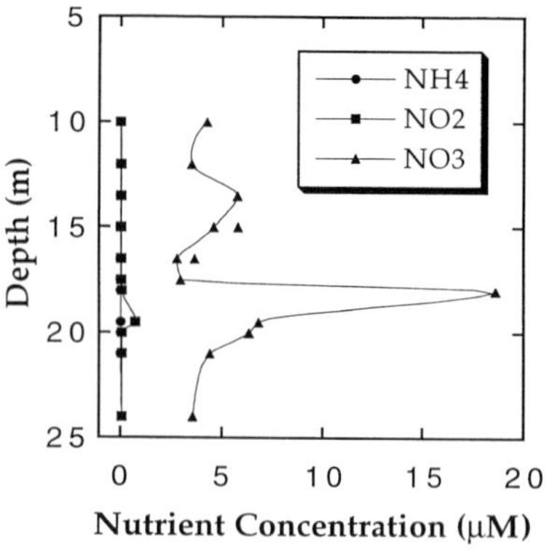

Fig. 18.4. Vertical profile of nutrient species of nitrogen from the Mayan Blue cave system.

A similar nitrate concentration was reported from the Mayan Blue halocline by Stoessell et al. (1989). Since abiological production of nitrate is negligible (Ward, 1986), the presence of a nitrate pulse can only be explained by nitrification. Second, corresponding to the increase in nitrate at the halocline, there was a minimum in dissolved oxygen (Fig. 18.2), the primary oxidant for nitrifying bacteria. These near-hypoxic levels of dissolved oxygen at the halocline provide optimum conditions for nitrification (Elkins et al., 1985). Third, the presence of a bacterial maximum at this depth in the water column of the Mayan Blue system has also been verified (Brett Dodson, pers. commun., 1996). Bacterial counts there ranged between 2.5 and 5.0 10^5 ml^{-1}. The bacteria, however, have not yet been classified. Compared to limnic and marine systems these counts are low, but because of inefficient mechanisms for carbon fixation by nitrifying bacteria (Ward, 1986) and possible grazing by pelagic crustaceans, this is to be expected.

The stable-isotope signature of some troglobitic crustaceans also supports the theory that chemoautotrophic production via nitrification is a possible food source within the anchialine cave ecosystem. Several individuals from two suspension-feeding species of Crustacea, the atyid shrimp *Typhlatya mitchelli* and the thermosbaenacean *Tulumella unidens,* had values for stable carbon isotopes that were "lighter" than any of the organic material measured. Considering that there is an enrichment of about 1‰ per trophic level transfer

(Peterson and Fry, 1987), it is impossible that these ^{13}C-depleted organisms consumed only the isotopically "heavier" PPOM and SPOM. Although there are no direct isotope data from nitrifying bacteria in this system, given the isotopic signature of the dissolved inorganic carbon, which ranged between −11 to −15‰ in the brackish portion of the water column, nitrifying bacteria should have $\delta^{13}C$ values between −35 and −45. This assumes a fractionation of −28 to −30‰, as reported by Peterson et al. (1980). Consumption of nitrifying bacteria would account for the unusually light values measured in *Tulumella unidens* and *Typhlatya mitchelli*.

Particulate organic matter entrained by the intruding seawater represents a fourth possible source of organic material. Its contribution, however, was probably negligible compared to the other sources. A sample of particulate organic matter collected below the halocline had a $\delta^{13}C$ value of −24.3. This number is nearly identical to the mean $\delta^{13}C$ value of particulate organic matter collected from the upper, brackish portion of the water column (−24.7‰) and within the range reported for marine waters (−22.0 to −24.4‰) (Williams and Gordon, 1970). Thus, it is impossible to determine the ultimate source of the seawater particulate organic matter based on isotope data alone. However, considering the lower particulate load in the seawater and the significant distance (6 km) of the collection site from the coast, the contribution from marine derived particulate organic matter is most likely minimal. It is doubtful that any appreciable amount of particulate organic matter was transported this far through the aquifer. Instead, the particulate organic matter found in the marine water was probably soil particulates that had settled through the upper, brackish portion of the water column. Finally, the scarcity of biota in the seawater reinforces the suggestion that marine particulate organic matter is a minor part of the organic pool. Only three troglobitic species are found exclusively below the halocline – the remipede *Speleonectes tulumensis,* and ostracods *Danielopolina mexicana* and *Spelaeoecia* n. sp. The remipedes are predatory carnivores and thus do not feed on particulate organic matter, while the ostracods are probable scavengers feeding on the available particulate organic matter as well as other organic matter.

Trophic classifications

The $\delta^{13}C$ values of the pelagic, troglobitic fauna from Mayan Blue varied as much as the identified sources of organic matter. For example, the three pelagic species (the shrimp *Typhlatya mitchelli,* the thermosbaenacean *Tulumella unidens* and the mysid *Antromysis cenotensis*) had mean $\delta^{13}C$ values distinctively different from one another and from the food sources (Fig. 18.5). The disparity between *Antromysis*

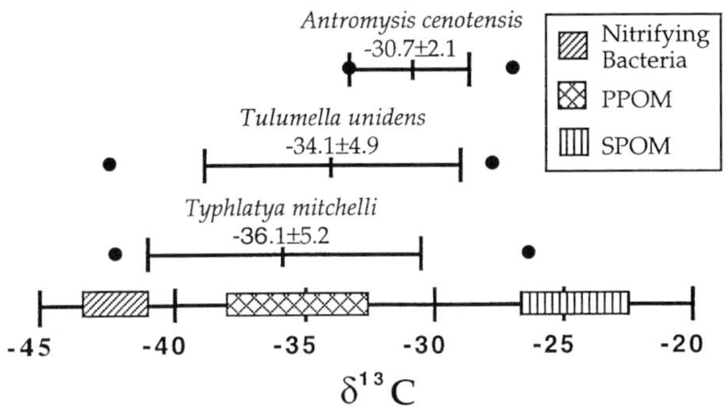

Fig. 18.5. Comparison of stable carbon isotope values of potential sources with those of fauna in the lower trophic level. For the sources, the boxes represent standard deviations of the PPOM and SPOM, and the potential range for the nitrifying bacteria. The mean values of the crustaceans are bracketed by the standard deviations and black dots represent the minimum and maximum measured values.

cenotensis, Tulumella unidens, and *Typhlatya mitchelli* suggests that each species utilized food sources in different proportions. A non-parametric Kruscal–Wallis test indicated that the mean $\delta^{13}C$ values of the crustacean populations may be different ($p = 0.15$). On that basis, the populations were clustered into two groups with the assigned names PPOM-dependent and SPOM-dependent (Table 18.1). A non-parametric

Table 18.1
Feeding groups identified by non-parametric Kruskal–Wallis and Wilcoxon Rank Sum Tests

Feeding group	Group mean of $\delta^{13}C$
PPOM-dependent	
Creaseria morleyi	−35.3 5.7‰; $n = 23$
Tulumella unidens	
Typhlatya mitchelli	
SPOM-dependent	
Amphipoda	
Antromysis cenotensis	−30.9 4.2‰; $n = 25$
Speleonectes tulumensis	
BOM-dependent	
Isopoda	−27.7‰; $n = 2$

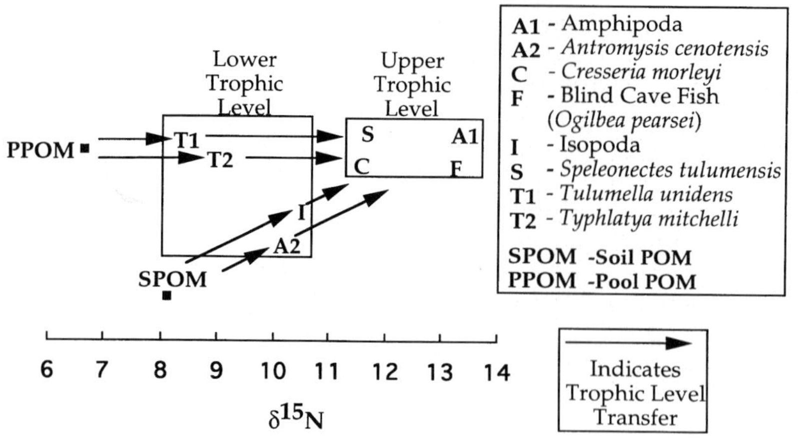

Fig. 18.6. Food-web hierarchy from the Mayan Blue cave system based on data for nitrogen stable isotopes. The lower trophic level consists of two pairs of crustaceans, each supported by independent food sources. The box for the upper trophic level contains members of the second and, possibly, third trophic levels.

Wilcoxon Rank Sum Testidentified these two groups as significantly different ($p=0.05$). These classifications imply different levels of dependence on each source, but should be used cautiously considering that there may be several trophic levels between crustaceans and either PPOM or SPOM.

The mean δ^{13}C values of the PPOM-dependent organisms (–35.3 5.7; $n=23$) were nearly identical to the mean values for PPOM (–35.3 2.6; $n=9$). This similarity and the high standard deviation suggests a strong dependence on PPOM, combined with flexibility in feeding on other sources. These additional food sources could be SPOM and chemoautotrophic bacteria. On the other hand, the δ^{13}C values of the SPOM-dependent organisms (–30.9 4.2; $n=25$) were intermediate between those for SPOM (–24.7 2.0; $n=7$) and PPOM, implying a greater dependence on the SPOM than the PPOM-dependent organisms displayed. The third group, dependent on benthic organic matter (BOM), was created to account for the benthic isopods. Although they were excluded from the statistical analysis because of the small sample number, they are considered a different feeding group since they occupy a separate niche. The BOM-dependent isopods had a mean δ^{13}C value of –27.7 1.5 ($n=2$), resembling the δ^{13}C values of the BOM (–24.1 2.8; $n=10$).

Trophic structure

The data for stable nitrogen isotopes allow us to define the hierarchy of the food web and support the trophic classifications discussed above (Fig. 18.6). Particulate organic matter originating from the pool (PPOM) had a mean δ^{15}N value of 6.7 ($n=2$), whereas that from the soil (SPOM) had a mean δ^{15}N value of 8.1 2.0 ($n=6$). The crustaceans *Tulumella unidens* and *Typhlatya mitchelli* had δ^{15}N values of 8.4 2.6 ($n=7$) and 9.1 1.7 ($n=10$), respectively. Meanwhile, *Antromysis cenotensis* had a mean δ^{15}N value of 10.0 1.1 ($n=14$). A single specimen of a BOM-dependent troglobitic isopod had a δ^{15}N value of 10.5. Assuming that the consumer is 2-3‰ "heavier" that its food source (DeNiro and Epstein, 1981), it appears that PPOM provided the majority of organic nitrogen for the PPOM-dependent crustaceans *Tulumella unidens* and *Typhlatya mitchelli,* while the SPOM provided the majority of organic nitrogen for the mysid *A. cenotensis* and the isopod.

The upper trophic level was occupied by predators and/or scavengers; thus their "higher" δ^{15}N value was expected. These species were the remipede *Speleonectes tulumensis* (11.7; $n=2$), the blind cave fish *Ogilbia pearsei* (13.3; $n=1$), two undifferentiated amphipod species (*Mayaweckelia cenoticola* and *Tuluweckelia cernua*) (13.6 0.6; $n=8$), and the paleamonid shrimp *Creaseria morleyi* (11.7; $n=2$). The data, although limited, also suggests that the amphipods and *O. pearsei* may have occupied a third trophic level. Amphipods have been observed scavenging on the exuviae of crustaceans, which could have come from crustaceans at the lower or higher trophic level crustaceans, or even other amphipods. This would explain why their δ^{15}N values were "heavy" relative to the other crustaceans. *Ogilbia pearsei* is the largest predator and

Fig. 18.7. Scanning electron micrographs of *Typhlatya mitchelli*: (a) lateral body view, arrow pointing to pereiopod; (b) setal tufts on propodus and dactylus of pereiopod, arrow pointing to setal tuft; (c) distal margin of pereiopod; (d) ventral view of oral region, arrow pointing to third maxilliped.

should be able to feed on any crustacean species in the food web. The parallels between diver observations of feeding behavior and what is suggested by the $\delta^{15}N$ data reinforces this idea, despite the limited data available.

ECOLOGY

Feeding mechanisms

Specialized feeding mechanisms may allow the crustaceans to feed selectively on the different forms of organic carbon available in the water column. This may explain why an isotope signature not corresponding with that of the particulate organic matter is reflected in the fauna. Two of the pelagic species, *Tulumella unidens* and *Typhlatya mitchelli,* have modified appendages which allow them to filter small, possibly even bacterial-sized, particles. Assuming that the PPOM (algae and other small suspended particles) and the SPOM (soil particles) differ in size, size-selective feeding behavior would allow the different species to feed on different organic sources. Even though the PPOM is present in such small quantities that it does not register in the isotopic signature of the total particulate organic matter in the cave, species selectively foraging on the small quantities of PPOM present could reflect the stable carbon isotope signal of the PPOM.

As with other atyid shrimps, *Typhlatya mitchelli* (Fig. 18.7a) has setal tufts on the distal end of its 1st and 2nd pereiopods which function like feather dusters to capture minute particles from the water (Fig. 18.7b) (Fryer, 1960). The individual setae have an array of filaments and saw-like structures that could ensnare and entangle bacteria and small particles (Fig. 18.7c). These structures may also be used for cleaning purposes. The maxillipeds are also highly setose (Fig. 18.7d), and may facilitate the transfer and sorting of particles from the pereiopods to the mouth.

The thermosbaenacean species *Tulumella unidens* is also equipped for collecting smaller particles. Thermosbaenacean feeding behavior is maxillary, and they feed entirely on detritus (Schram, 1986). As the maxillules of *Tulumella unidens* are highly setose, they should be capable of removing small particles from the water (M. Wicksten, pers. commun., 1995). Mysids like *Antromysis cenotensis* use their thoracic endopods to gather particles into a "food basket" (Schram, 1986). Given this feeding behavior, *A. cenotensis* would be expected to feed on larger particles than *Tulumella unidens* and *Typhlatya mitchelli.*

The adaptation to feed on the less abundant PPOM is an ecological advantage helping to explain why those that do so are more abundant than those that do not. Diver observations suggest that *Tulumella unidens* and *Typhlatya mitchelli* have the highest biomass in the Mayan Blue system. Because the algal material from the cenote and the chemoautotrophic bacteria are less degraded, they would be expected to be of higher nutritional value. During the decomposition of organic material, labile nitrogen-rich compounds such as amino acids and nucleic acids are preferentially degraded (Henrichs, 1993). Decomposed material with a high carbon/nitrogen ratio is less nutritious, since it provides less nitrogen for building protein. Pool particulate organic matter and chemoautotrophic bacteria presumably have higher nitrogen content than SPOM and, therefore, would be a preferred food source. It is no surprise, then, that *Tulumella unidens* and *Typhlatya mitchelli,* both adapted to filtering minute particles, are so successful in the anchialine environment.

Referring back to Fig. 18.5, the reason for the discrepancies in stable carbon isotopes between the three different species should now be more evident. *Typhlatya mitchelli,* the species seemingly most adept at gathering the small ^{13}C-depleted particles, was isotopically "lighter" than the other two species and the PPOM source. On the other hand, the thermosbaenaceans also have the ability to feed on smaller particles, but do not seem to be as efficient as *Typhlatya mitchelli,* as is manifest by their slightly "heavier" isotope signature. Finally, the "basket feeding" mysids appear to rely more on the larger SPOM particles, which had the most ^{13}C-enriched isotope values.

Niche partitioning

The apparent differential utilization of food resources by troglobitic Crustacea at the lower trophic level is known as "niche partitioning". Niche partitioning occurs when the available resources in a habitat are partitioned among different species employing specialized feeding strategies (Roughgarden, 1974). This is to be distinguished from niche separation or competitive exclusion where the available habitat is partitioned among the community members (Culver, 1970). For example, amphipods and isopods in an Appalachian cave-stream community exhibited niche separation by competitively excluding other species from particular areas in the stream-bed where they fed (Culver, 1970). On the other hand, because the members of the anchialine community each feed on a different fraction of the resource pool, several species coexist within the same niche.

Comparative ecology of other cenotes

With the exception of Cenote Angelita, both upstream (heading inland) and downstream (heading toward the coast) cave passages extended away from the cenote pools investigated. This coupling between cenote and cave appears to influence the ecology of the cave. Primary production in the pool and groundwater exchange rates between the cave and pool determine how much PPOM enters the cave system from the pool. This variable input of PPOM between sites, in turn, seems to influence the stable carbon isotope signatures of the cave fauna and the structure of the community.

The cenotes, although similar in many respects, are notably different in others. For example, Temple of Doom Cenote, a classic "jug-shaped" cenote surrounded by dense forest, receives little sunlight, while the open pool cenotes – Naharone, Carwash and Mayan Blue – are exposed to intense solar radiation. As a result, primary production in Temple of Doom Cenote is slight compared to the others. However, even within the open-water pools, visually observable production varies because of different hydrology and nutrient inputs. Carwash Cenote appears the most productive, and Mayan Blue the least. Stable-isotope data measured from each of these sites allow one to speculate on how the different cenotes influence the local ecology of their surrounding caves.

Productivity and the residence time of the water in the pool influences how much organic material is entering the cave. Assuming similar SPOM inputs from site to site, cenotes with the highest productivity and groundwater exchange rates will contribute more PPOM to the downstream cave community, thus lowering the ratio of SPOM to PPOM (SPOM:PPOM). As a result of lower SPOM:PPOM in caves downstream from highly productive pools, one would expect a community rich with the PPOM-dependent feeders *Tulumella unidens* and *Typhlatya mitchelli.* Furthermore, an increase in food availability would stimulate the production of more ammonia because of grazing regeneration (Fasham et al., 1990). Oxidation of this ammonia by nitrifying bacteria would fuel higher *in situ* chemoautotrophic production, and may support a larger troglobite biomass.

Since PPOM and nitrifying bacteria are isotopically lighter than the SPOM, one would expect the $\delta^{13}C$ values of the fauna from caves near highly productive cenotes also to be "lighter". This was exactly the case for the four cave systems studied. In the caves associated with the "open pool" cenotes – Naharone, Carwash and Mayan Blue – the mean $\delta^{13}C$ values for the organisms were –40.3 5.8 ($n=2$), –37.7 6.3 ($n=5$) and –32.9 4.8 ($n=46$), respectively, while the mean $\delta^{13}C$ value for troglobites in the "jug-shaped" Temple of Doom Cenote was –26.7 2.6 ($n=7$). The $\delta^{13}C$ values of troglobites in caves adjacent to "open pool" cenotes resembled the PPOM from the Mayan Blue pool (–35.3 2.6), indicating that PPOM is also an important source of organic matter within these systems. Carwash Cenote had the highest pool primary production, so it would be expected that the transport and subsequent mineralization of this organic material should support more chemoautotrophic activity. This may account for the unusually light values of the Carwash cave fauna. On the other hand, the oligotrophic Temple of Doom Cenote contributed a minimal amount of organic material into the cave. At this site, SPOM was most likely the primary food source. Indeed, the $\delta^{13}C$ values of the fauna collected from Temple of Doom cave passages (–26.7 2.6; $n=7$) were very similar to the Mayan Blue cave POM (–24.7 2.0; $n=7$) which we hypothesized was primarily soil-derived.

The SPOM:PPOM also seem to control the structure of the troglobitic community. At the Carwash and Naharone sites where we hypothesized that SPOM:PPOM was low, over half of the organisms collected were *Typhlatya mitchelli.* These shrimps, as discussed previously, may specialize in feeding on the more abundant PPOM. However, at the Temple of Doom site where PPOM input was slight, over half of the troglobites collected were the SPOM feeders *Antromysis cenotensis.* This situation demonstrates how the abiotic environment of an ecosystem can control the diversity of the community.

Inferences on the ecology of other Caribbean sites

Troglobitic genera characteristic of Yucatan caves also inhabit anchialine caves from other sites in the Caribbean including the Bahamas, Dominican Republic and Cuba (Holsinger, 1989; Yager, 1994). These crustaceans include amphipods (*Bahadzia* spp.), thermosbaenaceans (*Tulumella* spp.), remipedes (*Speleonectes* spp.), and isopods (*Bahalana* spp.). Measurements of dissolved oxygen, salinity and temperature from Mayan Blue are comparable to those reported by Iliffe et al. (1984) from various Bermuda caves and by Yager (1994) from Cueva de los Carboneros, an anchialine cave in Cuba. Considering the parallels in habitat and community structure, it is reasonable to hypothesize that common biogeochemical processes occurring at these sites govern their ecology.

CONCLUSIONS

Over the past two decades, there has been a tremendous increase in our knowledge of anchialine faunal composition and biogeographic distribution. This fundamental information has provided a basis for initiating detailed ecological and biogeochemical studies. Through techniques such as stable isotope tracing, it has been possible to produce a conceptual model explaining the complexities of the organic cycling, nutrient biogeochemistry and community ecology of anchialine cave ecosystems.

During this study of Mexican anchialine caves, we found that the organic matter supporting the cave community originates from at least three separate sources. These sources are soil organic matter from the overlying jungle soils, freshwater algae from the cenote pool, and chemoautotrophic bacteria from the cave interior. Soil and algal organic matter are allochthonous, meaning that they are transported into the systems from an outside source, while the chemoautotrophic bacteria represent an autochthonous source (i.e., produced *in situ*). It was tentatively assumed that these cave systems were completely detrital in nature; thus, the discovery of chemoautotrophic production in this environment represents a novel finding. Chemoautotrophic production in aquatic caves has been observed (Sarbu and Popa, 1992; Sarbu et al., 1996), but this is the first report of this type of production in an anchialine system. Finally, we argue that particulate organic matter associated with intruding seawater is a relatively insignificant source in these systems.

Food resources at the base of the food web were partitioned among specialized suspension-feeding crustaceans at lower trophic levels, which supported crustaceans and fish at higher trophic levels. The trophic structure of the community appeared to be controlled by the proportionate contributions of soil particulate organic matter (SPOM) and pool particulate organic matter (PPOM). Environmental and faunal similarities at other locations throughout the Caribbean support the idea that the events we observed are common anchialine processes.

Because of the rapid domestic, commercial, industrial and agricultural growth occurring in the coastal tropics, a clear understanding of the groundwater ecology and biogeochemistry is imperative for protecting these precious and sensitive environmental resources. By understanding how these systems operate without human interference, one can more easily monitor the condition and "health" of the groundwater in these areas as development continues.

ACKNOWLEDGEMENTS

The authors would like to thank the following people for their creative, logistical and laboratory assistance: Mike Madden, Jim Coke, Elva Escobar, Javier Alcocer, Thomas Boutton, John Morse, Mary Wicksten, Clarence McMahon, Lynn Roelke, Rody Best and Tony White. This study was supported by grants from the EPA (CR 820669), The Texas Institute of Oceanography, Texas A&M at Galveston, The Richmond Area Speleological Society and The Cave Research Foundation.

REFERENCES

Back, W. and Hanshaw, B.B., 1974. Hydrogeochemistry of the Northern Yucatan Peninsula, Mexico with a section on Mayan water practices. In: A.E. Weidie (Editor), *Field Seminar on Water and Carbonate Rocks of the Yucatan Peninsula, Mexico.* New Orleans Geological Society, New Orleans, pp. 45–77.

Back, W., Hanshaw, B.B., Herman, J.S. and Van Driel, J.N., 1986. Differential dissolution of a Pleistocene reef in the groundwater mixing zones of coastal Yucatan, Mexico. *Geology,* 43: 137–140.

Boutton, T.W., 1991. Stable carbon isotope ratios of natural materials: II. Atmospheric, terrestrial, marine, and freshwater environments. In: D.C. Coleman and B. Fry (Editors), *Carbon Isotope Techniques.* Academic Press, New York, pp. 173–185.

Bowman, T.E., Garner, S.P., Hessler, R.R., Iliffe, T.M. and Sanders, H.L., 1985. Mictacea, a new order of Crustacea Peracarida. *J. Crustacean Biol.,* 5: 74–78.

Carlucci, A.F. and McNally, P.M., 1969. Nitrification by marine bacteria in low concentrations of substrate and oxygen. *Limnol. Oceanogr.,* 14: 736–739.

Checkly, D.M. and Entzeroth, L.C., 1985. Elemental and isotopic fractionation of carbon and nitrogen by marine, planktonic copepods and implications to the marine nitrogen cycle. *J. Plankton Res.,* 7: 553–568.

Culver, D.C., 1970. Analysis of simple cave communities: niche separation and species packing. *Ecology,* 54: 102–110.

Cunliffe, S., 1985. The flora and fauna of Sagittarius, an anchialine cave and lake in Grand Bahama. *Cave Sci.,* 12: 103–109.

Dando, P.R., Southward, A.J., Southward, E.C., Terwilliger, N.B. and Terwilliger, R.C., 1985. Sulfur-oxidizing bacteria and haemoglobin in the gills of the bivalve mollusc *Myrtea spinifera. Mar. Ecol. Prog. Ser.,* 23: 85–98.

de Angelis, M.A., Reysenbach, A. and Baross, J.A., 1991. Surfaces of hydrothermal vent invertebrates: Sites of elevated microbial CH_4 oxidation activity. *Limnol. Oceanogr.,* 36: 570–577.

DeNiro, M.J. and Epstein, S., 1981. Influence of diet on the distribution of nitrogen isotopes in animals. *Geochem. Cosmochem. Acta,* 45: 341–351.

Elkins, J.W., Lipschultz, F., Wofsy, F.C., McElroy, M.B., Codispoti, L.A. and Friederich, G., 1985. Dissolved N_2O and the appearance of N_2O minima in Peruvian waters during NITROP-85. *EOS,* 66: 1320.

Fasham, M.J.R., Ducklow, H.W. and McKelvie, S.M., 1990. A nitrogen-based model of plankton dynamics in the ocean mixed layer. *J. Mar. Res.,* 48: 591–639.

Felbeck, H., 1981. Chemoautotrophic potential of the hydrothermal vent tube worm, *Riftia pachyptila* Jones (Vestimentifera). *Science,* 213: 336–338.

Fogel, M.L. and Cifuentes, L.A., 1993. Isotope fractionation during primary productivity. In: M.H. Engel and S.H. Macko (Editors), *Organic Geochemistry.* Plenum Press, New York, pp. 73–98.

Fosshagen, A. and Iliffe, T.M., 1985. Two new genera of Calanoida and a new order of Copepoda, Platycopoida, from marine caves in Bermuda. *Sarsia,* 70: 345–358.

Fry, B., 1991. Stable isotope diagrams of freshwater food webs. *Ecology,* 72(6): 2293–2297.

Fry, B. and Sherr, E.B., 1984. $\delta^{13}C$ measurements as indicators of carbon flow in marine and freshwater ecosystems. *Contr. Mar. Sci.,* 27: 13–47.

Fryer, G., 1960. The feeding mechanism of some atyid prawns of the genus *Caridina. Trans. R. Soc. Edinburgh,* 64: 217–244.

Gundersen, K., 1966. The growth and respiration of *Nitrosocystis oceanus* at different partial pressures of oxygen. *J. Gen. Microbiol.,* 42: 387–396.

Haines, E. and Montague, C., 1979. Food sources of estuarine invertebrates analyzed using $^{13}C/^{12}C$ ratios. *Ecology,* 60: 48–56.

Hart Jr., C.W., Manning, R.B. and Iliffe, T.M., 1985. The fauna of Atlantic marine caves: evidence of dispersal while maintaining ties to deep water. *Proc. Biol. Soc. Washington,* 98: 288–292.

Henrichs, S.M., 1993. Early diagenesis of organic matter: The dynamics (rates) of cycling of organic compounds. In: M.H. Engel and S.H. Macko (Editors), *Organic Geochemistry.* Plenum Press, New York, pp. 101–117.

Holsinger, J.R., 1989. Preliminary zoogeographic analysis of five groups of crustaceans from anchialine caves in the West Indian region. *Proc. 10th Int. Congr. Speleol.,* 1: 25–26.

Iliffe, T.M., 1992. Anchialine cave biology. In: A.I. Camacho (Editor), *The Natural History of Biospeleology.* Museo Nacional de Ciencias Naturales, Madrid, pp. 615–636.

Iliffe, T.M., Jickells, T.D. and Brewer, M.S., 1984. Organic pollution of an inland marine cave from Bermuda. *Mar. Environ. Res.,* 12: 173–189.

Kennicutt, M.C., Burke Jr., R.A., MacDonald, I.R., Brooks, J.M., Denoux, G.J. and Macko, S.A., 1992. Stable isotope partitioning in seep and vent organisms: chemical and ecological significance. *Chem. Geol.,* 101: 293–310.

Kornicker, L.S. and Iliffe, T.M., 1989a. Ostracoda (Myodocopina, Cladocopina, Halocyridina) from anchialine caves in Bermuda. *Smithson. Contr. Zool.,* 475: 1–88.

Kornicker, L.S. and Iliffe, T.M., 1989b. Troglobitic Ostracoda (Myodocopa: Thaumatocyprididae) from anchialine pools on Santa Cruz Island, Galapagos Islands. *Smithson. Contr. Zool.,* 483: 1–38.

Kornicker, L.S. and Iliffe, T.M., 1995. Ostracoda (Halocypridina, Cladocopina) from anchialine lava tube in Lanzarote, Canary Islands. *Smithson. Contr. Zool.,* 568: 1–32.

Kornicker, L.S. and Yager, J., 1996. The troglobitic halocyprid Ostracoda of anchialine caves in Cuba. *Smithson. Contr. Zool.,* 580: 1–16.

Macko, S.A., 1993. The characterization of organic matter in abyssal sediments, pore water and sediment traps. In: G.T. Rowe (Editor), *Deep-sea Food Chains and the Global Carbon Cycle.* Kluwer Academic Publishers, Boston, pp. 325–338.

Macko, S.A. and Ostram, N.E., 1994. Pollution studies using stable isotopes. In: K. Lajtha and R.H. Michener (Editors), *Stable Isotopes in Ecology and Environmental Science.* Blackwell Scientific Press, Oxford, pp. 45–62.

Macko, S.A., Lee, W.Y. and Parker, P.L., 1982. Nitrogen and carbon isotope fractionation by two species of marine amphipods: laboratory and field studies. *J. Exp. Mar. Biol. Ecol.,* 63: 145–149.

Manning, R.B., Hart Jr., C.W. and Iliffe, T.M., 1986. Mesozoic relicts in marine caves of Bermuda. *Stygologia,* 2: 156–166.

Mohr, C.E. and Poulson, T.L., 1966. *The Life of the Cave.* McGraw-Hill, New York, 232 pp.

O'Leary, M.H., 1981. Carbon isotope fractionation in plants. *Phytochemistry,* 20: 553–567.

Parzefall, J. and Wilkens, H., 1975. Zur ethologie augenreduzierter tiere. Untersuchungen an *Muniodopsis polymorpha* Koelbel (Anomura, Galatheidae). *Ann. Speleol.,* 30(2): 325–335.

Peterson, B.J. and Fry, B., 1987. Stable isotopes in ecosystem studies. *Annu. Rev. Ecol. Syst.,* 18: 293–320.

Peterson, B.J., Howarth, R.W., Lipschultz, F. and Ashendorf, D., 1980. Salt marsh detritus: an alternative interpretation of stable carbon isotope ratios and the fate of *Spartina alterniflora. Oikos,* 34: 173–177.

Peterson, B.J., Howarth, R.W. and Garritt, R.H., 1985. Multiple stable isotopes used to trace the flow of organic matter in estuarine food webs. *Science,* 227: 1361–1363.

Peterson, B.J., Howarth, R.W. and Garritt, R.H., 1986. Sulfur and carbon isotopes as tracers of salt-marsh organic matter flow. *Ecology,* 67: 685–674.

Quay, P.D., Emerson, S.R., Quay, B.M. and Devol, A.H., 1986. The carbon cycle for Lake Washington – A stable isotope study. *Limnol. Oceanogr.,* 31: 596–611.

Rau, G.H., 1978. Carbon-13 depletion in a subalpine lake: Carbon flow implications. *Science,* 201: 901–902.

Rau, G.H., 1981. Low 15N/14N in hydrothermal vent animals: ecological implications. *Science,* 289: 284–285.

Richardson, L.L., Aguilar, C. and Nealson, K.H., 1988. Manganese oxidation in pH and O_2 microenvironments produced by phytoplankton. *Limnol. Oceanogr.,* 33: 352–363.

Roughgarden, J., 1974. Niche width: Biogeographic patterns among *Anolis* lizard populations. *Am. Nat.,* 108: 429–442.

Sarbu, S.M. and Popa, R., 1992. A unique chemoautotrophically based cave ecosystem. In: A.I. Camacho (Editor), *The Natural History of Biospeleology.* Museo Nacional de Ciencias Naturales, Madrid, pp. 637–663.

Sarbu, S.M., Kane, T.C. and Kinkle, B.K., 1996. A chemoautotrophically based cave ecosystem. *Science,* 22: 1953–1955.

Schram, F.R., 1986. *Crustacea.* Oxford University Press, Oxford, 606 pp.

Stock, J.H., Iliffe, T.M. and Williams, D., 1986. The concept of "anchialine" reconsidered. *Stygologia,* 2: 90–92.

Stoessell, R.K., Ward, W.C., Ford, B.H. and Schuffert, J.D., 1989. Water chemistry and $CaCO_3$ dissolution in the saline part of an open-flow mixing zone, coastal Yucatan Peninsula, Mexico. *Geol. Soc. Am. Bull.,* 101: 159–169.

Stout, J.D., Goh, K.M. and Rafter, T.A., 1981. Chemistry and turnover of naturally occurring resistant organic compounds in soil. In: E.A. Paul and J.N. Ladd (Editors), *Soil Biochemistry.* Marcel Dekker, New York, pp. 1–73.

Ward, B.B., 1986. Nitrification in marine environments. In: J.J. Prosser (Editor), *Nitrification.* IRL Press, Oxford, pp. 157–184.

White, J.R. and Roman, M.R., 1991. Measurement of zooplankton grazing using particles labeled in light and dark with [methyl-^{3}H] methylamine hydrochloride. *Mar. Ecol. Progr. Ser.,* 71: 45–52.

Williams, P.M. and Gordon, L.I., 1970. Carbon 13: Carbon 12 ratios in dissolved and particulate organic matter in the sea. *Deep Sea Res.,* 17: 19–27.

Yager, J., 1981. Remipedia, a new class of Crustacea from a marine cave in the Bahamas. *J. Crustacean Biol.,* 1: 328–333.

Yager, J., 1987. *Cryptocorynetes haptodiscus,* new genus, new species, and *Speleonectes benjamani* from anchialine caves in the Bahamas, with remarks on distribution and ecology. *Proc. Biol. Soc. Washington,* 100(2): 302–320.

Yager, J., 1994. *Speleonectes gironensis,* new species (Remipedia: Speleonectidae), from anchialine caves in Cuba, with remarks on biogeography and ecology. *J. Crustacean Biol.,* 14: 752–762.

Chapter 19

GEOLOGY AND BIOLOGY OF THE FRASASSI CAVES IN CENTRAL ITALY: AN ECOLOGICAL MULTI-DISCIPLINARY STUDY OF A HYPOGENIC UNDERGROUND KARST SYSTEM

Serban M. SARBU, Sandro GALDENZI, Marco MENICHETTI and Gabriele GENTILE

INTRODUCTION

The Frasassi caves are located in central Italy, in the Marche region on the Adriatic side of the Apennine mountains, 40 km from the coast. Karst systems develop on both sides of the Gola di Frasassi (Frasassi Gorge), a canyon 500 m deep and 2 km long, cut by the Sentino river from west to east (Fig. 19.1). Elevations range between 200 m at river level to 957 m at the top of the gorge (Fig. 19.2). More than 100 caves open into the Gola di Frasassi, and are believed to be of hypogenic origin (i.e., they resulted from the interaction between limestone and acidic waters ascending from a deep aquifer). Grotta Grande del Vento is the most popular show cave in Italy. Its largest room, Abisso Ancona (Fig. 19.1*h*; see also Fig. 19.3) and a few adjacent rooms are visited every year by approximately 350 000 tourists.

The vegetation in the gorge is bipartite: mediterranean thermophil species such as *Arbutus unedo, Ficus carica, Phillyrea latifolia, Pistacia terebinthus, Quercus ilex, Smilax aspera* and *Viburnum tinus,* are prevalent on the northern side of the gorge. Deciduous trees forming the *Scutellaria–Ostryetum* association are prevalent on the colder southern side. Plantation pine forests and pastures cover the top of the mountains surrounding the gorge. *Moehringia papulosa, Potentilla caulescens* and *Rhamnus saxatilis* are a few examples of plants growing on the limestone cliffs in the gorge (Ballelli et al., 1981). Bare rock outcrops are scattered in the cliffs along the gorge (Fig. 19.2).

The climate of the region is temperate sub-continental in the main valleys and apenninic continental in the mountains. The annual average temperature in the Gola di Frasassi is 13°C and decreases to 11°C on the top of the hills. The average amount of precipitation varies between 900 mm y^{-1} in the valley and 1100 mm y^{-1} in the mountains. Snowfall occurs occasionally in January and February, with an annual average of about 20 cm. Precipitation tends to reach maximal values in the late fall and winter, whereas the water balance shows a deficit in July and August. The wind regime consists of wet and cold winds from the northeast and warm winds from the southwest.

GEOLOGY

The Umbria–Marche Apennines were formed during the Upper Miocene tectonic compressive phase and represent a fold-and-thrust belt, trending northeast. The Frasassi karst system is developed in the external limb of an anticline bordered by north–south faults (Fig. 19.4). The subterranean karst mainly follows the fold-hinge zone where the bedding is sub-horizontal. The main tectonic fractures, accounting for about 85% of the discontinuities in the rock mass, form a system of cross and longitudinal extensional joints that originated during the last stage of the folding. These tectonic fractures are visible in the scarps of the gorge and in the cave passages; they had an important role in the development of the karst drainage and influenced the morphology of the large solutional voids (Fig. 19.3). A set of east–west faults and a northwest–southeast joint system influence the Grotta del Mezzogiorno–Grotta di Frasassi cave system (Fig. 19.5) as well as the surface hydrographic network. Waters rich in hydrogen

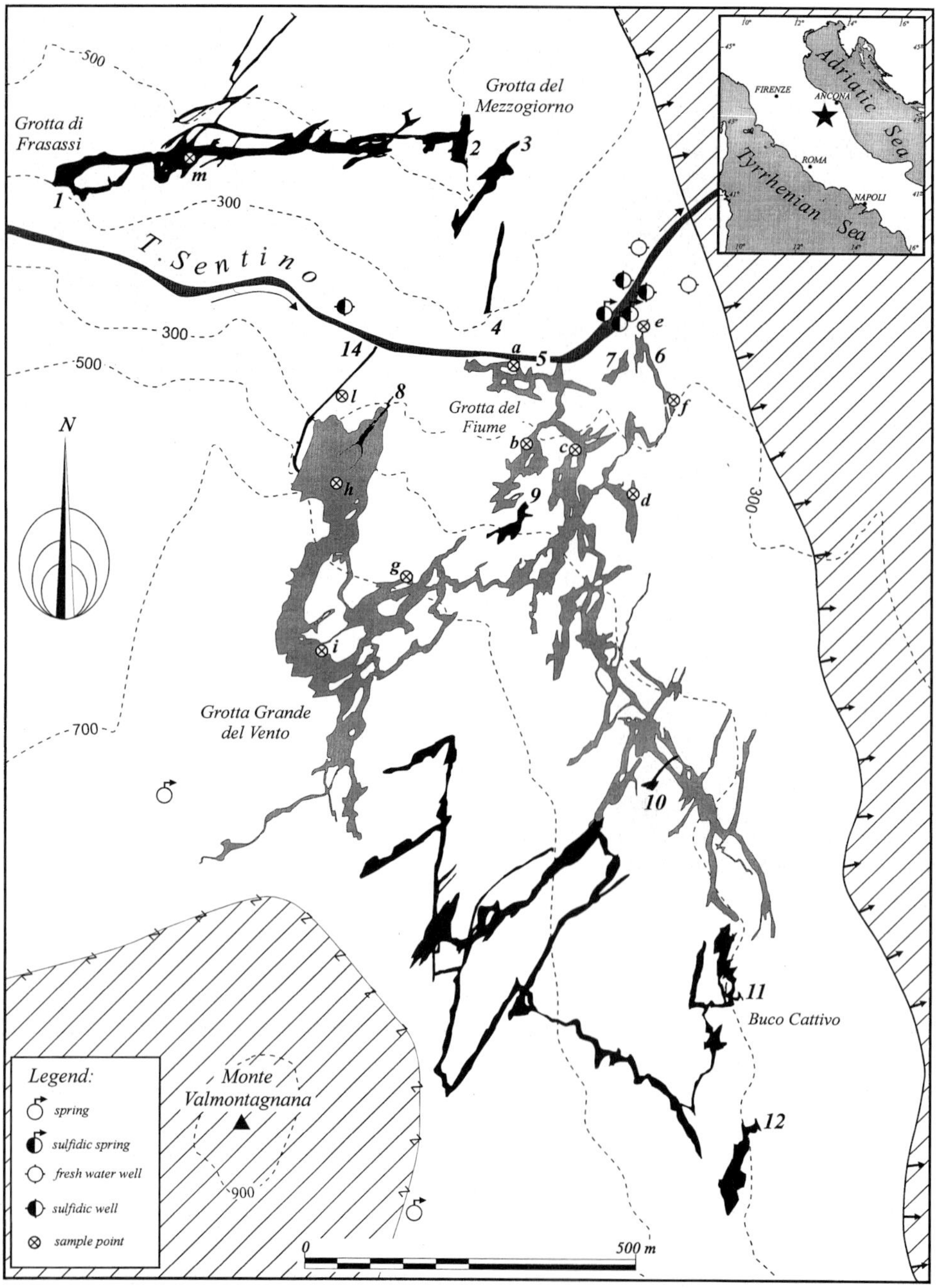

Fig. 19.1. Schematic plan of the Gola di Frasassi underground karst system. The areas with oblique shading represent outcrops of Cretaceous marls with low permeability, while the white area is the permeable limestone; the arrows indicate the dip of the geological contact. Caves developed at altitudes above 300 m are represented in black, those at lower altitudes in gray. The numbers are the cave entrances: *1*, Grotta di Frasassi; *2*, Grotta del Mezzogiorno; *3*, Buco del Colonnello; *4*, Grotta Verde; *5*, Grotta del Fiume; *6*, Grotta Sulfurea; *7*, Grotta Bella; *8*, Grotta Grande del Vento, natural entrance; *9*, Grotta del Paradiso; *10*, Buco del Diavolo; *11*, Buco Cattivo; *12*, Grotta dell'Infinito; *13*, artificial entrance of the Grotta Grande del Vento.
Sample points: *a*, entrance area of Grotta del Fiume; *b*, Lago Verde in Grotta del Fiume; *c*, Sala Guano in Grotta del Fiume; *d*, Ramo Sulfureo section of Grotta del Fiume; *e*, entrance area of the Grotta Sulfurea; *f*, deep sulfidic section of the Grotta Sulfurea; *g*, Laghi di Lucia; *h*, dripping water in Abisso Ancona; *i*, Lago dell'Orsa; *l*, sulfidic pool in a shaft intercepted by the artificial access tunnel; *m*, Sala Guano in Grotta di Frasassi.

Fig. 19.2. Oblique aerial view of Gola di Frasassi from the west. The vegetation on the slope of the gorge is mainly mediterranean thermophile species and deciduous trees. Jurassic limestone outcrops are visible on both sides of the Sentino.

sulfide ascend to the surface following the network of fractures related to the north–south fault system.

The Frasassi karst system is developed in Jurassic strata 900 m thick (Lower Lias, 190 10^6 years old). The limestone is very pure (over 99% calcium carbonate), and consists of cyclic sequences 4–5 m thick. The high intergranular porosity and the well developed network of fractures favored water flow within the limestone and the initial development of the cave system (Galdenzi and Menichetti, 1990; Menichetti, 1990).

The Jurassic carbonate overlies an Upper Triassic sequence, 1800 m thick, of anhydrites, gypsum, and dolomites, and is covered by a thin layer of nodular

Fig. 19.3. The largest room in the Frasassi caves: Abisso Ancona in Grotta Grande del Vento with a volume exceeding 1 million cubic meters (*h* in Fig. 19.1). Limestone strata are visible on the walls, and broken blocks are visible on the floor of the room.

limestone, interbedded with marls and by a thick layer of Lower Cretaceous limestone 130 10^6 years old (Fossa-Mancini, 1921; Parotto and Praturlon, 1975). This stratigraphic sequence, which forms an important karstifiable level in the overall Apennine chain, is delimited by an overlying marly rock formation of Lower Cretaceous age (120 10^6 years old), low in permeability.

HYDROGEOLOGY

The surface drainage network is dominated by the Sentino river, which flows perpendicular to the hinge of the Frasassi anticline structure. Superpositions[1], antecedences and regressions of river networks with respect to anticline structures are often present in the Apennine chain (Parotto and Praturlon, 1975).

The regional aquifers are hosted by the Jurassic limestone in the core of the anticline, while the recharge occurs in an area of $5\,km^2$ of surface karst. The groundwater supplies the springs located in the most important valleys and in the main fault zones. The average overall base-flow discharge is $22\,\ell\,s^{-1}\,km^{-2}$ (Bani et al., 1986). Aquicludes impeding water flow, represented by marly layers, isolate local perched aquifers. Groundwater flow in the vadose zone is controlled by conduit systems along fissures, while the network of tectonic fractures regulates water movement in the phreatic zone. The hydrodynamics of larger

[1] When geologists refer to rivers with antecedences, it means that the geologic structure that the river crosses is younger than the river itself – the river is said to maintain its course as the fold or fault, or whatever the structure is, is being formed. Superimpositions are drainages that were established on an earlier surface, so that the pattern of the river was kept because of the river's incision into a geologic structure.

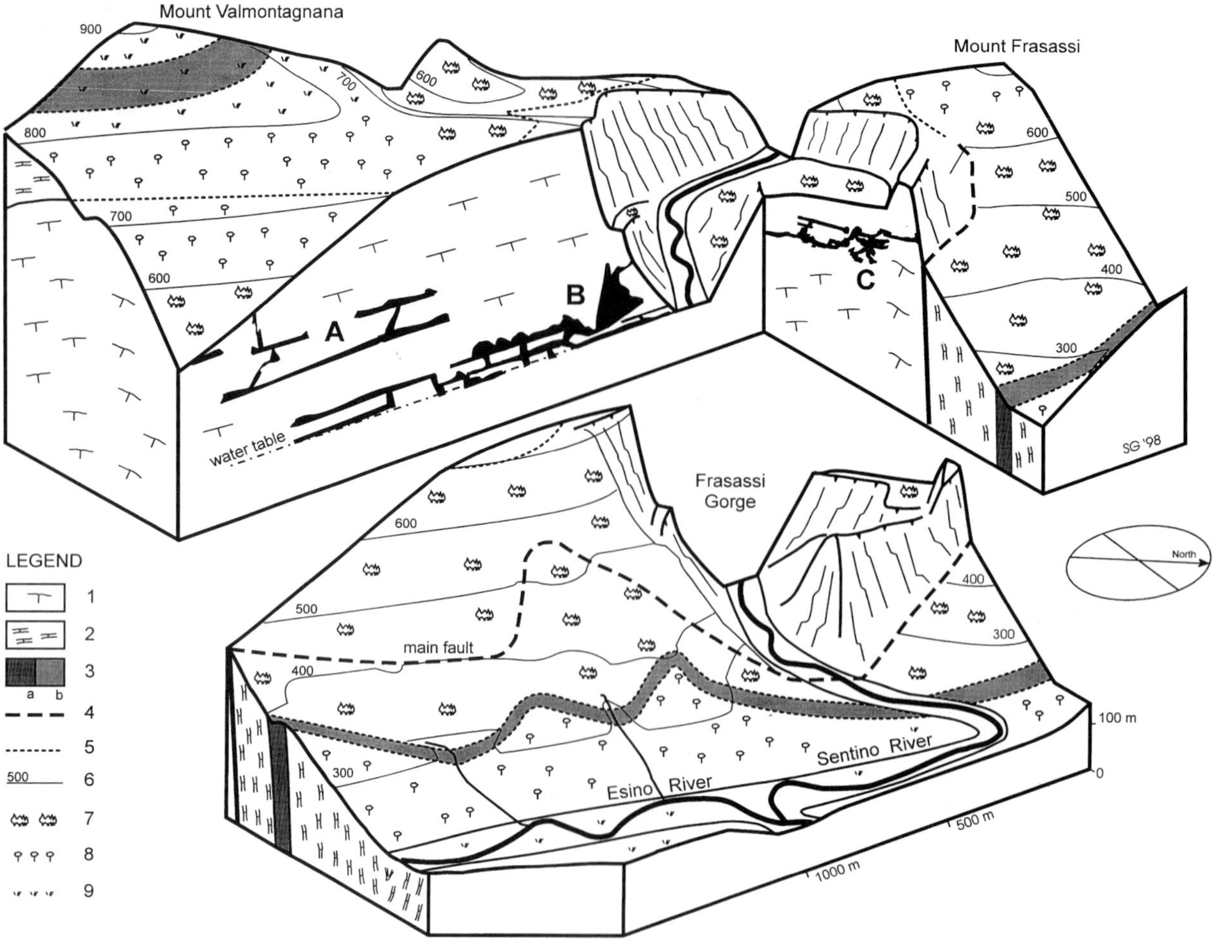

Fig. 19.4. Block diagram of the Frasassi karst system. (1) Jurassic limestone; (2) Cretaceous siliceous carbonates and marls; (3) Cretaceous marls: a, section; b, outcrops; (4) faults; (5) stratigraphic boundary; (6) contours; (7) broad-leaved woods and coppices; (8) pine plantations; (9) pastures. (A) Buco Cattivo; (B) Grotta del Fiume–Grotta Grande del Vento cave system; (C) Grotta del Mezzogiorno–Grotta di Frasassi cave system.

Fig. 19.5. The entrance of the Grotta di Frasassi (*1* in Fig. 19.1), with the church built in 1828 by the French architect Valadier for Pope Leone XII who was born in the Frasassi area.

springs is regulated primarily by the base flow, but occasional fast flow is evidence for input from the vadose zone.

Two types of groundwater are found in the Frasassi region: carbonate waters resulting from the infiltration of surface meteoric waters through the limestone, and mineralized waters that ascend to the surface after having encountered deep Triassic anhydrite rocks in their flow path (Fig. 19.6). The chemical content of the latter is primarily controlled by sodium and chloride ions and secondarily by sulfur ions (i.e., sulfate and sulfide). Sulfide can reach concentrations up to 0.4 mMol ℓ^{-1} in the mineralized aquifer (Menichetti, 1985; Sighinolfi, 1990; Tazioli et al., 1990).

Conductivity and temperature of water in the sul-

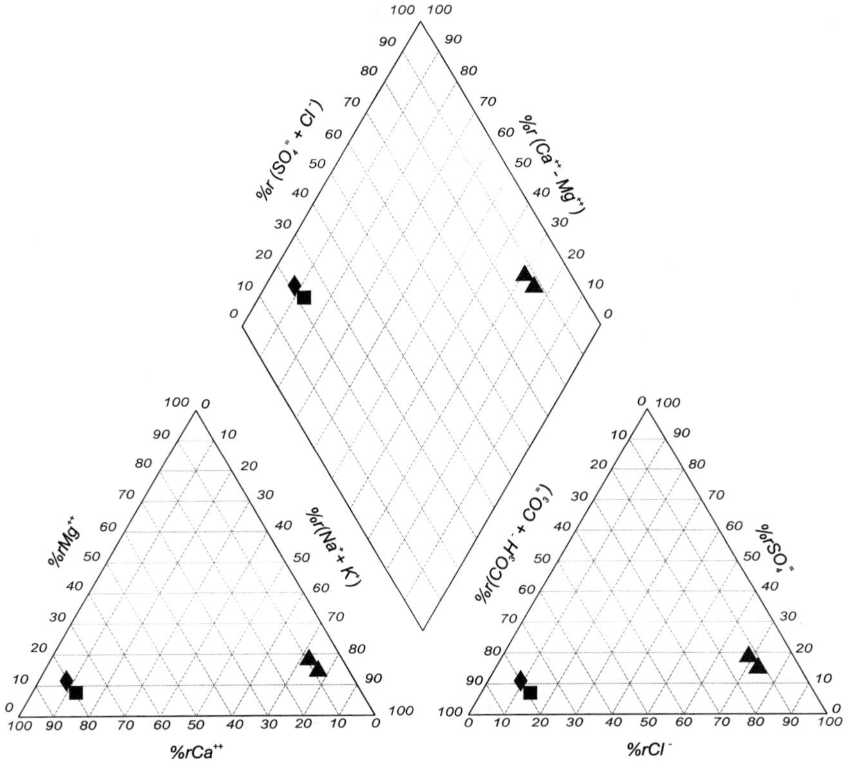

Fig. 19.6. Piper diagram of the waters in the Gola di Frasassi karst system. Triangles indicate the sulfidic waters, squares are waters from rimstone dams, and the diamonds represent fresh waters infiltrating from the surface.

Fig. 19.7. Sulfidic lake in the Grotta del Fiume (*d* in Fig. 19.1). Gypsum micro- and macro-crystals dot the limestone host rock. White organic matter and elemental sulfur are visible on the limestone rock at the air–water interface.

fidic stream in the Ramo Sulfureo section of Grotta del Fiume (Fig. 19.1*d*; see also Fig. 19.7) exhibit interesting variations related to the seasonal recharge. Here the water temperatures vary from 12.8 to 13.7°C (Fig. 19.8). The data show the lack of any significant shift in either the conductivity or the temperature peaks. In spring and summer the conductivity increases from 660 $\mu S\,cm^{-1}$ to 1000 $\mu S\,cm^{-1}$ due to seepage from the base flow in the sulfidic aquifer. During periods of abundant surface precipitation, especially during the fall, an increased input of fresh water rich in carbonate reaches the water table and dilutes the sulfuric mineralized water in the phreatic zone (Fig. 19.8). The variations in water temperature and conductivity in the sulfidic aquifer at Frasassi show the importance of the fast-flow recharge.

Stable isotope ratio analysis (SIRA) of the water suggests a meteoric origin for both the carbonate-rich and the mineralized waters, with a recharge area located at altitudes of 600–1000 m (Tazioli et al., 1990) (Fig. 19.9).

The tritium content of the water in the Frasassi area shows values of 1–13 Units of Tritium (U.T.) (Cigna, 1989; Tazioli et al., 1990); for the dripping water in the Grotta Grande del Vento (Fig. 19.1*h*), while the water in the sulfidic springs and in the underground sulfidic streams reaches 35–40 U.T. (Fig. 19.1*b,d*). The relationship between the total dissolved solids (T.D.S.) and the tritium content can delineate the underground flow

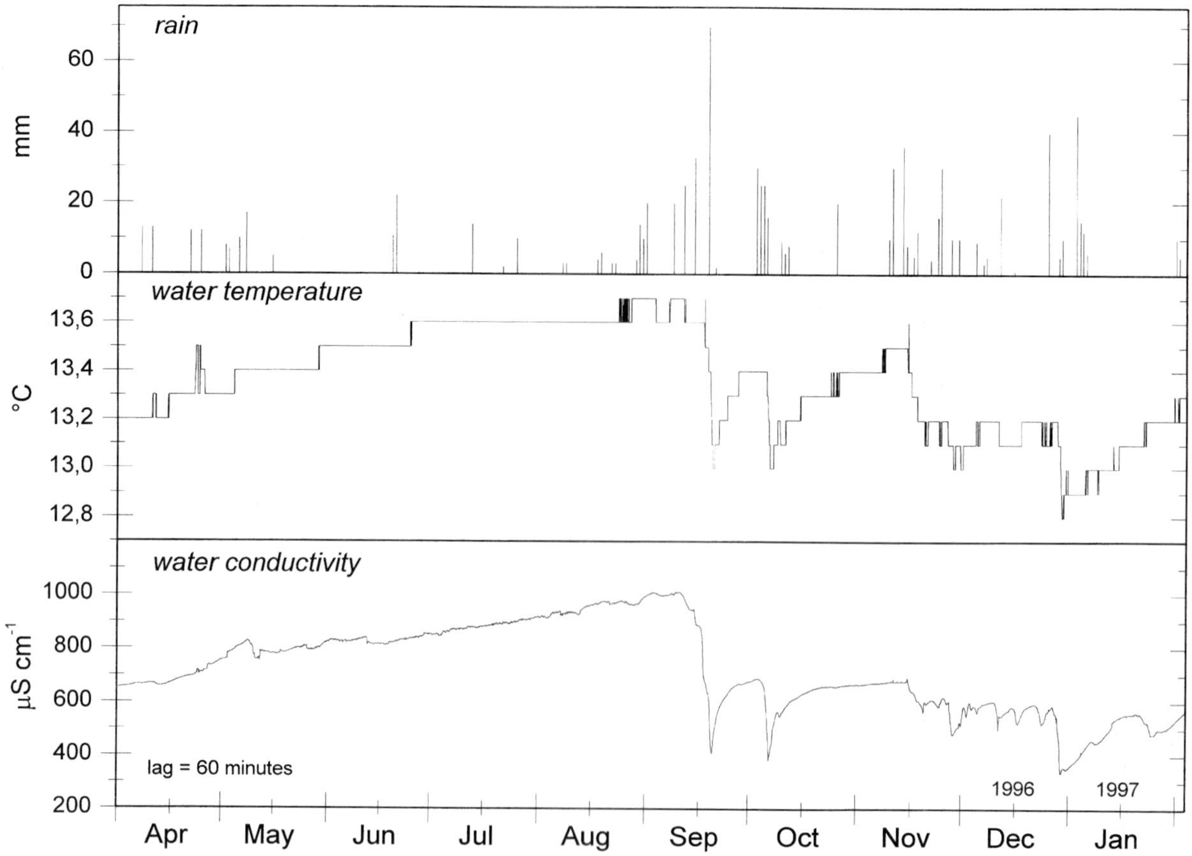

Fig. 19.8. Hourly records of conductivity and temperature of the water in the Ramo Sulfureo branch (*d* in Fig. 19.1) of the Grotta del Fiume compared with the surface precipitation (April 1996 through January 1997).

path from the surface to the vadose zone (Fig. 19.10), through a direct relationship between the age of the water and its degree of mineralization. The dripping water has a low level of mineralization, with a residence time in the aquifer of a few years. Some of this water reaches the deep Triassic evaporites in the Frasassi structure, where it becomes enriched in sodium, chloride, sulfate and sulfide. Eventually, the water rises to the surface after having been underground for a few decades.

KARST

The Frasassi karst system

The Frasassi caves represent the most interesting karst system in central Italy. Surface karst features (i.e., sinkholes and karren) are rare in this region, despite the lack of extensive surface drainage. Only a few shaft caves are present. On the other hand, the subsurface karst is well developed, consisting of a total of over 35 km of cave passages (Figs. 19.1, 19.4). Cave entrances open at elevations between 200 and 500 m. Horizontal passages and rooms of phreatic origin are the most common features, while shafts, meanders and other running-water features that develop in the vadose zone are quite rare.

The Grotta del Fiume and Grotta Grande del Vento (Fig. 19.1*5*,*8*), making up the largest cave system in the Gola di Frasassi, are developed in the southern wall of the gorge and consists of a total of over 20 km of cave passages located at altitudes between 200 and 360 m. With a volume that exceeds 1 000 000 m^3, Abisso Ancona is the largest room in the cave and one of the largest cave rooms in Europe (Fig. 19.3). The lowest sections of this cave are still active and contain several sulfidic lakes and streams. Other important caves, developed at altitudes of 400 to 500 m, are Buco Cattivo (5000 m long) (Fig. 19.1*11*) and the Grotta del

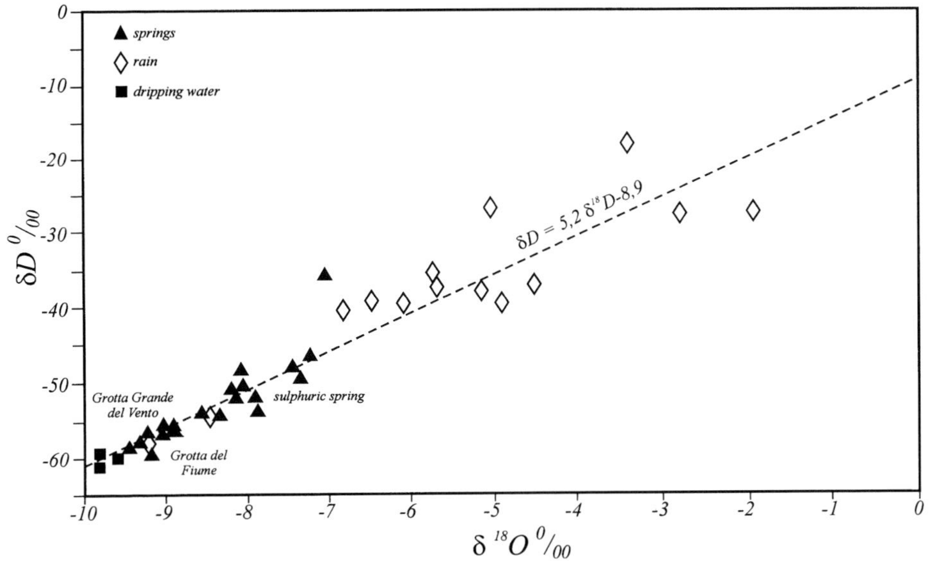

Fig. 19.9. Relationship between $\delta^{18}O$ (‰) and δD (‰) for the water in the Frasassi area. Data from dripping water in the Grotta Grande del Vento (*h* in Fig. 19.1) and the Grotta del Fiume (*b* in Fig. 19.1). Values of δ(‰) are calculated as $X = [(R_{sample} - R_{standard})/R_{standard}]\ \ 10^3$, where X is δD or $\delta^{18}O$, and R is deuterium/hydrogen, or $^{18}O/^{16}O$. The standard is: Standard Mean Ocean Water (SMOW) for hydrogen and oxygen. Precision of measurements was 0.2‰ or better.

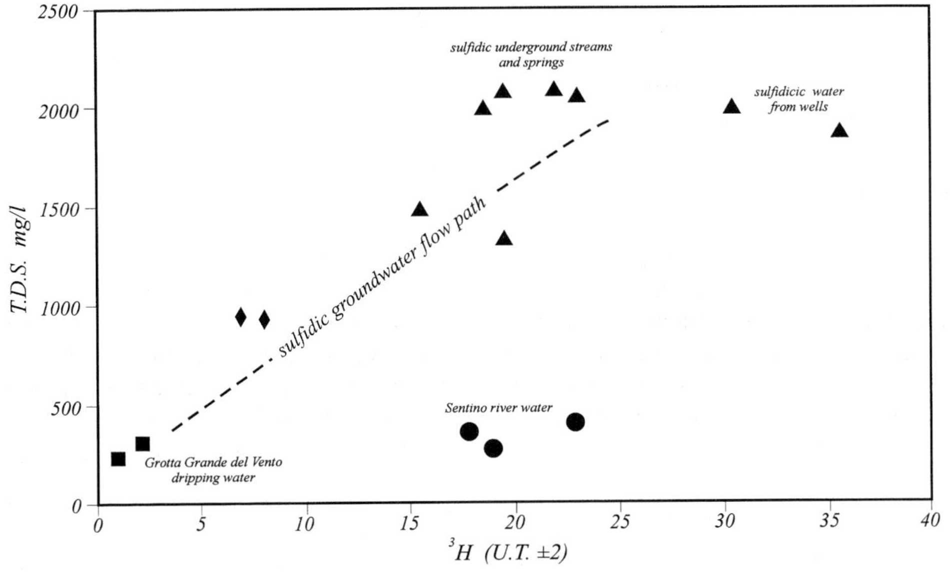

Fig. 19.10. Relationship between the total dissolved solids (T.D.S.) and ^3H (U.T.) for the groundwaters of the Frasassi area and the hypothetical subterranean flow path of the sulfidic groundwater.

Mezzogiorno–Grotta di Frasassi cave system (4500 m long) (Fig. 19.1*1*,*2*). Both are dry, well decorated caves. The latter represents the main cave system in the northern wall of the Gola di Frasassi Gorge and its two entrances are the largest and the most beautiful in the area (Fig. 19.3).

The Frasassi caves are developed in at least four main horizontal levels. The two lowest levels (e.g., most of the passages in Grotta del Fiume and Grotta Grande del Vento) correspond in elevation to the alluvial gravel terraces of Middle to Upper Pleistocene age in the nearby valleys, suggesting that the history of the

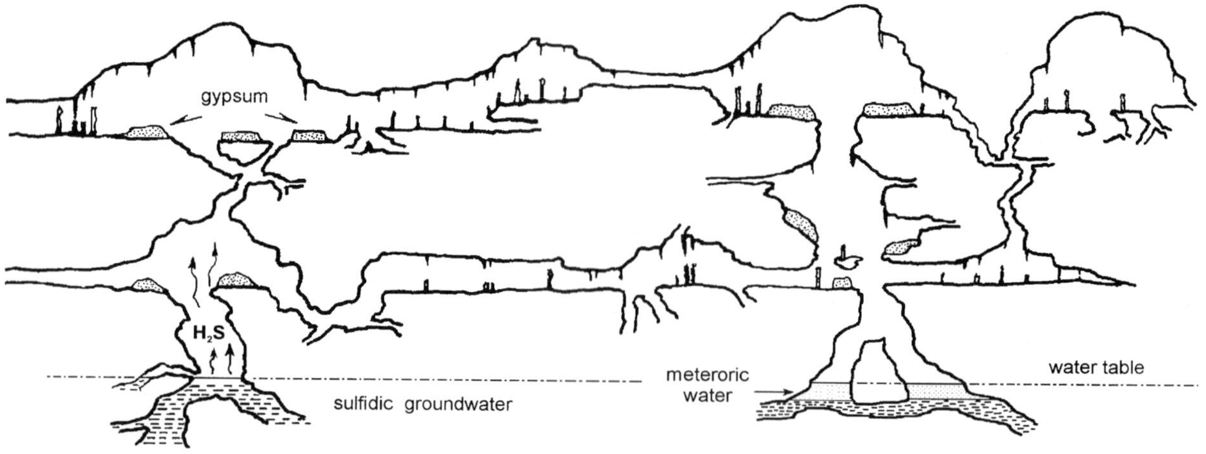

Fig. 19.11. Schematic morphology of cave passages in lower level of Frasassi. Each cave level consists of large chambers and passages connected by shafts and by narrow tubes and crevices of phreatic origin. At present, groundwater is often stratified with a layer of oxygenated water at the surface, which prevents hydrogen sulfide vapors from rising and protects the limestone walls from sulfuric acid corrosion.

cave can be related to the evolution of the surface hydrographic network. The development of the cave levels is believed to correspond to the deposition of surface terraces during the longest stages between the surface river down-cutting episodes of the Middle and Upper Pleistocene. Using the uranium series technique, Taddeucci et al. (1992) dated carbonate speleothems in Grotta del Fiume and Grotta Grande del Vento showing stalagmite ages of up to 200 000 years, and pointed out that changes in the surface climate are correlated with changes in the growth rate of the speleothems in the Grotta di Frasassi. This supports the hypothesis of a connection between the development of cave levels and alluvial terraces.

Each karst level shows a complex pattern, strongly influenced by faults and by the hydrogeological setting. Dendritic networks of passages with tributaries draining into a main stream are absent from the Frasassi caves. Instead, the size and the general morphology of the cave passages change without a clear rule. The caves exhibit a typical ramifying pattern, developed in overlying levels. The lower levels, ranging between altitudes of 200 and 300 m, were developed in morphological and hydrogeological settings similar to the present, as shown by the similarity between their morphology and that of currently active caves. The upper levels located at altitudes of 300 to 500 m show different characteristics and it has been suggested that they developed in different hydrogeological settings. Their origin, however, is still an open question, and research on this subject is in progress.

Hypogean features can best be recognized and studied in the Grotta del Fiume–Grotta Grande del Vento system, which has developed almost entirely in the lower karst levels (Fig. 19.11) (Galdenzi, 1990). Phreatic tubes, 1–10 m in diameter, are present in many parts of the cave. They are often anastomosing, forming complex mazes. Several large rooms resulted from the collapse of the limestone separating superimposed cave levels. Other large rooms show wide rounded ceilings and flat erosional rock floors where shafts and crevices open. Some of the shafts connect the different cave levels or provide access to the water table (Fig. 19.7). Typical phreatic features are widespread throughout the cave (Fig. 19.12). These include roof pendants and knife-edges resulting from the enlargement of nearby passages; spongeworks, a feature of diffuse corrosion, with many small interconnected passages giving the limestone a vacuolar appearance; and bubble trails – grooves on the cave walls produced by acidic gas bubbles rising towards the water table. Cupolas of different sizes are present on the walls and ceilings of the cave, both in the small passages and in the largest rooms. Silt and clay fills of residual origin are widespread in the cave (Bertolani et al., 1977). Except for the passages located very close to the gorge, which are affected by occasional severe floods of the Sentino river, allochthonous alluvial gravels are absent from the cave.

Sulfuric acid speleogenesis

The presence of gypsum deposits inside the Grotta del

Fig. 19.12. Sala Finlandia in the Grotta del Fiume–Grotta Grande del Vento karst system, a classic phreatic passage with cupolas of different sizes on the ceiling. Later dripstone formations (i.e., stalactites, stalagmites, and gours) develop in the cave passage after the lowering of the water table.

Fiume–Grotta Grande del Vento system is the most evident effect of the circulation of sulfidic groundwater. Until recently, the presence of gypsum and other minerals in the main caves of the region was considered of secondary importance, and the speleogenesis was thought to result from carbon dioxide enrichment of the drip water. However, as early as 1931, Principi suggested that a small cave in the Umbria–Marche Appennini resulted from interaction between sulfidic water and limestone. More recently, it was proposed that the origin of the main caves at Frasassi and in other areas in the Umbria–Marche Appennini was associated with the circulation of sulfidic groundwater through limestone (Galdenzi, 1990; Galdenzi and Menichetti, 1995). Similar speleogenic mechanisms have also been suggested for some caves in other parts of the world (Egemeier, 1981; Palmer, 1991; Hill, 1987; Ford and Williams, 1989; Forti et al., 1989; Sarbu and Kane, 1995).

The rising of water rich in hydrogen sulfide within the limestone increases the solutional aggressiveness which leads to limestone dissolution. The oxidation of hydrogen sulfide to sulfate (Eq. 19.1) occurs in the upper phreatic zone, near the watertable, in the presence of oxygen, and is mediated by sulfur-oxidizing microorganisms. Limestone dissolution (Eq. 19.2) takes place both in phreatic and vadose conditions:

$$H_2S + 2O_2 \rightarrow 2H^+ + SO_4^{2-} \quad (19.1)$$

$$2H^+ + SO_4^{2-} + CaCO_3 + H_2O \rightarrow CaSO_4\ \ 2H_2O + CO_2 \quad (19.2)$$

Gypsum produced in the phreatic zone is transported away in solution by groundwater. However, when formed above the water table, gypsum produces white, massive, finely grained replacement crusts on the cave walls and ceilings (Fig. 19.13). Later, seepage water undersaturated with respect to gypsum may remove the crusts, exposing small, rounded erosional pockets on the cave walls.

Fig. 19.13. White, finely grained gypsum, forming in some pockets on the limestone wall in the Grotta del Fiume–Grotta Grande del Vento karst system.

The rate of sulfuric acid speleogenesis was determined during an experiment carried out in the deep sulfidic sections of the Grotta del Fiume (Galdenzi et al., 1997). Limestone tablets (80 40 10 mm) were placed both in the sulfidic groundwater and above the water level in a cave room containing a fast-flowing sulfidic stream. Some tablets were removed and weighed after two and a half years, while the rest of the tablets were left in the cave for five years. All tablets exhibited evident surface corrosion, and the weight loss reached values up tp 20 mg cm^{-2} yr^{-1}. The surface of the

Fig. 19.14. Experimental limestone tablet exposed to H_2S vapors in the cave atmosphere for five years: left, replacement gypsum covering the surface of the tablet; right, the morphology of the tablet surface after the removal of the gypsum.

submerged tablets was uniformly corroded and did not exhibit gypsum deposition. The limestone surface was entirely covered by a microbial biofilm which appears to play an important role in the oxidation of hydrogen sulfide and, subsequently, in limestone corrosion. The rate of weight loss in the submerged tablets remained constant throughout the experiment. The tablets placed above the water level were completely covered with replacement gypsum crusts derived from the oxidation of hydrogen sulfide and subsequent reaction with the limestone (Fig. 19.14). Under the gypsum crust, the tablet surface was covered by a thin layer of organic matter, and was severely and quite irregularly corroded, exhibiting incipient erosional pockets (Fig. 19.14). In the emersed tablets, the rate of weight loss increased with time, reaching the highest values at the end of the experiment. Additionally, the weight loss was strongly influenced by the location of each tablet within the cave room and by the concentration of hydrogen sulfide in the cave atmosphere. This experiment confirmed the conclusions of the earlier morphological studies and pointed out the intense solutional aggressiveness of the sulfidic groundwater.

Yellow crusts of elemental sulfur, resulting from the oxidation of hydrogen sulfide, are present in some sections of the cave where intense evaporation of hydrogen sulfide takes place (e.g., in the proximity of sumps or sulfidic streams). Microbial mats consisting of sulfur-oxidizing microorganisms grow on the surface of the cave walls where elemental sulfur is present. Values for pH between zero and one were measured on the surface of these microbial mats, indicating that the sulfur-oxidizing microbiota secrete sulfuric acid ($2H^+ + SO_4^{2-}$) as a byproduct of their metabolic activity.

The hydrogeological setting has strongly influenced speleogenesis in the Frasassi area. The largest cave rooms occur where the thin, surficial marly cover permitted intense infiltration of oxygen-rich water from the surface (Fig. 19.11). The oxidation of hydrogen sulfide was intensified in these zones, causing an acceleration of limestone dissolution which generated large cave chambers and galleries. Gypsum deposition was more intense in some periods of the cave history, as shown by the presence of old gypsum replacement crusts, common in the upper, now dry cave passages, as well as by the existence of white, fine-grained, vast floor deposits of gypsum (over $1000\,m^3$) (Fig. 19.15) in the largest cave rooms (Galdenzi, 1990).

Fig. 19.15. Large deposits of micro-crystalline gypsum overlying phreatic clay deposits in Sala Duecento, in the Grotta Grande del Vento. These gypsum deposits were formed above the water table by gravity flow and accumulation of gypsum resulting from sulfuric acid limestone corrosion. The dripping water dissolves gypsum, forming deep and narrow furrows. Calcite stalagmites are presently forming in the passage.

The most intensive and widespread subaerial deposition of gypsum probably occurred in geomorphological and hydrogeological settings that favored widespread contact of the sulfidic groundwater with the oxygen-rich cave atmosphere. At present, deposition of gypsum is limited to the few narrow zones where the watertable can be reached. In most of these locations, a layer of fresh water often overlies the mineralized sulfidic groundwater (Fig. 19.16).

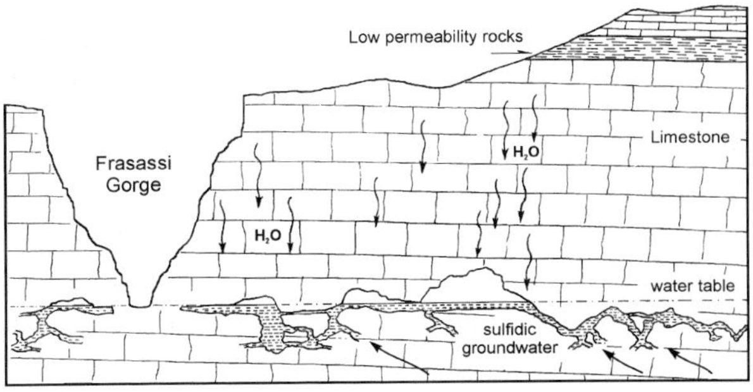

Fig. 19.16. Schematic model for cave development in the lower levels of the Frasassi caves. The solutional aggressiveness of the groundwater increases on account of non-homogeneous flow of descending meteoric water. Only small phreatic tubes, tapering inwards, develop under the thin marly cover.

The caves in the Gola di Frasassi are good examples of multi-loop caves, with development in three main phases:

(1) phreatic phase – at first, passages were developed in the shallow phreatic zone as a consequence of hydrogen sulfide oxidation favored by the mixing of the oxygen-rich infiltration waters with the mineralized groundwater;
(2) "watertable" phase – a slight lowering of the watertable and the gradual enlargement of cave passages enabled widespread contact between the groundwater and the cave atmosphere. During this phase, sulfuric acid corrosion of limestone occurred above the watertable, generating gypsum replacement crusts which cover the cave walls. This also led to the growth of the ceiling in the large cave rooms.
(3) vadose phase – as a consequence of the lowering of the watertable, many of the former phreatic and "watertable" passages became part of the vadose zone. Carbonate speleothems developed in the upper passages, while the formation of gypsum crusts was limited to deep shafts that reached sulfidic waters at the new water table.

Cave evolution in the Gola di Frasassi was clearly favored by the tectonic uplift and climate changes that occurred during the Pleistocene. Nevertheless, some smaller periods of positive shifts of the watertable during this time may also be hypothesized, and the true history of the cave may have been more complex.

TOURISM

Grotta Grande del Vento in the Gola di Frasassi was discovered in the early 1970s. The opening of an artificial access tunnel (Fig. 19.1*14*), and the construction of a concrete pathway, have made it possible for nine million tourists to visit the cave since 1974, with an average of 1000 visitors each day. An environmental monitoring network with twenty sensors for air and water temperature, air humidity, velocity of the air, carbon dioxide concentration and atmospheric pressure, as well as pluviometers, has been recording data at 30-minute intervals since 1989 (Menichetti et al., 1994).

The average air temperature in the cave varies between 13.5 and 13.9°C in the inner sections of the cave, and the humidity is almost 100%. In the tourist part (Fig. 19.1*h*), temperature increases of a few degrees take place during the summer season as a consequence of the influx of warm surface air through the artificial tunnel and the large number of tourists visiting the cave (in the month of August an average of 3700 persons visit the cave each day). Only in September and October does the underground air temperature reach the original values of the colder months. These cyclic temperature variations progressively decrease from the entrance to the deeper section of the cave.

The natural air circulation in the cave is characterized by a "chimney effect": in winter when the external air temperature is lower, air enters the cave through the lower entrance and leaves the cave through the upper entrance. The circulation is reversed during the warmer season, and the switch in the direction of air flow depends on the external weather conditions. The air velocity in the upper small natural cave entrance may reach $3\,m\,sec^{-1}$. The air flux is characterized by pulses alternating with stagnant periods depending on the external weather conditions (i.e., temperature, wind, and barometric pressure). In the artificial tunnel the air speed can exceed $1\,m\,sec^{-1}$, especially in the late morning when larger groups of tourists are present in the cave and when the access doors remain completely open.

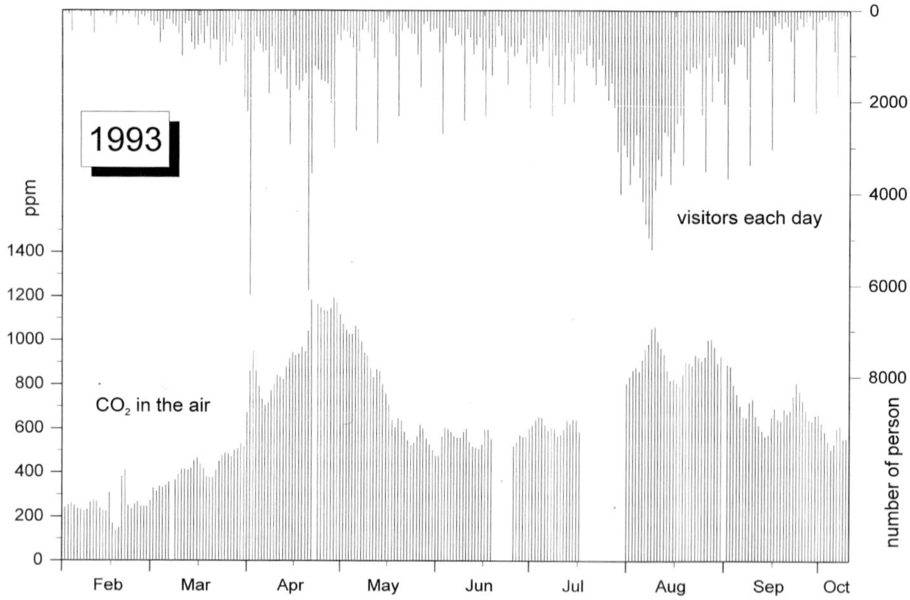

Fig. 19.17. Carbon dioxide concentration and the number of visitors entering the Grotta Grande del Vento. A total of 300 000 visitors entered the cave in the time interval shown in this figure (February 1 through October 15, 1993).

The carbon dioxide concentration in the cave atmosphere has both a natural and an anthropogenic component. The natural background, which varies from 600 to 1400 ppm, is related to the rising of sulfidic water and to seasonal biological cycles. Tourists' breath, as well the opening of the access doors, contributes significantly to variations in carbon dioxide concentration in the cave atmosphere. The carbon dioxide level recorded close to the tourist pathway in 1993 shows seasonal variations, with a maximum in summer exceeding 1000 ppm, and a minimum of about 400 ppm in winter (Fig. 19.17). Daily and hourly fluctuations of up to 1100 ppm are correlated with the number of visitors. Sometimes, especially in summer, it may take a few days for the cave to recover after a severe increase in the carbon dioxide concentration caused by natural and anthropogenic factors.

The heat flow generated by tourists and by the artificial lights installed along the visitors' path accounts for 86% of the total energy input into the cave. The energy output is represented mainly by cooling convection and radiation through the limestone walls (19%) and through the exchange of air (67%) through the natural and artificial entrances. The total estimated energy input is of 574 GJ while the energy output represents 566 GJ, leaving a substantial positive balance in the overall annual energy budget (Menichetti, 1996).

The temporal evolution of the underground natural and artificial non-stationary processes and the variables involved in the energetic budget are hard to predict. They are controlled by chaotic dynamic processes where small and apparently insignificant variations may be amplified in time. A strong effort is currently being made by speleologists and researchers, and by the Consorzio Frasassi[2] to protect the fragile underground ecosystem, mainly from the effects of warming, growth of algae, dust and trash that visitors carry into the caves. However, more resources must be aimed in this direction in the future.

FAUNA

The cave fauna of the hypogean karst system at Frasassi was the object of an in-depth, prospective scientific investigation (Bertolani et al., 1994) commissioned by the Commissione Scientifica of the Consorzio Frasassi as part of a comprehensive environmental impact study for the Gola di Frasassi and its caves. The purpose of the survey was to obtain detailed information on the hypogean faunal populations. Prior to this investigation,

[2] The public administration of the show caves in Frasassi.

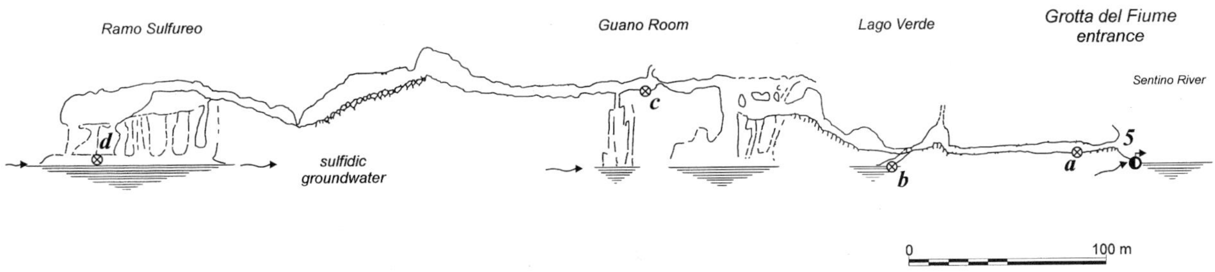

Fig. 19.18. Location of the collecting sites LAV, GUA, and RSO within the lower levels of the Grotta del Fiume–Grotta Grande del Vento cave system (schematic representation). Vertical scale exaggerated by 150%. The symbols *a, b, c, d,* and *5* have the same meaning as in Fig. 19.1.

data on the biology of this karst system was limited to reports on individual taxonomic groups. Fauna from the Frasassi caves was mentioned in scientific publications as early as 1939, when Stranero studied the geographic distribution of the troglobitic carabid beetle *Duvalius bensai.* Many other investigators have studied the fauna inhabiting the caves at Frasassi: Ruffo and Vigna Taglianti (1969), Vigna Taglianti (1970), Brignoli (1971, 1979, 1985), Cola and Freude (1974) and Karaman (1985), to mention a few.

The thorough investigation of 12 caves in the Gola di Frasassi by Bertolani et al. (1994) resulted in the identification of 57 species, of which half were troglophiles. Only two species were troglobites – the amphipod *Niphargus ictus* and the carabid beetle *Duvalius bensai lombardi*; both of these taxa are endemic to the Grotte di Frasassi. Bertolani et al. interpreted this low level of adaptation to cave-dwelling as a consequence of the young geological age of the caves, which suggested a relatively recent faunistic colonization. They concluded that the cave communities were relatively homogeneous and highly stable throughout the entire karst system at Frasassi. This homogeneity was considered to be the consequence of the geographic proximity of the caves to one another, and of their common origin. It was also noted that the homogeneity observed in the Frasassi caves extends to other caves in this region. Of the 26 species found by Bani (1984) in Grotta dei Cinque Laghi 45 km northeast of Frasassi in the Apennine chain, 16 are also present in the Grotte di Frasassi. An additional troglobitic species, not mentioned by Bertolani et al. (1994), is the spider *Centromerus pasquini,* which had earlier been reported from the Frasassi Caves by Brignoli (1979, 1985). This species is endemic to caves in northern Italy.

Following preliminary reports (Galdenzi, Menichetti, and Sarbu, unpublished) regarding the presence of rich and abundant invertebrate communities in the sulfidic sections of the caves, as well as the presence of abundant microbial communities in the sulfidic streams and on the limestone walls in close proximity to the sulfidic waters (Galdenzi et al., 1997), the Consorzio Frasassi in 1997 commissioned a biological investigation of the cave sections containing sulfidic streams or sulfidic vapors in the atmosphere. Non-sulfidic sections of the caves were also examined during this study for comparative purposes. Studies performed in other sulfidic caves, such as Movile Cave in Romania (Sarbu, 1990; Sarbu and Kane, 1995; see also Chapter 17, this volume) and submarine caves at Capo Palinuro in Italy (Southward et al., 1996), identified rich and abundant invertebrate communities based on autochthonous food produced by sulfur-oxidizing chemoautotrophic microorganisms.

Four cave passages containing sulfidic streams and lakes were investigated in the Grotte di Frasassi: the Ramo Sulfureo section of Grotta del Fiume (RSO) (Fig. 19.1*d*; Fig. 19.18*b*), the Lago Verde room in Grotta del Fiume (LAV) (Fig. 19.1*b*; Fig. 19.18*b*), the Laghi di Lucia in the Grotta del Fiume–Grotta Grande del Vento system (Fig. 19.1*g*), and the deep sulfidic section of Grotta Sulfurea (SUL) (Fig. 19.1*f*). Two non-sulfidic habitats were also investigated: a cave passage containing a large deposit of guano in the Grotta del Mezzogiorno–Grotta di Frasassi system (MEZ) (Fig. 19.1*m*), and a room in Grotta del Fiume which contains a small pile of guano (GUA) (Fig. 19.1*c*; Fig. 19.18*c*). In addition, samples were collected from several surface locations in the Gola di Frasassi and from the entrances of Grotta del Fiume (ENF) (Fig. 19.1*a*; Fig. 19.18*a*) and of Grotta Sulfurea (ENS) (Fig. 19.1*e*). The faunal investigations in the sulfidic sections of the Frasassi caves led to the discovery of 15 invertebrate species (Table 19.1).

Table 19.1
List of species of invertebrates found recently in the sulfidic sections of the Frasassi caves

ANNELIDA	
Oligochaeta	
Lumbriculidae	
Rhynchelmis n. sp.	troglobite [1]
MOLLUSCA	
Gastropoda	
Hydrobiidae	
Islamia pusilla	troglomorphic population
CRUSTACEA	
Isopoda	
Oniscidae	
Androniscus dentiger	troglophile
Amphipoda	
Gammaridae	
Niphargus ictus	troglobite [1]
ARACHNIDA	
Pseudoscorpiones	
Chthoniidae	
Chthonius n. sp.	troglobite [1]
Acarina	
Labidostomatidae	
Labidostoma cornuta	endogeous
Araneae	
Linyphiidae	
Centromerus pasquinii	troglobite [2]
Lepthyphantes pallidus	troglophile
Lepthyphantes n. sp.	troglobite [1]
Porrhomma n. sp.	troglobite [1]
Nesticidae	
Nesticus eremita	troglophile
INSECTA	
Collembola	
Neelipleona	
Neelus murinus	endogeous
Onichiuridae	
Deuteraphorura n. sp.	troglobite [1]
Coleoptera	
Carabidae	
Duvalius bensai lombardi	troglobite [1]
Pselaphidae	
Tychobythinus gladiator gladiator	endogeous

[1] Species endemic to the Frasassi caves.
[2] Species endemic to northern Italy.

Niphargus ictus (Amphipoda, Gammaridae) and *Islamia pusilla* (Gastropoda, Hydrobiidae), the only aquatic invertebrates found to date, were observed in all the sulfidic lakes and streams in the Frasassi caves. A small population of annelid worms belonging to a new species of *Rhynchelmis* (Oligochaeta, Lumbriculidae) was discovered on the wet surface of cave walls in the Ramo Sulfureo section of Grotta del Fiume, as well as in the deep sulfidic section of Grotta Sulfurea. Numerous specimens of *Androniscus dentiger* (Isopoda, Oniscidae) and *Nesticus eremita* (Araneae, Nesticidae) were found in all the sulfidic cave passages and in the proximity of the guano deposits. Several very small arthropods were collected from the cave walls and from the floor of the passages in the Ramo Sulfureo section: *Labidostoma cornuta* (Acarina, Labidostomatidae), an endogeous species; four new endemic troglobites: *Chthonius* n. sp. (Pseudoscorpiones, Chthoniidae), *Deuteraphorura* n. sp. (Collembola, Onichiuridae) and two new species of spiders *Porrhomma* n. sp. (Araneae, Linyphiidae) and *Lepthyphantes* n. sp. (Araneae, Linyphiidae); *Neelus murinus* (Collembola: Neelipleona), a species considered a trogloxene but which occurs in large numbers in the sulfidic sections of the cave; and an endogeous beetle *Tychobythinus gladiator gladiator* (Coleoptera, Pselaphidae). Two additional species of spiders were collected in the sulfidic section of Grotta Sulfurea: the troglophile *Lepthyphantes pallidus* (Araneae, Linyphiidae), and the troglobite *Centromerus pasquinii* (Araneae, Linyphiidae), a species endemic for northern Italy. A single specimen of *Duvalius bensai lombardi* (Coleoptera, Carabidae), endemic for the Frasassi caves, was collected in sulfidic habitats (LAV).

A total of nine troglobitic species were found in the sulfidic sections of the Frasassi caves, seven of which are endemic to the Frasassi karst system

(Table 19.1). Five of these species are new to science. *Androniscus dentiger, Centromerus pasquinii, Duvalius bensai lombardi, Nesticus eremita* and *Niphargus ictus* are among the few invertebrate species previously mentioned from the Frasassi caves.

ECOSYSTEM ANALYSIS

Sources of food within the Frasassi caves

The absence of food in the upper dry levels of the Grotta del Fiume–Grotta Grande del Vento system was thought to account for the complete lack of fauna in these passages (Bertolani et al., 1994). Early faunistic studies were primarily based on samples collected in the vicinity of the cave entrances, as well as on samples collected in cave passages containing guano deposits. In both cases, the fauna is feeding on allochthonous organic matter of photosynthetic origin, transported into the cave gravitationally (e.g., leaf litter, tree logs and branches), by water (e.g., plant debris, soil), or by bats (e.g., guano).

Recent research in the sulfidic sections of the Frasassi caves (*b,d,f,g,* Fig. 19.1) led to the discovery of rich and abundant invertebrate communities. The majority (60%) of the species discovered here are endemic, and the abundance of some of these populations (e.g., 10 to 50 specimens m^{-2}) is comparable to abundances of troglobites found in the sulfidic passages in Movile Cave, Romania (see Chapter 17 by Sarbu, Table 17.4, p. 335). Nine of the twelve invertebrate species found in the sulfidic passages of the Frasassi caves are predators.

The discovery of a rich microbiota in the sulfidic streams and lakes, as well as on the cave walls above the water surface, and the presence of hydrogen sulfide in the atmosphere, suggested that sulfur-oxidizing microorganisms are the primary producers in the subterranean ecosystems. Large populations of flagellates live in the sulfidic waters associated with the sulfur-oxidizing microorganisms. Preliminary incubations of samples of cave microbiota from the sulfidic streams with bicarbonate labeled with ^{14}C showed an uptake of radio-labeled inorganic carbon, indicating that organic matter (i.e., food for the cave community) is being produced autochthonously, *in situ,* in the deep recesses of the caves.

Microbial biofilms consisting of sulfur-oxidizing microorganisms grow on the surface of the thick gypsum crusts covering the limestone walls above the sulfidic streams. Values of pH between zero and one were measured on the mucous surface of these biofilms, indicating that the sulfur-oxidizing microbiota secrete sulfuric acid as a byproduct of their metabolic activity. Preliminary microbiological investigations of several samples of biofilm collected from the cave walls indicate the presence of acidophilic strains of *Thiobacillus thiooxidans* and *T. ferrooxidans. Thiobacillus thiooxidans* is typically found in acidic environments where it accounts for the production of copious amounts of sulfuric acid from reduced sulfur compounds. *Thiobacillus ferrooxidans* is able to derive energy from the oxidation of ferrous iron as well as reduced sulfur compounds. Mucous biofilms exhibiting very low pH values have recently been also reported from Cueva de Villa Luz, Chiapas, Mexico (Taylor, 1999).

Structure of the cave food-web

Stable isotope ratio analysis (SIRA) of various organic samples was done to determine the source of food used by various populations of cave and surface invertebrates. This is a valuable method for studying food webs, because organisms fractionate isotopes of carbon ($^{13}C/^{12}C$) and nitrogen ($^{15}N/^{13}N$) in predictable ways. Fractionation values for carbon isotopes of approximately –25‰ between inorganic carbon and the biomass of chemoautotrophs have previously been reported for sulfur-oxidizing microorganisms (Degens, 1969; Ruby et al., 1987; Vlasceanu et al., 1997). Consumers are typically enriched by approximately 1–2‰ in ^{13}C and by 3–5‰ in ^{15}N, relative to their food source (Paull et al., 1985; Ehleringer et al., 1986; Peterson and Fry, 1987). Stable isotope ratio analysis has recently been used to investigate the food-web structure of cave communities, and has led to the identification of autochthonous sources of food (Sarbu et al., 1996; Southward et al., 1996; Pohlman et al., 1997).

The stable isotope ratios of carbon in inorganic samples from Frasassi have been analysed at the Department of Geology, University of Michigan. The ratios of carbon and nitrogen stable isotopes in organic samples were analysed at the Institute for Marine

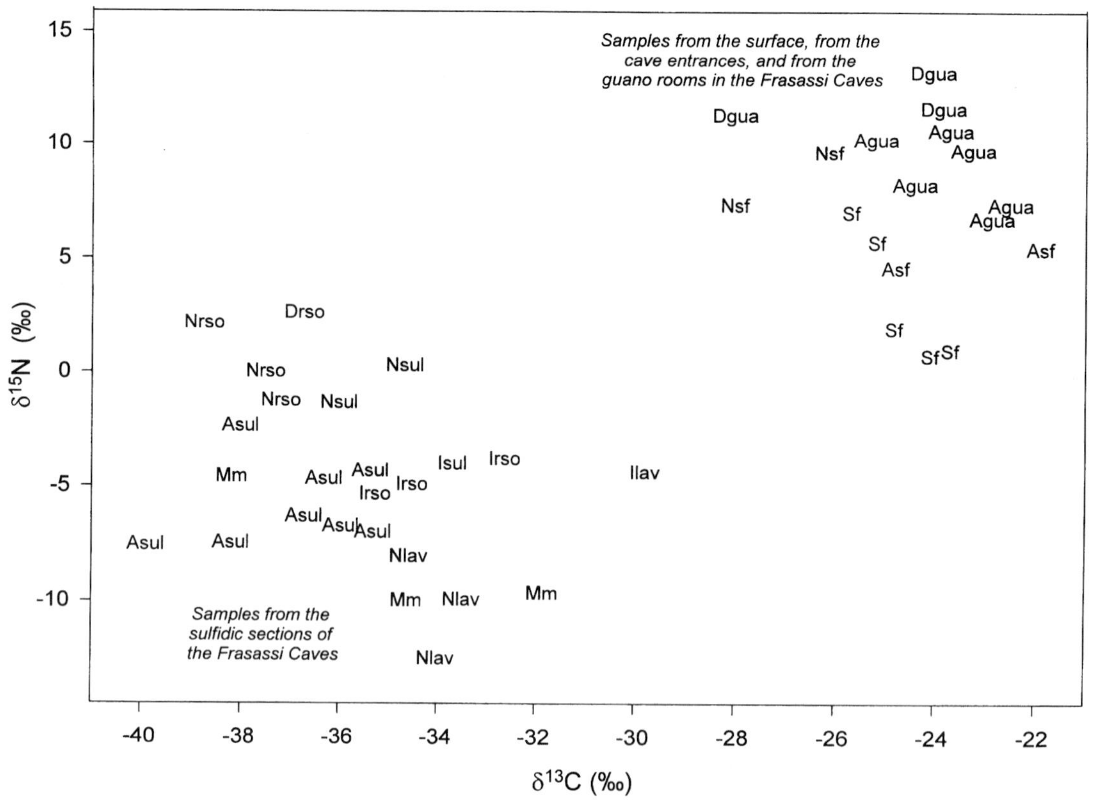

Fig. 19.19. Stable isotope ratios of carbon and nitrogen for organic samples. Agua, *Androniscus dentiger* from cave sections containing deposits of guano; Asf, *A. dentiger* from the surface, collected under leaf litter in the Gola di Frasassi; Asul, *A. dentiger* from sulfidic passages in the Grotta del Fiume–Grotta Grande del Vento and in Grotta Sulfurea; Dgua, *Duvalius bensai lombardi* from GUA; Drso, *Duvalius bensai lombardi* from RSO; Ilav, *Niphargus ictus* from LAV; Irso, *Niphargus ictus* from RSO; Isul, *N. ictus* from SUL; Mm, microbial mats from RSO; Nlav, *Nesticus eremita* from LAV; Nrso, *N. eremita* from RSO; Nsf, *N. eremita* collected at the surface, near the cave entrance; Nsul, *N. eremita* from SUL; Sf, surface fauna collected under leaf litter in the Gola di Frasassi;.

Science, University of Alaska at Fairbanks. Results are expressed as δ (delta) values[3].

Bicarbonate in the sulfidic waters in the Frasassi region is isotopically light with respect to carbon, values for $\delta^{13}C$ ranging between –7.5‰ and –9‰. Microbial mats (Mm, Fig. 19.19), consisting of sulfur-oxidizing bacteria, collected in the sulfidic stream in the Ramo Sulfureo section of the Grotta del Fiume–Grotta Grande del Vento system showed average $\delta^{13}C$ values of –35‰, indicating a fractionation of approximately –20‰. This is consistent with fractionation values known from chemoautotrophic sulfur-oxidizing microorganisms (Degens, 1969; Ruby et al., 1987; Vlasceanu et al., 1997). Surface leaf litter, which represents the food source for grazers living at the surface and in the cave entrance areas, has $\delta^{13}C$ values of –27‰, which are characteristic for plants using the Calvin cycle (O'Leary, 1988).

Samples of invertebrates collected in the sulfidic sections of the Frasassi caves (Fig. 19.19) are isotopically lighter in carbon and nitrogen than invertebrates from the surface that feed on leaf litter (Sf, Fig. 19.19). On the other hand, organisms feeding on bat guano in the dry sections of the caves are isotopically heavier than those feeding on leaf litter (Fig. 19.19).

Two species inhabiting sulfidic sections of the caves also occur at the surface and around deposits of guano, thus making possible a comparison between different populations of the same species. Samples of *Androniscus dentiger, Duvalius bensai lombardi,*

[3] Values of δ (‰) are calculated as $X = R_{\text{sample}} - R_{\text{standard}} / R_{\text{standard}} \quad 10^3$, where X is $\delta^{13}C$, δD, $\delta^{18}O$, or $\delta^{35}S$ and R is $^{13}C/^{12}C$, D/H, $^{18}O/^{16}O$, or $^{35}S/^{32}S$. Standards are Pee Dee belemnite (PDB) for carbon and air for nitrogen. Precision of measurements was 0.2‰ or better.

and *Nesticus eremita* were collected in the sulfidic passages (Asul, Drso, Nrso, Nsul, Nlav, Fig. 19.19), around deposits of guano in several parts of the cave (Agua, Dgua, Fig. 19.19), and at the surface (Asf, Nsf, Fig. 19.19). The populations feeding on photosynthetic food (i.e., leaf litter, bat guano) are isotopically heavier, both with respect to carbon and to nitrogen, than the populations collected in the sulfidic rooms. This shows that the sulfidic populations rely on a different source of food, such as the autochthonously synthesized chemoautotrophic food, present in large quantities in the sulfidic sections of the caves.

The stable isotope data indicate that *Niphargus ictus* (Irso, Fig. 19.19) appears to graze on microbial mats that cover the sediments in the sulfidic streams (Mm, Fig. 19.19). *Nesticus eremita* in the Ramo Sulfureo section of Grotta del Fiume (Nsul, Fig. 19.19) appears to feed on the abundant *Androniscus dentiger* (Asul, Fig. 19.19), as suggested by the stable isotope data. In the guano-based food webs, *A. dentiger* appears to represent the food for *Duvalius bensai lombardi* (Asul, Fig. 19.19). An interesting observation was that, for unknown reasons, *Nesticus eremita* collected from the Lago Verde room in Grotta del Fiume, is heavier with respect to carbon and lighter with respect to nitrogen than the population of the same species collected from the Ramo Sulfureo section of the cave.

POPULATION GENETICS

The data obtained from the analysis of carbon and nitrogen stable isotope ratios offer insight regarding the food-web structure within the cave community during the particular period when the analysis was performed. It is arguable that, if trophic relationships are stable over long periods of time, they should be consistent with the genetic structure of the organisms inhabiting this system. Thus, study of the genetic structure of the populations and the gene-flow patterns of these organisms represents an important complement to the stable isotope analysis.

The oniscidean isopod *Androniscus dentiger* is the species which is most abundant in different habitats in the Frasassi caves. Allozyme electrophoresis was used to investigate the genetic structure of several populations of *A. dentiger* (Gentile and Sarbu, unpublished observations). The sampling sites were the same as those used in the stable-isotope analysis.

Table 19.2
Genotype frequencies in the Frasassi caves. Values of F_{IS} [1] (Weir and Cockerham, 1984) for each of the 18 loci investigated

Locus	Cave systems [2,3]			
	MEZ	GUA	RSO	LAV
Ada	−0.027	0.686**	0.823**	0.709**
Aldo	–	–	–	–
Aph-1	0.511*	0.559**	0.517*	0.596**
Aph-2	–	–	–	–
Apk	–	–	–	–
Ca	–	−0.155	0.786**	−0.043
Dia	–	–	–	–
β*Gal*	0.349	0.786**	–	–
Got-1	–	–	–	–
Got-2	–	–	–	–
Gpi	–	–	–	–
Mdh-1	−0.150	−0.067	0.064	
Mdh-2	–	–	–	–
Mpi	–	0.087	0.197	0.654**
Pep-1	–	0.067	–	–
Pep-2	–	−0.011	–	–
Pgm	–	–	–	–
Pk	–	–	–	–

[1] The F_{IS} index provides an estimate of the difference between expected and observed heterozygosities. Negative values indicate an excess of heterozygotes, positive values a heterozygote deficency.

[2] MEZ, Grotta del Mezzogiorno–Grotta di Frasassi system; GUA, Sala Guano in the Grotta del Fiume; RSO, Ramo Sulfureo section of the Grotta del Fiume; LAV, Lago Verde room in the Grotta del Fiume.

[3] Significance of differences from zero is indicated by asterisks: $*$, $.05 > p > .01$; $**$, $p < .01$.

Some loci showed genotype frequencies deviating from Hardy–Weinberg equilibrium, generally attributable to heterozygote deficiencies (Table 19.2).

F_{IT}, F_{ST} and F_{IS} estimates[4] [0.564, 0.284 and 0.390, differing significantly from zero in each case ($p < 0.001$)] suggested that there is genetic variance both among and between populations. The major contribution to genetic variation among populations is

[4] *F* statistics are a valuable tool for partitioning the genetic diversity into its within-population and between-population components (Nei, 1987). The F_{IS}, F_{IT}, and F_{ST} statistics used here were proposed by Wright (1951). In his terminology, F_{IS} and F_{IT} are the correlations between the two uniting gametes relative to the subpopulation of a subdivided population, and relative to the total population, respectively. F_{ST} is the correlation between two gametes drawn at random from each subpopulation.

Table 19.3
Gene-flow (Nm) values[1] in the Frasassi caves

	MEZ[2]	GUA[2]	RSO[2]
GUA[2]	0.508		
RSO[2]	0.606	1.536	
LAV[2]	0.227	4.958	0.462

[1] Nm is an estimate of gene flow, where N is the effective population size, and m the migration rate. The estimate of Nm can be related to that of F_{ST} (Wright, 1951). We calculated F_{ST} by using θ (Weir and Cockerham, 1984). A gene flow value of $Nm=1$ represents the theoretical threshold beyond which gene flow due to genetic drift prevents independent differentiation of populations, if an insular model is assumed (Wright, 1931).

[2] MEZ, Grotta del Mezzogiorno–Grotta di Frasassi system; GUA, Sala Guano in the Grotta del Fiume; RSO, Ramo Sulfureo section of the Grotta del Fiume; LAV, Lago Verde room in the Grotta del Fiume.

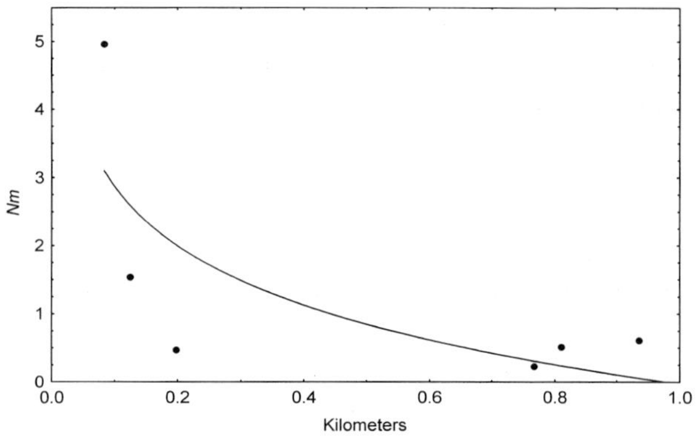

Fig. 19.20. Pairwise estimates of gene flow among all populations. Gene-flow values (Nm) are plotted against the corresponding distances (km). Gene flow is quite low even at very small geographic distances.

provided by the sample from the Grotta del Mezzogiorno (MEZ). In fact, after removing this sample, F_{ST} assumed its lowest value ($F_{ST}=0.180$).

Estimates of gene exchange between the colonies obtained from F_{ST} (Table 19.3) suggest that samples from Grotta del Mezzogiorno and from Grotta del Fiume represent independent gene pools. Such an outcome could be explained by the geographic distance between the two cave systems, as well as by the possibility that the river Sentino represents a geographic barrier for *Andronisus dentiger*.

Gene exchange among the populations within Grotta del Fiume is not homogeneous either. The RSO and LAV populations are nearly isolated, showing a very low level of gene flow between them. Conversely, both of these populations exchange genes with the GUA population. The population from the Ramo Sulfureo section (RSO) shows the highest degree of isolation. It is likely that restricted home ranges, small effective size of populations and the combined effect of drift and inbreeding may explain the general heterozygote deficit observed (Gentile and Sarbu, unpublished results).

Cave organisms are expected to exhibit a high level of genetic structuring. Numerous studies have shown a high degree of genetic fragmentation occurring at larger geographic scales (Crouau-Roy, 1989; Sarbu et al., 1993; Kane et al., 1994; Culver et al., 1995). What is less expected, is to have the large degree of genetic structure found within a single cave, Grotta del Fiume. In fact, F_{ST} values observed within the Grotta del Fiume are comparable in magnitude to values often observed in studies carried out over a large geographic area (Caccone, 1985; Sbordoni et al., 1985; Kane et al., 1992).

Presently, very few studies of the genetic structure of cave organisms have been carried out within a single cave (Sbordoni et al., see Chapter 24, this volume). Caves are often expected to be very homogeneous environments, as a result of their habitat stability. This may not necessarily be true, at least in instances where abundant localized trophic inputs may transform a cave into a very patchy habitat, constraining populations to the trophic resources despite the potential dispersal capability of individuals. This is likely what is occurring in the Grotta del Fiume.

Populations from this cave show heterozygote deficiency. This was not observed in two samples taken from MEZ at seven-month intervals, nor in a number of cave populations of *Androniscus dentiger* previously studied (Gentile and Sbordoni, 1998). This may be related to the remarkably different distribution of the abundant sources of food within the Grotta del Fiume compared to other caves. The distribution of *A. dentiger* in the Grotta del Fiume overlaps the distribution of the trophic sources. This pattern has not been observed in other caves where the scarcity of food, generally consisting of small inputs from the surface, forces the individuals to maximize exploration of the environment. Consistent with this scenario, gene exchange between populations inhabiting the Grotta del Fiume is less than would be expected from the small distances between the sampling sites (Fig. 19.20), in the absence of any apparent geographic barrier (Gentile and Sbordoni, 1998). This scenario is strongly supported by data for stable isotopes of carbon and nitrogen showing that troglobitic populations are strictly constrained by their food resources. The samples collected in the Sala Guano are isotopically heavy, indicating that guano is the only source of food they use. On the other hand, samples from the sulfidic rooms are isotopically light, indicating that they rely entirely on autochthonous chemoautotrophic food. This clear separation between the organisms from the sulfidic rooms and the organisms from the surface and from the Sala Guano (Fig. 19.19) is consistent with the low values of gene flow observed.

ACKNOWLEDGMENTS

The authors thank the Consorzio Frasassi for the logistic and financial support that made this study possible. Special thanks to the Marchean Speleologists for their help during the field work, and in particular to R. Bambini whose efforts are very much appreciated. The authors are grateful to the researchers who analyzed the samples and to those who identified the biological material collected in the Frasassi caves: R. Bernasconi (Gastropoda), M. Bertrand (Acarina), I. Weiss (Araneae), M. Dzwillo (Oligochaeta), R. Poggi (Pselaphidae), G. Gardini (Pseudoscorpiones), and R. Dallai and P.P. Fanciulli (Collembola). The stable isotope samples were analyzed at the Institute for Marine Science of the University of Alaska and the Department of Geological Sciences of the University of Michigan. The electrophoretic analysis was carried out at the Department of Biology, University Tor Vergata, Roma, Italy. Helpful comments for the preparation of the manuscript were received from R. Argano, P. Forti, D. Joyce, T.C. Kane, P. Palmer, and A. Summers-Engel. Special thanks to V. Sbordoni for contributing to this research with his criticism and financial support.

REFERENCES

Ballelli, S., Biondi, E., Cortini Pedrotti, C., Francalanciani, F., Orsomando, E. and Pedrotti, F., 1981. *Il Patrimonio Vegetazionale delle Marche.* Regione Marche, Assessorato dell'Urbanistica e dell'Ambiente, Ancona, carta floristica n. 18.

Bani, C., Bono, P. and Capelli, G., 1986. Schema idrogeologico dell'Italia Centrale. *Mem. Soc. Geol. Ital.,* 25: 991–1012.

Bani, M., 1984. *La Grotta dei Cinque Laghi.* Sezione Speleologico Citta di Castello, Urbania, pp. 1–214.

Bertolani, M., Garuti, G., Rossi, A. and Bertolani-Marchetti, M., 1977. Motivi di interesse mineralogico e petrografico nel complesso carsico Grotta Grande del Vento-Grotta del Fiume. *Le Grotte d'Italia,* 4: 109–144.

Bertolani, R., Manicardi, G.C. and Rebecchi, L., 1994. Faunistic study in the karst complex of Frasassi (Genga, Italy). *Int. J. Speleol.,* 32: 61–77.

Brignoli, P.M., 1971. Note su ragni cavernicoli Italiani (Araneae). *Fragm. Entomol.,* 7: 1–229.

Brignoli, P.M., 1979. Ragni d'Italia XXXI. Specie cavernicole nuove o interesanti (Araneae). *Quad. Mus. Speleol. "V. Rivera",* 5: 3–48.

Brignoli, P.M., 1985. Aggiunte e correzioni al Catalogo dei Ragni Cavernicoli Italiani. *Mem. Mus. Civ. Stor. Nat. Verona,* 4: 51–61.

Caccone, A., 1985. Gene flow in cave arthropods: a qualitative and quantitative approach. *Evolution,* 39: 1223–1235.

Cigna, A.A., 1989. Underground water dating by tritium measurements. *10th International Speleological Congress, Budapest, II.* Union Internationale de Spéleologie (U.I.S.), pp. 562–563.

Cola, L. and Freude, H., 1974. Prima nota sulla fauna dell grotte nella zona di Genga (Ancona). *Boll. Soc. Entomol. Ital.,* 106: 37–39.

Crouau-Roy, B., 1989. Population studies on an endemic troglobitic beetle: geographical patterns of genetic variation, gene flow and genetic structure compared with morphometric data. *Genetics,* 121: 571–582.

Culver, D.C., Kane, T.C. and Fong, D.W., 1995. *Adaptation and Natural Selection in Caves.* Harvard University Press, Cambridge, 189 pp.

Degens, E.T., 1969. Biochemistry of stable carbon isotopes. In: G. Elington and M.T.J. Murphy (Editors), *Organic Geochemistry, Methods and Resulsts.* Springer, New York, pp. 304–329.

Egemeier, S.J., 1981. Cavern development by thermal waters. *Natl. Speleol. Soc. Bull.,* 43: 31–51.

Ehleringer, J.R., Rundel, P.W. and Nagy, K.A., 1986. Stable isotopes in physiological ecology and food web research. *Trends Ecol. Evol.,* 1: 42–45.

Ford, D.C. and Williams, P., 1989. *Karst Geomorphology and Hydrology.* Unwin Hyman, London, 601 pp.

Forti, P., Menichetti, M. and Rossi, A., 1989. Speleothems and speleogenesis of the Faggeto Tondo Cave (Umbria, Italy). *10th International Speleological Congress, Budapest (I).* Union Internationale de Spéleologie (U.I.S.), pp. 74–76.

Fossa-Mancini, E., 1921. Geologia ed idrogeologia della Gola del Sentino nella Marca di Ancona. *G. Geol. Prat.,* 16: 37–76.

Galdenzi, S., 1990. Un modello genetico per la Grotta Grande del Vento. In: S. Galdenzi and M. Menichetti (Editors), *Il carsismo della Gola di Frasassi.* Instituto Italiano di Speleologia, pp. 123–142.

Galdenzi, S. and Menichetti, M. (Editors), 1990. *Il carsismo della Gola di Frasassi.* Instituto Italiano di Speleologia, 243 pp.

Galdenzi, S. and Menichetti, M., 1995. Occurrence of hypogenic caves in a karst region: examples from Central Italy. *Environ. Geol.,* 26: 39–47.

Galdenzi, S., Menichetti, M. and Forti, P., 1997. La corrosione di placchette calcaree ad opera di acque solfuree: dati sperimentali in ambiente ipogeo. *Proc. 12th Int. Congr. Speleol.,* 1: 187–190.

Gentile, G. and Sbordoni, V., 1998. Indirect methods to estimate gene flow in cave and surface populations of *Androniscus dentiger* (Isopoda, Oniscidea). *Evolution,* 52: 332–342.

Hill, C.A., 1987. Geology of Carlsbad Cavern and other caves in the

Guadalupe Mountains, New Mexico and Texas. *N. M. Bur. Mines Miner. Resour. Bull.,* 117: 1–150.

Kane, T.C., Barr Jr., T.C. and Badaracca, W., 1992. Cave beetle genetics: geology and gene flow. *Heredity,* 68: 277–286.

Kane, T.C., Culver, D.C. and Matthieu, J., 1994. Biotic fluxes and gene flow. In: J. Gibert, D.L. Danielopol and J.A. Stanford (Editors), *Groundwater Ecology.* Academy Press, London, pp. 245–270.

Karaman, G.S., 1985. New data of the genus *Niphargus* Sciodte (Fam. Niphargidae) in Italy and adjacent regions (contributions to the knowledge of Amphipoda 138). *Boll. Mus. Civ. Stor. Nat. Verona,* 12: 209–228.

Menichetti, M., 1985. Caratteristiche chimico-fisiche delle acque carsiche dell'Appennino umbro-marchigiano. In: Prhomos (Editor), *Atti Congr. Naz. Biospeleol.* Città di Castello, pp. 33–73.

Menichetti, M., 1990. Influenze tettonico-strutturali nello sviluppo del carsismo della Gola di Frasassi. In: S. Galdenzi and M. Menichetti (Editors), *Il carsismo della Gola di Frasassi.* Instituto Italiano di Speleologia, pp. 65–92.

Menichetti, M., 1996. Bilancio energetico di una grotta turistica: la Grotta Grande del Vento a Frasassi (An). In: *Proceeding of the International Symposium "Show caves and Environmental Monitoring", Cuneo.* Union Internationale de Speleologie (U.I.S.), pp. 211–223.

Menichetti, M., Cigna, A.A., Galdenzi, S. and Marinelli, G., 1994. Relazione tra flusso turistico e condizioni ambientali nella Grotta Grande del Vento a Frasassi (AN) XVII. In: *Congresso Nazionale di Speleologia Castelnuovo Garfagnana (Lu).* Union Internationale de Speleologie (U.I.S.), pp. 209–224.

Nei, M., 1987. *Molecular Evolutionary Genetics.* Columbia University Press, New York, 512 pp.

O'Leary, M.H., 1988. Carbon isotopes in photosynthesis. *BioScience,* 38: 328–336.

Palmer, A.M., 1991. Origin and morphology of limestone caves. *Geol. Soc. Am. Bull.,* 103: 1–25.

Parotto, M. and Praturlon, A., 1975. Geological summary of the Central Apennines. Consiglio Nazionale delle Ricerche. *Quad. Ric. Sci.,* 90: 257–311.

Paull, C.K., Jull, A.J.T., Toolin, L.J. and Linick, T., 1985. Stable isotope evidence for chemoautotrophy in an abyssal seep community. *Nature,* 317: 709–711.

Peterson, B.J. and Fry, B., 1987. Stable isotopes in ecosystem studies. *Annu. Rev. Ecol. Syst.,* 18: 293–320.

Pohlman, J.W., Iliffe, T.M. and Cifuentes, L.A., 1997. A stable isotope study of organic cycling and the ecology of an anchialine cave ecosystem. *Mar. Ecol. Prog. Ser.,* 155: 17–27.

Principi, P., 1931. Fenomeni di idrologia sotterranea nei dintorni di Triponzo (Umbria). *Grotte Ital.,* 5: 1–4.

Ruby, E.G., Jannasch, H.W. and Deuser, H.W., 1987. Fractionation of stable carbon isotopes during chemoautotrophic growth of sulfur-oxidizing bacteria. *Appl. Environ. Microbiol.,* 53: 1940–1943.

Ruffo, S. and Vigna Taglianti, A., 1969. Alcuni *Niphargus* delle acque sotteranee dell'Italia centro-meridionale e considerationi sulla sistematica del gruppo *orcinus* (Amphipoda, Gammaridae). *Mem. Mus. Civ. Stor. Nat. Verona,* 16: 1–29.

Sarbu, S.M., 1990. The unusual fauna of a cave with thermomineral waters containing H_2S, from southern Dobrogea, Romania. *Mem. Biospeol.,* 17: 191–195.

Sarbu, S.M. and Kane, T.C., 1995. A subterranean chemoautotrophically based ecosystem. *Natl. Speleol. Soc. Bull.,* 57: 91–98.

Sarbu, S.M., Kane, T.C. and Culver, D.C., 1993. Genetic structure and morphological differentiation: *Gammarus minus* (Amphipoda: Gammaridae) in Virginia. *Am. Midl. Nat.,* 129: 145–152.

Sarbu, S.M., Kane, T.C. and Kinkle, B.K., 1996. A chemoautotrophically based groundwater ecosystem. *Science,* 272: 1953–1955.

Sbordoni, V., Allegrucci, G., Cesaroni, D., Cobolli Sbordoni, M. and De Matthaeis, E., 1985. Genetic structure of populations and species of *Dolichopoda* cave crickets: evidence of peripatric divergence. In: V. Sbordoni (Editor), *Contact Zones and Speciation. Boll. Zool.,* 52: 139–156.

Sighinolfi, G.P., 1990. Chimismo ed origine delle acque del sistema ipogeo "Grotte di Frasassi" (Ancona) – implicazioni speleogenetiche ed ambientali. *Mem. Inst. Ital. Speol. Costacciaro-Perugia,* 4: 109–122.

Southward, A.J., Kennicutt II, M.C., Alcala-Herrera, J., Abbiati, M., Airoldi, L., Cinelli, F., Bianchi, C.N., Morri, C. and Southward, E.C., 1996. On the biology of submarine caves with sulfur springs: Appraisal of 13C/12C ratios as a guide to trophic relations. *J. Mar. Biol. Assoc. U.K.,* 76: 265–285.

Stranero, S.L., 1939. Sulla distributione geographica del *Duvalius bensai* Gestro (Coleopt. Carab.). *Boll. Soc. Entomol. Ital.,* 71: 105–107.

Taddeucci, A., Tuccimei, P. and Voltaggio, M., 1992. Studio geocronologico del complesso carsico "Grotta del Fiume–Grotta Grande del Vento" (Gola di Frasassi, AN) e implicazioni paleoambientali. *Quaternario,* 5: 213–222.

Taylor, M.R., 1999. A trip to the lighted house. *Natl. Speleol. Soc. N.,* 57: 36–48.

Tazioli, G.S., Cocchioni, M., Coltorti, M., Dramis, F. and Mariani, M., 1990. Circolazione idrica e chimismo delle acque sotterranee dell'area carsica di Frasassi nelle Marche. *Mem. Inst. Ital. Speol. Costacciaro-Perugia,* 4: 93–108.

Vigna Taglianti, A., 1970. Osservationi su alcuni *Duvalius* appenninici (Coleoptera, Carbidae). *Fragm. Entomol.,* 7: 45–54.

Vlasceanu, L., Popa, R. and Kinkle, B.K., 1997. Characterization of *Thiobacillus thioparus* LV43 and its distribution in a chemoautotrophically based groundwater ecosystem. *Appl. Environ. Microbiol.,* 63: 3112–3127.

Weir, B.S. and Cockerham, C.C., 1984. Estimating F-statistics for the analysis of population structure. *Evolution,* 38: 1358–1370.

Wright, S., 1931. Evolution in Mendelian populations. *Genetics,* 6: 111–123.

Wright, S., 1951. Theoretical structure of populations. *Ann. Eugen.,* 15: 323–354.

Part VI

Evolution

Chapter 20

CRITICAL REVIEW OF THE RELEVANT THEORIES OF THE EVOLUTION OF SUBTERRANEAN ANIMALS

David C. CULVER and Horst WILKENS

INTRODUCTION

Subterranean animal species are characterized by the absence or reduction of a series of features. Among these, eyelessness has long been especially fascinating to biologists since the time of Lamarck (1809) (Elliot, 1984). Unlike some aspects of evolution, such as the evolution of sex and the origin of evolutionary novelty, eyelessness, and regressive evolution in general have always appeared to be easy to explain. To an orthogeneticist (e.g., Vandel, 1964), eye loss in cave animals was a by-product of a phyletic line. To a neo-Lamarckian (e.g., Packard, 1888), eye loss was a classic example of the role of disuse. Indeed, the evolution of cave animals played a major role among North American neo-Lamarckians (Richardson and Kane, 1988), and terms like "regressive evolution" and "sensory compensation" retain a neo-Lamarckian connotation (Banister, 1985; Romero, 1985). Even Darwin (1872, p. 135) attributed eye loss of cave animals to disuse. To a neo-Darwinian (e.g., Poulson, 1963), eye loss in cave animals was the result of an evolutionary trade-off resulting from the selection for elaborated non-optic sensory structures. To neutral mutationists (Kosswig, 1960; Chakraborty and Nei, 1974), eye loss in cave animals was easily explained by the accumulation of selectively neutral, structurally reducing mutations made possible by the lack of stabilizing selection. Kosswig (1963) made many contributions to the study of regressive evolution. He put his emphasis, however, on the genetics of such processes, with the ultimate general aim of evaluating the role of selection in evolution.

In his review of cave biology, Barr Jr (1968) listed 12 hypotheses put forward to explain the reduction of biologically functionless features, like the evolution of eyelessness, and to this Christiansen (1992) added two more – and no doubt a thorough review would find even more. In a modern context and more particularly in a genetic context, theories that account for regressive evolution in cave animals fall into two groups (see Fong et al., 1995) – those that involve natural selection either directly or indirectly, and those that involve neutral mutation and genetic drift. Both of these share an emphasis on genetic change. Theories that emphasize other aspects of regressive evolution, such as allometric effects (Heuts, 1951) and ontogenetic timing (Magniez, 1985), must ultimately be based on genetic changes resulting from drift or selection.

We first present an overview of the two main theories and the objections raised to each. We then provide a detailed examination of the two best-studied cave organisms in this context – the Mexican cave fish *Astyanax fasciatus* and the Appalachian cave amphipod *Gammarus minus*.

Neutral mutation

For complex structures such as eyes and pigmentation, the phenotypic effect of most mutations will be to decrease the complexity or expression of the characters involved. For example, in Sexton and Clark's (1936) catalog of eye mutations of the amphipod *Gammarus chevreuxi,* only one of the dozens of mutations increases the complexity or size of the eye in any sense, and this one increases the amount of inter-ommatidial pigment. In surface environments, such mutations are selected against. That is, there is stabilizing selection for the maintenance of eye and pigment. In an aphotic environment such as caves, stabilizing selection on optic, and all other light-dependent features, is likely to be greatly reduced or absent. Structurally reducing mutations are not selected against, and some, due to chance alone, will increase in frequency in the population. This is in accord with the high variability of the biologically functionless vestigial structures in cave species in general (Culver, 1982). As Kimura and other theoretical population geneticists have shown (e.g., Kimura, 1983), some selectively neutral mutations will come to be fixed in a population. For a strictly neutral mutation, the probability that it will ever become fixed is

$$\frac{1}{(2N_e)}$$

where N_e is the effective population size. The expected time (in generations) to fixation of this selectively neutral mutation is

$$4N_e.$$

Much of the discussion of selectively neutral mutations has centred around mutations with no apparent phenotypic effects, such as mutations in pseudogenes and synonymous third-position mutations in exons (expressed sequences of the genome). What makes cave animals interesting in this context is the large morphological effect of mutations which may potentially be selectively neutral.

It has often been argued that there has not been enough time for neutral mutation and genetic drift to account for the amount of regressive evolution observed (Ludwig, 1942; Barr Jr, 1968). These arguments are based on the rate of loss of a single gene (or equivalently, the rate of fixation of a single gene); but, because eye and pigment systems are clearly polygenic (e.g., Wilkens, 1988), any realistic model of the time required must take this into account. When this is done, for most reasonable values of mutation rate, population size, and the extent of the polygenic system, neutral processes proceed at a rate rapid enough to explain the observed levels of regressive evolution.

An example of the kinds of rate calculations that are possible comes from Culver et al. (1995). The probability that the original allele at a frequency of 1 becomes lost from a population by generation t, with a mutation rate of μ, is

$$1 - e^{-\mu t}.$$

If time since isolation is only 50 000 generations, mutation rate is 10^{-6}, and a mutation at any one of 10 different loci can result in eye loss, then the probability that none of the original ten alleles is lost is only .006. Thus neutral mutation can explain eye loss in this hypothetical example.

Natural selection

If natural selection is involved in regressive evolution of cave animals, the question is not, as for neutral mutation, its efficacy and speed. Given large differences in fitness of organisms with different genotypes and correspondingly different eye phenotypes, selection can result in rapid loss of such biologically functionless structures. The question is: what could cause differences in fitness of different eye phenotypes in the absence of light? There seems to be little likelihood that eyes, much less all the other biologically functionless structures of cave animals, are directly disadvantageous, as Barr Jr (1968) pointed out. By direct disadvantage, we mean that the eye or any other functionless feature, in and of itself, results in increased mortality or decreased fecundity. A highly implausible example would be if an eye injury resulted in significantly increased mortality.

Three more or less indirect disadvantages for eyes and pigment systems have been put forward. Poulson (1963) and Sket (1985) suggested that eyes and pigment systems represent a metabolic inefficiency in a resource-scarce environment. Mutations that reduce useless structures then allow for an increase in other structures such as the lateral-line system in fish. The difficulty with this hypothesis is that it can neither account for regressive evolution in resource-rich subterranean environments such as Movile Cave (see Chapter 17, this volume) and many tropical caves (see Chapter 31, this volume), nor for certain behavioural features like aggressive behaviour or dorsal light reaction, which are subject to genetic reduction although they may not even have been manifested by the cave ancestral form in darkness (Wilkens, 1988; Hausberg, 1995).

Regal (1977) argued that natural selection would favour the reduction or elimination of unused pathways, a process he called "streamlining evolution". He noted that this could be accomplished by either eliminating genetic material from the genome or by "turning off" sections of desoxyribonucleic acid (DNA). While the hypothesis is interesting, it is difficult to test. Fong et al. (1995) have reviewed evidence for genome streamlining and its relevance to Regal's hypothesis.

Jones and Culver (1989) and Jones (1990) suggested

Fig. 20.1. Epigean and cave form of *A. fasciatus*.

that animals with eyes are at a disadvantage with respect to selection, not because of selection in respect of the eyes themselves but rather in respect of the function they perform – detection of visual stimuli. If visual stimuli and other stimuli (e.g., chemical) are not processed independently, then the presence of optic connections in the nervous system may interfere with extra-optic functions.

The mechanism of indirect selection against eyes in all three of the above scenarios is thought to involve pleiotropic effects (Barr Jr, 1968; Fong, 1989). This is the mechanism by which indirect selection may occur, rather than the cause.

CASE STUDY: *ASTYANAX FASCIATUS* (CHARACIDAE, TELEOSTEI)

Introduction

Astyanax fasciatus (Characidae) is a Neotropical freshwater fish which was able to colonize the North American continent northward to the Rio Grande system following the end of Pliocene, when the intercontinental land bridge between the Americas emerged. This common epigean fish has developed a series of hypogean populations in a rather restricted area in the northern Sierra Madre Oriental in the Mexican States San Luis Potosi and Tamaulipas (Fig. 20.1). They inhabit the sinkhole area of captured streams, the waters of which again resurge at the eastern base of the limestone escarpment. The degree of isolation and convergent evolution is for the larger part of cave populations still under dispute. At the moment, more than 20 forms are known with reduced eyes and pigment (Mitchell et al., 1977) (Fig. 20.2). Two of them, the Micos fish and the Chica fish, possess dark body pigmentation and variable externally visible eyes (Mitchell et al., 1977; Wilkens, 1988). In nature, as well as in the laboratory, epigean and cave forms may interbreed. Thus, a genetic analysis of regressive and constructive features, as well as that of the corresponding evolutionary processes, is possible.

Genetics

Genetic control of the manifestation and number of genes involved in the development of regressive structures in *A. fasciatus* may be monogenic or additively polygenic.

Monogenic inheritance

All regressive traits of monogenic inheritance observed in cave forms of *A. fasciatus* affect body coloration. All three types of body colours developed by this species have been found to be subject to deleterious mutations.

Melanin: Melanin pigment helps the fish to adjust body colour to the differing light or dark albedos of the underground. The melanin is contained in melanophore cells, which are able to perform both a physiological and a morphological colour change.

The melanin content of the melanophores of cave forms of *Astyanax* is affected by two independently segregating recessive genes. The "albino factor" is responsible for the total lack of melanin in the pigmentary epithelium of the eye as well as in all body melanophores (Sadoglu, 1955; Wilkens, 1970b).

The second factor, the "brown gene", simply diminishes the melanin content of these cells. This causes a brownish appearance of the eye pigmentary epithelium

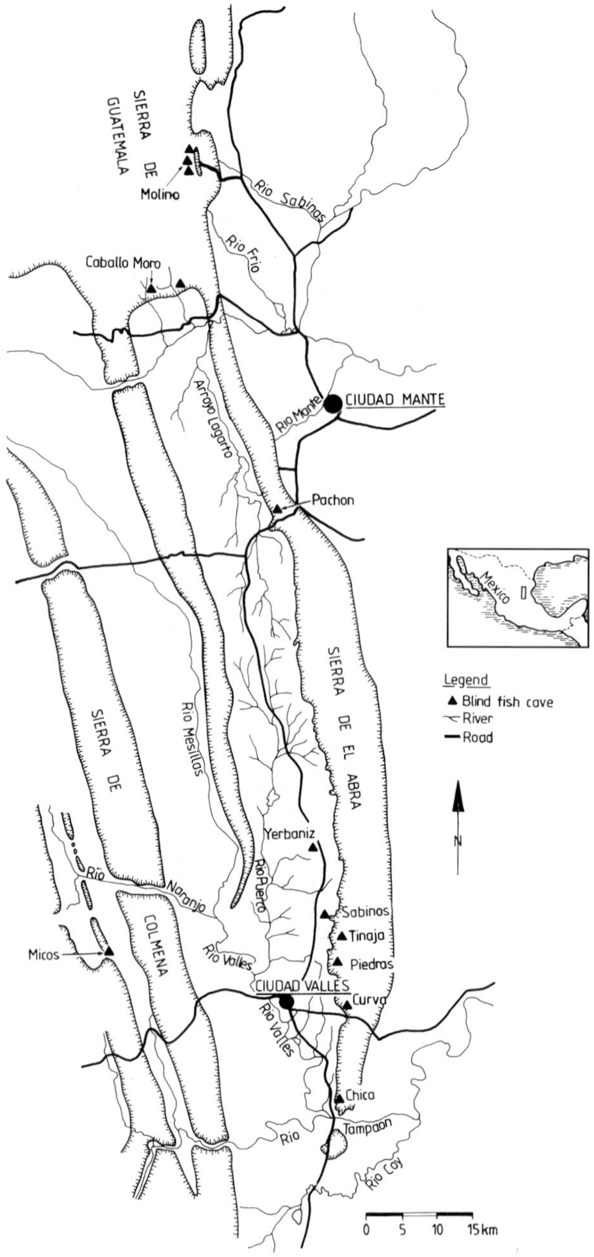

Fig. 20.2. Distribution of the *Astyanax* blind fish populations.

and the body melanophores (Sadoglu and McKnee, 1969; Wilkens, 1970b).

All cave populations are homozygote for the "brown gene". Only in the phylogenetically young Micos fish does it fail to occur. The "albino gene" has been found in three of nine forms studied, the Pachon, the Yerbaniz and the Molino fish. The populations are homozygous for it.

Guanine: The silvery appearance characteristic of most fishes is caused by guanine crystals contained in iridophore cells. The reflection of light by guanine crystals camouflages fish specimens against predators viewing them from below when they swim in the free water column.

Only two populations, the Piedras and the Molino fish, have yet been found to possess a gene responsible for guanine reduction. In the Piedras fish the scales totally lack guanine deposits, whereas there seems to be no reduction of guanine in the dorsal iridophores, a lateral dermal stripelike layer of iridophores, or in the peritoneum.

Xanthochromes: In the Piedras cave population, there are specimens of which the body has a homogeneous yellow colour, resulting from xanthochromes stored in the fat tissue. The "yellow gene" responsible is recessive, resulting from a single mutation reducing the ability to break down carotinoid pigments ingested with the food.

Polygenic inheritance

Except for the different types of body pigmentation all troglomorphic features hitherto analysed in cave forms of *A. fasciatus* show a polygenic basis. This is the case for regressive traits like eyes, number of melanophore cells, and the dorsal-light reaction (Wilkens, 1988; Langecker, 1993), which are biologically useless in darkness, as well as for regressive behavioural traits like aggression (Hausberg, 1995) or schooling, which are not expressed under such conditions, because of the absence of light as the signal-transmitting agent. The same expression is characteristic of constructive traits like the taste-bud density and feeding behaviour of cave fishes (Schemmel, 1974a,b, 1980).

Polygenic expression is similar for both regressive and constructive features. The polygenes show an additive manner of expression, each of them in principle contributing the same amount. However, in all these systems there is a threshold effect after a specific quantity of expression, when the degree of differentiation or the size and quantity of a specific character abruptly increase (Fig. 20.3). The identity of phenotypic expression of regressive as well as of constructive features can be explained by the fact that, in both cases, functioning non-rudimentary genetic systems are involved.

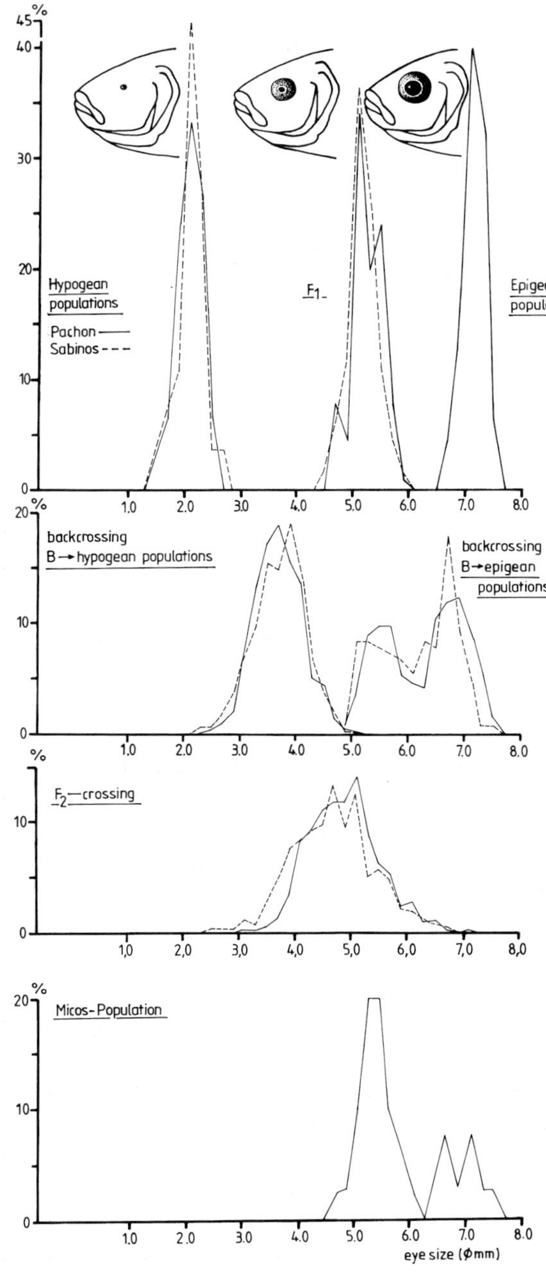

Fig. 20.3. Distribution of eye size in the epigean, the cave forms and their crossings. In the phylogenetically young Micos fish and the backcross the threshold effect develops at approximately an eye size of 6.0 mm.

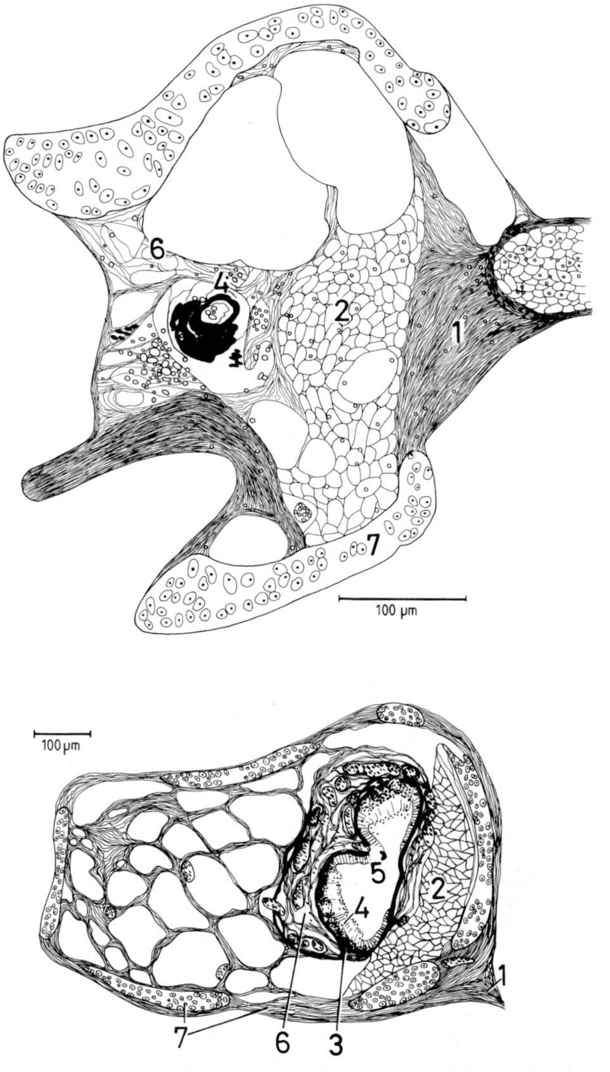

Fig. 20.4. Histological sections of rudimentary cave fish eyes showing the different degrees of differentiation. 1, connection to body surface; 2, *ligamentum annulare* contained in rudimentary eye chamber; 3, pigmentary epithelium; 4, retinal rudiment; 5, vitreous body; 6, chorioid; 7, sclera.

Eyes: The eye of *A. fasciatus* is the best-studied of all regressive features. In the adult cave forms, the most reduced eyes possess only a rudimentary sclera which encloses some undifferentiated cell material. Better-developed eyes show a retinal rudiment which has lost outer nuclear and plexiform layers. All such cave-fish eyes are externally no longer visible (Fig. 20.4). Crossing experiments have shown that the size of all particular structures like retina, lens, or pupillary opening are correlated with eye-ball size. The larger the eye-ball, the better each organ is developed. This is due to physiological interdependence among the structures during development. A large proportion of smaller-sized eyes have a rudimentary visual function: vision is disturbed or impossible because the retina is undifferentiated, the lens is not transparent, or they are incorrectly aligned (Wilkens, 1970a).

These disturbances are probably caused by the fact that the intricate pattern of physiological interdependence within the eye during development is out of balance. One of the reasons could be the mutation of structural genes, as has been shown for an opsin (eye protein) gene (Yokoyama et al., 1995). Where the disturbances mentioned are observed in crosses, they may also result from genetic incompatibility between epigean and cave populations.

The threshold effect is observed in the eye of the phylogenetically young Micos fish and in all backcrosses of the cave fish with the epigean form. In the F_2-cross, genetic incompatibility obviously inhibits the manifestation of this character. There is no threshold (Fig. 20.3).

Analysis of crossing experiments suggests that at least six genetic factors are responsible for eye reduction in the cave populations where the reduction is extreme (Wilkens, 1970a, 1988). The nature of these factors, which were called "eye genes", is still unknown. The analysis of defined genes and their expression in determining the process of eye regression at the molecular level has only just begun. Sequence comparison of several opsin genes from different epigean and cave-living populations revealed, besides many polymorphic alterations, only one cave-specific deletion which could possibly alter the properties of one of the opsin proteins (Yokoyama et al., 1995). The genes for structural proteins of the lens, *gamma*-s, as well as *alpha*-A-crystallin, are not active in cave specimens (Langecker et al., 1995; Behrens et al., 1998), although at least the *alpha*-A-crystallin does not show any mutations. The messenger ribonucleic acid (mRNA) for a red-like opsin could even be detected as being expressed for a short time during the development of photoreceptor cells in the retina of the larval blind form of *A. fasciatus* (Langecker et al. 1993).

Contrary to all other regressive features, eye reduction is only achieved after a process of ontogenetic regression. In the early larval eye almost all structures characteristic of this organ are developed. On this basis it has been proposed that certain structural eye-specific genes mentioned above may have been inactivated during the evolution of eye regression, but that defects in developmental-control genes are primarily responsible for the reduced-eye phenotype.

In order to elucidate a possible role of developmental-control genes, the expression of the master control gene of eye development, *Pax*-6, was investigated in eyes and brains of different larval stages of epigean and cave fish by *in situ* hybridization[1] (Behrens et al., 1997). It could be shown that pattern, strength and time course of *Pax*-6 expression was not altered in the tissues of the cave form when compared to the epigean fish. *Pax*-6 is even expressed in the highly degenerated eyes of late larval stages of the cave form. Comparative sequence analysis of the *Pax*-6 gene in the epigean and cave forms shows the complete integrity of cave-fish *Pax*-6. Probably this gene is not involved in the evolutionary process of eye degeneration in the cave fish. As the target gene of *Pax*-6, the *alpha*-A-crystallin gene is also unaffected by mutations, it is assumed that the genetic defect is located in the regulation cascade responsible for eye development below the *Pax*-6 and above the *alpha*-A gene (Behrens et al., 1998).

Melanophore system: The dark body coloration is not only dependent on the amount of melanin contained in the colour cells, but above all on the number of melanophores. In the phylogenetically old cave fish the number of melanophores has been reduced by 80% compared with the epigean form. As concerns the number of melanophores left, slight differences between the cave populations may be observed, which are genetically based. By crossing experiments it was found that one or at most two genetic factors in the epigean form have deteriorated in the cave fish as a result of mutations. The physiological colour change – that is, the ability of the colour cells to contract and expand pigment – still functions (Wilkens, 1970b).

The manifestation of the polygenic system responsible for the number of melanophores is characterized by the threshold effect mentioned above, which in this case shows interesting characteristics. Contrary to all other thresholds observed for polygenic structures in *Astyanax,* the phenotypic expression of this one is dependent on environment. It is only exhibited when the fish are kept over a black background. The threshold is the genetic basis of the morphological colour change – a well-known characteristic of fish in general (Wilkens, 1984).

Most interestingly, another *Astyanax* cave population, the phylogenetically young Micos fish which possesses the full complement of melanophores, exhibits non-functional morphological colour change in the

[1] A procedure based on molecular genetics in which hybridization of nucleic acid takes place within the cell or in tissue culture.

F_1 crossing with the cave forms in which pigmentation is extremely reduced. This does not happen in the F_1 cross between the latter cave forms and the epigean fish. This can be explained by the fact that the deterioration of the melanophore system has just started genetically in the Micos form, although it does not yet manifest itself phenotypically. It can be concluded that at least one polygene responsible for the melanophore system has already mutated (Wilkens, 1976).

As for the eyes, the reduction of the number of melanophore colour cells is probably also based genetically on deleterious mutations of developmental control genes. Because of their developmental integration into the formation of the neural crest, they also have an important inductive function. Thus, rudimentation may only proceed in small quantitative steps.

Conclusions

General remarks

The regressive evolution of complex structures has a quantitative character (Peters and Peters, 1968). Complex systems are not reduced by structural disturbance but by diminution in size. The analysis of eye regression suggests that it is caused by the loss of function of developmental control genes. Defective structural genes, which are exclusively responsible for the loss of certain features of body coloration, only play a minor role in the reduction of complex structures. This interpretation is proven by the ontogenetic regression of the eye, during which the larval visual organ develops all the particular structures characteristic of an intact eye. This is only possible because almost all genes responsible for eye formation are still functioning. Because of the inductive function of this organ in head formation, a complex structure is functionally and genetically integrated into the organism. This is also true for the melanophore system, the development of which is integrated into that of the neural crest. Mutations of structural genes that result in a defective development during early developmental stages are likely to have lethal effects because of severe malformations. Therefore the regressive evolution of a complex feature to some extent follows an orderly pattern. Catastrophic structural deterioration has lethal effects.

Due to this, regressive and constructive evolution are ruled by the same genetic principles. In both, developmental genes provide for a constructive or a regressive process by enlargement or a reduction in the overall size of an organ, whereas structural genes are mostly unaffected.

Drift and random fixation

The degree of genetic isolation and the amount of convergent evolution of the different *Astyanax* cave populations is still under study. Only for the Micos fish, which is geographically and hydrologically separated and phenotypically distinct, can an entirely independent evolution be assumed. Additionally, an evolution separate from the populations of the Sierra de El Abra can be assumed for those of the Sierra de Guatemala (Fig. 20.2) (Borowsky and Espinasa, 1997).

It has been shown that some of the different monogenic defective colour mutants are restricted to a single population, or only a few. For example, the guanineless factor and the yellow xanthochrome factor only occur in one population, while the albino gene has been found in three of them, the Pachon, the Yerbaniz and the Molino forms.

The polygenic systems also show genetic differences between the different cave forms. The reduction of the number of melanophores relies on at least two genetic factors in the Pachon population and on at least one genetic factor in the Sabinos cave fish. Crossing experiments between different eye-reduced cave forms have also proved that the genes responsible for defective eye formation in cave forms are in part non-allelic: F_1-hybrids between two different cave forms develop better differentiated eyes than are found in both parental cave forms (Wilkens, 1971). This phenomenon is probably due to the mutation of different developmental control genes in different populations. When combined in the F_1-generation, restitutive genetic interaction makes the eyes larger and better differentiated. Similar observations were made in the melanophore system: F_1 crossings between different cave populations may also show, parallel to the rudimentary eye, an enlarged number of melanophores when compared with their parental generation.

In general, it can be concluded that occurrence and fixation of mutant alleles happen by chance because the structures for which they provide are functionless. This is a demonstration that the process of regression of a certain structure is not driven by a specific directional selective force, in which case there should be no differences between the different cave populations in the distribution of mutant alleles.

Linkage and pleiotropy

It has often been claimed that the genetic linkage

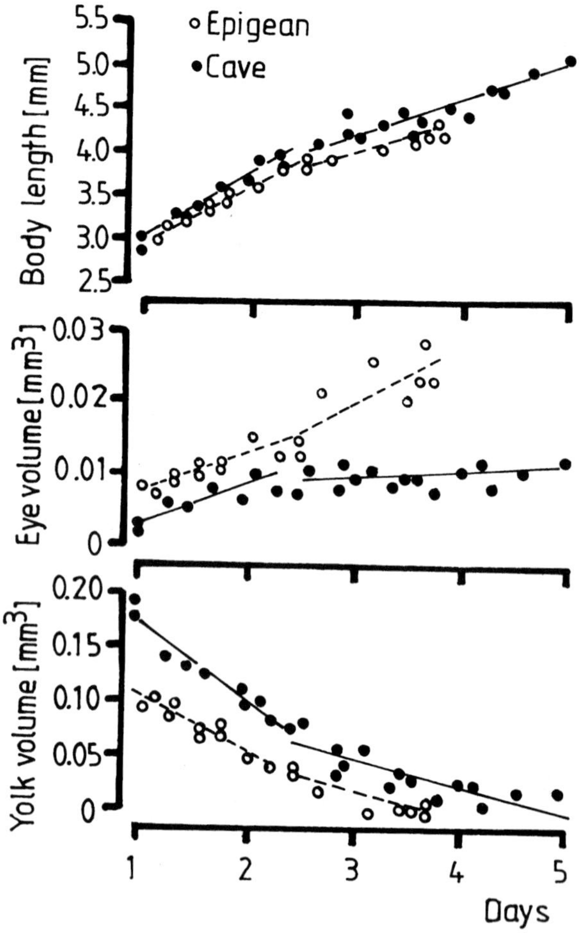

Fig. 20.5. Comparison of body and eye growth with egg-yolk consumption in larval epigean and cave *A. fasciatus* (after Hüppop, 1988).

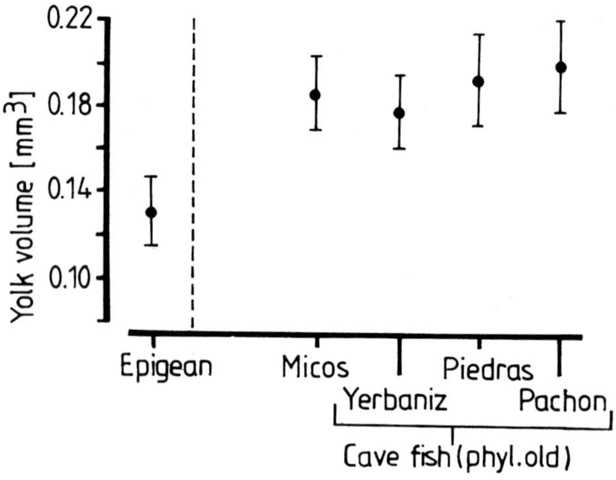

Fig. 20.6. Egg-yolk content in epigean and cave individuals of *A. fasciatus* (after Hüppop and Wilkens, 1991).

of a regressive with a constructive trait and pleiotropy should provide the indirect selective force for structural reduction. Energy economy has been especially emphasized in this context (Barr Jr, 1968). This hypothesis was developed on the assumption that the restricted food supply of many temperate caves is universal. However, it has been shown that reduction of eyes and pigment also occurs in food-rich tropical caves, as well as in chemoautotrophically based subterranean ecosystems without food limitation (see Chapters 17 and 31, this volume).

In *Astyanax* it can be demonstrated that, at least during early ontogeny, eye reduction in the cave-fish embryo has no influence either on the consumption of egg yolk or on growth, in comparison with the epigean fish (Fig. 20.5) (Hüppop, 1985, 1986, 1988; see also Chapter 8, this volume). Also, the phylogenetically long existence of rudiments of functionless structures contradicts the role of energy economy in evolution. Furthermore, much more efficient strategies against food scarcity have been developed in cave forms – fat storage, increased content of yolk in eggs, reduced metabolism (Hüppop, 1985; see also Chapter 8, this volume).

Another prominent argument for indirect selection is that of sensory compensation, whereby the constructive improvement of compensatory sensory organs, like taste buds, profits from the reduction of the eyes by the use of their sensory pathways and centres within the brain (Culver et al., 1995). This assumption, however, has the disadvantage of not explaining the reduction of non-sensory traits like pigmentation, or behavioural features like the dorsal light reaction or the aggressive behaviour.

In some cave forms of *Astyanax,* the possible influences of the linkage of a regressive character to a constructive one by pleiotropy have been investigated in some cases. In the Micos fish, the egg-yolk content is increased relative to the epigean form. Although eye size in this population is highly variable, this has no effect on the variability of egg-yolk content, which is the same as that of the other cave populations and of the epigean form (Fig. 20.6, Table 20.1). Furthermore, it was not possible to correlate eye size with the number of taste buds, both of which are variable in the Micos fish (Schemmel, 1974b). Finally, neither the guanineless nor the albino mutation seems to have any influence on constructive traits in these cave fish.

Reduction of non-manifested characters

In *A. fasciatus,* not only morphological but also

Table 20.1
Coefficients of variability of constructive and regressive features in the epigean population, different cave forms, and F_2 crosses[1]

Population or crossing	Eye	Pigment	Aggression	Taste buds	Yolk	Feeding behaviour
Surface	2.25	7.0	5.63	no data	7.9	4.4
Pachon cave	9.0	29.8	0	11.3	7.1	3.4
Sabinos cave	9.05	17.7	0	12.0	no data	3.4
Micos cave	12.0	7.9	42.3	26.0	8.0	no data
F_2 Pachon Epigean	15.6	32.4	no data	29.0	11.4	11.4
F_2 Sabinos Epigean	14.9	34.0	no data	24.0	no data	11.4

[1] Data from Hüppop (1988), Schemmel (1974a,b, 1980), Wilkens (1970a,b) and Hofmann (1990).

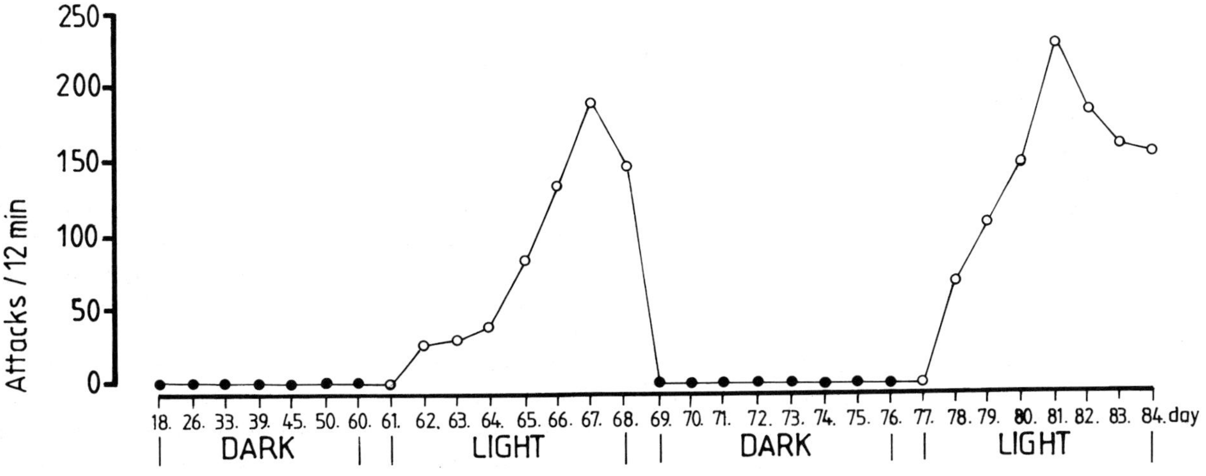

Fig. 20.7. Aggressive interactions of epigean *A. fasciatus* in light and darkness (after Hausberg, 1995).

behavioural traits, like schooling, the dorsal light reaction or aggressive behaviour, are reduced (Parzefall, 1986; Parzefall and Senkel, 1986; Langecker, 1993; Hausberg, 1995). For example, the epigean fish are highly aggressive, but aggression is released exclusively by optical stimuli. This is the reason why, when epigean specimens are placed in absolute darkness, not the slightest form of aggressive behaviour is observed (Fig. 20.7).

The cave fish have genetically lost this specific epigean manner of aggressive behaviour. This can be analyzed in the F_2 generation following crossings with the epigean fish, in which different individuals display different intensities of epigean aggressive behaviour. In the phylogenetically young Micos fish, which is polymorphic in respect of eye size, aggressive behaviour is also polymorphic: there are highly aggressive as well as completely non-aggressive individuals (Fig. 20.8). Furthermore, it could be shown that another light-

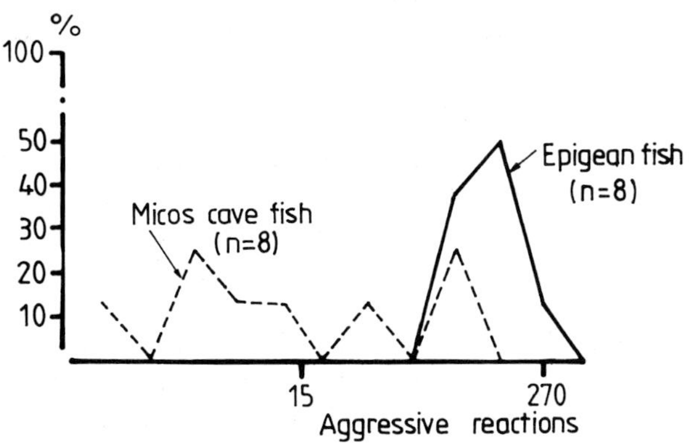

Fig. 20.8. Percentage distribution of the number of aggressive interactions within 10 minutes in the phylogenetically young Micos cave fish and the epigean fish (after Hofmann, 1990).

dependent behavioural trait, the dorsal light reaction, is also reduced in the cave forms (see Chapter 7).

It is concluded that traits not phenotypically manifested in darkness and thus not subject to selection in cave fish, nevertheless are reduced. The most plausible explanation of this phenomenon seems to be that of genetic deterioration and accumulation of mutant alleles due to mutation pressure.

Variability of regressive traits

Variability is characteristic of regressive structures. It can be observed in the functionless reduced wings of carabid beetles (Oertel, 1934; Krumbiegel, 1960) as well as in the eyes of cave-living animals. Studies have shown that variability may be environmentally modified, but above all is genetically based (Wilkens, 1988). In *Astyanax* the comparative measurement of ontogenetic eye growth shows that in the epigean population the variability of eye size does not change during development. In contrast, in the cave fish the variability increases during growth until it considerably surpasses that of the epigean fish (Fig. 20.9).

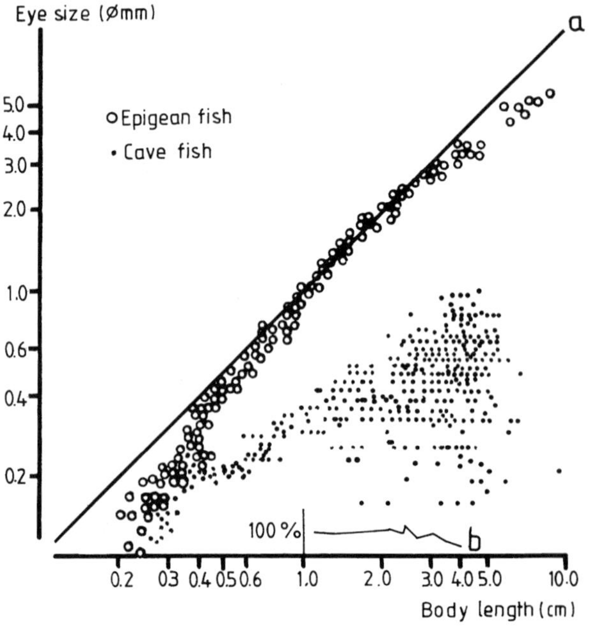

Fig. 20.9. Eye growth correlated with body size in epigean and phylogenetically old cave fish. (a) isometric correlation; (b) occurrence of a lens rudiment (after Wilkens, 1980).

A comparison of the variability of regressive and constructive traits in epigean and cave *Astyanax* clearly revealed that the variability of the eyes and the melanophore pigmentation is higher than that of the constructive traits yolk content, taste-bud density, and feeding behaviour (Wilkens, 1993) (Table 20.1). This can only be explained by the fact that in regressive structures stabilizing selection is absent, and consequently reducing mutations are not eliminated.

CASE STUDY: *GAMMARUS MINUS* (CRUSTACEA: GAMMARIDAE)

Geography and geographical genetics

The range of *Gammarus minus* forms a broad arc from southern Pennsylvania to western Kentucky and southern Indiana, with outlying populations from Missouri to eastern Oklahoma (Fig. 20.10). It is widespread in springs and spring brooks, but effectively is limited to carbonate springs with a pH greater than 7 (Glazier et al., 1992). There are numerous records of *G. minus* in caves throughout its range. Most of these cave populations show only slight morphological differentiation from spring populations (Holsinger and Culver, 1970). Some of these populations may be quite temporary, and are unlikely to be genetically differentiated from nearby spring populations (Culver et al., 1995). In two areas in its range (southwestern Virginia and south-central West Virginia), highly modified cave populations are found (Fig. 20.10). These highly modified cave populations, which Shoemaker (1940) described as the variety *tenuipes,* show striking morphological differences from the spring forms. The most obvious differences are greatly reduced eyes, a bluish rather than brownish colour, and a "slender and more delicate appearance" (Shoemaker, 1940) which is the result of appendages being longer and thinner. The number of ommatidia in the eye of *tenuipes* individuals averages two, while the number of ommatidia in the eye of spring individuals averages about 40. The morphological differences are extensive. Popa (1997) studied 44 appendage lengths and widths in two cave and two spring populations, one of each in each of two basins, and thus presumably independently evolved. He found that, in 38 of 44 cases, appendages were significantly longer or thinner in both sexes in both cave populations. In the other six cases, appendages were significantly longer or thinner in one sex in both cave populations. Furthermore, most of these differences appear to be genetically, rather than environmentally, determined. In a study of nine morphological characters, including six used by Popa, Fong (1989) found that morphological differences between the same cave and spring populations studied by Popa, reared in identical laboratory conditions, persisted. Genetic heritabilities (the percentage of overall phenotypic variation due to additive genetic variation) was significantly greater than zero in 34 of 36 cases, and averaged 72%. Heritabilities of eye characters (ommatidia number

Fig. 20.10. The left panel is a range map for the *Gammarus minus* species complex. The two darkened areas are the areas where cave populations of *G. minus* show extensive morphological modification, i.e., troglomorphy. The upper right panel shows detail of the distribution of the West Virginia populations studied and the lower right panel shows detail of the distribution of the Virginia populations studied. Numbering of the basins follows Culver et al. (1995). Squares indicate resurgence populations; circles, cave populations; triangles, karst-window populations.

and eye area) were in line with the general pattern, averaging 73%.

Given this background, three basic questions come to mind. First, why have populations of *Gammarus minus* with reduced eyes only evolved in two highly restricted parts of its range? Second, how many times did populations of *G. minus* with reduced eyes evolve? Third, what are the evolutionary causes of eye reduction?

To understand why populations with reduced eyes only evolved in two highly restricted parts of its range, it is necessary to begin with the unique characteristics of the subterranean environment where these populations of *G. minus* occur. The two areas of occurrence – Ward's Cove in southwestern Virginia and the Greenbrier Valley in south-central West Virginia – are unusual, if not unique, within the range of *G. minus* in the size of the caves and their integration into large subsurface drainage basins. Of the 50 caves in the United States with more than 17 km of known passages, nine are within the Greenbrier Valley. All of the basins where populations of *G. minus* with reduced eyes occur are greater than 10 km^2 in area. These basins, the subterranean equivalent of a drainage basin, contain a network of streams that resurge at a spring. The cave streams are separated from the resurgence by a flooded passage about 1 km in length, termed a "phreatic loop" by Ford and Williams (1989). The environment of the phreatic loop is quite different from that of the cave stream (an open-channel conduit). Water velocity in the phreatic loop is higher, turbulent flow more likely, and little sediment is likely to be present. Although the phreatic loop has not been sampled directly, it is unlikely that permanent populations of *Gammarus minus* occur there. Thus, there is at least a partial barrier between the putative ancestral spring populations and the cave populations. Likewise, there is a barrier between the surface and the cave streams at the upstream tips of the drainage basin. There are other large subsurface basins in the Greenbrier Valley that have *G. minus* populations, but they show no sign of eye reduction. These basins are all characterized by numerous sinking streams. Basins where populations with reduced eyes occur have few or no permanent surface streams. Most of the cave streams in these basins (Fig. 20.10) are formed by vertically percolating water rather than sinking streams (swallets). Thus, at both the upstream and downstream ends of basins where populations with reduced eyes occur, there are barriers between the surface and the cave stream. A final characteristic of cave streams where the populations of *G. minus* have reduced eyes is that food is not as scarce as in many other aquatic subterranean habitats. Finely divided leaf debris is often visible in streams, and Kevin Simon (pers. comm.), in a preliminary study of food resources, found that cave streams in Organ Cave (West Virginia), where the eyes of *G. minus* were reduced, had higher concentrations of coarse particulate organic matter (CPOM) than streams in the same cave which lacked *G. minus*.

Since populations with reduced eyes occur in six distinct subsurface basins (five in West Virginia and one in Virginia), the question arises as to how many times *G. minus* invaded caves,especially in West Virginia. All available morphological and genetic evidence is consistent with the hypothesis that *G. minus* invaded each basin separately, and the evidence is not consistent with the hypothesis that the eyeless populations in West Virginia represent a single invasion. A morphological example is the relationship between head length (x) and eye area (y) with the two fitted parameters a and b:

$$y = ax^b. \tag{20.1}$$

The allometric coefficients b were significantly different among cave populations from four basins (II, III, V, and VI in Fig. 20.10), ranging from 0.49 (negative allometry) to 1.33 (positive allometry), as shown in Fig. 20.11. Based on 13 enzyme systems encoded by 18 presumptive gene loci, a dendrogram (Fig. 20.12) based on the Unweighted Paired-Group Method using Arithmetic averages (UPGMA) of Sneath and Sokal (1973) also indicates that each basin was invaded separately. Resurgence populations, with one exception, are differentiated from upstream cave and karst-window populations. Upstream cave and karst-window populations tend to cluster together genetically. The exception cases are also informative. The cave population and the karst-window population of Basin I form a cluster genetically with the resurgence populations. These populations are also undifferentiated morphologically from the resurgence population. The only basin within which there is differentiation among cave and karst window populations is Basin III, which is also the largest (Fig. 20.10) and most complex. Differentiation between the two southernmost populations in the drainage reflects the presence of two parallel subterranean flows (east and west sides of the basin) which converge at the resurgence. The northern populations actually form a cluster with the cave population of Basin II. Subterranean drainage evolution has been such that the drainage divide between Basins II and III has been moving northward. Thus, northern populations of

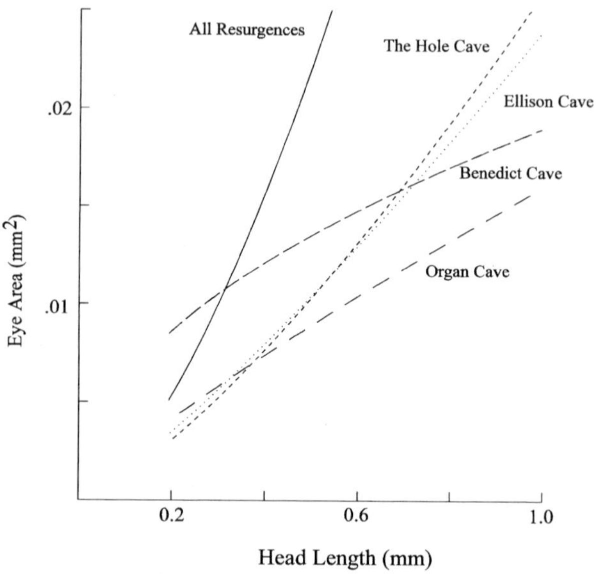

Fig. 20.11. Allometric curves ($y = ax^b$) of eye area and head length for four cave populations of *Gammarus minus* from four basins (Fig. 20.10): the Hole Cave–Basin II; Benedicts Cave–Basin III; Ellison Cave–Basin V; Organ Cave–Basin VI. Allometric curves for the four resurgence populations did not differ signifcantly.

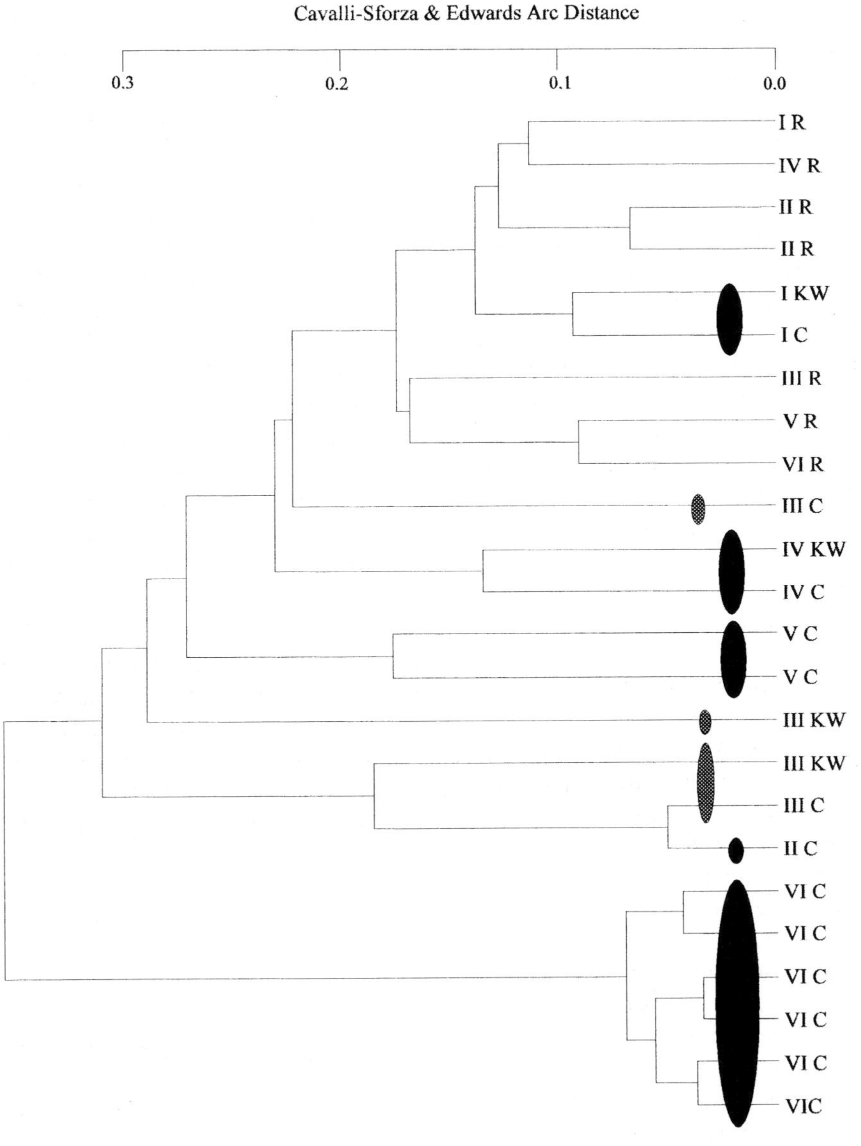

Fig. 20.12. UPGMA dendrogram using the arc distances of Cavalli-Sforza and Edwards (1967) for electrophoretic data for *Gammarus minus* populations in West Virginia. Numbering is as in Fig. 20.10. C, cave populations; R, resurgence populations; KW, karst-window populations. The Roman numerals indicate basins. Solid ellipses cover upstream populations in individual basins; hatched ellipses cover upstream populations in Basin III. Resurgence populations cluster together at the top of the dendrogram.

Basin III have probably been captured, hydrologically, from Basin II (Culver et al., 1995). Recent work by T.C. Kane and his colleagues at the University of Cincinnati with mitochondrial desoxyribonucleic acid (mtDNA) sequences of these populations supports the hypothesis of independent invasion of each basin. In addition, in some cases the genetic connection between resurgence and upstream cave and karst-window populations can still be detected.

Indirect measures of gene flow, based on Wright's F-statistics (Wright, 1978) suggest little or no gene flow between resurgence and upstream cave and karst-window populations and between upstream cave and karst-window populations in different basins. There is gene flow between populations within a basin (especially Basin VI) and between geographically adjacent resurgence populations (Culver et al., 1995).

Electrophoretic data also indicate that effective pop-

Table 20.2
Selection gradients (b) for antennal, eye, and head length characters for cave and spring habitats (from Culver et al., 1995)

Character	Habitat	Number of samples	Mean [1]	SD [1]	Skew [1]
Head Length	Cave	55	0.18**	0.19	0.37
	Spring	36	0.19**	0.22	0.52
Antennae	Cave	130	0.06*	0.21	0.72**
	Spring	90	0.06*	0.23	0.37
Eye	Cave	85	–0.08**	0.21	–1.09**
	Spring	54	0.06**	0.22	0.54

[1] Asterisks indicate means and skews significantly different from zero: * $0.01 < p < 0.05$; ** $P < 0.01$.
All distributions were skewed away from zero.

ulation sizes are large and that the populations never went through a genetic bottleneck. Heterozygosity levels are equivalent to those of surface-dwelling freshwater invertebrates. There were no differences in heterozygosity among caves (median = 0.087), resurgences (median = 0.086), and karst windows (median = 0.065). Assuming selectively neutral variation and genetic equilibrium, average heterozygosity (H) can be related to the effective population size (N_e) and the mutation rate (ν) to electrophoretically neutral alleles by:

$$H = \frac{4N_e\nu}{4N_e\nu + 1}. \qquad (20.2)$$

Assuming $\nu = 1 \quad 10^{-7}$ (Kimura, 1983), effective population sizes are approximately 10^5, well above the range where genetic drift predominates.

Causes of regressive evolution

Lande and Arnold (1983) developed multiple-regression models that allow the estimation of selection gradients for a particular morphological character. The vector of standardized directional-selection gradients, B, for the multiple regression

$$W_R = a + \sum b_i X_i, \qquad (20.3)$$

measures the correlation between fitness, W_R, and a character, X_i, when all other characters are held constant. If the characters included in the model are phenotypically uncorrelated with any unmeasured character, then B is a direct effect of selection on this character. In *G. minus,* two components of fitness are easily analyzed:

(1) Amplexus – the carrying of females by males prior to fertilization.
(2) Fecundity – the number of fertilized eggs carried in the external brood pouch by ovigerous females.

Both of these components of fitness may be affected by changes in the eye. The ability to find mates as well as food may be linked to the quality of extra-optic sensory structures, such as antennal length. In turn, the elaboration of extra-optic sensory structures at the neurological level may require a diminution of optic structures at that level. A series of characters were measured including eye area, ommatidia number, peduncle lengths of first and second antennae, and number of flagellar segments of first and second antennae. All measurements were size-corrected, and computational details have been given by Jones et al. (1992).

Directional-selection gradients were computed for 40 collections of *G. minus* from two caves and three resurgences. Overall, 49% of all multiple linear regressions and 16% of measured selection gradients were statistically significant (Table 20.2). Fourteen per cent of measured selection gradients for eyes were statistically significant. Directional selection on eyes tended to be negative in caves and positive in springs. A similar pattern obtains if only statistically significant selection gradients are considered. A nested analysis of variance (ANOVA) of significant eye selection with cave and springs as groups (habitats) and particular caves and springs as subgroups (sites) indicated that 94% of the variance was due to differences between groups and only 6% to differences within sites between groups (Table 20.3). We were unable to resolve the circumstances under which strong selection was likely.

How can the strong directional selection against eyes be explained? If the advantage of smaller eyes is simply that it makes larger antennae possible, through some form of energy economy or evolutionary trade-off,

Table 20.3
Nested analysis of variance of statistically significant selection gradients on eye characters (from Culver et al., 1995)

Source	Sum of squares	Degrees of freedom	Mean square	F	P
Between habitats[1]	1.487	1	1.487	20.34[2]	0.02
Sites[1] within habitats	0.244	4	0.061	0.84	*0.55*
Error	1.019	14	0.073		

[1] Habitats are caves and springs, and sites are particular caves or springs, e.g., Davis Spring.
[2] F for habitat is Mean square [Between habitats] / Mean square [Between sites within habitats].

then this effect should be reflected in direct selection on extra-optic sensory structures, but with negligible direct selection on eyes. This does not correspond to the observed selection gradients (Table 20.2). A more plausible explanation is that selection is not acting on any particular part of the anatomy (for example, the first antennal peduncle) but rather on the function it performs. For antennae and eyes the functions are detection of chemical and visual stimuli, respectively. These two sets of information are not processed independently in the central nervous system of the amphipod, either anatomically or physiologically (Bullock and Horridge, 1965). The relative size of brain regions in *G. minus* has changed in relation to the relative size of the sensory structures and the amount of stimuli they receive (Culver et al., 1995). The important point is that selection against eyes is occurring in cave populations. This does not mean that neutral mutation is unimportant, as will be seen below.

Before turning to evidence of neutral mutation, there is one other aspect of selection that is worth considering. Both selectionists and neutral mutationists agree that there should be a relaxation of stabilizing (balancing) selection in cave populations. We measured the extent of balancing selection by using the non-parametric regression technique of cubic splines (Schluter, 1988). This technique has the advantage of making no *a priori* assumptions about the shape of the selection curve, but the disadvantage of being computationally difficult. We studied a series of synthetic characters (principal components) in a large collection (>400 individuals) from Organ Cave (Basin VI in Fig. 20.10) and Davis Spring (Basin III in Fig. 20.10). The results, summarized in Table 20.4, indicate, as expected, that stabilizing selection is rare in cave populations. What was surprising is that it was also rare in the spring population, occurring in only one of nine cases.

The role of neutral mutation in the evolution of

Table 20.4
Summary of shape characteristics of selection curves for *Gammarus minus* from Davis Spring and Organ Cave[1]

Shape of curve	Davis Spring	Organ Cave
No Selection	3	3
Monotonic[2]	1	5
Stabilizing	1	0
Disruptive	1	0
Complex[2]	3	1

[1] Numbers are the number of cases out of a maximum of 9 for each population. See Culver et al. (1995) for details.
[2] "Monotonic" indicates curves with no internal minima or maxima, and correspond to directional selection. "Complex" shapes have more than one internal maximum or minimum.

reduced eyes in *Gammarus minus* is tied up with the question of time. Genetic data can be used to provide dates for the isolation of *G. minus* in caves. Nei (1987) estimated that the average rate of codon substitution detectable by electrophoresis was 10^{-7} per locus per year. From this rate and from Nei's electrophoretic distance, D, one can make an estimate of the time since isolation, t:

$$t = 5 \quad 10^{6} D. \tag{20.4}$$

For Organ Basin (Basin VI in Fig. 20.10), the best-studied basin, the estimate of time since isolation is 670 000 years ($D = 0.134$). Preliminary results of an ongoing study of Kane of 16SrDNA, a mitochondrial gene, give similar estimates. If the rate of nucleotide substitution is 10^{-8} per site per year, the time since divergence is

$$t = 5 \quad 10^{5} d^{*}, \tag{20.5}$$

where d^{*} is the percentage nucleotide divergence. Time since divergence in Organ Cave Basin is estimated at 450 000 years. Using this information, one can compare the observed rates of morphological evolution with the expected rates resulting from neutral mutation. Lynch

(1990) suggested a measure of evolutionary rate, Δ, that is independent of measurement scale:

$$\Delta = \frac{\mathrm{var_B}(\ln z)}{[t\ \mathrm{var_W}(\ln z)]}, \qquad (20.6)$$

where $\mathrm{var_B}(\ln z)$ and $\mathrm{var_W}(\ln z)$ are the observed between- and within-population variance components for log-transformed measures of morphology, z. The results are summarized in Table 20.5. Evolutionary rates were an order of magnitude greater for eye characters than for those of antennae, and all were lower than the expected neutral rate (Culver et al., 1995). While this difference in rates may be due to greater numbers of genes involved in the eye characters, it is interesting to speculate that it may in fact be due to the addition of neutral mutation to the rate of change of eye characters. Antennal characters, on the other hand, change largely as a result of selection. If this is so, then the rate of change of eye characters is largely determined by neutral mutation while the that for the antennae may be largely determined by selection not only on antennae but also on eyes.

Table 20.5
Evolutionary rates of divergence

Morphological traits [1]	Δ [2]	$\mathrm{var_b}(\ln z)$	$\mathrm{var_w}(\ln z)$
Size			
HL	$4.0\ 10^{-7}$	0.00248	0.01237
Eyes			
EN	$2.4\ 10^{-5}$	0.79983	0.06601
EA	$1.5\ 10^{-5}$	0.42515	0.05784
Antennae			
P1	$8.2\ 10^{-7}$	0.00890	0.02168
L1	$4.1\ 10^{-7}$	0.00544	0.02677
NI	$1.1\ 10^{-6}$	0.01165	0.02156
P2	$1.1\ 10^{-6}$	0.01315	0.02472
L2	$8.9\ 10^{-7}$	0.01472	0.03314
N2	$1.4\ 10^{-6}$	0.00996	0.01450

[1] Traits are head length (HL), eye ommatidia number (EN), surface area of compound eye (EA), peduncle length of first antenna (P1), flagellum length of first antenna (L1), number of flagellar segments of first antenna (N1), and the second antenna analogs (P2, L2, and N2).
[2] $\Delta = \mathrm{var_b}(\ln z)/[t\ \mathrm{var_w}(\ln z)]$, where $\mathrm{var_b}(\ln z)$ and $\mathrm{var_w}(\ln z)$ are the phenotypic variances between and within populations for log-transformed measures in populations from the Organ resurgence and Organ Cave. Time since divergence, t, was set at $5\ 10^5$ years.

REFERENCES

Banister, K.E., 1985. A subterranean population of *Garra barreimae* (Teleostei: Cyprinidae) from Oman, with comments on the concept of regressive evolution. *J. Nat. Hist.,* 18: 927–938.

Barr Jr., T.C., 1968. Cave ecology and the evolution of troglobites. In: Th. Dobzhansky, M.K. Hecht and W.C. Steere (Editors), *Evolutionary Biology, Vol. 2.* Plenum Press, New York, pp. 35–101.

Behrens, M., Langecker, T.G., Wilkens, H. and Schmale, H., 1997. Comparative analysis of Pax-6 sequence and expression in the eye development of the blind cave fish *Astyanax fasciatus* and its epigean conspecific. *Mol. Biol. Evol.,* 14(3): 299–308.

Behrens, M., Wilkens, H. and Schmale, H., 1998. Cloning of the αA-crystallin genes of a blind cave form and the epigean form of *Astyanax fasciatus*: a comparative analysis of structure, expression and evolutionary conservation. *Gene,* 216: 319–326.

Borowsky, R. and Espinasa, L., 1997. Antiquity and origins of troglobitic tetras, *Astyanax fasciatus. Proc. 12th Int. Congr. Speleology,* 1: 359–361.

Bullock, T. and Horridge, G., 1965. *Structure and Function in the Nervous System of Invertebrates, Vol. 2.* Freeman, San Francisco, 1719 pp.

Cavalli-Sforza, L.L. and Edwards, A.W.F., 1967. Phylogenetic analysis: models and estimation procedures. *Evolution,* 21: 550–570.

Chakraborty, R. and Nei, M., 1974. Dynamics of gene differentiation between incompletely isolated populations of unequal size. *Theor. Pop. Biol.,* 5: 460–469.

Christiansen, K.A., 1992. Cave life in the light of modern evolutionary theory. In: A. Camacho (Editor), *The Natural History of Biospeleology.* Museo Nacional de Ciencias Naturales, Madrid, pp. 453–478.

Culver, D.C., 1982. *Cave Life: Evolution and Ecology.* Harvard Univ. Press, Cambridge, MA, 223 pp.

Culver, D.C., Kane, T.C. and Fong, D.W., 1995. *Adaptation and Natural Selection in Caves: the Evolution of Gammarus minus.* Harvard Univ. Press, Cambridge, Mass.

Darwin, C., 1872. *On the Origin of Species by Means of Natural Selection, or the Preservation of Favoured Races in the Struggle for Life, 6th Edition.* John Murray, London, 454 pp.

Elliot, H., 1984. *Zoological Philosophy: an Exposition with Regard to the Natural History of Animals,* by J.B. Lamarck. University of Chicago Press, Chicago, translation.

Fong, D.W., 1989. Morphological evolution of the amphipod *Gammarus minus* in caves: quantitataive genetic analysis. *Am. Midl. Nat.,* 121: 361–378.

Fong, D.W., Kane, T.C. and Culver, D.C., 1995. Vestigialization and loss of nonfunctional characters. *Ann. Rev. Ecol. Syst.,* 26: 249–268.

Ford, D. and Williams, P., 1989. *Karst Geomorphology and Hydrology.* Unwin Hyman, London, 601 pp.

Glazier, D., Horne, M. and Lehman, M., 1992. Abundance, body composition and reproductive output of *Gammarus minus* (Crustacea: Amphipoda) in ten cold springs differing in pH and ionic content. *Freshwater Biol.,* 28: 149–163.

Hausberg, C., 1995. *Das Aggressionsverhalten von Astyanax fasciatus (Cuvier 1819) Characidae, Teleostei: Ontogenie, Genetik und Evolution bei der epigäischen und hypogäischen Form*. Thesis University Hamburg, 139 pp.

Heuts, M., 1951. Ecology, variation, and adaptation of the blind African cave fish *Caecobarbus geertsi* Boulenger. *Ann. Soc. R. Zool. Belg.*, 82: 155–230.

Hofmann, S., 1990. *Vergleichende Untersuchungen zum Aggressionsverhalten einer oberirdischen und einer unterirdischen Population von Astyanax fasciatus (Characidae, Pisces)*. Thesis University Hamburg, 73 pp.

Holsinger, J.R. and Culver, D.C., 1970. Morphological variation in *Gammarus minus* Say (Amphipoda: Gammaridae), with emphasis on subterranean forms. *Postilla*, 146: 1–24.

Hüppop, K., 1985. The role of metabolism in the evolution of cave animals. *Natl. Soc. Speleol. Bull.*, 47(2): 136–146.

Hüppop, K., 1986. Oxygen consumption of *Astyanax fasciatus* (Characidae, Pisces): A comparison of epigean and hypogean populations. *Environ. Biol. Fish.*, 17: 299–308.

Hüppop, K., 1988. *Phänomene und Bedeutung der Energieersparnis beim Höhlenfisch Astyanax fasciatus*. Thesis, University Hamburg, 42 pp.

Hüppop, K. and Wilkens, H., 1991. Bigger eggs in subterranean *Astyanax fasciatus*. *Z. Zool. Syst. Evolutionsforsch.*, 29: 280–288.

Jones, R., 1990. *Evolution of Cave and Surface Populations of the Amphipod Gammarus minus*. Ph.D. Thesis, Northwestern Univ., 221 pp.

Jones, R. and Culver, D.C., 1989. Evidence for selection on sensory structures in a cave population of *Gammarus minus* (Gammaridae). *Evolution*, 43: 688–693.

Jones, R., Culver, D.C. and Kane, T.C., 1992. Are parallel morphologies of cave organisms the result of similar selection pressures? *Evolution*, 46: 353–365.

Kimura, M., 1983. *The Neutral Theory of Molecular Evolution*. Cambridge Univ. Press, New York.

Kosswig, C., 1960. Darwin und die degenerative Evolution. *Abh. Verh. Naturwiss. Ver. Hamburg N.F.*, 4: 21–42.

Kosswig, C., 1963. Genetische Analyse konstruktiver und degenerativer Evolutionsprozesse. *Z. Zool. Syst. Evolutionsforsch.*, 1: 205–239.

Krumbiegel, J., 1960. *Die Rudimentation*. Gustav Fischer Verlag, Stuttgart, 141 pp.

Lamarck, J.B., 1809. *Philosophie zoologique: ou exposition des considérations relatives à l'histoire naturelle des animaux*. Chez Dentu, 2 Vols.

Lande, R. and Arnold, S., 1983. The measurement of selection on correlated characters. *Evolution*, 33: 402–416.

Langecker, T.G., 1993. Genetic analysis of the dorsal light reaction in epigean and cave-dwelling *Astyanax fasciatus* (Teleostei, Characidae). *Etho. Ecol. Evol.*, 3: 357–364.

Langecker, T.G., Schmale, H. and Wilkens, H., 1993. Transcription of the opsin gene in degenerate eyes of cave-dwelling *Astyanax fasciatus* (Teleostei, Characidae) and of its conspecific epigean ancestor during early ontogeny. *Cell Tissue Res.*, 273: 183–192.

Langecker, T.G., Wilkens, H. and Schmale, H., 1995. Developmental constraints in regressive evolution: studies of the expression of the gs-crystallin gene in the developing lens of cave-dwellling *Astyanax fasciatus* (Cuvier, 1819) (Teleostei, Characidae) by *in situ* hybridization. *J. Zool. Syst. Evol. Res.*, 33: 123–128.

Ludwig, W., 1942. Zur evoluorischen Erklärung der Höhlentiermerkmale durch Allelelimination. *Biol. Zentral.*, 62: 447–482.

Lynch, M., 1990. The rate of morphological evolution in mammals from the standpoint of neutral expectation. *Am. Nat.*, 136: 727–741.

Magniez, G., 1985. Regressive evolution in stenasellids (Crustacea, Isopoda, Asellota) of underground waters. *Natl. Speleol. Soc. Bull.*, 47: 118–122.

Mitchell, R.W., Russell, W.H. and Elliott, W.R., 1977. Mexican eyeless characin fishes, genus *Astyanax*: Environment, distribution, and evolution. *Spec. Publ. Mus. Texas Tech. Univ.*, 12: 1–89.

Nei, M., 1987. *Molecular Evolutionary Genetics*. Columbia Univ. Press, New York.

Oertel, R., 1934. Studien über Rudimentation, ausgeführt an den Flügelrudimenten der Gattung *Carabus*. *Z. Morph. Okol.*, 1: 38.

Packard, A.S., 1888. The cave fauna of North America, with remarks on the anatomy of the brain and the origin of the blind species. *Mem. Natl. Acad. Sci. USA*, 4: 1–156.

Parzefall, J., 1986. Behavioural ecology of cave-dwelling fishes. In: T.J. Pitcher (Editor), *The Behaviour of Teleost Fishes*. Croom Helm, London & Sydney, pp. 433–458.

Parzefall, J. and Senkel, S., 1986. Schooling behaviour in cavernicolous fish and their epigean conspecifics. *Cong. Int. Espeleol.*, 2(9): 107–109.

Peters, N. and Peters, G., 1968. Zur genetischen Interpretation morphologischer Gesetzmäßigkeiten der degenerativen Evolution. *Z. Morphol. Tiere*, 62: 211–244.

Popa, R., 1997. *The Dual Role of Convergent Natural Selection and Evolutionary History in the Morphology of Gammarus minus populations*. M.S. Thesis, American Univ., Washington, D.C., 160 pp.

Poulson, T.L., 1963. Cave adaptation in amblyopsid fishes. *Am. Midl. Nat.*, 70: 257–290.

Regal, P., 1977. Evolutionary loss of useless features: is it noise suppression? *Am. Nat.*, 111: 123–133.

Richardson, R. and Kane, T.C., 1988. Orthogenesis and evolution in the 19th century: the idea of progress in American neo-Lamarckism. In: M. Nitecki (Editor), *Evolutionary Progress*. Univ. Chicago Press, Chicago, pp. 149–167.

Romero, A., 1985. Can evolution regress? *Natl. Speleol. Soc. Bull.*, 47(2): 86–88.

Sadoglu, P., 1955. A mendelian gene for albinism in natural cave fish. *Experientia*, 13: 394–395.

Sadoglu, P. and McKnee, A., 1969. A second gene that affects eye and body color in Mexican blind cave fish. *J. Hered.*, 60(1): 10–14.

Schemmel, C., 1974a. Genetische Untersuchungen zur Evolution des Geschmacksapparates bei cavernicolen Fischen. *Z. Zool. Syst. Evolutionsforsch.*, 12: 169–215.

Schemmel, C., 1974b. Ist die cavernicole Micos-Population von *Astyanax mexicanus* (Characidae, Pisces) hybriden Ursprungs?. *Mitt. Hamb. Zool. Mus. Inst.*, 71: 193–201.

Schemmel, C., 1980. Studies on the genetics of feeding behaviour in the cave fish *Astyanax mexicanus f. anopichthys*. An example of apparent monofactorial inheritance by polygenes. *Z. Tierpsychol.*, 53: 9–22.

Schluter, D., 1988. Estimating the form of natural selection on a quantitative trait. *Evolution*, 42: 849–861.

Sexton, E. and Clark, A., 1936. A summary of the work on the amphipod *Gammarus chevreuxi* Sexton carried out at the Plymouth Laboratory (1912–1936). *J. Mar. Biol. Assoc. U.S.*, 21: 357–414.

Shoemaker, C., 1940. Notes on the amphipod *Gammarus minus* Say and description of a new variety *Gammarus minus* var. *tenuipes*. *J. Wash. Acad. Sci.*, 30: 388–394.

Sket, B., 1985. Why all cave animals don't look alike: a discussion of the adaptive value of reduction processes. *Natl. Speleol. Soc. Bull.*, 47: 78–85.

Sneath, P.H.A. and Sokal, R.R., 1973. *Numerical Taxonomy*. W.H. Freeman, San Francisco.

Vandel, A., 1964. *Biospéologie. La Biologie des Animaux Cavernicoles.* Gauthier-Villars, Paris, 619 pp.

Wilkens, H., 1970a. Beiträge zur Degeneration des Auges bei Cavernicolen, Genzahl und Manifestationsart. Untersuchungen an mexikanischen Höhlenfischen. *Z. Zool. Syst. Evolutionsforsch.*, 8: 1–47.

Wilkens, H., 1970b. Beiträge zur Degeneration des Melaninpigments bei cavernicolen Sippen des *Astyanax mexicanus* (Filippi). *Z. Zool. Syst. Evolutionsforsch.*, 8: 173–199.

Wilkens, H., 1971. Genetic interpretation of regressive evolutionary processes: Studies on hybrid eyes of two *Astyanax* cave populations. *Evolution*, 25(3): 530–544.

Wilkens, H., 1976. Genotypic and phenotypic variability in cave animals. Studies on a phylogenetically young cave population of *Astyanax mexicanus* (Fillippi). *Ann. Speleol.*, 31: 137–148.

Wilkens, H., 1980. Zur Problematik der Rudimentation, untersucht an der Ontogenie des Auges von Höhlenfischen (*Astyanax mexicanus*). *Z. Zool. Syst. Evolutionsforsch.*, 18: 232–238.

Wilkens, H., 1984. Zur Evolution von Polygensystemen: untersucht an ober- und unterirdischen Populationen des *Astyanax mexicanus* (Characidae, Pisces). *Fortschr. Zool. Syst. Evolutionsforsch. Beih.*, 3: 55–71.

Wilkens, H., 1988. Evolution and genetics of epigean and cave *Astyanax fasciatus* (Characidae, Pisces). Support for the neutral mutation theory. In: M.K. Hecht and B. Wallace (Editors), *Evolutionary Biology*, 23: 271–359.

Wilkens, H., 1993. Neutrale Mutationen und evolutionäre Fortentwicklung. *Z. Zool. Syst. Evolutionsforsch.*, 31: 98–109.

Wright, S., 1978. *Evolution and Genetics of Populations, Vol. 4, Variability Within and Among Populations.* University of Chicago Press, Chicago.

Yokoyama, S., Meany, A., Wilkens, H. and Yokoyama, R., 1995. Initial mutational steps toward loss of opsin gene function in cavefish. *Mol. Biol. Evol.*, 12: 527–532.

Chapter 21

ECOLOGICAL DERIVATION, COLONIZATION, AND SPECIATION

John R. HOLSINGER

OVERVIEW

Evolutionary theories that have been advanced to explain why and how surface-dwelling organisms invade and colonize subterranean environments (caves, interstices, shallow underground compartments, phreatic waters, etc.) and ultimately become troglobites (stygobites) generally encompass three phenomena: preadaptation of epigean ancestors, invasion and colonization of caves by founder populations, and speciation. Each phenomenon will be examined in some detail in this chapter.

The evolution of cave faunas poses for students of evolutionary biology a number of challenging questions, most of which are central to issues currently being debated by evolutionists. For example, how important is exaptation, defined by Gould and Verba (1982) as "features that now enhance fitness but were not built by natural selection for their current role", in equipping putative epigean ancestors for invasion and colonization of hypogean environments? Is this the most important aspect of preadaptation? What role do troglophiles play in the evolution of troglobites? Do they provide an intermediate stage in the evolutionary transition from epigean ancestor to cave-limited species? and do their ecological and evolutionary roles differ between temperate and tropical caves?

Do troglobites evolve from organisms that invade caves under constraints imposed by fluctuating climates, and then become isolated in underground refugia when populations of surface relatives become extinct, as suggested by the climatic-relict model, or do they evolve from ancestors that actively invaded caves to exploit new niches, subsequently transforming into troglobites without a period of physical isolation from surface relatives as suggested by the adaptive-shift model?

Is speciation in cave animals invariably allopatric as stressed in the climatic-relict model (Barr, 1968; Sbordoni, 1982), or can it be parapatric or even sympatric as implied in the adaptive-shift model of Howarth (1973, 1981, 1987) or in the scenarios of Wilkens and Hüppop (1986)? Do any of the three founder-effect or founder-induced speciation models (Carson and Templeton, 1984) offer a plausible explanation for the origin of cave species? and, if so, which one is the most acceptable given currently available data? Barr's (1968) application of Mayr's (1970) genetic-revolution model has been the most popular to date, but its universality is questionable, especially with respect to the origin of troglobites in tropical caves (Chapman, 1986a; Howarth, 1980, 1987; Bellés, 1991; Ashmole, 1993; Peck and Finston, 1993).

How do levels of genetic variability in cave species compare with those of epigean species? According to Sbordoni (1982), heterozygosity in troglobitic species increases with evolutionary time, so that genetic variability can be greater in small populations of old troglobites than in considerably larger populations of epigean relatives. At other scales, however, Culver (1982) noted that differences in heterozygosity and polymorphism are inconsistent both among different cave species and between cave and surface species.

Are troglobitic species composed of groups of reproductively isolated populations that meet the definition of biological species (Mayr, 1970)? or are most of them described without any real knowledge of reproductive isolation and therefore more likely to be evolutionary (phylogenetic) species (Wiley, 1978)? Christiansen (1992) believed that very few cave animals, as currently described, meet the requirements for biological species, but he also cautioned that species models should be tested against the data for a given group before conclusions are reached about what kind of species it is. Sbordoni (1982) and Caccone (1985) have shown that some cave species are reproductively isolated and therefore by definition qualify as biological species; but for many others there is simply a lack of definitive information on reproductive isolation.

Finally, how strongly differentiated morphologically are troglobitic species, and how common are cases where genetic differentiation has outpaced morphological change to the point where sibling species exist that can be separated only through biochemical analysis? Gel electrophoresis has become an effective tool for analyzing genetic differences, both among populations and between species of cave animals (Culver, 1982; Sbordoni, 1982; Kane et al., 1991). According to allozyme studies by Laing et al. (1976a,b), Sbordoni (1982) and Culver et al. (1995), which covered a variety of taxa living in caves, the occurrence of local populations of troglobites that are highly differentiated genetically but lack corresponding morphological differentiation may be common.

PREADAPTATION AND THE STATUS OF TROGLOPHILES

There is a consensus among biospeleologists that troglobites are ultimately derived from preadapted epigean propagules or founders that live in either terrestrial or aquatic environments. It is widely accepted that troglobite faunas evolve from adaptatively radiating epigean species, which invade the subterranean realm through fissures, mesocavernous spaces, interstices, stream beds, swallet holes, springs, and other openings giving access to subterranean environments. Preadaptation is a prerequisite for successful colonization of subterranean environments, and virtually all workers agree that colonizing populations had achieved some degree of preadaptation for subterranean life prior to the actual invasion of hypogean environments (Vandel, 1964; Barr, 1968; Culver, 1982; Sbordoni, 1982; Wilkens and Hüppop, 1986; Boutin and Coineau, 1990; Notenboom, 1991; Ashmole, 1993; Holsinger, 1994). Much of the preadaptation is probably in the form of exaptations (Gould and Verba, 1982; Christiansen, 1992), which may be resident in the genomes of founding organisms and thus provide them with the rudimentary mechanisms necessary for survival in a new, subterranean adaptive zone, where selective pressures are markedly different from those on the surface (Howarth, 1980; Rouch and Danielopol, 1987; Ashmole, 1993).

Preadaptations may be morphological, physiological, or behavioral. Ashmole (1993) called attention to several probable preadaptations (exaptations) in surface animals closely related to putative ancestors of troglobites, including sensitivity to substrate-borne vibrations in cixiid plant hoppers and spiders, and use of non-visual cues in foraging and detection of predators by certain fishes and primitive insects. Other adaptations, such as photo-negativity, locomotion in or on the substrate and absence of pelagic larvae, are mentioned by Parzefall (1986). Notenboom (1991) suggested that the absence of highly specialized feeding habits and mouthpart morphology in hadzioid amphipods may be preadaptive. Other good examples are given by Vandel (1964), Barr (1968), Rouch and Danielopol (1987), Wilkens (1988), and Boutin and Coineau (1990); continued study will undoubtedly reveal many more.

Speciation rates across the epigean–hypogean interface determine how quickly preadapted epigean organisms are transformed into troglobites, and whether or not there is a significant preliminary or intermediate troglophile (or stygophile) stage in the evolutionary transition from surface ancestors to true cave forms. By definition, troglophiles are facultative species that are able to complete their life cycle within a cave but may also occur in ecologically suitable habitats outside caves. Thus, the assumption follows that troglophiles are strongly preadapted for life in caves and often appear to be morphologically and ecologically intermediate between epigean ancestors and fully evolved (troglomorphic) troglobites. The potentially important role played by troglophiles in the evolution of troglobites was discussed in some detail by Barr (1967, 1968). Certain species of troglophiles have been referred to in the past as "troglobites *in statu nascendi*" (Barr, 1968).

More recently Barr (1985) suggested an evolutionary scheme for troglobitic trechine beetles (Carabidae: Trechinae) in which the hypothetical, preadapted intermediate stage would be an edaphobite (deep soil inhabitant). Under a constraint hypothesis, such as the climatic-relict model, colonizers would first become established in caves as troglophiles or their equivalent (e.g., edaphobites, etc.) and later become troglobites after all contact and gene exchange with surface populations had been eliminated (Barr, 1967, 1968, 1985; Sbordoni, 1982). However, under an adaptive-shift paradigm, in which troglobites are believed to evolve very quickly from preadapted invaders, the occurrence of an intermediate troglophile stage is improbable (Howarth, 1981).

It should also be stressed that the ultimate fate of many troglophiles, or edaphobites for that matter, living in caves remains unclear, and that numerous troglophiles are apparently well adapted for existence in subterranean systems and will probably remain at this level of adaptation indefinitely. The same analogy holds for the status of edaphobites, which when found caves are often regarded as troglophiles, or sometimes even confused with true troglobites because of their regressive features (e.g., loss of eyes and pigment) (Sket et al., 1994). Whereas some troglophiles, edaphobites, or even habitual trogloxenes may represent various intermediate stages between epigean and troglobitic species and could conceivably become troglobites under the influence of certain environmental changes, such as extinction of surface populations by climatic changes, others are apparently well adapted at their present level of existence (Howarth, 1981; Culver, 1982; Holsinger and Culver, 1988; Kane et al., 1994). The occurrence of successful, well-adapted non-troglobitic cavernicoles suggests that many troglophiles are not merely "intermediate" steps in the evolution of troglobites (Howarth, 1981; Culver, 1982; Chapman, 1986b; Holsinger and Culver, 1988; Notenboom, 1991; Christiansen, 1992).

Moreover, many of the edaphobites presently found in caves probably occur there through accidental penetration or chance dispersal into open cave passages from deep soil habitats, and are not in the process of colonizing caves and evolving into troglobites. Although it will be emphasized below that deep soil habitats are an important extension of the deep leaf litter environments of forest floors, and thus serve both as a transition zone between surface litter and subterranean passages and a source of preadapted troglobite ancestors, it is unlikely that the majority of edaphobites found in caves are shifting into a new adaptive zone and evolving into troglobites.

The ecological and/or evolutionary status of troglophiles may also differ between caves in tropical and temperate climates, and many species previously categorized as troglophiles in tropical caves may in fact be cave-limited and technically speaking, troglobites. Chapman (1980, 1984, 1986a,b) pointed out that troglophiles in tropical caves are often well adapted to the cave environment despite showing little or no obvious troglomorphy, and that these species are actually cave-limited even though they do not possess all of the regressive features that commonly characterize temperate-zone troglobites. Primarily on the basis of these observations, Chapman (1986b) contended that a distinction between troglobite and troglophile can seldom be made empirically and is of little practical value anyway, especially since these terms cannot possibly encompass all of the many different kinds of subterranean faunas that have been recognized in recent years. This reasoning may also apply to edaphobites that normally live in the deep soil but sporadically occur in caves. Many of them, including species of millipedes, mites, collembolans and beetles, display regressive features and are sometimes difficult to distinguish from troglobites. To resolve the problem, Chapman (1986b) suggested either a radical reform of the conventional Schiner–Racovita system (e.g., usage of troglobite, troglophile, trogloxene as defined by Barr, 1967, 1968) or discarding it in favor of a system that emphasizes the ecological role that an organism plays in a cave environment rather than how troglomorphic it is. The problem Chapman described is real, and finding a solution for it is a major challenge for students of biospeleology.

ORIGIN AND DERIVATION OF TROGLOBITES

The origin and derivation of aquatic and terrestrial cave species are considered separately below, inasmuch as there are important differences between the two with respect to modes of origin, routes of invasion, habitats, and dispersal potential.

Terrestrial troglobites

A variety of subterranean environments, including caves and numerous smaller mesocaverns in limestone (karst) and lava (basalt), the *milieu souterrain superficiel* (MSS) (= shallow underground environment), and even mine adits (Uéno, 1977), have been colonized by founder species from the surface. In temperate regions troglobites are derived primarily from putative epigean ancestors inhabiting moist humus, organic litter and moss on the floors of mesic upland and montane forests and the underlying soil or endogenous zone that extends to the greatest depth to which the roots of vegetation may penetrate (Vandel, 1964; Barr, 1968, 1985; Jefferson, 1976; Peck, 1981a; Juberthie, 1984;

Barr and Holsinger, 1985; Eberhard, 1992). Thus, it would appear that a deep, cool, moist forest-floor litter and the underlying soil provide the ideal environment for a highly diverse and distinctive cryptozoic fauna of small invertebrates, consisting largely of arthropods (primarily arachnids, millipedes, and insects). This diverse assortment of hydrophilic organisms is the primary source for the founder populations that ultimately colonize subterranean spaces and evolve into terrestrial troglobites.

In tropical regions terrestrial troglobites are derived from epigean inhabitants of the organic litter of rainforest floors or root systems that extend into cracks and crevices in the underlying bedrock (Howarth, 1993). In addition to the taxonomic orders commonly well represented in temperate-region caves, such as Araneae, Pseudoscorpionida, Opilionida, Diplopoda, Diplura, Collembola, Orthoptera, Coleoptera, and Diptera, tropical caves may also be inhabited by Amblypygida, Schizomida, Blattodea, Heteroptera, and Homoptera.

The origin of much of the remarkably diverse tropical terrestrial troglobite faunas in the tropics – of lava caves in the Hawaiian, Canary and Galapagos islands, and Undara in the Chillagoe region of northeastern Australia – is attributed to the invasion of numerous mesocavernous spaces in relatively recent lava flows (Howarth, 1987, 1988; Oromi et al., 1991; Ashmole, 1993; Peck and Finston, 1993). In tropical volcanic regions, such as Hawaii, corridors of undisturbed rainforest or kipukas lying adjacent to recent lava flows are very important sources of founding propagules. In a relatively short period of time after volcanic activity has ceased, propagules may disperse from protected "islands" of vegetation into the numerous mesocavernous spaces of the adjacent young basalt. As Howarth (1987) pointed out, a new lava flow contains an immense complex of mesocavernous spaces extending from the surface and providing numerous avenues for organisms to follow underground. However, not all Hawaiian troglobites have evolved from ancestors living on the floors of rainforests. Some of them, like rock crickets (*Caconemobius*) and lycosid spiders, may also be derived from epigean ancestors inhabiting wet rocky substrates (Howarth, 1987).

The importance of moist, forest-floor litter environments as the primary source for epigean ancestors of terrestrial troglobites has been noted for many cavernous areas of the world. Excellent examples include the Appalachians and Interior Low Plateaus of the eastern and southeastern United States (Barr, 1968; Peck, 1981a; Holsinger and Culver, 1988), parts of the southwestern United States (Peck, 1980, 1981b), the big island of Hawaii (Howarth, 1987), the Ribeira Valley of southeastern Brazil (Trajano, 1995), parts of southern Europe (Jeannel, 1943; Vandel, 1964), the Pyrénées of southern France (Juberthie, 1984), the Mulu caves of Sarawak (Chapman, 1984), central and western Tasmania (Eberhard, 1992), and the Cape Range karst in the northwest of Western Australia (Humphreys, 1993a).

To further emphasize the importance of a forest-floor litter fauna in the evolution of troglobite faunas, Peck and Lewis (1978) compared the biodiversity of terrestrial cave faunas of the [greater] Appalachian region of the eastern United States with the Ozark region of the central United States. They concluded that the comparative poverty of terrestrial troglobites in the Ozarks can be attributed to the inability of this region to support substantial montane forests and litter faunas through the periods of time needed for the development of a large, regionally distinct litter fauna.

The prolonged absence of forests in karst regions is typically reflected in impoverished troglobite faunas. In most such regions, forests that once occurred there have since been eliminated by extreme drought or glaciation. Massive disruption, displacement or outright destruction of cool, moist forest-floor environments would drastically reduce the resident litter fauna, leading ultimately to extermination of most or all of the potential ancestors of a cave fauna in these regions. For example, extensive karst regions of the northern United States, Canada and northern Europe, which were covered by ice during the last major glaciation (Wisconsin or Würm advance), lack terrestrial troglobite faunas at present, although they are inhabited by trogloxenes and some troglophiles of very recent origin (Peck, 1988). Similarly, arid, unforested karst or volcanic regions in the western and southwestern United States (e.g., southeastern New Mexico, parts of Arizona, southeastern Oregon) and parts of southern Australia (excluding Tasmania) have many caves but very few troglobites. The increasing aridity in these areas during the Quaternary has resulted in extirpation of either forests and their litter faunas, or cave faunas, or both, depending on the biogeographical history of the area. The scarcity of troglobites in cavernous areas of southern Australia has been attributed to the increasing aridity of the continent, which in turn caused dehydration of cave systems and extirpation of preexisting cave faunas (Moore, 1964; Hamilton-Smith,

Table 21.1
Comparison of the two principal theories for the origin of terrestrial troglobites

Model	Area(s) where developed	Principal mechanisms suggested	Type of speciation	Evidence
Climatic Relict	Temperate karst in North America, Europe and Japan	Climatic changes precipitate invasion under constraint; isolation of founders in caves	Allopatric	Numerous relicts in caves
Adaptive Shift	Tropical lava caves in Hawaiian Islands	Founders colonize newly developing food niches; no isolation from surface relatives	Parapatric	Few or no relicts in caves; species pairs respectively epigean and hypogean

1969; Eberhard, 1992). In contrast, however, the arid tropical Cape Range Karst (W.A., Australia contains an exceptionally rich, highly troglomorphic terrestrial cave fauna (Humphreys, 1993a; see also Chapter 22, this volume).

Based on observations in Brazilian karst regions, Trajano (1995) suggested that aridity has approximately the same inimical effect on a regional troglobite fauna as that of glaciation. When conditions become extreme enough to interrupt energy flow into caves, because of extensive penetration either of drought in desert regions or of frost and ice in glaciated regions, existing cave faunas may have become much less diverse or even have been obliterated. This explains why caves in recently glaciated areas, as well as those in very dry areas, have impoverished troglobite faunas. Their faunas were either extirpated in the recent past, or are under great stress at present because of aridity and dehydration.

Two principal theories have been proposed to explain cave colonization and the evolution of terrestrial troglobites (Table 21.1): (a) the climatic-relict model; and (b) the adaptive-shift model. Although these theories are often regarded as competing alternative hypotheses, they are best regarded as different explanations for related phenomena operating under different geological and/or climatological conditions. In view of the considerable differences in geological structure, climate, and biota among the various cave and karst regions of the world, it is unrealistic to believe that a single model for the origin of terrestrial troglobites will apply universally. As Peck and Finston (1993) pointed out, a newly emerging pattern suggests major differences between the obligate cave faunas of old continental land-masses, those of younger oceanic islands, and those of tropical continental regions. These differences may be due more to the age of the respective faunas than to any other factor.

The **climatic-relict model** (also called the relictual or Pleistocene-effect model) (Holsinger, 1988; Ashmole, 1993; Peck and Finston, 1993) was until recently the most widely accepted model for the evolution of terrestrial troglobites in temperate latitudes. It was originally proposed to explain the evolution of troglobites in the unglaciated karst regions of North America, Europe, and Japan, and relied heavily on the effects of fluctuating climates in the Pleistocene (Vandel, 1964).

In its original form, the model suggested that during glacial maxima in areas just south of the ice sheets, and probably also on mountains in more southerly latitudes, cool climates favored wide distribution of invertebrates inhabiting moist, forest-floor litter at low temperatures. In the periglacial climate beyond glacial margins, caves and forest floors were inhabited by similar species. Later, as glaciers retreated, these species became progressively restricted to the cool, moist interiors of caves, sinkholes, deep wooded ravines, and cool forest floors at higher altitudes. Ultimately, with climatic warming and concomitant changes in the composition of surface vegetation during interglacial periods, the surface populations of these species became extinct and most of the surviving populations were those living in caves. Continued geographic isolation, no longer compromised by occasional gene exchange between hypogean and epigean populations, promoted genetic isolation, accelerated speciation, and influenced the development of a troglomorphic facies among the newly evolving troglobites.

Subsets of the climatic-relict model have been developed recently to reconcile the discoveries of diverse troglobitic faunas in tropical regions with important new evidence for significant climatic changes in the tropics during the Tertiary (Humphreys, 1993b) and Quaternary. Prior to the early 1970s, terrestrial troglobites were thought to be rare in the tropics, and tropical climates were believed to have remained relatively stable throughout the Pleistocene. During the last three decades, however, a large number of terrestrial troglobites have been discovered and catalogued from many tropical areas (e.g., in Australia, Brazil, the Canary Islands, Cuba, the Galapagos Islands, Hawaii, Jamaica, Mexico, Papua New Guinea and Sarawak) (Reddell, 1981; Howarth, 1973; Bellés, 1991; Peck and Finston, 1993), and strong evidence has accumulated that clearly points to significant climatic fluctuations in much of the world during the Pleistocene (Lynch, 1988; Peck and Finston, 1993; deMenocal, 1995).

Whereas, earlier, the climatic-relict model was applied exclusively to the origin of troglobites in temperate-climate karst regions, the newer subsets attempt to relate the occurrence of relict troglobite faunas in tropical karst areas to major climatic changes that affected these areas in the past. Humphreys (1993a) in Australia and Trajano (1995) in Brazil have made strong cases for the profound effects of extended droughts and loss of rainforests on the isolation of relict cave faunas in tropical karst regions. According to Humphreys (1993a,b) the rich troglobite fauna of the semi-arid Cape Range karst in the northwest of Western Australia has affinities mostly with the faunas of rainforest litter of northern and northeastern Australia, and appears to be a relict fauna that was isolated by increasing aridity and retreat of rainforests from this area during the late Tertiary. Preliminary observations by Trajano (1995) suggest that a karst area in the Ribeira Valley of southeastern Brazil, which has changed from a corridor of open vegetation in the late Pleistocene to forests more recently, has a much more diverse troglobite fauna than a karst area in central Brazil, which has been continuously forested throughout this period of time. The comparatively high number of troglobites in the Ribeira Valley is believed to represent a relict cave fauna that was isolated during a period of aridity coincident with the Wisconsin glaciation in the Northern Hemisphere.

The **adaptive-shift model** (also called the local habitat shift model) (Ashmole, 1993; Peck and Finston, 1993), which has become more widely accepted in the last few years, was originally proposed by Howarth (1973, 1981) to explain the evolution of troglobites in the lava caves of the tropical Hawaiian Islands. It has since been used as an explanation for the evolution of troglobites in other volcanic regions (the Canary and Galapagos islands: Oromi et al., 1991; Peck and Finston, 1993), in tropical karst areas (Papua New Guinea and Sarawak: Chapman, 1986a), and for troglobitic spiders in temperate-climate karst areas in Bosnia, Croatia and Slovenia (Deeleman-Reinhold, 1981). It suggested that troglobites evolve from preadapted ancestors that actively invade caves to exploit new niches, subsequently transforming into troglobites without a period of physical isolation from surface relatives. Accordingly, troglobites are derived from the speciating native fauna of a region, and their evolution is regarded as a continuing process, rather than episodic as suggested by the climatic-relict theory. This model was originally based on extensive biological data from Hawaiian lava caves, which are geologically younger and have a food supply (primarily root exudates) different from those in most limestone caves in karst areas. Many of the lava-cave troglobites observed by Howarth, especially cixiid plant hoppers, crickets, and thread-legged bugs (*Nesidiolestes ana*), are morphologically closely similar to species living on the surface near the caves, thus providing strong evidence that putative epigean relatives have not necessarily become extinct concurrently with the evolution of their troglobite sister species. It follows then, that under this theory caves do not serve as refugia against the vicissitudes of climatic change, and that most of the troglobites evolving there are not relicts, at least in the short run.

Two potential weaknesses of the adaptive-shift model have been mentioned, though neither appears to be serious. First is the implication that speciation under the model is not allopatric, but instead is parapatric or sympatric. There is increasing evidence, however, to support divergence-with-gene-flow speciation (Rice and Hostert, 1993) and other non-allopatric speciation processes, such as disruptive selection (Wilkens and Hüppop, 1986; Ashmole, 1993; Peck and Finston, 1993) or other forms of evolutionary divergence (Coyne, 1992), and earlier arguments against adaptive shift on the grounds that it was not an allopatric model are no longer convincing. It seems likely that some kind of disruptive selection, such as that suggested by Ashmole (1993) and Peck and Finston (1993) for certain terrestrial arthropods, or by Wilkens and Hüppop (1986) for cave fishes, can occur at the epigean–hypogean interface, where significant selective pressures may change over a very short distance.

Second is the contention that the adaptive-shift theory is applicable only to the unique fauna of tropical volcanic regions, where, in comparison with limestone karst in temperate and some tropical regions, troglobites are of recent origin and few if any are relicts, and caves are geologically much younger. For most temperate karst regions, it was assumed earlier that animals invaded caves to survive climatic changes primarily associated with glaciation, and not to exploit novel food niches and seek additional living space. However, it has become obvious that there are a number of reasons why animals invade caves, and the type of cave and its geological age may be just two of several important factors involved in cave invasion.

Bellés (1991) listed three reasons why animals invade caves: survival (caves serve as immediate refugia against surface stress); opportunism (caves offer virgin space for colonization); and convenience (caves act as escape valves when competition occurs in surface habitats). He also pointed out that all three modes of colonization can be well exemplified in temperate regions that suffered the direct impact of Pleistocene glaciation, but they can just as easily be postulated for other situations, which have nothing to do with climatic revolutions. Thus, it is logical to assume that a form of opportunism, similar to that which was suggested as the reason for invasion of lava caves in Hawaii, could have also occurred many times in temperate karst regions. This is not to say that glaciation and climatic fluctuations did not have a significant impact on the evolution of troglobites in temperate regions, but, as Howarth (1973, 1981) pointed out, periodic changes in Pleistocene climates also probably obscured evidence of the early history of troglobite evolution and distribution.

The complex biogeographical, ecological, geological, and climatological changes that have occurred in most karst areas throughout the Quaternary make an accurate reconstruction of the evolutionary history of their cave faunas very difficult. In some cave regions with troglobite relicts, it is possible that the initial invasion of caves had little or nothing to do with subsequent climatic changes that led to the extinction of the close surface relatives of the cave fauna and resulted in the caves becoming refugia. In other regions, however, troglobite relicts probably evolved in close accordance with the scenario suggested by the climatic-relict hypothesis. The presence of both relict and non-relict species in the troglobite faunas of the Canary and Galapagos islands (Oromi et al., 1991; Peck and Finston, 1993) suggests that both the climatic-relict and adaptive-shift models may apply to the evolution of obligate cave faunas in some regions.

Aquatic troglobites

Aquatic troglobites (stygobionts) may be derived from marine, brackish or freshwater ancestors, depending on the environmental conditions under which the founding populations were living at the time of invasion of subterranean waters. Presumably a majority of potential founders, whether limnic or marine, are associated with benthic sediments containing organic matter (Coineau and Boutin, 1992; Holsinger, 1994). This is true for almost all invertebrates, and most vertebrates (fishes and salamanders) as well. Putative marine ancestors occur in a variety of unconsolidated macroporous sediments composed of sand, coral rubble, etc., and in rock crevices, in the littoral zones of tropical and warm-temperate seas along the margins of continents or around islands (Rouch and Danielopol, 1987).

In comparison, putative freshwater ancestors are commonly found in leaf litter, moss, gravels and shallow fissures, which furnish secluded habitats in the substrates of woodland seeps, sinkhole ponds, swamps, sloughs, small streams, and spring outflows. These cryptic habitats are often characterized by organically enriched benthic sediments. They are inhabited by a variety of Tricladida, Gastropoda, Oligochaeta and a impressive diversity of Crustacea that include Ostracoda, Copepoda, Isopoda, Amphipoda, and Decapoda, together with small fishes (e.g., Amblyopsidae, Clariidae, Cyprinidae, Ictaluridae, Pimelodidae) and salamanders (Plethodontidae). These same aquatic groups are also well represented in subterranean freshwater aquifers.

Four evolutionary pathways are envisaged for the evolution of stygobionts (Table 21.2):

(a) **thalassostygobionts** (marine or brackish-water stygobionts) evolve directly from epigean marine ancestors through active invasion of sediments and crevicular habitats, and adaptive shifts or vicariance;
(b) **limnostygobionts** (freshwater stygobionts) evolve from marine ancestors by stranding during regression of marine embayments;
(c) limnostygobionts may also evolve from marine ancestors by active invasion of freshwater aquifers along interstitial routes and by means of adaptive shifts; and
(d) limnostygobionts may also evolve from epigean freshwater ancestors through surface to subterranean invasions, assisted by stream capture, spring failure and depression of the groundwater table.

Table 21.2
Summary of theories for the origin of stygobionts (aquatic troglobites)

Type of transition	Model	Mechanism	Areas of occurrence	Resulting stygobiont
Marine–Marine	Step 1 of 2-step model	Adaptive shift, sea level changes	Interstitial and crevicular habitats in littoral zones	Thalassostygobiont
Marine–Limnic	Step 2 of 2-step model (= regression model)	Stranding during marine regressions	Former embayments; emerging islands; coastal zones	Limnostygobiont
Marine–Limnic	Active invasion	Adaptive shift	Coastal zones	Limnostygobiont
Limnic–Limnic	Active or passive invasion, or combination of these	Stream capture, spring failure, depression of groundwater table, etc.	Inland (continental) freshwaters	Limnostygobiont

According to the two-step models proposed by Boutin and Coineau (1990), Notenboom (1991), and Coineau and Boutin (1992), thalassostygobionts, including the smaller interstitial species of unconsolidated sediments (coarse sands and gravels) and the comparatively larger species of cracks, crevices, fissures, etc. (crevicular habitats), are derived directly from preadapted epigean marine ancestors. The critical first step of both these models takes place in the marine environment, and involves the vertical migration of preadapted benthic organisms into interstitial and/or crevicular habitats. Presumably this sets the stage for all subsequent evolution of stygobionts in the marine environment.

The "epigean" benthic habitats of marine waters do not appear to differ materially from the deeper ones inhabited by marine stygobionts, except that they may be less remote and ecologically more variable. Because a mixture of epigean and stygomorphic species is not uncommon in some of these sediments, the vertical penetration of deeper sediments from the shallower ones by preadapted organisms may be relatively easy for some groups, as is suggested by both the Boutin–Coineau and Notenboom models.

Marine–brackish-water stygobionts that commonly occur in interstitial or anchialine waters in coastal and insular areas, and have strong morphological affinities with other marine taxa, appear to have evolved directly from "epigean" marine ancestors through active invasion of sediments and crevicular habitats, and adaptive shifts or vicariant processes involving changes in sea level. Although many of these species are of relatively recent origin, as suggested by their close phylogenetic affinities with "epigean" forms, some, like the exclusively stygobiont remipede crustaceans, may be much older, and their origins are obscure (Holsinger, 1988; Yager and Humphreys, 1996).

Iliffe et al. (1984) proposed a deep-sea hypothesis for the origin of some of the stygobionts living in the marine–brackish-water of anchialine caves, but evidence compiled by Stock (1986) favors a shallow-water origin for almost all members of this fauna. The shallow-water theory has since been reinforced and corroborated by the findings of Danielopol (1990) and Holsinger (1992). Moreover, a preadapted benthic ancestry for many groups of stygobiont amphipod and isopod crustaceans is well documented in the studies of Boutin and Coineau (1990), Notenboom (1991), and Holsinger (1994). Although a deep-sea origin for some thalassostygobionts cannot be ruled out, it seems more likely in view of recent observations that a majority are derived from shallow-water ancestors.

Limnostygobionts have a dual origin, inasmuch as some of them appear to have evolved directly from marine ancestors, whereas others have probably evolved from "epigean" limnic ancestors by active or passive invasion of subterranean groundwater aquifers from surface waters. The derivation of limnostygobionts from marine ancestors may involve the stranding of founder populations in gradually freshening groundwaters during marine regressions in accordance with the regression model (Stock, 1977, 1980; Holsinger and Longley, 1980), or it may occur by way of active invasion of inland freshwater aquifers along a continuous interstitial pathway, followed by adaptive shift.

Marine relict crustaceans living in subterranean freshwater habitats provide the best evidence for the stranding phenomenon. These stygobiont organisms, which inhabit continental or insular groundwaters, have strong morphological affinities with marine taxa and were presumably derived from marine ancestors following the regression of marine embayments (Stock,

1980; Holsinger, 1994). The pattern is clearly indicated by the distribution of many crustacean taxa (especially amphipods and isopods), which occur in subterranean limnic habitats on parts of continents that were last covered by shallow marine waters in the Late Cretaceous or early to middle Tertiary, or on islands, such as those in the West Indies, that were slowly uplifted from the sea by tectonics during approximately the same time periods.

Following the regressive phase, marine regressions may fragment distributions into isolated pockets, which in turn will lead to the isolation of founders in a given area. According to the two-step model of Notenboom (1991), regressions are much more important in step 2 than in step 1. Step 2 is vicariant and leads to isolation of founders by stranding.

Ward and Palmer (1994) suggested that anchialine water provides one of two hypogean pathways for evolutionary invasions of continental waters by marine forms, the other being the marine psammolittoral, which is discussed below. Anchialine pathways, consisting of coastal caves and crevicular habitats, furnish numerous hypogean voids for invasion and occupancy along the margins of continents inundated by marine embayments, and on islands that are slowly emerging from the sea. Stranding would have to follow colonization of anchialine waters before the evolutionary transition from marine ancestor to limnostygobiont was complete. Thus, for the evolution of many limnostygobionts with a marine ancestry, the mixohaline water of anchialine caves has probably provided an important transitional zone during marine regressions (Holsinger, 1994).

Other limnostygobionts probably have evolved from marine ancestors by active invasion of freshwater aquifers along interstitial routes and adaptive shifts. The marine psammolittoral, including both sands (thalassopsammon) and gravels (thalassopsephon), provides habitats from which meiofaunal groups may invade hypogean fresh waters along continuous interstitial pathways leading from marine littoral zones into the sediments of inland freshwater aquifers (Delamare-Deboutteville, 1960; Vandel, 1964; Hutchinson, 1967; Pennak, 1968; Husmann, 1971; Botosaneanu, 1986; Ward and Palmer, 1994). However, some of the interstitial organisms that occur in the thalassopsammon are meiobenthic and therefore not considered here. Invertebrate groups that are believed to have invaded limnic groundwaters from the interstitial waters of marine sands and gravels are listed by Ward and Palmer (1994: fig. 2). However, for many of these species it is not known whether they ultimately adapted to limnic waters by adaptive shift or were stranded.

Rouch and Danielopol (1987) introduced the term "active colonization" to cover transitions from marine to limnic environments by organisms migrating through sediments in coastal areas and crossing the salinity barrier in a variety of ways. They proposed this as the primary mode of origin of limnostygobionts from marine relatives, stating that it is an active model rather than the "passive" regression model proposed earlier by Stock. Whereas the Rouch–Danielopol model provides a plausible explanation for the origin of some limnostygobionts, as already suggested above, it by no means replaces Stock's regression model. Stranding (regression model) is a better explanation for the origin of a much larger number of taxa and has the added advantage of being a vicariant process. Botosaneanu and Holsinger (1991) suggested that the origin of virtually all the marine-derived stygobionts of the Mediterranean and greater Caribbean–Gulf of Mexico regions can be explained by some form of the regression model.

Theoretically, another route of direct (active) invasion of hypogean fresh waters from marine sources is through submarine karst springs emitting mixohaline water (Notenboom, 1991). Although such springs may provide a dispersal avenue for invasion and active colonization of groundwaters of mixed salinities by marine organisms, it is questionable whether or not the immigrants would become limnostygobionts prior to a marine regression followed by stranding. The invasion of anchialine habitats is perhaps similar to that of coastal karst springs, since the water in both of these habitats is generally mixohaline. The conversion of either anchialine caves or coastal springs into freshwater habitats during a marine regression or island emergence would involve changes in salinity and conceivably result in stranding before their inhabitants were fully limnic.

The derivation of limnostygobionts directly from epigean freshwater ancestors through surface to subterranean invasions is apparently a very common phenomenon, and may well account for more than half of all limnostygobionts. The close phylogenetic relationship of many limnostygobionts with epigean freshwater taxa which have presumably lived in continental freshwater habitats for millions of years and lack close taxonomic alliance with any marine forms is strong evidence for their freshwater ancestry

(Holsinger, 1988, 1994). However, in certain regions a subset of limnostygobionts may exist whose epigean ancestry is historically not as old. Sket (1986) pointed out that certain limnostygobionts, such as sphaeromatid isopods, living in the Dinaric karst region of Slovenia and Croatia, have strong taxonomic affinities with marine organisms, but were derived from epigean freshwater ancestors which had evolved earlier from marine ancestors, possibly in the Miocene. It is also likely that certain species of the amphipod genus *Gammarus,* which inhabit freshwater caves in the Holarctic region, are derived from "epigean" ancestors, which in turn were probably derived from relatively recent marine immigrants (Holsinger, 1986, 1994).

Preadapted epigean ancestors probably invade and colonize subterranean groundwater habitats by penetration through the substrates of small streams and ponds, upstream migration into caves from springs, and diversion underground when a surface stream is captured into an underground channel (Holsinger and Culver, 1988; Holsinger, 1988, 1994). Both spring failure and subterranean stream capture, along with changes in stream gradients and flow patterns resulting from regional uplifting and depression of groundwater tables, are relatively common hydrological phenomena in karst areas, and are believed to have significant impact on the evolution of stygobiont species.

Active invasion of hyporheic zones, *nappes phréatiques* and fissures in fractured rocks through the substrates of streams and ponds is probably a common mode of access to groundwater habitats, although the precise underlying mechanism responsible for the evolution of stygobionts in these shallow groundwaters is unclear. Under some conditions it may involve adaptive shift through disruptive selection at the epigean–hypogean interface, whereas under other conditions it may be a vicariant process that occurs when a prolonged drought or depression of the ground water table results in eliminating surface populations and isolating the founding propagules in subterranean waters.

Invasion of subterranean groundwaters by organisms through active upstream migration into caves from spring resurgences, and passive introduction of organisms into subterranean waters when surface streams in which they dwell are captured through sinkholes into underground channels, have been discussed in detail in recent papers (Mitchell et al., 1977; Culver, 1982; Sweet, 1982; Barr and Holsinger, 1985; Wilkens and Hüppop, 1986; Holsinger, 1988, 1994; Holsinger and Culver, 1988; Culver et al., 1995). It is widely accepted that these two pathways are among the most important routes of invasion of subterranean groundwater aquifers by preadapted organisms.

Preadapted organisms that have actively invaded and colonized a cave stream by upstream migration from a spring may become fully isolated from the spring population if the spring fails through prolonged drought or lowering of the water table. In either event, the cave stream will be diverted into a lower channel and no longer flow to the surface through the old spring outlet, thereby isolating the colonizers in the underground habitat. Organisms may initially invade cave streams to survive failure of their surface habitats during severe droughts, as suggested for plethodontid salamanders in central Texas (Sweet, 1982), or to escape changes in water temperature or predation, as suggested for *Gammarus minus* in the central Appalachians (Culver et al., 1995). Invasion of subterranean waters by way of stream capture in karst areas has been suggested as the mode of origin for the highly diverse stygobiont crayfish fauna of north-central Florida (Franz and Lee, 1982) and the troglomorphic, cave-limited populations of characid fishes (*Astyanax fasciatus*) in east-central Mexico (Mitchell et al., 1977; Wilkens and Hüppop, 1986).

SPECIATION IN TROGLOBITIC FAUNAS

Limited space here prevents all but brief examination and discussion of several common patterns of speciation in cave faunas. Detailed syntheses of data relevant to speciation have been provided in review papers by Sbordoni (1982), using genetic (biochemical) data, and by Barr and Holsinger (1985) in relation to biogeographical patterns. Aquatic and terrestrial cave faunas from both Europe and North America are analyzed in these reviews. Additional discussions of speciation have been provided by Juberthie (1989) for island faunas and by Stoch (1995) for groundwater crustaceans.

Speciation in cave faunas may occur at or just below the epigean–hypogean interface when immigrants from surface populations successfully colonize subterranean habitats, and in the subterranean realm when established cave faunas diversify. These two speciation events, which may be termed "Phase 1" and "Phase 2", differ in a number of important ways. In addition to dissimilarities in their timing and in the composition and size of founder populations, the various factors that promote isolation and initiate genetic changes differ completely between boundary areas at the epigean–hypogean interface and the deeper recesses of subterranean habitats.

The first type of speciation has received more attention than the second in previous discussions of speciation in cave animals, due in part to the fact that it involves the evolutionary changes that occur during invasion and colonization, which have been extensively discussed in the literature. While Phase 1 speciation would be allopatric under the climatic-relict model, it would conceivably be parapatric or even sympatric under the adaptive-shift model. With few exceptions, Phase 1 speciation models for aquatic faunas are allopatric, primarily because they are constructed around physically isolating or vicariant events such as stranding, spring failure, stream capture, etc. (Barr and Holsinger, 1985; Holsinger and Culver, 1988; Notenboom, 1991; see also pp. 405–406 above). One exception, however, may be the cave-limited populations of *Astyanax fasciatus* in Mexico, which, while they originated as a result of stream capture, are believed to be evolving by a kind of sympatric speciation involving disruptive selection (Wilkens and Hüppop, 1986).

Phase-2 speciation involves diversification of lineages in subterranean adaptive zones, and in terms of the number of species generated it is probably more important than the Phase-1 process. It is attributed to what Christiansen (1992) called "the capacity to evolve lineages of increasingly troglomorphic and competitive species within caves." Stoch (1995) referred to it as "diversification in [cave] adaptive zones," and suggested that a test can be performed to see if diversification increases in subterranean habitats (as opposed to epigean ones) by comparing the species richness of two sister groups, one epigean and the other hypogean. Most models proposed for Phase-2 speciation events are allopatric, and based on the premise that local (cave) populations become isolated with respect to gene flow. The phylogenies of these speciating groups should be amenable to cladistic analysis and therefore testable, provided that outgroups can be identified for the relevant taxa.

In general, observations on the distribution patterns of troglobites in Europe, North America and elsewhere (Decu and Iliffe, 1984; Barr and Holsinger, 1985; Ginet and Juberthie, 1988) suggest that invasion and colonization of subterranean environments and concomitant Phase-1 speciation entails (a) multiple invasion by a single, widespread, preadapted epigean ancestor, or (b) a single invasion by a preadapted epigean ancestor (Figs. 21.1, 21.2).

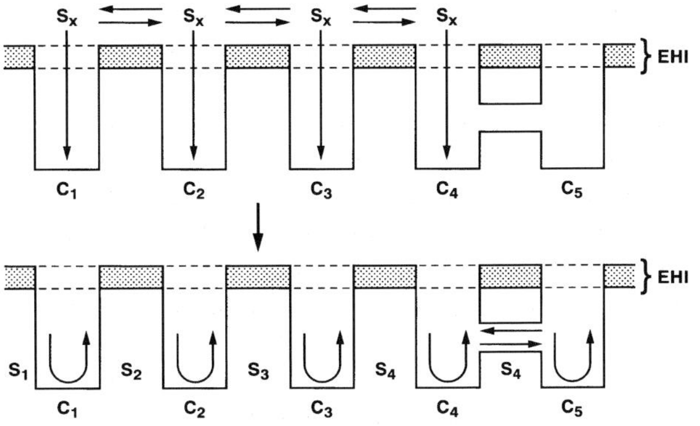

Fig. 21.1. Generalized scheme for multiple invasion and colonization of caves by a single, widespread, preadapted epigean ancestor. In the upper panel, a preadapted species (S_x) invades caves (C_1–C_4) over a broad area and newly founded populations are isolated from each other in separate caves. In the lower panel, troglobitic species (S_1–S_4) evolve in caves following speciation at or below the epigean–hypogean interface (EHI). Some species (e.g., S_4) subsequently expand their ranges by subterranean dispersal into previously uninhabited caves (C_5). The ancestral epigean species may or may not become extinct, depending on conditions at the surface and whether or not speciation is allopatric or parapatric. Arrows indicate gene flow between populations. Modified from Barr (1967, 1968) and Culver et al. (1995).

Multiple invasion may have two results depending on the extent of physical isolation that follows initial invasion. If invading populations are isolated from each other as they colonize subterranean habitats, then each founder deme theoretically has the potential to become a new cave species (viz. through a kind of peripheral isolation). Ultimately, the result would be a series of closely related congeneric species inhabiting adjacent caves – or, more likely, clusters of caves – in adjacent karst areas (Fig. 21.1). However, in some cases the species in these clusters may be genetically distinct but too poorly differentiated morphologically to be recognized as separate species by taxonomists. In some cases this similarity might be the result of convergent or parallel evolution, influenced by strong directional selection in the cave environment. In eastern North America, both the wide-ranging southern cave fish *Typhlichthys subterraneus* and the amphipod crustacean *Gammarus minus* are excellent cases in point. These two species are composed of numerous morphologically similar populations, many of which are differentiated genetically as revealed by electrophoretic analysis, which appear to have arisen through multiple invasions (Swofford, 1982; Culver et al., 1994, 1995).

Research by Culver and his co-workers (Culver et al., 1994, 1995) on the widely distributed, presumably troglophilic, amphipod *G. minus* in Appalachian caves and springs has revealed that this species consists of a complex of several species, which are recognizable biochemically but so far indistinguishable morphologically. The work on *G. minus* is particularly important because it has established the existence of a suite of undescribed allopatric sibling species, and offers a plausible explanation for the origin of these species from a common surface ancestor by independent invasions of cave streams. It is assumed that, following invasion, cave populations were physically isolated from surface populations through hydrological changes, and similar selection pressures in separate cave systems resulted in convergent morphologies.

The troglobitic millipede *Stygiochiropus communis,* which inhabits the Cape Range karst of Western Australia, is another example of a widespread cave species composed of morphologically identical but genetically distinct populations. Allozyme analysis by Humphreys and Shear (1993) indicated the occurrence of three distinct genetic provinces or clusters in areas bounded by geomorphological divides formed by deep erosional gorges. None of the clusters could be distinguished by morphological criteria. The study was unable to determine, however, whether this species complex resulted from multiple invasion of caves over a broad area, or from a restricted entry followed by dispersal, with subsequent isolation and genetic differentiation.

Simultaneous, independent invasions of separate cave or karst systems by a single ancestral species, leading to the evolution of a subset of new species, might be reflected on a cladogram by an unresolved polytomy. In contrast, successive independent invasions of separate cave systems that lead to speciation would more likely appear on a cladogram as a series of dichotomous branches. Thus, multiple invasion, whether simultaneous or successive, has the potential to produce monophyletic subsets of a number of closely related species, some of which may be nearly identical morphologically but differentiated genetically. Depending on the state of systematic knowledge for a given group, it might be possible to relate invasions to branching points on a cladogram generated from a phylogenetic analysis of the taxon under consideration.

If multiple invasion is followed by little or no physical subterranean isolation between founder populations, and members of these populations are able to disperse freely, and to maintain contact and gene exchange with each other, either underground after colonization, or by very limited dispersal through endogenous routes or the *milieu souterrain superficiel* near the surface, a single troglobitic species may result. Its range would correspond approximately to the geographic area of the original invasion. A number of widespread troglobites, such as linyphiid spiders in eastern North America (Barr and Holsinger, 1985), appear to fit this pattern, but there are other equally plausible interpretations of these distributions.

A second or even third invasion of a subterranean system by members of the same taxon (genus or family) that furnished the founders for the original colonization would greatly complicate the existing pattern, and prove difficult to decipher without a thorough knowledge of the taxonomic and phylogenetic relationships of the group. Some of the distributions described by Barr (1981, 1985) for troglobitic trechine beetles in parts of the southern Appalachians have been interpreted as good evidence for secondary and even tertiary invasions of the same cave systems by different epigean ancestors. The sympatric occurrence of representatives of different species groups of amphipods (*Stygobromus*) and isopods (*Caecidotea*), or of species in closely related genera (e.g., *Crangonyx* and *Stygobromus,* or *Caecidotea* and *Lirceus*), is additional evidence for secondary multiple invasion of subterranean habitats [examples have been given by Holsinger (1978, 1986); Lewis (1981); Holsinger and Culver (1988) and Culver et al. (1994)].

The invasion of a single subterranean system, such as a system or cluster of interconnected caves in contiguous karst, by a preadapted epigean ancestor, followed by the founding of a colonizing population, would result in a second pattern of Phase-1 speciation (Fig. 21.2). The eventual elimination of contact between the epigean and hypogean populations, either through extinction of the surface populations or through some kind of physical or ecological separation such as disruptive selection (Wilkens and Hüppop, 1986; Ashmole, 1993), would probably result in the evolution of a new cave species. If the epigean parental populations remain extant, a pair of sister species will result, one epigean and one hypogean. Ultimately, the difference between these species will depend on the amount of divergence that follows the split. Under

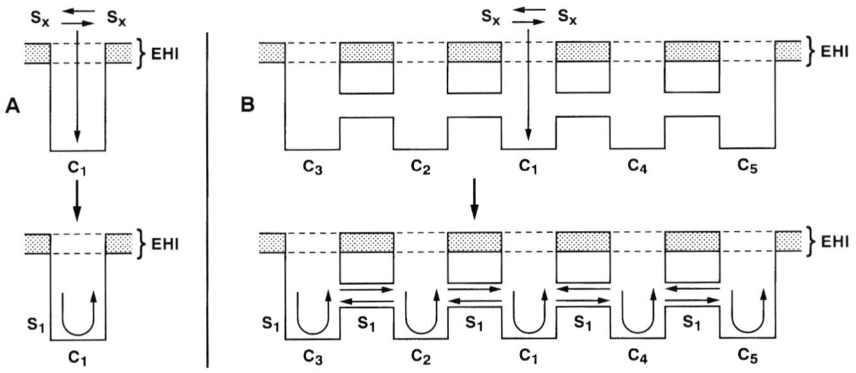

Fig. 21.2. Generalized scheme for the invasion and colonization of a single cave system by a preadapted epigean ancestor. In the upper panels of A and B, S_x is a preadapted, epigean species with a very limited geographic distribution; it invades and colonizes a single cave system (C_1). In the lower panels of A and B, a troglobitic species (S_1) evolves in caves following speciation at or below the epigean–hypogean interface (EHI) and either remains restricted to a single cave system (C_1) as shown in A, or disperses to other cave systems and becomes widely distributed as shown in B. The ancestral epigean species may or may not become extinct, depending on conditions at the surface and whether or not speciation is allopatric or parapatric. Arrows indicate gene flow between populations. Modified from Barr (1967, 1968) and Culver et al. (1995).

most conditions, the newly evolving cave species will probably diverge the most, inasmuch as it is likely to have been founded by a small, peripherally isolated population, which carried with it only a small amount of the original genetic variability of the parental species and therefore would be more prone to genetic drift. As shown in Fig. 21.2, a cave species derived through a single invasion may experience one of two fates: it will remain restricted to the area generally defined by its invasion and be recognized as a local endemic, or it will expand its range through subterranean dispersal and undergo subsequent diversification.

A large number of locally endemic troglobitic species whose ranges are confined to a small cluster of caves in a single karst area probably owe their present geographic restriction to single invasions, with very little dispersal following colonization. Numerous examples of "one-cave" endemics are found among trechine beetles, pseudoscorpions and crustaceans in the karst regions of eastern North America (Barr and Holsinger, 1985) and elsewhere, and their origin is apparently best explained by the single-invasion hypothesis.

Other cave species originate from single cave invasions, but ultimately expand their ranges by subterranean dispersal and come to occupy a broad area, encompassing many caves. The relatively widespread troglobitic trechine beetle *Neaphaenops tellkampfi* is an excellent case in point (Barr, 1979; Kane and Brunner, 1986). It is believed to have been derived through a single isolation from a surface-dwelling ancestor in southern Kentucky. Subsequently it dispersed from its place of origin and established many new cave populations, some of which are now locally isolated and constitute subspecies.

Phase-2 speciation is the multiplication of species in the subterranean realm, and involves diversification of lineages of organisms already adapted to subterranean environments. The high diversity of troglobites in many groups of cave animals, for example crustaceans (Stoch, 1995) and trechine carabid beetles (Barr, 1985; Kane and Brunner, 1986), as well as the large numbers of troglobitic species in exclusively subterranean genera of amphipods (e.g., *Bogidiella, Niphargus, Pseudoniphargus, Stygobromus*) and isopods (e.g., *Microcharon, Speocirolana, Stenasellus, Typhlocirolana*), are strong indications that speciation has continued to occur in subterranean faunas after their ancestors initially colonized the subterranean realm. It is therefore apparent that speciation in subterranean environments is an ongoing phenomenon, with the potential to generate numerous new species over time. However, surprisingly few details are available for this important phase of speciation; and, as pointed out above, most discussions of speciation in cave faunas to date have focused on the putative changes that accompany invasion and colonization.

In general, the speciation process in Phase 2 probably results from the dispersal of individuals through subterranean routes, followed by invasion and colonization of new habitat spaces, and eventual isolation from parental populations. Many studies suggest that isolation and differentiation occur underground. For example, an analysis of patterns of species distribution in the large, exclusively subterranean amphipod genus *Stygobromus* suggested that speciation in this group probably results in part from dispersal of organisms from one kind of groundwater habitat to another, followed by isolation of founder populations when previously occupied habitats are destroyed by lowered water tables, erosion, siltation, and other ongoing physical and/or ecological processes (Barr and Holsinger, 1985).

Populations of species in the predominantly European, genus *Niphargus* of subterranean amphipods are probably isolated by many of the same factors that affect *Stygobromus* (Sket, 1994). For example, *N. virei* is widespread in the groundwaters of eastern and southeastern France, but its range is disjunct between the Jura and Ardèche regions, having presumably been split by alpine glaciation. The species has failed to recolonize this area since glacier recession (Ginet and Decu, 1977; Kane et al., 1994). Populations of *N. longicaudatus* in central Italy are strongly differentiated genetically, and there is good evidence that some populations are separated by hydrological differences across short distances (Kane et al., 1994).

There are many ways in which populations of troglobitic organisms might become isolated. The most obvious physical barriers to dispersal include stratigraphic discontinuities, karst drainage divides, stream capture, deep base-level streams, destruction of cave passages by erosion, siltation, collapse, and flowstone (calcite) deposition, glaciation (as noted above), and a variety of other changes in geomorphology and hydrology (Barr and Holsinger, 1985; Stoch, 1995). Isolation may also be effected by vegetation and soil types not suitable for epigean dispersal, in cases where cave-limited species are able to disperse into shallow habitats near the surface, and some isolation may even result from the ecological requirements of the species themselves (Sbordoni, 1982; Caccone, 1985).

Dispersal and ultimately physical isolation of cave animals may be vertical as well as lateral; thus, a founding population may end up beneath rather than beside the ancestral one (see also Ashmole, 1993). Since new generations of caves are actively forming beneath old ones, it is a logical assumption that animals will move downward to occupy newly developing living space and available niches (Holsinger, 1988). Thus, the progressive isolation of populations in newly developing, lower-level passages may very well be the consequence of such activity, and is almost certain to play an important role in subterranean speciation.

The dispersal of small crustaceans, such as amphipods and isopods, through subterranean groundwaters from one kind of habitat to another is well documented (Holsinger, 1978; Culver et al., 1994). For example, in the Appalachian region amphipods in the genera *Crangonyx* and *Stygobromus* have been observed to pass from epikarstic waters of the subcutaneous zone or perched water table into cave drip pools via dripping water from cave ceilings (Holsinger, 1978). In the Organ Cave system in southeastern West Virginia, Culver and his co-workers (Culver et al., 1994) demonstrated that two species of stygobiont amphipods in the genus *Stygobromus,* and probably an isopod in the genus *Caecidotea,* have invaded small, first-order headwater streams from epikarstic waters in the overlying karst above the streams themselves. In contrast, the amphipod *Gammarus minus* has invaded the larger third-order streams in the same cave system by upstream migration directly from surface resurgences or springs. Although a majority of the invasions and re-invasions of different groundwater habitats by small crustaceans will not lead to permanent isolation of populations, it is likely that some such events will eventually fragment distributions and result in speciation.

ACKNOWLEDGMENTS

I am grateful to Drs. Fabio Stoch and Eleonora Trajano for allowing me to read and cite material from their unpublished manuscripts prior to publication and Dr. William F. Humphreys for making several suggestions that improved the manuscript. I also thank the many colleagues who have kindly exchanged valuable data and reprints on cave animals over the years and the Office of Publications and Graphics at Old Dominion University for assistance with preparation of the figures.

REFERENCES

Ashmole, N.P., 1993. Colonization of the underground environment in volcanic islands. *Mem. Biospeol.,* 20: 1–11.

Barr Jr., T.C., 1967. Observations on the ecology of caves. *Am. Nat.,* 101(922): 475–492.

Barr Jr., T.C., 1968. Cave ecology and the evolution of troglobites. *Evol. Biol.,* 2: 35–102.

Barr Jr., T.C., 1979. The taxonomy, distribution, and affinities of *Neaphaenops,* with notes on associated species of

Pseudanophthalmus (Coleoptera, Carabidae). *Am. Mus. Novit.*, 2682: 1–20.

Barr Jr., T.C., 1981. *Pseudanophthalmus* from Appalachian caves (Coleoptera: Carabidae): the *engelhardti* complex. *Brimleyana,* 5: 37–94.

Barr Jr., T.C., 1985. Pattern and process in speciation of trechine beetles in eastern North America (Coleoptera: Carabidae: Trechinae). In: G.E. Ball (Editor), *Phylogeny and Zoogeography of Beetles and Ants.* Junk, Dordrecht, pp. 350–407.

Barr Jr., T.C. and Holsinger, J.R., 1985. Speciation in cave faunas. *Annu. Rev. Ecol. Syst.,* 16: 313–337.

Bellés, X., 1991. Survival, opportunism and convenience in the processes of cave colonization by terrestrial faunas. In: J.D. Ros and N. Prat (Editors), *Homage to Ramon Margalef; or, Why There is Such Pleasure in Studying Nature. Oecol. Aquat.,* 10: 325–335.

Botosaneanu, L., 1986. General introduction. In: L. Botosaneanu (Editor), *Stygofauna Mundi.* Brill, Leiden, pp. 1–4.

Botosaneanu, L. and Holsinger, J.R., 1991. Some aspects concerning colonization of the subterranean realm – especially of subterranean waters: a response to Rouch and Danielopol, 1987. *Stygologia,* 6(1): 11–39.

Boutin, C. and Coineau, N., 1990. "Regression model," "modèle biphase" d'évolution et origine des micro-organisms stygobies interstitiels continentaux. *Rev. Micropaleontol.,* 33(3–4): 303–322.

Caccone, A., 1985. Gene flow in cave arthropods: a qualitative and quantitative approach. *Evolution,* 39(6): 1223–1235.

Carson, H.L. and Templeton, A.R., 1984. Genetic revolutions in relation to speciation phenomena: the founding of new populations. *Annu. Rev. Ecol. Syst.,* 15: 97–131.

Chapman, P., 1980. On the invertebrate fauna of caves of the Serrania de San Luis, Edo. Falcon, Venezuela. *Trans. Br. Cave Res. Assoc.,* 7(4): 179–199.

Chapman, P., 1984. The invertebrate fauna of the caves of Gunung Mulu National Park 1. *Sarawak Mus. J.,* 30: 1–18 (special issue).

Chapman, P., 1986a. Non-relictual cavernicolous invertebrates in tropical Asian and Australasian caves. *Commun. 9th Int. Congr. Speleol. (Barcelona, Spain),* 2: 161–163.

Chapman, P., 1986b. A proposal to abandon the Schiner–Racovitza classification for animals found in caves. *Commun. 9th Int. Congr. Speleol. (Barcelona, Spain),* 2: 179–182.

Christiansen, K.A., 1992. Biological processes in space and time: Cave life in the light of modern evolutionary theory. In: A.I. Camacho (Editor), *The Natural History of Biospeleology.* Monografias del Mueso National de Ciencias Naturales, Madrid, pp. 453–478.

Coineau, N. and Boutin, C., 1992. Biological processes in space and time: Colonization, evolution and speciation in interstitial stygobionts. In: A.I. Camacho (Editor), *The Natural History of Biospeleology.* Monografias del Mueso National de Ciencias Naturales, Madrid, pp. 423–451.

Coyne, J.A., 1992. Genetics and speciation. *Nature,* 355: 511–515.

Culver, D.C., 1982. *Cave Life, Evolution and Ecology.* Harvard Univ. Press, Cambridge, 189 pp.

Culver, D.C., Jones, W.K., Fong, D.W. and Kane, T.C., 1994. Organ Cave karst basin. In: J. Gibert, D.L. Danielopol and J.A. Stanford (Editors), *Groundwater Ecology.* Academic Press, New York, pp. 451–473.

Culver, D.C., Kane, T.C. and Fong, D.W., 1995. *Adaptation and Natural Selection in Caves.* Harvard University Press, Cambridge, 223 pp.

Danielopol, D.L., 1990. The origin of the anchialine cave fauna – the "deep sea" versus the "shallow water" hypothesis tested against the empirical evidence of the Thaumatocyprididae (Ostracoda). *Bijdr. Dierkd.,* 60(3/4): 137–143.

Decu, V. and Iliffe, T.M., 1984. A review of the terrestrial cavernicolous fauna of Romania. *Natl. Speleol. Soc. Bull.,* 45: 86–97.

Deeleman-Reinhold, C.L., 1981. Remarks on origin and distribution of troglobitic spiders. *Proc. 8th Int. Congr. Speleol. (Bowling Green, Kentucky),* 1: 305–308.

Delamare-Deboutteville, C., 1960. *Biologie des Eaux Souterraines Littorales et Continentales.* Hermann, Paris, 740 pp.

deMenocal, P.B., 1995. Plio-Pleistocene African climate. *Science,* 270: 53–59.

Eberhard, S.M., 1992. *The Invertebrate Cave Fauna of Tasmania: Ecology and Conservation Biology.* M.S. Thesis, University of Tasmania, 186 pp.

Franz, L.R. and Lee, D.S., 1982. Distribution and evolution of Florida's troglobitic crayfishes. *Bull. Fla. State Mus. Biol. Sci.,* 28: 53–78.

Ginet, R. and Decu, V., 1977. *Initiation à la Biologie et à l'Écologie Souterraines.* Delarge, Paris, 345 pp.

Ginet, R. and Juberthie, C., 1988. Le peuplement animal des karst de France. Deuxième partie: éléments de biogéographie pour les invertébrés terrestres. *Karstologia,* 11–12: 61–71.

Gould, S.J. and Verba, E.S., 1982. Exaptation – a missing term in the science of form. *Paleobiology,* 8: 4–15.

Hamilton-Smith, E., 1969. Studies of the Australian cavernicolous fauna. *Proc. IV Int. Congr. Speleol. Yugoslavia (1965),* 4–5: 113–120.

Holsinger, J.R., 1978. Systematics of the subterranean amphipod Stygobromus (Crangonyctidae), Part III: species of the eastern United States. *Smithson. Contrib. Zool.,* 266: 1–144.

Holsinger, J.R., 1986. Zoogeographic patterns of North American subterranean amphipod crustaceans. In: R.H. Gore and K.L. Heck (Editors), *Crustacean Biogeography. Crustacean Issues 3.* A.A. Balkema, Rotterdam, pp. 85–106.

Holsinger, J.R., 1988. Troglobites: the evolution of cave-dwelling organisms. *Am. Sci.,* 76: 146–153.

Holsinger, J.R., 1992. Two new species of the subterranean amphipod genus *Bahadzia* (Hadziidae) from the Yucatan Peninsula region of southern Mexico, with an analysis of phylogeny and biogeography of the genus. *Stygologia,* 7(2): 85–105.

Holsinger, J.R., 1994. Pattern and process in the biogeography of subterranean amphipods. In: D.C. Culver and J.R. Holsinger (Editors), *Biogeography of Subterranean Crustaceans: the Effects of Different Scales. Hydrobiologia,* 287(1): 131–145.

Holsinger, J.R. and Culver, D.C., 1988. The invertebrate cave fauna of Virginia and a part of eastern Tennessee: Zoogeography and ecology. *Brimleyana,* 14: 1–162.

Holsinger, J.R. and Longley, G., 1980. The subterranean amphipod crustacean fauna of an artesian well in Texas. *Smithson. Contrib. Zool.,* 308: 1–62.

Howarth, F.G., 1973. The cavernicolous fauna of Hawaiian lava tubes. I. Introduction. *Pac. Insects,* 15: 139–151.

Howarth, F.G., 1980. The zoogeography of specialized cave animals: A bioclimatic model. *Evolution,* 34(2): 394–406.

Howarth, F.G., 1981. Non-relictual terrestrial troglobites in the tropic Hawaiian caves. *Proc. 8th Int. Congr. Speleol. (Bowling Green, Kentucky)*, 2: 539–541.

Howarth, F.G., 1987. The evolution of non-relictual tropical troglobites. *Int. J. Speleol.*, 16: 1–16.

Howarth, F.G., 1988. Environmental ecology of north Queensland caves: or why there are so many troglobites in Australia. In: L.M. Pearson (Editor), *Reprints of the 17th Biennial Congress "Tropicon"*. Australian Speleological Federation, Sydney, pp. 76–84.

Howarth, F.G., 1993. High-stress subterranean habitats and evolutionary change in cave-inhabiting arthropods. *Am. Nat.*, 142: S65–S77.

Humphreys, W.F., 1993a. Cave fauna in semi-arid tropical Western Australia: A diverse relict wet-forest litter fauna. *Mem. Biospeol.*, 20: 105–110.

Humphreys, W.F., 1993b. The significance of the subterranean fauna in biogeographical reconstruction: examples from Cape Range peninsula, Western Australia. In: W.F. Humphreys (Editor), *The Biogeography of Cape Range, Western Australia. Rec. West. Aust. Mus. Suppl.*, 45: 165–192.

Humphreys, W.F. and Shear, W.A., 1993. Troglobitic millipedes (Diplopoda: Paradoxosomatidae) from semi-arid Cape Range, Western Australia: systematics and biology. *Invertebrate Taxonomy*, 7: 173–195.

Husmann, S., 1971. Ecological studies on freshwater meiobenthon in layers of sand and gravel. In: N.C. Hulings (Editor), *Proceedings of the First International Conference on Meiofauna. Smithson. Contrib. Zool.*, 76: 161–169.

Hutchinson, G.E., 1967. *A Treatise on Limnology, Vol. 2*. Wiley, New York, 1115 pp.

Iliffe, T.M., Wilkens, H., Parzefall, J. and Williams, D., 1984. Marine lava cave fauna: composition, biogeography, and origins. *Science*, 225: 309–311.

Jeannel, R., 1943. *Les Fossiles Vivants des Cavernes*. Editions Gallimard, Paris, 321 pp.

Jefferson, G.T., 1976. Cave faunas. In: T.D. Ford and C.H.D. Cullingford (Editors), *The Science of Speleology*. Academic Press, London, pp. 359–421.

Juberthie, C., 1984. La colonisation du milieu souterrain: théories et modèles, relations avec la spéciation et l'évolution souterraine. *Mem. Biospeol.*, 11: 65–102.

Juberthie, C., 1989. Insularité et spéciation souterraine. *Mem. Biospeol.*, 16: 3–14.

Kane, T.C. and Brunner, G.D., 1986. Geographic variation in the cave beetle *Neaphaenops tellkampfi* (Coleoptera: Carabidae). *Psyche*, 93(3–4): 231–251.

Kane, T.C., Barr Jr., T.C. and Badaracca, W.J., 1991. Cave beetle genetics: geology and gene flow. *Heredity*, 68(1992): 277–286.

Kane, T.C., Culver, D.C. and Mathieu, J., 1994. Biotic fluxes and gene flow. In: J. Gibert, D.L. Danielopol and J.A. Standford (Editors), *Groundwater Ecology*. Academic Press, New York, pp. 245–270.

Laing, C.D., Carmody, R.G. and Peck, S.B., 1976a. How common are sibling species in cave inhabiting invertebrates? *Am. Nat.*, 110: 181–184.

Laing, C.D., Carmody, R.G. and Peck, S.B., 1976b. Population genetics and evolutionary biology of the cave beetle *Ptomaphagus hirtus*. *Evolution*, 30: 484–498.

Lewis, J.J., 1981. The subterranean *Caecidotea* of the Interior Low Plateaus. *Proc. 8th Int. Congr. Speleol., (Bowling Green, Kentucky)*, 1: 234–236.

Lynch, J.D., 1988. Refugia. In: A.A. Myers and P.S. Giller (Editors), *Analytical Biogeography*. Chapman and Hall, London, pp. 311–342.

Mayr, E., 1970. *Populations, Species, and Evolution*. Harvard Univ. Press, Cambridge, 453 pp.

Mitchell, R.W., Russell, W.H. and Elliott, W.R., 1977. Mexican eyeless characin fishes, genus *Astyanax*: Environment, distribution, and evolution. *Special Publ. Mus. Texas Tech Univ.*, 12: 1–89.

Moore, B.P., 1964. Present-day cave beetle fauna in Australia. A pointer to past climatic changes. *Helictite*, 3(1): 3–9.

Notenboom, J., 1991. Marine regressions and the evolution of groundwater dwelling amphipods (Crustacea). *J. Biogeogr.*, 18: 437–454.

Oromi, J.L., Martin, A.L. and Izquierdo, I., 1991. The evolution of the hypogean fauna in the Canary Islands. In: E.C. Dudley (Editor), *The Unity of Evolutionary Biology. Proceedings 4th International Congress of Systematic and Evolutionary Biology*. Dioscorides Press, Portland, Oregon, pp. 381–395.

Parzefall, J., 1986. Behavioural preadaptations of marine species for the colonisation of caves. *Stygologia*, 2(1/2): 144–155.

Peck, S.B., 1980. Climatic change and the evolution of cave invertebrates in the Grand Canyon, Arizona. *Bull. Natl. Speleol. Soc.*, 42: 53–60.

Peck, S.B., 1981a. The geological, geographical, and environmental setting of cave faunal evolution. *Proc. 8th Int. Congr. Speleol., (Bowling Green, Kentucky)*, 2: 501–502.

Peck, S.B., 1981b. The invertebrate fauna of the caves of the Uinta Mountains, northeastern Utah. *Great Basin Nat.*, 41: 207–212.

Peck, S.B., 1988. A review of the cave fauna of Canada, and the composition and ecology of the invertebrate fauna of caves and mines in Ontario. *Can. J. Zool.*, 66: 1197–1213.

Peck, S.B. and Finston, T.L., 1993. Galapagos Islands troglobites: The questions of tropical troglobites, parapatric distributions with eyed-sister species, and their origin by parapatric speciation. *Mem. Biospeol.*, 20: 19–37.

Peck, S.B. and Lewis, J.J., 1978. Zoogeography and evolution of the subterranean invertebrate faunas of Illinois and southeastern Missouri. *Bull. Natl. Speleol. Soc.*, 40: 39–63.

Pennak, R.W., 1968. Historical origins and ramifications of interstitial investigations. *Trans. Am. Microsc. Soc.*, 87(2): 214–218.

Reddell, J.R., 1981. A review of the cavernicole fauna of Mexico, Guatemala, and Belize. *Tex. Mem. Mus. Bull.*, 27: 1–327.

Rice, W.R. and Hostert, E.E., 1993. Laboratory experiments on speciation: What have we learned in 40 years? *Evolution*, 47(6): 1637–1653.

Rouch, R. and Danielopol, D.L., 1987. L'origine de la faune aquatique souterraine, entre le paradigme du refuge et le modèle de la colonisation active. *Stygologia*, 3(4): 345–372.

Sbordoni, V., 1982. Advances in speciation of cave animals. In: C. Barigozzi (Editor), *Mechanisms of Speciation*. Alan R. Liss, New York, pp. 219–240.

Sket, B., 1986. Evolution of some taxonomically, zoogeographically, or ecologically interesting finds in the hypogean waters of Yugoslavia (in the last decades). *Commun. 9th Int. Congr. Speleol., (Barcelona, Spain)*, 2: 126–128.

Sket, B., 1994. Distribution patterns of some subterranean Crustacea in the territory of the former Yugoslavia. In: D.C. Culver and J.R.

Holsinger (Editors), *Biogeography of Subterranean Crustaceans: The effects of Different Scales. Hydrobiologia,* 287: 65–75.

Sket, B., Habic, P. and Juberthie, C., 1994. Slovenia. In: C. Juberthie and V. Decu (Editors), *Encyclopaedia Biospeologia.* Société Biospéologie, Moulis-Bucarest, pp. 801–807.

Stoch, F., 1995. The ecological and historical determinants of crustacean diversity groundwaters, or: Why are there so many species? *Mem. Biospeol.,* 22: 139–160.

Stock, J.H., 1977. The taxonomy and zoogeography of the hadziid Amphipoda, with emphasis on the West Indian taxa. *Stud. Fauna Curaçao,* 55: 1–130.

Stock, J.H., 1980. Regression model evolution as exemplified by the genus *Pseudoniphargus* (Amphipoda). *Bijdr. Dierkd.,* 50: 104–144.

Stock, J.H., 1986. Deep sea origin of cave faunas: an unlikely supposition. *Stygologia,* 2(1/2): 105–111.

Sweet, S.S., 1982. A distributional analysis of epigean populations of *Eurycea neotenes* in central Texas, with comments on the origin of troglobitic populations. *Herpetologica,* 38: 430–444.

Swofford, D.L., 1982. *Genetic Variability, Population Differentiation, and Biochemical Relationships in the Family Amblyopsidae.* M.S. Thesis, Eastern Kentucky University, Richmond, 193 pp.

Trajano, E., 1995. Evolution of tropical troglobites: Applicability of the model of Quaternary climatic fluctuations. *Mem. Biospeol.,* 22: 203–209.

Uéno, S.-I., 1977. The biospeleological importance of non-calcareous caves. *Proc. 7th Int. Congr. Speleol. (Sheffield, England),* pp. 407–408.

Vandel, A., 1964. *Biospéologie. La Biologie des Animaux Cavernicoles.* Gauthier-Villars, Paris, 619 pp.

Ward, J.V. and Palmer, M.A., 1994. Distribution patterns of interstitial freshwater meiofauna over a range of spatial scales, with emphasis on alluvial river-aquifer systems. In: D.C. Culver and J.R. Holsinger (Editors), *Biogeography of Subterranean Crustaceans: the Effects of Different Scales. Hydrobiologia,* 287(1): 147–156.

Wiley, E.O., 1978. The evolutionary species concept reconsidered. *Syst. Zool.,* 17: 17–26.

Wilkens, H., 1988. Evolution and genetics of epigean and cave *Astyanax fasciatus* (Characidae, Pices). Support for the neutral mutation theory. *Evol. Biol.,* 23: 271–367.

Wilkens, H. and Hüppop, K., 1986. Sympatric speciation in cave fishes? Studies on a mixed population of epi- and hypogean *Astyanax* (Characidae, Pisces). *Z. Zool. Syst. Evolutionforsch.,* 24: 223–230.

Yager, J. and Humphreys, W.F., 1996. *Lasionectes exleyi,* the first remipede crustacean recorded from Australia and the Indian Ocean, with a key to the world species. *Invertebrate Taxonomy,* 10: 171–187.

Chapter 22

> A ... rich area of enquiry is the study of refugia, 'ecological museums' that in both the geological record and selected modern habitats appear to house archaic biotas. ... useful insights may well be obtained from habitats as diverse as deep-water marine faunas and terrestrial swamp floras. (Conway Morris, 1995, p. 293)

RELICT FAUNAS AND THEIR DERIVATION

W.F. HUMPHREYS

INTRODUCTION

Underground ecosystems may be buffered from the vicissitudes of climate through geological time (e.g., Stock, 1986a) and hence, perhaps more than any others, have retained a broad diversity of living relicts, both phyletic and distributional. The pioneering studies of cave life were made, predominantly, in the karst areas of Europe and North America, areas directly or closely associated with regions repeatedly covered by ice during the Pleistocene. It is not surprising, then, that the effects of glaciation came to dominate the debate about both the distribution and the origin of troglobitic animals. These were viewed as the remnants of surface populations that took refuge underground from the adverse climatic conditions, which eliminated surface populations (Barr Jr, 1968; Banarescu, 1975; Sbordoni, 1982; Barr Jr and Holsinger, 1985), and where, in addition, effete lineages could survive under conditions that protected them from both the changeable climate and competition from more vigorous animals (Jeannel, 1943; Vandel, 1965). Hence, the early history of biospeleology is replete with studies of remnant distributions and taxa, respectively relicts and relics.

So dominant was this theme – that glacial cycles were the driving force for the evolution of troglobites – that it was widely held that the tropics, where such effects would be minimal, would contain few troglobites (Vandel, 1965; Barr Jr, 1968, 1973; Mitchell, 1969; see especially Sbordoni et al., Chapter 24, this volume), a view now thoroughly dispelled (Howarth, 1980, 1987, 1988; Humphreys, 1993a,b,d).

Recent work on the processes driving the evolution and distribution of troglobites has increased considerably the scope, as well as the temporal and global scale, of biospeleology. Examples of para- and peri-patric speciation are known from both aquatic and terrestrial systems (Stock, 1980; Wilkens and Hüppop, 1986; Howarth, 1987) and hence, the ablation of surface populations may not be a prerequisite for troglogenesis. Compelling evidence for sympatric speciation of cavernicolous species has been provided for the Mexican cave fish *Astyanax fasciatus* (Pisces: Characidae) (Wilkens and Hüppop, 1986) – sympatric speciation occurs on the basis that individuals with more than average cave preadaptation have a selective advantage in the cave over the epigean relatives, most cave species developing within the distribution range of their ancestral form. Many cave-living species are characterized by the fact that they start cavernicolous evolution within the distributional area of their epigean ancestral species (Wilkens and Hüppop, 1986). Furthermore, convergent adaptation to the cave environment in a number of locations has been demonstrated in the amphipod *Gammarus minus* (Culver et al., 1995).

The discovery of troglobite faunas in the finer voids within fractured substrates (Uéno, 1977; Juberthie, 1984; Oromi et al., 1991), in the superficial underground compartment (Juberthie and Delay, 1981) – the submarine equivalent is Iliffe's (1986) crevicular habitat – and in lava tubes (Howarth, 1972, 1973; Oromi et al., 1991), has both extended the type and area of study, and the range of plausible hypotheses for their dispersal and evolution. Similarly, the distributional debate has now been set in a global context. Rich tropical troglobite faunas have been discovered; and it has been hypothesised that plate tectonics have been involved in the distribution of troglobites, both in islands (Manning et al., 1986) and in continental systems (Stock, 1994). These changes in perception have provoked some of the most vigorous debates in biospeleology (Iliffe et al., 1984; Stock, 1986b; Rouch and Danielopol, 1987; Danielopol, 1990; Botosaneanu and Holsinger, 1991), in which the firm positions taken are likely to be resolved by diverse theories (Peck, 1990; Botosaneanu and Holsinger, 1991: p. 34), some even of hybrid character (see Humphreys, Chapter 30, this volume).

In many debates on relicts, confusion arises because the distinction is not clearly drawn between the process of regional vicariance and those resulting in the movement of species into the hypogean realm (e.g., Botosaneanu and Holsinger, 1991: p. 16). This is irrespective of whether the ancestors were driven underground [refugial – under constraint (cf. Botosaneanu and Holsinger, 1991)] or drawn underground [active colonization (Juberthie, 1984, 1989; Rouch and Danielopol, 1987)], or even whether the process depended on stochastic events (Wilkens and Hüppop, 1986; Fong and Culver, 1994). In this chapter I concentrate on the means by which the distributions of relict species arise; in doing so I necessarily draw selectively from an extensive literature, often citing only synoptic papers.

Darwin (1859: chapter 4) recognized the importance of vicariance and, in a brief discourse into cave life (chapter 5), he expected that conservative habitats, such as caves, which exert a low selective pressure, would be richer in "living fossils" than was known at that time. He also recognized many of the concepts important to biospeleology, namely (using descriptive terms) refugia, preadaptation, troglomorphies, parapatric speciation, regressive evolution, relics and relicts. From the inception of modern biospeleology it was recognized that Recent subterranean troglobites are relicts of pre-Quaternary faunas (Racovitza, 1907).

DEFINITIONS, SCALE, AND DISPERSAL ABILITY

In evolutionary terms a refuge is a region in which certain types or suites of organisms are able to persist during a period in which most of the original geographic range becomes uninhabitable because of climatic change (Morton et al., 1995). Continuing climatic change may result in the development of **refugia** – characterized by an accentuated long-term stability of various environmental parameters – which allow the survival of a species, through fragmentation of the refuge. This fragmentation may itself be the vicariant event, namely the division of the biota or taxon through the development of a natural barrier. Such relictual faunas are especially widespread in the subterranean world, and are represented by ecogeographic **relicts** (separated from a parent population by some vicariant event – French *relictes*) and phyletic **relics** (the last survivors of an ancient radiation: Newman, 1991 – French *reliques*). It should be noted that, while a lineage may be old, many species are likely to have evolved after the lineage invaded the hypogean environment (Wägele, 1990; Holsinger, 1994), especially in aquatic systems.

It is necessary to distinguish between the processes causing the separation of the fauna, which leads to their recognition as relicts, from those that result in the taxa entering the hypogean realm in the first place. While the same processes may be involved, their contributions must in most cases remain unproven; once a relict pattern has been established and the surface populations are, by definition, lost, one can only speculate about the processes causing the populations to move underground. I draw examples from terrestrial and aquatic faunas.

In northwestern Australia a rich terrestrial troglobite fauna occurs in humid caves in arid Cape Range, the affinities of which lie with the litter fauna of temperate and tropical rainforest now several thousand kilometres distant (Humphreys, 1993a,c). While the vicariant event was probably the loss of rainforest resulting from the onset of aridity (see Humphreys, Chapter 30, this volume), the fauna was not necessarily driven (drawn?) underground by the onset of aridity but may already have actively colonized the underground – that is, the colonization of the subterranean realm was not necessarily 'under constraint' (Botosaneanu and Holsinger, 1991).

Recent analyses of tropical cave biota have provided evidence for widespread parapatric, even sympatric, speciation (e.g. Wilkens and Hüppop, 1986; Howarth, 1987; Hoch and Howarth, 1989a, 1993); this has been argued also for glaciated regions (Rouch and Danielopol, 1987). For example, in the Hawaiian archipelago, the islands of which vary in age from 1 to 6 Ma (million years), colonization of emergent land led to explosive adaptive radiations. These resulted in swarms of closely related species exploiting different habitats, including the hypogean habitat (e.g., *Oliarus* planthoppers: Hoch and Howarth, 1993), in which troglobites independently evolved on each island, in at least ten different radiating groups (Howarth, 1987).

Howarth categorized 27 of the 43 species of troglobites present, and considered that speciation in 22% was sympatric (related species occur in the same cave), 48% parapatric (related species occur in neighbouring habitats), and 30% allopatric (related species are geographically isolated). Similarly in Australia a great diversity of *Solonaima* and *Undarana* planthoppers occurs, with closely related epigean and cavernicolous species (Hoch and Howarth, 1989a,b).

A widely disjunct, but recognizable and predictable, community inhabits anchialine waters (inland groundwater affected by marine tides and with limnic and marine influence, and usually with limited aerial connection) of northwestern Australia, the Canary Islands and the Caribbean region (Holsinger, 1989, 1992; Yager, 1987a,b, 1994; Humphreys and Adams, 1991; Poore and Humphreys, 1992; Bruce and Humphreys, 1993; Humphreys, 1993b; Knott, 1993; Yager and Humphreys, 1996). This specialized, relic and relictual fauna comprises mainly obligate stygobiont taxa containing several species congeneric between the three regions. While this distribution is consistent with the hypothesis that movement of tectonic plates was the vicariant event, and thus that the genera were already present in cave habitats, possibly as early as the Jurassic, certainly by the Middle Cretaceous (Yager and Humphreys, 1996; see also Humphreys, Chapter 30, this volume), this adds nothing to the understanding of the events that resulted in their entering, and becoming restricted to, the underground realm.

Scale

The vicariance event necessary for relictualization can occur at any scale, from the level of adjacent unconnected caves, to the global scale. Of the vicariant agents, some are physical (e.g. geological, stratigraphic, geomorphological, fluvial, stream capture, spring failure, perched water tables, plate tectonics, and the relative changes in level between land and sea that result from many processes), some are environmental (e.g., temperature, humidity, anoxia) and some temporal or merely stochastic. Clearly the effectiveness of these factors as barriers, or even their existence, depends on the biological attributes of the species (e.g., rivers may be barriers to terrestrial but not to aquatic troglobites: Barr Jr and Holsinger, 1985), and the temporal scale (e.g., plate-tectonic events are irrelevant to Quaternary relictualization).

Dispersal ability

Despite the occurrence of troglobites in the superficial underground compartment (Juberthie and Delay, 1981), such animals have very poor dispersal abilities, as is evident, *inter alia,* by the polytypic genera with high endemism (Strayer, 1994), the low frequency of sympatry of congeners (Notenboom, 1991), the lack of easily dispersed stages (e.g., larvae: Stock, 1980), low reproductive potential, the production of relatively large offspring and the long period of brood care (K-strategists: Stock, 1986a). Because of poor dispersal abilities, troglobites are excellent palaeogeographical monitors (Sbordoni, 1982; Danielopol, 1980).

MECHANISMS OF RELICTUALIZATION

Historical

The classic view of the evolution of troglobites is, in essence, that climatic instability in the epigean environment, especially the temperature and humidity changes associated with the onset and retreat of Pleistocene glaciations, led to the extinction of surface populations of species inhabiting soil and litter, mostly in the boreal forests. This disrupted gene flow between the various troglophilic populations, and permitted adaptation to cave conditions in the now relictual cave populations (Barr Jr, 1968; Sbordoni, 1982; Barr Jr and Holsinger, 1985).

The relatively dry climates of the interglacials caused epigean cryptozoic terrestrial lineages to retreat to the wetter conditions to be found at higher altitudes and latitudes. Hence, the extinction of the epigean populations permitted cave adaptation to proceed at lower latitudes and altitudes. This thesis, with aridity as the driving mechanism, has been extended globally to any area where the Pleistocene glaciations would have had climatic implications. For example, changes in the altitude range of vegetation zones associated with the onset of aridity that accompanied the glacial periods are evident in the Grand Canyon (Peck, 1980) and the Galapagos Islands (Peck, 1990).

In aquatic lineages, the onset of colder conditions during the glacial periods resulted in thermophilic lineages seeking refuge in groundwater (winter-warm conditions), while psychrophilic lineages colonized groundwater as the climate warmed and dried during interglacials (Gibert et al., 1994). Repeated cycles of glaciation eliminated surface populations and allowed cave populations to speciate.

The presence of cave-adapted species on tropical oceanic islands, and widely within the lowland tropics, has suggested that the extinction of surface populations is not necessary for the evolution of troglobites. In both northeast Australia and Hawaii closely related surface relatives of troglobites still exist, hence, rather than being relicts isolated in caves by changing climate, such troglobites seem to have become highly specialized in order to exploit resources within the system of interconnected medium-sized voids (Howarth, 1987; Hoch and Howarth, 1989b).

Terrestrial

On a global scale, the dipluran genus *Litocampa* contains 32 species in the Americas, Europe and Africa, of which 20 species are confined to caves in the United States where no epigean forms are known. The distribution and perceived retention of plesiomorphic characters suggests that the genus predates the fragmentation of Pangaea in the Mesozoic (Ferguson, 1981).

The eastern United States of America has been the focus of intense biospeleological study, and provides well-worked examples of terrestrial relictualization. There, for example, the trechine beetles (Carabidae) are the most widely represented and abundant group of terrestrial troglobites, the genus *Pseudanophthalmus* alone containing 240 species, and another seven genera containing from one to five species each. There are no obvious ancestors for the cave genera in North America, but there are oculate relatives in Eurasia. This distribution pattern suggest that the trechines of North America are relicts of formerly widespread epigean faunas (Barr Jr and Holsinger, 1985).

The evolution of cave beetles of the subfamily Bathysciinae is a good European example of relictualization of a terrestrial lineage (Juberthie, 1988). Just two examples from the extant taxa may be considered – the separation of *Batinoscelis,* known from southern Sardinia, from the species complex of *Speonomus delarouzeei* found in the eastern Pyrenees. The ancestors of the Bathysciinae inhabited tropical forest at low altitude (Jeannel, 1952) – they are currently found under relict tropical forest at an elevation of 1000 m in Tenerife (Oromi et al., 1986) in the mesovoid shallow substratum (Juberthie, 1988). During the Oligocene (35 million years ago) the Corsico-Sardinian plate abutted the Pyrénées (which then had a tropical climate), and the range of bathysciine cave beetles included southern Sardinia and the Pyrenees (Fig. 22.1a). Palaeomagnetic evidence shows that, with the opening of the western Mediterranean, the Corsico-Sardinian plate broke free in the Miocene (25–20 million years ago), rotating anti-clockwise as it moved to its present position near Italy, and isolating the Sardinian beetle population in the south of the island (Fig. 22.1b). Other populations were isolated by marine transgressions into the Rhône and Saône Valleys during the Pliocene (Juberthie, 1988). Estimates of genetic distance suggest that the genera separated about 25 million years ago (Sbordoni, 1982), in broad agreement with the palaeogeographic evidence [one may note that Sbordoni (1982) used a conversion factor between genetic distance and age that is about twice that of Juberthie (1988)]. Hence, the current distribution is explained by the close proximity of southern Sardinia and the Pyrenees during the Oligocene.

There are good-quality palaeoclimate data for the Pyrenees, and they have been used to interpret the speciation there of the *Speonomus delarouzeei* complex. Juberthie (1988) assembled a wealth of palaeoclimatic, behavioural, ecological and electrophoretic evidence to support his thesis that the three species arose owing to the effects of successive Pleistocene glaciations. During glacial periods the forests retreated to lower altitudes to be replaced by steppe, while during interglacials the forest zone migrated to higher altitudes. These repeated events served both to permit movement of the beetles through altitude, and to isolate the populations, permitting their adaptation to different thermal conditions (Juberthie, 1988).

Aridity

Analysis of pollen and pack-rat middens, for example, has allowed determination of the altitudinal migration of forest associated with climatic cycles over wide areas of the earth, not simply on the margins of glaciated regions. The association of these migrations with the altitudinal occurrence of troglobitic animals has been used to support the climatic-refuge hypothesis for the origin of troglobites (Peck, 1980, 1981b, 1990; Juberthie, 1988). Peck (1984) developed a coherent hypothesis associating speciation events (branching in the cladogram) in the *Ptomaphagus hirtus* species complex of leiodid beetles inhabiting North American caves with particular dry periods, using a combination of studies on speleothems, allozymes and hybridization (Fig. 22.2).

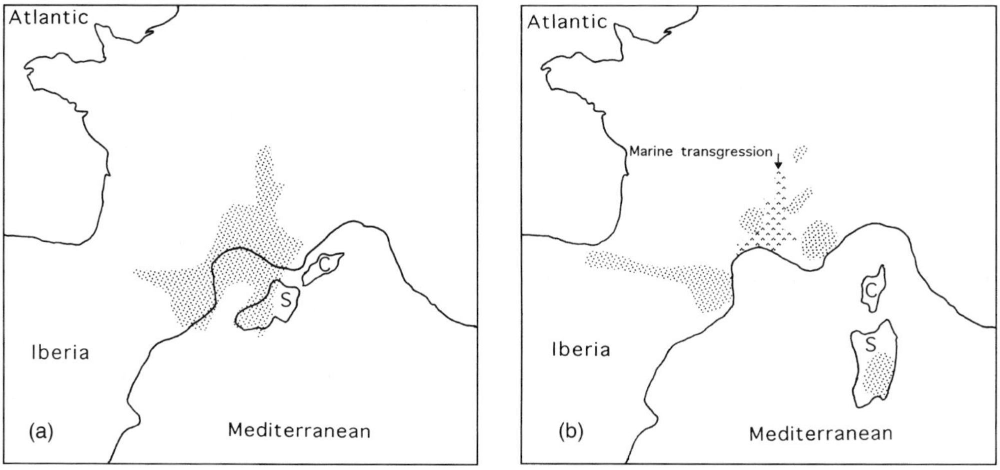

Fig. 22.1. Reconstruction of the western Mediterranean in the (a) Oligocene (35 million years ago) and (b) Late Miocene (5 million years ago), showing the fragmentation of the range of bathysciine cave beetles (dots). Marine transgression is depicted only in (b) and only for the Rhône and Saône valleys. The maps are greatly simplified from sequential plots in Juberthie (1988). C, Corsica; S, Sardinia.

In Puerto Rico the community of terrestrial troglophiles found in the arid tropical zone of the south coast is similar to that found in the moist forest zone of the north coast. The 43% similarity is primarily due to species requiring a moist environment which, on the north coast, are also to be found outside caves in leaf litter of the moist forest; this suggests that the northern habitat once extended over the southern part of the island (Peck, 1981a). An analogous argument has been applied to the community resemblance between arid-zone caves on the western seaboard of Australia and lowland rainforest on the east coast of the continent, thousands of kilometres away (Humphreys, 1993b,c).

Survival beneath continental ice sheets

There are several well-founded examples involving both terrestrial species (mites: Zacharda and Pugsley, 1988) and aquatic species (amphipods and isopods: Holsinger, 1978; Holsinger et al., 1983) where caves seem to have served as deep, subglacial refugia for some preglacial fauna, during periods when continental ice sheets removed surface populations.

Aquatic

Stranding

The stranding of faunas due to the marine regressions resulting from eustatic, orogenic or tectonic events, especially during the Late Tertiary, has been a major focus for theory on the origin of aquatic hypogean faunas as well as their relictualization. Although many studies have focused on the Late Tertiary (e.g., Por, 1986; Stock, 1986a, 1990, 1994; Boutin and Coineau, 1987), others have recognized the likely ancient vicariance in many stygofaunas (e.g., Boutin and Cals, 1985a,b; Boutin and Coineau, 1987; Rouch and Danielopol, 1987). The occurrence and distribution of aquatic hypogean species is commonly explained by marine regressions having forced their ancestors underground and stranded them inland (Stock, 1977, 1980; Wagner, 1994). The distinction between the processes driving the evolutionary process (troglogenesis) and those driving the distribution of species (vicariance, relictualization) is not always clearly established. Alternative explanations, for which there is wide-ranging evidence, include suggestions that colonization is an active process (Juberthie, 1984; Rouch and Danielopol, 1987), and that speciation may not be vicariant, but rather peripatric, parapatric, even sympatric (see pp. 418–419 above). Furthermore, much of the fauna must have already been hypogean long before the occurrence of the stranding events described. For example, there are higher taxonomic categories (classes, orders, families) whose entire membership is highly troglomorphic and for which non-troglomorphic members are unknown, including the remipedes and most genera of Thermosbaenacea (Newman, 1991). Many of these taxa contain species congeneric on either side of the Atlantic – some also in northwestern Australia – and are thus likely to have been dispersed by plate tectonics and sea-floor spreading starting in the Mesozoic. Despite this, the

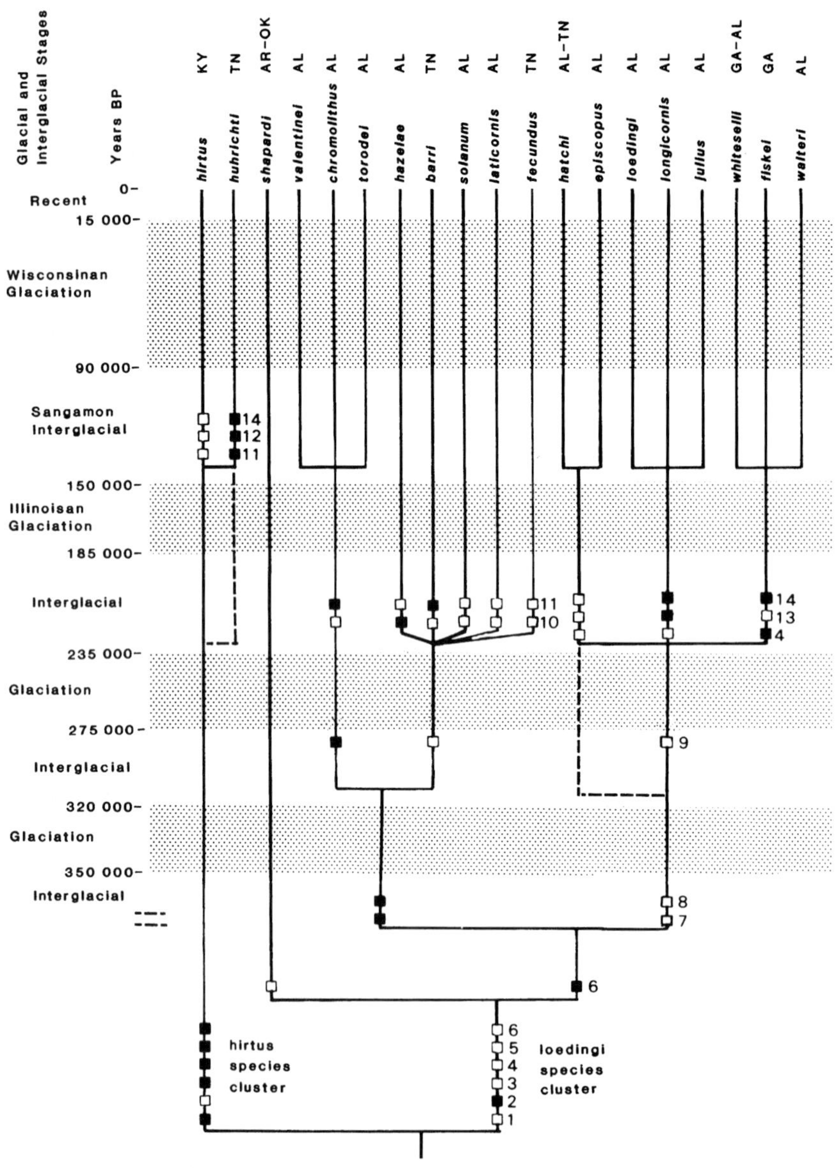

Fig. 22.2. Cladogram of the *Ptomaphagus hirtus* species group of cavernicolous leiodid beetles in the southeastern USA. The distributions are indicated by the state abbreviations (AL, Alabama; AR, Arkansas; GA, Georgia; KY, Kentucky; OK, Oklahoma; TN, Tennessee). Apomorphic character states are shown by solid squares, plesiomorphic character states by open squares. Isotope studies of cave stalagmites were used to determine the time and temperature of the glacial and interglacial periods. Branching times were determined by a combination of allozyme and hybridization studies. The lineage splits and character origin are shown at the most recent likely times for the events. From Peck (1984), with permission.

distribution of these forms is commonly explained by invoking changes in sea level during the Late Tertiary, especially during the Miocene (Stock, 1986a, 1990).

In essence the 'regression model' (Stock, 1977) is an explanation of stranding: marine transgressions allow marine stygobionts to expand their range by invading new areas and habitats, and subsequent marine regressions isolate the lineages inland where they adapt to the decreasing salinities (Fig. 22.3). Although the distributions of hadziid amphipods in the Caribbean (Stock, 1977) and *Pseudoniphargus* (Amphipoda) in Europe (Fig. 22.3) are consistent with this model, Stock (1980) recognized that the model did not account

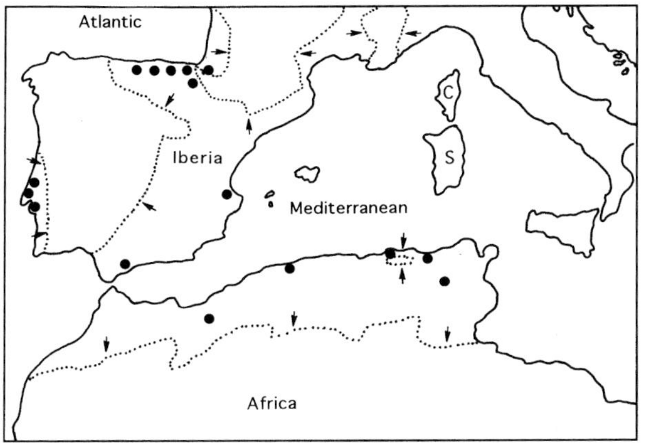

Fig. 22.3. The actual localities of *Pseudoniphargus* (points) in continental waters of Europe with the position (dotted line) of the maximum marine incursion from the Eocene to the present (arrows point inland). They are found in a wide range of habitats – island and continental, from marine/polyhaline to fresh-water at 1000 m altitude. All localities of *Pseudoniphargus* were on dry land in the Pliocene. During the Miocene most species were located on the edge of land and sea. The populations in northern Spain were submerged only during the Eocene. Much simplified from sequential plots in Stock (1980).

for the presence of congeneric species of obligate stygobionts on either side of the Atlantic.

Stock's (1977) model has been modified and extended by Boutin and Coineau (1990: *Modèle biphase*; Fig. 22.4), Notenboom (1991: Two-step model; Fig. 22.5), and Holsinger (1994: Three-step model). The similar '*biphase*' and the 'two-step' models were independently derived from different exemplars. The first step included many factors, such as active dispersal of generalized animals, a vertical transition from surface marine waters to an interstitial or crevicular littoral habitat – presumably aided by some kind of adaptive shift (*sensu* Howarth, 1987) – and passive processes, such as the isolation of interstitial freshwater animals from their marine or littoral ancestral populations. Subsequently these models essentially follow Stock's regression model.

The three-step model of Holsinger (1994) describes three levels along a continuum of hypogean adaptation, from each of which various vicariant and invasional events can proceed. This complex of models is applicable to invasion from both freshwater and marine sources and thus, at the same time, is both a generalization of, and an extension to, the two-step model. The model invokes processes of troglogenesis (direct invasion by preadapted taxa, as well as passive processes such as stranding from marine or fresh waters) and relictualization (vicariance by stream

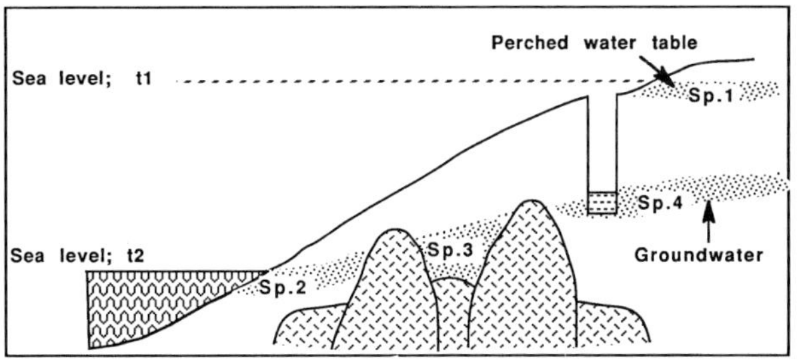

Fig. 22.4. Hypothetical origin of continental stygal species stranded inland by marine regression. Species 1 was stranded in a perched water table at the position of the beach during maximum marine transgression (t_1); species 2 followed the regressing beach to time t_2; species 4 followed the lowering groundwater level and is accessible in a well; and species 3 became isolated by impermeable strata. After Boutin and Coineau (1990).

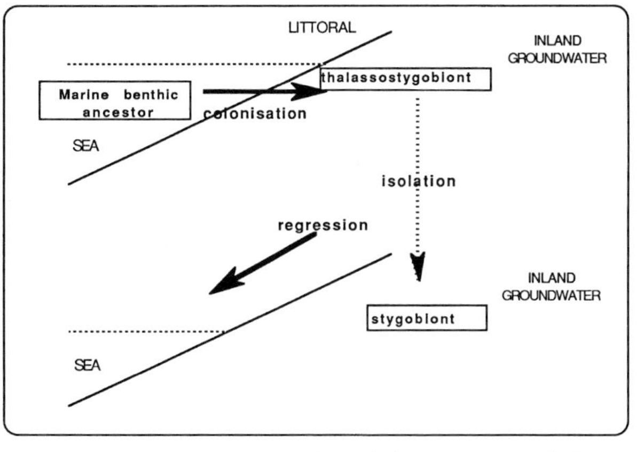

Fig. 22.5. Schematic representation of the two-step evolutionary model; of marine lineages of inland stygobionts. After Notenboom (1991).

capture, spring failure, changes in water table and marine regression).

Plate tectonics and Tethyan relicts

While Poulson and White (1969) proposed that some species had been relictual in caves for 100 milion years, Barr Jr and Holsinger (1971) argued strongly that isolation in caves was not earlier than the Late Pliocene (1.5–2.0 million years ago). The age of relictualization of hypogean fauna has always been contentious, and will remain so until adequate distributional and phyletic data, from molecular, morphological and fossil material, are analysed appropriately.

The distributions of many higher taxa of obligatory hypogean animals, because their powers of dispersal are otherwise very limited (see above, p. 419), provide evidence that the distribution of some relict faunas occurs through rafting on tectonic plates by means of seafloor spreading. As no order of terrestrial troglobites is restricted to the hypogean habitat, the most compelling evidence comes from aquatic groups, among which some Crustacea need to be considered: obligatory freshwater groups (e.g., Syncarida), or anchialine taxa (e.g., Thermosbaenacea and Remipedia).

Freshwater groups: On account of their worldwide distribution in fresh waters and absence from marine habitats, Syncarida are considered to be Pangaean (Schminke, 1974; Boutin and Coineau, 1987), as is the amphipod family Bogidiellidae (Karaman, 1982; Stock, 1977, 1983), although with a more restricted distribution. Their biogeography is explicable if they entered fresh water before the Triassic and were subsequently dispersed by tectonic-plate rafting.

It needs to be borne in mind that the apparent affinities of rare taxa can change substantially with the addition of small amounts of information; an example of this follows. For three decades the peracarid crustacean order Spelaeogriphacea was known from a single location and species – *Spelaeogriphus lepidops,* described from a freshwater stream in a cave located at an altitude of about 900 m on Table Mountain in South Africa (Gordon, 1957), and a fossil of Carboniferous age from a shallow-water marine deposit in the Atlantic Provinces of Canada. The discovery of *Potiicoara brasiliensis* from the Mato Grosso, Brazil, and *Mangkurtu mityula* from northwestern Australia, fulfilled Schram's (1974) prediction of the Gondwanan biogeographic affinities of the order. However, in 1997 two new fossil sites changed the complexion of the biogeography of the order. Modern looking species, close to *Spelaeogriphus,* were recognized from freshwater lacustrine deposits from the Lower Cretaceous of Spain and the Jurassic of China. Hence, it would appear that the order Spelaeogriphacea has Pangaean affinities (Poore and Humphreys, 1998), like the Syncarida discussed above.

There are numerous troglobitic amphipods which, owing to their limited powers of dispersal, are locally endemic and thus provide many examples of relicts, both phylogenetic and distributional (Holsinger, 1991). For example, the crangonyctids, which are exclusively freshwater and lack close marine relatives, probably had a Laurasian origin prior to the separation of North America and Eurasia in the Jurassic. Five of eight described genera are exclusively troglobitic (82% of the species), and two of the genera, *Crangonyx* and *Synurella,* occur both in Europe and North America (Holsinger, 1986).

Wägele (1990) considered that at least three groups of isopods (Aselloidea, Calabozoida, Phreatoicidea and probably the Protojaniridae) probably existed before the opening of the Atlantic. The wide distribution of Stenasellidae in subterranean freshwater habitats also suggests a broad Laurasian distribution prior to opening of the Atlantic, as well as connection between present-day Bermuda and older continental fragments (Wägele, 1990); this is an example of how the distribution of relictual taxa may have a contribution to make to the earth sciences, in that they often suggest a history at variance with current geological interpretation (Stock, 1986a,c; Notenboom, 1991).

Tethys Sea: Tethys came into existence only after the Mesozoic fragmentation of the Triassic Pangaea (Por, 1986), when it became a seaway separating the continents of Gondwana and Laurasia, and it persisted from the Triassic until the late Eocene (that is, from 200 million to 40 million years ago: Smith and Briden, 1977). Cosmopolitan elements of the Tethyan marine biota girdled the earth during the Mesozoic and through the Cretaceous (from 100 to 60 million years ago) via currents and oceanic islands of Panthalassa in the epicontinental seas (see Newman, 1991; Fig. 22.3).

The Mediterranean area is often considered to epitomize Tethys, but this part of it became landlocked, and almost completely dried out for several million years in the late Miocene (6 million years ago), leaving a series of hypersaline lakes (Por, 1986). This event, the Messinian Salinity Crisis, left only stygal habitat available for animals after marine regression (Botosaneanu and Holsinger, 1991). In consequence the circum-Mediterranean subterranean waters contain relicts (tropical elements) of the last pre-Messinian marine transgressions of the Miocene (~11 million years ago: Dimentman and Por, 1991).

Sphaeromidae are typically marine isopods, but species in the genera *Caecosphaeroma* and *Monolistra* (Monolistrini) are relicts in freshwater cave environments of southern Europe, from France to Bosnia-Herzegovina. Their distribution has been related to the location of deep, long-lasting, marine incursions during the Miocene, and their isolation and evolution linked with the Messinian Salinity Crisis. Estimates of the age of divergence using genetic-distance data support the coincidence of the salinity crisis and some of the speciation events (e.g., Sbordoni et al., 1980; Sbordoni, 1982).

Tethyan relicts: The term "Tethyan relicts" is variously used in the literature. Por (1986) referred to "Tethyan relicts" when dealing with subterranean species which have no congenerics in the open sea, and which appear in areas that were covered by the Eocene–Miocene Tethys. While accepting vicariance by plate tectonics, and the existence of congeneric species on either side of the Atlantic, he considered that the "classical case of **Miocenic** [my emphasis] amphi-atlantic patterns are those of the Thermosbaenacea, . . . [and] the hadziid amphipods." (p. 73). Similarly, Stock (1986b) stated "both [stygobiont and deep-water taxa] evolved after the Oligo–Miocene anoxia crisis in the Atlantic . . . ", although he subsequently accepted the likely role of plate tectonics in distributions (Stock, 1994).

Although there is a greater diversity of Tethyan relicts in the Indo-Pacific than in the Atlantic (Newman, 1991), several higher taxonomic groupings of Crustacea, namely the Remipedia, the Thermosbaenacea, the Spelaeogriphacea and the Mystacocarida, were thought to have exclusively circum-Atlantic distribution patterns. These are especially convincing as relictual distributions, since all members are blind, have life histories seemingly not conducive to long-range dispersal, and are restricted to hypogean habitats – anchialine, marine interstitial brackish, and freshwater. Furthermore, many contain species congeneric across the Atlantic, for instance *Speleonectes* (Remipedia); *Halosbaena* and *Monodella* (Thermosbaenacea); *Curassanthura* (Isopoda); *Gevgeliella, Psammogammarus, Pseudoniphargus, Spelaeonicippe* and *Stygogidiella* (Amphipoda).

The argument for vicariance due to the opening of the Atlantic is convincing, as their distribution on the reconstructed Triassic land-masses would be relatively modest (Fig. 22.6). However, the Remipedia (Yager and Humphreys, 1996), Thermosbaenacea (Poore and Humphreys, 1992), Spelaeogriphacea (Poore and Humphreys, 1998), and Mystacocarida (Bartsch, 1993) are now known from Western Australia. This Australian fauna contains species congeneric with members of the classically Tethyan relict fauna found on either side of the North Atlantic (*Halosbaena*: Canaries and Caribbean), or with otherwise widely disjunct distributions (*Haptolana*: Cuba, Somalia; *Lasionectes*: Turks and Caicos; *Liagoceradocus*: Canaries and Pacific (Bradbury and Williams, 1996); *Danielopolina*: North Atlantic and Pacific (Baltanas and Danielopol, 1995). In consequence, the easy acceptance of a rather constrained distribution pattern represented on a reconstructed Triassic land-mass (Fig. 22.6), gives way to a renewed broad geographic range, much broader even than the current, non-reconstructed, amphiatlantic distribution. Furthermore, many being obligate stygofauna, the genera may have been present in the stygofauna before the isolation of the Australian plate (at least 100 million years ago: Cande and Mutter, 1982).

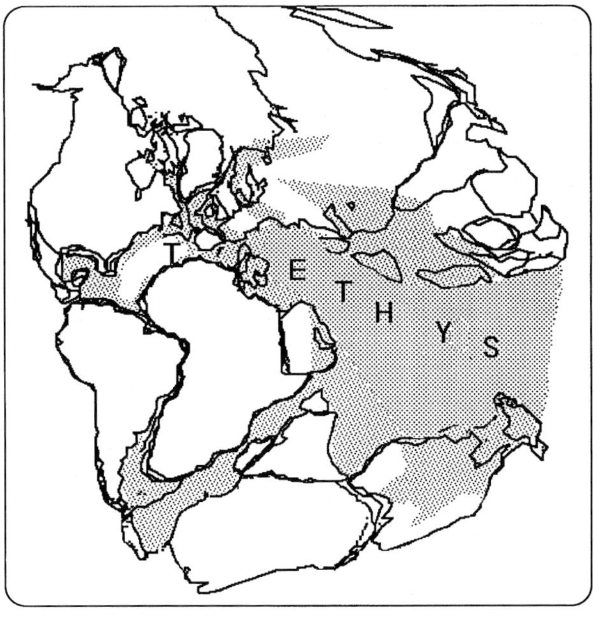

Fig. 22.6. The relative location of the continental plates and the extent of the epicontinental seas in the Jurassic (151 million years ago), when the Tethys Sea first connected the present Caribbean, Mediterranean and Indian Ocean regions. Even with this reconstruction the genera *Lasionectes* (Remipedia), *Halosbaena* (Thermosbaenacea) and *Haptolana* (Isopoda: Cirolanidae) (Remipedia), known from northwestern Australia, have widely disjunct distributions, variously including Somalia, the Canary and Caribbean islands. The plate map was prepared by the technique of Scotese and Denham (1988).

The distribution of remipedes and their associated fauna can be explained by cave colonization along the shores of Tethys during the Triassic and Jurassic 225 to 160 million years ago (Cals and Boutin, 1985), and subsequent dispersal by the break-up of Pangaea in the Mesozoic (Wilkens et al., 1986) through sea-floor spreading and continental drift (Stock and Longley, 1981; Iliffe et al., 1984; Hart et al., 1985; Wägele and Brandt, 1985). During the early Cretaceous the edge of the plate now lying in northwestern Australia formed the eastern shore of Greater Tethys (Howarth, 1981).

Fossil ancestors of *Danielopolina* occur in the fillings of submarine fissures, showing that they had already colonized crevicular systems by the Late Jurassic (Aubrecht and Kozur, 1995). The Australian species of *Danielopolina,* an anchialine species, constitutes a sister group to the remaining eight species, which include a deep-sea species, *D. carolynae,* from the Atlantic; this refutes (Baltanas and Danielopol, 1995) a deep-sea ancestry for the genus (Boxshall, 1989), and is inconsistent with the general model for the colonization of the anchialine habitat during the Tertiary proposed by Stock (1986a,b). Similarly, the Australian species of *Halosbaena* (Thermosbaenacea) constitutes a sister group for the other two species in the genus, known from the Canary Islands and the Caribbean (Wagner, 1994; H.P. Wagner, pers. comm.).

Together, these analyses indicate that the Australian populations underwent vicariance before those on either side of the Atlantic, which would support the proposed sequence in the break-up of Pangaea – that the Atlantic started to open up ~170 million years ago eastern and western Gondwana had almost separated by the Upper Jurassic (145 million years ago: fig. 13.9 in Howarth, 1981), were well separated by the Lower Cretaceous (~125 million years ago: fig. 13.12 in Howarth, 1981), and by the Upper Cretaceous (~80 million years ago – probably even by the Middle Cretaceous, 100 million years ago: Cande and Mutter, 1982) Australia was not even joined by epicontinental seas to land-masses other than Antarctica (fig. 13.15 in Howarth, 1981). This concordance is consistent with the hypothesis that the genera rafted on their respective tectonic plates, in which case there has been remarkable morphological conservatism following a very ancient vicariant event.

Oceanic islands, deep-sea ancestry and anoxia

The presence of congeneric species of obligate cavernicoles on continental land-masses, as well as on islands on mid-oceanic ridges, has always presented problems in explaining the overall distribution of the fauna (e.g., Sket and Iliffe, 1980; Iliffe et al., 1983, 1984; Hart et al., 1985; Wilkens et al., 1986). Indeed, some taxa are so widespread as seemingly to defy simple explanation. For example, some genera belonging to higher taxa of apparently monophyletic lineages of obligate subterranean freshwater inhabitants – for instance, the Bathynellacea at a continental level (Schminke, 1974) – are present on widely separated mid-oceanic-ridge islands and on continental land-masses. The anthurid isopod *Stygocyathura* occurs widely in the Caribbean region and the western Pacific (Wägele et al., 1987); other examples have been given by Hart et al. (1985). These distributions are consistent with the hypothesis that the isolation of the lineages on oceanic islands (and hence also continental vicariance) is associated with the mid-oceanic ridges and the fragmentation of the continents.

To account for the vicariant distribution of apparently ancient relict fauna and its presence on oceanic islands it has been suggested that a connection may be made, between the deep-sea and anchialine cave fauna, through suitable habitats in submerged rock, namely the crevicular system (Maciolek, 1983; Hart et al., 1985). Recent findings have suggested that many of these groups dispersed by sea-floor spreading when the continents fragmented (Hart et al., 1985; Boxshall, 1989), and that some were isolated on islands, or in crevicular systems associated with the mid-oceanic ridges (Hart et al., 1985; Boxshall, 1989), but there is no direct evidence for such crevicular habitats except in very shallow waters. There are, for example, many common elements on Bermuda, the Bahamas and Ascension, all of which have a common origin in the mid-Atlantic ridge in the late Cretaceous (110 million years ago: Manning et al., 1986). Boxshall (1989) suggested that a continuity of the crevicular habitat through geological time along the mid-oceanic ridges may have permitted the dispersal of misophrioid copepods – e.g., *Expansophria* and *Speleophria* both having species in anchialine habitats on Atlantic and Pacific islands.

Boxshall (1989) has provided the most compelling evidence for a deep-sea ancestry for misophrioid copepods from inland marine caves (Fig. 22.7), suggesting vicariance events in three dimensions – temporal, horizontal (movement from hot spots by sea-floor spreading) and vertical (rising of islands). In this context, it is worth noting that sulphide-oxidizing mixotrophic bacteria of the genera *Beggiatoa* and *Thiothrix,* which may be primary producers using sulphide-based chemosynthesis (Brigmon et al., 1994; cf. Kane et al., 1994), occur widely in anchialine and other groundwaters (Yager, 1991; Brigmon et al., 1994), as well as in deep-sea hydrothermal vents (Belkin and Jannasch, 1989). However, as occurred with *Danielopolina* (Baltanas and Danielopol, 1995), a single new observation could reverse the argument.

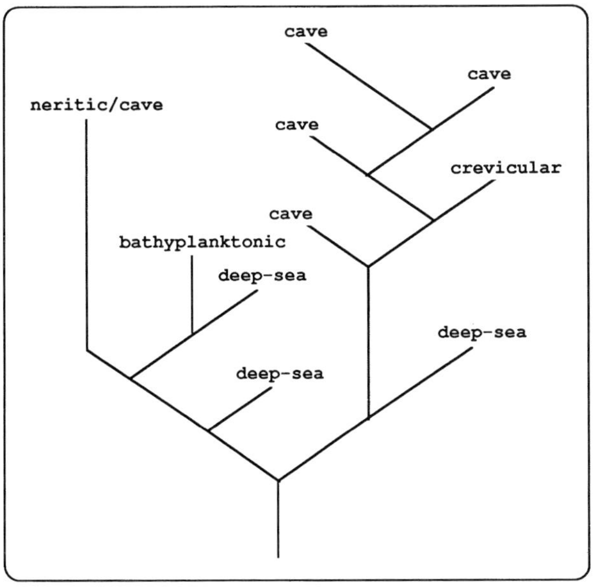

Fig. 22.7. Habitat cladogram which supports the deep-sea ancestry of the genera of misophrioid copepods. After Boxshall (1989).

Data from a number of taxa are also consistent with this hypothesis (Iliffe et al., 1983, 1984; Hart et al., 1985; Wilkens et al., 1986; Boxshall, 1989; Iliffe, 1990). However, it is not consistent with the distribution of *Bahadzia* (Hadziidae) (Holsinger, 1992), or thaumatocyprid ostracods generally (Danielopol, 1990), and for the genus *Danielopolina* has been both proposed (Wilkens et al., 1986; Boxshall, 1989) and refuted (Baltanas and Danielopol, 1995).

Stock (1986a,b) strongly criticized the concept of deep-sea ancestry for the anchialine fauna and sought the origin of many stygobionts in the Oligocene and Miocene. He argued that the cooling of the oceans and the associated periods of anoxia in mid-ocean waters led to the extinction of the deep-sea Tethyan fauna after the Eocene, so that the abyssal animals must have been derived subsequently from shallow-water lineages. Hence, the presence of similar taxa on either side of the Atlantic, for example, points to a common origin, but not a derivation from deep biotas.

Stock (1986a,b) offered no explanation for amphiatlantic (or more widespread) congeners, the distribution of which cannot be explained by dispersal along deep shelf waters. However, under the constraints of his hypothesis for a Tertiary origin of much of the anchialine fauna, it would need to have dispersed through the deep oceans, and could, perhaps, have followed the surface of the anoxic zone, as do many extant anchialine forms. Stock himself suggested that animals that prefer, or are capable of withstanding, oligoxic conditions would have found optimal habitats in the deep shelf waters just above the anoxic layer. Many taxa in anchialine habitats, including remipedes, thermosbaenaceans, amphipods and isopods, are found in the highest densities in, or are even restricted to, waters very deficient in oxygen ($<1\,\mathrm{mg}\,\ell^{-1}$: Holsinger and Yager, 1985; Stock, 1986a, 1994; Yager and Humphreys, 1996). "There are ... intervals of geological time where a non-uniformitarian approach that goes beyond familiar appeals to climatic change and sea-level fluctuations may yield rich rewards" (Conway Morris, 1995, p. 293).

It should be noted, however, that periods of mid-oceanic anoxia ($<0.2\,\mathrm{mg}\,\ell^{-1}$ oxygen) were not especially associated with the Eocene, but occurred regularly throughout the Mesozoic, with a major anoxic episode in the Albian (~100 million years ago: Fischer and Arthur, 1977). Furthermore, while the oceans were anoxic at intermediate depths, the deep seas were oxic at the same time (Fischer and Arthur, 1977; Kaiho, 1991), and for at least 90 million years (Childress, 1995), and these processes then, as now (Kennett and Ingram, 1995), occurred synchronously in all tropical oceans (Kaiho, 1991). Stock (1994) subsequently accepted the likely role of plate tectonics in the broad disjunctions seen, for example, in *Namanereis hummelincki* (Polychaeta), *Curassanthura* and *Ingolfiella* (Isopoda), *Metaniphargus* and *Pseudoniphargus* (Amphipoda) and *Procaris* (Macrura). However, *Procaris* occurs on Christmas Island (Indian Ocean), which is not associated with a mid-oceanic ridge (Humphreys and Eberhard, 1998).

TRACK ANALYSIS

One of the tenets of vicariance biogeography is that the present biotic distribution resulted from the fragmentation of earlier continuous ranges of putative ancestors. As the same eco-geographic isolating events should affect disparate taxa in similar ways, these taxa should exhibit similar (covariant) distribution patterns – the generalized tracks of Croizat et al. (1974). As relictualization seems to be such a dominant theme in the story of hypogean faunas, one would expect generalized tracks to be a common theme. Holsinger (1989) used them in his analysis of the group of genera which occur almost exclusively in anchialine caves in the Caribbean area, where they occur beneath a halocline in water with a very low oxygen content – *Danielopolina* (Thaumatocyprididae: Ostracoda), *Speleonectes* (Remipedia), *Tulumella* (Thermosbaenacea), *Bahalana* (Cirolanidae: Isopoda), *Bahadzia* (Hadziidae: Amphipoda) and *Agostocaris* (Agostocarididae; Decapoda). The broadly overlapping ranges of these genera, and their shared areas of endemism, suggests that the same fragmentation of an earlier continuous range of putative ancestors was caused by the same isolating events. At a higher taxonomic level (ostracods, remipedes, thermosbaenaceans, cirolanid isopods, gammarid amphipods and atyid shrimps), this community structure is characteristic (Holsinger, 1989, 1992; Yager, 1994) and has led to successful prediction of other members in northwestern Australia (Poore and Humphreys, 1992; Wagner, 1994) involving the genera *Danielopolina, Halosbaena, Haptolana, Lasionectes* and *Liagoceradocus.*

PROSPECT

It is generally assumed that the colonization of the hypogean realm is one-way, consistent with a traditionalist view that it houses "archaic biotas that have been displaced by competitive pressures" (Conway Morris, 1995, p. 293).

If many lineages were already present in the hypogean realm at the time that Pangaea fragmented, then some lineages could have become hypogean in association with the major global climate change at the end of the Palaeozoic; from 270 million to 200 million years ago, the atmospheric oxygen concentration fell from ~36% to 15%, with a concomitant four fold increase in atmospheric carbon dioxide (Conway Morris, 1995). In any case the hypogean fauna would subsequently have survived a number of anoxic episodes in the mid-ocean levels throughout the Mesozoic and Cainozoic (Fischer and Arthur, 1977), as well as the mass extinctions at the end of the Cretaceous which are associated with meteorite impacts. A major impact – Chicxulub on the Yucatán Peninsula – deposited a thick layer of ejecta throughout the Caribbean (fig. 9 in Hilderbrand, 1993), a stronghold of troglobitic fauna. The thick sequence of carbonate and sulphate rocks at the impact site increased the potential for disruptive climatic effects, and the resulting acid rain may have caused the extinction of both terrestrial and aquatic fauna (Hilderbrand, 1993). However, the massive carbonate deposits, in which most subterranean fauna are found, would neutralize the effects of acid rain, and would have increased further, over and above the protection provided by their inhabiting subterranean systems, the chance that some lineages may have survived such a catastrophic event; one wonders whether they could have re-emerged to populate the non-hypogean world and re-acquire epigean traits, in violation of Dollo's and Arber's laws (Lincoln et al., 1982) of the irreversibility of evolution.

Such a counterintuitive proposition finds support elsewhere; for example:

(1) A number of epigean species of Australian melitid and crangonyctoid amphipods are thought to have an hypogean ancestry in view of their lack of eyes (J.H. Bradbury, personal communication, 1996);
(2) *Niphargus pectencoronatae* is a limnic immigrant into anchialine habitats (Sket, 1996);
(3) The fact that the freshwater isopod genus *Microcerberus* is found on either side of the Atlantic Ocean supports the early divergence of the stygofauna (>100 million years ago). Moreover, importantly, Wägele et al. (1995) have concluded that the coastal mesopsammal taxa are derived from freshwater taxa and not, as is postulated in most models, *vice versa.*
(4) Culver et al. (1995, fig. 8.4) have provided strong evidence that some karst-window populations of *Gammarus minus* in West Virginia have evolved from cave populations, and, in the process, have re-acquired compound eyes, in apparent violation of Dollo's law. However, it is now recognized that eyes can evolve quickly (Osorio et al., 1996), and that the evolution of morphological complexity may commonly be reversed (see Gould, 1996).

The view expressed by Gibert et al. (1994) – that groundwater systems are not sparsely populated by rare relict forms, but are dynamic ecosystems responding to a plethora of ecological processes similar to those occurring on the surface – may have more widespread relevance in the hypogean realm.

ACKNOWLEDGEMENT

I thank Caroline Lawrence for drawing many of the figures.

REFERENCES

Aubrecht, R. and Kozur, H., 1995. *Pokornyopsis* (Ostracoda) from submarine fissure fillings and cavities in the Late Jurassic of

Czorsztyn Unit and the possible origin of the Recent anchialine faunas. *Neues Jahrb. Geol. Palaeontol. Abh.*, 196: 1–17.

Baltanas, A. and Danielopol, D.L., 1995. Cladistic analysis of *Danielopolina* species (Ostracoda: Thaumatocyprididae) and the origin of anchialine fauna. *Mitt. Hamb. Zool. Mus. Inst.*, 92: 315–324.

Banarescu, P., 1975. *Principles and Problems of Zoogeography.* Nolit, Belgrade, 212 pp.

Barr Jr., T.C., 1968. Cave ecology and the evolution of troglobites. In: T.H. Dobzhansky, M.K. Hecht and W.C. Steere (Editors), *Evolutionary Biology, Vol. 2.* Plenum Press, New York, pp. 35–102.

Barr Jr., T.C., 1973. Refugees of the Ice Age. *Nat. Hist.*, 1973(May): 26–35.

Barr Jr., T.C. and Holsinger, J.R., 1971. Biogeography of troglobites. *Natl. Speleol. Soc. Bull.*, 33: 115–118.

Barr Jr., T.C. and Holsinger, J.R., 1985. Speciation in cave faunas. *Annu. Rev. Ecol. Syst.*, 16: 313–337.

Bartsch, I., 1993. *Ctecheilocaris echra,* a new mystacocarid (Crustacea) from Western Australia. In: F.E. Wells, D.I. Walker, H. Kirkman and R. Lethbridge (Editors), *The Marine Flora and Fauna of Rottnest Island, Western Australia,* Vol. 1. Western Australian Museum, Perth, pp. 105–111.

Belkin, S. and Jannasch, H.W., 1989. Microbial mats at deep-sea hydrothermal vents: new observations. In: Y. Cohen and E. Rosenberg (Editors), *Microbial Mats: Physiological Ecology of Benthic Microbial Communities.* American Society for Microbiology, Washington, pp. 16–21.

Botosaneanu, L. and Holsinger, J.R., 1991. Some aspects concerning colonisation of the subterranean realm – especially of subterranean waters: a response to Rouch & Danielopol. *Stygologia,* 6: 11–39.

Boutin, C. and Cals, P., 1985a. Stygobiologie et corrélations biostratigraphiques transatlantiques: L'apport de la connaissance des Mystacocarides et des Thermosbaenacés actuels (Crustacés des eaux souterraines interstitielles). *Rev. Fac. Sci. Marrakech, Sect. Sci. Terre,* 183: 111–126.

Boutin, C. and Cals, P., 1985b. Importance en biogéographie évolutive de la découverte d'un crustacé phréatobie, *Modella atlantomaroccana* n. sp. (Thermosbaenacea) dans la plaine alluviale de Marrakech (Maroc atlantique). *C.R. Acad. Sci. Paris, Sér. III,* 300(7): 267–270.

Boutin, C. and Coineau, N., 1987. *Iberobathynella* (Crustacea, Syncarida, Bathynellacea) sur le continent africain. Implications paléobiogéographiques. *C.R. Acad. Sci. Paris, Sér. III,* 304(13): 355–357.

Boutin, C. and Coineau, N., 1990. "Regression Model", "Modèle Biphase" d'évolution et origine des micro-organismes stygobies interstitiels continentaux. *Rev. Micropaléontol.*, 33: 303–322.

Boxshall, G.A., 1989. Colonisation of inland marine caves by misophrioid copepods. *J. Zool. London,* 219: 521–526.

Bradbury, J.H. and Williams, W.D., 1996. Two new species of anchialine amphipod (Crustacea: Hadziidae: *Liagoceradocus*) from Western Australia. *Rec. West. Aust. Mus.*, 17: 395–409.

Brigmon, R.L., Martin, H.W., Morris, T.L., Bitton, G. and Zam, S.G., 1994. Biogeochemical ecology of *Thiothrix* spp. in underwater limestone caves. *Geomicrobiol. J.*, 12: 141–159.

Bruce, N.L. and Humphreys, W.F., 1993. *Haptolana pholeta* sp. nov., the first subterranean flabelliferan isopod crustacean (Cirolanidae) from Australia. *Invertebr. Taxon.*, 7: 875–884.

Cals, P. and Boutin, C., 1985. Découverte au Cambodge, domaine ancien de la Tethys orientale, d'un nouveau 'fossile vivant' *Theosbaena cambodjiana* n.g., n. sp. (Crustacea, Thermosbaenacea). *C.R. Acad. Sci. Paris, Sér. III* 300(8): 337–340.

Cande, S.C. and Mutter, J.C., 1982. A revised identification of the oldest sea-floor spreading anomalies between Australia and Antarctica. *Earth Planet. Sci. Lett.*, 58: 151–160.

Childress, J.J., 1995. Are there physiological and biochemical adaptations of metabolism in deep-sea animals? *Trends Ecol. Evol.*, 10: 30–36.

Conway Morris, S., 1995. Ecology in deep time. *Trends Ecol. Evol.*, 10(7): 290–294.

Croizat, L., Nelson, G. and Rosen, D.E., 1974. Centers of origin and related concepts. *Syst. Zool.*, 23: 265–287.

Culver, D.C., Kane, T.C. and Fong, D.W., 1995. *Adaptation and Natural Selection in Caves: the Evolution of Gammarus minus.* Harvard University Press, Cambridge, MA, 189 pp.

Danielopol, D.L., 1980. An essay to assess the age of the freshwater interstitial ostracods of Europe. *Bijdr. Dierkd.*, 50: 243–291.

Danielopol, D.L., 1990. The origin of the anchialine cave fauna – the "deep sea" versus the "shallow water" hypothesis tested against the empirical evidence of the Thaumatocyprididae (Ostracoda). *Bijdr. Dierkd.*, 60: 137–143.

Darwin, C., 1859. *On the Origin of Species by Means of Natural Selection.* Murray, London, 524 pp.

Dimentman, C. and Por, F.D., 1991. The origin of the subterranean fauna of the Jordan–Dead Sea Rift Valley: new data. *Stygologia,* 6: 155–164.

Ferguson, L.M., 1981. *Systematics, evolution, and zoogeography of the cavernicolous campodeids of the genus Lithocampa (Diplura: Campodeidae) in the United States.* Ph.D. Dissertation. Virginia Polytechnical Institute and State University, Blacksburg, VA.

Fischer, A.G. and Arthur, M.A., 1977. Secular variations in the pelagic realm. *Soc. Econ. Paleontol. Mineral. Spec. Publ.*, 25: 19–50.

Fong, D.W. and Culver, D.C., 1994. Fine scale biogeographic differences in the crustacean fauna of a cave system in West Virginia, USA. *Hydrobiologia,* 287: 29–37.

Gibert, J., Stanford, J.A., Dole-Olivier, M.J. and Ward, J.V., 1994. Basic attributes of groundwater ecosystems and prospects for research. In: J. Gibert, D.L. Danielopol and J.A. Stanford (Editors), *Groundwater Ecology.* Academic Press, London, pp. –40.

Gordon, I., 1957. On *Spelaeogriphus,* a new cavernicolous crustacean from South Africa. *Bull. Br. Mus. Nat. Hist. Zool.*, 5: 31–47.

Gould, S.J., 1996. *Full House: the Spread of Excellence from Plato to Darwin.* Harmony Books, USA.

Hart, C.W., Manning, R.B. and Iliffe, T.M., 1985. The fauna of Atlantic marine caves: evidence of dispersal by sea floor spreading while maintaining ties to deep waters. *Proc. Biol. Soc. Washington,* 98: 288–292.

Hilderbrand, A.R., 1993. The Cretaceous/Tertiary boundary impact (or the dinosaurs didn't have a chance). *J.R. Astron. Soc. Canada,* 87: 77–118.

Hoch, H. and Howarth, F.G., 1989a. Six new cavernicolous cixiid planthoppers in the genus *Solonaima* from Australia (Homoptera: Fulgoridea). *Syst. Entomol.*, 14: 377–402.

Hoch, H. and Howarth, F.G., 1989b. The evolution of cave-adapted cixiid planthoppers in volcanic and limestone caves in North

Queensland, Australia (Homoptera: Fulgoroidea). *Mém. Biospéol.*, 16: 17–24.

Hoch, H. and Howarth, F.G., 1993. Evolutionary dynamics of behavioral divergence among populations of the Hawaiian cave-dwelling planthopper *Oliarus polyphemus* (Homoptera: Fulgoroidea: Cixiidae). *Pac. Sci.*, 47(4): 303–318.

Holsinger, J.R., 1978. Systematics of the subterranean amphipod genus *Stygobromus* (Crangonyctidae). II. Species of the eastern United States. *Smithson. Contrib. Zool.*, No. 266, 144 pp.

Holsinger, J.R., 1986. Holarctic crangonyctid amphipods. In: L. Botosaneanu (Editor), *Stygofauna Mundi: A Faunistic, Distributional, and Ecological Synthesis of the World Fauna Inhabiting Subterranean Waters (including the Marine Interstitial).* E.J. Brill, Leiden, pp. 535–549.

Holsinger, J.R., 1989. Preliminary zoogeographic analysis of five groups of crustaceans from anchialine caves in the West Indian region. In: *Proceedings of the 10th International Congress of Speleology, Budapest,* Vol. 2, pp. 25–26.

Holsinger, J.R., 1991. What can vicariance biogeographic models tell us about the distributional history of subterranean amphipods. *Hydrobiologia,* 223: 43–45.

Holsinger, J.R., 1992. Two new species of the subterranean amphipod genus *Bahadzia* (Hadziidae) from the Yucatan Peninsula region of southern Mexico, with an analysis of phylogeny and biogeography of the genus. *Stygologia,* 7: 85–105.

Holsinger, J.R., 1994. Pattern and process in the biogeography of subterranean amphipods. *Hydrobiologia,* 287: 131–145.

Holsinger, J.R. and Yager, J., 1985. A new genus and two new species of subterranean amphipod crustaceans (Hadziidae) from the Bahamas and Turks and Caicos Islands. *Bijdr. Dierkd.*, 55: 283–294.

Holsinger, J.R., Mort, J.S. and Recklies, A.D., 1983. The subterranean crustacean fauna of Castleguard Cave, Columbia Icefields, Alberta, Canada, and its zoogeographic significance. *Arct. Alp. Res.*, 15: 543–549.

Howarth, F.G., 1972. Cavernicoles in lava tubes on the island of Hawaii. *Science,* 175: 325–326.

Howarth, F.G., 1973. The cavernicolous fauna of Hawaiian lava tubes, 1. Introduction. *Pac. Insects,* 15: 139–151.

Howarth, F.G., 1980. The zoogeography of specialized cave animals: a bioclimatic model. *Evolution,* 34: 394–406.

Howarth, F.G., 1987. The evolution of non-relictual tropical troglobites. *Int. J. Speleol.*, 16: 1–16.

Howarth, F.G., 1988. Environmental ecology of north Queensland caves: or why there are so many troglobites in Australia. In: L. Pearson (Editor), *17th Biennial Conference, Australian Speleological Federation Tropicon Conference, Lake Tinaroo, Far North Queensland, 27–31 December 1988.* Australian Speleological Federation, Cairns, pp. 76–84.

Howarth, M.K., 1981. Palaeogeography of the Mesozoic. In: L.M.R. Cooks (Editor), *The Evolving Earth.* Oxford University Press, Oxford, pp. 197–200.

Humphreys, W.F., 1993a. Cave fauna in semi-arid tropical Western Australia: a diverse relict wet-forest litter fauna. *Mém. Biospéol.*, 20: 105–110.

Humphreys, W.F., 1993b. Stygofauna in semi-arid tropical Western Australia: a Tethyan connection? *Mém. Biospéol.*, 20: 111–116.

Humphreys, W.F., 1993c. The significance of the subterranean fauna in biogeographical reconstruction: examples from Cape Range peninsula, Western Australia. *Rec. West. Aust. Mus., Suppl.,* 45: 165–192.

Humphreys, W.F. (Editor), 1993d. The biogeography of Cape Range, Western Australia. *Rec. West. Aust. Mus. Suppl.,* 45: 1–248.

Humphreys, W.F. and Adams, M., 1991. The subterranean aquatic fauna of the North West Cape peninsula, Western Australia. *Rec. West. Aust. Mus.,* 15: 383–411.

Humphreys, W.F. and Eberhard, S.M., 1998. *Assessment of the ecological values and management options for cave use on Christmas Island: Project 97/002,* Report prepared for Parks Australia North. Western Australian Museum, Perth, 134 pp.

Iliffe, T.M., 1986. The Zonation Model for the evolution of aquatic faunas in anchialine caves. *Stygologia,* 2: 2–9.

Iliffe, T.M., 1990. Crevicular dispersal of marine cave faunas. *Mém. Biospéol.,* 17: 93–96.

Iliffe, T.M., Hart Jr., C.W. and Manning, R.B., 1983. Biogeography and the caves of Bermuda. *Nature,* 302: 141–142.

Iliffe, T.M., Wilkens, H., Parzefall, J. and Williams, D., 1984. Marine lava cave fauna: composition, biogeography and origins. *Science,* 225: 309–311.

Jeannel, R., 1943. *Les Fossiles Vivants des Cavernes.* Gallimard, Paris, 321 pp.

Jeannel, R., 1952. L'évolution souterraine dans la région méditerranéenne et sur les montagnes du Kivu. *Notes Biospéol.,* 7: 7–13.

Juberthie, C., 1984. La colonisation du milieu souterrain; theories et modeles, relations avec la speciation et l'evolution souterraine. *Mém. Biospéol.,* 11: 65–102.

Juberthie, C., 1988. Paleoenvironment and speciation in the cave beetle complex *Speonomus delarouzeei* (Coleoptera, Bathysciinae). *Int. J. Speleol.,* 17: 31–50.

Juberthie, C., 1989. Insularité et speciation souterraine. *Mém. Biospéol.,* 16: 3–14.

Juberthie, C. and Delay, B., 1981. Ecological and biological implications of the existence of a Superficial Underground Compartment. *Proceedings of the 8th International Congress of Speleology, Bowling Green, Kentucky,* 6: 203–206.

Kaiho, K., 1991. Global changes of Paleogene aerobic/anaerobic benthic foraminifera and deep-sea circulation. *Palaeogeogr. Palaeoclimatol. Palaeoecol.,* 83: 65–85.

Kane, T.C., Sarbu, S.M. and Kinkle, B.K., 1994. Chemoautotrophy: methodological approaches, geological implications, and a case study from Romania. In: I.D. Sasowsky and M.V. Palmer (Editors), *Breakthroughs in Karst Geomicrobiology and Redox Geochemistry Symposium. Special Publication 1.* Karst Waters Institute, Charles Town, WV, pp. 38–39.

Karaman, G.S., 1982. Critical remarks to the recent revisions of the *Bogidiella* group of genera with some study of some taxa (Fam. Gammaridae). *Poljopr. Sumarst. Titograd,* 28: 31–57.

Kennett, J.P. and Ingram, B.L., 1995. A 20,000-year record of ocean circulation and climate change from the Santa Barbara basin. *Nature (London),* 377: 510–514.

Knott, B., 1993. Stygofauna from Cape Range peninsula, Western Australia: Tethyan relicts. *Rec. West. Aust. Mus. Suppl.,* 45: 109–127.

Lincoln, R.J., Boxshall, G.A. and Clark, P.F., 1982. *A Dictionary of Ecology, Evolution and Systematics.* Cambridge University Press, London, 298 pp.

Maciolek, J.A., 1983. Distribution and biology of Indo-Pacific insular hypogeal shrimps. *Bull. Mar. Sci.*, 33: 606–618.

Manning, R.B., Hart, C.W. and Iliffe, T.M., 1986. Mesozoic relicts in marine caves of Bermuda. *Stygologia*, 2(1/2): 156–166.

Mitchell, R.W., 1969. A comparison of temperate and tropical cave communities. *Southwest. Nat.*, 14(1): 73–88.

Morton, S.R., Short, J. and Barker, R.D., 1995. *Refugia for biological diversity in arid and semi-arid Australia. Biological diversity series*, paper No. 4. Biological Diversity Unit, Department of the Environment, Sport and Territories, Canberra, 171 pp.

Newman, W.A., 1991. Origins of southern hemisphere endemism, especially among marine crustacea. *Mem. Queensl. Mus.*, 31: 51–76.

Notenboom, J., 1991. Marine regressions and the evolution of ground dwelling amphipods (Crustacea). *J. Biogeogr.*, 18: 437–454.

Oromi, P., Medina, A.L. and Tejedor, M.L., 1986. On the existence of a superficial underground compartment in the Canary Islands. *Actas IX Congr. Int. Espeleol., Barcelona*, 2: 147–151.

Oromi, P., Martín, J.L., Medina, A.L. and Izquierdo, I., 1991. The evolution of the hypogean fauna in the Canary Islands. In: E.C. Dudley (Editor), *The Unity of Evolutionary Biology*, Vol. I. Dioscorides Press, Portland, OR, pp. 380–395.

Osorio, D., Averof, M. and Bacon, J.P., 1996. Arthropod evolution: great brains, beautiful bodies. *Trends Ecol. Evol.*, 10: 449–454.

Peck, S.B., 1980. Climatic change and the evolution of cave invertebrates in the Grand Canyon, Arizona. *Natl. Speleol. Soc. Bull.*, 42: 53–60.

Peck, S.B., 1981a. Zoogeography of invertebrate cave faunas in southwestern Puerto Rico. *Natl. Speleol. Soc. Bull.*, 43: 70–79.

Peck, S.B., 1981b. The geological, geographical, and environmental setting of cave faunal evolution. In: *Proceedings of the 8th International Congress of Speleology, Bowling Green, Kentucky*, Vol. 2, pp. 501–502.

Peck, S.B., 1984. The distribution and evolution of cavernicolous *Ptomaphagus* beetles in the southeastern United States (Coleoptera; Leiodidae; Cholevinae) with new species and records. *Can. J. Zool.*, 62: 730–740.

Peck, S.B., 1990. Eyeless arthropods of the Galapagos Islands, Ecuador: composition and origin of the cryptozoic fauna of a young, tropical, oceanic archipelago. *Biotropica*, 22: 366–381.

Poore, G.C.B. and Humphreys, W.F., 1992. First record of Thermosbaenacea (Crustacea) from the Southern Hemisphere: a new species from a cave in tropical Western Australia. *Invertebr. Taxon.*, 6: 719–725.

Poore, G.C.B. and Humphreys, W.F., 1998. First record of Spelaeogriphacea from Australasia: a new genus and species from an aquifer in the arid Pilbara of Western Australia. *Crustaceana*, 71: 721–742.

Por, F.D., 1986. Crustacean biogeography of the Late Middle Miocene Middle Eastern landbridge. In: R.H. Gore and K.L. Heck (Editors), *Crustacean Biogeography*. A.A. Balkema, Rotterdam/Boston, pp. 69–84.

Poulson, T.L. and White, W.B., 1969. The cave environment. *Science*, 165: 971–981.

Racovitza, E.G., 1907. Essai sur les problèmes biospéologiques. *Arch. Zool. Exp. Gén.*, 6: 371–488.

Rouch, R. and Danielopol, D.L., 1987. L'origine de la faune aquatique souterraine, entre le paradigme du refuge et le modèle de la colonisation active. *Stygologia*, 3: 345–372.

Sbordoni, V., 1982. Advances in speciation of cave animals. In: *Mechanisms of Speciation*. A.R. Liss, New York, pp. 219–240.

Sbordoni, V., Caccone, A., De Matthaeis, E. and Cobolli Sbordoni, M., 1980. Biochemical divergence between cavernicolous and marine Sphaeromidae and the Mediterranean salinity crisis. *Experientia*, 36: 48–50.

Schminke, H.K., 1974. Mesozoic intercontinental relationships as evidenced by bathynellid crustacea (Syncarida: Malacostraca). *Syst. Zool.*, 23: 157–164.

Schram, F.R., 1974. Paleozoic Peracarida of North America. *Fieldiana: Geol.*, 33: 95–124.

Scotese, C.R. and Denham, C.R., 1988. *Terra Mobilis: Plate Tectonics for the Macintosh*. Earth in Motion Technologies, Houston, TX.

Sket, B., 1996. The ecology of anchihaline caves. *Trends Ecol. Evol.*, 11: 221–225.

Sket, B. and Iliffe, T.M., 1980. Cave fauna of Bermuda. *Int. Rev. Gesamten Hydrobiol.*, 65: 871–882.

Smith, A.G. and Briden, J.C., 1977. *Mesozoic and Cenozoic Paleocontinental Maps*. Cambridge University Press, Cambridge, 63 pp.

Stock, J.H., 1977. The taxonomy and zoogeography of the Hadziid Amphipoda, with emphasis on the West Indian taxa. *Stud. Fauna Curaçao, Amsterdam*, 55: 1–130.

Stock, J.H., 1980. Regression model evolution as exemplified by the genus *Pseudoniphargus* (Amphipoda). *Bijdr. Dierkd.*, 50: 105–144.

Stock, J.H., 1983. Discovery of a Bogidiellid amphipod crustacean in inland waters of the East Indian Archipelago: *Bogidiella (Medigidiella) sarawacensis* n. sp. *Crustaceana*, 44: 198–204.

Stock, J.H., 1986a. Caribbean biogeography and a biological calendar of geological events. In: R.H. Gore and K.L. Heck (Editors), *Crustacean Biogeography – Crustacean Issues*, Vol. 4: 195–203.

Stock, J.H., 1986b. Deep sea origin of cave faunas: an unlikely supposition. *Stygologia*, 2: 105–111..

Stock, J.H., 1986c. Two new amphipod crustaceans of the genus *Bahadzia* from 'blue holes' in the Bahamas and some remarks on the origin of the insular stygofaunas of the Atlantic. *J. Natl. Hist.*, 20: 921–933.

Stock, J.H., 1990. Insular groundwater biotas in the (sub)tropical Atlantic: a biogeographical synthesis. *Atti Convegni Lincei, Accad. Naz. Lincei, Roma*, 85: 695–713.

Stock, J.H., 1994. Biogeographic synthesis of the insular groundwater fauna of the (sub)tropical Atlantic. *Hydrobiologia*, 287: 105–117.

Stock, J.H. and Longley, G., 1981. The generic status and distribution of *Monodella texana* Maguire, the only known north American Thermosbaenacean. *Proc. Biol. Soc., Washington*, 94: 569–578.

Strayer, D.L., 1994. Limits to biological distributions in groundwater. In: J. Gibert, D.L. Danielopol and J.A. Stanford (Editors), *Groundwater Ecology*. Academic Press, London, pp. 287–310.

Uéno, S.-I., 1977. The biospeleological importance of non-calcareous caves. In: *Proceedings of the 7th International Speleological Congress, Sheffield*, pp. 407–408.

Vandel, A., 1965. *Biospeleology: The Biology of Cavernicolous Animals*, translated by B.E. Freeman. Pergamon Press, Oxford, 524 pp.

Wägele, J.W., 1990. Aspects of the evolution and biogeography of stygobiontic Isopoda (Crustacea, Peracarida). *Bijdr. Dierkd.*, 60: 145–150.

Wägele, J.W. and Brandt, A., 1985. New west Atlantic localities for the stygobiont panthurid *Curassanthura* (Crustacea, Isopoda, Anthuridae) with descriptions of *C. bermudensis* n. sp. *Bijdr. Dierkd.,* 55: 324–330.

Wägele, J.W., Coleman, O. and Hosse, U., 1987. Two new hypogean species of *Cyathura* from Melanesia (Crustacea: Isopoda: Anthuridae): further Tethyan relicts? *Stygologia,* 3: 89–106.

Wägele, J.W., Voelz, N.J. and Vaun McArthur, J., 1995. Older than the Atlantic Ocean: discovery of a freshwater *Microcerberus* (Isopoda) in North America and erection of *Coxicerberus,* new genus. *J. Crustacean Biol.,* 15(3): 733–745.

Wagner, H.P., 1994. A monographic review of the Thermosbaenacea (Crustacea: Peracarida). *Zool. Verh.,* 291: 1–338.

Wilkens, H. and Hüppop, K., 1986. Sympatric speciation in cave fishes? Studies on a mixed population of epi- and hypogean *Astyanax* (Characidae, Pisces). *Z. Zool. Syst. Evolutionsforsch.,* 24: 223–230.

Wilkens, H., Parzefall, J. and Iliffe, T.M., 1986. Origin and age of the marine stygofauna of Lanzarote, Canary Islands. *Mitt. Hamb. Zool. Mus. Inst.,* 83: 223–230.

Yager, J., 1987a. *Cryptocorynetes haptodiscus,* new genus, new species, and *Speleonectes benjamini,* new species, of Remipede crustaceans from anchialine caves in the Bahamas, with remarks on distribution and ecology. *Proc. Biol. Soc. Washington,* 100: 302–320.

Yager, J., 1987b. *Speleonectes tulumensis,* n. sp. (Crustacea: Remipedia) from two anchialine cenotes of the Yucatan Peninsula, Mexico. *Stygologia,* 3: 160–166.

Yager, J., 1991. The Remipedia (Crustacea): recent investigations of their biology and phylogeny. *Verh. Dtsch. Zool. Ges.,* 84: 261–269.

Yager, J., 1994. *Speleonectes gironensis,* new species (Remipedia: Speleonectidae), from anchialine caves in Cuba, with remarks on biogeography and ecology. *J. Crustacean Biol.,* 14: 752–762.

Yager, J. and Humphreys, W.F., 1996. *Lasionectes exleyi,* sp. nov., the first remipede crustacean recorded from Australia and the Indian Ocean, with a key to the world species. *Invertebr. Taxon.,* 10: 171–187.

Zacharda, M. and Pugsley, C.W., 1988. *Robustocheles occulta* sp. n., a new troglobitic mite (Acari: Prostigmata: Rhagidiidae) from North American caves. *Can. J. Zool.,* 66: 646–650.

Chapter 23

EVOLUTIONARY RATES AND PHYLOGENETIC AGE IN SOME STYGOBIONTIC SPECIES

Claude BOUTIN and Nicole COINEAU

Evolution of subterranean species, as in other animals, can be studied from different points of view – intraspecific or interspecific molecular evolution, physiological, ethological, ecological and finally morphological evolution. The last is the most common approach because the observation of morphological differences is the easiest way to characterize different species and is the basis of taxonomy. Taxonomy, including recognition, description and classification of species, is necessarily the most fundamental and the most important requirement of any study of biodiversity and evolution. So it is not surprising that the morphology of subterranean species is more often considered than other characteristics by biospeleologists interested in the evolution of subterranean animals. On the other hand, the evolutionary history of lineages leading to the present subterranean species was always marked by one important event – the change of environment associated with the vertical transition from surface to subterranean biotopes (Boutin and Coineau, 1990; Notenboom, 1991; Coineau and Boutin, 1992). This new environment initiates a new kind of selection pressure. Consequently, even if molecular changes are generally considered by biologists as the most reliable time-dependent markers of evolution, morphological characteristics of subterranean species, gradually developed after the transition to the new biotope, can also be used as relevant evolutionary markers, from the time of the vertical transition. Thus, a number of morphological characteristics of species will be considered here as good indicators of subterranean evolution, and morphological differences between species as indicators either of the duration or of the rate of the evolution of subterranean species, even if this certainly does not mean that all evolutionary traits of subterranean animals may be described in terms of morphological changes.

The structure of this contribution will be as follows. First, some speculative considerations will be presented concerning the comparative molecular and morphological evolutionary rates of organisms in general, both in stable and changing environments. The remainder of the chapter emphasizes the thalassoid limnostygobionts – that is, freshwater stygobionts whose ancestors were marine. A brief review of different classes of stygobionts (Botosaneanu, 1986) allows to delineate the successive stages in the evolution of the ancestral lineages of each kind of stygobionts. Then an idealized model of colonization of continental groundwaters by successive lineages of stygobionts will be presented. The main phylogenetic traits, the age, and the biogeography of the different lineages are considered, with a special emphasis on the relation between the phylogenetic age of a lineage and the palaeogeographical history of the region where the species occur. Finally, four case studies are presented. The first is that of the metacrangonyctid amphipods on which the model is based, and which has already been used in a geological perspective. The second is that of the cirolanid isopods, which also fit well with the model. The last two case studies – the amphipod *Pseudoniphargus* and the isopod *Microcharon* – are briefly mentioned, as they include groups of species which exhibit evolutionary stages related with the phylogenetic age of each lineage and the historical geology of the areas where they occur.

MOLECULAR AND MORPHOLOGICAL EVOLUTION IN SUBTERRANEAN FAUNAS: SOME SPECULATIVE CONSIDERATIONS

When considering changes at the molecular scale, it is very likely that the rate of evolutionary changes which occurred in subterranean species may not differ significantly from those generally observed in surface species. It is also conceivable and postulated, even if not demonstrated, that the resulting molecular variability in subterranean and surface species is also similar. This hypothesis might very usefully be tested by studying simultaneously the molecular polymorphism of species which have colonized caves or other subsurface biotopes for a long time, and that of the more closely related surface species. To date, a few molecular studies only (Sbordoni et al., 1979) have been performed with such an aim in view. The results are not always supported by those proceeding from the comparison of genetic variability in other troglophilic or troglobiontic models [e.g., North American *Gammarus minus* (Culver et al., 1995) or amblyopsid fishes (Swofford, 1982)]. From a recent synthesis it appears that "cave and spring dwelling populations do not differ in levels of genetic variability" in *Gammarus minus* and that "the troglobite *Niphargus longicaudatus* appear no more differentiated than the troglophile *G. minus*" (Kane et al., 1994). However, *G. minus* as well as *Astyanax fasciatus* and possibly *N. longicaudatus* likely colonized subterranean waters recently, so that it still seems difficult to get generally valid and clear conclusions until studies are available comparing more ancient troglobites with their nearest surface relatives.

A possible alternative and challenging hypothesis is that, after their ancestors have lived for a long time (geologically speaking) in the subterranean realm, the extant subterranean species exhibit a higher molecular polymorphism than their closely related surface species. This difference between the two kinds of organisms seems conceivable even if the rate of occurrence of mutations was sensibly the same in the two groups of species during the same period, as suggested by the Neutral Theory of molecular evolution (Thorpe, 1982; Kimura, 1983; Wilkens, 1988). This second hypothesis needs to be corroborated, for instance by sequencing one or several suitable informative macromolecules. It is well known that different proteins, and their structural genes, evolve at different rates (Wilson et al., 1977; Thorpe, 1982) and the analysis has necessarily to concern parts of molecules where only low constraints allow the maximum rate of retention of point genetic changes. Indeed it can be supposed *a priori* that the present subterranean lineages have generally accumulated and retained a higher rate of neutral molecular changes than their surface relatives, because the subterranean realm is generally characterized by environmental conditions that have been more stable during millions or tens of millions of years than surface habitats, except in some particular biotopes such as littoral marine interstitial habitats or the alluvium of active rivers. This relative temporal stability of subterranean habitats, postulated for a long time by many authors, has been demonstrated in many particular cases and constitutes one of the generally accepted paradigms of biospeleology (Vandel, 1964; Poulson and White, 1969; Barr Jr and Kuehne, 1971; Ginet and Decou, 1977; Culver, 1982; Danielopol and Wouters, 1992; Gibert et al., 1994a; Culver et al., 1995).

However, there are probably very few laws, rules or paradigms concerning living creatures without one or more exceptions; after analysing *Niphargus virei* as a case of demographic transition in the subterranean domain, Turquin (1984) concluded: "Le milieu souterrain, pour un animal à longue durée de vie, ne peut être considéré comme un environnement constant [For a long-lived animal, the subterranean habitat cannot be considered as a constant environment]". In fact even in the deepest parts of caves, aquatic and often terrestrial fauna may be exposed to seasonal fluctuations in water supply by percolation or flooding which induce noticeable changes in trophic resources and other physicochemical parameters (Barr Jr, 1968; Camacho, 1992; Gibert et al., 1994b; Culver et al., 1995). Nevertheless, the stability of the subterranean environment, especially in continental groundwaters, is generally higher than that of surface biotopes, and this difference may have some evolutionary implications.

When environmental conditions are stable and quiet for a long time, for instance in some subterranean biotopes such as the phreatic waters of a large alluvial basin, successive generations may gradually accumulate a large proportion of the neutral mutations which occurred during the long history of the lineage (Fig. 23.1A). These neutral mutations contribute to increasing the molecular polymorphism of the present subterranean populations and species.

On the other hand, if the ancestors of a present surface species have been once or several times greatly reduced in numbers by ecological crisis (important climatic changes, rapid immigration of predators, destruction or disappearance of most of the favourable biotopes), the surviving species has often descended from a very small part of the ancestral population which have by chance overcome the crisis. In this case, by a founder effect, the descendent populations

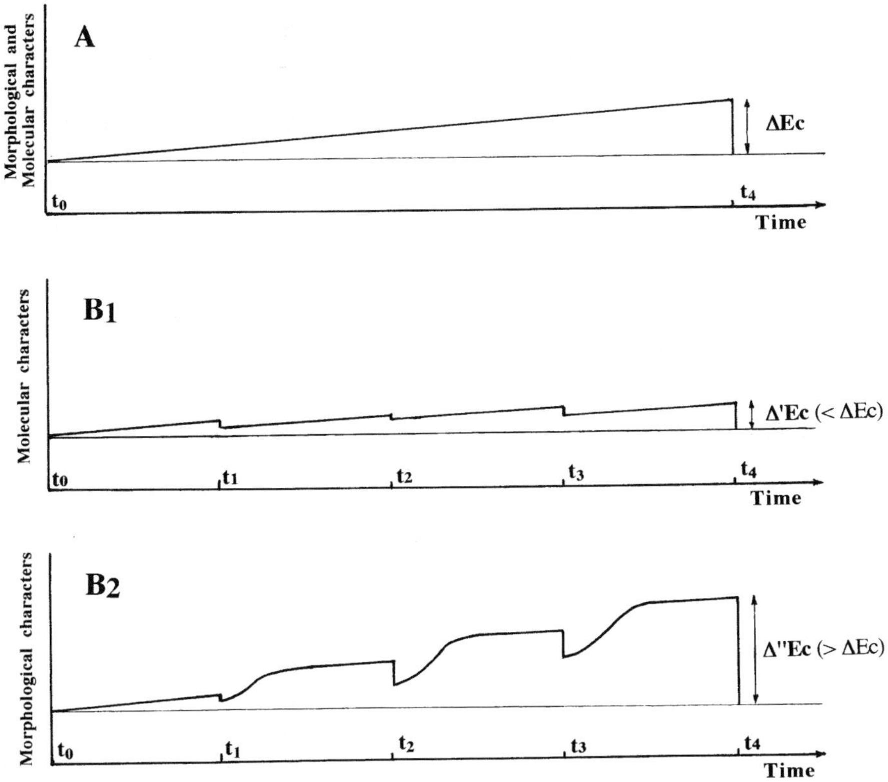

Fig. 23.1. Schematic diagram illustrating the differences of evolutionary rates in species living in a relatively constant environment (A) compared with species living in an unstable environment (B) subject to repeated crises at times t_1, t_2 and t_3. In species A (for example a continental and interstitial limnostygobiont), the rate of molecular and morphological change is constant and low. In species B (for example a littoral thalassostygobiontic species, or a surface species), the rate of molecular change is the same, but part of molecular diversity is lost during each crisis, and finally during the period t_0–t_4, the molecular evolutionary change ($\Delta' Ec$) may be smaller than in species A. However the morphological evolutionary change of species B ($\Delta'' Ec$) is more important than in species A, because during and immediately after each crisis the new environmental selective forces act more strongly on surviving populations, and new phenotypes appear or become dominant.

may have retained only a small part of the genetic diversity of the ancestral population living before the crisis. And when this kind of event occurred several times during the history of a lineage, a great part of the molecular polymorphism gradually accumulated during the quiet periods between the crisis is probably lost (Fig. 23.1B_1).

On the other hand, the new environmental conditions may result in a new type of selective pressure on the remaining diversity of populations after each crisis. Consequently some phenotypes and corresponding genotypes will be eradicated. But recent studies show that environmental stress characterizing the crisis periods could have an influence on the activity of genetic systems controlling the mutation rate and hence the variability of populations (Taddei et al., 1996, 1997a,b). When the environment changes repeatedly, the rate of mutation is higher, the selective pressure also changes, and as a result evolution occurs at a higher rate (Taddei et al., 1997a). This genetic process was observed first in Bacteria and is now considered as general in Eucaryota (Taddei et al., 1996). Thus, a lineage repeatedly affected by a number of crises will likely evolve physiologically and morphologically more rapidly than a related lineage which has lived in a relatively stable environment where the selective pressure on different phenotypes does not change during the same period, and where the mutation rate remains roughly constant (Fig. 23.1B_2). Biologists looking for the causes of different rates in evolution generally acknowledge that a great part of the observed variability in evolutionary rates no doubt arises because of different rates of ecological changes (Arthur, 1982); however the consequences and the importance of environmental changes are rarely considered when the pattern of the evolutionary change of a morphological character is tentatively simulated and figured, for instance in the simulation of Arthur (1984); therefore the resulting patterns of variation proposed (Arthur, 1984, p. 212), do not resemble that of Fig. 23.1B_2, but look like that of Fig. 23.1A.

In fact, it now seems possible to conceive that an animal taxon which led a subterranean life, for instance in continental groundwaters, for a long time (from several millions to hundreds of millions of years) would probably exhibit a relatively greater molecular evolution but relatively less physiological and morphological evolution (Fig. 23.1A) than a closely related taxon still living up to the present in surface habitats or in some unstable coastal interstitial habitats (Fig. 23.1B). Finally and somewhat paradoxically, the rate of molecular evolution in subterranean animals might seem higher and the rate of morphological evolution lower than that of evolution in closely related surface animals or in a species living in an unstable marine environment (Figs. 23.1B_1, 23.1B_2). Generally speaking, Arthur (1982) considered that, even if both processes were driven by mutations and natural selection, "the rate of evolution of the primary structure of proteins would be expected to be more uniform than the morphological evolutionary rate".

We think that the comparative study of related species, some of them living for a fairly long time in subterranean waters and some others still living in surface waters, would be very desirable, and that interstitial crustaceans would probably provide exceptionally relevant material for such an evolutionary analysis. The rate of molecular evolution in subterranean species, compared to that of surface species of the same group, could certainly be appreciated by estimation of the number of point mutations detectable when analysing the genetic diversity in the same part of the genome in the two groups of species. In order to test such a hypothesis, a comparison of the genetic differences between some surface marine species with that of related species which lived for a long time in a quiet interstitial biotope – such as the phreatic waters of an ancient plain – and also with that of some related species still living in a more changeable biotope – such as the marine littoral interstitial milieu – is needed.

The rate of physiological evolution is probably more difficult to study, because usually the ecology and physiology of the extant subsurface "known species" are poorly documented, and generally that of fossil species are even less well known. However, it would probably be possible to estimate, from simple laboratory tests and from field and biogeographic observations, some evolutionary changes in the range of salinity or temperature which can be endured by different species of a genus of subterranean crustaceans now living in fresh or brackish waters, but related with marine or limnic species. A significant advance in this field was already obtained by the research group of Danielopol, who studied the differing ability of adaptation to low concentrations of dissolved oxygen in different surface- and subterranean-dwelling species of asellid isopod crustaceans (see a review in Danielopol et al., 1994).

Conversely and in principle, morphological evolution in subterranean species is probably more easy to detect than in surface species or within different subterranean lineages evolved in different environments. It is possible to measure or to estimate the degree of regressive or constructive evolution of some traits which differ in a surface species (or population) and a closely related subterranean species (or population), belonging to the same family, genus or species (Wilkens, 1973, 1992; Peters, 1988). It is also possible, when aiming at the estimation of these evolutionary rates more globally, to consider the number of apomorphic features present in the subterranean taxa and not in the surface-dwelling species; finally the relationships and the rate of evolution in different lineages of a subterranean monophyletic group can be visualized by the structure of a phylogenetic cladogram. In this respect, recent studies dealing with some stygobiontic species of marine origin now living in continental groundwaters provide interesting data and conclusions regarding also the age of the different lineages (Boutin and Coineau, 1991; Boutin, 1993a,b, 1994a,b).

ORIGIN, EVOLUTION AND DIVERSITY OF STYGOBIONTS

Stygobionts are either marine or freshwater subterranean species (Botosaneanu, 1986). Marine stygobionts may occur in interstitial waters of littoral or marine sediments; they include the burrowing species of the endobenthos and the interstitial meiobenthos inhabiting marine sediments at different depths, and also the endopsammic fauna of the littoral zone, living in water-saturated sands, beneath the beaches or near the shore line. Other marine stygobionts may live in sea-water of marine caves or in brackish waters of anchihaline caves (Wilkens and Parzefall, 1974; Iliffe et al., 1984; Stock et al., 1986; Iliffe, 1992). All these species, like their ancestors, always occur in salt (or brackish) waters and are still geographically and ecologically linked to the marine environment. Collectively, they are called **thalassostygobionts** (Botosaneanu and Holsinger, 1991; Notenboom, 1991; Coineau and Boutin, 1992; Holsinger, 1994a). Most thalassostygobionts result from the adaptive evolution of marine epibenthic or free-swimming pelagic species, which successfully colonized the marine or coastal sediments or marine and anchihaline caves. Nevertheless, some rare thalassostygobionts are derived from continental freshwater species secondarily returned to the sea; some examples are known in different zoological groups of protozoans such as Arcellidae (Rhizopoda, Thecamoebea), or metazoans such as some mites (Arthropoda, Arachnida, Acarina).

Continental aquifers, in karstic limestones or other fissured rocks as well as in unconsolidated sediments, may also harbour different kinds of freshwater subterranean species which may be called **limnostygobionts** (Coineau and Boutin, 1992; Holsinger, 1994a).

If one excludes the rare exceptions mentioned below, thalassostygobionts generally are derived from ancestral marine epigean species. In contrast, the origin of limnostygobionts is double and two kinds of limnostygobionts have to be distinguished and considered separately. A number of limnostygobionts (like some species of the amphipod genus *Gammarus* or of the isopod genus *Proasellus,* as also a number of copepods) are closely related to species of the same genus or family living in surface freshwaters of lakes or rivers. They have no very close relatives in marine biotopes and because of their surface freshwater origin they may be called **limnicoid limnostygobionts** (Coineau and Boutin, 1992; Boutin, 1993a). A number of other limnostygobionts, such as hadzioid amphipods or cirolanid isopods, have no (or very few) closely related species in surface fresh waters and belong to taxonomic groups well known in marine habitats; they may be called **thalassoid limnostygobionts** (Notenboom, 1991; Coineau and Boutin, 1992; Boutin, 1993a). In fact, they are considered as derived from coastal marine populations which colonized the subterranean fresh waters either by the way of marine caves connected with continental karsts or, more often, through the littoral interstitial biotopes where salt and fresh groundwaters come into contact and mix together. Thus, the thalassoid limnostygobionts are descended either from interstitial or from troglobiontic littoral thalassostygobionts.

Necessarily, evolutionary changes which made possible the ecological transition from the marine to the continental subterranean habitat occurred along a shore line, and the marine regressions which marked a number of periods during geological times (or the emergence of oceanic islands) were favourable periods for this colonization, as pointed out by Stock (1977, 1980), who proposed the "Regression model evolution" in order to explain the origin of limnostygobionts from marine ancestors. Later on, this widely accepted model was redescribed elsewhere with more detail as a "Two-step model" (Boutin and Coineau, 1990; Notenboom, 1991; Coineau and Boutin, 1992), and even as a "three-step model" (Holsinger, 1994a). This "Multi-step model" of evolution and colonization implies a number of evolutionary, ecological and biogeographical changes, of different nature, which may have occurred one after the other (the occurrence of one change does not imply necessarily that the next one will happen and the different changes may sometimes occur in different order). In fact, the complete history of a limnostygobiontic group of species generally includes a great number of events from the time when the distant ancestral populations lived in open marine waters up to the present subterranean freshwater species. In any case there are always a minimum of two main biogeographic and ecological changes: one, named the "vertical transition" (Boutin and Coineau, 1991; Boutin, 1993a), which leads from a free swimming or epibenthic species to a stygobiontic species; and the other, named the "horizontal transition" by the same authors, which leads from a marine to a freshwater species.

When considering limnicoid limnostygobionts, it is clear that the "horizontal transition" – the conquest of continental surface freshwaters by marine ancestral populations which succeeded in colonizing the rivers – occurred first, and that the "vertical transition" – the sinking of surface freshwater populations into sediments and subterranean waters – occurred subsequently. It is also probable that such transitions may take place at every period of the history of a continent. Within a given taxonomic group, such transitions, especially the second, may occur several times in different regions or even in the same hydrographic basin. Such examples of repeated colonization of subterranean waters are given for freshwater ostracods by Marmonier and Danielopol (1988); for copepods by Rouch and Danielopol (1987); for cave crustaceans such as the amphipod *Gammarus minus* by Kane and Culver (1991) and Culver et al. (1995); and for cave fishes by Wilkens (1992). Thus, the occurrence of limnicoid stygobionts in a given area is rarely informative about the history of the area. Conversely, the chronology of the geological events which have marked the palaeogeography of a given area generally provides no precise information about the age of colonization of subterranean waters by surface freshwater species and consequently about the age of the extant limnicoid stygobiontic species.

In contrast, for the thalassoid limnostygobionts, even though the "vertical transition" (from epibenthic to interstitial life in marine sediments) may have occurred repeatedly, this is far from being the case for the "horizontal transition". This ecological and biogeographical transition of populations to phreatic continental waters – presently located far from the marine shores – is generally correlated with a withdrawal of the sea which occurred during a period of marine regression (or during the emergence of an island) generally well dated by geologists. Alternatively, it has been suggested that the marine ancestors of limnostygobiontic crustaceans such as the cirolanid isopods now living in groundwater in the Mediterranean region, could have actively colonized the continental subterranean waters through coastal karsts (Botosaneanu et al., 1985; Botosaneanu and Notenboom, 1989, 1992). But this scenario seems unlikely when one considers the distribution of many species known in Morocco. Many species of *Typhlocirolana,* and of the related genus *Marocolana,* live very far from any karst, and often at hundreds of kilometres from marine shores. In the light of the fact that stygobiontic interstitial peracarids have very poor active dispersion abilities (Coineau, 1984; Holsinger, 1991, 1994a; Coineau and Boutin, 1992; Boutin, 1993a), and in the light of the geological history of the areas where they occur in North Africa, it seems much more likely (Boutin, 1993a,b) that the conquest of continental subterranean waters by cirolanids occurred through the littoral interstitial zone during the periods of marine regression, according to the classical "regression model" proposed by Stock (1980). The second step of the "two-step evolutionary model" (Boutin and Coineau, 1990), previously considered as the main part of the regression model (Stock, 1977, 1980), seems to be the most plausible scenario leading to the evolutionary pattern observed at present in most (if not all) thalassoid limnostygobionts occurring far from the seas. One noteworthy aspect of this scenario is that, using the available geological data, it provides a possibility of analyzing both the biogeographic and the phylogenetic patterns of a group of thalassoid limnostygobionts, leading to dating of the different species or lineages of a genus or of a family.

EVOLUTIONARY RATES AND PHYLOGENETIC AGE OF THALASSOID STYGOBIONTS

When a family of stygobionts presently includes one coastal thalassostygobiontic species and a great number of related thalassoid limnostygobiontic species, it may happen that the cladogram of the family exhibits a small number of monophyletic groups of species. In the cladogram represented in Fig. 23.2, for example, there are only five lineages numbered from 1 to 5. The four groups of freshwater species, L1 to L4, are more and more remote from the original form, but less than Sp5, the sole still coastal species. When the areas successively covered by the past transgressive seas are included within each other, as in Fig. 23.2, an interesting relation generally appears between the rank of a lineage in the cladogram and the geographic location of the stations where the different species of the lineage occur. In the case represented in Fig. 23.2, some localities, such as the station numbered 4, may harbour up to four species belonging to the four different lineages of limnostygobionts. Conversely, in others, such as Stations 7 and 8, representatives of only one lineage occur (Table 23.1). In stations numbered 5, representatives of the three lineages L1, L2 and L3 can occur together (Table 23.1). During the last marine regression the thalassostygobionts which lived in the SLT3 shores of Fig. 23.2 colonized first the continental groundwaters, resulting in lineage L3. Later on, the more ancient limnostygobionts (already settled further inland and belonging to lineages L1 and L2) recolonized the area by passive downstream dispersal. They gradually drifted down by the movement of groundwater and also by the rivers which, surprisingly, play an important role in the passive and downstream dispersal of interstitial stygobionts living in the sediments of the valley (Boutin et al., 1992; Boutin, 1993a). In a case such as the one represented in Fig. 23.2, it is clear that the species which belong to lineages L1, L2, L3 or L4 have been living in continental groundwaters since the period of marine regressions which followed the T1, T2, T3 or

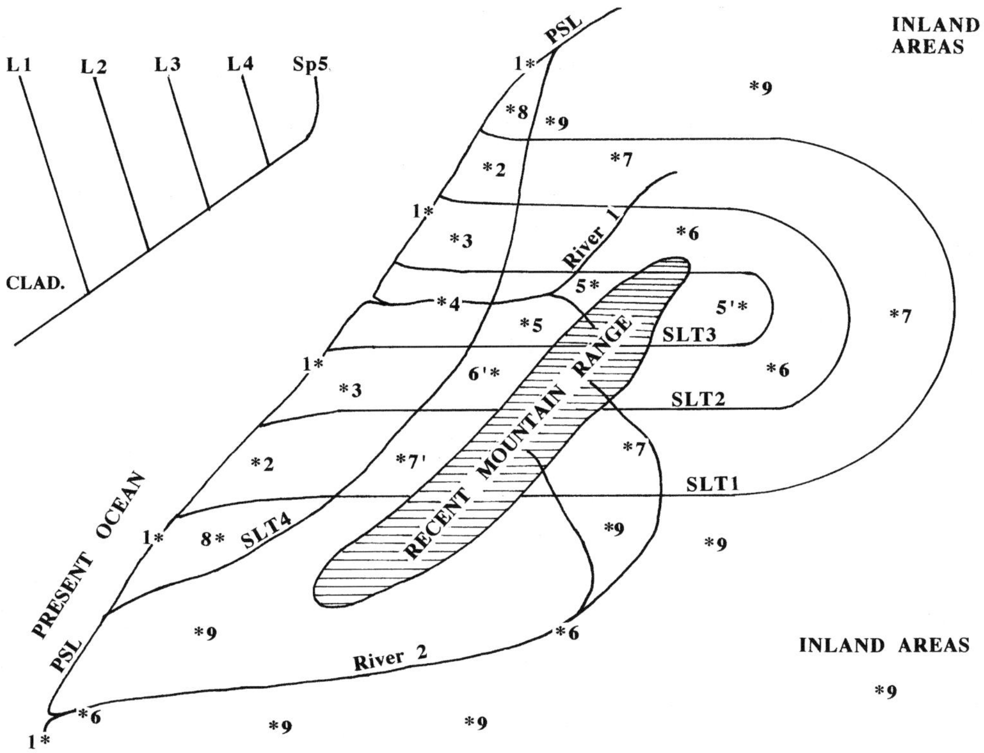

Fig. 23.2. Schematic diagram illustrating the idealized biogeographic and phylogenetic patterns of a group of thalassoid limnostygobionts and the concept of phylogenetic age of lineages or species. SLT1 to SLT4 are the past shore lines at the maxima of transgressive periods T1 to T4. After these 4 successive marine transgressions the sea receded up to a place near to PSL, the present shore line. The mountain range formed subsequently in time T5. In stations numbered 1 to 8 one or several species of the stygobiontic group considered occur, but none in stations numbered 9. Characteristics and number of possible species in each station are given in the text and in Table 23.1. The schematic cladogram of the group (above, left) exhibits one thalassostygobiontic coastal species Sp5 and 4 monophyletic groups (or lineages) of limnostygobiontic species L1 to L4, whose ancestors have colonized the groundwaters during the T1 to T4 regressions, respectively.

T4 transgressions. Thus, the moment of colonization by the ancestral littoral thalassostygobionts originating each lineage can be dated from the geological data, and one can know the duration of life of the stygobionts in continental waters, and thus the age of the lineages.

Comparison of morphological differences between different species belonging to the same lineage (and descended from the same ancestral thalassostygobiontic littoral population) may give an idea of the evolutionary rate in continental interstitial biotopes, in the local palaeogeographical and ecological conditions which prevailed in the area. For example, Stations 7 and 7′ in Fig. 23.2 harbour only one species each, and these two species, belonging to lineage 1, are very similar because they are derived, by vicariance, from the same ancestral population, which was fragmented when a recent mountain range was uplifted in time T5. The two limnostygobiontic species living in Stations 7 and 7′ have the same origin and the same age, and exhibit some differences which have appeared between the time T5 and the present. In the same way the species occurring in stations 6 and 6′, or in stations 5 and 5′, differentiated after the formation of the mountain range formed during the period T5. In fact, if sufficiently old, each local orogenic event, each water-parting formed between two hydrographic basins, can interrupt the genic fluxes between limnostygobiontic populations and so result in the origin of new vicariant species belonging to the same lineage. Thus, the stygobionts inhabiting the two stations numbered 3 may also include vicariant species belonging to L1 and L2, the speciation having occurred during the T3 transgression. Generally, these vicariant species resulting from random mutations occurring in isolated populations already living in continental groundwater differ slightly from each other by some minor apomorphic traits, the rate of morphological evolution in phreatic waters being very low (Boutin and Coineau, 1991).

Table 23.1
Main characteristics of stygobiont diversity in the taxonomic group considered in Fig. 23.2 [a]

Station number in Fig. 23.2	Number of possible species [a]	Groups of species possibly occurring in the station	Age of colonization by ancestors in the station
1	1	Sp5	Probably more ancient than T1
2	2	L1, L4	T1, then T4 regressions
3	3	L1, L2, L4	T1 & T2, then T4 regressions
4	4	L1, L2, L3, L4	T1, T2 & T3, then T4 regressions and later
5 & 5′	3	L1, L2, L3	T1 & T2, then T3 regressions and later
6 & 6′	2	L1, L2	T1, then T2 regressions
7 & 7′	1	L1	T1 regression
8	1	L4	T4 regression
9	0	None	

[a] The possible number of species belonging to different lineages which may occur in the same area. The number of species actually present in a station may sometimes be higher when several vicariant species belonging to the same group occur, or more often lower when the local ecological conditions or the lack of recent dispersal prevents the establishment of some species.

On the other hand, comparison of evolutionary traits which characterize one lineage such as L2 with those which characterize its sister clade (the group L3-L4-sp5) leads one to consider the apomorphic traits exhibited by the more derived clade (and not by the first one) as having appeared in the ancestral populations when they were still living in the littoral biotopes (between periods T2 and T3), even though during the same period the ancestors of the first group L2 were already settled in continental groundwaters. It is thus possible to observe that, during the same period, morphological evolution is more important in the littoral, changing and unstable, biotopes than in the quieter continental phreatic biotopes. This difference between the evolutionary rates in continental and in coastal environments has been emphasized by Boutin and Coineau (1991) and by Boutin (1993a,b) in thalassoid limnostygobiontic crustaceans. Several authors such as Lamotte (1988) have already focused attention on the importance of fortuitous events in the change of selective forces acting on population polymorphism and driving morphological evolution. On the other hand, the selective effect on the evolutionary rates of different kinds of stresses, particularly of changes in salinity, temperature and granulometry which characterize the coastal marine biotopes, have been well established (Parsons, 1987). Figure 23.1 symbolically represents this difference between evolutionary rates in biotopes such as continental groundwaters (Fig. 23.1A) and coastal groundwaters (Fig. 23.1B_2), in relation with their stability or instability. These evolutionary differences certainly explain the divergences between the lineages and the structure of the cladogram.

Finally, the age of a lineage, including species living in continental groundwaters, may be established when the phylogenetic relationships within the species of the group in question, the distribution of the different species (descriptive biogeography) and also the historical geology of the area are known. Thus, a regional monographic study of the group is the basis for further studies, especially studies of isolated species which may occur out of the main domain of evolution and differentiation (speciation) of the group.

In fact, the phylogenetic age of a group of species (the date of the origin of the lineage in a cladogram) is the age at which colonization of the area by the thalassostygobiontic ancestors of the lineage considered took place. Thus, when the study of the group has already been performed in the main part of its area, the occurrence of a new species reported from another region, after including this species in the cladogram of the group, may provide information on the date of colonization by the thalassobiontic ancestral populations in the region where the new species occurs. When a species belonging to lineage L3 for instance (Fig. 23.2) is reported from another region, for example in a sedimentary basin located far to the south or north of the area represented in Fig. 23.2, or in groundwaters of an oceanic island off the continent where the group has been studied, the occurrence of this species means that the part of the continent where it occurs has been flooded by the sea during times T3, or that the island has emerged (totally or partly) during the T3 regression. In this case, if the geological history of the region where the new species occurs is incompletely known, the occurrence of the thalassoid limnostygobiont may be used as a palaeobiogeographic argument for the palaeogeographic reconstruction (Boutin, 1994b).

A CASE STUDY: THE METACRANGONYCTID AMPHIPODS

Historical biogeography of Metacrangonyctidae

Metacrangonyctidae are small-sized amphipods (2 to 5 mm long); the family (Boutin and Messouli, 1988) includes some 40 species in two genera, inhabiting subterranean waters of the southern and eastern Mediterranean countries and of some nearby Atlantic or Mediterranean islands. *Metacrangonyx* is the most widely distributed genus. It is known from the Canary Islands to the Levant (Israel). Some species live in Mediterranean islands, in Algeria, and in the Sinai; others, some 30 species, are reported from Morocco (Messouli, 1988; Boutin, 1994a). The genus *Longipodacrangonyx* includes only seven species from Morocco, which differ from *Metacrangonyx* by 12 apomorphic traits. Except for one species (which lives in groundwater and in the anchihaline waters of a cave in the Balearic Islands), all populations of Metacrangonyctidae presently known live in fresh groundwater and can be captured by filtration of pumped groundwater or, in ordinary wells, using traps or nets. Some species only occur in islands or near the coast; most of the others live in the interior of the African continent, often hundreds of kilometres from the shores.

The marine – pelagic or epibenthic – epigean ancestors of the metacrangonyctid amphipods were likely widespread in the waters of the Tethys Ocean during the Upper Palaeozoic or the Early Mesozoic. The colonization of shallow or coastal sediments probably occurred along the southern shores of the Tethys during the Jurassic. This vertical transition (Boutin and Coineau, 1990) was an active extension of the initial niche or the active conquest of a new niche (Rouch and Danielopol, 1987). After the opening and widening of the North and Central Atlantic Ocean the coastal populations of amphipods were separated into two groups – the metacrangonyctid amphipods in the Eastern Atlantic and Mediterranean countries, and the weckeliid Hadziidae in the Western Atlantic (Boutin, 1993a, 1994a). Subsequently the interstitial populations established along the North African shores penetrated inland during the marine transgressions and colonized the groundwaters during the Mesozoic or Cenozoic marine regressions. In Morocco a very important marine transgression occurred during the Cenomano–Turonian period of the Upper Cretaceous. Another transgression, less extensive than the latter, occurred during the Senonian and the shore lines fluctuated until the terminal Cretaceous. Some limited marine gulfs penetrated inland in the Morocco during the Eocene. Further transgressions resulted from slight fluctuations of shore lines and affected only the littoral parts of the country (Choubert and Faure-Muret, 1962; Michard, 1976; Piqué, 1995).

Age of the different lineages in Metacrangonyctidae

The phylogenetic relationships within the Metacrangonyctidae have been extensively studied by Messouli (1988, 1994) and by Boutin (1994a). No less than 60 morphological characters, variable within the family, were used; half of them concerns the shape of the body or of some appendage segments and the other half are chaetotaxic traits (i.e., involving the number and distribution of spines and setae on appendages). It was possible to determine the character polarity by considering as plesiomorphic the character states observed in the marine species of the related family Hadziidae and by considering the character polarity already recognized in other interstitial amphipods [e.g., by Stock (1981) for the bogidiellids]. The cladograms show four lineages (Fig. 23.3A). The more evolved lineage includes the seven species of the genus *Longipodacrangonyx*. The genus *Metacrangonyx* seems more heterogeneous, as it is divided into three lineages: the "more primitive" *spinicaudatus* and *remyi* groups, each of them including seven species, and a more derived lineage, the *panousei* group, including 18 species (Boutin, 1994a). As the clade formed by the *panousei* group and the *Longipodacrangonyx* group has none of the apomorphic traits present in either of the two more primitive groups, the origin of these two primitive groups cannot be separated and the cladogram displays a trifurcation. Although some authors assumed that a multifurcation in a cladogram reveals an incomplete discriminative analysis of relationships, it is now accepted that such a multifurcation may express

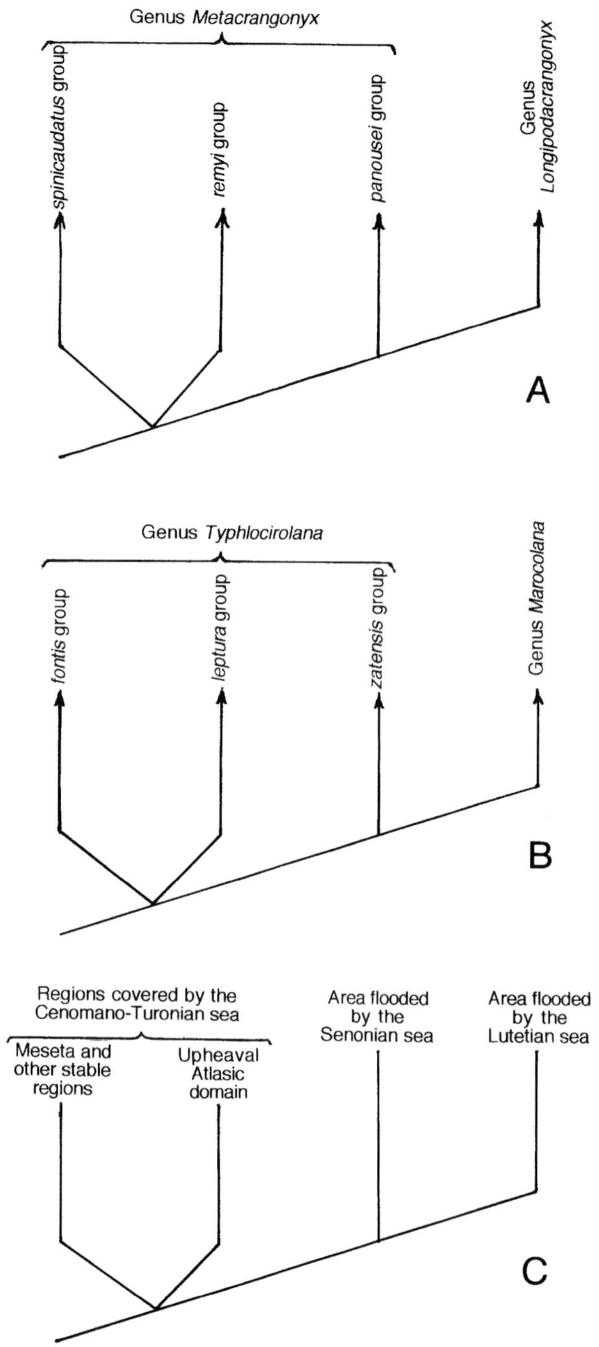

Fig. 23.3. Phylogenetic relationships among the four lineages of the metacrangonyctid Amphipoda (A), and among the four lineages of cirolanid Isopoda of the *Typhlocirolana* group (B) occurring in the eastern part of the Mediterranean basin. C is an area cladogram of the lineages of metacrangonyctid and cirolanid Crustacea occurring in Morocco: the regions where they occur are put in place of the names of the four groups of species forming each family. Modified from Boutin (1993a).

an evolutionary reality. Purvis and Garland (1993), following Maddison (1989), clearly distinguished between two kinds of polytomies: the "soft" and the "hard" polytomies; "soft" polytomies may be resolved in theory by the use of additional characters in parsimony analysis, whereas "hard" polytomies cannot because they "represent true cases of multiway speciation events". The phylogeny of Metacrangonyctidae is an interesting example with respect to this question, as will be shown after a biogeographic analysis of the distribution of different species.

To date, the Metacrangonyctidae are known from several hundred stations in Morocco. Species belonging to the *spinicaudatus* group and to the *remyi* group both occur over an extensive part of the country which was flooded by the large Cenomano–Turonian marine transgression. The species of the *spinicaudatus* group are located in the plains and in the regions which were orogenically stable after the Turonian regression until the Present; the species of the *remyi* group occur in areas which were subsequently uplifted, such as the southwestern plateaus of Atlantic Morocco, or strongly overturned such as the Haut Atlas mountains. In these mountainous regions, sedimentary layers including the aquifers were generally eroded or shifted, and the extreme changes in the ecological environment due to the shift of aquifers from a superficial layer to a deeper one seems to explain the peculiar apomorphic traits of the *remyi* group. Thus, it is likely that the *remyi* lineage has the same origin as the *spinicaudatus* group – related to the Turonian regression – but nevertheless exhibits particular apomorphic traits absent from all the other lineages.

When compared with the *spinicaudatus* and *remyi* groups, the *panousei* group of species exhibits some important chaetotaxic reductions, especially in the number of spines still present on the external surface of the body and appendages, and so is better adapted for moving and living in the interstitial milieu. Its biogeographic range is less extensive than that of the two primitive lineages and it is limited to the area covered by the Senonian marine gulfs. Thus, the ancestors of the *panousei* group of species likely were still littoral thalassostygobionts from the Turonian to the Senonian, and the age of the lineage may be dated from the Senonian.

As for the species of the more derived group, they form the genus *Longipodacrangonyx*. They are known from a small number of stations only, and all these stations are located within an area included inside the boundaries of the Eocene palaeo-gulfs. Furthermore, most known stations are in places where the last marine transgression occurred during the Eocene. So one can consider that metacrangonyctid amphipods which occupied interstitial habitats on the marine shores of the Atlantic Ocean and Mediterranean Sea during the Palaeocene and the Eocene colonized the inland biotopes during the Lutetian regression (Middle Eocene).

Finally, it seems clear that the two more "primitive" lineages of Metacrangonyctidae settled in continental groundwaters during the Turonian, that a more "evolved" lineage of *Metacrangonyx,* the *panousei* group, colonized the continental groundwaters during the Upper Senonian or the terminal Cretaceous, and that the more "evolved" Metacrangonyctidae, which are the fourth and last lineage forming the genus *Longipodacrangonyx,* colonized the continental groundwaters during the Lutetian (Fig. 23.3A). Consequently, the occurrence of Metacrangonyctidae in a region means that the region where its ancestors colonized the fresh groundwaters emerged either during the Turonian, the Senonian or the Lutetian, according to the lineage to which the species belongs.

Many thalassoid limnostygobiontic crustaceans presently inhabiting continental groundwaters belong to genera or groups that include some thalassostygobiontic species inhabiting the littoral subterranean waters. This is the case for the amphipods of the genus *Pseudoniphargus* and for the family Bogidiellidae, for the microcerberid isopods, and also for the Thermosbaenacea. In some other interstitial groups similarly widespread in continental groundwaters, such as the isopod *Microcharon,* littoral and marine species living in the marine sediments far from the shores still exist today. But this is not the case in the Metacrangonyctidae. In spite of intensive searching, no marine or littoral thalassostygobiontic species are presently known, and no continental Metacrangonyctidae more "evolved" than the species of genus *Longipodacrangonyx* have been found in the regions of Atlantic Morocco which were flooded by marine transgressions after the Eocene. That means that the littoral metacrangonyctids probably disappeared during the terminal Eocene.

The four lineages of metacrangonyctid amphipods differ from each other by clear morphological traits, so that any species can easily be identified with one of the lineages (Messouli, 1988; Boutin, 1994a). Thus, when a new species is discovered in a newly investigated region, the inclusion of this new species in one lineage of Metacrangonyctidae provides, from the phylogenetic data (the phylogenetic age of the lineage), information on the age of the colonization of the area by the metacrangonyctid ancestral populations, and therefore on the period of emergence of the land in the region considered.

Some studies using the stygofauna in a geological context have already been performed, leading to a different view of the age and origin of the Canary Islands (Boutin, 1994b) and of the age of the first permanent emergence of land in Israel and in the Sinai (Boutin, 1997). In Fuerteventura, one of the oldest Canary Islands, as in the Levant countries (Sinai and Israel), the metacrangonyctid amphipods presently living in fresh groundwaters belong to the *Metacrangonyx panousei* lineage, indicating a definitive emergence of a part of these territories during the terminal Cretaceous, and not during the mid-Cenozoic as is generally believed by geologists.

Age of different species within a lineage of Metacrangonyctidae

For the phylogenetic age of different species of Metacrangonyctidae, the problem is quite different. By "age of a species" we mean the age of the ancestral population which evolved to the present species, to the exclusion of any other related species. If two species belonging to the same lineage appear very similar and very closely related in a cladogram, the age of these two "sister species" is the date of the environmental event responsible for the interruption of genic fluxes within the common ancestral population and the evolutionary divergence leading to these two extant and separated species. Generally, this age cannot be deduced from the place of the species in a cladogram. Other data, concerning the local palaeogeography or the geological history of the region has to be taken into account. In a study concerning the origin and the history of a small monophyletic group of 6 species belonging to the *panousei* lineage, Boutin et al. (1992) had to consider no less than four different geological events: one marine transgression responsible for the separation of the area first occupied by the common ancestral population during the Eocene, then a post-Lutetian orogenic stage, then one of the subsequent orogenic stages which occurred in the Oligocene, the Miocene or the Plio-Villafranchian, and finally Quaternary erosion due to the rivers which cut the slopes of the Haut Atlas and originated separated hydrographic basins.

Thus, the establishment of the phylogenetic age of species may be useful for the evolutionary study of a stygobiontic group, but generally this assessment requires local palaeogeographical data and so cannot be used for palaeogeographic studies themselves.

PHYLOGENETIC AGE OF THE STYGOBIONTIC CIROLANID ISOPODS FROM MOROCCO AND THE MEDITERRANEAN BASIN

The phylogenetic and biogeographic patterns of the genera *Typhlocirolana* and *Marocolana* will be briefly presented because of a noteworthy resemblance with those of metacrangonyctid amphipods (Fig. 23.3B). The limnostygobiontic cirolanids belonging to the *Typhlocirolana* group *sensu* Wägele (1989) presently include three genera: *Typhlocirolana* Racovitza, 1905, *Turcolana* Argano and Pesce, 1980 and *Marocolana* Boutin, 1993b.

The first genus is widely distributed from Morocco to the Levant in regions flooded by the marine transgressions during the Cretaceous; it occurs also in some Mediterranean islands. Discovered first in Algeria, this genus has been split into a number of species occurring in many stations located in different regions of Morocco (Boutin, 1993a,b). The evolutionary traits of the genus *Typhlocirolana* lead to three lineages being distinguished. The two more primitive are also the more widely distributed on lands that emerged during the Turonian, and the most derived seems limited to lands that emerged during the Senonian regression. *Marocolana* and *Turcolana* are sister genera characterized by a number of significant apomorphic traits which did not appear in the genus *Typhlocirolana*. They are known from a small number of stations only; *Marocolana* occurs in some stations of Morocco which emerged definitively with the recession of the Eocene marine palaeo-gulfs; *Turcolana* occurs in Eastern Mediterranean countries, especially in Turkey, Rhodos and Israel, in regions which emerged partly during the Eocene or Early Cenozoic (Boutin, 1993a,b, 1997).

In a word, the biogeography and the phylogenetic pattern of the Cirolanidae and that of the Metacrangonyctidae, established from the data from Morocco, where the two groups are widely distributed, and the palaeogeography is well known, are very similar (Fig. 23.3). The phylogenetic age of the four lineages in cirolanids is probably the same as those of the four lineages of the metacrangonyctid amphipods. If it so happens that an area under study harbours both a metacrangonyctid and a cirolanid belonging to homologous lineages, the age of these lineages, as established in Morocco, may be informative about the date of the colonization of groundwaters by the ancestors of these stygobionts – that is to say, the date of the first definitive emergence of the area where they presently occur.

Finally, the age and the early stages of the palaeogeography of some Mediterranean islands will need revision. The fact that a species of *Metacrangonyx* and a species of *Typhlocirolana* belonging to the more ancient lineages of each group occur in the same Balearic island supports the hypothesis of a first definitive emergence of a part of the island during the Turonian.

PHYLOGENETIC AGE IN OTHER INTERSTITIAL THALASSOID LIMNOSTYGOBIONTS

Taxonomic, phylogenetic and biogeographic studies in other groups of interstitial crustaceans of marine origin have also resulted in the recognition of several phylogenetic lineages which differ by their evolutionary traits within these groups. Owing to the particular species richness of the stygobiontic fauna of Morocco and the knowledge of the palaeogeography of the different regions of this country, the Moroccan fauna, and also that of the Iberian peninsula, are particularly favourable for the identification of phylogenetic lineages and the establishment of their age in different groups.

For example, the amphipod genus *Pseudoniphargus* shows a Tethyan origin, and is known mainly from the western Mediterranean regions and also from various islands of the North Atlantic. This genus was specially studied in a phylogenetic context by Stock (1980), Boutin and Coineau (1988), Notenboom (1990, 1991) and Coineau and Boutin (1996). The phylogenetic analysis of the genus shows five main lineages (Fig. 23.4). The most primitive lineage and the most derived one occur in Spain, where especially the genus split into numerous species. In Morocco a number of species recently discovered are distributed along the southern piedmont of the Rif mountains where their ancestors colonized the groundwaters during the late Miocene. Consideration of the palaeogeography of the countries of the Mediterranean Basin where the genus occurs and particularly of the palaeo-shores of the

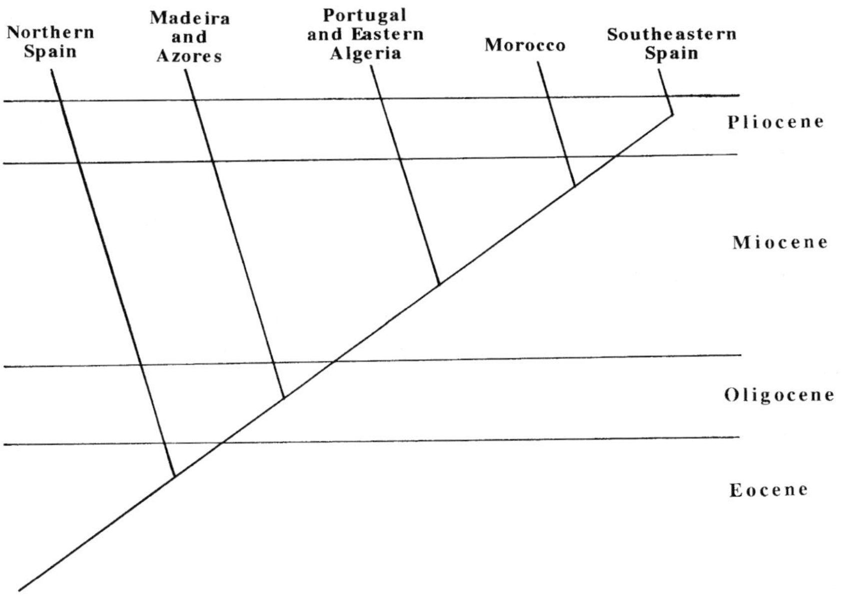

Fig. 23.4. Simplified area cladogram of the main groups of species in the genus *Pseudoniphargus*, and dating of the colonization of continental groundwaters by the coastal ancestral populations of each lineage, from the Eocene to the Pliocene. Modified from Boutin and Coineau (1988).

Cenozoic seas resulted in a reduced area cladogram (Fig. 23.4) including the date of the main cladogenetic event originating each lineage, from the Eocene to the Pliocene.

The strictly interstitial microparasellid isopod *Microcharon* is widely distributed. The genus includes some marine and some coastal thalassostygobiontic species and a great number of continental species living in groundwaters. About 70 species occur in southern Europe, the Middle East and North Africa. Coineau (1994) and Boulanouar et al. (1995) showed that, within one region – e.g., in Maghreb – species belonging to different lineages occur in areas with different geological history and palaeogeography. In Morocco, some species inhabiting the groundwaters of the Marrakesh Province have retained many plesiomorphic traits and belong to primitive lineages dating from Turonian, Senonian or Eocene, according as they are located in regions where the last marine embayment occurred during these periods. Other species, more derived and present in the south of the Rif mountains, belong to the Eastern Mediterranean group, and are dated from Late Miocene (their thalassostygobiontic ancestors colonized the fresh groundwaters during the late Tortonian and the Messinian at the end of the Oligocene, as the Peri-Rifian Channel filled and disappeared). Thus, in *Microcharon* as in other thalassoid interstitial crustaceans, it appears that the most primitive lineages include species developed from ancestral populations that settled in continental groundwaters during the more ancient marine regressions, and that conversely the species belonging to the more derived lineages developed from colonization by ancestral coastal populations during more recent regressions.

A number of studies published during the last decade concern the historical biogeography of different groups of thalassoid stygobiontic crustaceans, either in Central and North America (Stock, 1986; Holsinger, 1986, 1994b) or in the Mediterranean Basin (Boutin, 1993a,b, 1994a,b; Notenboom, 1988, 1990, 1991; Coineau, 1994) or through the world (Holsinger, 1991, 1993). In most of these cases, one of the conclusions of the studies emphasizes that the origin of stygobiontic lineages in continental groundwaters is more ancient than previously supposed. Several groups, usually considered as "stranded" by the Pliocene or Miocene seas, appear to have settled in groundwaters since the early Cenozoic, and more often since the Cretaceous. Another conclusion of these studies, when considering different families of stygobiontic crustaceans inhabiting the groundwaters for a long period, is that the evolutionary rate in continental groundwater populations seems to be very different from one group to another. For instance in the Thermosbaenacea the coastal populations which colonized the continental groundwaters of Morocco and Somalia during the Turonian regression, just after the large Mid-Cretaceous marine transgression, evolved during some 90 million years and at present constitute *Monodella atlantomaroccana* and *M. somala,* which are morphologically very similar (Boutin, 1993a). By way of contrast, the continental Bogidiellidae (Amphipoda), also occurring in the same regions and probably related with the same marine regressions, are now morphologically so different that their separation into two genera, *Afridiella* and *Maghrebidiella,* was justified (Stock, 1981; Diviaco and Ruffo, 1985; Boutin, 1993a). Thus, the evolutionary rate in stygobiontic interstitial lineages appears to be dependent both on the ecological instability of the environmental conditions and on some intrinsic evolutionary potentialities of the taxa.

DISCUSSION AND CONCLUSIONS

From the above examples it can be concluded that simultaneously studying the taxonomy, the cladistics and the biogeography of a whole stygobiontic group may be very fruitful, particularly if the study is performed in a region including the birthplace of the group, and where the different phylogenetic lineages occur. After these different phylogenetic lineages have been well defined and their origins dated, subsequent discovery of new species belonging to the group under study allows one to include this new species in a line of the cladogram and thus to date the colonization of the geographic area by its ancestors. If the new species occurs inside the first studied area, or in another region the historical geology of which being already sufficiently known, the new species and its distribution area generally provide a corroboration of the evolutionary scenario elaborated for the group, and so contribute to ensure the exactness and the accuracy of the evolutionary model already developed. Conversely, if the new species occurs in a continental region, or an island, the origin of which is poorly documented by geological data, it is possible to use the biological and the geological data in the reverse order. The inclusion of the new species in a lineage which is already known makes possible the dating of a geological event related to the colonization of the area by the ancestral stygobiontic populations.

However, it is clear that only a part of the aquatic subterranean fauna can be used for such a purpose. This chapter has focused on the continental interstitial thalassoid limnostygobionts because it is generally impossible or very difficult to assess with any precision the age of the colonization of the subterranean biotopes by the ancestors of the present limnicoid stygobionts. From the surface fresh waters, the "vertical transition" (active colonization of sediments or subterranean waters) by epibenthic populations may happen at any period, without relation to a particular geologic event. As for the strictly troglobiontic stygobionts (living in caves), even thalassoid ones cannot usually be used, or are not so easily usable, as interstitial species for several reasons:

(1) Groundwater occurs almost everywhere, but karst areas account for about 10 to 20% of the land area, and some karst areas have few caves. Therefore, they are in general not especially favourable for establishing the distribution areas of stygobiontic species.

(2) In karstic areas where only a few caves are penetrable, an effective sampling of the karstic fauna from the surface is possible but it requires heavy means such as the drilling of deep wells and the use of air-lift pumps (Gibert et al., 1994b; Malard et al., 1994). Thus, it is generally not possible to study a great number of stations or to prospect vast regions extensively, and troglobiontic stygobionts are consequently not very favourable subjects for palaeobiogeographic and evolutionary reconstructions.

(3) Finally, a number of evolutionary characteristics of interstitial stygobionts, in particular the elongation of body shape, the reduction of body size and chaetotaxy (bristle abundance), are generally not observed in troglobionts, because selective forces acting in the evolution of stygobionts are very different in caves and in interstitial biotopes (Henry and Magniez, 1983; Kane et al., 1994). The evolutionary history of troglobionts is not marked by the important change in the direction of the environmental pressures which operate when an ancestral surface species colonizes sediments and subterranean waters in interstitial milieus, and evolutionary rates probably do not differ greatly between littoral populations and continental stygobionts living in caves. Thus, is it probably more difficult to perform a reliable phylogenetic analysis in a troglobiontic group of species, and to establish a relation between some morphological traits and the phylogenetic age of the lineage to which a species belongs.

Consequently, it seems likely that thalassoid interstitial limnostygobionts are more suitable material than others for the study of the phylogenetic age of lineages of stygobionts.

Although particularly interesting, the above method may have nevertheless some limits and even may seem somewhat uncertain in some cases. The uncertainty may be related to the reality of the phylogenetic reconstruction, and thus concern the structure of the cladogram; it may also concern the dating of the main divergences between the lineages displayed in the cladogram, and thus the phylogenetic age of the lineages.

Concerning the phylogenetic reconstruction, uncertainty is never completely absent because, in all cases, a cladogram is only a phylogenetic hypothesis. Even when there are many reasons to consider this hypothesis as the more probable one, the cladogram is only the best estimate of the true phylogeny. Reasons for uncertainty in the more parsimonious cladogram may result from polytomies which may be due either to an incomplete resolution of the phylogenetic analysis, or to a polytomic pattern of evolution. Conversely, the uncertainty may proceed from the unwarrantable grouping of species in one monophyletic group because of apparent but false synapomorphies due to a number of homoplasies.

Several authors, such as Felsenstein (1985), Maddison (1989), and Purvis and Garland (1993), have proposed some tests aiming at the discrimination between the different kinds of polytomies. These tests proposed for differentiating the "soft", the "firm" and the "hard" polytomies are not easy to apply to the case of interstitial stygobionts, because polytomies have to be considered as "firm" or "almost hard" if the cladogram includes "many characters per taxon" or if the polytomy "is based on molecular distance analyses using full data matrices" or if the "fossil evidence shows the radiation to have been rapid" (Purvis and Garland, 1993). This is never or rarely the case for interstitial crustaceans. Conversely, a polytomy must be treated as "soft" when it "comes from a standard taxonomy featuring taxonomic levels" (Purvis and Garland, 1993), or from an incomplete matrix of molecular data, or when the ratio of informative characters to taxa is low, which is often the case in stygobiontic crustaceans. Although most authors (Felsenstein, 1985; Purvis and Garland, 1993) assume that "hard" polytomies are very much the exception in nature, it may happen in the interstitial crustaceans that related species or groups of species differ by a few characters only, but are nevertheless "hard" polytomies resulting from simultaneous speciations. This evolutionary pattern is far from being rare in thalassoid stygobionts because, during each regression of the sea, several marine gulfs were formed along the same shoreline along a continental margin and might be the place of simultaneous local colonizations of the new continental groundwaters by interstitial crustaceans previously thalassostygobiontic in coastal sediments. In such a case the cladogram generally displays polytomies resulting from the simultaneous origin of the different continental species (or subsequently of groups of species) descending from the same ancestral coastal population. This kind of polytomous pattern has already been reported by Notenboom (1988) in the Iberian species of the genus *Pseudoniphargus*. It appears also in some cladograms of the Metacrangonyctidae presented by Boutin (1994a), Messouli (1994) and in Fig. 23.3 above. Finally "hard polytomies" are not very rare in thalassoid stygobionts, and if any polytomy is considered with a special attention, the "hard polytomies" are generally identifiable and explainable, not by a statistical test and consideration of the character matrix alone, but rather in considering the geographic distribution of species and the geological history of their area.

Another reason for uncertainty in phylogenetic reconstruction could be the high degree of homoplasy resulting, in many interstitial stygobionts, from a parallel evolution. As in the genus *Pseudoniphargus* (Notenboom, 1990, 1991), homoplasy is also important in the Metacrangonyctidae. Messouli (1994) noted up to 25 of the 32 characters used in a matrix a subject to some degree of homoplasy. In addition, the resulting cladogram displayed a very low Consistency Index[1] (0.40). In fact, after performing the vertical transition, all interstitial populations undergo a similar selective pressure from the new biotopes, and all species, belonging to the same clade or not, tend to have a more elongated body, a smaller size and a reduced number of external bristles. Thus, similar apomorphies may happen independently (Sket, 1985). Fortunately, this adaptive evolution operated at an

[1] The Consistency Index (Kluge and Farris, 1969) ranges between 0 and 1, and reflects inversely the number of homoplasies (evolutionary changes occurring several times in different lineages), which in turn is related inversely to the reliability of the phylogenetic reconstruction.

appreciable rate in the littoral biotopes and at a very low rate in the continental biotopes, except in some particular periods when the distribution areas were subject to orogenic upheaval. Nevertheless, even in this last case, the direction of the evolutionary changes seems different in coastal and in continental biotopes – at least in the two groups specially studied from this point of view, the metacrangonyctid amphipods and the cirolanid isopods. Thus, for a species or group exhibiting possible parallel evolution in some traits, it is generally possible, by considering the whole set of evolutionary traits and also the biogeographic data, to discriminate for a particular synapomorphy between homoplasy due to similar selection and homology due to common ancestry. It is then possible either to keep the homoplasies in the cladogram or to eliminate the corresponding characters, this last choice being the surest solution when some doubt remains. For example, Boutin (1994a), searching for a reliable cladogram of the lineages of the metacrangonyctid amphipods, first analysed 60 morphological characters present in different states within the family. In the first cladogram, some characters were present in different states in the same clades, or were visibly homoplasic, exhibiting the same derived state in several species of different lineages. These characters were eliminated, as well as autapomorphies which did not contribute to the structure of the tree. Finally the "reduced cladogram" so obtained, reproduced in Fig. 23.3A, was inferred from a small matrix of only 16 characters which differ between the lineages, but which are in the same state in all the species of one lineage and which change only once in the cladogram, and so vouch for the reliability of the phylogenetic reconstruction.

In estimating the evolutionary rates and phylogenetic age of stygobionts, even from a reliable cladogram, another possible cause of inaccuracy may result from the method used for dating each bifurcation of the cladogram. Currently, a number of authors prefer reconstructions and estimations based on molecular data. It is clear that the comparative molecular analysis may provide a great number of data and that the analysis of a large data matrix performed with the help of a computer may provide objective results less debatable than a traditional cladogram based on a small number of morphological characters which may differ by their nature, their evolutionary significance and their independence or linkage. It is sure also that the Molecular Clock (Zuckerland and Pauling, 1965) and its use for dating the divergences is very attractive. However, a problem is that different molecules, proteins for instance, evolve at different rates (Fitch and Langley, 1976; Wilson et al., 1977; Fitch, 1980; Thorpe, 1982), and that the rate of the Clock, mainly when one considers long periods, depends on the molecule chosen and also on the organism. For example the classical "Genetic Distance" proposed by Nei (1972) and always used nowadays, was considered by him to vary by 1.0 per 5 million years, but other authors estimated that such a variation of 1.0 occurs in 18 million years (Yang et al., 1974) or in 20 million years (Sarich, 1977). Thus, even if the method used for calculation of the Nei Distance by different authors is not always exactly the same (Thorpe, 1982); from the literature it seems that the rate of the Molecular Clock may be variable and probably depends on the animal model studied. It is also known that the evolutionary rates estimated from different molecules are not necessarily correlated (Thorpe, 1982).

More recently, it is not rare to see the publication of different conflicting molecular phylogenies either within a small group of related species [for instance, the phylogenies of rodents derived from comparisons of their nuclear and their mitochondrial desoxyribonucleac acid (nDNA and mDNA)] or within a large clade such as the Arthropoda. Thus, it seems clear that a judicious approach would probably involve search for agreement between results provided by different and independent methods. At the same time, a number of authors emphasize noticeable differences between the dating of lineage ages using the Molecular Clock, and the dating of the same lineages on the basis of palaeontological data. For example, Lamotte (1994) quoted the duration of 5 million years obtained from Molecular Clock since the separation of the lineage of Man from that of the African apes, whereas palaeontological data provide evidence for 15 million years since the separation. Generally speaking, it is clear however that the molecular data are especially useful when they lead to conclusions corroborated by other methods, either in phylogenetic reconstructions or in dating lineage origins.

The approach proposed here for the estimation of the phylogenetic ages and evolutionary rates of species of thalassoid limnostygobionts is based on a cladistic analysis using morphological characters and also on the biogeography and the age of dated geological events. Cladograms obtained from molecular sequencing can also be used similarly in the future.

Some other results already obtained in the same field display a significant difference in evolutionary rate and in the orientation of morphological evolution in interstitial crustaceans living in either the littoral or the continental biotopes. From the phylogenetic age of different lineages it now seems possible to use thalassoid limnostygobionts for dating of a marine regression or an island emergence (Stock, 1995, 1996).

Thus, the study of this type of subterranean crustaceans opens a new field of research for stygobiology. This kind of study will be more informative with regard to the evolutionary rates when combined with comparative estimation of the molecular evolution in different lineages of groups already studied on the basis of morphological features.

ACKNOWLEDGEMENTS

This work has been supported by grants from the Ministère de l'Éducation Nationale de l'Enseignement Supérieur et de la Recherche (Paris): Programme ACC-SV nº7, "Systématique et Biodiversité: Diversité fonctionnelle et Évolution en milieu souterrain", from the French and Moroccan Ministères des Affaires Etrangères: "Action intégrée de coopération interuniversitaire franco-marocaine 93/643: Faune aquatique souterraine du Maroc", and by the CNRS GDR 88 "Evolution des Organismes microscopiques et environnement".

REFERENCES

Argano, R. and Pesce, G.L., 1980. A cirolanid from subterranean waters of Turkey (Crustacea, Isopoda, Flabellifera). *Rev. Suisse Zool.*, 87(2): 439–444.

Arthur, W., 1982. A developmental approach to the problem of variation in evolutionary rates. *Biol. J. Linn. Soc.*, 18: 243–261.

Arthur, W., 1984. Variation in evolutionary rates. In: *Mechanisms of Morphological Evolution: A Combined Genetic, Developmental and Ecological Approach.* Wiley, Chichester, pp. 204–219.

Barr Jr., T.C., 1968. Cave ecology and the evolution of troglobites. *Evol. Biol.*, 2: 35–102.

Barr Jr., T.C. and Kuehne, R.A., 1971. Ecological studies in the Mammoth Cave system of Kentucky. II: The ecosystem. *Ann. Speleol.*, 26: 47–96.

Botosaneanu, L. (Editor), 1986. *Stygofauna Mundi. A Faunistic, Distributional and Ecological Synthesis of the World Fauna Inhabiting Subterranean Waters (including the Marine Interstitial).* E.J. Brill/W. Blackhuys, Leiden, 740 pp.

Botosaneanu, L. and Holsinger, J.R., 1991. Some aspects concerning colonization of the subterranean realm – especially of subterranean waters: a response to Rouch and Danielopol, 1987. *Stygologia,* 6(1): 11–39.

Botosaneanu, L. and Notenboom, J., 1989. Eastern Mediterranean freshwaters stygobiont Cirolanids (Isopoda, Cirolanidae) with description of three new species. *Zool. Jahrb. Abt. Syst.,* 116: 1–19.

Botosaneanu, L. and Notenboom, J., 1992. The stygobiont Cirolanidae (Isopoda) of Israel and Sinai. *Isr. J. Zool.*, 37: 213–224.

Botosaneanu, L., Boutin, C. and Henry, J.-P., 1985. Deux remarquables Cirolanides stygobies nouveaux du Maroc et de Rhodes. Problématique des genres *Typhlocirolana* Racovitza, 1905 et *Turcolana* Argano et Pesce, 1980 (Isopoda). *Stygologia,* 1(2): 186–207.

Boulanouar, M., Yacoubi, M., Messouli, M. and Coineau, N., 1995. Un nouveau *Microcharon* (Isopoda, Janiroidea) du Maroc. Origine et biogéographie historique. *Contrib. Zool.*, 65(1): 53–64.

Boutin, C., 1993a. *Biogéographie historique des Crustacés Malacostracés stygobies du Maroc,* Thèse de doctorat. Université Claude Bernard, Lyon I, 263 pp.

Boutin, C., 1993b. Biogéographie historique des Crustacés Isopodes Cirolanidae stygobies du groupe *Typhlocirolana* dans le bassin méditerranéen. *C.R. Acad. Sci., Sci. Vie Paris,* 316: 1505–1510.

Boutin, C., 1994a. Phylogeny and biogeography of metacrangonyctid amphipods in North Africa. *Hydrobiologia,* 287(1): 49–64.

Boutin, C., 1994b. Stygobiology and historical geology: the age of Fuerteventura (Canary Island), as inferred from its present stygofauna. *Bull. Soc. Geol. France,* 165(3): 273–285.

Boutin, C., 1997. Stygobiologie et géologie historique: l'émersion des terres de Méditerranée orientale datée à partir des Amphipodes Metacrangonyctidae (Micro-crustacés souterrains). *Geobios, M.S.*, 21: 67–74.

Boutin, C. and Coineau, N., 1988. *Pseudoniphargus maroccanus* n. sp. (subterranean amphipod) the first representative of the genus in Morocco. Phylogenetic relationships and paleobiogeography. *Crustaceana,* Suppl. 13: 1–19.

Boutin, C. and Coineau, N., 1990. "Regression model", "Modèle biphase" d'évolution et origine des microorganismes stygobies interstitiels continentaux. *Rev. Micropaléontol.*, 33(3/4): 303–322.

Boutin, C. and Coineau, N., 1991. Instabilité des conditions environnementales et vitesse de l'évolution. L'exemple des micro-crustacés souterrains d'origine marine dans les pays méditerranéens. *Bull. Inst. Geol. Bassin d'Aquitaine,* 50: 63–69.

Boutin, C. and Messouli, M., 1988. *Metacrangonyx gineti* n. sp. d'une source du Haut-Atlas marocain et la famille des Metacrangonyctidae n. fam. (Crustacés Amphipodes stygobies). *Vie Milieu,* 38(1): 67–84.

Boutin, C., Messouli, M. and Coineau, N., 1992. Phylogénie et biogéographie évolutive d'un groupe de Metacrangonyctidae, Crustacés Amphipodes stygobies du Maroc. II – Cladistique et paléobiogéographie. *Stygologia,* 7(3): 159–177.

Camacho, A.I., 1992. A classification of the aquatic and terrestrial subterranean environment and their associated fauna. In: A.I. Camacho (Editor), *The Natural History of Biospeleology. Mus. Nac. Cienc. Nat. C.S.I.C., Madrid, Monogr.* 7: 57–103.

Choubert, G. and Faure-Muret, A., 1962. Évolution du domaine atlasique marocain depuis les temps paléozoïques. In: *Volume jubilaire dédié à P. Fallot, Bull. Soc. Geol. France,* Hors-série 1: 447–454.

Coineau, N., 1984. Quelques aspects de l'adaptation des Crustacés Isopodes, Amphipodes et Syncarides à la vie dans le milieu interstitiel. *Mém. Biospeol.,* 11: 7–15.

Coineau, N., 1994. Evolutionary biogeography of the microparasellid isopod *Microcharon* (Crustacea) in the Mediterranean Basin. *Hydrobiologia,* 287(1): 79–96.

Coineau, N. and Boutin, C., 1992. Biological processes in space and time. Colonization, evolution and speciation in interstitial stygobionts. In: A.I. Camacho (Editor), *The Natural History of Biospeleology, Mus. Nac. Cienc. Nat. C.S.I.C., Madrid, Monogr.,* 7: 423–451.

Coineau, N. and Boutin, C., 1996. Age and origin of the stygobiontic amphipod *Pseudoniphargus* (Crustacea) in Morocco, with the description of three new species. *Boll. Mus. Civ. Stor. Nat. Verona* [1993(1996)] 20(2): 503–520.

Culver, D.C., 1982. *Cave Life. Evolution and Ecology.* Harvard University Press, Cambridge, MA, 189 pp.

Culver, D.C., Kane, T.C. and Fong, D.W., 1995. *Adaptation and Natural Selection in Caves. The Evolution of Gammarus minus.* Harward University Press, Cambridge, MA, 223 pp.

Danielopol, D.L. and Wouters, K., 1992. Evolutionary (paleo)biology of marine interstitial Ostracoda. *Geobios,* 25(2): 207–211.

Danielopol, D.L., Creuzé des Châtelliers, M., Mösslacher, F., Pospisil, P. and Popa, R., 1994. Adaptation of crustacea to interstitial habitats: a practical agenda for ecological studies. In: J. Gibert, D.L. Danielopol and J.A. Stanford (Editors), *Groundwater Ecology.* Academic Press, San Diego, CA, pp. 217–243.

Diviaco, G. and Ruffo, S., 1985. Nuovi Bogidiellidi delle acque sotterranee africane (Crustacea, Amphipoda). *Monit. Zool. Ital.,* 7: 135–148.

Felsenstein, J., 1985. Phylogenies and the comparative method. *Am. Nat.,* 125: 1–15.

Fitch, W.M., 1980. Estimating the total number of nucleotide substitutions since the common ancestor of a pair of homologous genes: comparison of several methods and three beta haemoglobin messenger RNA's. *J. Mol. Evol.,* 16: 153–209.

Fitch, W.M. and Langley, C.H., 1976. Evolutionary rates in proteins: neutral mutations and the molecular clock. In: M. Goodman and R.E. Tashian (Editors), *Molecular Anthropology.* Plenum, New York, pp. 197–219.

Gibert, J., Stanford, J.A., Dole-Olivier, M.J. and Ward, J.V., 1994a. Basic attributes of groundwater ecosystems and prospects for research. In: J. Gibert, D.L. Danielopol and J.A. Stanford (Editors), *Groundwater Ecology.* Academic Press, San Diego, CA, pp. 8–40.

Gibert, J., Vervier, P., Malard, F., Laurent, R. and Reygrobellet, J.-L., 1994b. Dynamics of communities and ecology of karst ecosystems: example of three karsts in eastern and southern France. In: J. Gibert, D.L. Danielopol and J.A. Stanford (Editors), *Groundwater Ecology.* Academic Press, San Diego, CA, pp. 425–450.

Ginet, R. and Decou, V., 1977. *Initiation à la Biologie et à l'Écologie Souterraine.* Ed. Univ. J.P. Delarge, Paris, 345 pp.

Henry, J.P. and Magniez, G., 1983. Introduction pratique à la systématique des organismes des eaux continentales françaises. 4 – Crustacés Isopodes (principalement Asellotes). *Bull. Soc. Linn. Lyon,* 52: 319–358.

Holsinger, J.R., 1986. Zoogeographic patterns of North American subterranean amphipod crustaceans. In: R.H. Gore and K.L. Heck (Editors), *Crustacea Issues, 4, Crustacean Biogeography.* Balkema, Rotterdam, pp. 85–106.

Holsinger, J.R., 1991. What can vicariance geographic models tell us about the distributional history of subterranean amphipods? *Hydrobiologia,* 223: 43–45.

Holsinger, J.R., 1993. Biodiversity of subterranean amphipod crustaceans: global patterns and zoogeographic implications. *J. Nat. Hist.,* 27: 821–835.

Holsinger, J.R., 1994a. Pattern and process in the biogeography of subterranean amphipods. *Hydrobiologia,* 287: 131–145.

Holsinger, J.R., 1994b. Biogeographic and ecological implications of newly discovered populations of the stygobiont isopod crustacean *Antrolana lira* Bowman (Cirolanidae). *J. Nat. Hist.,* 28: 1047–1058.

Iliffe, T.M., 1992. Anchialine cave biology. In: A.I. Camacho (Editor), *The Natural History of Biospeleology. Mus. Nac. Cienc. Nat. C.S.I.C., Madrid, Monogr.,* 7: 613–636.

Iliffe, T.M., Wilkens, H., Parzefall, J. and Williams, D., 1984. Marine lava cave fauna: composition, biogeography and origins. *Science,* 225: 309–311.

Kane, T.C. and Culver, D.C., 1991. The evolution of troglobites *Gammarus minus* (Amphipoda: Gammaridae) as a case study. *Mém. Biospeol.,* 18: 3–14.

Kane, T.C., Culver, D.C. and Mathieu, J., 1994. Biotic fluxes and gene flow. In: J. Gibert, D.L. Danielopol and J.A. Stanford (Editors), *Groundwater Ecology.* Academic Press, San Diego, CA, pp. 245–270.

Kimura, M., 1983. *The Neutral Theory of Molecular Evolution.* Cambridge University Press, New York, 472 pp.

Kluge, A.G. and Farris, J.S., 1969. Quantitative phyletics and the evolution of the anurans. *Syst. Zool.,* 18: 1–32.

Lamotte, M., 1988. Phénomènes fortuits et évolution. In: *L'Évolution dans sa Réalité et ses diverses Modalités.* Colloque International, Fondation Singer-Polignac, Paris, 1985, pp. 241–268.

Lamotte, M., 1994. *Théorie Actuelle de l'Évolution.* Hachette, Paris, 444 pp.

Maddison, W.P., 1989. Reconstructing character evolution on polytomous cladograms. *Cladistics,* 5: 365–377.

Malard, F., Gibert, J., Laurent, R. and Reygrobellet, J.-L., 1994. A new method for sampling the fauna of deep karstic aquifers. *C.R. Acad. Sci. Paris, Sci. Vie,* 317: 955–966.

Marmonier, P. and Danielopol, D.L., 1988. Découverte de *Nannocandona fava* Eckman (Ostracoda Candoninae) en Basse Autriche, son origine et son adaptation au milieu interstitiel. *Vie Milieu,* 38: 35–48.

Messouli, M., 1988. *Les Crustacés Amphipodes Souterrains du Groupe Metacrangonyx. Répartition, Systématique et Phylogénie,* Thèse (Ph.D.). Université de Marrakech, Morocco, 234 pp.

Messouli, M., 1994. *Evolution, Phylogénie et Biogéographie Historique des Metacrangonyctidae, Crustacés Amphipodes Stygobies du Nord de l'Afrique et des Régions Voisines,* Thèse (Doctorat d'État). Université de Marrakech, Morocco, 312 pp.

Michard, A., 1976. *Éléments de Géologie du Maroc,* Notes et Mémoires du Service géologique du Maroc. Rabat, 408 pp.

Nei, M., 1972. Genetic distance between populations. *Am. Nat.,* 106: 283–292.

Notenboom, J., 1988. Phylogenetic relationships and biogeography of the groundwater dwelling Amphipod genus *Pseudoniphargus*

(Crustacea), with emphasis on the Iberian species. *Bijdr. Dierk.*, 58: 159–204.

Notenboom, J., 1990. Introduction to Iberian groundwater amphipods. *Limnetica,* 6: 165–176.

Notenboom, J., 1991. Marine regressions and the evolution of groundwater dwelling amphipods (Crustacea). *J. Biogeogr.,* 18: 437–454.

Parsons, P.A., 1987. Evolutionary rates under environmental stress. *Evol. Biol.,* 21: 311–347.

Peters, N., 1988. Quantitative Aspekte der regressiven Evolution – dargestellt am Beispiel der Augenrückbildung bei Höhlenfischen. *Z. zool. Syst. Evolutionsforsch.,* 28: 430–441.

Piqué, A., 1995. *Géologie du Maroc. Les Domaines Régionaux et leur Évolution Structurale.* Pumag, Marrakech, 284 pp.

Poulson, T.L. and White, W.B., 1969. The cave environment. *Science,* 165: 971–981.

Purvis, A. and Garland Jr., T., 1993. Polytomies in comparative analyses of continuous characters. *Syst. Biol.,* 42(4): 569–575.

Racovitza, E.G., 1905. *Typhlocirolana moraguesi* n.g., n.sp. Isopode aquatique cavernicole des grottes du Drach (Baléares). *Bull. Soc. Zool. Fr.,* 20: 72–80.

Rouch, R. and Danielopol, D.L., 1987. L'origine de la faune aquatique souterraine entre le paradigme du refuge et le modèle de la colonisation active. *Stygologia,* 3(4): 345–372.

Sarich, V.M., 1977. Rates, sample sizes and the neutrality hypothesis for electrophoresis in evolutionary studies. *Nature,* 265: 24–28.

Sbordoni, V., Cobolli Sbordoni, M. and De Matthaeis, E., 1979. Divergenza genetica tra populazioni e specie ipogee ed epigee di *Niphargus* (Crustacea, Amphipoda). *Lav. Soc. Ital. Biogeogr. N.S.,* 6: 329–351.

Sket, B., 1985. Why all cave animals don't look alike: a discussion of the adaptative value of reduction process. *Natl. Speleol. Soc. Bull.,* 47: 78–85.

Stock, J.H., 1977. The taxonomy and zoogeography of the Hadziid Amphipoda, with emphasis on the West Indian taxa. *Stud. Fauna Curaçao,* 55: 1–30.

Stock, J.H., 1980. Regression model evolution as exemplified by the genus *Pseudoniphargus* (Amphipoda). *Bijdr. Dierk.,* 50(1): 105–144.

Stock, J.H., 1981. The taxonomy and zoogeography of the family Bogidiellidae (Crustacea, Amphipoda), with emphasis on the West Indian taxa. *Bijdr. Dierk.,* 51(2): 345–374.

Stock, J.H., 1986. Carribean biogeography and a biological calendar for geological events. In: R.H. Gore and K.L. Heck (Editors), *Crustacean Biogeography,* Vol. 4 of *Crustacean Issues.* Balkema, Rotterdam–Boston, pp. 195–203.

Stock, J.H., 1995. Age of limnic and stygobiont Amphipoda of the Madeira Archipelago and La Gomera, Canary Islands. *Pol. Arch. Hydrobiol.,* 42(4): 409–415.

Stock, J.H., 1996. Biogeography and evolutionary scenario of aquatic organisms in Macronesia. *Bol. Mus. Mun. Funchal,* Suppl. 4 (dated 1995, published 1996): 729–745.

Stock, J.H., Iliffe, T.M. and Williams, D., 1986. The concept "anchihaline" reconsidered. *Stygologia,* 2: 90–92.

Swofford, D.L., 1982. *Genetic variability, population differentiation, and biochemical relationships in the family Amblyopsidae,* Thesis. Eastern Kentucky University, Richmond, 230 pp.

Taddei, F., Matic, I. and Radman, M., 1996. Du nouveau sur l'origine des espèces: Le paradigme darwinien revu par l'analyse moléculaire des bactéries. *La Recherche,* 291: 52–59.

Taddei, F., Matic, I. and Radman, M., 1997a. Mutagenèse et Adaptation. *Bull. Soc. fr. Génétique (Médecine/Sciences),* 13: 1–6.

Taddei, F., Radman, M., Maynard-Smith, J., Toupance, B., Gouyon, P.H. and Godelle, B., 1997b. Role of mutator alleles in adaptive evolution. *Nature,* 387: 700–702.

Thorpe, J.P., 1982. The molecular clock hypothesis: Biochemical evolution, genetic differentiation and systematics. *Annu. Rev. Ecol. Syst.,* 13: 139–168.

Turquin, M.J., 1984. Un cas de transition démographique dans le milieu souterrain. *Verh. Int. Verein. Theor. Angew. Limnol.,* 22: 1751–1754.

Vandel, A., 1964. *Biospéologie: La Biologie des Animaux Cavernicoles.* Gauthier-Villars, Paris, 619 pp.

Wägele, J.W., 1989. Evolution and phylogenetisches System der Isopoda. *Stand Forsch. Neue Erkenntn. Zool.,* 140: 1–262.

Wilkens, H., 1973. Über das phylogenetische Alter von Höhlentieren. Untersuchungen über die cavernicole Süßwasserfauna Yukatans. *Z. zool. Syst. Evolutionsforsch.,* 11: 49–60.

Wilkens, H., 1988. Evolution and genetics of epigean and cave *Astyanax fasciatus* (Characidae, Pisces). Support for the Neutral Mutation Theory. *Evol. Biol.,* 23: 271–367.

Wilkens, H., 1992. Neutral mutations and evolutionary progress. In: A.I. Camacho (Editor), *The Natural History of Biospeleology, Mus. Nac. Cienc. Nat. C.S.I.C., Madrid, Monogr.,* 7: 401–422.

Wilkens, H. and Parzefall, J., 1974. Die Ökologie der Jameos del Agua (Lanzarote). Zur Entwicklung limnischer Höhlentiere aus marinen Vorfahren. *Ann. Speleol.,* 29: 419–434.

Wilson, A.C., Carlson, S.S. and White, T.J., 1977. Biochemical evolution. *Annu. Rev. Biochem.,* 46: 573–639.

Yang, S.Y., Soulé, M. and Gorman, G.C., 1974. *Anolis* lizards of the Eastern Caribbean: a case study in evolution. I – Genetic relationships, phylogeny and colonization sequence of the *roquet* group. *Syst. Zool.,* 23: 387–399.

Zuckerland, E. and Pauling, L., 1965. Evolutionary divergence and convergence in proteins. In: V. Bryson and H.J. Vogel (Editors), *Evolving Genes and Proteins.* Academic Press, New York, pp. 97–166.

Chapter 24

POPULATION GENETIC STRUCTURE, SPECIATION AND EVOLUTIONARY RATES IN CAVE-DWELLING ORGANISMS

Valerio SBORDONI, Giuliana ALLEGRUCCI and Donatella CESARONI

INTRODUCTION

Cave organisms offer an unrivalled opportunity to investigate the process of population structuring and genetic divergence in space and time. Depending on the organism and the habitat features, it is possible to outline virtually the entire evolutionary process starting from an ancestral panmictic population, passing through fragmentation of reduced metapopulational units which may have been subjected to extinction and recolonization processes, leading to the accumulation of genetic divergence, and finally speciation. The different phases of this process have a strong geographical component. In fact, the distinct phylogeographic structure associated with discontinuous geographic distribution often shown by cave-dwelling organisms strongly suggests that fragmentation through habitat vicariance is by far the prevailing mechanism promoting evolutionary divergence and speciation in the cave habitat.

The model assumes fragmentation of a widespread epigean species and isolation of small populations in cave habitats, through interruption of gene flow mainly caused by local extinction of epigean demes. Genetic divergence is subsequently driven by genetic drift and, to a variable extent, selection. As a consequence, cavernicolous sister taxa at the rank of species or even genus are allopatric and almost completely isolated from each other.

Extrinsic causes for fragmentation are often detectable both in karst areas and in volcanic islands as tectonic, palaeoclimatic and vegetational changes, or lava-flow events. In many cases these phenomena can be dated to a reasonable accuracy. An account of several mechanisms of fragmentation of cave faunas at different time scales is given by Humphreys in Chapter 22 of this volume. However, fragmentation of populations can be perceived at much smaller scales, even within a single cave, where extrinsic factors represented by habitat and resource patchiness interact with biological properties of organisms like fecundity, dispersal ability and behavioural cues in the choice of an optimal habitat.

Present-day spatial patterns shown by relictual cave organisms reflect historical relationships between populations and species more clearly than in the case of epigean organisms. The geographic distribution and variation of the latter is often complicated by recurrent episodes of extinction, recolonization, and secondary contacts, eventually leading to complex patterns of introgression and hybridization. To some extent processes of these kinds also concern cave faunas. While the great majority of cave-adapted species are relicts, many are not, showing range overlap or parapatry with their presumed epigean relatives. A major challenge for evolutionary biospeleologists in the near future is to ascertain the nature of these contacts, whether they lead to primary genetic clines, with genetic differentiation occurring across an ecotone, or are merely secondary contacts between gene pools that had already diverged in allopatry.

In this paper we shall discuss relevant data on the genetic structure of cave-dwelling populations in order to improve and refine description of the process of genetic divergence and speciation in the cave habitat at different scales in space, and attempt to provide a time scale for these processes.

TROPICAL CAVE FAUNAS AND THE REFUGIA HYPOTHESIS

One important cause for fragmentation of populations

within a species is their limited survival in refugial habitats that are geographically isolated from each other. In the Northern Hemisphere, Pleistocene climatic changes affected to a very large extent the whole distribution of vegetation types, leading to alternate relictualization of one or other kind of biota. These effects are so well documented from pollen records (Huntley and Birks, 1983) that there is little place for questioning their role as factors of isolation and evolution of both animals and plants in a variety of environments. Also, there is little doubt that these changes strongly affected the present distribution and diversity of cave faunas in temperate regions (Jeannel, 1943; Vandel, 1964; Peck, 1973; Juberthie, 1988). Although other kinds of cave relicts were recognised in the early biospeleological literature (see Vandel, 1964), emphasis was on the glacial relict hypothesis in order to account for the occurrence of troglobites in the unglaciated regions of Europe, North America and Japan, where most of the troglobites known at that time were recorded. However, the refuge hypothesis could not account for the growing amount of records of troglomorphic animals in tropical caves and in lava tubes of volcanic archipelagos (Chapman, 1986; Howarth, 1987; Rouch and Danielopol, 1987).

However, there are two major points, raised in the context of criticism of the "refuge paradigm", which, instead of weakening the theory, seem to provide valuable elements to widen its scope. The first is the mode of colonization of the subterranean habitat, which can be either a constrained retreat to survive climatic shifts or competition, or an active occupation of available habitat to exploit its resources (Rouch and Danielopol, 1987; Bellés, 1991). All these modalities are clearly compatible with the subsequent isolation and fragmentation of populations in caves, since colonization and relictualization are distinct processes, usually occurring successively.

The second point is the relevance claimed for the "recent discovery of tropical troglobites" as undermining the climatic refugium hypothesis (Howarth, 1987). In fact, the occurrence of varying proportions of troglobites in tropical cave communities from different areas has long been recognized by biospeleologists since the sixties (Mitchell, 1969; Peck, 1974, 1976; Sbordoni and Argano, 1972; Sbordoni et al., 1973, 1977, 1980a; Orghidan et al., 1973, 1977; Reddell, 1981; and references therein). Since these pioneering investigations the number of known tropical troglobites has increased to over a thousand species, representing roughly one-fifth of the known troglobitic species (Peck and Finston, 1993). Rather than threatening the climatic-refuge hypothesis, these discoveries appear to deepen existing knowledge on the uneven geographical and taxonomic distribution of tropical troglobites on both islands and continental areas, thus enhancing the fit of data to this model and providing an extensive data set to improve one's understanding.

As applied to the tropical rainforests the theory of refugia (Haffer, 1969, 1982) has been challenged by some scientists questioning whether strong climatic changes occurred in the tropics during the Pleistocene (Broecker, 1995). However, recent analyses of noble-gas ratios in ancient groundwaters in Brazil provided fresh evidence that temperatures at the last glacial maximum were around 5.4°C cooler than now (Stute et al., 1995). Moreover, studies on changes in dust flux in the global atmosphere suggest that tropical lowlands were more arid during the last glacial period than today (Thompson et al., 1995).

The effects of these changes on the cave faunas have begun to be documented by ongoing studies in Western Australia (Humphreys, 1993), Brazil (Trajano, 1995), and Chiapas, Mexico (Sbordoni, 1994, and unpublished). In this last region (including Tabasco and Guatemala), around 140 caves have been studied in widely different bioclimatic conditions, from the tropical rainforests, deciduous forests and shrub forests in the lowlands to the alpine steppe and cloud forests in the highlands at altitudes up to 3300 m. The zoological data so far collected (Sbordoni et al., 1977; see also Reddell, 1981) indicate that cave communities from different areas differ markedly not only in taxonomic composition but also in the relative occurrence of troglomorphic species, as evaluated by the "specialization index" (Sbordoni et al., 1977). Interestingly, the degree of troglomorphy of cave communities is highest in areas characterized by mountain steppe or shrub and deciduous forests, where the opportunity for migration and genetic exchange between populations from different caves is limited by relatively dry climate and the correspondingly xeric vegetation occurring on the surface. In contrast, it is very low or absent in most of the rainforests. It appears, therefore, that the degree of evolutionary specialization of cave communities in a given area is related to their degree of past or present isolation from other caves in the same area. Preliminary investigations in caves of the Selva Lacandona (Chiapas, Mexico), thought to be a typical rainforest refugium (Toledo, 1982), showed the lowest specialization index, with no terrestrial troglobites, probably because the wet, constant climate allowed constant migration and genetic exchange between cave populations and their relatives living outside. On the other hand, more recently developed rainforests, such as the Selva del Mercadito in the Malpaso lake area of Chiapas, Mexico, host highly developed troglomorphic cave communities as a result of the harsh, drier climate and vegetation during the last glacial period. These findings support the refugia hypothesis and corroborate recent studies by Humphreys (1993) and Trajano (1995), who have clearly recognized the profound effects of Pleistocene bioclimatic changes in the evolution of troglobitic relict cave faunas in Western Australia and Brazil, respectively (see also Holsinger, Chapter 21, this volume). To some extent this explanation also applies to aquatic faunas, as exemplified by the evolution of cave fishes in Somalia (*Barbopsis devecchii* and *Phreatichthys andruzzii*) in relation to the patterns of desertification in that region (Cobolli Sbordoni et al., 1996).

In conclusion, it is hard to deny the prevalent relictual character of both terrestrial and aquatic cave communities not only in temperate karst regions, so strongly affected by periodical climate changes in the Pleistocene, but also in the tropics where the alternation of wet and dry phases in climate strongly shaped hydrological regimes and cycles in forest expansion and retreat.

SPECIATION MODELS

In a recent review article, Coyne (1992) remarked that speciation is still a poorly understood process but "despite persistent controversy, evidence still favours the neodarwinian view that species usually arise as the by-products of evolution in geographically isolated populations." Previous reviews of genetic and systematic data on speciation of cave organisms (Sbordoni, 1982; Barr Jr and Holsinger, 1985) strongly supported this view, which was first presented as a verbal model by Barr Jr (1967, 1968).

There has been a tendency in the recent biospeleological literature to minimize the importance of habitat fragmentation and its consequences for gene flow, speciation and cave adaptation in favour of "dynamic" models like local habitat shifts promoting parapatric speciation (Howarth, 1980, 1987; Peck and Finston, 1993), or even sympatric divergence (Wilkens and Hüppop, 1986) of cave populations from their presumed source populations. These speculations have thrown fresh insight into the evolutionary debate in biospeleology, promoting a fine-tuning of arguments such as the diversity of subterranean habitats, the modes of colonization and the role of biological peculiarities of organisms (Juberthie, 1984; Chapman, 1986; Rouch and Danielopol, 1987; Botosaneanu and Holsinger, 1991; Danielopol and Rouch, 1991; Howarth, 1993). However, the alleged critical situations stimulating these thoughts have still not been investigated genetically. Whether these ideas represent correct interpretations of facts should be determined by proper empirical testing, requiring at least the analysis of the genetic structure of the populations involved and the assessment of their strict monophyly.

Such analyses are unfortunately lacking for most of the critical situations where an epigean species lives in parapatry or even sympatry with the cave-adapted species presumably derived from it. Therefore the hypothesis that these situations are the product of a secondary contact of populations already diverged in allopatry cannot be refuted. Based on allozyme and morphometric data, this appeared to be the most tenable explanation for the sympatric occurrence of two related crayfish species showing different degrees of troglomorphy in a Mexican cave (Sbordoni et al., 1988; Allegrucci et al., 1992a; Cesaroni et al., 1992). Similar explanations could apply to many instances of eyeless vs. eyed species pairs reported for lava tubes and nearby epigean environments in Hawaii (Howarth, 1991), the Canary Islands (Martin et al., 1989) and the Galápagos Archipelago (Peck and Finston, 1993). The same hypothesis cannot be rejected as an explanation for the heterogeneous assemblage of troglomorphic and epigean fish in Micos Cave studied by Wilkens and Hüppop (1986). This population shows discontinuous variation in eye size. In this situation, the mechanism of disruptive selection hypothesized as promoting sympatric speciation could well be viewed as a reinforcement mechanism at work between a cave population of *Astyanax fasciatus* already diverged from its epigean relatives, and new colonizing waves of these relatives which only recently reinvaded the cave following changes in hydrologic regimes. This could be tested by estimating the degree of overall genetic divergence between the two "sub-populations" at gene markers different from those involved in eye development.

In summary, since all of presently available genetic data concerning the accumulation of genetic differentiation of cave populations and species from their cave-adapted sister taxa and/or their presumed epigean relatives support the idea that species arise as by-product of geographic isolation, we accept this hypothesis as the simplest and most tenable explanation for the origin of cave-dwelling species. Parapatric divergence and speciation across cave–surface ecotones is a challenging possibility. The sympatric model appears much weaker when applied to cave evolution and considered outside the usual context of host–parasite relationships.

In contrast, the geographic speciation model represents a sufficiently adaptable framework to incorporate improvements in knowledge of habitats, biology and ecology of organisms, and to fine-tune the temporal and geographical scale of events.It should be remarked that at the local scale, and in presence of strong selection pressures, the theoretical boundary between allopatric and parapatric divergence can be very narrow. In summary, the relative importance of the genetic system, drift, and selection in promoting divergence between gene pools can be expected to affect both the amount and the rate of evolutionary changes in a wide range of situations – from tropical to temperate caves, from lava tubes to limestone cave systems, from islands to mainland.

GENETIC STRUCTURE, GENE FLOW AND GEOGRAPHIC SCALING

The study of spatially and genetically structured populations, such as populations of cave organisms, has the potential to increase understanding of the evolutionary role of ecological, environmental, and geographical factors in determining continuous or disjunct distributional patterns. In terms of evolutionary genetics, the interpretation of the genetic structure in fragmented relict populations is strictly dependent on levels of migration and dispersal ability of individuals. This implies that it is necessary to assess the nature of the extrinsic factors that could modify the extent of gene flow between nearby populations, such as the extent of genetic connection or discontinuity between populations. In the last few years powerful methodologies have been implemented to analyse population structure at different hierarchical levels of geographical organization, and to compare the behaviour of different descriptors. Investigating the opportunity for actual or potential gene exchange between populations requires careful genetic analysis. The emerging results may help to disclose historical relationships between populations and to formulate hypotheses on processes generating such spatially structured population systems.

In particular, the geographic dimension has priority in studying the genetic structure of populations when the basic questions for understanding microevolutionary processes are:

(1) What is the extent of genetic connectivity between populations at different geographic scales and under what conditions could it change?
(2) What is the limit at which a geographic barrier becomes effective in preventing gene flow and hence in starting the passive accumulation of genetic divergence between conspecific populations?
(3) What is the spatial scale of the metapopulation, and what is the spatial scale for the ultimate separation (genetic discontinuity) between populations?
(4) To what extent is genetic structure, at varying geographic scale, influenced by selective factors or barriers against dispersal or isolation by distance?

Cave organisms appear particularly suitable for this kind of investigation because caves exemplify the process of habitat fragmentation, starting from more or less continuously distributed populations. They host relict populations at different steps of isolation, depending upon the biological properties and/or the history of the organisms involved. The occurrence of similar habitat characteristics in both natural and artificial caves provides an unusual opportunity to study populations and metapopulations of the same species with different colonization times and opportunities for colonization.

A comparative study of the gene-flow levels between populations of several species of cave-dwelling arthropods showed that the patterns are consistent with the dispersal abilities and the ecological requirements of each species (Caccone, 1985). In particular, differences in habitat specialization seem to be more important in determining population structure than the availability of routes for dispersal. In fact, epigean and troglophilic species are characterized by high to moderate levels of gene flow, whereas troglobitic species have low levels of gene flow.

To deepen insight on the scaling of these processes, we introduce two model situations involving both epigean and cave-dwelling rhaphidophorid crickets and nesticid spiders, and compare the results with

Table 24.1
Data on genetic structure and gene flow among populations of cave and epigean organisms

Populations	"Window" scale (km)	Average distance (km)	F_{ST}	*Nm*
Dolichopoda laetitiae				
GDP (6 subpopulations)	0.1 0.05	0.05	0.021	11.65
CER (5 subpopulations)	0.4 0.05	0.1	0.016	15.38
PSC (4 subpopulations)	1 0.1	0.5	0.024	10.17
3 populations	20 3	12.3	0.057	4.14
5 populations	60 15	31	0.091	2.50
14 populations	300 16	136	0.249	0.75
Euhadenoecus fragilis				
2 populations	34	34	0.177	1.16
3 populations	60 30	48	0.556	0.20
4 populations	120 50	74	0.687	0.11
5 populations	270 50	129	0.724	0.10
6 populations	380 50	179	0.726	0.09
Euhadenoecus puteanus				
2 populations	43	43	0.027	9.01
3 populations	145 45	106	0.057	4.14
4 populations	280 145	183	0.138	1.56
6 populations	460 220	226	0.160	1.31
8 populations	600 220	265	0.186	1.09
Nesticus eremita				
4 populations	15 12	14	0.060	3.92
5 populations	40 20	22	0.175	1.18
7 populations	125 60	62	0.231	0.83
9 populations	200 60	104	0.248	0.76
12 populations	680 85	227	0.336	0.49
Niphargus longicaudatus				
2 populations	0.2	0.2	0.115	1.92
2 populations	6	6	0.121	1.82
4 populations	55 30	40	0.203	0.98
7 populations	90 80	63	0.201	0.99

vicariant taxa in Italy and the eastern United States. Cricket populations of *Dolichopoda laetitiae* (Sbordoni et al., 1985; Cesaroni et al., 1997) in peninsular Italy, and *Euhadenoecus fragilis* and *E. puteanus* (Caccone and Sbordoni, 1987) distributed in the eastern United States respectively, were studied by means of allozyme polymorphism. Levels of gene flow between populations were also evaluated for *Nesticus* cave spiders from the Italian peninsula by means of allozyme polymorphism (Cesaroni et al., 1981) and from the southern Appalachians (USA) on the basis of nucleotide sequence data in mitochondrial desoxyribonucleic acid (mtDNA) (Hedin, 1997).

In all these organisms, the more or less enhanced degree of population fragmentation is chiefly the result of a gradual reduction of gene flow between populations caused by climatic changes that occurred in the Pleistocene. Therefore, levels of gene flow determined on the basis of the present patterns of genetic differentiation are supposed to reflect the historical gene flow that occurred between cave and epigean populations living in a continuum of wet or mesophilous woody environments.

The amount of genetic structuring within populations of *Dolichopoda* was evaluated by means of the *F* statistics (Wright, 1965). F_{ST} values were calculated for multiple combinations of populations at different geographic scales, and used in turn to estimate the levels of gene flow among populations by using the relation F_{ST} $1/(1+4Nm)$. The product *Nm*, where *N* is the population size and *m* is the migration rate between populations, is considered a direct estimate of gene flow, since it represents the number of individuals exchanged between populations per generation. On the basis of polymorphic allozyme loci, F_{ST} and *Nm* values were calculated for different groups of populations by considering geographic "windows" (rectangular areas with established dimensions) of progressively greater size. In this context, examples from *D. laetitiae* populations probably are useful to illustrate the analysis of population structuring at different geographic scales (Table 24.1, Fig. 24.1). The size of the geographic "window" for population sampling varied from a minimum where a genetically structured population was subsampled within an area 100 m long inside the labyrinthine Piane cave [Grotta delle Piane (Umbria, Italy) – GDP] to a maximum involving 14 populations in a geographic "window" 300 160 km^2 in central Italy. Details on scales of geographic "windows" and population structure are listed in Table 24.1. The diagram in Fig. 24.1 illustrates the *Nm* values calculated for six increasingly wider geographic "windows" by using only loci presumed to be neutral. The population in Piane cave (GDP), as already outlined, is the minimum scale to be considered, since it represents the internal subdivision of a single population into six subpopulations sampled

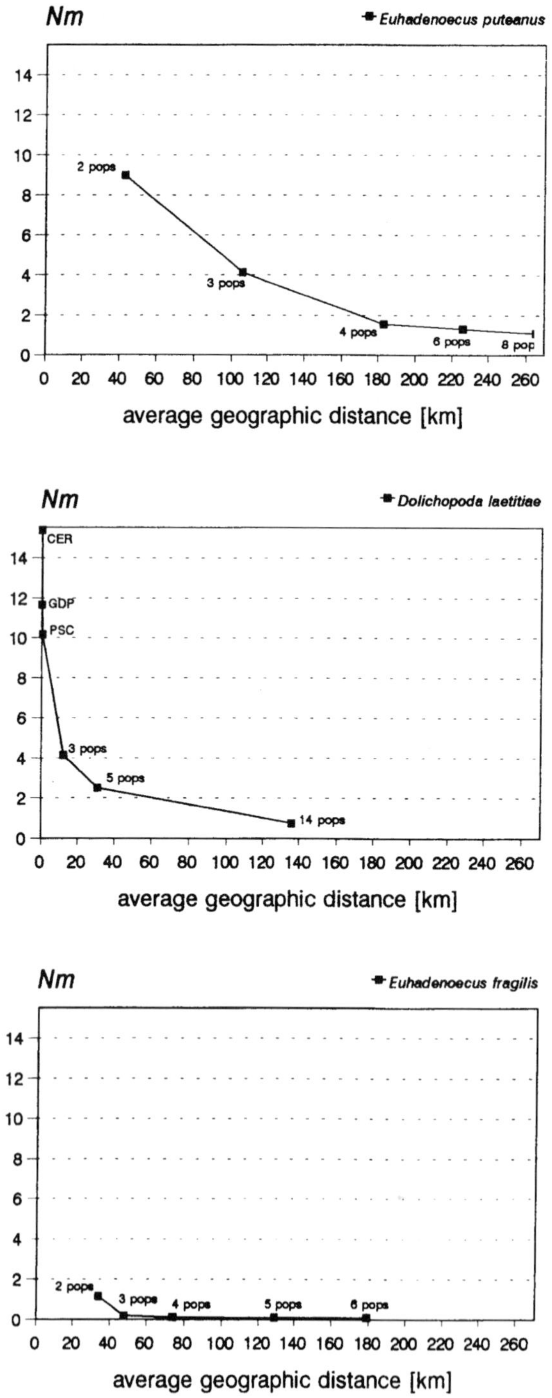

Fig. 24.1. Diagram showing the trend of the estimates of gene flow (*Nm*, as measured by Wright's F_{ST}) between groups of populations which are included in progressively wider geographic "windows". The spatial structuring of the cave cricket *Dolichopoda laetitiae* in Italy, and the forest-dwelling cricket *Euhadenoecus puteanus* and the cave cricket *Euhadenoecus fragilis* in the eastern USA, are compared. Details on genetic and geographic data are reported in Table 24.1.

in nearby but distinguishable sections of the cave (unpublished data). The spatially structured population living in the tombs of the Etruscan necropolis of Cerveteri near Rome (CER) (Sansotta, 1993; and unpublished) represents the second step of geographic scaling; the partitioning involves five groups of tombs. The third level of scaling is a complex of four nearby cave populations between which the average distance is about 500 m; they were sampled in the recently colonized area of Poscola cave (Grotta di Poscola – PSC) (Bernardini et al., 1997) in the Monti Lessini in Veneto (northern Italy). These three situations showed *Nm* values ranging approximately from 15 to 10 (Table 24.1, Fig. 24.1). This result suggests an effective gene flow, although varying in extent, between subpopulations at distances of up to half a kilometre. Grouping of populations progressively more distant produces lower and lower *Nm* values (Fig. 24.1). Comparisons among different geographic "windows" for *D. laetitiae* indicate that F_{ST} and *Nm* values depend essentially on the proximity between populations. The smooth decrease of the line indicates that gene-flow effects on patterns of genetic differentiation among populations are appreciable, and that populations as distant as 50 km might have exchanged migrant individuals recently. In fact, *Nm* values of about two to three should probably be interpreted as due to a residual effect of historical gene flow. However, at smaller scales the extent of gene flow varies between populations sampled in areas with different climatic conditions. In fact, analogous studies on other *Dolichopoda* species, namely *D. geniculata* and *D. schiavazzii* in central Italy and *D. linderi* in the eastern Pyrenees, indicate that the opportunity for gene flow between populations is dependent on the habitat structure surrounding the caves (Di Russo, 1993; Allegrucci et al., 1997; Cesaroni et al., 1997). These studies provided quite concordant evidence that groups of populations from caves surrounded by xeric vegetation, in coastal Mediterranean habitats, are genetically more structured than groups of populations living in montane mesophilous forests, which show, at the same scale, higher values of *Nm*. This result is also supported by several records of *Dolichopoda* in pitfall traps set in mesophilous montane woods, which indicate active migration of individuals outside caves (unpublished data).

Interesting observations can be made if the population genetic structure of *Dolichopoda laetitiae* is compared with that of *Euhadenoecus fragilis* and

E. puteanus. Both the latter species are found in caves in the eastern United States. *Euhadenoecus puteanus* commonly occurs outside caves in forest habitats from Pennsylvania to Tennessee, while *E. fragilis* is a strictly cavernicolous species inhabiting a patchy limestone area extending from West Virginia to Tennessee. Data on allele frequencies at 41 loci were available for nine populations of *E. puteanus* and six populations of *E. fragilis* (Caccone and Sbordoni, 1987). Caccone (1985), using Slatkin's methods (1981, 1985), had previously estimated an overall high level of gene flow between populations of *E. puteanus*. The eight population samples of *E. puteanus* here examined are distributed within an area twice as wide as those sampled for *D. laetitiae* and *E. fragilis*, and the average geographic distances among population samples are also larger. However, if one compares the overall relationship between geographic distances and *Nm* estimates in *D. laetitiae* and *E. puteanus* (Fig. 24.1) within the overlapping span, one can observe a clear difference in absolute values and slope of the two relationships. At the same geographic scale the extent of gene flow among populations of *E. puteanus* is higher than among *Dolichopoda* populations. The available samples of *E. fragilis* do not allow comparison with *Dolichopoda* in the smallest geographical "windows" (Fig. 24.1). However, the degree of population structuring is much greater than in either *Dolichopoda* or *E. puteanus*. No gene flow is detectable at any scale, suggesting that population fragmentation and geographical isolation have already proceeded to a very large extent. Interestingly, the troglomorphic features are much more marked in *E. fragilis* than in *E. puteanus*, *Dolichopoda laetitiae* being intermediate both in the degree of leg elongation and in the rate of pigment reduction. This corroborates the idea that gene flow in epigean populations prevents or retards adaptation to the cave habitat, and that isolation is an important prerequisite for the evolution of troglomorphic adaptations.

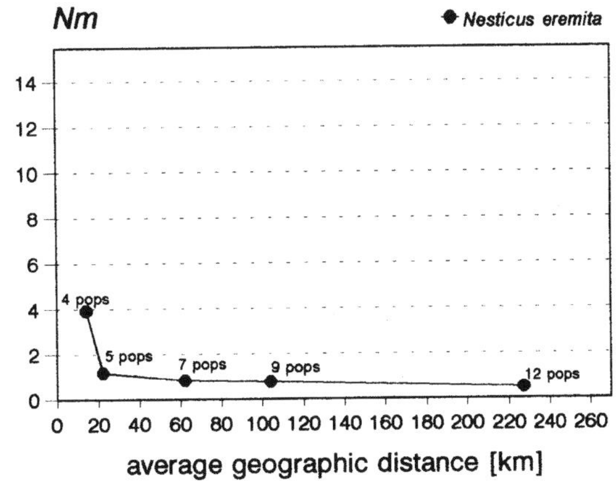

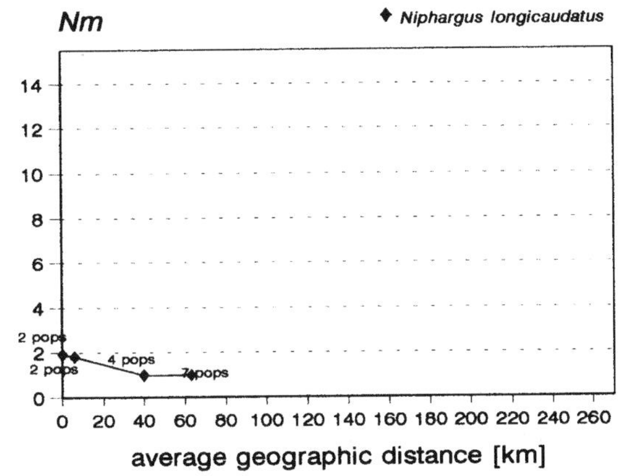

Fig. 24.2. Diagram showing the trend of the estimates of gene flow (*Nm*, as measured by Wright's F_{ST}) between groups of populations which are included in progressively wider geographic "windows". Data for population samples of the cave spider *Nesticus eremita* and the amphipod *Niphargus longicaudatus* are shown. Details on genetic and geographic data are reported in Table 24.1.

The European *Nesticus eremita* is a widespread troglophilic species colonizing not only natural limestone caves, but also artificial caves, cellars and some epigean habitats, just like *D. laetitiae.* Analysis of allozyme polymorphisms was carried out for 13 populations distributed from north-western to central Italy (Cesaroni et al., 1981). There was a highly significant correlation between genetic and geographic distances between populations, suggesting an important role of historical gene flow in shaping the patterns of genetic differentiation between populations. Previous estimates of gene-flow levels carried out by Slatkin's method suggested high to moderate levels of gene flow occurring between these spider populations, with estimates of *Nm* higher than those for cave-dwelling crickets in the genera *Euhadenoecus* and *Hadenoecus*, and for the forest-dweller *Euhadenoecus puteanus* (Caccone, 1985; Caccone and Sbordoni, 1987). Following the method used for crickets, we recalculated *Nm*, as measured by Wright's F_{ST}, for progressively increasing geographic "windows" (Table 24.1, Fig. 24.2). The smallest "window" includes four populations sampled in an area 15 12 km^2. As expected, *Nm* values increase as the "window" scale decreases, and migration is actually detectable only between very close populations. The relation between *Nm* estimates and geographic distance between populations appears very regular for this spider species, and there is a trend similar to that observed for *D. laetitiae.* It can be hypothesised that gene flow is negligible between populations of *N. eremita* when the average distance between populations is greater than 20–30 km. This outcome indicates this *Nesticus* species, individuals of which possibly can be airborne by silken cobwebs (i.e., ballooning), have the ability to disperse. The comparison to cricket species indicates that the dispersal ability for *N. eremita* is lower than for *Euhadenoecus puteanus,* perhaps less than for *Dolichopoda laetitiae,* but undoubtedly higher than for *Euhadenoecus fragilis.* For comparative purposes we report also the results of a study, carried out by analysing variation in mtDNA nucleotide sequences (Hedin, 1997), on the genetic structure of several populations of allopatric species of cave and epigean spiders belonging to the *Nesticus tennesseensis* complex in the southern Appalachian region (eastern USA). The results of this study demonstrated the occurrence of sequence monophyly within populations and large sequence divergence between conspecific populations. Such a result should be expected if, in the past, a large interbreeding population was fragmented into several subpopulations, and, subsequent to fragmentation, the exchange of migrants among these populations became very close to zero. All Appalachian *Nesticus* species studied by Hedin appeared to be characterized by similar and essentially complete population subdivision, regardless of differences in the degree of cave dependence (i.e., epigean, troglophilic or troglobitic species). These results led Hedin to conclude that limited dispersal capabilities combined with narrow physiological tolerances resulted in discontinuous structures and relatively small sizes in both cave and epigean populations of *Nesticus.* Although the population structure of the American species was studied through variation in mtDNA sequences, comparison with the results for *N. eremita* populations, based on allozyme data, leads to quite clear conclusions. The gene-flow patterns emerging in Italian and American populations of *Nesticus* are different, with higher levels of historical and/or present gene flow between nearby populations in the former than in the latter. *Nesticus* species can differ in their degree of specialization to the cave life as well as in dispersal ability, which in this case seems to play an important role in shaping pattern of gene flow. Nevertheless, the genetic structure of the species examined appears to be largely the result of historical processes in both cases. However, as already reported by Hedin (1997), other more technical reasons, concerning differences between study methods using mtDNA and allozymes (maternal haploid inheritance of mtDNA, possibly male-biased dispersal rates, etc.), might be important in explaining such a strong discordance.

In aquatic organisms, hydrologic relationships appear to reflect population connectivity and gene flow much better than geographic distance. In several instances, population samples from different caves belonging to a same groundwater aquifer may show no trace of genetic differentiation, suggesting that they belong to a single panmictic gene pool. Examples include several crustaceans, *Typhlocirolana moraguesi* from Majorca (Caccone et al., 1986), *Troglocaris anophthalmus* from different caves in the karst around Trieste (Cobolli Sbordoni et al., 1990) and *Spelaeomysis bottazzii* from Apulia (De Matthaeis et al., 1982), and the Somali cave fish *Barbopsis devecchii* (Cobolli Sbordoni et al., 1996). In other situations, however, strong population subdivision occurs even over very short geographic distances. Examples include the amphipods *Niphargus longicaudatus* in central Italy (Sbordoni et al., 1979; see Table 24.1 and Fig. 24.2) and *Gammarus minus* in West Virginia (Culver et al., 1995), and the amblyopsid fish *Typhlichthys subterraneus* in the southeastern United States (Swofford, 1982).

The case studies examined here demonstrate that different cave species, even if closely related, can show different levels of isolation, fragmentation and dispersal, depending on both extrinsic obstacles to gene flow and intrinsic properties of the organisms. The extent to which extrinsic and intrinsic factors interact can be illustrated by comparing the genetic structure of two prey–predator pairs of species in two distinct areas of the Mississippian Plateau in the eastern United States (Kane et al., 1992), namely the cave cricket *Hadenoecus subterraneus* (Caccone and Sbordoni, 1987) and the carabid beetle *Neaphaenops tellkampfi* (Kane and Brunner, 1986) which is a predator of *Hadenoecus* eggs in the Mississippian Plateau I, and *H. cumberlandicus* (Caccone and Sbordoni, 1987) and its egg predator *Darlingtonea kentuckensis* (Kane et al., 1992) in the Mississippian Plateau II. These two regions show different extents of fragmentation in karst coverage, the Mississippian Plateau II being much more dissected into relatively small limestone fragments

(Barr Jr and Holsinger, 1985). These differences may be expected to affect migration via underground routes to a greater extent than via surface routes, since substantial forest cover was almost continuous in both regions, at least in historical times. Results from allozymic analyses showed a similar overall degree of genetic structuring in the two cave cricket species (average values of F_{ST} were 0.58 and 0.46 in *H. subterraneus* and *H. cumberlandicus*, respectively). On the other hand, while *N. tellkampfi* also showed similar values ($F_{ST}=0.56$), *Darlingtonea* displayed much higher degrees of genetic structuring ($F_{ST}=0.96$, $Nm=0.01$). Genetic distances between *Darlingtonea* populations suggested that some of them have already evolved into reproductively isolated species, and that population fragmentation and divergence have been strongly affected by extrinsic barriers.

ARE CAVE-LIMITED POPULATIONS GOOD SPECIES?

Cave populations may differ in a number of traits, yet may not be regarded as different species. One of the best known examples is the fish *Astyanax fasciatus*, with populations differing strongly in their pigmentation and eye development (Wilkens, 1988). In *Dolichopoda*, conspecific populations from different caves also show consistent differences in behaviour, ecology and life-cycle parameters (Carchini et al., 1991; Di Russo et al., 1994; De Pasquale et al., 1995). On the other hand, almost identical features can be observed in cave populations which, because of their large degree of genetic divergence, can only be regarded as good species.

Dealing with the species problem in cave-dwelling populations is certainly a hard task. The biological-species concept shows severe limitations, indeed, when applied to allopatric populations. One possible way to bypass these limitations is to approach the species problem by means of multidimensional tools, where reproductive isolation is just one of several character sets to be considered (Sbordoni, 1993). From this point of view, cave populations are good species when, compared to other cave populations, they represent consistently discrete genetic units, and/or morphological and ecological units. Like most species, cave species are primarily described and identified by morphological characters. However, detectability of taxonomically sound differences for species discrimination varies among taxa according to the degree of morphological complexity and systematic knowledge. In cave insects like trechine and leiodid beetles, thoroughly studied for nearly a century, a higher number of features have been scored, even in very fine structures in the endophallus, and offer a better ground for morphological discrimination than in oniscidean isopods or nematodes. While in certain taxa two closely related species can be distinguished independently in respect of a number of characters, in others the occurrence of a diagnostic character between two species may be an accident rather than an expected feature.

SIBLING SPECIES

In troglobitic species, a large amount of genetic differentiation at structural loci may not necessarily be accompanied by divergence at the morphological level. Consequently the occurrence of local populations that are genetically differentiated but do not show corresponding morphological differentiation may be common, probably as a result of convergent evolution in similar environments. Several examples of morphologically close but genetically differentiated cave taxa have been reported (Laing et al., 1976; Sbordoni et al., 1979; Delay et al., 1980; Caccone et al., 1986; Juberthie, 1988; Cobolli Sbordoni et al., 1990).

In Table 24.2 average values and ranges for genetic distance [*D* of Nei (1972, 1978)] are reported for several cavernicolous taxa. Values between populations are compared with those between sibling and morphologically differentiated species.

The study of allozyme variation of two populations of cave millipedes (*Scoterpes copei*) enabled Laing et al. (1976) to discover the existence of two morphologically close species, coming from two distinct caves. On the basis of ten gene loci, they estimated a genetic distance equal to 1.576, a value comparable to those recorded for pairs of morphologically differentiated species in other cave arthropods (Table 24.2; Sbordoni, 1982), or even different genera in epigean organisms (Thorpe, 1983).

Allozyme electrophoresis was also instrumental in the discovery of cryptic species within the Pyrenean bathysciine beetle *Speonomus delarouzeei*. Eight cave populations ranging in altitude from 470 to 1190 metres on the south-western side of Mont Canigou (eastern Pyrenees) were studied. Data for genetic distance, based on 12 allozyme loci, suggested the existence of

Table 24.2
Nei's genetic distances[a] between populations and species of cave organisms and their relatives

	Number of loci	Comparisons between populations			Comparisons between sibling species			Comparisons between morphologically differentiated species			References
		N^b	Mean	Range	N^b	Mean	Range	N^b	Mean	Range	
ARACHNIDA											
Araneae											
Nesticus, three species	21	78	0.092	0.006–0.294				27	0.902	0.467–1.126	Cesaroni et al. (1981)
CRUSTACEA											
Amphipoda											
Crangonyx antennatus	8	21	0.091	0.032–0.153							Dickson et al. (1979)
Gammarus minus group	19										Culver et al. (1995)
-resurgence populations		36	0.021	0.002–0.005							
-cave and karst window populations		36	0.089	0.003–0.165							
-cave populations		15	0.005	0.001–0.008							
-resurgence vs cave and karst window		81	0.079	0.021–0.159[3]							
-resurgence vs cave		54	0.140	0.117–0.171[3]							
-cave vs cave and karst window		54	0.188	0.098–0.277[3]							
Niphargus, seven species	18	28	0.221	0.055–0.458	27	0.435	0.247–0.725	36	0.698	0.333–1.192	Sbordoni et al. (1979)
Isopoda											
Androniscus, four species	19							12	0.749	0.300–1.539	Gentile (1994); Gentile and Allegrucci (1997)
Androniscus dentiger group	19	87	0.224	0.005–0.481	283	0.541	0.118–0.710				Gentile (1994); Gentile and Allegrucci (1997)
Monolistra, three species	22							3	0.752	0.551–0.859	Sbordoni et al. (1980b)
Oritoniscus paganus group	14				1	0.327		4	0.700	0.693–0.707	Cobolli Sbordoni et al. (1995)
Spelaeomysis bottazzii	12	6	0.063	0.011–0.116							De Matthaeis et al. (1982)
Stenasellus, three species	15	2	0.205	0.167–0.243				8	1.359	1.182–1.590	Messana et al. (1995)
Trichoniscus, two species	15							10	1.397	1.196–1.839	Cobolli Sbordoni et al. (1997)
Trichoniscus pusillus group	15	3	0.105	0.035–0.141	42	0.751	0.308–1.745				Cobolli Sbordoni et al. (1997)

continued on next page

Table 24.2, *continued*

	Number of loci	Comparisons between populations			Comparisons between sibling species			Comparisons between morphologically differentiated species			References
		N^b	Mean	Range	N^b	Mean	Range	N^b	Mean	Range	
Troglocaris, two species	32							10	0.828	0.820–0.840	Cobolli Sbordoni et al. (1990)
Troglocaris anophthalmus group	32	13	0.001	0.000–0.005	32	0.228	0.139–0.357				Cobolli Sbordoni et al. (1990)
Typhlocirolana moraguesi group	18	1	0.000		2	0.691					Caccone et al. (1986)
DIPLOPODA											
Chordeumatida											
Scoterpes copei group	10				1	1.576					Laing et al. (1976)
INSECTA											
Orthoptera											
Ceuthophilus gracilipes	26	28	0.001	0.000–0.004							Cockley et al. (1977)
Dolichopoda, nine species	24	77	0.074	0.000–0.218				309	0.614	0.121–1.123	Allegrucci et al. (1992b, 1997) and unpublished data
Euhadenoecus, four species	41	72	0.077	0.002–0.368				203	0.517	0.256–0.634	Caccone and Sbordoni (1987)
Hadenoecus, five species	41	76	0.106	0.003–0.300				222	0.531	0.311–0.767	Caccone and Sbordoni (1987)
Troglophilus, two species	17	4	0.060	0.015–0.109				6	0.374	0.307–0.467	Sbordoni et al. (1981)
Coleoptera											
Bathysciola derosasi group	13	6	0.068	0.012–0.169	4	0.304	0.245–0.356				De Matthaeis et al. (1983)
Darlingtonea kentuckensis	11	45	0.299	0.000–0.978[3]							Kane et al. (1992)
Duvalius, two species	20	1	0.028					2	1.150	1.130–1.170	Sbordoni (1982)
Neaphaenops tellkampfi	12	10	0.029	0.005–0.071							Giuseffi et al. (1978)
Orostygia, two species	12							1	0.947		Sbordoni et al. (1980c)
Pseudanophthalmus, three species	11	24	0.024	0.003–0.117				54	0.139	0.003–0.264	Kane et al. (1992)
Ptomaphagus hirtus	13	15	0.219	0.004–0.478							Laing et al. (1976)
Speonomus delarouzeei, four species	12	11	0.063	0.002–0.157	17	0.451	0.333–0.588	2	2.418	2.294–2.542	Delay et al. (1980) and unpublished data
Speonomus, three species	18				3	0.817	0.430–1.020				Crouau-Roy (1988)
Speonomus hydrophilus	17	253	0.035								Crouau-Roy (1989)

continued on next page

Table 24.2, *continued*

	Number of loci	Comparisons between populations			Comparisons between sibling species			Comparisons between morphologically differentiated species			References
		N^b	Mean	Range	N^b	Mean	Range	N^b	Mean	Range	
OSTEICHTHYES											
Characiformes											
Astyanax mexicanus	17										Avise and Selander (1972)
-cave populations		3	0.140	0.027–0.208							
-surface populations		15	0.006	0.002–0.014							
-cave vs surface		18	0.142	0.062–0.218							
Cypriniformes											
Barbopsis devecchii	30	3	0.0003	0.000–0.001				9	0.728	0.708–0.761	Cobolli Sbordoni et al. (1996)
Phreatichthys andruzzii	30	3	0.033	0.009–0.053							
AMPHIBIA											
Urodela											
Proteus anguinus	40										Sket and Arntzen (1994)
-cave populations		3	0.233	0.140–0.560[3]							
-cave vs surface (*P. anguinus parkelj*)		2	0.325	0.230–0.410[c]							

[a] Nei (1972, 1978).
[b] Number of pair-wise comparisons.
[c] Likely to include distance values between two or more species.

three species, distributed in different climatic zones, separated by very short geographic distances (Delay et al., 1980; Cobolli Sbordoni et al., 1983). Studies on mating behaviour corroborated this view (Juberthie-Jupeau, 1988), and preliminary morphological data justified rehabilitation of the old name *Speonomus brucki* to distinguish the montane group of populations from those inhabiting caves in the Mediterranean zone, which included the typical *S. delarouzeei*. New records in the superficial underground compartment (MSS, Milieu Souterrain Superficiel; see also "Mesovoid Shallow Substratum" in Chapter 1 of this volume) and preliminary taxonomic studies suggested the existence of three additional species, and a pattern of speciation was hypothesized bringing together behavioral, physiological, ecological, geological, palaeoecological, and genetic information (Juberthie et al., 1984; Juberthie, 1988).

The situation of amphipods in the *Niphargus longicaudatus* group (Sbordoni et al., 1979) is also rather complex. The study of 18 allozyme loci in eleven population samples from central Italy revealed high genetic differentiation, reaching the species level. Even geographically close populations may show high divergence values. The genetic distance between two populations in the Abruzzi Apennines is 0.12, indicating an almost complete absence of migration in spite of the geographical closeness of the two caves, which are separated by a distance of only 300 metres.

Another case of sibling species concerns cave-dwelling isopods belonging to the genus *Typhlocirolana* (Caccone et al., 1986). This genus includes six species discontinuously distributed in underground freshwater systems throughout the southern Mediterranean coasts. The discovery of a Sicilian population of *Typhlocirolana* morphologically close to *T. moraguesi* from the Balearic Islands raised two issues: the taxonomic status of the Sicilian population, and the evolutionary history leading to the present distribution of *Typhlocirolana* species, which are thought to be derived from a *Cirolana*-like marine ancestor. The study of 18 allozyme loci revealed a high value of genetic distance, $D = 0.815$, between the Sicilian sample and two from Majorca, strongly suggesting their status as separate species.

The case of *Typhlocirolana* does not appear to be exceptional, since most genetic studies of cave-dwelling isopods, both aquatic and terrestrial, have revealed the occurrence of high genetic distances even between morphologically indistinguishable populations, as reported from studies of the genera *Oritoniscus* (Cobolli Sbordoni et al., 1995), *Stenasellus* (Messana et al., 1995) and *Trichoniscus* (Cobolli Sbordoni et al., 1997).

A complex situation in the genus *Androniscus* (Isopoda) has been thoroughly investigated. Twenty-eight populations from central Italy referable to *A. dentiger* have been analysed for allozymic variation at 19 gene loci. Morphological characters currently used in *Androniscus* taxonomy did not reveal any differences between the populations assayed. On the other hand, the study of allozyme variation revealed a high degree of genetic differentiation between the populations, highlighting the existence of two groups of populations (group A and group B), basically separated by the low valley of the Tiber river. The populations in group A proved to be genetically homogeneous, whereas those in group B were separated by high genetic distances, often comparable with those observed between morphologically well-differentiated species.These results strongly suggests that *A. dentiger* might be considered as a complex of sibling and cryptic species (Gentile, 1994; Gentile and Allegrucci, 1997; Gentile and Sbordoni, 1998).

The troglobitic cave shrimp *Troglocaris anophthalmus* is widely distributed in groundwater systems of the Italian, Slovenian and Croatian karst. The study of variation at 32 gene loci of seven population samples in the Italian karst highlighted the existence of three sharply discriminated allozyme forms with Nei's genetic distances D varying from 0.15 to 0.35 (Cobolli Sbordoni et al., 1990).While one of the three forms (Form C) distributed in the Val Rosandra area (south-east of Trieste) is strictly allopatric, the remaining two show some overlap. Form A in the Gorizia karst and Form B in the karst north-west of Trieste coexist in three caves. In these cases of sympatry there is no trace of gene flow between them, since no heterozygote was detected at any of the loci fixed for alternative alleles. Morphologically the three forms are very close and could not be distinguished. Thus, on the basis of the reproductive barriers detected at least between Forms A and B, they can only be regarded as sibling species. The most reasonable hypothesis to explain sympatry of these two species is secondary contact between populations which had evolved in allopatry in previously separated groundwater systems. This interpretation was strongly supported by tectonic and hydrogeological evidence, which also account for present mixing of flooding groundwaters from the Trieste and Gorizia karsts in caves where sympatric *Troglocaris* populations were recorded (Cobolli Sbordoni et al., 1990).

Interestingly enough, the occurrence of sibling species in *Troglocaris* seems to be reflected in its main predator, the well-known cave salamander *Proteus anguinus*. Recent studies (Sket and Arntzen, 1994) have documented a high degree of allozymic differentiation between cave populations from the Postojna and Sticna areas. In fact, genetic distance between troglomorphic populations from these two Slovenian localities (Nei's $D = 0.49$) was even higher than between the Sticna population and a recently discovered black, non-troglomorphic population (*Proteus anguinus parkelj*, $D = 0.23$). These preliminary findings, and the wide distribution area of this salamander in the Dinaric Karst, strongly suggest that the taxon *Proteus anguinus* may consist of a cluster of different species, with at least some of the troglomorphic forms morphologically indistinguishable from each other.

EVIDENCE OF REPRODUCTIVE ISOLATION

Most cave-related species are allopatric, and therefore it is usually difficult to test the occurrence of reproductive isolation between them. However, situations of secondary contact can be extremely useful in understanding the evolutionary outcome of allopatric divergence through estimates of hybridization and introgression rates. In such circumstances it is sometimes possible to make inferences on the tempo and mode of speciation.

Situations like that previously outlined for *Troglocaris anophthalmus* are not frequent, but provide unequivocal evidence of the completion of speciation at relatively low levels of genetic differentiation. Indeed, allozymic data clearly show the absence of any hybridization or even introgression between the two species, suggesting that they are reproductively isolated by either prezygotic barriers or hybrid inviability.

Two undescribed species of *Procambarus* crayfish belonging to the *P. mirandai* species group inhabit the subterranean stream running in the Cueva de los Camarones in Chiapas (Mexico). Individuals ranging from dark, robust, large-eyed forms with an epigean mien to pale, comparatively slender, microphthalmic ones resembling related troglobites were found, but only the extremes could be distinguished visually. Evolutionary relationships among individuals were investigated morphometrically by means of multivariate analysis on 14 body dimensions. These analyses revealed a significant discontinuity between the "light" species and the "dark" species, mainly determined by the shape of rostrum, chelae and telson (Allegrucci et al., 1992a). The same samples were assayed electrophoretically, and individual patterns at 23 gene loci were obtained. High levels of heterozygosity in both the "light" and the "dark" species and a clear discrimination between the two gene pools were revealed (Cesaroni et al., 1992). Although no F1 individual was detected between the two forms, the analysis of individual multilocus profiles showed the occurrence of a few individuals genetically intermediate, with allozymic variants from both species. At the morphometric level, some of these individuals showed intermediate features, while other were aberrant phenotypes (Allegrucci et al., 1992a). The occurrence of alternative alleles in the two gene pools allowed the quantification of patterns of introgression, which revealed asymmetric gene flow between the two species. Moreover, differential levels of introgression in subsamples within the surface-like species were found. Most introgressed individuals came from the inner section of the cave where the two species were greatly mixed. The existence of clearly discontinuous variation both in morphometric and allozymic characters appears to reflect a past history of allopatric divergence between these two *Procambarus* species. The hypothesis that these two crayfish populations have become isolated because of lack of connection between cave and surface waters was supported by geological data which indicated changes in the patterns of drainage in the karst area around the cave. A subsequent interconnection could have allowed a secondary contact and some introgression between the two crayfish species in the same cave. The persistence of a hybrid zone seems to be due to the ecological discontinuity, particularly food availability, observed between the inner and outer sections of the cave.

In *Dolichopoda* hybridization studies were performed both in nature and in the laboratory (Allegrucci et al., 1982; Sbordoni et al., 1987, and unpublished). The occurrence of inter-specific hybridization between *D. baccettii* and *D. schiavazzii* has been revealed under natural conditions. These species cohabit in a small artificial cave in the Mount Argentario promontory in Tuscany, where *D. schiavazzii* has probably been introduced in historical times. Out of 164 specimens in three samples collected in the late 1970s, 15 hybrids were identified by their heterozygous pattern at the

Phi locus, which is fixed for alternative alleles in these two species. Both larval and adult hybrids were scored, with a sex ratio near to unity (Allegrucci et al., 1982). No additional hybrids were found in a recent sampling from the same cave. In the laboratory, crosses between these two species (Nei's $D=0.539$: Allegrucci et al., 1992b) produced a few hatched eggs which resulted in malformed hybrids (unpublished data). Several other kinds of crosses have been performed in our laboratory in order to test the relationship between the degree of reproductive isolation between populations and their genetic distance. Preliminary results on the *D. laetitiae–D. geniculata* complex showed that both egg mortality and the percentage of hatched eggs were related to the genetic distance between parental populations (Sbordoni et al., 1987). Moreover, populations of these two species have been the subject of a release experiment, carried out by introducing a large sample of *D. laetitiae* from an artificial cave near Rome into a small natural cave already inhabited by *D. geniculata*. The two parental populations were fixed for alternative alleles at the Pgm locus, with an overall Nei's genetic distance of 0.216 (Sbordoni et al., 1985). During the following three years of regular sampling, out of 326 assayed specimens only eleven F1 hybrids were recorded, including both larval and adult specimens of both sexes, indicating that development could be completed normally (Allegrucci et al., 1982).

On the whole, these results suggest that no consistent premating reproductive barrier is at work in these crickets. In principle, there is no reason to suspect that anti-hybridization mechanisms have evolved under any selection pressure, since *Dolichopoda* species are strictly allopatric, as are single populations. Therefore reproductive isolation is expected to evolve neutrally as a by-product of the genetic divergence; affecting post-mating rather than pre-mating mechanisms. Moreover, repeated laboratory and field observations suggest that in *Dolichopoda* courtship behaviour is very simple, mainly relying on olfactory cues.

However, there is no reason to exclude the possibility that, in animals with more complex models of courtship behaviour, pre-mating isolating mechanisms may evolve as a by-product of overall genetic divergence. This seems to be the situation in the *Speonomus delarouzeei* complex of cave beetles, where reproductive isolation has been investigated to a rather detailed extent by means of crossbreeding experiments (Juberthie-Jupeau and Cazals, 1985; Sbordoni and Menichetti, unpublished) and by the analysis of mating behaviour (Juberthie-Jupeau and Cazals, 1984; Juberthie-Jupeau, 1988). The existence of at least three species, as revealed by allozyme genetic distance, was confirmed by evidence of reproductive isolation driven by atypical matings which resulted in a complete lack of sperm transfer. Moreover, Juberthie-Jupeau (1988) showed the occurrence of distinct behavioural patterns of mating, involving different numbers of steps. These behavioural differences have probably been established as a by-product of genetic divergence between populations which became isolated at least since the beginning of the Pleistocene.

NEUTRAL VERSUS ADAPTIVE DIVERGENCE

A basic question in speciation studies is the relative role of natural selection and genetic drift in shaping evolutionary divergence. Several lines of evidence seem to support the idea that most of the genetic divergence observed in isolated cave populations is the randomly produced outcome of the combined play of mutation pressure and small effective population size. Even in the absence of divergent selection pressures, it would be expected that, given sufficient time, reproductive isolation would occur as a by-product of the overall changes accumulated in the genome. The common occurrence of sibling species in the cave habitat strongly supports this view. However, differences in morphology between sister taxa may reflect either evolutionary time (i.e., time elapsed since isolation from a common ancestor) or differences in selection pressures. It is therefore informative to investigate the behaviour of different genes or characters in order to get a better picture of the evolutionary scenario produced by the interplay of environment, history and evolutionary changes. The comparative study of geographic variation, based on different data sets, either biological features or ecological parameters, represents a powerful tool for uncovering mechanisms underlying these processes.

The rather extensive amount of biological information available on the cave crickets *Dolichopoda* spp. allowed us to investigate both geographic variation in the *D. laetitiae–D. geniculata* complex and phylogenetic relationships among five species from central Italy (*D. aegilion*, *D. baccettii*, *D. geniculata*, *D. laetitiae* and *D. schiavazzii*), comparing different character sets.

Patterns of geographic variation were studied by

means of the kriging analyses and Mantel tests (Cesaroni et al., 1997). The kriging method (Matheron, 1969, 1970) is an interpolation approach widely employed in geostatistics, which permits the production of contour maps. The Mantel test (Mantel, 1967) is a non-parametric procedure, based on randomization, which generates statistical significance levels for correlational measures of similarity between distance matrices, without assumptions of specific probability distributions. Four different character sets were compared in the *D. laetitiae–D. geniculata* complex, using these methods. A first data set included allozyme polymorphisms, which in cave organisms seemed to evolve at rather constant rates, since genetic distances provided times for cladogenetic events in remarkable agreement with palaeogeographic evidence (Sbordoni et al., 1990; Allegrucci et al., 1992b). A second data set described the adult male epiphallus structure. The shape of the epiphallus was chosen because this structure is one of the major diagnostic characters on which taxonomic relationships among *Dolichopoda* species have traditionally been based (Baccetti, 1982). Its importance as a taxonomically informative character depends on the relatively safe assumption that its evolution is not influenced by environmental factors as much as other morphological traits. The third data set was represented by another morphological character, the metatibia length, which can be regarded as a synthetic descriptor of the level of adaptation of *Dolichopoda* populations to the cave environment, probably reflecting phenotypic plasticity and/or short-term adaptation to local environments rather than the isolation of gene pools (Allegrucci et al., 1987; Sbordoni et al., 1991). As a synthetic descriptor of the environmental parameters, the mean annual cave temperature was used. Temperature is considered a relevant factor affecting evolution in the cave habitat. Previous studies on *Dolichopoda* suggested a plausible role of temperature in determining morphological differences between populations (Allegrucci et al., 1987; Sbordoni et al., 1987), and as a selective agent on allozyme polymorphisms at specific loci (Sbordoni et al., 1976).

Data sets for allozymes and epiphallus shape produced similar patterns of geographic variation within the *Dolichopoda laetitiae–D. geniculata* species complex. Such resemblance in trend led us to the hypothesis that these two biological markers behave quite similarly in following historical pathways between populations. This suggested that geographic variation in both markers reflects the historical reduction of gene flow between populations, starting from the most ancient events causing speciation in the study area (Cesaroni et al., 1997). On the other hand, selective pressures controlling the shape of genital structures, which are probably under polygenic control, may be relatively homogeneous across the range of taxa. In allopatry, in fact, there is no ground for supposing that diversifying selection is acting on these structures.

The results of kriging analyses and Mantel tests also indicated that metatibia length and mean annual cave temperature vary in a quite similar fashion across the study area. Such agreement in spatial patterns suggests that variation in leg elongation is almost entirely under the control of an environmental gradient, synthetically described by the mean cave temperature. This last parameter reflects both latitudinal and especially altitudinal gradients, which in turn track the climate outside the cave. Two explanations for concordant variation in *Dolichopoda* leg elongation and cave temperature would appear plausible. Longer legs, and correlated characters, might be a direct response to local and/or warmer climatic conditions, the result of phenotypic plasticity. Alternatively, better adaptation to cave life, as evidenced by leg elongation, can be enhanced by the interruption of gene flow between cave and surface populations. This is in turn influenced by the type of vegetation outside caves and, to a large extent, by temperature.

Populations of the *D. laetitiae–D. geniculata* species complex provided a case study which was favourable for decoupling the environmental and the historical phylogenetic components in shaping geographic variation.

The phylogenetic relationships among five species of *Dolichopoda* from central Italy (*D. aegilion*, *D. baccettii*, *D. geniculata*, *D. laetitiae* and *D. schiavazzii*) were investigated using three different sets of molecular measures: allozyme polymorphisms, single-copy DNA–DNA hybridization (Sbordoni et al., 1991; Allegrucci et al., 1992b) and restriction length polymorphisms of mitochondrial DNA (Venanzetti et al., 1993). The same group of species were analysed by means of three different morphological data sets: body and appendage morphology, epiphallus morphometry, and egg chorion ultrastructure (Sbordoni et al., 1991). The latter, as well as epiphallus structure, was chosen because of the hypothesis of reduced selective constraints on this character, compared with other morphological traits and differences among species

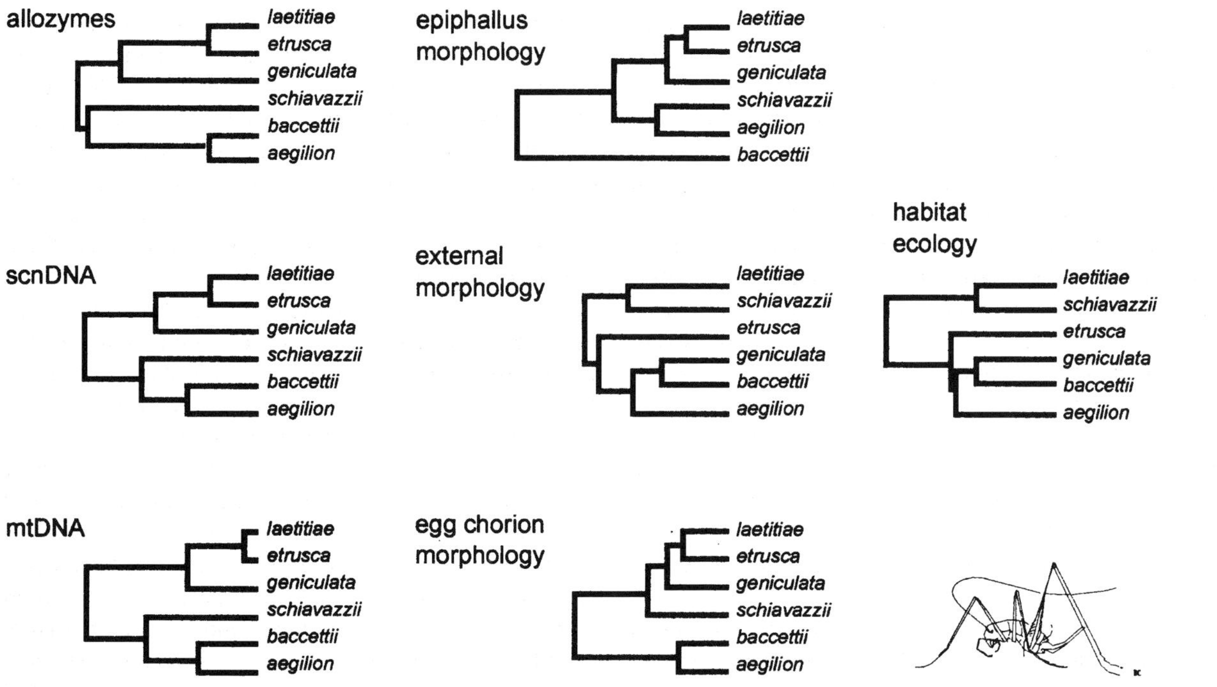

Fig. 24.3. Comparison among results deriving from molecular, morphological and ecological data available for six populations of *Dolichopoda* belonging to five species (*etrusca* is a subspecies of *D. laetitiae*). The trees from unweighted pair-group analysis using arithmetic averages [UPGMA (Sneath and Sokal, 1973)] based on distance matrices derived from seven character data sets outline the relationships among populations.

might be expected to evolve as a by-product of the overall divergence following geographic isolation.

Figure 24.3 illustrates the trees produced from the different character sets, using the Unweighted Pair-Group Method (UPGMA) of Sneath and Sokal (1973). The topology of the dendrograms is not the same for all types of data. Congruence of any pair of trees was evaluated from original distance matrices from which trees were derived, by means of Mantel tests. While all molecular data sets lead to similar tree topologies, with groupings in agreement with traditional taxonomy, the branch lengths differ. Morphological data sets were more or less dissimilar from each other and from those based on molecular data. The most contrasting tree is produced by the data from body and appendage morphology, confirming, as previously outlined, that these characters are largely influenced by environmental effects (Sbordoni et al., 1991; Cesaroni et al., 1997). In order to test this hypothesis we investigated the degree of correspondence between the morphometric tree and a tree obtained from a set of habitat descriptors, including cave age and temperature, scores for abundance of bat guano and scores for climate and vegetation type outside the caves. The branching pattern of morphometric data was highly congruent with that of habitat descriptors, supporting, at a different magnification, what we basically inferred from the geographic structure of the data.

However, epiphallus morphology, like the morphometrics of body and appendages, was not correlated with any data matrix, either morphological or molecular. On the other hand, egg chorion distances were correlated with molecular data to a statistically significant extent. The different behaviour of the data set for the epiphallus in geographic variation and in genetic relationship is clearly due to the different "windows" used in the two cases (Sbordoni et al., 1991).

These contrasting results obtained in geographic variation and in phylogenetic study on *Dolichopoda* cave crickets are not surprising, since characters evolve at different rates and are subjected to different forces, producing either constant or variable evolutionary rates.

Factors affecting morphological evolution and correlations among biological and physical measures of distance between populations have also been investigated in *Gammarus minus* (Culver et al., 1995:

pp. 149–155). A highly significant correlation was found between the genetic distance matrix and that for hydrologic distance, but not between the former and distance matrices for geography and habitat. This result suggested that the relevant parameter for understanding the genetic structure of populations of these aquatic organisms in caves is hydrologic distance, a measure of subsurface connectivity. Since habitat distance reflects the structure of the selective environment, Culver et al. also concluded that the lack of correlation between habitat and genetic distances indicated that selection does not influence allozyme variation. On the other hand, a strong dependence of morphological distance on habitat distance was found; this could be interpreted as the outcome of selection, purely environmental effects, or gene–environment interaction.

MOLECULAR CLOCKS, SPECIATION AND ADAPTATION

One major question addressed by evolutionary theory concerns the time of divergence of living taxa. In particular, when information from stratigraphy and palaeontology is not available, the possibility of dating the speciation events by molecular data is very attractive. The concept that molecules can serve as molecular clocks of evolution has been strongly debated since its introduction by Zuckerkandl and Pauling (1965). The idea of a universal clock ticking at similar rates in all organisms has been challenged, and a number of studies have shown that evolutionary rates exhibit heterogeneities of several types [for reviews see Li (1993) and Avise (1994)]. Nucleotide substitution rates vary dramatically within gene regions and among divergent taxonomic groups in both nuclear and mitochondrial genomes [for a review see Li and Graur (1991)]. These differences are probably due to different metabolic rates (Thomas and Beckenbach, 1989; Avise et al., 1992; Martin et al., 1992), differences in DNA repair efficiency (Britten, 1986), differences in nucleotide generation times (Martin and Palumbi, 1993), and differences in organismal generation time (Wu and Li, 1985; Li et al., 1996). On the basis of these results, there is a general consensus that the molecular-clock hypothesis in comparisons among distantly related taxa should be used only with extreme caution (see also Hillis et al., 1996).

On the other hand, comparisons among closely related species have been generally accepted. Several authors have suggested the existence of "local" molecular clocks [for a review see Li (1993)]. Rates of evolution for a particular gene among closely related species with similar life histories, metabolic rates and generation times, are likely to be stable. According to this hypothesis, predictions of time can be made by calibrating the evolution rate separately for each gene in each taxonomic group (Hillis et al., 1996). Once the molecular-clock hypothesis has been evaluated, the next problem is comparison of the molecular clock against absolute time. Approximate calibrations have been made of distance estimates based on data from electrophoresis, immunology and single-copy DNA (scDNA) hybridization, using taxa where palaeontological and biogeographical information provided trustworthy divergence dates between pairs of taxa. Electrophoretic studies in a number of different organisms indicate that, for small values of D ($D < 1$), genetic distances seem to be linearly related to time by the equation

$$t \text{ (in years)} = 5 \quad 10^6 D$$

(Nei, 1975). This relationship has been widely examined and discussed by several authors both from a theoretical and an empirical point of view [for a review see Nei (1987)]. In several organisms, the clock does not seem to track time since divergence accurately. This could be because of lineage-specific differences in rates, and/or strong differences in evolutionary rates of different coding regions that make up the pool of allozymic variants studied. Another possible source of bias could be poor assessment of the historical vicariant relationship between sister taxa to be studied. Notwithstanding these objective limitations of allozymic variation in the study of molecular rates of evolution, in cave organisms genetic distances based on allozymic surveys provide times of divergence that are in remarkable agreement with times of divergence based on biogeographical data (Sbordoni et al., 1990). Cave faunas are excellent palaeogeographic monitors, because cave populations have no opportunity to disperse over long distances, and have a definite origin in time that can often be dated back to specific biogeographic events such as glaciations, xerothermic phases and salinity crises. One such example is represented by *Monolistra* (Sbordoni et al., 1980b). The occurrence of Monolistrini (isopods of the typically marine family Sphaeromidae) in continental Europe is a remarkable example of relict distribution. Allozymic variation in proteins encoded by 20 loci was analysed in three troglobite species of *Monolistra* and two epigean species of *Sphaeroma* (*S. serratum*, a marine species, and *S. hookeri*, a brackish-water species). High levels of heterozygosity and calculations of divergence time (7 10^6 yr) based on genetic distance support the hypothesis that *Monolistra* diverged from its *Sphaeroma*-like marine ancestor during the Messinian Age (late Miocene, from 6.7 to 5.2 10^6 yr ago) in connection with the Mediterranean salinity crisis.

A different situation is represented by the occurrence in Sardinia of cave beetles of the genus *Speonomus* endemic to the Pyrenees (Jeannel, 1924). The present distribution seems to be due to an earlier connection between the Pyrenees and the Sardinian–Corsican plate, which rotated towards the Italian peninsula in the early Miocene to reach its present location (Alvarez, 1973; Cherchi and Montadert, 1982). According to evidence from palaeomagnetic and deep-sea drilling sources, the two areas separated 29 million years ago. The study of genetic divergence between *S. lostiai* in Sardinia and *S. delarouzeei* in the Pyrenees highlighted a value for genetic distance of 2.418, which, calibrated with albumin immunological distance for 60% of fast-evolving loci, gave an estimate of approximately 25 10^6 yr for the time of divergence (Sbordoni et al., 1990).

Newts belonging to the genus *Euproctus* present a distribution similar to that of *Speonomus*, with three vicariant species distributed in the Pyrenees (*E. asper*, with both surface and cave populations), and in fresh water outside caves in Corsica (*E. montanus*) and Sardinia (*E. platycephalus*). Values for genetic distance, obtained on the basis of 29 allozyme loci, corrected and weighted for fast and slowly evolving loci (Sbordoni et al., 1990), gave estimates of divergence time in good agreement with the geological dating of the disjunction events. In particular, estimated divergence times were respectively 25.2 10^6 yr between *E. asper* and *E. montanus*, 27.6 10^6 yr between *E. asper* and *E. platycephalus*, and 7.8 10^6 yr between the two insular species (Sbordoni et al., 1990). These three species and three other species belonging to the same family (Salamandridae), *Pleurodeles waltl*, *Triturus carnifex* and *T. vulgaris* were also analysed for sequence divergence in segments of three mitochondrial DNA genes, encoding 12S and 16S ribosomal RNA and cytochrome b. Particular attention will be paid to the results obtained from the genus *Euproctus*. On the basis of time estimates for tectonic events that led to the separation of the three land masses and of the divergence times obtained by allozyme data, sequence data were used to estimate rates of evolution for the three gene fragments and to investigate whether they conform to the rate-constancy hypothesis (Caccone et al., 1994, 1997). Assuming that these genes evolved linearly over time, and that gene flow was interrupted simultaneously with vicariant events, the time of divergence estimated by molecular distances was then compared with the divergence times based on the geological estimates. In particular, ratios of molecular divergence were compared with the ratios of geological time divergence. The ratios were very similar, indicating that the molecular-clock hypothesis cannot be rejected (Caccone et al., 1997). Moreover, on the basis of geological events, it was possible to calculate absolute rates of evolution for ribosomal and cytochrome b genes, and to compare them with rates for the same regions in other salamandrids and other vertebrates. Rates for the ribosomal genes (per million years since the last common ancestor) were about 0.38% for all substitutions and 0.10% for transversions only. Rates for the cytochrome b gene were 0.77% for all substitutions and 0.19% for transversions only, indicating that the cytochrome b gene evolves nearly twice as fast as the ribosomal genes. Ribosomal mtDNA rates are comparable to those reported for other vertebrates, but cytochrome b gene rates are lower in salamanders than in other ectotherms by a factor of 3 to 7.

HOW MUCH TIME IS NECESSARY TO SPECIATE?

As discussed by Sbordoni (1982), it is possible, on the basis of experimental data, to establish a threshold value of genetic distance at which reproductive isolation is likely to occur between cave populations. Although conclusions of this kind should be taken with caution because of the large variance expected in estimates of genetic distance from different organisms, samples of loci, and even laboratories, one can reach a rough generalization that this value seems to be as low as 0.2. The relationship between the allozymic genetic distance between two populations and the time since divergence has been discussed in the previous section, and we emphasise that vicariant histories of cave organisms are more reliable than for most organisms not dwelling in caves. Speciation events are inferred to be strictly allopatric, and mostly determined

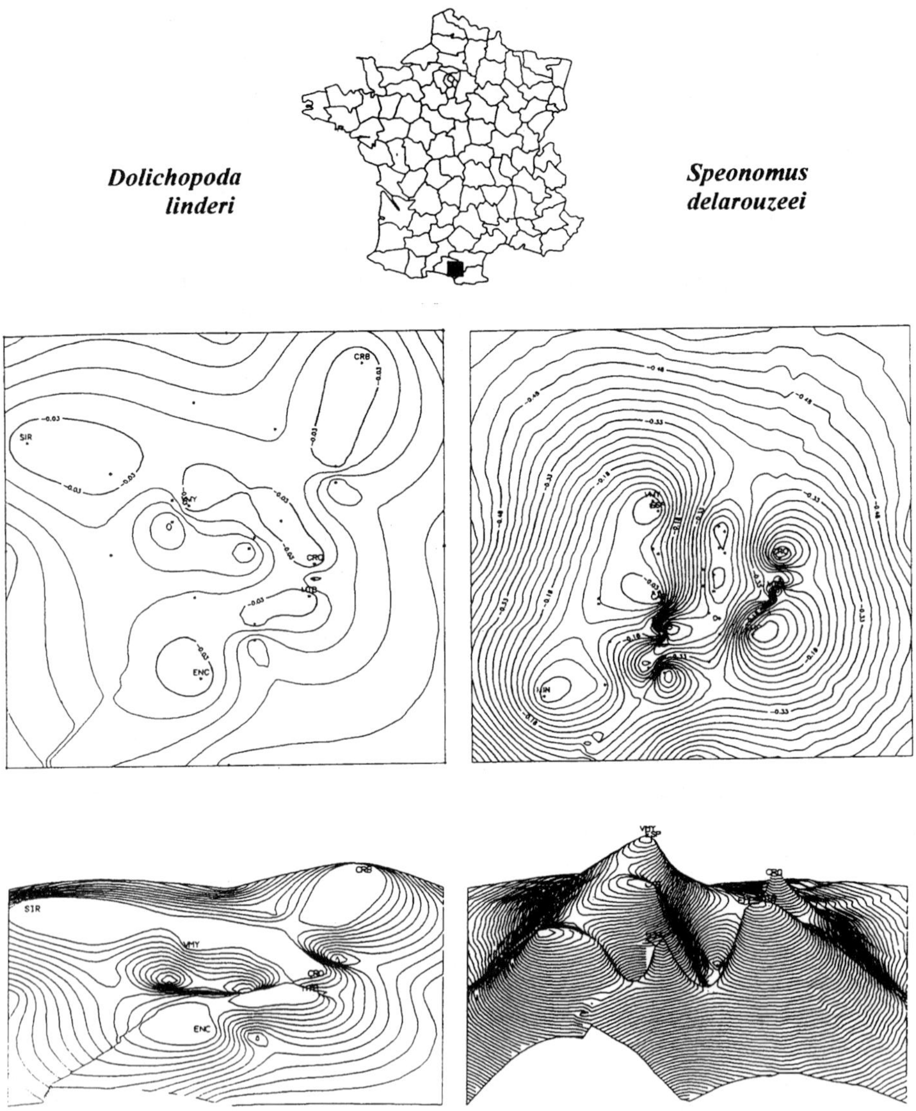

Fig. 24.4. Topographic contour maps and three-dimensional surface maps deriving from kriging interpolation of the genetic distances between populations and species of *Dolichopoda linderi* (on the left) and *Speonomus delarouzeei* complexes (on the right) inhabiting the Mont Canigou area (Pyrénées Orientales, France). Difference in height between valleys and peaks is proportional to the estimate of genetic distance between two populations. *Dolichopoda* and *Speonomus* data are reported on the same scale.

by isolation of different populations in isolated cave systems (Sbordoni, 1982; Barr Jr and Holsinger, 1985). When time estimates from geological data are available and accurate, one could attempt to estimate the time necessary for speciation.

An example of this type is represented by a species of cave cricket belonging to the genus *Dolichopoda*. The divergence of the insular species *D. aegilion* from its continental sister taxon *D. baccettii* can be dated back at least to the time when the island Giglio separated from the continent, following an earlier connection. This occurred during a marine transgression in the Günz–Mindel interglacial, 700 000 to 1 000 000 years ago (Lipparini, 1976; Mazzanti, 1983). By using Nei's (1975) relationship between time of divergence and genetic distance, we obtain an estimate of $t = 795\,000$ 405 000, which is consistent with the estimate from geological data (Allegrucci et al., 1992b). During this time *D. aegilion* and *D. baccettii* achieved morphological differentiation in body and appendage proportions, as well as in the shape of the epiphallus. Crossing experiments in progress show that, as in other *Dolichopoda* species, *D. aegilion* and *D. baccettii* are separated by post-zygotic reproductive barriers affecting zygote and hybrid viability, yet not by pre-zygotic behavioural cues. In captivity they freely mate despite substantial differences in the epiphallus morphology. Probably the two species do not represent an extreme example of fast divergence, and we argue that the time needed to speciate in isolation can be even shorter. On this basis we could predict that *Dolichopoda* populations evolving in isolation for 500 000 years are likely to have accumulated sufficient divergence in their gene pools, mostly by random processes, to achieve reproductive isolation.

This prediction can be tested in another instance where independent estimates of isolation are available. *Dolichopoda linderi* is a species widely distributed on Mont Canigou, at the eastern-most end of the Pyrenees. An electrophoretic study of seven cave populations sampled at different elevations in that area demonstrated a high level of genetic similarity between five populations located either on the northern slope of Mont Canigou or in the montane area, mostly characterized by mesic vegetation (Di Russo, 1993). Two populations markedly differed from the others, with Nei's average genetic distances of $D=0.15$–0.16. This is not surprising for the first population, Voleurs, which is geographically distant and separated from the Canigou populations by two rivers, la Tet and l'Agly. The second population however, in the Grotte de Montbolo, located on the southeastern slope of Mont Canigou, is only 2.5 km distant from another cave, Crouanques, which has a population genetically related to the main group of *D. linderi*. No obvious physical barrier occurs between these two caves. They are at similar altitudes (520–610 m) in a woody area dominated by *Quercus ilex* and *Q. pubescens*. Considering the possibility that these crickets can disperse outside caves for short distances, these two populations are almost parapatric, with a potential for gene flow. However, estimates both of genetic distance ($D=0.12$) and of gene flow ($F_{ST}=0.217$; $Nm=0.9$) suggest that they already have achieved reproductive isolation. The degree of genetic differentiation achieved between these two populations can only be understood in the light of historical factors, namely palaeoecology which, in the eastern Pyrenees, has been thoroughly investigated by Jalut (1974). Montbolo Cave lies in a south-facing limestone outcrop, which behaved as a mesic forest refugium during the coldest periods of the Pleistocene, when most of the surrounding area was characterized by a steppe vegetation dominated by *Artemisia*, a condition not favourable for *Dolichopoda*. Probably as a result of this situation, a small population unit of the ancestral *D. linderi* stock could survive there, evolving in isolation, during the last two glaciations, when most of the populations were obliged to follow the retreat of the mesic woodland. Interestingly enough, this history is remarkably similar to that of the *Speonomus delarouzeei* species complex, where the relative homogeneity of populations in the montane zone of Mont Canigou is in contrast to the genetic diversity of low-altitude populations, including those inhabiting Montbolo (*S. delarouzeei*) and Crouanques (*S. charlottae*), which evolved as separate species (Delay et al., 1980; Juberthie, 1988). However, *Dolichopoda* and *Speonomus* differ in the amount of genetic distance achieved between species, suggesting that speciation in the latter started about the beginning of Pleistocene, whereas the Montbolo population of *Dolichopoda* probably started to diverge in isolation only in the middle Pleistocene, around 600 000 years B.P. Figure 24.4 compares the evolutionary history of *Dolichopoda* and *Speonomus* in the Mont Canigou area.

ACKNOWLEDGEMENTS

We are grateful to all persons concerned with us in a long-term research program on cave biology. In particular, we wish to thank G. Gentile and C. Di Russo for their permission to include in this chapter some data from their Ph.D. theses and still unpublished papers. These studies have been supported by grants from the Italian Research Council (CNR) and Ministry of University and Scientific and Technologic Research (MURST).

REFERENCES

Allegrucci, G., Caccone, A., Cesaroni, D., Cobolli Sbordoni, M., De Matthaeis, E. and Sbordoni, V., 1982. Natural and experimental interspecific hybridization between populations of *Dolichopoda* cave crickets. *Experientia,* 38: 96–98.

Allegrucci, G., Cesaroni, D. and Sbordoni, V., 1987. Adaptation and speciation of *Dolichopoda* cave crickets (Orth. Rhaph.): geographic variation of morphometric indices and allozyme frequencies. *Biol. J. Linn. Soc.*, 31: 151–160.

Allegrucci, G., Baldari, F., Cesaroni, D., Thorpe, R.S. and Sbordoni, V., 1992a. Morphometric analysis of interspecific and microgeographic variation of crayfish from a Mexican cave. *Biol. J. Linn. Soc.*, 47: 455–468.

Allegrucci, G., Caccone, A., Cesaroni, D. and Sbordoni, V., 1992b. Evolutionary divergence in *Dolichopoda* cave crickets: a comparison of single copy DNA hybridization data with allozymes and morphometric distances. *J. Evol. Biol.*, 5: 121–148.

Allegrucci, G., Minasi, M.G. and Sbordoni, V., 1997. Patterns of gene flow and genetic structure in cave-dwelling crickets of the Tuscan endemic, *Dolichopoda schiavazzii* (Orthoptera, Rhaphidophoridae). *Heredity,* 78: 665–673.

Alvarez, W., 1973. The application of plate tectonics to the Mediterranean Region. In: D.H. Tarling and S.K. Runcorn (Editors), *Implications of Continental Drift to the Earth Sciences.* Academic Press, New York, pp. 893–908.

Avise, J.C., 1994. *Molecular Markers, Natural History and Evolution.* Chapman and Hall, New York and London, 512 pp.

Avise, J.C. and Selander, R.K., 1972. Evolutionary genetics of cave-dwelling fishes of the genus *Astyanax. Evolution,* 26: 1–19.

Avise, J.C., Bowen, B.W., Lamb, T., Meylan, A.B. and Bermingham, E., 1992. Mitochondrial DNA evolution at a turtle's pace: evidence for low genetic variability and reduced microevolutionary rate in the testudines. *Mol. Biol. Evol.*, 9: 457–473.

Baccetti, B., 1982. Ortotteri cavernicoli italiani (Notulae orthopterologicae XXXVI). *Lav. Soc. Ital. Biogeogr. Verona 1978 (n.s.),* 6: 195–200.

Barr Jr., T.C., 1967. Observations on the ecology of caves. *Am. Nat.*, 101: 475–492.

Barr Jr., T.C., 1968. Cave ecology and the evolution of troglobites. *Evol. Biol.*, 2: 35–102.

Barr Jr., T.C. and Holsinger, J.R., 1985. Speciation in cave faunas. *Annu. Rev. Ecol. Syst.*, 16: 313–337.

Bellés, X., 1991. Survival, opportunism and convenience in the processes of cave colonization by terrestrial fauna. In: J.D. Ross and N. Prot (Editors), *Homage to Roman Margalef; or Why there is such pleasure in studying nature, Oecol. Aquat.*, 10: 325–335.

Bernardini, C., Di Russo, C., Rampini, M., Cesaroni, D. and Sbordoni, V., 1997. A recent colonization of *Dolichopoda* cave crickets in the Poscola cave (Orthoptera Rhaphidophoridae). *Int. J. Speleol.*, 25(1996): 15–31.

Botosaneanu, L. and Holsinger, J.R., 1991. Some aspects concerning colonization of the subterranean realm – especially of subterranean waters: a response to Rouch & Danielopol, 1987. *Stygologia,* 6: 11–39.

Britten, R.J., 1986. Rates of DNA sequence evolution differ between taxonomic groups. *Science,* 231: 1393–1398.

Broecker, W.S., 1995. Cooling the tropics. *Nature,* 376: 212–213.

Caccone, A., 1985. Gene flow in cave arthropods: a qualitative and quantitative approach. *Evolution,* 39: 1223–1235.

Caccone, A. and Sbordoni, V., 1987. Molecular evolutionary divergence among North American cave crickets. I. Allozyme variation. *Evolution,* 41: 1198–1214.

Caccone, A., Allegrucci, G., Cesaroni, D., Cobolli Sbordoni, M., De Matthaeis, E., La Rosa, G. and Sbordoni, V., 1986. Genetic variability and divergence between cave dwelling populations of *Typhlocirolana* (Crustacea, Isopoda) from Majorca and Sicily. *Biochem. Syst. Ecol.*, 14: 215–221.

Caccone, A., Milinkovitch, M.C., Sbordoni, V. and Powell, J.R., 1994. Molecular biogeography: using the Corsica–Sardinia microplate disjunction to calibrate mitochondrial rDNA evolutionary rates in mountain newts (*Euproctus*). *J. Evol. Biol.*, 7: 227–245.

Caccone, A., Milinkovitch, M.C., Sbordoni, V. and Powell, J.R., 1997. Mitochondrial DNA rates and biogeography in European newts (genus *Euproctus*). *Syst. Biol.*, 46: 126–144.

Carchini, G., Di Russo, C. and Sbordoni, V., 1991. Contrasting age structures in cave cricket populations: patterns and significance. *Ecol. Entom.*, 16: 305–314.

Cesaroni, D., Allegrucci, G., Caccone, A., Cobolli Sbordoni, M., De Matthaeis, E., Di Rao, M. and Sbordoni, V., 1981. Genetic variability and divergence between populations and species of *Nesticus* cave spiders. *Genetica,* 56: 81–92.

Cesaroni, D., Allegrucci, G. and Sbordoni, V., 1992. A narrow hybrid zone between two crayfish species from a Mexican cave. *J. Evol. Biol.*, 5: 643–659.

Cesaroni, D., Matarazzo, P., Allegrucci, G. and Sbordoni, V., 1997. Comparing pattern of geographic variation in cave crickets by combining geostatistic methods and Mantel tests. *J. Biogeogr.*, 24: 419–431.

Chapman, P., 1986. Non-relictual cavernicolous invertebrates in tropical Asian and Australasian caves. *Communications of the 9th International Congress of Speleology, Barcelona, Spain,* 2: 161–163.

Cherchi, A. and Montadert, L., 1982. Oligo-Miocene rift of Sardinia and the early history of the western Mediterranean basin. *Nature,* 298: 736–739.

Cobolli Sbordoni, M., Delay, B., De Matthaeis, E. and Sbordoni, V., 1983. Divergenza genetica e barriere al flusso genico tra popolazioni cavernicole di *Speonomus delarouzeei* (Col. Cat.). *Atti XII Congresso Nazionale Entomologia, Roma 1980,* 2: 247–252.

Cobolli Sbordoni, M., Mattoccia, M., La Rosa, G., De Matthaeis, E. and Sbordoni, V., 1990. Secondary sympatric occurrence of sibling species of subterranean shrimps in the karst. *Int. J. Speleol.,* 19: 9–27.

Cobolli Sbordoni, M., Taiti, S., De Matthaeis, E. and Ketmaier, V., 1995. Genetic relationships among populations of the *Oritoniscus paganus* group (Crustacea, Isopoda, Oniscidea) from Corsica and Giannutri (Tuscan Archipelago). *Mem. Biospeol.,* 22: 13–16.

Cobolli Sbordoni, M., De Matthaeis, E., Mattoccia, M., Berti, R. and Sbordoni, V., 1996. Genetic variability and differentiation of hypogean Cyprinid fishes from Somalia. *J. Zool. Syst. Evol. Res.*, 34: 75–84.

Cobolli Sbordoni, M., Ketmaier, V., De Matthaeis, E. and Taiti, S., 1997. Biochemical systematics and evolutionary relationships in the *Trichoniscus pusillus* complex (Crustacea, Isopoda, Oniscidea). *Heredity,* 79: 463–472.

Cockley, D.E., Gooch, J.L. and Weston, D.P., 1977. Genetic diversity in cave dwelling crickets (*Ceuthophilus gracilipes*). *Evolution,* 31: 313–318.

Coyne, J.A., 1992. Genetics and speciation. *Nature,* 355: 511–515.

Crouau-Roy, B., 1988. Genetic variability and differentiation in a species complex of troglobitic beetles based on isozyme data. *Biochem. Syst. Ecol.,* 16: 303–310.

Crouau-Roy, B., 1989. Population studies on an endemic troglobitic beetle: geographical patterns of genetic variation, gene flow and genetic structure compared with morphometric data. *Genetics,* 121: 571–582.

Culver, D.C., Kane, T.C. and Fong, D.W., 1995. *Adaptation and Natural Selection in Caves. The Evolution of Gammarus minus.* Harvard University Press, Cambridge, MA, 223 pp.

Danielopol, D.L. and Rouch, R., 1991. L'adaptation des organismes au milieu aquatique souterrain. Réflexions sur l'apport des recherches écologiques récentes. *Stygologia,* 6: 129–142.

De Matthaeis, E., Colognola, R., Sbordoni, V., Cobolli Sbordoni, M. and Pesce, L., 1982. Genetic differentiation and variability in cave dwelling and brackish water populations of Mysidacea (Crustacea). *Z. zool. Syst. Evolutionsforsch.,* 20: 198–208.

De Matthaeis, E., Cobolli Sbordoni, M., Mattoccia, M., Allegrucci, G., Caccone, A., Cesaroni, D., Rampini, M. and Sbordoni, V., 1983. Struttura genetica di popolazioni cavernicole di *Bathysciola derosasi* (Col. Cat.): bottleneck e divergenza genetica. *Atti XII Congresso Nazionale Entomologia, Roma 1980,* 2: 253–254.

De Pasquale, L., Cesaroni, D., Di Russo, C. and Sbordoni, V., 1995. Trophic niche, age structure and seasonality in *Dolichopoda* cave crickets. *Ecography,* 18: 217–224.

Delay, B., Sbordoni, V., Cobolli Sbordoni, M. and De Matthaeis, E., 1980. Divergences génétiques entre les populations de *Speonomus delarouzeei* du Massif du Canigou (Coleoptera, Bathysciinae). *Mem. Biospeol.,* 7: 235–247.

Di Russo, C., 1993. *Adaptations et divergence évolutive des populations des Dolichopodes des Pyrénées Orientales,* Thèse de Doctorat en Ecologie et Biologie des Populations. Université Paul Sabatier de Toulouse, France, 70 pp.

Di Russo, C., Carchini, G. and Sbordoni, V., 1994. Life history variation in *Dolichopoda* cave crickets. In: H.V. Danks (Editor), *Insect Life-Cycle Polymorphism.* Kluwer Academic Publishers, Dordrecht, pp. 205–226.

Dickson, G.W., Patton, J.C., Holsinger, J.R. and Avise, J.C., 1979. Genetic variation in cave-dwelling and deep sea organisms, with emphasis on *Crangonyx antennatus* (Crustacea, Amphipoda) in Virginia. *Brimleyana,* 2: 119–124.

Gentile, G., 1994. *Struttura genetica di popolazioni e variazione geografica nell'isopode terrestre Androniscus dentiger dell'Italia Centrale (Isopoda, Oniscidea, Trichoniscidae),* Tesi di Dottorato di Ricerca in Biologia Evoluzionistica. Università Tor Vergata di Roma, Italia, 50 pp.

Gentile, G. and Allegrucci, G., 1997. Geographic variation and genetic relationships in populations of the *Androniscus dentiger* complex from Central Italy (Isopoda, Oniscidae, Trichoniscidae). *Int. J. Speleol.,* 26: 47–62.

Gentile, G. and Sbordoni, V., 1998. Indirect methods to estimate gene flow in cave and surface populations of *Androniscus dentiger* (Crustacea, Isopoda). *Evolution,* 52: 432–442.

Giuseffi, S., Kane, T.C. and Duggleby, W.F., 1978. Genetic variability in the Kentucky cave beetle *Neaphaenops tellkampfi* (Coleoptera, Carabidae). *Evolution,* 32: 679–681.

Haffer, J., 1969. Speciation in Amazonian forest birds. *Science,* 165: 131–137.

Haffer, J., 1982. General aspects of the refuge theory. In: G.T. Prance (Editor), *Biological Diversification in the Tropics.* Columbia University Press, New York, pp. 6–26.

Hedin, M.C., 1997. Molecular phylogenetics at the population/species interface in cave spiders of the southern Appalachians (Araneae: Nesticidae: *Nesticus*). *Mol. Biol. Evol.,* 14: 309–324.

Hillis, D.M., Mable, B.K. and Moritz, C., 1996. Applications of molecular systematics: The state of the field and a look to the future. In: D.M. Hillis, C. Moritz and B.K. Mable (Editors), *Molecular Systematics,* 2nd edition, Sinauer, Sunderland, MA, pp. 513–543.

Howarth, F.G., 1980. The zoogeography of specialized cave animals: a bioclimatic model. *Evolution,* 34: 394–406.

Howarth, F.G., 1987. The evolution of non-relictual tropical troglobites. *Int. J. Speleol.,* 16: 1–16.

Howarth, F.G., 1991. Hawaiian cave faunas: macroevolution on young islands. In: E.C. Dudley (Editor), *The Unity of Evolutionary Biology,* Vol. 1. Dioscorides, Portland, OR, pp. 285–295.

Howarth, F.G., 1993. High-stress subterranean habitats and evolutionary change in cave-inhabiting arthropods. *Am. Nat.,* 142: S65–S77.

Humphreys, W.F., 1993. Cave fauna in semi-arid tropical western Australia: a diverse relict wet-forest litter fauna. *Mem. Biospeol.,* 20: 105–110.

Huntley, B. and Birks, H.J., 1983. *An Atlas of Past and Present Pollen Maps for Europe: 0–13000 Years Ago.* Cambridge University press, 667 pp.

Jalut, G., 1974. *Evolution de la Végétation et Variations Climatiques durant les Quinze Derniers Millénaires dans l'Extrémité Orientale des Pyrénées,* Thèse de Doctorat d'Etat, Toulouse.

Jeannel, R., 1924. Monographie des Bathysciinae. Biospeleologica L. *Arch. Zool. Exp. Gen. Paris,* 63: 1–436.

Jeannel, R., 1943. *Le Fossiles Vivants des Cavernes.* Gallimord, Paris, 321 pp.

Juberthie, C., 1984. La colonisation du milieu souterrain; théories et modèles, relations avec la spéciation et l'évolution souterrain. *Mem. Biospeol.,* 11: 65–102.

Juberthie, C., 1988. Paleoenvironment and speciation in the cave beetle complex *Speonomus delarouzei* (Coleoptera, Bathyscinae). *Int. J. Speleol.,* 17: 31–50.

Juberthie, C., Delay, B., Juberthie-Jupeau, L., Durand, J.P., Escola, O., Bellés, X. and Blas, M., 1984. Etude du complexe *Speonomus delarouzei. Mem. Biospeol.,* 11: 141–164.

Juberthie-Jupeau, L., 1988. Mating behaviour and barriers to hybridization in the cave beetle *Speonomus delarouzei* complex (Coleoptera, Catopidae, Bathyscinae). *Int. J. Speleol.,* 17: 51–64.

Juberthie-Jupeau, L. and Cazals, M., 1984. Accouplement et comportement sexuel chez un Bathysciinae souterrain *Speonomus delarouzei* Fairm. *Behav. Processes,* 9: 147–155.

Juberthie-Jupeau, L. and Cazals, M., 1985. Isolement reproducteur entre deux populations ayant valeu d'espéces dans le complexe *Speonomus delarouzei. C.R. Acad. Sci. Paris,* 301: 393–396.

Kane, T.C. and Brunner, G.D., 1986. Geographic variation in the cave beetle *Neaphenops tellkampfi* (Coleoptera: Carabidae). *Psyche,* 93: 231–251.

Kane, T.C., Barr Jr., T.C. and Badaracca, W.J., 1992. Cave beetle genetics: geology and gene flow. *Heredity,* 68: 277–286.

Laing, C., Carmody, R.G. and Peck, S.B., 1976. How common are sibling species in cave-inhabiting invertebrates? *Am. Nat.,* 110: 184–189.

Li, W.H., 1993. So, what about the molecular clock hypothesis? *Curr. Opin. Genet. Dev.,* 3: 896–901.

Li, W.H. and Graur, D., 1991. *Molecular Evolution.* Sinauer, Sunderland, MA, 284 pp.

Li, W.H., Ellsworth, D.L., Krushkal, J., Chamg, B.H.J. and Hewett-Emmett, D., 1996. Rates of nucleotide substitution in primates and rodents and the generation-time effect hypothesis. *Mol. Phylogenet. Evol.*, 5: 182–187.

Lipparini, T., 1976. Per la storia del popolamento delle isole dell'Arcipelago Toscano. *Lav. Soc. It. Biogeogr.*, 5: 13–25.

Mantel, N., 1967. The detection of disease clustering and a generalized regression approach. *Cancer Res.*, 27: 209–220.

Martin, A.P. and Palumbi, S.R., 1993. Body size, metabolic rate, generation time and the molecular clock. *Proc. Natl. Acad. Sci. USA,* 90: 4087–4091.

Martin, A.P., Naylor, J.P. and Palumbi, S.R., 1992. Rate of mitochondrial DNA evolution is slow in sharks compared to mammals. *Nature,* 357: 153–155.

Martin, J.L., Izquierdo, I. and Oromi, P., 1989. Sur les relations entre les troglobies et les espèces épigées des Iles Canaries. *Mem. Biospeol.,* 16: 25–34.

Matheron, G., 1969. Le krigeage universel. *Cah. Cent. Morphol. Math.*, 1: 1–83.

Matheron, G., 1970. La théorie des variables régionalisées et ses applications. *Cah. Cent. Morphol. Math.*, 5: 1–212.

Mazzanti, R., 1983. Il punto sul quaternario della fascia costiera e dell'arcipelago di Toscana. *Boll. Soc. Geol. Ital.,* 52: 13–25.

Messana, G., Cobolli Sbordoni, M., De Matthaeis, E., Ketmaier, V. and Argano, R., 1995. A preliminary essay on the biogeography of Italian stenasellids. *Mem. Biospeol.,* 22: 109–114.

Mitchell, R.W., 1969. A comparison of temperate and tropical cave communities. *Southwest. Nat.,* 14: 73–88.

Nei, M., 1972. Genetic distance between populations. *Am. Nat.*, 106: 283–292.

Nei, M., 1975. *Molecular Population Genetics and Evolution.* North-Holland, Amsterdam, 288 pp.

Nei, M., 1978. Estimation of average heterozygosity and genetic distance from a small number of individuals. *Genetics,* 89: 583–590.

Nei, M., 1987. *Molecular Evolutionary Genetics.* Columbia University Press, New York, 512 pp.

Orghidan, T., Núñez Jiménez, A., Botosaneanu, L., Decou, V., Negrea, S. and Viña Bayés, N. (Editors), 1973. *Résultats des Expéditions Biospéologiques Cubano-Roumaines à Cuba,* Vol. 1. Editura Academiei Republicii Socialiste România, Bucarest, 424 pp.

Orghidan, T., Núñez Jiménez, A., Decou, V., Negrea, S. and Viña Bayés, N. (Editors), 1977. *Résultats des Expéditions Biospéologiques Cubano-Roumaines à Cuba,* Vol. 2. Editura Academiei Republicii Socialiste România, Bucarest, 420 pp.

Peck, S.B., 1973. A systematic revision and the evolutionary biology of the *Ptomaphagus* (*Adelops*) beetles of North America (Coleoptera; Leiodidae; Catopinae), with emphasis on cave-inhabiting species. *Mus. Comp. Zool. Bull.,* 145: 29–162.

Peck, S.B., 1974. The invertebrate fauna of tropical American caves, Part II: Puerto Rico, an ecological and zoogeographic analysis. *Biotropica,* 6: 14–31.

Peck, S.B., 1976. The invertebrate fauna of tropical American caves, Part III: Jamaica, an introduction. *Int. J. Speleol.,* 7: 303–326.

Peck, S.B. and Finston, T.L., 1993. Galapagos island troglobites: the questions of tropical troglobites, parapatric distributions with eyed-sister-species, and their origin by parapatric speciation. *Mem. Biospeol.,* 20: 19–37.

Reddell, J.R., 1981. *A Review of the Cavernicole Fauna of Mexico, Guatemala and Belize, Texas Mem. Mus. Univ. Texas Bull.* 27, University of Texas at Austin.

Rouch, R. and Danielopol, D.L., 1987. L'origine de la faune aquatique souterraine, entre le paradigme du refuge et le modèle de la colonisation active. *Stygologia,* 3: 345–372.

Sansotta, A., 1993. *Grilli Cavernicoli e Tombe Etrusche: Struttura Genetica di una Popolazione in Rapporto alla Eterogeneità dell'Habitat,* Tesi di Dottorato di Ricerca in Biologia Evoluzionistica. Università Tor Vergata di Roma, 84 pp.

Sbordoni, V., 1982. Advances in speciation of cave animals. In: C. Barigozzi (Editor), *Mechanisms of Speciation.* Liss, New York, pp. 219–240.

Sbordoni, V., 1993. Molecular systematics and the multidimensional concept of species. *Biochem. Syst. Ecol.,* 21: 39–42.

Sbordoni, V., 1994. Cave systems of Chiapas (Mexico): a theatre for evolutionary plays and the shaping of biodiversity. In: *XI Int. Symp. of Biospeleology, 28.VIII–2.IX.1994, Firenze,* Abstracts pp. 13–14.

Sbordoni, V. and Argano, R., 1972. Introduction; caves studied during the first mission to Mexico (1969). In: *Subterranean Fauna of Mexico, Parte I, Acc. Naz. Lincei, Quaderno N.,* 171: 5–21.

Sbordoni, V., Argano, R. and Zullini, A., 1973. Biological investigations on the caves of Chiapas (Mexico) and adjacent countries: introduction. In: *Subterranean Fauna of Mexico, Parte II, Acc. Naz. Lincei, Quaderno N.,* 171: 5–45.

Sbordoni, V., De Matthaeis, E. and Cobolli Sbordoni, M., 1976. Phosphoglucomutase polymorphisms and natural selection in populations of the cave cricket *Dolichopoda geniculata. Z. zool. Syst. Evolutionsforsch.,* 14: 292–299.

Sbordoni, V., Argano, R., Vomero, V. and Zullini, A., 1977. Ricerche sulla fauna cavernicola del Chiapas (Messico) e delle regioni limitrofe: grotte esplorate nel 1973 e nel 1975. Criteri per una classificazione biospeleologica delle grotte. In *Subterranean Fauna of Mexico, Parte III. Acc. Naz. Lincei, Quaderno N.,* 171: 5–74.

Sbordoni, V., Cobolli Sbordoni, M. and De Matthaeis, E., 1979. Divergenza genetica tra popolazioni e specie ipogee ed epigee di *Niphargus* (Crustacea, Amphipoda). *Lav. Soc. Ital. Biogeogr. N.S.,* 6: 1–23.

Sbordoni, V., Vomero, V., Argano, R. and Zullini, A., 1980a. The cave fauna of Chiapas (Mexico): tropical and temperate cave communities and their evolution. In: J.I. Furtado (Editor), *Tropical Ecology and Development, Part 1.* The International Society of Tropical Ecology, Kuala Lumpur, Malaysia, p. 168.

Sbordoni, V., Caccone, A., De Matthaeis, E. and Cobolli Sbordoni, M., 1980b. Biochemical divergence between cavernicolous and marine Sphaeromidae and the Mediterranean salinity crisis. *Experientia,* 36: 48–49.

Sbordoni, V., Allegrucci, G., Caccone, A., Cesaroni, D., Cobolli Sbordoni, M. and De Matthaeis, E., 1980c. A preliminary report of the genetic variability in troglobitic Bathysciinae: *Leptodirus hohenwarti* and two *Orostygia* species (Col. Catopidae). *Fragm. Entomol.,* 15: 327–336.

Sbordoni, V., Allegrucci, G., Caccone, A., Cesaroni, D., Cobolli Sbordoni, M. and De Matthaeis, E., 1981. Genetic variability and divergence in the cave populations of *Troglophilus cavicola* and *T. andreinii* (Orth. Rhaph.). *Evolution,* 35: 226–233.

Sbordoni, V., Allegrucci, G., Cesaroni, D., Cobolli Sbordoni, M. and De Matthaeis, E., 1985. Genetic structure of populations and species of *Dolichopoda* cave crickets: evidence of peripatric divergence. In: V. Sbordoni (Editor), *Genetics and Ecology in Contact Zones of Populations, Boll. Zool.*, 52: 95–114.

Sbordoni, V., Allegrucci, G., Caccone, A., Carchini, G. and Cesaroni, D., 1987. Microevolutionary studies in Dolichopodinae cave crickets. In: B. Baccetti (Editor), *Evolutionary Biology of Orthopteroid Insects.* Horwood, Chichester, UK, pp. 514–540.

Sbordoni, V., Allegrucci, G., Baldari, F. and Cesaroni, D., 1988. Evolutionary genetics and morphometrics of a cave crayfish population from Chiapas (Mexico). *Int. J. Speleol.*, 17: 65–81.

Sbordoni, V., Caccone, A., Allegrucci, G. and Cesaroni, D., 1990. Molecular island biogeography. In: *Biogeographical Aspects of Insularity, Atti Conv. Lincei,* 85: 55–83.

Sbordoni, V., Allegrucci, G. and Cesaroni, D., 1991. A multidimensional approach to the evolution and systematics of *Dolichopoda* cave crickets. In: G.M. Hewitt, A.W.B. Johnston and J.P.W. Young (Editors), *Molecular Techniques in Taxonomy,* NATO ASI Series, Vol. H 57. Springer, Berlin, pp. 171–199.

Sket, B. and Arntzen, J.W., 1994. A black, non-troglomorphic amphibian from the karst of Slovenia: *Proteus anguinus parkelj* n. ssp. (Urodela: Proteidae). *Bijdr. Dierk.*, 64: 33–53.

Slatkin, M., 1981. Estimating levels of gene flow in natural populations. *Genetics,* 99: 323–335.

Slatkin, M., 1985. Rare alleles as indicator of gene flow. *Evolution,* 39: 53–65.

Sneath, P.H.A. and Sokal, R.R., 1973. *Numerical Taxonomy.* W.H. Freeman, San Francisco.

Stute, M., Forster, M., Frischkorn, H., Serejo, A., Clark, J.F., Schlosser, P., Broecker, W.S. and Bonani, G., 1995. Cooling of tropical Brazil (5°C) during the last glacial maximum. *Science,* 269: 379–383.

Swofford, D.L., 1982. *Genetic Variability, Population Differentiation, and Biochemical Relationships in the Family Amblyopsidae,* Thesis. Eastern Kentucky University, Richmond.

Thomas, W.K. and Beckenbach, A.T., 1989. Variation in salmonid mitochondrial DNA: Evolutionary constraints and mechanisms of substitution. *J. Mol. Evol.*, 29: 233–245.

Thompson, L.G., Mosley-Thompson, E., Davis, M.E., Lin, P.N., Henderson, K.A., Cole-Dai, J., Boltzan, J.F. and Liu, K.B., 1995. Late glacial stage and Holocene tropical ice core records from Huscarian, Peru. *Science,* 269: 46–50.

Thorpe, J.P., 1983. Enzyme variation, genetic distance and evolutionary divergence in relation to levels of taxonomic separation. In: G.S. Oxford and D. Rollinsons (Editors), *Protein Polymorphism: Adaptive and Taxonomic Significance.* Academic Press, London, pp. 13–152.

Toledo, V.M., 1982. Pleistocene changes of vegetation in tropical Mexico. In: G.T. Prance (Editor), *Biological Diversification in the Tropics.* Columbia University Press, New York, pp. 93–111.

Trajano, E., 1995. Evolution of tropical troglobites: applicability of the model of quaternary climatic fluctuations. *Mem. Biospeol.*, 22: 203–209.

Vandel, A., 1964. *Biospéologie. La Biologie des Animaux Cavernicoles.* Ed. Gauthiers Villars, Paris, 619 pp.

Venanzetti, F., Cesaroni, D., Mariottini, P. and Sbordoni, V., 1993. Molecular phylogenies in *Dolichopoda* cave crickets and mtDNA rate calibration. *Mol. Phylogenet. Evol.*, 2: 275–280.

Wilkens, H., 1988. Evolution and genetics of epigean and cave *Astyanax fasciatus* (Characidae, Pisces): support for the Neutral Mutation Theory. *Evol. Biol.*, 23: 271–367.

Wilkens, H. and Hüppop, K., 1986. Sympatric speciation in cave fishes? Studies on a mixed population of epi- and hypogean *Astyanax* (Characidae, Pisces). *Z. zool. Syst. Evolutionsforsch.*, 24: 223–230.

Wright, S., 1965. Interpretation of population structure by F-statistics, with special regard to systems of mating. *Evolution,* 19: 395–420.

Wu, C.I. and Li, W.H., 1985. Evidence for higher rates of nucleotide substitution in rodents than in man. *Proc. Natl. Acad. Sci. USA,* 82: 1741–1745.

Zuckerkandl, E. and Pauling, L., 1965. Evolutionary divergence and convergence in proteins. In: V. Bryson and H.J. Vogel (Editors), *Evolving Genes and Proteins.* Academic Press, New York, pp. 97–166.

Part VII

Ecosystem Studies and Studies of Particular Cave Systems

VII.1. Ecosystem studies (parafluvial aquifers)

Chapter 25

A GROUNDWATER ECOSYSTEM IN THE DANUBE WETLANDS AT WIEN (AUSTRIA)

Dan L. DANIELOPOL, Peter POSPISIL, Jorge DREHER, Friederike MÖSSLACHER, Peter TORREITER, Margot GEIGER-KAISER and Amara GUNATILAKA

INTRODUCTION

Groundwater ecology, as a research field, started in Europe during the first part of this century with descriptive studies, mainly on subterranean animals (see Danielopol, 1982; Gibert, 1992; Gibert et al., 1994). In Austria, groundwater ecological studies were initiated by Spandl (1926). He showed *inter alia* that the groundwater in porous sediments along the Donau (Danube), in the Wien (Vienna) area, did not display the thermal stability typical of other subterranean environments like cave waters. He was able to show, through field and laboratory observations, that exclusively subterranean crustaceans of the genus *Niphargus* (Amphipoda), considered closely adapted to their stenothermic environment, could easily live under eurythermic conditions. Subsequently, Löffler (1960, 1964) related the structure of subsurface aquatic assemblages occurring in unconsolidated sediments to the abiotic (physical and chemical) environment. Such habitats and faunal assemblages were later described for the alluvial sediments of the Danube around Vienna (Danielopol, 1976, 1983).

During the last two decades, groundwater ecological research has become increasingly interdisciplinary (Gibert, 1992). It has been proposed that research on the Danube in Austria, both fundamental (ecological) and applied (environmental), should integrate descriptive studies on the subsurface organisms and their habitats into a framework of the structure and functioning of ecosystems (Fig. 25.1: Danielopol, 1980; Danielopol et al., 1991). Groundwater organisms are able to react to parameters in their habitats, and to modify their local environmental conditions. Both structural and functional changes of subsurface habitats and organisms are related to the dynamics of the underlying groundwater ecosystem.

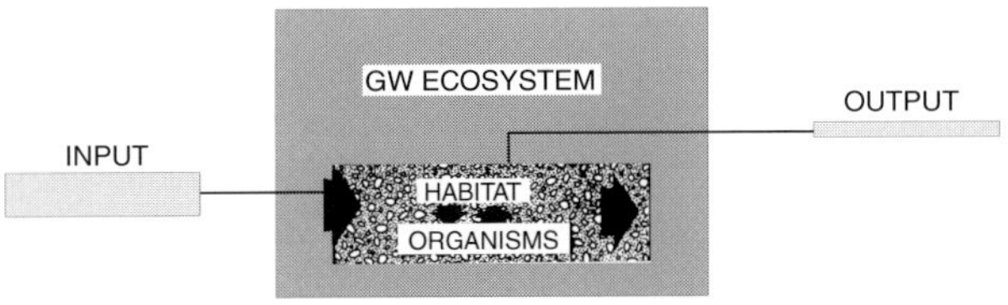

Fig. 25.1. Schematic presentation of an ecosystem (arrows suggests how matter and/or energy advance through a subsurface system).

This chapter summarizes a long-term project on a groundwater ecosystem located in the wetlands of the Danube, at Vienna. The investigations were initiated in 1975 and continued until 1982 with studies on various superficial gravel habitats of the so-called Lobau-A area, within the backwater system Eberschüttwasser, in the Lobau wetlands (Danielopol, 1980, 1983, 1989, 1991; Pospisil, 1994a, 1994b). Between 1991 and 1993 we investigated a subsurface area called the Lobau-C aquifer, which is located close to the Eberschüttwasser (Danielopol et al., 1992, 1997; Gunatilaka et al., 1994; Pospisil, 1994a,b; Pospisil et al., 1994; Negrea and Pospisil, 1995; Mösslacher, 1994, 1997a,b; Mösslacher et al., 1996; Dreher et al., 1997; Rouch and Danielopol, 1997). Here we deal with conceptual approaches used to define and/or describe the Lobau-C groundwater ecosystem, and further present relevant structural and functional aspects of this ecosystem. In this context, an attempt is made to evaluate the role of meio- and macro-organisms in the functioning of a subsurface system at a meso-scale – that is, from metres to hundreds of metres in length.

Besides field description of the ecosystem, information is presented on experimental simulations on micro-

and macro-organisms, either in the field or under laboratory conditions. This approach helps to a better understanding of complicated and dynamic features of the Lobau-C ecosystem. It is expected that the information presented here (the conceptual and methodological approach being realistic and innovative) will permit a deeper insight into the structure and functioning of groundwater ecosystems in wetland landscapes.

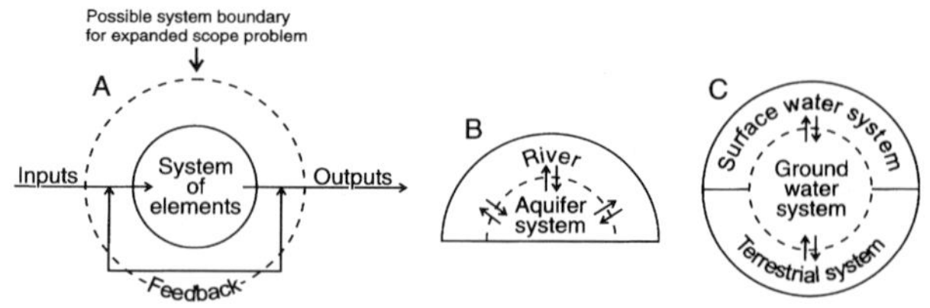

Fig. 25.2. (A) General system presentation (redrawn from Meta Systems Inc., 1975); (B) river–aquifer system; (C) connectivity of groundwater system with lotic and terrestrial (riparian) systems.

CONCEPTUAL MODELS

Tansley (1935) used the term "ecosystem" for complex systems formed by organisms and their surrounding environment. The boundaries of such systems were not precisely defined, but he mentioned (Tansley, 1935, p. 300) that what is peculiar to such systems is their internal organic cohesion ("Some of the systems are more isolated in nature, more autonomous than others. They all show organisation, which is the inevitable result of the interactions and consequent mutual adjustment of their components."). The difficulties with Tansley's concept became apparent when ecologists tried to delineate the boundaries of an ecosystem. Theoretically the interactions between organisms and their habitats form an ecosystem, as depicted in Fig. 25.1. Because organisms are in a dynamic state – they actively select their habitats, they adapt to them and/or are able to modify them – the practical designation of ecosystem boundaries is a very difficult task.

Odum (1969: p. 262) expanded the Tansleyan definition by considering the ecosystem to be an open system through which energy and matter flows and/or is cycled ("The ecosystem ... is considered to be a unit of biological organisation made up of all of the organisms in a given area, that is community, interacting with the physical environment so that a flow of energy leads to characteristic trophic structure and material cycles within the system."). It is credited to E. Odum that an ecosystem should be defined as a functional unit of the landscape (Likens and Bormann, 1985).

Groundwater ecologists (e.g. Rouch, 1977; Danielopol, 1983; Dole-Olivier et al., 1994; Gibert et al., 1994; Culver et al., 1995) adopted the Odum definition of the ecosystem. This latter allowed them in a satisfactory way to integrate empirical observations, at both population and community level, within a higher hierarchical framework.

Danielopol (1980), inspired by a conceptual model of hydrologists (Fig. 25.2A), proposed to consider the groundwater aquifer in the wetlands of the lower terrace of the Danube, in the Vienna area, as a large ecosystem. For this aquifer one can identify the input and output zones through which water and organic matter infiltrate within the underground porous system. Theoretically, such an alluvial aquifer displays feedback regulatory capacities through the water exchanges between the Danube and the aquifer reservoir as well as through the circulation of groundwater within the subsurface system. It was suggested (Danielopol, 1983; Pospisil and Danielopol, 1990; Danielopol et al., 1991) that the concept of aquifer as ecosystem could be useful not only for fundamental ecological research but also for planning groundwater protection activities.

More recently, Danielopol et al. (1997) showed that one could approach the description of the groundwater life within a Danube aquifer using the more general concept of the Global Aquifer/River System (GARS) developed by Castany (1985) for hydrogeological and water-management purposes (Fig. 25.2B). In alluvial plains of large rivers, like the Danube, a closely interactive system is formed through the constant exchange between surface (lotic) and subsurface water of the alluvial aquifer. The biological activity which develops within the aquifer is modulated by the water exchanges and/or dynamics, of the GARS. The idea of describing a subterranean ecosystem as a river/aquifer system was adopted independently by Stanford and Ward (1993) to describe the hyporheic corridor system of a large river in North America. More recently the concept was used by Rouch et al. (1997) for the meso-scale ecosystem they investigated along the Nert brook, in the Pyrenees, southern France.

Groundwater systems which exist in wetlands of large rivers are interconnected not only with the surface lotic system but also with the riparian (terrestrial) system (Fig. 25.2C). The boundary between the surface-water domain and the subsurface groundwaters can be identified close beneath the river-bed surface as the area where major flux discontinuities occur. The boundary

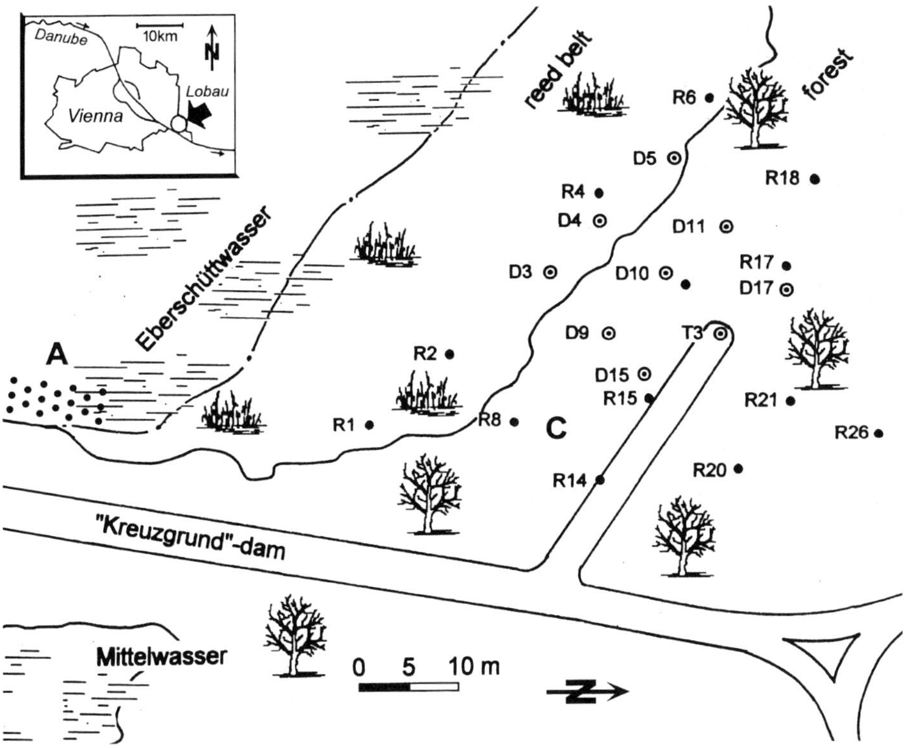

Fig. 25.3. The Lobau study area: A, alluvial sediments investigated during the period 1975–1981; C, the Lobau-C ecosystem (black dots indicate locations of mini-piezometers; encircled black dots indicate multilevel wells).

between terrestrial and groundwater systems is generally represented by the groundwater table (Freeze and Cherry, 1979). The view that a groundwater system has to be studied within the framework of its connectivity with both aquatic and terrestrial (riparian) systems has recently been emphasized in various studies (Gibert et al., 1997; Trémolières et al., 1997; Vervier et al., 1997; Ward et al., 1998). The groundwater ecosystem Lobau-C, which is described here, fits this latter model well.

GEOGRAPHICAL LOCATION

The Lobau-C groundwater ecosystem is situated in the southeastern part of the city of Vienna (Fig. 25.3). It is an area of about 900 m^2 located in the Lobau wetlands between the Eberschüttwasser and Mittelwasser backwaters. The Lobau wetlands, on the left side of the Danube, forms the forefront of the Marchfeld plain (Fig. 25.4A). Before the beginning of the regulation of the braided channel system of the Danube, at the end of the last century, the Lobau-C area was one of the Danube river banks (Fig. 25.4B) within the floodplain woods. During this century many of the old arms of the Danube river, being isolated from the main aggraded channel, clogged rapidly and were transformed into shallow backwaters. The Lobau-C area belongs to the Eberschüttwasser system (Fig. 25.5) which communicates with the Danube during high water stands. At low water level the water is retained in this system by a series of weirs. To protect the human population against floods in the inner part of the Marchfeld plain a double dike system was build around the Lobau wetlands (Fig. 25.5).

The alluvial sediments of the Marchfeld plain form a large aquifer which is recharged by water which infiltrates from the Danube and from the Bisamberg hills (Fig. 25.4A). Within the lower terrace of the Danube (the stippled area of Fig. 25.4A) the direction of groundwater flow is more or less parallel to the Danube river. During low water stands in the Danube, the inner part of the Marchfeld flows towards the Danube. In the alluvial sediments of the Lobau wetlands, where the study site is located, an important aquifer has developed which is used for the supply of drinking water to the

A

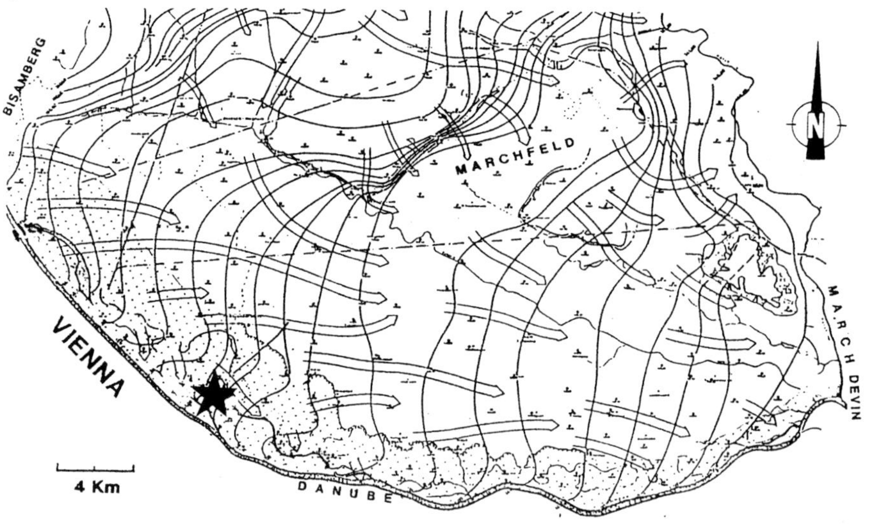

B

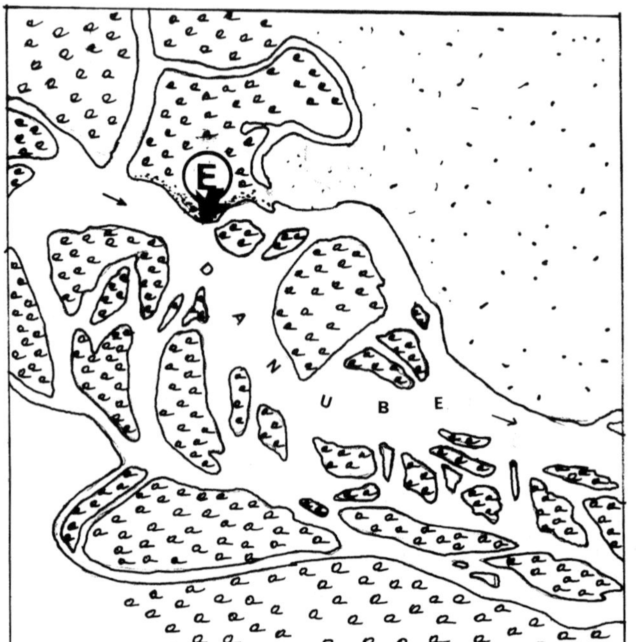

Fig. 25.4. (A) Hydrogeological map of the Marchfeld aquifer (with isopiestic lines and main subsurface flow directions) and the Danube wetlands (stippled area); the location of the Lobau study area is indicated with a star. (B) The Danube wetlands before the river regulation. the encircled E indicates the approximate location of Eberschüttwasser and the location of the Lobau-C area. (Redrawn from Danielopol, 1983.)

city of Vienna (see wells W1–W5 in Fig. 25.5). The environmental protection of this area against chemical pollution and landscape degradation has become a major concern for the Viennese environmentalists and water managers during the last three decades (Schiemer et al., 1992). Since 1996 the Lobau was integrated into the Danube Wetlands National Park.

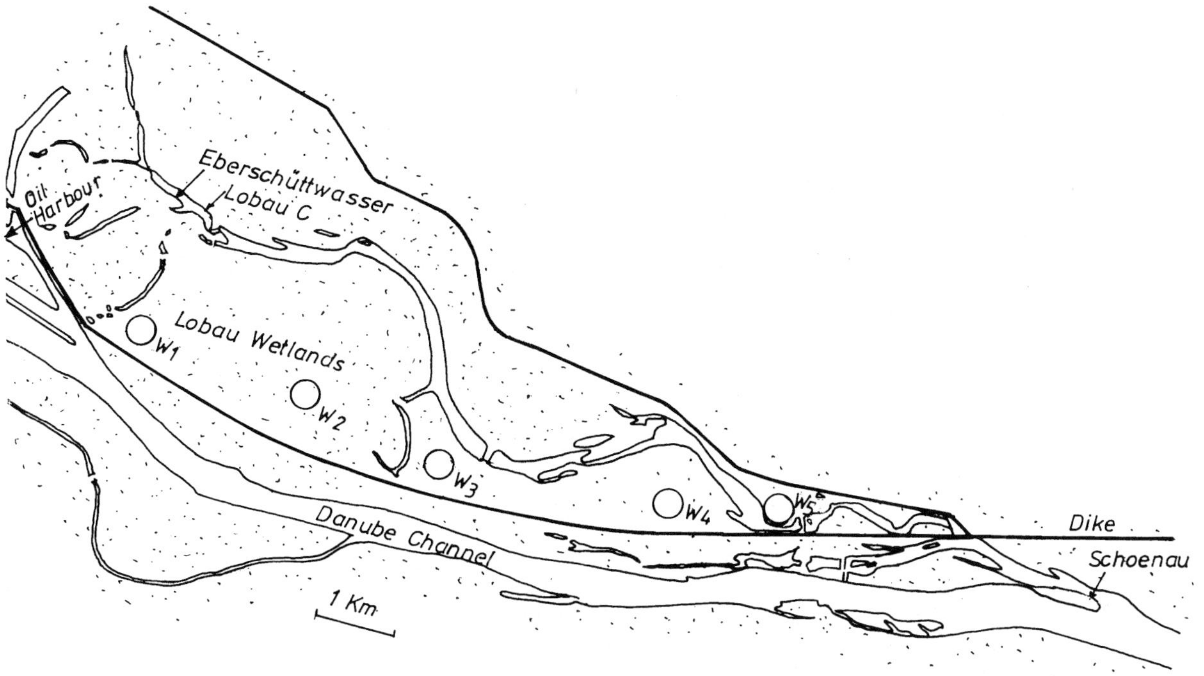

Fig. 25.5. Lobau wetlands with the location of water wells and the Lobau-C area; W1–W5, water wells for the drinking water supply of the city of Vienna; heavy lines, the dike system protecting the Lobau and the Marchfeld area against catastrophic Danube floods.

METHODOLOGICAL ISSUES

As a pilot study for the present description we investigated various parts of the Lobau wetlands qualitatively (Danielopol, 1983; Pospisil, 1994a,b). The area called Lobau-C by Pospisil (1994a) has the advantage that it is easy to identify the main sectors where surface water infiltrates into the aquifer and/or discharges from the subsurface sediments into the surface-water system (Dreher et al., 1997). Therefore it forms a miniature aquifer in which one can describe structural and functional aspects of a groundwater ecosystem.

Groundwater, sediment and interstitial-dwelling organisms were monitored seasonally from May 1991 until September 1993, and sampled through both simple and multilevel wells (Fig. 25.3) mainly from the upper part of the aquifer – that is, the first 4–6 m of depth (see Appendix). These wells were made either from Plexiglas tubes 5 cm in internal diameter (Fig. 25.6C) with 0.5 cm holes (the D wells in Fig. 25.3), or from metal tubes with the same specifications (the T well in Fig. 25.3). The simple piezometers were metal tubes 2.5 cm in internal diameter, the lower 10 cm of which is perforated with 0.5 cm holes (the R wells in Fig. 25.3). The wells were driven into the sandy-gravel alluvial sediments with a power hammer, and protected at the surface with an additional plastic tube (Fig. 25.6B–D).

We placed the 5 cm wells in a regular 3 3 grid, with 9 m spacing. Thirteen mini-piezometers were randomly distributed within the whole area of 900 m^2 (Fig. 25.3). Finally, five other mini-piezometers were placed in one row as controls (Dreher et al., 1997: fig. 2). The 5 cm wells, having lengths between 5 and 18 m (Pospisil, 1994a), were sampled over discrete depth ranges using a double-packer sampling device (Danielopol and Niederreiter, 1987) and the mini-piezometers (3–4 m long) were sampled with a Bou–Rouch pump (Pospisil, 1992, 1994a; Mösslacher, 1994).

The structure of the sediment and the presence of micro- and macro-organisms were observed and/or monitored with a mini-video camera in the field through the transparent Plexiglas tubes of the 5 cm wells (Fig. 25.7; Niederreiter and Danielopol, 1991).

Water levels were recorded mechanically with two limnigraphs for the surface water and one for the groundwater (Fig. 25.6A). Additionally, the piezometric level in the wells was measured weekly by hand over a two-year period, 1991–1993 (Dreher et al., 1997).

Hydraulic conductivities were measured (Dreher et

Fig. 25.6. (A) View of the Eberschüttwasser and the reed belt in front of the Lobau-C area; (B) ramming of multilevel well in alluvial sediments using a power hammer; (C) perforated Plexiglas tube for a multilevel well; (D) the Lobau-C area during autumn 1991; arrows point to wells D11 and D17.

al., 1997) using the infiltration-rate method in the mini-piezometers (Rouch, 1992) and the double-packer method in the 5 cm wells (Obermann, 1976). Seepage rates of surface water into the subsurface sediments along the Eberschüttwasser were measured by the infiltrometer method of Lee (1977).

Fig. 25.7. (A) The video camera (arrow) and the TV monitoring set-up for subsurface exploration; (B–F) images of sedimentary layers in several boreholes: B, well D3, 1 m deep; C, well D17, 2 m deep; D, well D3, 8 m deep; E, well D3, 4 m deep; F, well D11, 4 m deep; a, channels through fine sediments dug by macroorganisms like *Cyphon palustris*; b, microbial biofilm; c, fine sediment displaying reducing chemical conditions. Scale bar is 42 mm.

Dissolved oxygen was recorded *in situ* either in multilevel wells with the modified double-packer device (Danielopol and Niederreiter, 1987) or in the mini-piezometers by hand using the technique of Danielopol (1983). During periods when the aquifer is well oxygenated the oxygen values measured in multilevel wells are about 0.5–2 mg ℓ^{-1} higher than in nearby mini-piezometers (Pospisil et al., 1994). Total dissolved organic carbon (DOC) was measured using a Shimadzu TOC–5000 carbon analyser after the water was filtered through Whatman glass-fibre filters (GF/F) previously sterilized by heat. The input of particulate organic carbon to the groundwater system is not quantified; however, qualitative observations exist. Exploratory analyses were done on the DOC in order to differentiate between the easily assimilable labile DOC (e.g. carbohydrates) and the refractory carbon.

Total bacterial counts (TBC) were obtained from 100 ml unfiltered water samples using the DAPI[1] staining method and epifluorescence microscopy (Porter and Feig, 1980). Metabolically active bacteria (MAC) from the same samples were counted using the INT-Formazan technique [1] (Zimmermann et al., 1978). An electron transport system (ETS) activity is characteristic of respiring cells. The microbial ETS activity was quantified using the protocol of Obst and Holzapfel-Pschorn (1988), sequentially incubating 100 ml unfiltered water samples and/or sediment for one hour at 37°C and five hours at each of 20°C and 12°C. The highest potential metabolic activity of the heterotrophic microbial communities, expressed as a percentage of substrate turnover (i.e. an INT-Tetrazolium salt as electron acceptor) was obtained for the 37°C incubation. The overall significant correlation of total bacterial counts with the amount of ETS activity at the three different incubation temperatures indicates a very wide temperature tolerance of the groundwater microbial populations inhabiting the Lobau-C area (Danielopol et al., 1992).

Meio- and macro-organisms were collected from 5-litre samples of groundwater and sediment extracted with a Bou–Rouch pump, either fixed to the double-packer for the 5 cm ID wells, or directly with this pump fixed onto mini-piezometers (Pospisil, 1992). Samples for biological and/or chemical analysis were taken seasonally between May 1991 and September 1993. Faunistic samples were sieved through a 100 μm net and subsequently the taxa (especially Crustacea) were identified to species level where possible.

Experimental manipulations which mimic ecosystem processes in the field were done with slow-filtration "module" columns (Fig. 25.21, below) as described by Torreiter et al. (1994). The advantage of these columns is the opportunity to analyse the sediment quantitatively at different depths. Oxygen was monitored *in situ* in these columns at three different depths (Fig. 25.21) using the micro-syringe module of the "Orbisphere" fitted to an oxymeter (Torreiter et al., 1994). Sediment cages for microbial colonization were constructed from tubular stainless steel sieves (4 cm internal diameter, 20 cm length, 1 mm mesh) closed with stoppers at the ends. Cages filled with various types of sterile sediments were exposed in the field below the groundwater table for two months. Finally, the movements of meio- and macro-organisms through the interstitial space of sediments under laboratory conditions were observed using an aquarium filled with sediment and video-camera techniques. The data were analysed from video tapes (Danielopol et al., 1994).

THE LOBAU-C ECOSYSTEM

Structural aspects

Landscape features

The Lobau-C area is a typical wetland within the Danube backwater system. Close to Eberschüttwasser (Figs. 25.3, 25.6A,B) there is a newly formed belt of reeds (*Phragmites communis*). A new wetland area resulted from terrestrialization within the last hundred years. Proceeding northwards, the elevation of the terrain increases from about 149 m above sea level at the Eberschüttwasser margin to 151 m a.s.l. inland towards the Kreuzgrund dam and well T3 (Fig. 25.3). This higher area forms an older terrace and it is covered with a meadow forest of mixed *Cornus, Fraxinus* and *Populus* (Fig. 6D). The Kreuzgrund dam separates the Eberschüttwasser basin from the adjacent Mittelwasser basin (Fig. 25.3) which is at an elevation about 0.3 m lower.

Hydrological outline

The Lobau-C area is flooded partially every year, and completely during exceptionally high water stands in

[1] DAPI = 4′,6-diamidino-2-phenylindole, used for nucleic acid stains of micro-organisms.
INT = 2(*p*-iodophenyl)-3-(*p*-nitrophenyl)-5-phenyl tetrazolium chloride.

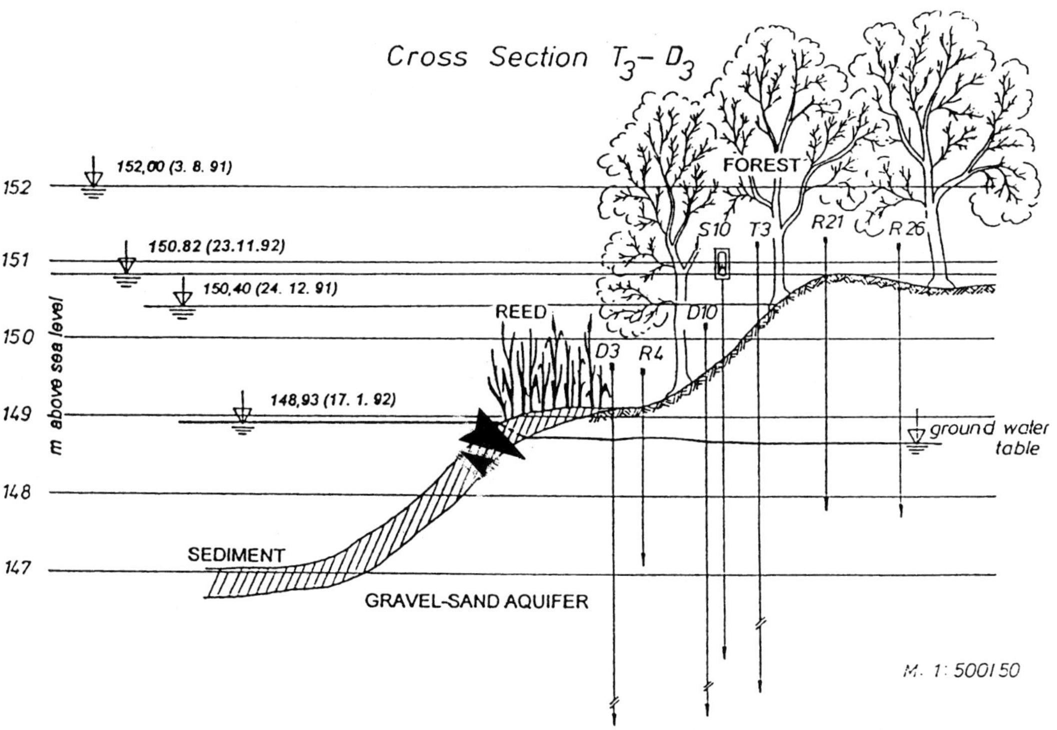

Fig. 25.8. Cross-section through the Lobau-C landscape with various water levels and the position of several wells including the limnigraph S10; black heavy arrows indicate area of infiltration or discharge (from or into the Eberschüttwasser and the aquifer).

the Danube (Fig. 25.8), every 2–4 years (Danielopol, 1983).

The groundwater table in the reed-belt zone lies close to the surface water level of the Eberschüttwasser and follows its fluctuations, while in the hinterland zone the groundwater table is about 1.5 m below the surface. The changes of the water level in the Eberschüttwasser and the Lobau-C local aquifer depend mainly on the water-level fluctuations of the regional Marchfeld aquifer (Fig. 25.4A) and in the Danube River. The high water fluctuations in the Danube reach the Eberschüttwasser with a reduced amplitude and a time lag. The mean annual precipitation in this area is 560 mm. Duc to the connection of the backwater system to the Danube main channel, the river floods also play an important role for the Lobau wetlands. During such events, and for a short period of time afterwards, the groundwater/surface water system of the Lobau-C area is inverted and the groundwater discharges into the Eberschüttwasser. Surface water from the Mittelwasser infiltrates into the aquifer Lobau-C, whereas for most of the year the margin of this latter backwater is an exfiltrating zone in the study area. When the flood recedes, the water level in the Eberschüttwasser becomes higher than the nearby groundwater system, and seepage occurs from the Eberschüttwasser into the groundwater system (see below, p. 490), which continues to flow towards the Mittelwasser owing to the steep hydraulic gradient (Dreher et al., 1997). When very high water in the Danube transgresses into the upper part of the backwater system, the Lobau-C aquifer can discharge even through the soil surface, as occurred in July 1991 (Danielopol et al., 1992).

Sediment heterogeneity and hydraulic conductivity

The soil layer is very thin, ranging from few centimctres in the middle of the forest to 10–20 cm in the reed-belt fringe. The alluvial sediments are layered, alternating between well-sorted gravel and stones loosely packed, and gravel-sandy and sandy-silty layers (Figs. 25.7, 25.9: Danielopol, 1983; Pospisil, 1994a,b). Generally one can divide the sedimentary column of 5–8 m depth into three units (Fig. 25.9): the uppermost horizon A, about 2–3 m deep, dominated by loosely packed gravel corresponding to the sediment of the former Danube channel; a B horizon 2–5 m deep consisting of very heterogeneous coarse sediment layers with good permeability alternating with fine

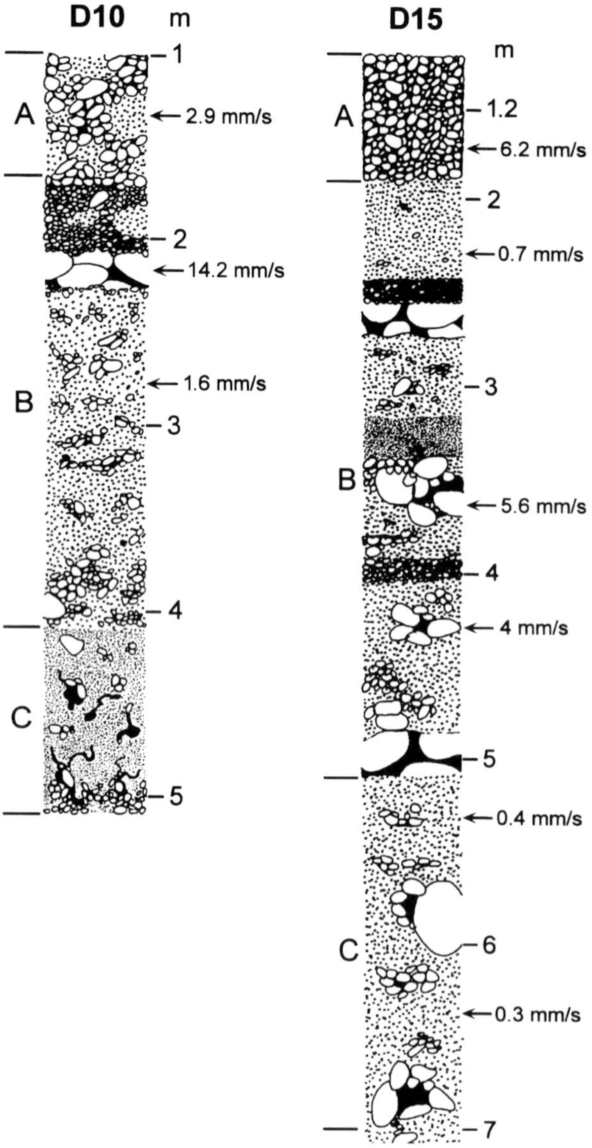

Fig. 25.9. Cross-sections of boreholes D10 and D15; (A–C) stratigraphic horizons; arrows indicate permeability values at selected depths.

grained deposits with low hydraulic conductivity; finally, a C horizon dominated by sandy-silty sediment mixed with gravel and stones.

The hydraulic conductivity varies according to the sediment heterogeneity. The average permeability in this shallow aquifer is estimated to be 0.8 mm s^{-1}, with extreme measured values ranging from 3.6 µm s^{-1} to 5 cm s^{-1} (Fig. 25.9).

The infiltration rate in front of the reed belt is about 1.1 mm s^{-1}, while in the area closer to the Kreuzgrund dam (e.g., close to the R1 well) the seepage rates are 10–20 times higher (Dreher et al., 1997). The bottoms of the Eberschüttwasser and Mittelwasser surface-water basins are largely filled with fine silty, organogenic sediments which clog the interstitial spaces of the coarser superficial layer; the filtration zones of the Lobau-C aquifer occur mainly on the upper margin of the basins (see arrows close to the reed belt in Fig. 25.8). The seepage of surface water into the aquifer is highly dependent on the water level in the Eberschüttwasser (Fig. 25.10A). The surface water largely infiltrates into the subsurface alluvium when the Eberschüttwasser level is more than 149.0 m above sea level (Fig. 25.10A). During floods and rainy periods, the local aquifer is also recharged by infiltration of water through the whole surface area. The groundwater flow-rate is estimated to vary between 1.6 and 4.8 µm s^{-1} for a mean permeability of 0.8 mm s^{-1} and a hydraulic gradient of 2 to 6 mm m^{-1}(Dreher et al., 1997).

Water temperature and concentration of dissolved oxygen

Groundwater temperature in the study area is strongly dependent on the annual atmospheric temperature variation near the shore area where the unsaturated layer is thinnest. Figure 25.10B shows that the groundwater temperature at 3.5–4 m below the soil surface around site D10 closely follows the temperature variation in the surface-water system. Variations of more than 15°C are registered, with a maximum of 20°C in summer and a minimum of 2–3°C during winter (see Appendix, p. 505, and also Pospisil, 1994a; Mösslacher, 1994). During summer and autumn when the aquifer recharge by surface water is small, the differences between the subsurface and the surface water become significantly larger than during the period when the aquifer is, hydrologically, strongly connected to the backwaters (Figs. 25.16B, D, below).

The oxygen concentration mainly depends on the infiltration rate of surface water into the subsurface system, as well as on the water temperature and the biological activity within the sediment (e.g. microbial respiration). During the period when the aquifer is being recharged, the Eberschüttwasser near-shore areas generally record higher oxygen concentrations (Fig. 25.16B, below), as occurred, for example, during flood events of March and November 1992 (Figs. 25.10A, B; see also Appendix, p. 505). During

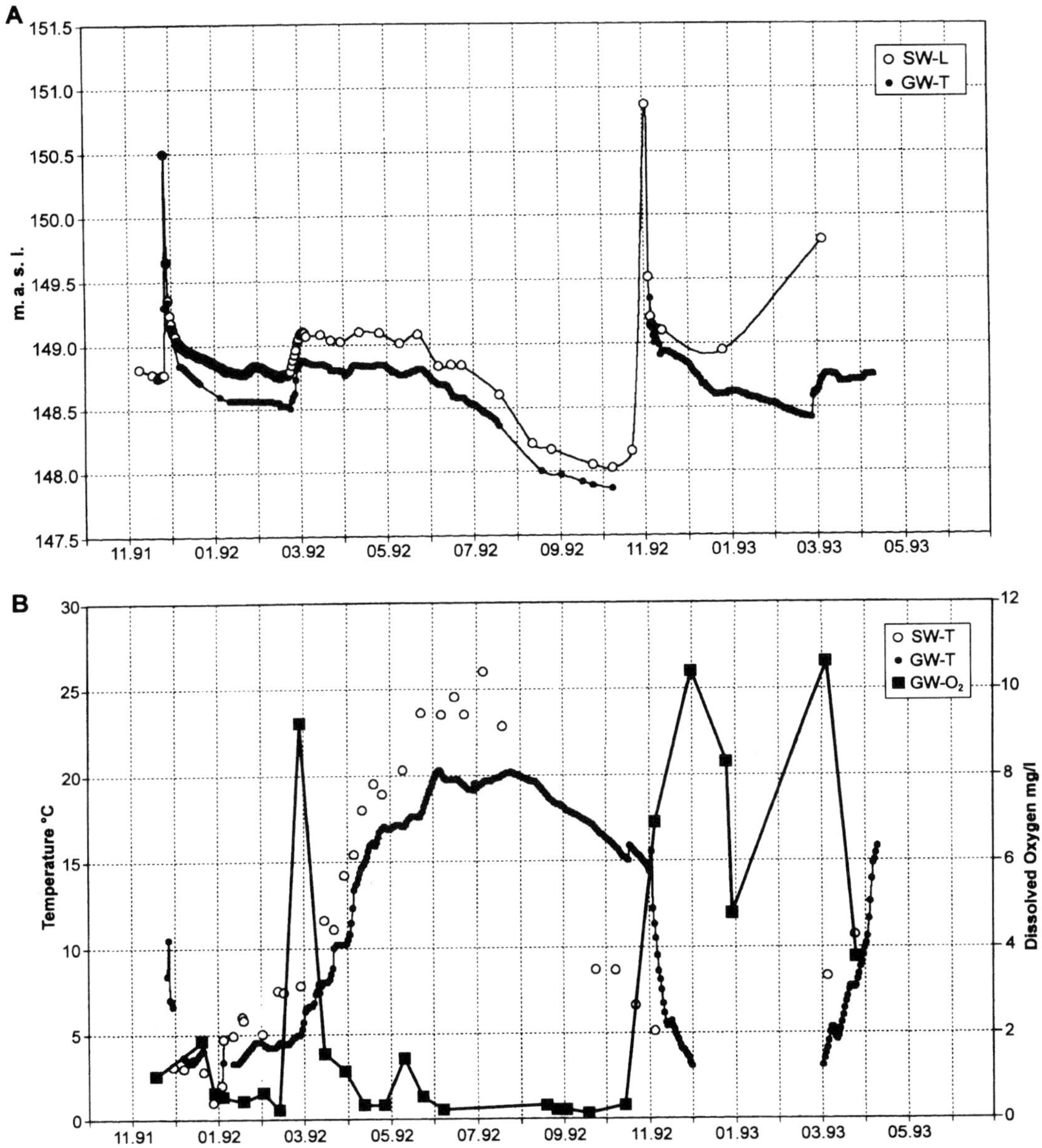

Fig. 25.10. (A) Hydrographs of the Eberschüttwasser water level (SW-L) and groundwater table (GW-T) at well S10. (B) Variation of temperature and oxygen through 1991–1993: surface water temperature of Eberschüttwasser (SW-T); groundwater temperature (GW-T) and oxygen (GW-O_2) at site D10 (3.5 m depth).

winter and spring the groundwater has higher oxygen concentrations than during the summer and autumn, when the water temperature and the microbial activity rise substantially. During these latter periods the aquifer becomes largely dysaerobic – that is, the oxygen concentrations drop drastically, below 1 mg ℓ^{-1}, and generally to around 0.1–0.01 mg ℓ^{-1} (Fig. 25.10D), giving rise to an anoxic micro-niche (Fig. 25.7E).

Total organic carbon, microbial counts and microbial activities

The riparian vegetation deposits a large amount of particulate organic carbon (POC), which accumulates at the surface of the soil (Figs. 25.6B–D, 25.7A) and which is further transported by the infiltrating water into the subsurface sediment layers during rainy periods and flood events. Large accumulations of particulate

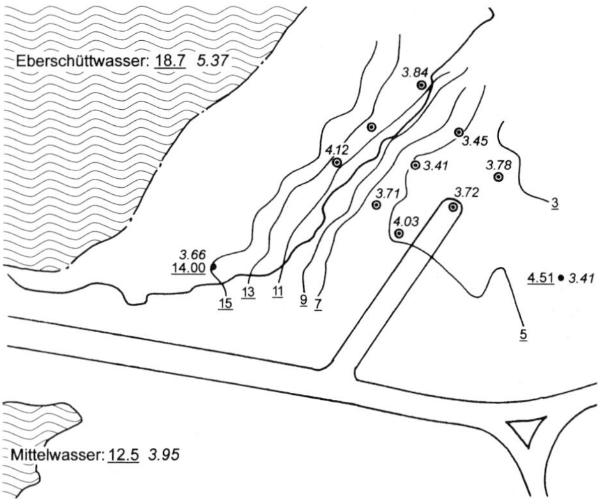

Fig. 25.11. Spatial distribution of DOC concentration and ETS activity (underlined numbers) in and around the Lobau-C system; median values calculated from the seasonal samples during one-year cycle, October 1992–September 1993; isolines for spatial distribution of ETS activity obtained by block kriging (Isaaks and Srivastava, 1989); black points, mini-piezometer wells; encircled points, multilevel wells.

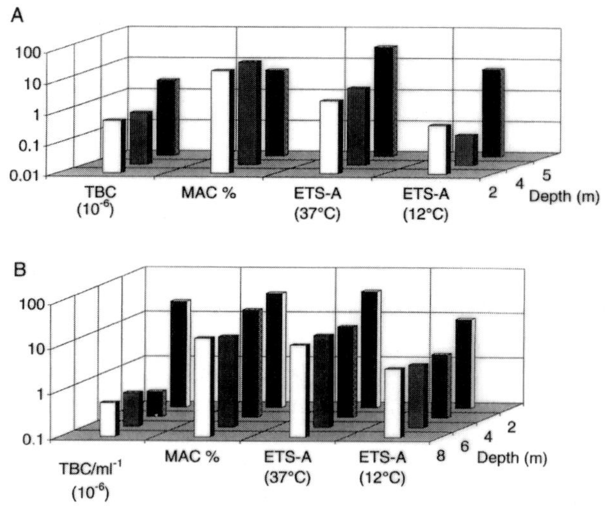

Fig. 25.12. Microbial densities and activities in the groundwater around wells D10 (A) and D15 (B) sampled on January 17, 1992; TBC, total bacterial counts; MAC, metabolically active (viable) bacteria as a percentage of total bacterial counts; ETS-A, electron transfer system activity at 37°C/1 h and at 12°C/5 h as percentage substrate turnover.

organic matter are visible in the uppermost two metres of sediments, especially in the reed-belt zone (e.g. around well D3). Fine particulate and dissolved organic matter commonly infiltrates from the surface water into the groundwater system. The *Phragmites* belt in Eberschüttwasser (Figs. 25.3, 25.6A,B) is considered to be a major producer of DOC, which infiltrates into the local aquifer (Danielopol et al., 1992). A detailed carbohydrate analysis of *Phragmites* litter collected during autumn shows high concentrations of C_5 and C_6 sugars (Gunatilaka et al., 1994).

The distribution of DOC within the study area is very heterogeneous (Danielopol et al., 1992) and hence only median values are presented here. In the Eberschüttwasser the annual median value for DOC during the period October 1992–September 1993 (calculated from four seasonal samples) was 5.37 mg ℓ^{-1} (Fig. 25.11), whereas in the samples taken from the multilevel wells (2 and 4 m layers) and two mini-piezometers (R1 and R26) the median values are lower, between 3.41 mg ℓ^{-1} and 4.12 mg ℓ^{-1}. Interestingly enough, in the Mittelwasser, where generally groundwater discharges, the median values of DOC were closer to those in the subsurface system. Detailed analysis of DOC from well D3 at a depth of 2 m during the spring of 1991 showed that the samples contained appreciable amounts of dissolved carbohydrates (nearly 50%), especially xylose, fructose, sucrose, glucose and a number of minor unidentified C_5 sugars (Gunatilaka et al., 1994). All these components are easily assimilable by microbes. In late autumn, measured concentrations of sugar (fructose and glucose) were much lower, indicating that they had been readily assimilated.

Heterotrophic and/or chemo-organotrophic microorganisms develop in the Lobau-C subsurface layer. Among the latter group one should mention the microaerophilic *Gallionella*. Bacteria densities in groundwater of the Lobau-C area are about 1 10^6 cells ml^{-1} (Fig. 25.12). High values were measured in water samples from sandy layers, for instance, in samples during January 1992 at a depth of 5 m at site D10 and 2 m at D15 (Fig. 25.9). Lower values were registered for the same date in samples from sediments where gravel and pebbles occur in high quantities, for instance at depths of 2 and 4 m at site D10 and at 4, 6 and 8 m at site D15. Microbial counts show that only 5–35% of bacteria were metabolically active (Figs. 25.12A,B).

The potential activity of heterotrophic microbial communities, expressed as the activity of their electron transport system (ETS) when incubated at 37°C, parallels the pattern showed for the total bacterial counts (Figs. 25.12A,B; Danielopol et al., 1992). Much lower ETS activity is recorded when the samples are incubated at 12°C, and this situation parallels the decrease of microbial activity with the lowering of the temperature in the field. Danielopol et al. (1992) showed that the mean values of ETS activity for groundwater samples collected in January 1992, when the water temperature was 3°C, were only 40% of those recorded in September when the water temperature was 18°C (Fig. 25.11). Even higher ETS values are found in surface water as compared to the groundwater, within which ETS activity decreases steadily as one proceeds inland from the Eberschüttwasser reed belt.

Sedentary microbial communities attached to sediment grains commonly display higher activities than those living in the interstitial water (Griebler and Slezak, 1997). As an example, the median ETS value of seasonal samples (November 1992–September 1993) at site R1 was 21% for the attached microbial and 14% for the water bacteria (Mösslacher, 1994). A significant correlation was found between the ETS activity of attached and nonattached microbial communities in the aquifer investigated (Mösslacher et al., 1996). Using the video-camera technique, thick microbial biofilms were visible filling the interstitial spaces of the sediment, for instance at site D11 at a depth of 4 m (Fig. 25.7F). The growth of microbial biofilms (Fig. 25.7D) may be stimulated by movements of macro-organisms causing bioturbation of the sediments.

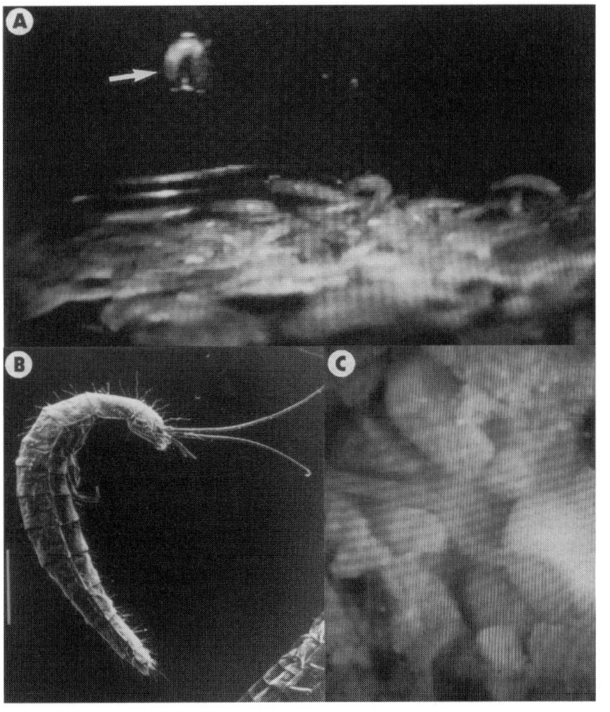

Fig. 25.13. Microorganisms from Lobau-C area: (A) larvae of *Cyphon palustris* (Coleoptera: Helodidae) in borehole T3, swimming at the groundwater surface and/or moving above it in the unsaturated sediment layer (arrow); (B) scanning electron micrograph of a *C. palustris* larva (scale bar: 2 mm); (C) an oligochaete (Tubificidae, gen. sp. indet.) around the borehole D3.

Faunistic diversity and spatial distribution

The fauna living within the Lobau-C groundwater aquifer is less diverse than those occurring in the benthic habitats of the Eberschüttwasser (Danielopol, 1983; Mösslacher, 1994, 1997b; Pospisil, 1994a,b; Negrea and Pospisil, 1995; Rouch and Danielopol, 1997). The interstitial-dwelling macrofauna is represented by only a few insect larvae, oligochaetes, gastropods and crustaceans (Amphipoda and Isopoda: see Appendix, p. 505). Benthic habitats of the Eberschüttwasser harbour various species of insects, like Ephemeroptera, Coleoptera and Diptera (Ceratopogonidae, Chironomidae) (Danielopol, 1983). In contrast to the surface-water benthic habitats in the Lobau-C aquifer, only one coleopteran species, *Cyphon palustris* (a helodid) occurs both in the reed-belt fringe (at the D3 and D5 sites) and also inland (at the D9, D10, D11 and T3 sites) during the winter and spring (Klausnitzer and Pospisil, 1991). *Cyphon* larvae, which are about 1.2 cm in length (Figs. 21.13A,B), accumulate in high numbers (tens of specimens per well) at the groundwater table, but are also found in numbers at depths of several metres in the sediment, where they leave visible trails (Fig. 25.7D); in some samples up to 60 specimens were caught. Within this area gastropods (*Bythiospeum* aff. *geyeri*), amphipods (*Niphargus* sp.) and oligochaetes occur rarely, and only in low numbers (1–10 specimens). *Bythiospeum* aff. *geyeri* is a small blind and unpigmented snail about 5 mm long, a typical hypogean-dwelling animal (Haase, 1995) occurring only in the deeper sediment layers, far from the surface water, where there is very little organic detritus (e.g. at the T3 and D15 sites at 8–12 m depth). In the laboratory it was noticed that this species grazes on the microbial biofilm which develops on decaying leaves (D.L. Danielopol, unpublished). Tubificid oligochaetes were observed especially along the reed-belt fringe which is rich in organic matter (e.g. at the D3 site in the upper 2 m layer; Fig. 25.13C).

Proasellus slavus vindobonensis (Isopoda) is a small stygobite species 3.5 mm in length, widely distributed within the Lobau-C aquifer (Fig. 25.14). Danielopol et al. (1994) and Mösslacher (1994) showed that this species is well adapted to live in a subterranean

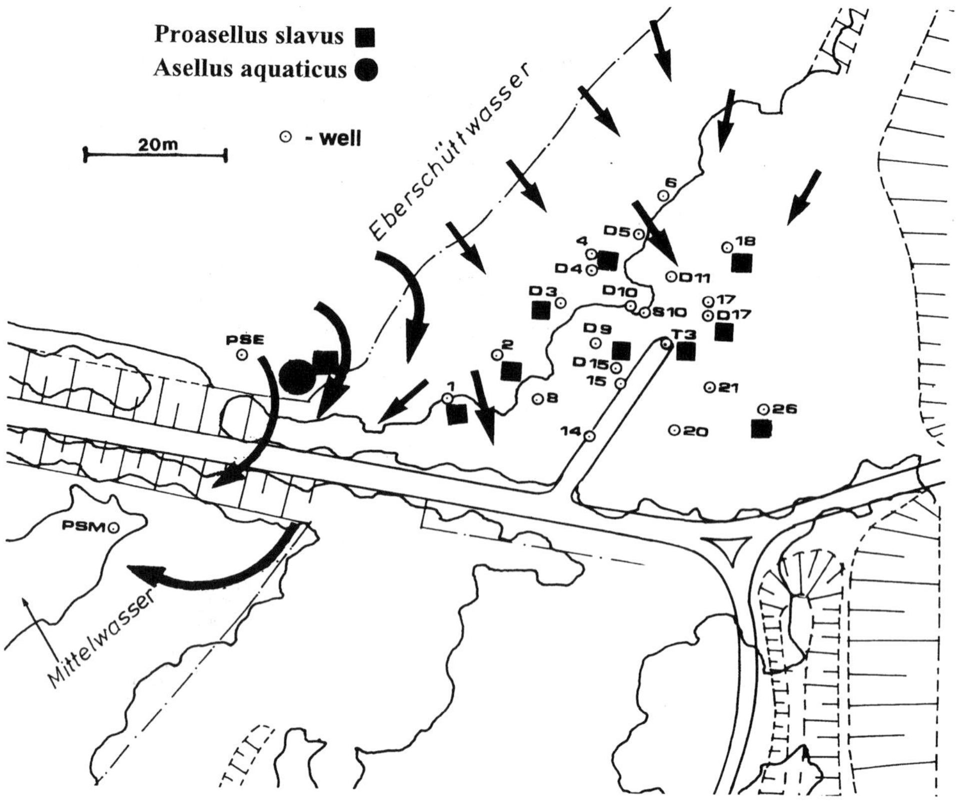

Fig. 25.14. Location of the epigean isopod *Asellus aquaticus* in the shallow alluvial sediments of Lobau-A area and of the stygobite *Proasellus slavus vindobonensis* in the Lobau-C aquifer; heavy arrows point schematically to the main pathways of the subsurface water circulation; numbers and letters are symbols for the various wells.

environment with low energetic resources, having great mobility and showing reduced feeding time and tolerance of low oxygen concentrations. At site T3, in the hinterland zone of the study area, *P. slavus vindobonensis* occurs throughout the year, during both normoxic and dysaerobic periods, in the upper part of the aquifer (0.1–0.5 m below the water table). Winter and spring are favourable for the extension of the isopods into the deeper layers (Table 25.1), because of the better oxygenation of the aquifer. The abundance of this species is always low (1–11 specimens in a sample of five liters). By comparison, the epigean *Asellus aquaticus* colonizes shallower interstitial habitats in the Lobau-A area (Fig. 25.3) which is well oxygenated and rich in organic detritus (Fig. 25.14; Danielopol, 1983).

Table 25.1
Seasonal changes in frequency (number of individuals in a five-litre sample) of *Proasellus slavus vindobonensis* in well T3

Date	Depth of sampling			Groundwater table (m)
	2 m	4 m	6 m	
May 1991	4	2	0	1.5
July 1991	0	0	0	1.5
September 1991	4	0	0	1.5
October 1991	11	0	0	1.8
January 1992	3	9	0	1.5
April 1992	9	9	3	1.5
July 1992	1	0	0	1.7
October 1992	8	0	0	2.4
January 1993	1	8	–	1.7
April 1993	11	1	1	1.6

Groundwater-dwelling meiofaunal groups (animals less than 2 mm in body size) are more diverse and abundant than macrofauna (Pospisil, 1994b). Ostracoda and Copepoda are the most common and abundant groups, while Rotatoria, Nematoda, Archiannelida, Turbellaria, Acari, Tardigrada and Cladocera are less common (Appendix, p. 505; see also Mösslacher, 1994; Pospisil, 1994a).

The various meiofaunal species are very patchily distributed, and the animal clumps change as the individuals move through the interstitial space. The

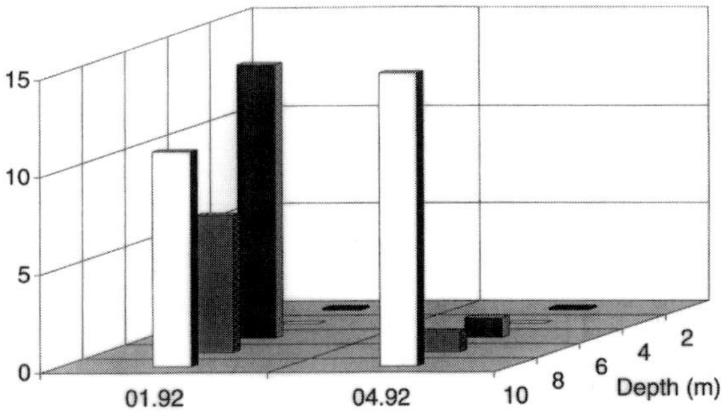

Fig. 25.15. Frequency (number of specimens in a five-litre sample) of harpacticoids (*Parastenocaris fontinalis*) caught at various depths around well D15 during January and April 1992.

copepod *Parastenocaris fontinalis* is a minute and filiform harpacticoid less than 1 mm long which was found at three sites: D15, various depths, D10 at 5 m and D3 at 2 m. Figure 25.15 shows that clumps of *P. fontinalis* were deeper (6–10 m) during winter and spring in very sandy sediments, but later samples did not contain the species at all.

Besides typical hypogean-dwelling animals, the Lobau-C aquifer also harbours epigean species which colonize the subsurface waters abundantly. Representatives of the former group are widely dispersed through the aquifer during the whole year (Figs. 25.16A,C).

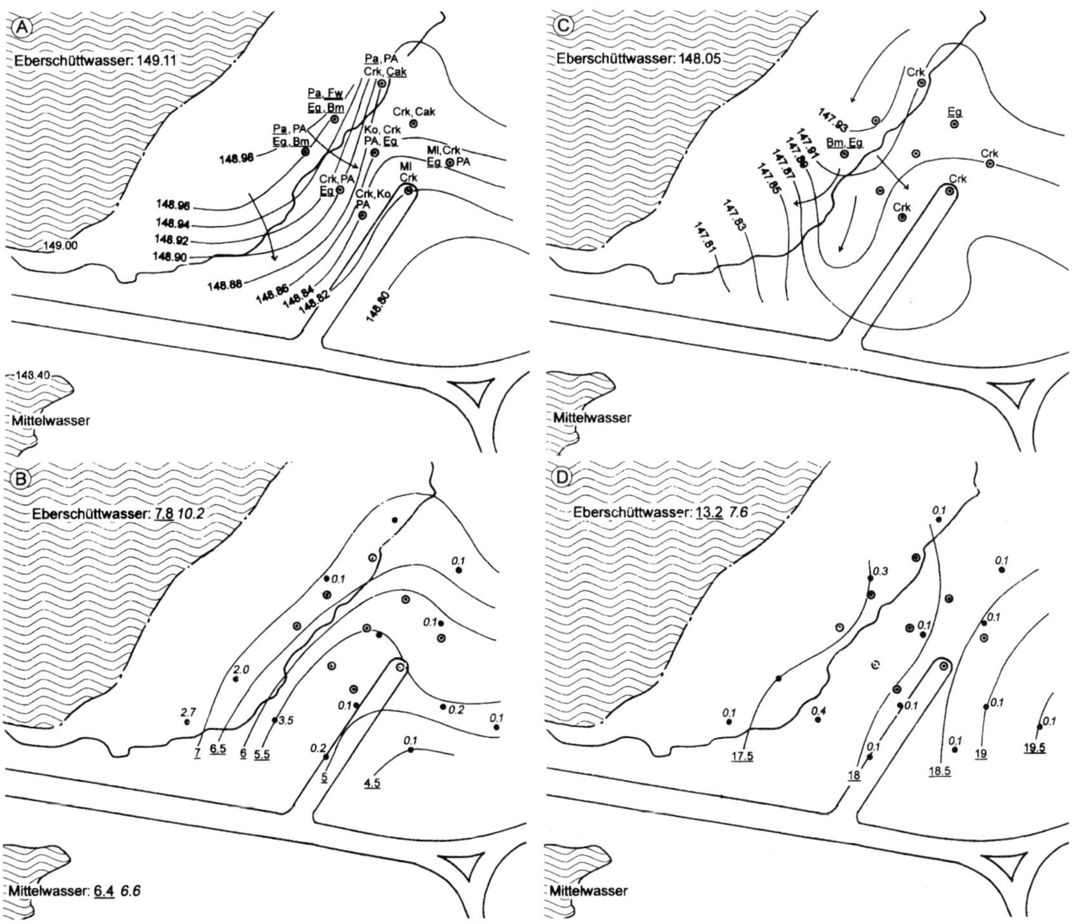

Fig. 25.16. Groundwater maps during two different seasons showing subsurface water flow (arrows; isopiestic lines drawn by hand) at different water levels of the Eberschüttwasser, as well as the distribution of meiofauna (letters for harpacticoid and ostracod species), the oxygen concentration (mg ℓ^{-1}) and water temperature (underlined values, in °C; isotherms obtained by block kriging): (A) Spring (March 25, 1992); hydrological maps with faunistic information. (B) Spring (March 25, 1992); temperature and oxygen concentration. (C) Autumn (October 10, 1992); hydrological map with faunistic information. (D) Autumn (October 10, 1992): temperature and oxygen concentrations. Black points, minipiezometer wells; encircled points, multilevel wells. Meiofauna: Ostracoda: Cak, *Candonopsis kingsleii*; Crk, *Cryptocandona kieferi*; Fw, *Fabaeformiscandona wegelini*; Ko, *Kovalevskiella* sp.; Mi, *Mixtacandona laisi vindobonensis*; Pa, *Pseudocandona albicans*. Harpacticoida: Bm, *Bryocamptus minutus*; Eg, *Elaphoidella gracilis*; PA, *Parastenocaris* (various species). Epigean species have a single underline, ecotonal species a double underline; others are exclusively hypogean.

During winter and spring, when the aquifer is better oxygenated, epigean species occur not only at the margin of the Eberschüttwasser where there are greater energetic resources, but also at more central sites (Fig. 25.16A). Summer and autumn are periods when the aquifer is largely dysaerobic and only a few species, mainly hypogean, occur (Fig. 25.16B). *Fabaeformiscandona wegelini* is a so-called ecotonal hypogean species (Danielopol et al., 1997). It is a hypogean animal commonly occurring in very superficial riverbed habitats rich in organic matter (Rogulj et al., 1993). In the study area it occurs further inland (Rouch and Danielopol, 1997, fig. 11) only when the environment is well oxygenated.

Densities of copepods and ostracods vary between several and 200 individuals in five-litre samples of groundwater (Figs. 25.17, 25.18). Very high densities were noticed generally just below the surface (0.1–0.5 m deep: Figs. 25.17A,C), particularly of the harpacticoid *Elaphoidella gracilis* (e.g. 275 individuals in the 5-litre sample at site D3: Fig. 25.17A). Low densities of groundwater meiofauna groups were noticed after exceptionally strong floods like those of July 1991 (Fig. 25.17B). In this case the fauna of the aquifer was partly flushed out, and started to recover one month later (Danielopol et al., 1992; Pospisil, 1994a,b). Also during the summer and/or autumn when the aquifer is largely dysaerobic the abundance and diversity of meiofauna decrease dramatically. Figure 25.18 shows the densities of cyclopoid copepods at six sites in the study area with low oxygen conditions during July 1992, and in normoxic conditions during spring 1992 and winter 1993. It is very likely that, during periods of low oxygen concentrations in the deeper part of the aquifer, mobile interstitial animals, such as copepods (Cyclopoida and Harpacticoida), migrate upwards and concentrate below the water table.

Not much is known about the feeding habits of the meio- and macro-fauna of the Lobau-C study area. Ostracods, the harpacticoid copepods, isopods (Mösslacher, 1994) and oligochaetes (e.g. tubificids) ingest fine sediment and organogenic detritus containing microbial biomass. *Cyphon palustris* feeds on plant detritus, while *Bythiospeum* aff. *geyeri* is a browser of bacterial biofilms, as mentioned above.

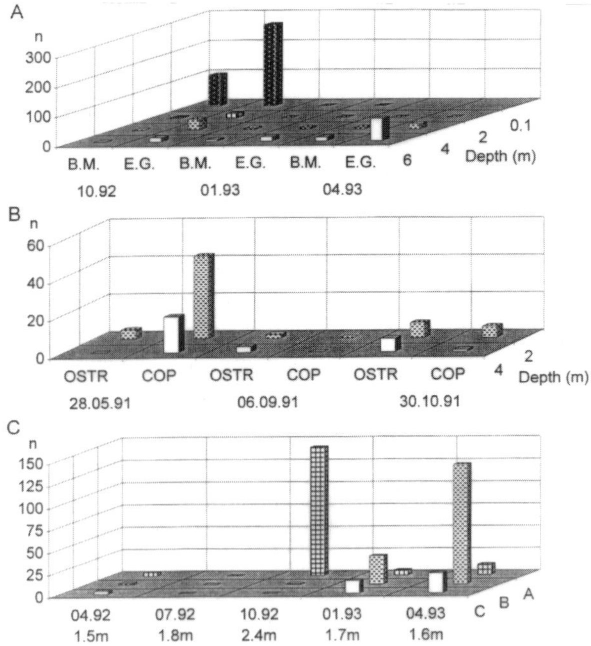

Fig. 25.17. Temporal and spatial distribution of various crustacean taxa in the Lobau-C area; *n*, number of individuals in a five-litre sample. (A) Frequency of two harpacticoid species (in five-litre samples) around the D3 borehole (B.M., *Bryocamptus minutus*, E.G., *Elaphoidella gracilis*). (B) Frequency of Ostracoda and Copepoda (Cyclopoida and Harpacticoida) from cumulated samples (five litres each) from the D3, D10 and T3 wells at 2 and 4 m depth. (C) Distribution of Cyclopoida around well D17 at various standardized depths (A, 2 m; B, 4 m; C, 6 m) below the soil surface; numbers below indicate the date of sampling and the water level in metres.

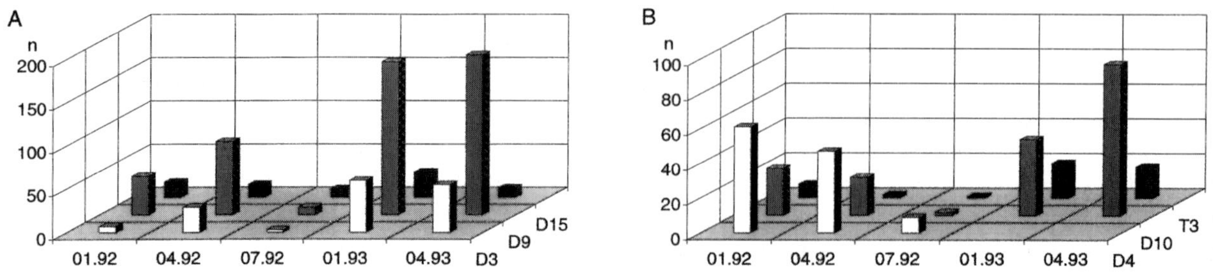

Fig. 25.18. Seasonal distribution of Cyclopoida at various sites of the Lobau-C aquifer; *n*, frequency of animals in five-litre samples, cumulated values. (A) 2 m depth; (B) 4 m depth.

Cyclopoid copepods and amphipod species of the genera *Acanthocyclops*, *Diacyclops* and *Niphargus* feed on a wide range of food types, plant detritus, fine sediment, and living and/or dead animals.

Functional aspects

Hydrological evolution

The Lobau-C sedimentary system can be visualized as a large open tank which on one side stores water while on the other side allows the water to flow. The water transfers matter and energy from epigean habitats into the aquifer, and on through the subsurface interstitial space. Because this local aquifer is an open system, water discharges again to the surface. During winter and spring, the main input zone, or recharge area of the aquifer, is along the Eberschüttwasser (Fig. 25.16A). The main subsurface pathways are located close to the fringe of the reed belt and the Kreuzgrund dam; the water discharges into the Mittelwasser and generally has higher concentrations of oxygen and DOC. During the summer and part of the autumn the water levels in the backwaters are lowered; those of the Eberschüttwasser drop to less than 149 m above sea level and the aquifer slowly becomes disconnected from this surface-water system. The aquifer now is fed by the poorly oxygenated groundwater of the inland Marchfeld area (Fig. 25.16C); this water either flows parallel to the margin of the Eberschüttwasser or from the back towards the Mittelwasser. Evidence for these two types of water dynamics comes not only from the interpretation of hydrographs but also from the examination of thermal profiles and the electric conductivity of the water. The groundwater originating from the Eberschüttwasser maintains the thermal and electric-conductivity signature of its source, and the electric-conductivity values of surface and subsurface waters are similar, between 450 and 500 µS (Fig. 25.16B; see also Appendix, p. 505). The groundwater has a different signature during the summer and autumn; the aquifer temperature is higher than that of the Eberschüttwasser, and the thermal gradient is reversed (Fig. 25.16D). Warm water from the hinterland heats the Lobau-C aquifer. The electric-conductivity values of the Eberschüttwasser drop gradually during the same period, reaching 328 µS in October, while those of the groundwater remain unchanged as compared with those of the previous period (Appendix, p. 505).

The second recharge area of the Lobau-C system is through the soil surface, and it occurs when surface water partly or entirely covers the area (Fig. 25.8), rapidly replenishing the aquifer and broadly recharging it with fine particulate organic matter, oxygen and epigean-dwelling organisms.

Dynamics of energetic resources

Oxygen, as electron acceptor, and organic carbon, as electron donor, are key energetic resources for organisms. Figures 25.10 and 25.16B show that, with the infiltration of highly oxygenated water from Eberschüttwasser during spring-time, the aquifer becomes better oxygenated along the main flow pathways, along the fringe of the reed belt and the Kreuzgrund dam, while the inner part of the aquifer remains poorly oxygenated (e.g. around well D17: see Appendix, p. 505). From Fig. 25.10B one can see that high oxygen concentrations in the inland well D10 drop rapidly, within several days or a few weeks. Figure 25.16B shows the strong oxygen consumption within the first few metres of the subsurface strata after the infiltration of surface water into the aquifer. An oxygen deficit of more than $6\,\mathrm{mg}\,\ell^{-1}$ was found between the Eberschüttwasser and the closest piezometers (R1, R2, R4, R8), several metres inland (Fig. 25.10), mainly on account of biological oxygen consumption. Because the subsurface water flow in this area during flood is strongly unidirectional, the oxygen consumption appears to be reduced from the groundwater interfaces with soil water and surface water towards the inner and deeper parts of the aquifer. This can be inferred from the spatial oxygen distribution and the general trend of the microbial activity as expressed by the ETS (Fig. 25.11). The microbial activity during late autumn and early winter decreases in parallel with the groundwater temperature, while the oxygen condition of the aquifer improves (Danielopol et al., 1992). During the summer–autumn stagnation period the aquifer becomes largely dysaerobic or even anaerobic (Fig. 25.16D). Rapid infiltration of surface water during floods (March 1992: Fig. 25.10B) again increases the oxygen concentration of the aquifer within a few hours or days.

The dynamics of the organic matter in the Lobau-C aquifer were not investigated, but qualitative and detailed analysis showed that part of the dissolved organic carbon (e.g. carbohydrates) close to the water surface at the reed-belt fringe is recycled or assimilated. Large quantities of particulate organic matter are visible in the same area and below the uppermost soil

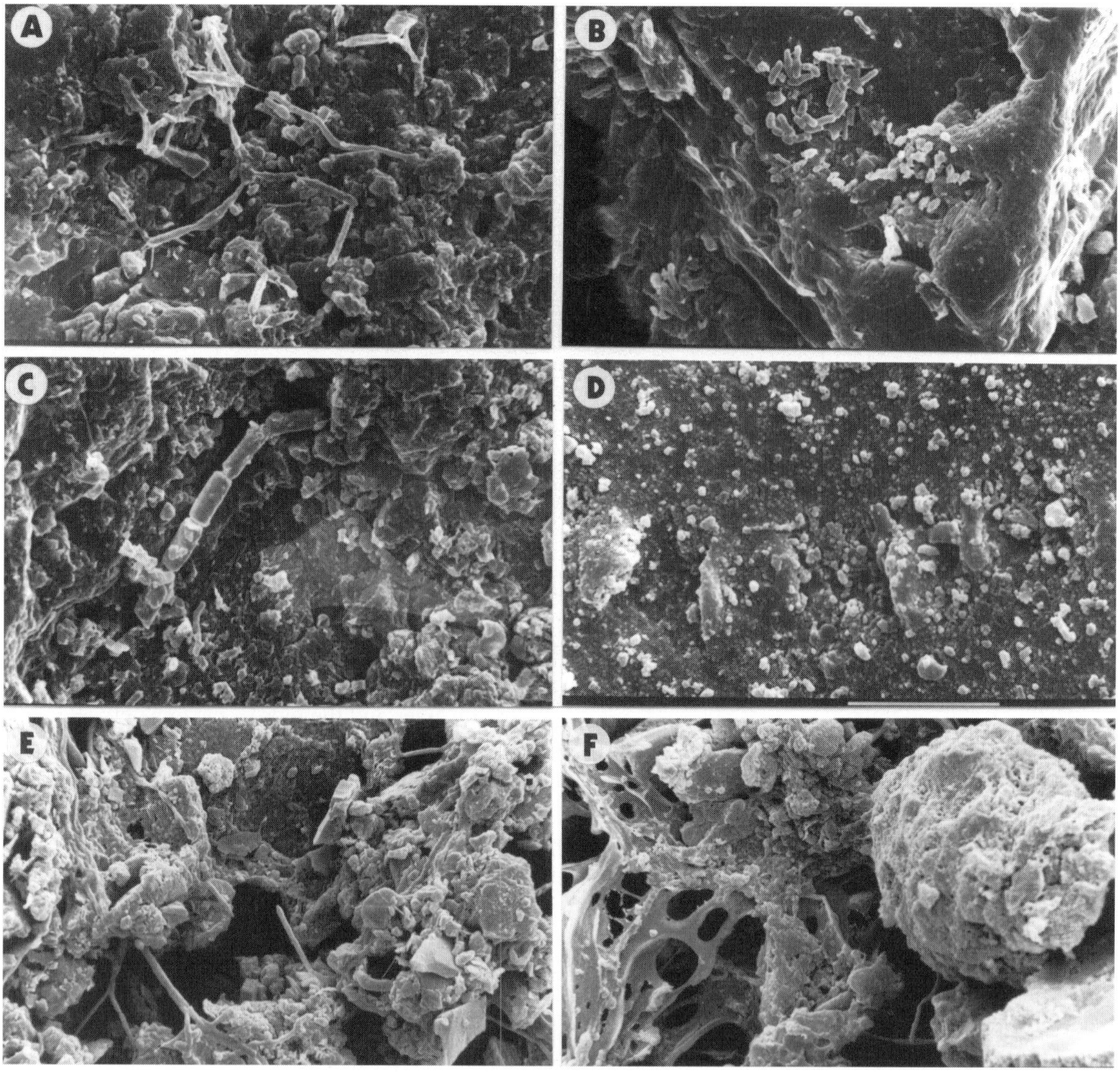

Fig. 25.19. Microbial colonization of sediments exposed in groundwater (see text), scanning electron micrographs; scale bar 10 μm. (A–C) Rod-shaped and filamentous microbes fixed onto rough surfaces of sand grains. (D) Coccoid bacteria on glass-bead surfaces. (E, F) Filamentous bacteria and bacterial exopolymers.

strata. These are precisely the zones where meio- and macro-fauna generally accumulate.

Activity of interstitial meio- and macro-fauna

As previously mentioned, many interstitial animals in this groundwater system are generalist feeders, ingesting organic detritus, microbial biomass and fine mineralogenic sediment, or even are carnivores, feeding on dead and/or living animals. High animal densities in the upper sediment layers of the surface-water system (the upper 0.5 m layer of sediment) were recorded at the site Lobau-A (Danielopol, 1983). Up to 200 oligochaetes were found in three-litre samples of interstitial water and sediment at one site (A1). The abundant populations of macro- and meiofauna in this superficial layer at the Lobau-A site contributed to the oxygen consumption within the sediment, and the oxygen concentration in the deeper layers (1.2–4 m deep) was much lower than values at a depth of 0.5 m, reaching dysaerobic levels; removal of oxygen by organisms in successively deeper strata causes sequential changes in the microenvironment (Danielopol, 1983). In this study we investigated only the distribution of the fauna in the deeper layers of the aquifer and at least one metre from the margin of the surface-water system in the Eberschüttwasser. We noticed that, at a micro-scale (millimetres to centimetres), feeding and movement of macro-organisms like *Cyphon palustris*, *Niphargus* sp. or *Proasellus slavus vindobonensis* change the sediment structure and possibly influence the oxygen conditions, as the fine silty sediments are bioturbated or pelletized by such organisms (Danielopol, 1989; Danielopol et al., 1997). The channels made in fine sediments by some oligochaetes and insect larvae allow, most probably, better micro-oxygenation around the tubes and facilitate the development of bacterial biofilms (Fig. 25.7D). However, during the Lobau-C study such effects were seldom observed because the populations of meio- and macro-organisms were commonly at low densities within large sectors of this aquifer.

Experimental simulation

Microbial colonization of sterile sediments in the field

Standardized exposure of sediment in wells, using cages, is a better way to compare and interpret field data obtained from mixed sediments or from mixtures of sediment and water. In order to observe the microbial assemblages in the substrate during the colonization process, we exposed cages filled with three different artificial substrates (sand with grain size less than 2 mm; fine glass beads 2 mm diameter; and larger glass beads 3 mm diameter) at a depth of 4 m in the groundwater for two months. The most prolific microbial growth occurred on the sandy substrate, where a diverse microbial assemblage was visible (Figs. 25.19A–C), with patchy accumulations of microbes on rough surfaces of sand grains and diffuse distribution of filamentous bacteria. On the glass beads (Fig. 25.19D) there were fewer attached microbes, mainly coccoid bacteria. Total bacterial counts and the proportion of metabolically active cells were higher in the sand (Fig. 25.20). These data are concordant with those obtained in the field, and confirm that sandy sediments with irregular grain shapes having a large surface support higher microbial densities and a higher metabolic activity than coarser sediments with smoother surfaces.

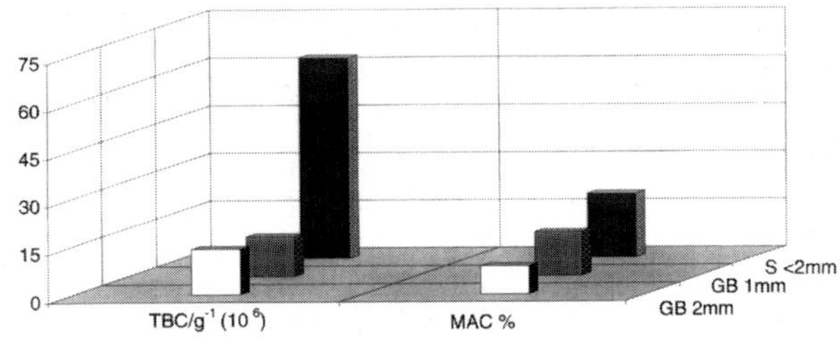

Fig. 25.20. Microbial densities and activities of various sediments exposed to groundwater (see text); TBC, total bacterial counts; MAC, proportion of metabolically active (viable) bacteria; S, sand; GB, glass beads of 1 and 2 mm diameter.

Laboratory experiments with slow-filtration columns

Three series of experiments with two parallel slow-filtration columns (50 cm long, 5 cm internal diameter: Fig. 25.21) were conducted in March 1993 to give

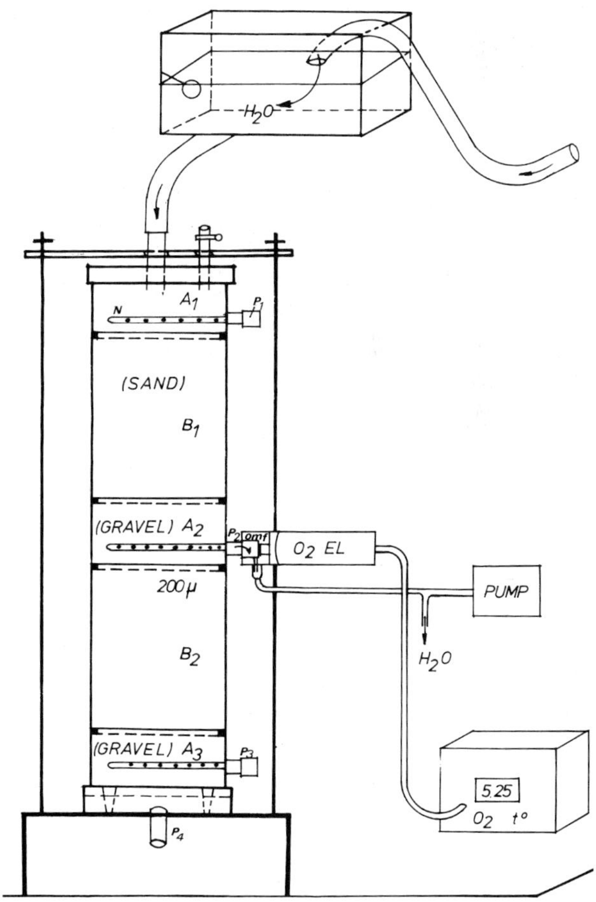

Fig. 25.21. The "module filter column": A1–A3, B1, B2, short and long modules (see text); P1–P3, ports for water extraction; P4, outflow; N, perforated needle on which the oxygen microflow module (omf) is fitted; EL, polarographic oxygen sensor.

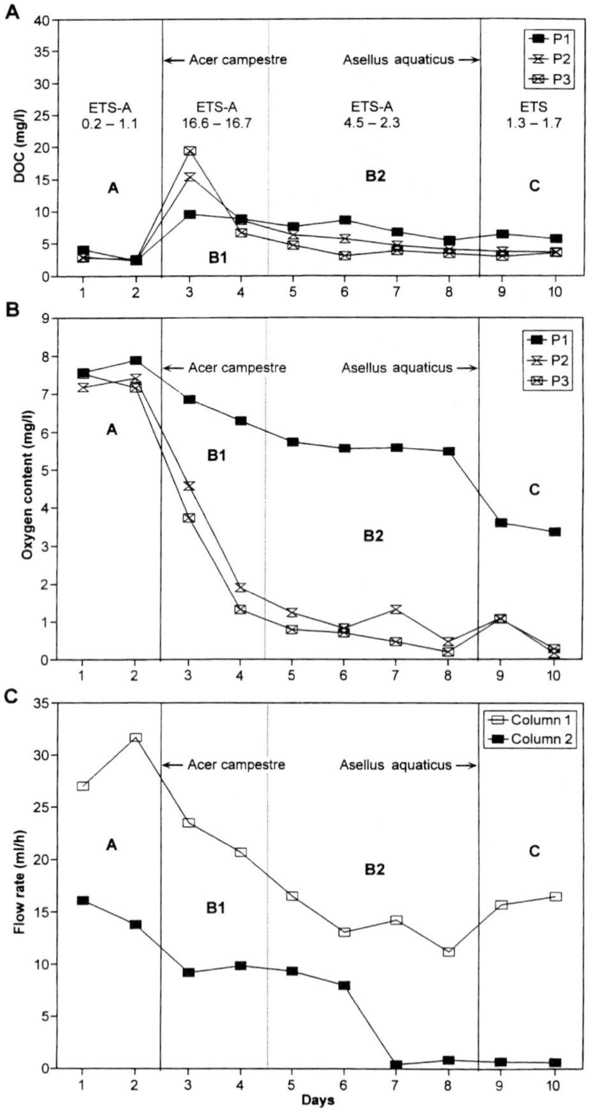

Fig. 25.22. Environmental changes in two slow filtration columns (see text). P1–P3, sampling points (see Fig. 25.21). A, B1, B2, C, successive experimental phases (see text for details): (A) dynamics of the dissolved organic carbon (DOC) and microbial activity (ETS-A, expressed as substrate turnover rate per hour at 37°C) in Column 1; (B) changes in the oxygen content in Column 1; (C) changes in the water seepage at the outflow P4 for both columns.

additional information on the dynamics of organic matter in through-flow sediment systems. Litter in the form of round dried leaves of *Acer campestre* (diameter 1 cm, 2 mg d.w.), previously leached in water for 24 h, were placed in Module A1 (4.5 cm height) to form a layer 1–2 mm thick. Medium-grained sand (200–600 μm) was added to Modules B1 and B2, each

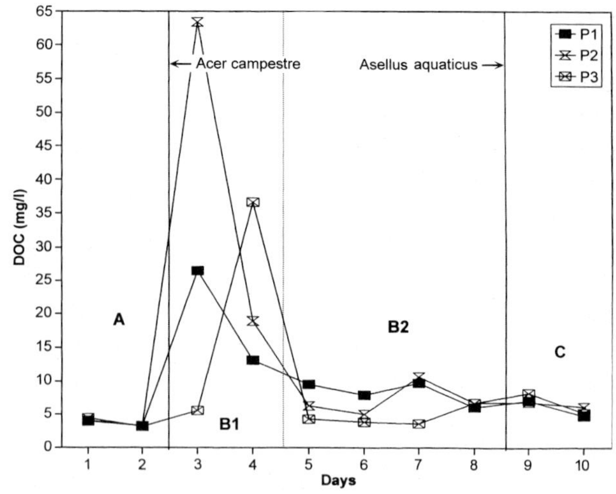

Fig. 25.23. Parallel generation of dissolved organic carbon in Column 2 (compare with Fig. 25.22).

19 cm long, while Modules A2 and A3 (same height as A1) were filled with fine gravel.

The sediments, heat-sterilized at 450°C for 2 h, were perfused with lake water from the Mondsee. Dissolved oxygen and dissolved organic carbon were monitored daily, from water extracted at ports P1–P3 (Fig. 25.21). Water flow rates were monitored daily for 16 days and integrated samples for ETS activity were taken at port P4. The water flow rate differed in the two columns, being higher in Column 1 (Fig. 25.22C). Figures 25.22 and 25.23 display relevant information for 10 days divided into three main phases (the first four and last two days of the runs are not shown). During Phase A (Fig. 25.22A) concentration of DOC and the ETS activity remained low (DOC between 2 and 4 mg ℓ^{-1}). After insertion of *Acer* leaves (Phase B) a dramatic increase in both DOC concentration and ETS activity occurred for a short period of time (Phase B1). In Column 1 maximum DOC values in the lower sediment layer were attained after 24 hours, suggesting that DOC was leached from the leaves for a short period and moved through the sediment downwards. In Column 2 with a lower flow rate, the maximum peak of the leached DOC lasted for two days (Fig. 25.23). During Phase B2, both DOC concentration and ETS activity remain at an intermediate value, suggesting that bacterial biofilms covering the leaves and/or the sediment grains are still active. The examination of the dissolved oxygen (DO) curves for Column 1 (Fig. 25.22B) and the changes in the water flow rates in both columns (Fig. 25.22C) give additional support to this interpretation. During

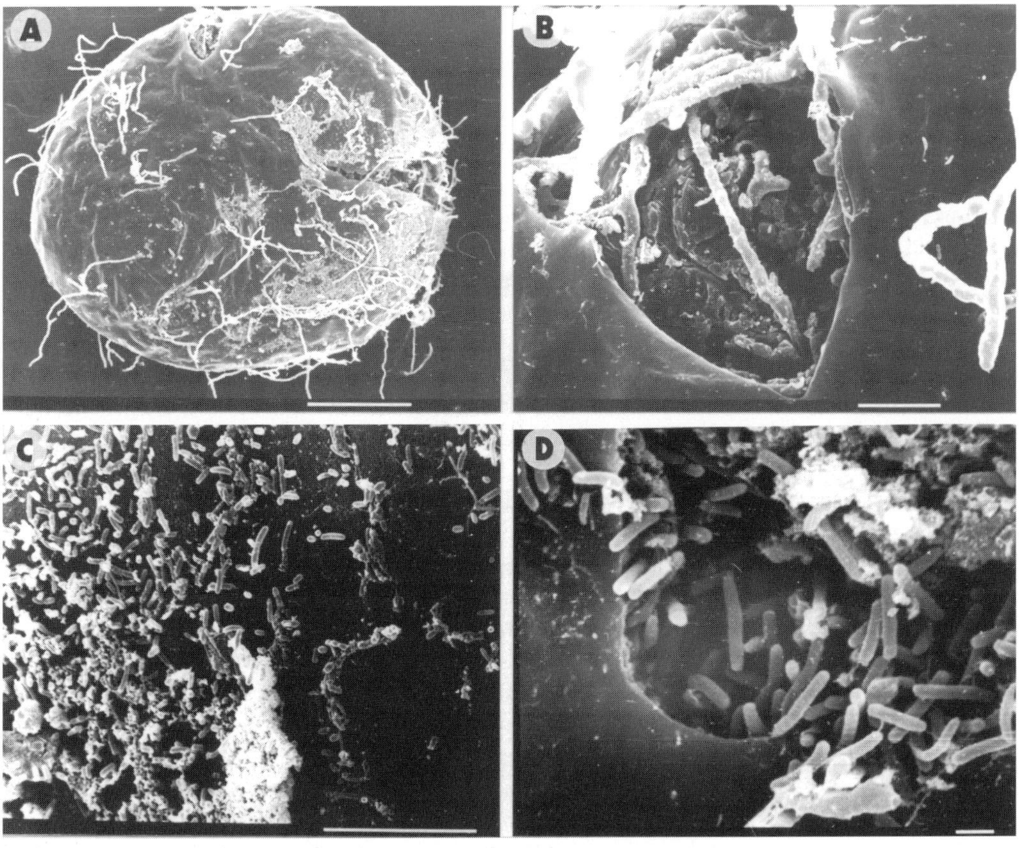

Fig. 25.24. Scanning electron micrographs of the microbial colonization of an isopod egg laid by *Asellus aquaticus* in the slow-filtration Column 2: (A) General view; scale bar 100 μm. (B, D) Penetration of filamentous and rod-shaped bacteria inside the egg through the broken wall (detail of A); scale bars 10 μm (B), 1 μm (D). (C) Accumulation of coccoid and rod-shaped microbes on the egg surface (detail of A); scale bar 10 μm.

Phase A, oxygen concentrations remain high throughout the column, but a rapid lowering of the dissolved oxygen content occurred in the sediment (Phase B1), leading to a dysaerobic Phase B2. The water flow decreased in both columns, most probably because the growth of the biofilm slowly clogged the interstitial spaces, especially in Column 2 in which the flow rate decreased to less than 1 ml h^{-1} by the end of Phase B2 (Fig. 25.22C).

The addition to each column of 25 specimens of adult isopods (*Asellus aquaticus*) did not lead to any significant increase in the amount of DOC leached from the leaves or any increase of ETS activity over the following two days (Figs. 25.22A, 25.23). This suggests that, even if the isopod fed on the leaves, the amount of DOC and bacterial activity were not stimulated within this short-term exposure. The only positive effect is seen in Column 1 (Fig. 25.22C), where the water flow rate increased after the addition of isopods; this effect is apparently due to the mechanical disturbance of the biofilm by the isopods.

Sediments sampled at the end of various experiments and examined with the scanning electron microscope showed high accumulations of organic detritus and diverse microbial types in the upper 2 cm layer of Module B1. In the deeper layers of sediment, like the upper part of Module B2, there was an accumulation of bacterial exopolymers (a mucopolysaccharide matrix) (Figs. 25.19E,F), which evidently contribute to the diminution of the rate of water flow. Organic matter which is not refractory, such as the proteins of isopod tissues, allows the development of dense microbial colonies (Figs. 25.24A–D) within a few days.

The aerial respiration of *Cyphon palustris* larvae

High numbers of *C. palustris* were repeatedly observed at the groundwater table, but only seldom at greater depths (see above), and we investigated whether this species is able also to live within the unsaturated zone or at least in the capillary fringe layer. We observed *Cyphon* larvae with a video camera in an aquarium filled half with coarse sediment and half with water. Anal respiration at the water surface was detected, and the larvae were able to move outside the water-saturated sediment and continue active into the unsaturated zone, at the capillary fringe.

DISCUSSION AND CONCLUSIONS

Groundwater systems, either in karstified rocks or in unconsolidated clastic sediments, were perceived by many ecologists as having rather simplified structures and functional patterns, as compared to analogues in the epigean environment (Culver, 1982; Gibert et al., 1990, 1994; Chapelle, 1992). The present documentation of a small-scale groundwater ecosystem stresses the inherent methodological difficulties for accurate descriptions, owing to the structural complexity of groundwater systems in alluvial sediments colonized by biota with strongly aggregated distributions.

Here we discuss several results of the present "Lobau Project" which may be considered innovative and useful for the development of other groundwater ecological projects.

(1) As mentioned above (p. 482), groundwater ecosystems have generally been defined loosely either as conceptual frameworks or landscape units. Here we started from an abstract view of what a groundwater ecosystem could be. We further delineated the ecosystem on the basis not only of its intrinsic characteristics but also of the relationships existing between this system and those of the surrounding wetlands, both aquatic and terrestrial. For descriptive purposes it was an advantage to use a small-scale groundwater area as an ecosystem which had partly recognizable boundaries, where one could identify the main input and output zones of the system. The approach therefore is holistic, and at a reduced scale: the river–aquifer and terrestrial–aquifer relationships are visible. We expect that principles of features and/or functioning similar to those which have been observed in the Lobau-C aquifer occur and/or operate within larger groundwater systems too (at a scale of hundreds of metres to kilometres). Here it was documented that the most active zones of a groundwater system are at the contacts with both surface aquatic and terrestrial systems. It can be postulated that within large landscape groundwater systems, like the Marchfeld aquifer, the most biologically active zones are located in similar areas – for instance, along the river banks of the Danube. Therefore such "sensitive areas" need to be better investigated and/or protected. This observation has important practical implications for the water-management and landscape-protection policies of Niederösterreich.

(2) One innovative aspect of this project, as compared to other ecosystem studies, such as the groundwater studies of the Rhône floodplains (Dole-Olivier et al., 1994) or those of the Baget karstic system in the Pyrenees (Rouch, 1986), was to use laboratory simulations in order better to understand field ecosystem processes. In this way we noticed that some of the ecological processes, like the transport of dissolved organic matter through the sedimentary system, proceed at very fast rates (hours or days), and therefore our field sampling program for the study of such parameters (seasonal sampling) was more or less inappropriate. A suitable strategy should sample groundwater following hydrological changes to the system, as discussed by Rouch (1986) and Gibert et al. (1994).

(3) Odum's classic ecosystem approach (McIntosch, 1985) required large interdisciplinary teams investigating with equal intensity a large number of ecosystem parameters in order to follow fluxes of matter and energy through the system. This 'big science' approach became impractical because of restricted funds. It is more realistic to adopt a minimalist approach (Danielopol and Marmonier, 1992) to identify and study key parameters and their impact on the whole ecosystem; one of these, in the context of groundwater ecosystem research, is dissolved oxygen. Beside organic carbon, oxygen is one of the most important energetic resources for subsurface organisms, and the dynamics of oxygen within groundwater ecosystems deserves more careful investigations in the future.

(4) Whereas many ecologists working either on groundwater or lotic systems have concentrated on riverbed sediments (Gibert et al., 1990, 1997; Brunke and Gonser, 1997), we have shown here that the terrestrial–aquatic boundary layer is another highly active peripheral zone of groundwater ecosystems (Fig. 25.25). Our documentation on

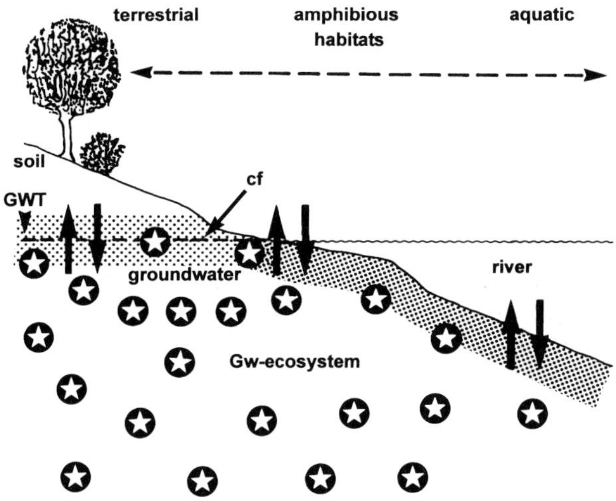

Fig. 25.25. A groundwater ecosystem with its peripheral zones, connected to riparian systems. Stippled areas: fine mesh, transitional zone between surface (lotic) and "deep" subsurface habitats; coarse mesh, transitional zone between subsurface habitats occurring in unsaturated and saturated sediment layers. GWT, groundwater table; cf, capillary fringe above the GWT. Heavy arrows indicate the exchange of matter and organisms between the different habitats and ecological systems; stars symbolize intensive sampling activity for groundwater ecological purposes.

the accumulation of meio- and macro-organisms near the groundwater table, below and possibly above it, gives additional support to the proposition of Madsen and Ghiorse (1993) that the boundary of the subsurface ecosystems on the terrestrial side lies in the unsaturated zone – at least at the capillary fringe level (Fig. 25.25).

(5) Chapelle (1992) and Gibert et al. (1994) noted that the food chains in groundwater systems differ from those of surface-water systems, being shorter and less diverse. This pattern, apparently, applies to the Lobau-C system where the meio- and macro-organisms observed are in many cases omnivores, feeding mainly on fine sediments, bacteria and/or plant detritus. Animals with high energetic requirements, like the ostracod *Cypridopsis vidua* for which we have evidence that it feeds selectively on diatoms (Mbahinzireki et al., 1991; Roca and Danielopol, 1991), do not penetrate deep into the aquifer. We found in the Lobau-C system a few facultative carnivores – for instance, amphipods of the genus *Niphargus*. However, in the future more information is needed on the exact characterization of the feeding preferences of the secondary producers of this system, a topic which was successfully addressed for hyporheic fauna by Schmid and Schmid-Araya (1997).

(6) It has been suggested repeatedly that groundwater animals are able to process sedimentary material within groundwater systems, and that in this way they participate actively to the recycling of energy and matter (Husmann, 1978; Danielopol, 1984, 1989; Schminke and Glatzel, 1988). Hence it has been questioned (Ward et al., 1998) whether subsurface interstitial metazoans could be considered as ecosystem engineers. Laboratory data presented here, as well as those of Merle (1994, and unpublished) and of Torreiter et al. (1994) demonstrate that environmental changes occurring at reduced scales (millimetre–centimetre) are induced *inter alia* by the activities of benthic and interstitial-dwelling animals, through their movement, feeding and respiration. These activities cause cascading effects within the deeper sedimentary layers as epitomized by the oxygen distribution in the sediment. Danielopol et al. (1997) doubted that feeding and movement in the low-density crustacean populations within the Lobau-C aquifer could impact substantially on the macroscopic functioning of this large (metre-scale) system, by, for example, improving the subsurface water flow through the sediments. This topic certainly deserves future scrutiny.

(7) In our opinion, the importance of groundwater meio- and macro-organisms should be seen in their capacity to reflect various aspects of the abiotical and/or biotical environment where they occur; in other words, they are useful as environmental descriptors (i.e. ecological indicators), an approach already documented for cave fauna by Poulson (1992). We have shown here (see also Danielopol et al., 1992; Pospisil, 1994a; Pospisil et al., 1994; Mösslacher et al., 1996; Mösslacher, 1997a,b) that, within the subsurface area of the Lobau-C aquifer, higher species richness and/or abundance were, in general, positively related to the groundwater sectors and/or seasonal periods when well oxygenated surface water penetrated at higher rates into the aquifer. Many of the groundwater-dwelling species display flexible ecological strategies (FES), especially energetic strategies (Fig. 25.26) – as, for example, has been found in the isopod *Proasellus slavus vindobonensis* (Danielopol et al., 1994). The groundwater system also supports narrowly specialized animals. In the subsurface sediments

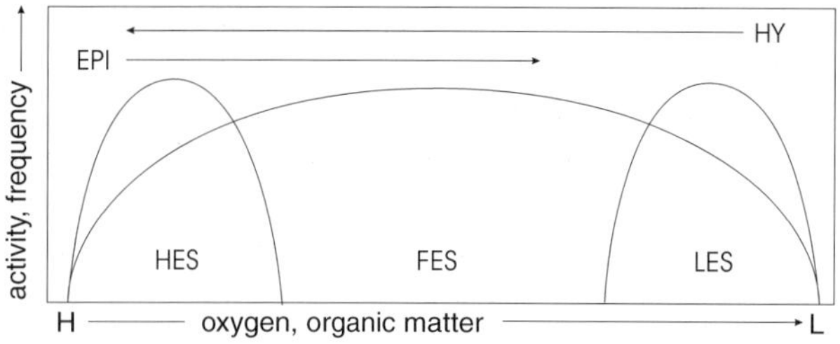

Fig. 25.26. Relative limits of tolerance for epigean (EPI) and hypogean (HY) organisms. H, L, high and low values of the environmental parameter; HES, LES, adaptive strategies for high and low energy requirements; FES, flexible adaptive strategies. The HES and LES curves characterize specialist taxa, while FES applies to generalists. Arrows starting from HY and EPI suggest the general spatial extension of the two animal groups, the hypogeans and the epigeans, into a groundwater system; the vertical axis indicates the biological activity and faunal abundance.

below the Eberschüttwasser backwater within the Lobau-A area (Danielopol, 1983, 1991) one finds species that display strategies adapted to their high energetic requirements, like the isopod *Asellus aquaticus* (Danielopol et al., 1994; Mösslacher and Creuzé des Châtelliers, 1996), the ostracod *Cypridopsis vidua*, and the harpacticoid *Canthocamptus staphylinus* (for this latter species and its ecological requirements, see Sarvala, 1979). Therefore, such animals, specialized to exploit habitats with high energetic resources (Fig. 25.26), indicate by their occurrence subsurface areas with temporarily or permanently enriched energetic supplies, such as the area around well D3 where *C. staphylinus* was observed in the upper two-metre layer. One should mention also the opposite alternative – oligotrophic animals displaying life strategies adapted to low energetic resources, such as the stygobitic mollusc *Bythiospeum* aff. *geyeri* found within deeper layers of the Lobau aquifer. Such organisms (Fig. 25.26) may characterize stable and energetically poor groundwater zones. Furthermore, ongoing ecotoxicological experiments (Mösslacher, 1997a) suggest the possible use of groundwater crustaceans for ecotoxicological surveillance of polluted groundwater habitats, as suggested elsewhere (Notenboom and Boessenkool, 1992; Notenboom et al., 1994). We consider that this research direction is a promising avenue for future groundwater ecological projects.

(8) Danielopol (1989) and Pospisil and Danielopol (1990) suggested that groundwater fauna should be integrated into the environmental protection and planning for National Parks because of its high cultural value, arguing that the study of subsurface organisms can offer both intellectual and aesthetic satisfaction.

Because one cannot protect everything, it has been suggested (Rouch and Danielopol, 1997) that priority be given to "groundwater hot-spot areas" where biota are well known through long-term research. The Lobau-C area in the National Park Donau-Auen of Niederösterreich is an excellent example, with priority to protect the boundary layers in the groundwater system which are most ecologically active (Danielopol et al., 1997).

(9) The concept of an ecosystem as a complex organized structure which processes energy and matter and allows their flow through the system (Odum, 1969) could be represented by the metaphorical image of a "powerful machine". Culver et al. (1995) used for the characterization of a subsurface ecosystem the metaphorical representation of an "ecological theatre"; that metaphor also fits the Lobau-C groundwater ecosystem, perhaps as an "opera house". In such a model the groundwater which modulates the distribution of matter and energy through the system could be compared to the opera music, and the meio- and macrofauna would be the equivalent of the artists who interpret (react ecologically) in subtle ways to the music that they perceive. Once groundwater ecologists have the conceptual skill and technical facilities to observe the subterranean life within the "opera house" (the aquifer) they may become active spectators, enjoying both the structure and the functioning of such a complicated system.

ACKNOWLEDGEMENTS

Useful information and/or material support for this contribution was provided also by various colleagues: C. Merle, M. Creuzé des Châtelliers (Université Claude Bernard, Lyon), R. Niederreiter (Uwitec, Mondsee), I. Bals (Orbisphère, Genève), C. Griebler, H. Ployer, H. Höllerer (Limnologisches Institut, Mondsee), K.H. Schminke (Universität Oldenburg), M. Haase (Universität Basel), G. Magniez and J.-P. Henry (Université de Bourgogne, Dijon), P. Marmonier (Université de Savoie, Le Bourget du Lac), R. Rouch, A. Mangin, M. Bakalowicz (Laboratoire Souterrain, CNRS, Moulis). The electron scanning micrographs were produced at the Universität Wien (Biocentrum) with the help of T. Loser and Prof. W. Klepal. I. Gradl and H. Ployer (Limnologisches Institut, Mondsee) helped to edit the manuscript. We are indebted to Dr. W.F. Humphreys, of the Western Australian Museum, Perth, and to the Editor-in-Chief, Dr. David W. Goodall, for editorial improvements. Permission to use the Lobau area for this groundwater project was given by the Forstverwaltung Lobau (Mrs. Placho, Mr. Schreckeneder, Dipl. Ing. G. Haubenberger). The project was funded by the "Fonds zur Förderung der wissenschaftlichen Forschung", grants P 7881 and P11149 (Vienna). One of us (D.L.D.) received financial support from the Österreichische Akademie der Wissenschaften to visit various laboratories abroad dealing with similar topics.

Appendix A. Ecological data for the Lobau ecosystem

Ecological data for the nine multilevel wells of the Lobau ecosystem and the surounding surface-water systems Eberschüttwasser and Mittelwasser are tabulated below.
The following abbreviations are used:
- D3, D5, . . . , T3: the names of the wells, as in Fig. 25.3
- m.a.s.l.: metres above the level of the Adriatic Sea.
- GWT (b.s.m.): depth of groundwater table below the soil surface, in metres.

Note that at wells D15, D17 and T3, the 20/10/92 samples were taken at a depth of 2.5 m instead of 2 m, because of the exceptionally low groundwater level in the aquifer; no data are available for well D4 after July 1992.

Well D3, Altitude 149.15 m.a.s.l.

Date	16/1/92		27/4/92		20/7/92		20/10/92		18/1/93		14/4/93	
GWT (b.s.m.)	0.4		0.3		0.6		1.2		0.5		0.4	
Depth (m)	2	4	2	4	2	4	2	4	2	4	2	4
Conductivity (μS at 25°C)	486	488	468	471	461	461	493	493	472	464	493	493
Oxygen (mg ℓ^{-1})	5.4	4.3	0.1	0.1	0.1	0.1	0.1	0.1	6.2	2.8	4.8	5.2
Temperature (°C)	3.0	3.1	15.7	15.4	20.4	20.4	16.4	16.3	1.0	5.3	8.4	8.6
Ostracoda	0	2	0	17	0	0	0	0	0	0	0	17
Cladocera	0	1	0	0	0	0	0	0	0	0	0	1
Cyclopoida	1	6	1	28	1	2	2	3	0	60	0	55
Harpacticoida	0	0	2	0	0	2	17	39	0	38	0	23
Isopoda	0	0	1	10	1	0	0	0	0	0	0	0
Amphipoda	0	1	0	0	0	0	0	0	0	0	0	0
Insect larvae	0	0	0	0	0	0	0	0	0	0	0	0

Well D5, Altitude 149.10 m.a.s.l.

Date	16/1/92		27/4/92		20/7/92		20/10/92		18/1/93		14/4/93	
GWT (b.s.m.)	–		0.3		0.6		–		0.5		0.4	
Depth (m)	2	4	2	4	2	4	2	4	2	4	2	4
Conductivity (μS at 25°C)		469	480	455	504	–	–	483	523	475	478	
Oxygen (mg ℓ^{-1})	–	–	0.1	0.1	0.1	0.1	–	–	6.2	2.8	3.4	4.2
Temperature (°C)	–	–	15.6	14.2	17.5	12.1	–	–	1.0	5.3	8.1	8.2
Ostracoda	–	–	6	0	1	0	–	–	0	0	1	1

Well D5, Altitude 149.10 m.a.s.l. (*cont'd*)

Date	16/1/92		27/4/92		20/7/92		20/10/92		18/1/93		14/4/93	
GWT (b.s.m.)	–		0.3		0.6		–		0.5		0.4	
Depth (m)	2	4	2	4	2	4	2	4	2	4	2	4
Cladocera	–	–	0	0	0	0	–	–	0	0	0	0
Cyclopoida	–	–	17	21	1	0	–	–	0	0	1	1
Harpacticoida	–	–	0	2	0	0	–	–	0	0	0	
Isopoda	–	–	0	0	0	0	–	–	0	0	0	0
Amphipoda	–	–	0	0	0	0	–	–	0	1	0	0
Insect larvae	–	–	0	0	0	0	–	–	0	3	0	0

Well D9, Altitude 149.15 m.a.s.l.

Date	16/1/92		27/4/92		20/7/92		20/10/92		18/1/93		14/4/93	
GWT (b.s.m.)	0.7		0.7		0.9		1.6		0.8		0.7	
Depth (m)	2	4	2	4	2	4	2	4	2	4	2	4
Conductivity (μS at 25°C)	484	484	472	473	464	468	503	501	500	496	511	513
Oxygen (mg ℓ^{-1})	2.85	1.1	0.1	0.1	0.1	0.1	0.1	0.1	5.4	6.1	4.1	4.5
Temperature (°C)	3.2	4.1	12	11.1	19.1	18.0	17.0	17.0	0.5	0.6	7.7	7.7
Ostracoda	0	0	1	2	0	0	0	0	0	0	1	0
Cladocera	0	0	0	0	0	0	0	6	0	0	0	0
Cyclopoida	36	19	71	14	9	0	21	2	109	78	96	89
Harpacticoida	1	1	3	0	0	0	0	1	1	2	1	7
Isopoda	0	0	0	0	0	0	0	0	0	0	0	0
Amphipoda	6	2	0	0	0	0	0	0	0	0	1	2
Insect larvae	0	0	0	0	0	0	0	0	0	8	0	2

Well D10, Altitude 149.61 m.a.s.l.

Date	16/1/92		27/4/92		20/7/92		20/10/92		18/1/93		14/4/93	
GWT (b.s.m.)	0.9		0.8		1.1		1.7		1.0		0.9	
Depth (m)	2	4	2	4	2	4	2	4	2	4	2	4
Conductivity (μS at 25°C)	487	488	475	475	456	460	486	512	504	509	510	515
Oxygen (mg ℓ^{-1})	0.4	0.4	1.1	0.8	0.2	0.1	0.2	0.1	4.7	4.5	3.7	3.2
Temperature (°C)	3	3.1	11.6	11.5	19.1	18.0	16.3	16.5	0.8	0.9	7.5	7.4
Ostracoda	0	1	0	8	0	5	0	0	5	39	0	5
Cladocera	0	0	0	0	0	0	0	0	0	0	0	0
Cyclopoida	1	26	1	21	0	0	3	0	0	0	2	86
Harpacticoida	0	0	0	1	0	0	0	0	0	1	0	1
Isopoda	0	0	0	0	0	0	0	0	0	0	0	0
Amphipoda	0	0	0	4	0	0	0	0	0	1	0	1
Insect larvae	0	0	0	0	0	0	0	0	0	5	0	0

Well D11, Altitude 149.15 m.a.s.l.

Date	16/1/92		27/4/92		20/7/92		20/10/92		18/1/93		14/4/93	
GWT (b.s.m.)	–		0.9		1.1		1.8		1.1		1.0	
Depth (m)	2	4	2	4	2	4	2	4	2	4	2	4
Conductivity (μS at 25°C)	–	–	475	478	463	467	439	451	508	511	505	500
Oxygen (mg ℓ^{-1})	–	–	0.3	0.1	0.3	0.1	0.1	0.1	2.9	2.1	2.7	1.3
Temperature (°C)	–	–	11.1	10.9	18.6	18.5	17.2	17.2	2.0	2.1	7.8	7.9
Ostracoda	–	–	0	5	3	2	0	0	0	0	0	0
Cladocera	–	–	0	0	0	0	0	0	0	0	0	0
Cyclopoida	–	–	4	2	0	0	0	1	2	1	2	5
Harpacticoida	–	–	0	0	0	0	0	0	1	1	0	0
Isopoda	–	–	0	0	0	0	0	0	0	0	0	0
Amphipoda	–	–	0	0	0	0	0	0	0	0	0	0
Insect larvae	–	–	0	0	0	0	0	0	0	0	0	0

Well D15, Altitude 149.95 m.a.s.l.

Date	16/1/92		27/4/92		20/7/92		20/10/92		18/1/93		14/4/93	
GWT (b.s.m.)	1.3		1.2		1.5		2.1		1.4		1.3	
Depth (m)	2	4	2	4	2	4	2	4	2	4	2	4
Conductivity (μS at 25°C)	482	485	473	474	477	477	499	501	527	528	520	511
Oxygen (mg ℓ^{-1})	2.5	1.8	0.9	0.8	0.2	0.3	0.2	0.1	7.2	1.8	5.4	4.4
Temperature (°C)	3.5	4.5	10.4	10.3	19.3	18.7	17.5	17.7	0.9	2.1	6.6	5.8
Ostracoda	3	1	3	1	3	0	1	0	0	0	1	0
Cladocera	0	0	0	0	0	0	0	0	0	0	0	0
Cyclopoida	13	3	6	9	9	1	48	0	6	23	9	2
Harpacticoida	0	0	0	0	0	0	0	0	0	1	0	0
Isopoda	8	0	0	0	0	0	0	0	0	0	0	0
Amphipoda	0	0	0	0	0	0	1	0	0	0	0	0
Insect larvae	0	0	0	0	0	0	0	0	0	0	0	0

Well D17, Altitude 150.31 m.a.s.l.

Date	16/1/92		27/4/92		20/7/92		20/10/92		18/1/93		14/4/93	
GWT (b.s.m.)	–		1.5		1.8		2.4		1.7		1.6	
Depth (m)	2	4	2	4	2	4	2	4	2	4	2	4
Conductivity (μS at 25°C)	–	–	475	477	458	457	438	431	445	472	503	515
Oxygen (mg ℓ^{-1})	–	–	0.1	0.1	0.1	0.1	0.1	0.1	6.8	0.9	4.0	0.4
Temperature (°C)	–	–	12.3	10.2	19.6	19.0	17.7	17.6	0.8	1.8	7.9	7.7
Ostracoda	–	–	6	1	2	0	1	0	0	0	0	0
Cladocera	–	–	0	0	0	0	0	0	0	0	0	0
Cyclopoida	–	–	4	2	0	0	144	1	6	32	12	134
Harpacticoida	–	–	0	0	0		0	0	0	2	0	5
Isopoda	–	–	1	0	0	0	0	0	0	0	0	0
Amphipoda	–	–	4	0	0	0	0	0	0	0	0	0
Insect larvae	–	–	0	0	0	0	0	0	0	0	0	0

Well T3, Altitude 150.24 m.a.s.l.

Date	16/1/92		27/4/92		20/7/92		20/10/92		18/1/93		14/4/93	
GWT (b.s.m.)	1.5		1.5		1.7		2.4		1.7		1.6	
Depth (m)	2	4	2	4	2	4	2	4	2	4	2	4
Conductivity (μS at 25°C)	491	490	476	475	467	466	467	478	491	482	510	509
Oxygen (mg ℓ^{-1})	1.8	1.4	0.6	0.4	0.1	0.1	0.1	0.1	8.6	8.4	5.3	5.2
Temperature (°C)	3.7	3.6	11.5	11.6	19.6	19.6	16.8	16.9	1.1	1.0	7.8	7.9
Ostracoda	3	0	0	0	0	0	1	0	0	0	2	0
Cladocera	0	0	0	0	0	0	9	0	0	0	0	0
Cyclopoida	8	0	0	2	1	0	0	1	20	0	17	1
Harpacticoida	0	0	0	0	0	0	0	0	0	0	0	0
Isopoda	3	9	9	9	0	0	8	0	1	8	11	1
Amphipoda	0	1	1	1	0	0	3	0	1	0		0
Insect larvae	0	0	0	0	0	0	0	0	0	0	0	0

Well D4, Altitude 149.12 m.a.s.l.

Date	16/1/92		27/4/92		20/7/92	
GWT (b.s.m.)	0.4		0.3		–	
Depth (m)	2	4	2	4	2	4
Conductivity (μS at 25°C)	483	483	467	468	455	460
Oxygen (mg ℓ^{-1})	5.2	4.8	0.2	0.2	0.1	0.1
Temperature (°C)	2.9	2.9	16.0	15.8	20.4	20.1
Ostracoda	0	0	3	10	0	0
Cladocera	1	3	0	0	0	0
Cyclopoida	0	61	6	41	2	7
Harpacticoida	0	63	0	4	0	0
Isopoda	0	0	0	0	0	0
Amphipoda	0	1	0	0	0	0
Insect larvae	0	0	0	0	0	0

Eberschüttwasser (ESW) and Mittelwasser (MW)

Date	16/1/92		27/4/92		20/7/92		20/10/92		18/1/93		14/4/93	
	ESW	MW	ESW	MW	ESW	MW	ESW	MW	ESW	MW	ESW	MW
Conductivity (μS at 25°C)	495	494	464	476	417	434	328	–	528	509	490	508
Oxygen (mg ℓ^{-1})	–	–	12.3	3.8	6.1	2.5	6.1	–	12.0	12.0	10.0	8.0
Temperature (°C)	3.1	4.7	19	14.3	23.6	21.9	10	–	4.7	4.9	10.7	9.5

REFERENCES

Brunke, M. and Gonser, T., 1997. The ecological significance of exchange processes between rivers and groundwater. *Freshwater Biol.*, 37: 1–33.

Castany, G., 1985. Liaisons hydrauliques entre les aquifères et les cours d'eau. *Stygologia*, 1: 1–26.

Chapelle, F.H., 1992. *Ground-water Microbiology and Geochemistry.* Wiley, New York, 424 pp.

Culver, D.C., 1982. *Cave Life; Evolution and Ecology.* Harvard University Press, Cambridge, MA, 189 pp.

Culver, D.C., Kane, T.C. and Fong, D.W., 1995. *Adaptation and Natural Selection in Caves.* Harvard University Press, Cambridge, MA, 223 pp.

Danielopol, D.L., 1976. The distribution of the fauna in the interstitial habitats of riverine sediments of the Danube and the Piesting (Austria). *Int. J. Speleol.*, 8: 23–51.

Danielopol, D.L., 1980. The role of the limnologist in groundwater studies. *Int. Rev. ges. Hydrobiol.*, 65: 777–791.

Danielopol, D.L., 1982. Phreatobiology reconsidered. *Pol. Arch. Hydrobiol.*, 29, 2: 375–386.

Danielopol, D.L., 1983. Der Einfluß organischer Verschmutzung auf das Grundwasser-Ökosystem der Donau im Raum Wien und Niederösterreich. *Bundesminist. Gesundheit Umweltschutz, Forschungsber.*, 5/83: 5–160.

Danielopol, D.L., 1984. Ecological investigations on the alluvial sediments of the Danube in the Vienna area; a phreatobiological project. *Verh. Int. Verein. Limnol.*, 22: 1755–1761.

Danielopol, D.L., 1989. Groundwater fauna associated with riverine aquifers. *J. North Am. Benthol. Soc.*, 8: 18–35.

Danielopol, D.L., 1991. Spatial distribution and dispersal of interstitial Crustacea in alluvial sediments of a backwater of the Danube at Vienna. *Stygologia*, 6: 97–110.

Danielopol, D.L. and Marmonier, P., 1992. Aspects of research on groundwaters along the Rhône, Rhine and Danube. *Regulated Rivers*, 7: 5–16.

Danielopol, D.L. and Niederreiter, R., 1987. A sampling device for groundwater organisms and oxygen measurements in multilevel monitoring wells. *Stygologia*, 3: 252–263.

Danielopol, D.L., Pospisil, P. and Dreher, J., 1991. Ecological basic research with potential application for groundwater management. In: H.P. Nachtnebel and K. Kovar (Editors), *Hydrological Basis of Ecologically Sound Management of Soil and Groundwater, Proc. Vienna Symp.*, IAHS Press, Wallingford, IAHS Publication 202, pp. 215–228.

Danielopol, D.L., Dreher, J., Gunatilaka, A., Kaiser, M., Niederreiter, R., Pospisil, P., Creuzé des Châtelliers, M. and Richter, A., 1992. Ecology of organisms living in a hypoxic groundwater environment at Vienna (Austria); methodological questions and preliminary results. In: J. Simons and J.A. Stanford (Editors), *Proc. 1st Int. Conf. on Ground Water Ecology.* U.S. E.P.A., A.W.R.A., Bethesda, MA, pp. 79–90.

Danielopol, D.L., Creuzé des Châtelliers, M., Mösslacher, F.,

Pospisil, P. and Popa, R., 1994. Adaptation of Crustacea to interstitial habitats: A practical agenda for ecological studies. In: J. Gibert, D.L. Danielopol and J.A. Stanford (Editors), *Groundwater Ecology.* Academic Press, San Diego, CA, pp. 217–243.

Danielopol, D.L., Rouch, R., Pospisil, P., Torreiter, P. and Mösslacher, F., 1997. Ecotonal animal assemblages; their interest for groundwater studies. In: J. Gibert, J. Mathieu and F. Fournier (Editors), *Groundwater/Surface Water Ecotones.* Cambridge University Press, Cambridge, pp. 11–20.

Dole-Olivier, M.J., Marmonier, P. and Creuzé des Châtelliers, M., 1994. Interstitial fauna associated with the alluvial floodplains of the Rhône River (France). In: J. Gibert, D.L. Danielopol and J.A. Stanford (Editors). *Groundwater Ecology.* Academic Press, San Diego, CA, pp. 313–346.

Dreher, J., Pospisil, P. and Danielopol, D.L., 1997. The role of hydrology in defining a groundwater ecosystem. In: J. Gibert, J. Mathieu and F. Fournier (Editors), *Groundwater/Surface Water Ecotones.* Cambridge University Press, Cambridge, pp. 119–126.

Freeze, R.A. and Cherry, J.A., 1979. *Groundwater.* Prentice-Hall, Englewood Cliffs, NJ, 604 pp.

Gibert, J., 1992. Groundwater ecology from the perspective of environmental sustainability. In: J. Simons and J.A. Stanford (Editors), *Proc. 1st Int. Conf. on Ground Water Ecology.* U.S. E.P.A., A.W.R.A., Bethesda, MA, pp. 3–14.

Gibert, J., Dole-Olivier, M.J., Marmonier, P. and Vervier, P., 1990. Surface water–groundwater ecotones. In: R.J. Naiman and H. Décamps (Editors), *The Ecology and Management of Aquatic–Terrestrial Ecotones. Man and Biosphere Series,* 4. Parthenon Publications, London, pp. 199–225.

Gibert, J., Stanford, J.A., Dole-Olivier, M.J. and Ward, J.V., 1994. Basic atttributes of groundwater ecosystems and prospects for research. In: J. Gibert, D.L. Danielopol and J.A. Stanford (Editors). *Groundwater Ecology.* Academic Press, San Diego, CA, pp. 7–40.

Gibert, J., Fournier, F. and Mathieu, J., 1997. The groundwater/surface water ecotone perspective: state of the art. In: J. Gibert, J. Mathieu and F. Fournier (Editors), *Groundwater/Surface Water Ecotones.* Cambridge University Press, Cambridge, pp. 3–8.

Griebler, C. and Slezak, D., 1997. Dimethylsulfoxide (DMSO) Reduktion. Ein neues Verfahren um die mikrobielle Aktivität in aquatischen Systemen zu bestimmen. In: M. Dokulil (Editor), *32. Konf. d. IAD, Limnologische Berichte – Donau 1997, 1 – Wissenschaftliche Referate.* Ö. N. Donauforschung d. IAD, Wien, pp. 123–126.

Gunatilaka, A., Dreher, J. and Richter, A., 1994. Dissolved organic carbon dynamics in a groundwater environment. *Verh. Int. Verein. Limnol.,* 25: 1420–1425.

Haase, M., 1995. The stygobiont genus *Bythiospeum* in Austria: a basic revision and anatomical description of *B.* cf. *geyeri* from Vienna (Caenogastropoda, Hydrobiidae). *Am. Malacol. Bull.,* 11: 123–137.

Husmann, S., 1978. Die Bedeutung der Grundwasserfauna für biologische Reinigungsvorgänge im Interstitial von Lockergesteinen. gwf - *Wasser Abwasser,* 119: 293–302.

Isaaks, E.H. and Srivastava, R.M., 1989. *Applied Geostatistics.* Oxford University Press, New York, 561 pp.

Klausnitzer, B. and Pospisil, P., 1991. Larvae of *Cyphon* sp. (Coleoptera, Helodidae) in groundwater. *Aquat. Insects,* 13: 161–165.

Lee, D.R., 1977. A device for measuring seepage flux in lakes and estuaries. *Limnol. Oceanogr.* 22: 140–147.

Likens, G.E. and Bormann, H., 1985. An ecosystem approach. In: G.E. Likens (Editor), *An Ecosystem Approach to Aquatic Ecology; Mirror Lake and its Environment.* Springer, New York, pp. 1–8.

Löffler, H., 1960. Die Entomostrakenfauna der Ziehbrunnen und einiger Quellen des nördlichen Burgenlandes. *Wiss. Arb. Burgenland,* 24: 1–32.

Löffler, H., 1964. 3. Beitrag zur Kenntnis der Entomostraken burgenländischer Brunnen und Quellen (südl. Burgenland). *Wiss. Arb. Burgenland,* 31: 156–169.

Madsen, E.L. and Ghiorse, W.C., 1993. Groundwater microbiology: subsurface ecosystem processses. In: T.E. Ford (Editor), *Aquatic Microbiology.* Blackwell Scientific Publishers, Cambridge, MA, pp. 167–213.

Mbahinzireki, G., Uiblein, F. and Winkler, H., 1991. Microhabitat selection on ostracods in relation to predation and food. *Hydrobiologia,* 222: 115–119.

McIntosch, R.P., 1985. *The Background of Ecology; Concept and Theory.* Cambridge University Press, Cambridge, 383 pp.

Merle, C., 1994. *Etude Éxperimentale de l'Activité des Oligochaetes Tubificidae sur les Écoulements d'Eau et les Flux d'Oxygène et de Nitrate au Niveau de l'Interface Eau–Sédient,* Master Thesis, Univ. Lyon 1, Villeurbanne, 34 pp.

Meta Systems Inc., 1975. *Systems Analysis in Water Resources Planning.* Water Information Center, Port Washington, NY, 393 pp.

Mösslacher, F., 1994. *Beobachtungen eines Grundwasserökosystems im Freiland (Lobau, Wien) und experimentelle Untersuchungen über die Respirometrie, Freß- und Bewegungsaktivität einiger Süßwasserisopoden (Crustacea, Asellidae),* Master Thesis, University of Vienna, 80 pp.

Mösslacher, F., 1997a. Ein Vorschlag für die zusätzliche Verwendung von Crustaceen zur Qualitätskontrolle von Grundwässern. In: M. Dokulil (Editor), *32. Konf. d. IAD, Limnologische Berichte – Donau 1997, 1 – Wissenschaftliche Referate.* Ö. N. Donauforschung d. IAD, Wien, pp. 449–452.

Mösslacher, F., 1997b. Subsurface dwelling crustaceans as indicators of environmental conditions in an alluvial aquifer. *Environ. Sci. Forum,* 96: 119–140.

Mösslacher, F. and Creuzé des Châtelliers, M., 1996. Physiological and behavioural adaptation of an epigean and a hypogean dwelling population of *Asellus aquaticus* (L.) (Crustacea, Isopoda). *Arch. Hydrobiol.,* 138: 187–198.

Mösslacher, F., Pospisil, P. and Dreher, J., 1996. A groundwater ecosystem study in the "Lobau" wetland (Vienna), reflecting the interactions between surface water and ground water. *Arch. Hydrobiol., Suppl. 113, Large Rivers,* 10: 451–455.

Negrea, S. and Pospisil, P., 1995. Contribution à la connaissance des Cladocères des eaux souterraines du Danube à Vienne. *Ann. Limnol.,* 31: 169–178.

Niederreiter, R. and Danielopol, D.L., 1991. The use of mini-videocameras for the description of groundwater habitats. *Mittblatt. Hydrogr. Dienstes Österreich,* 65/66: 85–89.

Notenboom, J. and Boessenkool, J.-J., 1992. Acute toxicity testing with the groundwater copepod *Parastenocaris germanica* (Crustacea). In: J. Simons and J.A. Stanford (Editors), *Proc. 1st Int. Conf. on Ground Water Ecology.* U.S. E.P.A., A.W.R.A., Bethesda, MA, pp. 301–309.

Notenboom, J., Plénet, S. and Turquin, M.J., 1994. Groundwater contamination and its impact on groundwater animals and ecosystems. In: J. Gibert, D.L. Danielopol and J.A. Stanford (Editors). *Groundwater Ecology*. Academic Press, San Diego, CA, pp. 477–504.

Obermann, P., 1976. Möglichkeiten der Anwendung des Doppelpakers in Beobachtungsbrunnen der Grundwassererkundung. *Bohrtech. – Brunnenbau-Rohrleitungen*, 27: 93–96.

Obst, U. and Holzapfel-Pschorn, A., 1988. *Enzymatische Tests für die Wasseranalytik*. R. Oldenbourg Verlag, München, 86 pp.

Odum, E., 1969. The strategy of ecosystem development. *Science*, 164: 262–270.

Porter, K.G. and Feig, Y.S., 1980. The use of DAPI for identifying and counting aquatic microflora. *Limnol. Oceanogr.*, 25: 943–947.

Pospisil, P., 1992. Sampling methods for groundwater animals of unconsolidated sediments.In: A.I. Camacho (Editor), *The Natural History of Biospeleology*. Monogr. Mus. Nac. Cienc. Nat. 7, Madrid, pp. 107–134.

Pospisil, P., 1994a. *Die Grundwassercyclopiden der Lobau in Wien (Österreich): faunistische, taxonomische und ökologische Untersuchungen*, Ph.D. Thesis. University of Vienna, 220 pp.

Pospisil, P., 1994b. The groundwater fauna of a Danube aquifer in the wetland Lobau at Vienna, Austria. In: J. Gibert, D.L. Danielopol and J.A. Stanford (Editors). *Groundwater Ecology*. Academic Press, San Diego, CA, pp. 347–366.

Pospisil, P. and Danielopol, D.L., 1990. Vorschläge für den Schutz der Grundwasserfauna im geplanten Nationalpark "Donauauen" östlich von Wien, Österreich. *Stygologia*, 5: 75- 85.

Pospisil, P., Danielopol, D.L. and Dreher, J., 1994. Measuring dissolved oxygen in simple and multilevel wells. In: J.A. Stanford and H.M. Valett (Editors). *Proc. 2nd. Int. Conf. Ground Water Ecology*. U.S. E.P.A., A.W.R.A., Herndon, VA, pp. 57–64.

Poulson, T.L., 1992. Case studies of groundwater biomonitoring in the Mammoth Cave region. In: J. Simons and J.A. Stanford (Editors), *Proc. 1st Int. Conf. Ground Water Ecology*. U.S. E.P.A., A.W.R.A., Bethesda, MA, pp. 331–340.

Roca, J.R. and Danielopol, D.L., 1991. Exploration of interstitial habitats by the phytophilous ostracod *Cypridopsis vidua* (O.F. Müller): experimental evidence. *Ann. Limnol.*, 27: 243–252.

Rogulj, B., Danielopol, D.L., Marmonier, P. and Pospisil, P., 1993. Adaptive morphology, biogeographical distribution and ecology of the species group *Mixtacandona hvarensis* (Ostracoda, Candoninae). *Mem. Biospeol.*, 20: 195–207.

Rouch, R., 1977. Considérations sur l'écosystème karstique. *C.R. Acad. Sc. Paris, Ser. D*, 284: 1101–1103.

Rouch, R., 1986. Sur l'écologie des eaux souterraines dans le karst. *Stygologia*, 2: 352–398.

Rouch, R., 1992. Caractéristiques et conditions hydrodynamiques des écoulements dans les sédiments d'un ruisseau des Pyrénées; implications écologiques. *Stygologia*, 7: 13–25.

Rouch, R. and Danielopol, D.L., 1997. Species richness of Microcrustacea in subterranean freshwater habitats. Comparative analysis and approximate evaluation. *Int. Rev. ges. Hydrobiol.*, 82: 121–145.

Rouch, R., Mangin, A., Bakalowicz, M. and D'Hulst, D., 1997. The hyporheic zone: hydrogeological and geochemical study of a stream in the Pyrenees mountains. *Int. Rev. ges. Hydrobiol.*, 82: 357–378.

Sarvala, J., 1979. A parthenogenetic life cycle in a population of *Canthocamptus staphylinus* (Copepoda, Harpacticoida). *Hydrobiologia*, 62: 113–129.

Schiemer, F., Pokorny, J., Gätz, N., Pospisil, P. and Christoff-Dirry, P., 1992. Limnologische Gesichtspunkte bei der Beurteilung von Augewässerdotationen. *Österr. Wasserwirtschaft*, 44: 300–306.

Schmid, P.E. and Schmid-Araya, J.M., 1997. Predation on meiobenthic assemblages: resource use of a tanypod guild (Chironomidae, Diptera) in a gravel stream. *Freshwater Biol.*, 38: 67–91.

Schminke, H.K. and Glatzel, Th., 1988. Besonderheiten und ökologische Rolle der Grundwassertiere. *Z. Dtsch. Geol. Ges.*, 139: 383–392.

Spandl, H., 1926. Die Tierwelt der unterirdischen Gewässer. Speläologische Monographien 11. Späleologisches Institut Verlag, Wien, 235 pp.

Stanford, J.A. and Ward, J.V., 1993. An ecosystem perspective of alluvial rivers: connectivity and the hyporheic corridor. *J. North Am. Benthol. Soc.*, 12: 48–60.

Tansley, A.G., 1935. The use and abuse of vegetational concepts and terms. *Ecology*, 16: 284–307.

Torreiter, P., Pitaksintorn-Watanamahart, P. and Danielopol, D.L., 1994. The activity of oligochaetes in relation to their ecological role in slow filtration columns. In: J.A. Stanford and H.M. Valett (Editors), *Proc. 2nd. Int.Conf. Ground Water Ecology*. U.S. E.P.A., A.W.R.A., Herndon, VA, pp. 85–94.

Trémolières, R., Carbiener, R., Eglin, I., Robach, F., Roeck, U. and Sanchez-Perez, J.-M., 1997. Surface water/groundwater/forest alluvial ecosystems: functioning of interfaces. The case of the Rhine floodplain in Alsace (France). In: J. Gibert, J. Mathieu and F. Fournier (Editors), *Groundwater/Surface Water Ecotones*. Cambridge University Press, Cambridge, pp. 91–101.

Vervier, P., Valett, M.H., Hakenkamp, C.C. and Dole-Olivier, M.J., 1997. Contribution of the groundwater/surface water ecotone concept to our knowledge of river ecosystem functioning. In: J. Gibert, J. Mathieu and F. Fournier (Editors), *Groundwater/ Surface Water Ecotones*. Cambridge Univ. Press, Cambridge, pp. 238–242.

Ward, J.V., Bretschko, G., Brunke, M., Danielopol, D.L., Gibert, J., Gonser, T. and Hildrew, A.G., 1998. The boundaries of river systems: the metazoan perspective. *Freshwater Biol.*, 40: 531–569.

Zimmermann, R., Itturiaga, R. and Becker-Birck, J., 1978. Simultaneous determination of the total number of aquatic Bacteria and their number involved in respiration. *Appl. Env. Microbiol.*, 36: 926–935.

Chapter 26

RHÔNE GROUNDWATER SYSTEMS

P. MARMONIER, M. CREUZÉ DES CHÂTELLIERS, M.J. DOLE-OLIVIER, S. PLÉNET and J. GIBERT

INTRODUCTION

Few groundwater systems associated with a river have been studied for such a long time and with such a high sampling effort as the river Rhône. In Europe, the only comparable set of studies is available for the Danube (Danielopol, 1984, 1989; Pospisil, 1994). The interest for underground habitats and fauna of the river Rhône arose during the 1970s; extensive studies emphasized the composition and spatial heterogeneity of groundwater fauna at the scale of the entire alluvial valley (Gibert et al., 1977, 1981; Seyed-Reihani, 1980). Other studies contributed to the precise description of the "vertical dimension" of the river in both the active channel and the alluvial plain (Dole, 1983a,b, 1984). During the same period, equivalent surveys of surface water systems were performed, which demonstrated that water exchanges between groundwater and surface systems have a key importance for the dynamics of river side-channels. More recently, research moved toward two major topics: (1) the study of abiotic and biotic processes that take place in the surface-water–groundwater ecotones; and (2) the effect of natural and anthropogenic disturbance on shallow interstitial and deep groundwater communities.

This chapter will emphasize the recent progress achieved in the study of the interstitial compartment of the river Rhône; it will be presented through vertical, longitudinal, and transverse (lateral) dimensions. From upstream to downstream, the processes that take place in groundwater habitats vary because of changes in river hydrology (disturbance regime), in nutrient content (characteristics of the river basin, inputs of fertilizers), and in faunal composition (linked to local history and ecological constraints). In the same way, from the main channel to the floodplain, the hydrology of surface water (flowing or stagnant water-bodies), the grain size of the sediment (gravelly or sandy sediments), and the availability of nutrients (relative importance of riparian vegetation and organic-matter decomposition) induce changes in fauna composition, assemblage dynamics, and energetics of the groundwater system. In these longitudinal and transverse dimensions, two major topics will be considered: (1) nutrient dynamics in interstitial habitats (focusing on exchanges and processes); and (2) dynamics of the interstitial communities (focusing on their composition, structure, and temporal changes). Finally, the recent progress of research on the effect of anthropogenic disturbances of groundwater systems (such as chronic pollution and river regulation) will be considered.

GEOGRAPHICAL CHARACTERISTICS OF THE RIVER RHÔNE

The Rhône (Fig. 26.1) is a large, sub-montane river, with a high specific discharge (~30 $\ell\ s^{-1}\ km^{-2}$: Bravard, 1987) and a large drainage basin (98 000 km^2). The river flows between eastern calcareous relief, a northern sedimentary basin, and a western crystalline massif combined with a calcareous plateau. Two major zones can be distinguished in the French part of the Rhône: (1) the Haut-Rhône from Genève to Lyon, with a moderate discharge (between 340 $m^3\ s^{-1}$ at Geneva and 600 $m^3\ s^{-1}$ at Lyon) and a strong gradient (0.50 ‰), and (2) the Bas-Rhône from Lyon to the Mediterranean Sea, with a higher discharge (between 1030 $m^3\ s^{-1}$ just downstream of Lyon, and 1700 $m^3\ s^{-1}$ approaching the Mediterranean) but with a similar slope. The river Saône (410 $m^3\ s^{-1}$ and a 30 600 km^2 drainage basin), the largest tributary of the Rhône, comes from the north-eastern part of France, reaches the Rhône at Lyon, and strongly modifies its hydrology: the river

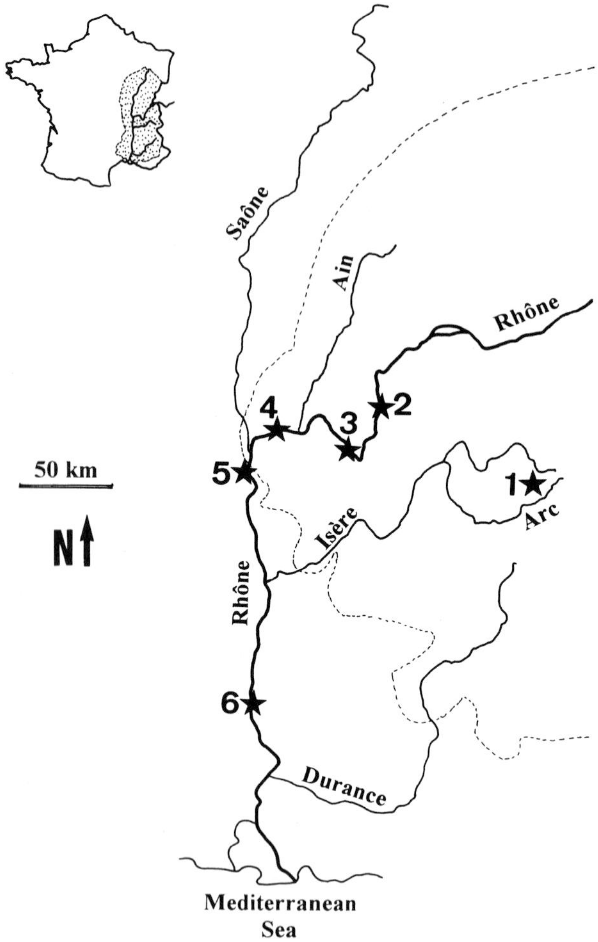

Fig. 26.1. The Rhône river basin, its location in France (top), and detailed map with major tributaries. The stars indicate study sites: (1) the Vanoise area between the Isère and Arc rivers; (2) the Chautagne sector; (3) the Brégnier sector; (4) the Jons sector (see detailed map in Fig. 26.4); (5) the Grand Gravier sector; (6) the Donzère sector. Dotted line: limit of the Pleistocene glacial maximum.

then changes from a glacial regime to a pluvial-oceanic regime. Two other large tributaries come from the Alps: the river Isère (350 $m^3 s^{-1}$ and 12 000 km^2) and the river Durance (220 $m^3 s^{-1}$ and 14 000 km^2) downstream of Lyon.

In the mountainous part of the catchment basin, interstitial habitats and fauna have still been only poorly studied; the only available data come from the upper part of the river Isère (and of its main tributary, the river Arc: Fig. 26.1) in the Parc National de la Vanoise. In this part of the French Alps, rivers flow on crystalline substratum with poorly developed lateral aquifers; habitats available for interstitial fauna are reduced.

Most of the studies have taken place in the foothill sector, where the valley is wide enough and the river develops large floodplains with a mosaic of abandoned channels and contrasted aquatic, semiaquatic, and terrestrial habitats, providing numerous slow-flowing and lentic patches for limnic organisms (Bravard et al., 1986; Amoros et al., 1987). A large part of these abandoned channels are fed by groundwater and represent valuable localities for studying interstitial fauna. The major aquifers studied are located just upstream or downstream of Lyon, in sedimentary basins composed of gravelly deposits (sometime thicker than 15 m) of high permeability (10^{-2} to 10^{-3} $m s^{-1}$).

FROM MOUNTAIN BROOKS TO LARGE ALLUVIAL VALLEY: THE LONGITUDINAL DIMENSION

Longitudinal gradients of benthic organisms are well known (Botosaneanu, 1988; Vannote et al., 1980; Sedell et al., 1989), but few studies deal with longitudinal gradients of interstitial fauna along large rivers. The most complete study is that of Ward and his colleagues (Ward and Voelz, 1990, 1994; Ward et al., 1994) on the South Platte River along an extensive elevation gradient (around 2000 m). The most striking conclusion of this study was the more-or-less complete lack of longitudinal gradient in the interstitial fauna: site-specific hydraulic conditions and substratum properties seemed to be more important than elevation in structuring interstitial communities. For the moment, such a large range of altitudes in the samples is not available for the Rhône basin; foothill sectors have been much more frequently sampled than high-mountain brooks. Despite this disequilibrium, this chapter will present (1) the first results available on interstitial assemblages of high-mountain brooks in the Vanoise sector and (2) some examples of longitudinal distribution of particularly well studied groups such as Ostracoda and Microparasellidae.

The high-mountain brooks of the Vanoise area

Low-order streams represent for all catchment basins the largest part of the lotic systems. In the Rhône basin, they are mostly mountain brooks of the Alps region. The only available data on interstitial habitat and fauna of these brooks come from the Vanoise area (Fig. 26.1). During the summer of 1994, six stations located in three alpine brooks were studied: two stations

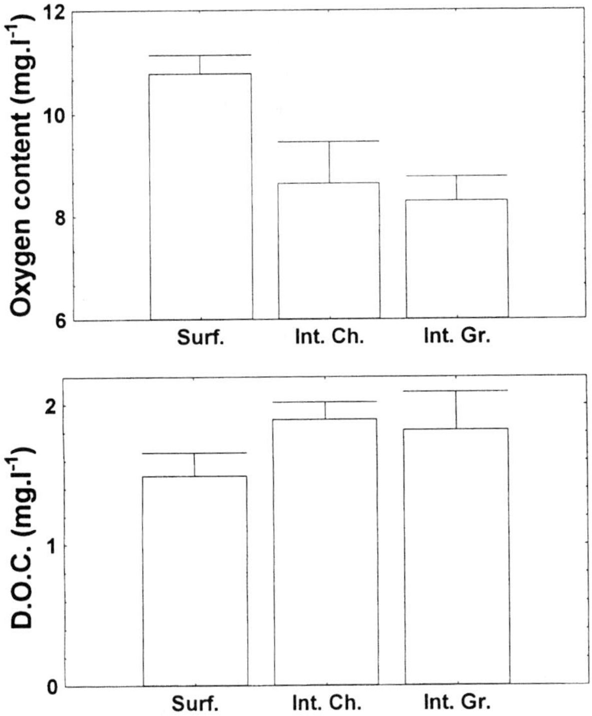

Fig. 26.2. Content of oxygen and dissolved organic carbon (DOC) of surface (Surf.) and interstitial waters from 40 cm deep in the channel (Int. Ch.) and from 40 cm deep in the gravel bar (Int. Gr.) of Vanoise brooks during summer 1993 (mean values SE for $n = 6$). Between-group differences were significant for oxygen content (Analysis of Variance: $p < 0.05$).

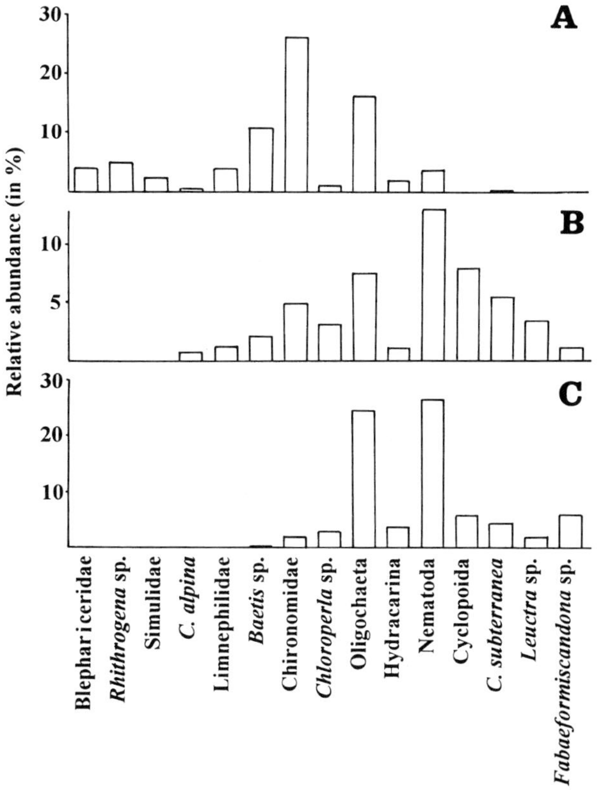

Fig. 26.3. Distribution of aquatic invertebrates (expressed in %) in different habitats of Vanoise brooks; (A) benthic layer, (B) interstitial habitat (40 cm deep) of the channel, and (C) interstitial habitat of gravel bars.

in the river Isère (at altitudes of 2056 and 890 m), two in the river Doron (at 1710 and 1560 m), and two in the river Arc (at 2046 and 1674 m). At each station, three sampling points were chosen according to the sampling design proposed by Pennak and Ward (1986): one sample in the benthic layer, one sample in the interstitial habitat of the channel at a depth of 40 cm in the sediment, and one sample at a depth of 40 cm in the sediment of a lateral gravel bar (at 3 or 5 m from the channel). A hand net and a Bou–Rouch pump (Bou, 1974; Bou and Rouch, 1967) were used for sampling the benthic layer and the interstitial habitat, respectively.

Despite the occurrence of large glaciers near these sampling stations, in most cases sediments were unclogged (except one station located on river Doron). Water temperature and specific conductance were not significantly different between interstitial and surface samples, demonstrating good exchanges of water between these two layers. In the same way, total organic-matter content of the sediment was low (with a mean content of 2.2 mg g^{-1} of dry sediment) at the three sampling points. In contrast, oxygen content significantly decreased from surface to groundwater, and dissolved organic carbon (which varied between 1 and 2 mg ℓ^{-1}) increased slightly from surface to groundwater (Fig. 26.2). The activity of interstitial biofilms would explain these spatial variations, oxygen consumption being associated with degradation of organic matter to dissolved organic carbon (DOC).

No altitudinal pattern could be observed in the composition of interstitial assemblages, but marked differences occurred between the three types of sampling point. The taxonomic richness of macroinvertebrates in these three alpine brooks was low and decreased from the surface to groundwater: 21 taxa were recorded in samples from the benthic layer, 14 taxa in the interstitial habitat of the channel, and 15 taxa in the interstitial habitat of the gravel bars. Most of the taxa had limited distribution (Fig. 26.3): (1) the ephemeropteran *Rhithrogena* sp. and the dipteran families Blepharoceridae and Simuliidae were restricted to the benthic layer;

(2) the planarian *Crenobia alpina*, the ephemeropteran *Baetis* sp., and the trichopteran family Limnephilidae were limited to surface and interstitial habitats of the channel; (3) most of the micro-crustaceans (such as the Cyclopoida and two ostracods *Cavernocypris subterranea* and a new species of the genus *Fabaeformiscandona*) and the plecopteran *Leuctra* sp. were limited to interstitial habitats of the channel and of the gravel bar. This faunal gradient is likely due to the decrease of water velocity and oxygen content from benthic to interstitial habitats. *Rhithrogena* sp., the Blepharoceridae and the Simuliidae have strictly rheophilous larvae; the planarian *Crenobia alpina* and the Ephemeroptera of the genus *Baetis* are generally considered as oxyphilous taxa. On the other hand, the ostracod *Cavernocypris subterranea* is a stygobite, and the ostracod *Fabaeformiscandona* sp. and the plecopteran *Leuctra* sp. are stygophiles for which environmental conditions are optimal in sediment interstices.

This first study of the interstitial compartment of alpine brooks located in the Rhône catchment provides results quite similar to those of Pennak and Ward (1986), with a strong spatial gradient at the scale of the station. This faunal gradient consists of a decrease in taxonomic richness and changes in fauna composition between the benthic layer, the interstitial habitat of the channel, and the interstitial habitat of the gravel bar.

Upstream–downstream faunal gradient

Upstream–downstream faunal gradients can be studied at two major spatial scales: at the scale of the entire river (500 to 1000 kilometres) and at the scale of a river section (from 5 to 20 km long). In the Rhône river these gradients are well documented at the scale of a river section. For example, longitudinal distribution of fauna has been studied in three regulated channels of the river Rhône: Brégnier; the Canal de Miribel (Jons sector); and Donzère (Fig. 26.1). In these sections, faunal distribution was mainly influenced by the hydrogeology of the sector and the geomorphology of river sediments (Creuzé des Châtelliers, 1991a,b). For example, in the Canal de Miribel, interstitial assemblages were dominated by hypogean fauna in areas fed by strong groundwater inputs, and by epigean organisms in areas with low groundwater influence (Creuzé des Châtelliers and Marmonier, 1990; Creuzé des Châtelliers and Reygrobellet, 1990).

Upstream–downstream distribution of interstitial fauna is not as well documented at the scale of the entire river, or for all the taxa. For Ostracoda, a total of 489 samples were available throughout the Rhône river basin, with 25 samples from high-mountain brooks (in the Vanoise sector, between altitudes of 2056 and 890 m) and 464 from five different sectors of the main channel of the Rhône itself (from 300 to 150 m: Fig. 26.1). The ostracods extracted from this set of samples can be readily ascribed to 4 distinct species groups (Table 26.1).

- The first group consists of two species found only in the high-mountain area; they are stygobites or stygophiles (*Cavernocypris subterranea* and *Fabaeformiscandona* sp.). These two species can be considered as strict high-mountain interstitial dwellers for the Rhône basin, even if one of them (*Cavernocypris subterranea*) is regularly collected in low-altitude springs in northern Europe (Marmonier et al., 1989).
- The second group consists of 13 species recorded only from the upstream area of the river (i.e. mountain brooks from the Vanoise, and the Chautagne, Brégnier, and Jons sectors of the Rhône itself). The most frequently collected epigean species were *Candona candida, Eucypris pigra, Pseudocandona albicans, Pseudocandona lobipes,* and *Pseudocandona marchica.* They are mainly spring-dwellers or typical of small streams and cold water; they can be considered as the typical ostracod fauna of the upper reaches of large rivers (Marmonier and Creuzé des Châtelliers, 1992). Three hypogean taxa also belong in this group: *Cryptocandona kieferi, Fabaeformiscandona wegelini* and *Pseudocandona triquetra. Cryptocandona kieferi* and *P. triquetra* only occurred in the Jons sector, while *F. wegelini* was common in the three submontane sectors. The occurrence of this ostracod fauna [with both psychrophilous crenobionts (living in springs) and stygobionts] may be explained by the montane (Vanoise sector) or sub-montane (Chautagne, Brégnier, and Jons sectors) characteristics of the river here – that is, low organic-matter content, oligotrophic water, and low pollution levels.
- The third group consists of seven species absent from mountain brooks, but widely distributed all along the remaining length of the river, such as *Cypridopsis vidua, Limnocythere inopinata* and *Potamocypris* cf *variegata.* They are phytophilic, lentic, or eurytopic species and can be considered as the typical ostracod fauna of large rivers (Marmonier and Creuzé des Châtelliers, 1992).

Table 26.1
Longitudinal distribution of ostracods in interstitial habitats of the Rhône and the Vanoise area[1]

Species	Springs and brooks, Vanoise area $n=25$	Rhône				
		Chautagne sector $n=94$	Brégnier sector $n=93$	Jons sector (Miribel canal) $n=186$	Grand Gravier sector $n=30$	Donzère sector $n=61$
Fabaeformiscandona sp	32					
Cavernocypris subterranea	44					
Eucypris pigra	4	2.1		1.0		
Pseudocandona marchica	4	8.5	10.7	8.6		
Cryptocandona vavrai			3.2			
Candona neglecta			2.1	6.4		
Psychrodromus olivaceus		1.0				
Pseudocandona lobipes		3.2	11.8	1.6		
Pseudocandona albicans		10.6	2.1	45.7		
Fabaeformiscandona wegelini[2]		12.7	16.1	14.5		
Pseudocandona sarsi				0.5		
Cyclocypris ovum				4.3		
Pseudocandona triquetra[2]				0.5		
Cryptocandona kieferi[2]				21.5		
Candona candida		5.3	3.2	29.0	3.3	
Herpetocypris reptans		4.2		1.6	6.6	
Cypridopsis vidua		8.5	2.1	14.5	56.0	34.6
Prionocypris zenkeri		3.2		2.1		1.6
Limnocythere inopinata		2.1	3.2	0.5	30.0	32.7
Cypria ophthalmica			1.0	5.9	10.0	1.6
Potamocypris variegata			1.0	2.7	36.6	40.9
Pseudocandona zschokkei[2]				60.7	26.6	6.5
Pseudocandona belgica[2]					6.6	
Isocypris beauchampi					36.6	9.8
Ilyocypris bradyi					30.0	4.9
Ilyocypris decipiens						1.6
Pseudocandona pratensis						1.6

[1] Frequencies are expressed in percent; n, number of samples.
[2] Hypogean species.

– The fourth group is made up of five species recorded exclusively in the two downstream sectors. *Isocypris beauchampi* was the most frequent, while *Ilyocypris decipiens* and *Pseudocandona pratensis* were only observed in the Donzère sector. They are mainly lentic and eutrophic species reputed to prefer muddy bottoms (*P. pratensis*) or vegetation (*I. beauchampi*). These species develop large populations in downstream sectors where eutrophication, increased water temperature, and decreased water depth lead to the development of periphytic algae on the stony bottom; they can colonize interstitial habitats from this surface layer. Two hypogean species are present in this fourth group: *Pseudocandona zschokkei* (found in the Jons, Grand Gravier and Donzère sectors) and *P. belgica* (found exclusively in the Grand Gravier sector).

This example of upstream–downstream distribution

underlines the longitudinal succession of fauna that occur in the Rhône system. This succession is not progressive from one sector to the other, but seems to be subject to two marked changes: (1) between mountain sites and the Chautagne sector, corresponding to a change in river slope, and (2) between the Jons and Grand Gravier sectors, corresponding to a change in the river hydrology at the confluence with the river Saône.

In contrast with epigean species, no hypogean ostracod had a wide distribution through all the sectors studied. Recent ecological characteristics are apparently not the key factors that explain the longitudinal distribution of hypogean ostracods (and many other stygobite groups). The major factor that may explain their distributional pattern is the quaternary history of the river basin and the overall limit of ice during the last glaciation (Fig. 26.1). The valley of the Haut-Rhône was deglaciated between 15 000 and 11 000 BP (Mandier, 1988; Monjuvent, 1988). In this historical context, *Cavernocypris subterranea* (a species frequent in Northern Europe at all elevations) and *Fabaeformiscandona* sp. may be considered as psychrophilous interstitial organisms that restricted their area of dispersion with the glacial retreat. *Fabaeformiscandona wegelini,* a hypogean species widely distributed in the Haut-Rhône, is abundant in the uppermost sediment layer, and may have recolonized these sectors during the Post-Glacial period. In contrast, *Cryptocandona kieferi, Pseudocandona belgica, P. triquetra* and *P. zschokkei,* which are deep groundwater dwellers, may not have recolonized the deglaciated area; they are still limited to downstream sections of the Rhône (from Lyon to Donzère).

The same observation can be made for the microparasellid *Microcharon reginae* (Dole and Coineau, 1987). The genus *Microcharon* has a marine origin, and may have colonized interstitial freshwater habitat during the Eocene and Pliocene. *Microcharon reginae* is abundant in the backwaters of the Jons sector, which represents the northernmost location of this genus in France. The destructive effect of glaciation may explain the lack of *M. reginae* in upper sectors of the Rhône (such as the Brégnier or Chautagne sectors).

The Jons sector harbours a very rich hypogean fauna with *M. reginae,* five hypogean ostracod species (of the seven present in the Rhône basin), and many other rare hypogean taxa (see examples below, pp. 521, 522). This richness can be explained by (1) the high number of abandoned river arms in this sector, which provide a wide range of groundwater–surface-water interfaces; (2) the thickness in this area of the aquifer feeding the abandoned channels; and (3) the location of the sector itself, close to the last glacial limit which permitted post-glacial recolonization (Ginet, 1982). Thus, the Jons sector represents a "hot-spot" for groundwater biodiversity at the scale of the Rhône system, and it must be protected.

Microdistribution of interstitial fauna in the active channel of the Rhône

The upstream–downstream distributional pattern presented above must not obscure the heterogeneity of the distribution of interstitial fauna in the Rhône channel. Interest in the faunal heterogeneity at this small spatial scale was recently increased (Boulton et al., 1992; Hakenkamp et al., 1993) because distribution of groundwater fauna may reflect the spatial heterogeneity of interstitial metabolism and nutrient transformation.

In the Rhône, two relevant factors explain the spatial distribution of interstitial biota: (1) the depth into the sediments; and (2) the direction of water exchanges between surface systems and groundwater (i.e. upwelling versus downwelling). The vertical distribution of fauna has been studied both in the main channel of the river (Dole, 1983a) and in abandoned channels in the alluvial plain (Seyed-Reihani, 1980; Dole, 1984). In most cases, the epigean fauna was more abundant and diversified near the surface than below a depth of 1 m; ubiquitous stygobites were abundant both in shallow interstitial water and in deep groundwater, and strict phreatobites were more abundant deep in the sediment (Dole, 1984, 1985). When the horizontal distribution of fauna was investigated at the scale of a few metres, the composition of interstitial assemblages was found to differ between areas fed by surface-water infiltration (downwellings) and areas fed by groundwater outflows (upwellings). Epigean fauna was generally dominant in downwelling zones and stygofauna more abundant in upwelling zones (Marmonier and Dole, 1986; Marmonier, 1988; Dole-Olivier and Marmonier, 1992a,b).

An example of the distribution of interstitial fauna is presented in Table 26.2. Vertical distribution of interstitial fauna was studied in a regulated channel of the river Rhône (the Canal de Miribel, Jons sector: Fig. 26.1) in August and September 1987 (Dole-Olivier and Marmonier, 1992a). Epigean Crustacea (ostracods and amphipods) and insect larvae (e.g. Trichoptera, Coleoptera, Diptera) were dominant in shallow inter-

Table 26.2
Vertical distribution of interstitial fauna in a downwelling and an upwelling zone in the Canal de Miribel (river Rhône)[1,2]

Fauna	Downwelling		Upwelling	
	−50 cm	−100 cm	−50 cm	−100 cm
Candona candida	0.5			
Esolus parallelepipedus	0.8			
Stenelmis canaliculata	0.3			
Chloroperla sp.	0.5			
Leuctra cf *fusca*	0.3			
Polycentropus sp.	0.3			
Sialis lutaria	0.3			
Ceratopogonidae	0.5			
Gammarus fossarum	9.3	0.8		
Bidessus minutissimus	14.5	1.0		
Habrophlebia sp.	0.8	0.3		
Chironominae	2.5	2.3		
Tanytarsini	3.0	2.5		
Orthocladiinae	0.8	1.0		
Tanypodinae	0.8	4.0		
Gastropoda		3.5		
Ilyocypris sp.		0.3		
Cypridopsis vidua		0.3		
Nematoda	34.0	5.5	0.5	0.5
Oligochaeta	10.0	46.0		0.5
Hydra sp.	4.5	1.5		0.3
Hydracarina	1.0	0.8		0.3
Cladocera	3.3	18.8	26.8	0.5
Harpacticoida	16.3	7.8	0.5	
Cyclopoida	805.8	190.3	91.8	4.5
Fabaeformiscandona wegelini[3]	0.3			1.0
Pseudocandona albicans	6.0			0.5
Pseudocandona marchica	1.5		0.5	
Niphargopsis casparyi[3]	12.5	0.3	33.5	16.0
Niphargus rhenorhodanensis[3]	7.5	13.0	38.3	7.3
Leuctra major			1.5	0.3
Pseudocandona zschokkei[3]		0.5		0.3
Proasellus walteri[3]		4.0		0.3
Salentinella sp.[3]				1.3
Cryptocandona kieferi[3]				0.5

[1] See location in Fig. 26.4.
[2] Observations were made in August and September, 1987; figures are mean abundance in 10 litres.
[3] Hypogean species.

stitial habitat of the downwelling zone. Ubiquitous stygobites (e.g. *Niphargus rhenorhodanensis*) were more abundant at a depth of 100 cm in downwelling zones and 50 cm in upwelling zones. Strict phreatobites were more abundant in deep sediment in both downwelling and upwelling zones (e.g. *Proasellus walteri, Pseudocandona zschokkei*), or only in deep sediment of upwelling zones (e.g. *Cryptocandona kieferi, Salentinella* sp.). This example, and related studies, underline the necessity of including hydrologic processes in research designs that will help future cross-system comparisons (Marmonier and Dole, 1986; Hakenkamp et al., 1993).

THE RIVER AND ITS ALLUVIAL PLAIN: THE TRANSVERSE DIMENSION

Depending on the geological characteristics of the region, large rivers can develop large floodplains around their main channel. Hence, by moving in the plain and leaving many side arms and former meanders, they generate a large number of different aquatic systems and wetlands (Schumm, 1977; Bravard et al., 1986; Amoros et al., 1987). Side arms are progressively isolated from the main channel by sediment deposits at their upstream end, and remain open at the downstream end. At this stage of the succession the old channel is regularly flooded, and good interstitial conditions may occur in the upstream part of the arm, where springs emerge (Gibert et al., 1977, 1981). When the old channel is completely isolated (when the downstream end is closed by sediment deposits) and progressively abandoned by surface water even during floods, groundwater flow may be maintained and a well-developed interstitial habitat can be found in spring-fed areas (Reygrobellet and Dole, 1982; Dole, 1983a). Finally, eutrophication of surface water and lowering of the groundwater table induces succession of the surface aquatic biotopes to wetlands and terrestrial habitats. In this latter case, groundwater inputs become irregular and the interface between surface water and groundwater is slowly modified to a transition zone between groundwater and soil with poor interstitial habitats (Marmonier et al., 1992).

In this very complex geological and geomorphological background, the study of the Rhône alluvial plain has been oriented to understand two major problems: (1) the inputs of matter and energy from the alluvial valley to the main channel through the groundwater system; and (2) the spatial heterogeneity of groundwater characteristics and interstitial fauna in the alluvial plain.

Dynamics of inputs from groundwater alluvial plain to river main channel

The main channel of a river is not isolated from the other water bodies of the alluvial plain. Many studies that deal with nutrient or faunal exchanges between backwaters and the main channel through surface water flow have been published (Cellot and Bournaud, 1986; Cellot and Rostan, 1993; Smock, 1994), but research dealing with groundwater inputs remains rare. Such exchanges can be measured only under particular hydrological conditions when surface water does not hide groundwater outflows. An estimation of solutes, nutrient, and microbial inputs to the river Rhône from the groundwater milieu was performed during maintenance works on a dam in the Chautagne sector (Fig. 26.1: Boissier et al., 1996). These maintenance works induced a marked decrease in the level of surface water in the area studied. Inputs to the river were compared between two types of groundwater: (1) the interstitial water of gravel bars; and (2) the groundwater of the alluvial plain. Eight springs (created by the decrease in river level) were investigated: four groundwater-fed springs, located at the bottom of the bank under the forest on the alluvial plain, and four interstitial-fed springs located at the downstream end of a gravel bar.

During the artificial and partial drought period associated with the maintenance works, the inflows of groundwater from the alluvial plain represented an average input of 0.77 0.24 mg ℓ^{-1} of nitrogen (of which 93.3% was nitrate), 0.018 0.014 mg ℓ^{-1} of total phosphorus (of which 42.2% was orthophosphate), 3.56 0.39 mg ℓ^{-1} of silica, 2.31 0.70 mg ℓ^{-1} of dissolved organic carbon (DOC, of which 47% was biodegradable), and (7.3 3.7) 10^4 bacteria per ml (of which 8.8% were active).

There was a marked spatial heterogeneity between the two types of groundwater that flow into the river (Boissier et al., 1996). Concentrations of calcium and alkalinity were higher in bank springs than in gravel-bar springs. The solute concentrations measured in the bank springs were characteristic of the surrounding mountain geology (high calcium content and high alkalinity). In the interstitial water flowing out from the gravel bars, the content of sulphate, sodium, nitrogen, and phosphorus was significantly higher. The higher nitrogen content in gravel-bar interstitial water than in groundwater from the alluvial plain may be explained by oxidation of the organic matter trapped in the sediment of the gravel bar. This oxidation induced the formation of nitrate, which flowed out of the bar and was detected in the springs. The weak development of gravel-bar vegetation (regularly disturbed by floods) was not sufficient to have had a significant effect on nitrate concentrations in spring water.

A high temporal variability was also observed during this three-day study. Oxygen, pH, the biodegradable fraction of the DOC, and bacterial concentrations varied with time and seemed to be linked to the biological activity of the groundwater biofilm. In both types of groundwater, the pH decreased from 8.1 on the first day to 7.7 on the third day, and the oxygen content from 11 to 4 mg ℓ^{-1}. This decrease may be explained by microbial degradation of organic matter in the groundwater inducing at the same time oxygen consumption, carbon dioxide release (thus a decrease of the pH), DOC fluctuation, and bacterial drift.

Direct studies of animal drift from groundwater to river channel are also very rare, because they are difficult to perform: inputs of groundwater are generally diffused through bed sediments. The occurrence of strict phreatobites (generally sampled in the alluvial plain) in the shallow interstitial habitats of the active river channel is not inevitably a consequence of interstitial drift, as an active colonization of the subsurface layer (–50 cm) is also possible after a long period of low water (Marmonier and Dole, 1986).

Spatial heterogeneity of nutrient content in the groundwater of the alluvial plain

Nutrient flow from groundwater to surface systems also plays an important role in the development of biofilms, microphytes, and macrophytes in river backwaters (Carbiener et al., 1988; Bornette and Amoros, 1991; Bornette and Large, 1995). As a consequence, the local heterogeneity of groundwater nutrient content may influence the functioning of secondary arms and abandoned arms fed by springs. This spatial heterogeneity was studied in the alluvial plain of the Rhône at Jons. Interstitial habitat was sampled at depths of 50 and 100 cm in the sediments, in April, July, and November, at seven stations distributed from east to west in this sector (Fig. 26.4); two stations (M1 and M2) were located in the eastern part of the plain near the confluence between the Ain and the

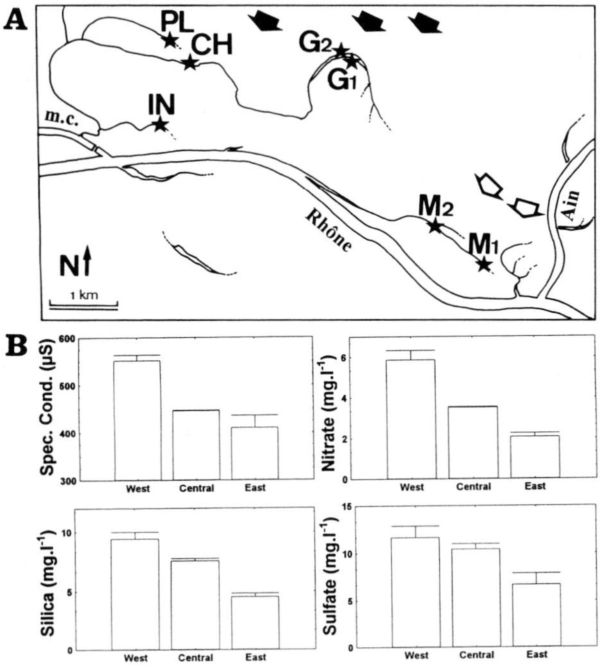

Fig. 26.4. Spatial heterogeneity of chemical characteristics of alluvial-plain groundwater. (A) Locality plan of the Jons sector: stars, sampling stations; open arrows, Ain river parafluvial groundwater; solid arrows, groundwater of the Dombes plateau; m.c., upstream part of the Canal de Miribel. (B) Chemical characteristics for eastern (M1 and M2), central (G1 and G2), and western (PL, CH and IN) stations. Mean (SE; $n = 12$ or 18) solute contents for two depths (50 and 100 cm) and three dates (April, July, and November 1987). In the analyses of variance, between-group differences were significant ($p < 0.001$) for specific conductance, silica and nitrate, and also for sulphate ($p < 0.05$).

Rhône, two (G1 and G2) in the central part of the plain, and three (IN, CH, and PL) in the western part of the plain at the foot of the Dombes plateau [area of corn (*Zea mays*) culture].

Specific conductance, and the concentrations of silica, sulphate, and nitrate, were significantly different between the three sectors of the plain. Stations located in the eastern part of the plain were characterized by low content of solutes and nutrients; they were strongly influenced by the underflow of the river Ain. Stations located in the western part of the plain were characterized by high content of solutes and nutrients: they were fed by groundwater flowing under the cultivated Dombes plateau. The stations located in the central part of the plain displayed intermediate characteristics.

This example clearly underlines the importance of local heterogeneity for inputs of solutes and nutrients in river backwaters; in most cases groundwater inputs in an abandoned channel maintain oligotrophic conditions in the water body, but under particular hydrogeologic or anthropogenic conditions groundwater inputs can be an additional factor for eutrophication of river backwaters.

Spatial heterogeneity of hypogean fauna in the groundwater of the alluvial plain

Studying both vertical and transverse heterogeneity of interstitial assemblages at the scale of an entire sector of the Rhône alluvial plain, Dole (1983a) observed that the distribution of groundwater fauna varies markedly along a cross-section through the alluvial plain. This result can be illustrated by the distribution of hypogean fauna sampled in three stations of the Jons alluvial plain (in April and July 1987, at depths of 50 and 100 cm in the sediment: Marmonier et al., 1992). Despite the lack of difference between the physico-chemical characteristics of interstitial water (these three stations are all located in the western part of the Jons alluvial plain: Fig. 26.4), important differences were observed in their faunal composition (Table 26.3). In the first one (IN), located close to the main channel, hypogean taxa only represented 24% of the total taxonomic richness, and consisted of crustaceans generally abundant in the main channel (*Niphargus rhenorhodanensis*) or of organisms widely distributed in secondary arms (*Hauffenia minuta, Microcharon reginae, Niphargus pachypus,* or *Salentinella* sp.). In the last station (PL), located far from the main channel, the percentage of hypogean taxa increased (34%) with the occurrence of the strictest phreatobites (such as *Bathynella* sp. or *Niphargus renei*). An intermediate situation was observed in the CH station, where hypogean taxa represented 25%, including some strict phreatobites such as *Troglochaetus beraneki* (Archiannelida).

The explanation of such differences may be found in the hydrological characteristics of these three stations, and particularly in their relation to floods: Station IN is regularly flooded during high-water periods (each winter), CH is rarely flooded (once or twice in each decade), and PL was never flooded during the 1980s. In conclusion, the distance to the active channel corresponds to a gradient of physical stability (Dole and Chessel, 1986), which structured the distribution of hypogean fauna from species resistant to instability (e.g. *N. rhenorhodanensis*) to species very sensitive to hydrological disturbances (e.g. *Bathynella* sp.).

Table 26.3
Spatial distribution of interstitial fauna in the Jons alluvial plain[1]

Fauna	Station PL[2]	Station CH[2]	Station IN[2]
Niphargus renei	0.5 (1.0)		
Niphargopsis casparyi	0.5 (1.0)		
Bathynella sp.	0.5 (0.6)		
Moitessieria lineolata	7.25 (4.0)		
Troglochaetus beraneki	0.75 (1.5)	0.5 (1.0)	
Parastenocaris sp.		0.25 (0.5)	
Bythiospeum diaphanum	17.8 (14.2)		4.25 (2.2)
Hauffenia minuta	5 (3.4)	0.5 (0.6)	20.5 (15.2)
Salentinella sp.	9.5 (13.7)	0.25 (0.5)	0.25 (0.5)
Microcharon reginae	15.8 (14.9)	4 (4.2)	1 (1.4)
Niphargus pachypus	4.25 (2.1)	12.3 (11.4)	10.3 (18.5)
Niphargus kochianus			0.75 (1.5)
Niphargus rhenorhodanensis			2.25 (1.5)
% of species hypogean	34%	25%	24%

[1] Mean abundance (and standard deviations, $n=4$) for depths of 50 and 100 cm, in April and July 1987. The two depths were combined.
[2] See location of the three stations in Fig. 26.4.

This gradient of physical stability also corresponds to different stages in the succession that exists between newly abandoned arms and old oxbow lakes. The end of this succession corresponds to a complete terrestrialization of the water body, with lowering of the water table, reduction of exchanges between surface water and groundwater, size reduction of sediment pores (with partial clogging), and an increase of temperature variability. An example of an interstitial community living in an astatic[1] pond is given below (p. 528). In this type of semi-aquatic system, interstitial assemblages are dominated by epigean fauna (with organisms adapted to desiccation); the abundance and diversity of stygobite fauna declines. Interstitial habitats of oxbow lakes with permanent surface water and groundwater inflow (such as stations PL, G1, or G2: Fig. 26.4) represent "hot spots" for groundwater biodiversity at the scale of the alluvial plain.

TEMPORAL DYNAMICS OF INTERSTITIAL ASSEMBLAGES

Temporal changes in running-water systems are related to two major causes: seasonal cycles and random hydrological disturbances. These two types of temporal fluctuations are discussed for the interstitial community because they have large consequences for the productivity and the stability of surface-water systems. The study of seasonal variation of interstitial assemblages poses the problem of a "nursery zone" permitting the early development of some benthic species (Puig et al., 1990) in the same way that the study of the effects of hydrological variation poses the problem of a "refuge zone" during periods of high water (Williams and Hynes, 1974). This section provides some examples of temporal variations linked to seasonal changes and hydrological disturbances.

Seasonal variations of interstitial assemblages

Seasonal variations are particularly marked for insect larvae, because adults generally leave the aquatic system (except some Coleoptera that have aquatic adults). For example, larvae of the plecopteran *Leuctra* cf *fusca* were abundant during winter in the interstitial habitats of the Canal de Miribel (Jons sector: Fig. 26.1) and rare during summer and autumn because adults emerged during this period (Marmonier, 1988). In the same canal, the abundance of Chironomidae varied greatly according to the seasons (Fig. 26.5). Tanypodinae, Orthocladiinae and Tanytarsini were more abundant during early and late summer (July and September 1984) than during autumn and winter (November 1983 and February 1984), because most species had finished their aquatic life before autumn. In the Canal de Miribel, the interstitial habitat can be considered as a "nursery zone" allowing some benthic species to develop during the early stages of aquatic life.

[1] An astatic pond is one with marked changes in water level during the year.

Table 26.4
Seasonal variation of the species richness and the mean abundance of ostracod assemblages in the Canal de Miribel (partially fed by groundwater) and the Donzère regulated channel (fed by surface water)[1]

	Canal de Miribel (France)		Donzère regulated channel	
	Summer ($n = 34$)	Winter ($n = 32$)	Summer ($n = 30$)	Winter ($n = 28$)
Total species richness	10	7	11	6
Mean abundance (S.E.)	1.9 (0.6)	3.7 (1.8)	361 (246)	0.37 (0.26)

[1] For location of the sectors see Fig. 26.1.

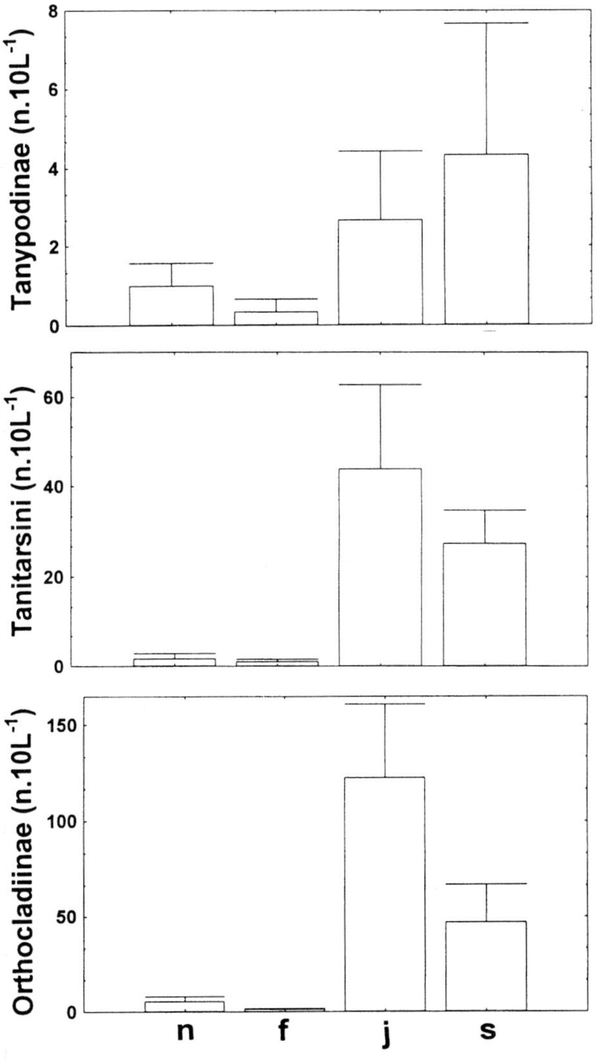

Fig. 26.5. Seasonal variations of chironomid abundance at a depth of 50 cm in the sediments at a station mostly fed by surface-water infiltration, in November 1983 (n), February 1984 (f), July 1984 (j), and September 1984 (s). Mean (SE; $n = 3$) abundances for ten-liter samples. Between-date differences were significant for Tanypodinae and Orthocladiinae (Analysis of variance, $p < 0.05$).

The effect of the seasonal cycle on interstitial communities is not comparable in all sectors of the Rhône (Creuzé des Châtelliers, 1991a). For example, temporal variations of ostracod assemblages were studied in two different sectors (Creuzé des Châtelliers and Marmonier, 1993): the Canal de Miribel in the Jons sector where large groundwater inflows occur from well-developed aquifers, and the regulated channel of the Rhône in the Donzère sector (Fig. 26.1), where groundwater inflows are weak and the sediment interstices are mostly fed by surface water. In both sectors, species richness decreased between summer and winter (Table 26.4); the rare species present during summer disappeared in winter (e.g., *Herpetocypris reptans, Pseudocandona marchica* and *P. sarsi* in the Canal de Miribel; *Ilyocypris decipiens, I. monstrifica, Prionocypris zenkeri* and *Pseudocandona pratensis* in Donzère). The species that remained during winter were those that were most abundant during summer (e.g. *Cypria ophthalmica, Cypridopsis vidua*) or stygobites (*Pseudocandona zschokkei*). In contrast, seasonal variations in ostracod abundance differed markedly according to the sector (Table 26.4). In Miribel, the mean abundance was not significantly different between summer and winter, while in Donzère abundance decreased markedly. Such differences in seasonal cycle between different sectors of the same river are surprising. Hydrological characteristics of the two sectors may explain them. Interstitial habitats of the Donzère channel are fed only by surface water, and are consequently subject to great variation in temperature. In contrast, variation in the temperature of interstitial water temperature in the Canal de Miribel is buffered by high groundwater inflow.

In conclusion, these two examples underline the fact that seasonal cycles in interstitial assemblages are more evident for amphibionts (such as insects) than for organisms that spend all their life in water (such as ostracods). These strictly aquatic organisms may clearly react to the thermic effects of groundwater inputs.

Effect of floods on interstitial assemblages

Hydrological events have a major role in controlling stream metabolism and the dynamics of stream patchiness (Resh et al., 1988). Indeed, spatial organization of resources and biota depends on hydrology, whose effects can be positive or negative in relation to flood amplitude and duration; short events (e.g. storms) may provide nutrient inputs that enhance productivity, large hydrological events may be considered as disturbances because they scour the bed sediment. Interstices of sediment provide a large spatial extent to the stream, which may be an important "refuge" for benthic organisms during hydrological disturbances (Williams and Hynes, 1974).

The effect of floods on interstitial assemblages has been studied in many stations in the Canal de Miribel (Marmonier and Dole, 1986; Marmonier and Creuzé des Châtelliers, 1991; Dole-Olivier and Marmonier, 1992c). For example, in the study of Marmonier and Creuzé des Châtelliers (1991) interstitial fauna were sampled in a groundwater upwelling zone on nine occasions between June 27 (just after a flood of 340 $m^3 s^{-1}$) and September 14, 1984 (after more than two months with a constant discharge of 30 $m^3 s^{-1}$: Table 26.5). The composition and structure of assemblages varied during the study period.

Just after the flood, many epigean organisms occurred, which are typically absent or extremely rare in upwelling zones. They included planktonic organisms (*Bosmina coregona, B. longirostris, Daphnia hyalina,* and the Cyclopoida and Calanoida) or strictly benthic animals (*Chloroperla* sp., *Gammarus* sp., *Hydropsyche* sp.). These species might be brought passively by water flow into the sediments or might have actively colonized the sediment interstices. Sediment interstices play a role of "refuge" or "storage zone" for these planktonic and benthic organisms. During the low-water period, the first group of species disappeared. The assemblage was dominated by benthic species that are well known to colonize interstitial habitats easily (Chironomidae, *Leuctra* cf *fusca,* Gastropoda) and by stygobites (*Alona phreatica,* Niphargidae, *Parastenocaris* sp.). These stygobites were more abundant at the end of the low-water period than just after the flood; they seemed to colonize this station actively after it had been defaunated by the flood. Rare taxa also appeared during the low-water period, most of them being good colonizers of the interstices (*Bidessus* sp., Chironomidae, *Esolus parallelepipedus, Pseudorthocladius* sp.) or stygobites (*Proasellus walteri, Pseudocandona zschokkei*).

Temporal dynamics observed in upwelling areas varied with hydrological characteristics of the station. In downwelling zones (where surface water infiltrates into the sediment), large floods induce a "wash-out" effect on interstitial fauna (Marmonier and Creuzé des Châtelliers, 1991). In this case, the interstitial layer was progressively recolonized by both epigean benthic animals (*Gammarus* sp.; Marmonier and Dole, 1986) and stygobites that developed populations in downwelling areas (such as *Fabaeformiscandona wegelini* and *Niphargus rhenorhodanensis*).

The contrasting reactions to floods in upwelling and downwelling areas of the same river underline the importance of patchiness in river sediments. Spatial heterogeneity of the interstitial habitat has to be considered in relation to its role as a "refuge" or "storage" zone (Marmonier and Creuzé des Châtelliers, 1991). In the same way, amplitude and duration of floods are of key importance to the response of interstitial fauna to hydrological disturbances. This temporal heterogeneity must be integrated into the design of studies dealing with refuge hypotheses (Dole-Olivier and Marmonier, 1992c; Dole-Olivier et al., 1997).

Anthropogenic effects

The Rhône valley has been altered by human activity along most of its length; for example, 20 hydroelectric dams were built between Genève and the Mediterranean Sea. The Rhône also passes through the city of Lyon, representing an industrial area of three million people. Anthropogenic effects on the Rhône river system can be observed at numerous levels: increase of organic matter content, toxic pollution, destruction of natural alluvial plains, impoundment. For interstitial habitat and groundwater fauna, two major topics were investigated during the past decade: effects of water pollution, and effects of river regulation (mostly dam building). In this section, we will consider the recent progress in the study of the effects of pollution (by chemicals, and by organic matter), the role of interstitial habitats, and the impact of river regulation on the abandoned channels of the alluvial plain.

Table 26.5
Composition of interstitial assemblages[1] sampled just after a flood (June 27) and during the low-water period (July 6 to September 14)[2]

Fauna	27/6	6/7	14/7	2/8	14/8	21/8	29/8	7/9	14/9
Bosmina logirostris	4								
Bosmina coregoni	68								
Daphnia hialina	6								
Chloroperla sp.	1								
Hydropsyche sp.	1								
Gammarus sp.	1								
Calanoida	8	1							
Gastropoda	4	1						2	1
Orthocladiinae	57	11	1			1		2	
Tanytarsini	9	2	9			1			
Hydra sp.			1	2	1	1			
Leuctra cf *fusca*	3	4	2	5	1	6	5	3	3
Hydracarina						12		21	9
Parastenocaris sp.[3]	16	8	2	16	36	12	24	100	104
Alona phreatica[3]	8	12	11	84	248	89	409	221	85
Niphargidae[3]	44	37	37	157	179	51	19	58	74
Pseudorthocladius sp.		1							
Bidessus sp.		1							
Esolus parallelepipedus		1		1		1			
Chironomini			1						
Corynoneura sp.	1		1				1		
Proasellus walteri[3]			2					1	
Baetis sp.			1					1	
Ceratopogonidae								1	
Pseudocandona albicans								1	2
Pseudocandona zschokkei[3]									1
Iliocriptus sordidus									1
Taxonomic richness[4]	21	16	15	11	10	13	9	14	14

[1] Abundance in 10 litre samples.
[2] The station studied was located in the upstream part of the Miribel canal at a depth of 50 cm in the sediments.
[3] Hypogean species.
[4] Nematoda, Oligochaeta, Cyclopoida, and Harpacticoida (except the genus *Parastenocaris*) were not identified or presented in this table, but are included in the taxonomic richness.

Chemical pollution of Rhône sediments

In a recent study (Plénet et al., 1995), concentrations of heavy metals in the sediment were compared at six stations. As an example, results obtained in the shallow interstitial habitats of two stations (at a depth of 15 cm in the sediment) will be presented. One station was located just upstream of Lyon (Station 4 in Fig. 26.1) and the other downstream, in the Grand Gravier sector (Station 5 in Fig. 26.1). Though there were no significant differences in copper concentration, marked differences could be observed between the concentrations of chromium, lead, nickel and zinc at the two stations (Table 26.6). The contaminated site (Station 5) has lower taxonomic richness (especially for macrocrustaceans and insects) and a lower diversity index than the reference site (Station 4). In contrast, some taxa, such as the Lumbriculidae, Tubificidae,

Table 26.6
Heavy metal content[1] of sediments in the river Rhône

Location	Pb	Cu	Zn	Ni	Cr
A[2]					
Station 4 (upstream of Lyon)	61.66 (11.06)	50.66 (29.29)	72.66 (13.05)	70.00 (17.77)	144.33 (33.08)
Station 5 (downstream of Lyon)	95.33 (12.58)	83.00 (33.64)	165.66 (21.93)	167.00 (28.84)	257.66 (41.74)
Significance of difference[3]	**	n.s.	**	**	**
B[4]					
Site A	34.55 (18.51)	60.94 (51.20)	100.77 (58.59)	93.22 (41.09)	131.72 (71.56)
Site B	36.88 (11.83)	48.77 (29.37)	115.94 (32.44)	64.38 (31.94)	82.55 (41.86)
Site C	64.83 (21.39)	154.61 (199.49)	241.22 (151.36)	94.72 (66.41)	140.50 (104.01)
Significance of differences[3]	**	*	**	*	**
C[5]					
−20 cm	49.0 (18.66)	130.05 (161.19)	172.77 (103.59)	109.55 (61.71)	158.33 (103.18)
−50 cm	45.88 (22.94)	91.72 (139.32)	175.22 (147.14)	80.22 (39.98)	108.77 (63.39)
−100 cm	41.38 (25.21)	42.55 (30.23)	109.94 (68.53)	62.55 (39.75)	87.66 (47.99)
Significance of differences[3]	n.s.	**	*	**	*

[1] Mean concentrations (with standard deviation) expressed in $\mu g\,g^{-1}$ of dry fine sediments.
[2] At each station, $n=3$.
[3] Significance based on analysis of variance: n.s., $p>0.1$; *, $0.1>p>0.05$; **, $p<0.05$.
[4] At Station 5, in the Grand Gravier sector (see Fig. 26.1). $n=18$ at each site.
[5] At three depths at Station 5, in the Grand Gravier sector. $n=18$ at each depth.

some species of Naididae, and Cladocera, seem to be more tolerant of metallic pollution and were abundant in polluted sites. Many physical characteristics of the river change between the sites upstream and downstream of Lyon (e.g., hydrology, sediment grain size), nevertheless, pollution may explain a part of the faunal changes observed.

The metal content in sediments downstream of Lyon is heterogeneous. Local characteristics of sediments (grain size, organic matter content) and hydrology (direction of water exchanges between surface and groundwater) may influence the metal content of the sediments. The patchy distribution of heavy metals in the interstitial habitats of the Rhône was studied in the Grand Gravier sector (Station 5 in Fig. 26.1: Gibert et al., 1995). The Grand Gravier site is one of the major pumping areas for drinking water for the southwestern part of Lyon (13 000 $m^3\,day^{-1}$). Most of the water pumped (70–90%) comes from the river channel (Ille et al., 1991). In this area, the river regularly receives industrial and domestic waste water from the city upstream. In contrast with the background pollution of surface water, an efficient bank filtration results in good groundwater quality (wells are located ~80 m from the river).

In this sector, we sampled fine sediment in three contrasting stations (Table 26.6; Gibert et al., 1995). Station A is located in a downwelling area with coarse sediments; Station B is an area with coarse sediments where the direction of water flux may change with time (alternately groundwater upwelling and surface water downwelling); and Station C is characterized by low water exchanges through a thick layer of fine sand. Sediment was sampled in January and May 1992 by pumping, at three points (from river to the shore) and from three depths at each point (depths of 20, 50, and 100 cm in the sediments). Metal concentrations were measured in the fine fraction of the sediment. Concentrations of copper, lead and zinc were significantly higher at Station C than at Stations A and B; concentrations of chromium and nickel were significantly higher at Stations A and C than at Station B (Table 26.6). With respect to the vertical dimension, concentrations of chromium, copper, nickel and zinc were significantly higher near the surface than at −100 cm in the sediment (Table 26.6). These results show that heavy metals accumulate near the surface (−20 and −50 cm), in deposits of fine sediment (Station C) or in downwelling zones (Station A). River sediment represents an efficient filter for metals flowing in lotic systems, and plays an important role in groundwater protection. Finally, interstitial fauna diversity and density were significantly lower at Station C than at the two other stations. Together with grain size and direction of water flow, metal pollution may explain the faunal characteristics of Station C.

These two recent studies provide a constructive basis for discussing the impact of metals on interstitial communities (composition, diversity, and abundance). The next step may consist of proposing indicators of the biological integrity of the river interstitial layer, for

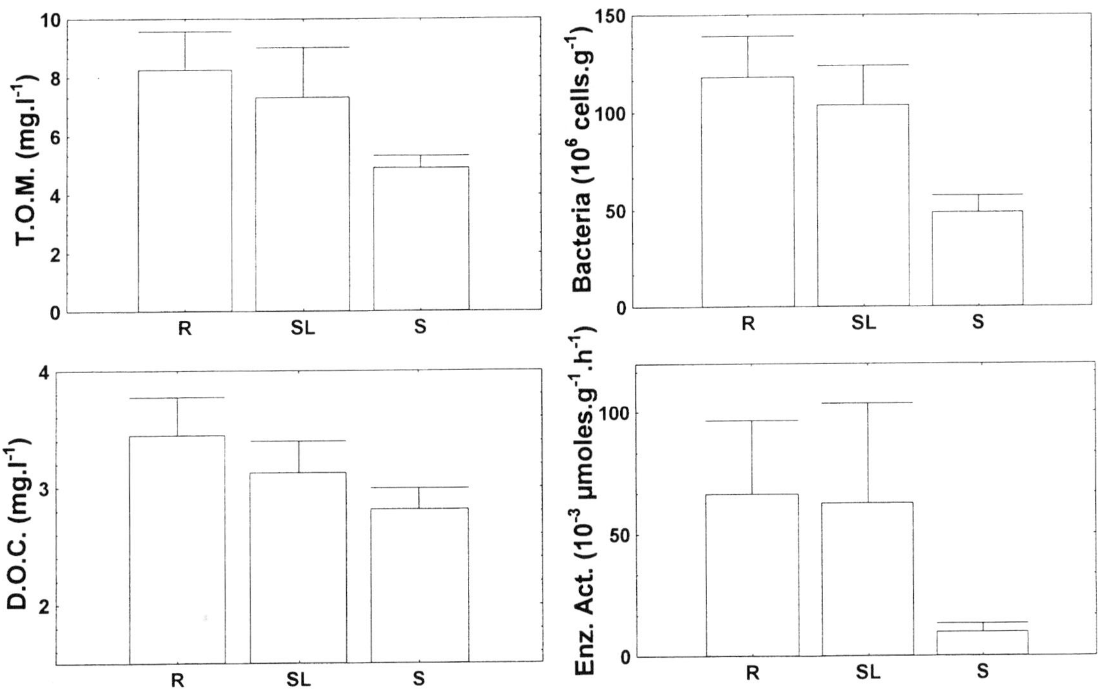

Fig. 26.6. Total Organic Matter (TOM) content of the sediments, Dissolved Organic Carbon (DOC) content of interstitial water, and bacterial abundance and hydrolytic enzymatic activity (measured with FDA degradation) of biofilm fixed on fine sediment of the Rhône river in the Grand Gravier sector. Three sampling points were studied: at 1.5 m in the river (R), just at the shoreline (SL), and at 1.5 m on the shore (S). Mean values (SE; $n = 12$) were calculated for three depths (20, 50, and 100 cm in the sediments) and four dates (January, May, August, and November 1992). Differences between sampling points were significant for bacterial abundances and enzymatic activity (analysis of variance, $p < 0.05$), with $p < 0.1$ for TOM.

biomonitoring metal contamination of both interstitial habitats and deep groundwater systems.

Role of interstitial habitat in groundwater protection

Artificial groundwater recharge based on the induced infiltration of surface water into aquifers through river-bed sediments is a common method of increasing natural groundwater resources. In the case of the surface water in polluted river systems, the efficiency of bed sediments in retaining pollutants is of key importance for aquifer protection.

The efficiency of bed sediment in the elimination of organic matter as it moves from surface water to groundwater (bank filtration) was studied immediately downstream of Lyon (Grand Gravier sector: Marmonier et al., 1995). Spatial and temporal variations in the organic-matter content of sediment, in the concentrations of dissolved organic carbon (both the biodegradable and the refractory fractions), and biofilm development (both bacterial abundance and microbial enzymatic activity) were surveyed in the first metre of sediment (in Station A, see p. 526). The study area was fed most of the year by surface-water inflow (downwelling), because of groundwater pumping. Samples were collected four times during 1992 (January, May, August, and November), at times when discharge was below the mean annual discharge. Samples consisted of one litre of interstitial water and fine sediments pumped with a Bou–Rouch pump at three points (at 1.5 m into the river, on the shoreline, and on the shore 1.5 m from the shoreline) and at three depths (20, 50, and 100 cm into the sediments) for each sampling period.

The major result of this study is the observation of decreasing gradients (1) from surface water to deep sediments and (2) from the river to the shore in most of the cases for the four variables (Marmonier et al., 1995; Fig. 26.6). The highest content of total organic matter in the sediment was observed near the surface water, demonstrating the physical filtration of infiltrating water. Dissolved organic carbon (DOC) also decreased with depth and distance to the river, but with contrasting behavior for the different fractions. The refractory fraction (RDOC) decreased in concentration whatever the season; this decrease was probably due to physical processes, such as adsorption on fine mineral particles. On the other hand, the biodegradable fraction (BDOC) decreased in concentration during the warm season when microbial enzymatic activity was high. This observation is supported by the negative correlation observed between enzymatic activity and concentration of biodegradable organic carbon. Finally, bacterial abundance did not vary according to the season. Microbial activity, rather than bacterial abundance, is the more relevant parameter for understanding the vertical stratification of dissolved organic matter in river sediments.

In conclusion, a simple model can be proposed describing how organic matter may be removed from infiltrating surface water (bank filtration) in a downwelling zone: particulate organic matter coming from the river surface water is trapped and stored in the sediment (mainly in the upper shallow interstitial habitat), where abundant bacterial populations partially convert this organic matter to dissolved organic carbon for their metabolic needs. When microbial activity is low, some of this biodegradable dissolved organic carbon is not assimilated, and slowly migrates downward with water infiltrating the bank. During periods with high microbial activity, the biodegradable components of dissolved organic carbon are rapidly assimilated by microbial populations. The refractory components are probably adsorbed physically onto fine mineral particles year-round. Physical and biological processes together make this first metre of sediments adjacent to the river an efficient filter to remove infiltrating organic matter.

Effect of river regulation on the interstitial community

Most of the dams built on the Rhône consist of a retention canal (or reservoir) limited by a dam that diverts most of the river discharge to an adjacent hydroelectric power station through a head-race canal. The old active channel becomes a by-pass section receiving a minimum low-water discharge (that varies between 10 and 30 $m^3 s^{-1}$). During spates or floods, the channel serves as a spillway for the adjacent hydroelectric facility, and its discharge can increase very rapidly within a few hours, whereas during the low discharge period it drains the surrounding aquifer. This type of regulation induces three major physical disturbances that modify the composition and dynamics of interstitial fauna: (1) in the reservoir, the decrease of water flow induces sedimentation of fine particles and total or partial clogging of sediments; (2) in the bypass section, floods induce erosion of the bed sediments and modification of water exchanges between surface water and groundwater; (3) in the alluvial plain, a decrease of the water table results in a partial or total terrestrialization of backwaters.

The effect of erosion on interstitial communities of the by-pass channel has been studied in the Canal de Miribel (Creuzé des Châtelliers and Reygrobellet, 1990; Creuzé des Châtelliers and Marmonier, 1990). In the eroded part of this canal, the level of surface water decreases and lateral inputs of groundwater increase; as a consequence, density and diversity of hypogean organisms increase in the shallow interstitial assemblages. In contrast, in the sedimentation areas, the

Table 26.7
The hypogean fauna of two stations located in the Méant abandoned channel [1,2]

Species	Station M1 [1]	Station M2 [1]
Niphargus kochianus	0.7 (1.5)	
Salentinella sp.	1.2 (1.8)	
Proasellus sp.	0.5 (1.0)	
Microcharon reginae	23.5 (34.2)	0.7 (0.9)
Niphargus pachypus		5.5 (6.8)
Hauffenia minuta		41.5 (46.8)
Bythiospeum diaphanum		5.7 (7.3)
Moitessieria lineolata		44 (39.5)
Niphargus gallicus		0.2 (0.5)
% hypogean taxa	17%	30%

[1] Station M1 is located in the astatic upstream zone and Station M2 is located in the permanent-water central zone (see location of the stations in Fig. 26.4).
[2] The figures are mean abundance (and standard deviation) for combined samples from depths 50 and 100 cm in April and July 1987 ($n = 4$).

reverse phenomenon occurs and the relative abundance of hypogean organisms decreases markedly.

In the alluvial plain, lowering of the water table particularly affects the upstream end of the abandoned arms. The Méant old channel (Jons sector: Fig. 26.1) is not directly influenced by the hydroelectric dam (located downstream), but lowering of the water table in this part of the alluvial plain makes it a good example of an abandoned channel modified by river regulation. Two stations were studied in April and July 1987 (at depths of 50 and 100 cm in the sediments): Station M1, located at the upstream end of the Méant channel, is subject to partial drought at the end of the summer; Station M2, located in the central part of the Méant channel, is permanent. Despite the lack of differences in the chemical characteristics of their interstitial water (they are both located in the eastern part of the Jons alluvial plain; see Fig. 26.4), hypogean assemblages sampled in these two stations were very different (Table 26.7). Station M1 is characterized by a low density and a low species richness of stygobites, while in Station M2 stygobites are more abundant and represent a third of the total taxonomic richness. Physical disturbance induced in the alluvial plain by the erosion of the river channel and the lowering of the water table has noticeable consequences on abandoned arms, with particularly great consequences for the stygofauna.

CONCLUSION

In most cases, intensive studies of the dynamics of interstitial assemblages were located in small or medium-size rivers (Palmer, 1990; Bretschko, 1991; Boulton and Stanley, 1995). The Rhône and the Danube are the only large European rivers that have been intensively studied in the past twenty years. Three major results have arisen from the large set of studies on the Rhône. Firstly, groundwater fauna in the Rhône river basin is diverse (38 hypogean species have been identified: Dole-Olivier et al., 1994). Secondly, the spatial heterogeneity of the fauna is mainly influenced by the hydrogeology of river sectors (characteristics of the alluvial-plain aquifer, direction of water exchanges between surface water and groundwater) and the geomorphology of river sediments (location of an abandoned arm in the alluvial plain, grain size of the bottom sediments: Dole-Olivier et al., 1993). Understanding of the relative importance of these factors (and their interactions at different scales) would provide a good framework for patterns of faunal distribution. Thirdly, temporal dynamics of interstitial assemblages is mainly governed by seasonal changes and river hydrology (rhythm and intensity of floods: Dole-Olivier et al., 1997). Nevertheless, many research topics need to be developed in the future.

Future research must focus on three major topics: (1) the vertical dimension of the Rhône system must be studied in the sediments to greater depths (down to −10 or −20 m) in the main channel, at the shore-line, and on the alluvial plain (up to the foot of the lateral plateau) because most of the ecological processes (such as organic-matter decomposition, pollution control, or faunal dynamics) are not limited to the shallow interstitial habitats, but continue in deep groundwater; (2) the longitudinal dimension must include high-mountain interstitial habitats (in the Alps) and the delta to the Mediterranean Sea (in Camargue). Valuable results could be obtained under these contrasts of climate and unusual ecological context (effect of glaciers and marine waters on interstitial habitats and fauna); (3) the description of biological indicators for assessment of river interstitial quality and for biomonitoring pollution of both shallow interstitial habitats and deep groundwater systems.

The final goal of many recent hydrobiological studies has been to build a general approach integrating groundwater (and its interface with surface system) into running-water theory. The knowledge accumulated for the river Rhône and its groundwater system constitutes a relevant contribution to this goal.

ACKNOWLEDGEMENTS

Studies included in this chapter were supported by grants from the Lyonnaise des Eaux-Dumez (Le Pecq, France) and the Program for Interdisciplinary Research on the Environment (PIR Environnement) within the U.R.A. CNRS 1974 Laboratory. We are grateful to specialists who helped us with the identification of macroinvertebrates (Prof. Ginet for amphipods, Dr Bernasconi for molluscs). Finally, we wish to thank E. Dova and C. Claret for their help in the field in Vanoise and K. Simon (Virginia Polytechnic Institute) for editing the English manuscript.

REFERENCES

Amoros, C., Roux, A.L., Reygrobellet, J.-L., Bravard, J.-P. and Pautou, G., 1987. A method for applied ecological studies of fluvial hydrosystems. *Regulated Rivers,* 1: 17–36.

Boissier, J.M., Marmonier, P., Claret, C., Fontvieille, D. and Blanc, P., 1996. Comparison of solutes, nutrients, and bacteria inputs from two types of groundwater to the Rhône river during an artificial drought. *Hydrobiologia,* 319: 65–72.

Bornette, G. and Amoros, C., 1991. Aquatic vegetation and hydrology of a braided river floodplain. *J. Veget. Sci.,* 2: 497–512.

Bornette, G. and Large, A.R.G., 1995. Groundwater–surface water ecotones at the upstream part of confluences in former river channels. *Hydrobiologia,* 310: 123–137.

Botosaneanu, L., 1988. Zonation et classification biologique des cours d'eau; développement récents, alternatives, perspectives. *Atti XV Congr. Naz. Ital. Entomol.* S.G.P. Communication, Rome, pp. 33–61.

Bou, C., 1974. Recherches sur les eaux souterraines. 25 – Méthodes de récolte dans les eaux souterraines interstitielles. *Ann. Speleol.,* 29: 611–619.

Bou, C. and Rouch, R., 1967. Un nouveau champ de recherche sur la faune aquatique souterraine. *C.R. Acad. Sci. Paris,* 265: 369–370.

Boulton, A.J. and Stanley, E., 1995. Hyporheic processes during flooding and drying in a Sonoran desert stream. II faunal dynamics. *Arch. Hydrobiol.,* 134: 27–52.

Boulton, A.J., Valett, H.M. and Fisher, S.G., 1992. Spatial distribution and taxonomic composition of the hyporheos of several Sonoran desert streams. *Arch. Hydrobiol.,* 125: 37–61.

Bravard, J.-P., 1987. *Le Rhône du Léman à Lyon.* La manufacture, Lyon, 451 pp.

Bravard, J.-P., Amoros, C. and Pautou, G., 1986. Impact of civil engineering works on the successions of communities in a fluvial system. A methodological and predictive approach applied to a section of the Upper Rhône River, France. *Oïkos,* 47, 92–111.

Bretschko, G., 1991. Bed sediments, groundwater and stream limnology. *Int. Verein. Theor. Angew. Limnol.,* 24: 1957–1960.

Carbiener, R., Trémolière, M., Ortscheit, A. and Klein, J.P., 1988. Les associations végétales, biorévélatrices des échanges hydrologiques eaux de surface–eaux souterraines. In: H. Kobus and L. Zilliox (Editor), *Contamination des Eaux Souterraines par les Nitrates. Incidences de l'Agriculture sur la Qualité des Eaux Souterraines et Mesures de Protection, Colloq. Franco-Allemand,* 71: 175–200.

Cellot, B. and Bournaud, M., 1986. Modifications faunistiques engendrées par une faible variation de débit dans une grande rivière. *Hydrobiologia,* 135: 223–232.

Cellot, B. and Rostan, J.C., 1993. Dissolved organic carbon in the upper Rhone: the influence of side-arms? *Regulated Rivers,* 8: 391–397.

Creuzé des Châtelliers, M., 1991a. *Dynamique de Répartition des Biocénoses Interstitielles du Rhône en Relation avec des Caractéristiques Géomorphologiques (Secteurs de Brégnier-Cordon, Miribel-Jonage et Donzère-Mondragon),* Thèse. Université Lyon I, 160 pp.

Creuzé des Châtelliers, M., 1991b. Geomorphological processes and discontinuities in the macrodistribution of the interstitial fauna. A working hypothesis. *Int. Verein. Theor. Angew. Limnol.,* 24: 1609–1612.

Creuzé des Châtelliers, M. and Marmonier, P., 1990. Macrodistribution of Ostracoda and Cladocera in a by-pass channel: exchange between superficial and interstitial layers. *Stygologia,* 5: 17–24.

Creuzé des Châtelliers, M. and Marmonier, P., 1993. Ecology of benthic and interstitial ostracods of the Rhône River, France. *J. Crustacean Biol.,* 13: 268–279.

Creuzé des Châtelliers, M. and Reygrobellet, J.-L., 1990. Interactions between geomorphological processes, benthic and hyporheic communities: first results on a by-passed canal of the french Upper-Rhône River. *Regulated Rivers,* 5: 139–158.

Danielopol, D.L., 1984. Ecological investigations on the alluvial sediments of the Danube in the Vienna area – a phreatobiological project. *Int. Verein. Theor. Angew. Limnol.,* 22: 1755–1761.

Danielopol, D.L., 1989. Groundwater fauna associated with riverine aquifers. *J. North Am. Benthol. Soc.,* 8: 18–35.

Dole, M.J., 1983a. *Le Domaine Aquatique Souterrain de la Plaine Alluviale du Rhône à l'Est de Lyon: Ecologie des Niveaux Supérieurs de la Nappe,* Thèse. Université Lyon I, 184 pp.

Dole, M.J., 1983b. Le domaine aquatique souterrain de la plaine alluviale du Rhône à l'est de Lyon. 1. Diversité hydrologique et biocenotique de trois stations représentatives de la dynamique fluviale. *Vie Milieu,* 33: 219–229.

Dole, M.J., 1984. Structure biocénotique des niveaux supérieurs de la nappe alluviale à l'est de Lyon. *Mem. Biospeol.,* 11: 17–26.

Dole, M.J., 1985. Le domaine aquatique souterrain de la plaine alluviale du Rhône à l'est de Lyon. 2. Structure verticale des peuplements des niveaux supérieurs de la nappe. *Stygologia,* 1: 270–291.

Dole, M.J. and Chessel, D., 1986. Stabilité physique et biologique des milieux interstitiels. Cas de deux stations du Haut Rhône. *Ann. Limnol.,* 22: 69–81.

Dole, M.J. and Coineau, N., 1987. L'Isopode Microcharon (Crustacea, Isopoda) abondant dans les eaux interstitielles de l'est lyonnais. *M. reginae* n. sp., écologie et biogéographie. *Stygologia,* 3: 200–216.

Dole-Olivier, M.J. and Marmonier, P., 1992a. Patch distribution of interstitial communities : prevailing factors. *Freshwater Biol.,* 27: 177–191.

Dole-Olivier, M.J. and Marmonier, P., 1992b. Ecological requirements of stygofauna in an active channel of the Rhône river. *Stygologia,* 7: 65–75.

Dole-Olivier, M.J. and Marmonier, P., 1992c. Effects of spates on interstitial assemblages structure. Disturbance–perturbation relationship, rate of recovery. *Hydrobiologia,* 230: 49–61.

Dole-Olivier, M.J., Creuzé des Châtelliers, M. and Marmonier, P., 1993. Repeated gradients in subterranean landscape – Example of the stygofauna in the alluvial floodplain of the Rhône River (France). *Arch. Hydrobiol.,* 127: 451–471.

Dole-Olivier, M.J., Marmonier, P., Creuzé des Châtelliers, M. and Martin, D., 1994. Interstitial fauna associated with the alluvial floodplain of the Rhône River (France). In: J. Gibert, D.L. Danielopol and J.A. Stanford (Editors), *Groundwater Ecology,* Academic Press, New York, pp. 313–346.

Dole-Olivier, M.J., Marmonier, P. and Beffy, J.L., 1997. Response of invertebrates to lotic disturbance: is the hyporheic zone a patchy refugium? *Freshwater Biol.,* 37: 257–276.

Gibert, J., Ginet, R., Mathieu, J., Reygrobellet, J.-L. and Seyed-Reihani, A., 1977. Structure et fonctionnement des écosystèmes du Haut Rhône français. IV. Le peuplement des eaux phréatiques, premiers résultats. *Ann. Limnol.,* 13: 83–97.

Gibert, J., Ginet, R., Mathieu, J. and Reygrobellet, J.-L., 1981. Structure et fonctionnement des écosystèmes du Haut Rhône français; IX: analyse des peuplements de deux stations phréatiques alimentant des bras morts. *Int. J. Speleol.*, 11: 141–158.

Gibert, J., Marmonier, P., Vanek, V. and Plénet, S., 1995. Water flows and sediment characteristics in river bank: relationship with heavy metals and invertebrate community dynamics. *Can. J. Fish. Aquat. Sci.*, 52: 2084–2097.

Ginet, R., 1982. Structure et fonctionnement des écosystèmes du Haut Rhône français: XXIV. Les Amphipodes des eaux interstitielles en amont de Lyon. *Pol. Arch. Hydrobiol.*, 29: 231–237.

Hakenkamp, C.C., Valett, H.M. and Boulton, A.J., 1993. Perspectives on the hyporheic zone: integrating hydrology and biology. Concluding remarks. *J. North Am. Benthol. Soc.*, 12: 94–99.

Ille, C., Suais, M.F., Durbec, A., Gaillard, B. and Laffont, M., 1991. Caractérisation du pouvoir épurateur des berges d'un fleuve. Application à l'Ile du Grand Gravier (Rhône). *Hydrogeologie*, 4: 283–289.

Mandier, C., 1988. *Le Relief de la Moyenne Vallée du Rhône au Tertiaire et au Quaternaire. Essai de Synthèse Paléogéographique.* B.R.G.M. Editions, Orléans, 230 pp.

Marmonier, P., 1988. *Biocénoses Interstitielles et Circulation des Eaux dans le Sous-Écoulement d'un Chenal Aménagé du Haut Rhône Français*, Thèse. Université Lyon I, 269 pp.

Marmonier, P. and Creuzé des Châtelliers, M., 1991. Effects of spates on interstitial assemblages of the Upper Rhône River. Importance of spatial heterogeneity. *Hydrobiologia*, 210: 243–251.

Marmonier, P. and Creuzé des Châtelliers, M., 1992. Biogeography of the benthic and interstitial living ostracods (Crustacea) of the Rhône River (France). *J. Biogeogr.*, 19: 693–704.

Marmonier, P. and Dole, M.J., 1986. Les Amphipodes des sédiments d'un bras court-circuité du Rhône: logique de répartition et réaction aux crues. *Sci. Eau*, 5: 461–486.

Marmonier, P., Meisch, C. and Danielopol, D.L., 1989. A review of the genus Cavernocypris Hartmann (Ostracoda, Cypridopsinae): systematics, ecology and biogeography. *Bull. Soc. Nat. Luxemb.*, 89: 221–278.

Marmonier, P., Dole-Olivier, M.J. and Creuzé des Châtelliers, M., 1992. Spatial distribution of interstitial assemblages in the floodplain of the Rhône river. *Regulated Rivers*, 7: 75–82.

Marmonier, P., Fontvieille, D., Gibert, J. and Vanek, V., 1995. Distribution of dissolved organic carbon and bacteria at the interface between the Rhône River and its alluvial aquifer. *J. North Am. Benthol. Soc.*, 14, 382–392.

Monjuvent, G., 1988. La déglaciation rhodanienne entre les moraines internes et le val du Bourget. *Géol. Alpine*, 64: 61–104.

Palmer, M.A., 1990. Temporal and spatial dynamics of meiofauna within the hyporheic zone of Goose Creek, Virginia. *J. North Am. Benthol. Soc.*, 9: 17–25.

Pennak, R. and Ward, J.V., 1986. Interstitial faunal communities of the hyporheic and adjacent groundwater biotopes of a Colorado mountain stream. *Arch. Hydrobiol., Suppl.*, 74: 356–396.

Plénet, S., Gibert, J. and Marmonier, P., 1995. Biotic and abiotic interactions between surface and interstitial systems in rivers. *Ecography*, 18: 296–309.

Pospisil, P., 1994. The groundwater fauna of a Danube aquifer in the wetland "Lobau" at Vienna, Austria. In: J. Gibert, D. Danielopol and J.A. Stanford (Editors), *Groundwater Ecology.* Academic Press, New York, pp. 347–366.

Puig, M.A., Sabater, F. and Malo, J., 1990. Benthic and hyporheic fauna of mayflies and stoneflies in the Ter River basin (NE-Spain). In: I.C. Campbell (Editor), *Mayflies and stoneflies.* Kluwer Academic Publishers, Dordrecht, pp. 255–258.

Resh, V.H., Brown, A.V., Covich, A.P., Gurtz, M.E., Li, H.W., Minshall, G.W., Reice, S.R., Sheldon, A.L., Wallace, J.B. and Wissmar, R., 1988. The role of disturbance in stream ecology. *J. North Am. Benthol. Soc.*, 7: 433–455.

Reygrobellet, J.-L. and Dole, M.J., 1982. Structure et fonctionnement des écosystèmes du Haut Rhône français, XVII. Le milieu interstitiel de la "lône du Grand Gravier"; premiers résultats hydrologiques et faunistiques. *Pol. Arch. Hydrobiol.*, 29: 485–500.

Schumm, J.A., 1977. *The Fluvial System.* Wiley Interscience, New York, 338 pp.

Sedell, J.R., Richey, J.E. and Swanson, F.J., 1989. The river continuum concept: a basis for the expected ecosystem behaviour of very large rivers? In: D.P. Dodge (Editor), *Proc. Int. Large River Symp., Can. Spec. Publ. Fish. Aquat. Sci.*, 106: 49–55.

Seyed-Reihani, A., 1980. *Etude Écologique du Milieu Interstitiel Lié au Fleuve Rhône en amont de Lyon*, Thèse. Université Lyon I, 70 pp.

Smock, L.A., 1994. Movement of invertebrates between stream channels and forest floodplains. *J. North Am. Benthol. Soc.*, 13: 524–531.

Vannote, R.L., Minshall, G.W., Cummins, K.W., Sedell, J.R. and Cushing, C.E., 1980. The river continuum concept. *Can. J. Fish. Aquat. Sci.*, 37: 130–137.

Ward, J.V. and Voelz, N.J., 1990. Gradient analysis of interstitial meiofauna along a longitudinal stream profile. *Stygologia*, 5: 93–99.

Ward, J.V. and Voelz, N.J., 1994. Groundwater fauna of the South Platte River system, Colorado. In: J. Gibert, D. Danielopol and J.A. Stanford (Editors), *Groundwater Ecology.* Academic Press, New York, pp. 391–423.

Ward, J.V., Voelz, N.J. and Poff, N.L., 1994. Gradient analysis of zoobenthos community structure along a mountain stream continuum. *Int. Verein. Theor. Angew. Limnol.*, 25: 1462–1464.

Williams, D.D. and Hynes, H.B.N., 1974. The occurrence of benthos deep in the substratum of a stream. *Freshwater Biol.*, 4: 233–256.

Chapter 27

KARST ECOSYSTEMS IN THE RHÔNE RIVER BASIN

J. GIBERT, F. MALARD, M.J. TURQUIN and R. LAURENT

INTRODUCTION

On a global scale, karst lands cover extensive parts of the landscape – about 30% of the European and French land surfaces. Karstic groundwater represents 25% of the total water supply in France, thus providing an essential resource for a substantial part of the population (Action de Coopération Scientifique et Technique 65, 1995). Research on karst aquifers in France, being vital resources to the French economy, has focused on the study of groundwater flow and contaminant transport. However, progress made in the field of groundwater ecology has clearly shown that karst systems also play a major role in sustaining biological diversity at the landscape scale (Gibert et al., 1994a).

Interest in cave habitats and life in massifs of the Pyrenees, Jura and Alps started at the end of the nineteenth century, but extensive biogeographical studies really began during the middle of the twentieth century (see also Geoffroy, 1993; Ginet, 1993; Turquin, 1993; Juberthie and Ginet, 1994). Long-term ecological research in the Rhône river basin (south-eastern France) has been carried out in a high diversity of karsts from different geographical areas (e.g., Jura, Pre-Alps, Ardèche), climatic zones (mediterranean and continental), geomorphological contexts (plain, plateau, mountain) and geological facies. Elsewhere in the world, only a few karst systems have been studied for such a long period of time and with such a high sampling effort. Present-day knowledge results from the collection of data for approximately 220 caves and springs in the Rhône river basin, published in a series of more than 130 papers. First, research focussed on the description of organisms; second, on the ecology of caves (Culver, 1982; Chapman, 1993); and third, on the ecology of karst systems (Rouch, 1977). Therefore, conceptions of the organization of subterranean life have changed radically as research evolved from cave exploration to ecosystem appraisal.

In the context of current hypotheses in subterranean ecology, this chapter is a first attempt to review studies of karst ecosystems in the Rhône river basin. Based mainly on research conducted by students of the Laboratoire d'Hydrobiologie et Ecologie Souterraines at the Université de Lyon, we summarize the evolution in knowledge and approach in research on karstic ecology and emphasize significant advances.

FROM CAVE EXPLORATION TO ECOSYSTEM APPRAISAL

The geographical and geological theatre

The karstic domain of the Rhône river basin covers 40 000 km^2, approximately one-third of the total basin area. It encompasses a variety of karsts having different structural characteristics (e.g. folded or tabular karsts), geomorphologic features (e.g. mountain or plateau karsts), bioclimatic conditions, and clastic covering and inheritances (Fig. 27.1, Table 27.1) (for more details see Nicod, 1995).

Mountain karsts developed mainly in folded calcareous structures (except in the Jura mountains) composed of thick Mesozoic carbonate series. High-alpine karsts (altitude ranging from 1800 to 2500 m) are dominated by Urgonian and/or Senonian limestone and dolomite. Runoff (2500 mm) and solution rate (100 m^3 km^2 y^{-1}) are high, and developed karstification creates deep speleological networks (deeper than 1000 m). Upland karsts are extremely diverse, but three subtypes can be structurally distinguished by their unequal extent of ante-Quaternary inheritance and the role of glacial and

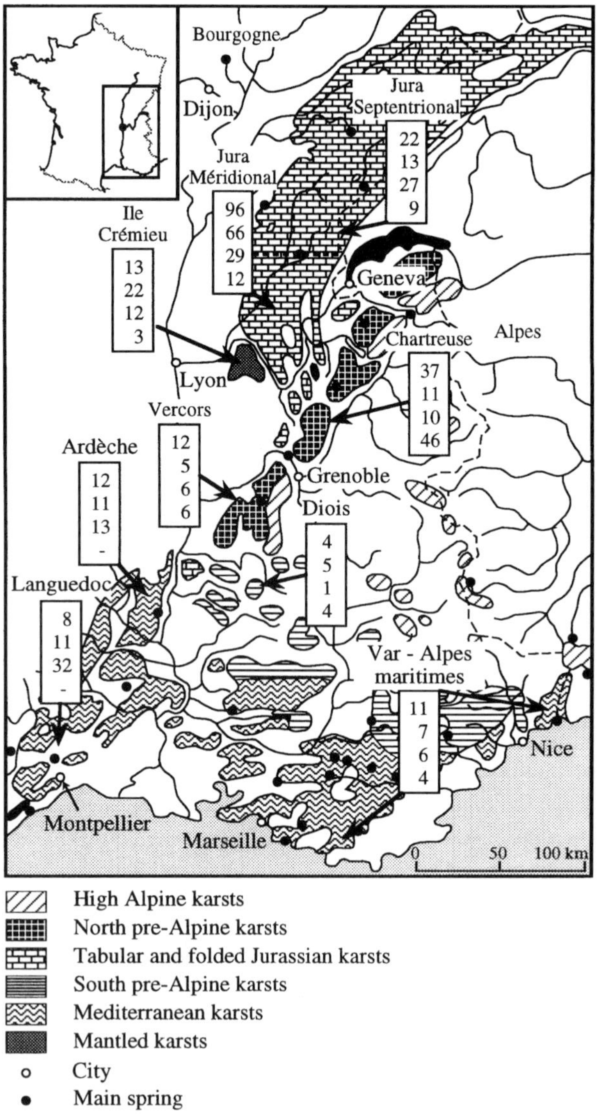

Fig. 27.1. Location of different karst units in the Rhône basin. For each unit, the number of sites investigated (caves or springs), publications, and the number of species of stygobites (aquatic fauna), and troglobites (terrestrial fauna) is indicated in the columns, running from top to bottom.

nival phenomena during the Pleistocene. Northern pre-Alpine karsts are characterized by the thickness of the calcareous series (especially the Barremian formation of the reef facies), the incidence of Quaternary glaciation such as ice caps on high areas and thick glaciers in valleys, and high solution rates ($150\,m^3\,km^2\,y^{-1}$). Folded and tabular Jurassic karsts that developed in Dogger and Malm formations are separated by a thick marl interstratification. Major karstic forms include dolines, large synclinal poljes, excavation of cross valleys, and large subhorizontal speleological networks. The different glacial stages during the Pleistocene led to the expansion of extensive moraine deposits in depressions. The "Ile Crémieu" (southwestern Jura), which operated as an under-glacial karst during the Quaternary period, is the largest French karst covered by morainic deposits. Southern pre-Alpine karst are scattered because of tectonic movements, variations of facies, and excavation of canyons during the Miocene period.

Mediterranean karsts that developed in a series of middle and upper Jurassic and lower Cretaceous (Urgonian facies) limestone and dolomite have low solution rates ($5–27\,m^3\,km^2\,y^{-1}$). Karstification is ancient; bauxite deposits that formed during the middle Cretaceous were preserved in some paleokarsts. These karsts have been affected by Quaternary regressions and transgressions (especially the Messinian regression). The Languedoc and Ardèche plateaus were shaped during the Cenozoic period by successive gradations that now persist as paleoforms. For example, the distribution of avens in the Ardèche plateau is related to the location of paleo-terraces.

First faunal investigations in caves

As early as the nineteenth century, entomologists already recognized the subterranean fauna of the karstic systems of the Rhône river basin, but specialists in groundwater fauna were rare at this time [see the review by Turquin (1993)]. In 1875, Bedel and Simon published a list of European cave arthropods containing many terrestrial species but only a few aquatic crustaceans. For example, several species of Coleoptera collected in caves of the Vercors were described by Abeille de Perrin (1869, 1875) and Bedel (1878, 1883): *Anophthalmus delphinensis* (now called *Duvalius*), *Pholeuon dapsoides* (now called *Cytodromus*), *Adelops tarissani* (now called *Royerella*), and *Anophthalmus gounellei* (now called *Trichaphaenops*). Concerning the groundwater fauna, in addition to a publication by Nicolas (1882) on the mollusc *Avenionia* collected in a well in Avignon (southeastern France), amphipods of the genus *Niphargus* were sampled in some Jura caves by Viré and described by Chevreux (1896). For example, *Niphargus virei* was known from some caves in the northern and southern Jura and in the Ardèche.

However, until the beginning of the twentieth century, biospeleological studies were more or less isolated, and the literature only provided some inventories

Table 27.1
Main features of different types of karsts in the Rhône river basin

Karst type	Geographical location	Massifs	Cover	Glacial influence	Nival influence	Runoff (mm)	Solution rate ($m^3\ km^{-2}\ y^{-1}$)
Mountain karsts							
highland	High Alpine	Désert de Platé	bare	present	present	2500	100
upland	North pre-Alpine	Vercors, Diois, Chartreuse	forested	past	present	1500	150
upland	Jura	North Jura	forested	past	present	1500	95
		South Jura (Ile Crémieu)				1000	80
upland	South pre-Alpine	Vaucluse	forested	–	present	700	60
Mediterranean karsts		Ardèche	forested and bare			400	27
		Languedoc					5
		Alpes-Basses				350	27

Table 27.2
Numbers of taxa in several caves of the Rhône river basin, and publications dealing with them

Location		Number of publications	Terrestrial taxa	Troglobitic taxa	Aquatic taxa	Stygobitic taxa
Grotte d'Hautecourt	Jura	45	89	10	2	2
Grotte de Rappe	Jura	3	41	5	16	16
Grotte de Corveissiat	Jura	5	3	3	11	10
Tunnel de Drom	Jura	8	57	5	7	6
Dorvan system	Jura	30	118	11	110	16
Karst de Foussoubie	Ardèche	6	–	–	97	11
Site Expérimental du Triadou	Languedoc	4	–	–	53	32

of caves with their associated fauna. Biospeleological research in France, and especially in the Rhône river basin, was organized by Jeannel, Chappuis, and Racovita up to 1940; their contributions have been reviewed by Balazuc (1993), Geoffroy (1993) and Réveillet (1993). The list of cavernicolous species for the basin increased as a result of the work of Sollaud (1936) and Ginet (1953a,b, 1956, 1961a,b; Ginet and Genest, 1954). Ginet founded the Laboratoire d'Hydrobiologie et Ecologie Souterraines at the Université de Lyon in 1960, where research was initially devoted to the biology, systematics and biogeography of the stygobitic amphipod *Niphargus* and aquatic and terrestrial animals of other systematic groups collected in caves, sinks or karstic springs, primarily in southeastern France (Jura, Alps and Cévennes).

Systematic inventory of cave animals

Not all the karstic massifs of the Rhône river basin have been been studied to the same degree, and not all systematic groups have received the same attention. Among mountain karsts, only the Jurassic karsts (northern Jura, southern Jura and Ile Crémieu) and the northern pre-alpine karsts (Vercors, Chartreuse, Diois) have been studied. Research in Mediterranean regions has been restricted to karsts of the Ardèche, Languedoc and Low Alps. Figure 27.1 indicates for each massif the number of sites investigated (caves or springs), publications, stygobites (aquatic fauna), and troglobites (terrestrial fauna). Table 27.2 provides similar information for particular sites (caves or karst systems) that were studied in detail. The number of species for each area is underestimated because only those species collected during our investigations were taken into consideration. Moreover, sampling effort was considerably higher in the southern Jura (96 sites sampled and 66 papers) than in other massifs.

Several important conclusions emerge from the systematic inventory of cave animals. First, cave dwellers are found in all karsts, but there are large differences in species number among massifs. For example, caves of the Ile Crémieu harbour only 3 to 5 troglobitic species, whereas approximately 50 troglobites are known from the Vercors. Second, although karstic areas of the Jura and Vercors show climatic, geographical, and geological similarities, they differ markedly from a biological point of view. Indeed, the scarcity of terrestrial troglobites in the Jura caves strongly contrasts with the richness of communities in the Vercors (Bouvet and Turquin, 1986). Third, in any large structural unit, for example the Jura mountains, a relatively small geographical area such as the "Ile Crémieu" may have species-poor communities of stygobites and troglobites (Fig. 27.1).

Karst ecosystem studies

During the second half of the twentieth century, the development of various sampling methods allowed rapid progress in groundwater ecology (Mathieu et al., 1991). Because of the inaccessibility of karstic systems, faunal sampling has essentially been carried out at the natural accesses to the karst – the caves and springs. Various techniques (sampling by sight, traps and artificial substrate: see Camacho, 1992) have been used to sample different microhabitats in caves and subterranean streams. As it became evident that caves were not the main habitat of stygobites, drifting animals were collected at springs by filtering water during floods. Such a methodological approach, initiated by Rouch et al. (1968) in the Baget karst (Ariège, France) and used afterwards in several karst systems of the basin (Gibert, 1986; Barthélémy, 1984; Turquin, 1984), considered the aquifer as a "black box". The structure of drifting communities at the outlets of a drainage basin was used to define the main biological characteristics of karstic systems. Using such a sampling technique, Rouch (1977) demonstrated that the physical sub-units of a karst also corresponded to biological components, and defined for the first time the concept of karstic ecosystems. However, because many aquifers lack natural accesses or suitable springs for filtering, a new method for sampling the fauna of deep karstic groundwater was recently developed by Malard et al. (1994a). Faunal pumping is achieved by an air-lift in deep uncased wells that intersect fractures of different hydraulic conductivities.

COMMUNITY ORGANIZATION: SPATIAL AND TEMPORAL DISTRIBUTION PATTERNS WITHIN KARST ECOSYSTEMS

The karst system: structural and functional aspects

Karstic ecological studies carried out in the Rhône river catchment have mainly relied on the hydrogeological representation of the karst system proposed by Mangin (1975) (Fig. 27.2). The karst system is primarily a physical unit defined as "the area at the level from which the flow of karstic type becomes organized to constitute a unit of drainage" (Mangin, 1994). It consists of three different zones: the drainage area, the unsaturated zone, and the saturated zone. The **drainage area** may include non-karstic terrains where surface runoff occurs. Then, surface runoff recharges the karst aquifer by diffuse or point infiltration. The **unsaturated zone** includes in its uppermost part a subsurface zone which extends to a depth of 3 to 10 m below the ground surface (Williams, 1983). Due to the decompression of the rock, the action of climatic factors and the infiltration of water enriched with carbon dioxide, the subsurface zone is highly fissured and karstified. It rapidly absorbs rainwater, whose infiltration with depth is limited because the number and size of fractures in the infiltration zone decrease with depth. Thus, infiltrating water is stored in a shallow zone called the epikarst aquifer (Mangin, 1975). This perched aquifer limits surface runoff, delays the infiltration of water (delayed infiltration *versus* direct infiltration) and promotes karstification at depths by storing groundwater. Water circulation in the unsaturated zone is mainly vertical and occurs rapidly within a relatively small number of large subvertical fractures (fast infiltration) or slowly in small-sized fissure networks (slow and diphasic infiltration, "diphasic" implying a mixture of air and water). Subvertical water flow may be followed by underground runoff ("the subterranean river zone") because the saturated zone develops mainly in the vicinity of springs. The **saturated zone** is made up of a drainage network which drains during low-water periods and recharges a series of independent annex systems during floods. The drainage network determining the transmissive function of the aquifer corresponds to the preferential flow path of groundwater, in which the flow velocity may be high. In contrast, the annex systems that determine the

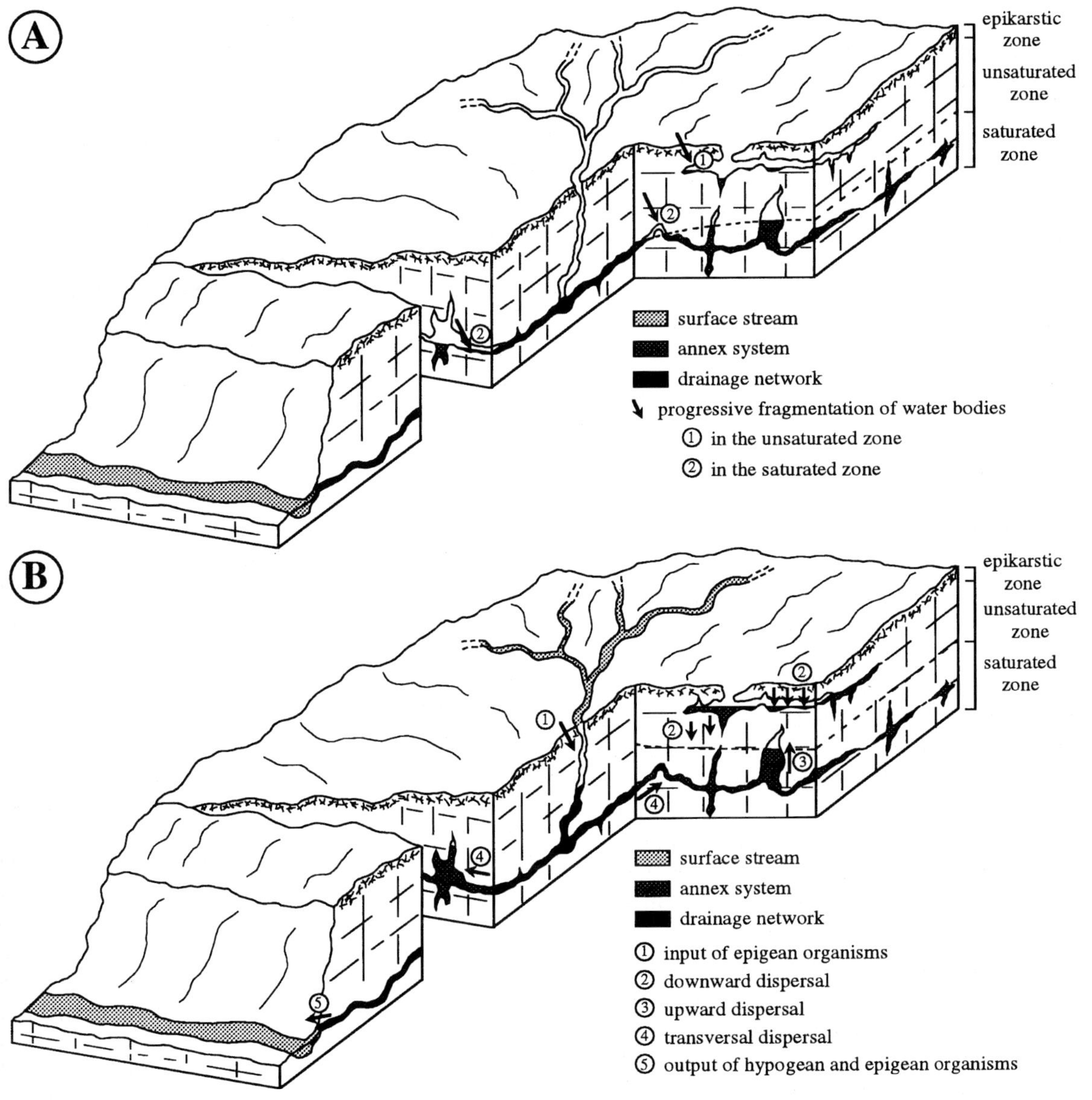

Fig. 27.2. Hydrogeological representation of a karst system showing (A) fragmentation of water bodies during low water and (B) fluxes of organisms during floods.

storage function of the aquifer are zones of low flow velocity. Because they may include large cavities, annex systems do not correspond strictly to small-sized fissure networks.

The karst ecosystem, as shown in Fig. 27.2, encompasses a wide range of terrestrial and aquatic habitats at different scales of space. For example, at a macroscale, one may distinguish between the communities of the epikarstic aquifer, the infiltration zone and the saturated zone. At a mesoscale, specific animal assemblages inhabit different parts of the saturated zone (i.e. the drainage network and the annex systems). Finally, fractures having different sizes, physico-chemical characteristics, and connections with the surface environment constitute a mosaic of microhabitats in the saturated zone. Whatever the scale considered, exchanges of organic matter (signifying energy supply) and fluxes of aquatic organisms between habitats are regulated by water flow. Except in the entrance zone of caves, where cryptogams may develop, supply of organic matter to the terrestrial and aquatic organisms is flow-dependent and varies with cave location within the karst system (i.e. whether the caves have active water flow or not).

External events affect the habitats and their communities at various time scales. In the long term, events such as glaciation or changes in the local base level (the lowest level of karstification of the limestone) are likely to affect the whole karst biotope and biocenosis. At the scale of a yearly hydrological cycle, one may observe changes in the community structure of the whole saturated zone; but moderate and short-term floods affect only some parts of the saturated zone (i.e. the drainage network and downstream annex systems). Thus, the spatial and temporal heterogeneity of animal communities has been perceived at various scales in karst ecosystems of the Rhône basin. Research has been carried out to determine distributional patterns and responses of fauna to ecological factors of the karstic subterranean milieu.

Spatial distribution patterns of fauna

Subterranean fauna of studied palearctic karsts results from the convergence of several driving forces of unequal importance. A whole set of factors leading to the diversity of world-wide subterranean taxa [see the review by Stoch (1995)] could probably be mentioned to explain the present-day composition and distribution of species in the Rhône basin. The review of all these factors is beyond the scope of this chapter, thus we present only those historical and ecological determinants that are believed to be of primary importance in this basin.

Regional scale: the karstic highway

The Rhône river basin is ideally located between the Mediterranean Sea (origin of many stygobites), the Rhine valley, and the North Sea, thus making possible the flow of ancient seas (transgressions). Both sides of the basin are occupied by limestone areas, which now lie only 400 km away from the ancient hercynian mountain the Massif Central. This 285-million years old massif remained above sea-level when many coastlands were submerged. Thus, the permanent existence of emerged lands and the repeated invasion of seas during the Tertiary were propitious to the invasion of the underground environment by troglobites and stygobites originating either from faunas with tropical affinities during the Lower Tertiary (Paleocene, Eocene, Oligocene) or from temperate faunas during the Upper Tertiary (Miocene, Pliocene). However, great differences have been observed between the gamma diversity (regional species richness) of different karstic systems that have been studied intensively (Fig. 27.1, Table 27.2).

The subterranean terrestrial fauna of the karstic Rhône river basin probably was derived from Tyrrhenian stock (Jeannel, 1926). This is exemplified by the high richness of Coleoptera in Vercors (more than 30 species or subspecies belonging to the genera *Cytrodomus, Diaprisus, Duvalius, Isereus, Royerella* and *Trichaphaenops*) and Collembola (about 16 species belonging to the genera *Arrhopalites, Isotomurus, Mesachorutes, Onychiurus, Pseudosinella, Schaefferia, Sminthurus* and *Tomocerus*: Gisin, 1953) (Fig. 27.3). High diversity in the Vercors massif is due to a high level of endemism that likely reflects a high rate of geographic speciation in a highly fragmented subterranean habitat. Probably flourishing during the Tertiary in the entire basin, the subterranean fauna was deeply affected during the Ice Ages, especially in the northern part of the basin (Bouvet and Turquin, 1986). Indeed, the subterranean Jura harbours only three species of troglobitic Coleoptera (*Royerella villardi, Trichaphaenops cerdonicus* and *T. sollaudi*) and seven species of Collembola (*Arrhopalites, Onychiurus, Pseudosinella, Schaefferia* and *Tomocerus*). Glacier extension also resulted in discontinuous species ranges within any particular massif. For example, within the Jura mountains, the lower diversity noted earlier in the Ile Crémieu may reflect particular paleoclimatic perturbations that affected this area during the Quaternary. This karst was entirely covered by an ice cap, and then morainic deposits filled the fissures and prevented post-glacial colonization.

Evolutionary invasion of karstic groundwater in the Rhône basin has occurred for many millions of years and resulted in a highly diversified stygobitic fauna. For example, the bathynellids (Syncarida) probably colonized groundwater prior to the breakup of Pangaea during the Jurassic (Schminke, 1981). *Stenasellus buili* [(Stenasellidae; Isopoda), found in karsts of the Languedoc region] is thought to have originated from a Mesogean ancestor that probably invaded groundwater during the Mid-Cretaceous (Magniez, 1981). However, the majority of groundwater taxa are thalassoid stygobites of Tertiary origin (i.e. they were derived from marine ancestors that directly colonized groundwater). This is likely to be the case for the isopod family Microparasellidae (Coineau, 1994; Dole and Coineau,

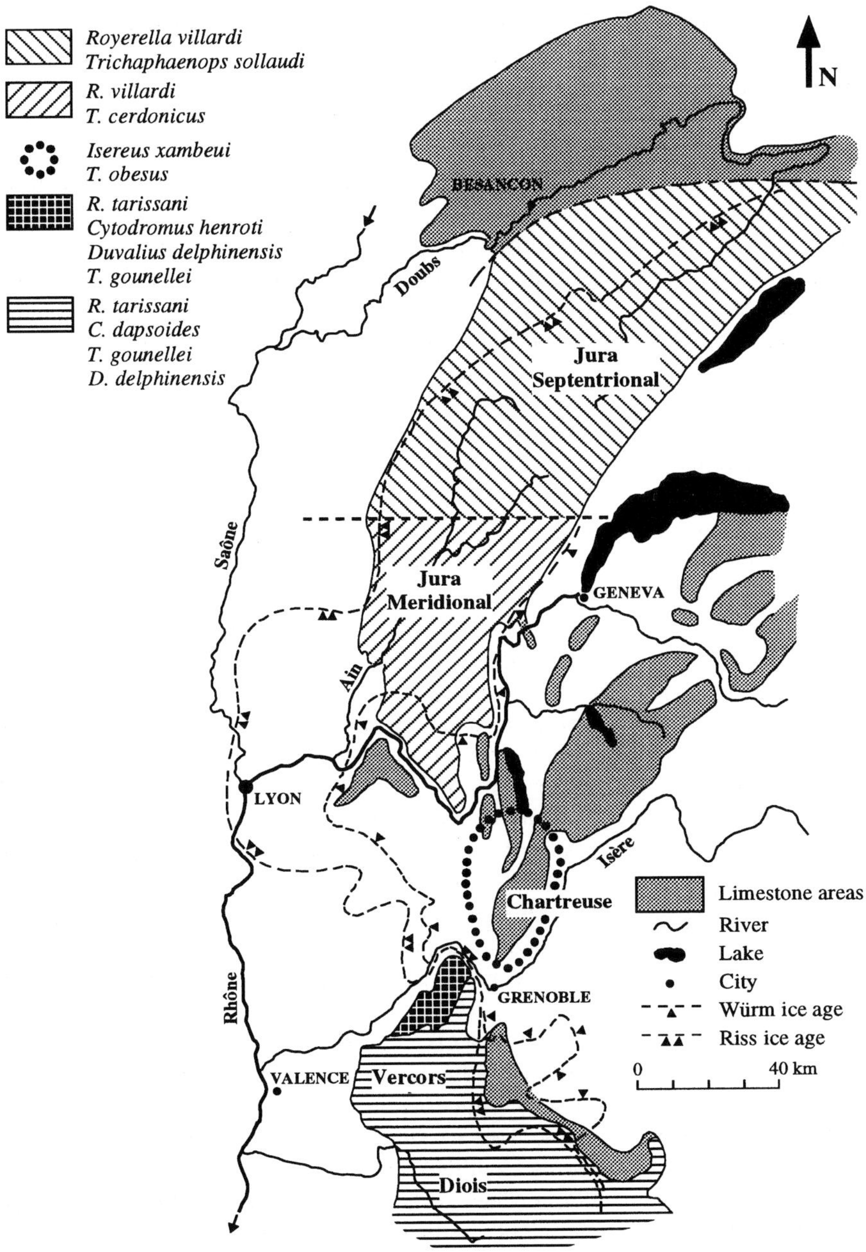

Fig. 27.3. Distribution map of troglobitic Coleoptera in the Jura and Vercors massifs (modified after Bouvet and Turquin, 1986).

1987), and the amphipod families Niphargidae (Ginet, 1983; Karaman and Ruffo, 1986), and Salentinellidae (Notenboom, 1991). Finally, some taxa are limnicoid stygobites (i.e. derived from marine ancestors that invaded surface water before colonizing groundwater) such as representatives of the isopod family Asellidae (Henry, 1976) or the harpacticoid Canthocamptidae (Rouch, 1994).

Differences in faunal composition between different karst systems have resulted from the combination of several historical factors that influenced the migration and extinction rates of aquatic species. The glacial history of the Ile Crémieu also probably caused its species-poor aquatic and terrestrial composition (no Coleoptera of the subfamilies Bathycinae or Trichinae) compared to the diverse communities encountered in the northern and southern Jura. However, glacial influence cannot explain the relatively low number of stygobites in the Vercors, which is known to harbour a diverse terrestrial fauna (see above). Aquatic habitat diversity is believed to be poor in well-karstified systems of this region, which usually have a deep unsaturated zone (with rapid vertical circulation) and a weakly-developed saturated zone. Limestone aquifers of the Languedoc are among the most diversified karst systems of the world, which include the Edwards Aquifer in Texas (Longley, 1981), the Mammoth Cave in Kentucky (Poulson, 1992), the Dinaric karst in Slovenia (Sket, 1996) and Movile Cave in Romania (Sarbu and Popa, 1992; see also Chapter 17, this volume). For example, Malard et al. (1996a) found 32 stygobitic species in only four wells of the Lez karst system; in one well there were 24 species (Table 27.2). Such a high diversity is primarily believed to be related to three factors: the proximity of the Mediterranean Sea, which promoted invasion of groundwater by thalassoid stygobites during the Tertiary; higher homeostasis due to large system size (500 km^2 for the Lez system according to Karam, 1989); and high habitat diversity in plateau karsts that contain extensive saturated zones.

Depending on their ecological tolerances and preferences, stygobites have been affected by historical changes. Therefore, differences in faunal composition are also related to species traits. For example, *Niphargus rhenorhodanensis* is a generalist widely distributed in karstic and interstitial groundwater of the Rhône basin. This species, which suffered from historical perturbations such as glaciation, was able to recolonize karst systems by way of the hyporheic zone of large rivers (interstitial highway). *Niphargus virei* also is a generalist, but a typical karstic species which colonizes almost all karst basins except those located in the southern Jura and Dauphiné regions. This hiatus in its distribution has been attributed to the destructive action of glaciers that spread during the Würm period (Ginet, 1953b; Ginet and Juberthie, 1987). In southern France (Languedoc and Corbières regions) *N. virei* is usually associated with *Faucheria faucheri, Sphaeromides buili* (Isopoda), *S. raymondi* and *Troglocaris inermis* (Decapoda). These thermophilic species may have undergone a reduction in their distribution in more northerly sites. This hypothesis is reinforced by the fact that the cirolanid *S. raymondi* is known as a locally endemic species in karsts of the Ardèche region, where, despite intensive research along the river, it was found in only two spring caves.

Spatial heterogeneity within karst systems

The distribution patterns of animals have been examined across a range of spatial scales in different karst systems of the Rhône basin [see Gibert et al. (1994b) for a more comprehensive view on this aspect]. At the scale of the karst system, the diversity and abundance of stygobitic species were shown to increase from the epikarst down to the saturated zone in the Dorvan massif (Gibert, 1986, 1988). Only four species were collected in water percolating in the topmost layers of the massif, seven species in the spring draining the epikarstic aquifer and 14 species at the outlet of the saturated zone. Changes in the structure of the harpacticoid community along the lateral dimension have been nicely documented for the Baget karst (Ariège, France) (Rouch et al., 1993). In this karst system, the main channels were mainly colonized by *Parapseudoleptomesochra subterranea*, whereas the assemblages in the lateral subsystems were dominated by *Ceuthonectes gallicus* (Hillère subsystem) and *Nitocrella gracilis* (Peyrère subsystem).

Mesoscale investigations carried out in several caves of the basin revealed a patchy distribution of the fauna which was related to environmental heterogeneities, either climatic as in the Grotte de Donzère (Turquin and Van Poucke, 1978), or geomorphological as in the Grotte d'Aromas (Turquin et al., 1976) and Grotte de Glandieu (Reygrobellet et al., 1975). The Grotte de Donzère has several entrances distributed over a 100 m altitudinal gradient causing air movement ranging from gentle winds to strong draughts. Entrance fauna spread all over the cave, but species are distributed along steep gradients. For example, *Scoliopteryx libatrix* resides in the breeze zone, while other Lepidoptera such as *Apopestes* and *Orneodes* only inhabit calmer walls. The Grotte d'Aromas is a horizontal gallery unevenly filled with bedded deposits, probably of Pleistocene glaciofluvial origin. Extensive areas of dry sandy sediments are interrupted only in places where vadose rivulets produce patches of loam and organic matter. This short cave harbours three species of *Niphargus* found only in particular "highlands" within this sandy desert. *Niphargus kochianus* enters the cave only during rainy periods and then disappears; *N. rhenorhodanensis* is restricted to vadose trickles; and *N. virei* is found in only one place, a fractured dome-pit which allows this large species living in the saturated zone to wander into the cave.

At a microscale, fractures having different sizes, physico-chemical characteristics, and connections with

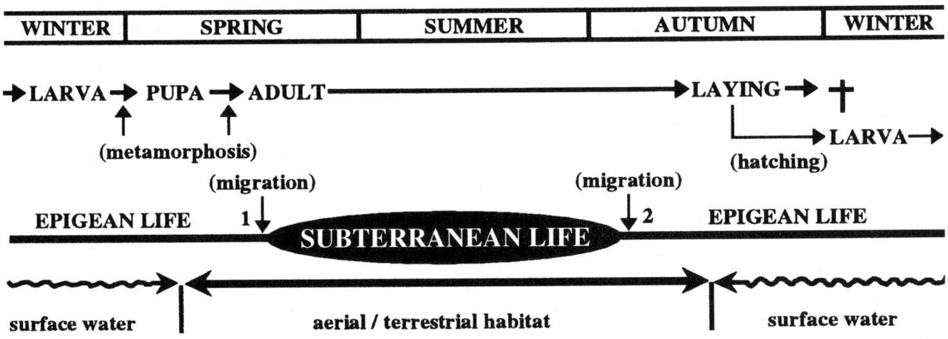

Fig. 27.4. Biological cycle of subtroglophilic Trichoptera belonging to the *Stenophylax* group (modified after Ginet and Decu, 1977).

the surface environment also constitute a dynamic mosaic of habitats in the saturated zone of a karst aquifer. In the Lez system, Malard and Simon (1997) found different groundwater assemblages in closely spaced wells intersecting fractures, with yields ranging from 150 to less than $2\,m^3\,h^{-1}$. In wells intersecting large conductive fractures characterized by fluctuating groundwater flow, invertebrate assemblages were dominated by ubiquitous stygophilic cyclopoids (e.g., *Acanthocyclops venustus, Diacyclops languidoides*) but also contained stygobitic molluscs (e.g. *Moitessiera rollandiana*) and ostracods (e.g. *Mixtacandona* sp.). In contrast, wells intersecting poorly karstified fissures with more stable flow conditions were colonized by stygobitic amphipods (*Niphargus pachypus, Salentinella juberthiae*), syncarids (Bathynellidae) and oligochaetes (*Henlea* sp., *Rhyacodrilus subterraneus*).

Temporal distribution patterns of fauna

Fluxes of water, organisms and organic matter entering the subterranean environment show both quantitative and qualitative temporal changes. Therefore, in spite of the ecological inertia of the subterranean environment, epigean seasonal rhythms can influence the biological cycle of many stygobitic organisms.

Seasonal dynamics

Seasonal rhythms are exerted in different ways depending on the kind of organisms and location within the karst system. They have a deep influence on threshold communities living in the entrances to caves or pot-holes. Indeed, strong seasonal changes in the entrance communities of several caves of the southern Jura – including a five-fold increase in total abundance from winter to summer – were reported by Turquin and Bouvet (1977). High abundance of the summer community is due to the input of trogloxene macrofauna, an increase in the number of troglophiles (e.g. the dipterans *Phora* and *Sciara*), and the arrival of troglobites such as the coleopteran *Royerella villardi* or the Dipluran *Plusiocampa sollaudi*.

The biological cycle of subtroglophilic Trichoptera (six species of *Stenophylax*, five species of *Micropterna* and two species of *Mesophylax*) which constitute a significant component of cave-entrance communities was studied in detail by Bouvet (1977, 1994). The aquatic phase, lasting from October to May in surface streams, includes five larval stages and two metamorphoses (nymphose and imago moults). In May, adults of golden caddis flies (*Stenophylax*) migrate from streams to cave entrances (Fig. 27.4). They spend the summer (in diapause) clinging to the walls and roof of caves, where they can sometimes be found 500 m from the cave entrance. This diapause in the subterranean environment, which provides favourable conditions with little changes in humidity and temperature, is required for the maturation of ovaries. A similar cycle is observed for some Lepidoptera (e.g. *Scoliopteryx, Triphoea*), but their underground diapause occurs during winter (Bouvet et al., 1974).

Input of terrestrial invertebrates into streams and pools of caves also shows a seasonal trend. In the pools of the Grotte d'Hautecourt, this "manna" (i.e. food supply to aquatic fauna) is two or three times higher (in number of individuals) in winter and spring than in summer and autumn (Turquin, 1975a). In contrast, the transit of terrestrial fauna in the main stream of the epikarstic zone of the Dorvan karst is higher during the summer period (Gibert, 1986).

Most populations of troglobites and stygobites show a continuous reproductive potential. For example, breeding females of *Niphargus* are found all the year in karst systems of the Rhône basin. However, a laying rhythm can be observed, which is particular to each

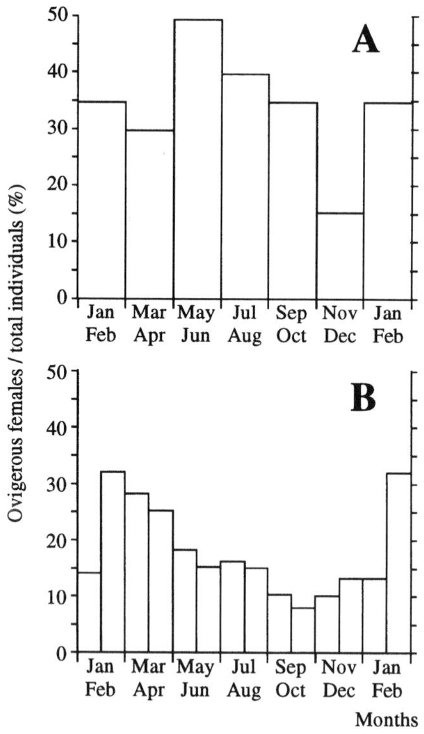

Fig. 27.5. Seasonal changes in the percentage of ovigerous females of: (A) *Niphargus virei* living in the flowing-water zone of a karst; (B) *N. rhenorhodanensis* living in pools of caves. (Modified after Ginet, 1960 and Turquin, 1975b).

species and also is dependent on habitat constraints. Karstic populations of *Niphargus rhenorhodanensis* and *N. virei* that are exposed to fluctuating hydrologic fluxes (floods) show maximum egg-laying in summer and a decrease in autumn (Fig. 27.5A). In contrast, populations of *Niphargus rhenorhodanensis* and *Asellus cavaticus* living in pools, fossil caves or interstitial aquifers have a higher number of ovigerous females in late winter or mid spring (Ginet, 1960; Henry, 1965; Turquin, 1975b) (Fig. 27.5B). For all these populations continuation of an internal rhythm is unlikely, but the existence of a seasonal trigger such as the "manna" or "flood factor" may act very rapidly between two reproduction phases (Mathieu and Turquin, 1992).

Effects of floods

Responses of aquatic communities to floods have been studied in detail in several karst ecosystems of the Rhône basin: the Drom system (Turquin, 1986), the Dorvan system (Gibert, 1986), the Foussoubie system in the Ardèche (Vervier, 1988) and the Lez system (Malard, 1995). The extent to which a flood affects karst communities depends on intrinsic characteristics of the flood (e.g. intensity, duration, spatial extent), specific features of the karst system (e.g. poorly or well karstified system, openness of the system with regard to epigean biological fluxes), the hydrological stage of the system (e.g. high or low water periods), and the type of organism (e.g. planktonic or benthic forms). Because generalization is difficult, the results presented in this section mainly concern the case of a flood occurring after a relatively long period of low water in a well-karstified and open karst system. Within the framework of a systems approach, we will consider the inputs and outputs of organisms as well as fluxes of organisms within karst ecosystem (cf. Fig. 27.2).

- **Habitat connectivity during low-water period.** During low-water periods, subsurface water bodies undergo fragmentation and spatial reduction. The absence of surface-water inputs results in a poor connection between water bodies of the unsaturated zone (cave water or epikarstic aquifer) and groundwater. As groundwater drainage proceeds, hydraulic connections between regions of the saturated zone also become weaker. Such a low connectivity is partly responsible for the spatial heterogeneity observed in karst communities; as biological communities become isolated they respond increasingly to locally unique ecological factors.
- **Input of epigean organisms.** Colonization of karstic underground water is achieved by epigean forms in several places (springs, beds of perched, effluent or influent streams, and sinkholes) during low-water and high-water periods. However, highest input of epigean organisms into karst aquifers probably occurs during flood events through the sinking of surface streams. There is a massive migration of a highly diverse epigean fauna at point inlets that are often directly connected to the drainage network. For example, drifting fauna at the main outlet of the Foussoubie karst system, which is partially fed by surface water of a sinking stream, include many insect larvae belonging to Plecoptera, Ephemeroptera and Diptera (Vervier and Gibert, 1991). Décamps and Laville (1975) studied drifting fauna in the intermittent Lachein stream that sinks into the Baget karst at a point called the "Perte de la Peyrère". During a flood, they estimated that 18 000 to 25 000 epigean invertebrates (Oligochaeta, Ephemeroptera, Plecoptera, Coleoptera, Diptera and Trichoptera) were introduced into the aquifer every hour. For the same sinkhole, Rouch (1979) estimated that 7 to 14 million epigean harpacticoids enter

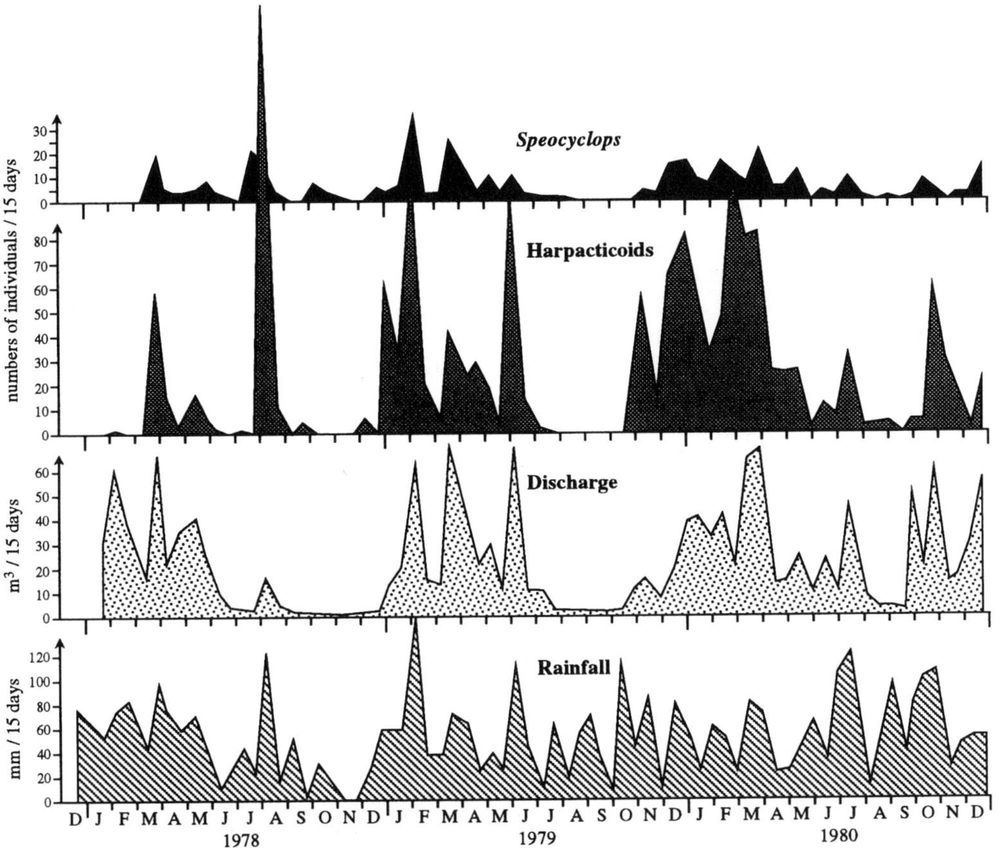

Fig. 27.6. Drift rate of cyclopoids (*Speocyclops* cf. *proserpinae*) and harpacticoids (*Limocamptus echinatus, Ceuthonectes serbicus* and *Elaphoidella cavatica*) in water percolating in the epikarstic zone of the Dorvan karst system (Cormoran Cave) (modified after Gibert, 1986).

the aquifer each year. Most epigean invertebrates do not pass from sinkholes to springs; rather, they are trapped within the aquifer [90% for the epigean invertebrates studied by Décamps and Laville (1975) and 70% for the epigean harpacticoids studied by Rouch (1982)]. However, they do not survive in the aquifer, and probably constitute a significant source of food for the autochthonous fauna. Indeed, Rouch and Carlier (1985) showed that, on a yearly basis, the bulk of the drifting epigean harpacticoids entering the system does not modify the structure of the epigean harpacticoid community living permanently in the aquifer.

– **Fluxes of organisms within the karst aquifer.** Increased subsurface water flow primarily leads to a reduction of terrestrial habitats. On account of percolation, subvertical or subhorizontal runoff, and flooding by rising of groundwater tables, many terrestrial organisms are drowned. Turquin (1975a) showed that terrestrial invertebrates falling into water pools of the Hautecourt cave constituted a major source of food for *Niphargus* (Amphipoda). Energetic input due only to drowning of Diptera was estimated at 125.10^4 J per year for $90\,m^2$ of pools (Turquin, 1981a). At the scale of the aquifer, it is likely that terrestrial fauna drifting into subterranean habitats during a flood turn into particulate and dissolved organic matter, and may subsequently be used by collectors and silt-feeding forms.

Greater hydrologic flux also results in greater biological exchange between disconnected subsurface water bodies. Dispersal of organisms may occur along three distinct pathways: downward, upward and lateral.

– **Downward dispersal** is associated with subvertical runoff and percolation through the unsaturated zone. Study of drifting aquatic fauna in percolating water of the Dorvan system (epikarstic zone) showed (Fig. 27.6) an increase in the drift rate of cyclopoids (*Speocyclops* cf. *proserpinae*) and harpacticoids (*Ceuthonectes serbicus, Elaphoidella cavatica* and *Limocamptus echinatus*) with increasing discharge rate (Gibert, 1986). Drifting fauna also included nematodes, oligochaetes and amphipods, but drift density usually remained very low (4.4 invertebrates m^{-3}). Although drift in percolating water may seem insignificant from a quantitative point of view, this downward dispersal allows renewal of pool communities in caves lacking substantial water circulation (Delay, 1968).

- **Upward dispersal** occurs when there are spectacular fluctuations in the groundwater table (tens of metres) during floods. Abrupt raising of the groundwater table followed by a rapid decline frequently occurs in well-karstified systems. Such events are likely to promote the dispersal of invertebrates in the unsaturated zone. Numerous individuals of *Niphargus virei* (Amphipoda) are found in pools of the Drom system (southern Jura) following floods. Turquin (1981b) suggested that offspring of *N. virei,* whose teguments are hydrophobic, could be dispersed by floating. She interpreted the summer period of reproduction as a strategy for dispersal of juveniles during groundwater recharge induced by autumn floods. Figure 27.7 shows the size–frequency histograms of *N. virei* for three samples obtained by filtering water of an intermittent artesian spring (Gueux system, Côte-d'Or, France) (Malard et al., 1997). Filtering was carried out while the spring flowed for the first time in autumn. The samples were collected at the beginning of the flood from the outlet of a small fracture whose maximum discharge did not exceed 3.5 ℓ s^{-1}. In three samples, the small size classes (i.e. with body length ranging from 0.2 to 0.8 cm) represented the major component of the total number of individuals (52, 71 and 65% for the first, second and third samples, respectively).
- **Lateral dispersal** of organisms occurs during the reversal of groundwater flow between the transmissive and capacitive parts of the saturated zone. This kind of dispersal has been clearly illustrated during a two-year sampling survey carried out in several wells of an experimental site of 500 m^2 in the Lez system (Malard et al., 1994b, 1996a). In wells intersecting the thinly fissured parts of the saturated zone, high abundances of epigean and hypogean invertebrates were observed after the floods. At the scale of the karst system, a similar process may

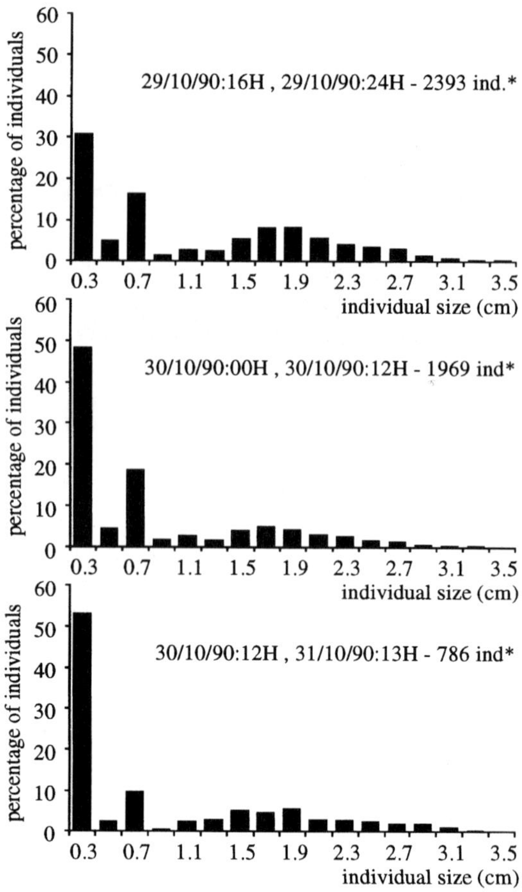

* sampling period and number of individuals (ind.)

Fig. 27.7. Length frequency distribution of *Niphargus virei* (hypogean amphipod) in three consecutive samples collected during the first autumn flood of an intermittent artesian spring (Gueux karst system, Côte-d'Or, France): (A) October 29, 16.00 to midnight, 2393 individuals; (B) October 30, midnight to noon, 1969 individuals; (C) October 30, 12.00 to 13.00 hrs, 786 individuals.

operate between the drainage network and annex systems. However, increases in biological exchange between regions of the saturated zone do not imply that aquatic communities are spatially homogeneous. For example, hypogean harpacticoid communities in mesohabitats of the Baget karst had distinctive characteristics which remained stable through time (Rouch and Bonnet, 1976a, 1977). Ecological factors specific to each mesohabitat probably act as a stabilizing force that maintained a stable community structure.

In most cases, dispersal of organisms should not be considered as the result of a constraint induced by current. Forced dispersal is limited to

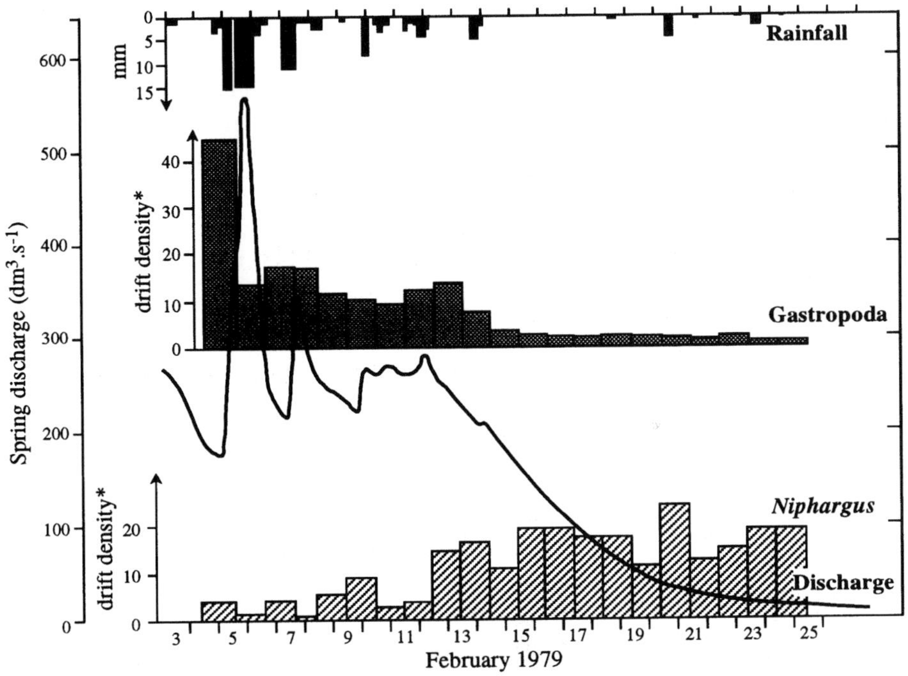

Fig. 27.8. Drift density (individuals per 1000 m^3) of gastropods (*Bythiospeum diaphanum*, *Carychium minimum* and *Hauffenia minuta*) and *Niphargus rhenorhodanensis* (hypogean Amphipoda) at the main outlet of the Dorvan karst system (modified after Gibert and Laurent, 1982).

zones with high flow velocities. Animals drifting in slowly percolating water are not swept away by the current, and the upward drift of *Niphargus virei* at Gueux spring is likely to be behavioral. Indeed, Ginet (1960) showed that the mean dislodgement velocity of *N. virei* was as high as $0.5\,\mathrm{m\,s^{-1}}$. At the Pissoir spring (Dorvan system), drift of *N. rhenorhodanensis* during low flow periods showed a nycthemeral rhythm (higher drift density during night) (Gibert, 1986). Dispersal of organisms during floods is primarily a passive process (i.e. animals do not use their locomotory organs), which probably corresponds to an explorative strategy.

– **Output of organisms.** Animal drift at the outlet of karst systems has been well studied by several authors (Rouch, 1970; Gibert and Laurent, 1982; Turquin, 1986; Vervier and Gibert, 1991). Drift involves epigean and hypogean invertebrates, but drift patterns vary among taxa. Gibert and Laurent (1982) studied the drift of gastropods (*Bythiospeum diaphanum, Carychium minimum* and *Hauffenia minuta*) and *Niphargus rhenorhodanensis* at the main outlet of the Dorvan karst system (Fig. 27.8). Drift density was positively correlated with discharge for gastropods ($r=0.82$; $p<0.01$) but negatively correlated for *Niphargus* ($r=-0.78$; $p<0.01$). Drift was "catastrophic" for gastropods but behavioural for *Niphargus*. Whatever the mode of drift, karst ecosystems lose a large number of organisms during floods. Rouch (1986) estimated that drift caused an annual loss of several million individuals for the harpacticoid community of the Baget system. A long-term study (from 1973 to 1979) of the drifting population of *N. virei* in the Drom system showed that excessive drift due to a long series of floods caused the near extinction of the population in 1977 (Turquin, 1981c, 1984, 1986; Turquin and Barthélémy, 1985). However, in 1979 the population was expanding (juveniles less than one-year old represented 70% of the population) as a result either of increased reproduction rates or of the arrival of juveniles from slow-moving water zones of the aquifer. This example illustrates that passive dispersal could be a costly strategy for *N. virei.* Turquin (1981c) pointed out that egg production in *N. virei* (60 eggs per clutch) could compensate for loss of individuals during drift. However, she also suggested that such a compensatory mechanism may not be sufficient in karst aquifers having rapid vertical circulation; *N. virei* is usually absent from such systems (karst aquifers of Chartreuse and Vercors).

Aquatic karst communities isolated in different habitats with few interconnections during low-water periods are likely to be in a climax stage (see for example Rouch and Bonnet, 1976b, p. 40). During floods, increases in biological flux between these habitats modify, at least temporarily, the community structure characterizing this climatic stage. In the short term, therefore, floods maintain karst communities in a state of dynamic equilibrium. Finally, although the effects of floods in karst systems may seem comparable with those observed in river flood plains (Ward and Stanford, 1995), there exists at least one major difference: floods do not modify the structural characteristics of the karstic milieu.

ECOTONES BETWEEN SURFACE AND SUBTERRANEAN SYSTEMS

The definition of exchanges or interactions between ecological systems that can influence nutrient cycling and community dynamics is important in understanding the functioning of these systems (Holland, 1988; Naiman and Décamps, 1990; Hansen and Di Castri, 1992). Ecotones also are known to affect energy flow between adjacent systems, to act upon landscape processes, and to improve landscape connectivity. Karst processes linked to limestone fissuring, dissolution activity, and flows along fractures and bedding-plane partings create a range of contact zones and modules of openness with the epigean world, ranging from large cave entrances and shafts to tiny fissures. These ecotones depend on speleogenesis, the nature of the overlying soil and the extent of passage obliteration by rainfall and weathering products (Turquin and Bouvet, 1977). They constitute spatially fixed boundaries on a short time scale, which separate periodically lighted and continuously dark environments. Change in the matrix also implies that the underground compartment can act as a filter for abiotic and biotic fluxes. The diversity, abundance and spatio-temporal distribution of ecotonal communities depend upon the size and distribution of penetrable spaces, organic-matter transport, and mobility of invertebrates.

In the Rhône basin, surveys of terrestrial ecotones such as cave or pot-hole entrances, were carried out as early as the 1970s, but interaction zones between surface and karstic groundwater systems such as springs or sinks were investigated only during the last 15 years (Gibert et al., 1990; Vervier et al., 1992). The nature and effect of exchanges between the surface and underground environments differ markedly in these two types of ecotones, and thus are presented separately.

Terrestrial ecotones

There is a distinct gradient of abiotic factors – mainly light, temperature and humidity – from cave or pot-hole entrances to deep karstic zones (Fig. 27.9). Three different communities are distributed along this gradient, namely, an epigean community in the vicinity of the entrance, a hypogean community in the deep zone of the karst, and a transitional or ecotonal community (entrance community with parietal fauna). The latter, usually having a higher species richness and evenness than the adjoining communities (Culver and Poulson, 1970; Turquin and Bouvet, 1977), has been studied in detail in several caves and pot-holes of the Jura and Pre-Alps: the Tunnel de Drom (Bouvet et al., 1972), the Dorvan system (Gibert et al., 1975), the Grotte de la Scierie (Bourne, 1976), the Grotte de la Bouna (Bourne, 1977), the Gouffre d'Antona (Bouvet and Turquin, 1976), the Puits de Gros Chêne (Turquin and Bouvet, 1977), and the Gouffre de Lent (Gibert et al., 1981). In the entrance zones of horizontal and vertical caves, organic matter is transported by different vectors. In the former, organic input is essentially due to the active penetration of epigean organisms (biological vector), whereas in the latter food supply depends on the fall of plant debris and animals (gravity vector) and transport by surface runoff of living and organic matter (water vector). Pot-holes can thus be considered as "ecological traps" because they have a higher ratio of trogloxenes to troglophiles plus troglobites (e.g. 1.43 for the Gouffre d'Antona) than in horizontal caves (e.g. 0.84 for the Tunnel de Drom) (Turquin and Bouvet, 1977).

The parietal fauna is a fluctuating set of species that colonize the intermediate zone, sometimes 200 to 500 m from the cave entrance. It is composed of a variety of trogloxenes, subtroglophiles, troglophiles and troglobites. Representatives of the subtroglophilic fauna, mainly belonging to the Arachnida, Coleoptera, Diptera, Lepidoptera, Trichoptera and Hymenoptera, show a patchy distribution that is closely linked to the morphologic and microclimatic factors of the cave entrance. Indeed, they usually do not feed in the cave

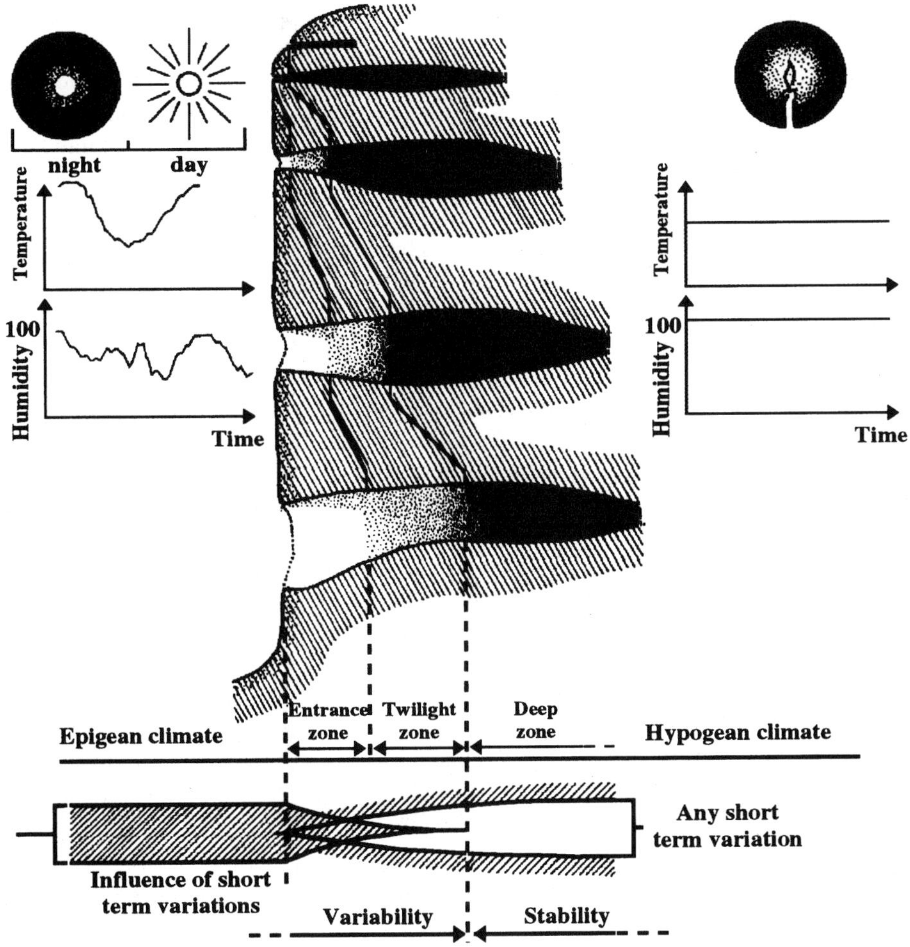

Fig. 27.9. Topographic variations of contact zones between surface and underground environments in relation to the size of karst openings, i.e., from a single fissure to a large cave entrance (modified after Ginet and Decu, 1977).

and benefit from the environmental conditions of various subterranean microhabitats during a part of their biological cycle (diapause). Therefore, the composition of the subtroglophilic fauna exhibits a well-defined seasonal periodicity, winter diapause species such as the butterfly *Triphosa dubitata* and the harvestman *Amelinus aurantiacus*, and summer diapause species such as the dipteran *Limonia nubeculosa* and Trichoptera of the *Stenophylax* group being alternately dominant. The intermediate zone of caves also constitutes a temporary refuge for many trogloxenes and troglophiles during unfavourable climatic periods. Moreover, some troglophilic and troglobitic predators such as the spider *Meta menardi* or the beetle *Quedius mesomelinus* can be particularly abundant in the ecotonal zone inhabited by easy prey. Spatial limits of ecotonal communities (i.e. elasticity) also change through time. In the Lent shaft, during summer there is a downward movement of purely trogloxene species (e.g. Coleoptera and Hymenoptera) and episodic upward movements of troglophiles (e.g. the dipteran *Triphleba aptina*) and troglobites (e.g. the coleopteran *Trichaphaenops cerdonicus* and the dipluran *Plusiocampa sollaudi*).

Aquatic ecotones

Water is the component which is most frequently exchanged between surface and underground environments. Contact zones between these two adjacent patches are diverse, ranging from a network of partly clogged fissures in the epikarstic zone to large openings. For example, karstic sinks and springs are highly permeable ecotones with a large-meshed boundary. The rates and directions of fluxes of matter and energy can change dramatically in a sinkhole, which may successively drain a surface stream, function as a spring, and finally dry up. In such contact zones, ecological contrast, amplitude of change of biological and physical variables, and stochasticity of time-series patterns, are the highest within the geohydraulic continuum. Flow characteristics (e.g. duration, frequency and timing of floods), strength of surface-water and groundwater interactions, and retention times, influence the composition and spatio-temporal distribution of fauna. The biotic interactions that result in the occurrence of a mixed community of epigean and hypogean taxa can therefore be used to delineate the limits of the groundwater/surface-water ecotone and identify its functional features (Gibert et al., 1990).

Results obtained in the Foussoubie karstic system (Ardèche) showed that the location and extent of the interaction zone between surface water and groundwater vary annually depending on the hydrological stage (Vervier and Gibert, 1991). During high-water periods, the relatively low ratio of bicarbonate to calcium and the high relative abundance of epigean insects demonstrate that abiotic and biotic interactions between the sinking Ruisseau de la Planche and the aquifer extend to the lower part of the massif down to the Source de Foussoubie. In this particular system, the interaction zone extended 5 km downstream from the physical boundary, showing that aquatic ecotones in karst systems may exhibit a high degree of elasticity. Biotic and abiotic interactions decreased progressively during flood recessions and were no longer detectable at the Source de Foussoubie during low-water periods.

At different time scales, karstic springs are seen as dispersion sites for hypogean and epigean species. During flood events, drifting hypogean taxa such as the amphipod *Niphargus rhenorhodanensis* may colonize the interstitial habitats of spring brooks (Mathieu and Essafi, 1991), whereas during low-water periods benthic forms such as *Gammarus fossarum* (Amphipoda) migrate upstream to some extent and settle in the karst system (Gibert, 1986). On a geological timescale, karstic springs can be seen as the convergence point of several colonization routes. This "cross-road effect" results in the occurrence of diverse communities composed of taxa belonging to different biogeographical areas. For example, oligochaete communities collected in several karstic spring brooks located at the interface between the Jura karsts and the Rhône floodplain included a wide range of species known from: (1) karstic Jura caves and springs (e.g. *Rhyacodrilus balmensis*); (2) the hyporheic zone of the river Rhône (e.g. *Rhyacodrilus amphigenus*); and (3) European groundwater (e.g. *Rhyacodrilus falciformis, Trichodrilus diversisetosus* and *T. strandi*) (Chafiq et al., 1992, Chafiq and Gibert, 1993, 1996). The importance of karstic springs as preferential routes of colonization of karst aquifers was also emphasized by Culver et al. (1995) for the freshwater amphipod *Gammarus minus* and by Danielopol and Bonaduce (1990) for marine ostracods of the subfamily Pseudolimnocytherinae.

Several researchers also have studied the movement of nutrients and organic matter, and the dynamics of epigean and hypogean communities or populations, in various spring ecotones of the basin (Vervier et al., 1986; Mathieu and Essafi, 1991; Chafiq et al., 1992; Essafi et al., 1992; Mathieu et al., 1992; Plénet et al., 1992; Chafiq and Gibert, 1993, 1996; Plénet and Gibert, 1995; Malard et al., 1997). For example, changes in the abundance and structure of interstitial populations of *N. rhenorhodanensis* and *Proasellus valdensis* (Isopoda) in karst outflows were related to the flow regime and geomorphologic characteristics of the springs. At the Verna spring in the southern Jura, the subterranean flow occurs at the base level of the massif, and thin layers of sediments provide little space for colonization. The interstitial hypogean populations, consisting of a few individuals only, are severely affected by the spate, and have a long recovery time (about seven months). In contrast, at overflow sites (e.g. the Pissoir spring) where alluvial deposits provide potential refuge, populations are abundant and recover rapidly after spates (Mathieu and Essafi, 1991; Essafi et al., 1992; Mathieu et al., 1992).

The edge effect that results in a higher species richness in the interaction zone is rarely observed for surface-water/groundwater ecotones. At "Le Château spring" (Ile Crémieu, southern Jura) the taxonomic richness was higher on the surface than underground (Plénet and Gibert, 1995). A faunal gradient was observed between the two adjacent systems, and three types of taxa were distinguished: strictly hypogean taxa restricted to the subterranean area, ubiquitous epigean and hypogean taxa distributed in both systems, and strictly epigean taxa collected only in surface water (Table 27.3). In another Jura spring (Source Bleue), the greatest biodiversity also was observed in the surface environment (13 species). Groundwater

Table 27.3
Spatial distribution of epigean and hypogean taxa in a karstic spring in the Jura[1]

Taxa	Distance from physical boundary (m)									
	Underground environment					Surface environment				
	−25	−15	−10	−5	−1	1	5	10	15	25
Proasellus valdensis[2]		+								
Proasellus* cf. *valdensis[2]				+						
Haplophthalmus sp.				+						
Niphargus rhenorhodanensis[2]	+		+				+			
Cyclocypris ovum	+									+
Diptera (Nymphe)			+			+			+	+
Diptera (Larvae)	+			+	+	+	+	+	+	+
Nemoura sp.					+		+	+	+	+
Nematoda			+				+	+	+	+
Hydracarina					+				+	+
Oligochaeta	+	+	+	+	+	+	+	+	+	+
Gastropoda						+		+	+	+
Hirudinea						+				+
Gammarus fossarum							+	+	+	+
Synagapetus dubitans (Larvae)							+	+	+	+
Elmis aenea (Larvae)							+	+	+	+
Elmis aenea (Adult)							+	+	+	+
Limnephilidae (Trichoptera)							+			+
Stenophylax (*lateralis* or *sequax*)							+	+	+	
Bivalvia								+	+	+
Plectronemia conspersa								+	+	+
Eucyclops serrulatus								+	+	+
Gammarus pulex								+	+	
Porcellio spinicornis									+	
Trichoniscus sp.									+	
Lype reducta									+	
Cyclopoïda									+	+
Synagapetus dubitans (Nymph)									+	+
Paracyclops fimbriatus										+
Harpacticoïda										+

[1] Data on presence/absence were collected during five sampling visits to the Source du Chateau (île Crémieu). After Plénet and Gibert (1995).
[2] Hypogean taxa are in bold type. The plus (+) sign indicates that a taxon was present at that point.

contained only two or three stygobitic species, and the transition zone in the vicinity of the physical boundary had an intermediate diversity ranging from five to nine species (Vervier et al., 1986). However, this intermediate biodiversity is not always found for all systematic groups and all springs (Gibert et al., 1990). In fact, community dynamics are a function of correspondence between environmental conditions and the ecological status of the organisms (i.e. the life-history traits), and reflect the collective response to a set of conditions occurring in the space and time window of interest.

HUMAN ALTERATIONS OF KARST COMMUNITIES

Many human activities are likely to affect aquatic and terrestrial communities of karst ecosystems, but few data are available to show the nature and extent of such effects. Among the various anthropogenic inputs, subsurface water contamination has become an environmental problem of increasing magnitude, and is probably the primary threat to karst communities. Three categories of contaminants are usually distinguished (Notenboom et al., 1994): toxic compounds, nitrogen and phosphorus compounds (nutrients for microorganisms), and organic matter used by heterotrophic organisms. Up to now, most studies have concerned only the sensitivity of karst communities to organic pollution (Holsinger, 1966; Branstetter, 1975; Culver et al., 1992; Iliffe et al., 1984). Further, because of the inaccessibility of the subterranean karst environment, investigations were limited mainly to caves. However, methodological approaches have been developed with various success in karst systems in the Rhône basin. In the Jura and Ain regions, Barthélémy (1984) attempted to demonstrate a relationship between the composition of drifting communities and the level of organic contamination (i.e. ammonium, chloride, nitrate and phosphate concentrations) for 16 springs. Several taxa were distributed along temperature, salinity or oxygen gradients but no biological indicator of pollution could be identified (Carrel et al., 1986). Richert (1990) sampled the benthos at the same springs and calculated a biological index for each spring brook, but failed to detect a significant correlation between the index and the level of groundwater contamination. Because of contaminant dilution within the aquifer, the biological impact of pollution is difficult to detect at the outlets of karst systems. Finally, sampling has been completed in deep wells intersecting contaminated regions of karst aquifers (Malard et al., 1994a). This last method has encouraging results that are described in the first example below. A second example illustrates the adverse effects of organic contamination of infiltration water on subterranean terrestrial fauna.

Example 1: Organic contamination and changes in groundwater communities of the Lez spring basin. The Lez spring aquifer, extending over 200 km^2 and belonging mainly to the Middle–Upper Jurassic and Lower Cretaceous formations, lies between the Hérault and Vidourle rivers in southern France (Fig. 27.10). Much of the groundwater flowing through this aquifer drains toward the Lez spring. This spring is pumped at a discharge rate of approximately 1300 $\ell\ s^{-1}$ to provide drinking water for the city of Montpellier. From 1991 to 1993, groundwater invertebrates were sampled in eight deep observation wells (depth ranging from 46 to 90 m) at three experimental sites. Groundwater quality at one site was severely impacted by the infiltration of secondary effluent from a malfunctioning sewage treatment plant. For example, during a period of high effluent discharge, the conductivity of the groundwater was 1246 $\mu S\ cm^{-1}$, chloride concentration was 132.5 mg ℓ^{-1} and the faecal coliform value was 35 000 CFU per 100 ml. Full details on the hydrogeological, physico-chemical and bacteriological characteristics of these sites have been given by Drogue (1980), Malard et al. (1994c) and Malard and Chapuis (1995).

Major differences were observed between the composition of the invertebrate communities of the pristine and impacted sites (Fig. 27.9). An increase in species richness at the sewage-polluted site was due essentially to a higher number of oligochaete and insect species. Indeed, 83 taxa – 36 crustaceans, 27 oligochaetes, 16 insects and 4 molluscs – were collected at this site, whereas the 53 taxa identified at the pristine sites consisted of 36 crustaceans, 12 oligochaetes, 2 insects, and 3 molluscs. Stygobite richness was depressed at the sewage-polluted site (15 stygobites and 55 stygoxenes), while as many as 32 stygobites and only 13 stygoxenes were collected at the pristine sites. Invertebrate populations were dominated (66%) by two ubiquitous species (the cyclopoid *Paracyclops fimbriatus* and the harpacticoid *Attheyella crassa*) and two polysaprobes (the cyclopoid *Macrocyclops albidus* and the oligochaete *Tubifex tubifex*). In addition, the groundwater fauna exhibited a patchy distribution that was related to the heterogeneous movement of sewage-polluted water in the aquifer (Malard et al., 1996a). For example, well W8 had the highest relative abundance of oligochaetes (mostly *Tubifex tubifex*), because it intersected an enlarged bedding joint that carried much of the sewage-polluted water.

Example 2: Organic contamination and changes in terrestrial communities of the Grotte de Rappe. Because input of organic matter into the unsaturated zone is flow-dependent, changes in the organic content of infiltrating waters are likely to affect both the aquatic fauna and terrestrial communities. Results of long-term studies for several caves of the Jura mountains clearly support this idea; we present those of the Grotte de Rappe here (Turquin, 1973; Malard et al., 1994d; Turquin and Crague, 1994). This cave is part of a low mountain karst area of Jurassic limestone; it is 2500 m long, and extends in a SW–NE direction (Fig. 27.11).

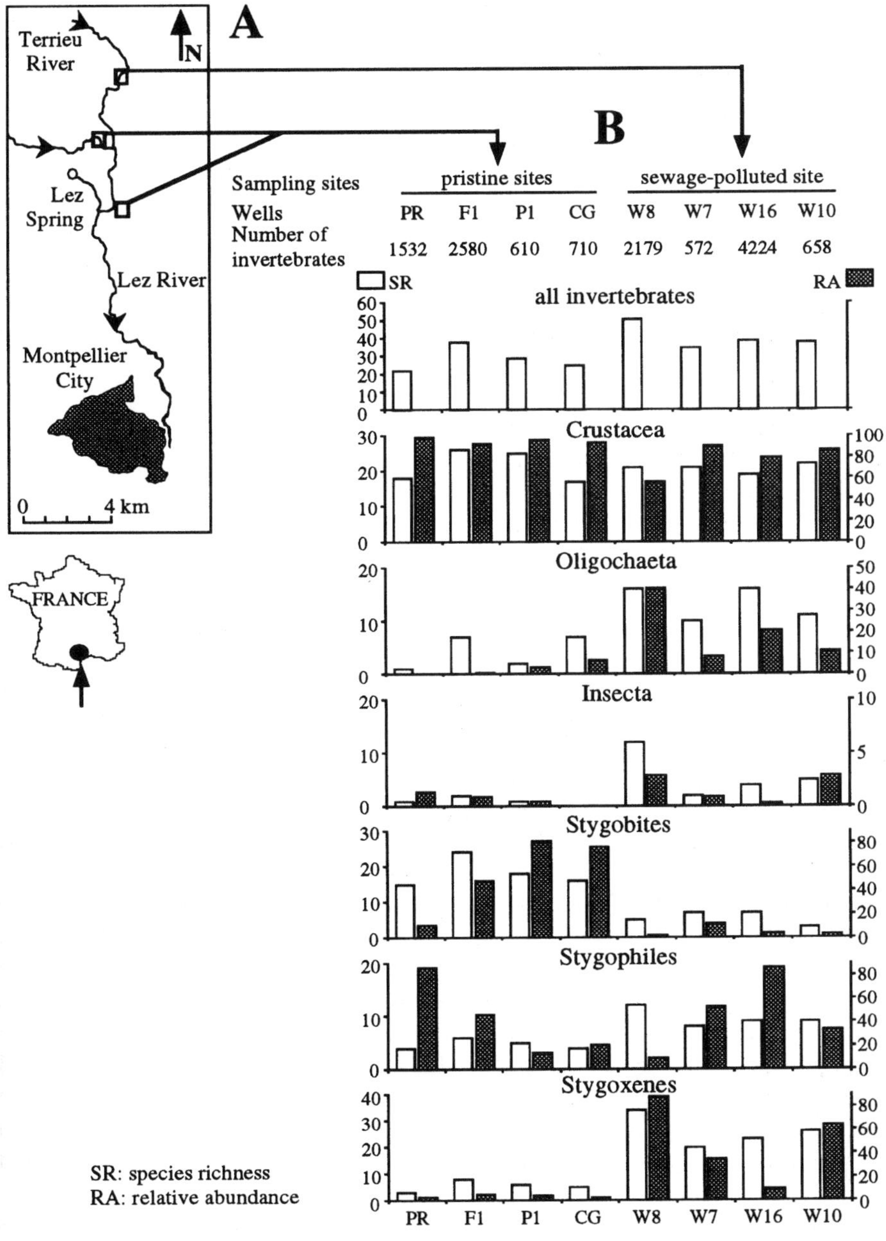

Fig. 27.10. Differences in composition and structure of invertebrate assemblages occurring in pristine and sewage-polluted sites of the Lez karst system: (A) location of the three limestone study sites; (B) total number of invertebrates collected from 1991 to 1993 in 8 wells, species richness, and relative abundance (ratio of the number of individuals of a given group to the total number of individuals of all groups) of crustaceans, oligochaetes, molluscs, and insects, as well as those of stygobites, stygophiles, and stygoxenes. (Modified after Malard et al., 1996b.)

It comprises three superposed galleries, the lowest one reaching the groundwater table at a depth of about 40 m below the ground surface. Extensive sampling of the terrestrial cave fauna was carried out in 1970 and 1992 by means of attractant traps placed in different galleries of the cave. During the 22 years that separated these two sampling surveys, human activity on the land surface had increased considerably. In 1992, pollution sources included septic tanks, a landfill, a timber mill, and a highway department salt storage facility. Because

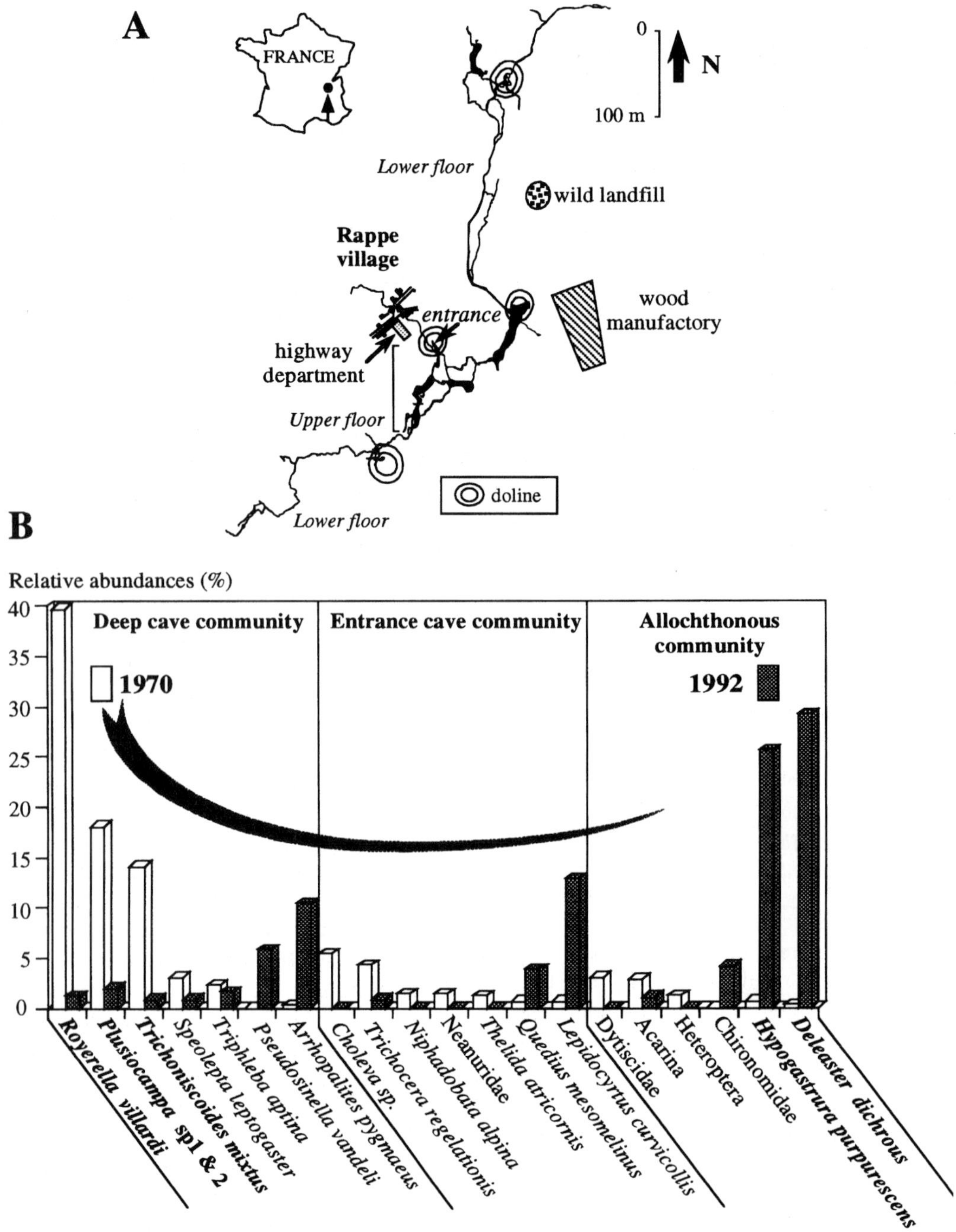

Fig. 27.11. Long-term changes in the structure of the terrestrial communities of the Rappe Cave (Ain, France): (A) map of the cave showing locations of the major sources of pollution; (B) relative abundances of the taxa collected in 1970 and 1992 (modified after Malard et al., 1994d).

there was no sewage treatment plant, raw sewage was poured into the ground and reached the cave.

Concentrations of dissolved constituents in percolating water indicated multiple sources of contamination, including domestic effluent (250 faecal coliforms per 100 ml), salt leachate (409.4 mg ℓ^{-1} chloride) and landfill leachate [150.5 mg ℓ^{-1} dissolved organic carbon (DOC)]. Changes in the quality of the infiltrating water resulted in the loss of native cave fauna and their replacement by trogloxenous taxa (Fig. 27.11). In 1970, terrestrial assemblages were dominated by four species belonging to the deep-cave community [*Trichoniscoides mixtus* (Isopoda); *Plusiocampa* sp. 1 and *Plusiocampa* sp. 2 (Diplura); *Royerella villardi* (Coleoptera)]. These four species had low relative abundance in 1992, whereas two allochthonous species (the springtail *Hypogastrura purpurescens* and the opportunist epigean staphylinid beetle *Deleaster dichrous*) constituted 55% of the total assemblage in the cave. Contamination of vadose water not only depressed the hypogean component of the cave, but also induced structural changes in the three different kinds of assemblages that comprised the cave community. For example, all species that were common in the deep-cave, entrance and allochthonous communities in 1970 were poorly represented in 1992.

These two case studies show that karst communities are highly sensitive to organic pollution. Locally, high concentrations of contaminants in the unsaturated and saturated zones occur as a result of rapid downward movement of organically enriched water through large fractures, with little attenuation. Oxidation of organic matter induces greater oxygen consumption, which rapidly exceeds the replenishment of oxygen by advection. In the worst case, anoxic conditions prevail continuously and the groundwater fauna disappears. Although oxygen concentrations may remain above critical levels, an increase in food resources in ground water (particulate organic matter, or the number of primary consumers) promotes the colonization of the subterranean environment by epigean species of detritus feeders and carnivores. This could be a direct relation to detritus feeders or an indirect consequence of the increase in the number of primary consumers for carnivorous species. Because food webs are less complex than in surface environments, a change at only one trophic level can rapidly affect the whole community. Moreover, stygobites also may be unable to compete with ubiquitous invaders. Finally, many authors also suggest that the resilience of karstic subterranean ecosystems to contamination events is low (Hobbs, 1988; Culver et al., 1992; Tercafs, 1992; Malard et al., 1996a). Even when the source of organic pollution is eliminated, subsequent elution during rainy periods of contaminants stored in the unsaturated zone promotes the potential for residual contamination. A decrease in organic-matter availability usually results in the progressive disappearance of saprobiont species, although ubiquitous taxa still may remain dominant.

Groundwater contamination is one of several causes of alternation of karst ecosystems. Another major threat to species diversity is modification of the hydrological regime of karst aquifers resulting from increased use of groundwater resources. This is a worldwide process, whose adverse consequences to karst communities have been little studied by groundwater ecologists. For example, excessive pumping from the 200 km^2 karst system of the Lez spring near Montpellier induced a lowering of the groundwater table by about 10 m during periods of low groundwater recharge. In this karst system, excessive pumping also resulted in an artificial fragmentation of mesohabitats by promoting the hydraulic disconnection of different regions of the saturated zone (Karam, 1989). In the Edwards Aquifer (Texas), one of the earth's most diverse groundwater ecosystems, excessive pumping also leads to degradation in water quality by the encroachment of highly saline water (Longley, 1981, 1992). On a long-term basis, excessive pumping may irreversibly affect the diversity of karst communities. A high-discharge pumping test of the Baget aquifer conducted for only five days induced changes in the harpacticoid community that were still noticeable one year after the test (Rouch et al., 1993). Submergence of karst aquifers following dam construction also modifies groundwater flow regimes (Karanjac and Günay, 1980; Milanovic, 1986). Poulson (1992) observed that construction of three dams and subsequent regulation of groundwater flow in the Green River (Mammoth Cave, Kentucky) promoted siltation and decreased food availability in the Styx–Echo River area, which resulted in the decline of the cave shrimp *Palaemonias ganteri*. Changes in landscape may alter evapotranspiration rates and thus infiltration rates in karst areas, and ultimately affect karst communities by modifying moisture content and food availability in the unsaturated zone.

The special nature of karstic groundwater communities underscores the importance of maintaining their biological integrity and diversity. The subterranean fauna contains many phylogenetic relicts which are true living witnesses of very ancient lineages. Because many species show high levels of endemism, the risk of species extinction due to anthropogenic disturbances is especially high. Bearing in mind the increasing rate at which human beings are altering karst ecosystems, many groundwater species are likely to fill the shelves of the extinct species room in the

Galerie de l'Evolution in Paris if strong measures are not taken to manage underground ecosystems. Debate on whether one should strive for "nondegradation" or "differential protection" in protecting groundwater resources makes sense if one simply considers the economic role of groundwater (Durenberger and Leavitt, 1988). Groundwater ecologists should tackle the problem in another way by stressing the importance of maintaining biodiversity, and thereby ecosystem function in protecting groundwater quality (Job and Simons, 1994).

SUBTERRANEAN KARST ECOLOGY IN THE RHONE BASINS: EVALUATION AND PERSPECTIVES

Karstic regions of the Rhône basin have been studied more intensively than any other French karstic areas (with the exception, perhaps, of Pyrenean karsts). During the past 30 years, considerable effort has been spent in (1) investigating a variety of karst systems, (2) examining distribution patterns of animals across a wide range of spatial and temporal scales, and (3) maintaining a multi-disciplinary perspective (i.e. ecology, systematics, biology including physiology, genetics, hydrogeology). Partly because of this three-tiered strategy, research in karst ecology within the basin underwent two major changes. At the beginning of the 1970s, cave explorations were gradually replaced by ecosystem studies, because caves were no longer considered as the main habitat of subterranean animals. At the end of the 1980s, the narrative–inductive approach that prevailed for many years in subterranean ecology was abandoned in favour of a more hypothetico-deductive approach. Studies were then completed in the context of current concepts in general ecology, such as patch dynamics (Townsend, 1989), the habitat templets (Southwood, 1977) and ecotones (Holland, 1988).

It is especially important to maintain a broad spatial and temporal perspective, as well as a multidisciplinary approach, in karstic subterranean ecology. Future research should test the relevance of basic concepts in ecology to karstic systems. For example, karst systems and their fauna offer novel opportunities to address the effects of habitat fragmentation and decreased ecological connectivity on biodiversity. They also constitute suitable material for developing conceptual models that could evaluate the combined effect of historical and ecological determinants on biological diversity. Indeed, the present-day diversity of troglobites and stygobites in the Rhône basin is the result of several historical factors such as marine transgressions (affecting colonization rates), deepening of valleys (affecting speciation rates), and glaciations (affecting extinction rates) and ecological determinants such as habitat heterogeneity, environmental stability and food resources (Stoch, 1995). For example, depending on the adaptability of a species, a change in sea level resulting in the fragmentation of a large system may lead to allopatric speciation (increase in species diversity) or to extinction (decrease in species diversity) associated with the lower homeostasis of smaller systems.

Abundant biological data have been collected in several French karst areas (e.g. the Champagne region, Côte-d'Or, Jura, Vercors, Chartreuse, Ardèche, Languedoc, Corbières, Pyrenees) and could be used to address various hypotheses regarding causes of subterranean species diversity. Efforts should be made to collect and analyse this available information to elucidate general trends and pose new questions that could serve as important guidelines for future studies.

Lastly, scientists must recognize the importance of transferring basic knowledge on subterranean ecology to resource managers and decision makers. Indeed, research in karstic subterranean ecology has direct ramifications relating to the conservation and protection of karst landscapes and their associated fauna, the management of groundwater contamination, and the supply of drinking water. Valuable results already have been obtained in the field of groundwater biomonitoring [e.g. the identification of biomonitors – see Notenboom et al. (1994) and Malard et al. (1996b)]. Research efforts should be encouraged in the field of groundwater pollution ecology and groundwater ecotoxicology.

ACKNOWLEDGEMENTS

Studies included in this chapter were supported by the Program for Interdisciplinary Research on the Environment (Programme Interdisciplinaire de Recherche en Environnement, CNRS, et programme MAB/UNESCO sur les Ecotones) within the Laboratoire d'Ecologie des Eaux Douces et des Grands Fleuves ESA CNRS 5023, and by grants from the ACC-SV N°7 Ministère de la Recherche et de la Technologie "Systématique et Biodiversité". We thank everyone who helped us in the field, especially D. Martin (Université de Lyon 1). Christopher T. Robinson (EAWAG/ETH, Dübendorf –

Department of Limnology) provided valuable help for editing the English manuscript.

REFERENCES

Abeille de Perrin, E., 1869. Description d'*Anophthalmus* nouveaux. *Ann. Soc. Entomol. France,* 9(4): 401–409.

Abeille de Perrin, E., 1875. Diagnoses de Coléoptères nouveaux.*Ann. Soc. Entomol. France,* 5(5): 213–216.

Action de Coopération Scientifique et Technique 65, 1995. *Karst groundwater protection. Final Report,* EUR 16547 EN, 466 pp.

Balazuc, J., 1993. Petite histoire de la biospéologie dans la région Causses-Cévennes. *Spelunca-Mémoires,* 7, 295–297.

Barthélémy, D., 1984. *Impact des Pollutions sur la Faune Stygobie Karstique: Approche Typologique sur Seize Émergences des Départements de l'Ain et du Jura,* Thesis. Université Lyon 1, 181 pp.

Bedel, L., 1878. Diagnose d'une nouvelle espèce française de Coléoptères du genre *Adelops. Ann. Soc. Entomol. France,* 8(4): 74–75.

Bedel, L., 1883. Diagnose d'une nouvelle espèce française de *Trechus* anophthalme. *Ann. Soc. Entomol. France,* 3(6): 53–54.

Bedel, L. and Simon, E., 1875. Liste générale des articulés cavernicoles de l'Europe. *J. Zool.,* 4: 69 pp.

Bourne, J.D., 1976. Notes préliminaires sur la distribution spatiale de *Meta menardi, Triphosa dubitata, Triphosa sabaudiata, Nelima aurantiaca* et *Culex pipiens* au sein d'un écosystème cavernicole. *Int. J. Speleol.,* 8(3): 253–267.

Bourne, J.D., 1977. Mise en évidence de groupements temporaires de la faune pariétale dans un tunnel artificiel en fonction de l'humidité et des mouvements d'air. *Rev. Suisse Zool.,* 84: 527–539.

Bouvet, Y., 1977. *Conditions de Vie des Trichoptères Subtroglophiles (Insectes, Limnephilidae) leurs Réactions aux Variations des Facteurs du Milieu,* Thesis, Université Lyon 1, 162 pp.

Bouvet, Y., 1994. Trichoptera. In: C. Juberthie and V. Decou (Editors), *Encyclopaedia Biospeologica,* Tome I. Société de Biospéologie, Moulis/Bucarest, pp. 327–332.

Bouvet, Y. and Turquin, M.J., 1976. Influence des modules d'ouverture du karst vers l'extérieur sur la répartition et l'abondance de son peuplement. *Bull. Mens. Soc. Linn. Lyon,* 45(7): 245–257.

Bouvet, Y. and Turquin, M.J., 1986. Les peuplements cavernicoles comparés des karsts du Vercors et du Jura: origine, causalité des différences observées. *Atti Conv. Int. Carso di Alta Montagna,* 2: 326–344.

Bouvet, Y., Turquin, M.J. and Michalon, E., 1972. Etude des biocoenoses du tunnel artificiel de Drom (Ain). *Ann. Speleol.,* 27: 563–574.

Bouvet, Y., Turquin, M.J., Bornard, C., Desvignes, S. and Notteghem, P., 1974. Quelques aspects de l'écologie et de la biologie de *Triphosa* et *Scoliopteryx* Lépidopères cavernicoles. *Ann. Speleol.,* 29(2): 229–236.

Branstetter, J., 1975. A reconnaissance investigation of groundwater pollution in a limestone terrane, Horse Cave, Kentucky. *Kentucky Caver,* 9(3): 39–41.

Camacho, A.I., 1992. Sampling the subterranean biota cave (aquatic environment). In: A.I. Camacho (Editor), *The Natural History of Biospeleology.* Museo Nacional de Ciencias Naturales, Madrid, pp. 136–168.

Carrel, G., Barthélémy, D., Auda, Y. and Chessel, D., 1986. Approche graphique de l'analyse en composantes principales normée: utilisation en hydrobiologie. *Acta Œcologica,* 7(2): 189–203.

Chafiq, M. and Gibert, J., 1993. Storage and dynamics of organic matter in different springs of small floodplain streams. *Hydrobiologia,* 251: 199–209.

Chafiq, M. and Gibert, J., 1996. Interactions between solutes, nutrients, and the biofilm in sediments of floodplain karstic and alluvial springs. *Int. Rev. ges. Hydrobiol.,* 81(3): 399–416.

Chafiq, M., Gibert, J., Marmonier, P., Dole-Olivier, M.J. and Juget, J., 1992. Spring ecotone and gradient study of interstitial fauna along two floodplain tributaries of the river Rhône, France. *Regulated Rivers,* 7(1): 103–115.

Chapman, C., 1993. *Caves and Cave Life.* The New Naturalist Library, Harper Collins Publishers, 224 pp.

Chevreux, E., 1896. Sur un Amphipode d'eau douce *Niphargus virei* nov. sp., provenant des grottes du Jura. *Bull. Mus. Hist. Nat. Paris,* 2: 136–137.

Coineau, N., 1994. Evolutionary biogeography of the Microparasellid isopod *Microcharon* (Crustacea) in the Mediterranean basin. *Hydrobiologia,* 287: 77–93.

Culver, D.C., 1982. *Cave Life: Evolution and Ecology.* Harvard University Press, Cambridge, MA, 189 pp.

Culver, D.C. and Poulson, T.L., 1970. Community boundaries: faunal diversity around a cave entrance. *Ann. Speleol.,* 25(4): 853–860.

Culver, D.C., Jones, W.K. and Holsinger, J.R., 1992. Biological and hydrological investigation of the Cedars, Lee County, Virginia, an ecologically significant and threatened karst area. In: J.A. Stanford and J.J. Simons (Editors), *Ground Water Ecology.* American Water Resource Association, Bethesda, MD, pp. 281–290.

Culver, D.C., Kane, T.C. and Fong, D.W., 1995. *Adaptation and Natural Selection in Caves. The Evolution of Gammarus minus.* Harvard University Press, Cambridge, MA, 223 pp.

Danielopol, D.L. and Bonaduce, G., 1990. The colonization of subsurface habitats by the Loxoconchidae Sars and the Psammocytheridae Klie. In: R. Whatley and C. Maybury (Editors), *Ostracoda and Global Events.* Chapman & Hall, London, pp. 437–453.

Décamps, H. and Laville, H., 1975. Invertébrés et matières organiques entrainés lors des crues à l'entrée et à la sortie du système karstique du Baget. *Ann. Limnol.,* 11(3): 287–296.

Delay, B., 1968. Données sur le peuplement de la zone de percolation temporaire. *Ann. Speleol.,* 24(3): 705–716.

Dole, M.J. and Coineau, N., 1987. L'isopode *Microcharon* (Crustacea, Isopoda) abondant dans les eaux interstitielles de l'est lyonnais. *M. reginae* n. sp. écologie et biogéographie. *Stygologia,* 3(3): 200–216.

Drogue, C., 1980. Essai d'identification d'un type de structure de magasins carbonatés, fissurés. Application à l'interprétation de certains aspects du fonctionnement hydrogéologique. *Mem. Soc. Geol. France,* 11: 101–108.

Durenberger, S.D. and Leavitt, B.R., 1988. Two views on groundwater strategy. *J. Water Pol. Control Fed.,* 60: 594–596.

Essafi, K., Mathieu, J. and Beffy, J.L., 1992. Spatial and temporal variations of *Niphargus* populations in interstitial aquatic habitat at the karst/floodplain interface. *Regulated Rivers,* 7(1): 83–92.

Geoffroy, J.J., 1993. De la caverne aux écosystèmes souterrains:

grande et petite histoire de la biospéologie. *Spelunca-Memoires,* 17: 277–284.

Gibert, J., 1986. Ecologie d'un système karstique jurassien. Hydrogéologie, dérive animale, transit de matières, dynamique de la population de *Niphargus* (Crustacé Amphipode). *Mém. Biospeol.,* 13: 1–379.

Gibert, J., 1988. Functional sub-units of an exsurgence karstic system, and exchanges with the surface environment. Reflections on the characterization of natural aquatic groundwater ecosystems. *Int. Verein. Theor. Angew. Limnol.,* 23(2): 1090–1096.

Gibert, J. and Laurent, R., 1982. L'écosystème karstique du massif de Dorvan (Torcieu, Ain, France). IV. La dérive d'invertébrés hypogés aquatiques au niveau de l'exutoire principal du massif. *Pol. Arch. Hydrobiol.,* 29(2): 471–483.

Gibert, J., Laurent, R., Mathieu, J. and Reygrobellet, J.-L., 1975. Biospéléologie du Département de l'Ain (France). Contribution à l'étude des biocénoses cavernicoles de la région de Torcieu. *Mém. Doc.,* 1: 21–46.

Gibert, J., Laurent, R., Mathieu, J. and Reygrobellet, J.-L., 1981. Ecological studies of openings into underground karst. I. The shaft wall fauna of an entrance pit (Gouffre de Lent, Ain, France) first results. In: B.F. Beck (Editor), *Proc. 8th Int. Congr. of Speleology, Bowling-Green, USA.* National Speleological Society, pp. 228–233.

Gibert, J., Dole-Olivier, M.J., Marmonier, P. and Vervier, P., 1990. Surface water–groundwater ecotones. In: R.J. Naiman and H. Décamps (Editors), *Ecology and Management of Aquatic–Terrestrial Ecotones.* Partenon Publishing, London, pp. 199–225.

Gibert, J., Danielopol, D.L. and Stanford, J.A. (Editors), 1994a. *Groundwater Ecology.* Academic Press, San Diego, CA, 571 pp.

Gibert, J., Vervier, P., Malard, F., Laurent, R. and Reygrobellet, J.-L., 1994b. Dynamics of communities and ecology of karst ecosystems: example of three karsts in eastern and southern France. In: J. Gibert, D.L. Danielopol and J.A. Stanford (Editors), *Groundwater Ecology.* Academic Press, San Diego, CA, pp. 425–450.

Ginet, R., 1953a. Faune cavernicole du Jura et des chaînes subalpines dauphinoises. I. Crustacés aquatiques. *Notes biospéol.,* 8(2): 185–198.

Ginet, R., 1953b. Contribution à la connaissance de la faune cavernicole du Jura méridional. Influence des glaciations quaternaires. In: B. Gèze (Editor), *Actes 1° Congr. Int. de Spéléologie, Paris,* Vol. 3. Imprimerie Artières, Millau, France, pp. 125–130.

Ginet, R., 1956. Faune cavernicole du Vercors et du Diois. II. Stations prospectées en 1954 et 1955. *Bull. Mens. Soc. Linn. Lyon,* 25(2/3): 57–64; 86–88.

Ginet, R., 1960. *Ecologie, Éthologie, et Biologie de Niphargus (Amphipodes, Gammaridés Hypogés),* Thesis. Université Lyon 1, 254 pp.

Ginet, R., 1961a. Crustacés Amphipodes des eaux souterraines du Var et des Alpes-Maritimes. *Spelunca-Memoires,* 1: 57–58.

Ginet, R., 1961b. Faune cavernicole du Jura Méridional et des Chaînes subalpines dauphinoises. II. Contribution à la connaissance des Invertébrés. *Ann. Speleol.,* 16(3): 303–325.

Ginet, R., 1983. Les *Niphargus* (amphipodes souterrains) de la région de Lyon (France) – observations biogéographiques, systématiques et écologiques. *Mem. Biospeol.,* 10: 179–186.

Ginet, R., 1993. Historique résumé des connaissances sur le genre *Niphargus* en France (Crustacé amphipode des eaux souterraines). *Spelunca-Memoires,* 17: 289–292.

Ginet, R. and Decu, V., 1977. *Initiation à la Biologie et à l'Écologie Souterraine.* J.-P. Delarge, Paris, 345 pp.

Ginet, R. and Genest, L.C., 1954. Faune cavernicole du Vercors et du Diois. I. Stations prospectées pendant la campagne 1953. *Bull. Soc. Linn. Lyon,* 23(2/3): 47–50; 73–80.

Ginet, R. and Juberthie, C., 1987. Le peuplement animal des karsts de France (Eléments de biogéographie souterraine pour les Invertébrés). Première partie: la faune aquatique. *Karstologia,* 10: 43–51.

Gisin, H., 1953. Collemboles cavernicoles du Jura, des Préalpes savoyardes, du Valais et du Tessin. *Stalactite,* 2: 11–24.

Hansen, A.J. and Di Castri, F., 1992. Landscape boundaries. Consequences for biotic diversity and ecological flows. *Ecol. Stud.,* 9: 4–452.

Henry, J.-P., 1965. Contribution à l'étude de la biologie d'*Asellus cavaticus* Leydig. *Int. J. Speleol.,* 1: 279–287.

Henry, J.-P., 1976. *Recherches sur les Asellidae Hypogés de la Lignée Cavaticus (Crustacea, Isopoda, Asellota),* Thesis. Université Dijon, 270 pp.

Hobbs, H.H., 1988. Gasoline pollution in an Indiana cave. *Natl. Speleol. Soc. Bull.,* 29.

Holland, M.M., 1988. Scope/MAB technical consultations on landscape boundaries: report of a Scope/MAB workshop on ecotones. *Biol. Int., Special Issue,* 17: 47–106.

Holsinger, J.R., 1966. A preliminary study of the effects of organic pollution of Banners Corner Cave, Virginia. *Int. J. Speleol.,* 2: 75–89.

Iliffe, T.M., Jickells, T.D. and Brewer, M.S., 1984. Organic pollution of an inland marine cave from Bermuda. *Mar. Environ. Res.,* 12: 173–189.

Jeannel, R., 1926. *Faune Cavernicole de la France, avec une Étude des Conditions d'Existence dans le Domaine Souterrain.* Editions Lechevalier, Paris, 334 pp.

Job, C.A. and Simons, J.J., 1994. Ecological basis for management of groundwater in the United States: statutes, regulations, and a strategic plan. In: J. Gibert, D.L. Danielopol and J.A. Stanford (Editors), *Groundwater Ecology.* Academic Press, San Diego, CA, pp. 523–540.

Juberthie, C. and Ginet, R., 1994. Historique de la biospéologie – France. In: C. Juberthie and V. Decu (Editors), *Encyclopaedia Biospeologica.* Société de Biospéologie, Moulis, France, pp. 665–692.

Karam, Y., 1989. *Essais de Modélisation des Écoulements dans un Aquifère Karstique. Exemple de la Source du Lez (Hérault – France),* Thesis, Université Montpellier 2, 286 pp.

Karaman, G.S. and Ruffo, S., 1986. Amphipoda: *Niphargus* group (Niphargidae *sensu* Bousfield, 1982). In: L. Botosaneanu (Editor), *Stygofauna Mundi.* E.J. Brill, Leiden, pp. 515–534.

Karanjac, J. and Günay, G., 1980. Dumanli Spring, Turkey – the largest karstic spring in the world? *J. Hydrol.,* 45: 219–231.

Longley, G., 1981. The Edwards aquifer: earth's most diverse groundwater ecosystem? *Int. J. Speleol.,* 11: 123–128.

Longley, G., 1992. The subterranean aquatic ecosystem of the Blacones fault zone Edwards aquifer in Texas – threats from overpumping. In: J.A. Stanford and J.J. Simons (Editors), *Ground Water Ecology.* American Water Resources Association, Bethesda, MD, pp. 291–300.

Magniez, G., 1981. Biogeographical and paleobiogeographical problems in Stenasellids (Crustacea Isopoda Asellota of underground waters). *Int. J. Speleol.*, 11: 71–81.

Malard, F. and Chapuis, R., 1995. Temperature logging to describe the movement of sewage-polluted surface water infiltrating into a fractured rock aquifer. *J. Hydrol.*, 173: 191–217.

Malard, F. and Simon, K., 1997. Sampling in wells for describing ecological patterns at a microscale in karst aquifers. In: I.D. Sasowsky, D.W. Fong and E.L. White (Editors), *Conservation and Protection of the Biota of Karst.* Karst Water Institute, Charles Town, West Virginia, pp. 46–55.

Malard, F., Gibert, J., Laurent, R. and Reygrobellet, J.-L., 1994a. A new method for sampling the fauna of deep karstic aquifers: *C.R. Acad. Sci. Paris,* 317: 955–966.

Malard, F., Reygrobellet, J.-L., Mathieu, J. and Lafont, M., 1994b. The use of invertebrate communities to describe groundwater flow and contaminant transport in a fractured rock aquifer. *Arch. Hydrobiol.*, 131: 93–110.

Malard, F., Reygrobellet, J.-L. and Soulié, M., 1994c. Transport and retention of fecal bacteria at sewage-polluted fractured rock sites. *J. Environ. Quality,* 23: 1352–1363.

Malard, F., Crague, G., Turquin, M.J. and Bouvet, Y., 1994d. Monitoring karstic ground water: the practical aspect of subterranean ecology. *Theor. Appl. Karstol.*, 7: 115–126.

Malard, F., Mathieu, J., Reygrobellet, J.-L. and Lafont, M., 1996a. Biomonitoring groundwater contamination: application to a karst area in southern France. *Aquat. Sci.*, 58(2): 158–187.

Malard, F., Plénet, S. and Gibert, J., 1996b. The use of invertebrates in ground water monitoring: a rising research field. *Ground Water Monit. Remed.*, 16(2): 103–113.

Malard, F., Turquin, M.J. and Magniez, G., 1997. Filter effect of karstic spring ecotones on the population structure of the hypogean amphipod *Niphargus virei.* In: J. Gibert, J. Mathieu and F. Fournier (Editors), *Groundwater/Surface Water Ecotones – Biological and Hydrological Interactions and Management Options.* International Hydrology Series and Cambridge University Press, Paris, pp. 42–50.

Malard, M., 1995. *Contribution à l'Étude Biologique de la Qualité des Eaux Souterraines Karstiques: Application à un Site Atelier Nord-Montpelliérain (Bassin de la Source du Lez),* Thesis, Université Lyon 1, 400 pp.

Mangin, A., 1975. Contribution à l'étude hydrodynamique des aquifères karstiques. Constitution et fonctionnement des aquifères karstiques. *Ann. Speleol.*, 30(1): 21–124.

Mangin, A., 1994. Karst hydrogeology. In: J. Gibert, D.L. Danielopol and J.A. Stanford (Editors), *Groundwater Ecology.* Academic Press, San Diego, CA, pp. 43–67.

Mathieu, J. and Essafi, K., 1991. Variations des effectifs de populations interstitielles de *Niphargus* (Amphipodes stygobies) à l'interface karst/plaine alluviale. *C.R. Acad. Sci. Paris,* 312(3): 489–494.

Mathieu, J. and Turquin, M.J., 1992. Biological processes at the population level. II. Aquatic populations – *Niphargus* (stygobiont amphipod) case. In: A.I. Camacho (Editor), *The Natural History of Biospeleology.* Museo Nacional de Ciencas Naturales, Madrid, pp. 263–293.

Mathieu, J., Marmonier, P., Laurent, R. and Martin, D., 1991. Récolte du matériel biologique aquatique souterrain et stratégie d'échantillonnage. *Hydrogéologie,* 3: 187–200.

Mathieu, J., Essafi-Chergui, K. and Culver, D.C., 1992. Variations in the structure of stygobiont crustacean populations (*Niphargus rhenorhodanensis* and *Proasellus valdensis*) within the sediments of a karst outflow. *Hydrobiologia,* 231: 41–49.

Milanovic, P., 1986. Influence of the karst spring submergence on the karst aquifer regime. *J. Hydrol.*, 84: 141–156.

Naiman, R.J. and Décamps, H., 1990. *The Ecology and Management of Aquatic–Terrestrial Ecotones.* Vol. 4 of Man and the Biospheres Series. Parthenon Publishing, London, 316 pp.

Nicod, J., 1995. Carte géomorphologique des karsts de France. Notice d'une maquette au 1/3000000 environ. *Karstologia,* 25: 21–34.

Nicolas, H., 1882. Quelques notes sur le genre *Aveniona,* nouveau mollusque découvert dans les puits et les eaux souterraines du sous-sol d'Avignon. *Mem. Acad. Vaucluse,* 159.

Notenboom, J., 1991. Marine regressions and the evolution of groundwater dwelling amphipods (Crustacea). *J. Biogeogr.*, 18: 437–454.

Notenboom, J., Plénet, S. and Turquin, M.J., 1994. Groundwater contamination and its impact on groundwater animals and ecosystems. In: J. Gibert, D.L. Danielopol and J.A. Stanford (Editors), *Groundwater Ecology.* Academic Press, New York, pp. 477–504.

Plénet, S. and Gibert, J., 1995. Comparison of surface water/groundwater interface zones in fluvial and karstic systems. *C.R. Acad. Sci. Paris,* 318: 499–509.

Plénet, S., Gibert, J. and Vervier, P., 1992. A floodplain spring: an ecotone between surface water and groundwater. *Regulated Rivers Res. Manage.*, 7: 93–102.

Poulson, T.L., 1992. Assessing groundwater quality in caves using indices of biological integrity. In: U.S. EPA and the Association of Ground Water Scientists and Engineers (Editors), *Proc. 3rd Conf. on Hydrogeology, Ecology, Monitoring, and Management of Groundwater in Karst Terranes.* U.S. EPA and AGWSE, Nashville, TN, 495–511.

Réveillet, P., 1993. Essai historique de la biospéologie du Vercors et du Diois. *Spelunca-Memoires,* 17: 299–302.

Reygrobellet, J.-L., Mathieu, J., Laurent, R., Gibert, J. and Renault, Ph., 1975. Répartition du peuplement par rapport à la géomorphologie de la grotte de la cascade de Glandieu (Ain). *Spelunca-Memoires,* 8: 195–204.

Richert, C., 1990. *Faisabilité d'un Indice Biologique de Qualité des Eaux Souterraines Karstiques,* Thesis. Université Lyon 1, 18 pp.

Rouch, R., 1970. Le système karstique du Baget. I. Le phénomène d'"hémorragie" au niveau de l'exutoire principal. *Ann. Speleol.*, 25(3): 665–709.

Rouch, R., 1977. Considérations sur l'écosystème karstique. *C.R. Acad. Sci. Paris,* 284: 1101–1103.

Rouch, R., 1979. Le système karstique du Baget. VIII. La communauté des Harpacticides. Les apports au sein du système par dérive catastrophique. *Ann. Limnol.*, 15(3): 243–274.

Rouch, R., 1982. Le système karstique du Baget. XIII. Comparaison de la dérive des Harpacticides à l'entrée et à la sortie de l'aquifère. *Ann. Limnol.*, 18: 133–150.

Rouch, R., 1986. Sur l'écologie des eaux souterraines dans le karst. *Stygologia,* 2(4): 352–398.

Rouch, R., 1994. Copepoda. In: C. Juberthie and V. Decu (Editors), *Encyclopaedia Biospeologica.* Société de Biospéologie, Moulis, France, pp. 105–111.

Rouch, R. and Bonnet, L., 1976a. Le système karstique du Baget.

V. La communauté des Harpacticides. Sur la constance du peuplement. *Ann. Speleol.,* 31: 43–53.

Rouch, R. and Bonnet, L., 1976b. Le système karstique du Baget. IV. Premières données sur la structure et l'organisation de la communauté des Harpacticides. *Ann. Speleol.,* 31: 27–41.

Rouch, R. and Bonnet, L., 1977. Le système karstique du Baget. VII. La communauté des Harpacticides. Structure du peuplement hypogé pendant les crues. *Ann. Limnol.,* 13(3): 251–259.

Rouch, R. and Carlier, A., 1985. Le système karstique du Baget. XIV. La communauté des Harpacticides. Evolution et comparaison des structures du peuplement épigé à l'entrée et à la sortie de l'aquifère. *Stygologia,* 1(1): 71–92.

Rouch, R., Juberthie-Jupeau, L. and Juberthie, C., 1968. Essai d'étude du peuplement de la zone noyée d'un karst. *Ann. Speleol.,* 23: 716–733.

Rouch, R., Pitzalis, A. and Descouens, A., 1993. Effets d'un pompage à gros débit sur le peuplement des crustacés d'un aquifère karstique. *Ann. Limnol.,* 29: 15–26.

Sarbu, S.M. and Popa, R., 1992. A unique chemoautotrophically based cave ecosystem. In: A.I. Camacho (Editor), *The Natural History of Biospeleology.* Museo Nacional de Ciencas Naturales, Madrid, pp. 637–666.

Schminke, H.K., 1981. Adaptation of Bathynellacea (Crustacea, Syncarida) to life in the interstitial ("Zoea Theory"). *Int. Rev. ges. Hydrobiol.,* 66(4): 575–637.

Sket, B., 1996. Biotic diversity of hypogean habitats in Slovenia and its cultural importance. In: A. Cimerman and N. Gunde-Cimerman (Editors), *International biodiversity seminar, ECCO XIV Meeting.* Slovenian National Commission for UNESCO, Ljubljana, pp. 59–74.

Sollaud, E., 1936. Sur quelques formes endémiques de la faune cavernicole du Jura. Livre jubilaire E.L. Bouvier, Paris, pp. 317–322.

Southwood, T.R.E., 1977. Habitat, the templet for ecological strategies? *J. Anim. Ecol.,* 46: 337–365.

Stoch, F., 1995. The ecological and historical determinants of crustacean diversity in groundwaters, or: why are there so many species? *Mem. Biospeol.,* 22: 139–160.

Tercafs, R., 1992. The protection of the subterranean environment: conservation principles and management tools. In: A.I. Camacho (Editor), *The Natural History of Biospeleology.* Museo Nacional de Ciencas Naturales, Madrid, pp. 483–523.

Townsend, C.R., 1989. The patch dynamics concept of stream community ecology. *J. North Am. Benthol. Soc.,* 8(1): 36–50.

Turquin, M.J., 1973. Une biocoenose cavernicole originale pour le Bugey: le Puits de Rappe. In: *Comptes Rendus du 96ème Congrès National des Sociétés Savantes, 1971, Toulouse, France,* Vol. 3, pp. 235–256.

Turquin, M.J., 1975a. Incidenses des biocénoses terrestres sur le rythme de ponte de l'Amphipode troglobie *Niphargus. Bull. Soc. Zool. France,* 100(2): 169–176.

Turquin, M.J., 1975b. Cycle biologique et rythme de ponte chez les Crustacés aquatiques troglobies. *C.R. Acad. Sci. Paris,* 280: 2033–2036.

Turquin, M.J., 1981a. La biocénose terrestre cavernicole: apport énergétique pour la communauté aquatique souterraine. *Mem. Biospeol.,* 8: 9–16.

Turquin, M.J., 1981b. The tactics of dispersal of two species of *Niphargus* (perennial troglobite Amphipoda). In: B.F. Beck (Editor), *Proc. 8th Int. Congr. of Speleology, Bowling Green (Kentucky),* pp. 353–354.

Turquin, M.J., 1981c. Profil démographique et environnement chez une population de *Niphargus virei* (Amphipode, Troglobie). *Bull. Soc. Zool. France,* 106(4): 457–466.

Turquin, M.J., 1984. Un cas de transition démographique dans le milieu souterrain. *Int. Verein. Theor. Angew. Limnol.,* 22: 1751–1754.

Turquin, M.J., 1986. Mortalité et stock de l'amphipode *Niphargus virei,* dans le système karstique de Drom. *Bull. Mens. Soc. Linn. Lyon,* 55(9): 293–304.

Turquin, M.J., 1993. Les citations françaises antérieures à 1900 dans les ouvrages de biospéologie. *Spelunca-Memoires,* 17: 285–288.

Turquin, M.J. and Barthélémy, D., 1985. The dynamics of a population of the troglobitic amphipod *Niphargus virei* Chevreux. *Stygologia,* 1(1): 109–117.

Turquin, M.J. and Bouvet, Y., 1977. Energy flow and faunistical distribution inside karst; the influence of module of openness. In: T.D. Ford (Editor), *Proc. 7th Int. Congr. of Speleology, Sheffield, UK.* Hawthornes, Nottingham, pp. 406–407.

Turquin, M.J. and Crague, G., 1994. Impact de la gestion des eaux usées sur la biocénose cavernicole, la zone noyée et la source du système karstique de Rappe (Neuville-sur-Ain, France). *Bull. Mens. Soc. Linn. Lyon,* 63(9): 337–348.

Turquin, M.J. and Van Poucke, S., 1978. Le Rhône et le peuplement cavernicole du defile de Donzère. Réflexions sur le choix des terrains d'étude en écologie souterraine. In: Féderation Française de Spéléologie (Editors), *Actes du XIIème Congrès National de Spéléologie, Grasse, 1976.* Comité de Spéléologie des Alpes-Maritimes, pp. 156–169.

Turquin, M.J., Bouvet, Y., Renault, P. and Pattee, E., 1976. Essai de corrélation entre la géomorphologie d'une cavité et la répartition spatiale de son peuplement actuel. In: R. Gigon (Editor), *Actes du 5ème Congrès National de Spéléologie, Interlaken, 1974.* Courvoisier S.A., Suisse, pp. 46–60.

Vervier, P., 1988. *Hydrologie et Dynamique des Peuplements Aquatiques Souterrains: Comparaison de Deux Systèmes Karstiques des Gorges de l'Ardèche,* Thesis, Université Lyon 1, 219 pp.

Vervier, P. and Gibert, J., 1991. Dynamics of surface water/ groundwater ecotones in a karstic aquifer. *Freshwater Biol.,* 26: 241–250.

Vervier, P., Gibert, J., Laurent, R. and Couturaud, A., 1986. Plongée souterraine et hydrobiologie dans la Source Bleue, Montperreux, Doubs. Utilisation de substrats artificiels pour étudier la faune des réseaux noyés. *Spelunca,* 22: 37- 42.

Vervier, P., Gibert, J., Marmonier, P. and Dole-Olivier, M.J., 1992. A perspective on the permeability of the surface freshwater–groundwater ecotone. *J. North Am. Benthol. Soc.,* 11(1): 93–102.

Ward, J.V. and Stanford, J.A., 1995. Ecological connectivity in alluvial river ecosystems and its disruption by flow regulation. *Regulated Rivers Res. Manage.,* 11: 105–119.

Williams, P.W., 1983. The role of the subcutaneous zone in karst hydrology. *J. Hydrol.,* 61: 45–67.

VII.2. Studies of particular cave systems

Chapter 28

ICE CAVES IN TEMPERATE REGIONS

Gheorghe RACOVITA

INTRODUCTION

A first overview of the temperature of karst caves resulted from measurements made over nearly three decades by Emil Racovita and René Jeannel, the two remarkable coordinators of the program "Biospeologica". Jeannel (1926) concluded that, in general, the temperature of a cave is equal to the average annual temperature of the external environment in the geographical location in question. This rule is supported, within each climatic region, by linear correlation between altitude and underground temperature, but only when this temperature refers to the stable meroclimate (*sensu* G. Racovita, 1975) of the deep cave environment.

In consequence, caves in temperate regions opening at a high altitude may have glacial temperatures and yet contain cavernicolous species. The few studies conducted mostly concern Coleoptera, and the classical study is of the leptodirine *Isereus xambeui* from Trou de Glas (in the French Alps) – which is said to live at a constant temperature between 0°C and 2°C – and the trechine *Arctaphaenops angulipennis*, living in similar temperatures in the Austrian Alps (Vandel, 1964). These species are often cited in support of arguments about the wide thermal tolerance of troglobitic species. However, the biology of the two species is practically unknown and the discussion that follows suggests that the supposition that they have a wide thermal tolerance is unfounded.

The cases mentioned above probably represent caves close to the altitudinal limit beyond which the climatic conditions are unsuitable for available lineages of animals to occupy ecological niches freed after the last glacial episode (Peck, 1980; Juberthie, 1984). However, there are other caves – and these are equally important from an ecological point of view – that preserve perennial speleothems of ice despite opening at medium altitudes. While at such altitudes the external climate is too moderate to cause the formation of a glacial topoclimate, nevertheless such climates may occur as a result of the thermocirculation within caves, generally involving a single opening leading to a downward-sloping cave. Such caves have a seasonal and bidirectional ventilation pattern, with aerodynamic exchange with the outside occurring only in winter, and in consequence have a constant accumulation of cold air in the cave (Trombe, 1952; Andrieux, 1971; G. Racovita, 1975).

Caves at medium altitudes only develop glacial topoclimates if other conditions are also met, such as the volume of the cave relative to the opening; the ratio required depends in turn on the outside climate (G. Racovita and Serban, 1990). As a result, there are very few mid-altitude glacial caves; one of the most intensely studied caves of this type is the Scarisoara glacier cave (Ghetarul de la Scarisoara, Romania) which is to serve here as the exemplar, on the basis of which the ecology of the terrestrial fauna inhabiting glacial caves may be closely examined.

GHETARUL DE LA SCARISOARA

The cave is situated at an altitude of 1165 m in the Bihor mountains, the central orographic unit of the Apuseni mountains (Transylvania) on the southwestern border of the Scarisoara karst plateau in the Ariesul Mare basin. Together with the cave Pojarul Politei, it represents the oldest level of karstification in the Ghetar–Ocoale karst system, probably of Neogene age (Rusu et al., 1970).

The cave was surveyed by Schmidl (1863), and was studied in detail by E. Racovita (1927), who explained

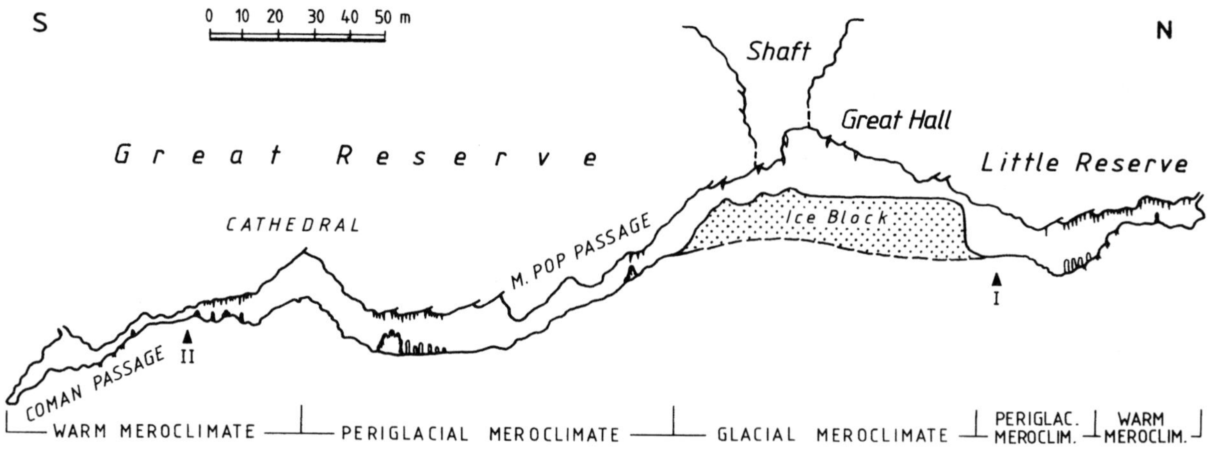

Fig. 28.1. Longitudinal profile of the Ghetarul de la Scarisoara (modified after Rusu et al., 1970). The black triangles mark the positions of the stations for beetle counts.

the processes by which ice speleothems are formed and maintained, and who drew attention to the special importance of this cave for both its ecological and its other attributes.

The cave is of simple structure (Fig. 28.1), extending 700 m horizontally and 105 m vertically. The shaft in the median section of the cave, 60 m in diameter, descends to the floor 50 m below, where there is a large block of permanent ice. This block is almost parallelopipedic in shape, has a volume of 75 000 m^3 and is at least 3500 years old (Rusu et al., 1970). The upper surface of this block forms the nearly horizontal floor of the Sala Mare ("Great Hall"), while at the edges one can descend into three other large halls, the Rezervatia Mica ("Little Reserve") to the north, the Biserica ("Church") to the north-west and the Rezervatia Mare ("Great Reserve") to the south. The Rezervatia Mare [including the Galeria Maxim Pop ("Maxim Pop Passage") and the Catedrala ("Cathedral")], continues as a narrow, steeply descending gallery – the Galeria Coman ("Coman Passage"), leading to the deepest point in the cave. In each of the three large halls, peripheral to the ice block, massive and isolated stalagmites of permanent ice are found (Fig. 28.2), which give place peripherally to calcite speleothems.

The spatial distribution of the speleothems is closely related to the meroclimate structure of the cave (G. Racovita, 1974, 1994), which consists of several zones, arranged in centrifugal succession distinguished principally by their temperature. The first is the **glacial zone** – coinciding with the position of the massive ice block – in the Great Hall where the temperature

Fig. 28.2. Permanent ice speleothems in the Biserica ("Church") of the Ghetarul de la Scarisoara.

falls below 0°C (down to −10°C) as a result of the activation of bidirectional ventilation, but is stabilized at 0.5°C in summer, when the thermocirculation is restricted to an exclusively underground convection cell. Next is the **periglacial zone** within the three lateral halls which are characterized by similar seasonal variation in temperature, but have less rigorous winter values – they coincide with the extent of the massive and isolated ice speleothems. The sequence ends in a **warmer zone** disposed in the extreme parts of the cave where the temperature is not affected by aerodynamic exchange with the outside, but remains above 0°C throughout the year – usually between 2° and 5°C – and consequently contains only calcite speleothems.

As a whole, the cave offers very few trophic resources, and these are limited to the segments of clay which occur sporadically in the calcite areas, and sparse bat excrement. However, adjacent to the ice block are tree-trunks which have fallen down the shaft; but this zone is practically azoic owing to the low temperature.

Preliminary data on the troglobitic fauna

Within the Ghetarul de la Scarisoara, it is notable that the components of the underground terrestrial community are influenced by strong microhabitat selection. The community includes troglophilic and troglobitic representatives of three groups: araneids (*Nesticus racovitzai* and *Troglohyphantes racovitzai*), springtails (*Onchopodura crassicornis*, *Onychiurus* spp. and *Tomocerus minor*), and leptodirine beetles (*Pholeuon* (s. str.) *proserpinae glaciale*). The first two groups are found only sporadically, and the community is dominated by the beetles which, being quite numerous, are amenable to ecological research.

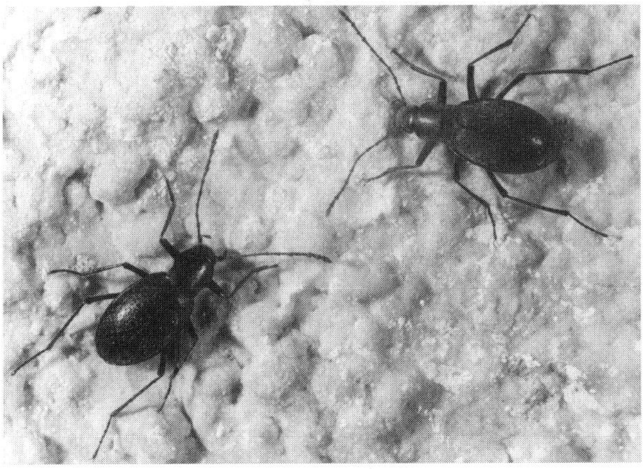

Fig. 28.3. The leptodirine beetle *Pholeuon proserpinae glaciale.*

The leptodirine beetle *Ph. p. glaciale* (Fig. 28.3) was described from Ghetarul de la Scarisoara (Jeannel, 1923) and is endemic to the karst plateau of Scarisoara, where it also occurs in two neighbouring caves, Pojarul Politei and the Ghetarul de sub Zgurasti. The first two are only separated by an infill of 4–5 m at the end of the Coman Passage and they are climatically similar, Pojarul Politei not having a glacial topoclimate. Ghetarul de sub Zgurasti is 3 km distant and, despite its name ("ghetar" in Romanian means "ice cellar" in English), it contains no permanent ice, and the temperature where the beetles are found is normally about 8°C (Jeannel and Racovita, 1929, p. 468). *Ph. proserpinae* is present in the superficial underground compartment ("MSS") in the valley of the Ordâncusa – the eastern limit of the Scarisoara karst plateau – but in small numbers, so that it is seemingly accidental in this community (G. Racovita, 1984); it is so sparse that the subspecific status is uncertain.

When *Ph. p. glaciale* was first found in the Ghetarul de la Scarisoara, only the Great Hall and the Church were known; hence, the beetles were in the periglacial meroclimate where they were seen to move slowly on the bare walls, rarely on the ice and then only in the autumn (August–October) when the air temperature was just above freezing (Jeannel and Racovita, 1929, p. 440). Having no information on the effect of winter on the biology of these beetles, E. Racovita (1927, p. 103) confined himself to stating that "... la basse température, pourvu qu'elle ne soit pas de façon permanente sous 0°C, n'empêche pas la colonisation et la formation de faunes et flores spéciales [low temperature, provided that it is not constantly below 0°C, does not prevent the colonization and the formation of a special fauna and flora]". Jeannel (1943, p. 58) was less circumspect, declaring that the life cycle of the beetle must occur at temperatures of 0.1° to 0.8°C, even though it avoids freezing in winter by withdrawing into the network of joints in the karst massif. His view prevailed, and *Ph. p. glaciale* became the convincing example of the capacity of troglobites to live in extreme thermal conditions (Vandel, 1964).

Subsequent observations made in newly discovered parts of the cave – the Rezervatia Mare and the Rezervatia Mica – demonstrated that the usual habitat of the beetles was in the **warm zone** where the temperature remained above freezing year-round, and that their presence in the frozen zone was accidental. Thus this correction must be made to the initial image of subsistence conditions in order to explain the reserve expressed towards the similar cases mentioned.

Despite long-term baiting, *Ph. p. glaciale* could not be found again in the Biserica (G. Racovita, 1969), possibly owing to increased tourist activity in the area. It has been established experimentally, using a sample coming for the Rezervatia Mare, that the optimal temperature for the subspecies is 4.3–4.8°C (M. Serban, personal communication, 1968), comparable to the temperature of its normal environment. Hence, there is no reason to consider that *Ph. proserpinae glaciale* is especially adapted to temperatures that are close to the lower limit of activity of leptodirines, a limit established at 0°C for *Speonomus* (Glaçon, 1952).

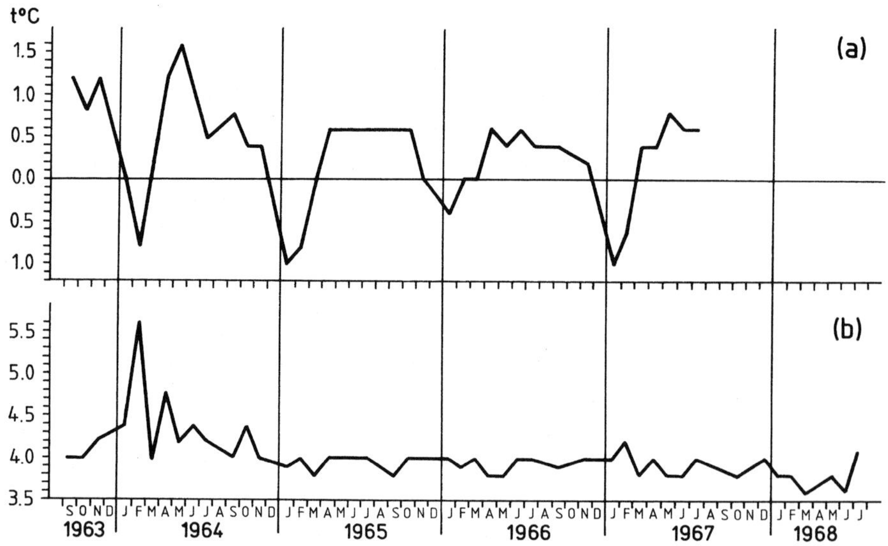

Fig. 28.4. Temperature variation (a) in the periglacial meroclimate in the Rezervatia Mica and (b) in the stable meroclimate in the Galeria Coman. The value of 5.6ºC, recorded in February 1964, is obviously an anomaly; it was probably caused by diminution of the thermocirculation as a consequence of a sudden major change in barometric pressure.

The dynamics of cave populations of *Ph. p. glaciale*

These conclusions were confirmed by a long-term demographic study from September, 1963 through July, 1968. These initial data were subsequently used comparatively in a study of the ecology of troglobitic leptodirines in the Apuseni mountains (G. Racovita, 1980). This study recorded the fluctuations in the adult beetles represented by "groups of individuals" (*sensu* Delay, 1975) from different cave habitats, which, when combined, constituted the cave population. The study was conducted by periodic examination of the size of these groups [by the method of Cabidoche (1968)], a process facilitated by concentrating individuals at "numbering stations" by means of long-term baits. Two bait stations were installed in the Ghetarul de la Scarisoara (Fig. 28.1); one in the **periglacial part** of the Rezervatia Mica, in a pocket situated in the eastern part of this chamber, and a second in the **warm zone** of the Rezervatia Mare immediately behind the Galeria Coman. The number of beetles was estimated every 30 days, at which times thermo-hygrometric measurements were also made.

According to the meroclimatic structure of the cave, these two habitats experience different physical conditions, mostly with respect to air temperature. The first station (Fig. 28.4a) is subject to obvious seasonal temperature variation, having negative temperatures in

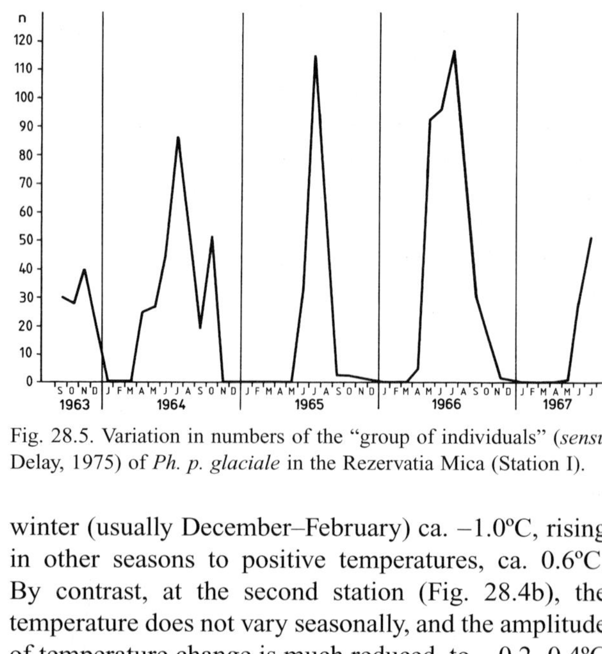

Fig. 28.5. Variation in numbers of the "group of individuals" (*sensu* Delay, 1975) of *Ph. p. glaciale* in the Rezervatia Mica (Station I).

winter (usually December–February) ca. −1.0ºC, rising in other seasons to positive temperatures, ca. 0.6ºC. By contrast, at the second station (Fig. 28.4b), the temperature does not vary seasonally, and the amplitude of temperature change is much reduced, to 0.2–0.4ºC around the average temperature of 4.0ºC.

This difference affects the biology of the beetles in the two parts of the cave. At the first station (Fig. 28.5), soon after the air temperature rise above the critical value of 0ºC, the beetles progressively increase in number, reaching a maximum in summer and diminishing again in autumn; the beetles completely disappear when the temperature falls below 0ºC.

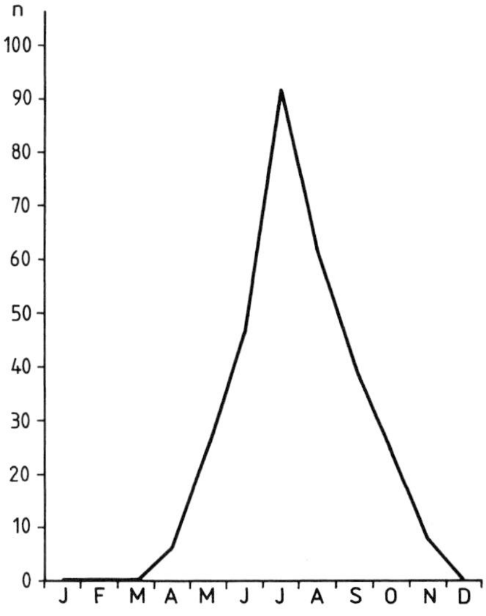

Fig. 28.6. Average number of beetles recorded in a month for the "group of individuals" in the Rezervatia Mica.

As shown by the number of beetles recorded each month (Fig. 28.6), the dynamics of the individuals in this periglacial part of the cave have two distinct phases: presence during the warm periods, and absence in winter, the succession of these phases being repeated in each annual cycle. Hence, the size of the group of individuals depends on the seasonal variation in the air temperature, a point that will be discussed later.

At the second station, the beetles are permanently present and at quite a high density. Here the dynamics of the group are quite different (Fig. 28.7), with a much less obvious seasonal periodicity; the progressive variation in numbers found at Station I is replaced by a seemingly irregular variation in numbers. The common element remaining is the considerable reduction in the number of beetles seen in the autumn. These fluctuations in population size occur in a microenvironment of very high physical constancy.

I consider that these heterogeneous fluctuations in numbers are better explained by considering the average variation in the numbers of individuals (Fig. 28.8); as well as a reduction in numbers in the autumn (October–November), an earlier oscillation occurs in the period January–July. Hence, in contrast to the population cycles at Station I, at Station II the annual population cycle is diphasic and not coincident with those seen in the meroclimate cycles in the chamber; thus, they are likely to have a different cause.

The irregular part of the curve characterizing the first of the two phases (Fig. 28.8) suggests that the size of the group of individuals could be influenced, during this period of their annual cycle, by a self-adjustment phenomenon, so that it may be under the control of biotic factors, such as population density. According to the model established by MacArthur and Connel (1966), this hypothesis is verified when there is a significant correlation between the absolute number of beetles observed in a month and the subsequent change

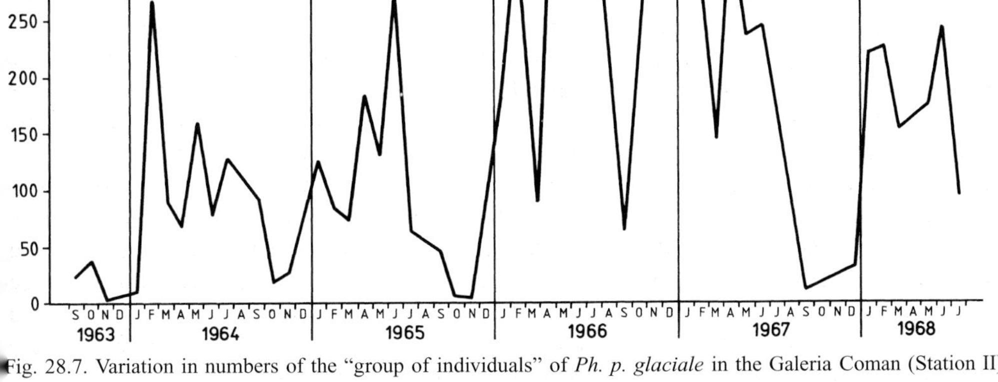

Fig. 28.7. Variation in numbers of the "group of individuals" of *Ph. p. glaciale* in the Galeria Coman (Station II).

in numbers (the relative numbers). Such a relationship can be seen in these data ($r = 0.98$; $t = 9.20$; $p < 0.01$), and it is characterized by a straight-line regression:

$$\Delta N = -1.86N + 352,$$

from which the mathematical model of the natural phenomenon can be constructed. The parameter that characterizes this model is the absolute value of the coefficient a, which is equal to the tangent of the angle α made by the regression line with the abscissa axis (G. Racovita, 1973). It this case $1 < a < 2$, implying $45° < \alpha < 63°$, and the model results in oscillations that become progressively smaller, asymptotically reaching a stable state – the size of the group. Comparison of the actual data with the model shows a close fit (Fig. 28.9), the stable state being respectively 186 and 189 individuals. It is concluded that, in the stable meroclimate in the Coman Passage, the density of beetles is controlled, for the first part of the annual cycle, by self-adjusting processes amongst the adults.

The considerable reduction in the number of beetles during the second part of the annual cycle is probably the result of interference of a seasonal reproductive maximum. This supposition is derived from the knowledge that the reduction in numbers is accompanied by a major change in the age structure of the population (many individuals have just hatched). The proportion of immature adults reaches 59% of the adult population in autumn, compared with 10 to 25% at other times. As sexual maturity in leptodirines is not reached until one year after the imago emerges (Deleurance-Glaçon, 1963), in the latter part of the year the cave community consists mostly of non-reproductive individuals.

There is no reason to reject Jeannel's (1943) scenario according to which, especially in the case of the most specialized underground beetles, oviposition and larval development most commonly take place in the joint network of the karst. Hence, the major changes in population size can be interpreted as a major migration, from the cave habitat ("speleic", *sensu* Decu et al., 1991) to the fissure network ("cleitric"), of those adults participating in reproduction. This results in selective reduction in the group of individuals living in the cave, and implicitly modification of the age structure.

This phenomenon, obvious in the context of the most favourable meroclimate of the Rezervatia Mare, can also be found among individuals living in the Rezervatia Mica, but here it is masked by the dominant effect of the temperature variations. Here, the population numbers in the group start to decline under conditions similar to those that occur when the maximum numbers are found in the Rezervatia Mare, and long before the air temperature drops below values suitable for the beetles. As this diminution in numbers occurs at the same time as that found in beetles of the Rezervatia Mare, which are part of the same population, it is likely that it results from the same cause.

A second demographic study undertaken in the Rezervatia Mare, from August 1982 to February 1986, showed that the dynamics of this group had not changed qualitatively in the intervening 20 years – the phase of self-adjustment and the autumnal minimum again proved dominant characteristics of the population cycle (G. Racovita, 1987).

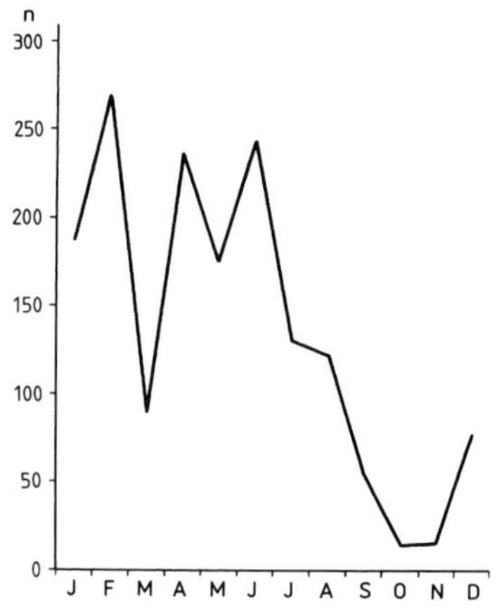

Fig. 28.8. Average number of beetles recorded in a month for the "group of individuals" in the Galeria Coman (on the ordinate, median values).

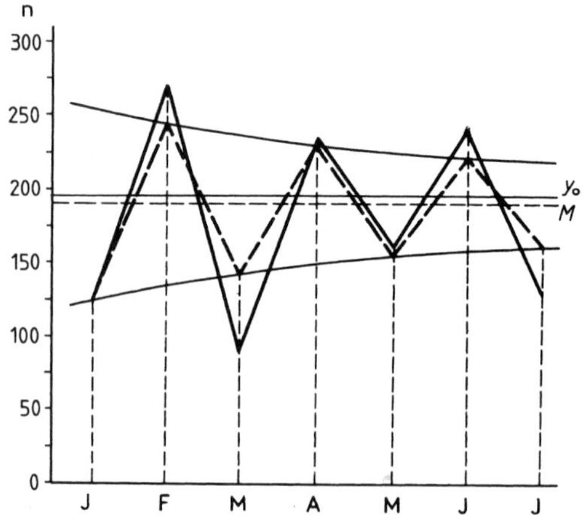

Fig. 28.9. Comparative diagram showing the actual fluctuations (dashed line) and the mathematical model of the self-adjustment of the number of beetles (solid line) in the "group of individuals" in the Galeria Coman: y_0, theoretical equilibrium value; *M*, average number of individuals.

The significance of quantitative variations in the population of *Ph. p. glaciale*

In summary, the fluctuations in the beetle populations found in the different cave habitats are controlled by physical factors of the environment, especially air temperature, in variable meroclimate conditions, but by cenotic factors – the self-adjustment of the density and ecological periodicity in the reproductive cycle – in meroclimate conditions that are stable enough to correspond constantly with the ecological optimum. Hence, the dynamics of the *Ph. p. glaciale* population in the Ghetarul de la Scarisoara cave support Dajoz's (1974) model of the determinism of the size of natural populations.

It is now established that fluctuations in cave populations occur, not through modification of birth or death rates, but through migrations of the insect between the cave and the joint network (Juberthie, 1969; Delay, 1978; G. Racovita, 1980; Rusdea, 1992). Delay (1978) contended that cavities suitable for human entry (caves) were not physically distinct from the smaller joints with which they connect, and thus that they do not really constitute distinct biotopes. However, the example from the Ghetarul de la Scarisoara discussed here suggests that this thesis should be modified – that it would be more appropriate to consider the karst cavities and the joint network **as different ecological entities lacking clear delimitation.**

Indeed, in the extreme habitat characteristic of the periglacial parts of the cave, biotope selection is so severe that it imparts to the troglobitic community a composition entirely different from that expected in the entire karst massif. It is thus remarkable that in the Pojarul Politei cave the leptodirines are represented not only by *Pholeuon,* but also by *Drimeotus* (*Bihorites*) *laevimarginatus*. While the density of the latter is considerably lower, this would likely result from the competition that usually occurs between the two genera (Jeannel, 1923). It should be noted that, if the presence of the beetles in the cave was not a periodic extension of their usual habitats, then it would be difficult, perhaps impossible, to determine the cause of this extension because it could not then be related to the physical context or to the trophic resources in the periglacial parts of the cave.

Without having special physiological adaptation, *Ph. p. glaciale* inhabits an environment at the limits of ecological tolerance, and generally the same applies in similar glacial situations; this indicates that the underground domain is not a simple "shelter", and that its colonization does not always represent a means of survival of lineages in the state of "sénilité phylétique [phyletic senility]" (Vandel, 1964). Instead, it is an active process, in the evolution of which all free ecological niches are occupied. If these free ecological niches are not optimal, they should at least provide conditions compatible with the ecological potential of those species susceptible to this process (G. Racovita, 1981).

This conclusion relates to the classical image of the structure and extension of the underground domain as conceived by E. Racovita (1907) in which the main elements were the accessible cavities and the joint network. This conclusion was confirmed by the discovery of the superficial underground compartment (Juberthie et al., 1980), a discovery that altered the dominant paradigm in biospeleology. This resulted in a reappraisal of the thesis that cave populations are completely and definitely isolated from epigean communities (Juberthie, 1984), an idea stemming from the "shelter medium" function assigned to caves.

REFERENCES

Andrieux, C., 1971. Contributions à l'étude du climat des cavités naturelles des massifs karstiques. IV – Influence de la morphologie des galeries, de la forme et de l'orientation des ouvertures sur le climat souterrain. *Ann. Speleol.,* 26(1): 5–30.

Cabidoche, M., 1968. Biocénose cavernicole de la Salle de la Verna (Gouffre de la Pierre Saint-Martin). Methode d'étude en milieu naturel. *Ann. Speleol.,* 24(1): 667–688.

Dajoz, R., 1974. *Dynamique des Populations.* Masson, Paris, 301 pp.

Decu, V., Racovita, G. and Vaczy, C., 1991. Sur la nomenclature des milieux souterrains, edaphiques et des annexes directes du sol. *Trav. Inst. Speol. "E. Racovitza",* 30: 63–67.

Delay, B., 1975. Étude quantitative de populations monospécifiques de coléoptères hypogés par la méthode des marquages et recaptures. *Ann. Speleol.,* 30(1): 195–206.

Delay, B., 1978. Milieu souterrain et écophysiologie de la reproduction et du développement des coléoptères Bathysciinae hypogés. *Mem. Biospeol.,* 5: 1–340.

Deleurance-Glaçon, S., 1963. Recherches sur les coléoptères troglobies de la sous-famille des Bathysciinae. *Ann. Sci. Nat. Zool.,* 5(1): 1–172.

Glaçon, S., 1952. La variation de l'activité de *Speonomus diecki* Saulcy en fonction de la température et du sexe. *Bull. Soc. Zool. France,* 77(1): 36–44.

Jeannel, R., 1923. Etude préliminaire des coléoptères aveugles du Bihor. *Bull. Soc. Sc. Cluj,* 1: 411–472.

Jeannel, R., 1926. *Faune Cavernicole de la France, avec une Etude des Conditions d'Existence dans le Domaine Souterrain.* P. Lechevallier, Paris, 334 pp.

Jeannel, R., 1943. *Les Fossiles Vivants des Cavernes*. Gallimard, Paris, 321 pp.

Jeannel, R. and Racovita, E., 1929. Enumération des grottes visitées, 1918–1927 (7ème serie). *Arch. Zool. Exp. Gen.*, 68(2): 293–608.

Juberthie, C., 1969. Etude écologique des larves de *Speonomus infernus* subsp. *infernus* dans la grotte de Sainte-Catherine. *Ann. Speleol.*, 24(3): 563–577.

Juberthie, C., 1984. La colonisation du milieu souterrain; théories et modèles, relations avec la spéciation et l'évolution souterraine. *Mem. Biospeol.*, 11: 65–102.

Juberthie, C., Delay, B. and Bouillon, M., 1980. Extension du milieu souterrain en zone non calcaire: description d'un nouveau milieu et de son peuplement par les coléoptères troglobies. *Mem. Biospeol.*, 7: 19–52.

MacArthur, R.H. and Connel, J.H., 1966. *The Biology of Populations*. Wiley, New York, 230 pp.

Peck, S.B., 1980. Climatic change and the evolution of cave invertebrates in the Grand Canyon, Arizona. *Natl. Speleol. Soc. Bull.*, 42: 53–60.

Racovita, E., 1907. Essai sur les problèmes biospéologiques. *Arch. Zool. Exp. Gen.*, 6: 371–488.

Racovita, E., 1927. Observations sur la glacière naturelle dite "Ghetarul de la Scarisoara". *Bull. Soc. Sci. Cluj*, 3: 75–108.

Racovita, G., 1969. Recherches sur l'écologie du Bathysciinae troglobionte *Pholeuon proserpinae glaciale* Jeann. In: *5 Int. Congr. of Speleology, Stuttgart*, 4, B 14: 1–6.

Racovita, G., 1973. Quelques significations de la variation numérique des populations de coléoptères cavernicoles. *Trav. Inst. Spéol. "E. Racovitza"*, 12: 9–29.

Racovita, G., 1974. Sur la structure méroclimatique des cavités souterraines. *Theor. Appl. Karstol.*, 1: 123–130.

Racovita, G., 1975. La classification topoclimatique des cavités souterraines. *Trav. Inst. Spéol. "E. Racovitza"*, 14: 197–216.

Racovita, G., 1980. Etude écologique sur les Bathysciinae cavernicoles. *Mem. Biospeol.*, 6: 1–199.

Racovita, G., 1981. Aspects évolutifs de l'écologie des coléoptères cavernicoles. *Mem. Biospeol.*, 7: 119–125.

Racovita, G., 1984. Etude faunistique et écologique du milieu souterrain superficiel de la Vallée d'Ordancusa (Monts du Bihor). *Trav. Inst. Spéol. "E. Racovitza"*, 23: 27–37.

Racovita, G., 1987. Variations à longue échéance dans la dynamique d'une population troglobie. *Studia Univ. "Babas-Bolyai" Cluj-Napoca, Biol.*, 32(1): 39–51.

Racovita, G., 1994. Bilan climatique de la grotte glacière de Scarisoara (Monts du Bihore, Roumanie), dressé sur dix annees d'observations. *Trav. Inst. Spéol. "E. Racovitza"*, 33: 107–158.

Racovita, G. and Serban, M., 1990. Interprétation en vue d'une reconstitution paléoclimatique des particularités structurales et dynamiques du dépôt de glace pérenne de la Grotte de Scarisoara. *Studia Univ. "Babas-Bolyai" Cluj-Napoca, Geol.*, 35(2): 47–61.

Rusdea, E., 1992. Stabilisierende Selektion bei microphthalmen Höhlentieren: untersuchungen zur tageszeitlichen Aktivitätsverteilung und Populationsdynamik von *Laemostenus schreibersi* (Küster) (Carabidae). *Mem. Biospeol.*, 19: 1–110.

Rusu, T., Racovita, G. and Coman, D., 1970. Contributions à l'étude du complexe karstique de Scarisoara. *Ann. Speleol.*, 25(2): 383–408.

Schmidl, A.A., 1863. *Das Bihar-Gebirge an der Grenze von Ungarn und Siebenburgen*. Verl. Förster u. Bartelmus, Wien, 442 pp.

Trombe, F., 1952. *Traité de Spéléologie*. Payot, Paris, 376 pp.

Vandel, A., 1964. *Biospéologie. La Biologie des Animaux Cavernicoles*. Gauthiers-Villars, Paris, 619 pp.

Chapter 29

THE SEGEBERGER HÖHLE – A PHYLOGENETICALLY YOUNG CAVE ECOSYSTEM IN NORTHERN GERMANY

Anne IPSEN

INTRODUCTION

The Segeberger Höhle shows several characteristics not found in other caves. It originated within a salt dome, rather far northwards at a latitude of 54°. The area where it is found was covered by Pleistocene glaciers until about 10 000 years BP. Furthermore the cave has an extremely isolated position, as a result of its geological peculiarity. The diluvial sediments covering this formerly glaciated part of Germany normally are not apt to give rise to caves.

Although the salt dome forms a low hill approximately 90 m in height, the existence of this cave was not discovered until 1913. First biological explorations only took place several years later (Mohr, 1929). An important discovery was that of the occurrence of a troglobitic beetle, *Choleva septentrionis holsatica* (Fig. 29.1). This species is endemic to this cave system.

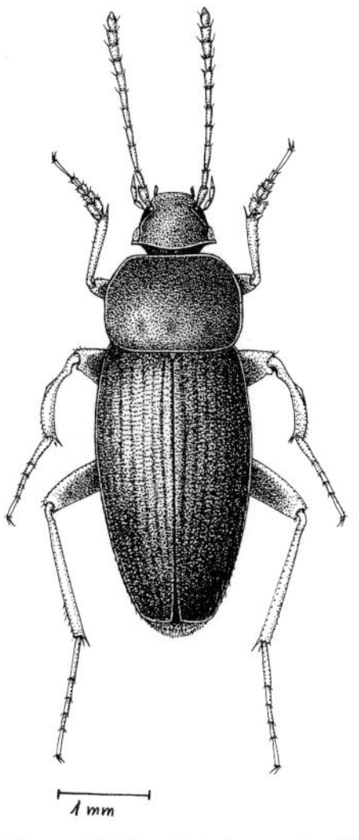

Fig. 29.1. Male of *Choleva septentrionis holsatica* (drawing U. Frerichs).

GEOLOGY AND ORIGIN OF THE SEGEBERGER HÖHLE

Geology and morphology of the cave

The cave was formed within the cap of a salt dome. As is typical, this cap mainly consists of gypsum ($CaSO_4\ 2H_2O$). The cave is situated 50 m below the top of the gypsum hill of 90 m elevation, which contains a diffuse system of cracks and crevices (Fig. 29.2) (Gripp, 1931; Teichmüller, 1948; Regge and Hahn, 1978).

It is usually assumed that the cave system underneath the top of the hill originated below the groundwater level. Probably the water level in the cave was very

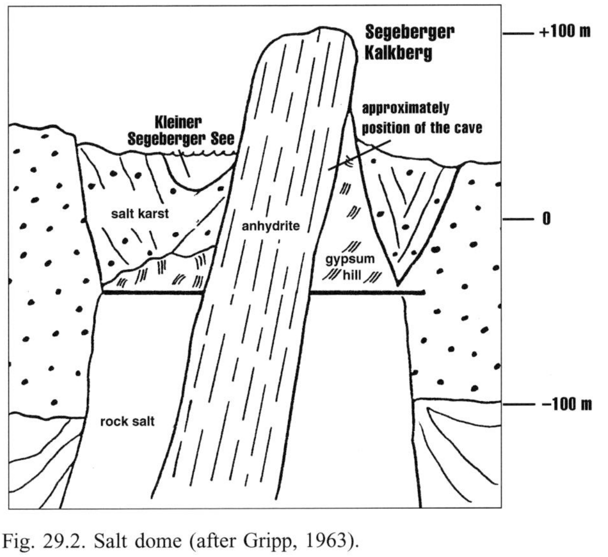

Fig. 29.2. Salt dome (after Gripp, 1963).

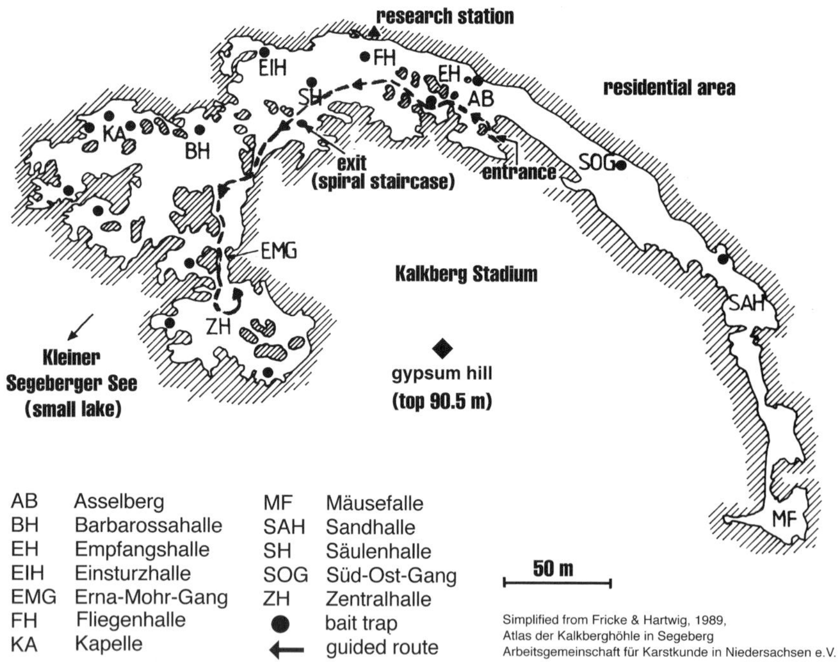

Fig. 29.3. Ground plan of the Segeberger Höhle, 90.5 m above sea level (modified after Fricke et al., unpublished).

Fig. 29.4. Cave passage (photo Ipsen).

recently lowered. In the middle of the 19th century the water level of a small lake (Kleiner Segeberger See), situated about 60 meters away from the cave, another salt dome formation (doline), was artificially lowered by about 10 to 15 m (Regge and Hahn, 1978). This may have drained the Segeberger Höhle as well.

Gripp (1931) explained the cave drainage differently. According to him the water penetrated into artificial shafts of the salt mine. These mine shafts were dug into the hill in 1870 to mine salt. The water pressure was too high and the shafts collapsed. The penetrating water might be the reason for the destruction of the passage between the Kleiner Segeberger See and the cave. Mud and rock debris blocked the connecting clefts and cracks.

The formation of the cave is unique in northern Germany. The cave consists of about twelve larger rooms which are connected by a system of narrower passages which are nearly 2 km long in total. The biggest hall is 40 m in diameter and up to 10 m high, the smallest one is only a few meters in length and 5 m high. Numerous dead ends are characteristic for the cave. On average, the cave floor lies about 50 m underneath the top of the gypsum hill. The floor is covered by clay, sand, loose coarse gravel and big stones (Figs. 29.3, 29.4.). The cavities were formed by leaching and erosion. They have horizontal ceilings and V-shaped walls. This is caused by the fact that the degree of water saturation with gypsum is highest at the bottom of the crevices and lowest at the upper groundwater level. The large cavities have their characteristic trough-like form as a result of the different solutional activities in the upper and lower water layers (Fig. 29.5). The water originally came from the small lake mentioned above. It had been seeping through the clefts and cracks into the gypsum hill. In addition, fractures have transformed the edges of the hollows (Hagel, 1978). Today, a system of clefts

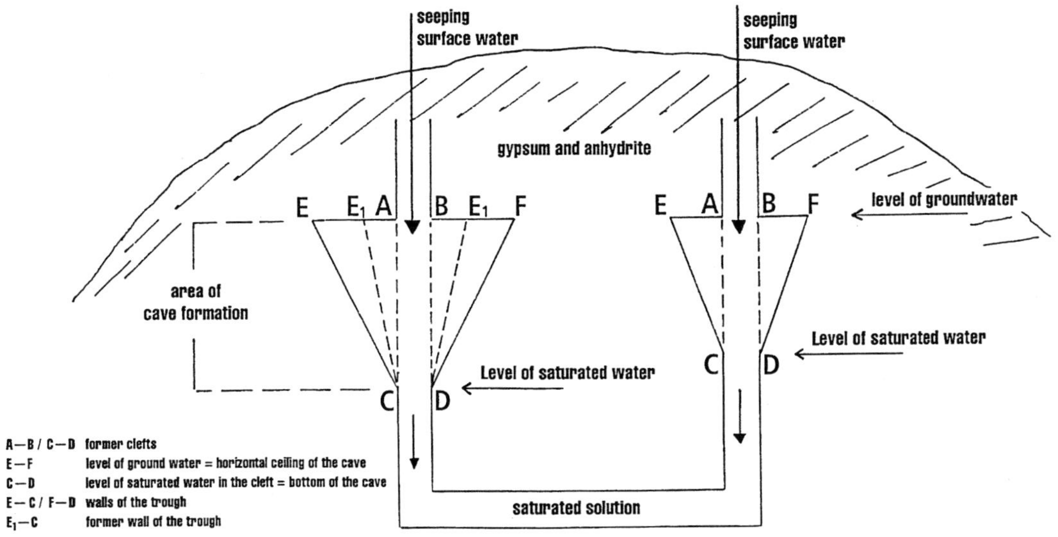

Fig. 29.5. Scheme of the origin of gypsum cave passages (modified after Gripp, 1931).

and cracks within the upper parts of the gypsum cap still exists.

Cave climate

The cave climate is characterized by a constant temperature of 9°C and saturated humidity throughout the year. Only close to the artificial entrances do both these factors vary, depending on the outside weather (Mohr, 1930, 1937; Heun, 1955; Kugelschafter, 1994; Ipsen, 1999). Air movement is low in all sections of the cave.

THE CAVE COMMUNITY

The fauna of the cave consists of about 30 species which can be divided into four different ecological groups (Table 29.1).

Troglobites

The most numerous species and the only troglobite is the beetle *C. s. holsatica,* which is endemic to the cave.

Distribution and relationship of *C. s. holsatica*

C. s. holsatica was first described by Benick and Ihssen (Benick, 1937). The beetle belongs to the family Cholevidae. Species of this family often live in burrows of rodents, under roots of trees, or in other natural hollows like caves. They are necrophagous and saprophagous (Freude et al., 1971).

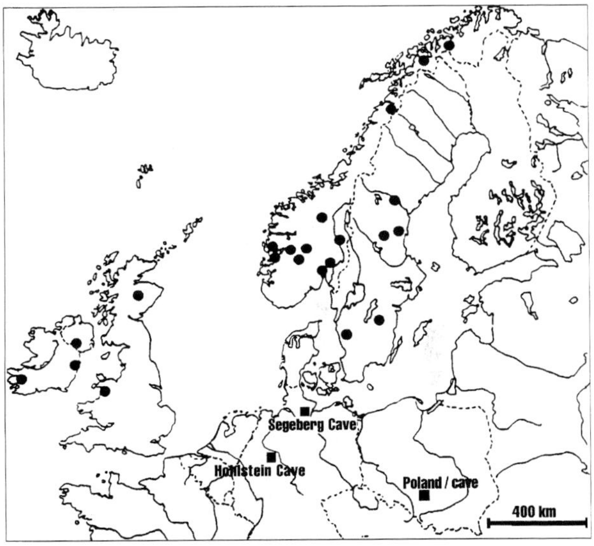

Fig. 29.6. Geographical distribution of *C. s. holsatica* and its closest relatives in Northern Europe. Solid circles, *C. s. septentrionis*; squares, *C. s. holsatica* from the Segeberg Cave (Germany), *C. gracilenta* from Sokola Góra (Poland) and *C. s. sokolowskii* from the Hohlstein Cave (Germany). Modified after Schilthuizen (1990).

Today the closest known relatives of *C. s. holsatica* are *Choleva lederiana* and *Choleva septentrionis septentrionis*. They occur in Scandinavia. According to Schilthuizen (1990) *C. s. septentrionis* is also found in Great Britain and in Ireland (Fig. 29.6). The species belong to the *Choleva agilis* (Illiger) group.

Table 29.1
Fauna of the Segeberger Höhle

(1) Troglobites	
Coleoptera	
	Choleva septentrionis holsatica
(2) Troglophiles	
Mollusca	
	Oxychilus cellarius
Arachnida	
	Meta menardi
	Clubionidae
Crustacea (Isopoda)	
	Oniscus asellus
	Porcellio scaber
Chilopoda	
	Lithobius forficatus
Collembola	
	(various species)
Insecta (Coleoptera)	
	Quedius mesomelinus
	Tachinus rufipes
Insecta (Diptera)	
	Fungifora unipunctata
	Triphleba hyalinata
(3) Seasonal Visitors	
Insecta (Coleoptera)	
	Catops sp.
Insecta (Diptera)	
	Culex pipiens
Mammalia (Chiroptera)	
	Myotis bechsteini
	Myotis brandti
	Myotis dasycmene
	Myotis daubentoni
	Myotis myotis
	Myotis nattereri
	Plecotus auritus
(4) Accidentals	
Oligochaeta	
	Lumbricidae
Mammalia (Rodentia)	
	Rattus norvegicus

Studies of the family suggest that *C. s. septentrionis* and also *Choleva agilis,* another surface species of this family, originally were distributed all over Germany, and that the geographical isolation of the Segeberg cave form and other cave forms in Poland and Romania occurred relatively late in the postglacial period. Thus, thinking in phylogenetic terms, the colonization of the caves took place only recently.

In 1996 *Choleva septentrionis sokolowskii,* which is closely related to *C. s. holsatica* was caught in the Hohlsteinhöhle in the Teutoburger Wald in central Germany (Ipsen and Tolasch, 1997; Ipsen, 1999) (Fig. 29.6.). This fact supports the theory that *C. s. holsatica* is a glacial relict, and that *C. s. septentrionis* migrated northwards with the retreating ice.

Morphology of *C. s. holsatica*

C. s. holsatica is 4.7–5.6 mm long, and slender in form (Fig. 29.7). As is characteristic of troglobite animals, it shows special adaptations to cave life, like lengthened extremities, in comparison to surface relatives such as *Choleva agilis* (Benick, 1950b). It is larger than most cholevids. Head, pronotum and elytrae are dark brown to black. Legs, antennae and mouthparts are uniformly covered with hairs. The body also is covered with appressed yellow pubescence. The surface of the head is finely and sparsely punctate, with fine distinct microsculpture (Fig. 29.8). The antennae are long and slender; all segments are longer than wide. The wings are well developed. The elytrae have not grown together, and have sharp tips. *C. s. holsatica* has not been observed to fly. The male metatrochanter is short with a long tooth (Freude et al., 1971). Compared to its relative *C. agilis,* its elytrae have less cutin and pigmentation (Benick, 1950b).

The eyes of *C. s. holsatica* are smaller than those of *C. agilis* (Mohr, 1929; Benick, 1950b; Heun, 1955). The average number of ommatidiae is 185 (range 156–215) (Fig. 29.9). The average number of ommatidia in the eyes of *C. agilis* is 230 (Benick, 1950b).

Fig. 29.7. Male of *C. s. holsatica,* approximately 5 mm long (photo Ipsen).

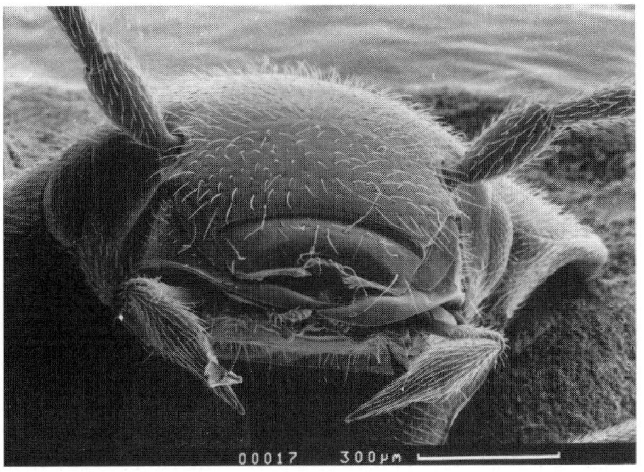

Fig. 29.8. Head of *C. s. holsatica* (photo Ipsen).

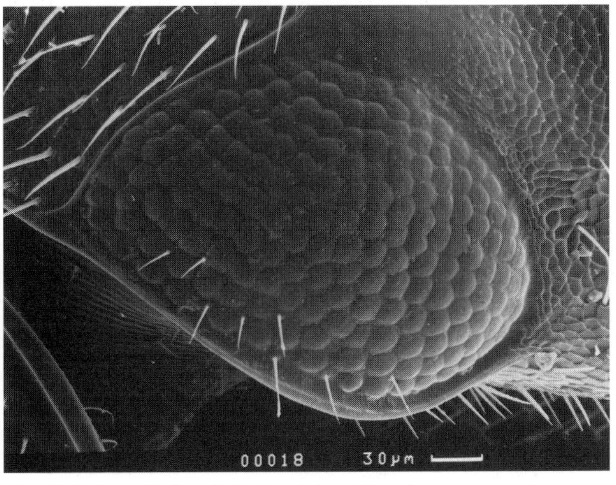

Fig. 29.9. Eye of *C. s. holsatica* (photo Ipsen).

Life cycle of *C. s. holsatica*

In the laboratory the development of the eggs of *C. s. holsatica* takes approximately 27 days. The ontogeny is rather difficult to observe because the eggs are not transparent. They are placed by the females singly or in groups of two to four on the ground, or directly in bat carcasses. The eggs are comparatively yolk-rich; they are extremely variable in size, but usually have a diameter of 0.5–1.0 mm. The females produce between 27 and 34 eggs within 4 weeks. Egg size is not correlated with the female body size.

Only a few days before hatching the larvae can be seen within the egg, which at this developmental stage has become transparent. After hatching, the larvae pass three different stages of development. The average duration of the first and second stage is 10 days each, the third takes 12–15 days. Molting lasts only 4 minutes. At the end of the third stage the larvae become slow and motionless. They stop feeding and build a pupa cocoon consisting of soil. This cocoon is open. In case no material can be found or it is dry and hard, the pupa lies directly on the ground. It remains dormant for 26–29 days. The pupa is very sensitive to light and every kind of disturbance.

Ecology of *C. s. holsatica*

C. s. holsatica is a stenotherm species. A test in a temperature-gradient apparatus, in which the beetles could choose their range of preference, showed that they prefer 9°C, a temperature which is typical of the cave. They need a high humidity of almost 100% to develop, otherwise the larvae cannot hatch and they die. Experiments show that the beetles' phototactic response is negative. Whether *C. s. holsatica* is able to discern shapes remains unclear. The beetle feeds mostly on bat guano and bat carcasses.

In the laboratory the beetles did not synchronize their activity rhythm. At any one time some of them were active searching for food, while others rested at the bait and agglomerated into groups of 4–5 individuals. This behavior corresponds to that observed in the cave (Fig. 29.10).

Population size and seasonality of *C. s. holsatica*

The beetles stay in parts of the cave which are absolutely dark and where the floor is relatively dry. They prefer sections with large numbers of bats (mainly *Myotis* spp.) hibernating during the winter season.

The population size is about 3500 beetles, as estimated by a capture–recapture-method based on the Lincoln index (Jolly, 1965). Most of the beetles are found in the cave hall called "Kapelle" (2000 specimens). This hall is uninfluenced by climatic variations from outside. The ground of the hall is sandy and relatively dry. During winter approximately 5000 bats roost on the ceiling and in the clefts. Floor and ceiling have many clefts, cracks and dead ends in which the beetles hide.

There is an obvious annual rhythm in the number of adult beetles counted in the cave during each visit. This phenological rhythm seems to depend on the presence of bats (Fig. 29.11). Similar results were found earlier (Heun, 1955). In October and November only a few adult beetles were observed, but many larvae. The number of adults increases up to the middle of December and decreases again in February. In March

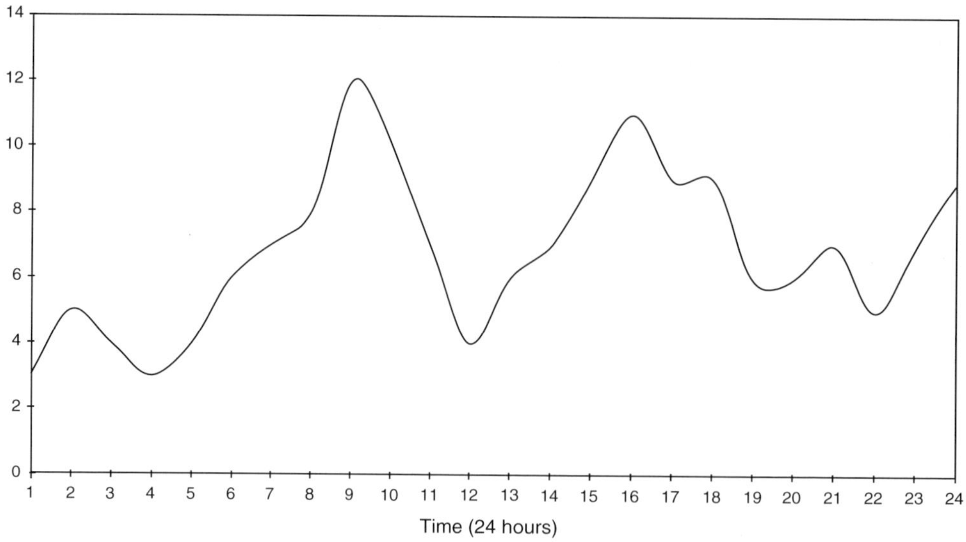

Fig. 29.10. Activity of *C. s holsatica.*

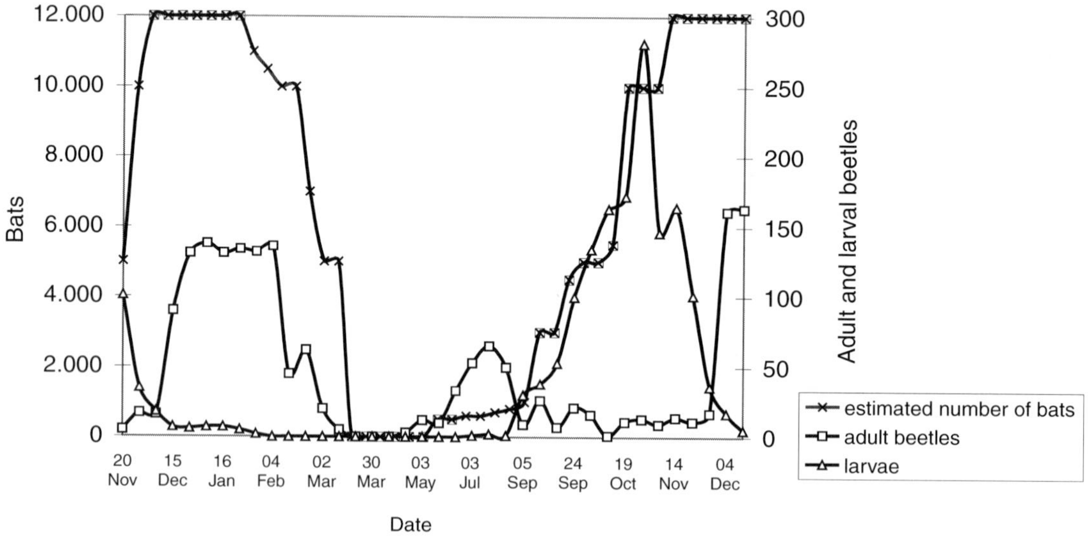

Fig. 29.11. Phenology of adult and larval *C. s. holsatica* and bats in the Segeberger Höhle (observation time Nov. 1994 to Dec. 1995; number of bats staying in the cave according to Kugelschafter (1994), beetles and their larvae found on 16 bait traps distributed over the cave according to Ipsen).

and April when the bats have left the cave only a few beetles were counted. Only in July and August could beetles be observed in copula. Larvae were seen in the cave from August until the end of December. The first newly hatched adult beetle was observed on November 20, a time when most bats have entered. In the section investigated, the beetles reproduce only once a year. In the laboratory with constant food supply, also, only one generation per year develops. The reason for the decrease of the number of beetles observed in February is not clear.

Heun (1955) postulated a period of dormancy from February till July as a result of the absence of bat guano. However, as the beetles also feed on bat carcasses which are found in the clefts all the time, her theory does not seem to be correct. This was shown by food-preference experiments. Having the choice between guano, carcasses and cheese, the beetles prefered fresh carcasses. Two weeks after the death of the bat, the carcass becomes less attractive than the cheese. The author's interpretation is that the beetles hide within the clefts and stay on the upper level of the cave which is not accessible to humans. Marking beetles during the whole spring and summer shows that some specimens stay in the low-lying parts of the cave all the time and do not move into the cracks. The beetles die in autumn after reproduction.

Systematic position and phylogenetic age of *C. s. holsatica*

The systematic position of *C. s. holsatica* is still under dispute. Benick (1950b) suggested that specimens of *C. s. septentrionis* migrated with the retreating ice-belt northward, and that some of them colonized the cave in Segeberg. According to him they stayed here as a relict of the glacial period. Benick (1950b) argued from similar morphological characters in both forms. The existence of *C. s. sokolowskii* (Ipsen and Tolasch, 1997) in central Germany some 350 km further south in the Teutoburger Wald supports his theory. If Benick's hypothesis is correct and if this previously unknown form has colonized the cave in the Teutoburger Wald earlier than *C. s. holsatica* the Segeberger Höhle, it is possible and probable that its troglobitic adaptations are more advanced.

The time of cave colonization by *C. s. holsatica* and other terrestrial animals in Segeberg, and their phylogenetic age, cannot be determined exactly. It is unknown at what time exactly the solution process of the salt-dome cap started, but it cannot be excluded that this might have been already during the last interglacial. The process of colonization by animal species, however, can only have started after the glaciers had been melting in post-Pleistocene times about 20 000 years BC. At that time the formation of the cave began. Crevices allowing bats and beetles to enter the system surely existed at earlier times. From a biological point of view it seems unlikely that this was as late as the 19th century, when the accessible part of the cave finally drained because the groundwater level fell (Gripp, 1913, 1963).

Consequently the phylogenetic age of *C. s. holsatica* is considered to lie between 15 000 and 20 000 years (Benick, 1950b; Gripp, 1913). That means that it is a relatively young species.

The ecological behavior of an animal and its morphological characters in many cases indicate its phylogenetic age. In cave species often the selective force for life-history adaptation is a low food supply (Culver, 1982; Culver et al., 1995). *C. s. holsatica,* as a phylogenetically young cave beetle, still shows the typical life history of an unspecialized surface species. Specialized cholevid beetles with a long trophilic history, for example *Speonomus hydrophilus* or *S. longicornis* (Bathysciinae), which are endemic to the Grotte Bastardech in the Pyrénées (Ariège, France) (Juberthie et al., 1980a,b) live far from the entrances. In that part of the cave food is so rare and highly dispersed that normal larvae, having limited mobility, would starve (Poulson and White, 1969). In the course of evolution these species have essentially dispensed with the larval stage. The mobile adult searches widely for food and, at irregular intervals, lays only a few large eggs each of which directly hatches into a mature last-instar larva. All intermediate larval stages are no longer developed. Immediately after hatching, the mature larva constructs its pupal case without feeding; it is smooth and has reduced sensory and feeding structures (Juberthie, 1969; Poulson and White, 1969). This example shows that troglophilic adaptations of old cave beetles are more advanced than that of *C. s. holsatica* (Gnaspini 1992, 1993).

The larva of *C. s. holsatica* is mobile and shows no reduction of larval characters. Compared to *C. agilis* the number of eggs is reduced (Benick, 1950a) – the average number of eggs of *C. agilis* is 60–70, and they are placed in groups of 5–10 (Heun, 1955; Koch, 1989; Topp, 1990). The egg shells of *C. s. holsatica* are not transparent and do not prevent desiccation. There is no pupa cocoon, which is closed and could protect the larvae against any kind of disturbance. Contrary to other epigean cholevids, *C. s. holsatica* has only one reproductive phase per year – that means only one generation developed each year – and the population size is relatively small. This could be because it is adapted to a low adult mortality or a low food supply. Surface cholevids produce two generations, one in spring and one in autumn, depending on the food supply and the abiotic factors temperature and humidity. During the winter season the adults fall into a period of dormancy and in summer, while the temperature is too high, they reduce their activity (Freude et al., 1971; Ruzicka, 1993; Topp, 1994).

The annual rhythm in the phenology of the beetle is different from that of surface forms. Its reproductive behavior is correlated with the migration of the bats into the cave. Summarizing, it is possible to name characters which indicate that *C. s. holsatica* is ecologically well adapted to cave life (temperature preference, small number of ommatidia, life cycle, annual rhythm in the phenology depending on the bats). On the other hand, the morphological characters are a sign of the young phylogenetic age of this cave form (eye reduction on a low level, wings developed: Wilkens, 1988; Ipsen, 1999). These hints coincide with the hypothesis about the age of the cave system and its evolution.

Troglophiles

The largest group of species living in the cave is that of troglophiles. These are animals which are able to spend their whole life in the cave and to propagate here. All of them are characterized by a low population density (Table 29.1).

There is a small population of the mollusc *Oxychilus cellarius* in the cave (max. 50 specimens). They have no natural predator in the cave, and feed on bacteria and fungi. Specimens of different ages can be found in all parts of the system.

The araneid spider *Meta menardi* often lives in caves. In Segeberg it is concentrated just below the entrance and in some other halls. Only small numbers, up to 25 adult specimens, can be counted on each visit. They feed mostly on the isopods *Oniscus asellus* and *Porcellio scaber*.

Oniscus asellus and *Porcellio scaber* are found in low numbers (1–10 specimens per visit) through the whole cave. Some specimens show less pigmentation, but it is not sure whether this is genetically based or just caused environmentally by darkness. However, when the animals are kept under light conditions the pigmentation does not become darker.

One to three specimens of the chilopod *Lithobius forficatus* were observed in the cave on each visit. It is a predator on *C. s. holsatica* and on the isopods (Mohr, 1937; Heun, 1955).

The abundance of dipteran larvae (Phoridae and Sarcophagidae) is quite high (10–100 specimens counted per visit). They occur in the whole cave. Collembola exist in high abundance and with several species in the cave (Schmitz, 1937).

A large number of Acari live in the cave floor (Rack, 1962), and others on the bats and on the beetle larvae.

Seasonal visitors

Approximately 15 000–20 000 bats hibernate during the winter in the cave (Kugelschafter, 1994; Table 29.1). They belong to the species *Myotis bechsteini, M. brandti, M. dasycmene, M. daubentoni, M. myotis* (Fig. 29.12), *M. nattereri* and *Plecotus auritus* (Fig. 29.13). The most abundant species are *Myotis daubentoni* (7000) and *Myotis nattereri* (8000).

Fig. 29.12. Hibernating specimens of *Myotis myotis* (about 100 mm long) (photo Tolasch).

Bat activity during wintertime does not depend only on the outside temperature. Internal rhythms may also be decisive factors (Fig. 29.14). The bats which stay in the Segeberger Höhle come from all over Central Europe. The largest distance of migration observed was 1500 kilometers. This specimen came from Scandinavia. Most bats stem from the coast of the North Sea, which is about 100 km away, or from the closer environment of Segeberg. Guano analysis has shown that the bats mostly feed on Chironomidae, Hymenoptera, Lepidoptera and Trichoptera. The most important place where the bats hunt at night is the shore of a large lake nearby (Großer Segeberger See).

Another seasonal guest is the dipteran *Culex pipiens,*

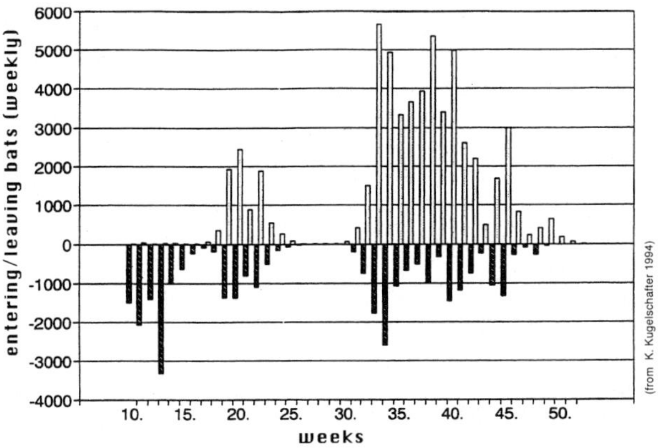

Fig. 29.13. Bat activity in the Segeberger Höhle in 1993 (from K. Kugelschafter, 1994). Dark columns, weekly numbers of departures of bats from the cave; light columns, weekly numbers of arrivals of bats at the cave.

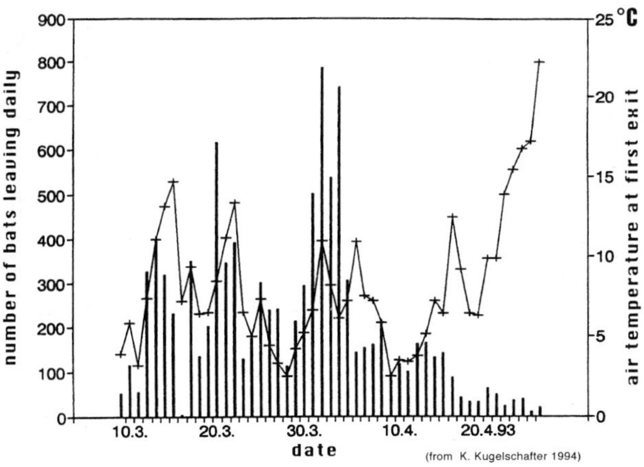

Fig. 29.14. Correlation of outside temperature and number of bats leaving the cave in spring 1993.

which stays near the entrances in large numbers. In spring it leaves the cave because it cannot reproduce there. The beetles *Quedius mesomelinus* and *Tachinus rufipes* (Staphylinidae) are seasonal guests, too. They can be found in small numbers in the entrance section of the cave. They are predators of *C. s. holsatica.*

Accidentals

This group includes animals which accidentally enter the cave (Table 29.1) and occur as single specimens. They can live here only for a short time and do not reproduce. They include surface forms like rats (*Rattus norvegicus*), lumbricids, snails and spiders.

FOOD WEB

The food web is quite simple (Fig. 29.15). The most important resource is the guano dropped by the bats beneath their roosts. The bat carcasses which are predominantly found in the deep clefts are also part of the food basis. Cave beetles rely exclusively on bat guano and carcasses. The dipterans *Fungifora unipunctata, Triphleba hyalinata* and the mollusk *Oxychilus cellarius,* and the isopods *Oniscus asellus* and *Porcellio scaber,* as well as mites and collembolans, also depend on this nourishment. Predators do not play an important role. The spider *Meta menardi* feeds on isopods. An undetermined spider of the family Clubionidae seems to catch collembolans and small dipterans. *C. s. holsatica* is obviously only predated on by the rare troglophile beetles *Quedius mesomelinus* and *Tachinus rufipes,* and the chilopod *Lithobius forficatus.*

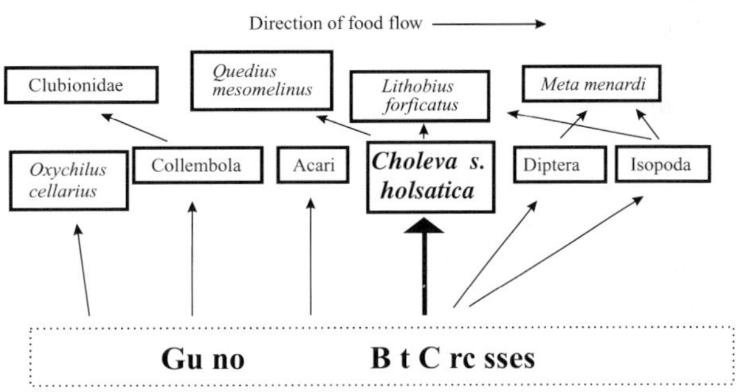

Fig. 29.15. Food web of the Segeberger Höhle.

HUMAN IMPACT

The Segeberger Höhle was only discovered in 1913 by children at play. Already six years after the discovery the cave was made accessible to tourists. Since that time thousands of people visit the cave yearly. The reason for its popularity is its uniqueness in northern Germany. The high attraction for tourists had several consequences for the animals living in the cave system. The cave has had artificial lighting since 1919. Until 1980 the whole cave was illuminated. The lights are used during guided tours, and caused some parts of the cave to become uninhabitable for animals adapted to darkness. Furthermore, the artificial light affects the ecosystem because the abiotic factors change in the surroundings of the bulbs. Algae (*Lampenflora*) which normally do not occur in the cave system can grow there. As a consequence, terrestrial animals like snails may reproduce because they feed on these plants.

Two artificial entrances were provided. Near these entrances the abiotic factors temperature and humidity vary in dependence on the weather outside the cave. These changes disturb the ecological equilibrium in that part of the cave. For construction of paths, parts of the floor were covered with gravel. Experiments show that this gravel functions as a barrier; the distribution of the beetles might be split into smaller isolated populations. These smaller populations are more endangered because the beetles cannot take refuge in other parts of the cave. Retreat in case of danger or changing conditions is not possible.

Tourists who are visiting the cave in a large number leave garbage in the entrance area. This garbage lures animals like rats which accidentally run into the cave. They cannot find any food here and unable to leave the cave again they are forced to feed on beetles (own observation). These new predators are a serious threat for the endemic beetle.

Already in the 1930s the quarry above the cave, which is a result of salt-mining in the 19th century, was rebuilt as an open-air theater. At least 250 000 people visit the so-called Kalkberg Stadion every year. The masses of people disturb the delicate cave ecosystem – for example, the music causes erosion. Until the early 1990s sewage was spilled directly into parts of the cave. At present terrestrial animals are hardly to be found at this place. Erosion and washing-out of the soil diminished the amount of substrate. In consequence of this, the beetles also retreated (Mohr, 1929; Heun, 1955; Ipsen, 1999).

REFERENCES

Benick, L., 1937. Die Käfer der Segeberger Höhle. *Schr. Naturw. Ver. Schleswig-Holst.*, 22(1): 146–176.

Benick, L., 1950a. Vom Kleintierleben in der Segeberger Höhle. *Die Heimat,* 57: 138–142.

Benick, L., 1950b. Die Reliktnatur der *Choleva holsatica* L. Bck. et G. Ihss. aus der Segeberger Höhle und die Möglichkeit ihrer Altersbestimmung. In: *Neue Ergebnisse und Probleme der Zoologie (Klatt-Festschrift).* Zool. Anz., Ergänzungsbd. zu Bd. 145, Leibzig, pp. 46–56.

Culver, D.C., 1982. *Cave Life, Evolution and Ecology.* Harvard University Press, Cambridge, MA.

Culver, D.C., Kane, T.C. and Fong, D.W., 1995. *Adaptation and Natural Selection in Caves.* Harvard University Press, Cambridge, MA.

Freude, H., Harde, K.W. and Lohse, G.A., 1971. *Die Käfer Mitteleuropas.* Vol. 1–4, Krefeld.

Gnaspini, P., 1992. Bat Guano Ecosystems, a new classification and some considerations with special references to neotropical data. *Mem. Biospeol.,* 19.

Gnaspini, P., 1993. Brazilian Cholevidae (Coleoptera), with emphasis on cavernicolous species. IV. Adelopsis (Iutururuca) biology and description of larvae. *Mem. Biospeol.,* 20: 91–99.

Gripp, K., 1913. Über den Gipsberg in Segeberg und die in ihm vorhandene Höhle. *Beiheft Jahrb. Hamb. Wiss. Anst. XXX (1912),* 6: 35–51.

Gripp, K., 1931. Neues über die Entstehung der Höhle im Gipsberg zu Segeberg. *Die Heimat,* 41: 234–237.

Gripp, K., 1963. Neues über den Gipsberg und die Höhle zu Segeberg. *Heimatkd. Jahrb. Kreis Segeberg,* 97–107.

Hagel, J., 1978. Der Segeberger Kalkberg in der Literatur. In: *Heimatkd. Jahrb. Segeberg.* Stuttgart.

Heun, C., 1955. Biologie und Ökologie von *Choleva holsatica* Ihss. & Ben. *Entomol. Mitt. Zool. Staatsinst. Zool. Museums Hamburg,* 7: 196–223.

Ipsen, A., 1999. *Biologie, Ökologie und Systematik van höhlenbewohnenden Käfern der Gattung* Choleva *(Coleoptera, Cholevidae) in Norddeutschland.* Ad fontes, Hamburg, p. 126.

Ipsen, A. and Tolasch, T., 1997. *Choleva septentrionis sokolowskii* n. ssp., eine neue cavernicole Unterart von *Choleva septentrionis* Jeannel aus Mittelschweden (Col. Cholevidae). *Entomol. Nachr. Ber.,* 41: 167–171.

Jolly, G.M., 1965. Explicit estimates from capture–recapture data with both death and immigration (stochastic model). *Biometrica,* 52: 225–247.

Juberthie, C., 1969. Etude écologique des larves de *Speonomus infernus* dans la grotte Sainte-Catherine. *Ann. Spéléol.,* 24: 563–577.

Juberthie, C., Delay, B. and Bouillon, M., 1980a. Extension du milieu souterrain en zone non-calcaire: description d'un nouveau milieu et de son peuplement par les Coléoptères troglobie. *C.R. Acad. Fr.,* 290: 49–52.

Juberthie, C., Delay, B. and Ruffat, G., 1980b. Variations biométriques entre différentes populations de *Speonomus hydrophilus* en relation avec leur situation géographique (col. Bathysciinae). *Mem. Biospeol.,* 7: 249–266.

Koch, K., 1989. *Die Käfer Mitteleuropas,* Vols. 1, 7. Goecke und Evers Verlag, Krefeld.

Kugelschafter, K., 1994. *Untersuchungen zur Bedeutung der Segeberger Kalkberghöhle und angrenzender Naturbiotope für Fledermäuse.* Gutachten im Auftrag des MNUL Schleswig-Holstein, Gießen.

Mohr, E., 1929. Biologische Untersuchungen in der Segeberger Höhle. *Nat. Ver. Bd. XIX,* 1: 1–25.

Mohr, E., 1930. Die Höhle von Segeberg (Holst.) und ihre Bewohner. *Mitt. Höhlen- und Karstforschung,* 81–89.

Mohr, E., 1937. Neue biologische Untersuchungen in der Segeberger Höhle. *Nat. Ver. Bd. XXII,* 1: 116–145.

Poulson, T.L. and White, W.B., 1969. The cave environment. *Science,* 165: 971–980.

Rack, G., 1962. Die Milben der Segeberger Höhle. *Heimatkd. Jahrb. Kreis Segeberg,* 146–167.

Regge, J. and Hahn, H., 1978. *Der Segeberger Kalkberg und seine Höhlen.* Verlag C.H. Wäser, Bad Segeberg.

Ruzicka, J., 1993. Three new species of *Choleva* (Coleoptera:

Leiodidae: Cholevinae) from the Caucasus and Turkey, with a key to species of the *cisteloides* group. *Eur. J. Entomol.,* 90: 337–348.

Schilthuizen, M., 1990. A revision of *Choleva agilis* (ILLIGER, 1798) and related species (Coleoptera: Staphylinoidea: Cholevidae), Zool. Med. Leiden, 64(10): 121–153.

Schmitz, H., 1937. *Triphleba* (s. str.) *morae* n. sp. aus der Segeberger Höhle. *Nat. Ver.,* 22(H1): 177–178.

Teichmüller, R., 1948. Das Oberflächenbild des Salzdomes von Segeberg in Holstein. *Z. Dtsch. Geol. Ges.,* 98.

Topp, W., 1990. Selection for an optimal monovoltine lifecycle in an unpredictable environment. Studies on the beetles *Catops nigrigans* Spence (Col. Catopidae). *Oecologia,* 84: 134–141.

Topp, W., 1994. Seasonal time partitioning and polymorphism in the development cycles of sympatric Staphylionidae (Coleoptera) living in unstable environments. In: H.V. Danks (Editor), *Insect Life-Cycle Polymorphism.* Netherlands, pp. 277–312.

Wilkens, H., 1988. Evolution and genetics of epigean and cave *Astyanax fasciatus* (Caracidae, Pisces). Support for the Neutral Mutation Theory. *Evol. Biol.,* 23: 271–367.

Chapter 30

THE HYPOGEAN FAUNA OF THE CAPE RANGE PENINSULA AND BARROW ISLAND, NORTHWESTERN AUSTRALIA

W.F. HUMPHREYS

INTRODUCTION

Rich troglobite communities have recently been found to inhabit caves in tropical limestones and lavas of both oceanic islands and continental areas (Howarth, 1973, 1980, 1982, 1988; Hoch and Howarth, 1989a,b, 1993; Howarth and Stone, 1990; Humphreys, 1993a–d). In addition, plate tectonics is increasingly invoked to explain the distribution of both island (Manning et al., 1986) and continental (Stock, 1994) troglobites, especially those associated with Tethys. These changes in perception have re-invigorated debates in biospeleology (see Humphreys, Chapter 22, this volume).

A priori, the arid northwest of Australia might be thought to be peripheral to these debates – not only is it far removed from the classical Tethyan regions centered on the Mediterranean and on either side of the Atlantic, but arid zones were considered to be sparsely populated with troglobites owing to lack of water (Howarth, 1980) and low food input from xeric plant communities (Peck, 1978). Hence, the discovery of the troglobite communities, amongst the world's richest, in the arid Cape Range region adds another level to the re-invigorated debates concerning the origin of troglobites.

This chapter addresses the nature of the karst area on the Cape Range peninsula of Western Australia and includes references to the geologically related Barrow Island lying 170 km to the northeast on the North West Shelf.

REGIONAL SETTING

Location

The Cape Range peninsula lies just within the tropics of north-western Australia (Figs. 30.1, 30.2) in the arid zone that occupies much of Central and Western Australia. It lies more that 1200 km from the nearest wet temperate forest in southwestern Australia, and more than 3000 km from the lowland rainforests of eastern Australia.

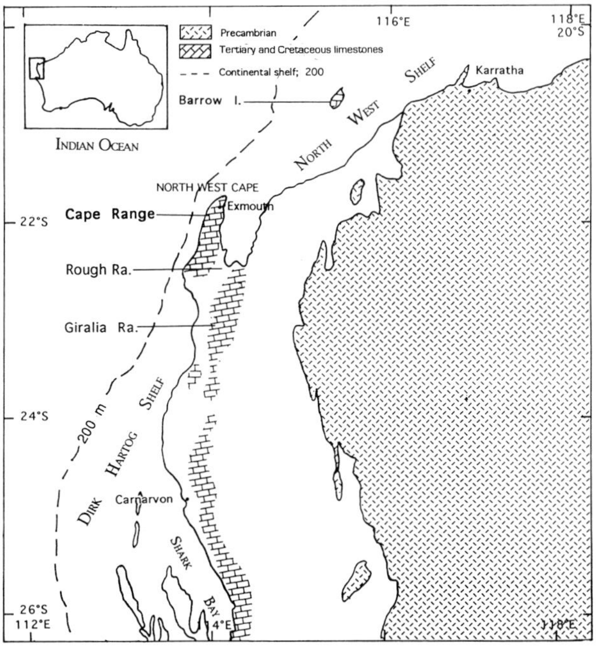

Fig. 30.1. The regional setting of the Cape Range peninsula and Barrow Island. The Precambrian Pilbara Craton has not undergone major regional deformation since 2400 Ma ago (Trendall, 1990) and has been continuously emergent for more than 600 Ma. This is fringed to the west by middle Mesozoic deposits (~100 Ma old). A series of anticlines of Cretaceous and Tertiary limestones occur further to the west. The cavernous Tulki Limestone is known only from Cape Range, and the Cape Range (*sensu stricto*) troglobite fauna is restricted to it. Compiled from Hocking et al. (1987) and Wyrwoll et al. (1993).

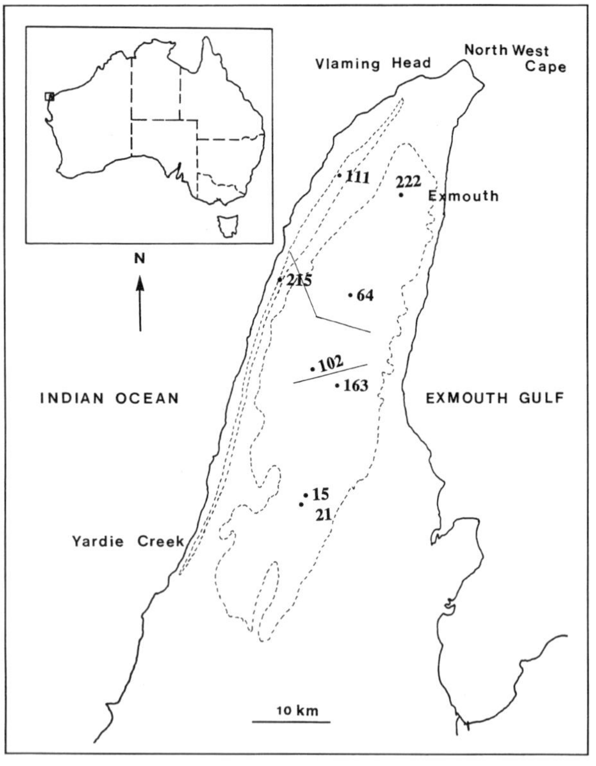

Fig. 30.2. The Cape Range peninsula. The dotted lines show the extent of the surface exposure of the cavernous Tulki Limestone, and the numbered points denote the location of caves mentioned in the text. The straight lines indicate the approximate boundaries between of the genetic provinces determined from allozyme electrophoresis of the millipede *Stygiochiropus communis* (see Fig. 30.5).

Geomorphology

Cape Range lies in the Carnarvon Basin, which is bounded in the east by a Precambrian craton – the Pilbara Craton, which has been continuously emergent for more than 600 Ma (million years). It is a faulted, gently folded Phanerozoic basin containing up to 7 km of Palaeozoic sediments covered by a Mesozoic veneer which thickens to the north and west. The break-up of Gondwana occurred here in the mid-Jurassic when rapid subsidence formed a transgression–regressive couplet 3 km thick. In the Late Cretaceous carbonate deposition occurred on the marine shelf and continental slope (Hocking et al., 1987).

Tectonic activity during the Cenozoic has resulted in the domination of the Cape Range–Exmouth Gulf region by a series of anticlines in marine limestones exposed at the surface at the Cape, Rough and Giralia Ranges (Fig. 30.1). Geologically the Cape Range Group comprises a sequence of marine, mainly carbonate, sediments comprising the Mandu (Late Oligocene), Tulki and Trealla (Middle [?Late] Miocene) Limestones (Wyrwoll et al., 1993). The Tulki-like Poivre Formation of Middle Miocene age occurs on Barrow Island (McNamara and Kendrick, 1994).

A series of emerged and warped marine erosion terraces cut into Tulki Limestone on the western side of Cape Range up to an altitude of ~60 m (van de Graaff et al., 1976), the lowest forming a continuous coralline bench up to 1.7 km wide. Warping of the upper terraces has been considered to result from tectonic uplift during the Quaternary (van de Graaff et al., 1976), whereas the lowest terrace (Tantabiddi Member) belongs to the last interglacial of the Late Pleistocene (132 000–127 000 years BP: Kendrick et al., 1991). However, recent fossil evidence suggests that the next highest terrace (Jurabi Member) is not younger than late Pliocene (>1.8 Ma BP: Wyrwoll et al., 1993).

Cape Range attains an altitude of about 314 m and is deeply dissected on the dip slopes. Erosion has removed much of the Trealla Limestone and several of the deeper gorges have cut through to the Mandu Limestone. Nearly all caves in the range occur in Tulki Limestone (Humphreys, 1993d: frontispiece).

The west coast of Cape Range is the closest point on the Australian continent to the outer edge of the continental shelf (Kendrick et al., 1991), but to the northeast extends the broad, shallow North West Shelf (Fig. 30.1), which is contiguous with a series of shelves stretching around northern Australia to New Guinea; the shelf also broadens southwards towards Shark Bay.

Caves

Currently, 416 caves and 13 man-made wells are known from the Cape Range. (R.D. Brooks, personal communication, 1998). Within the Cape Range *sensu stricto,* most of the caves are vertical, reaching depths of up to 80 m (C-102), and only one cave has significant lateral extent (~5 km, C-163). Permanent water can be reached in only four of the caves within Cape Range proper, at an altitude of ~180 m above sea level, and only one outflow cave is known (C-64).

On the coastal plain several caves and sinkholes, a number of wells and numerous bores permit access to the groundwater. Marine tides are apparent in water bores up to 3.5 km inland (Forth, 1973) indicating open-conduit flow of the groundwater within the coastal plain. Such conduits form at the water table, especially by mixture–corrosion[1], and are thought to have developed in karstified Tulki Limestone as it dips beneath sea level. Owing to Pleistocene sea-level fluctuations, the limestone is likely to be cavernous to depths of more than 100 m below sea level – divers have penetrated caves to 32 m below sea level, deeper than the sea over much of the North West Shelf.

Barrow Island has only three caves of biospeleological interest, permanent water being reached in only one, but bores permit access to the groundwater and the mesocaverns above.

Groundwater

The general hydrogeological model for oceanic islands has been applied to the ground water of Cape Range peninsula (Allen, 1993), namely, following the Ghyben–Herzberg principle, a freshwater lens overlies salt water, so that locally a wedge of salt water intrudes under the fresh water contained in the limestone – a zone of mixing occurs between them, broadening towards the coast. In the Exmouth area the fresh-water–salt-water transition is at ~5 km (Martin, 1990), exceptionally far inland (cf., 0.5 km on the coral island of Niue: Jacobson and Hill, 1980), but this is to be expected as the greater the transmissivity of the limestone and the more arid the area, the further inland the wedge should penetrate. In the Exmouth area the inland limit to the salt-water interface appears to be controlled by the presence below the water table of solution cavities and channels (Martin, 1990).

Climate

Annual evaporation (3219 mm) exceeds the sparse precipitation (median = 280 mm; 10th percentile = 100 mm; 90th percentile = 500 mm) by more than an order of magnitude. The mean daily temperature is ~27°C; but the mean monthly maximum temperature exceeds 35°C for four months of the year, and the mean relative humidity is persistently low (53% at 09.00; 33% at 15.00: Vine et al., 1988; Humphreys, 1991a).

Rainfall in the region results from at least four processes – tropical cyclones, monsoons, northwesterly epitropical flows, and frontal events (Gentilli, 1972) – and this results in very low predictability, even for the driest inhabited continent (Humphreys et al., 1989: table 1), in both the seasonal distribution and the between-year variation in rainfall (Fig. 30.3). The very high temperature, low humidity, and high evapo-transpiration excess make it a difficult place for cryptozoic animals.

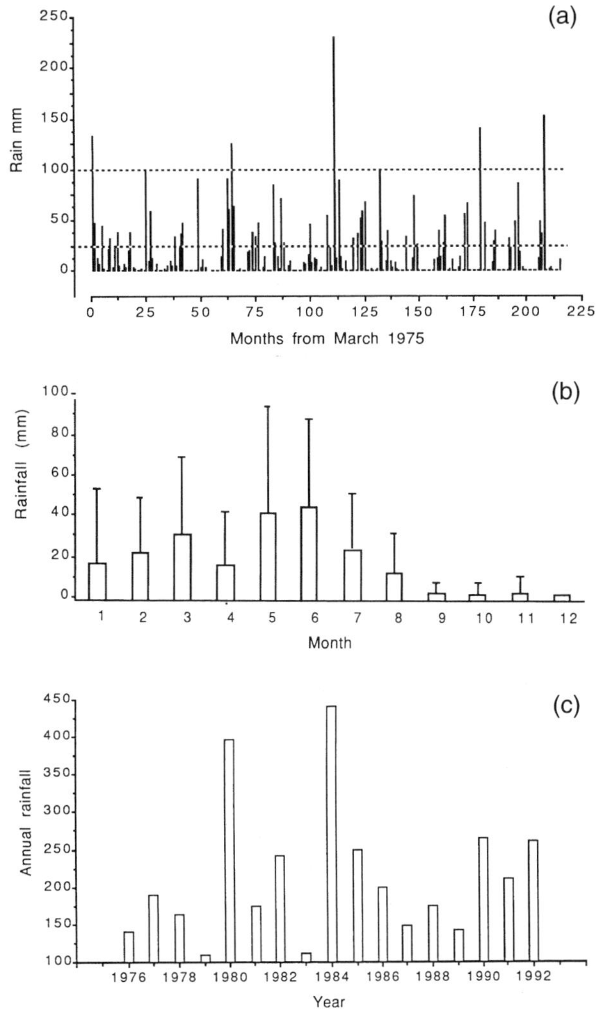

Fig. 30.3. (a) Monthly rainfall (mm) at Learmonth from March 1975. Broken lines at 25 mm and 100 mm are, respectively, the amount of rainfall required for surface flow for minor recharge of some caves, and the amount of rainfall required to flood the major caves deeply. (b) Mean monthly rainfall (mm) (standard deviation) at Learmonth (January = 1). (c) Annual rainfall at Learmonth, 1976–1992.

Rainfall necessary to flood caves

As the caves are recharged with organic matter

[1] See Glossary, p. 6.

primarily carried by the influx of water, the intensity, frequency and periodicity of rainfall will also determine that of the energy recharge.

The karst terrain precludes surface run-off except after exceptional rainfall (Figs. 30.3, 30.4). No recharge is likely to occur following a single rainfall of less than about 25 mm – such rainfall, and hence minor recharge of some caves, occurs on average once every 5.4 months, but with very low predictability (Humphreys et al., 1989). Major flows of water sufficient to flood caves deeply are likely to occur once in 4.7 years (single falls of rain >150 mm: Humphreys et al., 1989) – rainfalls of more than 100 mm occur about once every three years (Fig. 30.3).

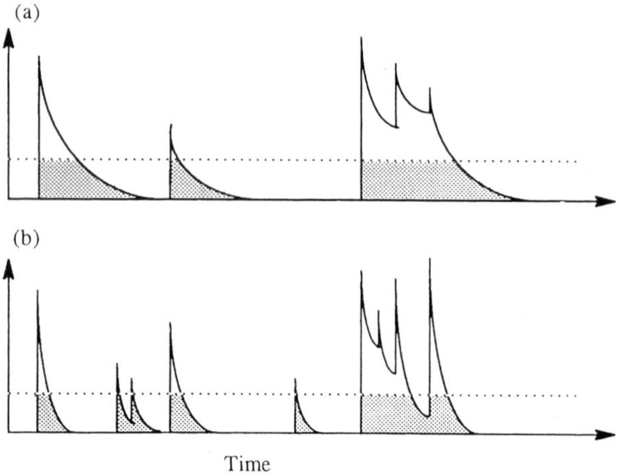

Fig. 30.4. Hypothetical availability and depletion of resources in Cape Range caves: (a) Slow-drying (small entrance) cave with a small catchment; (b) fast-drying cave with large catchment. The fine horizontal line denotes the resource level (energy or water) below which the populations are eliminated from the cave (shaded areas). Population responses would lag following the influx of new resources. Caves with large catchments (b) flood and are recharged more frequently. Slow-drying caves (a) maintain populations for longer after flooding. After Humphreys (1991c, 1993c).

Vegetation

Essentially a widespread arid-zone flora covers the limestone hills, and the adjacent sand and coastal plains. The limestone ranges are dominated by *Eucalyptus* over *Triodia* and the lower slopes by *Acacia* over *Triodia*. The coastal plains are vegetated by grasslands, low shrublands, or by sandplain heath over *Triodia* (Keighery and Gibson, 1993).

BIOSPELEOLOGICAL HISTORY

The biological diversity of the subterranean fauna of the Cape Range area has been recognized only recently.

Aquatic fauna

The first anchialine fauna, considered to be a freshwater fauna (Knott, 1993) until recently (Humphreys, 1993b, 1994), was reported from the west side of the peninsula (Whitley, 1945; Mees, 1962), and later from the east side (Cawthorn, 1963) of the peninsula. Two species of blind fishes *Milyeringa veritas* (Whitley, 1945) and *Ophisternon candidum* (Mees, 1962) and two of atyid shrimps, *Stygiocaris lancifera* and *S. stylifera* (Holthuis, 1960) remained the only aquatic fauna reported until the 1990s, when many more taxa were described and other biospeleological studies commenced: these covered distribution, environment and genetic relations of the fauna (Humphreys and Adams, 1991; Humphreys, 1993b,d); Thermosbaenacea (Poore and Humphreys, 1992); Cirolanidae (Bruce and Humphreys, 1993); Mollusca (Slack-Smith, 1993); recognition of a Tethyan connection (Humphreys, 1993b; Knott, 1993); broad-scale distributions and recognition of anchialine habitat (Humphreys, 1994); trophic relations (Humphreys and Feinberg, 1995); *Danielopolina* (Ostracoda: Thaumatocyprididae: Baltanas and Danielopol, 1995; Danielopol et al., 2000); amphipods (Barnard and Williams, 1995); cyclopoid copepods (Pesce et al., 1996a,b); and Remipedia (Yager and Humphreys, 1996). Representatives of the fauna were later found on Barrow Island on the North West Shelf, 170 km to the northeast (Bruce and Humphreys, 1993), together with the recognition of the great diversity of the amphipod stygofauna on Barrow Island (Bradbury and Williams, 1996a,b, 1997).

Terrestrial fauna

Arachnids (Schizomida and a troglobitic spider) were collected (Main, 1980) from an unidentified cave (Humphreys, 1990a) in Cape Range in 1962. However, the diversity of the terrestrial troglobitic fauna in Cape Range and its status as a relict wet-forest fauna was not appreciated until the late 1980s (Humphreys, 1989), when sustained work commenced on the terrestrial fauna; this was followed by an ever-increasing amount of work in the range, including studies of the Schizomida (Harvey, 1988, 1992; Vine et al., 1988; Humphreys et al., 1989; Humphreys, 1990b), pseudoscorpions (Harvey, 1991, 1993), spiders

(Deeleman-Reinhold, 1993; Millidge, 1993), arachnids and myriapods (Harvey et al., 1993), millipedes (Shear, 1992; Humphreys and Shear, 1993; Shear and Humphreys, 1996), terrestrial isopods (Dalens, 1992, 1993), insects (Roth, 1991, 1995; Sturm and Smith, 1993; Hoch, 1993), Mollusca (Slack-Smith, 1993), ecophysiology of cave arthropods (Humphreys and Collis, 1990; Humphreys, 1991c), biological and other surveys (Humphreys, 1991b,d, 1993a,c,d), experimental ecology of troglobite populations (Humphreys, 1991c), and isozyme variation (Humphreys, 1991d; Humphreys and Adams, 1991; Adams and Humphreys, 1993).

Distinct faunas were later found on the Cape Range coastal plain (Humphreys, 1994; Harvey and Humphreys, 1995; Shear and Humphreys, 1996), and on Barrow Island (Humphreys, 1991e; Hoffman, 1994), and subsequently taxa common to these two areas were recognized (Adams and Humphreys, 1993; Harvey and Humphreys, 1995; Humphreys, 1999).

FOOD SUPPLY

Primary production

No data are available for the primary productivity of hummock grasslands comprising *Triodia* spp. in Australia (G. Griffin, pers. comm., 1995). Nonetheless, plant biomass and productivity is low in deserts and highly correlated with rainfall (Noy-Meir, 1985), which is itself unpredictable (see above, p. 583). The effects of variability in rainfall are exacerbated because the ratio of production to rainfall (P/R) decreases below an annual rainfall of 100 mm (cf., Fig. 30.3c), and there is a threshold rainfall below which production is zero (Noy-Meir, 1985). Hence, the resource base for hypogean communities is very limited, variable and unpredictable.

Food sources

Large trogloxene populations (e.g., bats and crickets) and the abundance of roots (e.g., in lava tubes), which characterize many cave systems (Harris, 1973; Norton et al., 1975; Kane et al., 1975; Studier and Lavoie, 1990), are lacking in Cape Range caves. In consequence, both the fauna and the energy inputs that are characteristically associated with them are absent from Cape Range.

These caves vary widely in food supply. Many have no obvious source of food other than organic carbon within the cave soil (Humphreys et al., 1989; Humphreys, 1990b; Humphreys and Adams, 1991); others receive irregular aseasonal influx of vegetable matter with flood waters (Vine et al., 1988; Humphreys et al., 1989), while others on the coastal plain contain dense mats of tree roots proliferating in the groundwater. Quantities of vegetable material, sometimes large, are washed into the caves after exceptional and unpredictable rainfall and, if they enter humid areas, serve as a major food supply for long periods; otherwise, they dry out and become unsuitable as food resources (Humphreys, 1991c).

Caves in the Cape Range rarely contain drip-water, and then it is mostly very localized following heavy rain. Hence, the percolation of fine particulate and dissolved organic matter into caves must be extremely restricted or even non-existent.

Seasonality

Owing to the unpredictable climate there is no consistent seasonality associated with rainfall, energy influx, or plant root activity, etc., nor are seasonal cues to be had from inputs by bats, cave crickets or other trogloxenes, as is found elsewhere.

Transport to caves

Owing to the aridity of the surface, primary productivity is low and the mechanisms to transport it to the hypogean system unreliable. However, these factors are offset to some extent because old tropical karst areas in arid climates are largely devoid of soil cover (Poulson and Culver, 1969), so that percolation to deep karst areas will be unimpeded – in consequence a superficial underground compartment [the MSS of Juberthie and Delay (1981); see also Chapter 1, this volume] is likely to be absent from the region.

Owing to these factors the hypogean realm in the Cape Range is largely dependent on the influx into caves of water-borne organic material. The energy entering a given cave is strongly determined by physical factors unrelated to amount and seasonality of the rainfall – it depends *inter alia* on catchment size and porosity, and on how fine-scale surface morphology directs drainage (Humphreys, 1991c).

The caves dry at rates that are a function, *inter alia,* of their size, water storage capacity (mainly in the cave soils), the opening size relative to cave volume, and characteristic ventilation. In addition, the drying rate is strongly seasonal owing to tropical winter effects (cf., Howarth, 1980). Dry caves do not support troglobites in Cape Range, and those occasionally drying become temporarily unsuitable for troglobites, probably by affecting food availability (Humphreys, 1991c). In consequence the occupation of caves, as opposed to mesocavernous spaces (*sensu* Howarth, 1983), is cyclical, determined by the influx pattern (Fig. 30.4).

In essence, some caves may receive inflow after relatively low rainfall (every few months) but then dry rapidly (over months). In contrast, other caves take inflow only after heavy rainfall (every few years) but thereafter dry very slowly, over several years [Fig. 30.4; see also Humphreys et al. (1989) and Humphreys (1991d)]. In consequence there is no level of energy input into caves on Cape Range that can be regarded as characteristic, either over the area or through time. Individual caves may, on average, have a higher energy input than others, but all experience periods of relative abundance and scarcity depending on their characteristics.

Transport within the matrix

As it is a tropical karstland largely devoid of soil, rainfall would be expected to infiltrate rapidly to the groundwater. In consequence, fine particulate and dissolved organic carbon entering the system, other than by flooding, is likely to bypass the caves, either entering the groundwater or becoming lodged within the rock matrix; hence this is where the bulk of the fauna is likely to occur.

Owing to the large evaporation excess, there is a tendency for water within the more superficial parts of the karst to move upwards, with a net downward movement only after exceptional rainfall. The resulting bidirectional movement of water in the phreatic zone is likely to strand organic carbon somewhere within the limestone where it serves as a food resource. The depth at which this equilibrium occurs is variable both in time (depending on the momentary vectors of inputs and outputs) and in space (depending on the local characteristics of the karst). In consequence, the troglobitic fauna is likely to need to move widely to harvest this temporally and spatially variable resource.

Tree roots

Tree roots penetrate some caves on the coastal plain (C-215) in quantity, and to depths of over 27 m in the range (e.g., C-21); they are roots of *Ficus* and probably *Brachychiton,* as found in Queensland (Howarth, 1988). Within a humid cave tree roots frequently do not reach the floor, but the fine rootlets serve as foci for condensation from which the tree probably absorbs water. Roots are not a major resource in most Cape Range caves – the caves may be too deep, unlike those in North Queensland (Howarth, 1988). Tree roots are used by planthoppers of the families Meenoplidae (*Phaconeura* spp.) and Cixiidae (gen. indet.), the former being attended by an undescribed species of ant (*Paratrechina* sp.) with much reduced eyes – it is one of few ants known that are especially adapted for cave life (S. Shattuck, pers. comm.).

On the coastal plain, root mats form in the underground water (cf., Jasinska and Knott, Chapter 15, this volume), although no characteristic fauna is known (W.F. Humphreys, unpublished; E. Jasinska, pers. comm.). Such root mats probably occur at the water table throughout the coastal plain – on which *Ficus* are common even though no cavities may be apparent – and provide a widespread habitat. The thermosbaenacean *Halosbaena tulki* was first found in such a habitat (Poore and Humphreys, 1992).

Accidentals

Probably owing to the aridity of the surface, accidentals appear to be relatively sparse in the Cape Range caves. However, they make a substantial contribution to the aquatic food chain on the coastal plain (Humphreys and Feinberg, 1995).

Trophic interactions: aquatic

The presence of characteristic red mud in the digestive tracts, and the deep layers of faecal pellets, suggests that thermosbaenaceans, atyid shrimps and harpacticoid copepods forage in the flocculent cave mud, presumably for microbiota.

The Cape Range atyids occur in a wide variety of habitat and salinity, being found in groundwater, caves and anchialine pools (Humphreys, 1994). Anchialine atyids occupy a more complex habitat (Bailey-Brock and Brock, 1993) than their more typical freshwater relatives (Fryer, 1977) for *Halocaridina rubra* does not rely only on filter-feeding, but also scrapes the benthic algal/cyanobacterial crust, ingesting Cyanobacteria, diatoms and other algae (Bailey-Brock and Brock, 1993).

Examination of the food of the two blind fishes showed them to be opportunistic feeders, taking specialized stygofauna (*Halosbaena, Stygiocaris*), freshwater fauna inhabiting the open pools (larval Odonata, Diptera and Trichoptera), and invertebrates accidentally entering the water (oniscoid isopods, cockroaches, ants). For *Milyeringa veritas* the number of prey items were as follows: cockroaches 26%, isopods 42%, *Stygiocaris* 5% (?10%), ants 11%.

Trophic interactions: terrestrial

The presence of characteristic red mud in the digestive tracts of *Stygiochiropus* spp. and the accumulation of faecal pellets, suggests that the millipedes process the cave soil for microbiota; they also forage directly on wet vegetable matter (Humphreys, 1991c). Oniscoid isopods (several species) are primarily associated with moist vegetable matter.

The prey are poorly known; the following associations have been observed. *Draculoides vinei* eats individuals of the same species, *Stygiochiropus,* and armadillid and philosciid isopods. *Draculoides vinei* is eaten by the undescribed genus of miturgid spider, which also itself eats *Stygiochiropus. Trichocyclus septentrionalis* (Pholcidae) eats weevils (*Myllocerus* sp.). An undescribed species of *Forsterina* sp. (Desidae) eats other individuals of the same species.

Baits

Trophic clues can be found from the acceptance of baits. In the terrestrial system a variety of baits (cheese, fish, sweet potatoes) were largely unsuccessful in attracting troglobites. In contrast, the provision of moist dead vegetable matter slowly attracts a succession of species (Humphreys, 1991c; W.F. Humphreys and R.D. Brooks, unpublished). In contrast, baits (cheese, fish, etc.) were successful in attracting many species of the aquatic cave fauna.

SPECIES RICHNESS OF CAVE FAUNA

The karst region contains at least 55 species of highly troglomorphic animals, of which 33 belong to terrestrial taxa (Table 30.1) and 22 to aquatic taxa (Table 30.2). In the terrestrial realm there are a large number of other species whose survival in arid Cape Range is probably dependent on caves (Harvey et al., 1993), although overt troglomorphies may be weakly developed; a number of taxa in the anchialine system are similarly of uncertain status (Table 30.3).

Of the terrestrial troglobites, all the species and many genera (>7) are endemic to the Cape Range Formation (Table 30.1), while amongst the aquatic troglobites the genera *Milyeringa* and *Stygiocaris,* and three genera of amphipods, are apparently endemic to the region (Table 30.2), most of the remaining being congeneric with species inhabiting anchialine caves on both sides of the North Atlantic (Humphreys, 1993b, 1994; Knott, 1993).

Stygobiont amphipods are diverse (Table 30.2), and on Barrow Island just 11 specimens yielded a diversity comparable to that of the Edwards Aquifer in the United States (total diversity = 2.197; generic diversity = 0.687; within-genus diversity = 1.513: Holsinger and Longley, 1980). Available data (Holsinger, 1994) indicate that Barrow Island alone contains ~1% of all known stygobiont amphipods.

Considerable genetic diversity occurs in those recognized species that have been examined, on the coastal plain and, especially, within the Cape Range (Adams and Humphreys, 1993). The borders of the major genetic provinces are associated with the deeper gorges that cut through the cavernous Tulki Limestone into the underlying non-cavernous Mandu Limestone (Humphreys and Shear, 1993) – as, for example, in the millipede *Stygiochiropus* (Fig. 30.5).

As the region is remote and has been examined only recently, many more species surely await discovery. The Cape Range karst is much smaller than many others that have been examined in Australia, and yet the richness of this fauna is notable (Tables 30.4 and 30.5), in part because of the juxtaposition of terrestrial, freshwater and anchialine faunas.

In a world context, for comparison, 43 species of troglobite have been reported from the Hawaiian archipelago, with no more than 24 species on any one island (Howarth, 1987), and 40 species were found in the Galapagos archipelago (Peck, 1990), whereas the well-worked Appalachians area of the United States, which is ~20 times the area of the Cape Range, contains 140 species of troglobites, of which 42 are aquatic and 98 terrestrial (Holsinger and Culver, 1988). Mexico, Guatamala and Belize combined contained

Table 30.1
Some terrestrial troglobitic fauna known from caves in northwestern Australia

Major taxa	Genera and species [1]	Locations
CHELICERATA		
Schizomida [2]	*****Draculoides** vinei*	Cape Range
	**Draculoides bramstockeri* Harvey and Humphreys	coastal plain of Cape Range peninsula; Barrow Island
	**Draculoides* sp. nov.	coastal plain of Cape Range peninsula
	**Bamazomus* sp. nov.	coastal plain of Cape Range peninsula
Pseudoscorpionida		
Hyidae	*****Hyella** humphreysi*	Cape Range
	**Hyella* sp. nov.	coastal plain of Cape Range peninsula
Chthoniidae	**Tyrannochthonius butleri*	Cape Range
Opilionida		
Assamiidae	**Anjolus* sp. (undescribed)	Cape Range
Phalangodidae	*Gen. indet. [3]	coastal plain of Cape Range peninsula
Araneae		
Hahniidae	*Gen.	coastal plain of Cape Range peninsula; Cape Range
Desidae	*Gen. & sp. indet. (Harvey et al., 1993)	?coastal plain of Cape Range peninsula; Cape Range
Ctenidae	***Gen.** & sp. nov. 1 (Harvey et al., 1993)	Cape Range
Pholcidae	*Trichocyclus septentrionalus*	
Symphytognathidae	**Anapistula troglobia*	Cape Range
Theridiidae	**Steatoda* sp. 1. (Harvey et al., 1993)	Cape Range
	**Pholcomma* sp. (Harvey et al., 1993)	Cape Range
Linyphiidae	**Dunedinia occidentalis*	Cape Range
	**Chthiononetes tenuis*	Cape Range
MYRIAPODA		
Polyzoniida		
Siphonotidae	*Gen. & sp. indet.	Cape Range
Polydesmida [4]		
Paradoxosomatidae	*****Stygiochiropus** communis*	Cape Range
	**Stygiochiropus isolatus*	Cape Range
	**Stygiochiropus sympatricus*	Cape Range
	**Stygiochiropus peculiaris*	coastal plain of Cape Range peninsula
Spirobolida		
Pachybolidae	*****Speleostrophus** nesiotes*	Barrow Island
INSECTA		
Diplura		
Japygidae [5]	*Indjapyx* n. sp. 1	Cape Range
	Indjapyx n. sp. 2	Cape Range
	Gen. nov. & sp. nov. 1	Cape Range
	Gen. nov. & sp. nov. 2	Cape Range
New family or sub-family		?coastal plain of Cape Range peninsula; Cape Range

continued on next page

Table 30.1, *continued*

Major taxa	Genera and species [1]	Locations
Thysanura		
Nicoletiidae: Atelurinae	∗*Trinemura troglophila*	Cape Range
Blattodea	∗*Nocticola flabella*	Cape Range
	∗*Nocticola* sp. nov.	Barrow Island
Orthoptera		
Nemobiinae	∗***Ngamarlanguia*** *luisae*	Cape Range
Coleoptera		
Curculionidae: Polydrosinae	*Myllocerus* sp. 1 [6]	Cape Range
Hemiptera		
Meenoplidae	∗*Phaconeura proserpina*	coastal plain of Cape Range peninsula
	∗*Phaconeura* sp. nov. 1 (H. Hoch, pers. com.)	coastal plain of Cape Range peninsula
	∗*Phaconeura* sp. nov. 2	Cape Range
Cixiidae	∗?gen.	Cape Range
CRUSTACEA		
Isopoda		
Philosciidae	∗Gen. & spp. indet. >2	coastal plain of Cape Range peninsula; Cape Range

[1] Those marked ∗ exhibit extreme troglomorphies; the remainder are possibly restricted to cave environments; genera in **bold** are endemic to the area, as are all determined species.
[2] Only genus of Schizomida endemic to Australia.
[3] First blind Phalangodidae in Australia, though usual in European caves (G. Hunt, pers. comm., 1994).
[4] *Antichiropus humphreysi* and *Boreohesperus capensis* are also found in caves but are well adapted to outside conditions.
[5] J. Pagés, pers. comm., 1994–95.
[6] *Myllocerus* sp. 2 is found outside caves.

317 species of troglobites, most being found in the Mexican states of Tamaulipas and San Luis Potosí with 51 and 48 species of troglobites, respectively (Reddell, 1981).

Hypogean communities are noted for the low frequency of sympatry of congeners (Notenboom, 1991), and in the eastern United States no two congeneric millipedes are known to be sympatric, and the few cavernicolous pseudoscorpions that coexist do so only in pairs (Barr Jr and Holsinger, 1985).

The Cape Range is unusual in the number of coexisting taxa and congeners – four species of pseudoscorpions coexist in C-15, two species of schizomids coexist in C-215, and congeneric species occur sympatrically in C-111 (the millipede *Stygiochiropus:* Humphreys and Shear, 1993), and in the western seaboard of the Cape Range (the shrimp *Stygiocaris:* Humphreys and Adams, 1991).

This troglobite fauna is almost devoid of beetles, as has been noted for other tropical areas (Vandel, 1965; Reddell, 1981: p. 217; see also Deharveng, Chapter 31, this volume); this is in sharp contrast to the diverse beetle fauna associated with temperate caves in both northern and southern hemispheres (Peck, 1978; Barr Jr and Holsinger, 1985; Juberthie, 1988; Eberhard et al., 1991).

DEGREE OF REGRESSIVE EVOLUTION

Taxonomists commonly comment on the high degree of troglomorphy seen in species from the Cape Range, and this general high level of regressive evolution is in accord with the apparent great age of this fauna (see next section). "*Nocticola flabella* Roth is probably the most troglobitic cockroach ever known for its remarkable depigmentation and very slender appendages" (Izquierdo and Oromi, 1994).

The millipede genus *Stygiochiropus,* endemic to the Cape Range, contains four species that are strongly modified for cave life, with greatly elongated legs and

Table 30.2
The composition of the stygofauna of northwestern Australia[1]

Major taxon	Genus and species[2]	Locality
Pisces: Eleotridae	***Milyeringa*** *veritas*	coastal plain bordering Cape Range; Barrow Island
Pisces: Synbranchiformes	*Ophisternon candidum*	coastal plain bordering Cape Range
Decapoda: Atyidae	***Stygiocaris*** *lancifera*	coastal plain bordering Cape Range
	Stygiocaris stylifera	coastal plain bordering Cape Range; Barrow Island
Isopoda: Cirolanidae	*Haptolana pholeta*	coastal plain bordering Cape Range; Barrow Island
Thermosbaenacea	*Halosbaena tulki*	coastal plain bordering Cape Range; Barrow Island
Amphipoda: Melitidae	***Nedsia*** *douglasi*	coastal plain bordering Cape Range
	N. fragilis, N. humphreysi, N. hurlberti, N. macrosculptilis, N. sculptilis, N. straskraba, N. urifimbriata	Barrow Island
	Norcapensis *mandibulis*	Cape Range
Amphipoda: Bogidiellidae	***Bogidomma*** *australis*	Barrow Island
Amphipoda: Hadziidae	*Liagoceradocus subthalassicus*	Barrow Island
	Liagoceradocus branchialis	coastal plain bordering Cape Range
Ostracoda: Halocyprida	*Danielopolina kornickeri*	coastal plain bordering Cape Range
Syncarida: Bathynellacea	*Atopobathynella* sp. nov. (H.K. Schminke, pers. comm.)	Barrow Island
Copepoda: Harpacticoida	?	coastal plain bordering Cape Range; Cape Range; Barrow Island
Copepoda: Cyclopoida	*Metacyclops mortoni*	coastal plain bordering Cape Range
	Microcyclops varicans	coastal plain bordering Cape Range; Cape Range
	Apocyclops dengizicus	coastal plain bordering Cape Range
	Halicyclops longifurcatus	coastal plain bordering Cape Range
	Halicyclops spinifer	coastal plain bordering Cape Range; Cape Range
Remipedia: Nectiopoda	*Lasionectes exleyi*	coastal plain bordering Cape Range
Turbellaria	?	coastal plain bordering Cape Range; Barrow Island

[1] About half of this fauna has been found since 1991.
[2] Genera in **bold** are endemic to the area, as are all the species save some copepods.

antennae, total lack of pigment, brittle cuticle, and narrow, elongated segments with short angular, spine-like paranota (Humphreys and Shear, 1993).

Amongst the pseudoscorpions, *Hyella humphreysi* has elongate appendages and is of very large size (Harvey, 1993); *Austrochthonius easti* has lost the posterior eyes and has reduced anterior eyes; *Tyrannochthonius butleri* has no eyes and is pale in colour. Many of the troglobitic spiders and the cave cricket are entirely eyeless. Comparative work on congeners is usually impossible as no closely related surface species, even congeners, are known.

All the aquatic species are depigmented and lack eyes, with a remarkable exception – the bogidiellid amphipod is the only member of the family known which has eyes (Bradbury and Williams, 1996a), despite it having been collected down a borehole 50 m deep towards the centre of Barrow Island.

Table 30.3
High-level taxonomic groupings of taxa collected in open water bodies[1]

Taxon	Notes	Authority[2]
Protista (sundry)	*Euplotes* sp.; *Paramecium* sp.	Knott (1993)
Rotatoria		
Turbellaria		
Nematoda		
Polychaeta	Syllidae (two species)	G. Hartmann-Schröder, pers. comm.
Oligochaeta		
Mollusca	*Iravadia* (*Iravadia*) sp.? *I. ornata*[3]	Slack-Smith (1993)
Ostracoda		
Amphipoda: Aoridae	*Grandidierella* n. sp.	Myers and Lowry, pers. comm.
Acarina: Hydracarina	*Coaustraliobates* sp.	Harvey, pers. comm.
Ephemeroptera		
Odonata		
Hemiptera		
Diptera: Chironomidae	*Kiefferulus intertinctus*	Knott (1993)
Coleoptera		
Pisces: Poeciliidae	*Poecilia reticulata* [feral guppies!]	

[1] These taxa may not constitute part of the stygofauna proper. However, a number of stygofaunal species are also known from the same habitats.
[2] The references denote the source of determinations.
[3] Status ambiguous (cf. Slack-Smith, 1993).

Table 30.4
The number of troglobites and stygobionts in four Tasmanian karst regions and New South Wales, compared with the Cape Range

	Precipitous Bluff[1]	Exit Cave[1]	Vanishing Falls[1]	Kubla Khan[1]	New South Wales[2,3]	Cape Range
Troglobites	15	15	14	11	9	41
?Troglo/stygobites	4	1	3	2	8 taxa	>10
Stygobionts	3	7	5	0	8	22

[1] Locations in Tasmania. Data from Knott (1993), after Eberhard (1992).
[2] Data from A.P. Spate and S.M. Eberhard (pers. commun., 1995).
[3] Forty-eight karst areas combined; numbers given are of genera.

AGE OF CAVE SYSTEMS AND CAVE FAUNA

Detailed geological and palaeoclimatic data, which are widely available for much of Europe (Juberthie, 1988) and North America (Peck, 1981, 1984) are unavailable for northwestern Australia, primarily because of the great age of the landscape. The age of the cave systems in Cape Range is unknown, but there are a number of clues which are pursued in the following paragraphs. It should be stressed, however, that the age of a cave system may not provide any clue as to the age of the fauna – either phylogenetically or geographically – because both aquatic and terrestrial troglobites may move widely through the interstices on a scale which may be local (Juberthie, 1983; Uéno, 1987; Howarth, 1981; Oromi et al., 1991; Humphreys, 1993c) or regional (Humphreys, 1993b,c, 1994); the system may possibly be in geologico-temporal continuity on a large, even global scale through plate tectonics and sea-floor spreading (Hart et al., 1985; Boxshall, 1989).

The cave systems must have formed after the Miocene limestones (>25 Ma ago) were thrust above sea level. Recent evidence suggests that the uplift of the Cape Range anticline had been completed by no later than the late Pliocene (>1.8 Ma BP) and possibly by the late Miocene (>5 Ma BP: Wyrwoll et al., 1993). Clearly

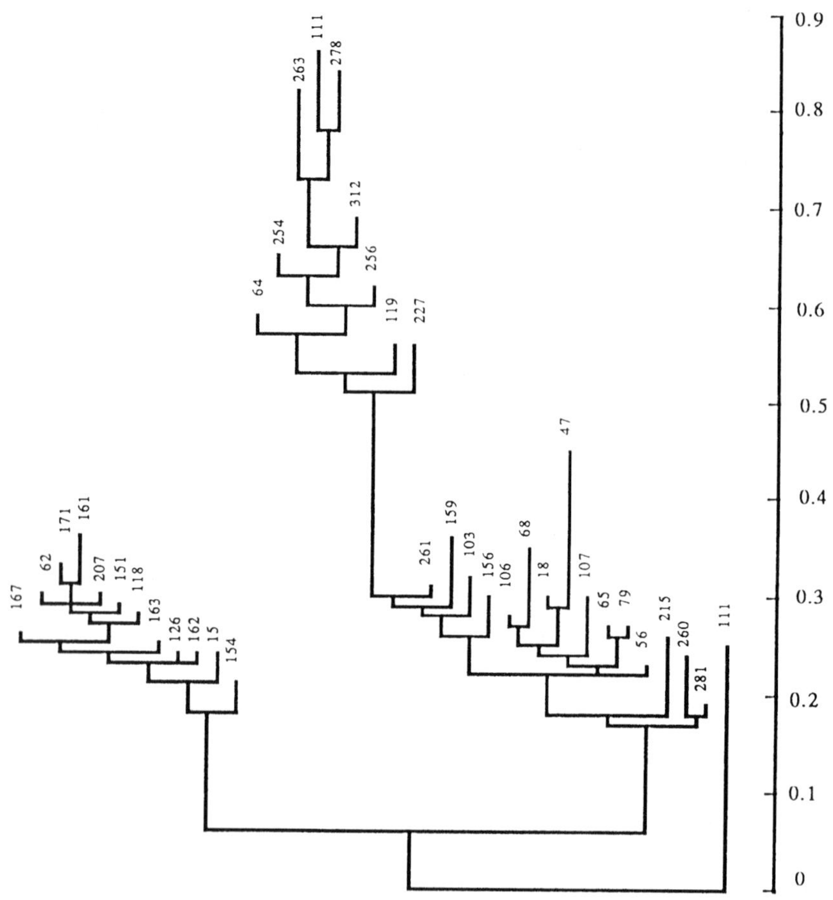

Fig. 30.5. Tree of the troglobitic millipede populations *Stygiochiropus communis* from the 38 caves numbered, produced by a Distance Wagner Analysis on the matrix of Rogers' Genetic Distance (Humphreys, 1991d) rooted on sympatric *S. sympatricus* from C-111.

Table 30.5
Number of troglobitic species of arachnids and myriapods in five regions of Australia [1]

Region	Number of species
Nullarbor	8
Chillagoe	14
Tasmania [2]	34
New South Wales [3]	8+
Cape Range	23 [4]

[1] Data from Harvey et al. (1993).
[2] Thirty karst regions combined.
[3] Forty-eight karst regions combined; the figure given is of genera.
[4] Updated from Harvey et al. (1993).

the limestones would have been emergent much earlier than this, as they currently reach an altitude of 311 m despite erosion having stripped the Trealla Limestone off more than 95% of the area of the range (Humphreys 1993d: frontispiece). The deeply dissected landscape, the caves and the cave morphology indicate that the caves formed at a time when the area was much wetter than at present.

Limited examination of speleothems has dated material up to 184 000 BP (D. Ford, pers. comm.); but the caves are much older – the location of some samples (e.g., hanging floors) shows that the caves were well developed at the time the speleothems were produced. Dates covering the period 170 000–80 000 years ago (D. Smith, pers. comm.) show growth rates of 0.72 $\mu m\ yr^{-1}$ (Humphreys, 1991d), an order of magnitude slower than any recorded by Shaw (1993: appendices 14 and 15) or Ford and Williams, 1989: table 8.5), suggesting that there was no major wet episode in that time interval.

The surface biota of the Cape Range region is largely

that found in the broad arid zone on the western border of which the Cape Range lies (see papers in Humphreys, 1993d); but the flora gives some evidence of a wetter past (Keighery and Gibson, 1993), with some tropical relicts. In contrast, the subterranean terrestrial fauna has clear affinities with the leaf-litter fauna found on the floor of wet forests, both temperate and tropical (Harvey et al., 1993; Humphreys, 1993a,c). Hence, study of the palaeoclimate and palaeovegetation may provide indicators as to the age of the fauna.

Palaeoclimatic data are sparse for northwestern Australia, hence one needs to rely unduly on data from further afield, in eastern and southern Australia – the affinities of the cave fauna provide the main evidence that there were extensive wet forests in the region from the Eocene onwards. While plants can establish rainforest patches by propagule dispersal (Kenneally et al., 1991), there is no evidence that this is the case with physico-chemically sensitive cryptozoic invertebrates – these are often found in rainforest as separate species with small ranges (e.g., the Schizomida: Harvey, 1992). Hence, it is argued, the caves were colonized at a time when Australia was widely blanketed with complex rainforest.

Complex rainforest occurred on the western shoulder of Australia during the Eocene (Truswell, 1990), but global drying in the late Miocene led to massive contraction of the rainforest (Kemp, 1981). While subsequent palaeovegetation data are lacking for Western Australia, the continental trends (Truswell, 1990) suggest that by the Pliocene there had been a progressive retreat of both complex rainforest and closed forest communities to the eastern and southwestern seaboards of the continent, as is the case today (Fig. 30.6). The regional rainforest would have progressively fragmented (Webb and Tracey, 1981) through the Miocene and Pliocene owing to the easterly wind pattern (Bowler, 1982). Along river valleys forest persisted (Truswell, 1990), providing corridors of suitable habitat between the Neogene Cape Range Formation and older landmasses (see also Wyrwoll et al., 1993). Hence, by the late Miocene extensive rainforest must have been eliminated from the Cape Range area, but possibly would have persisted until later in the deep gorges of the range, permitting genetic exchange between cave areas (Humphreys, 1993c).

Ages of vicariance in troglobites estimated from figures for genetic distance are often in broad agreement with the palaeogeographic evidence (Delay et al., 1980; Sbordoni, 1982; Juberthie, 1988). Data for genetic distance are available for a number of taxa within the Cape Range (Table 30.6). Data for melitid amphipods on the coast and those inhabiting fresh cave waters at ~200 m altitude within the Cape Range suggests that the elevation of the range could have occurred ~7 Ma ago (Table 30.6). There is evidence from both the strongly troglomorphic congeneric species within the genera *Stygiocaris* (shrimps) and *Stygiochiropus* (millipedes), that by 4.3–4.0 Ma ago they were already inhabiting the subterranean realm, and that this realm was fragmenting, permitting speciation (Table 30.6). Hence both the aquatic and terrestrial components of the fauna were present and speciating in the Late Miocene/Upper Pliocene.

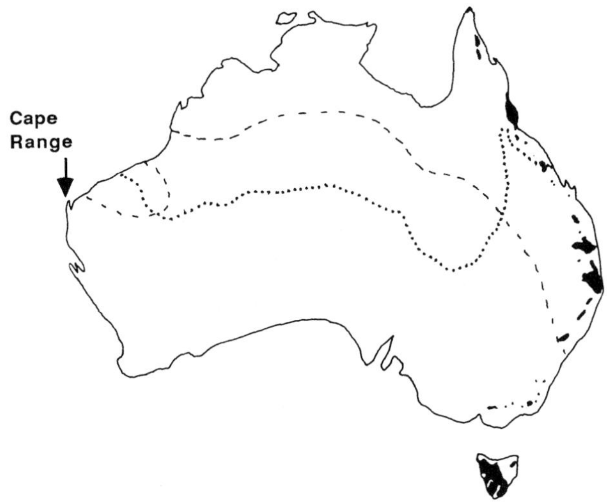

Fig. 30.6. The current distribution of extensive coastal rainforest (black) and the southerly limit (dashed lines) of remnant coastal rainforest elements in contemporary Australia (after Webb and Tracey, 1981). The former occurs in an arc through eastern Australia, with a distribution similar to the Pliocene limits of forest with *Nothofagus*. The dotted line shows the known northerly extent of forest with *Nothofagus* in the Eocene; data from further north are mostly lacking. By the middle Eocene (45 Ma ago), complex rainforest was restricted to riverine and lacustrine areas across the southern half of Australia and was absent by the early Miocene (20 Ma ago). Through the Pliocene, forest with *Nothofagus* retreated to the eastern seaboard (Truswell, 1990: fig. 7). After Humphreys (1993c).

The genetic provincialization of the cave fauna occurred during the Pleistocene – within the Cape Range from 2.6 to 0.8 Ma ago (Humphreys and Shear, 1993), as indicated by the data for *Stygiochiropus communis* (Table 30.6). From 1.3 to 0.6 Ma ago the perched water bodies became separated, isolating populations of melitid amphipods within the Range (Table 30.6), and possibly indicating continued drying. Differentiation

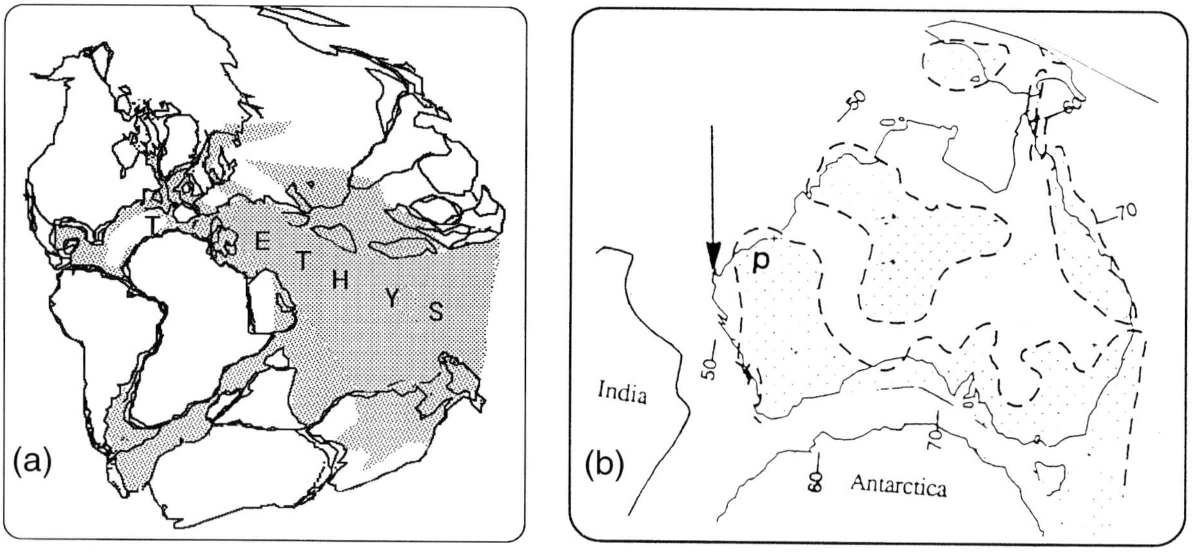

Fig. 30.7. (a) The relative location of the continental plates in the Jurassic (151 Ma ago) when the Tethys Sea first connected the present Caribbean, Mediterranean and Indian Ocean regions. Even with this reconstruction the genera *Halosbaena* (Thermosbaenacea), *Haptolana* (Isopoda: Cirolanidae) and *Lasionectes* (Remipedia), known from the Cape Range area, have widely disjunct distributions variously including Somalia, the Canary Islands and Caribbean islands. The plate map was produced by the technique of Scotese and Denham (1988). (b) The current outline of Australia superimposed on the emergent parts of the Australian plate (dashed lines) in the early Cretaceous (118 Ma ago), when shallow seas surround the Pilbara Craton (p); the arrow shows the location of the present Cape Range. The numbers on the map show the latitude. Continental margins after Veevers et al. (1991).

of populations of the amphipods, *Draculoides vinei* and *S. communis* between various caves within the broadly defined genetic provinces continued through the Pleistocene (550 000–100 000 BP: Table 30.6).

There is general accord between the geological, fossil and biological evidence as to the age of emergence of the Cape Range. However, for the anchialine fauna the evidence is quite different. While our "ignorance about the time of colonization of cave waters is complete" (Culver, 1982: p. 141), the presence of species in the anchialine fauna congeneric with obligatory troglobites in the North Atlantic (e.g., *Halosbaena, Haptolana,* and *Lasionectes*) suggests that the fauna had colonized caves prior to the disruption of Pangea in the Mesozoic (Fig. 30.7). In this case a spatial and temporal continuity of crevicular habitat must be sought to explain their occurrence in Cape Range, which is of Miocene age. This must be sought between the Mesozoic formations that fringe the Precambrian Pilbara Craton (Fig. 30.1) and their current location in Miocene strata; this continuity needs to be sought in the following geological context.

The Permian Callytharra Formation, which is karstic in places and up to 244 m deep, occurs widely at depth, extending from Onslow in the north to Wooromel at the foot of the Cape Range peninsula, and from the coast to well inland (Hocking et al., 1987). In the lower Mesozoic there were successive waves of sedimentation, giving marine silt and clay stones, alluvial fans and fluvial sandstones, followed by shallow marine and coastal sandstone (Jurassic). During the Cretaceous there was a progradation of a deltaic complex over marine clays followed by calcareous pelagic deposition formed on the extensive low-energy shelf deepening towards the northwest. The advent of calcareous deposition was part of a worldwide event caused by changed oceanic-circulation patterns following continental break-up, which continued through the remainder of the Cretaceous. Such carbonate deposition continued in the northern Carnarvon Basin throughout the Tertiary. The subsequent uplift of the area by reverse thrust may be the result of compensation for the break-up, adjustment and foundering of the continental margin.

CONCLUDING REMARKS

The hypothesis that regions lacking forest throughout the Pleistocene, such as parts of the American southwest and Australia, have a very depauperate terrestrial

Table 30.6
The evidence supporting a proposed sequence of events in the Cape Range area[1]

Event	Period	Date	Evidence
1000 years ago			
N.W. Shelf inundated[2]	Holocene	8–0	calculated from Chappell and Thom (1977)
Wetter period	Pleistocene	14–11	U/Th speleothem date, C-125[4]
E and W coast populations isolated	Pleistocene	25	*D* between *Stygiocaris stylifera* populations[3,5]
Differentiation of populations	Pleistocene	30	*S. stylifera*[5]
N. W. Shelf inundated[2]	Pleistocene	85–80	calculated from Chappell and Thom (1977)
Differentiation[6]; southern province	Pleistocene	100	*D* between millipede populations[5,7,14]
N.W. Shelf inundated[2]	Pleistocene	111–99	calculated from Chappell and Thom (1977)
Wetter period	Pleistocene	128	*Anadara trapezia* distribution[9]
N.W. Shelf inundated[2]	Pleistocene	135–116	calculated from Chappell and Thom (1977)
Differentiation in central province	Pleistocene	165	*D* between amphipods in range[5,7,10,15]
Differentiation of populations	Pleistocene	200	*D* between *Draculoides* populations[7]
Atyid populations isolated	Pleistocene	205	*D* between *S. lancifera* populations[7,10]
Wetter period	Pleistocene	219–68	U/Th speleothem date, C-126 and C-163[4]
N.W. Shelf inundated[2]	Pleistocene	222–216	calculated from Chappell and Thom (1977)
Wetter period	Pleistocene	225	*Anadara trapezia* distribution[9]
Differentiation[6]; central populations	Pleistocene	400	*D* between millipede populations[5,7]
E and W coast populations isolated	Pleistocene	315	*D* between cave fish[11] populations[5,7,10,12]
Differentiation[6]; northern populations	Pleistocene	350	*D* between millipede populations[5,7]
Barrow Island isolated from Cape Range	Pleistocene	380	*D* between *S. stylifera* populations[5,7,13]
Schizomid populations diverge	Pleistocene	450	*D* within *Draculoides vinei* populations[5,7]
Within-province differentiation	Pleistocene	550	*D* within millipede genetic provinces[5,7,14]
Northern cave waters isolated	Pleistocene	650	*D* between amphipods in range[5,7,10,15]
Southern caves isolated from rest	Pleistocene	800	*D* between millipede populations[5,7,14]
1 000 000 years ago			
Wet forest lost from gorges	Pleistocene	4.3–1.6	Deduced from genetic distances above
Southern cave waters isolated	Pleistocene	1.3	*D* between amphipods in range[5,7,10,15]
Northern cave areas isolated	Pleistocene	1.6	*D* between millipede genetic provinces[5,7]
Cape Range fully emergent	Pliocene	>2	Marine terraces; Wyrwoll et al. (1993)
Southern cave areas isolated	Pliocene	2.2	*D* between millipede genetic provinces[5,7,14]
Shrimps speciate	Pliocene	4.0	*D* between congeneric atyid shrimps[5,7]
Millipedes speciate	Pliocene	4.3	*D* between congeneric millipedes[5,7,14]
Anticline uplift	Miocene/Pliocene	7	*D* between amphipod genera[5,7,10,15]
Indian and Pacific Oceans isolated[16]	Late Miocene	7	Kennett et al. (1985)
Anticline uplift	Late Miocene	>7	Malcolm et al. (1991)
Closed forest with *Nothofagus*	Miocene	>7	Truswell (1990)
Cooling sea and land	Miocene	8–6	MacFadden and Cerling (1994)
Indian and Atlantic Oceans separate[17]	Middle Miocene	14–12	Kennett (1982)
Pilgrammuna Formation formed	Middle Miocene	~17	Hocking et al. (1987)
Trealla Limestone formed	Middle Miocene	~17	Hocking et al. (1987)

continued on next page

Table 30.6, *continued*

Event	Period	Date	Evidence
Increased aridity[18]	Early Miocene	~22	Circum-Antarctic circulation
Tulki Limestone formed	Early Miocene	~22	Hocking et al. (1987)
Mandu Limestone formed	Late Oligocene	~28 onwards	Hocking et al. (1987)
Complex rainforest in vicinity	Eocene	>38	Truswell (1990)
Tethys Seaway ends	Late Eocene	40	Smith and Briden (1977)
Carbonate deposition continues	Cainozoic	65 onwards	Hocking et al. (1987)
Epicontinental sea connection cut	Middle Cretaceous	~100	Cande and Mutter (1982)
Tethys encircles Pilbara craton	Cretaceous	118	Veevers et al. (1991)
NW Australia is E. shore of Tethys	Early Cretaceous	~120	M.K. Howarth (1981)
E. and W. Gondwana well separated	Cretaceous	125	M.K. Howarth (1981)
Eastern Gondwana fragments	Cretaceous	133–96	Veevers et al. (1991)
Advent of calcareous deposition[19]	Cretaceous-		Hocking et al. (1987)
Extensive low energy shelf	Cretaceous		Hocking et al. (1987)
E. and W. Gondwana separated	Upper Jurassic	145	M.K. Howarth (1981)
Onset of seafloor spreading	Jurassic	155	Hocking et al. (1987)
Tethys Seaway starts	Triassic	200	Smith and Briden (1977)
Shallow marine environments start	Triassic	~200	Hocking et al. (1987)
Deltaic progradation	Mid Triassic	210	Hocking et al. (1987)
Marine transgression; local seaways	Early Triassic	220	Hocking et al. (1987)
Broad marine shelf	Permian	250	Hocking et al. (1987)
Clear water carbonate deposition[8]	Early Carboniferous	340	Hocking et al. (1987)
Pilbara Craton emergent	Precambrian	>600	Hocking et al. (1987)

[1] The purpose of this Table is to provide a heuristic focus for the timing of biogeographically pertinent events in the area. Expanded from Humphreys (1993c).
[2] Defined as Barrow Island being separated from the mainland when the sea level rises above a level more than 20 m below the present level.
[3] "*D*" is Nei's (1987) genetic distance.
[4] The ages include the 95% confidence intervals for the date. The dating was conducted by Dr D. Smith, Department of Inorganic Chemistry, University of Melbourne. Owing to low uranium content the error bounds are large.
[5] Adams and Humphreys (1993), unpublished.
[6] By inference, caves isolated.
[7] Calculated from the average Nei's genetic distance (D) assuming $D = 1$ corresponds to five million years of divergence (Nei, 1987).
[8] Moogooree Limestone.
[9] Kendrick et al. (1991).
[10] Humphreys and Adams (1991).
[11] *Milyeringa veritas.*
[12] Between populations on the east and west coast of the Cape Range peninsula.
[13] Between populations on Barrow Island and the Cape Range peninsula.
[14] Humphreys and Shear (1993).
[15] The populations within the Cape Range *sensu stricto* are not congeneric with those on the coastal plain or Barrow Island (J. Bradbury, pers. comm., 1995).
[16] Diversion of North Equatorial Current from the Indian Ocean to the North Pacific.
[17] Tropical connection only.
[18] The opening of a seaway between South America and Antarctica in the Early Miocene permitted the West Wind Drift and hence the establishment of the circum-Antarctic circulation pattern; Antarctica became refrigerated and the temperature gradient between the equator and the poles intensified, creating the arid regions where they are currently found (see Main, 1993).
[19] Part of a worldwide event caused by changed oceanic-circulation patterns following continental break-up (Hocking et al., 1987).

cave fauna (Peck, 1980) is not supported by the Cape Range data.

Howarth (1988) suggested that tropical cave faunas were associated with parts of cave systems free from draughts, where the air was rich in carbon dioxide; there is no evidence for such requirements by the diverse and highly troglomorphic fauna of the Cape Range area. Indeed, some species, such as *Stygiochiropus communis* and *S. isolatus,* will move into sunlight at the cave threshold if the humidity is high (e.g., C-222) and moisture in the air flowing out of the cave condenses there (e.g., C-15).

The history of the Cape Range area is complex, and as yet poorly documented. However, a plurality of factors appear to have been involved in the evolution of the fauna of the Cape Range peninsula (see Table 30.7), as has been suggested for some other regions (Peck, 1990; Botosaneanu and Holsinger, 1991: p. 34).

Table 30.7
Vicariance and the presence of relictual and relic taxa in the subterranean fauna of the Cape Range peninsula

Item [1]	Terrestrial	Groundwater
Vicariant event	aridity	plate tectonics
	eustatic	eustatic
	orogenic	orogenic
Relictual taxa		
global	Hyidae	many stygofauna spp.
regional	Schizomida	amphipods
local	*Stygiochiropus*	amphipods in Cape Range
Relic taxa	e.g., Remipedia, Thermosbaenacea, Atyidae	
Refugium	caves	freshwater
	crevicular habitats	anchialine

[1] See definitions in Glossary, p. 6.

ACKNOWLEDGEMENTS

The numerous taxonomists who have been involved in this work are acknowledged in the more specific publications. I thank Julianne Waldock for her continued support, and Caroline Lawrence for drawing some figures. The field work in Cape Range has been variously funded by the Western Australian Museum, Australian National Parks and Wildlife Service, The National Estate Grants Scheme, The Butler Trust fund at the Western Australian Museum, and numerous volunteers. Victualling and logistics on Barrow Island were provided by Western Australian Petroleum Pty Ltd.

REFERENCES

Adams, M. and Humphreys, W.F., 1993. Patterns of genetic diversity within selected subterranean fauna of the North West Cape peninsula, Western Australia: systematic and biogeographic implications. *Rec. West. Aust. Mus. Suppl.,* 45: 145–164.

Allen, A.D., 1993. Outline of the geology and hydrogeology of Cape Range, Carnarvon Basin, Western Australia. *Rec. West. Aust. Mus. Suppl.,* 45: 25–38.

Bailey-Brock, J.H. and Brock, R.E., 1993. Feeding, reproduction, and sense organs of the Hawaiian anchialine shrimps *Halocaridina rubra* (Atyidae). *Pac. Sci.,* 47: 338–355.

Baltanas, A. and Danielopol, D.L., 1995. Cladistic analysis of *Danielopolina* species (Ostracoda: Thaumatocyprididae) and the origin of anchialine fauna. *Mitt. Hamb. Zool. Mus. Inst.,* 92: 315–324.

Barnard, J.L. and Williams, W.D., 1995. The taxonomy of freshwater Amphipoda (Crustacea) from Australian freshwaters: Part 2. *Rec. Aust. Mus.,* 47: 161–201.

Barr Jr., T.C. and Holsinger, J.R., 1985. Speciation in cave faunas. *Annu. Rev. Ecol. Syst.,* 16: 313–337.

Botosaneanu, L. and Holsinger, J.R., 1991. Some aspects concerning colonisation of the subterranean realm – especially of subterranean waters: a response to Rouch & Danielopol. *Stygologia,* 6: 11–39.

Bowler, J.M., 1982. Aridity in the late Tertiary and Quaternary of Australia. In: W.R. Barker and P.J.M. Greenslade (Editors), *Evolution of the Flora and Fauna of Arid Australia.* Peacock Publications, Frewville, South Australia, pp. 35–45.

Boxshall, G.A., 1989. Colonisation of inland marine caves by misophrioid copepods. *J. Zool. (London),* 219: 521–526.

Bradbury, J.H. and Williams, W.D., 1996a. Freshwater amphipods from Barrow Island, Western Australia. *Rec. Aust. Mus.,* 48: 33–74.

Bradbury, J.H. and Williams, W.D., 1996b. Two new species of anchialine amphipod (Crustacea: Hadziidae: *Liagoceradocus*) from Western Australia. *Rec. West. Aust. Mus.,* 17: 395–409.

Bradbury, J.H. and Williams, W.D., 1997. The amphipod (Crustacea) stygofauna of Australia: description of new taxa (Melitidae, Neoniphargidae, Paramelitidae), and a synopsis of known species. *Rec. Aust. Mus.,* 49: 249–341.

Bruce, N.L. and Humphreys, W.F., 1993. *Haptolana pholeta* sp. nov., the first subterranean flabelliferan isopod crustacean (Cirolanidae) from Australia. *Invertebr. Taxon.,* 7: 875–884.

Cande, S.C. and Mutter, J.C., 1982. A revised identification of the oldest sea-floor spreading anomalies between Australia and Antarctica. *Earth Planet. Sci. Lett.,* 58: 151–160.

Cawthorn, P., 1963. Discovery of subterranean freshwater fauna on the eastern side of North West Cape. *West. Aust. Nat.,* 8: 129–132.

Chappell, J. and Thom, B.G., 1977. Sea levels and coasts. In: J. Allen, J. Golson and R. Jones (Editors), *Sunda and Sakul: Prehistoric*

Studies in Southeast Asia, Melanesia & Australia. Academic Press, London, pp. 275–291.

Culver, D.C., 1982. *Cave Life: Evolution and Ecology.* Harvard University Press, Cambridge, MA, 189 pp.

Dalens, H., 1992. Oniscidea (Crustacea, Isopoda) from caves of Cape Range in Western Australia. I. The genus *Buddelundia. Rec. West. Aust. Mus.,* 16: 87–102.

Dalens, H., 1993. Two new genera of terrestrial isopods (Crustacea: Isopoda: Oniscidea) from north-western Western Australia. *Rec. West. Aust. Mus.,* 16: 257–267.

Danielopol, D.L., Baltanas, A. and Humphreys, W.F., 2000. *Danielopolina kornickeri* sp. n. (Ostracoda: Thaumatocypridoidea) from a western Australian anchialine cave – morphology and evolution. *Zool. Scr.,* 29: 1–16.

Deeleman-Reinhold, C.L., 1993. *Trichocyclus septentrionalis* n. sp. a new species of cave dwelling pholcid spider from northwestern Australia (Araneae: Pholcidae). *Rec. West. Aust. Mus.,* 16: 323–329.

Delay, B., Sbordoni, V., Cobolli Sbordoni, M. and De Matthaeis, E., 1980. Divergences génétique entre les populations de *Speonomus delarouzeei* du Massif du Canigou (Coleoptera, Bathysciinae). *Mém. Biospéol.,* 7: 235–247.

Eberhard, S.M., 1992. *The invertebrate cave fauna of Tasmania: ecology and conservation biology, M.Sc. Thesis.* Department of Zoology, University of Tasmania, Hobart, 184 pp.

Eberhard, S.M., Richardson, A.M.M. and Swain, R., 1991. *The Invertebrate Cave Fauna of Tasmania.* Zoology Department, University of Tasmania, 174 pp.

Ford, D.C. and Williams, P.W., 1989. *Karst Geomorphology and Hydrology.* Unwin Hyman, London, 601 pp.

Forth, J.R., 1973. Exmouth water supply. *Geol. Surv. West. Aust. Annu. Rep.,* 1972: 11–15.

Fryer, G., 1977. Studies on the functional morphology and ecology of the atyid prawns of Dominica. *Philos. Trans. R. Soc. London B,* 277: 57–128.

Gentilli, J., 1972. *Australian Climatic Patterns.* Nelson, Melbourne, 285 pp.

Harris, J.A., 1973. Structure and dynamics of a cave population of the guano mite, *Uroobovella coprophila* (Wolmersley). *Aust. J. Zool.,* 21: 239–275.

Hart, C.W., Manning, R.B. and Iliffe, T.M., 1985. The fauna of Atlantic marine caves: evidence of dispersal by sea floor spreading while maintaining ties to deep waters. *Proc. Biol. Soc. Washington,* 98: 288–292.

Harvey, M.S., 1988. A new troglobitic schizomid from Cape Range, Western Australia (Chelicerata: Schizomida). *Rec. West. Aust. Mus.,* 14: 15–20.

Harvey, M.S., 1991. The cavernicolous pseudoscorpions (Chelicerata: Pseudoscorpionida) of Cape Range, Western Australia. *Rec. West. Aust. Mus.,* 15: 487–502.

Harvey, M.S., 1992. The Schizomida (Chelicerata) of Australia. *Invertebr. Taxon.,* 6: 77–129.

Harvey, M.S., 1993. The systematics of the Hyidae (Pseudoscorpionida). *Invertebr. Taxon.,* 7: 1–32.

Harvey, M.S. and Humphreys, W.F., 1995. Notes on the genus *Draculoides* Harvey (Schizomida: Hubbardiidae), with the description of a new troglobitic species. *Rec. West. Aust. Mus. Suppl.,* 52: 183–189.

Harvey, M.S., Gray, M.R., Hunt, G.S. and Lee, D.C., 1993. The cavernicolous Arachnida and Myriapoda of Cape Range, Western Australia. *Rec. West. Aust. Mus. Suppl.,* 45: 129–144.

Hoch, H., 1993. A new troglobitic planthopper species (Homoptera: Fulgoroidea: Meenoplidae) from Western Australia. *Rec. West. Aust. Mus.,* 16: 393–398.

Hoch, H. and Howarth, F.G., 1989a. Six new cavernicolous cixiid planthoppers in the genus *Solonaima* from Australia (Homoptera: Fulgoroidea). *Syst. Entomol.,* 14: 377–402.

Hoch, H. and Howarth, F.G., 1989b. The evolution of cave-adapted cixiid planthoppers in volcanic and limestone caves in North Queensland, Australia (Homoptera: Fulgoroidea). *Mém. Biospéol.,* 16: 17–24.

Hoch, H. and Howarth, F.G., 1993. Evolutionary dynamics of behavioral divergence among populations of the Hawaiian cave-dwelling planthopper *Oliarus polyphemus* (Homoptera: Fulgoroidea: Cixiidae). *Pac. Sci.,* 47: 303–318.

Hocking, R.M., Moors, H.T. and van de Graaff, W.J.E., 1987. Geology of the Carnarvon Basin, Western Australia. *Geol. Surv. West. Aust. Bull.,* 133: 1–289.

Hoffman, R.L., 1994. Studies on spiroboloid millipeds. XVIII. *Speleostrophus nesiotes,* the first known troglobitic spiroboloid milliped, from Barrow Island, Western Australia (Diplopoda: Pachybolidae: Trigoniulinae). *Myriapodologia,* 3: 19–24.

Holsinger, J.R., 1994. Pattern and process in the biogeography of subterranean amphipods. *Hydrobiologia,* 287: 131–145.

Holsinger, J.R. and Culver, D.C., 1988. The invertebrate cave fauna of Virginia and a part of eastern Tennessee: zoogeography and ecology. *Brimleyana,* 14: 1–164.

Holsinger, J.R. and Longley, G., 1980. The subterranean amphipod crustacean fauna of an artesian well in Texas. *Smithson. Contrib. Zool.,* 308: 1–62.

Holthuis, L.B., 1960. Two new species of Atyid shrimps from subterranean waters in N.W. Australia (Decapoda: Natantia). *Crustaceana,* 1: 47–57.

Howarth, F.G., 1973. The cavernicolous fauna of Hawaiian lava tubes, 1. Introduction. *Pac. Insects,* 15: 139–151.

Howarth, F.G., 1980. The zoogeography of specialized cave animals: a bioclimatic model. *Evolution,* 34: 394–406.

Howarth, F.G., 1981. Non-relictual terrestrial troglobites in the tropics: Hawaiian caves. In: *Proceedings of the 8th International Congress of Speleology, Bowling Green, Kentucky,* Vol. 2, pp. 539–541.

Howarth, F.G., 1982. Bioclimatic and geologic factors governing the evolution and distribution of Hawaiian cave insects. *Entomol. Gen.,* 8: 17–26.

Howarth, F.G., 1983. Ecology of cave arthropods. *Annu. Rev. Entomol.,* 28: 365–389.

Howarth, F.G., 1987. The evolution of non-relictual tropical troglobites. *Int. J. Speleol.,* 16: 1–16.

Howarth, F.G., 1988. Environmental ecology of north Queensland caves: or why are there so many troglobites in Australia. In: L. Pearson (Editor), *17th Biennial Conference, Australian Speleological Federation Tropicon Conference, Lake Tinaroo, Far North Queensland, 27–31 December 1988.* Australian Speleological Federation, Cairns, pp. 76–84.

Howarth, F.G. and Stone, F.D., 1990. Elevated carbon dioxide levels in Bayliss Cave, Australia: implications for the evolution of obligate cave species. *Pac. Sci.,* 44: 207–218.

Howarth, M.K., 1981. Palaeogeography of the Mesozoic. In

L.M.R. Cooks (Editor), *The Evolving Earth.* Oxford University Press, Oxford, pp. 197–220.

Humphreys, W.F., 1989. *The status of relict cave fauna of Cape Range, Western Australia, especially the Schizomida,* Report to the Australian National Parks and Wildlife Service, Canberra, 104 pp.

Humphreys, W.F., 1990a. The location of Spiral Cave, Cape Range. *West. Caver,* 30: 36–37.

Humphreys, W.F., 1990b. The biology of a troglobitic schizomid (Chelicerata: Arachnida) from caves in the semi-arid Cape Range, Western Australia. *Acta Zool. Fenn.,* 190: 181–186.

Humphreys, W.F., 1991a. Biological research into the Cape Range karst area, North West Cape peninsula, Western Australia. In: S. Brooks (Editor), *Cave Leeuwin – Proceedings of the 18th Biennial Conference of the Australian Speleological Federation Inc., Margaret River, Western Australia. 30 December 1990–5 January 1991.* Australian Speleological Federation Inc., Nedlands, Western Australia, pp. 6–14.

Humphreys, W.F., 1991b. *Survey of caves in Cape Range, North West Cape Peninsula, Western Australia,* Report to the Australian Heritage Commission and the Western Australian Heritage Committee, 178 pp.

Humphreys, W.F., 1991c. Experimental reactivation of pulse driven populations in a terrestrial troglobite community. *J. An. Ecol.,* 60: 609–623.

Humphreys, W.F., 1991d. *The cave fauna of Cape Range: specific analyses,* Report to the Australian National Parks and Wildlife Service, Canberra, 132 pp.

Humphreys, W.F., 1991e. Troglobites on Barrow Island, Western Australia. *West. Caver,* 31: 11–14.

Humphreys, W.F., 1993a. Cave fauna in semi-arid tropical Western Australia: a diverse relict wet-forest litter fauna. *Mém. Biospéol.,* 20: 105–110.

Humphreys, W.F., 1993b. Stygofauna in semi-arid tropical Western Australia: a Tethyan connection? *Mém. Biospéol.,* 20: 111–116.

Humphreys, W.F., 1993c. The significance of the subterranean fauna in biogeographical reconstruction: examples from Cape Range peninsula, Western Australia. *Rec. West. Aust. Mus. Suppl.,* 45: 165–192.

Humphreys, W.F. (Editor), 1993d. The Biogeography of Cape Range, Western Australia. *Rec. West. Aust. Mus. Suppl.,* 45: 1–228.

Humphreys, W.F., 1994. *The subterranean fauna of the Cape Range coastal plain, Northwestern Australia,* Report to the Australian Heritage Commission and the Western Australian Heritage Committee, 202 pp.

Humphreys, W.F., 1999. The distribution of the Australian cave fishes. *Rec. West. Aust. Mus.,* 19: 469–472.

Humphreys, W.F. and Adams, M., 1991. The subterranean aquatic fauna of the North West Cape peninsula, Western Australia. *Rec. West. Aust. Mus.,* 15: 383–411.

Humphreys, W.F. and Collis, G., 1990. Water loss and respiration of cave arthropods from Cape Range, Western Australia. *Comp. Biochem. Physiol.,* 95A: 101–107.

Humphreys, W.F. and Feinberg, M.N., 1995. Food of the blind cave fishes of northwestern Australia. *Rec. West. Aust. Mus.,* 17: 29–33.

Humphreys, W.F. and Shear, W.A., 1993. Troglobitic millipedes (Diplopoda: Paradoxosomatidae) from semi-arid Cape Range, Western Australia – systematics and biology. *Invertebr. Taxon.,* 7: 173–195.

Humphreys, W.F., Adams, M. and Vine, B., 1989. The biology of *Schizomus vinei* (Chelicerata: Schizomida) in the caves of Cape Range, Western Australia. *J. Zool. (London),* 217: 177–201.

Izquierdo, I. and Oromi, P., 1994. Dictyoptera – Blattaria. In: C. Juberthie and V. Decu (Editors), *Encyclopaedia Biospeologica.* Société de Biospéologie, Moulis (France) and Bucarest, pp. 295–300.

Jacobson, G. and Hill, P.J., 1980. Hydrogeology of a raised coral atoll, Niue Island, South Pacific. *J. Aust. Geol. Geophys.,* 5: 271–278.

Juberthie, C., 1983. Le milieu souterrain: tendue et composition. *Mém. Biospéol.,* 10: 17–65.

Juberthie, C., 1988. Paleoenvironment and speciation in the cave beetle complex *Speonomus delarouzeei* (Coleoptera, Bathysciinae). *Int. J. Speleol.,* 17: 31–50.

Juberthie, C. and Delay, B., 1981. Ecological and biological implications of the existence of a Superficial Underground Compartment. *Proc. 8th Int. Congr. Speleol., Bowling Green, Kentucky,* 6: 203–206.

Kane, T.C., Norton, R.M. and Poulson, T.L., 1975. The ecology of predacious troglobitic beetles, *Neaphaenops tellkampfii* (Coleoptera: Carabidae, Trechinae). I. Seasonality of food input and early life history stages. *Int. J. Speleol.,* 7: 45–54.

Keighery, G. and Gibson, N., 1993. Biogeography and composition of the flora of the Cape Range peninsula, Western Australia. *Rec. West. Aust. Mus. Suppl.,* 45: 51–85.

Kemp, E.M., 1981. Tertiary palaeogeography and the evolution of Australian climate. In: A. Keast (Editor), *Ecological Biogeography of Australia.* Dr W. Junk, The Hague, pp. 31–49.

Kendrick, G.W., Wyrwoll, K.H. and Szabo, B.J., 1991. Pliocene – Pleistocene coastal events and history along the western margin of Australia. *Quat. Sci. Rev.,* 10: 419–439.

Kenneally, K.F., Keighery, G.J. and Hyland, B.P.M., 1991. An ecological survey of tropical rainforests in Western Australia: background and methods. In: N.L. McKenzie, R.B. Johnston and P.G. Kendrick (Editors), *Kimberley Rainforests.* Surrey Beatty and Sons Pty Limited, Chipping Norton, New South Wales, pp. 93–131.

Kennett, J.P., 1982. *Marine Geology.* Prentice-Hall, New York, 813 pp.

Kennett, J.P., Keller, G. and Srinivasan, M.S., 1985. Miocene planktonic foraminiferal biogeography and paleoceanographic development of the Indo-Pacific region. *Geol. Soc. Am. Mem.,* 63: 197–236.

Knott, B., 1993. Stygofauna from Cape Range peninsula, Western Australia: Tethyan relicts. *Rec. West. Aust. Mus. Suppl.,* 45: 109–127.

MacFadden, B.J. and Cerling, T.E., 1994. Fossil horses, carbon isotopes and global change. *Trends Ecol. Evol.,* 9: 481–486.

Main, A.R., 1993. Synthesis and prospects. *Rec. West. Aust. Mus. Suppl.,* 45: 243–248.

Main, B.Y., 1980. Record of the arachnid order Schizomida in Western Australia. *West. Aust. Nat.,* 14: 236–237.

Malcolm, R.J., Pott, M.C. and Delfos, E., 1991. A new tectono-stratigraphic synthesis of the North West Cape area. *APEA J.,* 1991: 154–176.

Manning, R.B., Hart, C.W. and Iliffe, T.M., 1986. Mesozoic relicts in marine caves of Bermuda. *Stygologia,* 2: 156–166.

Martin, M.W., 1990. *Exmouth town water supply investigation report and recommendations for future work, hydrogeology report No. 1990/36,* Western Australian Geological Survey, 12 pp.

McNamara, K.J. and Kendrick, G.W., 1994. Cenozoic molluscs and echinoids of Barrow Island, Western Australia. *Rec. West. Aust. Mus. Suppl.,* 51: 1–50.

Mees, G.F., 1962. The subterranean fauna of Yardie Creek station, North West Cape, Western Australia. *J.R. Soc. West. Aust.,* 45: 24–32.

Millidge, A.F., 1993. Three new species of the spider family Linyphiidae from Australia (Araneae). *Rec. West. Aust. Mus.,* 16: 211–219.

Nei, M., 1987. *Molecular Evolutionary Genetics.* Columbia University Press, New York, 512 pp.

Norton, R.M., Kane, T.C. and Poulson, T.L., 1975. The ecology of predacious troglobitic beetles, *Neaphaenops tellkampfii* (Coleoptera: Carabidae, Trechinae). II. Adult seasonality, feeding and recruitment. *Int. J. Speleol.,* 7: 55–64.

Notenboom, J., 1991. Marine regressions and the evolution of ground dwelling amphipods (Crustacea). *J. Biogeogr.,* 18: 437–454.

Noy-Meir, I., 1985. Desert ecosystem structure and function. In: M. Evenari, I. Noy-Meir and D.W. Goodall (Editors), *Hot Deserts and Arid Shrublands,* Vol. 12A of *Ecosystems of the World.* Elsevier, Amsterdam, pp. 93–103.

Oromi, P., Martin, J.L., Medina, A.L. and Izquierdo, I., 1991. The evolution of the hypogean fauna of the Canary Islands. In: E.C. Dudley (Editor), *The Unity of Evolutionary Biology,* Vol. 1. Dioscorides Press, Portland, OR, pp. 380–395.

Peck, S.B., 1978. New montane *Ptomaphagus* beetles from New Mexico and zoogeography of southwestern caves (Coleoptera; Leiodidae; Catopinae). *Southwest. Nat.,* 23: 227–238.

Peck, S.B., 1980. Climatic change and the evolution of cave invertebrates in the Grand Canyon, Arizona. *Natl. Speleol. Soc. Bull.,* 42: 53–60.

Peck, S.B., 1981. The geological, geographical, and environmental setting of cave faunal evolution. *Proc. 8th Int. Congr. Speleol., Bowling Green, Kentucky,* 2: 501–502.

Peck, S.B., 1984. The distribution and evolution of cavernicolous *Ptomaphagus* beetles in the southeastern United States (Coeloptera; Leiodidae; Cholevinae) with new species and records. *Can. J. Zool.,* 62: 730–740.

Peck, S.B., 1990. Eyeless arthropods of the Galapagos Islands, Ecuador: composition and origin of the cryptozoic fauna of a young, tropical, oceanic archipelago. *Biotropica,* 22: 366–381.

Pesce, G.L., De Laurentiis, P. and Humphreys, W.F., 1996a. Copepods from ground waters of Western Australia. I. The genera *Metacyclops, Mesocyclops, Microcyclops* and *Apocyclops* (Crustacea: Copepoda: Cyclopidae). *Rec. West. Aust. Mus.,* 18: 67–76.

Pesce, G.L., De Laurentiis, P. and Humphreys, W.F., 1996b. Copepods from ground waters of Western Australia. II. The genus *Halicyclops* (Crustacea: Copepoda: Cyclopidae). *Rec. West. Aust. Mus.,* 18: 77–85.

Poore, G.C.B. and Humphreys, W.F., 1992. First record of Thermosbaenacea (Crustacea) from the Southern Hemisphere: a new species from a cave in tropical Western Australia. *Invertebr. Taxon.,* 6: 719–725.

Poulson, T.L. and Culver, D.C., 1969. Diversity in terrestrial cave communities. *Ecology,* 50: 153–158.

Reddell, J.R., 1981. A review of the cavernicole fauna of Mexico, Guatamala, and Belize. *Tex. Mem. Mus. Univ. Tex. Austin Bull.,* 27: 1–327.

Roth, L.M., 1991. A new cave-dwelling cockroach from Western Australia (Blattaria: Nocticolidae). *Rec. West. Aust. Mus.,* 15: 17–21.

Roth, L.M., 1995. New species and records of cockroaches from Western Australia (Blattaria). *Rec. West. Aust. Mus.,* 17: 153–161.

Sbordoni, V., 1982. Advances in speciation of cave animals. In: C. Barrigozzi (Editor), *Mechanisms of Speciation.* A.R. Liss, New York, pp. 219–240.

Scotese, C.R. and Denham, C.R., 1988. *Terra Mobilis: Plate Tectonics for the Macintosh.* Earth in Motion Technologies, Houston, TX.

Shaw, T.R., 1993. *History of Cave Science: the Exploration and Study of Limestone Caves to 1900,* 2nd edn. Sydney Speleological Society, Sydney, 338 pp.

Shear, W.A., 1992. A new genus and two new species of millipedes from Cape Range, Western Australia (Diplopoda, Polydesmida, Paradoxosomatidae). *Rec. West. Aust. Mus.,* 15: 777–784.

Shear, W.A. and Humphreys, W.F., 1996. A new *Stygiochiropus* from a North West Cape (Western Australia) coastal plain cave (Diplopoda, Polydesmida, Paradoxosomatidae). *Rec. West. Aust. Mus.,* 17: 447–449.

Slack-Smith, S.M., 1993. The non-marine molluscs of the Cape Range peninsula, Western Australia. *Rec. West. Aust. Mus. Suppl.,* 45: 87–107.

Smith, A.G. and Briden, J.C., 1977. *Mesozoic and Cenozoic Paleocontinental Maps.* Cambridge University Press, Cambridge, 63 pp.

Stock, J.H., 1994. Biogeographic synthesis of the insular groundwater fauna of the (sub)tropical Atlantic. *Hydrobiologia,* 287: 105–117.

Studier, E.H. and Lavoie, K.H., 1990. Biology of cave crickets, *Hadenoecus subterraneus,* and camel crickets, *Ceuthophilus stygius* (Insecta: Orthoptera): metabolism and water economies related to size and temperature. *Comp. Biochem. Physiol.,* 95A: 157–161.

Sturm, H. and Smith, G.B., 1993. New bristle tails (Meinertellidae, Machiloidea, Archaeognatha, Insecta) from Australia. *J. Aust. Entomol. Soc.,* 32: 233–240.

Trendall, A.F., 1990. Pilbara Craton, introduction. Geology and mineral resources of Western Australia. *Geol. Surv. West. Aust. Mem.,* 3: 128–194.

Truswell, E.M., 1990. Australian Rainforest: the 100 million year record. In: L.J. Webb and J. Kikkawa (Editors). *Australian Tropical Rainforest, Science – Values – Meaning.* CSIRO, Melbourne, pp. 7–22.

Uéno, S.-I., 1987. The derivation of terrestrial cave animals. *Zool. Sci.,* 4: 593–606.

van de Graaff, W.J.E., Denman, P.D. and Hocking, R.M., 1976. Emerged Pleistocene marine terraces on Cape Range, Western Australia. *Geol. Surv. West. Aust. Annu. Rep.,* 1975, 63–70.

Vandel, A., 1965. *Biospeleology: The Biology of Cavernicolous Animals,* translated by B.E. Freeman. Pergamon Press, Oxford, 524 pp.

Veevers, J.J., Powell, C.McA. and Roots, S.R., 1991. Review of seafloor spreading around Australia. I. Synthesis of the patterns of spreading. *Aust. J. Earth Sci.,* 38: 373–390.

Vine, B., Knott, B. and Humphreys, W.F., 1988. Observations on the environment and biology of *Schizomus vinei* (Chelicerata: Schizomida) from Cape Range, Western Australia. *Rec. West. Aust. Mus.*, 14: 21–34.

Webb, L.J. and Tracey, J.G., 1981. The rainforests of northern Australia. In: R.H. Groves (Editor), *Australian Vegetation.* Cambridge University Press, Cambridge, pp. 67–101.

Whitley, G.P., 1945. New sharks and fishes from Western Australia. *Aust. Zool.*, 11: 34–37.

Wyrwoll, K.H., Kendrick, G.W. and Long, J.A., 1993. The geomorphology and Late Cenozoic geological evolution of the Cape Range – Exmouth Gulf region. *Rec. West. Aust. Mus. Suppl.*, 45: 1–23.

Yager, J. and Humphreys, W.F., 1996. *Lasionectes exleyi,* sp. nov., the first remipede crustacean recorded from Australia and the Indian Ocean, with a key to the world species. *Invertebr. Taxon.*, 10: 171–187.

Chapter 31

THE CAVE FAUNA OF SOUTHEAST ASIA. ORIGIN, EVOLUTION AND ECOLOGY

Louis DEHARVENG and Anne BEDOS

INTRODUCTION

For two decades, an increasing number of speleological expeditions have been organized to explore the numerous karsts and huge underground systems of Southeast Asia. Biospeleological investigations have followed at a much slower pace, and the cave fauna still remains virtually unknown in most karstic areas of Burma (Myanmar), Laos, Vietnam, southern China, Java, Kalimantan, central and eastern Sulawesi, the Moluccas (Malukku) and the Philippines. At a country-wide scale, only in Thailand has the karst been fairly well surveyed. The situation is even worse regarding the biology and ecology of cave species, where no progress can be expected in the near future. Yet Southeast Asia hosts much of the world's biodiversity, and several key biological problems might greatly benefit from the study of its cave fauna:

(1) The karsts of Southeast Asia occupy large areas, in continuity with the huge subtropical/temperate limestone areas of China. As such, they constitute the best available model on earth for the study of latitudinal patterns of cave biodiversity.
(2) These karsts are very diverse in their climatic conditions and age, and are highly and variously fragmented, particularly in the Sunda islands, giving unique opportunities to test virtually any model of colonization and speciation.
(3) The environmental stressfulness of caves, particularly high levels of carbon dioxide, have been pointed out recently as a possibly major clue in the evolution of subterranean lineages, but data are still limited to a few subtropical lava tubes (Howarth, 1993). The largest set of "bad-air" caves in the tropics (i.e., caves where the air has a high concentration of carbon dioxide) has been located in northern Thailand (Deharveng, 1987a), with a rich troglobitic fauna and wide natural gradients of carbon dioxide content.
(4) Highly troglomorphic fauna is increasingly being collected in tropical caves, but Southeast Asia is of special interest for several terrestrial groups. For instance, the speciose lineages of *Troglopedetes* (Collembola) or *Diestrammena* (crickets) offer large clines of adaptation to cave life from epigean to strictly troglobitic species, with wide series of morphological modifications. Reconstructing phylogenies in these non-relict taxa could be a most efficient way to address the problem of the troglobitization process and of the origin of troglobites.
(5) While the Neotropics, India and tropical Africa are relatively homogeneous biogeographically, Southeast Asia is not, and this is one of its most attractive biological feature. The fauna of Southeast Asia is open to northern Palearctic influence through China, meets the Indian fauna in Burma (Myanmar), and is introgressed by Austral lines in the south. Isolated in a buffered and low-competition environment, troglobitic faunas of the region are likely to have retained some vestiges of ancient ecogeographical events, which could have been erased from the more exposed epigean communities.

Some of these questions are addressed in this chapter on the basis of the few available data, while others are open avenues for future research. After an introduction about biological exploration in the region, we shall describe four large subterranean systems which have been extensively sampled for their fauna. Faunal diversity will be next examined in relation to that of other tropical and temperate areas. The relations between energy input, environmental parameters and the organization of subterranean communities will be considered, and some basic problems related to the origin and the evolution of troglobites in Southeast Asia will be discussed.

THE BIOLOGICAL EXPLORATION OF SOUTHEAST ASIAN CAVES

In 1914, Jeannel and Racovitza, commenting on a large collection of cavernicolous fauna from tropical Africa, observed that a number of troglobites were included in their material: "A l'inverse des troglophiles, ces troglobies sont nombreux en tant qu'espèces, mais très rares comme individus [In contrast to troglophiles, these troglobites include numerous species, but are rare as individuals]". They suggested that true terrestrial troglobites (in the sense of troglomorphic species) were in fact limited to the deepest parts of caves, "loin de la population du guano [far away from the guano population]". These observations were overlooked by subsequent biospeleologists, including Jeannel himself, and the opinion prevailed that troglobites, at least terrestrial troglobites, were absent, or very rare, in tropical caves (Vandel, 1964: p. 36). The recent discovery of a well-diversified troglobitic fauna in the tropics was therefore considered as a major advance in the field of cave biology (Howarth, 1983). Two decades of investigations have largely confirmed this view. Not only have highly troglomorphic species now been found in the tropics, but evidence is accumulating that rich cave-restricted faunas are present in most tropical caves, be they ancient karstic systems (Peck, 1975; Chapman, 1976; Sbordoni et al., 1980; Decu, 1981), or geologically recent lava tubes (Howarth, 1973).

The Batu caves of Malaya were among the first investigated in the tropics for their cave life (Ridley, 1898). Extensive researches were carried out on this underground system over several decades, and the specimens collected there still constitute the largest single collection of cave organisms ever made in a Southeast Asian cave (McClure et al., 1967; Bullock, 1965). Data were also gathered at the beginning of the century on the fauna of the Farm caves near Moulmein in southern Burma and to a lesser extent of the caves of the Malayan State of Jalor, now southern Thailand (Annandale et al., 1913). Most of the animals of the Batu caves, the Farm caves or the Jalor caves, as well as those of different caves of Southeast Asia reported by Boutin (1971), were however not troglobites, or at least did not display marked cave-dependent morphological characters in the sense of Christiansen (1961). According to recent literature about tropical faunas, they would be classified today, rather, as guanophiles, guanobites or troglophiles: "no species found in the caves of Burma and the Malay peninsula is so highly specialized in correlation with this mode of life as are certain species found in the caves of Europe and North-America" (Annandale et al., 1913). Thus, these first investigations failed to bring any positive information about the presence and the importance of terrestrial troglobitic species in Southeast Asia, and rather supported the general opinion that they were absent from tropical caves. It was not until the exploration of the Mulu caves in Sarawak, northern Borneo (Chapman, 1982a), that the existence of a well-diversified fauna of troglobites was established for a lowland cave of Southeast Asia.

The delay of biospeleologists in discovering troglomorphic species in tropical caves is clearly related to the difficulty of access to oligotrophic caves (Howarth, 1983), and to the remoteness of the oligotrophic habitats to which these species are restricted (Jeannel and Racovitza, 1914). Howarth (1980) has analysed in detail why habitats suitable for terrestrial troglobitic species are less frequent in tropical than in temperate countries. The relevant factors he recognized were related to the tropical climate, which directly induces higher evaporation rates in caves [because of both the overall higher temperatures and the diurnal fluctuations in temperature outside the cave – the "winter effect" of Howarth (1980); see also Barr (1968)], and indirectly leads to a more open structure of the cave systems (because of higher rates of solution and erosion). Another most important feature is the abundance of non-troglobite competitors (Jeannel and Racovitza, 1914). Non-troglobite species are more numerous in tropical than in temperate caves; they enter much deeper into the cave, and may saturate habitats otherwise favourable to troglobites, because of the similar climatic conditions in and outside caves during the year, and the large food availability from bats, swiftlets and flood detritus. It is therefore not surprising that the full richness of tropical caves has only been discovered when careful attention was given to the more inaccessible oligotrophic habitats which host the troglobitic fauna. In Southeast Asia, most biospeleological discoveries were made as a result of

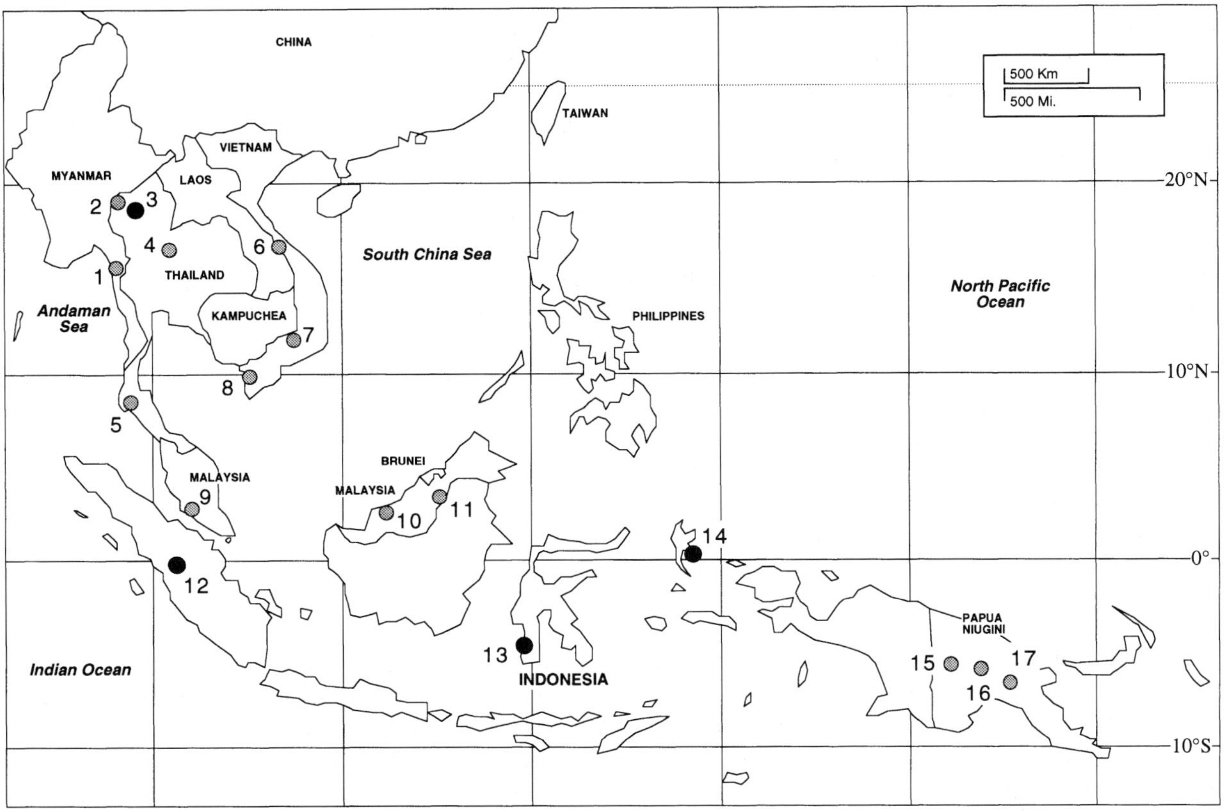

Fig. 31.1. Caves and cave areas of Southeast Asia studied for their fauna. Black circles, reference cave systems; grey circles, other important caves. **Myanmar (Burma)**: 1, Farm caves. **Thailand**: 2, Nam Lang karst; 3, Tham Chiang Dao; 4, Tham Kubio–Tham Phulu system; 5, Phangnga caves. **Vietnam**: 6, Ke Bang karst; 7, lava tubes north of Ho Chi Minh City; 8, Hon Chong caves (Deharveng and Le Cong Kiet, 2000). **Malaysia**: 9, Batu caves; 10, Mulu caves; 11, Niah caves (Chapman, 1984). **Indonesia**: 12, Ngalau Surat, Sumatra; 13, the Gua Salukkan Kallang/Towakkalak system (SKT), Maros karst, Sulawesi; 14, Batu Lubang cave, Halmahera. **Papua New Guinea**: 15, Telefomin caves (Chapman, 1976); 16, Muller Range (Smith, 1980); 17, Elimbari (Deharveng, 1982).

extensive speleological explorations in recent years (Chapman, 1976, 1982a, 1984; Deharveng et al., 1986; Deharveng, 1987b; Deharveng and Leclerc, 1989; Stone, 1983).

In spite of efforts during the last three decades, few underground systems have been thoroughly investigated, considering the extent of limestone and basaltic areas in Southeast Asia. Information about lava-tube fauna is limited to a few small caves in southern Vietnam (Deharveng and Le Cong Kiet, 2000). Sparse information is available for many limestone caves, but significant amounts of data about cave faunas are available only for 17 large systems or cave groups (Fig. 31.1). Among them, the Papuan sites are high in elevation (above 1500 m), and are more temperate than tropical in climate and fauna (Chapman, 1976; Deharveng, 1981, 1982). In contrast, the remaining caves are all located at low elevation (less than 600 m). On the map, except for northern Thailand, the caves studied are 400 to 1700 km apart, thus affording only a sparse coverage of Southeast Asia (Fig. 31.1). These 14 lowland underground systems have been very unevenly sampled. Except for Tham Chiang Dao and Tham Kubio–Tham Phulu, aquatic collections from most sites have been insufficient. The Mulu caves were extensively sampled for vertebrates and the larger arthropods, but microarthropods were not recorded. Flying arthropods associated with guano have been accurately collected only in the Batu caves; but very few troglobites have been reported from these caves, or from the Farm caves – probably because of a lack of suitable habitats, as well as the emphasis put on guano habitats during the biological surveys. The faunas of Batu Lubang, the Gua Salukkan Kallang/Towakkalak system and Ngalau Surat are relatively well known, except for guano habitats and vertebrates. The other caves have been less intensively studied.

Although several papers include comments on the ecology of cave fauna in Southeast Asia (Boutin, 1971; Stone, 1983; Whitten et al., 1984, 1987; Deharveng and Leclerc, 1989), only five deal in depth with ecological aspects of invertebrate cave assemblages: Bullock (1965) for the guano communities of Batu caves, Chapman (1980b, 1982a, 1983) for the terrestrial macroarthropods of Mulu caves and Gibert (1987) for the aquatic fauna of Chiang Dao cave. Autecological and biological data about cave species are even rarer, except for a few troglophilic or guanophilic species such as the spider *Liphistius batuensis* of the Batu caves (McClure et al., 1967) or the Amblypygida *Myodalis* sp. of Cambodia (Boutin, 1969). A few studies have also been devoted to bat ecology (McClure et al., 1967) and to bat parasites biology (Dunn, 1965).

THE SYSTEMS STUDIED

Particular attention will be given in this study to four large underground systems recently investigated. Each comprises a wide range of habitats, from guano piles to low-energy habitats. All have been extensively sampled, with a parallel sampling of soil and aquatic habitats outside, allowing a reasonable assignment of cave species to ecological categories of troglobite or troglophile.

Tham Chiang Dao (TCD) (Fig. 31.2)

This system 5.2 km long is situated in northern Thailand, at 19°20′N and at an altitude of 455 m

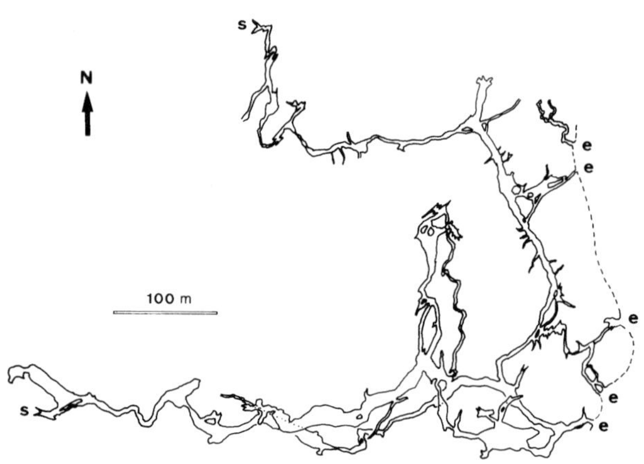

Fig. 31.2. Tham Chiang Dao (Thailand); e, entrance; s, sump.

(Deharveng and Brouquisse, 1986). The origin of the underground stream running in a short portion of the cave is not known; the absence of organic debris on the banks is noticeable, and suggests internal drainage rather than large external stream sinks for the waters, although a few streamlets which sink in the remote western part of the Chiang Dao mountain may contribute to the flow.

Bats are relatively abundant in Tham Chiang Dao, with guano scattered in all galleries, and a few large piles locally. Troglobites and guanobites are abundant in many parts of the cave. A detailed ecological analysis of its aquatic fauna is available (Gibert, 1987).

Ngalau Surat (Sumatra) (Fig. 31.3)

This cave in central Sumatra, located on the equator (0°26′S) at an altitude of about 400 m, was formed by the sinking of the Sangki river into the Gunung Seribu limestone range. It consists in a gallery six kilometres long, with the river running on the floor, and over 1000 metres of side passages (Deharveng and Bedos, 1996; and unpublished data).

Terrestrial and aquatic fauna were sampled during three expeditions (Deharveng and Bedos, 1995; and unpublished data). Several troglobitic species were

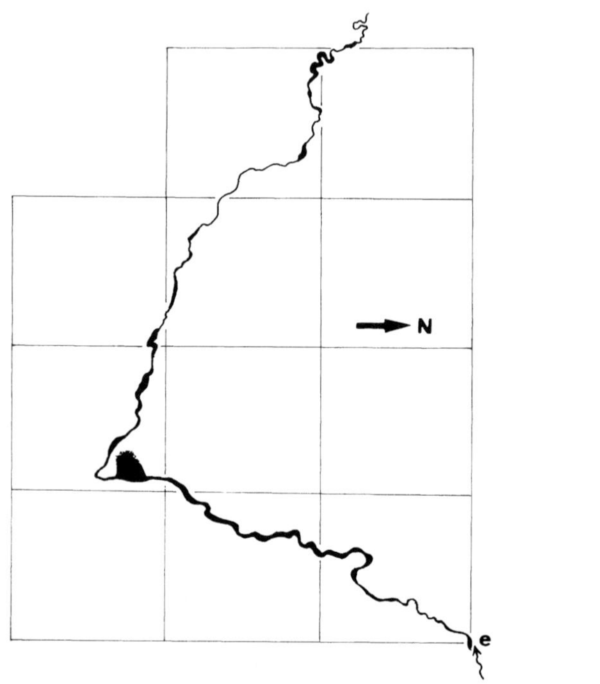

Fig. 31.3. Ngalau Surat (Sumatra); each square side is 0.5 km long; e, entrance.

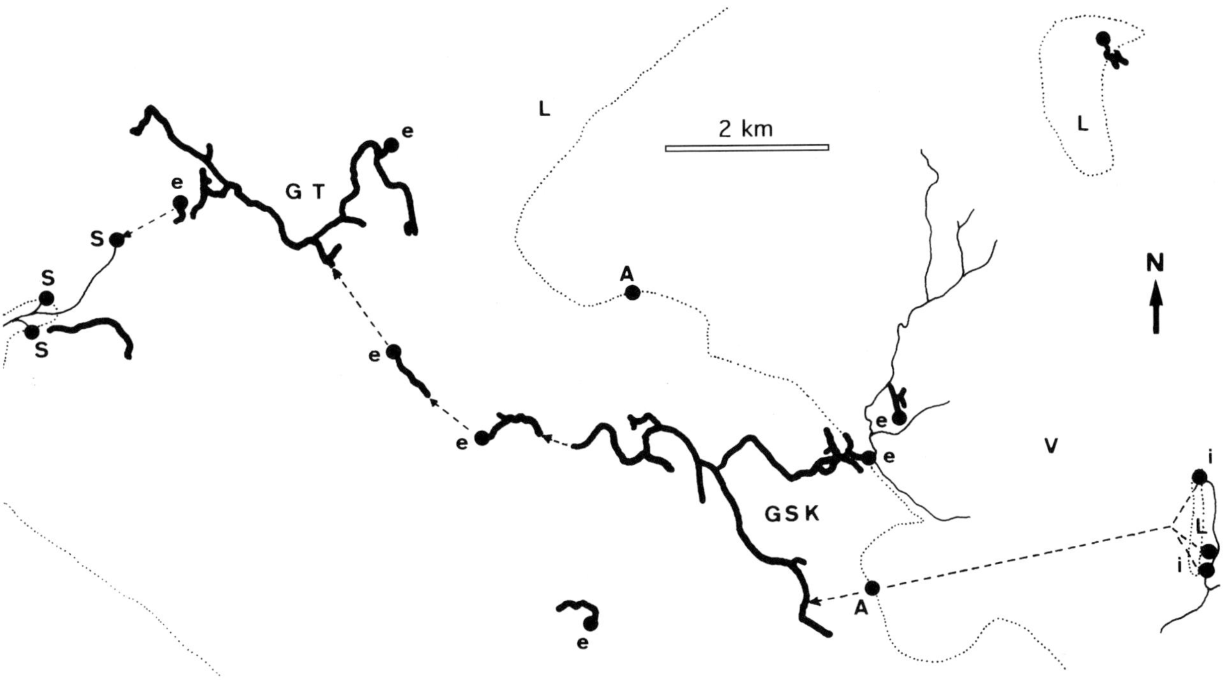

Fig. 31.4. The SKT (Gua Salukkan Kallang/Towakkalak) system in the Maros karst (southern Sulawesi); GT, Gua Tanette; GSK, Gua Salukkan Kallang; A, aven; e, entrance; i, impassable sink; S, spring; L, limestone; V, volcanic rocks.

collected; guanobites were abundant on guano deposits throughout the cave.

The Gua Salukkan Kallang/Towakkalak system (SKT) (Fig. 31.4)

The huge underground system of Gua Salukkan Kallang or Towakkalak lies between 40 and 260 m of altitude in the Maros karst of southern Sulawesi, at 5°01′S (Brouquisse and Rigal, 1987; Rigal, 1987; Bedos et al., 1994).

A speleological survey and a biospeleological survey of this system were conducted in parallel during five expeditions. The large underground river of the system runs through five different caves: gua Wattanang (length: 440 m), gua Tanette (length: 9.7 km), gua Batu Neraka (length: 750 m), gua Lubang Kabut (length: 1145 m), gua Salukkan Kallang (length: 12.5 km). It emerges in an unexplored sump. Large amounts of flood debris along the river banks, and bat and swiftlet guano, scattered along the underground galleries or locally in large patches, provide abundant trophic resources exploited by a rather rich fauna.

Batu Lubang (Fig. 31.5)

Batu Lubang, on the remote island of Halmahera

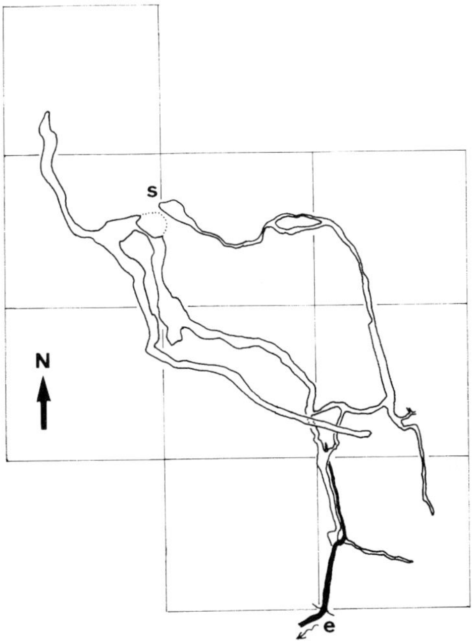

Fig. 31.5. Batu Lubang (Halmahera); each square side is 0.5 km long; e, entrance; s, sump.

(Moluccas, Indonesia), lies on the equator (0°29′N) at an altitude of less than 10 m. Over eight and a half kilometres have been surveyed, mostly in huge galleries (Brouquisse and Brouquisse, 1992) through which the stream forming them has ceased to flow ("fossil galleries"). Access to the subterranean stream is limited to the entrance zone and to a sump two kilometres inside. The terrestrial fauna was extensively sampled during a twelve-day period. The troglobite fauna is diversified, and the large piles of guano in the first 700 metres of the cave host huge invertebrate populations (Deharveng, 1989).

FAUNAL DIVERSITY

Species richness and taxonomic diversity

Local diversity

The number of species recorded in the four systems studied ranges from 72 to 93 (Tables 31.1 to 31.4), with a large proportion of taxa still unidentified and several groups not analysed at the species level. These numbers have however limited significance, because the non-troglobitic fauna (mostly guanophilic species), which has been sampled very unequally from cave to cave, provides the largest contribution to this richness. Non-troglobitic species range from 66.6 to 81.5% of the total number of species in the study sites (Table 31.4). This large proportion of non-troglobitic species is probably the most conspicuous characteristic of tropical cave faunas in Southeast Asia. Moreover, this ratio is probably an underestimate because non-troglobitic microarthropods are frequently overlooked. For instance, as a result of different collecting techniques in high-energy habitats and of different possibilities of identifying specimens, 24 Collembola, and no Diptera are recorded to species or genus from the Gua Salukkan Kallang/Towakkalak system (Table 31.2), as against 64 Diptera and only two Collembola from the Batu caves (McClure, 1965).

Between-cave comparison

Considering species richness for troglobites makes more sense, as this category has been thoroughly sampled in the study sites. As indicated by the small number of species found once only, no significant increase in species richness may be expected from further sampling effort in the four reference systems. The analysis of this troglobitic fauna shows that the overall number of troglobitic species hosted by each system is similar, except for the SKT system in Sulawesi which is richer in both terrestrial and aquatic species (Table 31.4). The richness of this system seems to result from the presence of some unusual additional taxa (blind crabs, Atyidae, the caraboid beetle *Mateuius troglobioticus*). The taxonomic groups which represent the core of the troglobitic fauna are the same in the four systems studied (Tables 31.1–31.3 and 31.5). Each of these groups has also a similar number of troglobitic species. For instance, the numbers of troglobitic Collembola (the most speciose group in Southeast Asian caves) are two (Batu Lubang), three (TCD and Ngalau Surat) and four (SKT), with the lowest richness in the least-studied cave (Batu Lubang).

Similarity in troglobite richness does not imply taxonomic similarity. For instance, the troglomorphic species of Collembola belong to cf. *Coecobrya* (Entomobryidae) in central Thailand, to *Troglopedetes* (Paronellidae) in Tham Chiang Dao, to *Cyphoderopsis* (Paronellidae) in Ngalau Surat, to *Pseudosinella* (Entomobryidae) in the SKT system and Batu Lubang. The same is found among troglomorphic spiders, which often belong to different families from one site to another (Table 31.2).

Habitat parameters and biodiversity

Chapman (1983) tested species diversity in the Mulu caves against habitat stability, predictability, substrate heterogeneity (an estimate of environmental diversity) and rate of food input (an estimate of productivity). Of these parameters, only predictability was found to be correlated with biodiversity. This agrees with the findings of Christiansen and Bouillon (1978) in the temperate caves of the Pyrenees. However, Chapman's work suffers from sampling heterogeneity (Chapman, 1980b: p. 141) and high subjectivity in evaluating the level of the parameters studied in the different habitats. Predictability in particular was estimated by the mode of food supply and flooding along a 1–4 scale, with 3 and 4 the only relevant levels (Chapman, 1983: p. 212), the most flood-prone habitats being the least predictable. Chapman concluded that unpredictable environmental perturbations, such as flooding, tend to lower biodiversity by suppressing "the development of a diverse climax community". In contrast, our results suggest that flooding promotes species richness by bringing together many taxa from various outside habitats, but also by attracting some cave-restricted and guanobitic species in Ngalau Surat

Table 31.1
The aquatic fauna of four cave systems in Southeast Asia

Fauna	Cave system			
	Tham Chiang Dao	Ngalau Surat	Gua Salukkan Kallang/Towakkalak system	Batu Lubang
PLATHELMINTHES: Turbellaria		Tricladida gen. sp. [1,4]	*Dugesia leclerci* [2]	
NEMATODA	Dorylaimidae gen. sp. *Tobrilus* sp.			
ANNELIDA: Oligochaeta	*Marionina* sp.			
MOLLUSCA: Gastropoda	gen. sp.	gen. sp.	Thiaridae gen. sp.	
ARTHROPODA				
Arachnida: Actinedida	gen.sp.			
Crustacea				
Ostracoda	gen. sp.			
Copepoda	Cyclopoidea gen. sp. *Elaphoidella* cf *grandidieri* *Tropocyclops prasinus* cf *guwana*	gen. spp. [5]	gen. spp.	gen. sp.
Syncarida	Parabathynellidae gen. sp. [2]			
Isopoda: Stenasellidae		*Stenasellus* sp. [2]		
Amphipoda: Bogidiellidae		Bogidiellidae gen. sp. [2]	Bogidiellidae gen. sp. [2]	*Bogidiella deharvengi* [2]
Decapoda: crabs		gen. sp.	*Cancrocaeca xenomorpha* [2] *Parathelphusa* cf *celebensis*	*Currothelphusa asserpes* [3] *Sesarmops* cf *impressum* [3]
Decapoda: shrimps	cf *Macrobrachium* gen. sp.		Atyidae gen.sp. [2] cf Atyidae gen. sp. [4] *Macrobrachium lar*	cf *Macrobrachium* gen. sp.
CHORDATA: Pisces		*Puntius* cf *binotatus*	Eleotridae gen. sp. [4] *Glossogobius* sp. [4]	

1 "gen. sp." : one species of an unidentified genus.
2 Blind troglomorphic species.
3 Troglomorphic species, also found in terrestrial habitats.
4 Microphthalmic troglomorphic species.
5 "gen. spp." several species of one or several unidentified genera.

Table 31.2
The terrestrial fauna of four cave systems in Southeast Asia

Fauna	Cave system			
	Tham Chiang Dao	Ngalau Surat	Gua Salukkan Kallang/Towakkalak system	Batu Lubang
ANNELIDA				
Oligochaeta		Enchytraeidae gen.sp. Lumbricidae gen.sp.	Enchytraeidae gen.sp. Lumbricidae gen.sp.	Enchytraeidae gen.sp.
Hirudinea		gen.sp.[1,4]		
MOLLUSCA: Gastropoda	gen.sp.	gen.spp.[2,3]	gen.sp.	
	Subulinidae gen.sp.[3]			Subulinidae gen.sp.[3]
ARTHROPODA: Arachnida				
Scorpionida				gen.sp.[3]
Pseudoscorpionida	*Megachernes* cf *grandis*[3] *Tyrannochthonius* sp.[4]	gen.sp.[4]	gen.sp.[4]	gen.sp.
Schizomida	*Schizomus* sp.[3,4]	*Schizomus* sp.[3,4]	*Schizomus* sp.[3,4]	
Palpigradi	*Eukoenenia* cf *lyrifer* *Eukoenenia thais*[4]		*Eukoenenia maros*[4] *Prokoenenia celebica*	
Amblypygida		gen.sp.[3]	gen.sp.[3]	gen.sp.[3]
Araneae		gen.spp.[5]	gen.spp.[4]	gen.sp.
Ctenidae			gen.sp.[4]	
Dipluridae				gen.sp.[5]
Gnaphosidae	*Scotophaeus* sp.[4]			gen.sp.[4]
Heteropodidae	*Heteropoda* sp.[3] *Panaretus* sp.[3]	*Heteropoda* sp.[3]	*Heteropoda* sp.[3]	*Heteropoda* sp.[3]
Mygalomorpha				gen.sp.[4]
Nesticidae	*Howaia mogera*			
Ochyroceratidae	*Althepus tibiatus* *Speocera minutissima*		*Psiloderces leclerci* *Speocera caeca*[4]	
Pholcidae	*Spermophora* sp.[4]			*Spermophora* sp.[4]
Tetrablemmidae	*Indicoblemma* sp.			
Theridiidae	*Pholcomma* sp. *Theridion rufipes*			

continued on next page

Table 31.2, *continued*

Fauna	Cave system			
	Tham Chiang Dao	Ngalau Surat	Gua Salukkan Kallang/Towakkalak system	Batu Lubang
Opiliones	*Bandona palpalis* *Paratakoia* sp.[5]	gen.sp.[4]	gen.sp.[5]	Dibuninae gen.sp.[5] *Dibunus pseudobiantes*
Acarina: Gamasida	gen.spp.[3]	gen.spp.[3]	gen.spp.[3]	gen.spp.[3]
Acarina: Uropodida	gen.sp.[3]	gen.sp.[3]	gen.sp.[3]	gen.sp.[3]
Acarina: Actinedida	gen.spp.	gen.spp.	gen.spp.	gen.spp.
Rhagidiidae			gen.sp.	gen.sp.
Trombidiidae	gen.sp.[4]	gen.sp.[4]	gen.sp.[4]	gen.sp.[4]
Acarina: Oribatida	gen.spp.	gen.spp.	gen.spp.	gen.spp.
ARTHROPODA: Crustacea				
Ostracoda			gen.sp	gen.sp.
Isopoda Oniscidea				
Armadillidae	*Cubaris* sp.[3,5]	gen.sp.[3,5]	gen.sp.[3,5]	gen.sp.[3,5]
Philosciidae	gen.sp.[4]	gen.sp.[4]	gen.sp.[4]	*Leucophiloscia* sp.[4]
Tricho/Styloniscidae		Trichoniscidae gen.sp.[4]	Styloniscidae gen.sp.[4]	*Clavigeroniscus* sp.[3,5]
ARTHROPODA: Myriapoda				
Diplopoda		gen.sp.[4]	gen.sp.	
Glomeridesmida		gen.sp.[4]	gen.sp.	
Stemmiulida				Stemmiulidae gen.sp.[3,4]
Julida		Glyphiulidae gen.sp.[3,5] Julida gen.sp.	?Glyphiulidae gen.sp.[3,5]	
Polydesmida	gen.sp. *Dyomerothrix gremialis*[4]		gen.sp.[4]	gen.sp.[4] Pyrgodesmidae gen.sp.[4]
Craspedosomatida		gen.sp.	?Craspedosomatida gen.sp.[4]	*Metopidiothrix* sp.
Pauropoda			Pauropodidae gen.sp.	
Symphyla	gen.sp.	gen.sp.	gen.sp.	gen.sp.
Chilopoda				
Geophilomorpha		gen.sp.[3]		gen.sp.[3]
Scolopendromorpha			gen.sp.[3]	*Ethmostigmus platycephalus*[3]
Scutigeromorpha		Scutigeridae gen.sp.[3]		Scutigeridae gen.sp.[3]

continued on next page

Table 31.2, *continued*

Fauna	Cave system			
	Tham Chiang Dao	Ngalau Surat	Gua Salukkan Kallang/Towakkalak system	Batu Lubang
ARTHROPODA: Insecta				
Protura			gen.sp.	
Collembola	cf. Table 31.3	cf. Table 31.3	cf. Table 31.3	cf. Table 31.3
Diplura	Japygidae gen.sp. [4]		*Lepidocampa hypogaea* [4] *Lepidocampa weberi borneensis*	*Lepidocampa* sp.
Microcoryphia				*Machilellus* sp. [3]
Pterygota				
Ephemeroptera			gen.sp.	
Dictyoptera	gen.sp. [3]	gen.sp. [3]		gen.sp. [3]
Nocticolidae	gen.sp. [4]	gen.sp. [4]	gen.sp. [4]	gen.sp. [4]
Isoptera			gen.sp.	
Plecoptera		gen.sp.		
Orthoptera	Rhaphidophoridae gen.sp. [3]	*Rhaphidophora* cf *oophaga*	*Rhaphidophora ?heinrichi* [3]	Rhaphidophoridae gen.sp. [3]
Coleoptera	gen.spp.	gen.spp.	gen.spp.	gen.spp.
Aderidae	gen.sp. [3]	gen.sp. [3]		
Caraboidea		gen.sp.	*Mateuius troglobioticus* [3,5]	*Clivina* sp.
Catopidae	Cholevinae gen.sp. *Ptomaphaginus* sp. [3]	*Ptomaphaginus* sp. [3]		
Corticariidae				gen.sp.
Curculionidae				gen.sp.
Histeridae			gen.sp.	gen.sp.
Merophysiidae	*Holoparamecus* sp. [3]			
Pselaphidae	gen.sp.		gen.sp.	Euplectini gen.sp.
Ptiliidae			gen.sp.	gen.sp.
Scydmaenidae		gen.sp.		
Sphaeridiidae				gen.sp.
Staphylinidae	gen.spp. [3] Oxytelinae gen.sp.	gen.spp. [3]	gen.spp. [3]	gen.spp. [3]

continued on next page

Table 31.2, *continued*

Fauna	Cave system			
	Tham Chiang Dao	Ngalau Surat	Gua Salukkan Kallang/Towakkalak system	Batu Lubang
Coleoptera (*cont'd*)				
Tenebrionidae			gen.sp.	
Trichoptera		gen.spp.[3]		
Lepidoptera Tineoidea	*Crypsithyris spelaea*[3] *Tinea antricola*[3] *Wegneria cerodelta*[3]	Tineoidea gen.spp.[3]	Tineoidea gen.spp.[3]	*Tinea porphyropa*[3]
Diptera	gen.spp.[3]	gen.spp.[3] Psychodidae gen.sp.[3]	gen.spp.[3] Psychodidae gen.sp.[3]	gen.spp.[3] Psychodidae gen.sp.[3]
Hymenoptera	gen.sp. Formicidae gen.sp.[3]	Formicidae gen.sp.[3]	gen.sp. Formicidae gen.sp.[3]	gen.sp. Formicidae gen.sp.[3]
Psocoptera	*Belapha* sp. *Liposcelis bostrychophilus* *Liposcelis entomophilus* *Pachytroctes* sp. *Perientomum* sp. *Psocathropos microps*		gen.sp.	
Thysanoptera	gen.sp.		gen.sp.	
Hemiptera	Heteroptera gen.sp. Cicadelloidea gen.sp.	Emesinae gen.sp.[3]	Heteroptera gen.sp. Aphididae gen.sp.	Heteroptera gen.sp. *Mesovelia horvathi*
CHORDATA: Reptilia		*Elaphe* sp.	*Elaphe* sp.	
CHORDATA: Aves		Apodidae gen.sp.	Apodidae gen.sp.	Apodidae gen.sp.
CHORDATA: Mammalia				
Insectivora	*Crocidura* sp. *Suncus murinus murinus*			
Chiroptera	*Hipposideros armiger armiger* *Hipposideros lylei* *Hipposideros* sp. *Pipistrellus javanicus paterculus* *Rhinolophus pusillus szechwanus* *Rhinolophus* sp.	gen.spp.	gen.spp.	*Dobsonia crenulata* *Hipposideros bicolor saevus*

[1] "gen. sp." : one species of an unidentified genus.
[2] "gen. spp." several species of one or several unidentified genera.
[3] Guanophilic or guanobitic species.
[4] Blind or, if belonging to a blind epigeous lineage, with elongate appendages (Arthropoda) or without pigment (leeches).
[5] Microphthalmic troglomorphic species.

Table 31.3
Collembolan species in the four underground systems studied

Fauna	Cave system			
	Tham Chiang Dao	Ngalau Surat	Gua Salukkan Kallang/ Towakkalak system	Batu Lubang
Troglobites				
Troglomorphic	*Pseudosinella chiangdaoensis* *Troglopedetes fredstonei*	*Cyphoderopsis* sp.1	*Coecoloba* sp. *Pseudosinella* sp.	*Pseudosinella* sp.1
Slightly troglomorphic	*Arrhopalites chiangdaoensis*	*Onchopodura* sp. *Pararrhopalites* sp.	*Onchopodura* sp. *Pararrhopalites* sp.	*Pseudosinella* sp.2
Guanobites-troglobites	*Acherontiella colotlipana* *Coecobrya guanophila* *Troglopedetes leclerci*	*Acherontiella* sp. *Cyphoderopsis* sp.2	*Willemia* cf *buddenbrocki*	*Acherontiella* sp. *Cryptopygus* sp. *Lepidonella* sp.
Guanophiles-troglophiles	*Folsomina onychiurina*	*Folsomina onychiurina*	*Folsomina onychiurina*	*Xenylla yucatana*
Others	*Arrhopalites anulifer* *Coecobrya similis* *Entomobrya* sp. *Folsomides exiguus* *Lepidocyrtus* sp. *Mesaphorura* sp. *Paronellidae* sp. *Protaphorura* sp. *Pseudosinella* sp. *Seira* sp.	*Arrhopalites* sp. *Folsomides centralis* *Folsomides exiguus* *Isotomiella nummulifer* *Isotomiella symetrimucronata* Lobellini gen.sp. [1] *Megalothorax* sp. *Mesaphorura* sp. *Neelus* sp. *Paranura* sp. *Pronura* sp. *Protaphorura* sp.1 *Protaphorura* sp.2 *Pseudosinella* sp. Sminthuridae gen.sp. *Willemia* sp.	*Brachystomella* sp. *Blasconura* sp. *Folsomides centralis* *Folsomides exiguus* *Friesea* sp. *Isotomiella* cf *annae* *Isotomiella nummulifer* *Isotomiella symetrimucronata* *Lepidocyrtus* sp. Lobellini gen.sp. *Megalothorax* sp. *Mesaphorura* sp. *Paleonura* sp. *Paranura* sp. *Protaphorura* sp. Pseudachorutinae gen.sp. *Sminthurinus* sp. *Sphaeridia* sp.	*Ascocyrtus* sp. *Isotomiella* cf *annae* *Mesaphorura* sp. *Neelus* sp. *Prabhergia* sp. *Pseudachorutes* cf *longisetus*

[1] "gen. sp.", one species of an unidentified genus.

Table 31.4
Numbers of troglobitic, stygobitic and guanobitic species in the four underground systems studied

Species numbers	Cave system							
	Tham Chiang Dao		Ngalau Surat		SKT [1]		Batu Lubang	
	N	%	*N*	%	*N*	%	*N*	%
Total number of species	92		74		93		72	
Number of terrestrial troglobitic species [2]	15	16.30	17	22.97	21	22.58	15	2.83
Number of stygobitic species	1	1.09	3	4.05	7	7.53	1	1.39
Number of guanobitic species [2]	23	25.00	24	32.43	17	18.28	24	33.33

[1] SKT, Gua Salukkan Kallang/Towakkalak system.
[2] Schizomida, Armadillidae, *Clavigeroniscus* sp., Glyphiulidae, Stemmiulidae and *Mateuius troglobioticus* are counted as troglobites and as guanobites; troglo-guanobite Collembola are not considered.

Table 31.5
Species richness (troglobites and troglomorphic guanobites) in underground systems

Taxa	Southeast Asia				France	China (Hunan):
	Tham Chiang Dao	Ngalau Surat	Gua Salukkan Kallang/Towakkalak system	Batu Lubang	(Haute-Garonne): Grotte de la Buhadère	Feihu Dong
PLATHELMINTHES						
Turbellaria	0	1	1	0	0	1
ANNELIDA						
Hirudinea	0	1	0	0	0	0
ARTHROPODA: Arachnida						
Pseudoscorpionida	1	1	1	0	0	0
Schizomida	1	1	1	0	0	0
Palpigradi	1	0	1	0	0	0
Araneae	2	>2	3	4	1	2
Opiliones	1	1	1	1	0	1
Acarina: Rhagidiidae	0	0	0	0	1	1
Acarina: Trombidiidae	1	1	1	1	0	0
ARTHROPODA: Crustacea						
Syncarida	1	0	0	0	0	0
Isopoda: Stenasellidae	0	1	0	0	1	0
Isopoda: Armadillidae	1	1	1	1	0	0
Isopoda: Philosciidae	1	1	1	1	0	0
Isopoda: Styloniscidae	0	0	1	1	0	0
Isopoda: Trichoniscidae	0	1	0	0	0	0
Amphipoda	0	1	1	1	1	1
Decapoda: crabs	0	0	1	0	0	0
Decapoda: shrimps	0	0	2	0	0	2
ARTHROPODA: Myriapoda						
Diplopoda	1	3	3	3	1	2
Chilopoda	0	0	0	0	0	1
ARTHROPODA: Insecta						
Collembola	3	3	4	2	6	2
Diplura	1	0	1	0	0	1
Dictyoptera	1	1	1	1	0	0
Orthoptera	0	0	0	0	0	2
Coleoptera	0	0	1	0	5	3–4
CHORDATA: Pisces	0	0	2	0	0	1
Total number of species	16	20	28	16	16	20–21
Total number of groups	13	15	19	10	7	13

and the SKT system. In fact, the local absolute values of food input and humidity, rather than their patterns of temporal change, obviously control most of the diversity fluctuations, at least in oligotrophic habitats, because troglobites are able to cope with irregular supplies of resources: they usually respond more or less rapidly to organic baits placed in any cave microhabitat, provided that humidity is high enough. At least, substrate heterogeneity, through the nature and amount of its organic content, has a subsidiary role in the Chapman study, because of the broad diet of most cave animals, and hence their lack of trophic specialization.

Biodiversity of cave species in Southeast Asia *versus* other regions of the world

Biological diversity will be considered here at two scales: the regional scale, and the local scale. In evaluating regional diversity, it is necessary to consider the effect of the endemic, vicariant species of each karstic system. Patterns may be misleading since what is measured by regional species richness is to a large extent proportional to the number of karstic systems studied in the area. Comparison of local diversities at species level is practically and biologically more meaningful, as local diversity is dependent on the ecological diversity of habitats, on the stock of lines present in the region and on their level of speciation, but not (at least not directly) on the number of geographically independent sites of speciation.

Comparison with other tropical cave faunas

The lack of detailed taxonomic or ecological information concerning epigean tropical invertebrates is a serious drawback for inter-regional comparisons of hypogean faunas. It results in uncertainty about the ecological status of the species recorded, with a risk of mis-evaluating the troglobitic species richness of the caves or regions studied. To minimize this problem for the reference cave systems, we coupled the faunistic survey of the caves with that of surface soil and water habitats. As this approach is not followed in most other studies of the literature, however, comparison are only indicative.

Regional diversity of Southeast Asia compared to other tropical cave areas: In a recent overview of the tropical cave faunas of the world, Peck and Finston (1993) recorded the highest diversity of troglobites in Mexico + Guatemala + Belize [~317 recorded species according to Reddell (1981), with 85% above 300 m] and Hawaii [50 terrestrial and 30 anchialine troglobites, but only 43 terrestrial, according to Howarth (1987)]. Ninety troglobitic species were listed from Cuba by Decu and Juberthie (1994) and probably a similar number is known from Australia (Howarth, 1988; Humphreys, 1993). Thailand has more than 94 troglobitic species (76 terrestrial and 18 aquatic: Deharveng and Bedos, 2000), including the 12 cave species of the Collembolan genus *Troglopedetes* (Deharveng and Gers, 1993). The three reference systems studied in Indonesia (Ngalau Surat, SKT and Batu Lubang) have 64 troglobitic species altogether; at least ten other troglobites are known from the country, which amounts to 74 species for Indonesia. Thailand and Indonesia should therefore be among the top-ranking in any list of tropical troglobite diversity, particularly for lowland troglobitic faunas. However, any cave sampling of a newly explored karst in these areas provides several new species of troglobitic invertebrates, and current species richness merely reflects sampling effort. Age, extent and fragmentation of the karsts also differ largely from country to country, preventing straightforward extrapolations from available data. All one can say is that the richest cave biodiversity (Mexico + Guatemala + Belize) falls in the area that has been most studied, with over 1000 caves sampled, while the country of Southeast Asia with the highest cave biodiversity, Thailand, is also by far the country most investigated in the region for its cave fauna.

Local diversity of Southeast Asia compared to other tropical caves: Recent studies, particularly in the Neotropics, make cave-to-cave comparisons possible. Peck (1971) listed 67 invertebrate species from Chilibrillo cave (Panama), of which three were doubtful terrestrial troglobites. In the Jumandi cave of Ecuador, among 22 invertebrate species only one troglobite was found (Peck, 1985). In contrast to these low numbers, in the caves of Serrania de San Luis (Venezuela) Chapman (1980a) found 11 troglobites, of which nine were terrestrial, among the 95 invertebrate species collected. The karstic areas of Mexico and Central America listed by Peck and Finston, 1993, [after Reddell (1981)] host 0 to 14 terrestrial troglobites and 0 to 10 stygobites. Bayliss Cave in North Queensland (Australia) hosts 17 troglobites, all terrestrial (Howarth and Stone, 1990). Biodiversity of troglobites in Southeast-Asian caves (16 to 28 species in the reference systems studied, Table 31.5) therefore ranges among the highest known values for tropical caves, with the caveat that group coverage varies and troglobite definition is slightly different among different authors. A second interesting feature is that large, well-studied caves of the Neotropics may be very poor in troglobites; this is never the case in Southeast Asia.

Comparison with temperate cave faunas

Taxonomic composition: Cave faunas of Southeast Asia, both aquatic and terrestrial, show important and constant differences from those of temperate caves.

Aquatic communities, as is usual in subterranean waters, are dominated by Crustacea. However, the dominant groups in Southeast Asia are generally not the same as in temperate countries. Decapoda (shrimps and crabs) are frequent as stygobitic elements (up to three species in a single system, e.g. the SKT system) and constant as troglophilic elements (crabs and microphthalmic shrimps). They are much less common in temperate caves. Conversely, with only one family (Bogidiellidae) in caves of Southeast Asia, amphipods are much less diverse than in temperate subterranean waters (Eberhard et al., 1991; Holsinger, 1993). In addition, though present in some areas of North America, temperate China and Japan, stygobitic fishes are mainly tropical, and particularly diversified in Southeast Asia.

Flying vertebrates (bats and swiftlets) nesting underground are constant cave-dwellers and the most important food suppliers for cave invertebrates in Southeast Asia; they are not more diversified than in temperate areas (at least North Temperate), but much more abundant. Sixteen species (13 bats and three swiftlets: Chapman, 1985) have been recorded from the caves of the Mulu karst, and as many as 10 species roost in the Batu caves (nine bats and one swiftlet: McClure, 1965). They are also more diversified trophically, including insectivorous bats, insectivorous swiftlets and frugivorous bats, whereas only the first are encountered in temperate caves.

Regarding terrestrial invertebrates, temperate cave communities are generally dominated by insects; in the caves of tropical Southeast Asia insects are generally outnumbered in richness and abundance by other arthropod groups, particularly arachnids. A more detailed account is given below.

(1) The first characteristic of caves of Southeast Asia compared to the temperate zone is a high diversity of arachnids at all taxonomic levels. Several groups present in Southeast Asia are absent from temperate caves (Amblypygida, Opilioacari, Schizomida and many families of Araneae), while none is restricted to the latter. This results in a profound difference in the trophic organisation of communities, as most Arachnids (other than certain mites) are active predators.

(2) Isopods are also more diverse than in temperate caves, with typically two or three species in each cave when suitable habitats are present: one more or less guanophilic armadillidiid, one troglomorphic philosciid and sometimes one troglomorphic styloniscid or trichoniscid. A temperate cave usually hosts one or sometimes two species of Trichoniscidae (or of Styloniscidae, their southern counterparts: Eberhard et al., 1991).

(3) Giant guanophilic Chilopoda (Scutigeridae, Scolopendromorpha), absent in temperate caves, are regular predators in caves in Southeast Asia.

(4) Diplopoda, though very insufficiently studied, are obviously more diversified than those of temperate caves.

(5) Collembola, though represented by a number of troglobitic species in Southeast Asia, are less diversified than in Mediterranean areas or in the southeastern United States, because of the absence of troglobites of several families, notably Hypogastruridae, Onychiuridae and Tomoceridae.

(6) Cockroaches, including guanophilic species and troglobitic Nocticolidae, are constant inhabitants of Southeast Asia caves, whereas Blattodea are exceptional in temperate caves.

(7) The rich troglomorphic fauna of Coleoptera observed in many temperate areas (Japan, Europe, North America) and apparently present in the highland caves of Papua New Guinea (Chapman, 1976; Smith, 1980), does not exist in lowland Southeast Asia. Only three troglomorphic species of Caraboidea (Deuve, 1988) have been found there, restricted to the most seasonal areas of northern and central Thailand, where they are scarce. *Mateuius troglobioticus*, a slightly troglomorphic carabid (eyes and pigment reduced) of the SKT system, *Eustra honchongensis* of the Hon Chong caves (Vietnam), and several undescribed species of oculated, but long-appendage Pselaphidae of northwestern Thailand (Besuchet, 1986) are the only other lowland troglobitic species of Coleoptera.

Regional diversity of cave faunas in Southeast

Asia *versus* temperate areas: The scarcity of information about the cave biodiversity of the main Southeast Asia regions contrasts with the large amount of data accumulated for years about regional cave faunas of temperate areas. Comparisons are therefore untrustworthy, but the following observations are of some interest.

(1) Considering large taxonomic units (orders and families), the differences listed in the previous section reflect a higher group diversity in Southeast Asia than in temperate caves, particularly for the guano fauna [as already noted for Cuban caves by Decu (1981)].

(2) At the species level, available data suggest a high level of endemism among Southeast Asia troglobites (and possibly among certain guanobites: Boutin, 1971). Although this probably applies to many groups (particularly terrestrial Isopoda and Diplopoda), only the case of *Troglopedetes* (Collembola) has been partially documented. A number of species of this genus, of which 12 have already been described (Deharveng and Gers, 1993), are restricted to caves in western Thailand. The average distribution range of a species is clearly smaller in *Troglopedetes* than in *Pseudosinella* and *Sinella*, their main temperate counterparts. *Troglopedetes* has more species in a given area, for a similar level of karst fragmentation and range of elevation. This difference could result either from a higher potential for speciation in *Troglopedetes*, or from a greater fragmentation and isolation of the underground habitats in the area. The tropical climate and pedogenesis induce a general filling-in of crevices and screes inside and around the karstic units, which, together with the absence of cryofragmentation of the rocks, results in a poor development of the Superficial Underground Compartment[1] outside the karstic units (Deharveng, 1987c). This may lead to smaller units available for colonization, in support to the hypothesis of high habitat fragmentation. But the tropical climate may also promote a more rapid evolution as a result of higher temperatures and shorter life cycles (Rohde, 1992). Detailed studies of the microgeographic distribution patterns of speciose underground lineages in the tropics, like those made by Christiansen and Culver in temperate karsts of North America (1968, 1969) or by Eberhard in Tasmania (1991), are clearly needed to establish if cave speciation is actually more active in the tropics independent of indirect (i.e. pedogenetic) influences, and more generally which environmental parameters may be involved in the evolution of tropical organisms.

(3) The patterns of generic diversity are very different in tropical *versus* temperate cave fauna. Generic diversity roughly mirrors the relative importance of the different taxonomic groups, which has been discussed above. Region-to-region differences in generic richness in a given climatic zone are only documented for Collembola. In this group, five troglobitic genera (*Arrhopalites*, *Onychiurus*, *Pseudosinella*, *Sinella* or *Coecobrya*, and *Tomocerus*) dominate the troglobitic mesofauna in Holarctic caves. In most of these genera occurrence is sympatric, but the Entomobryidae show clear allopatric dominance. In the European and Mediterranean regions the only troglomorphic Entomobryidae are species of *Pseudosinella*; in North America both *Pseudosinella* and *Sinella* occur, whereas in the Asiatic portions of the Palearctic only *Coecobrya* and *Sinella* occur. Conversely, even in a much smaller area like Thailand, the troglobitic fauna is dominated by one or two genera which are not the same in the different karstic areas and remain parapatric in their distributions: *Cyphoderopsis* in the south, cf. *Coecobrya* in the centre and *Troglopedetes* in the northwest (Deharveng and Leclerc, 1989). Other tropical troglobites have not been analysed to this degree, and the general applicability of the pattern observed for collembolan genera in Southeast Asia (low diversity inside the regions, high diversity between the regions) needs to be examined in other groups.

Local diversity of cave faunas in Southeast Asia *versus* temperate areas: Information about biodiversity of temperate caves is scattered among a huge amount of papers, but the number of caves that have been thoroughly sampled for a large array of invertebrates remains surprisingly low; and often they are small caves with a limited number of habitats. We shall use here original data for a more relevant comparison.

We sampled two temperate caves in a way similar to that which we used for Southeast Asian caves: the

[1] Elsewhere called "Mesocavernous Shallow Substratum (MSS)" – see, for instance, Chapter 1 of this volume.

Grotte de La Buhadère in the central Pyrenees (France) and Feihu Dong in Hunan (central China). La Buhadère is a narrow, oligotrophic cave over one kilometre long located in one of the richest biospeleological area of Europe (massif d'Arbas). Feihu Dong is a huge system already 14 km long, with a large variety of mostly oligotrophic habitats. The Grotte de La Buhadère is at the lowest diversity level (16 troglobites) of the caves studied (Table 31.5). The number of troglobites of the Chinese cave is only outnumbered by the SKT system. Most temperate cave assemblages are however not so diverse as those of La Buhadère and Feihu Dong. D.C. Culver (personal communication) made available a list of species richness for 132 caves of West Virginia, a region moderately rich in troglobites. Of these caves, 96 had less than five troglobites and only 8 had more than 10; the richest cave hosted 15 troglobites – less than the lowest richness of the four cave communities of Southeast Asia studied here (16 troglobites in Batu Lubang and Tham Chiang Dao). In temperate Australia [Tasmania (Eberhard et al., 1991: p. 112) and New South Wales (Eberhard and Spate, 1995: p. 29)] figures are similar to those for West Virginia; in both areas, 11 troglobitic species represented the richest recorded cave fauna.

The number of phyletic groups and their relative species richness are very different between Southeast Asia and temperate regions (Table 31.5), as already noted for regional diversity. In Southeast Asia, spiders and several smaller groups usually replace Coleoptera and Collembola: there are more groups with troglobites, and less species per group on average, with Feihu Dong being intermediate in that respect.

The limited evidence available indicates therefore that cave biodiversity is usually higher in Southeast Asia than in northern or southern temperate regions. However, with one documented site only for the whole Chinese karst, the largest on earth, it would be premature to draw any firm conclusion about latitudinal patterns of alpha-diversity.

ENERGY INPUT AND THE ORGANIZATION OF SUBTERRANEAN COMMUNITIES

Energy input into caves

Biological diversity has been frequently related to productivity or heterogeneity (Brown, 1981). As primary productivity in caves is almost always nearly zero and cave habitat heterogeneity is low compared to epigean habitats, the low species richness of underground habitats compared to epigean ones is not unexpected; however, the absence of primary producers in caves is strongly compensated in the wet tropics, notably in Southeast Asia, by the high energy input derived from flying vertebrates and flood detritus; and the trophic parameter food availability remains at the base of species distribution in these habitats. The importance and the nature of energy input have been used by Sbordoni et al. (1977) to propose a biological classification of caves, and by Chapman (1982a) to define the habitats of the Mulu caves. A generalized typology of food resources would therefore be particularly useful for comparing the energetic status of different habitats.

In aquatic habitats, biota feed on exogenic organic matter either dissolved or carried by streams in the form of coarse and fine particles.

Five major sources of food are listed by Culver (1982) and Holsinger and Culver (1988) for terrestrial habitats of Virginia caves (bat guano, eggs and guano of cave crickets, microorganisms, mammalian faeces and dead animals, and plant detritus left by flooding). Most of these resources are essential also in tropical caves of Southeast Asia; but there cricket guano is a by-product of bat guano because most crickets, at least in the four systems of reference, remain in the cave for feeding. Additional sources of food are also present (swiftlet guano), while others are unimportant (faeces of mammals other than bats). Other energy supplies (tree roots, detritus falling into the caves by gravity, and invertebrates entering caves regularly or accidentally) are of marginal importance in those caves of Southeast Asia that have been studied so far.

We are finally left with three main categories of exogenous energy supply brought into caves by three main vectors: guano (from vertebrate faeces, urine, nests and corpses); detritus, mainly vegetable (brought by allochthonous streams); and dissolved organic matter (brought by percolating waters). Once underground, these resources support the whole trophic chain, in which they are integrated through an eventual transformation by microorganisms.

Scarcity of food, sometimes considered a general feature of temperate caves (Culver, 1970), occurs less frequently in the lowland caves of Southeast Asia investigated so far. There, fossil caves are regularly fed with guano, and active caves usually have flood detritus (and often guano as well), providing abundant, relatively stable and relatively predictable food sources for the fauna. In the remote and isolated parts of the underground systems, percolating groundwater may still bring dissolved organic matter, offering a kind of biotope suitable to troglobites, which are adapted to cope with low food resources.

Guano

Guano is an important food resource in the caves of Southeast Asia (Annandale et al., 1913), particularly in the marginal areas of the karsts where caves are smaller and more open. Often restricted to limited areas, the amount of guano usually declines with distance from the entrance. Its spatial distribution depends on local temperature and air movements, and its production is slightly season-dependent even in Southeast Asia (McClure et al., 1967). However, guano piles are long-term sources of food because colonial bats (more than swiftlets, which may change nesting area over the years) use the same areas for very long periods, with guano decomposition taking many years (Decu, 1981).

Guano is available in two different forms: (1) in large amounts it constitutes the basic source of energy for highly diverse and crowded invertebrate communities; (2) in scattered faeces, it is the commonest available resource for troglobites (Jeannel, 1926: pp. 61–62).

The ecological status of invertebrate species associated with guano in Southeast Asia

Guano piles are the archetype of high-energy habitats because of their high organic-matter content and the density of invertebrates they support. According to Poulson (1972: p. 58), species living in guano "have wide food and habitat niches"; in other words, there would be no specialized guanobite fauna. In Southeast Asia, it is true that some of the dominant guano species are able also to colonize vegetable detritus in caves (some Collembola, and several zoophagous species like Amblypygida or Schizomida), but most are absent from energy-rich habitats outside caves, like forest litter. They are not generalists, in spite of their apparently wide trophic range, with a few exceptions like the parthenogenetic collembolan *Folsomina onychiurina*, which succeeds equally well in forest soil. Moreover, key taxa such as Glyphiulidae (Diplopoda) or Aderidae (Coleoptera) seem to be restricted to guano, being only exceptionally found in flood detritus, the other energy-rich habitat in caves. In fact, most guano species of Southeast Asia are specialized to exploit definite kinds of high-energy resources under special microclimatic constraints [guanobite–troglobite in the sense of Decu (1981) or Gnaspini (1992)], while oligotrophic habitats are occupied by the troglobitic species *sensu stricto*.

Guano assemblages

The abundance of guano is the most striking characteristic of many caves in Southeast Asia. Patterns of guano deposits vary from cave to cave, and inside a single cave, in relation to the distribution, the biology and the abundance of guano providers (swiftlets, colonial frugivorous bats, colonial and non-colonial insectivorous bats, crickets), which are themselves strongly dependent on microclimate and the spatial configuration of the site. Given the large number of bat species which may occur in a single cave [nine species in the Batu caves (McClure, 1965); seven species in Taronggo cave, Sulawesi (Whitten et al., 1987)], it is likely that competition for space also plays an important role. All these factors are certainly influential, but their importance for predicting guano distribution has never been tested in the field, and the way the guano communities exploit this multiform source of food has rarely been studied in Southeast Asia. Reports by McClure et al. (1967) and Chapman (1982a, 1985), and the data of the present study shed some light on this last point, however.

The animal communities of guano, where arthropods are by far the dominant and most speciose group, are similarly structured in the four reference sites, as shown in the generalized food chains of Figs. 31.6 and 31.7. More detailed interspecific relationships have been suggested by McClure (1965) for the Batu caves and by Chapman (1982a, 1985) for the Mulu caves, but they remain largely speculative (Bullock, 1965).

A major characteristic of the guano communities in Southeast Asian caves is their tripartite organization, which can be easily seen in the Chapman food webs.

Basic food supply is provided by the faeces, urine and corpses of flying vertebrates and the associated fungi and bacteria. These support two well-defined and largely independent communities: the giant arthropod community (GAC) and the meso- and micro-invertebrate community (MIC). In the food chains given by McClure (1965) these two assemblages seem to be intermixed; but this merely results from unchecked specific diets assigned to the giant arthropods.

The giant arthropod communities of the four reference systems include one primary consumer (a

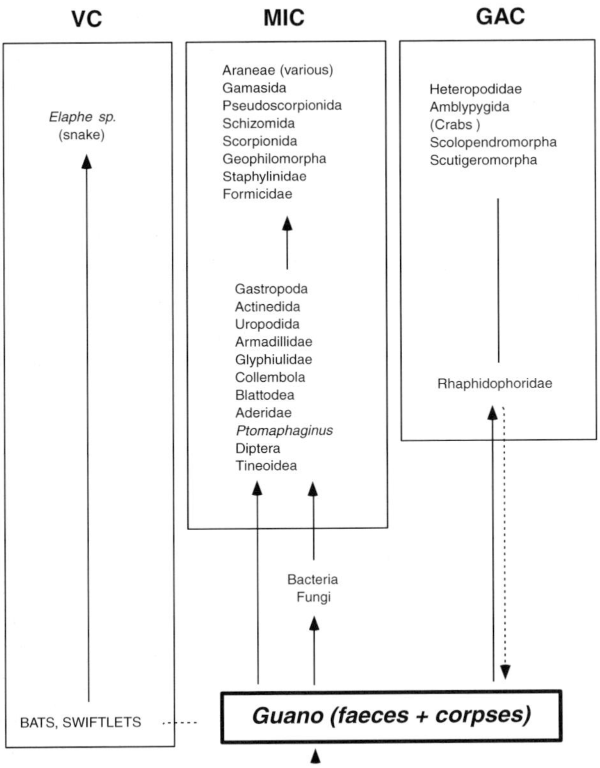

Fig. 31.6. A generalized food web of guano communities in Southeast Asia caves, parasitic taxa excluded. Dotted lines: main sources of guano. VC, vertebrate community; GAC, giant arthropod community; MIC, meso- and micro-invertebrate community.

large rhaphidophorid cricket, always occurring at high density) which is preyed on by four or five predators. Crickets ingest a huge quantity of organic matter without apparent choice. One of the largest, *Rhaphidophora oophaga*, sometimes listed as an eater of swiftlet eggs (Bullock, 1965), is probably omnivorous (Chapman, 1982a: p. 146). Three species of Rhaphidophoridae differing in size have been recognized in the Mulu caves by Chapman, and two, differing in the degree of eye reduction, may be present in caves of central Thailand (Deharveng and Delnatte, 1988). Predators are giant Chilopoda and Arachnida, more rarely crabs (observed as cricket predators in Batu Lubang). The highly vagile giant arthropods have often a parietal localisation, and may be found far from guano deposits. They are pigmented and oculated, but most, except some Rhaphidophoridae (Chapman, 1982a) and Amblypygida (Boutin, 1969: p. 167) which may leave the cave at night, seem to be restricted to caves. Rhaphidophorid faeces and corpses complement the vertebrate guano in all caves, making

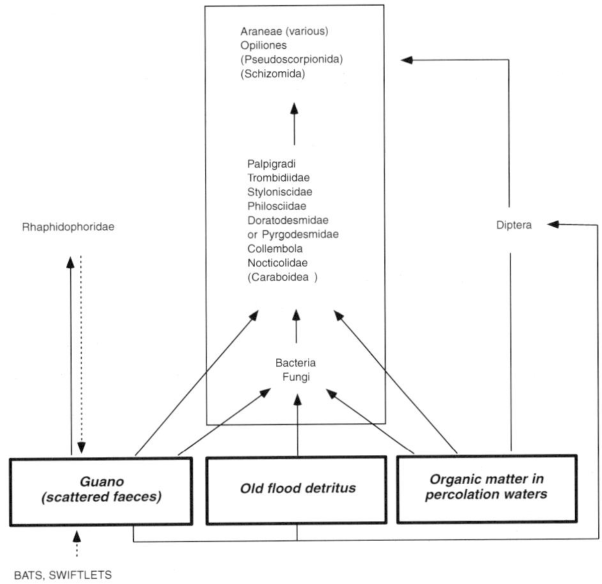

Fig. 31.7. A generalized food web of oligotrophic communities in Southeast Asia caves, parasitic taxa excluded. Groups in parentheses are present only rarely. Dotted lines: main sources of guano.

guano resources available for the meso- and micro-invertebrate guanobites and providing scattered faeces exploited by troglobites.

Primary consumers of the meso- and micro invertebrate community usually live in very large populations made up of fairly large-size species (glyphiulid millipedes, cockroaches, tineid moths, or woodlice), and a number of smaller invertebrates (Gastropoda, Acarina, Collembola, Coleoptera, Diptera). Glyphiulidae are characteristic of Southeast Asia guano habitats. Tineoidea are represented by innumerable larvae foraging in guano piles, with several species frequently coexisting (three in Tham Chiang Dao: Deharveng, 1987b). Among medium-sized primary consumers, Diptera and Staphylinidae may be incredibly numerous; Staphylinidae have never been studied, but Diptera are highly diversified: 64 species in 20 families are listed for the Batu caves, with 11 species for the genus *Psychoda* alone (McClure et al., 1967). These abundant populations of primary consumers are preyed upon by numerous arachnids, Geophilomorpha, sometimes ants or even small scorpions (Batu Lubang). The overall richness of such micro- and meso-invertebrate assemblages has never been evaluated in any one underground system of Southeast Asia, but is likely to be much higher than that of troglobite assemblages.

Unlike the giant arthropods, micro- and meso-invertebrates are found in, on or near the guano deposits, on account either of reduced vagility (for the non-flying species, which include all the predators except winged ants) or of higher stenotopy (for cockroaches and flying species: Coleoptera, Diptera, Lepidoptera). Most of these species have eyes, but a trend towards eye and pigment reduction is noticeable among several lines [*Troglopedetes* (Collembola), *Cubaris*, Glyphiulidae (Diplopoda)] and may possibly even occur within a single population (*Ptomaphaginus boutini* in a Cambodian cave: Boutin, 1971). Eyeless species belong to epigean lines which are already eyeless (such as the Collembola *Coecobrya guanophila* and *Folsomina onychiurina*).

Regional and local variability of guano associations

Among the variations observed in the composition of guano communities in Southeast Asia, two are particularly striking and deserve some comments: the changes in the balance between Blattodea and Glyphiulidae among the mesoinvertebrates, and the presence or absence of Amblypygida.

Glyphiulid Diplopoda and, to a lesser extent, cockroaches are regular inhabitants of guano in Southeast Asia. When present, they tend to dominate the invertebrate assemblage. Both can coexist in the same cave, but not on the same patch of guano, suggesting strong competition at their contact zone. In Ngalau Pangian (Sumatra), *Pycnoscelus* cockroaches live in huge populations on the guano of a small passage where the temperature is over 25°C; they are completely absent in the main gallery (temperature <23°C), where Glyphiulidae are dominant on guano. More generally, cockroaches tend to replace Glyphiulidae in the warmest caves of Southeast Asia. Light is a second influential factor (McClure et al., 1967). Among the four food webs published for guano communities in Southeast Asia (McClure et al., 1967; Chapman, 1982a), two diverge widely from the general pattern of Fig. 31.6: the "bat community food nexus" of Mulu and the "food chains of cavern A" of the Batu caves. They lack the giant arthropods, but host *Pycnoscelus* cockroaches and peculiar micro- and meso-invertebrates such as trogid beetles. A remarkable feature is that both these food webs were described from the twilight zone. We encountered assemblages of this type in the Hon Chong caves of southern Vietnam under similar light conditions. Warmth and light clearly favour cockroaches against glyphiulid millipedes.

Large Amblypygida are regular predators in the guano communities of Southeast Asia. They are however absent when Rhaphidophoridae, their usual prey, are absent – for instance in the case of the twilight guano communities discussed above. They are also absent in some regions in spite of the presence of Rhaphidophoridae, as in northern Thailand. In that case, the 2000 mm annual rainfall isohyet was found a good predictor of the distribution limits of the group (Deharveng and Leclerc, 1989). Amblypygid activity is dependent on humidity, as shown in Cambodia caves by Boutin (1969); but, as permanently saturated atmosphere may be found in most underground systems of Southeast Asia, it is likely that other parameters control the geographical distribution of these arachnids. Temperature could be one of these factors. Large Amblypygida are typically lowland inhabitants, present in most caves studied below 300 m in Cambodia (Boutin, 1969), Thailand (Deharveng and Bedos, 1988; Leclerc et al., 1988, Leclerc and Dalger, 1988), southern Vietnam (Deharveng and Le Cong Kiet, 2000), Sulawesi (Bedos et al., 1990), the Moluccas (Deharveng, 1989), Nias island and Java (unpublished data). In contrast, Amblypygida are absent in caves above 300 m: Tham Chiang Dao (550 m, Table 31.2), Ngalau Surat (400 m, Table 31.2; only a small species is present in the cave), Sagada caves in Luzon (over 1300 m, unpublished data). They also are absent below 300 m in the colder areas of the region: the Ke Bang karst of central Vietnam (Deharveng and Le Cong Kiet, 2000) and the caves of central Thailand (Deharveng and Delnatte, 1988). In further support of this limit to their distribution, Amblypygida have not been recorded from caves beyond 20°N in Southeast Asia, and the same seems to be true for latitudes south of about 20°S in the Neotropics (Trajano and Gnaspini-Netto, 1991).

Guano from different sources may be exploited by different communities in Brazil (Gnaspini, 1992: p. 136). It was suggested that the same could be true for Southeast Asia (Chapman, 1982a: p. 156; Deharveng and Leclerc, 1989). The only study supporting this view in the region was in the Mulu caves, where Chapman (1982a) found two types of species assemblages, one on swiftlet guano [and on guano of *Miniopterus* and *Myotis* (Chapman, 1982a: p.158)], and the other on bat guano. However, differences in light intensity were associated with the type of guano in these examples, preventing an evaluation of the relative influence of these two factors. In Indonesia, two of the systems studied (the SKT system, and Batu Lubang) hosted swiftlets, frugivorous bats and insectivorous bats, and one (Ngalau Surat) had swiftlets and insectivorous bats. In these caves, no clear-cut difference was detected between bat and swiftlet guano, while in Batu Lubang frugivorous and insectivorous bat guano hosted different assemblages. There, a huge colony of frugivorous bats roosts at 400–600 m inside the cave, providing a continuous rain of urine and droppings which covers the rocks and the ground; from 600 m onwards, this damp guano is progressively replaced by that of insectivorous bats and swiftlets, with obvious changes in community composition (disappearance of the collembolans *Cryptopygus* sp. and *Lepidonella* sp., appearance of scorpions). However, these modifications could, once again, also be related to decreasing light intensity. Available observations are therefore not conclusive, and the source of the guano does not yet emerge as a major factor structuring guano communities.

Temporal variations in guano assemblages

Fluctuations of abundance in a number of guanophilic and troglophilic species of the Batu caves (Malaya) have been followed for nearly two years (McClure et al., 1967). Significant changes were observed in several cases, but not clearly related to seasonal fluctuations in bat populations (which experience a heavy mortality from March to May in the Batu caves according to Bullock, 1965). Moreover, many species showed erratic peaks of abundance at various times of the year, with high interannual differences (McClure et al., 1967: p. 423). This pattern suggests that the guano invertebrates probably reproduce more or less independently of outside climatic fluctuations.

Daily migrations are typical of several trogloxenic species, which feed outside at night and use the cave as shelter during the day: bats, swiftlets, but also some crickets [in the Mulu caves (Chapman 1982a: p. 153), and in the SKT system] and Amblypygida [in Cambodian caves (Boutin, 1969: p. 167)]. The extent of these daily moves is poorly documented however, and many populations of these last two groups are found far inside caves where they probably live permanently. More limited changes may be caused by microclimatic variations. Amblypygida have thus been shown to move from place to place inside Cambodian caves, in reaction to local fluctuations of air humidity (Boutin, 1969).

Flood detritus

Streams, particularly during flooding, may bring huge amounts of organic matter into the deep parts of a cave. Such a food resource is patchy and season-dependent, because (1) it is limited to cave systems having stream sinks, (2) it is restricted to a usually narrow, well-delimited strip along the allochthonous streams of the cave, (3) it roughly declines from the entrance to the deepest part of the cave and (4) it is strongly related to seasonal flooding. However, because the deposit of vegetable detritus depends on the spatial topography of the river edges, it generally occurs in the same areas each year, and its decomposition may take several years; it may provide a constant source of food during very long periods.

Floods or ordinary flow of exogenic streams bring into caves not only enormous quantities of organic matter, but also a large number of living organisms, which will intervene in the local communities as predators or competitors, and may colonize the available cave habitats for a while. This process, as pointed out by Hawes (1939), has ecological and evolutionary implications of prime importance. Sadly, while the guano habitat has been studied in Southeast Asia only rarely, flood detritus has not been studied at all, with the exception of the "Shallow" habitat recognized by Chapman (1982a) in the Mulu caves, a special tower-karst cave habitat with flood sediment and roots as food supply.

Flood detritus has a clear zonal distribution in each cave system, but its patchy and heterogeneous nature prevents it from supporting a well-defined food web.

Local zonation parallels the stream flow. It starts with the recent detritus near the stream, and ends with the rotten detritus of the oldest and largest flood episodes. A flood-line can usually be recognized, which separates the ordinary flood detritus (deposited every year during the rainy season) from the exceptional flood detritus. Only the latter habitat has a significant number of troglobites, whereas the former mainly hosts outside species, the proportion decreasing with the time since the last flood.

A second zonation pattern results from the decrease in the amount of detritus with increasing distance from the point where the stream enters. The proportion of non-troglobites decreases accordingly, more or less irregularly following the configuration of the stream and of the cave.

We found a great richness in flood detritus in Ngalau Surat and the SKT system, where guanophiles, troglobites, troglophiles and outside species may coexist. In addition, a few species seem to have a predilection for this habitat, like the troglobitic Philosciidae of Ngalau Surat. Schizomida (guanophiles) and Nocticolidae (troglobites) are also frequently observed there. This is in sharp contrast with the observation of Chapman in Mulu (1982a, 1983), that there is a poor fauna in his "Shallow" habitat: he found that most of the giant arthropod species of the guano were present, but none of the local troglobites. Two things explain this apparent discrepancy: (1) the "Shallow" habitat contains only superficial or shallow parts of small marginal caves (Chapman, 1982a), whereas flood detritus as considered here is also found in the deepest parts of large caves; (2) microarthropods are not listed in the "Shallow" species assemblage, whereas they make the largest part of the fauna extracted from flood detritus. In fact, the apparent patterns of richness change drastically according to sampling methods, and the resulting difference in the taxonomic or functional groups collected.

Parietal communities

The rich temperate parietal community is largely inactive and seasonal (aestivating and hibernating species: Jeannel, 1926: pp. 41–44), and differs therefore fundamentally from the tropical parietal assemblages (Ginet and Decou, 1977: p. 105). There, most species present on the cave walls belong in fact to guano assemblages (particularly the highly vagile giant arthropods), exploiting the nearby guano piles or preying on detritivorous guanobites. Some Rhaphidophoridae which feed outside the cave during the night are the only regular counterpart of the normal parietal fauna of temperate caves. They form abundant populations in many caves of the Maros karst in Sulawesi.

Parietal assemblages in tropical caves are usually species-poor. Only recently was a rich parietal fauna discovered in Southeast Asia. This community is located on slightly overhanging rocks along the subterranean river of Ngalau Surat (Sumatra), several tens of metres from guano accumulations and at over half a kilometre from the entrance. Large numbers of leeches, spiders (two oculated species), small Amblypygida, microphthalmic Opiliones, oculated Armadillidae, larvae of Keroplatidae, and Emesinae were active on this apparently atrophic substrate. Most species are not or hardly cave-adapted, and (except the Armadillidae and leeches) probably prey on the numerous trogloxenic insects (Ephemeroptera, Trichoptera, Diptera) which continuously hatch from the stream waters. Cave walls, which in most cases are azoic, or support only some guanobitic macroarthropods, may therefore host a diversified and unusual fauna in certain conditions.

Dense monospecific groupings of invertebrates have also been observed occasionally on the walls of several caves (epigean Orthoptera in Tham Pha Tai, Thailand; Staphylinidae in Gua Tanette, a cave of the SKT system in Sulawesi). Most impressive are the nurseries of giant Thelyphonida reported in a deep cave of the Thakhek karst during the Laos 1996 expedition (Brouquisse and Vacquié, personal communication). The cause of such groupings is unknown, but it might correspond to some kind of resting period, recalling that of temperate parietal species.

Terrestrial troglobitic fauna and oligotrophic habitats

How troglobite species benefit from guano resources

It has been noted that troglobites in tropical caves often rely on scattered guano of swiftlets (Chapman, 1982a), but are absent from large piles of guano. Thus, the distribution type rather than the nature of guano may be the factor controlling the presence of troglobites. Our observations corroborate this view in the four caves we studied; however, scattered guano more frequently originated from bats and crickets than from swiftlets. The large crickets are to feed on bat or swiftlet guano, then to disperse rather far from the food source. The large *Rhaphidophora* spp. of Ngalau Surat or the SKT system may thus move several tens of metres in a short time. This is one way by which troglobites indirectly benefit from vertebrate guano in places where vertebrates are absent.

A second indirect food supply is provided by the flying guanophilic Coleoptera and Diptera. These insects, often in huge populations in guano piles, are preyed upon by guanophilic predators in the area covered by guano, but also by strict troglobites such as spiders at some distance from the guano.

Troglobites may at least approach the edges of the guano piles, where organic matter is continuously washed away by seeping water, giving a much less energy-rich food source.

Guano accumulation therefore determines a variable peripheral zone where organic matter is progressively more scattered or less concentrated ("satellite communities" of Chapman, 1980b: p. 142), providing increasingly favourable habitats to troglobites.

The troglobite habitats in Southeast Asia caves

Troglobitic species are by far the best known of cave species, and reliable species lists are available for several Southeast Asia caves. In contrast, ecological study of them has been limited to one paper produced by Chapman (1982a).

Chapman (1982a) has characterized four associations in the Mulu caves, of which one, the "SCAT" association on scattered faeces, contains only troglobitic species, whereas they are completely absent from the three others. The SCAT association is restricted to dark, oligotrophic, damp habitats with scattered droppings of swiftlets as the main resource, where evaporation rate is near zero. Some SCAT species may live on bog guano rich in organic matter which is not exploited by the usual guano fauna, but this has not been confirmed outside the Mulu karst. Only in Ngalau Surat does guano washed by percolating waters, and forming a thin damp layer on calcite surfaces, host troglobitic species.

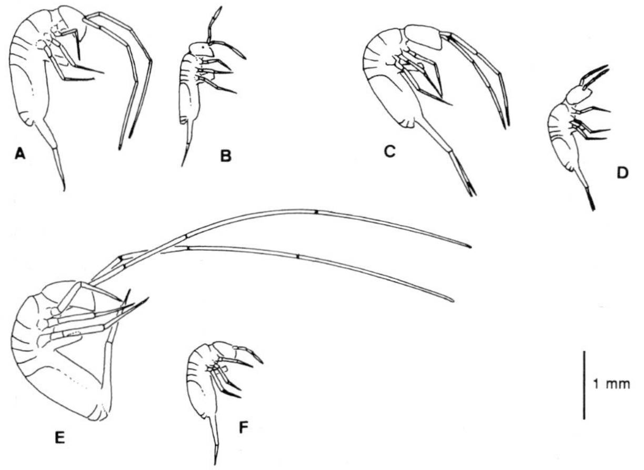

Fig. 31.8. Habitus of some troglomorphic species of Collembola in Southeast Asia, compared to that of their epigean or guanobitic relatives: (A, B) cf *Coecobrya* spp. (undescribed) of the Tham Kubio–Tham Phulu system from oligotrophic habitats (A) and from guano (B, microphthalmic). (C, D) *Troglopedetes* of Tham Pha Mon (Nam Lang karst, Thailand) from oligotrophic habitats (C, *T. longicornis*) and living in relatively high-energy habitats (D, *T.* cf. *maffrei*); (E, F) *Cyphoderopsis* spp. (undescribed) of Ngalau Surat (Sumatra) from deep in the cave (E) and from outside litter (F).

Oligotrophic habitats are the domain of the troglobitic terrestrial fauna in the caves of Southeast Asia that we have studied, as they are in Mulu and probably in all the caves of the world. Food resources of these habitats include the guano-related supply examined in the previous section (scattered faeces of swiftlets, bats and crickets, organic matter of waters seeping from guano piles, invertebrates flying out from the guano), but also organic matter brought through the soil by percolating waters, old flood detritus, and sometimes roots. In spite of the different nature and origin of these nutrient sources, the terrestrial troglobitic fauna is rather similar in all oligotrophic habitats for a given underground system, provided that humidity is high. The amount rather than the nature of the food supply seems to control the presence of troglobites and the composition of their assemblages.

Troglobitic faunas have fewer species than guano communities. Giant arthropods are absent, probably because of the low density of potential prey. Most primary consumers are insects, while all predators are arachnids, mainly small spiders. Most of these species exhibit some degree of troglomorphy, particularly the Nocticolidae and the Collembola (Fig. 31.8).

Aquatic habitats and fauna

Terrestrial and aquatic habitats are not as clearly separated in caves as they are in epigean environments (Jeannel, 1926: p. 39). The Stenasellidae of Ngalau Surat are sometimes found crawling on the wet substrate from puddle to puddle for a short distance. The large crabs of Batu Lubang are active both in water and outside it. The most remarkable case of amphibian life concerns the isopod *Thailandoniscus annae* discovered in a cave of southern Thailand. This species, unlike nearly all other species of the group, spends most of its life under water, leaving it only for moulting (Dalens, 1989).

The diversity of aquatic communities

When one examines the aquatic fauna of the caves of Southeast Asia, the overall impression is of a strikingly low diversity. Stygobitic species and groups are much less numerous than terrestrial troglobites, and there is no counterpart in underground waters for the rich terrestrial guano communities. Cave fish and crabs, the only potential predators, are often scarce or absent. Rarely more than three different taxa represent the primary-consumer level in any given underground system: two Crustacea (often one bogidiellid and one *Stenasellus* species) and one planarian (Table 31.1). Frequently, the number of taxa is even more reduced – a syncarid is for instance the unique stygobitic mesoarthropod in Tham Chiang Dao; cave fish are lacking in Ngalau Surat; fishes and stenasellid isopods have not been found in Batu Lubang. On the other hand, the SKT system has no isopod, but two cave fishes, one highly troglomorphic crab, and two shrimps, one blind and one microphthalmic. This low richness of stygobites cannot be however extrapolated to microinvertebrates, which have not been studied in the caves of Southeast Asia except Tham Chiang Dao (Gibert, 1987).

Stygobite assemblages are also characterized by a strong local heterogeneity. In Mulu, Chapman (1980b) cited one or two stygobites per cave, while the total fauna of the karst amounted to four species. At the habitat-unit level (one pond or one streamlet), a maximum of three taxa coexist in any of the four reference systems. In Ngalau Surat, only a few puddles, for instance, contain the bogidiellid, the *Stenasellus* and the planarian. While richer, the fauna of the SKT system (seven stygobitic species) has a low site diversity, with generally only one species per habitat unit. Most other caves we have sampled in the region fit this overall pattern (low site diversity, with large microgeographic differences), suggesting a strong ecological segregation among stygobites. The only exception is the system Tham Kubio–Tham Phulu of central Thailand, where a single puddle of two square metres may host as many as six different stygobitic species, all in relatively large numbers: a thermosbaenacean (probably a species of *Theosbaena*), a bogidiellid (*Aequigidiella aquilifera*), a stenasellid (*Stenasellus deharvengi*), a large planarian (*Dugesia deharvengi*), a blind dytiscid (*Siamoporus deharvengi*), and a large worm (*Haplotaxis glandularis*). The puddles of the Tham Kubio–Tham Phulu system are windows on phreatic waters. Whether such a richness could be encountered in similar ecological situations remains to be seen, but this example demonstrates how premature it would be to attempt to evaluate the biodiversity status of the whole Southeast Asian stygobitic fauna at the present time.

The fauna of Tham Chiang Dao

The low richness of tropical underground waters is exemplified by the study of the fauna of Tham Chiang Dao by Gibert (1987). She featured six characteristics of the aquatic communities of this cave: (1) the faunal density is lower than that of temperate caves; (2) the taxonomic richness is low, and the faunistic spectrum markedly variable from place to place; (3) the taxonomic groups are the same as those of temperate regions; (4) a large proportion of epigean species are present in karstic waters, in spite of the apparent absence of colonization from upstream since the water is presumed to be of endogenous origin; (5) a continuum exists between cave and interstitial communities; (6) Oligochaeta and small species are largely dominant. (1) and (6) were related to local conditions (important filling in, diffuse water alimentation), but the other points may well represent general characteristics of the aquatic faunas of tropical caves.

Aquatic habitats and their faunas

Like terrestrial habitats, the different aquatic habitats in Southeast Asian caves can be categorized by their energy supply. High-energy habitats include those where energy is provided by guano and those where energy is brought in by exogenic streams. At the opposite end, low-energy habitats comprise the water bodies fed by percolating groundwater, filtered stream water or phreatic groundwater.

Eutrophic water, in contact with guano accumulations, was sampled at several locations in the SKT system and Ngalau Surat. They hosted no specialized fauna. Dipteran larvae and copepods, probably non-stygobitic, were often abundant even in the deepest parts of the caves (SKT), whereas stygobitic species were only locally present in Ngalau Surat. In this cave, clean and guano-floored pools occur side by side at one kilometre from the entrance. The latter hosted on average many fewer individuals of *Stenasellus* than the former, and lacked Bogidiellidae and planarians; on the other hand, another site two km farther had guano-rich pools with many planarians and *Stenasellus*. The guano-floored pools of the Mulu caves, but rarely the clean pools, host the stygobitic crab *Cerberusa caeca* (Chapman, 1980b). The existence of different kinds of eutrophic water may best explain these contrasting observations, but detailed analyses are lacking.

Allochthonous streams which have not been filtered through boulders, gravels or sands are largely or totally dominated by allochthonous fauna (fishes, molluscs, insect larvae), and devoid of stygobitic taxa, as observed in the SKT system, Batu Lubang and Ngalau Surat. The absence of stygobitic species in such epigean-dominated waters is probably the result of competitive exclusion rather than narrow ecological requirements of the stygobites. Breeding of various stygobites (*Nemacheilus* sp. from Tham Hud, Nam Lang karst, Thailand; *Thailandoniscus annae* from Tham Poung Chang, Phangnga, southern Thailand; and

Stenasellus sp. from Ngalau Surat) has shown that they remain active through a large range of temperature (18ºC–23ºC at least), which indicates a weak thermal stenotopy, already emphasized by Vandel (1964: p. 367) for temperate troglobites. The competitive exclusion hypothesis is further supported by data on local species distribution. In Gua Tanette (the SKT system), epigean fish are abundant in several kilometres of the underground river, whereas cave-adapted fish were only collected in a small inlet and in a pond far from the main stream. The remarkable stygobitic crabs of the SKT system (*Cancrocaeca xenomorpha*) are limited to a few sections of the same river where outside competitors are absent. Stygobitic fish of Thailand are found in the dark zones of the caves devoid of epigean fish, with one noticeable exception, *Nemacheilus* sp., which occurs up to the twilight zone of Tham Hud (Maffre and Rigal, 1986: p. 43), in a very small temporary streamlet devoid of any other species of fish.

Percolating waters, endogenic inlets and phreatic waters provide the most favourable habitats for stygobitic species in the Mulu caves (Chapman, 1980b), as well as in all the systems we explored in Southeast Asia. In these habitats however, faunas have a very patchy distribution, and most of the apparently suitable biotopes are devoid of any conspicuous fauna. For instance, Bogidiellidae were collected only once under small cascades in the SKT system; both there and in Batu Lubang, active sampling in similar biotopes failed to produce any additional records.

ADAPTATION TO CAVE LIFE AND ORIGIN OF THE TROGLOBITES OF SOUTHEAST ASIA

Cave life in temperate areas is associated with a large array of spectacular morphological, physiological and biological adaptations which have long intrigued biospeleologists. It was not before the 1970s that similar features were established as widespread among tropical troglobites. Meanwhile, Howarth (1973) showed that adaptation to cave life might occur in very short time. The classical "relict hypothesis" of troglobite origin was apparently unable to account for these two facts, and different alternative models were proposed (Chapman, 1982b; Howarth, 1987; Rouch and Danielopol, 1987; Peck and Finston, 1993). We shall address here three points central to all of them, in the light of the troglobitic fauna of Southeast Asia: the relict state, troglomorphy and regressive evolution.

Relict and non-relict species in the caves of Southeast Asia

The idea that, once a certain degree of troglobitic adaptation has been reached, reverse evolution can no longer proceed is implicitly illustrated in the troglobitization models of Chapman, 1982b, or Peck and Finston, 1993. From simple refuges or potential habitats in the short term, caves therefore turn into ecological traps in the evolutionary process (Christiansen, 1961: p. 300; Vandel, 1964: p. 566). Once adapted to cave life, taxa largely escape ecological changes on the surface (bioclimatic fluctuations or water recession) which may lead to the local extinction of their epigean populations (for troglophiles) or relatives (for troglobites). **Relicts** are these species which have lost their epigean relatives in the region where they are established, while **phyletic relicts** ("reliques" of French authors) have lost all their epigean relatives (Jeannel, 1943; Botosaneanu and Holsinger, 1991). Although not a phenomenon restricted to caves, the relict state reaches by far its highest expression in underground habitats, and feeds most of the recent debates around subterranean evolution.

Contrary to the claim of Peck and Finston (1993: p. 28), a number of non-relict cave species have long been recognized in temperate countries (Jeannel, 1943: p. 22; Vandel, 1964: pp. 189, 272). They were however considered as unimportant (and uninteresting?) compared to relict troglobites. Conversely, in the tropics, Leleup (1965: pp. 153, 154) stressed repeatedly that troglobites, including those of Southeast Asia, were few, non-relict, and weakly troglomorphic. The discovery of predominantly non-relict but well diversified and often highly troglomorphic species in the caves of Hawaii led Howarth (1973, 1987) to reconsider the question of the relict state and to propose new hypotheses about speciation and adaptation to cave life. Subsequent observations were however in conflict, with non-relict (in Southeast Asia: Deharveng and Leclerc, 1989: p. 105), relict (in Australia: Humphreys, 1993) and mixed (in Galápagos: Peck, 1990) troglobitic fauna being recorded from different tropical regions of the world. The emerging pattern indicates a large dominance of non-relict troglobites in the humid and aseasonal tropics, with the proportion of relicts increasing with drought and seasonality.

A variety of ecological events may lead to the relict state. Five categories of relict taxa have been recognized among troglobites by Vandel (1964: p. 554): thermophilic (relicts of warmer bioclimates), temperate (relicts of colder bioclimates, including glacial relicts), hygrophilic (relicts of wetter bioclimates), marine (relicts of marine littoral environment: Chappuis, 1956), and bathyal (a puzzling category including species related to deep-sea taxa, and not relevant for the troglobites known in Southeast Asia).

Thermophilic relicts are particularly numerous in temperate caves, while hygrophilic relicts are found in the caves of most dry areas, including tropical ones like the Cape Range in Australia (Humphreys, 1993). Though these two categories could theoretically exist in the tropics, attesting to ancient bioclimatic events that might have eliminated ancient tropical fauna from the region, none has been recognized so far among the troglobites of Southeast Asia. Temperate relicts might be expected to occur more frequently, but only the guanophilic Opilioacari of southern Thailand and possibly the *Trechiama* carabid of central Thailand belong to this category. The only probable marine relict species is the stygobitic crab (Hymenosomatidae) of the SKT system, while all other cave crabs collected so far in Southeast Asia belong to freshwater lines and are not relicts. In Parabathynellidae, Bogidiellidae and Stenasellidae, geographically widespread but at present relegated to subterranean waters, the shift to cave life is a very ancient event which probably occurred in freshwater ancestors (Wägele, 1990); these taxa cannot be ascribed to any of the Vandel's categories. The same uncertainty, but rooted in insufficient taxonomic knowledge of the groups, concerns the probably relict collembolan *Pseudosinella chiangdaoensis* of Tham Chiang Dao and the isopod *Thailandoniscus annae* of southern Thailand. Thus, remarkable relict taxa exist in Southeast Asia, but their functional importance is usually negligible as they are rare and scattered in a few subterranean sites of the region.

Most of the troglobitic taxa of Southeast Asia are probably not relicts. They belong to lines often widespread in the caves and epigean habitats of the region. The dominant troglobitic Collembola (*Pseudosinella* of Sulawesi, cf. *Coecobrya*, *Cyphoderopsis* and *Troglopedetes* of Thailand) are closely related to species abundant in the local soil fauna. The troglomorphic palpigrade *Eukoenenia thais* of Tham Chiang Dao has strong affinities with the less troglomorphic cave species *Eukoenenia lyrifer*, and with the soil species *E. siamensis* found in the same region (Condé, 1992). The primitive liphistiid spiders of Thailand, Malaya and Sumatra include both non-cave and cave species (though the latter are not troglomorphic) (Boutin, 1971). The blind and the microphthalmic Atyidae of the SKT system are probably related to macrophthalmic species present in surface waters of southern Sulawesi.

There remain a number of cave species in Southeast Asia which cannot be assigned to relict or non-relict categories. They belong to lineages not yet taxonomically studied, or insufficiently looked for in surface habitats. That is the case for many Arachnida, terrestrial Isopoda and Diplopoda. Given the large number of species concerned, the suspected non-relict status of the largest part of the cave fauna of Southeast Asia still needs confirmation.

Troglomorphy and regressive evolution among Southeast Asia troglobites

Adaptation to oligotrophic habitats in caves involves a number of features (Jeannel, 1926; Vandel, 1964: p. 571), grouped by Christiansen (1961: p. 288) under the term cave-dependent or troglomorphic characters. Only morphological traits have been documented for troglobites in Southeast Asia. They include regressive evolution of some characters and progressive evolution of others, with possibly partial links between these two trends in the form of sensorial compensation.

In most lineages, eyes, pigment and wings are, when present, the main organs subject to regressive evolution. This process of reduction [Anophthalmy, Depigmentation, Apterism – the "ADA" of Laneyrie (1960: p. 94)] occurs widely among temperate troglobitic species, which usually reach the ultimate level of regression – that is, the complete loss of wings, eyes and pigment. Leleup (1965: p. 124) pointed out the close relation of "ADA" among terrestrial arthropods with (1) habitat stability; (2) high humidity; (3) richness in carbon dioxide; (4) darkness; and (5) stable and relatively low temperature. The rarity of "ADA" in tropical species of warm hypogean habitats which Leleup observed led him to consider low temperatures as a prerequisite to cave adaptation. The picture seems to be different in Southeast Asia where "ADA" is the usual status of many hypogean species in all areas investigated, including lowland caves and deep soil. In this respect, the result and the extent of regressive evolution do not differ between caves of the temperate zone and those of Southeast Asia.

The regression towards "ADA" together with progressive traits define troglomorphy. In arthropods, the latter include claw elongation, appendage elongation, size increase and development of sensorial receptors. Spectacular troglomorphic species are known in temperate caves, particularly among relict troglobites (Jeannel, 1943). On the other hand, tropical troglobites are considered to rarely exhibit such advanced evolution (Leleup, 1965). In Southeast Asia however, a number of highly troglomorphic species have been described in recent years among a variety of groups, from Coleoptera (*Ozaenaphaenops* spp. from northern Thai caves) to Palpigradi (*Eukoenenia thais* from Tham Chiang Dao) and aquatic crabs (*Cancrocaeca xenomorpha* from the SKT system). Tropical troglobites, as they are usually non-relict, could be expected at least to exhibit lower levels of troglomorphy than temperate troglobites. That is not the case in Collembola in Southeast Asia, the only group examined so far in this respect. In the most troglomorphic species of the genus *Troglopedetes* in Thailand, claw elongation is weaker, but antenna elongation is stronger than in the most troglomorphic *Pseudosinella,* its counterpart in western temperate caves. More generally, the elongation of antennae in the troglobitic species of Collembola compared to their outside relatives is particularly important in Southeast Asia (Fig. 31.8), and much greater than in temperate caves. Similar trends towards larger size or appendage elongation are obvious for several other groups, from spiders (Deeleman-Reinhold, 1995: pp. 20, 35 for the genus *Psiloderces*) to aquatic Crustacea (Magniez, 1991: p. 99 for *Stenasellus* spp.). In summary, very high levels of troglomorphy are not less frequent in Southeast Asia than among temperate troglobites.

CONCLUSIONS AND PERSPECTIVES

Study of the biospeleology of Southeast Asia is at its very beginning. Though the numbers of recorded troglobitic species and genera have dramatically increased in the last fifteen years, only a very small proportion of karsts have been investigated so far, and published work is still mostly restricted to taxonomic descriptions. On existing data, however, the general characteristics of the fauna of Southeast Asia caves appear to be those of many caves in the wet tropics: an overwhelming importance of guano as a food resource, dominance of non-relict troglobites, a large proportion of Arachnida and a low proportion of troglobitic Coleoptera. The few caves sampled extensively all exhibit a high taxonomic richness, with numerous cave-restricted taxa, important cave-to-cave differences and very high levels of endemism. Troglomorphy is widespread among many lineages, and may reach levels as high as in the temperate zone. The functional organization of these diverse communities is poorly understood, and the biology of their key species is unknown in most cases. Given these gaps, faunistical survey of the subterranean habitats of Southeast Asia remains a basic need before any sound hypotheses can be advanced on their origin and their regional evolution.

ACKNOWLEDGEMENTS

This work has been supported by a grant from Ministère de l'Education Nationale, de l'Enseignement Supérieur et de la Recherche (programme ACC-SV No. 7 Systematique et Biodiversité: "Diversité Fonctionnelle et Evolution en Milieu Souterrain"). We thank David Culver for providing useful original data about West Virginia cave biodiversity. We are also grateful to K. Christiansen for very useful comments about the text, and improvement of the English style.

REFERENCES

Annandale, N., Coggin-Brown, J. and Gravely, F.H., 1913. The limestone caves of Burma and the Malay peninsula. *J. Asiat. Soc. Bengal,* 9: 391–424.

Barr Jr., T.C., 1968. Cave ecology and the evolution of troglobites. *Evol. Biol.,* 2: 35–102.

Bedos, A., Deharveng, L., Deharveng, L., Leclerc, P., Rigal, D. and Solier, P., 1990. *Expéditions Maros 88 – Maros 89, Rapport Spéléologique.* APS, Toulouse, 51 pp.

Bedos, A., Brouquisse, F., Deharveng, L., Leclerc, P. and Rigal, D., 1994. 4. Grandes cavités du karst de Maros. In: APS (Editor), *Indonésie 92, Rapport Spéléologique.* APS, Toulouse, pp. 39–45.

Besuchet, C., 1986. 27 – Coléoptères Psélaphides. In: APS (Editor), *Expédition Thaï-Maros 85.* APS, Toulouse, p. 199.

Botosaneanu, L. and Holsinger, J.R., 1991. Some aspects concerning colonisation of the subterranean realm – especially of subterranean waters: a response to Rouch and Danielopol, 1987. *Stygologia,* 6(1): 11–39.

Boutin, C., 1969. Premières observations sur un Arachnide Amblypyge cavernicole du Cambodge (*Myodalis* sp.). *Ann. Fac. Sci. Phnom Penh,* 2: 161–180.

Boutin, C., 1971. Observations biospéologiques en Asie du Sud-Est. *Ann. Fac. Sci. Phnom Penh,* 4: 157–186.

Brouquisse, F. and Brouquisse, R., 1992. 3. Halmahera: le réseau de Batu Lubang. In: APS (Editor), *Expédition Indonésie 90, Rapport Spéléologique et Scientifique.* APS, Toulouse, pp. 27–32.

Brouquisse, F. and Rigal, D., 1987. 6. Résultats spéléologiques Sulawesi. In: APS (Editor), *Expédition Thaï-Maros 86, Rapport Spéléologique et Scientifique.* APS, Toulouse, pp. 47–74.

Brown, J.H., 1981. Two decades of homage to Santa Rosalia: toward a general theory of diversity. *Am. Zool.,* 21: 877–888.

Bullock, J.A., 1965. The ecology of Malaysian caves (and a note on the faunistic list from Batu caves). *Malay. Nat. J.,* 19(1): 57–64.

Chapman, P., 1976. The British New Guinea Speleological expedition 1975 – b) Speleobiology. *Trans. Br. Cave Res. Assoc.,* 3(3+4): 192–203.

Chapman, P., 1980a. The invertebrate fauna of caves of the Serrania de San Luis, ed. Falcon, Venezuela. *Trans. Br. Cave Res. Assoc.,* 7(4): 179–199.

Chapman, P., 1980b. The biology of caves in the Gunung Mulu National Park, Sarawak. *Trans. Br. Cave Res. Assoc.,* 7(3): 141–149.

Chapman, P., 1982a. The ecology of caves in the Gunung Mulu National Park, Sarawak. *Trans. Br. Cave Res. Assoc.,* 9(2): 142–162.

Chapman, P., 1982b. The origin of troglobites. *Proc. Univ. Bristol Spelaeol. Soc.,* 16(2): 133–141.

Chapman, P., 1983. Species diversity in a tropical cave ecosystem. *Proc. Univ. Bristol Spelaeol. Soc.,* 16(3): 201–213.

Chapman, P., 1984. Invertebrates in Niah Great Cave, Batu Niah National Park, Sarawak. *Cave Sci.,* 11(2): 89–91.

Chapman, P., 1985. Cave frequenting vertebrates in the Gunung Mulu national park, Sarawak. *Sarawak Mus. J.* 1985: 101–115.

Chappuis, P.A., 1956. Sur certaines reliques marines dans les eaux souterraines. *Comm. 1er Congr. Int. Spéléologie, Paris, 1953,* III. CNRS, pp. 47–53.

Christiansen, K.A., 1961. Convergence and parallelism in cave Entomobryidae. *Evolution,* 15: 288–301.

Christiansen, K.A. and Bouillon, M., 1978. An evolutionary and ecological analysis of the terrestrial Arthropods of caves in the Central Pyrenees. *NSS Bull.,* 40: 103–117.

Christiansen, K.A. and Culver, D.C., 1968. Geographical variation and evolution in *Pseudosinella hirsuta. Evolution,* 22: 237–255.

Christiansen, K.A. and Culver, D.C., 1969. Geographical variation and evolution in *Pseudosinella violenta* (Folsom). *Evolution,* 23(4): 602–621.

Condé, B., 1992. Palpigrades cavernicoles et endogés de Thaïlande et des Célèbes (1ère note). *Rev. Suisse Zool.,* 99(3): 655–672.

Culver, D.C., 1970. Analysis of simple cave communities I. Caves as islands. *Evolution,* 24: 463–474.

Culver, D.C., 1982. *Cave Life. Evolution and Ecology.* Harvard University Press, Cambridge, MA, 188 pp.

Dalens, H., 1989. Sur un nouveau genre d'Oniscoide "aquatique" provenant du sud-est Asiatique: *Thailandoniscus annae. Spixiana,* 192(1): 1–6.

Decu, V., 1981. Quelques aspects de la biospéologie tropicale résultant des expéditions biospéologiques cubano–roumaines à Cuba. In: *Résultats des Expéditions Biospéologiques Cubano–Roumaines à Cuba,* Vol. 3. Editura Academiei Republicii Socialiste România, Bucarest, pp. 9–16.

Decu, V. and Juberthie, C., 1994. Cuba. In: C. Juberthie and V. Decu (Editors), *Encyclopaedia Biospeologica, Tome I.* Société de Biospéologie, Moulis/Bucarest, pp. 459–475.

Deeleman-Reinhold, C.L., 1995. The Ochyroceratidae of the Indo-Pacific Region (Aranea). *The Raffles Bulletin of Zoology,* Suppl. 2, 103 pp.

Deharveng, L., 1981. The fauna of caves. In: *Papua New Guinea. Spelunca,* Suppl. 3, pp. 38–39.

Deharveng, L., 1982. Observations préliminaires sur les Collemboles cavernicoles de Papouasie Nouvelle-Guinée. *Mem. Biospeol.,* 8: 69–75.

Deharveng, L., 1987a. 9 – Nouvelles données sur le gaz carbonique des sols et des cavités de Thaïlande et de Sulawesi. In: APS (Editor), *Expédition Thaï-Maros 86, Rapport Spéléologique et Scientifique.* APS, Toulouse, pp. 97–110.

Deharveng, L., 1987b. 10 – Programme zoologique: bilan général et principaux résultats. In: APS (Editor), *Expédition Thaï-Maros 86, Rapport Spéléologique et Scientifique.* APS, Toulouse, pp. 111–116.

Deharveng, L., 1987c. Cave Collembola of South-East Asia. *Korean J. Syst. Zool.,* 3(2): 165–174.

Deharveng, L., 1989. 4. La faune souterraine de Batu Lubang. In: APS (Editor), *Expédition Batukarst 88, Rapport Spéléologique et Scientifique.* APS, Toulouse, p. 199.

Deharveng, L. and Bedos, A., 1988. 7. Le sud: provinces de Ranong, Chumphon, Surat Thani et Phangnga. In: APS (Editor), *Expéditions Thai87–Thai88, Rapport Spéléologique et Scientifique.* APS, Toulouse, pp. 79–114.

Deharveng, L. and Bedos, A., 1995. 6. La faune souterraine de Sumatra Barat. In: APS (Editor), *Expédition Sumatra 93, Rapport Spéléologique et Scientifique.* APS, Toulouse, pp. 41–44.

Deharveng, L. and Bedos, A., 1996. Nouvelles explorations au Gunung Seribu (Sumatra). *Spelunca,* 61: 5–6.

Deharveng, L. and Bedos, A., 2000. Thailand. In: C. Juberthie and V. Decu (Editors), *Encyclopaedia Biospeologica, Tome III.* Société de Biospéologie, Moulis/Bucarest, in press.

Deharveng, L. and Brouquisse, F., 1986. 4. Le massif du Doï Chiang Dao, les karsts de l'est de Chiang Mai. In: APS (Editor), *Expédition Thaï-Maros 85, Rapport Spéléogique et Scientifique.* APS, Toulouse, pp. 23–30.

Deharveng, L. and Delnatte, E., 1988. 5. Le centre-est: provinces de Phetchabun, Khon Kaen et Chaiyaphum. In: APS (Editor), *Expéditions Thai87–Thai88, Rapport Spéléologique et Scientifique.* APS, Toulouse, pp. 41–63.

Deharveng, L. and Gers, C., 1993. Ten new species of *Troglopedetes* Absolon, 1907 from caves of Thailand (Collembola, Paronellidae). *Bijdr. Dierk.,* 63(2): 103–113.

Deharveng, L. and Le Cong Kiet, 2000. Vietnam. In: C. Juberthie and V. Decu (Editor), *Encyclopaedia Biospeologica, Tome III.* Société de Biospéologie, Moulis/Bucarest, in press.

Deharveng, L. and Leclerc, P., 1989. Recherches sur les faunes cavernicoles d'Asie du Sud-Est. *Mem. Biospeol.,* 16: 91–110.

Deharveng, L., Leclerc, P., Lebreton, B., Besson, J.P. and Gibert J., 1986. Biologie, 19. Considérations générales et catalogue. In: APS (Editor), *Expédition Thaï-Maros 85, Rapport Spéléologique et Scientifique.* APS, Toulouse, pp. 164–173.

Deuve, T., 1988. Coléoptères Caraboidea des milieux souterrains de l'Asie du Sud-Est. In: APS (Editor), *Expéditions de l'APS en Asie du Sud-Est, Travaux Scientifiques – 1.* APS, Toulouse, pp. 17–20.

Dunn, F.L., 1965. Gua Anak Takun: ecological observations. *Malay. Nat. J.,* 19(1): 75–87.

Eberhard, S.M. and Spate, A., 1995. *Cave Invertebrate Survey: Towards an Atlas of New South Wales Cave Fauna.* Departmen of Urban Affairs and Planning and the Australian Heritage Commission, 174 pp.

Eberhard, S.M., Richarson, A.M.M. and Swain, R., 1991. *The Invertebrate Cave Fauna of Tasmania.* Zoology Department, University of Tasmania, 115 pp.

Gibert, J., 1987. 11. Le système karstique du Doi Chiang Dao (Thaïlande). Peuplements aquatiques souterrains, répartition, relations entre le milieu karstique et le sous-écoulement de l'exutoire. In: *Expédition Thaï-Maros 86, Rapport Spéléologique et Scientifique.* APS, Toulouse, pp. 117–128.

Ginet, R. and Decou, V., 1977. *Initiation à la Biologie et à l'Écologie Souterraines.* J.P. Delarge, Paris, 345 pp.

Gnaspini, P., 1992. Bat guano ecosystems: a new classification and some considerations with special references to neotropical data. *Mem. Biospeol.,* 19: 135–138.

Hawes, R.S., 1939. The flood factor in the ecology of caves. *J. Anim. Ecol.,* 8(1): 1–5.

Holsinger, J.R., 1993. Biodiversity of subterranean amphipod crustaceans: global patterns and zoogeographic implications. *J. Nat. Hist.,* 27: 821–835.

Holsinger, J.R. and Culver, D.C., 1988. The invertebrate cave fauna of Virginia and a part of Eastern Tennessee: zoogeography and ecology. *Brimleyana,* 14: 1–162.

Howarth, F.G., 1973. The cavernicolous fauna of Hawaiian lava tubes. 1. Introduction. *Pac. Insects,* 15(1): 139–151.

Howarth, F.G., 1980. The zoogeography of specialised cave animals: a bioclimatic model. *Evolution,* 34(2): 394–406.

Howarth, F.G., 1983. Ecology of cave Arthropods. *Annu. Rev. Entomol.,* 28: 365–389.

Howarth, F.G., 1987. The evolution of non-relictual tropical troglobites. *Int. J. Speleol.,* 16: 1–16.

Howarth, F.G., 1988. Environmental ecology of North Queensland caves: or why there are so many troglobites in Australia. In: L. Pearson (Editor), *17th Biennial Conf. Australian Speleological Federation, Cairns,* pp. 76–84.

Howarth, F.G., 1993. High-stress subterranean habitats and evolutionary change in cave-inhabiting arthropods. *Am. Nat.,* 142(Supp): 65–77.

Howarth, F.G. and Stone, F.D., 1990. Elevated carbon dioxide levels in Bayliss Cave, Australia: implications for the evolution of obligate cave species. *Pac. Sci.,* 44(3): 207–218.

Humphreys, W.F., 1993. The significance of the subterranean fauna in biogeographical reconstruction: examples from Cape Range peninsula, Western Australia. in: W.F. Humphreys (Editor), *The Biogeography of Cape Range, Western Australia.* Western Australian Museum, Perth, pp. 165–192.

Jeannel, R., 1926. *Faune Cavernicole de la France.* Lechevalier, Paris, 334 pp.

Jeannel, R., 1943. *Les Fossiles Vivants des Cavernes.* Gallimard, Paris, 321 pp.

Jeannel, R. and Racovitza, E.G., 1914. Biospeologica XXXIII. Enumération des grottes visitées 1911–1913 (5ème série). *Arch. Zool. Exp. Gen.,* 53: 325–558.

Laneyrie, R., 1960. Résumé des connaissances actuelles concernant les coléoptères hypogés de France. *Ann. Soc. Entomol. France,* 129: 89–149.

Leclerc, P. and Dalger, D., 1988. 8. Le sud: province de Yala. In: APS (Editor), *Expéditions Thai87–Thai88, Rapport Spéléologique et Scientifique.* APS, Toulouse, pp. 115–120.

Leclerc, P., Giani, N. and Bouguenec, V., 1988. 6. Le sud: provinces de Kanchanaburi, Ratchaburi, Phetchaburi et Prachuap Khiri Khan. In: APS (Editor), *Expéditions Thai87–Thai88, Rapport Spéléologique et Scientifique.* APS, Toulouse, pp. 64–78.

Leleup, N., 1965. La faune entomologique cryptique de l'Afrique intertropicale. Musée Royal de l'Afrique Centrale. *Ann. Sci. Zool.,* 141: 1–186.

Maffre, L. and Rigal, D., 1986. 6. Le karst de la Nam Lang. In: APS (Editor), *Expédition Thaï-Maros 85, Rapport Spéléologique et Scientifique.* APS, Toulouse, pp. 38–50.

Magniez, G., 1991. Bons et mauvais caractères taxonomiques: exemple des Stenasellides d'Extrême Orient. *Mem. Biospeol.,* 18: 99–104.

McClure, H.E., 1965. Microcosms of Batu Caves and a list of species collected at Batu Caves. *Malay. Nat. J.,* 19(1): 65–74.

McClure, H.E., @Boo-Liat Lim and Winn, S.E., 1967. Fauna of the dark cave, Batu Caves, Kuala Lumpur, Malaysia. *Pac. Insects,* 9(3): 399–428.

Peck, S.B., 1971. The invertebrate fauna of tropical American caves, part 1: Chilibrillo cave, Panama. *Ann. Speleol.,* 26(2): 423–437.

Peck, S.B., 1975. The invertebrate fauna of tropical American caves, III: Jamaica, an introduction. *Int. J. Speleol.,* 7: 303–326.

Peck, S.B., 1985. The invertebrate faunas of tropical American caves, part 6: Jumandi cave, Ecuador. *Int. J. Speleol.,* 14: 1–8.

Peck, S.B., 1990. Eyeless Arthropods of the Galapagos islands, Ecuador: composition and origin of the cryptozoic faunas of a young, tropical, oceanic archipelago. *Biotropica,* 22: 366–381.

Peck, S.B. and Finston, T.L., 1993. Galapagos Islands troglobites: the questions of tropical troglobites, parapatric distributions with eyed-sister-species, and their origin by parapatric speciation. *Mem. Biospeol.,* 20: 19–37.

Poulson, T.L., 1972. Bat guano ecosystems. *Natl. Soc. Speleol. Bull.,* 34(2): 55–59.

Reddell, J.R., 1981. Review of the cavernicole fauna of Mexico, Guatemala and Belize. *Tex. Mem. Mus. Bull.,* 27: 1–327.

Ridley, H.N., 1898. Report on caves in the Malay peninsula. *Rep. Br. Assoc. Adv. Sci., Bristol,* pp.572–582.

Rigal, D., 1987. Gua Salukkan Kallang, karst de Maros, Célèbes Sud, Indonésie. *Spelunca,* 28: 32–39.

Rohde, K., 1992. Latitudinal gradients in species diversity, the search for the primary cause. *Oikos,* 65(3): 514–527.

Rouch, R. and Danielopol, D.L., 1987. L'origine de la faune aquatique souterraine, entre le paradigme du refuge et le modèle de la colonisation active. *Stygologia,* 3(4): 345–372.

Sbordoni, V., Argano, R., Vomero, V. and Zullini, A., 1977. Subterranean fauna of Mexico, part III – Ricerche sulla fauna cavernicola del Chiapas (Messico) et delle regione limitrofe: grotte esplorate nel 1973 e nel 1975. Criteri per una classificazione bioespeleologica delle grotte. *Cuad. Accad. Naz. Lincei,* 171(3): 5–74.

Sbordoni, V., Vomero, V., Argano, R. and Zullini, A., 1980. The cave fauna of Chiapas (Mexico): tropical and temperate cave communities and their evolution. In: J.I. Furtado (Editor), *Tropical Ecology and Development; Proc. 5th Int. Symp. on Tropical Ecology, April 1979, Kuala Lumpur,* Part I, pp. 687–692.

Smith, G.B., 1980. Biospeleology. In: J.M. James and H.J. Dyson (Editors), *Caves and Karsts of the Muller Range.* Atea 78, Newtown, Australia, pp. 121–129.

Stone, F.D., 1983. Biogeography and ecology of the caves of Thailand: preliminary observations. *Natl. Soc. Speleol. Bull.*, 45(1): 5.

Trajano, E. and Gnaspini-Netto, P., 1991. Notes on the food-webs in caves from Southeastern Brazil. *Mem. Biospeol.,* 18: 75–79.

Vandel, A., 1964. *Biospéologie. La Biologie des Animaux Cavernicoles.* Gauthier-Villars, Paris, 619 pp.

Wägele, J.W., 1990. Aspects of the evolution and biogeography of stygobiontic Isopoda. *Bijdr. Dierk.,* 60 (3&4): 145–150.

Whitten, A.J., Damanik, S.J., Anwar, J. and Hisyam, N., 1984. *The Ecology of Sumatra.* Gadjah Mada University Press, Yogjakarta, 583 pp.

Whitten, A.J., Mustapha, M. and Henderson, G.S., 1987. *The Ecology of Sulawesi.* Gadjah Mada University Press, Yogjakarta, 777 pp.

Chapter 32

ECOLOGY AND SPECIES DISTRIBUTION OF THE MONTE CORONA LAVA TUNNEL ON LANZAROTE (CANARY ISLANDS)

Thomas M. ILIFFE, Jakob PARZEFALL and Horst WILKENS

INTRODUCTION

Significance of the cave and its fauna

Lanzarote, the easternmost of the Canary Islands, is of volcanic origin. Particularly noteworthy is the presence on Lanzarote of one of the world's longest lava tubes, located near the northern end of the island. This lava tube begins at the base of the volcano Monte Corona, continues 7 km down the island slope, and finally crosses the coastline where it is inundated by the Atlantic Ocean. This last section of the lava tube ranks as the world's longest underwater volcanic cave.

Lanzarote possesses only a very shallow freshwater lens. This is due to the absence of higher mountains, resulting in a very low annual rainfall. On other islands within this archipelago, higher rainfall generates increased freshwater flow into the ocean so that groundwater at the shoreline is brackish. In contrast, seawater penetrates deeply into the groundwater of Lanzarote. This is probably facilitated by the volcanic composition of the bottom, which provides a crevicular substrate. Consequently, tides and water of fully marine salinity are present in the groundwater to some distance inland from the coast, as well as in the lava tube.

The first published reports on this lava tube consisted of a description of the Jameos del Agua, the coastal section of the tube, and its fauna (Koelbel, 1892; Calman, 1904; Harms, 1921). Intensive studies and important biological discoveries, as well as the first diving exploration of the submarine tunnel began in the 1970's (Wilkens and Parzefall, 1974). During the 1980's, continued diving explorations yielded further new species, limited to the submarine sections of the Túnel de Atlantida (see below), as well as the complete exploration and mapping of the cave (Exley, 1983; Iliffe et al., 1984; Méndo and Ortega, 1988). The seaward end of the Túnel de Atlantida was first reached by Swiss cave diver Oliver Isler in 1986 (Isler, 1989).

A particularly rich and diverse endemic fauna inhabits the marine sections of the lava tube. Many of these cave-limited species show eye and pigment reductions characteristic of troglobites. The first and best-known invertebrate to be described from this cave was the galatheid crab *Munidopsis polymorpha* (Koelbel, 1892). Later diving explorations led to the discovery of primitive relict species such as the remipede *Speleonectes ondinae*. While many of these anchialine troglobites show close taxonomic relationships with deep-sea forms, others are closely related to species inhabiting limestone caves on the opposite side of the Atlantic Ocean (Iliffe et al., 1984; Wilkens et al., 1986). Furthermore, the Jameos del Agua and Túnel de Atlantida constitute the most significant anchialine cave biotope in the Eastern Atlantic region, as well as the most important submarine lava tube in the world, from the faunistic point of view.

Most studies of anchialine caves have dealt with those formed in limestone rock such as those in Bermuda (Sket and Iliffe, 1980; Iliffe et al., 1983; Iliffe, 1994), the Bahamas (Juberthie and Iliffe, 1994), and the Yucatan Peninsula of Mexico (Iliffe, 1993). The presence of a very similar and closely related fauna inhabiting volcanic caves raises interesting questions pertaining to their origins and modes of dispersal.

HABITAT DESCRIPTION

Lava tube formation and age

The Lanzarote lava tube was formed during an eruption

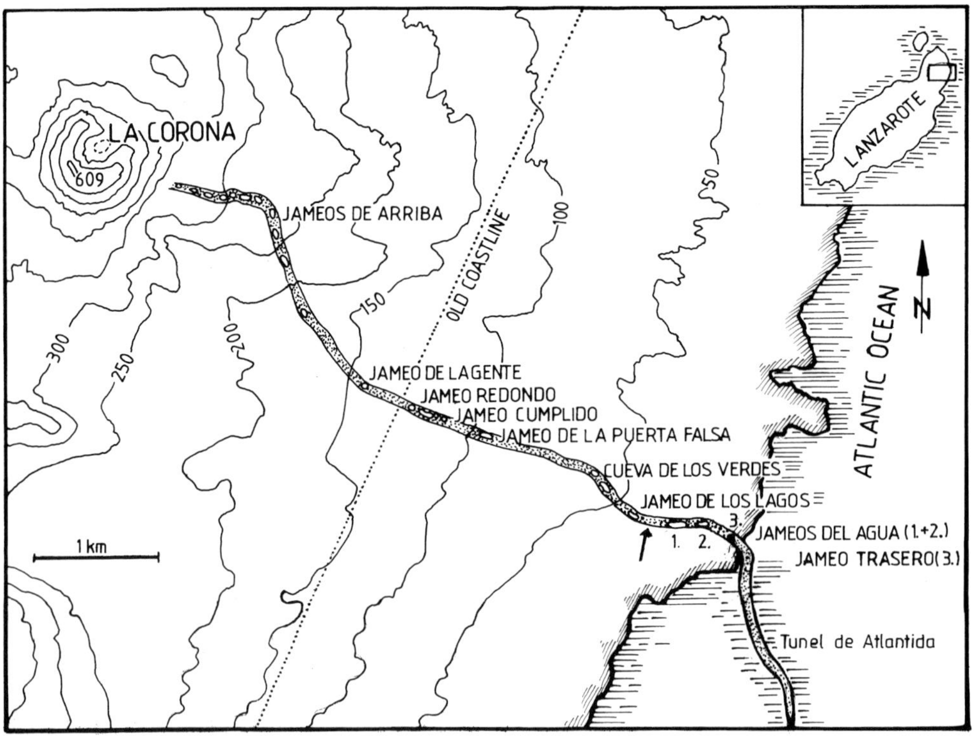

Fig. 32.1. The lava tube of the volcano Monte Corona (1, Jameo Grande; 2, Jameo Chico; arrow indicates how far the seawater penetrates in the lava tube).

of the volcano Monte Corona, possibly occurring as recently as two to three thousand years ago (Bravo, 1964; Macau, 1965). It extends from the volcano over 7 km to the shoreline and continues for a farther 1.4 km into the sea, reaching a depth of 50 m below sea level at its terminus. As a result of variations in the amount of lava produced during phases of the eruptive sequence, several levels of galleries have developed within the tube (Fig. 32.1).

In various places, the lava roof of the tube has broken down producing openings to the surface called "Jameos". They provide access to intact sections of the tube that are named after them. The tube can be divided into terrestrial and marine sections. The terrestrial upper part, near the base of the volcano, is accessible through the Jameos de Arriba and the Jameo de la Puerta Falsa. This section ends with the Cueva de los Verdes, which today is a commercial tourist attraction. Seawater partially or totally floods the lower marine parts of the tube, beginning some 1000 m inland from the coast in a completely dark section of the tube called Cueva de los Lagos. The touristically operated Jameos del Agua extends between the latter and the wholly submarine Túnel de Atlantida (Figs. 32.1, 32.2).

The lava tube is situated in a lava field, called Malpais de la Corona, which originated during the eruption. Lava making up the Malpais is of a very low density, so that sea water can penetrate beneath this lava surface reaching up to the old coastline situated more than 3 km inland. Semi-diurnal temperature and humidity variations in the Cueva de los Verdes section of the lava tube may be produced by fluctuations in the phreatic level induced by the oceanic tides (Vieira et al., 1991). The vegetation of the Malpais consists of a sparse xerophytic flora.

Cueva de los Lagos

This part of the lava tube is the most inland section into which marine groundwater penetrates. It consists of a series of shallow pools 1–2 m deep, extending over a horizontal distance of about 150–200 m. Although the water is calm in the pools, there is significant tidal fluctuation. At the seaward end of this section, the tube becomes submerged and extends underwater in the direction of the Jameos del Agua. As these pools are

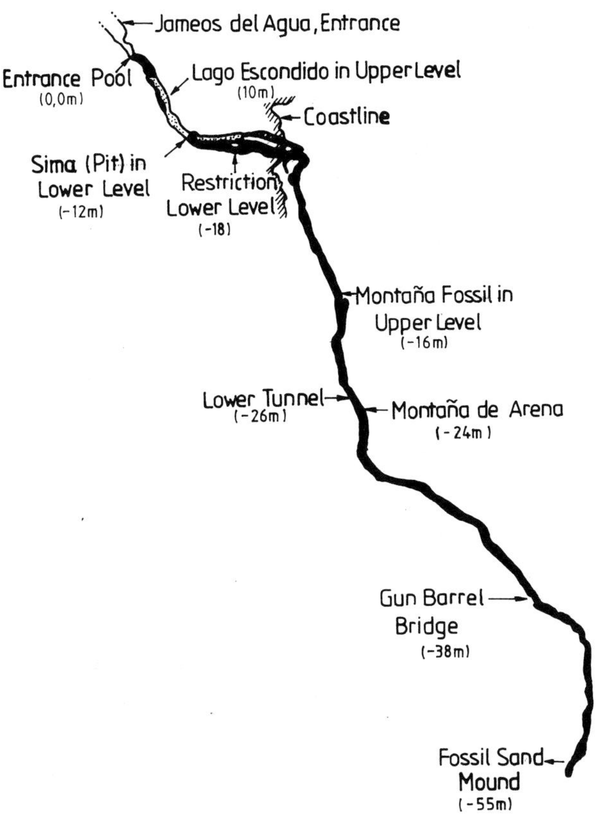

Fig. 32.2. Túnel de Atlantida – the submerged part of the lava tube (water depths in brackets).

situated about 300 m from the entrance, the Jameo de los Lagos, they remain in absolute darkness and thus qualify as a true cave habitat.

The most common species found here is the gastropod *Littorina saxatilis.* Additionally, the polychaete *Gesiella jameensis,* the gastropod *Peringiella epidaurica,* the ostracod *Danielopolina wilkensi,* the remipede *Speleonectes ondinae,* the copepods *Expansophria dimorpha, Muceddina multispinosa* and *Speleophriopsis canariensis,* the mysid *Heteromysoides cotti* and the amphipod *Spelaeonicippe buchi* have been observed, although *Munidopsis polymorpha* has not (Wilkens et al., 1993; Jaume and Boxshall, 1996a).

Jameos del Agua

The Jameos del Agua is a section of the lava tube near the coast that has been used as a tourist attraction since the 1970's. Here, a salt-water lake, 50 m long by 10 m wide and 8 m deep, is bounded by breakdown mounds beneath two collapse entrances, Jameo Chico and Jameo Grande (Fig. 32.4). This lake is naturally illuminated by limited sunlight from these two entrances. On account of the significant terrestrial input of light and nutrients, this lake can be characterized as an anchialine biotope.

The low intensity of light is just sufficient to permit the growth of diatoms, but no green or brown algae are able to develop. Due to limited algal diversity, the marine faunal spectrum able to colonize this lake is restricted. Some species, however, develop high population densities. A large population of the galatheid crab *Munidopsis polymorpha* is present (Figs. 32.5, 32.6). Also abundant are the mysid *Heteromysoides cotti* and the amphipod *Parhyale multispinosa* (= *P. hawaiiensis*). *Munidopsis polymorpha* and *H. cotti* show the same distribution pattern – that is, they are found feeding or swimming, respectively, over the diatom-covered rocks. *Parhyale multispinosa* is limited to the eulitoral, where it occurs in great numbers. Other species are found in lower numbers. These include the amphipod *Liagoceradocus acutus* and the isopod *Curassanthura canariensis,* which inhabit coarse lava gravel; the polychaete *Gesiella jameensis,* the copepod *Oromiina fortunata* and the amphipod *Spelaeonicippe buchi,* which are occasionally found swimming in the water column; and the polychaetes *Fauveliopsis jameoaquensis*, *Macrochaeta* sp. and *Mesonerilla diatomeophaga,* which inhabit the coarse black sand and gravel bottom of the pool.

The sedentary echiuran worm *Bonellia viridis*, first reported from here by Harms (1921), is another inhabitant of the Jameos del Agua. A comparison of the abundance of specimens observed today with the situation in the 1970's reveals an increase in the population density. A dense network of worm proboscises now covers the bottom of the anchialine pool.

Túnel de Atlantida

A second, but smaller, pool lies on the seaward side of the breakdown pile below the Jameo Chico. Apart from artificial lights that illuminate this pool during the day, it remains in total darkness. From this pool, the Túnel de Atlantida, a submerged extension of the tube, continues seaward for 1415 m (Figs. 32.1–32.3). The first 150 m of underwater passage averages about 5–8 m high and 6–10 m wide at water depths to 12 m. The "Sima" or pit at 150 m horizontal penetration connects the initial shallower section of the tube with the gradually descending main level. Directly above

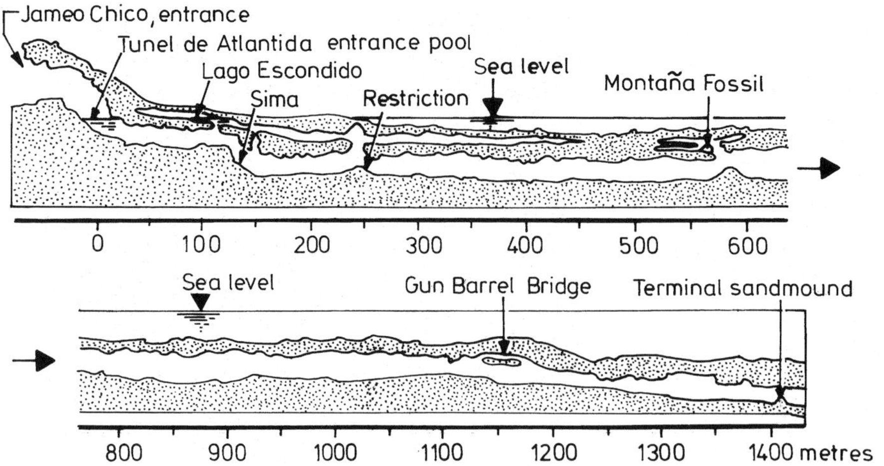

Fig. 32.3. Extended profile view of the Túnel de Atlantida. The scale bar beneath the figure indicates the distance in meters from the entrance pool.

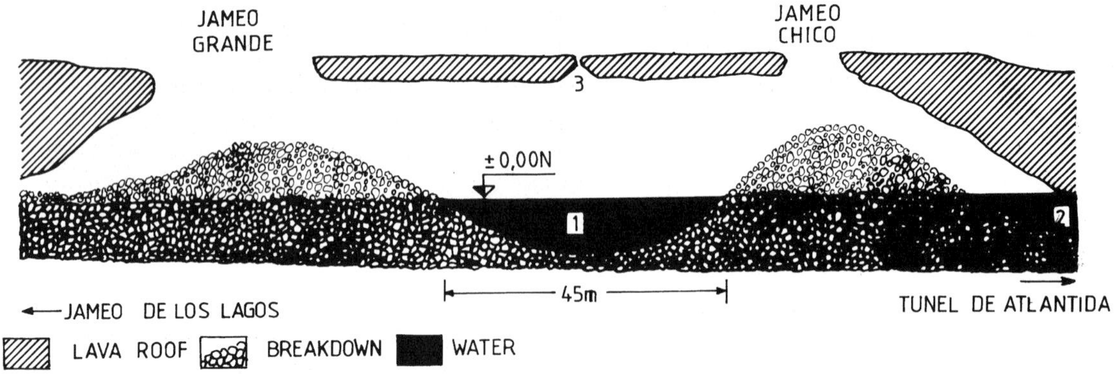

Fig. 32.4. Section of the Jameos del Agua (1, anchialine lake; 2, entrance pool of the Túnel de Atlantida; 3, blowhole).

the Sima is one of three connections to an upper level lying immediately above the main lower level. Heading inland, the upper level emerges above sea level at the Lago Escondido or Hidden Lake that can only be reached by diving. This lake room extends inland as a dry passage ending near a collapse sinkhole on the surface. In the opposite direction, the upper level bisects a major collapse chamber that connects the two levels and reaches just above sea level at the ceiling. Beyond this chamber, the upper level becomes lower and eventually pinches out. In the lower level, breakdown boulders from the collapse chamber have created a narrow restriction that divers must pass through in order to reach the more distant sections of the underwater cave.

At a point below the present coastline, both upper and lower levels take a sharp dogleg turn to the right. This may indicate that the lava flow forming the cave occurred at a time when sea level was similar to present elevations, causing the tube to be deflected upon reaching the sea.

Common species in the water column of the Túnel de Atlantida include the archiannelid *Protodrilus* sp.; the polychaete *Gesiella jameensis*; the ostracods *Danielopolina phalanx, D. wilkensi* and *Eupolycope pnyx*; the remipede *Speleonectes ondinae*; the copepods *Dimisophria cavernicola, Expansophria dimorpha* and *Palpophria aestheta*; the thermosbaenacean *Halosbaena fortunata*; and the amphipods *Liagoceradocus acutus* and *Spelaeonicippe buchi*. It is noteworthy that such a rich and diverse fauna occurs primarily in the water column, which is characterized by exceptional transparency and a lack of visible particulate matter. This obviously raises questions as to the food sources for these species. An unidentified species of archiannelid polychaete is abundant in gravel at the bottom

Fig. 32.5. *Munidopsis polymorpha.* (Photo: H. Wilkens)

of the entrance slope. *Munidopsis polymorpha* and the polychaete *Syllis cornuta* have also been found in this same locale.

Montaña de Arena and farther reaches of the Túnel de Atlantida

Another upper level of the tube at 550 m horizontal penetration contains a partially lithified carbonate sand mound, the Montaña Fossil. At one time, sand must have entered the cave from a hole in the ceiling that later became plugged. A lower tunnel at 700 m penetration ends in a sand plug. This sand comes from the main level, where it originates from a hole in the ceiling 20 cm in diameter, still hydrologically active. This opening is situated above a mound of loose sand 15 m high, the Montaña de Arena. It is the only site in the cave where sea water directly enters.

Sponges, hydroids, and other encrusting invertebrates are present around the hole, although absent in all other parts of the cave. Holothurians, echinoids and ophiuroids have also only been observed near this tidal vent. The ostracod *Eusarsiella bedoyai* and the shrimp *Stenopus spinosus* have been collected from the Montaña de Arena, and are probably troglophiles or accidentals.

Beyond the Montaña de Arena, the cave continues for another 700 m before ending. At its termination, the ceiling turns down to meet the floor and the tunnel pinches out at a water depth of 57 m. A final fossil sand mound is present at this point. Visual observations and limited collections from sections of the cave beyond the Montaña de Arena suggest that faunal abundance varies inversely with distance from the island.

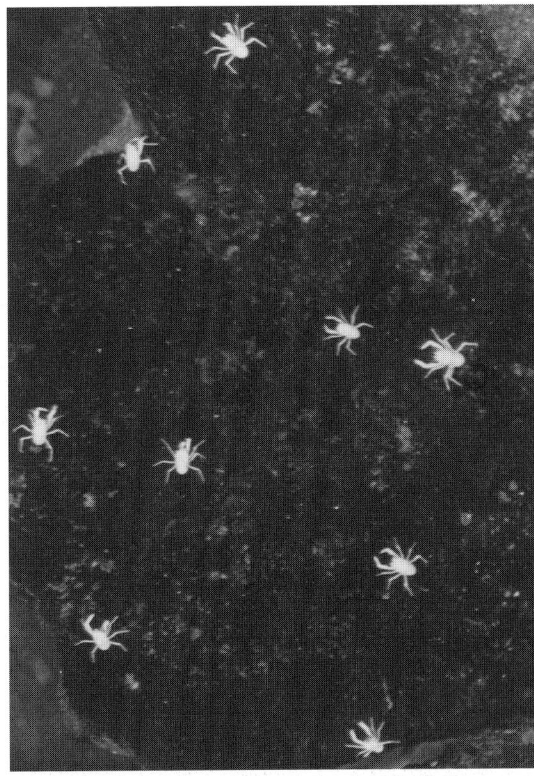

Fig. 32.6. *Munidopsis polymorpha* grazing on diatom-covered rocks. (Photo: H. Wilkens)

Crevicular habitats

In both terrestrial and submarine sections of the lava tube, there is close connection with the groundwater crevicular habitat. This was demonstrated by sampling in artificial wells that supply seawater for salt production in salinas. In this fully marine habitat, many of the characteristic species inhabiting the lava tube, such as *Danielopolina* spp., *Heteromysoides cotti, Liagoceradocus acutus* and *Munidopsis polymorpha,* were also found. Furthermore, *H. cotti, M. polymorpha,* and the nocturnally active *Parhyale multispinosa* were collected from several small inland tidal pools connected with the groundwater crevicular system (Wilkens et al., 1986, 1993).

Currents, tides and water quality

Tides in the entrance pool of the Túnel de Atlantida have a range of 1.5 m (Vieira et al., 1991) and occur about 1.5 hours after the corresponding tides in the open ocean. Strong tidal currents are evident in all parts of the cave, particularly at the base of the Sima and at the restriction. Salinity is slightly lower than that of the open sea (35.5 to 35.6 g ℓ^{-1} in the cave, *versus* 36.6 to 37.0 g ℓ^{-1} in littoral waters). Cave water temperatures remain almost constant year-round at 18°C. The water is exceptionally transparent, with very low levels of apparent particulate matter. Artificial light is present only at the entrance of the Túnel de Atlantida, but this is extinguished within the first few tens of meters. The floor, walls, and ceiling of the cave consist primarily of bare rock, breakdown, or solidified lava flow. A white carbonate submarine cement covers the upper surface of exposed rocks on the floor of the cave. Other than the sand mounds, little loose sediment is present.

ANCHIALINE FAUNA

Annotated species list

General description of the lava-tube fauna

Taxonomically, crustaceans and polychaetes constitute the main components of the lava-tube fauna. However, only five of the 13 reported polychaete species, namely *Fauveliopsis jameoaquensis, Gesiella jameensis, Macrochaeta* sp., *Protodrilus* sp., and *Speleobregma lanzaroteum,* lack eyes and pigment and thus are likely to be troglobitic. All other phyla occur in lower numbers, e.g., two gastropod species (Fig. 32.7). Characteristically, most crustacean species have reduced eyes and pigment. Only *Parhyale multispinosa* and *Heteromysoides cotti* have functional eyes and brown body coloration. The eyes of *M. polymorpha* are partially functional vestiges.

Cestum veneris (Cestidae, Ctenophora). Open-ocean, planktonic species, such as the ctenophore *C. veneris*, can occasionally be carried far into the cave by tidal currents (Harms, 1921).

Macrochaeta sp. (Acrocirridae, Polychaeta). This undescribed species is so far represented by only a single specimen. It is blind, depigmented and translucent. It was collected from coarse sand at 3 m depth within the Jameos del Agua (Núñez et al., 1997).

Protodorvillea kefersteini (Dorvilleidae, Polychaeta). The widely distributed carnivorous polychaete *P. kefersteini* was collected from fine sand at the Montaña de Arena (García-Valdecasas, 1985).

Fauveliopsis jameoaquensis (Fauveliopsidae, Polychaeta). This eyeless, depigmented, minute species was collected from coarse black sand at 5 m depth within the Jameos del Agua (Núñez et al., 1997).

Mesonerilla diatomeophaga (Nerillidae, Polychaeta). *Mesonerilla diatomeophaga* feeds on benthic diatoms growing on gravel and rocks in the Jameos del Agua (Núñez et al., 1997). It is related to *Leptonerilla prospera* from anchialine caves in Bermuda. Tiny archiannelid polychaetes have also been found to be locally abundant in gravel substrate from the Túnel de Atlantida and in sand at the Montaña de Arena. It is not known if these specimens belong to *M. diatomeophaga* or possibly another species.

Paradoneis lyra (Paraonidae, Polychaeta). *Paradoneis lyra* has a wide geographic distribution, being known from the northeastern and northwestern Atlantic as well as the Mediterranean (Núñez et al., 1997). It is present in the Jameos del Agua and the Montaña de Arena. It is probably accidental.

Gesiella jameensis (Polynoidae, Polychaeta). This monospecific polynoid probably has a deep-sea origin (Wilkens et al., 1986). It occurs in all aquatic parts of the lava tunnel, where it is typically observed swimming or drifting in open water of the cave.

Protodrilus sp. (Protodrilidae, Polychaeta). This archiannelid is a relatively common inhabitant of the water column in the Túnel de Atlantida and Cueva de los Lagos.

Speleobregma lanzaroteum (Scalibregmatidae, Polychaeta). *Speleobregma lanzaroteum* is apparently troglobitic, being eyeless and depigmented. The thickened areolated epidermis, common among scalibregmatids, is reduced in this species (Bertelsen, 1986).

Exogone (Parexogone) gambiae (Syllinae, Polychaeta). This species, previously known only from the Mediterranean Sea, was reported for the first time in the Atlantic by Núñez et al. (1997).

Miscellania dentata (Syllinae, Polychaeta). *Miscellania dentata* was found in association with diatom-covered gravel (Núñez et al., 1997). It had previously been reported from the Canary Islands and the northwestern Mediterranean.

Syllis cornuta (Syllinae, Polychaeta). The globally distributed polychaete *S. cornuta* is present in macro-

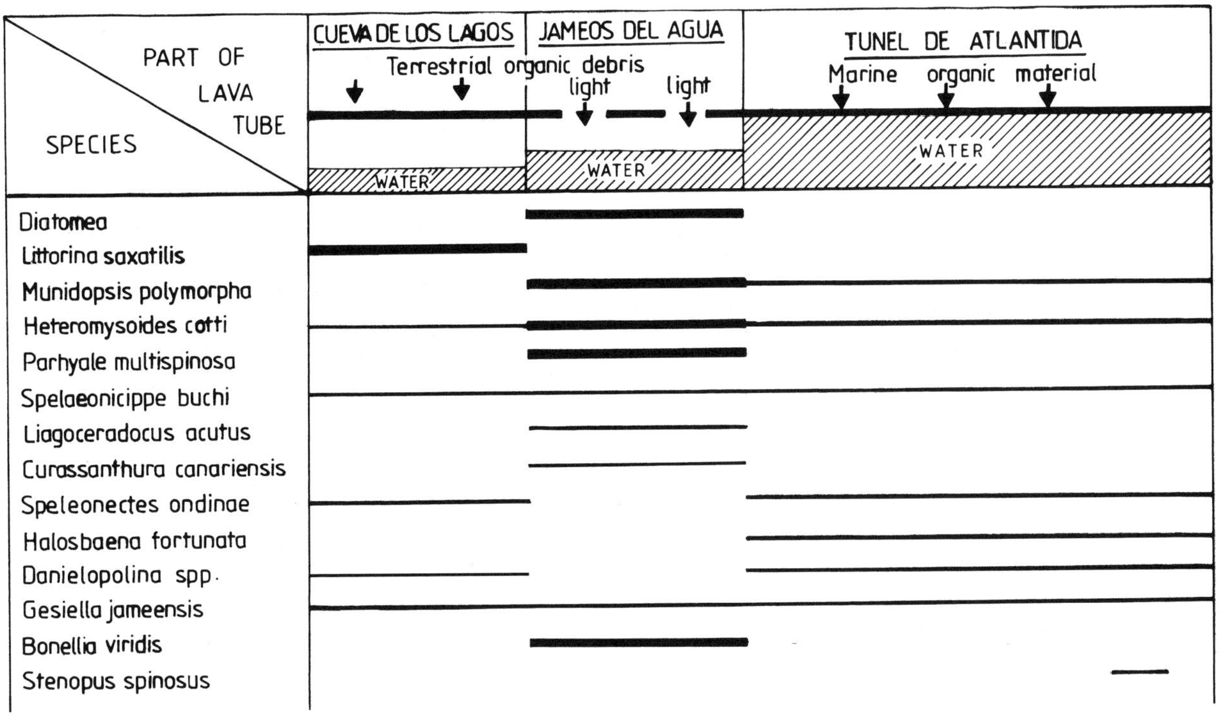

Fig. 32.7. Energy sources and distribution of species in the different parts of the lava tube. The line thickness reflects the relative frequency of the species.

porous gravel in the Jameos del Agua lake and at the entrance to the Túnel de Atlantida.

Syllis garciai (Syllinae, Polychaeta). This species is known from the northeastern and northwestern Atlantic and the western Mediterranean. It was reported from sand and diatom-covered gravel in the Jameos del Agua (Núñez et al., 1997).

Syllis truncata cryptica (Syllinae, Polychaeta). This species is known from the northeastern Atlantic, Mediterranean and Red Sea. It was found in gravel pores from the Jameos del Agua (Núñez et al., 1997).

Bonellia viridis (Echiura). *Bonellia viridis* is restricted to the anchialine Jameos del Agua lake and the artificially lighted entrance pool of the Túnel de Atlantida. It occurs there in very large numbers.

Littorina saxatilis (Littorinidae, Gastropoda). *Littorina saxatilis* is the most abundant species in the Cueva de los Lagos lakes.

Peringiella epidaurica (Rissoidae, Gastropoda). *Peringiella epidaurica* is found in the Cueva de los Lagos (Wilkens et al., 1993).

Paramisophria reducta (Arietellidae, Copepoda). *Paramisophria reducta* from the Jameos del Agua lava tube and *P. galapagensis* from an anchialine fissure on Santa Cruz in the Galapagos Islands are the first records of cavernicolous arietellid copepods. These two species may be derived from neritic hyperbenthic ancestral forms that first invaded caves (Ohtsuka et al., 1993).

Oromiina fortunata (Cyclopinidae, Copepoda). *Oromiina fortunata* occurs in the Jameos del Agua. *Oromiina* possesses the ancestral state for the Cyclopoida in many characters (Jaume and Boxshall, 1997).

Muceddina multispinosa (Cyclopinidae, Copepoda). *Muceddina multispinosa* has also been found to inhabit marine caves along the coast of Sardinia and the Balearic Islands (Jaume and Boxshall, 1996b).

Dimisophria cavernicola (Misophriidae, Copepoda). Misophrioid copepods occur primarily in deep-sea or anchialine cave habitats, suggesting a link between these two marine environments (Boxshall, 1989). All misophrioids were collected in plankton tows from the water column of the Túnel de Atlantida. This monospecific genus resembles, but is distinct from two other anchialine genera, *Speleophria* and *Palpophria* (Boxshall and Iliffe, 1987).

Expansophria dimorpha (Misophriidae, Copepoda). The genus *Expansophria* contains four species: *E. apoda* from a limestone sinkhole in Palau, *E. dimorpha* from the Túnel de Atlantida, *E. galapagensis* from volcanic fissures in the Galapagos Islands, and *E. sarda* from a limestone cave in Sardinia (Boxshall and Iliffe, 1987, 1990; Jaume and Boxshall, 1996a). Morphologically, *E. dimorpha* very closely resembles *E. galapagensis*.

Palpophria aestheta (Misophriidae, Copepoda). *Palpophria aestheta* is the only species known from this genus. It is probably a raptorial predator, with maxillules, maxillae and maxillipeds adapted for grasping and manipulating prey (Boxshall and Iliffe, 1987).

Speleophriopsis canariensis (Misophriidae, Copepoda). *Speleophriopsis canariensis* from the Jameo de los Lagos has three stygobiont, anchialine congeners – *S. balearicus* from the Balearic Islands (Mediterranean), *S. campaneri* from Palau (Western Pacific) and *S. scottodicarloi* from Bermuda (Western Atlantic) (Jaume and Boxshall, 1996a). The circumtropical distribution of species in this genus in the entire region occupied by the former Tethys Sea suggests that they are relicts of an ancient (late Mesozoic) and once widely distributed shallow-water fauna (Jaume and Boxshall, 1996a). The distribution of these species involved vicariant processes associated with the motion of tectonic plates including the opening of the Atlantic (~120 10^6 years BP) and the closure of the Tethys Sea (~20 10^6 years BP).

Unidentified copepodid IV (Misophriidae, Copepoda). This immature specimen could not be fully described but it clearly represents a fifth misophrioid genus from this cave. Although these genera differ considerably in morphology, they might be derived from a common ancestor that colonized separate and older cavernicolous habitats on the island (Boxshall and Iliffe, 1987). Following allopatric speciation and subsequent generic separation in isolated caves, dispersal with rising sea levels could have led to their colonization of this newly formed cave.

Stygocyclopia balearica (Pseudocyclopiidae, Copepoda). This species, previously reported only from anchialine caves in the Balearic Islands, was found in Cueva de los Lagos and the Túnel de Atlantida (Damià Jaume, pers. commun.).

Neoechinophora karaytugi (Superornatiremidae, Copepoda). *Neoechinophora karaytugi,* in the recently described harpacticoid copepod family Superornatiremidae, has three congeners inhabiting anchialine caves in Bermuda and one from anchialine caves in the Balearic Islands (Huys, 1996; Jaume, 1997). The amphi-Atlantic/Mediterranean distribution of these species is interpreted to indicate a Tethyan origin (Jaume, 1997).

Stenopus spinosus (Stenopodidea, Natantia). Specimens of this conspicuous, red-colored shrimp have been found only at the Montaña de Arena. It is probably a troglophile.

Munidopsis polymorpha (Galatheidae, Anomura). The galatheid crab *M. polymorpha* is the most conspicuous species in the inundated parts of the lava tube (Figs. 32.5, 32.6) (Koelbel, 1892; Fage and Monod, 1936). It is the only species within this genus that can be observed and collected in shallow water (Calman, 1904; Wilkens and Parzefall, 1974). All other species of *Munidopsis* are typically distributed at depths of more than 2000 m.

The crab has a maximum length from rostrum to telson of 3 cm. Typically, it is completely pale, but, depending on the carotene content of its food, a yellowish coloration may be present. The eye stalks are reduced to tiny vestiges, so that *M. polymorpha* is incapable of vision. Light perception, however, is well developed.

Munidopsis polymorpha picks up food with the chelae of the first pereiopods. As a result of a larger food supply, the crab reaches its highest population density in the Jameos del Agua. The distribution pattern achieved here resembles a net with the mesh size regulated by aggressive behavior (Fig. 32.6) (Parzefall and Wilkens, 1975). The feeding area defended by an individual is determined by the length of its second antennae. These are steadily moved around in circular fashion. When contact with another individual occurs, usually the smaller one will try to escape or aggressive behavior is displayed.

Reproduction is characterized by maternal care. Large, yolk-rich eggs, numbering from one to ten, with a mean of three, are carried beneath the female pleon. Eggs are continuously cleaned by the fifth pereiopods. The hatching larva, a zoea, still contains yolk. It drops to the substrate but is not able to perform local movements to hide or take refuge from predators. Only after several molts does a tiny subadult develop, which begins activities typical of the adult crabs (Wilkens et al., 1990).

Liagoceradocus acutus (Gammaridea, Amphipoda). The genus *Liagoceradocus* contains four other species, *L. lonomaka*, a blind anchialine species from brackish lava pools in the Hawaiian Islands, *L. pusillus* from marine algae in the Caroline Islands, and *L. branchialis* and *L. subthalassicus* from brackish limestone caves in

Western Australia. It is present in the Jameos del Agua and the Túnel de Atlantida.

Spelaeonicippe buchi (Pardaliscidae, Amphipoda). *Spelaeonicippe buchi* has one congener, *S. provo*, from an anchialine cave in the Caicos Islands (Wilkens et al., 1986). These species belong to the primarily deep-sea family Pardaliscidae. *Spelaeonicippe buchi* is present in all aquatic parts of the lava tube.

Parhyale multispinosa (Talitroidea, Amphipoda). This amphipod is endemic to the Jameos del Agua. It is closely related to the cosmopolitan species *P. hawaiiensis*. It does not show specific adaptations to subterranean life, and is abundant in the Jameos del Agua lake and at the beginning of the Túnel de Atlantida. Furthermore, it was discovered in some anchialine pools north of the Jameos del Agua, in the Malpais de la Corona (Wilkens et al., 1993).

Halophiloscia canariensis (Oniscoidea, Isopoda). *Halophiloscia canariensis* inhabits supralitoral rocks bordering the entrance pool of the Túnel de Atlantida.

Curassanthura canariensis (Paranthuridae, Isopoda). Along with *C. canariensis*, the genus *Curassanthura* contains two other species, *C. bermudensis* in anchialine caves in Bermuda and *C. halma* from marine groundwaters in Curaçao (Wilkens et al., 1986). *Curassanthura canariensis* is found within gravel at the edge of the Jameos del Agua lake, close to the Jameo Grande (Wägele, 1985).

Heteromysoides cotti (Heteromysini, Mysidacea). *Heteromysoides cotti* has affinities to the sponge-dwelling mysid, *H. spongicola*, from Cuba. This mysid is present in the water column from all parts of the submerged lava tube, but is most abundant in the Jameos del Agua.

Halosbaena fortunata (Halosbaenidae, Thermosbaenacea). *Halosbana fortunata* is the first fully marine thermosbaenacean to be reported. It has two congeners, *H. acanthura*, which inhabits coastal groundwater in Curaçao and *H. tulki* from a low-salinity limestone cave in Western Australia (Wagner, 1994). All specimens were collected from the water column within the Túnel de Atlantida.

Eupolycope pnyx (Polycopinae, Ostracoda). The genus *Eupolycope* contains eight species, with only *E. pnyx* inhabiting caves (Kornicker and Iliffe, 1995). All other species in the genus are found in open marine environments in the Northern Hemisphere. Specimens of *E. pnyx* were collected from the water column in the Túnel de Atlantida.

Eusarsiella bedoyai (Sarsiellidae, Ostracoda). This is the second species of *Eusarsiella* to be described from anchialine caves. *Eupolycope pnyx* has been collected from six anchialine caves in Bermuda on ten different occasions over a five-year period (Kornicker and Iliffe, 1989), whereas *E. bedoyai* has been present in two collections from the Montaña de Arena on different dates. Most of the 50+ species from this genus inhabit sandy substrates in shallow waters on the continental shelf. *Eusarsiella bedoyai* and *E. pnyx* are probably troglophilic species living in habitats linking the seabed and cave environments (Baltanas, 1992).

Danielopolina phalanx (Thaumatocyprididae, Ostracoda). While *D. carolynae* was collected from a depth of 3459 m in the South Atlantic, all other *Danielopolina* species inhabit anchialine caves. These include cave species in the Bahamas, Cuba, Galapagos Islands, Jamaica, Yucatan, and Western Australia, along with two species from the Túnel de Atlantida. The Lanzarote lava tube is the only location containing two species of *Danielopolina*. *Danielopolina phalanx* has so far only been collected from the water column in the Túnel de Atlantida.

Danielopolina wilkensi (Thaumatocyprididae, Ostracoda). Specimens of *D. wilkensi* were collected from Cueva de los Lagos and the Túnel de Atlantida (Kornicker and Iliffe, 1995), as well as from wells 16 km away from the lava tube (Wilkens et al., 1986). One specimen was collected in a baited trap placed at a depth of 24 m in the Túnel de Atlantida. The guts of most specimens contained minute amber or brown organic particles. No recognizable fragments of animals were identifiable. Kornicker and Iliffe (1995) concluded that this species apparently feeds upon detritus.

Speleonectes ondinae (Speleonectidae, Remipedia). Besides *S. ondinae*, other species in this exclusively anchialine genus include *S. benjamini*, and *S. lucayensis* from several islands in the Bahamas, *S. gironensis* from Cuba, and *S. tulumensis* from Yucatan. *Speleonectes ondinae* has been collected from the Túnel de Atlantida and from Cueva de los Lagos. It is always observed actively swimming in the water column, and even in captivity it retains this behavior. Remipedes are most abundant in the large collapse chamber connecting upper and lower levels located about 250 m into the Túnel de Atlantida. Remipedes are believed to be raptorial predators.

Ecological and biogeographic relationships

Ecologically, the fauna of the lava tube can be subdivided into endemic troglobites showing pronounced reduction of eyes and pigment, troglophiles maintaining permanent populations within the cave but also living in suitable habitats outside, and accidentals that can survive but not reproduce and proliferate within the cave. At least 24 troglobitic species inhabit the cave: *Curassanthura canariensis, Danielopolina phalanx, D. wilkensi, Dimisophria cavernicola, Expansophria dimorpha, Fauveliopsis jameoaquensis, Gesiella jameensis, Halophiloscia canariensis, Heteromysoides cotti, Liagoceradocus acutus, Macrochaeta* sp., *Mesonerilla diatomeophaga, Muceddina multispinosa, Munidopsis polymorpha, Neoechinophora karaytugi, Oromiina fortunata, Palpophria aestheta, Paramisophria reducta, Protodrilus* sp., *Spelaeonicippe buchi, Speleobregma lanzaroteum, Speleonectes ondinae, Speleophriopsis canariensis* and *Stygocyclopia balearica.* In addition, another as yet undescribed species, a misophrioid copepodid, is likely to be a troglobite. Other copepods are still under study, but are likely to include troglobitic representatives. Seven species of troglophiles inhabit the cave: *Bonellia viridis, Eupolycope pnyx, Eusarsiella bedoyai, Parhyale multispinosa, Peringiella epidaurica, Stenopus spinosus,* and *Syllis cornuta.* Accidental species consist of plankton swept into the cave by tidal currents and normal marine species surviving only in the area of direct sea water input at the Montaña de Arena. Plankton include the ctenophore *Cestum veneris,* Foraminifera, and cnidarians, while marine species include unidentified sponges, echinoids, holothurians, ophiuroids, flatworms, gastrotichs, priapulids, nematodes, and molluscs (García-Valdecasas, 1985). Several ubiquitous species of polychaetes with normal pigment and eyes that inhabit the lava tube include *Exogone (Parexogone) gambiae, Miscellania dentata, Paradoneis lyra, Protodorvillea kefersteini, Syllis cornuta, S. garciai*, and *S. truncata cryptica* (Núñez et al., 1997). These are probably accidental species, able to survive in only some sections of the cave such as the Jameos del Agua.

Examination of the biogeographic affinities of the lava tube fauna shows two main patterns. First, a sizable number of anchialine genera present in Lanzarote have amphi-Atlantic distributions or even have representatives from the Pacific. Amphi-Atlantic genera include the ostracod *Danielopolina*, the remipede *Speleonectes,* the thermosbaenacean *Halosbaena,* the mysid *Heteromysoides,* the isopods *Curassanthura* and *Halophiloscia,* and the amphipod *Spelaeonicippe.* Furthermore, some genera, including the copepods *Expansophria*, *Paramisophria* and *Speleophriopsis* and the amphipod *Liagoceradocus,* as well as *Danielopolina* and *Halosbaena,* also occur in anchialine habitats in the Pacific. Second, many anchialine species from Lanzarote belong to taxa that primarily inhabit the deep sea. The fauna with deep-sea associations include the polychaete *Gesiella*, the copepods *Dimisophria, Expansophria, Palpophria* and *Speleophriopsis,* and the galatheid *Munidopsis,* along with *Danielopolina* and *Spelaeonicippe.*

It has been suggested that the amphi-Atlantic distribution of species results from cave colonization during an early stage in the formation of the Atlantic followed by dispersal by sea-floor spreading and continental drift (Iliffe et al., 1984; Hart et al., 1985; Wägele, 1985). Some genera including *Danielopolina* and *Expansophria* which are present in the Galapagos Islands may have entered the Pacific before the closure of the Panama land bridge (Iliffe, 1991). The presence of some anchialine taxa such as *Danielopolina, Expansophria,* and *Liagoceradocus* in the western Pacific presents more of a problem. Particularly noteworthy has been the discovery of numerous anchialine taxa, originally thought to be limited to or known primarily from the Atlantic and Caribbean, in caves in the Cape Range of Western Australia (Humphreys and Adams, 1991; Humphreys, 1993). The wide range and overlapping distributions of such taxa suggest similar dispersal mechanisms and simultaneous colonization of the cave habitat during the Tethyan period.

A deep-sea origin of some anchialine taxa has been suggested by Hart et al. (1985), Manning et al. (1986) and Iliffe (1990). Cladistic analysis of misophrioid copepods indicates that species inhabiting both Atlantic and Pacific caves are descended from deep-sea ancestors (Boxshall, 1989). However, a deep-sea origin of cave faunas has been deemed unlikely by Stock (1986) and Danielopol (1990).

In any case, the Lanzarote lava-tube fauna is obviously of considerable age, indicating that these species must have existed in alternate habitats on Lanzarote before the formation of the lava tube. The island is underlain by much older rocks that date back to the Tertiary Period. As large lava tubes are relatively short-lived and not altogether common, it is most likely that the present-day anchialine fauna survived in a crevicular habitat on Lanzarote prior to the formation of the cave they now inhabit. This is substantiated by the presence of many lava-tube species in groundwater accessible through artificial wells on Lanzarote (Wilkens et al., 1986).

Species distribution and food supply

The marine fauna inhabiting the lava tube is not homogeneously dispersed. Species distribution and population density vary in the diverse sections and are mainly influenced by trophic factors. Each of the three main sections of the lava tube appear to have separate and distinct sources of food input (Fig. 32.7).

The submarine Túnel de Atlantida is mainly supplied by faunal and algal material, usually small planktonic organisms, carried in by the tidally exchanging waters. Harms (1921) observed one specimen of the ctenophore *Cestum veneris* as far inside as the Jameos del Agua. Furthermore, the occurrence of planktonic cnidarians has been repeatedly reported.

Although planktonic organisms may drift far up the lava tube, the main trophic source for Cueva de los Lagos is probably organic debris derived from the overlying lava field. Seasonal rainfall and tidal movements are believed to wash organic particles down from the surface to the crevicular groundwater system surrounding the lava tube. The presence of this particulate matter may explain the occurrence of gastropods.

In contrast to the other parts of the lava tube, the Jameos del Agua is characterized by an enriched food supply. Here, daylight from two large ceiling openings penetrates to the anchialine lake, enabling primary production to occur. In addition, the prevailing strong winds blow a significant amount of organic material directly into the Jameos del Agua lake from the surface of the lava field.

As a result of these greater food resources, some species including *Bonellia viridis, Heteromysoides cotti, Munidopsis polymorpha* and *Parhyale multispinosa* have attained very high densities in the Jameos del Agua. The latter three species are benthic and littoral, feeding on detritus and diatoms. In addition, the polychaete *Mesonerilla diatomeophaga* feeds on the carpet of diatoms covering rocks in the Jameos del Agua.

Bonellia viridis and *P. multispinosa* occur exclusively in the Jameos del Agua and the entrance pool of the Túnel de Atlantida. Conversely, other species seem to be restricted to the Cueva de los Lagos and/or the Túnel de Atlantida. Supplemental to food availability, light plays an important role in determining the occurrence of species inhabiting the darker, shaded habitats in the Jameos del Agua.

Munidopsis polymorpha and *P. multispinosa* avoid bright light and are primarily nocturnal. The degree of scotophilia characteristic of these species enables them to be active in the twilight of the Jameos del Agua. In contrast to these forms, all other lava-tube species have reduced eyes and pigment, and appear to be extremely scotophilic.

ACKNOWLEDGEMENTS

Diving expeditions to the Túnel de Atlantida in 1983, 1992 and 1994 by T. Iliffe were supported by grants from the National Science Foundation (BSR-8215672), National Academy of Sciences, National Geographic Society and the Texas Institute of Oceanography. We thank the Cabildo Insular de Lanzarote for providing invaluable assistance to our scientific studies. This chapter is dedicated to the memory of cave diver Sheck Exley, a member of the 1983 and 1992 expeditions.

REFERENCES

Baltanas, A., 1992. *Eusarsiella bedoyai* (Myodocopida, Sarsiellidae), a new ostracode species from a marine lava cave in the Canary Islands. *Bijdr. Dierkd.*, 61: 251–255.

Bertelsen, R.D., 1986. *Speleobregma lanzaroteum,* a new genus and species of Scalibregmatidae (Polychaeta) from a marine cave in the Canary Islands. *Proc. Biol. Soc. Washington,* 99: 375–379.

Boxshall, G.A., 1989. Colonization of inland marine caves by misophrioid copepods. *J. Zool.*, 219: 521–526.

Boxshall, G.A. and Iliffe, T.M., 1987. Three new genera of misophrioid copepods (Crustacea) from anchialine caves on Indo-West Pacific and North Atlantic Islands. *Zool. J. Linn. Soc.,* 91: 223–252.

Boxshall, G.A. and Iliffe, T.M., 1990. Three new species of misophrioid copepods from oceanic islands. *J. Nat. Hist.,* 24: 595–613.

Bravo, T., 1964. *El volcán y el malpaís de la Carona. La "Cueva de los Verdes" y los "Jameos".* Publicaciones de Cabildo Insular de Lanzarote, Arrecife, 31 pp.

Calman, W.T., 1904. On *Munidopsis polymorpha* Koelbel, a cave dwelling marine crustacean from the Canary Islands. *Ann. Mag. Nat. Hist.,* 10: 127–131.

Danielopol, D.L., 1990. The origin of anchialine cave fauna – the "deep sea" versus the "shallow water" hypothesis tested against empirical evidence of the Thaumatocyrididae (Ostracoda). *Bijdr. Dierkd.,* 60: 137–143.

Exley Jr., S., 1983. Lanzarote Volcanic Cave Expedition. *Explor. J.,* 61(3): 118–123.

Fage, L. and Monod, T., 1936. La faune marine du Jameo de Agua, lac souterrain de l'isle de Lanzarote (Canaries). *Arch. Zool. Exp.,* 78: 97–113.

García-Valdecasas, A., 1985. Estudio faunístico de la cueva submarina "Túnel de la Atlántida", Jameos del Agua, Lanzarote. *Nat. Hisp.,* 27: 1–56.

Harms, W., 1921. Das rudimentäre Sehorgan eins Höhlendekapoden, *Munidopsis polymorpha* Koelbel. *Zool. Anz.*, 52: 101–115.

Hart Jr., C.W., Manning, R.B. and Iliffe, T.M., 1985. The fauna of Atlantic marine caves: Evidence of dispersal by sea floor spreading while maintaining ties to deep waters. *Proc. Biol. Soc. Washington,* 98: 288–292.

Humphreys, W.F., 1993. Stygofauna in semi-arid tropical Western Australia: A Tethyan connection?, *Mém. Biospéol.*, 20: 111–116.

Humphreys, W.F. and Adams, M., 1991. The subterranean aquatic fauna of the North West Cape peninsula, Western Australia. *Rec. West. Aust. Mus.*, 15: 383–411.

Huys, R., 1996. Superornatiremidae fam. nov. (Copepoda, Harpacticoida) – an enigmatic family from North Atlantic anchihaline caves. *Sci. Mar.*, 60: 497–542.

Iliffe, T.M., 1990. Crevicular dispersal of marine cave faunas. *Mém. Biospéol.*, 17: 93–96.

Iliffe, T.M., 1991. Anchialine cave fauna of the Galápagos Islands. In: M.J. James (Editor), *Galápagos Marine Invertebrates.* Plenum Press, New York, pp. 209–231.

Iliffe, T.M., 1993. Fauna Troglobia Acuática de la Península de Yucatán. In: S.I. Salazar-Vallejo and N. Emilia Gonzáles (Editors), *Biodiversidad Marina y Costera de México.* Commision Nacional para la Biodiversidad y CIQRO, México, pp. 673–686.

Iliffe, T.M., 1994. Bermuda. In: C. Juberthie and V. Decu (Editors), *Encyclopaedia Biospeologica,* Vol. 1. Société de Biospéologie, Moulis (France) and Bucarest, pp. 417–424.

Iliffe, T.M., Hart Jr., C.W. and Manning, R.B., 1983. Biogeography and the caves of Bermuda. *Nature,* 302: 141–142.

Iliffe, T.M., Wilkens, H., Parzefall, J. and Williams, D., 1984. Marine lava cave fauna: composition, biogeography, and origins. *Science,* 225: 309–311.

Isler, O., 1989. 1986 international expedition to the Tunel de la Atlantida. *Caves & Caving,* 45: 16–21.

Jaume, D., 1997. First Record of Superornatiremidae (Copepoda, Harpacticoida) from Mediterranean waters, with description of three new species from Balearic anchihaline caves. *Sci. Mar.,* 61: 131–152.

Jaume, D. and Boxshall, G.A., 1996a. The persistence of an ancient marine fauna in Mediterranean waters – new evidence from misophrioid copepods living in anchihaline caves. *J. Nat. Hist.*, 30: 1583–1595.

Jaume, D. and Boxshall, G.A., 1996b. Two new genera of cyclopinid copepods (Crustacea) from anchihaline caves on Western Mediterranean and Eastern Atlantic Islands. *Zool. J. Linn. Soc.*, 117: 283–304.

Jaume, D. and Boxshall, G.A., 1997. Two new genera of cyclopinid copepods (Cyclopoida: Cyclopinidae) from anchialine caves of the Canary and Balearic Islands, with a key to genera of the family. *Zool. J. Linn. Soc.*, 120(1): 79–101.

Juberthie, C. and Iliffe, T.M., 1994. Bahama Islands. In: C. Juberthie and V. Decu (Editors), *Encyclopaedia Biospeologica,* Vol. I. Société de Biospéologie, Moulis (France) and Bucarest, pp. 447–458.

Koelbel, K., 1892. Beiträge zur Kenntniss der Krustaceen der Kanarischen Inseln. *Ann. K.K. Naturhist. Hofmus.*, 7, 105 pp.

Kornicker, L.S. and Iliffe, T.M., 1989. Ostracoda (Myodocopina, Cladocopina, Halocypridina) from anchialine caves in Bermuda. *Smithson. Contrib. Zool.*, 475: 1–88.

Kornicker, L.S. and Iliffe, T.M., 1995. Ostracoda (Halocypridina, Cladocopina) from anchialine lava tube in Lanzarote, Canary Islands. *Smithson. Contrib. Zool.*, 568: 1–32.

Macau, F., 1965. Tubos volcánicos en Lanzarote "La Cueva de los Verdes". *An. Estud. Atlant.*, 11.

Manning, R.B., Hart Jr., C.W. and Iliffe, T.M., 1986. Mesozoic relics in marine caves of Bermuda. *Stygologia,* 2: 156–166.

Méndo, A. and Ortega, L., 1988. Retorno a la prehistoria. *Geo,* 14(Marzo 1988): 9–24; 116–117.

Núñez, J., Ocaña, O. and Brito, M.C., 1997. Two new species (Polychaeta: Fauveliopsidae and Nerillidae) and other polychaetes from the marine lagoon cave of Jameos del Agua, Lanzarote (Canary Islands). *Bull. Mar. Sci.*, 60: 252–260.

Ohtsuka, S., Fosshagen, A. and Iliffe, T.M., 1993. Two new species of *Paramisophria* (Copepoda, Calanoida, Arietellidae) from anchialine caves on the Canary and Galapagos Islands. *Sarsia,* 78(1): 57–67.

Parzefall, J. and Wilkens, H., 1975. Zur Ethologie augenreduzierter Tiere. Untersuchungen an *Munidopsis polymorpha* Koelbel (Anomura, Galatheidae). *Ann. Speleol.*, 30: 325–335.

Sket, B. and Iliffe, T.M., 1980. Cave fauna of Bermuda. *Int. Rev. Gesamten Hydrobiol.*, 65: 871–882.

Stock, J.H., 1986. Deep sea origin of cave faunas, an unlikely supposition. *Stygologia,* 2: 105–111.

Vieira, R., Van Ruymbeke, M., Fernández, J., Arnoso, J. and Toro, C., 1991. The Lanzarote underground laboratory. *Cah. Cent. Eur. Geodyn. Seismol.*, 4: 71–86.

Wägele, J.W., 1985. On the Tethyan origin of the stygobiont Anthuridea *Curassanthura* and *Cyathura* (*Stygocyathura*), with a description of *Curassanthura canariensis* n. sp. from Lanzarote (Crustacea, Isopoda). *Stygologia,* 1: 258–269.

Wagner, H.P., 1994. A monographic review of the Thermosbaenacea (Crustacea: Peracarida). *Zool. Verh.*, 291: 1–338.

Wilkens, H. and Parzefall, J., 1974. Die Ökologie der Jameos del Agua (Lanzarote). Zur Entwicklung limnischer Höhlentiere aus marinen Vorfahren. *Ann. Spéléol.*, 29: 419–434.

Wilkens, H., Parzefall, J. and Iliffe, T.M., 1986. Origin and age of the marine stygofauna of Lanzarote, Canary Islands. *Mitt. Hamb. Zool. Mus. Inst.*, 83: 223–230.

Wilkens, H., Parzefall, J. and Ribowski, A., 1990. Population biology and larvae of the anchialine crab *Munidopsis polymorpha* (Galatheidae) from Lanzarote (Canary Islands). *J. Crustacean Biol.*, 10(4): 667–675.

Wilkens, H., Parzefall, J., Ocaña, O. and Medina, A.L., 1993. La fauna de unos biotopos anquialinos en Lanzarote (I. Canarias). *Mém. Biospéol.*, 20: 283–285.

Part VIII

Conservation

Chapter 33

CONSERVATION OF CAVE COMMUNITIES IN AUSTRALIA

Elery HAMILTON-SMITH and Stefan EBERHARD

INTRODUCTION

Despite a high level of interest in caves during the 19th century, and some promising early beginnings in the study of cave biology in Australia (e.g., Higgins and Petterd, 1883; Urquhart, 1889; Rainbow, 1904, 1905, 1907; Lea, 1910), research on cave fauna was extremely limited until about 1970. Until very recently, Australian cave communities have been perceived as being depauperate compared with the well-studied communities of Europe and North America, especially with regard to the representation of taxa that display a high degree of troglomorphism.

A higher proportion of troglobitic species, perhaps comparable with those in some Northern Hemisphere countries, was recognized only in Tasmania, with its more humid climatic history and episodes of glaciation. However, during the last decade in particular this traditional perception has been overturned by dramatic discoveries of highly diverse communities containing very diverse and ancient troglomorphic taxa.

It must be appreciated that most Australian karst areas have not been subject to any thorough or systematic biological examination, and that much of the known fauna remains undescribed. This in itself is a major impediment to conservation planning and management. Further, some cave communities have been lost entirely, and a great many more have been seriously degraded or are threatened by human activities.

This chapter describes the biophysical characteristics of the various karst environments and their cave communities, then details major conservation issues and the nature of impacts threatening these communities, and concludes with a discussion of potential conservation strategies.

OVERVIEW

There are a great many similarities between the cave communities of the Australian mainland and those elsewhere, but there are also some fundamental differences in the systematic composition of the fauna, due at least in part to the unique geologic and climatic history of the continent. In the prehistoric past Australia was part of the southern super-continent Gondwana, and was attached to Antarctica. These lands shared a Gondwanan flora and fauna, and when the final separation between them occurred (45 million years ago), both lands were well watered and supported cool temperate and subtropical forests (White, 1994). As Australia separated from Antarctica and drifted northwards, the continent began to dry out, so that its forests contracted and were replaced with arid-adapted vegetation.

The formation of the Antarctic ice cap 15 million years ago saw the beginning of a series of marked climatic fluctuations that have greatly stressed the Australian (and other Gondwanan) flora and fauna. Warm and wet interglacial periods alternated with very dry, cool and windy glacial stages, but only a small area of the Eastern Highlands and Tasmania were subject to extensive ice cover. The cyclic fluctuation of the climate in the late Cainozoic, superimposed upon a generally increasing and spreading aridity over the last 2.4 million years, provided conditions under which subterranean refugia played an important role (Flannery, 1994).

Many Australian karst landscapes are quite ancient, and may have been available for colonization by animals over a very long period of time. Palaeokarst features of Devonian age (350 to 400 million years ago) are known in New South Wales for instance, although subsequent episodes of burial, marine transgressions, flooding, sedimentation and major climate changes may well have eliminated faunas from that time (Osborne and Branagan, 1988). The occurrence of highly modified stygobiontic taxa belonging to morphologically conservative groups, such as the Anaspidacea, imply that subterranean freshwater environments have been continuously available for a very long period, possibly since the end of the Palaeozoic era. The distribution of remipede Crustacea and the fauna associated with them can be explained by cave colonization along the shores of the Tethys Seaway during the Triassic and Jurassic (225–160 million years ago), whereas, during the early Cretaceous, the plate that is now Western Australia formed the eastern shore of Greater Tethys (Yager and Humphreys, 1996). Bats have been present in Australia since the Miocene, and they are found in the cave fossil record of some two million years ago. Available methods of dating cave sediments are generally restricted to the last few millions of years, but the occurrence of isolated caves of probable phreatic origin situated high above valley floors of the present-day landscape suggest great ages for such sites.

The composition of Australian cave communities at the higher systematic levels is broadly similar to that of other continents, but frequently involves taxa that are uniquely southern in origin, such as the phreatoicidean isopods and anaspidacean syncarids. The communities naturally contain a high proportion of taxa with Gondwanan affinities, although others, such as the Remipedia, have Pangean or Tethyan affinities. Like cave faunas elsewhere, many taxa are phylogenetic or distributional relicts. There is a very high degree of endemism, with many taxa displaying vicariant distribution patterns. Other obligate cave taxa have extant relatives dwelling in nearby surface habitats. Many of the cave-dwelling taxa are derived from the terrestrial or aquatic habitats found within mesic forest environments. Some of the obligate terrestrial cave taxa from temperate regions are derived from only weakly hygrophilous groups, and many are only slightly troglomorphic. The majority of stygobionts have freshwater origins, although some have a marine origin. Troglophiles and trogloxenes make up a substantial portion of the diversity within many cave communities. Bats are widespread in continental Australian caves, and their guano supports a rich assemblage of guanophilic invertebrates.

CAVE-FAUNA PROVINCES OF AUSTRALIA

Although the proportionate area of carbonate rock is less in Australia than the world average (Gillieson and Spate, 1999), there are about 500 discrete localities documented (Fig. 33.1). In addition, there are cave and groundwater communities established within non-carbonate rocks such as basalt, sandstone, quartzite, acid igneous rocks, unconsolidated sediments and others. These include lava tunnels, sea caves, boulder and talus caves, aeolian caves, piping caves beneath silcrete crusts, and interstitial environments.

Karst, pseudokarst and groundwater localities are distributed across the Australian continent, which encompasses climate types ranging from tropical through to temperate. We have used regional climates, which reflect the surface vegetation and animal communities, as a basis for delineating four major cave-fauna provinces within Australia. In most cases, these regions also each have a dominant pattern of karst landforms. As defined by Bridgewater (1987) and depicted in Fig. 33.1, these provinces are: a tropical-climate zone (II); a subtropical dry-climate zone (III); a transitional zone with winter rain (IV); and a warm-temperate zone extending to a warm-temperate/tropical transitional climate zone (V, II–V). The biophysical characteristics of the karst areas and cave communities within each of these provinces are described below.

Tropical-climate province

The tropical-type climate occurs in the northern part of Australia, as delineated by a line running roughly east/west and intersecting the coast at Broome (latitude 18°S Western Australia) and north of Rockhampton (latitude 22°S Queensland). It is characterized by some seasonality in mean daily temperature, and monsoonal rainfall concentrated in the summer months. This province includes about a dozen major karst areas distributed across Western Australia, the Northern Territory, and Queensland. The major rock types include Precambrian dolomites, Cambrian limestones and Devonian and other Palaeozoic limestones. All of these comprise what Jennings (1969) described as the 'seasonally humid' tropical karsts of Australia. Pseudokarst features are developed in sandstones and quartzites, and there are extensive lava tunnels in Cainozoic basalts at Undara in North Queensland. The topography of the karsts is expressed as a peneplain at Katherine and on the Barkly Tableland; ranges and

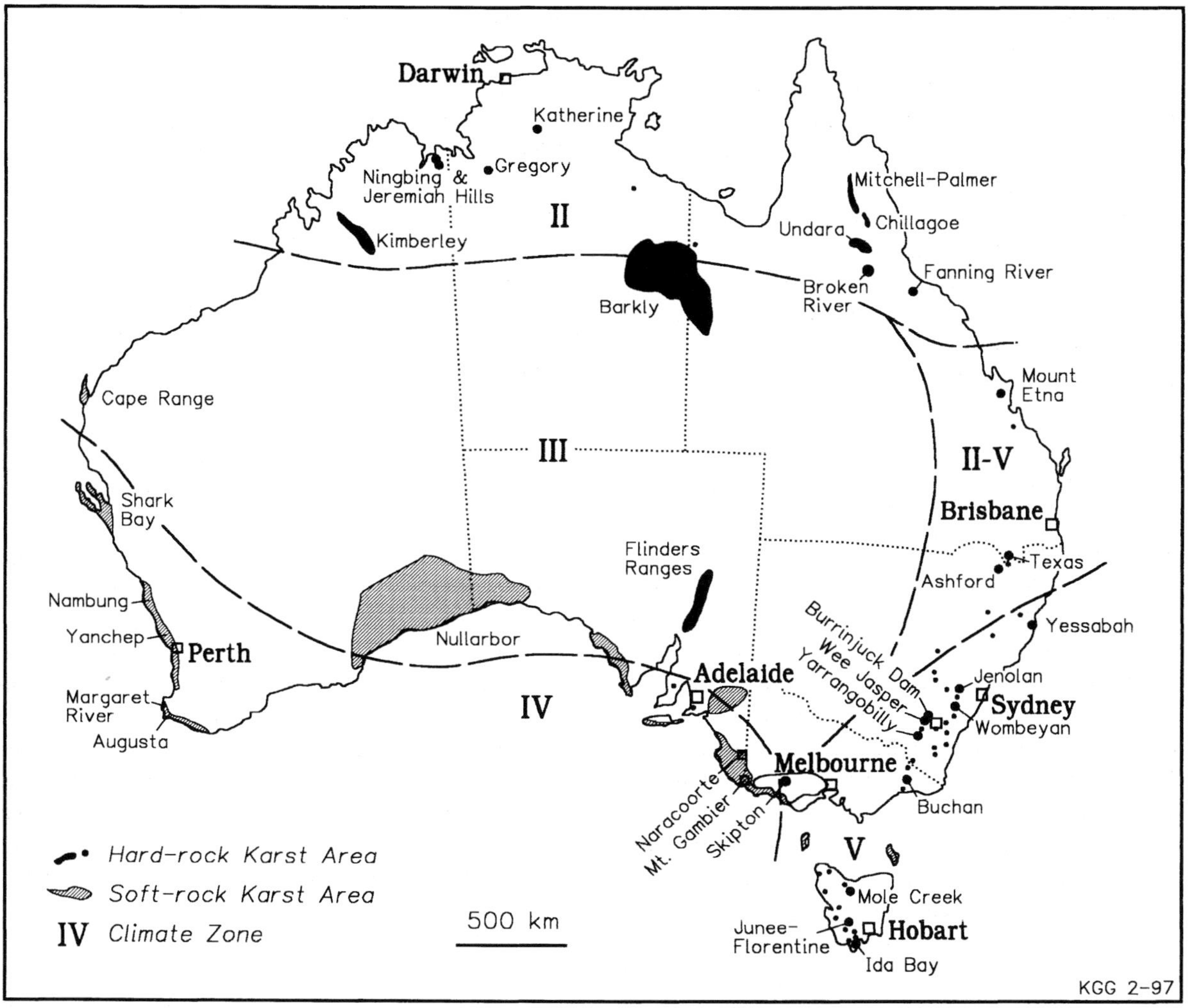

Fig. 33.1. Climatic zones and karst areas in Australia. (Source: K.G. Grimes, after Bridgewater, 1987).

dissected blocks in the Kimberley; escarpments at Colless Creek; rounded stepped hills at Gregory River; and isolated towers at Chillagoe and the Mitchell–Palmer area.

Cave development consists of predominantly horizontal passages, often in joint-controlled three-dimensional mazes with numerous daylight holes and other surface openings and with some vertical or inclined passages descending to the water table. Cave climates are warm to hot and humid. The surface vegetation types include tropical evergreen woodlands, semi-arid grasslands and shrublands, and occasional patches of remnant rainforest.

Some limited studies have been undertaken in the Barkly region and Katherine by Hamilton-Smith, and work by others has recently commenced at Gregory River, the Kimberley, and Fanning River. Extensive research has been undertaken on invertebrate populations at Chillagoe and in the lava tunnels at Undara, where 40 species of predominantly terrestrial troglobites have been recorded (Howarth, 1988). At these sites, studies on speciation and reductive evolutionary trends in planthoppers (Fulgoroidea: Cixiidae and Meenoplidae), where closely related congeners dwell in nearby surface habitats, has lent supportive evidence to Howarth's adaptive-shift hypothesis for the evolution of cave species. (See Howarth, 1987, 1993; Hoch and Asche, 1988; Stone 1988; Hoch and Howarth, 1989a,b; Howarth and Stone, 1990).

Many of the tropical cave systems consist of spacious

interconnected passages with multiple entrances, so they are poorly insulated from daily and seasonal climate changes on the surface. With the relatively arid tropical winter (Howarth, 1980), much of the interior environment of these caves consists of an extended transition zone, which correspondingly influences the distribution of cave fauna. The faunal communities of the transition zone are relatively depauperate, and generally consist almost entirely of non-troglomorphic species. Even the guanophile communities are often of low diversity, the taxa including oniscoid isopods; scutigerid centipedes; various families of spiders; cockroaches; reduviid bugs; and beetles belonging to the families Carabidae, Staphylinidae, Tenebrionidae, and Trogidae. The origins and affinities of many of the terrestrial cavernicolous taxa lie with a tropical rainforest fauna. Terrestrial troglobites are found in the deeper, poorly ventilated recesses that often have elevated concentrations of carbon dioxide and depressed concentrations of oxygen. The so-called "stagnant-air" zone first described by Howarth and colleagues from lava tunnels at Undara contains a diverse community of highly troglomorphic taxa. Remarkably, 24 of the 51 resident species in Bayliss Cave are troglobites (Malipatil and Howarth, 1990).

Stygobiontic taxa are distributed throughout the tropical province. They have freshwater origins and are generally relictual. They include amphipods, copepods, atyid decapods, and a flabelliferan isopod of uncertain affinities (Williams, 1964; Dumont and Maas, 1985; Bruce, 1992; Wilson and Ponder, 1992). Vertebrates entering the caves include pythons and other snakes, swiftlets (*Aerodramus spodiopygius*), and many species of bats.

Subtropical dry-climate province

The subtropical dry-climate zone occurs across much of Central Australia, and is characterized by very low rainfall, high daytime temperatures in summer, and low winter minima. Given the size of this area, carbonate rocks are poorly represented, with karst development retarded by the arid climate, although the Nullarbor Plain is one of the largest areas of continuously exposed limestone in the world. There is no one dominant karst land-form; karsts are excavated in narrow bands of Cambrian dolomites and limestones in the Flinders Ranges of South Australia, an immense plain of relatively soft and unmetamorphosed Tertiary limestones on the Nullarbor, and a tectonically complex orogenic tertiary limestone at Cape Range and on Barrow Island in Western Australia. The karst landscape is expressed as flat plains, islands, and inland ranges, and as a range with a fringing coastal plain (Cape Range). Cape Range and Barrow Island are faunistically distinct from other areas, in that the terrestrial biota is relictual from the tropical zone. Cave development includes vertical potholes, extensive horizontal passages and mazes, water-filled passages, and anchialine habitats. Small caves resulting from piping of unconsolidated materials from below the broken silcrete or laterite crusts of this region are numerous but little studied. The surface vegetation includes evergreen woodland, semi-arid grassland and shrublands.

Cape Range with the related Barrow Island comprises perhaps the most comprehensively researched karst region in Australia, and from a biological perspective is certainly the most significant. A total of 66 troglobite or stygobiont species have been, or are being, described from 35 families and 46 genera (Humphreys, personal communication, 1996). These taxa include at least 24 aquatic species and some 42 terrestrial species. The terrestrial fauna is primarily a relict wet-forest litter fauna, whilst the aquatic fauna has Tethyan affinities (Humphreys 1993; see also Chapter 30, this volume). The remarkable stygobiontic fauna includes Australia's only two troglobitic fish, together with a remarkable assembly of crustaceans – atyids, melitid and other amphipods, an isopod, a bathynellacean, copepods, the first thermosbaenacean and the first remipede from the Southern Hemisphere.

Considerable collecting has been undertaken on the Nullarbor Plain, but only 15 species of terrestrial troglobites have been recorded from the entire area. These species, which appear to be distributional relicts dating from times of more humid climate, are highly troglomorphic nevertheless, and include two cockroaches, an isopod, a scolopendrid, two carabid beetles, a mygalomorph spider and several other spiders that have congeneric cave-dwelling relatives located in distant eastern and western margins of the continent. The Nullarbor caves and dolines shelter troglophilic and trogloxenic invertebrates, many of which are associated with guano of the chocolate wattled bat (*Chalinolobus morio*). They include harvestmen, mites, pseudoscorpions, spiders, isopods, rhaphidophorid crickets, cockroaches, Diptera, and beetles belonging to the families Anisotomidae, Carabidae, Dermestidae, Histeridae, Lathridiidae, Ptinidae,

Scarabaeidae, Staphylinidae, Tenebrionidae and Trogidae (Richards, 1971).

In general, only micro-organisms have been collected from the many submerged cave passages of the Nullarbor, and investigation of these has only recently commenced. Enormous clusters of colonial microflora are present, including at least sulphate-reducing bacteria; sulphide-oxidizing species may also be present (James, 1996). The general absence of metazoans is difficult to explain given the apparently suitable habitat offered by this large aquifer. The occurrence of a melitid amphipod within Nurina Cave, situated at a low level near the coast, probably relates to a stranding event during a Pleistocene marine transgression, although this cave also has an unusually diverse population of terrestrial troglomorphs. Elsewhere in this subtropical dry-climate province, there are stygobiontic amphipods recorded from the Flinders Ranges, and atyid shrimps from the Gibson Desert, as well as molluscs and Crustacea associated with groundwater and mound-spring habitats.

The terrestrial cave fauna of the Flinders Ranges is sparse and poorly known, but includes carabid and tenebrionid beetles. The lack of studies of the caves underlying the duricrust in the deserts of the region has already been noted. However, these caves are used by many vertebrates to shelter from extreme temperatures, and so make possible an extension of range into areas where they would not otherwise survive. Research may yet reveal an important and diverse faunal community in these sites (M. Denny, personal communication, 1996).

Transitional-zone province with winter rain

The transitional zone with winter rain is characterized by almost complete summer drought; but this province receives cyclonic rains during winter. This climate type is experienced in two disjunct regions: the southwestern corner of Western Australia, and the lower southeast of South Australia and the southwestern corner of Victoria.

There are a number of important karst areas within both these regions. In Western Australia, there is a strip of coastal plains and dune ridges composed of intensely karstified Pleistocene aeolian calcarenite. Similar karsts in Miocene limestones overlain with Pleistocene calcarenites occur in the Otway Basin of Victoria and South Australia, as well as on the Eyre Peninsula and Kangaroo Island, with very small areas of Cambrian dolomites and limestones on Yorke Peninsula and in the Adelaide Hills, all in South Australia. In all of these areas, the rock is unmetamorphosed and relatively soft, with some aeolian calcarenites still in the process of lithification. Thus, this area is characterized by what Jennings (1968) has termed 'syngenetic karst'. Cave types generally comprise extensive horizontal passage networks (including mazes at or below the water table), often with extensive collapse caverns formed in the soft limestones, and sometimes with limited development of vertical shafts. There are also sea caves, cenotes and vauclusian springs. The surface vegetation is temperate forests, grasslands and shrublands, with some semi-arid grasslands and shrublands.

Few terrestrial troglobites are known, and the fauna is dominated by troglophiles or guanophile communities associated with the bat species *Chalinolobus morio* in Western Australia and *Miniopterus schreibersii* in South Australia and Victoria (Hamilton-Smith 1968, 1972). The taxa include isopods, mites, harvestmen, spiders, pseudoscorpions, cockroaches, cave crickets, psocopterans, many dipterans, tineid moths, and beetles from the families Anisotomidae, Carabidae, Histeridae, Jacobsoniidae, Lathridiidae, Ptinidae and Staphylinidae. Amongst the few troglomorphs are planthoppers (Meenoplidae) from Nambung (Western Australia) which are congeneric with species from the Cape Range, Barrow Island, the Kimberley and Chillagoe.

The aquatic communities within this province are remarkably diverse; for instance, in Yanchep Caves (Western Australia) unusual communities in tree root mats are found containing up to 25 aquatic species (Jasinska and Knott, 1991; see also Chapter 15, this volume). In the Otway Basin, the Mount Gambier cenotes and springs are very rich in aquatic plant life and associated epigean invertebrates and fishes. They also harbour a stygobiontic amphipod (Stock and Iliffe, 1990) and a species of syncarid (Zeidler, 1985).

A great diversity of stromatolitic Cyanobacteria restricted to karst occur in both the Mount Gambier district (Hallam and Thurgate, 1992; Thurgate, 1995) and Western Australia, at Shark Bay, Lake Thetis, Lake Clifton, Lake Richmond and several lakes on Rottnest Island (McNamara, 1992).

Province of warm-temperate and warm-temperate/tropical transitional climates

This province essentially coincides with the Eastern Highlands, which extend virtually the entire length of the eastern Australian seaboard, including the island of Tasmania. The warm-temperate type of climate has no strongly defined winter, with year-round rainfall, and encompasses southeastern New South Wales, central and eastern Victoria, and Tasmania. The region of northern New South Wales and southeastern Queensland is a transition zone between the warm-temperate and tropical types of climates, with higher temperatures and some monsoonal influence.

There are about 200 discrete karst localities scattered along the length of the Eastern Highlands, east coast and western slopes of the highlands, plus an additional 300 localities within Tasmania. The majority of these are excavated in Palaeozoic limestones, and comprise relatively small areas of impounded karst. In Tasmania, Precambrian dolomites are also an important karst-bearing rock type, and in western Victoria there are extensive Tertiary basalt plains containing lava tunnels. The topographic setting of the karsts is predominantly as impounded hill-flank fluviokarsts, but includes residual hills, plains, riverine, alpine and coastal contexts. The surface vegetation varies widely across the region, with subtropical wooded vegetation in the north which may be evergreen or (only on the karsts) deciduous, temperate forests, grasslands and shrublands in New South Wales and Victoria, and evergreen forests (including rainforest), heathlands and grasslands in Tasmania. From the faunistic perspective, Tasmania with its more humid climatic history and periods of glaciation is distinctive in its higher proportion of troglomorphic species. There are a great variety of cave types represented within this province, including active stream caves, dry caves, horizontally and vertically developed systems, springs, sea caves, and lava tunnels.

The cavernicolous fauna has been extensively, but still incompletely, sampled in Tasmania and New South Wales, whereas the Victorian and Queensland faunas are less well known. A large number of terrestrial and aquatic cave-obligate taxa occur, many with Gondwanan affinities and some with Pangean affinities. Tasmanian cave communities rank amongst the richest within Australia's temperate zone, with a high systematic diversity overall and a high proportion of cave-limited taxa, which may be a consequence of this island's more humid climatic history combined with episodes of Pleistocene glaciation (Eberhard et al., 1991). At the familial level, the overall richness of New South Wales cave communities is only slightly less than Tasmania (115 families from New South Wales compared with 130 families from Tasmania), although Tasmania holds approximately double the number of clearly troglobitic genera (34 in Tasmania compared with 17 in New South Wales). The terrestrial troglobites from Tasmania are often more highly troglomorphic than their counterparts in continental Australia, which suggests that they have undergone a longer period of isolation in caves. The terrestrial cave communities in New South Wales are notable for their abundance of apparently cave-limited taxa that display negligible troglomorphism, or so-called second-level troglophiles (*sensu* Hamilton-Smith, 1972). The harvestmen and pseudoscorpions have Tasmanian and Gondwanan affinities (Beier, 1967, 1968; Hunt, 1972).

Some individual cave systems in Tasmania contain more than 70 resident species, of which 15 or more are troglobites. The typical terrestrial invertebrates encountered are triaenonychid harvestmen, chthoniid pseudoscorpions, several families of spiders, stylioniscid isopods, symphylids, diplopods, springtails, rhaphidophorid crickets, carabid beetles, and Diptera (including glow-worms). All of these groups are represented in the cave communities of eastern continental Australia. Cave-dwelling bats do not occur in Tasmania, but various species, of which *Miniopterus* spp. are by far the most numerous, occupy many of the caves in New South Wales, Victoria and southeastern Queensland, whilst their guano supports a distinctive assemblage of guanophiles including mites, pseudoscorpions, spiders, diplopods, chilopods, springtails, psocopterans, cockroaches, tineid moths, hymenopterans, many dipterans, and beetles belonging to the families Anisotomidae, Carabidae, Cryptophagidae, Histeridae, Jacobsoniidae, Lathridiidae, Pselaphidae, Ptinidae, Staphylinidae and Tenebrionidae. Additional terrestrial taxa that commonly occur in the more northerly localities of the province include scutigerid centipedes, reduviid bugs, fulgoroid planthoppers, and trogid beetles. Mount Etna and nearby sites in Central Queensland have a transitional character, in that guanophiles similar to those found further south occur here, together with subtropical forms such as scutigerid centipedes, reduviid bugs, fulgoroid planthoppers and trogid beetles (Hamilton-Smith, 1970a).

The communities of stygobionts in Tasmania and New South Wales consist of distributional and/or phylogenetically relict taxa such as anaspidacean syncarids, phreatoicidean and janirid isopods, crangonyctoid amphipods, hydrobiid gastropods, and paludicolan flatworms. Both regions are comparable in terms of their

Fig. 33.2. The Mt. Etna Quarry destroyed an unknown number of caves. (Photograph D. & J. Vavryn.)

overall species richness, their degree of troglomorphy and significance to zoogeographic and evolutionary studies (Eberhard and Spate, 1995b).

THREATS TO AUSTRALIAN CAVE COMMUNITIES

Ever since European settlement of Australia a little over 200 years ago, the natural environment has endured a legacy of neglect and abuse. This has been the result of a colonial ideology fostering wholesale exploitation of the country's natural resources. Many karst and cave communities have suffered the brunt of Australian cultural tradition, but the last decade or so has witnessed a dramatic improvement in awareness, as environmental issues have figured prominently within the social and political arena. Indeed, there is cause for optimism in the future outlook as a number of major conservation battles have been recently fought, and won, with karst issues being a key ingredient; but there is little room for complacency, as some highly significant sites have already been lost whilst others remain under imminent threat.

Owing to a paucity of earlier baseline data, and until recently a general lack of impact-related studies undertaken in Australia, it has proved difficult to quantify and predict precisely the effects of various human activities on cave fauna. Nevertheless, there are numerous examples where human activities have caused the local extinction of populations of cave-dwelling bats and invertebrates, or where they have been seriously degraded or compromised. A great many sites remain highly vulnerable to a range of threatening processes, including quarrying and flooding, land clearance (deforestation) and agricultural activities (crop-raising and stock raising) and forestry. The nexus between land clearance and agricultural activities is the most widespread and insidious threat, affecting cave microclimates, nutrient inputs, hydrological regimes, and water-quality parameters such as sedimentation, nutrient enrichment and pollutants. A further threat to cave communities is direct disturbance caused by cave visitors, guano miners, and activities in cave management and research. The Australian experience of these threats is described below.

Quarrying and flooding

A number of caves, along with their resident fauna, have been obliterated through quarrying activities, whilst others have been inundated following the construction of water impoundments. Some of these sites were the focus of high-profile conservation campaigns finally settled in the law courts, but decisions have fallen both ways in recent times. A determined conservation campaign was waged over some 25 years at

Mount Etna (Fig. 33.2) in Queensland, where quarrying of limestone for cement has removed a number of significant caves (Sprent, 1970; Hamilton-Smith and Champion, 1975; Bonyhardy, 1993). Eventually, during the course of lengthy legal proceedings, speleologists resorted to occupying caves in order to prevent further blasting. This phase of the battle ended when the quarry company illegally blasted the entrance of one of the occupied caves (Speaking Tube Cave), which was an important roost site of the endangered ghost bat, *Macroderma gigas*, and finally destroyed it in 1989. The high cost of legal proceedings prevented their final determination, but eventually in 1991, largely as a result of a change in government, the site was included in a new National Park.

This experience does not stand alone; a number of sites were lost before the importance of karst and cave communities was recognized beyond the speleological fraternity. Texas Caves in Queensland, for example, were the type locality of an endemic troglobitic silverfish, *Trinemura russendenensis* (Smith and Shipp, 1978), and other undescribed invertebrates, but they were inundated by waters of the Glenlyon Dam (Archer, 1978). In New South Wales, construction of the Burrinjuck Dam during the 1930s flooded the type locality of the horseshoe bat, *Rhinolophus megaphyllus* (Ryan, 1965). In 1993, a South Australian quarrying company, with the apparent support of the State Mines Department, used explosives to destroy the entrance of a highly significant cave system at Sellicks Hill. This event resulted in extensive legal and political action, but the future of the site remains unresolved (Buswell, 1994). Other karsts have been inundated at Dartmouth in Victoria, Canberra in the Australian Capital Territory, and Lorinna and Moina in Tasmania. Quarrying of karst has occurred, and is still occurring, at Flowery Gully in Tasmania, Lilydale in Victoria, Marulan in New South Wales, and elsewhere.

However, there have been a number of successful conservation campaigns against quarrying. During the 1970s, the Colong karst in New South Wales and the Precipitous Bluff karst in Tasmania were both spared from quarrying proposals as a result of concerted efforts by speleologists and other interested groups. In the late 1980s cave archaeological sites on Tasmania's Franklin River provided a key piece of evidence that helped to save the river valley from being dammed. In 1992 a limestone quarry at Yessabah in New South Wales ceased operations following legal challenge by a speleologist who successfully demonstrated that he held a 'vested interest' in the caves. His case was enhanced by the fact that one of the caves threatened was a major bat-roosting site. Cave faunal values were a significant criterion in the decision to include Exit Cave within the boundaries of the World Heritage area in southwestern Tasmania. Then, in 1992, evidence of the deleterious impacts of quarrying upon this fauna was a major consideration in the decision to close the quarry concerned.

Many other limestone quarries are not known to have impacted upon caves (Gillieson 1989, pp. 48–49), and there are large resources of non-cavernous limestones scattered throughout the country. Nevertheless, a number of highly significant karsts continue to be placed under considerable pressure for mining development, owing to their geographical proximity to population centres or to the limestone quality, as well as other economic and political considerations. At the time of writing, a proposal for quarrying over part of the highly significant Cape Range karst (see Humphreys, Chapter 30, this volume) is under consideration by the Western Australian government. At least, in this case geological review has led to a limitation being placed upon the area which might be quarried. Regrettably, further protection of this region, or any other cave faunal site, is hindered by the universal problem of inadequate recognition being accorded to invertebrate communities by political and administrative decision-makers (Hill and Michaelis, 1988).

Land clearance and agricultural activities

Land clearance (deforestation), agricultural activities (crop-raising and stock-raising) and forestry are the most common and widespread threats facing karst biota in Australia. The impacts resulting from these activities are often subtle and difficult to quantify, because much of the degradation occurred long before any biota had been recorded. The impacts which potentially affect the biota include alteration to cave microclimates (including moisture, carbon dioxide and other parameters) and increased nutrient inputs from both grazing animals and artificial fertilizers, as well as changes to hydrological regimes and water quality, particularly sedimentation and pollutants.

It is very likely that a substantial proportion of the cave communities within New South Wales, Victoria, southwestern Western Australia and southeastern South Australia have undergone a marked alteration of composition or structure as a result of widespread land-clearance activities with their resultant wide range of environmental impacts. Of the approximately 120 karst areas in New South Wales and Victoria, some 60% have been subject to removal or major modification of their native vegetation cover. This has clearly caused a loss of biodiversity associated with removal of the native vegetation, and has reduced the input into caves of food, such as leaves, twigs and accidental fauna. Deforestation has increased the desiccation of cave environments, particularly around entrances, resulting in depletion of the twilight and entrance-zone sub-communities. It has altered water-table levels and caused changes in water quality and flow regime which impact upon the aquatic communities.

The extent of these impacts can best be demonstrated from extensive studies undertaken in South Australia on the effect of exotic pine (*Pinus radiata*) plantations on groundwater levels and atmospheric moisture regimes. As obvious examples, the water level in Sheathers Cave dropped by approximately one metre over a five-year period following the establishment of a pine plantation above the cave, whereas, conversely, the level in the nearby Mount Burr Cave rose significantly, probably to approximately the same extent, when the pine plantation above it was destroyed by wildfire (Grimes et al., 1995, p. 16). A significant increase in humidity, which resulted in the regeneration of previously dormant speleothems, occurred at Naracoorte Caves following the removal of pines. Similarly, caves beneath pine plantations at Yarrangobilly in New South Wales are less humid and contain less biota than those beneath indigenous forests (New South Wales, 1983).

Observations have indicated a general decline in the numbers of cave-dwelling bats in central and southeastern New South Wales (Hall and Dunsmore, 1974) and, anecdotally, elsewhere. Throughout New South Wales there are many caves that previously had extensive use by large numbers of bats, but which now are used only infrequently by smaller numbers, or not at all. The evidence includes the absence of bats from large numbers of caves that contain piles of bat guano and roof-staining from bat urine. The decline may be attributed, at least in part, to widespread deforestation, which has severely restricted the area of suitable foraging habitat available to the insectivorous bats. Use of inappropriate insecticides may lead to the accumulation of residues in bats and consequent mortality (Dunsmore et al., 1974), although there is no direct Australian evidence that this may lead to mortality of cave-dwelling species.

Local extinction of guano-dependent invertebrates has occurred at sites that are no longer used by bats. One of the few well-documented examples involves the disappearance, between 1866 and 1895, of a large maternity colony of the common bent-wing bat (*Miniopterus schreibersii*) from Mount Widderin Cave in Victoria (Simpson and Smith, 1964). The bats probably abandoned the cave as a result of the clearing of native scrubland, which also caused more rapid percolation of water into the cave. In turn, the loss of the bats impacted upon the invertebrate fauna of the cave, which then came to include less than half the number of taxa recorded from other *Miniopterus* maternity sites in the region (Hamilton-Smith, 1968).

The effects of agricultural and forestry activities, and industrial and metropolitan development, upon stream ecosystems has been the subject of a number of studies within Australia, and clearly these activities can result in major physico-chemical and ecological disturbances. Of these, agriculture and forestry impinge most widely on Australian karst areas, and the impacts derived from quarrying may extend some distance downstream from the initial site of disturbance. The price of agricultural prosperity has been a substantial blight upon the natural integrity of karst environments in continental Australia, and many Tasmanian karst areas are located within forestry concessions subject to logging operations involving clear-felling.

A variety of hydrological impacts consequent upon deforestation, agricultural or other land-use activities threaten aquatic cave and groundwater communities. The most apparent and widespread threat involves the lowering of water tables through groundwater abstraction for irrigation purposes, watering of stock, or human consumption. Other hydrological impacts include changes to flow regimes and water quality, particularly sedimentation, nutrient enrichment and pollutants.

Lowering of the water table in caves is acutely evident in Western Australia in the Augusta-Margaret River region, and at Yanchep where a uniquely rich community associated with underwater mats formed from tree roots is at risk (Jasinska and Knott, 1991; see also Chapter 15, this volume). The Cape Range stygofauna is also threatened by water abstraction.

Changes in flow regime and water quality become evident following alteration or removal of the native vegetation cover. The changes include an increase in

Fig. 33.3. Rubbish dumping at Earls Cave, South Australia October 1971. (Photograph R. Sweet, provided by F. Aslin.)

water yield, thus exacerbating flooding in caves (Kiernan, 1988). More rapid run-off after a precipitation event increases the rapidity in rise and peak discharge of streams flowing into caves. Previously perennial stream courses may become ephemeral or intermittent. A limestone-quarrying operation caused significant adverse impacts to aquatic cave-dwelling fauna at Ida Bay in Tasmania. Changes in flow regime and water quality (including sedimentation, eutrophication, and toxins) are responsible for causing the extinction of fauna in Bradley Chesterman Cave (Eberhard, 1995).

A number of recent Australian studies of surface waters (e.g. Doeg and Milledge, 1991; Growns and Davis, 1994; O'Connor and Lake, 1994) implicate sedimentation as a primary cause of stream degradation and faunal decline, and this appears to be the case in underground streams also. In Exit Cave at Ida Bay in Tasmania, for example, sedimentation originating from the limestone quarry altered stream habitats and restricted the distribution of hydrobiid snails. During February 1992, when the quarry was operational, a study showed that snail abundance was significantly lower in sediment-affected streams than in control streams. After closure of the quarry operation in October 1992, a rehabilitation program has sought to minimize further environmental degradation of the cave system by restoring natural in flow regimes and limiting the further influx of sediment. Continued monitoring of the snail populations now indicates similar abundance in the sediment-affected and the control sites (Eberhard, 1995). Since European settlement, many other stream-cave systems in eastern Australia have been subject to increased rates of sediment deposition as a result of land-use activities within their catchments (Gillieson, 1989).

The protection of water resources in general, and karst groundwater resources in particular, is an acute problem within rural communities throughout Australia. Ignorance and disregard of the nature of karst hydrological systems have severely compromised water resources and subterranean fauna in some localities. These impacts have been most often caused by improper rubbish disposal in sinkholes, and poor management of pollutants. Municipal disposal sites at Mole Creek in Tasmania and Buchan in Victoria both drain directly into karst aquifers. On one occasion, some 300 000 litres of waste milk was dumped in a cave at Allensford in Victoria (White, 1976).

An extreme example is the dumping of thousands of sheep carcasses in Earls Cave, in the Mount Gambier region of South Australia, and other dumping at this site over a long period (Aslin, 1972a; see also Fig. 33.3). One study in the region identified 26 other caves, 21 industrial drainage bores and a variety of other natural depressions as potential points of entry to the aquifer for pollutants (Aslin, 1972b). The extent of the problem is indicated by the nitrate pollution of this aquifer, which was measured as ten times in excess of safe levels (Waterhouse, 1973). However, there is evidence of impacts to the biota occurring much earlier than this. The first European settlement in the region was at a site in Mount Gambier now known as the Town Cave, selected because of its ready access to good water. Woods (1862, pp. 359–360) described this cave, saying that the water in the bottom of the cave was "... so deep as to give it, clear as it is, a deep sea-blue tint." He then added "The water is full of a cypris and cyclops, the shells of which seem to strew the bottom. There is also much conferva, a shrimp-like brachiopod and a minute paludina ...". The site later became a drain for the waste of the growing town, and today the water level is lower, with accumulations of silt and rubbish; the water is dirty and foul-smelling, and the aquatic community described by Woods has disappeared.

Nutrient enrichment may displace stygobiont communities and facilitate colonization of underground waters by epigean taxa. This has occurred at Bradley Chesterman Cave in Tasmania, although petroleum hydrocarbons have also been implicated in elimination of the original community at this site (Eberhard, 1995). Pollution from sewage has been identified at both Mole Creek in Tasmania and Buchan in Victoria, although action has been taken to eliminate the problem at the latter site. Subsidence in a limestone area also ruptured a sewage main in Tasmania (Kiernan, 1988, pp. 9–10). Eutrophication is the major threat to stromatolite colonies in karst lakes and cenotes in Western and South Australia (McNamara, 1992; Hallam and Thurgate, 1992; Thurgate, 1995). Run-off laden with agricultural fertilizers has increased the concentration of phosphates in Lake Clifton, Western Australia, ten-fold. The resultant algal blooms have stifled continuing growth of the stromatolite colonies by reducing available light levels, as well as directly smothering them (McNamara, 1992, pp. 23–24).

Human visitors

Direct impacts to cave communities have been caused by human visitors, including guano miners, cave managers and tourists, recreational cavers, and scientists.

Mining of bat guano has long ceased in Australia, but during the last century large quantities were removed from caves throughout eastern Australia. Doubtless these operations caused major disturbance to resident bat populations at the time, as well as to the associated guano communities of invertebrates. At Naracoorte in South Australia for example, guano miners cut an access hole through the roof of the Bat Cave that is the maternity site for the regional population of the bent-winged bat, *Miniopterus schreibersii*. The resultant ventilation and lowering of temperature would have made the cave unsuitable as a maternity site. Caves that provide the right microclimatic conditions for breeding in this species are few and far between, and there are only a handful of maternity sites known throughout the entire range of the species in Australia (Dwyer and Hamilton-Smith, 1965). Although it is not known what happened to the bat population at the time, the hole was sealed on the cessation of mining and the bats have occupied the cave ever since. The cave, now located in a World Heritage area, has more recently been accorded a very high level of protection.

Traditionally, cave managers concerned with developing caves for tourism purposes have seen their conservation responsibility as being confined to the preservation and display of the most beautiful speleothems, and until recently scant regard was given to the habitat requirements of the fauna. Cave microclimates have been significantly altered by the excavation of new entrances and connecting tunnels, or the enlargement of existing entrances. Additionally, entrances have been gated or sealed-off entirely, thus inhibiting access for bats and other trogloxenes, and altering the input of nutrients such as leaf litter and accidental animals. At a number of sites, cave entrance gates have subsequently been modified to allow the free passage of bats, or to restore the original microclimatic conditions. The Alexandra Cave at Naracoorte, for example, was discovered by digging open the entrance, thus allowing entry to rhaphidophorid crickets which established a flourishing colony and created considerable interest amongst visitors. Later it was realized that the free air flow permitted through the wooden gate placed on the new entrance had desiccated many speleothems, so a new air-tight door was installed to further their natural regeneration. This succeeded, but resulted of course in loss of the cricket colony. In this example, the crickets are found widely throughout the region, and the loss of one cave is not a matter of concern. However, in the Northern Territory, a badly designed gate resulted in the destruction of the largest recorded population of the rare golden horseshoe bat, *Rhinonycteris aurantius*.

The metabolism of large numbers of human visitors affects cave microclimates by increasing levels of carbon dioxide, moisture and temperature. Visitors also introduce lint, skin flakes, soil particles, food scraps, microflora, and fungi. These provide a rich food source which benefits some cavernicolous species, but may also permit colonization by adventitious species. This phenomenon is evident at Jenolan Caves in New South Wales, where considerable populations of springtails dwell alongside the tourist paths. The growth of algae, mosses and other vegetation near artificial lighting (*Lampenflora*) may also encourage colonization of the dark zone by invertebrates normally restricted to the twilight zone.

Other foreign materials that tend to accumulate in tourist caves include soot from carbide and other early combustible lighting sources, pitch from old electrical wiring seals, verdigris from scraps of electrical wiring or from coins placed in 'wishing wells', broken light globes, scrap metal and timber (Bonwick and Ellis, 1985). Old and disused timber left dumped in caves attracts cavernicolous species, and clean-up programs, although well-intentioned, can cause mortality to many individuals, and perturbation of the ecological balance established in the faunal community when the timber is removed.

Recreational caving activities have caused widespread and profound disruption to bat populations, as colonies that are repeatedly disturbed may abandon their cave sites and not return to them; this has occurred at Cotter Cave in the Australian Capital Territory and elsewhere. Even minor disturbance of bats at their overwintering sites and maternity roosts may result in significant fatalities (Hamilton-Smith, 1970b). Caving activities have increased in popularity over the last 25 years especially, and may be part of the cause for the decline in bat populations throughout New South Wales and Victoria during this period, although hard data are lacking. Most cave visitors are simply unaware of the disturbance they may cause, but occasionally bats are deliberately killed.

Floor-dwelling invertebrates in many caves are also widely under threat from recreational cavers who inadvertently trample them underfoot. A side effect of intensive trampling is the compaction of soft floor sediment, which may render it less suitable as invertebrate habitat. In extensively trampled caves the area of substrate inhabited by invertebrates has been greatly reduced, and the animals are confined to untrampled substrates alongside the walls, or in accessible crevices, and may even become extinct. This phenomenon is evident in many popular caving areas. Regrettably, the residual invertebrate population in Mount Widderin Cave at Skipton in Victoria has been almost totally exterminated by opening of the cave to large parties of visitors who have been allowed to trample and compact the whole floor area (Hamilton-Smith, 1968; Spate and Hamilton-Smith, 1991). Similarly, conduct of guided parties through a cave at Mount Etna in Queensland resulted in destruction of the root mat that had provided the habitat for planthoppers (Cixiidae and Meenoplidae). In this case, the management agency argued that the limited root destruction would not impact upon the surface vegetation, and so permitted the tours to occur; the fauna was not even considered (D. & J. Vavryn, personal communication, 1995). Other microhabitats that are vulnerable to damage from trampling are pools and small watercourses, which often contain a distinctive and specialized fauna.

Speleological activities such as digging can alter cave microclimates and the distribution of the fauna. There is one documented instance of a cave which, when first dug into, contained a high relative humidity and many invertebrates. Shortly after opening of the entrance, the cave dried out and the fauna disappeared (Hamilton-Smith, 1970b).

The final direct human threat to cave fauna involves the activities of scientific researchers. At least part of the decline in numbers of cave-dwelling bats during the 1960s and 1970s may be attributed to the pressure of specimen-collecting and banding programs. The former activity is known to have decimated local populations on at least two occasions with minimal scientific return (Hamilton-Smith, 1970b). At the same time, although participants in the banding scheme took particular care to minimize the disturbance they caused, and this research yielded much useful information on the ecology of the species (Simpson and Hamilton-Smith, 1965), there were occasional negative impacts. More recently, concern has been expressed about the possibility that invertebrate specimens may be over-collected. Slaney and Weinstein (1995) found that continuous trapping over a one-month period in Rope Ladder Cave in Queensland caused a significant decline in numbers of invertebrate taxa. They recommended that the general ecology of organisms in a cave be investigated before any intensive sampling is carried out, and that, where possible, sampling with replacement should be used.

STRATEGIES IN CONSERVATION

Present understanding of the nature of karst landscapes emphasizes the importance of the interrelationships between environmental conditions prevailing on the surface and those underground. We argue that this is fundamental to the development of effective strategies for the conservation of subterranean faunal communities. This approach may involve, for instance, the management of water-catchment areas which extend well beyond the geological boundaries of the karst outcrop. Aside from highly publicized conservation campaigns which have been waged in attempts to conserve imminently threatened sites, the conservation strategies that have been applied in Australia to date include: legislative protection of threatened species and development of recovery plans for them; the protection of areas of sensitive habitat at individual cave sites; habitat restoration; the location of karstlands within National Parks and other protected areas; and public awareness campaigns. Each of these strategies is discussed in turn below.

Species protection

Virtually all bat species are protected under legislation throughout the Australian region. The legislation has restrained the shooting of bats, but was not implemented until long after the widespread loss of bat foraging habitat through land-clearance activities had already occurred. The highest level of protective legislation is afforded to species that are considered endangered. In New South Wales, for example, the bent-winged bats (*Miniopterus* spp.) and some other cave-dwelling species are listed as vulnerable under Schedule 12 of the Threatened Species Conservation Act 1995. This means that forestry or development proposals must, in addition to undertaking an environmental impact assessment (EIA), also carry out a fauna-impact statement (FIS) if such a species is involved. This legislation potentially offers considerable security to listed species, but for some reason aquatic species are excluded from protection under the Act. Legislative protection of cave-dwelling invertebrates has generally lagged far behind that for vertebrates, although in 1971 a number of Tasmanian cave species were wholly protected under that state's Parks and Wildlife Act of 1970. Although it was a remarkable step forward at the time, a number of problems have become evident. When the legislation was enacted, the dozen or so species listed comprised, virtually, the known extent of Tasmania's cave fauna; despite the discovery of many additional taxa in the mean time, it is still widely and erroneously believed by park managers and academic biologists alike that all species of cave fauna are protected. A permit is required to collect any of the listed species, including those located outside national parks or state reserves. The Threatened Species Protection Act of 1995 incorporates a number of the recently described taxa, but the majority remain undescribed and therefore cannot be listed.

It could be further argued that some of the species are not threatened at all. The original proclamation was hurriedly instituted in reaction to fears of scientific over-collecting of one species which, ironically, is one of the most abundant and widely distributed of all cavernicoles. Although species protection of this kind may inhibit deliberate killing of individual animals, it cannot be invoked to prevent trampling, pollution or other destruction of habitat.

A more sophisticated approach is the development and implementation of recovery plans for endangered species. This approach is multi-faceted and often involves habitat restoration and protection, public education programs, and enhanced breeding. Currently a recovery plan for Australian bats is being developed, and will include attention to cave-dwelling species. A few plans have been prepared for non-cavernicolous invertebrates, but there are formidable problems involved, including lack of sufficient baseline information to identify species demanding such action, lack of research upon which to base plans, and the complexity of implementation (Yen and New, 1995). Nevertheless, this approach must be pursued whenever it appears to offer the opportunity for more adequate conservation.

A substantial portion of Australia's invertebrate cave taxa remain undescribed, and are likely to remain so in the foreseeable future. The lack of adequate taxonomic knowledge will continue to be a major stumbling block in gaining protection for individual species. One solution may be to seek protection of ecological communities, which is provided for under the Australian Endangered Species Protection Act 1992, and at the time of writing a proposal has been submitted which seeks listing of cave communities as Endangered Ecological Communities under the Act (Eberhard and Spate, 1995a). At the very least, such recognition can be used as a basis for further conservation action.

Habitat protection

The protection of areas of sensitive habitat within individual caves, coupled with an education program, has proved to be a pragmatic and successful conservation strategy which works much along the lines of a community recovery plan. Thus, in Mullamullang and Nurina Caves on the Nullarbor Plain, certain key habitat areas have been delineated with protective strings and signs which explain the reason for trying to exclude visitors from these areas (Poulter, 1991, 1994). Similarly, in Kubla Khan and Little Trimmer Caves in Tasmania, so-called substrate protection zones have been maintained by marking out pathways and "no-go" areas. This approach relies upon voluntary compliance by the cave's visitors, and its relative efficacy will depend on a number of factors such as the accessibility and popularity of the site, the presence of a dedicated management authority, and the type of people who visit the cave (e.g. speleologists vs. the general public). There are many caves where the impacts of trampling are ongoing, and these sites would greatly benefit from some form of habitat protection.

Similarly, bat conservation has been fostered by some land management agencies and caving groups. For example, a number of caves at Bungonia are closed annually by the New South Wales National Parks and Wildlife Service during critical periods of the bat breeding season and winter torpidity. Another example is at Anticline Cave in eastern Victoria, where a major maternity site used by the horseshoe bat *Rhinolophus megaphyllus* had been abandoned, and the species was no longer seen in the area for a period of almost 20 years. Cavers took the initiative, with the support of the landowner, in restricting visits to the cave; as a result the population has been re-established (Hamilton-Smith, 1991). A similar return by bent-winged bats to a cave at Jenolan has been recorded after the responsible manager had closed the cave to visitors, but its significance has not yet been fully assessed.

Habitat restoration

The importance of restoring previously damaged and degraded habitats is only now being realized. Both speleologists and tourism managers have been involved in aesthetic restoration (e.g., Bonwick and Ellis, 1985), and this clearly has some benefits in terms of habitat restoration. The experiences at Horse Cave in the United States (Elliott, Chapter 34, this volume) and Waitomo in New Zealand clearly demonstrate the potential importance of restoration. Restoration of climatic patterns or breaking-up of compacted floors are two examples where action is being taken at the time of writing, or might be taken, in Australia. Rehabilitation of the former quarry at Ida Bay in Tasmania has been undertaken with a full recognition of the impacts upon fauna, and, as noted above, the hydrobiid snail population is already returning to its former condition as a result of siltation control. Artificial fertilizers, which would have accelerated the surface revegetation program, have been excluded because of their negative impacts upon cave biota (Gillieson, 1995; see also Fig. 33.4).

Protected areas

Land reserves for the specific protection of caves, as opposed to cave communities *per se*, were first gazetted at Wombeyan in 1865, and then elsewhere later. Although some of these reservations were adequate, many were not, often only encompassing the cave entrance rather than the whole cave or karst system, and very rarely protecting the total catchment. Substantial portions of some reserved cave systems extend beneath private land, state forest, or other land tenure, thus compounding the task of management and conservation. In some instances this problem has been solved by further land acquisition, but it has often involved protracted negotiations to settle conflicts of interest.

Quite a number of important karst areas are located within National Parks or other conservation reserves, although, more often than not, their inclusion has been by fortuitous circumstance rather than deliberate intention. Consequently, some parts of the karst system, or its catchment, may not be included within the reserve boundaries. Such land protection is no panacea, and a range of other problems may arise, particularly in relation to fauna conservation. Lack of training for park managers in karst issues, combined with limited funding availability for karst-related conservation projects, is one of the more important. Many caves in National Parks are under great pressure from large numbers of recreational visitors, and in some parks this problem has been addressed by limiting access through institution of a permit system. Some caves have been closed entirely, and enforcement of the closure sometimes requires the installation and maintenance of

Fig. 33.4. Rehabilitation of the limestone quarry at Ida Bay, Tasmania. Note the use of filtering to prevent further entry of silt to underground waters. (Photograph D.S. Gillieson.)

entrance gates which in turn may interfere with access for bats.

Park authorities walk a difficult line between rigorous enforcement of conservation ideals on the one hand versus their need to maintain public and political legitimation on the other, and this may discourage strong action on conservation issues. There have been a number of incidents where park rangers have apprehended persons for contravening regulations, but the institution of legal proceedings has been halted at the ministerial level. Further, depending on the political climate at the time, conservation agendas may be discarded in favour of economic initiatives. Certainly financial problems have led to a shortage in management resources, particularly in personnel on the ground. Nevertheless, there has been a continuing improvement in the overall quality of cave management, aided by scientific research and monitoring, over the last 20 years.

A great many significant karst and cave communities are located on privately owned land, or they are located in areas of state forest subject to logging operations, especially in Tasmania. Only in recent years, the Forestry Department of the Tasmanian government has played a leading role in conserving karst resources within production forests by developing a forest-practices code which embraces proper recognition of karst. In addition, logging operations have been excluded within some sensitive karst catchments, and operations have been deferred in other areas pending further investigations of their significance and vulnerability. For karstlands and cave communities under private ownership there is no legislative mechanism in place which specifically ensures their conservation, although recent developments pertaining to land clearance and catchment protection within New South Wales may prove to be relevant. Unfortunately, many karst communities on private land are already seriously degraded. The future of these communities depends to a large extent upon the attitude of the landholder, which varies between individuals. Some caves and sinkholes continue to be utilized as rubbish dumps, whereas other landowners provide a high level of conservation management by excluding or limiting access to visitors. More often than not, however, this action is based on sheer convenience, or unsubstantiated fears of legal liability in case of a caving accident, rather than a genuine concern for conservation. The key to any conservation program is essentially one of enhanced understanding through education and interpretation at all levels, and insufficient attention has yet been given to the involvement of private landowners as partners in conservation, particularly in respect to karst areas.

International treaties are playing an increasing role in resource protection. The stromatolites of Shark Bay, together with caves in southwestern Tasmania, at Riversleigh in western Queensland, and Naracoorte in South Australia, fall within World Heritage areas. Most recently, a number of karst sites have also been recognized under the Ramsar Convention on Wetlands. These include Cape Range and several sites in the Kimberley of Western Australia, part of the Katherine karst in the Northern Territory, caves at Chillagoe and Undara in Queensland, and a number of sites in the Naracoorte–Mount Gambier region of South Australia (Australian Nature Conservation Agency, 1996).

Public education

It would appear that the future management and conservation of cave communities will rest both on a legislative footing and on better public recognition and understanding of the complexities of karst processes and karst environments (Eberhard and Spate, 1995a). The privatization of guiding services may, and often does, result in less emphasis on conservation and public education, but at Undara in Northern Queensland privatization has led to perhaps the best guiding practice in Australia. More generally, collaboration between cavers, academic researchers, and cave managers has proved to be a fertile starting point, from which there is a positive flow-on effect to the general public. The development of a Minimum Impact Code for caving has been adopted by some operators within the adventure-tourism industry (Webb, 1993). Another example is a pamphlet published by speleologists which has raised awareness of the plight of cave-dwelling bats in Victoria, and cooperation between the speleologists, a private landholder, and the state conservation department has enabled recovery of the Buchan population (Hamilton-Smith, 1991; Friends of Buchan Caves, 1993). Experience and expertise continue to evolve, and this is transmitted through the auspices of organisations such as the Australian Speleological Federation and the Australasian Cave and Karst Management Association.

REFERENCES

Archer, M., 1978. Modern fauna and flora of the Texas Caves area, Southeastern Queensland. *Mem. Queensl. Mus.,* 19: 111–119.

Aslin, F., 1972a. Caves of Australia, No. 6: Earls Cave. *Newsl. Aust. Speleol. Fed.,* 57: 15–16.

Aslin, F., 1972b. Preliminary report on potential pollution of underground water in and around Mt. Gambier. *Newsl. Aust. Speleol. Fed.,* 57: 17–18.

Australian Nature Conservation Agency, 1996. *A Directory of Important Wetlands in Australia,* 2nd edition. The Agency, Canberra, 964 pp.

Beier, M., 1967. Some Pseudoscorpionida from Australia, chiefly from caves. *Aust. Zool.,* 14: 199–205.

Beier, M., 1968. Some cave-dwelling Pseudoscorpionida from Australia and New Caledonia. *Rec. South Aust. Mus.,* 15: 757–765.

Bonwick, J. and Ellis, R., 1985. New caves for old: cleaning, restoration and redevelopment of show caves. *Cave Manage. Australasia,* 6: 134–153.

Bonyhardy, T., 1993. *Places Worth Keeping: Conservationists, Politics and the Law.* Allen and Unwin, Sydney, 192 pp.

Bridgewater, P.B., 1987. The present Australian environment: terrestrial and freshwater. In: G.R. Dyne and D.W. Walton (Editors), *Fauna of Australia.* Australian Government Publishing Service, Canberra, Vol. 1A, pp. 69–100.

Bruce, A.J., 1992. *Pycnisia raptor,* a new genus and species of predatory troglobitic shrimp (Crustacea: Decapoda: Atyidae) from Northern Australia. *Invertebrate Taxon.,* 6: 553–566.

Buswell, C., 1994. Sellick's Hill Quarry Cave. *Aust. Caver,* 136: 18–23.

Doeg, T.J. and Milledge, G.A., 1991. Effect of experimentally increasing concentrations of suspended sediment on macroinvertebrate drift. *Aust. J. Mar. Freshwater Res.,* 42: 519–526.

Dumont, H.J. and Maas, S., 1985. *Mesocyclops cuttacuttae* n. sp. from a cave in Northern Australia (Crustacea: Copepoda, Cyclopoidea). *Beagle,* 1: 115–122.

Dunsmore, J.D., Hall, L.S. and Kottek, K.H., 1974. DDT in the bent-winged bat inAustralia. *Search,* 5(3): 110–111.

Dwyer, P.D. and Hamilton-Smith, E., 1965. Breeding caves and maternity colonies of the bent-winged bat in southern Australia. *Helictite,* 4: 3–21.

Eberhard, S.M., 1995. Impact of a limestone quarry on aquatic cave fauna at Ida Bay in Tasmania. Paper presented at *11th Conf., Australasian Cave and Karst Management Association, Tasmania, May 1995.*

Eberhard, S.M. and Spate, A.P., 1995a. *Cave Communities.* Submission to the Endangered Species Unit, Australian Nature Conservation Agency, Canberra, 15 pp.

Eberhard, S.M. and Spate, A.P., 1995b. *Cave Invertebrate Survey: Toward an Atlas of NSW Cave Fauna.* Report to the National Parks and Wildlife Service of New South Wales, 112 pp.

Eberhard, S.M., Richardson, A.M.M. and Swain, R., 1991. *The Invertebrate Cave Fauna of Tasmania.* Zoology Department, University of Tasmania, Hobart, 174 pp.

Flannery, T., 1994. *The Future Eaters.* Reed Books, Melbourne 423 pp.

Friends of Buchan Caves, 1993. *Cave Dwelling Bats of Eas Gippsland.* Brochure, 6 pp. with poster on verso.

Gillieson, D.S., 1989. Effects of land use on karst areas in Australia In: D.S. Gillieson and D.I. Smith (Editors), *Resource Managemen in Limestone Landscapes: International Perspectives.* Departmen of Geography and Oceanography, University College, Australia Defence Force Academy, Canberra, pp. 43–60.

Gillieson, D.S., 1995. Rehabilitation of the Lune River Quarr Tasmanian Wilderness World Heritage Area, Australia. *Journ of the Australasian Cave and Karst Management Association,* 1 4–10.

Gillieson, D.S. and Spate, A.P., 1999. Caves and karst in Australia and New Guinea. In: Yuan Daoxian (Editor), *Global Karst Correlation.* VSP Press, Utrecht, Netherlands, pp. 229–254.

Grimes, K.G., Hamilton-Smith, E. and Spate, A.P., 1995. *South East Karst Province of South Australia.* Report on behalf of the Australasian Australasian Cave and Karst Management Association for the National Parks and Wildlife Service of South Australia, 110 pp.

Growns, I.O. and Davis, J.A., 1994. Effects of forestry activities (clearfelling) on stream macroinvertebrate fauna in South-western Australia. *Aust. J. Mar. Freshwater Res.,* 45: 963–975.

Hall, L.S. and Dunsmore, J.D., 1974. A survey of cave-dwelling bats. *Aust. Speleol. Fed. Newsl.,* 64: 9–10.

Hallam, N.D. and Thurgate, M., 1992. *The Biology of Ewens and Piccaninnie Ponds and Woolwash, Blacks and Gouldens Holes, South Australia.* Department of Ecology and Evolutionary Biology, Monash University, Melbourne, 79 pp.

Hamilton-Smith, E., 1968. The insect fauna of Mt. Widderin Cave, Skipton, Victoria. *Victorian Nat.,* 85: 294–6.

Hamilton-Smith, E., 1970a. Preliminary notes on the cavernicolous fauna of the Mt. Etna caves area, Central Queensland. In: J.K. Sprent (Editor), *Mount Etna Caves.* University of Queensland Speleological Society, St. Lucia, Queensland, pp. 65–71.

Hamilton-Smith, E., 1970b. Biological aspects of cave conservation. *J. Sydney Speleol. Soc.,* 14(7): 157–164.

Hamilton-Smith, E., 1972. The bat population of the Naracoorte Caves area. In: A. Goede and R. Cockerill (Editors), *Proc. Eighth Biennial Conf., Australian Speleological Federation.* Australian Speleological Federation, Sydney, N.S.W., pp. 66–75.

Hamilton-Smith, E., 1991. The return of the bats. *Nargun,* 23 (10): 81.

Hamilton-Smith, E. and Champion, R., 1975. *Mount Etna and the Caves: A Plan for Action.* Queensland Conservation Council, Brisbane, 147 pp.

Higgins, E.T. and Petterd, W.F., 1883. Description of a new cave-inhabiting spider, together with notes on mammalian remains from recently discovered caves in the Chudleigh district. *Pap. Proc. R. Soc. Tasmania,* 1882: 191–192.

Hill, L. and Michaelis, F.B., 1988. *Conservation of Insects and Related Wildlife,* Occasional Paper No. 13. Australian National Parks and Wildlife Service, Canberra, 40 pp.

Hoch, H. and Asche, M., 1988. Cave-dwelling planthoppers of Australia. In: L. Pearson (Editor), *Papers, 17th Biennial Conference, Australian Speleological Federation.* Australian Speleological Federation, Cairns, Qld, pp. 67–75.

Hoch, H. and Howarth, F.G., 1989a. Reductive evolutionary trends in two new cavernicolous species of a new Australian Cixiid genus (Homoptera: Fulguroidea). *Syst. Entomol.,* 14: 179–196.

Hoch, H. and Howarth, F.G., 1989b. Six new cavernicolous cixiid planthoppers in the genus *Solonaima* from Australia (Homoptera: Fulguroidea). *Syst. Entomol.,* 14: 377–402.

Howarth, F.G., 1980. The zoogeography of specialised cave animals. *Evolution,* 34: 394–406.

Howarth, F.G., 1987. The evolution of non-relictual tropical troglobites. *Int. J. Speleol.,* 16: 1–16.

Howarth, F.G., 1988. Environmental ecology of North Queensland caves: Why there are so many troglobites in Australia. In: L. Pearson (Editor), *Papers, 17th Biennial Conference, Australian Speleological Federation.* Australian Speleological Federation, Cairns, Qld, pp. 76–84.

Howarth, F.G., 1993. High-stress subterranean habitats and evolutionary change in cave-inhabiting arthropods. *Am. Nat.,* 142(Suppl.): S65–S67.

Howarth, F.G. and Stone, F.D., 1990. Elevated carbon dioxide levels in Bayliss Cave, Australia: Implications for the evolution of obligate cave species. *Pac. Sci.,* 44 (3): 207–218.

Humphreys, W.F. (Editor), 1993. The Biogeography of Cape Range, Western Australia. *Rec. Western Aust. Mus., Suppl.,* 45: 1–248.

Hunt, G.S., 1972. Notes on Australian cave harvestmen. In: A. Goede and R. Cockerill (Editors), *Proc. Eighth Biennial Conf., Australian Speleological Federation.* Australian Speleological Federation, Sydney, N.S.W., pp. 76–80.

James, J.M., 1996. Evidence for a "sulfureta" in Weebubbie Cave, Nullarbor Plain, Western Australia. In: E. Hamilton-Smith (Editor), *Abstracts, Karst Studies Seminar,* Naracoorte 9–11 February. Regolith Mapping, Hamilton, Vict., pp. 11–12.

Jasinska, E.J. and Knott, B., 1991. The importance of caves at Yanchep. In: S. Brooks (Editor), *Proc. Eighteenth Biennial Conf., Australian Speleological Federation.* Australian Speleological Federation, Nedlands, WA., pp. 55–57.

Jennings, J.N., 1968. Syngenetic karst in Australia. In: J.N. Jennings and P.W. Williams (Editors), *Contributions to the Study of Karst.* Department of Geography Publication G/5. Australian National University, Canberra, pp. 41–110.

Jennings, J.N., 1969. Karst of the seasonally humid tropics in Australia. In: O. Stelcl (Editor), *Problems of the Karst Denudation.* Institute of Geography, Brno, Czechoslovakia, pp. 149–158.

Kiernan, K., 1988. *The Management of Soluble Rock Landscapes.* Speleological Research Council, Sydney, 61 pp.

Lea, A.M., 1910. On some Tasmanian cave-inhabiting beetles. *Tasmanian Nat.,* 2: 53–58.

Malipatil, M.B. and Howarth, F.G., 1990. Two new species of *Micropolytaxus* Elkins from Northern Australia (Hemiptera: Reduviidae: Saicinae). *J. Aust. Entomol. Soc.,* 29: 37–40.

McNamara, K., 1992. *Stromatolites.* Western Australian Museum, Perth, 27 pp.

New South Wales, 1983. *Harvesting and Rehabilitation of Jounama Pine Plantation, Kosciusko National Park: Environmental Impact Statement.* National Parks and Wildlife Service and Forestry Commission, Sydney, N.S.W., 198 pp.

O'Connor, N.A. and Lake, P.S., 1994. Long-term and seasonal large-scale disturbances of a small lowland stream. *Aust. J. Mar. Freshwater Res.,* 45: 243–255.

Osborne, R.A.L. and Branagan, D.F., 1988. Karst landscapes of New South Wales. *Earth Sci. Rev.,* 25: 467–480.

Poulter, N., 1991. Cave rights for troglobites. In: S. Brooks (Editor), *Proc. Eighteenth Biennial Conf., Australian Speleological Federation.* Australian Speleological Federation, Nedlands, WA., pp. 15–24.

Poulter, N., 1994. Protecting caves from people II. *Aust. Caver,* 137: 17–26.

Rainbow, W.J., 1904. Studies in Australian Araneidae III. *Rec. Aust. Mus.,* 5: 326–336.

Rainbow, W.J., 1905. Studies in Australian Araneidae IV. *Rec. Aust. Mus.,* 6: 9–12.

Rainbow, W.J., 1907. Two new species of Australian Collembola. *Rec. Aust. Mus.,* 6: 313–314.

Richards, A.M., 1971. An ecological study of the cavernicolous fauna of the Nullarbor Plain, Southern Australia. *J. Zool. London,* 164: 1–60.

Ryan, R.M., 1965. The type locality of the Australian horsehoe bat, *Rhinolophus megaphyllus. Aust. J. Sci.,* 27: 259.

Simpson, K.G. and Hamilton-Smith, E., 1965. Third and fourth annual reports on bat banding in Australia. *CSIRO Aust.: Div. Wildlife Res. Tech. Pap.,* 9: 1–24.

Simpson, K.G. and Smith, G.T., 1964. Bat mandible from Mt. Widderin Cave, Skipton, Victoria. *Victorian Nat.,* 81: 78–79.

Slaney, D.P. and Weinstein, P., 1995. Conservation of cave fauna: more than just bats. Paper presented at *Conference on Invertebrate Biodiversity and Conservation, Melbourne, November 1995.*

Smith, G.B. and Shipp, E., 1978. A new species of cave-dwelling Nicoletid silverfish (Thysanura: Insecta) from the Texas Caves, Queensland. *Mem. Queensl. Mus.,* 19: 121–123.

Spate, A. and Hamilton-Smith, E., 1991. Caver's impacts – some theoretical and applied considerations. In: P. Bell (Editor), *Proc. Ninth Conf.* Australasian Cave and Karst Management Association, Margaret River, WA., pp. 20–30.

Sprent, J.K. (Editor), 1970. *Mount Etna Caves.* University of Queensland Speleological Society, Brisbane, 116 pp.

Stock, J.H. and Iliffe, T.M., 1990. A new Australian crangonyctid *Niphargus. Stygologia,* 5: 137–142.

Stone, F.D., 1988. The cockroaches of North Queensland caves and the evolution of tropical troglobites. In: L. Pearson (Editor), *Papers, 17th Biennial Conf., Australian Speleological Federation.* Australian Speleological Federation, Cairns, Qld., pp. 88–93.

Thurgate, M., 1995. *Sinkholes, Caves and Spring Lakes,* Occasional Paper no. 1. South Australian Underwater Speleological Society, Adelaide, 48 pp.

Urquhart, A.T., 1889. On two species of Aranea new to science from the Jenolan Caves, New South Wales. *Trans. N.Z. Inst.,* 22 236–239.

Waterhouse, J.D., 1973. Contamination of groundwater in the Moun Gambier area. *Miner. Resour. Rev.,* 139: 27–40.

Webb, R., 1993. Mimimal impact caving code – Fourth Draft. *Aust Caver,* 135: 24–25.

White, M.E., 1994. *After the Greening: The Browning of Australia* Kangaroo Press, Sydney, 288 pp.

White, N.J., 1976. Dumping of milk, Allansford, Western Victoria *Nargun,* 8(7): 11.

Williams, W.D., 1964. Subterranean freshwater prawns (Crustacea Atyidae) in Australia. *Aust. J. Mar. Freshwater Res.,* 15: 93–106.

Wilson, G.D.F. and Ponder, W.F., 1992. Extraordinary new subterranean Isopods (Peracarida: Crustacea) from the Kimberley Region, Western Australia. *Rec. Aust. Mus.,* 44: 279–298.

Woods, J.E. (Tenison), 1862. *Geological Observations in South Australia.* Longman, Green, Longman, Roberts and Green London, 404 pp.

Yager, J. and Humphreys, W.F., 1996. *Lasionectes exleyi,* sp. nov., the first remipede crustacean recorded from Australia and the Indian Ocean, with a key to the world species. *Invertebrate Taxon.* 10 171–187.

Yen, A.L. and New, T.R., 1995. Some problems affecting implementation of recovery programmes for threatened Australian non-marine invertebrate species. In: A. Bennett, G. Backhouse, and T. Clark (Editors), *People and Nature Conservation: Perspectives on Private Land Use and Endangered Species Recovery.* Royal Zoological Society of New South Wales, Sydney, pp. 9–14.

Zeidler, W., 1985. A new species of crustacean (Syncarida Anaspidacea: Koonungidae) from sinkholes and caves in the south east of South Australia. *Trans. R. Soc. South Aust.,* 109: 63–75.

Chapter 34

CONSERVATION OF THE NORTH AMERICAN CAVE AND KARST BIOTA

William R. ELLIOTT

INTRODUCTION

In this chapter I briefly review the biogeography of the North American cave and karst biota, then the many threats to the biota and the remedies that have been devised. I cite cases mainly from the United States of America, with a few examples from Canada, Mexico, Belize, and Panama (Fig. 34.1, overleaf).

NORTH AMERICAN CAVE BIOGEOGRAPHY

The study of cave biota in the United States began late in the 19th Century, and has been especially productive since the 1950s (Barr, 1968). The first biological study of a Mexican cave was in 1866, and important studies were made in 1932 in Yucatán. Published works on the Mexican cave fauna accelerated since the 1950s (Reddell, 1981).

For too long many biospeleologists believed that troglobites were rare in the tropics. However, it has become apparent that many troglobites exist in Mexico, Hawaii, and other tropical areas where the high biodiversity, complex geology, rugged terrain, and multiple isolating mechanisms provide many opportunities for speciation.

Peck (1997) tabulated the troglobitic fauna of the 48 contiguous states of the United States. There are at least 1307 troglobitic species in the United States, including 401 aquatic species (31%) and 906 terrestrial species (69%). Peck thought that more species were to be discovered, but doubted that there were 6000 troglobitic species, as predicted by Culver and Holsinger (1992). In southeastern Alaska, two troglobitic amphipods of the genus *Stygobromus* are known from island caves; one of these species (*S. quatsinensis*) also occurs in caves on Vancouver Island (British Columbia, Canada) (Aley et al., 1993). Castleguard Cave in Alberta, contains Canada's only other troglobite, *Stygobromus canadensis* (Holsinger et al., 1997).

Reddell (1981) reported about 317 troglobites from caves in Mexico, Guatemala, and Belize, and nearly 1952 species in all. Troglobites were about 16% of the total number of species, and about 25% of the troglobites were aquatic. Hundreds of new species awaited study and description at that time. The number of named troglobitic species today probably would be about 500, but no recent tabulation has been published (Elliott, 1994b).

Based on these studies and typical trends, I estimate that continental North America currently contains about 1800 named troglobitic species, of which 25 to 30% are aquatic. In addition, caves provide shelter for bats and numerous other vertebrates. Humans benefit from the feeding activities of bats, in controlling night-flying insects, pollinating plants, and re-seeding forests.

Karst environments, such as deep, moist sinkholes and limestone glades, sometimes have relictual plant species, such as the Hart's tongue fern *Phyllitis scolopendrium* var. *americanum*, a Pleistocene relict found in Ontario, Michigan, New York state, and cave entrances in Tennessee and Alabama (Evans, 1982). A high degree of endemism occurs in plant species in the limestone glades of the southeastern United States (Baskauf, 1997). The only known United-States locality for the tropical hay-scented fern, *Dennstaedtia globulifera*, is in Fern Cave, Texas (Elliott, 1994f). These patchy environments are just as vulnerable as the caves.

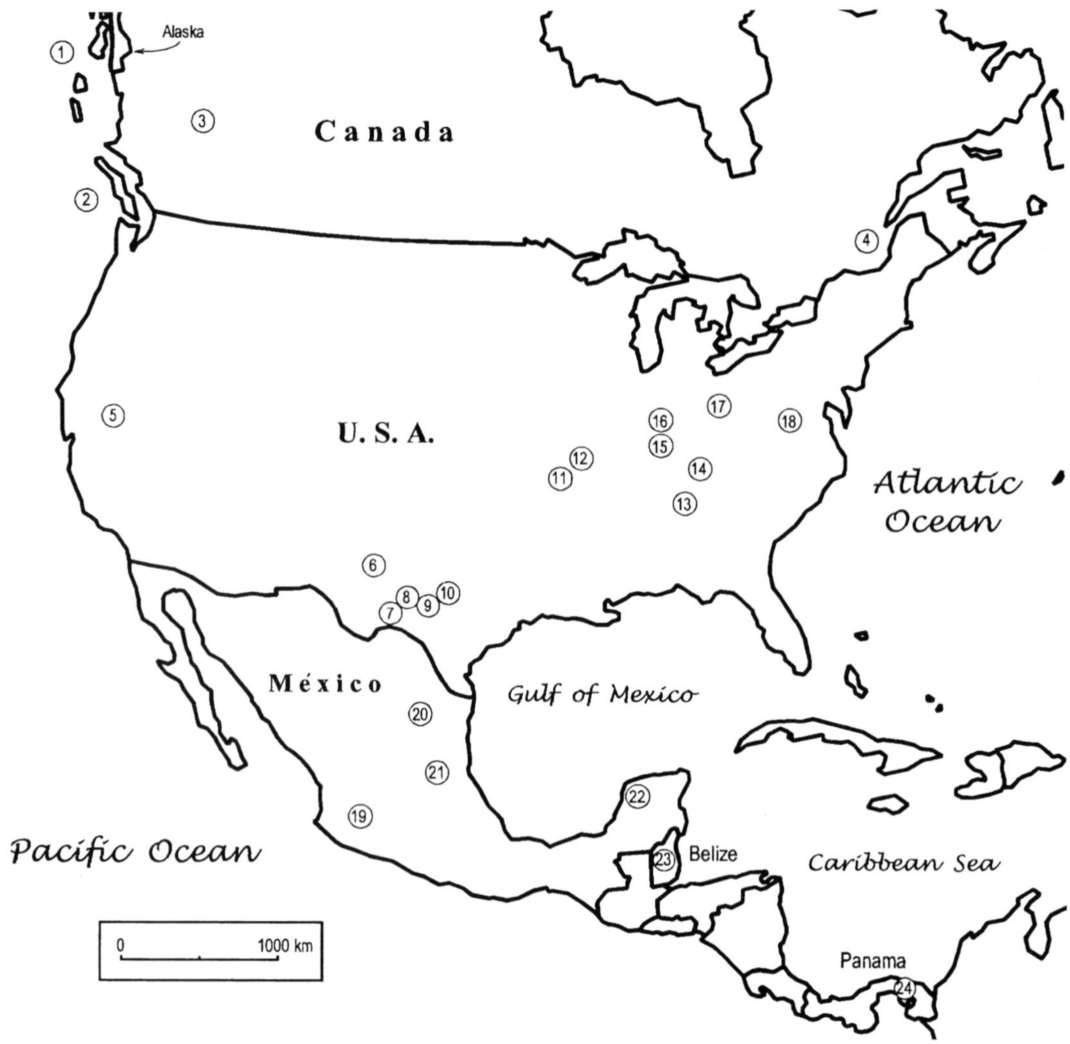

Fig. 34.1. Map of North America with selected locations discussed in text. (1) Southeast Alaska; (2) Vancouver Island; (3) Castleguard Cave, Alberta; (4) Fourth Chute Cave, Quebec; (5) McLean's Cave and Transplant Mine, California; (6) Carlsbad Cavern, New Mexico; (7) Marshall Bat Cave, Texas; (8) Valdina Farms Sinkhole, Texas; (9) San Antonio and San Marcos, Texas; (10) Austin, Texas; (11) Blanchard Springs Cavern, Arkansas; (12) Meramec Spring, Missouri; (13) Shelta Cave, Alabama; (14) caves at Cookeville, Tennessee; (15) Mammoth and Hidden River caves, Kentucky; (16) Blue River Karst, Indiana; (17) Cave Hill Cave, Ohio; (18) Virginia and West Virginia karst; (19) Mezcala, Jalisco; (20) Toxic Sink and Pozo del Cañon El Buey, Coahuila; (21) Sótano de Médico, San Luís Potosí; (22) Mérida, Yucatán; (23) Actun Chapat, Belize; (24) Chilibrillo Cave, Panama.

EXTINCT AND ENDANGERED SPECIES

Four or five North American troglobitic species are thought to be extinct, but it is likely that others became extinct before they could be discovered or described (Table 34.1). It is certain that local populations of invertebrates, fishes, salamanders and bats have been extirpated. Probably fewer than half of the United-States troglobites have been described, and in Mexico the proportion is even lower. Since some species are endemic to a single cave or a small cluster of caves, and many caves have been disturbed, filled, quarried, mined, or polluted, it is possible that some species have disappeared without it being known.

The Valdina Farms Sinkhole salamander, *Eurycea troglodytes*, may be extinct. The salamander's only known cave in Texas was used as a recharge well by the Edwards Underground Water District, which excavated a flood channel from a nearby creek to the entrance. In 1987 a large flood pulse cleaned out the

Table 34.1
Possibly extinct cave-dwelling species of North America

Species	Last year seen	Threats	Range[1]
Crustacea			
Bactrurus n.sp., amphipod	1963	sealed spring, pesticides	Indiana
Stygobromus lucifugus (=*subtilis*?) Dubious Cave amphipod	1882	?	Illinois
Insecta			
Pseudanophthalmus krameri, Kramer's cave beetle	1973?	?	Ohio
Amphibia			
Eurycea robusta, Blanco blind salamander	1951	?	Texas
Eurycea troglodytes, Valdina Farms Sinkhole salamander	1985?	recharge dam	Texas

[1] All species in this table were each known from a single cave only.

cave. Also lost were a colony of four million free-tailed bats, *Tadarida brasiliensis mexicana*, and a rare colony of the leaf-chinned bat *Mormoops megalophylla*. A follow-up study failed to find the salamander (Veni, 1987; Elliott, 1993a, 1994d).

Jordan Hall Spring, in the sub-basement of the biology department at Indiana University, was altered by construction and then poisoned by a termite treatment. The spring used to contain a unique isopod, *Caecidotea jordani* and two amphipods, *Crangonyx* n. sp. and *Bactrurus* n. sp., the latter being unique to that spring (Lewis, 1996a; John Holsinger, personal communication).

Hobbs (1997) reported that the carabid beetle *Pseudanophthalmus krameri* appears to be extinct or extremely rare. The beetle was known only from Cave Hill Cave, Ohio, which is in a mixed, mesophytic, second-growth forest with nearby fields, in a formerly glaciated area. The cave supports 23 terrestrial and aquatic species, and there were no obvious perturbations, although pesticide spraying could have occurred. Repeated visits and intense sampling yielded no individuals of *P. krameri*.

Some troglobites are extremely rare, so that it is difficult to prove or disprove whether they are extinct. For example, intensive surveys of the Blue River Bioreserve, Indiana, turned up a cave pseudoscorpion, *Kleptochthonius packardi*, which had not been seen for over 100 years (Lewis et al., 1997). The endangered blind Kentucky cave shrimp *Palaemonias ganteri* was thought to be extinct because it was not seen between 1967 and 1979, although it was seen again subsequently (Lisowski, 1982, 1983). There had been a reduction in sightings of cave fish, crayfish, and shrimps just after brine contamination from oilfields in 1961. In Mammoth Cave (near Bowling Green, Kentucky), the shrimp inhabits quiet, silt-bottomed pools with seasonal sediment deposition. This was altered by river dams, which caused back-flooding, back-ponding, siltation, temperature changes, and increased pollution input to streams in the Mammoth Cave system (Fig. 34.2). The normal levels of back-ponding were documented by Hovey in 1897 and 1909.

The rarity of the blindfish *Amblyopsis spelaea* in Mammoth Cave and its absence from adjacent areas to the north led to speculation that it was either introduced or decimated during the long period when blindfish were sold as curios. Poulson (1968) examined historical and scientific records, and found that most early records from Mammoth Cave were for *A. spelaea*, not *Typhlichthys subterraneus*, which also inhabits the cave. *Amblyopsis spelaea* was the dominant species in the Echo and Roaring River areas of Mammoth Cave around 1890, and it is still common in Roaring River. The present rarity of *A. spelaea* is probably related to silting and flooding associated with deforestation, forest fires, and the construction of Lock and Dam #6 on the nearby Green River (Poulson, 1968).

Table 34.2 details the twenty troglobitic species in the United States and one in Mexico that are on the lists of endangered or threatened species for the United States or for individual states. The Texas blind salamander, *Typhlomolge rathbuni*, was the first endangered species listed by the United States under the Endangered Species Conservation Act, a forerunner to the current Endangered Species Act. The salamander

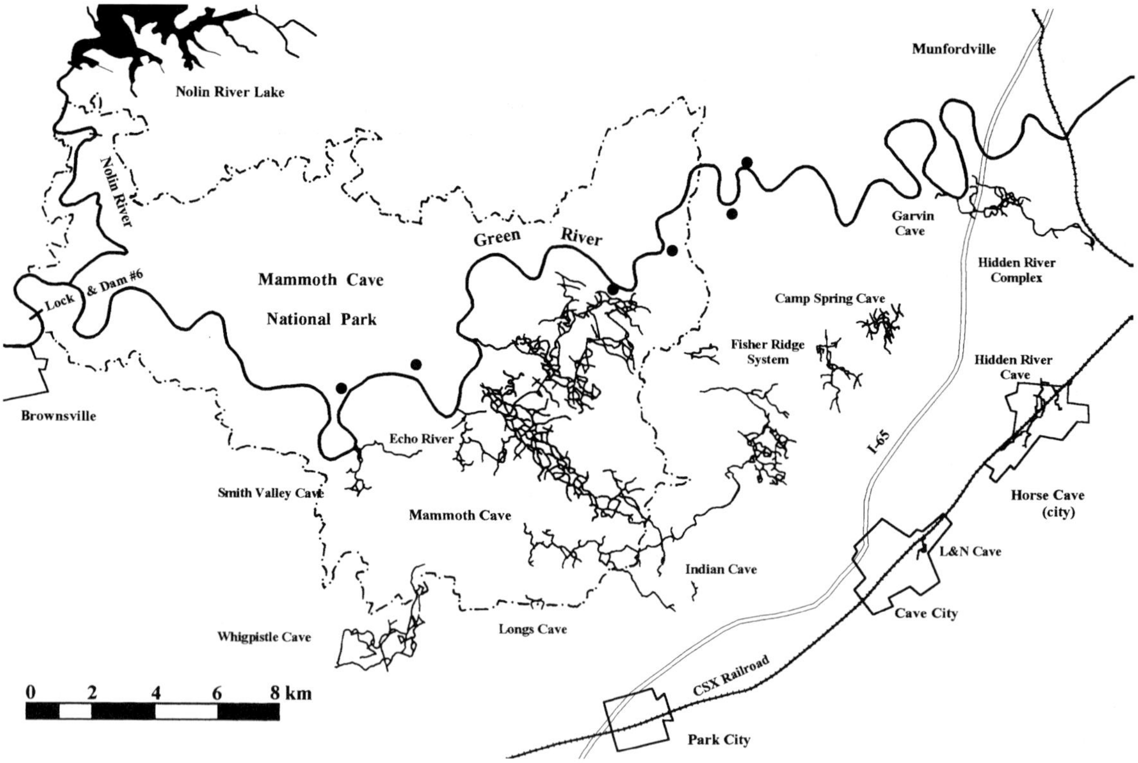

Fig. 34.2. The region around Mammoth Cave, Kentucky, depicting places mentioned in the text. The Green River flows to the west (left), and is impounded by Lock & Dam #6, which causes backflooding into some caves. Cave City and Horse Cave (city) are hydrologically connected to the Green River via Hidden River Cave and its downstream Complex.

was first seen in 1895 when an artesian well was drilled close to the San Marcos River. It is considered one of the most cave-adapted vertebrates in the world. Today it is also known in the nearby Ezell's Cave, which is protected by The Nature Conservancy, and two or three other localities in the vicinity. These localities are threatened by excessive pumping from the Edwards Aquifer, which serves as the sole drinking water source to more than 1.5 million people in central Texas (Longley, 1991).

The first seven Texas species in Table 34.2 all occur in two counties around Austin, and were listed in 1988 as endangered by urbanization, road-building, quarrying, and imported red fire ants (*Solenopsis invicta*). Subsequent studies have expanded the known range of some of the species, but all are still significantly threatened. Williamson County in Texas is currently the second fastest growing county in the United States. Since 1850, Austin (Travis County) has doubled in population about every 20 years (Reddell and Elliott, 1994).

Table 34.3 lists the six bats in the continental United States that are currently on the United States endangered species list. All are dependent on caves for part of their life cycle, and disturbance has been the major factor in their decline. Indiana bats, *Myotis sodalis,* have lost significant numbers, largely through disturbance of their hibernacula and improper gating, but also through loss of their typical summer habitat (under the bark of riparian trees).

My work in ten caves in Belize in 1992 and 1993 convinced me that insectivorous and frugivorous bats are often dependent on caves for roost sites in the tropics. In a cave called Actun Chapat, I found the frugivore *Artibeus jamaicensis* and the insectivores *Glossophaga* sp. and *Mormoops megalophylla* (unpublished data). In two other caves I found *Carollia* sp., a frugivore. "Ecotours" are increasing their visits to some of these caves. Disturbance of these cave roosts could lead to a decline in the reseeding of cut-over forest areas by bats.

McCracken (1989) questioned the utility of the IUCN *Red Data Book*, which in 1988 listed 4% of the world's bat species as endangered or threatened; he thought that this list gave an inaccurate and minimal

Table 34.2
Endangered and threatened cave and karst species of North America[1]

Species	Status[2]	Year listed	Range
Arachnida			
Neoleptoneta myopica, Tooth Cave spider	E	1988	Texas
Tartarocreagris texana, Tooth Cave pseudoscorpion	E	1988	Texas
Texella reddelli, Bee Creek Cave harvestman	E	1988	Texas
Texella reyesi, Bone Cave harvestman	E	1988	Texas
Crustacea			
Antrolana lira, Madison Cave isopod	T	1982	Virginia
Cambarus aculabrum, cave crayfish	E	1993	Arkansas
Cambarus zophonastes, cave crayfish	E	1987	Arkansas
Gammarus acherondytes, Illinois cave amphipod	E	1998	Illinois
Lirceus usdagalun, Lee County cave isopod	E	1992	Virginia
Palaemonetes cummingi, Squirrel Chimney Cave = Florida cave shrimp	T	1990	Florida
Palaemonias alabamae, Alabama cave shrimp	E	1988	Alabama
Palaemonias ganteri, Kentucky cave shrimp	E	1983	Kentucky
Insecta			
Batrisodes texanus, Coffin Cave mold beetle	E	1988	Texas
Rhadine persephone, Tooth Cave ground beetle	E	1988	Texas
Texamaurops reddelli, Kretschmarr Cave mold beetle	E	1988	Texas
Pisces			
Amblyopsis rosae, Ozark cavefish	T	1984	Arkansas, Missouri, Oklahoma
Prietella phreatophila, Mexican blindcat (catfish)	E	1970	Mexico (Coahuila)
Speoplatyrhinus poulsoni, Alabama cavefish	E	1977	Alabama
Amphibia			
Gyrinophilus palleucus, Tennessee cave salamander	ST (Tenn.)		Tennessee, Alabama, Georgia
Typhlomolge rathbuni, Texas blind salamander	E	1967	Texas
Typhlomolge robusta, Blanco blind salamander	SE	P 1995	Texas

[1] From the U.S.Fish and Wildlife Service, http://www.fws.gov/~r9endspp/endspp.html, and state web sites.
[2] E, endangered; T, threatened; P, subjects of citizen petitions; SE, State endangered list; ST, State threatened list.

assessment of our current extinction crisis. This was because the list largely reflects the ignorance of the status of bats in most parts of the world. Besides the six endangered United States bats, another 13 species in the United States rely substantially on caves; but this does not mean that the other 34 United States bat species are doing well.

Table 34.4 lists twelve species that are being seriously considered for listing by the United States Fish and Wildlife Service (USFWS). Some were the subject of citizen petitions, and some were promoted by the Service because of imminent land development, as in the case of *Cicurina wartoni* (Ruth Stanford, USFWS, personal communication). In 1996 the Service discontinued the C2 list, a sort of "pre-candidate list".

The United States Fish and Wildlife Service has received many petitions for additional listings of cave species, but has not acted on most of the petitions because of a lack of data, lack of funding and staff, or political pressure. Table 34.5 summarizes the 11 species which are the subject of petitions still being considered; for at least 64 others that were the subject of petitions, they were ruled "not substantial", "not warranted", "warranted but precluded", or "withdrawn" (Susan Lawrence, USFWS, personal communication). Frequently there was a lack of information in the

Table 34.3
Endangered North American bats and their use of cave roosts [1]

Species	Food preferences	Cave roost requirements	Range
Leptonycteris sanborni, Lesser (Sanborn's) long-nosed bat	agave nectar, pollen, insects	Year-round	southern Arizona, Mexico
Leptonycteris nivalis, Mexican long-nosed bat	agave nectar, pollen	Year-round	West Texas, Mexico
Myotis grisescens, Gray bat	insects	Year-round	Oklahoma, Kansas to Kentucky, then south to Florida
Myotis sodalis, Indiana bat	insects	Winter hibernacula	Vermont to Oklahoma, Michigan to Florida
Corynorhinus townsendii ingens, Ozark big-eared bat	moths	Year-round	Oklahoma, Arkansas (now absent from Missouri)
Corynorhinus townsendii virginianus, Virginia big-eared bat	moths	Year-round	Kentucky, Virginia, West Virginia

[1] Sources: Barbour and Davis (1969); McCracken (1989); U.S. Fish and Wildlife Service web site, http://www.fws.gov/~r9endspp/endspp.html.

Table 34.4
Candidates for the United States list of endangered species [1]

Species	Status and priority [2]	Threats	Range (and number of sites)
Mollusca			
Antrobia culveri, Tumbling Creek cave snail	C, 7	potential pollution	Missouri (1)
Arachnida			
Adelocosa anops, Kauai cave wolf spider or pe'e pe'e maka 'ole	C, 1	land development	Hawaii
Cicurina bandida, Spider	SC?	land development	Texas (1)
Cicurina cueva, Spider	SC?	land development	Texas (1)
Cicurina wartoni, Warton's cave spider	C, 2	land development	Texas (1)
Neoleptoneta microps, spider	SC?	land development	Texas (1)
Crustacea			
Spelaeorchestia koloana, Kauai cave amphipod	C, 1	land development	Hawaii
Stygobromus pecki, Peck's cave amphipod	PE [3]	overuse of aquifer	Texas (1)
Caecidotea filicispeluncae, isopod	SC	tree cutting, microclimatic changes	Ohio (1)
Caecidotea rotunda, isopod	SC	human visitation	Ohio (1), Indiana (4)
Insecta			
Pseudanophthalmus holsingeri, Holsinger's cave beetle	C, 2	gasoline, sewage, siltation	Virginia (1)
Pseudanophthalmus ohioensis, beetle	SC	small population, flooding, visitation	Ohio (1)

[1] U.S. Fish and Wildlife Service web site, http://www.fws.gov/~r9endspp/endspp.html.
[2] PE, formally proposed for the list of endangered species. C, species that probably will be listed as endangered (priority increases from 12 to 1). SC, species of concern (lowest status).
[3] Year of petition 1990.

Table 34.5
Cave species for which petitions by United States citizens for listing as "endangered" are currently being considered

Species	Year of petition	Range
Arachnida		
Cicurina (Cicurella) baronia, Robber Baron Cave spider	1992	Texas [1]
Cicurina (Cicurella) madla, Madla's Cave spider	1992	Texas [1]
Cicurina (Cicurella) venii, Cave spider	1992	Texas [1]
Cicurina (Cicurella) vespera, Vesper Cave spider	1992	Texas [1]
Neoleptoneta microps, Government Canyon Cave spider	1992	Texas [1]
Texella cokendolpheri, Robber Baron Cave harvestman	1992	Texas [1]
Crustacea		
Stygobromus clantoni, Clanton's Cave amphipod	1989	Kansas
Insecta		
Rhadine exilis, ground beetle	1992	Texas [1]
Rhadine infernalis infernalis, ground beetle	1992	Texas [1]
Rhadine infernalis ewersi, ground beetle	1992	Texas [1]
Batrisodes (Excavodes) venyivi, Helotes mold beetle	1992	Texas [1]

[1] All Texas species in this table are from Bexar County, the San Antonio area.

Table 34.6
Species wholly dependent on karst aquifers, which have been listed as "endangered" or "threatened", or have been proposed for listing [1]

Species	Status [2]	Year of listing or petition	Range
Insecta			
Heterelmis comalensis, Comal Springs riffle beetle	PE	1990	Texas
Stygoparnus comalensis, Comal Springs dryopid beetle	PE	1990	Texas
Pisces			
Cyprinodon diabolis, Devil's Hole pupfish	E	1967	Nevada
Cyprinodon elegans, Comanche Springs pupfish	E	1967	Texas
Cyprinodon bovinus, Leon Springs pupfish	E	1980	Texas
Etheostoma fonticola, fountain darter	E	1970	Texas
Amphibia			
Eurycea nana, San Marcos salamander	T	1978	Texas
Eurycea sosorum, Barton Springs salamander	E	1997	Texas
Liliopsida			
Zizania texana, Texas wild rice	E	1978	Texas

[1] U.S. Fish and Wildlife Service web site, http://www.fws.gov/~r9endspp/endspp.html.
[2] E, listed as "Endangered"; T, listed as "Threatened"; P, subject of citizens' petition.

petitions about population trends or threats. However, few of the species are under study because of little or no funding from any source.

Table 34.6 summarizes endangered "karst-dependent" species; these are not particularly cave-adapted, but are wholly dependent on the flow of karst springs. The San Marcos and Comal rivers in central Texas are fed by large springs from the Edwards Aquifer. The rivers support multiple endemic species, and contribute to the Guadalupe River, which feeds the San Antonio Bay estuary on the Gulf of Mexico coast, where there are important fisheries.

The process by which endangered species are listed in the United States is haphazard (Elliott, 1990, 1992a). Not only must there be adequate scientific knowledge to list a species, often there must be a political push in the form of a petition from conservationists and scientists. Consequently, unpopular or obscure species, even if they be declining, often are not listed because they are not championed by anyone. Although the cave species that are currently listed have multiple threats against them, other species may be in more danger than those already listed. Species on the list often serve, however imperfectly, as proxies that protect other species which should also be listed. Elliott (1991) analyzed the biogeography of more than 100 rare cave species in the Balcones Fault Zone of Texas. Travis County contains more than 30 troglobitic species endemic to that county, but only six of them were listed. His analysis selected a list of caves for the Balcones Canyonlands Conservation Plan that would protect all the endangered species as well as most of the other rare endemics, all of which are vulnerable. That plan now includes about 65 caves, half of which contain endangered species. However, lack of funding has imperiled even those caves (Elliott, 1997c).

The value of cave species to the public often is considered low; but cave species have potential scientific, practical, and educational value. To many, the size or intelligence of a creature is equated with its importance. Cave species may have good potential value to humans as "indicator species" in karst areas. That is, the decline of sensitive species because of pests or pollution may serve as a natural alarm for regulators and public health agencies. This is especially true of groundwater species, which may be affected by pollutants, pathogens, or nutrient stress over long distances. These contaminants can also affect people.

THREATS TO CAVE AND KARST COMMUNITIES

Declines in cave bat populations were noted as early as 1952 (Mohr, 1972). Studies of troglobites lagged behind because of the small number of biospeleologists. Poulson and Kane (1977) outlined several kinds of disturbance in cave ecosystems. They gave examples from Mammoth Cave, and with a flow chart they illustrated the results and interrelations of these disturbances. For a scientific management program, they outlined the elements of a biological inventory coupled with environmental measurements, taxonomic identifications, and evaluation of results. Many of the threats discussed below are so interrelated that researchers have difficulty determining which threats cause the greatest problems.

Hydrological threats

Water projects have caused many problems for cave biota. The Mammoth Cave system in Kentucky has the longest and best documented ecological history of any cave in North America. The Styx and Echo river areas in Mammoth Cave had an apparent decline of troglobites from the late 1800s to the 1910s or 1920s. In 1906, an impoundment (Lock and Dam #6) was installed on the Green River below Mammoth Cave. Green River naturally back-floods into the cave, but the levels are higher now than before this construction (Lisowski and Poulson, 1981; Lewis, 1982). About 1937, back-flooding into Bat Cave killed 300000 individuals of *Myotis sodalis* (Mohr, 1972). From the late 1950s to the 1970s, the cavefish and cave crayfish seen were all large, but in low numbers. In the 1970s the Nolin River, a tributary to Green River, was dammed below Mammoth Cave, and Green River was dammed upstream. Since then the maximum height of floods has decreased and the time to return to base-level flow after floods has increased. The Kentucky blind shrimp (*Palaemonias ganteri*) declined, and is now found farther upstream in the cave. Poulson (1996) concluded that toxins and organic enrichment, though present, were not the cause of the declines in the biota, which was probably due to siltation. Loss of nutrients also may have occurred (see below). The present rarity of *Amblyopsis spelaea* is probably related to silting and flooding associated with deforestation, forest fires, and the construction of Lock and Dam #6 (Poulson, 1968).

Lewis and Lewis (1980) studied *Caecidotea stygia* and another unnamed species of *Caecidotea* in Mammoth Cave. Base-level streams typically provide greater microhabitat diversity and perennial water supplies, but environmental disturbances have reduced habitat diversity and introduced physical and chemical conditions which cause animal communities to deteriorate. Isopods do not usually occur in Echo or Styx rivers, except where large breakdown slabs remain above the silt or where boards from old tourist trails have been discarded into the stream, creating artificial microhabitats harboring isopods. The once abundant crayfish *Orconectes pellucidus*, reported by Hay about 1902 in the River Styx, is no longer common, since the food source (*Caecidotea*) has disappeared.

A water-quality monitoring program in Mammoth Cave National Park from 1990 to 1993 provided baseline data on spatial and temporal changes, but no biological data were gathered. Two rivers and eight springs were sampled synoptically in natural woodlands, agricultural lands, and areas influenced by urban use and oil and gas exploration. The 36 parameters recorded at monthly intervals included discharge, turbidity, chloride, fecal coliform, organisms and triazine herbicides. The first 19 months of data demonstrated a strong correlation between land use in the drainage basin and water quality. In June 1990, Echo River Spring back-flooded from the Green River and received a large amount of flow from heavily used agricultural land, whereas its normal input is of low-turbidity water from parkland (Meiman, 1993).

In the late 1970s the New Melones Reservoir was built on the Stanislaus River in California. About 30 caves were inundated, including McLean's Cave, one of only two known localities for the troglobitic harvestman *Banksula melones*. The McLean's Cave community was transplanted to a nearby mine, but in a follow-up study *B. melones* was found in 18 caves and is now considered safe (Elliott, 1978, 1981). [See further discussion of this case under Ecological Transplantation below, p. 684].

The loss of recharge to and excessive pumping from a karst aquifer results in a gradual decline of water quantity and quality. The Devil's Hole pupfish (*Cyprinidon diabolis*) in Nevada is endangered by excessive pumping from the regional aquifer, which has reduced the fish to a tiny population at the bottom of a sinkhole (Elliott, 1981). Other karst pupfishes are also endangered (Table 34.6).

The Balcones Fault Zone portion of the Edwards Aquifer (Texas) supports more than 40 aquatic species. This unique community is in danger of severe impact from excessive pumping of groundwater, which will cause spring failures and dewatering of some of the system, and also cause encroachment of highly saline water into areas that now have high-quality water. The annual withdrawal by pumping of about 666 080 000 m^3 now approaches the annual average recharge of 774 627 000 m^3 (Longley, 1991). Four endangered animal species (the beetles *Heterelmis comalensis* and *Stygoparnus comalensis*, the fountain darter *Etheostoma fonticola*, and the salamander *Eurycea sosorum*), as well as the endangered Texas wild rice *Zizania texana*, and one threatened species (*Eurycea nana*), are dependent on this aquifer (see Tables 34.2 and 34.6).

Land development

The development of karst areas without regard to sensitive natural features has destroyed or filled numerous caves. In the area of Austin (Texas), caves are often searched out before the construction of roads and buildings because of endangered species protected by the United States Fish and Wildlife Service (USFWS). The excavation of sealed cave entrances during karst studies may lead to drying of the entrance zone. However, this may be offset by increased nutrient input from organic detritus, colonizing cave crickets, bats, and other species; but pests, such as fire ants, may also invade the caves more easily (see below, p. 678). Studies have shown that occasional feces deposited by visiting raccoons (*Procyon lotor*), who routinely use caves, enhances the population sizes of terrestrial cave communities; therefore the opening of sealed caves may be beneficial to cave communities if inputs and disturbance are not excessive (Elliott, 1994c; Elliott and Reddell, 1989; Reddell, 1991; Reddell and Elliott, 1991).

The mining of caves for saltpeter, bat guano, or other minerals can have a drastic effect on bat colonies and other fauna. Mexican free-tailed bats, *Tadarida brasiliensis mexicana*, have been disturbed by some guano mining in Texas, whereas other miners may have aided the colonies by mining-out rooms that might eventually have filled with guano. The better operations mine only in the winter when the bats are gone (Elliott, 1994d).

Quarrying and road building has completely destroyed many caves, but has also revealed some significant caves that were put to good use. Inner Space Cavern, near Georgetown, Texas, was discovered in 1963 during construction of Interstate Highway 35. It proved to be an important biological and paleontological cave, and was so attractive that it was developed as a show cave in 1966. The cave also contains two endangered species, but it is threatened by road spills, quarrying (and its possible effects on troglobite populations and groundwater), utilities, and encroaching residential developments (Elliott, 1994g).

The opening of large second entrances can severely alter the meteorology of a cave, causing bats to vacate it. Marshall Bat Cave in Texas lost its colony of free-tail bats after 1945, when a large shaft 40 m deep was dug into the rear of the cave to hoist out guano, causing too much ventilation and cooling of the cave (Elliott, 1994d). Free-tails require warmth and large caverns for maternity roosts (Herreid, 1963, 1967). Mammoth Cave in Kentucky probably harbored bats prehistorically, before the entrance was modified to block incursions of cold winter air. The National Park Service is currently trying to reinstate the natural temperature profile of the cave (Rick Olson, personal communication).

Sealing caves can be very harmful to cave fauna – even an imperfect gate can harm bats and other fauna (see below). Although there is little direct information on the effects of actually sealing a cave, I have observed that areas of caves that are naturally sealed by calcite, or truly entranceless caves that have been bored into, are rather depauperate. Poulson (1997) has noted the dry areas of Mammoth Cave caused by the sandstone cap-rock, which hinders percolation and food input. In the shallow caves of central Texas, those areas with soil creeping in from plugged sinkholes, or near tree roots, or with feces from crickets, bats or raccoons, have more abundant and diverse communities (Elliott and Reddell, 1989; Reddell and Elliott, 1991; Reddell, 1991). Some cave communities are highly dependent on cave crickets, which leave at night to scavenge on the surface; for example, different species of troglobitic beetles that prey on cricket eggs are found from Indiana to Alabama and Texas. Therefore, if entrances are sealed the input from cave crickets ceases unless access through small holes is still possible.

Caves that are sealed under pavement or buildings usually receive less infiltrating water, and may become barren. For example, Mayor Elliott Cave, which was entranceless, was discovered under a street and a house in Georgetown, Texas, in May, 1997. Investigation of the cave showed that the speleothem areas under the house slab foundation were dry, and there was no fauna. Passageways under a street received infiltration from irrigation of lawns and leaking street gutters. A moist soil slope from a sink or crevice in the yard contained a delicate, endangered, troglobitic harvestman, *Texella reyesi* (Elliott, 1997b).

Nutrient stress

Nutrient loss

Cases in which nutrient loss has resulted in noticeable effects in a cave have rarely been documented; one that is often cited is Shelta Cave, in Huntsville, Alabama. Shelta had the most diverse cave community known in the southeastern United States, but land development encroached on the cave in the 1960s, and the townspeople were concerned about youths entering the cave. The cave harbored a large colony of *Myotis grisescens*, the endangered gray bat. The National Speleological Society purchased the cave in 1967 to save it, and they moved their headquarters to a building nearby. The cave was gated in 1968 with a strong, cross-barred gate taken from an old jail. In hindsight, this gate was inappropriate for bats, and they abandoned the cave within two years. However, the urbanization of the area might also have doomed the colony. In 1981 a modern, horizontal-bar, "Tuttle style" door was put on the gate, but no bats returned to the cave (Hobbs and Bagley, 1989). *Myotis grisescens* usually does not tolerate even well-designed gates, especially for its maternity colonies (Robert Currie, personal communication).

Studies by Cooper (1969, 1975) elucidated the rich aquatic community in Shelta Cave, which included three species of extremely long-lived crayfish with low reproductive rates, a shrimp, cavefish, and numerous other troglobites. The aquatic system was dependent on food from bat guano, and it declined after the bats disappeared in 1970. The counts in Cooper's 1968–1969 census studies, when compared to data collected from 1985 to 1989 by Hobbs and Bagley (1989), showed that, after the bats left, crayfish decreased from a usual range of 49–250 (usually >100) to 0–10, cavefish decreased from "many" to 3–15, and the Alabama cave shrimp dropped from 1–25 to none (Hobbs, 1996). Hobbs and Bagley analyzed cave water samples and found small amounts of heptachlor epoxide ($0.5\,\mu g\,\ell^{-1}$ in 1987, $0.04\,\mu g\,\ell^{-1}$ in 1988). The insecticide probably was leaching into the groundwater from house foundations in the area treated against termites. The shrimp was known from two caves, and was listed as endangered in 1988 (Table 34.2), but has since been found in a river cave in the area (H.H. Hobbs, III, personal communication). It is still not certain whether the cave gate, which caused a nutrient loss by causing the bats to leave, was the primary cause of the community's decline, or if toxins and disturbance were also to blame. All of these factors probably played a role in the decline.

Poulson (1996) suggested that loss of fine particulate organic matter, caused by river damming, could have influenced the decline of aquatic fauna in Mammoth Cave.

Sometimes terrestrial invertebrate cave communities can survive well in urban areas. Elliott and Reddell in 1989 found that Bandit Cave, under a wooded lot in a residential area of Austin, Texas, still contained all of the species found there in the 1960s. The owner protects the cave, and raccoons still visit it, leaving feces. However, there are no real census data to evaluate the situation properly. It may be difficult to maintain aquatic communities and colonies of some bat species in urban areas.

Nutrient enrichment

Enrichment appears to be a much more common problem in caves than nutrient loss. Pasquarell and Boyer (1993) reported on weekly water samples taken in the fall of 1990 from four springs in Greenbrier County, West Virginia. Besides low levels of the herbicide atrazine being detected in all springs, mean nitrate levels were 13.6 and 10.8 mg ℓ^{-1} from two basins. Mean bacterial levels for the two basins were 101 and 139 fecal coliform colonies per 100 ml and 266 and 276 fecal *Streptococcus* colonies per 100 ml. Samples from nine cave stream sites had nitrates ranging from 13.4 to 63.7 mg ℓ^{-1}. Fecal coliforms ranged from 110 to 28 588 colonies per 100 ml. One cave, which had the highest levels of nitrate and fecal coliforms, receives flow from a sinkhole which is immediately adjacent to a feedlot on the surface. Species diversity in this last cave was substantially reduced as compared with the figures for other, less contaminated caves. Aley (1997) mentioned the odor of animal wastes from a hog-raising operation permeating the air of one of the largest caves in Missouri.

A study of three caves in Cookeville, Tennessee, suggested a relationship between degraded water quality and reduced biotic diversity. Capshaw and Ament caves, which receive sewage-contaminated runoff, were dominated by oligochaetes and dipterans, and had larger coefficients of variation for most water-quality parameters than City Spring Cave (Pride et al., 1988).

Karstic groundwater is the major water source in the Yucatán Peninsula, and had a major influence on the Maya culture. In Mérida some waste water is disposed of by deep-well injection, but its fate has not been traced. Pig farms and cattle ranches are another potential source of pollution, and use of fertilizers and pesticides threatens the local water supply in some areas. Solid waste is often dumped at the edges of towns or discarded into dry caves. Cenote Dzitya, near Mérida, was contaminated by a nearby pig farm, according to water chemistry and algal data. Preliminary data from 75 water bodies showed that undisturbed inland water bodies contained relatively low concentrations of nutrients (0.6–16.0 μmol $NO_3^- + NO_2^-$, 0.1–6.0 μmol NH_4^+, 0.2–1.8 μmol P, and 2.3–74 μmol Si), whereas culturally impacted ecosystems and coastal lagoons had evidence of enrichment (19–162 μmol $NO_3^- + NO_2^-$, 6–62 μmol NH_4^+, 2.5–13.8 μmol P, and 93–544 μmol Si). A pig farm (Agropecuaria Yucatán) was constructed above Cueva de El Pochote, which contained a unique cave fauna, including the fishes *Ogilbia pearsei*, *Ophisternon infernale*, the isopod *Creaseriella anops*, and the shrimps *Creaseria morleyi* and *Typhlatya pearsei* (Horst Wilkens, personal communication). Cenotes provide important habitat for stygobionts and other species, such as the threatened Morelet's crocodile, and provide drinking water for endangered mammals such as the jaguar (Brenner et al., 1995). No studies of pollution effects on cave species of this region have been published to date.

In July 1993 I observed many annelid worms in a stream in Whispering Canyon Cave, in the Tongass National Forest, southeastern Alaska, which may have reflected elevated nutrient loads from nearby logged areas (Aley et al., 1993). Problems caused by logging have been reported in similar cave areas on Vancouver Island (British Columbia, Canada). Many caves were choked with slash and sediments, and small buffer zones of uncut timber around caves were blown down by windstorms because they were inadequate in size and design (Blackwell and Associates, Ltd., 1995; Stokes, 1996; Paul Griffiths, personal communication).

Enrichment is a common problem, especially in show caves, where it is caused by lunch rooms, garbage, litter, sewage, cave lint, and lighting that promotes the growth of Cyanobacteria. These disturbances change the distribution and abundance of species within the cave. Examples from several prominent caves are given below.

Mammoth Cave, Kentucky. The cave amphipod, *Crangonyx packardi*, used to be common in Shalers Brook in Mammoth Cave, which at one time was enriched by sewage effluent from the Mammoth Cave

Hotel. The sewage problem has now abated and *C. packardi* is no longer there, but *Stygobromus vitreus*, which is more cave-adapted, is present (Lewis, 1984). The artificial Crystal Lake, near the Frozen Niagara Entrance, was created from damming a short shaft drain, and was used for boat rides. Barr and Kuehne (1971) reported that *Caecidotea stygia* and *Orconectes pellucidus* were there, but Lewis (1984) found neither and noted that, "Prying boards from the bottom of the lake uncovers a black layer of anaerobically produced *sapropel* with its characteristic accompanying odor, a habitat which does not seem conducive to cavernicoles."

Pollution from the flushing of a sewage lagoon used by the Job Corps in the summer of 1967 reduced the complexity of the cave community from that seen under unpolluted conditions in 1966. It also reduced the numbers of terrestrial organisms trapped per trap–day by a factor of more than ten at two stations near the headwaters of Eyeless Fish Trail. Organic content was 240% of the 1966 level. A "high-tide-line" of Cyanobacteria, like that associated with foam and coliform pollution in Keller Shafts, was marked (Poulson, 1967). The Job Corps Center was removed several years later. Attention was focused on potential sources of contamination. The United States Environmental Protection Agency issued an Environmental Impact Statement on Mammoth Cave. New waste-load allocations were to be issued for public treatment plants in the area. The National Park Service financed a regional sewage treatment plant (Poulson, 1997).

In Mammoth Cave most contaminant transfer occurs in the first flush of rainfall events. Farmlands on karst contribute non-point-source sediments, pesticides, herbicides, animal wastes, and bacterial loads to cave streams. Urban development brings sewage, solid waste, and leakage from buried storage tanks and pipelines. Unnatural sediment deposition, entry of exotic species, and changes in deposition of particulate organic matter from upstream threaten ecosystems at the water table or base level (Poulson, 1996, 1997).

The pathogen *Salmonella choleraesuis* was recovered in water from the Hawkins River and Owl Cave in Mammoth Cave National Park, and *Salmonella* sp. was found in Owl Cave sediment. Echo River in Mammoth Cave lacked *Salmonella*. A possible source of the bacteria was septic systems that had failed (Rusterholtz and Mallory, 1990).

Poulson (1992) developed community signatures for different kinds of pollutants, contrasting acute toxins, chronic toxins, organic enrichment, and siltation. "Indices of biological integrity" included the number of species and metrics for their well-being, and the presence and relative abundance of tolerant and intolerant species. Gross enrichment is usually from point sources, and can be detected as a slippery biofilm on rocks in cave streams, with stimulation of short-lived troglobites like flatworms and isopods.

In many show caves, plant growth around electric lights is a continual problem. Algae and moss protonemata had to be steam-cleaned from ceiling formations at Frozen Niagara in Mammoth Cave (Aley, 1997).

Enrichment sometimes comes from unusual sources. Lisowski et al. (1986) found that release of oil brines or sulphur water from oil and gas exploration could have major adverse impacts on the Mammoth Cave system. They found sulfur bacteria in Sulphur River in Parker's Cave, which is strikingly different from other cave streams in the region. The water contained high levels of sodium chloride (up to 1.6% Cl^-) and sulfide. The substrate of the stream was covered with a white mat-like material several centimeters thick. Two sulfur oxidizers, *Beggiatoa alba* and "*Thiothrix tenuissima*", were in the mat along with others. Water and mat samples contained a number of Protozoa in 12 genera, including flagellates, ciliates, and amoeboid forms. Two species of annelids, a cave snail, five species of collembolans, a cave carabid, a linyphiid spider *Phanetta subterranea*, and several species of mites were found. Water droplets on a spider's web were milky white because of rod-shaped bacteria, and had a pH of 0.13. Northup et al. (1997a) characterized the bacteria as sulfide oxidizers, and found phylogenetic relationships with communities of deep-sea thermal vents and tidal mud-flats.

Hidden River Cave, Kentucky. The Hidden River (Horse) Cave system was commercialized in 1916, the tourist industry co-existing with water pumping and hydroelectric generation. Hidden River Cave had a troglobitic fauna, including the cavefish *Typhlichthys subterraneus*, the crayfish *Orconectes pellucidus* and the isopod *Caecidotea bicrenata*. However, increasing groundwater contamination from indiscriminate sewage disposal led to closing of the cave's tourist operation in 1943. Creamery wastes, and sewage from Cave City and the City of Horse Cave, were introduced into the cave stream. In 1970 waste water from a chrome-plating factory was added to the sewage-plant effluent entering the cave. The troglobitic community was extirpated as a result of degradation of the cave. The South Branch of Hidden River Cave contained large numbers of red tubificid worms (*Tubifex* sp.) at the edges of stream pools, "sewage fungus" (characterized by the bacterium *Sphaerotilus natans*), nearly anaerobic conditions, low nitrates, high nitrites, and blackened iron sulfide. In 1983 the East Branch of Hidden River had recovered much of its natural character by the time it reached Hidden River Cave from the Cave City treatment plant. The water was nearly saturated with oxygen and probably supported troglobites. In 1989, new sewage-treatment facilities were opened and the flow of effluent to the cave stopped. By about 1995 the original animal community had substantially recolonized the section of the cave which had once been heavily polluted, from relatively unpolluted, upstream tributaries (Lewis et al., 1983; Lewis, 1996a). Such an outcome could not have occurred in many other hydrologic situations. Today, visitors can take an ecological tour into the cave and observe the troglobites.

Carlsbad Cavern, New Mexico. Except for the Bat Cave section, where a large colony of free-tail bats (*Tadarida brasiliensis mexicana*) resides, Carlsbad Cavern had a sparse fauna which became disturbed after the cave was opened to the public in 1923. Bailey (1928) observed that the Lunch Room in the cave was already in the 1920s leading to an invasion of the King's Palace area, the southern Big Room, and the lower cave by *Peromyscus leucopus* mice. Although this species commonly is found in cave entrance areas, where it feeds on crickets, the mice already were permanent residents deep in the cave, as evidenced by females with embryos or nursing young. Bailey found cave crickets and tourists' lunches in the mouse stomachs.

Studies of rhaphidophorid crickets in Carlsbad Cavern have shown unnatural distributions. *Ceuthophilus carlsbadensis* naturally occurs in the Bat Cave section. *Ceuthophilus longipes* now occurs far from the entrance in total darkness in the Sand Passage. In the Left Hand Tunnel (just past the Lunch Room) *C. carlsbadensis* and *C. longipes* occurred near reliable food sources such as trash containers. In the Left Hand Tunnel, a greater number of crickets was found directly behind the gate about 100 m from the Lunch Room, and also 300 m from there at a probable oviposition site. This appears to be an artificial-entrance situation, with the lighted Lunch Room serving as an *Ersatz* outdoors, complete with food, and the crickets hiding in a darkened area behind the gate, and ovipositing farther back. Any food dropped by park visitors attracts crickets to feed upon it. Rangers have reported crickets feeding on bat carcasses in the Queen's Chamber, and on human feces left on or near visitor trails (Northup and Kuper, 1987; Northup et al., 1987, 1989). The National Park Service instituted tighter housekeeping controls on the Lunch Room operators, and now removes the contents of all trash barrels by the end of the day to combat the large number of raccoons that invaded the cave at night (National Park Service, 1996; Dale Pate, personal communication). The National Park Service's new management plan (1996) would remove the facilities from the Lunch Room.

Aley et al. (1985) conducted studies on exotic plant growth in Carlsbad Caverns. They identified 26 species of algae, together with moss protonemata and two ferns (*Cyrtomium auriculatum* and *Dryopteris filix-mas*). There may have been a total of 100 to 200 species of algae growing in the cave. The growths in the cave included about 70% Cyanobacteria, 20% green algae, and 10% moss protonemata. Diatoms were present in about 25% of all clusters, and a few yellow-green algae were present. Many of the algal genera, and three of the algal species, in Carlsbad have also been found growing in total darkness in other caves. Algal growth, once abundantly established due to artificial lighting, will not quickly disappear if it is deprived of light (Aley et al., 1985). Light intensities were measured at a number of sites. In alcoves the minimum light intensity threshold for algal growth had a mean value of 17.2 lx (1.6 ft c) with a standard deviation of 7.5 lx. In non-alcove sites (most situations) the mean threshold was 47.4 lx (SD = 20.5 lx).

Cave lint studies more or less started in Carlsbad Caverns in connection with annual "lint camp" clean-ups by volunteers. Jablonsky et al. (1995) placed "seeded lint" (which looks different under ultraviolet light from normal lint) on the trail, and found that it moved up to 100 m down the trail; much of it moved to the edge or off the trail. Slides of lint contained synthetic and natural fibers, dirt, wood, insect parts, human hair, animal fur, fungus, processed tobacco, paper, and other things. Unidentified mites were seen in some of the samples (Pat Jablonsky, personal communication).

In Carlsbad Cavern, Elliott (1997a) studied old woodpiles, which probably were remnants of structures installed in the 1920s. Most of the wood was depleted of nutrients and supported little visible fungal or bacterial growth any longer. However, some wood was sodden, consisting of up to 80% water by weight, and served as a moist substrate. Some piles contained almost no invertebrates, while others were a haven for several species. Elliott recommended removing most of some of the piles over a period of time, leaving a small residue for the fauna to utilize. He studied Signature Pool near the trail, which is a microcosm of algae, eyed flatworms (*Phagocata* sp.) and eyed copepods which probably were introduced; the small community is driven by a light bulb hanging over the pool. In another part of the cave a sewage leak infiltrated into the cave and caused some localized fungal growths and swarms of fungus gnats, which supported a local community of spiders.

In another part of Carlsbad Caverns National Park, urine and feces left by cavers in Lechuguilla Cave contaminated some areas with bacteria, despite strict rules about carrying wastes out of the cave. The National Park Service was concerned about such microbiota altering the cave environment, possible effects on any cave-adapted species, and health effects. Allowing contaminated areas to "rest" for a few months often resulted in the disappearance of the exotic species, provided that visible residues were removed (Northup et al., 1997b).

A Midwestern Show Cave. In a bizarre case in 1993, earthworms (of two exotic European species, *Dendrobaena rubida* and *Eisiella tetraedra*) came out of the cave walls, and rocks fell out of the ceiling for an extended time. I was acting as consultant for the cave management, who thought that the earthworms were causing the problems, and that they should be exterminated. Systematic observation and sampling revealed that the earthworms were following infiltrating sewage into the cave from broken sewer lines and forgotten, leaking, septic tanks built on top of the cave. The epikarst in that area, though mantled with good soil, was highly transmissive. Dye traces proved that the cave was cross-connected to many input points on the surface, including septic systems. Bacteriological sampling showed the presence of *Escherichia coli*, *Salmonella*, and *Shigella* in many air and water samples from the cave. The earthworms reported were found crawling on wet surfaces in the cave, and were actually eating softer, marly beds of rock, which were probably laden with bacteria. There was little or no odor in the cave, but levels of airborne bacteria were high enough to cause concern about prolonged exposure. Clay banks in some areas were festooned with rich and colorful fungus gardens, which were inhabited by invading hothouse millipedes and isopods. No troglobites could be found in the cave. After these initial findings, the owners initiated intensive maintenance and repair of the septic system, and took steps to prevent infiltration of nutrients from the theme park above the cave. The earthworms were not killed, but were to be considered as "friends" who would eat up much of the contamination. After about two years of work, the management reported that, biologically, the cave was returning to normal, and that rock-fall (which probably was related to both high rates of water infiltrations and earthworm activity) had decreased to a low level (Elliott, unpublished data).

Exotic and pest species

As has been seen above, enrichment of caves with nutrients can bring sewage bacteria and fungi, red tubificid worms, earthworms, hothouse millipedes, isopods (sowbugs, woodlice), and other exotic species into the cave. In central Texas, imported red fire ants, *Solenopsis invicta*, began invading caves in 1988, not because of enrichment but because of the ongoing adaptation of this Brazilian pest to the southern United States. The multiple-queen colonies live in soil mounds, but the workers invade shallow caves in the summer, when the soil is hot and dry, and forage for moisture and food in the caves. They have been observed attacking and carrying off various troglobitic species and cave crickets. Some caves in the area of Austin (Texas) are so overrun with fire ants in the summer that it is unsafe to crawl into the caves. The ants retreat to the surface again in the autumn, but can reproduce year-round if the weather is mild. The best control method is to kill the mounds with boiling water. This is instantaneous, but labor-intensive; however, it avoids the problem of using insecticidal baits, which could be picked up by cave crickets as they forage on the surface at night (Elliott and Reddell, 1989; Elliott, 1990, 1991, 1992b, 1993a, 1993b, 1994b,d, 1997b; Reddell, 1991).

Chemical pollution

Cave life has been affected more often by water-borne pollutants than by solid waste. In this section I mostly discuss chemical pollutants, but waste from sewage, agriculture, or cavers can have toxic effects too, and they often go together.

Spent carbide from acetylene lamps was the most

common poison recognized by cavers in their environment. Until the 1960s many cavers dumped or buried their spent carbide in caves, but the practice was discouraged. Peck (1969) pointed out that the calcium hydroxide in spent carbide was poisonous to cave fauna, which he demonstrated in experiments with the cave beetle *Ptomaphagus hirtus*. Fortunately, carbide dumping and use is declining. I found no reports of any kills resulting from spent carbide, but the effects would be subtle. Batteries left by careless cave visitors can leak toxic materials, including mercury.

The most dramatic kill of cave biota yet reported was in November of 1981, when some 80 000 ℓ of liquid fertilizer (ammonium nitrate and urea) spilled at a pipeline break near Dry Fork Creek, Missouri. Dry Fork is a disappearing stream, and a recharge area for Meramec Spring, the state's third largest spring. Seven days following the break, dissolved oxygen at Meramec Spring, a distance of 21 km from the break site, dropped to less than 1 mg ℓ^{-1} for nine days, resulting in a loss of over 37 000 fish at a hatchery. Concentrations of ammonia and nitrate nitrogen in the spring were elevated for over 38 days. Aquatic organisms killed included >10 000 of the Salem cave crayfish *Cambarus hubrichti*, nearly 1000 of the southern cavefish *Typhlichthys subterraneus*, which had not previously been reported from the Meramec basin, and a small number of the Ozark blind salamander *Typhlotriton spelaeus*. Numerous other cave organisms killed included amphipods, isopods, and gastropods, but no attempt was made to quantify these losses (Crunkilton, 1985). A few years later cave divers observed some of these species in side passages or pockets in the spring system (Eugene Vale and Jo Schaper, personal communication). Baseline data and real follow-up data are lacking, so the long-term effects of this spill are still unknown. This incident points up the need for baseline data and for ongoing hydrological and biological monitoring.

Perhaps even more devastating than the Meramec Spring kill was the presumed sterilization by nitric acid of caves at the Indiana Army Ammunition Plant, near Charlestown, Clark County, Indiana. The plant was constructed during World War II for the production, shipping, and storage of nitrocellulose propellant. The caves of the Jenny Lind Run drainage received nitric acid at a pH of 2.3 as a waste product of the nitrocellulose production process, at a rate estimated at about 85 000 ℓ min^{-1}. Now, 50 years later, six species of troglobite have recolonized the nitric acid caves of Jenny Lind Run. The aquatic troglobitic isopod *Caecidotea stygia* has recolonized all of the caves and springs on Jenny Lind Run. Terrestrial troglobites inhabiting the caves include the millipede *Pseudotermia nefanda,* the dipluran *Litocampa* sp., the collembolan *Pseudosinella* sp., the fly *Spelobia tenebrarum* and the spider *Phanetta subterranea* (Lewis, 1996b). The true composition of the original fauna is not known.

Bats with insectivorous diets, long life, and long migrations are sensitive to biological accumulation of pesticides. Mexican free-tailed bats, *Tadarida brasiliensis mexicana*, in Carlsbad Cavern declined from 8.7 million in 1936 to only 200 000 in 1973, at least partly due to residues of dichlorodiphenyltrichloroethane (DDT) in their diets. This insecticide had been used extensively on irrigated cotton in the Pecos River Valley nearby. Studies showed that residues of DDT and DDE (dichlorodiphenyldichloroethylene, a metabolite) in the colony did not decline much after the insecticide was banned in the United States in 1972. After 1980 the residues declined slightly, but rose again in 1987–1988, perhaps due to usage of DDT in Mexico. Gray bats, *Myotis grisescens,* from two maternity caves in Franklin County, Missouri were killed by dieldrin, a metabolite of aldrin[1], in 1976–1978.

Residues of dieldrin and heptachlor epoxide, which had been used locally on corn (*Zea mays*), were found in guano and insect prey. One colony declined from 1800 in 1978 to none in 1979–1982. Dieldrin also killed gray bats in three caves in Boone County, Missouri (Clark, 1988). Peck (1974) reported that public health authorities fumigated Chilibrillo Cave in Panama to remove the bat colonies shortly after his survey there. The effect on invertebrates was not studied. Other factors have also reduced bat populations (see below, p. 681).

Sewage and gasoline leaking from a service station into groundwater poisoned parts of the Young–Fugate Cave System, Virginia. The cave is inhabited by two troglobitic crustaceans, and is the only known locality for the cave beetle *Pseudanophthalmus holsingeri* (Table 34.4). The service station was removed. Contaminants that have fouled karst systems in Virginia include spills or leaks of petroleum products, herbicides, sheep

[1] aldrin = (1R,4S,4aS,5S,8R,8aR)-1,2,3,4,10,10-hexachloro-1,4,4a,5,8,8a,hexahydro-1,4:5,8-demethanonaphthalene.
dieldrin = (1R,4S,4aS,5R,6R,7S,8S,8aR)-1,2,3,4,10,10-hexachloro-1,4,4a,5,6,7,8,8a-octahydro-6,7,epoxy-1,4:5,8-demethanonaphthalene.

Fig. 34.3. Used drums of chlordane and methamidophos in the entrance of Toxic Sink, Coahuila, Mexico. Photo by Peter S. Sprouse.

and cattle dip, solvents, fertilizers, sewage, milk, cream, and the leachate from improperly disposed waste materials (Hubbard, 1996; Hubbard and Balfour, 1993). In 1995, a resident crustacean was decimated in Wildcat Saltpeter Cave, Virginia, by diesel fuel from a leaking underground storage tank (David Hubbard, personal communication). Sawdust and bark caused a die-off of the crustaceans *Caecidotea recurvata*, *Crangonyx antennatus* and *Lirceus usdagalun* in Thompson Cedar Cave, Virginia (Culver et al., 1992).

Brine pollution from the Greensburg Oilfield upstream from Mammoth Cave began in 1958 and continued for several years. The chloride ion allowed Hendrickson (1961) to estimate the contribution of water from Green River to the cave through Styx River Spring and out again at Echo Spring – about 40% from Green River and 50% from local groundwater sources. There was a reduction in sightings of cavefish, crayfish, and shrimps just after the brine contamination. A massive kill of cave crayfish occurred in Hawkins River (Mammoth Cave) under Joppa Ridge in 1979, probably from a petroleum spill or leaking underground gasoline tank. An accident in 1980 involving two trucks spilled ink and cyanide; quick action by officials prevented large quantities from entering Hawkins River and killing all the aquatic life downstream to the Green River. At Mammoth Cave most contaminant transfer occurs in association with rainfall events (Poulson, 1996).

Slash and sedimentation from logging can choke cave entrances and alter aquatic communities. Logging and road-building practices in the karst areas of the Tongass National Forest, southeastern Alaska, caused some caves to receive elevated loads of sediment, sometimes including diesel fuel and other petroleum products, which could affect cave stream communities, salmon runs downstream, and drinking-water sources (Aley et al., 1993; Elliott, 1994a). Similar problems have occurred on the karst of Vancouver Island (British Columbia, Canada) (Blackwell and Associates, Ltd., 1995).

Waste is being dumped into caves in some parts of Mexico. The name of Sótano de Médico, near Tlamaya, San Luís Potosí, reflects the large amount of medical waste, including used syringes, that had been dumped into it (Minton, 1992). In July 1996 I personally saw used drums of the organochlorine pesticide chlordane[2] and the organophosphate pesticide methamidophos, both highly toxic, in Toxic Sink, Coahuila (Fig. 34.3).

Nearby, there was medical waste, including syringes, in Pozo del Cañon El Buey (Fig. 34.4). As mentioned above, fertilizers and pesticides threaten the local water supply in some areas of Yucatán.

In an experiment in a remote area of Mammoth

[2] chlordane = 1,2,4,5,6,7,8,8-octachlor-2,3,3a,4,7,7a-hexahydro-4,7-methanoindane. methamidophos = O,S-dimethyl phosphoramidothioate.

Fig. 34.4. A bag of medical waste, including syringes, at the bottom of Pozo del Cañon El Buey, Coahuila, Mexico. Photo by Peter S. Sprouse.

Cave, Rusterholtz (1989) dosed core samples of saturated sand with 5 μg of chlordane per g of sample for 30 minutes, and observed a 25.8% decrease in respiration. Cell counts were 6.2 10^5–1.5 10^6 cells per gram; over 50% of the total cells were actively respiring before treatment. This shows that even a very low concentration of pesticide can kill or suppress microbes in cave sediments.

Welding fumes can be very toxic, especially if the metal being welded contains zinc or other toxic elements. Welding in caves should be preceded by a study of air movements to see if contaminants will be naturally blown out of the cave entrance, or else a temporary exhaust system should be used (Elliott, 1995).

Nicotine is a powerful insecticide, which was extensively used to fumigate greenhouses before synthetic pesticides became popular (Feinstein, 1952). Howarth (1983) warned against the use of tobacco in caves because of its nicotine content; but there are about 4000 other harmful chemicals in tobacco smoke, including acrolein, formaldehyde, α-benzopyrene, and carbon monoxide. Elliott (1995) improvised an exhaust system for removing welding fumes during the construction of a bat gate in Gorman Cave, Texas; however, the system was overwhelmed by the workers, who smoked cigarettes in the cave, creating a cloud of particulates. Nitric oxide and carbon monoxide were within acceptable limits as defined by the United States Occupational Safety and Health Administration for workers, but equipment was lacking for measuring other contaminants from tobacco smoke. Jablonsky et al. (1995) found processed tobacco in cave lint, but this was not quantified.

Killing, over-collecting, and disturbance of fauna

Bats in caves have often been killed as a result of human action; a few examples are provided. As mentioned above, back-flooding from the dammed Green River into Bat Cave in Mammoth Cave National Park killed 300 000 individuals of *Myotis sodalis* (Mohr, 1972). In 1960, three boys intentionally killed 10 000 individuals of this species in Carter Caves State Park, Kentucky (Mohr, 1972). In 1961 an entire colony of *Myotis velifer* was killed or driven off from Chinaberry Cave, Williamson County, Texas, by an intruder using a .22 caliber rifle with rat-shot (Elliott, 1994d). Eagle Creek Cave, Arizona, housed a colony of Mexican free-tailed bats, *Tadarida brasiliensis mexicana*, numbering 25 to 50 million in 1964, but the population shrank to 600 000 by 1970 (Mohr, 1972), possibly because vandals disturbed the bats, using shotguns; DDT may have also been involved. In 1973 a developer filled Bear Cave, Bexar County, Texas, with large boulders, sand, and gravel, out of fear of liability because a person had been stuck in the entrance and had to be rescued; at least 20 000 bats (probably *Myotis velifer*) were roosting in the cave at the time (Elliott, 1994d). In 1983 in Walkup Cave, Hardeman County, Texas, local people used shotguns against bats, sprayed them with gasoline, and set them on fire. Walkup was an important roost for *Eptesicus fuscus pallidus*, *Myotis velifer* and *Corynorhinus townsendii pallescens* (Elliott, 1994d,e). In 1987, four men were convicted of shooting and crushing to death at least 66 (but probably several hundred) endangered Indiana bats, *Myotis sodalis*, in Thornhill Cave, Kentucky (Foster and MacGregor, 1987).

In Ontario, Canada, *Myotis lucifugus* and other species have declined in most of the unprotected caves and mines. In Fourth Chute Cave in Quebec, Canada, where the largest known population of *Myotis leibii* in eastern North America hibernated, commercialization eliminated the cold-air circulation, and the bats abandoned the cave (Mohr, 1972). Some time between 1993 and 1996, a colony of *Leptonycteris* sp. was killed or driven off by local people from a cave near Mezcala, Jalisco (Mexico). The people probably were after the common vampire bat, *Desmodus rotundus*, which inhabits another cave nearby (Mario Sgro, personal communication). Many other colonies of cave bats have been destroyed in Mexico in efforts to kill vampire bats, or as a result of guano and phosphate mining.

A general decline in many species of bat has caused great concern to bat biologists and conservationists. Bats are especially sensitive to human disturbance, even as slight as travel past their hibernating or nursery roosts. Continued disturbance can cause bats to abandon a cave. Disturbance during either summer or winter is critical, since only a few caves have microclimates suitable for colonial bats (Mohr, 1972; Poulson, 1976). Disturbance and arousal of hibernating bats can cause fat reserves to be used that would last them for 10 to 30 days (Brady, 1982). College classes and tourists at some show caves commonly visited bat roosts until the 1970s.

The banding of bat wings with bird leg-bands was widely used by bat biologists from 1932 to about 1960, but the practice ceased when biologists became aware of the high mortality rate from the bands, which tore wing membranes and injured bones. Ironically, the banding studies alerted biologists, as early as 1952, to the ongoing declines of cave bats, some of which was caused by banding and collecting, but more by habitat destruction, disturbance, and insecticides (Mohr, 1972). The steady recovery of *Myotis sodalis* in Bat Cave, Mammoth Cave National Park (Kentucky), was reversed after a biologist banded all 250 in 1971; the population dropped to only 68 in 1975. Biologists adopted lipped bands, and today they use tiny markers or radio transmitters glued to the fur.

Over-collecting and handling of animals was an early concern of biospeleologists. Sullivan (1956) cautioned against excessive collecting of specimens, and the National Speleological Society adopted a conservation policy that discouraged collecting by amateurs or for any commercial use. Over-collecting was feared to have caused the decline of the cavefish *Amblyopsis spelaea* in Mammoth Cave, specimens of which were sold as curios in the 19th Century (Poulson, 1968). Over-collecting can threaten small populations with low reproductive rates, as in the case of Shelta Cave, where some crayfish do not become sexually mature until they are over 40 years old. Generally, larger troglobites, such as cave-fish, salamanders, and crayfish, may be long-lived, have small population sizes, and reproduce slowly; therefore collecting should be highly restrained. Smaller, more abundant troglobites, such as some amphipods, isopods, millipedes, and beetles, may withstand collecting better. However, some small species are exceedingly rare, which should be taken into account. Culver (1982) emphasized the need for restraint in collecting by reporting that he had caused a severe decline in the populations of several cave isopods.

Over-use by visitors may cause compaction of substrates, with disturbance or loss of microhabitats for small, cryptic species. Many troglobites are thigmotactic – that is, they like to hide under rocks. The effect of substrate trampling on such species has not been adequately studied.

Isolation

The isolation of caves and karst areas by land development, road-building, utilities and quarries has rarely been addressed in the academic scientific literature, but is a common concern of applied scientists. The concerns are at several levels: potential alteration of hydrologic inputs to the cave; loss of nutrient inputs by making intervening areas unsuitable for trogloxenes like cave crickets and raccoons; and vegetational changes.

A large quarry near Inner Space Cavern, Williamson County, Texas, probably has destroyed many caves since 1963. This quarry may be creating a barrier to subterranean fauna in the narrow band of Edwards Limestone south of Inner Space Cavern. For several reasons, and as a hedge against the extirpation of some populations, the species recovery plan for seven terrestrial troglobites in that area calls for at least three cave preserves in each recognized "karst faunal zone", based on geology and recognizable cave faunas particular to karst blocks (Veni, 1992; O'Donnell et al., 1994).

MANAGEMENT OF CAVES AND KARST PRESERVES

Baseline faunal and ecological surveys

It has become axiomatic that baseline faunal surveys,

with simultaneous monitoring of ecological factors such as temperature, humidity, moisture, air movements, and nutrient inputs, are essential for understanding changes in cave faunas (Poulson and Kane, 1977; Elliott and Reddell, 1989; Perkins, 1990; Elliott, 1994c; Northup and Welbourn, 1995; Lewis, 1996a). Cave faunas change seasonally and in response to climate. Initial faunal surveys can be accomplished in several ways, including hand collecting, baiting, Berlese extraction, and pitfall trapping. Once a fauna list is compiled, census work in different seasons and years increases the range and value of the data set. Large fluctuations in some trogloxenic populations can occur on an annual time scale; for instance, cave cricket populations in some caves in central Texas dropped by half after a drought year in 1996 (Elliott, unpublished data). Historical data from the Mammoth Cave ecosystem over 100 years have been useful in understanding the current ecology of the cave. Comparative studies of different types of cave communities provide a better idea of how to manage a particular cave, such as a show cave that has been perturbed for many years.

Cave gating

Early cave gates were sometimes harmful to natural cave communities, especially bats; they were either too weak to exclude vandals, or too restrictive for bats and for air flow (Elliott, 1996b). MacGregor (1993) found that some early gates actually caused most of the decline in Indiana bats (*Myotis sodalis*) since the 1950s and 1960s. These gates often had crisscross bars or steel plates, which blocked air flow, and bats actually had to land on some of them to get through. Some gates had concrete sills or walls that projected into the passageway, restricting the flow of cold air along the floor, which changed the temperature profile of the cave. Indiana bats prefer very cold caves for hibernation, so they abandoned some caves that had become too warm. Better gates have allowed some of those populations to increase again. The standard bat gate now recommended by the American Cave Conservation Association, Bat Conservation International, the United States Fish and Wildlife Service, and the National Speleological Society is usually made of horizontal pieces of stiffened angle iron spaced at intervals of 15 cm, with vertical supports no closer than 1.2 m apart (Tuttle and Taylor, 1994; Elliott, 1996b). Some bat species do not tolerate gates at all, while some tolerate them better if the gate is built inside the cave instead of at the entrance, where there is more light and a higher risk of predators lurking (Robert Currie and Merlin Tuttle, personal communication). As mentioned above, *Myotis grisescens* vacated Shelta Cave after a gate was installed. *Tadarida brasiliensis mexicana*, the Mexican free-tailed bat, cannot tolerate gates because of its flight geometry and huge colonies.

Fig. 34.5. A strong cave gate on the entrance of Lakeline Cave, Texas; length about 2 m. The gate has shielded hinges and two locks for greater security. Note the animal access hole through the ledge, which provides access to raccoons and mice. Photo by William R. Elliott.

In Texas, many caves have been gated for the protection of endangered or rare troglobites. Such gates may have tight spacing of the angle iron bars, but have animal access holes at least 20 cm in diameter built into the edge to allow cave crickets, raccoons, and other fauna to pass through (Fig. 34.5). Conservationists have made so many innovations in cave-gate design and construction that the gating handbook of the National Speleological Society (NSS) (Hunt and Stitt, 1975) is being revised by the American Cave Conservation

Association, Bat Conservation International, the NSS, and the USFWS.

Helf et al. (1996) emphasized the importance of protecting and managing cave crickets at Mammoth Cave, and recommended bypasses for crickets around gates. Vale and Jones (1993) restored microclimates in Onondaga Cave, Missouri, by plugging artificial entrances which, though gated, allowed too much dry air into the cave.

Cave restoration

Cave restoration projects are beneficial in removing harmful materials, trash, and graffiti. Some precautions are needed to avoid stressing cave communities that may have colonized organic materials, particularly old woodpiles, which may be a haven for invertebrates (see p. 678 above). Elliott (1982) and Hubbard (1995) emphasized that wood should be examined by a biologist during clean-up projects, or left alone. Such woodpiles may have attracted a large population of invertebrates over decades of time, and to remove it suddenly may, in effect, remove a significant portion of the population from that area of the cave. Oftentimes such wood can be removed gradually, and a small residue kept to provide habitat for the remaining fauna (Elliott, 1997a). An important concept is to give some caves or microhabitats time to "rest".

In Blanchard Springs Cavern, Arkansas, Aley (1972) successfully killed algae, moss, and ferns growing near electric lights with a steam generator, without damaging speleothems. Aley et al. (1985) found that, in Carlsbad Cavern, lighting in moist alcove sites should be kept below 9.7 lx. In moist non-alcove sites lighting should be below 30.1 lx. Some studies were also conducted on chemical plant-control agents. The best agent for general plant control appears to be a 5.25% solution of sodium hypochlorite (bleach). Copper sulfate and calcium hypochlorite solutions have caused damage to cave features when tested in Carlsbad Cavern.

Ecological transplantation

Wildlife rescue operations have often been used for big game, but the concept has rarely been applied to cave animals. Ezell's Cave, Texas, had been sealed by the owner to keep out intruders, but the loss of the bat colony and its guano was thought to have contributed to a decline in the resident population of the salamander *Typhlomolge rathbuni*. In 1970 biologists twice attempted to re-establish the bat species *Myotis velifer* in Ezell's Cave. Bats were captured in northwestern Texas, but they did not stay in Ezell's Cave in central Texas for even a day. This species is adaptable and will readily colonize newly opened caves, but the trip and handling probably frightened them severely (Elliott, 1993a, 1994d).

In 1976, and again in 1997–1978, a threatened population of harvestmen (*Banksula melones*, a troglobitic opilionid) was transplanted from McLean's Cave, California, to a nearby mine (Elliott, 1978, 1981). This species was known from just two caves at the time, one of which was threatened by quarrying. The other site, McLean's Cave, was scheduled for complete inundation by the New Melones Reservoir. The United States Army Corps of Engineers, which was building the dam, supported the project so as to avoid this species being listed as "endangered". During the second, larger project, the cave population was intensively collected for three months, and the community of about 30 arthropod species transplanted to the Von Trump Mine, now called the "Transplant Mine". The mine was stocked like a terrarium with cave soil, rocks, and rotting wood from the surrounding area. The cave soil and fauna were brought in insulated containers to the mine and carefully placed in higher areas to avoid the minor flooding that sometimes occurs from seepage. A follow-up visit in 1979 revealed that the cave community was doing well, although the population ratios of some of the species had changed dramatically. Thomas Briggs (personal communication) returned to the mine in 1996 and found that *B. melones* was still there.

Transplanting cave fauna is an experimental technique, and it cannot seriously be considered as a good solution for endangered cave faunas. Despite the apparent success of the "New Melones Transplant", several philosophical and practical problems were pointed out by Elliott (1978, 1981). Assuming that the transplant would work, one must have a suitable "empty" habitat into which to transplant. In this case, the Transplant Mine had a cave-like microclimate, was only 50 years old, and was excavated in Calaveras Limestone, which is the major speleifer in the area. However, transplanting to a natural cave would present three problems:

Problem 1: If the cave had its own natural community, and if it did not already contain the endangered species to be transplanted, then a question arises as to whether the cave would be in any case suitable for that species.

Problem 2: If the cave were suitable, the transplanted species would have to compete with the existing community for resources, which might defeat the purpose of the transplantation.

Problem 3: Two cave communities would be altered or destroyed instead of one.

Ecological transplants have other potential problems, such as the need to replenish nutrients (the Transplant Mine has to be restocked with wood occasionally as it has no sinkhole-like entrance), and the need for long-term funding to monitor the transplant. Other ecologically more appropriate solutions may need to be considered. For example, ironically, a follow-up study in 1979 found that *Banksula melones* occurred in 18 caves in the area. Obviously, good faunal surveys of a karst area are needed before a questionable rescue effort should be considered.

Ecosystem management

The ultimate survival of a cave or karst community depends on the proper protection and management of the cave and the surrounding terrain. In the case of some endemic, terrestrial cave invertebrates, a one-hectare preserve may be sufficient as long as nutrient and water inputs are not altered and disturbances are minimal. However, often one must often reach beyond the immediate karst area to contributing catchments or, in the case of bats, to alternate roost habitats in planning for the survival of an ecosystem. With migratory bats, these considerations must be on a continental scale to insure success.

Many cave preserves that have been set aside only protect the entrance area, and not the entire catchment for the cave system. However, the need for ecosystem management, which considers the whole karst area, is often recognized. The trend by governments has been toward increasing aspirations for karst ecosystem management; but in practice this is very difficult financially and politically. In the Austin (Texas) area, even though seven terrestrial troglobites have been listed as endangered, the federal government has not bought any cave preserves for them, even though it has spent millions of dollars to purchase a national wildlife refuge nearby for the benefit of two endangered bird species. Sixty-five cave preserves that were planned will have to await funding through a local development fee collected by the county and city (Elliott, 1997c).

Some believe that caves cannot be managed, and that they should be left alone (except for occasional caving trips). Such a passive conservation ideal cannot work in developed areas, where drainage patterns are changed, native vegetation and faunas are gradually perturbed, nutrient and moisture inputs are altered, and pests invade. In managing cave preserves, it is important to be able to measure the results from time to time. Therefore, baseline surveys, census surveys, and written management plans are essential.

Karst groundwater issues are so important, economically and politically, that they move beyond the influence of the technical experts who best understand them (Elliott 1994d, 1996b). In Texas, the regulation of over-pumping of the Edwards Aquifer, one of the most important karst aquifers in North America, has been gradually improved only through a succession of state rulings and legislation, which were contradicted by federal court decisions. It is quite possible that the San Marcos and Comal Springs will become dry within the next decade, endangering many species and the human economy (Elliott, 1993a, 1996a).

The Karst Waters Institute, based in Charles Town, West Virginia, has begun an annual event to publicize the "Top Ten List of Endangered Karst Ecosystems". In the first edition 40 areas were nominated by biologists, and the top ten included terrestrial and aquatic cave communities in Georgia, Hawaii, Indiana, Kentucky and Texas, as well as Australia, Bermuda, the Canary Islands, France, and Vietnam (Karst Waters Institute, 1997).

GENERAL CONCLUSIONS

In general, the most dramatic declines in cave faunas have been caused by the direct disturbance and killing of bats, and massive kills of aquatic troglobites by water projects, sewage, and chemicals. Perhaps four North American cave species have become extinct as a result of human activities, and it is possible that other extinctions have occurred. Local extirpations of several species of bats, cave-fishes, and crustaceans have been documented. However, the gradual and inexorable decline of some cave communities over decades may go unnoticed because of a lack of baseline surveys and systematic monitoring. Nutrient stress is a problem that few cave biologists have studied, but the long series of records in Mammoth Cave and Carlsbad Cavern provide a few insights into this subtle process.

Although many cave-management plans have been devised across North America, and 20 species are under protection, it is apparent that many other species are just as threatened by human activities. It is more obvious than ever that the management of regional karst ecosystems and protection strategies should be encouraged, while the funding of small cave preserves to protect locally endemic species with a minimum investment is not forgotten.

ACKNOWLEDGEMENTS

I am grateful to the following people who have assisted me with literature, unpublished data, and suggestions: Tom Aley, Tom Briggs, James Cokendolpher, Ronald Crunkilton, David Culver, Robert Currie, Horton H. Hobbs III, John Holsinger, David Hubbard, Pat Jablonsky, Jim Kennedy, Kathleen Lavoie, Susan Lawrence, Rick Olson, Dale Pate, James R. Reddell, Jo Schaper, Mario Sgro, Peter S. Sprouse, Ruth Stanford, Merlin Tuttle, Daryl Ubick, Eugene Vale, George Veni, and Horst Wilkens. I thank the Texas Memorial Museum at The University of Texas at Austin, the Texas Speleological Survey and the Missouri Department of Conservation for their support.

NOTE ON LITERATURE

In writing this chapter I discovered a large amount of useful literature which is difficult to obtain. I have assembled an annotated bibliography, which is too large to include here, but I shall post it on my "Biospeleology" World Wide Web site, accessible through the Internet at

`http://www.utexas.edu/depts/tnhc/.www/biospeleology`

REFERENCES

Aley, T., 1972. Control of unwanted plant growth in electrically lighted caves. *Caves & Karst,* 14: 33–35.

Aley, T., 1997. Caves in crisis. In: *1997 Yearbook of Science and the Future.* Encyclopaedia Britannica, Inc., Chicago, IL, pp. 116–133.

Aley, T., Aley, C. and Rhodes, R., 1985. Control of exotic plant growth in Carlsbad Caverns, New Mexico. In: *Proc. 1984 Natl. Cave Management Symp., Missouri Speleol.,* 25(1–4): 159–171.

Aley, T., Aley, C., Elliott, W.R. and Huntoon, P.W., 1993. *Karst and cave resource significance assessment, Ketchikan Area, Tongass National Forest, Alaska,* Report of the Karst Resources Panel, 79 pp. + 43 pp. appendices.

Bailey, V., 1928. *Animal Life of the Carlsbad Cavern. Monographs of the American Society of Mammalogists,* Vol. 3. Williams & Wilkins, Baltimore, MD, 195 pp.

Barbour, R.W. and Davis, W.H., 1969. *Bats of America.* The University Press of Kentucky, Lexington, KY, 287 pp.

Barr Jr., T.C., 1968. Cave ecology and the evolution of troglobites. In: T. Dobzhansky, M.K. Hecht and W.C. Steere (Editors), *Evolutionary Biology,* Vol. 2. K. Holland, pp. 35–102.

Barr Jr., T.C. and Kuehne, R.A., 1971. Ecological studies in the Mammoth Cave ecosystems of Kentucky. II. The ecosystem. *Ann. Speleol.,* 26: 47–96.

Baskauf, C.J., 1997. Population genetic studies of plants endemic to karst, with an emphasis on the limetone glades of Tennessee. In: I.D. Sasowsky, D.W. Fong and E.L. White (Editors), *Conservation and Protection of the Biota of Karst, Symposium at Nashville, Tennessee, February 13–16, 1997,* Special Publication 3. Karst Waters Institute, Charles Town, WV, pp. 2–4.

Blackwell and Associates, Ltd., 1995. *Literature review of management of cave/karst resources in forest environments,* Report to Vancouver Forest Region, British Columbia, Canada, 19 pp.

Brady, J.T., 1982. The status of the Indiana bat (*Myotis sodalis*). In: R.C. Wilson and J.J. Lewis (Editors), *National Cave Management Symposia Proceedings, Carlsbad, New Mexico 1978 and Mammoth Cave, Kentucky 1980.* Pygmy Dwarf Press, Oregon City, OR, pp. 127–32.

Brenner, M., Medina Gonzalez, R. and Zetina Moguel, C., 1995. Water resources of the Yucatan Peninsula, Mexico: Special concerns and management priorities. *Land and Water,* November–December: 18–20.

Clark, D.R., 1988. Environmental contaminants and the management of bat populations in the United States. In: U.S. Fish & Wildlife Service (Editor), *Management of Amphibians, Reptiles, and Small Mammals in North America, Symposium, Flagstaff, Arizona, July 19–21, 1988.* U.S. Fish & Wildlife Service, Washington, D.C., pp. 409–413.

Cooper, J.E., 1969. Biological studies in Shelta Cave, Alabama. In: *5. Internationaler Kongress für Speläologie, Stuttgart, 1969, Abhandlungen,* Vol. 4: B1/1–8.

Cooper, J.E., 1975. Ecological and Behavioral Studies in Shelta Cave, Alabama, with Emphasis on Decapod Crustaceans, Ph.D. Dissertation. University of Kentucky, Lexington, KY, xvi+364 pp.

Crunkilton, R., 1985. Subterranean contamination of Maramec Spring by ammonium nitrate and urea fertilizer and its implication on rare cave biota. In: *Proc. 1984 Natl. Cave Management Symp., Missouri Speleol,* 25(1–4): 151–158.

Culver, D.C., 1982. *Cave Life. Evolution and Ecology.* Harvard University Press, Cambridge, MA, 189 pp.

Culver, D.C. and Holsinger, J.R., 1992. How many species of troglobites are there? *Natl. Speleol. Soc. Bull.,* 54: 79–80.

Culver, D.C., Jones, W.K. and Holsinger, J.R., 1992. Biological and hydrological investigation of The Cedars, Lee County, Virginia, an ecologically significant and threatened karst area. In: *First Int. Conf. on Ground Water Ecology.* U.S. Environmental Protection

Agency and American Water Resources Association, Washington, D.C., pp. 281–290.

Elliott, W.R., 1978. *Final Report on the New Melones Cave Harvestman Transplant.* U.S. Army Corps of Engineers, Sacramento, 62 pp.

Elliott, W.R., 1981. Damming up the caves. *Caving Int.,* 10: 38–41.

Elliott, W.R., 1982. An introduction to biospeleology, In: J. Hassemer (Editor), *Caving Basics.* National Speleological Society, Huntsville, pp. 100–108. 2nd edition: 1987, G.T. Rea (Editor).

Elliott, W.R., 1990. Endangered species, endangered caves. *Natl. Soc. Speleol. News,* 48: 225–231.

Elliott, W.R., 1991. *Endangered and Rare Karst Species in Travis County, Texas: Options for the Balcones Canyonlands Conservation Plan,* Report to Balcones Canyonlands Conservation Plan. U.S. Fish & Wildlife Service, Texas Parks & Wildlife Dept., Texas Nature Conservancy. 9 pp. + 12 pp. appendix.

Elliott, W.R., 1992a. Caves, endangered species, and biodiversity. *Am. Caves,* Spring/Summer: 19.

Elliott, W.R., 1992b. Fire ants invade Texas caves. *Am. Caves,* 5(Winter): 13.

Elliott, W.R., 1993a. Cave fauna conservation in Texas. In: D.L. Foster (Editor), *National Cave Management Proceedings, Bowling Green, Kentucky, October 23–26, 1991.* American Cave Conservation Association, Horse Cave, Kentucky, pp. 323–337.

Elliott, W.R., 1993b. *Fire Ants and Endangered Cave Invertebrates: A Control and Ecological Study,* Report to Texas Parks & Wildlife Department. 33 pp.

Elliott, W.R., 1994a. Alaska's forested karstlands. *Am. Caves,* 7(Winter/Spring): 8–12.

Elliott, W.R., 1994b. Biodiversity and conservation of North American cave faunas: An overview. In: B. Mixon (Editor), *1994 NSS Convention Program, Abstracts,* National Speleological Society, Huntsville, AL, p. 48.

Elliott, W.R., 1994c. *Community Ecology of Three Caves in Williamson County, Texas: A Three-Year Summary,* Report to Simon Development Co. Inc. Texas Parks & Wildlife Dept., and U.S. Fish & Wildlife Service, 46 pp.

Elliott, W.R., 1994d. Conservation of Texas caves and karst. In: W.R. Elliott and G. Veni (Editors), *The Caves and Karst of Texas, Convention Guidebook.* National Speleological Society, Huntsville, AL, pp. 85–97.

Elliott, W.R., 1994e. Conservation of western Oklahoma bat caves. *Oklahoma Underground,* 17: 44–53. Reprint of report to The Nature Conservancy, Oklahoma Chapter.

Elliott, W.R., 1994f. Fern Cave. In: W.R. Elliott and G. Veni (Editors), *The Caves and Karst of Texas, 1994 Convention Guidebook.* National Speleological Society, Huntsville, AL, pp. 287–289.

Elliott, W.R., 1994g. Inner Space Cavern. In: W.R. Elliott and G. Veni (Editors), *The Caves and Karst of Texas, 1994 Convention Guidebook.* National Speleological Society, Huntsville, AL, pp. 140–142.

Elliott, W.R., 1995. Air monitoring during construction of a cave gate. In: D.L. Pate (Editor), *Proc. 1993 Natl. Cave Management Symp., Carlsbad, New Mexico.* National Cave Management Symposium Steering Committee, Carlsbad, NM, pp. 45–51.

Elliott, W.R., 1996a. The Barton Springs salamander. *Am. Caves.* 9(1): 14–15.

Elliott, W.R., 1996b. The evolution of cave gating – How the philosophy and technology have changed. *Am. Caves,* 9(2): 9–15.

Elliott, W.R., 1997a. *A Survey of Ecologically Disturbed Areas in Carlsbad Cavern, New Mexico,* Report to Carlsbad Caverns National Park. 10 pp.

Elliott, W.R., 1997b. *Biological, Geological, and Engineering Aspects of Mayor Elliott Cave, Georgetown, Texas,* Report to Paul Price Associates and The City of Georgetown, Texas. 8 pp.

Elliott, W.R., 1997c. *The Caves of the Balcones Canyonlands Conservation Plan, Travis County, Texas,* Report to Travis County, Texas, Transportation and Natural Resources Department, Balcones Canyonlands Conservation Plan. 156 pp.

Elliott, W.R. and Reddell, J.R., 1989. *The Status and Range of Five Endangered Arthropods from Caves in the Austin, Texas, Region.* Austin Regional Habitat Conservation Plan, 100 pp.

Evans, A.M., 1982. The Hart's tongue fern – an endangered plant in cave entrances. In: R.C. Wilson and J.J. Lewis (Editors), *Natl. Cave Management Symp. Proc., Carlsbad, New Mexico 1978 and Mammoth Cave, Kentucky 1980.* Pygmy Dwarf Press, Oregon City, OR, pp. 143–145.

Feinstein, L., 1952. Insecticides from plants. In: *Insects, The Yearbook of Agriculture.* U.S. Department of Agriculture, pp. 222–229.

Foster, D. and MacGregor, J., 1987. Four Kentuckians convicted of killing Indiana bats, reward helps! *Devil's Advocate, Diablo Grotto Newsl., Berkeley, CA,* 20: 84.

Helf, K., Poulson, T.L. and Lavoie, K.H., 1996. Protection and management of the cave cricket (*Hadenoecus subterraneus*) at Mammoth Cave National Park. In: G.T. Rea (Editor), *Proc. 1995 Natl. Cave Management Symp., Spring Mill State Park, Mitchell, Indiana.* Indiana Karst Conservancy, Indianapolis, IN, pp. 155–160.

Hendrickson, G.E., 1961. *Sources of water in Styx and Echo rivers, Mammoth Cave, Kentucky,* U.S. Geological Survey Professional Paper, 424-D: D41–D43.

Herreid II, C.F., 1963. Temperature regulation of Mexican free-tailed bats in cave habitats. *J. Mammal.,* 44: 560–573.

Herreid II, C.F., 1967. Temperature regulation, temperature preference and tolerance, and metabolism of young and adult free-tailed bats. *Physiol. Zool.,* 40: 1–22.

Hobbs III, H.H., 1996. Impacts of surface perturbations in karst areas in southeastern United States: A biologist's perception. In: G.T. Rea (Editor), *Proc. 1995 Natl. Cave Management Symp., Spring Mill State Park, Mitchell, Indiana.* Indiana Karst Conservancy, Indianapolis, IN, p. 163.

Hobbs III, H.H., 1997. A biological assessment of five invertebrate stygobionts from southwestern Ohio. In: I.D. Sasowsky, D.W. Fong and E.L. White (Editors), *Conservation and Protection of the Biota of Karst, Symposium at Nashville, Tennessee, February 13–16, 1997,* Special Publication 3. Karst Waters Institute, Charles Town, WV, pp. 22–25.

Hobbs III, H.H. and Bagley, F.M., 1989. *Shelta Cave Management Plan.* Biological Subcommittee of the Shelta Cave Committee, National Speleological Society. 78 pp.

Holsinger, J.R., Carlson, K.R. and Shaw, D.P., 1997. Biogeographic significance of recently discovered amphipod crustaceans (*Stygobromus*) in caves of southeastern Alaska and Vancouver Island. In: *Proc. 12th Int. Congr. Speleology, Switzerland,* Vol. 3: 347–349.

Howarth, F.G., 1983. The conservation of Hawaii's cave resources. *Newsl. Cave Conservat. Manage.*, 2(1–2): 19–23.

Hubbard Jr., D.A., 1995. Cave conservation and cave clean-ups: Not always one in the same. *Natl. Speleol. Soc. News,* 53: 31.

Hubbard Jr., D.A., 1996. Contaminant case studies in Virginia Karst. In: *Convention Program, National Speleological Society, Huntsville, AL,* pp. 43–44.

Hubbard Jr., D.A. and Balfour, W.M., 1993. An investigation of engineering and environmental concerns relating to proposed highway construction in a karst terrane. *Environ. Geol.,* 22: 326–329.

Hunt, G. and Stitt, R.R., 1975. *Cave Gating, A Handbook.* National Speleological Society, Huntsvill, AL, 43 pp. 2nd edition: 1981, 60 pp.

Jablonsky, P., Kraemer, S. and Yet, B., 1995. Lint in caves. In: D.L. Pate (Editor), *Proc. 1993 Natl. Cave Manage. Symp., Carlsbad, New Mexico.* pp. 73–81.

Karst Waters Institute, 1997. *Top Ten List of Endangered Karst Ecosystems,* Press Release, 4 pp.

Lewis, J.J., 1982. Aquatic ecosystems and management problems in the Mammoth Cave area. In: R.C. Wilson and J.J. Lewis, (Editors), *Natl. Cave Management Symp. Proc., Carlsbad, New Mexico 1978 and Mammoth Cave, Kentucky 1980.* Pygmy Dwarf Press, Oregon City, OR, pp.a 73–76.

Lewis, J.J., 1984. Observations on aquatic communities in the historic section of Mammoth Cave. *Cave Res. Found. Annu. Rep.,* 1981: pp. 17–19.

Lewis, J.J., 1996a. Cave bioinventory as a management tool. In: G.T. Rea (Editor), *Proc. 1995 Natl. Cave Management Symp., Spring Mill State Park, Mitchell, Indiana.* Indiana Karst Conservancy, Indianapolis, IN, pp. 228–236.

Lewis, J.J., 1996b. The devastation and recovery of caves and karst affected by industrialization. In: G.T. Rea (Editor), *Proc. 1995 Natl. Cave Management Symp., Spring Mill State Park, Mitchell, Indiana.* Indiana Karst Conservancy, Indianapolis, IN, pp. 214–227.

Lewis, J.J. and Lewis, T.M., 1980. The distribution and ecology of two species of subterranean *Caecidotea* in Mammoth Cave National Park. *Cave Research Foundation Annual Report,* 1980. pp. 23–26.

Lewis, J.J., Lewis, T.M. and Eckstein, J., 1983. A biological reconnaissance of a polluted cave stream: The Hidden River Groundwater Basin. *Cave Res. Found. Annu. Rep.,* 1982: 9–10.

Lewis, J.J., Pursell, F.A. and Huffman, H., 1997. The biological inventory of caves of the Blue River Bioreserve. In: I.D. Sasowsky, D.W. Fong and E.L. White (Editors), *Conservation and Protection of the Biota of Karst, Symposium at Nashville, Tennessee, February 13–16, 1997,* Special Publication 3. Karst Waters Institute, Charles Town, WV, pp. 37–41.

Lisowski, E.A., 1982. The endangered Kentucky blind cave shrimp. In: R.C. Wilson and J.J. Lewis (Editors), *Natl. Cave Management Symp. Proc., Carlsbad, New Mexico 1978 and Mammoth Cave, Kentucky 1980.* Pygmy Dwarf Press, Oregon City, OR, pp. 138–142.

Lisowski, E.A., 1983. Distribution, habitat, and behavior of the Kentucky cave shrimp *Palaemonias ganteri* Hay. *J. Crustacean Biol.,* 3: 88–92.

Lisowski, E.A. and Poulson, T.L., 1981. Impacts of Lock and Dam Six on baselevel ecosystems in Mammoth Cave. In: T.L. Poulson (Editor), *Cave Research Foundation Annual Report, 1979.* Adobe Press, Albuquerque, NM, pp. 48–54.

Lisowski, E.A., Olson, R.A., Roy, W.R. and Thompson, D.B., 1986. Geochemistry and biology of Sulphur River, Parker's Cave, Kentucky. *Cave Res. Found. Annu. Rep.,* 1985: 26–27.

Longley, G., 1991. Threats to the subterranean aquatic ecosystem of the Balcones Fault Zone, Edwards Aquifer, in Texas. *Natl. Speleol. Soc. Bull.,* 53(2): 110 (abstract).

MacGregor, J., 1993. Responses of winter populations of the federal endangered Indiana bat (*Myotis sodalis*) to cave gating in Kentucky. In: D.L. Foster (Editor), *Natl. Cave Management Proc., Bowling Green, Kentucky, October 23–26, 1991.* American Cave Conservation Association, Horse Cave, KY, pp. 364–370.

McCracken, G.F., 1989. Cave conservation: Special problems of bats. *Natl. Speleol. Soc. Bull.,* 51: 49–51.

Meiman, J., 1993. The effects of recharge basin land-use practices on water quality at Mammoth Cave National Park, Kentucky. In: D.L. Foster (Editor), *Natl. Cave Management Proc., Bowling Green, Kentucky, October 23–26, 1991.* American Cave Conservation Association, Horse Cave, KY, pp. 105–115.

Minton, M., 1992. Mexico news. *AMCS Activities Newsl.,* 19: 5–21.

Mohr, C.E., 1972. The status of threatened species of cave-dwelling bats. *Natl. Speleol. Soc. Bull.,* 34: 33–47.

National Park Service, 1996. *Final General Management Plan/ Environmental Impact Statement.* Carlsbad Caverns National Park, NM, 287 pp.

Northup, D.E. and Kuper, R., 1987. Natural history of arthropods of Carlsbad Caverns emphasizing Rhaphidophoridae of the genus Ceuthophilus. *Cave Res. Found. Annu. Rep.,* 1986: 29–30.

Northup, D.E. and Welbourn, W.C., 1995. Conservation of invertebrates and microorganisms in the cave environment. In: D.L. Pate (Editor), *Proc. 1993 Natl. Cave Management Symp., Carlsbad, New Mexico.* National Cave Management Symposium Steering Committee, Carlsbad, NM, pp. 292–301.

Northup, D.E., Ziegler, W.S. and Ingham, K.L., 1987. Community structure of the arthropods of Carlsbad Cavern. *Cave Res. Found. Annu. Rep.,* 1987: 30–32.

Northup, D.E., Hardy, J.M. and Ingham, K.L., 1989. Arthropod species diversity in the Big Room and the environs. *Cave Res. Found. Annu. Rep.,* 1988: 39–40.

Northup, D.E., Angert, E., Reysenbach, A.L., Peek, A. and Pace, N., 1997a. Microbial communities in Sulphur River, Parker Cave: A molecular phylogenetic study. In: I.D. Sasowsky, D.W. Fong and E.L. White (Editors), *Conservation and Protection of the Biota of Karst, Symposium at Nashville, Tennessee, February 13–16, 1997,* Special Publication 3. Karst Waters Institute, Charles Town, WV, pp. 55–56.

Northup, D.E., Beck, K.M. and Mallory, L.M., 1997b. Human impact on the microbial communities of Lechuguilla Cave: Is protection possible during active exploration? *National Speleological Society Convention Program, Sullivan, MO,* p. 55.

O'Donnell, L., Elliott, W.R. and Stanford, R., 1994. *Endangered Karst Invertebrates (Travis and Williamson Counties, Texas): Recovery Plan.* U.S. Fish & and Wildlife Service, 30 pp.

Pasquarell, G.C. and Boyer, D.B., 1993. Water quality impacts of agriculture on karst conduit waters, Greenbrier County, WV. In: D.L. Foster (Editor), *Natl. Cave Management Proc., Bowling Green, Kentucky, October 23–26, 1991.* American Cave Conservation Association, Horse Cave, KY, pp. 72–78.

Peck, S.B., 1969. Spent carbide – a poison to cave fauna. *Natl. Speleol. Soc. Bull.,* 31(2): 53–54.

Peck, S.B., 1974. The invertebrate fauna of tropical American caves, Part II: Puerto Rico, An ecological and zoogeographic analysis. *Biotropica,* 6(1): 14–31.

Peck, S.B., 1997. Origin and diversity of the North American cave fauna. In: I.D. Sasowsky, D.W. Fong and E.L. White (Editors), *Conservation and Protection of the Biota of Karst, Symposium at Nashville, Tennessee, February 13–16, 1997,* Special Publication 3. Karst Waters Institute, Charles Town, WV, pp. 60–66.

Perkins, J.M., 1990. Winter report on *Plecotus townsendii* hibernacula survey. *Cave Res. Found. Annu. Rep.,* 1989: 39.

Poulson, T.L., 1967. Comparison of terrestrial cave communities. *Cave Res. Found. Annu. Rep.,* 1967: 27–32.

Poulson, T.L., 1968. Aquatic cave communities. *Cave Res. Found. Annu. Rep.,* 1968: 16–18.

Poulson, T.L., 1976. Management of biological resources in caves. In: D. Rhodes (Editor), *Proc. Natl. Cave Management Symp., Albuquerque, New Mexico, 1975.* Speleobooks, Albuquerque, NM, pp. 46–52.

Poulson, T.L., 1992. Assessing groundwater quality in caves using indices of biological integrity. *Cave Res. Found. Annu. Rep.,* 1991: 50–51.

Poulson, T.L., 1996. Research aimed at management problems should be hypothesis-driven: Case studies in the Mammoth Cave Region. In: G.T. Rea (Editor), *Proc. 1995 Natl. Cave Management Symp., Spring Mill State Park, Mitchell, Indiana.* Indiana Karst Conservancy, Indianapolis, IN, pp. 267–273.

Poulson, T.L., 1997. The Mammoth Cave Tour. In: I.D. Sasowsky, D.W. Fong and E.L. White (Editors), *Conservation and Protection of the Biota of Karst, Symposium at Nashville, Tennessee, February 13–16, 1997,* Special Publication 3. Karst Waters Institute, Charles Town, WV, pp. 113–117.

Poulson, T.L. and Kane, T.C., 1977. Ecological diversity and stability. In: T. Aley and D. Rhodes (Editors), *Proc. Natl. Cave Management Symp., Mountain View, Arkansas, 1976.* Speleobooks, Albuquerque, NM, pp. 18–21.

Pride, T.E., Harvey, M.J., Ogden, A.E. and Smith, W.P., 1988. Water quality and benthic community structure of caves receiving urban runoff. *Natl. Speleol. Soc. Bull.,* 53: 15.

Reddell, J.R., 1981. A review of the cavernicole fauna of Mexico, Guatemala, and Belize. *Texas Mem. Mus. Bull.,* 27. 327 pp.

Reddell, J.R., 1991. *Further Study of the Status and Range of Endangered Arthropods from Caves in the Austin, Texas, Region,* report on a study for the U.S. Fish & Wildlife Service. iv+178 pp.

Reddell, J.R. and Elliott, W.R., 1991. *Distribution of Endangered Karst Invertebrates in the Georgetown Area, Williamson County, Texas.* A report on a study for The City of Georgetown, 64 pp.

Reddell, J.R. and Elliott, W.R., 1994. The impact of urbanization on endemic cave fauna in Travis and Williamson counties, Texas. In: B. Mixon (Editor), *1994 NSS Convention Program, Abstracts.* National Speleological Society, Huntsvill, AL, p. 49.

Rusterholtz, K.J., 1989. Response of a model ground water microbial community to acute chemical stress. *Cave Res. Found. Annu. Rep.,* 1988: 46.

Rusterholtz, K.J. and Mallory, L.M., 1990. Detection and quantification of viable *Salmonella* sp. in flowing underground streams. *Cave Res. Found. Annu. Rep.,* 1989: 39.

Stokes, T., 1996. *Preliminary Problem Analysis of Cave/Karst Issues Related to Forestry Activities on Vancouver Island.* Report by Terra Firma Geoscience Services to Vancouver Forest Region, British Columbia. 13 pp.

Sullivan, G.N., 1956. Biological aspects of cave conservation. *Natl. Speleol. Soc. News,* 14: 30–31.

Tuttle, M.D. and Taylor, D.A.R., 1994. *Bats and Mines,* Resource Publication 3. Bat Conservation International, Austin, TX, 42 pp.

Vale, E. and Jones, R.H., 1993. It's an open and shut cave: Plugging artificial entrances at Onondaga Cave State Park. In: D.L. Foster (Editor), *Natl. Cave Management Proc., Bowling Green, Kentucky, October 23–26, 1991.* American Cave Conservation Association, Horse Cave, KY, pp. 129–131.

Veni, G., 1987. *Valdina Farms Sinkhole: Hydrogeologic and Biologic Evaluation,* Report for Edwards Underground Water District, San Antonio, TX, 103 pp.

Veni, G., 1992. *Geologic Controls on Cave Development and the Distribution of Cave Fauna in the Austin, Texas, Region,* Prepared for U.S. Fish & Wildlife Service. v+77 pp.

Chapter 35

CONSERVATION OF SUBTERRANEAN HABITATS AND SPECIES

Christian JUBERTHIE

INTRODUCTION

Awareness of the necessity for the conservation of caves and their fauna, on the basis of national or regional legislation is developed to different extents in different countries. The state of advancement of conservation and protective measures also differs from one country to another. The differences are due to:
- the level of development of the country;
- the degree of knowledge of the subterranean fauna and habitats;
- the level of activity and the number of speleo-clubs and biospeleologists;
- the richness of karst, caves, and other habitats;
- the level of pollution;
- the politics of conservation; and
- the legislation and its application.

A summary of the cave legislative policy in the different European countries has been published by the Council of Europe (Juberthie, 1995). Many authors have written about cave conservation, generally, in their own country, or caves of special interest (see References section below).

Today it is indispensable to develop and harmonize conservation strategies and measures, legislative aspects and management. The first very important point is to define the field concerned, the types of habitats, the kinds of species, and their vulnerability. Indeed, the subterranean habitat extends much more widely and is more diversified than caves alone, and its conservation cannot be reduced to the protection of caves and to a speleological approach.

DIVERSITY OF SUBTERRANEAN HABITATS AND SPECIES

Subterranean biodiversity

In the world, about 7800 species are known exclusively from subterranean habitats (troglobites in terrestrial habitats, and stygobites in aquatic habitats). Seventy orders include troglobites or stygobites, as follows:

Vertebrates: about 100 stygobite species are recorded (85 fishes and 14 amphibians).

All the other species are invertebrates. The two most numerous groups are Crustacea and Coleoptera.

Crustacea: around 2585 species [2300 aquatic (Remipedia, 10; Ostracoda, 115; Cladocera, 10; Copepoda, 690; Mysidacea, 17; Mictacea, 1; Isopoda aquatica, 360; Amphipoda, 740; Decapoda, 150; Thermosbaenacea, 35; Syncarida, 180), and 285 terrestrial isopods];

Insecta: 2500 species (Collembola, 250; Diplura-Campodeidae, 98; Diplura-Japygidae, 13; Thysanura, 17; Dictyoptera-Blattaria, 31; Dermaptera, 15; Psocoptera, 9; Hemiptera-Heteroptera, 4; Homoptera, 56; Coleoptera, 2000; Diptera, 22);

Arachnida: 1525 species (Scorpiones, 13; Pseudoscorpiones, about 330; Palpigradida, 28; Schizomida, 25; Ricinulei, 10; Amblypygi, about 10; Araneae, about 200; Opiliones, 82; Hydracarina, 600; Acari terrestria, 200);

Myriapoda: about 250 (Chilopoda, 55; Diplopoda, about 200; Symphyla, 2);

Other groups: 850 species (Hydrozoa, 1; Turbellaria, 165; Microturbellaria, 25; Nemertina, 3; Rotifera, about 100; Gastrotrichida, 10; Nematoda, 17; Polychaeta, 3; Oligochaeta *s.l.*, 115; Hirudinea, 14; Mollusca aquatic and terrestrial, 400; Onychophora, 2; Tardigrada, 10).

Presumably the number of troglophilic and stygophilic species which inhabit epigean and subterranean habitats is similar (about 7500). Thus the number of species inhabiting and reproducing in subterranean habitats is probably about 15 000. Of special interest are troglophylic bats because of their importance in the food chains in caves, and the oil bird "Guacharo" (*Steatornis caripensis*) in neotropical caves, because it is an endangered species.

Diversity of habitats

Until the middle of the 20th century, most biologists had a limited view of subterranean habitats, restricted to caves, which are the part of the hydrogeologically active or fossil systems that humans can enter. The discovery of other subterranean habitats which cannot be explored and studied with speleological techniques, such as subterranean populations in mesovoids or microvoids, cracks in rocks other than limestones, and in alluvial or colluvial layers, have greatly enlarged the diversity of habitats, and as a consequence the fields of biospeleology and biospeleological conservation. Therefore, areas for protection should be considered on the basis of the following principles:

(1) The distribution of subterranean species is not restricted to caves.
(2) The network of karstic cracks, mesovoids in the case of colluvial layers (MSS), voids of the interstitial aquatic habitats of the stream, phreatic aquifers frequently without caves, brackish and marine-water habitats in coastal caves, should be taken into account. A survey of the areas of species distribution is necessary.
(3) The resources of the subterranean system come from surface plant and animal production. Consequently, protection should include a surface zone to control the impact of agricultural and forestry practices and possible pollution.

The different subterranean habitats for conservation are as follows:

Karst caves

Caves represent the totality of the subterranean habitat only for trogloxenic bats and Orthoptera, and presumably for several species of stygobitic fishes.

The subterranean system (caves + network of active or fossil cracks)

Terrestrial and aquatic invertebrates colonize not only the caves but the network of cracks usually related to the caves. The size of populations is highest along the voids where surface food enters the system. The mark–recapture method demonstrates that in the temperate zone of the northern hemisphere only 5 to 10% of the population of troglobite beetles inhabits caves. The rest of the population dwells in the limestone cracks. Seasonally, a part of the crack dwellers migrate from the cracks to the caves in spring and summer, and return to cracks in autumn.

Volcanic subterranean habitat

In basaltic volcanoes, numerous endemic troglobitic and stygobitic species are known in lava tubes and mesovoids (MSS) of the lava flows. This type of habitat is well developed in Hawaii, Galápagos, the Canary Islands (with 130 subterranean endemic species!), Madeira, and in mainland volcanic areas (Fuji Jama, etc., in Japan, Korea, United States, etc.).

Pseudokarsts

Hypogean systems and habitats developed in rocks other than limestone and dolomite. In the old Precambrian shield (Guyana, Africa, the Baltic) most of the known caves are formed in primary rocks (gneiss, quarzite, diabase sills, and granite) and under the crust of altered primary rock (laterite) in the tropical zone. Numerous endemic species have been described from caves in quartzite and gneiss (for example from the Transvaal and the Cape Province in South Africa and from Venezuela), and from caves in sandstone (for example, caves in Guinea).

Mesocavernous Shallow Substratum (MSS) or Underground Superficial Compartment

The network of voids in the screes along the slopes of the valley in different types of rocks (limestone, schist, granite, granodiorite, etc.) in the mountains of the temperate zone and the network of cracks and mesovoids in volcanic layers, are inhabited by troglobitic or stygobitic invertebrates. Some species are the same as those in caves, others are endemic to the MSS. Usually soil and often forest cover the MSS. Consequently the conservation of areas with MSS presents some special problems because:

(1) the major part of the areas is outside the limestone and karst; and
(2) the conservation of MSS habitat needs to be associated with conservation of surface ecosystems such as forest or forest soil and their fauna.

Phreatic aquifers, alluvial aquifers, hyporheic "interstitial" habitat of surface streams

On the mainland 98.7% of fresh water is subterranean, and about 50.7% is inhabited by stygobitic and stygophilic species. The habitats include karstic aquifers; phreatic aquifers in valleys in all types of sediments and rocks; underground flow in surface streams (interstitial or hyporheic habitat); and the interstitial habitat on the beaches of freshwater lakes. Some species are similar to cave species, others are endemic, with morphological traits associated with life in interstices. These are perhaps the species most endangered by mineral and organic pollution, channelization, dams, and development of irrigation. This habitat is present in all latitudes, and frequently is locally destroyed.

Marine and anchialine coastal cave habitats

Stygophilic, blind and depigmented, marine or euryhaline species colonize the marine and anchialine caves of the coastal limestone or lava tube of volcanic islands. Some of these marine cave dwellers have a great zoological interest, such as the extraordinary Remipedia, a primitive group of Crustacea only discovered in 1981; or some Porifera considered as fossil twenty years ago.

Mines

Numerous mines and underground quarries in limestone and other rocks have communities composed of troglophilic species which came in through the entrance or with pieces of wood. Often, the deep zone of old artificial cavities in Europe is colonized by true troglobites such as Collembola, Diplura, and beetles. These artificial cavities have a great importance for bats which roost, hibernate and reproduce in them. In Europe, some of the biggest bat colonies are found in underground quarries such as those of Sint Pietersberg near Maastricht in the Netherlands, or the artificial caves of Miedyrzecz in Poland.

Heritage interest

The special interest in the conservation of subterranean species is due to the two following points:

(1) Some are "living fossils", species or groups which have disappeared from the surface waters and terrestrial habitats, some for millions of years, and which survive in caves, groundwater, and other subterranean habitats protected from surface cataclysms. They represent, in a sense, living zoological archives, especially for several groups of Crustacea (such as Syncarida, Thermosbaenacea, Isopoda, Stenasellidae, etc.), Porifera (Pharetronida and Hexatinellida) and some vertebrates, such as *Proteus* in Slovenia, Croatia and Bosnia-Herzegovina.

(2) Others are endemic species, sometimes rare, often endangered, which actively colonized the subterranean ecosystem. Their evolution and speciation continues, affected by paleoclimatic, tectonic, geomorphologic events, which isolate the populations in independent habitats, or generate new mosaics of subterranean habitats of MSS type, or new lava tubes quickly ready for colonization. Often there is a founder effect. In these cases, the colonization of the subterranean ecosytem is active, similar to the other surface ecosystems; but only preadapted species, with low rates of reproduction and metabolism, lucifuge habits, etc., can colonize these dark and food-scarce habitats.

Subterranean ecosystems, which differ from surface ecosystems with respect to climatic and biotic parameters, are among the most widespread on the mainland, with a large diversity of fauna – terrestrial, freshwater, brackish and marine.

VULNERABILTY OF THE HABITATS

The vulnerability of subterranean karstic habitats and species

The particular vulnerability of a large number of hypogean species comes from their low ecological vigour, and certainly from the small area of their territory, so that local pollution or modification of ecological conditions may bring about their disappearance. Many factors combine to place subterranean species at risk:

Destruction of habitats

Destruction of caves and their associated network of cracks by quarrying has been reported. For example, in England 73 of 360 caves were destroyed in the Mendip Hills. In Holland near Maastricht a part of the bat habitat of Sint Pietersberg was destroyed. In Italy about twenty caves were destroyed in the Geni Bolognesi region, and in Malaysia the well-known Batu Caves were endangered by a quarry project before steps were taken to conserve them. Intensive quarrying of gravel from stream-beds can destroy the interstitial habitat. Building of roads and highways across karstic areas can also destroy caves. Flooding of karstic caves by hydroelectric dams and reservoirs destroys the terrestrial subterranean habitats.

Excessive visitation of caves

Excessive visitation is a result of the considerable development of speleology, and group visits by adults and school children, and tourists. Group visits may serve as an inititation to nature and the cave environment, or as private, local or national recreational activities. This phenomenon has increased during the last two decades in many countries. For example, in 1985, 20 000 people visited sixty caves in Derbyshire (England). This may cause:

– destruction of biotopes and concretions by trampling and crawling;
– pollution of water and disappearance of aquatic populations as a result of toxic products left in caves (such as carbide and batteries);
– accumulation of non-biodegradable products which render the habitats unsuitable for subterranean species;
– an increase in the mortality of hibernating bats due to frequent awakening during severe and long winters. These disturbances cause bats to arouse prematurely from hibernation, and excessive consumption of their energy reserves, which the animals cannot restore;
– disturbance of bat nurseries, causing mortality of the young.

Vandalism

Concretions, stalactites, and stalagmites are the first objects destroyed in caves for a variety of reasons: destructive rage, collection, national or international marketing. An international black market tends to intensify destruction by providing financial support for the removal of all marketable speleothems. Numerous cases are reported from France, Italy, Morocco, Brazil, the United States, Korea, South-Africa, etc. Numerous cases of break-ins through entrance barriers have occurred and several of the strongest barriers have been attacked with explosives, often by irresponsible speleologists.

There have been clandestine excavations to collect fossil bones (for instance, recently skulls and canines of *Ursus spelaeus* in Pyrénéan caves) or prehistoric objects for the international black market.

Intensive trapping

Several factors may induce people, especially entomologists, to use and leave pitfall traps in caves for long periods of time: the search for rare species; the desire to complete a private collection with the rarest species; attempts to supplement collections at the request of certain museums; and the high market value of rare specimens, especially trechine beetles. This trapping is especially dangerous for carnivorous trechines, the numbers of which are very limited. Some of these species are very rare or known from only one or a few caves. In contrast, troglobitic saprophagous bathysciine beetles often have large populations. For numerous European species the discovery of the MSS habitat has greatly increased the number of sites from which they are known.

Show caves

The opening of caves to tourism has complex, negative and positive effects. The fact of controlling entrances and limiting movements to marked pathways may assure a better conservation of fauna in the galleries that are not visited than in caves with open access. On the other hand, in the parts of caves that are visited, new openings for tourist purposes may modify air circulation and change the microclimate sufficiently to cause the disappearance of the fauna. Lighting causes the disappearance of fauna in illuminated areas and growth of *Lampenflora.* Trechines and some depigmented Crustacea are especially sensitive to illumination, and disappear. However, if illumination is well planned and leaves dark regions in the galleries visited, at least a part of the fauna, such as Araneae, Collembola, Diplopoda, terrestrial Isopoda and some bathysciine beetles, may take refuge in these undisturbed areas. Some types of installations, such as concrete pathways and stairs, can destroy biotopes or make the cave too artificial for the fauna. Several cases have been reported of partial or total disappearance of the cave fauna in show caves (such as the Vadu Crisului cave in Romania and the Cango cave in South Africa).

At present, methods of estimating troglobitic populations are well developed and common practice for cave beetles and other terrestrial species. They can be used to estimate population size before modification and to measure the impact of the modifications and the tourist visits. Early studies in the Grotte de Sainte-Catherine in the Pyrénées showed the importance of climate (especially velocity of air flow, temperature and humidity of walls and cracks) on the seasonal presence or absence of blind *Aphaenops* beetles.

Long-term climatic and microclimatic studies in caves with prehistoric paintings in France (Lascaux, Niaux, Pech-Merle, and Gargas caves) have demonstrated the impact of tourist groups on microclimatic parameters (temperature, humidity) affecting the walls with paintings. Management of tourist groups and visits (number of persons in a group, time between visits, total numbers of visitors a day) has been initiated.

Chemical pollution

Chemical pollution is linked to industry, urbanization, and intensive agricultural practices (Fig. 35.1). I

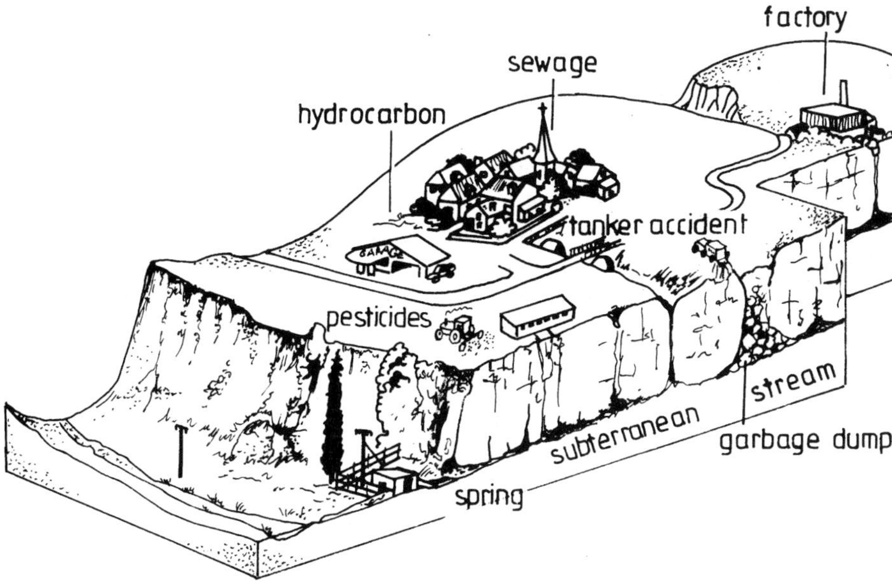

Fig. 35.1. Schematic diagram showing the different origins of pollution in karstic regions. (From the poster issued by the Fédération Française de Spéléologie: "Protégeons nos cavernes", modified.)

may be diffuse, accidental or chronic. Mainly, it affects aquatic habitats and aquatic fauna because water is the main agent of transport of pollution. It also affects bats, especially in countries with intensive agricultural practices. Pesticides used to control insect populations have had negative impacts on the populations of many bats. One may recall the drastic effect of dichlorodiphenyltrichloracetic acid (DDT) on several species. Two effects seem likely: direct poisoning of the bats; and reduction in the resource base of those bats which eat insects. The use of insecticides to treat wood in roofs also endangers the bats roosting in them.

The risk of accidental pollution is high. Pollution by hydrocarbons and other petroleum products is the most frequent kind of pollution [e.g., in river basins of France, Belgium, Indiana (USA), and in Slovenia], followed by pesticides, organic toxic products, and heavy metals (e.g., the Sloans Valley cave system in Kentucky, USA). The origins include tanker accidents, leakage of fuel, kerosene, motor oil and gasoline, toxic wastes from factories, and rubbish dumps in karst areas.

Non-point-source pollution is primarily due to intensive agricultural methods in many countries, with the use of high level of nitrates, chlorine-based organic herbicides and pesticides, proximity of areas with a high density of livestock, and salt runoff from highways. Cave-specific legislation is not necessary to solve these impacts.

Bacteriological pollution

The need to preserve the purity of drinking water became obvious at the beginning of the twentieth century, at which time health risks were recognized in karst areas, posed by pathogenic bacteria coming from dead animals discarded in pits and sink-holes. These bacteria are transported to springs and resurgences by karst water. It is an important human health problem, and laws have been promulgated to avoid this danger by forbidding the throwing of dead animals into sink-holes. However, decomposed carcasses of sheep and dogs have often been seen with thousands of scavenger troglobitic beetles, such as *Speonomus*. This sporadic type of pollution does not have a lethal impact on the species in subterranean communities, because these bacteria do not seem to be pathogenic for the cave invertebrate fauna.

Salinity change

Salinity changes in caves or phreatic aquifers near the coasts due to freshwater extraction may affect the quality of the underwater cave environment and change the composition of the communities. At present, there are no clear data on this potential impact.

Vulnerability of the interstitial (hyporheic) habitat

The main dangers for the interstitial habitat are:

(1) Accidental and diffuse pollution (see the preceding paragraph).
(2) Disappearance of gravel through exploitation.
(3) Excessive pumping of water for irrigation or storage in dams, which causes the reserve flow to be too low.
(4) Increase of water temperature in summer, with the formation of hypoxic zones unsuitable for subterranean species. In the hypoxic zones, only some rare subterranean species, adapted to life in a low oxygen environment, can survive.
(5) Dams and small-scale hydroelectric power stations, which cause a change in the stream profile, modifications in the distribution and quantity of surface and interstitial flows, and discharge of excessive amounts of silt and sand. The interstitial habitat is destroyed by tiny mineral particles coming from dams, which clog the voids within the gravel.

Vulnerability of bats

Caves, mines and subterranean quarries constitute only part of the habitat for bats. Caves are used to roost, to hibernate, and as maternity sites. The decline of several species is due to a series of factors: pesticide poisoning of their invertebrate prey in the past by DDT; human disturbance of hibernating clustered individuals during winter; human disturbance of nursery colonies; local vandalism, with destruction of numerous specimens hanging from ceilings; loss of habitat by closing of caves and mines; change in cave microclimate by new openings; change in landscape surface with decrease of plant diversity and consequently of the bats' prey; and bat capture for scientific studies. Several species are endangered, for instance *Rhinolophus ferrumequinum* and *R. hipposideros* in Europe.

CONSERVATION OF SPECIES

The best method of protecting the subterranean species is to protect their habitat. However, a list of selected subterranean species in each country permits the application of international, European and national legislation. It gives a legal basis for protecting the habitat, and for limiting the intensive trapping of species. Protection of troglophylic bats presents specific problems, on account of their life cycle, and their movements from one cave to another. They use different caves at different stages of their life cycle. Bats use numerous caves as roosting places during the warm season, when they exploit the invertebrate resources of the surface communities. They use a reduced number of caves for hibernation and maternity. Furthermore, the meta-population of bats is subdivided into colonies which move about roosting over a fairly extended region, so that hibernation and nursery caves may be located in different countries (e.g., France and Spain for the Pyrénéan region). The necessity to coordinate conservation policy is obvious.

LEGISLATIVE ASPECTS OF CONSERVATION

A conservation policy must be translated into the legislative field at different levels: international, national, and regional.

International conservation policy

Three world Conventions can be applied to subterranean species and habitats:

(1) The Washington Convention (3 March, 1973) to protect wild fauna and to regularize their international trade. Except for bats and some other rare vertebrates such as *Proteus* and the Guacharo, no subterranean species are listed.
(2) The Rio Convention (13 June, 1992) to protect biodiversity.
(3) The Ramsar Convention (2 February, 1971) to protect the "humid zone". Recently, in 1996, the Ramsar Convention was extended to the karst subterranean aquifers. It offers a new possibility to protect large karst areas of international importance from an ecological, zoological, limnological, hydrological, or aesthetic point of view.

European legislation

Four general European conventions and a specific recommendation (No. 36) may be used to protect species, habitats and a special type of cave (artificial caves of archaeological interest):

(1) The Bonn Convention (23 June, 1979) for the conservation of wild migratory species (it is relevant to troglophylic bats).

(2) The Bern Convention (19 September, 1979) for the conservation of wild fauna and natural habitats (Special Recommendation No. 36 concerns subterranean animals and habitats).
(3) The Directive Habitat CEE (1992–2004): "Natura 2000".
(4) The Malta Convention (16 January, 1992) to protect archeological heritage. It is also relevant to mines, quarries, and the artisanal and industrial techniques related to them.

Recommendations

Conservation of habitats (based on Recommendation No. 36)

Recommendation No. 36 on the conservation of the European underground habitats, from the Council of Europe, should be extended to the whole world. It recommends that different countries:

(A) Compile national inventories of subterranean habitats of high biological value, using the following selection criteria:
  (1) All types of subterranean habitats [including caves, lava tubes, Meso Shallow Substratum (MSS), hyporheic habitat in surface streams, alluvial phreatic aquifers, hydrothermal caves, anchialine caves, and marine caves];
  (2) Habitats characterized by the presence of species adapted to subterranean life. They are characterized by morphological, life-history and physiological traits, the chief being regression of the eyes, loss of pigment, K reproductive strategy, low metabolism, and disappearance of circadian rhythms of locomotory activity.
  (3) Habitats of biotic communities of considerable species richness which are representative of the various biogeographical and climatic regions of the world;
  (4) Habitats containing key species of great heritage and zoological interest, survivors of faunas which have disappeared from terrestrial and aquatic surface ecosystems ("living fossils");
  (5) Habitats with endemic species. Endemic species are so numerous that it is necessary to make a choice; priority should be given to species most representative of the biotic community of the region, or species with very narrow range;
  (6) Terrestrial habitats with very rare species or exceptional for the type of species present (for example: Cueva de los Guacharos in Venezuela, the most important cave with oil birds; the Waitomo caves in New Zealand, with the "Glow worm";
  (7) Uniqueness of the habitat. Besides the most common underground habitats, there are a small number of unusual habitats such as the hydrothermal habitat (for example: Movile Cave, a sulphurous habitat with endemic species; Baile Herculane cave in Romania, a tropical habitat in the temperate zone); caves with hot chambers due to bats and their guano with very high invertebrate biodiversity (for example: Cueva de las Majaes in Cuba); ice caves (for example ice caves with endemic beetles: Eisreisenvelt in Dachstein, Austria and Scarisoara in Romania; marine lava tubes (for example Jameos del Agua of Lanzarote, in the Canary Islands and Lucayan Caverns in the Bahamas with Remipedia;
  (8) Endangered habitats. These include primarily all the aquatic subterranean habitats, which have high vulnerability to chemical or organic pollution by surface water entering caves and phreatic aquifers. But they also include the terrestrial habitat, which can be destroyed by quarrying, excessive visiting, or excessive hunting and collecting;
  (9) Caves, quarries or mines which are used by bats (all species are protected) for winter hibernation or as a summer nursery; for a meta-population of species using many caves in a region, a list of all the hibernacula and breeding sites should be compiled in order to prepare a network of caves, mines, and quarries to be protected.
  (10) Habitats which have been the subject of very thorough ecological studies and can serve as a reference, or be used for long-term follow-up of populations, biotic communities and management, should be conserved for their scientific value.

Some complementary general criteria should be used to assess the relative importance of habitat, based on the criteria of the Habitat Directive "Natura 2000" of European Community":

(11) Degree of conservation of the elements of the habitat;

(12) Size of the species populations compared to other sites on the national territory;

(13) Degree of isolation of the population of the habitat in question from the general area of the species.

(B) Compile a list of the subterranean habitats protected, using the preceding criteria, with the nature of the protection (physical or biological) and the legislation. Indeed, many karst areas, hyporheic habitats of surface streams, and mines with bats, are included in various types of protected areas (National Parks, Natural Reservations, etc.) created for conservation of flora, birds, and aquatic mammals. However, invertebrates and especially cave invertebrates have not been considered, and, most often, there are not even inventories of the cave fauna. Thus, this recommendation is the primary step.

(C) Compile a list of the most endangered sites in each country.

(D) Compile a short list of natural subterranean habitats (e.g., from one to five) in each country of Europe, America, Africa, Asia, Oceania, and finally the world, for their fauna, their uniqueness, their biogeographical representativness and their biological diversity. The general survey of the world cave fauna "Encyclopaedia Biospeologica" (Juberthie and Decu, 1994, 1998), including 75 animal groups and 7725 species from 161 countries, which gives a strong base for conservation policy, should be used.

(E) It is necessary to manage habitats in order to protect wild invertebrate and vertebrate cave species. Accordingly, the compilation of inventories must be supplemented by studies of population dynamics, on a long-term basis, in order to discern the factors that affect population size and species diversity. For this purpose, it is necessary to choose reference sites in different countries where changes in communities and populations may be monitored.

Conservation of subterranean species

Subterranean invertebrate species requiring special conservation measures should be identified, and national or regional red lists of such species as candidates for conservation should be drawn up.

Conservation of troglophylic bats and their habitat

(A) Establishment of regional and national lists of the most threatened sites (hibernation and breeding sites in caves, mines and quarries). Selection of these sites should be based on their being permanent shelters for reproduction and breeding of young, on the significance of the population, and on their importance in the network of caves used in the course of migration.

(B) Regional maps (state, province, district, etc.) of subterranean sites used by each species of bat should be prepared.

(C) Regional and national maps of hibernation and nursery subterranean sites used by bats are needed.

(D) Populations should be monitored on a continuing basis, using standard methods, to estimate the number of bats in the reference sites and to determine the most endangered species.

(E) Long-term studies are needed of the causes of decreases in population size (changes in landscape, forestry and woodlands, conversion of land to agricultural production, and intensive agricultural practices which reduce abundance and diversity of insect prey).

(F) Conservation measures should be based on a regional network of the most endangered sites (hibernation and nurseries).

Specific conservation of caves and selection of caves for different uses (geological, mineralogical, archaeological and biological aspects)

The number of caves is very high; for instance, more than 130 000 caves are recorded in Europe. They have many different uses, so that there is no ecological or scientific need to conserve all caves. The main trends for the use and conservation strategy of caves should be to classify caves into different types according to their importance, and to apply the appropriate legislation. The types are:

(1) Show caves;

(2) Protected caves of interest for heritage, biodiversity, mineralogical, and cultural reasons, protected:
   - for conservation of species and their habitats (e.g., bats, cave beetle *Aphaenops*);
   - for mineralogical and palaeoclimatic purpose (e.g., speleothems, and as museums preserving evidence of past climatic events on the earth's surface);

  - for archaeology (e.g., human remains);
  - for prehistoric paintings and drawings;
  - for scientific research (such as monitoring of populations and habitats over a long term).

(3) Protected mines (interest in bats, or for cultural reasons).
(4) Caves for recreational and speleological use (the very large majority).
(5) Cave reserved for educational groups (discovery of the cave environment).

Education

Education and knowledge of the composition, diversity, and vulnerability of caves, subterranean habitats and species, and measures of conservation, are the bases of a conservation policy. It is becoming increasingly important to educate young people and tourists during visits to caves. National or regional natural parks, natural reserves and other conservation reserves, speleo-clubs and ecological associations are and will be important places to present general information on the subterranean environment as ecosystem.

The training of cave guides is an important way to transfer information to the visitors. In France the association of show caves has this type of educational program. The education of cavers is a necessity. In France, publications and posters are provided by the Fédération Française de Spéléologie. The U.I.S. (International Union of Speleology) has a conservation commission for the caves.

CONCLUSIONS

The conservation of subterranean species and habitats requires immediate measures and the application of world-wide, European or national conservation directives, and the establishment of specific protective measures or laws in the different countries.

REFERENCES [1]

Audetat, M., 1974. Problèmes de la protection des cavernes en Suisse. *Actes 5e Congr. nat. Spéléol. Interlaken. Stalactite, suppl.* 9: 198–201.

Barthélémy, D., 1984. *Impact des pollutions sur la faune stygobie karstique: approche typologique sur 16 emergences des départements de l'Ain et du Jura, Thèse.* Université de Lyon, 183 pp.

Black, G.P., 1964. The conservation of caves in Britain. *Stud. Speleol.,* 1(1): 16–21.

Bolner, K.T., Tardi, J. and Nemedi, L., 1989. Evaluation of the environmental impacts in Budapest's caves on the basis of the study of the quality of dripping water. In: *Proc. 10th Int. Congr. Speleology, Budapest,* Vol. 2. International Union of Speleology, pp. 634–639.

Bosnak, A.D. and Morgan, E.L., 1981. Acute toxicity of cadmium, zinc and total residual chlorine to epigean and hypogean isopods (Asellidae). *Natl. Speleol. Soc. Bull.,* 43: 12–18.

Boutin, C., 1984. Sensibilité à la pollution et répartition, de quelques espèces de Crustacés phréatobies à Marrakech (Maroc occidental). *Mem. Biospel.,* 9: 55–64.

Brady, J.T., 1983. Cave management for the endangered Indian bats (*Myotis sodalis*) and Gray bat (*Myotis grisescens*). *Natl. Speleol. Soc. News,* 41: 298–300.

Christinat, J.-L., 1987. Protection des cavernes dans le canton de Neuchâtel: le bilan après 5 ans. *Actes 8e Congr. Nat. Suisse Spéléol.,* 165–168.

Craven, S.A., 1989. Conservation at the Cango caves. *Bull. S. Afr. Speleol. Assoc.,* 30: 57–74.

Danielopol, D.L., 1989. Groundwater fauna associated with riverine aquifers. *J. Am. Benthol. Soc.,* 8(1): 18–35.

De Broyer, C., Thys, G. and Fairon, J., 1992. *Atlas du Karst Wallon. Province de Namur, Tome I.* Commission Wallonne d'Étude et de Protection des Sites Souterrains, Bruxelles, 246 pp.

Deltshev, C.D., 1989. Conservation problems of the Bulgarian cave fauna. In: *Proceedings of the 10th International Congress of Speleology, Budapest,* Vol. 3, pp. 778–779.

Duclaux, G. and Mouret, C. (Editors), 1989. Protection des cavernes et du milieu souterrain. *Spelunca,* 35: 1–100.

Edworthy, K.J., 1987. Groundwater contamination. A review of current knowledge and research activity. *Stygologia,* 3: 279–295.

Elliott, W.R., 1990. Endangered species and endangered caves. *Natl. Speleol. Soc. News,* 48: 225–231.

Ganter, J., 1989. Cave exploration, cave conservation. Some thoughts on compatibility. *Natl. Speleol. Soc. News,* 47: 249–253.

Gibert, J., Danielopol, D.L. and Stanford, A. (Editors), 1994. *Groundwater Ecology.* Academic Press, New York, 571 pp.

Habe, F., 1986. Skocjanske Jame. Exposition des motifs pour l'enregistrement dans la liste du patrimoine mondial. In: *Proc. 9th Int. Congr. Speleology, Barcelona,* Vol. 1. Comis. Organitz. 9th Int. Congr. Espeologia, Barcelona, pp. 233–234.

Hamilton-Smith, E., 1970. Biological aspects of cave conservation. *J. Sydney Speleol. Soc.,* 14: 157–164.

Hardwick, P. and Gunn, J., 1989a. The impact of agricultural operations on British caves. In: *Proc. 10th Int. Congr. Speleology, Budapest,* Vol. 1. International Union of Speleology, pp. 198–200.

Hardwick, P. and Gunn, J., 1989b. Cave management and conservation in Britain: an historical revue. In: *Proc. 10th Int. Congr. Speleology, Budapest,* Vol. 1. International Union of Speleology, pp. 196–197.

Hegedus, G., 1971. Cave closing as a conservation method. In: B.F. Beck (Editor), *Proc. 8th Int. Congr. Speleology, Bowling Green,* Vol. 2, pp. 401–402.

[1] As the present chapter is synthetic, the reference list serves mainly as a bibliographic guide: most of the publications are not referred to in the text.

Hobbs, H.H., 1988. Gasoline pollution in an Indiana cave. *Natl. Speleol. Soc. Bull.,* 46: 29.

Holsinger, J.R., 1966. A preliminary study of the effects of organic pollution of Banners Corner Cave, Virginia. *Int. J. Speleol.,* 2: 75–89.

Hubart, J.-M., 1973. Urgence d'une protection des cavernes et biotopes souterrains en Belgique. *Nat. Belg.,* 54: 141–154.

Huppert, G.N. and Thorn, J.B., 1989. Federal cave protection in the United States: the Federal Cave Resources Protection Act of 1988. In: *Proc. 10th Int. Congr. Speleology, Budapest,* Vol. 1. International Union of Speleology, pp. 188–189.

Iliffe, T.M., Jickells, T.D. and Brewer, M.S., 1984. Organic pollution of an inland marine cave from Bermuda. *Marine Environ. Res.,* 12, p. 173–189.

International Union of Speleology (Editor), 1989. Cave tourism. *Proc. Int. Symp. at the 170th anniversary of Postojna Jama (Yugoslavia).* Centr. Sci. Rs. S.A.Z.U. Postojna, 204 pp.

Juberthie, C., 1995. Underground habitats and their protection. *Nat. Environ.* (Council of Europe, Strasbourg), 72: 1–158.

Juberthie, C. and Decu, V. (Editors), 1994. *Encyclopaedia Biospeologica,* Vol. I. Société de Biospéologie, Moulis (France) and Bucarest, 834 pp.

Juberthie, C. and Decu, V. (Editors), 1998. *Encyclopaedia Biospeologica,* Vol. II. Société de Biospéologie, Moulis (France) and Bucarest, pp. 835–1373.

Kenrick, L.D., 1983. Education and secrecy in cave conservation. *Natl. Speleol. Soc. Bull.,* 45: 1–7.

Maire, R., 1979. Comportment de karst vis-à-vis des substances polluantes. *Ann. Soc. Geol. Belg.,* 102: 101–108.

McCracken, G.F., 1989. Cave conservation: special problems of bats, *Natl. Speleol. Soc. Bull.,* 51: 49–51.

McFarlane, D.A., 1988. Endangered cave species. *Natl. Speleol. Soc. News,* 46: 36–37.

Middelton, G.L., 1977. The conservation and management of caves in Tasmania. Australia. In: *Proc. 7th Int. Congr. Speleology, Sheffield.* Cave Research Asoociation, Somerset, England, pp. 311–314.

Ministère de l'Environnement (Editor), 1994. *Inventaire de la faune menacée en France.* Nathan ed./Mus. Nat. Hist. Nat., Paris, and W.W.F. France, 176 pp.

Ministère de l'Environnement (Editor), 1996. *La diversité biologique en France et programme d'action pour la faune et la flore sauvages.* Ministère de l'Environnement, Paris, 378 pp.

Novak, D., 1986. The pollution of the karstic ground water: the example of Krupa. In: *Proc. 9th Int. Congr. Speleology, Barcelona.* Comis. Organitz. 9th Congr. Int. Espeologia, Barcelona, pp. 87–89.

Pride, Th., Ogden, A. and Harvey, M., 1989. Biology and water quality of caves receiving urban runoff in Cookville, Tennessee, U.S.A. In: *Proc. 10th Int. Congr. Speleology, Budapest,* Vol. 1. International Union of Speleology, pp. 27–29.

Pugsley, C.W., 1984. Management of a biological resource – Waitomo Glowworm Cave, New Zealand. In: B.F. Beck (Editor), *Proc. 8th Int. Congr. Speleology, Bowling Green,* Vol. 2, pp. 489–492.

Ransome, R., 1990. *The Natural History of Hibernating Bats.* Christopher Helm, London, 235 pp.

Sket, B., 1983. Proposal for protection of some speleobiologically important regions. *Naš. Kraš Sarajevo,* 9: 123–127.

Stebbings, R.E., 1988. *Conservation of European Bats.* Christopher Helm, Bromley, Kent, 246 pp.

Stitt, R., 1981a. Underground wilderness. A conservation principle and a management tool. In: B.F. Beck (Editor), *Proc. 8th Int. Congr. Speleology, Bowling Green,* Vol. 2, pp. 185–186.

Stitt, R., 1981b. Cave conservation in the United States of America. An overview in 1981. In: B.F. Beck (Editor), *Proc. 8th Int. Congr. Speleology, Bowling Green,* Vol. 2, pp. 187–189.

Szekely, K., 1983. Cave conservation in Hungary. *Karst es Barlang,* Special Issue, pp. 87–89.

Tercafs, R., 1995. The protection of the subterranean environment. Conservation principles and management tools. In: A.I. Camacho (Editor), *The Natural History of Biospeology.* National Museum of Natural Sciences, Madrid, pp. 485–522.

Trajano, E., 1986. Vulnerabilidad dos troglobios pertubaqoes ambiantais. *Espeleo-Tema,* 15: 19–24.

Turquin, M.J., 1980. La pollution des eaux souterraines: incidences sur les biocenoses aquatiques. *Actes 1er Colloque National sur Protection des Eaux Souterraines, Besançon,* Vol. 2. Université Besançon, pp. 341–347.

Waltham, A.C., 1983. A review of karst conservation sites in Britain. *Stud. Speleol.,* 4: 85–92.

Wilde, K., 1985. An historical review of cave and karst conservation and management in New Zealand 1889–1985. *N. Z. Speleol. Soc. Bull.,* 7: 461–490.

Wilmut, J., Young, I., Davies, G., Long, H., Mellors, P.T., Heaver, S., Price, G., Cawthorn, R. and Jodson, D., 1985. The conservation of caves in Great Britain. *Stud. Speleol.,* 6(Special Issue): 4–39 (chap. I–VIII).

Wilson, J.E., 1983. The role of the American Conservation Cave Association in the cave conservation movement. *Natl. Speleol. Soc. Bull.,* 45: 1–8.

SYSTEMATIC LIST OF GENERA

All taxa mentioned in the text are here included in their taxonomic position. Division of genera is to the Family level in higher plants, to the Order level in arthropods and vertebrates, to the Phylum, Subphylum or Class elsewhere. Names of intermediate taxa mentioned are inserted immediately below the higher taxon to which they are ascribed, unless they are themselves divided. Within Class, Order or Family, lower taxa are listed in alphabetical order.

MONERA (PROKARYOTA)

ARCHAEA

Acidianus
Archaeoglobus
Methanobacterium
Methanococcus
Pyrobaculum
Pyrodictium
Pyrolobus
Stygiolobus
Sulfolobus
Thermoproteus

EUBACTERIA

Acetobacterium
Alcaligenes
Aquifex
Bacillus
Beggiatoa
Clostridium
Desulfobacter
Desulfococcus
Desulfotomaculum
Desulfovibrio
Escherichia
Gallionella
Hydrogenobacter
Leptospirillum
Nitrobacter
Nitrococcus
Nitrosococcus
Nitrosolobus
Nitrosomonas
Nitrospira
Pseudomonas
Salmonella
Shigella
Sphaerotilus
Streptococcus
Thiobacillus
Thiomicrospira
Thioploca
Thiothrix

FUNGI

LICHENES

Stereocaulon

PLANTS

PTERIDOPHYTA
FILICOPSIDA

Polypodiaceae
Dennstaedtia
Dryopteris
Phyllitis
Dryopteridaceae
Cyrtomium

PINOPHYTA (GYMNOSPERMAE)

Pinaceae
Pinus

MAGNOLIOPHYTA (ANGIOSPERMAE)
LILIOPSIDA (MONOCOTYLEDONES)

Liliaceae
Smilax
Poaceae
Phragmites
Spartina
Triodia
Zea
Zizania

MAGNOLIOPSIDA (DICOTYLEDONES)

Aceraceae
Acer
Anacardiaceae
Pistacia
Schinus
Asteraceae
Artemisia
Caprifoliaceae
Viburnum
Caryophyllaceae
Moehringia
Casuarinaceae
Casuarina
Cornaceae
Cornus
Ericaceae
Arbutus
Fagaceae
Nothofagus
Quercus
Mimosaceae
Acacia
Moraceae
Ficus
Myrtaceae
Agonis
Corymbia
Eucalyptus
Metrosideros
Oleaceae
Fraxinus
Phillyrea
Proteaceae
Banksia
Rhamnaceae
Rhamnus
Rosaceae
Potentilla
Salicaceae
Populus
Sterculiaceae
Brachychiton

ANIMALS

PROTOZOA
CILIATA

Euplotes
Paramecium
Trachelocerca

FORAMINIFERA

Marenda

CNIDARIA
HYDROZOA

Hydra
Psammohydra

CTENOPHORA

Cestum

PLATHELMINTHES

ARCHOOPHORA
Machrostomidae
Acantomacrostomum
CATENULIDA
Stenostomum
Dendrocoelidae
Dendrocoelum
Dugesiidae
Dugesia
Macrostomida
Macrostomum
Planariidae
Crenobia
Phagocata

BRYOZOA

GYMNOLAEMATA
Ctenostomata
Monobryoontidae
Monobryozoon

NEMATHELMINTHES
NEMATODA

Leptolaimidae
Chronogaster
Panagrolaimidae
Panagrolaimus
Rhabditidae
Protorhabditis
Tripylidae
Tobrilus

ROTIFERA

Habrotrochidae
Habrotrocha

ECHIURA

Bonellia

ANNELIDA
APHANONEURA

Aeolosomatidae
Aeolosoma

ARCHIANNELIDA

Protodrilidae
Protodrilus

POLYCHAETA

Acrocirridae
Macrochaeta
Capitellidae
Capitella
Dorvilleidae
Protodorvillea

Fauveliopsidae
Fauveliopsis
Nereididae
Namanereis
Nerillidae
Leptonerilla
Mesonerilla
Troglochaetus
Paraonidae
Paradoneis
Polynoidae
Gesiella
Pelagomacellicephala
Scalibregmatidae
Speleobregma
Syllinae
Exogone
Miscellania
Syllis

CLITELLATA

HIRUDINEA
Haemopidae
Haemopis
OLIGOCHAETA
Enchytraeidae
Henlea
Marionina
Haplotaxidae
Haplotaxis
Lumbricidae
Allolobophora
Dendrobaena
Eiseniella
Lumbriculidae
Rhynchelmis
Trichodrilus
Naididae
Pristina
Tubificidae
Antipodrilus
Phallodrilus
Rhyacodrilus
Tubifex

MOLLUSCA

BIVALVIA

Arcidae
Anadara
Mytilidae
Bathymodiolus
Mytilus
Calyptogena

GASTROPODA

Ellobiidae
Carychium
Hydrobiidae
Moitessiera
Avenionia
Bythiospeum
Hauffenia
Islamia
Iravadiidae
Iravadia
Littorinidae
Antrobia
Littorina
Moitessieriidae
Heleobia
Physidae
Physa
Rissoidae
Peringiella
Zonitidae
Oxychilus

ONYCHOPHORA

Peripatidae
Speleoperipatus

UNIRAMIA

DIPLOPODA

CALLIPODIDA
Abacionidae
Tetracion
CHORDEUMATIDA
Cleidogonidae
Cleidogona
Pseudotremia
Diplomaragnidae
Diplomaragna
Trichopetalidae
Scoterpes
Trichopetalum
Craspedosomatida
Metopidiotrichidae
Metopidiothrix
GLOMERIDA
Trachysphaeridae
Trachysphaera
JULIDA
Blaniulidae
Antrokoreana
Julidae
Archiboreoiulus
Trogloiulus
POLYDESMIDA
Cryptodesmidae
Peridontodesmella
Oniscodesmidae
Katantodesmus
Paradoxosomatidae
Antichiropus
Boreohesperus
Stygiochiropus
Epanerchodus
Vanhoeffeniidae
Antriadesmus
SPIROBOLIDA
Pachybolidae
Speleostrophus
SPIROSTREPTIDA
Cambalidae
Cambala
Nannolene
Pseudonannolenidae
Pseudonannolene

CHILOPODA

GEOPHILOMORPHA
Geophilidae
Clinopodes
Geophilus
LITHOBIOMORPHA
Lithobiidae
Lithobius
SCOLOPENDROMORPHA
Cryptopidae
Cryptops
Scolopendridae
Ethmostigmus
Scolopendra
SCUTIGERIMORPHA
Scutigeridae
Scutigera

SYMPHYLA

Scolopendrellidae
Symphynella

COLLEMBOLA

Arrhopalitidae
Arropalites
Brachystomellidae
Brachystomella
Cyphoderidae
Onchopodura
Onychiurus
Entomobryidae
Pseudosinella
Sinella
Ascocyrtus
Coecobrya
Entomobrya
Heteromurus
Lepidocyrtus
Seira
Hypogastruridae
Acherontides
Acherontiella
Hypogastrura
Mesachorutes
Schaefferia
Willemia
Xenylla
Isotomidae
Cryptopygus
Folsomia
Folsomides
Folsomina
Isotomiella
Isotomurus
Neanuridae
Blasconura
Coecoloba
Friesea
Paleonura
Paranura
Pronura
Neelidae
Megalothorax
Neelipleona
Neelus
Oncopoduridae
Oncopodura
Onichiuridae
Deuteraphorura
Mesaphorura
Onichiurus
Prabhergia
Protaphorura
Paronellidae
Cyphoderopsis
Lepidonella
Troglopedetes
Pseudachorutinae
Pseudachorutes
Sminthuridae
Arrhopalites
Pararrhopalites
Sminthurinus
Sminthurus
Sphaeridiidae
Sphaeridia
Tomoceridae
Tomocerus

DIPLURA

Campodeidae
Litocampa
Lepidocampa
Plusiocampa
Japygidae
Troglojapyx
Indjapyx

INSECTA

ARCHAEOGNATHA
Machilellus
THYSANURA
Lepismatidae
Ctenolepisma
Nicoletiidae
Cubacubana
Trinemura
EPHEMEROPTERA
Baetidae
Baetis
Habrophlebiidae
Habrophlebia
Heptageniidae
Rhithrogena
BLATTODEA
Blaberidae
Pycnoscelus
Blattellidae
Loboptera
Nocticolidae
Alluaudellina
Nocticola
PLECOPTERA
Chloroperlidae
Chloroperla
Nemouridae
Nemoura
Leuctridae
Leuctra
ORTHOPTERA
Gryllidae
Hymenoptila
Nemobiinae
Caconemobius
Ngamarlanguia
Speonemobius
Rhaphidophoridae
Ceuthophilus

Dolichopoda
Euhadenoecus
Hadenoecus
Rhaphidophora
Spelaiacris
Troglophilus
DERMAPTERA
Pygidicranidae
Anataelia
PSOCOPTERA
Lepidopsocidae
Perientomum
Liposcelidae
Belapha
Liposcelis
Troglotroctes
Pachytroctidae
Pachytroctes
Psyllipsocidae
Psocathropos
HEMIPTERA
Aphididae
Aphis
Dilachnus
Cixiidae
Cixius
Javesella
Oliarus
Solonaima
Undarana
Delphacidae
Notuchus
Meenoplidae
Meenoplus
Phaconeura
Mesoveliidae
Mesovelia
Nepidae
Nepa
Reduviidae
Nesidiolestes
MEGALOPTERA
Sialidae
Sialis
COLEOPTERA
Agoninae
Speogonium
Bathycinae
Batinoscelis
Josetekiella
Speonomus
Carabidae
Anophthalmus
Aphaenops
Benbidion
Eustra
Duvalius
Gastrogonum
Geotrechus
Mateuius
Mayaphaenops
Mexaphaenops
Neaphaenops
Ozaenaphaenops
Pseudanophthalmus
Pseudophthalmus
Rhadine
Trechiama
Trichaphaenops
Catopidae
Ptomaphaginus
Cholevidae
Bathysciola
Catops
Choleva
Cytodromus
Diaprisus
Dissochaetus
Orostygia
Royerella
Clivinidae
Clivina
Curculionidae
Myllocerus
Dryopidae
Stygoparnus
Dytiscidae
Bidessus
Sternopriscus
Siamoporus
Elmidae
Elmis
Esolus
Heterelmis
Stenelmis
Helodidae
Cyphon
Lampyridae
Photuris
Leiodidae
Adelops
Adelopsis
Ptomaphagus
Leptodirinae
Drirneotus
Isereus
Pholeuon
Leptodirus
Melyridae
Gietella
Ifnidius
Merophysiidae
Holoparamecus
Nitidulidae
Urophorus
Pselaphidae
Batrisodes
Decumarellus
Marellus
Texamaurops
Tychobythinus
Staphylinidae
Atheta
Deleaster
Domene
Medon
Quedius
Trechinae
Arctaphaenops
Darlingtonea
TRICHOPTERA
Glossosomatidae
Synagapetus
Hydropsychidae
Hydropsyche
Limnephilidae
Stenophylax
Polycentropidae
Plectronemia
Polycentropus
Psychomiidae
Lype
LEPIDOPTERA
Geometridae
Triphosa
Noctuidae
Apopestes
Scoliopteryx
Orneodidae
Orneodes
Tineidae
Tinea
Wegneria
Crypsithyris
DIPTERA
Chironomidae
Corynoneura
Kiefferulus
Pentapedilum
Polypedilum
Pseudorthocladius
Triphleba
Culicidae
Culex
Drosophilidae
Drosophila
Fanniidae
Fannia
Helomizidae
Thelida
Leptoceridae
Leptocera
Limoniidae
Limonia
Milichiidae
Pholeomyia
Mormotomyiidae
Mormotomyia
Muscidae
Psilochaeta
Mycetophilidae
Fungivora
Macrocera
Speolepta
Phoridae
Conicera
Dohrniphora
Megaselia
Phora
Tryphleba
Psychodidae
Psychoda
Sciaridae
Sciara
Sphaeroceridae
Spelobia
Syrphidae
Syrphus
Tipulidae
Niphadobata
Trichoceridae
Trichocera
HYMENOPTERA
Formicidae
Paratrechina
Solenopsis

CRUSTACEA
REMIPEDIA
ENANTIOPODA
Tesnusocaris
NECTIOPODA
Godzilliognomus
Lasionectes
Speleonectes
PHYLLOPODA
BRANCHIOPODA
CLADOCERA
Daphnidae
Alona
Daphnia
Bosminidae
Bosmina
MAXILLOPODA
OSTRACODA
Cladocopina
Polycopidae
Eupolycope
Cytheracea
Candinidae
Cyclocypris
Candona
Candonopsis
Cryptocandona
Cypria
Potamocypris
Prionocypris
Cyprididae
Cavernocypris
Cypridopsis
Eucypris
Fabaeformiscandona
Herpetocypris
Ilyocypris
Isocypris
Mixtacandona
Nannocandona
Pseudocandona
Psychrodromus
Cypridinidae
Skogsbergia
Darwinulidae
Darwinula
Entocytheridae
Sagittocythere
Halocyprididae
Deeveya
Spelaeoecia
Ilyocyprididae
Iliocriptus

Limnocytheridae
Kovalevskiella
Limnocythere
Sarsiellidae
Eusarsiella
Thaumatocyprididae
Danielopolina
Xestoleberididae
Xestoleberis

COPEPODA
CALANOIDA
Diaptomidae
Stygodiaptomus
Epacticeriscidae
Enantiosis
Erebonectes
CYCLOPOIDA
Cyclopidae
Acanthocyclops
Apocyclops
Diacyclops
Eucyclops
Halicyclops
Kieferiella
Macrocyclops
Metacyclops
Microcyclops
Neocyclops
Paracyclops
Speocyclops
Tropocyclops
Cyclopinidae
Muceddina
Oromiina
Pseudocyclopiidae
Stygocyclopia
HARPACTICOIDA
Ameiridae
Nitocrella
Arietellidae
Paramisophria
Canthocamptidae
Attheyella
Bryocamptus
Canthocamptus
Ceuthonectes
Elaphoidella
Limocamptus
Gelyellidae
Gelyella
Harpacticidae
Parapseudoleptomesochra
Parastenocaridae
Parastenocaris
Superornatiremidae
Neoechinophora
MISOPHRIOIDA
Misophriidae
Boxshallia
Dimisophria
Expansophria
Misophria
Palpophria
Speleophria
Speleophriopsis
PLATYCOPIOIDA
Platycopiidae
Antrisocopia
Nanocopia
MYSTACOCARIDA
Derocheilocaris

MALACOSTRACA
SYNCARIDA
Anaspidacea
Koonungidae
Koonunga
Psammaspididae
Anaspides
Parastygocaris
Psammaspides
Stygocaris
Bathynellacea
Bathynellidae
Bathynella
Transvaalthynella
Parabathynellidae
Acanthobathynella
Atopobathynella
Cteniobathynella
Hexabathynella
Iberobathynella
Leptobathynella
Notobathynella
Vandelibathynella
Palaeocaridacea
Palaeocarididae
Palaeocaris
ISOPODA
Anthuridae
Stygocyathura
Armadilliidae
Armadillidium
Cubaris
Asellidae
Asellus
Caecidotea
Lirceus
Proasellus
Atlantasellidae
Atlantasellus
Cirolanidae
Antrolana
Bahalana
Creaseriella
Faucheria
Haptolana
Marocolana
Speocirolana
Turcolana
Typhlocirolana
Janiridae
Iais
Jaera
Limnocytheridae
Gomphodella
Microcerberidae
Microcerberus
Protocerberus
Yvesia
Microparasellidae
Angeliera
Microcharon
Oniscidae
Androniscus
Oniscus
Oritoniscus
Thailandoniscus
Paranthuridae
Curassanthura
Philosciidae
Halophiloscia
Leucophiloscia
Platyarthridae
Trichorhina
Porcellionidae
Porcellio
Sphaeromadidae
Caecosphaeroma
Monolistra
Sphaeroma
Sphaeromides
Stenasellidae
Stenasellus
Styloniscidae
Clavigeroniscus
Trachelipidae
Trachelipus
Trichoniscidae
Caucasonethes
Haplophthalmus
Trichoniscoides
Trichoniscus
SPELAEOGRIPHACEA
Mangkurtu
Potiicoara
Spelaeogriphus
AMPHIPODA
Aoridae
Grandidierella
Bogidiellidae
Aequigidiella
Afridiella
Bogidiella
Bogidomma
Maghrebidiella
Medigidiella
Pseudingolfiella
Stygogidiella
Ceinidae
Austrochiltonia
Crangonyctidae
Bactrurus
Crangonyx
Stygobromus
Stygonyx
Synurella
Gammaridae
Gammarus
Psammogammarus
Hadziidae
Bahadzia
Liagoceradocus
Mayaweckelia
Tuluweckelia
Weckelia
Spelaeorchestia
Ingolfiellidae
Balcanella
Gevegiella
Gevgeliella
Ingolfiella
Trogloleleupia
Melitidae
Nedsia
Norcapensis
Metacrangonyctidae
Longipodacrangonyx
Metacrangonyx
Niphargiidae
Metaniphargus
Niphargopsis
Niphargus
Pontoniphargus
Pseudoniphargus
Paramelitidae
Hurleya
Pardaliscidae
Spelaeonicippe
Perthidae
Perthia
Salentinellidae
Salentinella
Talitridae
Parhyale
DECAPODA
Aeglidae
Aegla
Agostocarididae
Agostocaris
Astacidea
Cambarus
Procambarus
Troglocambarus
Atyidae
Halocaridina
Stygiocaris
Troglocaris
Typhlatya
Cambaridae
Orconectes
Galatheidae
Munida
Munidopsis
Hippolytidae
Barbouria
Yagerocaris
Hymenosomatidae
Cancrocaeca
Mictocarididae
Mictocaris
Mysidae
Antromysis
Heteromysoides
Mysis
Spelaeomysis
Nebaliidae
Speonebalia
Palaemonidae
Creaseria
Palaemonetes
Palaemonias
Macrobrachium

Palinuridae
Jasus
Cherax
Parathelphusidae
Parathelphusa
Penaeidae
Penaeus
Potamidae
Cerberusa
Procarididae
Procaris
Ventrocaris
Pseudothelphusidae
Typhlopseudothelphusa
Sesarmidae
Sesarmops
Stenopodidae
Stenopus
Sundathelphusidae
Currothelphusa
THERMOSBAENACEA
Halosbaenidae
Halosbaena
Theosbaena
Monodellidae
Monodella
Tethysbaena
Thermaobaenidae
Thermosbaena
Tulumellidae
Tulumella

CHELICERATA
ARACHNIDA

SCORPIONIDA
Chactidae
Alacran
Sotanochactas
Troglocormus
Troglotayosicus
Typhlochactas
PSEUDOCORPIONIDA
Cheiridiidae
Apocheiridium
Chelanops
Hesperochernes
Maxchernes
Megachernes
Troglochernes
Chthoniidae
Austrochthonius
Chthonius
Kleptochthonius
Troglochthonius
Tyrannochthonius
Hyidae
Hyella
Neobisiidae
Alabamocreagris
Balkanoroncus
Insulocreagris
Ronchus
Roncus
Tartarocreagris
Syarinidae
Hadoblothrus
Pseudoblothrus
Troglobisium
UROPYGI
Thelyphonidae
Mastigoproctus
SCHIZOMIDA
Hubbardiidae
Sotanostenochrus
Stewartpeckius
Tayos
Bamazomus
Draculoides
Protoschizomidae
Agastoschizomus
Protoschizomus
Schizomidae
Schizomus
AMBLYPYGI
Charinidae
Charinus
Phrynichidae
Phrynichus
Phrynidae
Paraphrynus
Phrynus
Tarantulidae
Myodalis
PALPIGRADI
Eukoeneniidae
Eukoenenia
Prokoeneniidae
Prokoenenia
RICINULEI
Ricinoididae
Cryptocellus
Pseudocellus
Ricinoides
OPILIONIDA
Assamiidae
Anjolus
Bandona
Dibunidae
Dibunus
Gonyleptidae
Goniosoma
Phalangiidae
Bunochelis
Phalangodidae
Amelinus
Banksula
Maiorerus
Paratakoia
Texella
ARANEAE
Barychelidae
Troglothele
Clubionidae
Agraecina
Desidae
Forsterina
Dictynidae
Cicurina
Dipluridae
Troglodiplura
Gnaphosidae
Scotophaeus
Hahniidae
Hahnia
Heteropodidae
Heteropoda
Panaretus
Leptonetidae
Neoleptoneta
Linyphiidae
Anthrobia
Centromerus
Chthiononetes
Dunedinia
Islandiana
Phanetta
Porrhomma
Loxoscelidae
Loxosceles
Lycosidae
Adelocosa
Lycosa
Mesothelae
Liphistius
Nemesiidae
Speloctеniza
Nesticidae
Howaia
Nesticus
Ochyroceratidae
Althepus
Psiloderces
Speocera
Pholcidae
Blechroscelis
Spermophora
Trichocyclus
Scytodidae
Scytodes
Symphytognathidae
Anapistula
Tetrablemmidae
Indicoblemma
Tetragnathidae
Meta
Theraphosidae
Spelopelma
Theridiidae
Lepthyphantes
Marianana
Pholcomma
Theridion
Troglohyphantes
Steatoda
ACARINA
Argasiidae
Ornithodoros
Halacaridae
Lobohalacarus
Soldanellonyx
Hydrachnellae
Wandesia
Hygrobatidae
Coaustraliobates
Macrochelidae
Macrocheles
Macrolaspis
Nicoletiellidae
Labidostoma
Omartacaridae
Neoguanolichus
Omartacarus
Rhagidiidae
Rhagidia

CHAETOGNATHA

Spadellidae
Paraspadella

ECHINODERMATA
HOLOTHURIA

Synaptidae
Leptosynapta

CHORDATA
VERTEBRATA

PISCES
Amblyopsiformes
Amblyopsidae
Amblyopsis
Chologaster
Speoplatyrhinus
Troglichthys
Typhlichthys
Cypriniformes
Balitoridae
Homaloptera
Indoreonectes
Nemacheilus
Noemacheilus
Oreonectes
Paracobitis
Schistura
Sundoreonectes
Triplophysa
Cobitidae
Protocobitis
Cyprinidae
Barbopsis
Barbus
Caecobarbus
Caecocypris
Carassius
Eilichthys
Garra
Gibbibarbus
Iranocypris
Phreatichthys
Puntius
Sinocyclocheilus
Typhlobarbus
Typhlogarra
Varicorhinus
Zaccarinia
Characiformes
Characidae
Anoptichthys
Astyanax
Stygichthys

Siluriformes
Clariidae
Clarias
Horaglanis
Uegitglanis
Siluridae
Pterocryptis
Silurus
Astrolepidae
Astroblepus
Ictaluridae
Prietella
Satan
Trogloglanis
Loricariidae
Ancistrus
Pimelodidae
Caecorhamdella
Caecorhamdia
Imparfinis
Phreatobius
Pimelodella
Rhamdia
Taunayia
Typhlobagrus
Trichomycteridae
Trichomycterus
Caecogilbia
Gymnotiformes
Sternopygidae
Eigenmannia
Scombroidea
Scombidae
Thunnus
Ophidiiformes
Bythitidae
Cyprinodontiformes
Poeciliidae
Mollienesia
Mollienisia
Poecilia
Cyprinodontidae
Cyprinodon

Synbranchiformes
Synbranchidae
Anommatophasma
Furmastix
Monopterus
Ophisternon
Pluto
Synbranchus
Perciformes
Eleotridae
Milyeringa
Oxyeleotris
Typhleotris
Gobiidae
Caecogobius
Glossogobius
Luciogobius
Percichthyidae
Bostockia
Percidae
Etheostoma
AMPHIBIA
Urodela
Plethodontidae
Ambystoma
Aneides
Bolitoglossa
Chiropterotriton
Desmognathus
Diestrammena
Eurycea
Gyrinophilus
Haideotriton
Hydromantes
Plethodon
Pseudoeurycea
Thorius
Typhlomolge
Typhlotriton
Proteidae
Hypochthon
Necturus
Proteus

Salamandridae
Euproctus
Pleurodeles
Triturus
REPTILIA
Squamata
Iguanidae
Anolis
AVES
Apodiformes
Apodidae
Aerodramus
Collocalia
Caprimulgiformes
Aegothelidae
Aegotheles
Strigiformes
Steatornithidae
Steatornis
Strigidae
Tyto
MAMMALIA
Insectivora
Soricidae
Crocidura
Suncus
Chiroptera
Furipteridae
Furipterus
Megadermatidae
Macroderma
Molossidae
Tadarida
Mormoopidae
Mormoops
Pteronotus
Natalidae
Natalus
Phyllostomidae
Anoura
Artibeus
Carollia
Chrotopterus
Desmodus

Glossophaga
Leptonycteris
Phylloderma
Platyrrhinus
Trachops
Pteropodidae
Dobsonia
Rhinolophidae
Hipposideros
Rhinolophus
Rhinonycteris
Vespertilionidae
Chalinolobus
Corynorhinus
Eptesicus
Miniopterus
Myotis
Pipistrellus
Plecotus
Rodentia
Cricetidae
Neotoma
Dasyproctidae
Agouti
Muridae
Peromyscus
Rattus
Carnivora
Canidae
Canis
Vulpes
Felidae
Felis
Mustelidae
Lutra
Procyonidae
Procyon
Ursidae
Ursus
Peropteryx
Proboscidea
Elephantidae
Loxodonta

AUTHOR INDEX[1]

[1] Page references to text are in roman type, to bibliographical entries in italics.

SYSTEMATIC INDEX

GENERAL INDEX [1]

[1] Page numbers in boldface type refer to the Glossary.

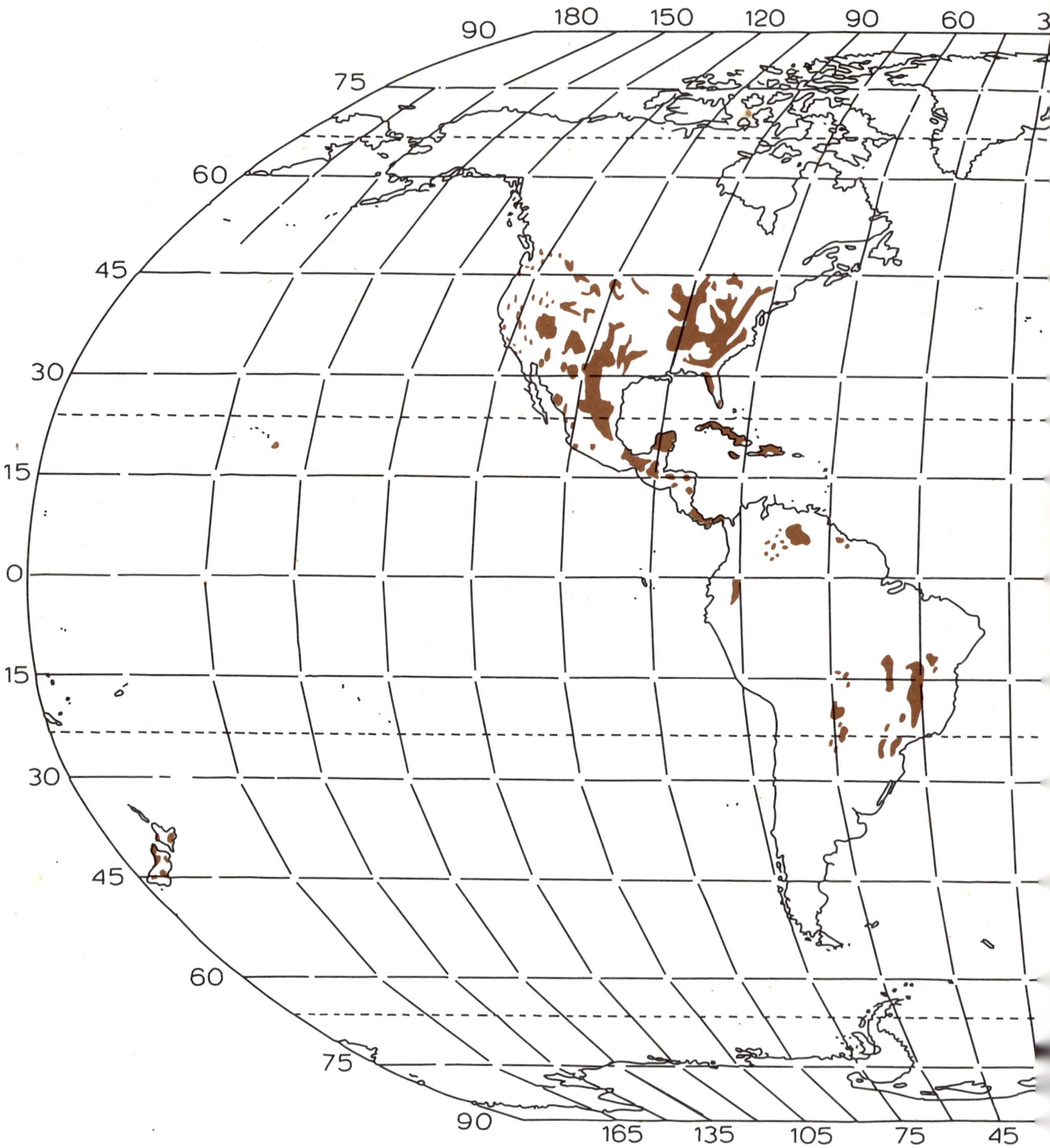
90
180
150
120
90
60
75
60
45
30
15
0
15
30
45
60
75
90
165
135
105
75
45